S. Prössdorf · B. Silbermann

Numerical Analysis for Integral and Related Operator Equations

Mathematische Lehrbücher und Monographien
Herausgegeben vom
Karl-Weierstraß-Institut für Mathematik, Berlin

II. Abteilung
Mathematische Monographien
Band 84

Numerical Analysis for Integral and Related Operator Equations

von S. Prössdorf und B. Silbermann

Numerical Analysis for Integral and Related Operator Equations

by Siegfried Prössdorf and Bernd Silbermann

Akademie Verlag

Autoren:

Prof. Dr. Siegfried Prössdorf
Karl-Weierstraß-Institut für Mathematik, Berlin

Prof. Dr. Bernd Silbermann
Technische Universität Chemnitz
Sektion Mathematik

Lektorat: Dipl.-Math. Renate Helle
Manuskriptbearbeitung: Dr. Reinhard Höppner
Herstellerische Betreuung: Barbara Hawelitschek

Die Deutsche Bibliothek-CIP-Einheitsaufnahme

Prössdorf, Siegfried:
Numerical analysis for integral and related operator equations / by Siegfried Prössdorf and Bernd Silbermann. — Berlin: Akad. Verl., 1991
(Mathematische Lehrbücher und Monographien: Abt. 2, Mathematische Monographien; Bd. 84)
ISBN 3-05-500696-8
NE: Silbermann, Bernd:; Mathematische Lehrbücher und Monographien / 02

ISBN 3-05-500696-8
ISSN 0076-5430

Erschienen in der Akademie Verlag GmbH, O-1086 Berlin (Federal Republic of Germany), Leipziger Str. 3—4
Gedruckt auf säurefreiem Papier

Gesamtherstellung: Maxim Gorki-Druck GmbH, O-7400 Altenburg
Bestellnummer: 9248
Printed in the Federal Republic of Germany

Preface

To the memory of Professor S. G. Mikhlin

About 15 years ago, when we were writing our booklet "Projection Methods and the Approximate Solution of Singular Equations" (1977) (that was the time when German native speakers still wrote in German, and so this little book is also in German), we were a long way from expecting that the numerical analysis of several classes of integral equations was on the point of entering a rapid stage of development, continuing until the present time and being certain to endure for a long time to come. This line of development was given decisive impetus by the increasing use of boundary element methods (BEM's, sometimes also referred to as boundary integral equation methods) in engineering and the natural sciences.

The essence of the methods alluded to may be described as follows. The solution to the boundary value problem under consideration is sought as an appropriately chosen integral over the boundary containing one or more unknown functions as well as a known fundamental solution of the differential equation to be solved. Inserting this ansatz (well known from potential theory for more than 100 years) into the boundary conditions leads to a boundary integral equation for the unknown function. The solution of the latter integral equation then provides the solution of the original boundary value problem in the form of an integral representation. The wide number of possible choices of boundary elements (e.g. of finite elements on the boundary as trial functions) and of discretizations of the boundary integral equations (e.g. collocation, Galerkin or quadrature methods) result in a whole variety of BEM's. Comparing BEM's with finite difference and finite element methods and considering the pros and cons, one will observe that BEM's take a series of advantages from the facts that they allow a reduction in the dimension of the problem by one unit and that they work equally well for both interior and exterior boundary value problems. Therefore, during the last decade BEM's have become a rather powerful and popular technique in engineering computations of boundary value problems arising from different fields of application.

Notwithstanding the more than one hundred year usage of boundary integral equation methods in the analytical theory of boundary value problems, going back to C. Neumann's pioneering work in 1877, the mathematically rigorous foundation and error analysis of BEM's has been started on and (at least for two-dimensional problems) has made fairly satisfactory progress recently. The reason for this delay lies, in the author's view, in the fact that boundary integral operators are in general neither integral operators of the form identity plus compact operator nor of the form identity plus an operator with small norm, so that the existing standard theories for the numerical analysis of second kind Fredholm integral equations cannot be applied. Boundary reduction rather leads to singular integral equations, convolution equations (of Wiener-Hopf or Mellin type), or even to pseudodifferential equations. For instance, solving the Dirichlet problem for the Laplace equation in a domain with corners by a double layer potential ansatz amounts to a convolution equation of Mellin type of the form 5.0(1) with the kernel 5.0(3).

The study of the equations we encounter when applying BEM's requires having recourse to a series of heavy guns from mathematical analysis. So it is not surprising that this peculiarity is shared by the numerical analysis of these equations. In our opinion, the profound investigation of a broad variety of approximation methods for solving such integral equations is a presentday problem of numerical analysis.

Due to the breadth and complexity of the questions touched upon above, it is impossible to cover all aspects of the matter by a single monograph. We therefore restrict our attention to the illumination of a few up-to-date methods and ideas, which, as we reckon, form an indispensable part of the numerical analysis of operator equations at present as well as in future. On the other hand, we are fully aware that the present book is nothing but a snapshot of what is going on and that its tone is set, moreover, by our own scientific interests.

The book is addressed to a wide audience of readers. We hope that both the mathematician interested in theoretical aspects of numerical analysis and the engineer wishing to see practically realizable recipes for computations will find a few suggestions. We tried to present the material in a form which allows any reader to go to the chapters or parts of the book he is interested in as quickly as possible. The interdependence table provides an overall view of the connection between the parts of this monograph.

Here now are some remarks concerning the contents of our book and methodological questions. First and foremost, Banach algebra techniques will play a dominant role in a series of chapters in this book. Some readers might consider the use of such tools for studying the stability of concrete approximation methods as out of the ordinary or even as an exotic extravagance. However, we wish to encourage these readers to overcome this psychological barrier and to read and see for themselves the power of such techniques for the investigation of very concrete approximation methods. For example, it will be seen that the Banach algebra approach is just the right device for conquering the afore-mentioned complications emerging from differential equations in domains with corners or singular integral equations with discontinuous coefficients. Chapter 7, which may be read independently of the remaining chapters, serves as both an introduction to and a demonstration of the power of Banach algebra techniques in numerical analysis; there this philosophy is exemplified by the reduction (= finite section) and collocation methods for singular integral operators on $\mathbf{L}^2(\mathbb{T})$ with piecewise continuous coefficients.

Moreover, Chapter 7 will also reveal that each concrete approximation method for a singular integral operator A is tied in with a certain operator valued function (depending on the approximation method under consideration) which owns many features of the symbol of singular integral and pseudodifferential operators known from classical analysis. The (loosely) summarised result will be that the approximation method for A is stable if and only if A itself and all values of the afore-mentioned operator valued functions are invertible. The approach developed in Chapter 7 is carried over to spline approximation methods in Chapters 10–13.

We here wish to express our hope that the notion of the "symbol" of approximation methods for operators will play an important and maybe even imperative role in the numerical analysis of the future.

Chapters 1, 2, and 6 contain the theoretical background of the book. Chapter 1 records some facts from functional analysis, including local principles, the notion of the convergence manifold, and a discussion of the Aubin-Nitsche lemma. Chapter 2 is devoted to the approximation properties of periodic functions (spline and trigonometric approximation in Sobolev and Hölder-Zygmund spaces) and is perhaps of independent interest. The subject of Chapter 6 is algebras of one-dimensional singular integral and pseudodifferential operators, the concept of locally strong ellipticity for singular integral operators with piecewise continuous coefficients, and the asymptotics of the solutions of singular integral equations on an interval.

Chapter 3 exemplifies the convergence manifold concept for Fredholm integral equations of the second kind. We there also introduce some important approximation methods for such equations, which will be thoroughly studied for much more general classes of integral and operator equations in the subsequent chapters.

Chapters 4 and 5 form an integrated whole, although they may be read independently of each other. In Chapter 4 we present a new variant of the well-known Gohberg-Feldman approach to convolution operators. Our version of this approach incorporates the considerable know-how on this topic we have gained during the last few years, and our emphasis is on those aspects of the matter which simplify the theory of the finite section method. Chapter 5 is a systematic discussion of approximation methods based on spline approximation for Wiener-Hopf integral equations and convolution equations of the Mellin type. As a point of fact, we mention that our technique of studying these approximation methods at long last rests on a reduction of the problems to the same problems for the finite section method.

Chapter 8 is a continuation of the theme started in Chapters 4 and 7 and concentrates on the finite section and collocation methods for singular integral equations on Hölder-Zygmund spaces.

Chapter 9 is an attempt to develop a detailed and mathematically rigorous theory of polynomial approximation methods for singular integral equations over an interval. It is this chapter in which we want also to present results suited to immediate computer implementation.

The concern of Chapter 10 is a general and uniform approach to the stability theory and error analysis of spline collocation and spline Galerkin methods on uniform meshes as well as of quadrature schemes and collocation procedures with trigonometric trial functions. The heart of our analysis is a localization technique for numerical schemes in which the matrices of finite systems are paired circulants.

Chapter 11 deals with approximation methods for singular integral equations of Cauchy and Mellin types on closed curves with corners and on a finite interval. In particular, we prove necessary and sufficient conditions for the stability of the piecewise constant collocation and of certain quadrature methods using simple quadrature rules.

In Chapter 12 we investigate Galerkin and collocation methods with splines for locally strongly elliptic singular integral equations on an interval using graded meshes. Moreover, modifications of the quadrature methods presented in Chapter 11 with a higher order of convergence are considered.

In the first part of Chapter 13 we study ε-collocation of pseudodifferential equations on closed curves in the scale of Sobolev spaces by smooth polynomial splines subordinate to uniform meshes. In the second part the nodal collocation and Galerkin methods for pseudodifferential equations using polynomial splines on arbitrary meshes are considered. In both cases we give necessary and sufficient stability conditions and prove optimal error estimates in Sobolev norms.

The circle of ideas we originally intended to cover with this book has essentially grown and been shaped by the influence of our (former) students and (subsequent) coworkers A. Böttcher, J. Elschner, R. Hagen, P. Junghanns, A. Rathsfeld, S. Roch, and G. Schmidt, and we are pleased to report that a major part of the present monograph is the result of fruitful team-work with these colleagues. Moreover, during the long years' work on this book we had several opportunities of discussing a series of the problems considered here with competent specialists; in this connection we especially acknowledge the useful advice of D. Arnold, M. Costabel, I. Gohberg, I. Graham, G. Hsiao, C. Johnson, N. Krupnik, W. McLean, G. Mastroianni, V. Mazya, E. Meister, S. Mikhlin, G. Monegato, C.-J. Nedelec, N. Nikolski, A. Pomp, J. Prestin, J. Saranen, I. Sloan, F.-O. Speck, E. Stephan, and W. Wendland. The stimulating conversations and the outstanding work of all these colleagues and friends have left their imprint on the present book. We wish to express our sincere thanks to all of them.

We are greatly indebted to Mrs. Scholz, Mrs. Tabbert, and Miss Ulrich for typing this monstrosity with remarkable care and expertise and to A. Böttcher for proofreading the entire manuscript. We furthermore express deep gratitude to the Akademie-Verlag Publishing House and to Dr. R. Höppner and his staff for the pleasant co-operation at all stages of our work on this project.

Last but not least, we thank our wives Roswitha Prössdorf and Lyudmila Silbermann for all the understanding they showed in the troublesome period of writing this book.

October 1990 Siegfried Prössdorf, Bernd Silbermann

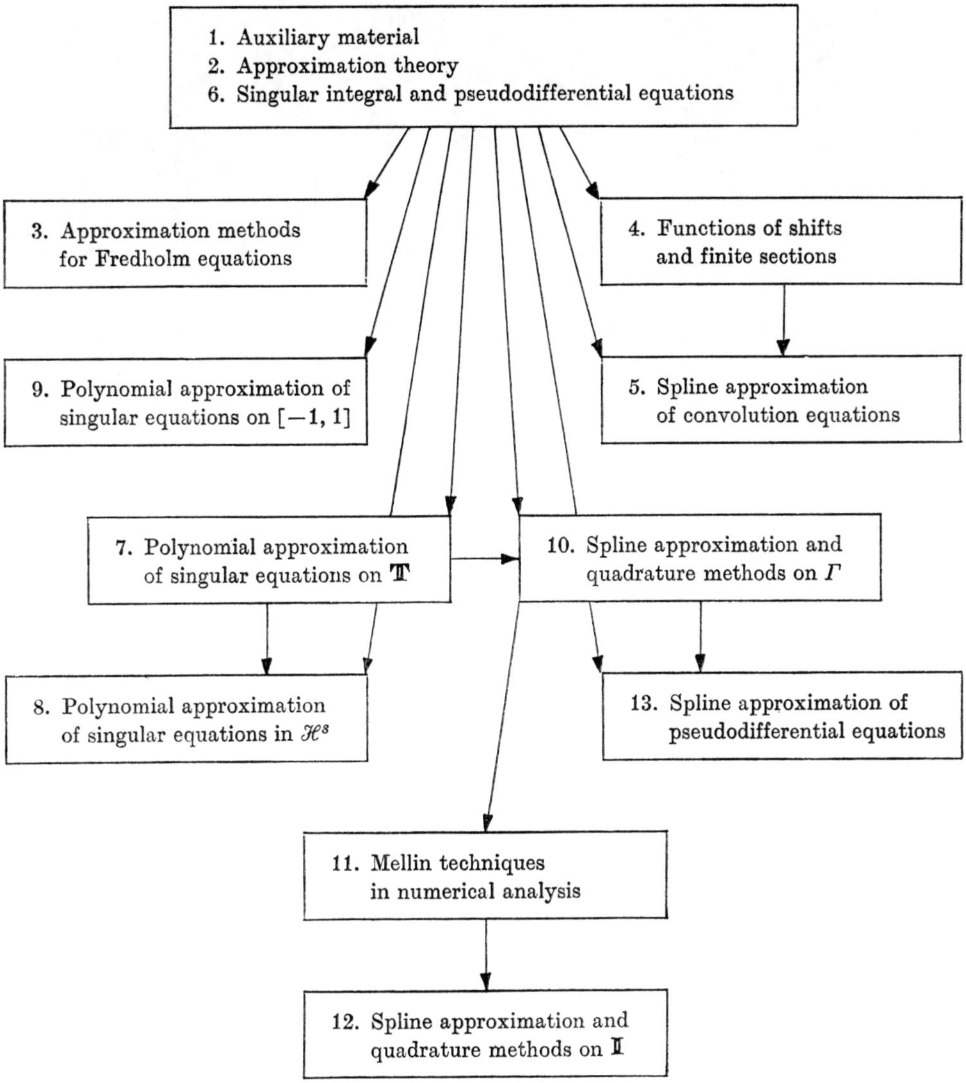
1. Auxiliary material
2. Approximation theory
6. Singular integral and pseudodifferential equations
3. Approximation methods for Fredholm equations
4. Functions of shifts and finite sections
9. Polynomial approximation of singular equations on [−1, 1]
5. Spline approximation of convolution equations
7. Polynomial approximation of singular equations on $\mathbb{T}$
10. Spline approximation and quadrature methods on Γ
8. Polynomial approximation of singular equations in $\mathcal{H}^s$
13. Spline approximation of pseudodifferential equations
11. Mellin techniques in numerical analysis
12. Spline approximation and quadrature methods on $\mathbb{I}$

Contents

Chapter 1
Auxiliary material

Operators

1.1. Bounded and compact operators. Let X and Y be Banach spaces. We denote by $\mathcal{L}(X, Y)$ the linear space of all (bounded and linear) *operators* from X to Y. We let $\mathcal{K}(X, Y)$ denote the collection of all linear *compact* operators from X into Y, and $\mathcal{K}_0(X, Y)$ refers to the set of all *finite-rank* operators from X into Y, i.e., F is in $\mathcal{K}_0(X, Y)$ if and only if $F \in \mathcal{L}(X, Y)$ and $\dim F(X) < \infty$. In the case $X = Y$ we shall write $\mathcal{L}(X) = \mathcal{L}(X, X)$, $\mathcal{K}(X) = \mathcal{K}(X, X)$, $\mathcal{K}_0(X) = \mathcal{K}_0(X, X)$. The identity operator on X will be denoted by I_X or simply by I.

An operator $A \in \mathcal{L}(X, Y)$ is called *invertible* if there is an operator $A^{-1} \in \mathcal{L}(Y, X)$ such that $A^{-1}Ax = x$ for all $x \in X$ and $AA^{-1}y = y$ for all $y \in Y$. A well-known theorem of Banach states that $A \in \mathcal{L}(X, Y)$ is invertible if and only if the *kernel* $\ker A$ of A is trivial and the *image* $\operatorname{im} A$ of A coincides with Y.

A sequence $\{A_n\}$ of operators $A_n \in \mathcal{L}(X, Y)$ is said to *converge* to an operator $A \in \mathcal{L}(X, Y)$

(a) *weakly*, if $f(A_nx) - f(Ax) \to 0$ for each $x \in X$ and each functional $f \in Y^*$, where Y^* denotes the dual space of Y, i.e., the space of all bounded linear functionals on Y;

(b) *strongly*, if $\|A_nx - Ax\|_Y \to 0$ for each $x \in X$; in this case we write also $A = \operatorname{s-lim}_{n\to\infty} A_n$;

(c) *uniformly*, if $\|A_n - A\| \to 0$, where, for $B \in \mathcal{L}(X, Y)$,

$$\|B\| := \sup_{\|x\| \leqq 1} \|Bx\|. \tag{1}$$

Let $\{A_n\}$ be a sequence of operators $A_n \in \mathcal{L}(X, Y)$. Then one has the following.

(d) If $A \in \mathcal{L}(X, Y)$, if $\|A_nx - Ax\|_Y \to 0$ as $n \to \infty$ for each x in a dense subset of X, and if $\sup_n \|A_n\| < \infty$, then A_n converges strongly to A.

(e) (Banach-Steinhaus theorem). If $\{A_nx\}$ is a convergent sequence in X for each $x \in X$, then $\sup_n \|A_n\| < \infty$, the operator A defined by $Ax = \lim_{n\to\infty} A_nx$ belongs to $\mathcal{L}(X, Y)$, and

$$\|A\| \leqq \liminf_{n\to\infty} \|A_n\|.$$

(f) If $A \in \mathcal{L}(X, Y)$, if $A_n \to A$ weakly, and if $K \in \mathcal{K}(Y, Z)$, then $KA_n \to KA$ strongly.

Equipped with the operator norm (1) the linear set $\mathcal{L}(X, Y)$ becomes a Banach space and $\mathcal{L}(X)$ a Banach algebra. Then $\mathcal{K}(X, Y)$ is a closed subspace of $\mathcal{L}(X, Y)$. In general, the subspace $\mathcal{K}_0(X, Y) \subset \mathcal{L}(X, Y)$ is not closed, but its closure is contained in $\mathcal{K}(X, Y)$. Notice the following implications:

$$A \in \mathcal{L}(X, Y), \quad K \in \mathcal{K}(Y, Z), \quad B \in \mathcal{L}(Z, V) \Rightarrow BKA \in \mathcal{K}(X, V),$$
$$A \in \mathcal{L}(X, Y), \quad K \in \mathcal{K}_0(Y, Z), \quad B \in \mathcal{L}(Z, V) \Rightarrow BKA \in \mathcal{K}_0(X, V).$$

In particular, $\mathcal{K}(X)$ is a closed two-sided ideal of $\mathcal{L}(X)$ and $\mathcal{K}_0(X)$ is a two-sided (but in general not closed) ideal of $\mathcal{L}(X)$.

For each operator $A \in \mathcal{L}(X, Y)$, the *adjoint* or *dual operator* $A^* \in \mathcal{L}(Y^*, X^*)$ is uniquely defined by

$$f(Ax) = (A^*f)\, x$$

for any $f \in Y^*$ and any $x \in X$. The mapping $A \to A^*$ is antilinear and normpreserving, i.e., one has $(\lambda_1 A_1 + \lambda_2 A_2)^* = \bar{\lambda}_1 A_1^* + \bar{\lambda}_2 A_2^*$ and $\|A\| = \|A^*\|$. Here $\bar{\lambda}$ denotes the complex adjoint to $\lambda \in \mathbb{C}$ ($\mathbb{C}$ being the field of complex numbers).

(g) $K \in \mathcal{K}(X, Y) \Leftrightarrow K^* \in \mathcal{K}(Y^*, X^*)$.

The following assertion is of great importance in numerical analysis.

(h) Let $\{A_n\} \subset \mathcal{L}(Z, V)$ converge strongly to $A \in \mathcal{L}(Z, V)$ and let $\{B_n^*\} \subset \mathcal{L}(Y^*, X^*)$ converge strongly to $B^* \in \mathcal{L}(Y^*, X^*)$. Then if $C \in \mathcal{K}(Y, Z)$,

$$\|A_n C B_n - ACB\| \to 0.$$

Proof. A little thought shows that it suffices to prove the assertion for $X = Y$, $B_n = I$.

Let S be the image of $\{\|x\| \leqq 1\}$ under the map C. Then S is relatively compact and for any $\varepsilon > 0$ there exist elements $s_1, \ldots, s_m \in S$ such that $\|y - s_i\| < \varepsilon$ for all $y \in S$ and for a suitable chosen i. Let $x_1, \ldots, x_m$ be elements from $\{\|x\| \leqq 1\}$ such that $Cx_i = s_i$, $i = 1, \ldots, m$. Then for $y \in \{\|x\| \leqq 1\}$

$$\begin{aligned}\|(A_nC - AC)\, y\| &\leqq \min\, (\|(A_n - A)\,(Cy - Cx_i)\| + \|(A_n - A)\, Cx_i\|)\\ &\leqq \varepsilon(\|A_n\| + \|A\|) + \min_i \|(A_n - A)\, Cx_i\|.\end{aligned}$$

The sequence $\{\|A_n\|\}$ is bounded by 1.1(e) and what results is that $\|A_nC - AC\| \to 0$ as $n \to \infty$. ∎

In fact, we showed that the stronge convergence is uniform on relatively compact sets. This is the key to the notion of collectively compact sets of operators. Recall that a set $B \subset \mathcal{L}(X, Y)$ is said to be *collectively compact* if the set $\{Af: A \in B, f \in S\}$ is relatively compact, where S denotes the unit ball in X. Clearly each element of a collectively compact set of operators is a compact operator. If $\{A_n\}_{n=1}^{\infty} \subset \mathcal{L}(X, Y)$ is collectively compact and $A = \text{s-lim}\, A_n$, then (obviously) $A \in \mathcal{K}(X, Y)$.

The following assertion is now obvious.

(k) Let $\{A_n\} \subset \mathcal{L}(Y, Z)$ converge strongly to the zero operator and let $\{B_n\} \subset \mathcal{L}(X, Y)$ be collectively compact. Then $\|A_nB_n\| \to 0$ as $n \to \infty$.

1.2. Unbounded operators. In what follows we have also to deal with linear unbounded operators. An unbounded operator A is (as a rule) defined only on a linear subset of the Banach space X. This subset is called the *domain* of the operator A and will be denoted by $\mathcal{D}(A)$. A linear operator $A: \mathcal{D}(A) \to Y$ is called *closed* if for any sequence $\{x_n\} \subset \mathcal{D}(A)$ such that $x_n \to x$ and $Ax_n \to y$ (in the norms of the spaces X and Y, respectively) one has $x \in \mathcal{D}(A)$ and $Ax = y$.

The famous closed-graph theorem of Banach states that a closed linear operator A defined on all of a Banach space X is bounded.

1.3. Fredholm operators. Let X and Y be Banach spaces and let $A \in \mathcal{L}(X, Y)$. We put

$$\ker A = \{x \in X: Ax = 0\}, \qquad \operatorname{im} A = \{y \in Y: \exists\, x \in X \text{ with } Ax = y\}.$$

The operator A is said to be *normally solvable* if im A is closed in Y. A normally solvable operator $A \in \mathcal{L}(X, Y)$ is called a *Φ_+-operator* (*Φ_--operator*) if $\dim \ker A < \infty$ $\big(\dim \operatorname{coker} A := \dim (Y/\operatorname{im} A) < \infty\big)$. If A is both a Φ_+- and Φ_--operator, then it is called a *Φ-operator* (or *Fredholm operator* or *Noetherian operator*) and the integer

$$\operatorname{ind} A = \dim \ker A - \dim \operatorname{coker} A$$

is referred to as the *index* of A. The collection of all Φ_+-operators from X to Y will be denoted by $\Phi_+(X, Y)$, and $\Phi_+(X, X)$ will be abbreviated to $\Phi_+(X)$. A similar definition is made for $\Phi_-(X, Y)$, $\Phi_-(X)$, $\Phi(X, Y)$, $\Phi(X)$.

A simple but nevertheless important example for a normally solvable operator is the *projection*, i.e., an operator $p \in \mathcal{L}(X)$ such that $p^2 = p$. Since $I - p$ is also a projection and $\operatorname{im} p = \ker (I - p)$, the closedness of im p follows from that of $\ker (I - p)$. Now we list up basic properties of Fredholm operators.

(a) For $A \in \mathcal{L}(X, Y)$ the following are equivalent:

(i) $A \in \Phi(X, Y)$.

(ii) There exist operators $R, L \in \mathcal{L}(Y, X)$ such that $AR - I_Y \in \mathcal{K}(Y)$, $LA - I_X \in \mathcal{K}(X)$.

(iii) There exists an operator $B \in \mathcal{L}(Y, X)$ such that $AB - I_Y \in \mathcal{K}_0(Y)$, $BA - I_X \in \mathcal{K}_0(X)$.

Any operator $B \in \mathcal{L}(Y, X)$ for which $AB - I_Y \in \mathcal{K}(Y)$, $BA - I_X \in \mathcal{K}(X)$ is called a *regularizer of A*.

(b) (Atkinson's theorem). Let X, Y, Z be Banach spaces and let $A \in \Phi(X, Y)$, $B \in \Phi(Y, Z)$. Then $BA \in \Phi(X, Z)$ and

$$\operatorname{ind} BA = \operatorname{ind} B + \operatorname{ind} A .$$

(c) $\Phi(X, Y)$, $\Phi_+(X, Y)$, $\Phi_-(X, Y)$ are open subsets of $\mathcal{L}(X, Y)$ and the mapping ind: $\Phi(X, Y) \to \mathbb{Z}$ ($\mathbb{Z}$ being the set of all real integers) is constant on the connected components of $\Phi(X, Y)$. If $A \in \Phi(X, Y)$ and $K \in \mathcal{K}(X, Y)$, then $A + K \in \Phi(X, Y)$ and $\operatorname{ind}(A + K) = \operatorname{ind} A$.

(d) Let X, Y, Z be Banach spaces and let $A \in \mathcal{L}(X, Y)$, $B \in \mathcal{L}(Y, Z)$. Then the following implications are true:

(i) $A \in \Phi_\pm(X, Y)$, $B \in \Phi_\pm(Y, Z) \Rightarrow BA \in \Phi_\pm(X, Z)$,

(ii) $BA \in \Phi_+(X, Z) \Rightarrow A \in \Phi_+(X, Y)$,

(iii) $BA \in \Phi_-(X, Z) \Rightarrow B \in \Phi_-(Y, Z)$,
(iv) $A \in \Phi(X, Y), BA \in \Phi(X, Z) \Rightarrow B \in \Phi(Y, Z)$,
(v) $B \in \Phi(Y, Z), BA \in \Phi(X, Z) \Rightarrow A \in \Phi(X, Y)$.

(e) Let $A \in \Phi_+(X)$ and let $K \in \mathcal{K}_0(X)$ be any projection of X onto $\ker A$. Then there is a $\delta > 0$ such that

$$\|Ax\| + \|Kx\| \geqq \delta \|x\| \quad \forall x \in X.$$

Vice versa, if for an operator $A \in \mathcal{L}(X)$ there exist operators $K_j \in \mathcal{K}(X)$ $(j = 1, \ldots, n)$ and $\delta > 0$ such that

$$\|Ax\| + \sum_{j=1}^{n} \|K_j x\| \geqq \delta \|x\| \quad \forall x \in X,$$

then $A \in \Phi_+(X)$. If in the last inequality all operators K_j are the zero operators, then A is said to be of *regular type*.

(f) We have $A \in \Phi_\pm(X, Y) \Leftrightarrow A^* \in \Phi_\mp(Y^*, X^*)$. Moreover, if $A \in \Phi(X, Y)$ then $\dim \ker A^* = \dim \operatorname{coker} A$, $\dim \operatorname{coker} A^* = \dim \ker A$, whence

$$\operatorname{ind} A^* = -\operatorname{ind} A.$$

(g) Assume that the Banach space X is continuously and densely embedded in the Banach space X_1, that there are operators $A \in \Phi(X)$, $A_1 \in \Phi(X_1)$ such that $A_1 \mid X = A$ and $\operatorname{ind} A = \operatorname{ind} A_1$. Then $\ker A = \ker A_1$.

Operator matrices and their determinants

1.4. Definitions. Given a linear space X, denote by X_N the linear space of column-vectors of length N with components from X and let $X_{N \times N}$ denote the linear space of $N \times N$ matrices with entries from X. If X is a Banach space, X_N can be made to become a Banach space on defining a norm in X_N by

$$\|(x_1, \ldots, x_N)^\mathsf{T}\|_{X_N} := \|x_1\|_X + \cdots + \|x_n\|_X \tag{1}$$

or by choosing any norm in X_N equivalent to that one. Every operator $A \in \mathcal{L}(X_N)$ may then be written as an operator matrix $A = (A_{ij})_{i,j=1}^N$, where $A_{ij} \in \mathcal{L}(X)$, that is, $\mathcal{L}(X_N)$ may be identified with $(\mathcal{L}(X))_{N \times N}$. It is easily seen that $F \in \mathcal{K}_p(X_N)$, $p = 0, \infty$, if and only if all entries F_{ij} of the matrix $F = (F_{ij})_{i,j=1}^N$ are in $\mathcal{K}_p(X)$. Here we used the notation $\mathcal{K}_\infty(X) := \mathcal{K}(X)$.

Let $A = (A_{ij})_{i,j=1}^N \in \mathcal{L}(X_N)$. The *determinant* $\det A$ of A is the operator in $\mathcal{L}(X)$ which is defined by

$$\det A = \sum_\sigma (-1)^{p(\sigma)} A_{N,\sigma(N)} \cdots A_{1,\sigma(1)}, \tag{2}$$

where σ ranges over all permutations of $\{1, \ldots, N\}$ and $p(\sigma)$ is the signum of the permutation σ. If the operators A_{ij} do not commute pairwise, then the order in which the factors in each item of the sum (2) are arranged is of significance. However, if we are only interested in whether $\det A$ is a Φ_+- or Φ_--operator and if the entries of A pairwise commute up to a compact operator, then the answer will not depend on the arrangement of the factors in the items of the sum (2).

In the theorems below $[A, B]$ denotes the *commutator* $AB - BA$.

1.5. Theorem. *Let X be a Banach space and let $A = (A_{ij})_{i,j=1}^{N} \in \mathcal{L}(X_N)$.*

(a) *If $[A_{ij}, A_{l,m}] \in \mathcal{K}(X)$ for $j \neq m$, then*
$\det A \in \Phi_-(X) \Rightarrow A \in \Phi_-(X_N)$,
$A \in \Phi_+(X_N) \Rightarrow \det A \in \Phi_+(X)$.

(b) *If $[A_{ij}, A_{lm}] \in \mathcal{K}(X)$ for $i \neq l$, then*
$\det A \in \Phi_+(X) \Rightarrow A \in \Phi_+(X_N)$,
$A \in \Phi_-(X_N) \Rightarrow \det A \in \Phi_-(X)$.

(c) *If the entries of A pairwise commute modulo compact operators, then $A \in \Phi(X_N)$ (resp. $\Phi_+(X_N)$, $\Phi_-(X_N)$) if and only if* $\det A \in \Phi(X)$ *(resp. $\Phi_+(X)$, $\Phi_-(X)$).*

This theorem says nothing about the connection between the index of A and that of $\det A$. Under the hypothesis of part (c) both equality and inequality of these indices are possible. The following theorem states sufficient conditions for equality.

1.6. Theorem (Markus-Feldman). *Let X be a Banach space, let $A \in \Phi(X_N)$ and suppose the entries of A pairwise commute modulo $\mathcal{K}_0(X)$. Then* $\det A \in \Phi(X)$ *and* $\operatorname{ind} A = \operatorname{ind} \det A$.

The proof of the Φ-partion (c) of Theorem 1.5 can be carried out purely algebraical using the following theorem, which is of its own interest. If A is a ring with identity e, then $A_{N \times N}$ endowed with natural algebraic operations is also a ring with identity element. Put $[a, b] = ab - ba$ $(a, b \in A)$.

1.7. Theorem (Krupnik). *Let $a = (a_{ij})_{i,j=1}^{N} \in A_{N \times N}$ and let $[a_{mk}, a_{pq}] = 0$ $(m, k, p, q = 1, \ldots, n)$. Then a is invertible in $A_{N \times N}$ if and only if* $\det a$ *is invertible in A.*

Banach algebras

1.8. Invertibility and spectrum. Let A be a Banach algebra over the complex field with identity e. An element $a \in A$ is said to be *left* (*right*, resp. *two-sided*) *invertible* in A if there exists an element $b \in A$ such that $ba = e$ ($ab = e$, resp. $ba = ab = e$). Two-sided invertible elements will be simply called *invertible*. The collection of all invertible elements of A will be denoted by GA. Note that GA is open and forms a group with respect to the multiplication in A. For $a \in A$, let $\exp a = \sum_{n \geq 0} a^n/n!$.

(a) Let A be a Banach algebra with identity e and let $a \in A$ be an element such that $\|a\| < 1$. Then $e - a \in GA$ and $(e - a)^{-1}$ is given by $(e - a)^{-1} = \sum_{n \geq 0} a^n$, the expansion converging in the norm.

(b) If A is commutative, then an invertible element a belongs to the connected component of GA containing the identity e if and only if a has a logarithm in A, i.e., if there exists a $b \in A$ such that $a = \exp b$.

The *spectrum* $\operatorname{sp} a$ (or $\operatorname{sp}(a)$) of an element $a \in A$ is the set of all $\lambda \in \mathbb{C}$ such that $a - \lambda e$ is not (two-sided) invertible in A. In order to emphasize that invertibility in A is meant, we shall sometimes write $\operatorname{sp}_A a$ instead of $\operatorname{sp} a$. For several applications the following result will often be useful.

(c) Let B be a closed subalgebra of A and suppose $e \in B$. If $a \in B$, then $\operatorname{sp}_B a$ is the union of $\operatorname{sp}_A a$ and a (possibly empty) collection of bounded connected components of the complement of $\operatorname{sp}_A a$.

1.9. Ideals. Let A be a Banach algebra. A closed subspace J of A is called a *closed left* (resp. *right*) *ideal* of A if $aj \in J$ (resp. $ja \in J$) for all $a \in A$ and $j \in J$. The *closed two-sided ideals* are the ideals which are both left and right. An ideal J of A is said to be *proper* if $J \neq A$.

A proper closed left (right, resp. two-sided) ideal J of A is called a *maximal* left (right, resp. two-sided) *ideal* if it is not properly contained in any other proper left (right, resp. two-sided) ideal of A.

(a) In a Banach algebra with identity every proper left (right, resp. two-sided) ideal is contained in some maximal left (right, resp. two-sided) ideal.

Maximal two-sided ideals will be simply referred to as *maximal ideals*. If J is a proper closed two-sided ideal of A, then the quotient algebra A/J is a Banach algebra under the norm $\|a + J\| := \inf\limits_{j \in J} \|a + j\|$.

(b) If A is commutative and has an identity and if J is a maximal ideal of A then the quotient algebra A/J is a field, i.e., every nonzero element of A/J has an inverse.

The *radical* $\mathfrak{R}(A)$ of A is the intersection of all maximal left ideals of A.

(c) If A has an identity, then $\mathfrak{R}(A)$ is a closed two-sided ideal of A and $\mathfrak{R}(A)$ coincides with the intersection of all maximal right ideals of A.

A Banach algebra with identity whose radical consists only of the zero element will be called *semisimple*.

1.10. The maximal ideal space. Let A be a commutative Banach algebra with identity e. A *multiplicative* linear functional on A is a continuous linear mapping $m: A \to \mathbb{C}$ which preserves multiplication ($m(ab) = m(a)\, m(b)$ for all $a, b \in A$) and takes the value 1 at e $(m(e) = 1)$. The kernel of m is the set of all $a \in A$ for which $m(a) = 0$. There is a one-to-one correspondence between the multiplicative linear functionals and the maximal ideals of A: the kernel of every multiplicative linear functional is a maximal ideal and every maximal ideal is the kernel of some (uniquely determined) multiplicative linear functional. Therefore no distinction between multiplicative linear functionals and maximal ideals will be made. We denote the set of all multiplicative linear functionals on A by $\mathbf{M}(A)$. The formula $\hat{a}(m) = m(a)$ $(m \in \mathbf{M}(A))$ assigns to each $a \in A$ a function $\hat{a}: \mathbf{M}(A) \to \mathbb{C}$. This function is called the *Gelfand transform* of a. Let $\hat{A}$ be the set of all functions $\hat{a}$, $a \in A$.

The *Gelfand topology* on $\mathbf{M}(A)$ is the coarsest (weakest) topology on $\mathbf{M}(A)$ that makes all functions $\hat{a} \in \hat{A}$ continuous. It is the topology induced on $\mathbf{M}(A)$ thought of as a subset of the dual space A^*, the latter space provided with the weak-star topology. Thus, an open neighborhood base of a point $m_0 \in \mathbf{M}(A)$ is formed by the sets

$$\begin{aligned} U_{a_1,\ldots,a_n;\,\varepsilon}(m_0) &= \{m \in \mathbf{M}(A): |\hat{a}_i(m) - \hat{a}_i(m_0)| < \varepsilon \text{ for } i = 1, \ldots, n\} \\ &= \{m \in \mathbf{M}(A): |m(a_i) - m_0(a_i)| < \varepsilon \text{ for } i = 1, \ldots, n\}, \end{aligned}$$

where $a_1, \ldots, a_n \in A$ and $\varepsilon > 0$. The set $\mathbf{M}(A)$ equipped with its Gelfand topology is called the *maximal ideal space* of A. Note that $\mathbf{M}(A)$ is a compact Hausdorff space and that $\hat{A}$ is a (not necessarily closed) subalgebra of $\mathbf{C}(\mathbf{M}(A))$, where $\mathbf{C}(X)$ denotes the algebra of all complex-valued continuous functions on the topological space X.

The mapping $\Gamma: A \to \mathbf{C}(\mathbf{M}(A))$, $a \mapsto \hat{a}$, will be referred to as the *Gelfand transform*. Notice that, in general, Γ is neither one-to-one nor onto. The kernel of Γ coincides with the radical of A. Thus, if A is semisimple, then the Gelfand transform is one-to-one and, hence, an (algebraical) isomorphism of A onto $\hat{A} \subset \mathbf{C}(\mathbf{M}(A))$. Therefore we shall then simply write a and A instead of $\hat{a}$ and $\hat{A}$. Finally, recall that if A is a (not necessarily semisimple) commutative Banach algebra with identity and if $a \in A$, then the range of $\hat{a}$ coincides with the spectrum $\text{sp}_A\, a$ and $\|\hat{a}\|_\infty \leqq \|a\|$, where $\|\hat{a}\|_\infty := \max_{m \in \mathbf{M}(A)} |\hat{a}(m)|$.

1.11. Singly generated algebras. A commutative Banach algebra A with identity e is said to be *singly generated* if there is an $a \in A$ such that the linear hull of the set $\{e, a, a^2, \ldots\}$ is dense in A,

$$A = \text{clos lin}\,\{e, a, a^2, \ldots\}.$$

In that case the maximal ideal space of A is naturally homeomorphic to $\text{sp}_A\, a$, where $\text{sp}_A\, a$ has the topology induced from the inclusion $\text{sp}_A\, a \subset \mathbb{C}$ and $\mathbb{C}$ is regarded as furnished with the usual topology.

C^*-algebras

1.12. Definitions. A mapping $a \mapsto a^*$ of a Banach algebra A into itself is called an *involution* on A if $(a^*)^* = a$, $(a + b)^* = a^* + b^*$, $(ab)^* = b^*a^*$, $(\lambda a)^* = \bar{\lambda} a^*$ for all $a, b \in A$ and $\lambda \in \mathbb{C}$. A Banach algebra A with an involution $a \mapsto a^*$ that satisfies $\|aa^*\| = \|a\|^2$ for all $a \in A$ is called a *C^*-algebra*. An element $a \in A$ is called *selfadjoint* if $a = a^*$. For such an element we have $\|a\| = \max_{\lambda \in \text{sp}\, a} |\lambda|$. If Y is a compact Hausdorff space and H is a Hilbert space, then $\mathbf{C}(Y)$ and $\mathcal{L}(H)$ are C^*-algebras.

1.13. Basic properties of C^*-algebras

(a) (Gelfand-Naimark). If A is a commutative C^*-algebra with identity, then the Gelfand transform is an isometrical star-isomorphism of A onto $\mathbf{C}(\mathbf{M}(A))$.

(b) (Gelfand-Naimark). If A is any C^*-algebra, then there exists a Hilbert space H such that A is isometrically star-isomorphic to some C^*-subalgebra of $\mathcal{L}(H)$.

(c) If A is a C^*-algebra with identity and B is a C^*-subalgebra of A containing the identity, then an element $a \in B$ is left (right, resp. two-sided) invertible in A if and only if it is so in B.

(d) Let A and B be C^*-algebras and let $\varphi: A \to B$ be an (algebraical) star-homomorphism. Then $\|\varphi(a)\| \leqq \|a\|$ for all $a \in A$ and the range of φ is closed in B. If, in addition, φ is one-to-one, then $\|\varphi(a)\| = \|a\|$ for all $a \in A$.

(e) Let A be a C^*-algebra and let J be a closed two-sided ideal of A. Then J is selfadjoint (that is, $J^* = J$) and A/J provided with the involution $(a + J)^* := a^* + J$ and the usual quotient norm is a C^*-algebra.

(f) Let A be a C^*-algebra, let B be a C^*-subalgebra of A, and let J be a closed two-sided ideal of A. Then $B + J$ is a C^*-subalgebra of A and the C^*-algebras $(B + J)/J$ and $B/B \cap J$ are isometrically star-isomorphic.

1.14. The algebra $\mathbf{C}(X)$. Let X be a compact Hausdorff space. Then $\mathbf{C}(X)$ equipped with norm $\|a\|_\infty = \max_{x \in X} |a(x)|$ is a C^*-algebra with respect to the involution $a(x) \mapsto \overline{a(x)}$. If F is a closed subset of X, then

$$I_F = \{a \in \mathbf{C}(X) : a(x) = 0 \text{ for all } x \in F\}$$

is obviously a closed (two-sided) ideal of A. Moreover, it can be shown that every closed ideal of A is of this form (see, e.g., NAIMARK [1, pp. 247–248]). In other words, there is a one-to-one correspondence between all closed subsets of X and all closed (two-sided) ideals of $\mathbf{C}(X)$.

Let F be a closed subset of X and let J be the collection of all continuous functions vanishing on a certain neighborhood (depending on the function) of F. Then J is an ideal in $\mathbf{C}(X)$ whose closure coincides with I_F. To see this note that for any point $x \in X$, $x \notin F$, there is a function in $\mathbf{C}(X)$ such that this function vanishes on F and takes the value 1 at x.

1.15. The algebra *$C^*(p, q, e)$*. Let A be a C^*-algebra and $p, q \in A$ be selfadjoint elements such that $p^2 = p$, $q^2 = q$. Denote by $C^*(p, q, e)$ the smallest closed C^*-subalgebra of A containing the elements p, q, e. Finally note that $\mathbf{C}_{N\times N}[0, 1] := (\mathbf{C}[0, 1])_{N\times N}$ is a C^*-algebra under the norm

$$\|a\| := \max_{t\in[0,1]} \left(\max_{i=1,\dots,N} \lambda_i(t)^{1/2} \right),$$

where $\lambda_i(t)$ are the eigenvalues of $(aa^*)(t)$. Let $\boldsymbol{p}, \boldsymbol{q}, \boldsymbol{e} \in \mathbf{C}_{2\times 2}[0, 1]$ be the elements defined by

$$\boldsymbol{q} = \begin{pmatrix} x & \sqrt{x(1-x)} \\ \sqrt{x(1-x)} & 1-x \end{pmatrix}, \quad \boldsymbol{p} = \begin{pmatrix} 1 & 0 \\ 0 & 0 \end{pmatrix}, \quad \boldsymbol{e} = \begin{pmatrix} 1 & 0 \\ 0 & 1 \end{pmatrix}.$$

1.16. Theorem. *Let A be a C^*-algebra and p, $q \in A$ be selfadjoint elements such that $p^2 = p$, $q^2 = q$ and $\operatorname{sp}(pqp) = [0, 1]$. Then the C^*-subalgebra $C^*(p, q, e)$ is star-isomorphic to the C^*-subalgebra $C^*(\boldsymbol{p}, \boldsymbol{q}, \boldsymbol{e})$ of $\mathbf{C}_{2\times 2}[0, 1]$ generated by $\boldsymbol{p}, \boldsymbol{q}, \boldsymbol{e}$.*

This theorem is of exceptional importance in the theory of singular integral operators and will be often used in what follows.

Local principles

1.17. Definitions. Let A be a Banach algebra with identity e. A subset $M \subset A$ is called a *localizing class* if

(i) $0 \notin M$,

(ii) for any $f_1, f_2 \in M$ there exists a third element $f \in M$ such that $f_j f = f f_j = f$ $(j = 1, 2)$.

Two elements $a, b \in A$ are said to be *M-equivalent* from the left (resp. from the right) if

$$\inf_{f\in M} \|(a - b) f\| = 0 \qquad \left(\text{resp. } \inf_{f\in M} \|f(a - b)\| = 0\right).$$

An element $a \in A$ is called *M-invertible* from the left (resp. from the right) if there are a $b \in A$ and an $f \in M$ such that $baf = f$ (resp. $fab = f$). A system $\{M_\tau\}_{\tau\in T}$ of localizing classes is said to be *covering* if from each choice $\{f_\tau\}_{\tau\in T}$ $(f_\tau \in M_\tau)$ there can be selected a finite number of elements $f_{\tau_1}, \dots, f_{\tau_m}$ whose sum is invertible in A.

Now suppose T is a topological space. Then a system $\{M_\tau\}_{\tau\in T}$ of localizing classes will be said to be *overlapping* if

(iii) each M_τ is a bounded subset of A;

(iv) $f \in M_{\tau_0}$ ($\tau_0 \in T$) implies that $f \in M_\tau$ for all τ in some open neighborhood of τ_0;
(v) the elements of $F = \bigcup_{\tau \in T} M_\tau$ commute pairwise.

Let $\{M_\tau\}_{\tau \in T}$ be an overlapping system of localizing classes. The *commutant* of F is the set $\operatorname{Com} F := \{a \in A : af = fa \ \forall f \in F\}$. It is clear that $\operatorname{Com} F$ is a closed subalgebra of A. For $\tau \in T$, let Z_τ denote the set of all elements in $\operatorname{Com} F$ which are M_τ-equivalent to zero both from the left and from the right. Note that Z_τ is a closed (by virtue of (iii)) two-sided ideal of $\operatorname{Com} F$ which does not contain the identity e (if $e \in Z_\tau$, then there are $f_n \in M_\tau$ such that $\|f_n\| \to 0$ as $n \to \infty$, and since there exist $g_n \neq 0$ in M such that $f_n g_n = g_n$, it follows that $\|f_n\| \geqq 1$, which is a contradiction). For $a \in \operatorname{Com} F$ let a^τ denote the coset $a + Z_\tau$ of the quotient algebra $\operatorname{Com} F/Z_\tau$.

Finally, recall that a function $f: Y \to \mathbb{R}$ given on a topological space Y is called *upper semi-continuous* at $y_0 \in Y$, if for each $\varepsilon > 0$ there is a neighborhood $U_\varepsilon \subset Y$ of y_0 such that $f(y) < f(y_0) + \varepsilon$ whenever $y \in U_\varepsilon$. The function f is said to be upper semi-continuous on Y if it is upper semi-continuous at each $y \in Y$. Equivalently, f is upper semicontinuous on Y if and only if $\{y \in Y : f(y) < \alpha\}$ is an open subset of Y for every $\alpha \in \mathbb{R}$. Notice that if Y is a compact Hausdorff space and $f: Y \to \mathbb{R}$ is a bounded upper semi-continuous function on Y, then there is a $y_0 \in Y$ such that $f(y_0) = \sup_{y \in Y} f(y)$.

1.18. Lemma. (a) *Let M be a localizing class, let $a, a_0 \in A$, and suppose a and a_0 are M-equivalent from the left (resp. from the right). Then a is M-invertible from the left (resp. from the right) if and only if a_0 is so.*

(b) *Let $\{M_\tau\}_{\tau \in T}$ be a system of localizing classes having property* 1.17(iii), *let $\tau \in T$ and $a \in \operatorname{Com} F$. Then a is M_τ-invertible in* $\operatorname{Com} F$ *from the left (resp. right) if and only if a^τ is invertible in* $\operatorname{Com} F/Z_\tau$ *from the left (resp. right).*

Proof. (a) Let a be M-invertible from the left. Then there are $b \in A$ and $f \in M$ such that $baf = f$. Since a and a_0 are M-equivalent from the left, there is a $g \in M$ such that $\|(a - a_0) g\| < 1/\|b\|$. Choose $h \in M$ so that $fh = gh = h$. Then

$$ba_0h = bah - b(a - a_0) h = bafh - b(a - a_0) gh = h - uh = (e - u) h,$$

where $u := b(a - a_0) g$, and because $\|u\| < 1$, we deduce that $e - u \in GA$. Thus, if we let $v = (e - u)^{-1} b$, then $va_0h = h$, which shows that a_0 is M-invertible from the left.

(b) Let a^τ be invertible in $\operatorname{Com} F/Z_\tau$ from the left. Then there is a $b \in \operatorname{Com} F$ such that $ba - e \in Z_\tau$. This implies that ba is M_τ-equivalent from the left to e, and from part (a) we deduce that ba and thus a is M_τ-invertible from the left. It can be shown similarly that a is M_τ-invertible from the right in case a^τ is invertible from the right. Conversely, if there are $b \in \operatorname{Com} F$ and $f \in M_\tau$ such that $baf = f$, then $(ba - e) f = 0$, hence $ba - e \in Z_\tau$, and thus $b^\tau a^\tau = e$. It can be shown similarly that a^τ is invertible from the right in case a is M_τ-invertible from the right. ■

The following theorem is the local principle of Gohberg and Krupnik.

1.19. Theorem. *Let A be a Banach algebra with identity, let $\{M_\tau\}_{\tau \in T}$ be a covering system of localizing classes, and let $a \in$* $\operatorname{Com} F$.

(a) *Suppose that, for each $\tau \in T$, a is M_τ-equivalent from the left (resp. right) to $a_\tau \in A$. Then a is left-invertible (resp. right-invertible) in A if and only if a_τ is M_τ-invertible from the left (resp. right) for all $\tau \in T$.*

(b) *Suppose the system* $\{M_\tau\}_{\tau\in T}$ *has property* 1.17(iii). *Then a is invertible* (*in* Com F) *from the left* (*resp. right*) *if* a^τ *is invertible from the left* (*resp. right*) *for all* $\tau \in T$. *Furthermore, if a is invertible* (*in* A) *from the left* (*resp. right*) *and one of its left inverses* (*resp. right inverses*) *belongs to* Com F *then* a^τ *is invertible from the left* (*resp. right*) *for all* $\tau \in T$. *The element a belongs to* GA *if and only if* $a^\tau \in G(\mathrm{Com}\, F/Z_\tau)$ *for all* $\tau \in T$.

(c) *If the system* $\{M_\tau\}_{\tau\in T}$ *is overlapping, then the mapping*

$$T \to \mathbb{R}^+, \qquad \tau \mapsto \|a^\tau\|$$

is upper semi-continuous.

Proof. (a) If a is left-invertible, then a is M_τ-invertible from the left for all $\tau \in T$ and hence, by Lemma 1.18(a), a_τ is M_τ-invertible from the left for all $\tau \in T$.

Conversely, suppose a_τ is M_τ-invertible from the left for all $\tau \in T$. It follows again from Lemma 1.18(a) that a is M_τ-invertible from the left for all $\tau \in T$. Thus, there are $b_\tau \in A$ and $f_\tau \in M_\tau$ such that $b_\tau a f_\tau = f_\tau$. Since $\{M_\tau\}_{\tau\in T}$ is covering, we can choose $f_{\tau_1}, \ldots, f_{\tau_m}$ so that $\sum_{i=1}^m f_{\tau_i}$ is in GA. Put

$$s = \sum_{i=1}^m b_{\tau_i} f_{\tau_i}.$$

Then

$$sa = \sum b_{\tau_i} f_{\tau_i} a = \sum b_{\tau_i} a f_{\tau_i} = \sum f_{\tau_i}$$

and it results that $(\sum f_{\tau_i})^{-1}\, s$ is a left-inverse of a.

(b) If a^τ is invertible from the left for all $\tau \in T$, then a is invertible in Com F and A from the left by virtue of Lemma 1.18(b) and part (a) of the present theorem. On the other hand, if a is invertible from the left and one of its left inverses belongs to Com F then it is obvious that a^τ is invertible from the left. If $a \in GA$ then clearly $a \in G(\mathrm{Com}\, F)$.

(c) Let $\tau_0 \in T$ and $\varepsilon > 0$. Choose a $z \in Z_{\tau_0}$ so that $\|a + z\| < \|a^{\tau_0}\| + \varepsilon/2$. Since z is M_{τ_0}-equivalent to zero from the left, there is an $f \in M_{\tau_0}$ such that $\|zf\| < \varepsilon/2$. From 1.17(iv) we deduce that $f \in M_\tau$ for all τ in some open neighborhood $U(\tau_0)$ of τ_0. Put $y = z - zf$. If $\tau \in U(\tau_0)$, then there exists a $g \in M_\tau$ such that $fg = g$ (1.17(ii)). Consequently, $yg = zg - zfg = zg - zg = 0$, and since $y \in \mathrm{Com}\, F$ (1.17(v)), it follows that $y \in Z_\tau$ for all $\tau \in U(\tau_0)$. Hence, $\|a^\tau\| \le \|a + y\|$ for $\tau \in U(\tau_0)$. Thus, if $\tau \in U(\tau_0)$, then

$$\|a^\tau\| - \|a^{\tau_0}\| < \|a + y\| - \|a + z\| + \varepsilon/2 \le \|y - z\| + \varepsilon/2$$

$$= \|zf\| + \varepsilon/2 < \varepsilon,$$

which proves the upper semi-continuity of $\tau \mapsto \|a^\tau\|$ at τ_0. ■

Now we are going to prove the local principle of Douglas mainly using Theorem 1.19.

1.20. Definitions. Let A be a Banach algebra with identity e. The *center* Cen A of A is the set of all elements $z \in A$ with the property that $za = az$ for all $a \in A$. Clearly, Cen A is a closed commutative subalgebra of A. Let B be a closed subalgebra of Cen A containing e. Thus, B is commutative, too. If $N \subset B$ is a maximal ideal of B, then J_N will denote the smallest closed two-sided ideal of A containing N, that is, $J_N = \mathrm{clos}_A \left\{\sum_{k=1}^m x_k a_k : m \in \mathbb{N}, x_k \in N, a_k \in A\right\}$.

Of course, it may happen that $J_N = A$. In case all underlying algebras are C^*-algebras, this situation cannot occur (this will be proved later). In case $J_N \neq A$ we denote by A_N the quotient algebra A/J_N and, for $a \in A$, by a_N the coset $a + J_N$.

1.21. Theorem. *Let A be a C^*-algebra with identity e and $B \subset$ Cen A be a C^*-subalgebra containing e.*

(a) *If $a \in A$, then a is left (right, resp. two-sided) invertible in A if and only if a_N is left (right, resp. two-sided) invertible in A_N for all $N \in \mathbf{M}(B)$.*

(b) *The mapping*

$$\mathbf{M}(B) \to \mathbb{R}^+, \qquad N \mapsto \|a_N\|$$

is upper semi-continuous.

(c) $\bigcap\limits_{N \in \mathbf{M}(B)} J_N = \{0\}$.

(d) $\|a\| = \max\limits_{N \in \mathbf{M}(B)} \|a_N\|$.

Proof. (a) and (b): By the Gelfand-Naimark theorem (1.13(a)) the C^*-algebra B is isometrically star-isomorphic to $\mathbf{C}(\mathbf{M}(B))$ under the Gelfand transform Γ. For $N_0 \in \mathbf{M}(B)$ let L_{N_0} be the collection of all elements b of B such that $(\Gamma b)\,(N) = 1$ for all N in a certain neighborhood of N_0 and Γb vanishes outside another neighborhood of N_0. Clearly, the system $\{L_N\}_{N \in \mathbf{M}(B)}$ forms a covering and overlapping system of localizing classes and Theorem 1.19 applies to any element of A. Obviously, the assertions (a) and (b) are proved as soon we have shown that the ideals Z_N defined in 1.17 coincide with the ideals J_N. Let $c \in Z_N$. Then there exists a sequence $\{f_n\} \subset L_N$ such that $\|f_n c\| \to 0$ as $n \to \infty$. Write $c = f_n c + (e - f_n)\, c$. Since $(e - f_n)\, c$ belongs to J_N the element c can be approximated in the norm by elements from J_N. Hence, $Z_N \subset J_N$. Conversely, let $c \in J_N$. Then c can be approximated in the norm by elements of the form $\sum\limits_{i=1}^{m} x_i a_i$, where $x_i \in B$, $a_i \in A$ and x_i vanishes on a certain neighborhood of N, $i = 1, \ldots, m$ (see Section 1.14). Obviously, we can find an element $f \in L_N$ such that $f\left(\sum\limits_{i=1}^{m} x_i a_i\right) = \sum\limits_{i=1}^{m} (f x_i)\, a_i = 0$. This proves that $\sum\limits_i x_i a_i \in Z_N$. Since Z_N is closed we get $c \in Z_N$ and $J_N \subset Z_N$. Since $e \notin Z_N$ we also have $e \notin J_N$.

(c) and (d): Since A is a C^*-algebra, (d) follows immediately for selfadjoint elements from part (a) of the present theorem. Since aa^* and a^*a are selfadjoint, (d) holds for these element and therefore for any $a \in A$. Now suppose $b \in \bigcap\limits_{N \in \mathbf{M}(B)} J_N$. Then $b_N = 0$ for all $N \in \mathbf{M}(B)$. By part (d) we have $b = 0$ and so we are done. ■

Convergence manifold and error estimates

1.22. Definitions and notation. The functional analysis of approximation methods for solving operator equations is a large field. It cannot be the aim of what follows to give a complete record of the methods and ideas which are of interest for this topic. We shall only point out some basic facts important for the numerical analysis of such operator equations considered in the following chapters. More special topics are also included in Chapter 4 (concerning mainly the finite section method and the so-called $\mathcal{P}$-convergence).

Let X and Y be Banach spaces and let $\{X_n\}_{n=0}^{\infty}$ and $\{Y_n\}_{n=0}^{\infty}$ be sequences of closed subspaces of X and Y, respectively. Suppose further that there is a sequence of (perhaps unbounded) projections $\{P_n\}_{n=0}^{\infty}$ with domain in Y and image space Y_n, that is $\mathcal{D}(P_n) \subset Y$, $P_n^2 = P_n$, and $\operatorname{im} P_n = Y_n$ for all n.

In order to solve approximately the equation

$$Ax = y \qquad (x \in X, y \in Y)$$

with $A \in \mathcal{L}(X, Y)$ the following notion is often very useful. Assume we are given a sequence $\{A_n\}$ of linear bounded operators acting from X_n into Y_n which are in a certain sense close to A and invertible for n large enough. If $y \in \bigcup_{n=0}^{\infty} \bigcap_{k=n}^{\infty} \mathcal{D}(P_k)$ then for sufficiently large n the equations $A_n x_n = P_n y$ are uniquely solvable and one may ask whether the sequence $\{x_n\}$ converges to a solution x of the equation $Ax = y$. By $\mathfrak{K}(A; \{A_n, P_n\})$ we denote the set of all $y \in Y$ such that

(i) $y \in \bigcup_{n=0}^{\infty} \bigcap_{k=n}^{\infty} \mathcal{D}(P_k)$,

(ii) the sequence of elements $x_n = A_n^{-1} P_n y$ converges in the norm of X to an element $x \in X$ as $n \to \infty$,

(iii) $Ax = y$.

The set $\mathfrak{K}(A; \{A_n, P_n\})$ is called *convergence manifold*. If A is invertible then the set $\mathfrak{K}(A; \{A_n, P_n\})$ is the maximal linear set on which the operators $A_n^{-1} P_n$ converge strongly to A^{-1}. If $P_n A|_{X_n} \in \mathcal{L}(X_n, Y_n)$ and $A_n := P_n A|_{X_n}$ then the underlying method for the approximate solution of the equation is usually called a *projection method*. The above operator sequence $\{A_n\}$ is said to be *stable* if the operators A_n are invertible for all n large enough and if there is a positive number C independent of n such that

$$\|A_n x_n\|_Y \geqq C \|x_n\|_X$$

for all $x_n \in X_n$.

1.23. Proposition. *Let the following conditions be fulfilled.*

(i) *The sequence $\{A_n\}$ is stable,*

(ii) $\|x_n - x\|_X \to 0$, $x_n \in X_n$, *implies* $\|A_n x_n - Ax\|_Y \to 0$ *as* $n \to \infty$,

(iii) *for each $x \in X$ there exists a sequence $\{x_n\}$ ($x_n \in X_n$) such that $\|x - x_n\|_X \to 0$ as $n \to \infty$.*

Then

(a) *The operator A is of regular type.*

(b) $\mathfrak{K}(A; \{A_n, P_n\}) = \Sigma_Y$,

where Σ_Y is the linear set of all elements $y \in \bigcup_{n=0}^{\infty} \bigcap_{k=n}^{\infty} \mathcal{D}(P_k)$ such that $\|P_n y - y\|_Y \to 0$ as $n \to \infty$.

(c) $\|x - x_n\| \leqq \inf_{v \in X_n} (\|x - v\| + \|A_n^{-1}\| \|A_n v - P_n y\|)$.

Proof. (a): For given $x \in X$ there is a sequence $\{x_n\}$, $x_n \in X_n$, such that $\|x - x_n\| \to 0$ as $n \to \infty$ (by (iii)). Passing to the limit we get from $\|A_n x_n\| \geqq C \|x_n\|$ the inequality $\|Ax\| \geqq C \|x\|$ for all $x \in X$.

(b): Let $y \in \Sigma_Y$, $Ax = y$ and $A_n x_n = P_n y$. We have (n large enough)

$$\|x - x_n\| \leqq \inf_{v \in X_n} (\|x - v\| + \|v - x_n\|) \leqq \inf_{v \in X_n} (\|x - v\| + \|A_n^{-1}\| \, \|A_n v - P_n y\|)$$
$$\leqq \inf_{v \in X_n} \left(\|x - v\| + \|A_n^{-1}\| \, (\|A_n v - Ax\| + \|y - P_n y\|)\right). \tag{1}$$

Condition (iii) gives that there is a sequence $\{v_n\}$, $v_n \in X_n$, such that $\|v_n - x\| \to 0$ as $n \to \infty$. Therefore, by condition (ii), $\|A_n v_n - Ax\| \to 0$ as $n \to \infty$. Now it is immediate that the right-hand side of (1) converges to zero as $n \to \infty$, and so we have proved that $\Sigma_Y \subset \mathfrak{K}(A; \{A_n, P_n\})$. Now suppose $y \in \mathfrak{K}(A; \{A_n, P_n\})$, $x_n = A_n^{-1} P_n y$ for n large enough. Then there is an $x \in X$ so that $\|x_n - x\| \to 0$ and $Ax = y$.

Hence, by (ii) we have $\|P_n y - y\| = \|A_n x_n - Ax\| \to 0$ as $n \to \infty$ and $\mathfrak{K}(A; \{A_n, P_n\}) \subset \Sigma_Y$.

(c): Immediate from (1). ∎

Now we are going to investigate the relations between the convergence manifold of the operators A and $A + T$, where T is compact or of small norm. Furthermore, we state estimates for the error $\|x - x_n\|$. To begin with we establish a result of rather technical character.

1.24. Lemma. *Let $B, B + S \in \mathcal{L}(X, Y)$ be invertible operators,* im $S \subset \mathfrak{K}(B; \{B_n, P_n\})$, *and suppose the operators $B_n, B_n + S_n \in \mathcal{L}(X_n, Y_n)$ to be invertible for n large enough (say for $n \geqq n_0$). For $y \in \mathfrak{K}(B; \{B_n, P_n\})$ put $x = (B + S)^{-1} y$ and $x_n = (B_n + S_n)^{-1} P_n y$. Then we have ($n \geqq n_0$)*

$$\|x - x_n\| \leqq \left(1 + \|(I_{X_n} + B_n^{-1} S_n)^{-1}\| \, \|B_n^{-1} P_n S\|\right) \|x - B_n^{-1} P_n B x\|$$
$$+ \|(I_{X_n} + B_n^{-1} S_n)^{-1}\| \, \|B_n^{-1}(P_n S - S_n) \, B_n^{-1} P_n B x\|. \tag{1}$$

Proof. Clearly,

$$\|x - x_n\| \leqq \|x - B_n^{-1} P_n B x\| + \|B_n^{-1} P_n B x - x_n\|$$
$$\leqq \|x - B_n^{-1} P_n B x\| + \|(I_{X_n} + B_n^{-1} S_n)^{-1}\| \, \|B_n^{-1} S_n B_n^{-1} P_n B x - B_n^{-1} P_n S x\|$$
$$\leqq \|x - B_n^{-1} P_n B x\| + \|(I_{X_n} + B_n^{-1} S_n)^{-1}\| \, \{\|B_n^{-1}(P_n S B_n^{-1} P_n B x - P_n S x)\|$$
$$+ \|B_n^{-1}(S_n - P_n S) \, B_n^{-1} P_n B x\|\},$$

and (1) is now obvious. ∎

1.25. Theorem. *Let $A, A + T \in \mathcal{L}(X, Y)$ be invertible operators such that*

(a) im $T \subset \mathfrak{K}(A; \{A_n, P_n\})$, $P_n T \in \mathcal{L}(X, Y)$ ($n \geqq n_0$),

(b) $\sup_{n \geqq n_0} (\delta_n + \varepsilon_n) < \delta$, *where*

$$\delta_n := \|A^{-1} T - A_n^{-1} P_n T\|, \qquad \varepsilon_n := \|A_n^{-1}(P_n T - T_n)\|_{\mathcal{L}(X_n)},$$
$$\delta := \left(\|(I + A^{-1} T)^{-1}\|\right)^{-1}.$$

(c) $A_n + T_n \in \Phi(X_n, Y_n)$ *and* ind $(A_n + T_n) = 0$ ($n \geqq n_0$).

Then $A_n + T_n$ is invertible for all sufficiently large n. If, in addition,

(d) $\varepsilon_n \to 0$ as $n \to \infty$,

then

$$\mathfrak{K}(A + T; \{A_n + T_n, P_n\}) = \mathfrak{K}(A; \{A_n, P_n\}),$$

and for $y \in \mathfrak{R}(A; \{A_n, P_n\})$, $(A + T)\, x = y$ *and* $(A_n + T_n)\, x_n = P_n y$ *the following error estimate holds:*

$$\|x - x_n\| \leqq (\delta - \delta_n - \varepsilon_n)^{-1} \{(\delta - \varepsilon_n + \|A^{-1}T\|)\, \|x - A_n^{-1}P_nAx\| + \varepsilon_n\, \|A_n^{-1}P_nAx\|\}. \tag{1}$$

Proof. For any $x_n \in X_n$ we get

$$\begin{aligned}\|x_n\| &\leqq \delta^{-1}\, \|(I + A^{-1}T)\, x_n\| \\ &\leqq \delta^{-1}\big(\|A^{-1}T - A_n^{-1}T_n\|_{\mathcal{L}(X_n,X)}\, \|x_n\| + \|(I_X + A_n^{-1}T_n)\, x_n\|\big) \\ &\leqq \delta^{-1}\big((\delta_n + \varepsilon_n)\, \|x_n\| + \|(I_X + A_n^{-1}T_n)\, x_n\|\big).\end{aligned}$$

What results is

$$\|(I_X + A_n^{-1}T_n)\, x_n\| \geqq (\delta - \delta_n - \varepsilon_n)\, \|x_n\|.$$

This together with (c) yields the invertibility of $I_{X_n} + A_n^{-1}T_n$ and thus that of $A_n + T_n$ for n large enough. Note that

$$\|(I_{X_n} + A_n^{-1}T_n)^{-1}\|_{\mathcal{L}(X_n)} \leqq c_n := (\delta - \delta_n - \varepsilon_n)^{-1}.$$

Therefore Lemma 1.24 applies. Further we have that the norms of $A_n^{-1}P_nT$ are bounded uniformly (by (b)). If $y \in \mathfrak{R}(A; \{A_n, P_n\})$ and $x = (A + T)^{-1}\, y$, then, by (a), Ax is also in $\mathfrak{R}(A; \{A_n, P_n\})$. Hence $\|x - A_n^{-1}P_nAx\| \to 0$ as $n \to \infty$. Finally note that

$$\|A_n^{-1}(P_nT - T_n)\, A_n^{-1}P_nAx\| \leqq \|A_n^{-1}(P_nT - T_n)\|\, \|A_n^{-1}P_nAx\|.$$

Putting these things together we deduce from 1.24(1) that $\|x - x_n\| \to 0$ as $n \to \infty$ and also get the estimate (1). In particular,

$$\mathfrak{R}(A; \{A_n, P_n\}) \subset \mathfrak{R}(A + T; \{A_n + T_n, P_n\}).$$

Conversely, let $y \in \mathfrak{R}(A + T; \{A_n + T_n, P_n\})$ and put $x = A^{-1}y$, $x_n = A_n^{-1}P_ny$. We have by (a) and the part already proved that $(A + T)\, x = y + Tx \in \mathfrak{R}(A + T; (A_n + T_n, P_n\})$. Obviously, Lemma 1.24 applies to the case $B = A + T$, $B_n = A_n + T_n$, $S = -T$, $S_n = -T_n$. By (b) the norms of $A_n^{-1}T_n$ are bounded uniformly. Therefore the norms of $\big(I_X - (A_n + T_n)^{-1}\, T_n\big)^{-1} = I_{X_n} + A_n^{-1}T_n$ are bounded uniformly, too. The inequalities

$$\begin{aligned}\|(A_n + T_n)^{-1}\, P_nT\| &\leqq \|(I_{X_n} + A_n^{-1}T_n)^{-1}\|\, \|A_n^{-1}P_nT\|, \\ \|(A_n + T_n)^{-1}\, (P_nT - T_n)\| &\leqq \|(I_{X_n} + A_n^{-1}T_n)^{-1}\|\, \|A_n^{-1}(P_nT - T)\|\end{aligned}$$

show that the norms $\|(A_n + T_n)^{-1}\, P_nT\|$ are uniformly bounded and $\|(A_n + T_n)^{-1} \times (P_nT - T)\|$ tends to zero. Similar to the first part of the theorem one obtains $\|x - x_n\| \to 0$ as $n \to \infty$. Thus,

$$\mathfrak{R}(A + T; \{A_n + T_n, P_n\}) \subset \mathfrak{R}(A; \{A_n, P_n\}). \blacksquare$$

1.26. Corollary. *Under the conditions of Theorem* 1.25, *the sequence* $\{A_n\}$ *is stable if and only if the sequence* $\{A_n + T_n\}$ *is so.*

Proof. Immediate from the previous proof. ∎

1.27. Remarks. 1°. If $\tilde{Y} \subset \mathfrak{R}(A; \{A_n, P_n\})$ is a Banach space continuously embedded in Y, $T \in \mathcal{K}(X, \tilde{Y})$, and $P_n \in \mathcal{L}(\tilde{Y}, Y)$ then conditions (a) are fulfilled and $\delta_n \to 0$ as $n \to \infty$.

2°. Let $A, A + T \in \mathcal{L}(X, Y)$ be invertible operators such that $A_n := A|_{X_n}$ maps X_n onto Y_n (then, clearly, A_n is invertible and $A_n^{-1} = A^{-1}|_{Y_n}$) and

(a) $\operatorname{im} T \subset \Sigma_Y$, $P_n T \in \mathcal{L}(X, Y)$ $(n \geqq n_0)$,
(b) $\delta_n := \|T - P_n T\| \leqq \delta_0 < \left(\|(I + A^{-1}T)^{-1}\| \, \|A^{-1}\|\right)^{-1}$ $(n \geqq n_0)$,
(c) $\varepsilon_n := \|P_n T - T_n\|_{\mathcal{L}(X_n, Y_n)} \to 0$ as $n \to \infty$,
(d) $A_n + T_n \in \Phi(X_n, Y_n)$ and $\operatorname{ind}(A_n + T_n) = 0$.

Then $\mathfrak{R}(A + T; \{A_n + T_n, P_n\}) = \Sigma_Y$, and for $y \in \Sigma_Y$, $x = (A + T)^{-1} y$ and $x_n = (A_n + T_n)^{-1} P_n y$,

$$\|x - x_n\| \leqq (\delta - \delta_n - \varepsilon_n)^{-1} \{(\delta - \varepsilon_n + \|A^{-1}T\|) \, \|Ax - P_n Ax\| + \varepsilon_n \|P_n Ax\|\} \, \|A^{-1}\|.$$

This follows immediately from Theorem 1.25 and $\mathfrak{R}(A; \{A_n, P_n\}) = \Sigma_Y$.

3°. Suppose now $P_n \in \mathcal{L}(Y)$ for all n and $P_n \to I_Y$ strongly. Put $A_n = P_n A|_{X_n}$, $T_n = P_n T|_{X_n}$. We shall write $A \in \Pi\{P_n\}$ if $\mathfrak{R}(A; \{A_n, P_n\}) = Y$. An immediate consequence of Proposition 1.23, the Banach-Steinhaus Theorem, and Theorem 1.25 is the following assertion:

(a) $A \in \Pi\{P_n\}$ *if and only if the sequence* $\{A_n\}$ *is stable. Moreover, if this condition is fulfilled then, for each* $y \in Y$, *the sequence* $x_n = A_n^{-1} P_n y$ *converges quasioptimally to the solution* $x = A^{-1} y$, *i.e.*

$$\|x - x_n\| \leqq c \inf_{v_n \in X_n} \|x - v_n\|.$$

(b) *Let* $T \in \mathcal{K}(X, Y)$, $A \in \Pi\{P_n\}$, *and let* $A + T$ *be invertible. Then* $A + T \in \Pi\{P_n\}$.
This assertion remains true if $T \in \mathcal{L}(X, Y)$ is an operator of sufficiently small norm.

Our next aim is to illuminate the role of collective compactness in the notion of convergence manifold.

1.28. Proposition. *Suppose*

(i) $\tilde{Y} \subset \mathfrak{R}(A; \{A_n, P_n\})$ *is a Banach space continuously embedded in* Y *and* $P_n \in \mathcal{L}(\tilde{Y}, Y)$.
(ii) *Let* $T, T_n \in \mathcal{L}(X, \tilde{Y})$, $T_n \to T$ *strongly and let* $\{T_n\}$ *be collectively compact. Suppose further* $\operatorname{im} T_n \subset Y_n$ *and denote the restriction of* T_n *on* X_n *also by* T_n.
(iii) *Assume* $A + T$ *to be invertible.*

Then $\mathfrak{R}(A; \{A_n, P_n\}) = \mathfrak{R}(A + T; \{A_n + T_n, P_n\})$. *Moreover, if* $x = (A + T)^{-1} y$ *and* $x_n = (A_n + T_n)^{-1} P_n y$ $(y \in \mathfrak{R}(A; \{A_n, P_n\})$, *then*

$$\|x - x_n\| \leqq C\left(\|x - A_n^{-1} P_n A x\| + \|A_n P_n (T - T_n) A_n^{-1} P_n A x\|\right). \tag{1}$$

Proof. First of all consider $A_n + T_n = A_n(I_{X_n} + A_n^{-1} T_n)$. By our assumptions $A_n^{-1} P_n \in \mathcal{L}(\tilde{Y}, X)$ and $A_n^{-1} P_n \to A$ strongly on $\tilde{Y}$. We now prove the invertibility of $I + A_n^{-1} T_n \in \mathcal{L}(X)$ for n large enough. Put $B_n = I - (I + A^{-1}T)^{-1} A_n^{-1} T_n$. Then $B_n(I + A_n^{-1} T_n) = I - (I + A^{-1}T)^{-1} (A_n^{-1} T_n - A^{-1} T) A_n^{-1} T_n = I - \Delta_n$. Since $A_n^{-1} T_n = A_n^{-1} P_n T_n$ converges strongly to $A^{-1} T$, the collective compactness of $\{T_n\}$ yields $\|\Delta_n\| \to 0$. Hence, for n large enough the operators $I - \Delta_n$ are invertible and $\dim \ker (I + A_n^{-1} T_n) = 0$.

Using that $I + A_n^{-1}T_n \in \Phi(X)$ is of index zero we get the invertibility of $I + A_n^{-1}T_n$ for n large enough (say $n \geqq n_0$), and $(I + A_n^{-1}T_n)^{-1} = (I - \Delta_n)^{-1} B_n$, $\sup_{n \geqq n_0} \|(I - \Delta_n)^{-1} B_n\| < \infty$. Now it is easy to see that, for $n \geqq n_0$, the operators $I_{X_n} + A_n^{-1}T_n \in \mathcal{L}(X_n, Y_n)$ are also invertible and that

$$\sup_{n \geqq n_0} \|(I_{X_n} + A_n^{-1}T_n)^{-1}\| < \infty.$$

Again Lemma 1.24 applies and for $x = (A + T)^{-1} y$, $x_n = (A_n + T_n)^{-1} P_n y$ $\big(y \in \mathfrak{R}(A; \{A_n, P_n\})\big)$ we get

$$\begin{aligned} \|x_n - x\| \leqq &\big(1 + \|(I_{X_n} + A_n^{-1}T_n)^{-1}\| \, \|A_n^{-1}P_nT\|\big) \|x - A_n^{-1}P_nAx\| \\ &+ \|(I_{X_n} + A_n^{-1}T_n)^{-1}\| \, \|A_nP_n(T - T_n) A_n^{-1}P_nAx\|. \end{aligned}$$

Because $Ax \in \mathfrak{R}(A; \{A_n, P_n\})$ and $\text{s-lim}\, T_n = T$, we obtain $\|x_n - x\| \to 0$ (as $n \to \infty$) and (1). Consequently,

$$\mathfrak{R}(A; \{A_n, P_n\}) \subset \mathfrak{R}(A + T; \{A_n + T_n, P_n\}).$$

The reverse inclusion $\mathfrak{R}(A + T; \{A_n + T_n, P_n\}) \subset \mathfrak{R}(A; \{A_n, P_n\})$ is now obvious. ■

An almost immediate consequence of the preceding proof is the following.

1.29. Theorem. *Let $T_n, T \in \mathcal{L}(X)$. Assume $T_n \to T$ strongly, and $\{T_n\}$ is collectively compact. Suppose $(I - T)^{-1}$ exists and define*

$$\Delta_n = (I - T)^{-1} (T_n - T) T_n.$$

Then $\|\Delta_n\| \to 0$. If $\|\Delta_n\| < 1$, then $(I - T_n)^{-1}$ exists,

$$\|(I - T_n)^{-1}\| \leqq \frac{1 + \|(I - T)^{-1}\| \, \|T_n\|}{1 - \|\Delta_n\|}$$

and

$$\|x_n - x\| \leqq \frac{\|(I - T)^{-1}\| \, \|T_ny - Ty\| + \|\Delta_n\| \, \|x\|}{1 - \|\Delta_n\|} \to 0, \tag{1}$$

where $x = (I - T)^{-1} y$, $x_n = (I - T_n)^{-1} y$.

Proof. We need only to prove the error estimate (1). But since $\|x_n - x\| = \|(I - T_n)^{-1}y - (I - T)^{-1} y\|$ and

$$\begin{aligned} (I - T_n)^{-1} - (I - T)^{-1} &= (I - \Delta_n)^{-1} \big(I + (I - T)^{-1} T_n\big) - (I - T)^{-1} \\ &= (I - \Delta_n)^{-1} [I + (I - T)^{-1} T_n - (I - T)^{-1} \\ &\quad + \Delta_n(I - T)^{-1}] \\ &= (I - \Delta_n)^{-1} [(I - T)^{-1} (T_n - T) + \Delta_n(I - T)^{-1}], \end{aligned}$$

we are done. ■

Stability in the sense of Mikhlin

1.30. Notation. The computation of both the operator $A_n \in \mathcal{L}(X_n, Y_n)$ and the right-hand side P_ny in the approximating equation

$$A_nx_n = P_ny \tag{1}$$

is, in general, accompanied with errors (e.g. rounding errors in the computation of scalar products, integrals etc.). So what we solve actually is not the equation (1), but a certain perturbed equation of the form

$$(A_n + T_n)\,\tilde{x}_n = P_n y + y_n. \tag{2}$$

Here one requires of course that $\tilde{x}_n \in X_n$, $y_n \in Y_n$, and $T_n \in \mathcal{L}(X_n, Y_n)$. However, the approximation method under consideration is stable in the following sense.

1.31. Theorem. *Let $A \in \mathcal{L}(X, Y)$ be an invertible operator and Z a Banach space continuously embedded in Y such that $Y_n \subset Z \subset \mathfrak{R}(A; \{A_n, P_n\})$. Furthermore, suppose $P_n|_Z \in \mathcal{L}(Z, Y)$, $A_n + T_n \in \Phi(X_n, Y_n)$ with* ind $(A_n + T_n) = 0$, *and let $y \in Z$.*

Then there exist positive constants p, q, γ which do not depend on n and y such that for $\|A_n^{-1}T_n\|_{\mathcal{L}(X_n, X)} < \gamma$ the following holds:

(i) *The equation* 1.30 (2) *has a unique solution for every $y_n \in Y_n$ $(n \geqq n_0)$.*

(ii) $\|\tilde{x}_n - x_n\| \leqq p\,\|y\|_Z\,\|A_n^{-1}T_n\| + q\,\|y_n\|_Z$.

Proof. The existence of a positive constant γ such that (i) holds is an immediate consequence of Theorem 1.25 (put $T = 0$ there). Further

$$\begin{aligned}\|\tilde{x}_n - x_n\| &\leqq \|(I_{X_n} + A_n^{-1}T_n)^{-1}\|\,\|x_n + A_n^{-1}T_n x_n - (I_{X_n} + A_n^{-1}T_n)\,\tilde{x}_n\| \\ &\leqq \|(I_{X_n} + A_n^{-1}T_n)^{-1}\|\,\|A_n^{-1}T_n x_n - A_n^{-1} y_n\|.\end{aligned}$$

Recall that the norms of $(I_{X_n} + A_n^{-1}T_n)^{-1}$ are uniformly bounded (see the proof of Theorem 1.25). Using that $x_n = A_n^{-1}P_n y$ we get, by our assumptions,

$$\|\tilde{x}_n - x_n\| \leqq p\,\|y\|_Z\,\|A_n^{-1}T_n\| + q\,\|y_n\|_Z. \quad \blacksquare$$

The Aubin-Nitsche lemma

The Aubin-Nitsche lemma is often referred to as "Nitsche's trick" when one is using finite element methods to solve boundary value problems for partial differential equations. Application of this lemma in appropriate Sobolev spaces gives an optimal convergence rate and the so-called superapproximation of the approximate solution to the Galerkin equations (Nitsche [1]; see also Ciarlet [1, p. 137]). This technique can also be applied to a wide class of integral equations, more precisely, to all strongly elliptic pseudodifferential equations (see Hsiao and Wendland [2]).

In this section we shall prove a certain abstract version of the Aubin-Nitsche lemma for operator equations in general Hilbert scales.

1.32. Definition. A family of abstract Hilbert spaces H^s $(s \in \mathbb{R},\ s \geqq 0)$ with scalar products $(\cdot, \cdot)_s$ and norms $\|u\|_s = \sqrt{(u, u)_s}$ $(u \in H^s)$ is called a *Hilbert scale* if it owns the following property:

(i) $H^s \subset H^t$ for $s > t$, i.e., H^s is a dense subset of H^t and $\|u\|_t \leqq \|u\|_s$ for each $u \in H^s$.

If we identify H^0 with its dual space, then H^0 may be identified with a subspace of the dual space $H^{-t} := (H^t)^*$, $t > 0$.

Indeed, each element $f \in H^0$ can be viewed as a functional on H^t. In fact, for any $u \in H^t$ we have

$$|(u, f)_0| \leqq \|f\|_0 \|u\|_t .$$

Since H^t is dense in H^0, the values of the functional $(u, f)_0$ $(u \in H^t)$ define the corresponding element $f \in H^0$ uniquely.

Hence, $H^0 \subset H^{-t}$ and the scalar product $(\cdot, \cdot)_0$ can be prolonged by

$$(u, f)_0 := f(u) \; \forall f \in H^{-t} \; \forall u \in H^t .$$

In particular,

$$\|u\|_t = \sup_{\|v\|_{-t}=1} |(u, v)_0| \; (u \in H^t, v \in H^{-t}) . \tag{1}$$

Moreover, the embedding $H^s \subset H^t$ holds for all $s, t \in \mathbb{R}$, $s > t$.

If $\{H^s\}$ $(s \in \mathbb{R})$ is a Hilbert scale and $A \in \mathcal{L}(H^s, H^t)$, then there exists the 0-adjoint operator $A^* \in \mathcal{L}(H^{-t}, H^{-s})$, which is connected with A by

$$(Au, v)_0 = (u, A^*v)_0 \; (u \in H^s, v \in H^{-t}) .$$

Let D be an unbounded positive operator with domain $\mathcal{D}(D)$ in the Hilbert space $\tilde{H}^0$ such that $\|u\|_{\tilde{H}^0} \leqq \|Du\|_{\tilde{H}^0}$ $\big(u \in \mathcal{D}(D)\big)$. Then $\mathcal{D}(D^s)$ $(s > 0)$ forms a Hilbert space H^s with scalar product $(u, v)_s := (D^s u, D^s v)_{\tilde{H}^0}$. If, for $s < 0$, H^s denotes the closure of $\tilde{H}^0$ with respect to the scalar product $(f, g)_s := (D^s f, D^s g)_{\tilde{H}^0}$, then the family $\{H^s\}$ $(s \in \mathbb{R})$ is a Hilbert scale in the sense of Definition 1.32 with $H^0 = \tilde{H}^0$.

For any $s, t \in \mathbb{R}$, H^{s-t} can be identified with the antidual of H^{s+t} with respect to the Hermitian form $(u, v)_s$ extended to $u \in H^{s-t}$, $v \in H^{s+t}$, and

$$\|u\|_{s-t} = \sup_{0 \neq v \in H^{s+t}} |(u, v)_s| / \|v\|_{s+t} . \tag{1'}$$

Moreover, $\{H^s\}$ possesses the *interpolation property*, that is, if $A \in \mathcal{L}(H^{s_k}, H^{t_k})$ and $\|A\|_{H^{s_k} \to H^{t_k}} \leqq C_k$, $k = 1, 2$ $(s_1 \leqq s_2, t_1 \leqq t_2)$, then $A \in \mathcal{L}(H^s, H^t)$ and

$$\|A\|_{H^s \to H^t} \leqq C_1^{1-\lambda} C_2^\lambda \tag{2}$$

for $s = (1 - \lambda) s_1 + \lambda s_2$, $t = (1 - \lambda) t_1 + \lambda t_2$ and $0 \leqq \lambda \leqq 1$.

Note that the Sobolev spaces $H^s(\Gamma)$ $(s \in \mathbb{R})$ form a Hilbert scale (cf. Chapter 2).

For a proof of these results we refer the reader to Berezanski [1].

Now let $\{H^s\}$ $(s \in \mathbb{R})$ be a Hilbert scale and $\{V_h\}$ $(h > 0)$ be a family of finite dimensional spaces satisfying

$$V_h \subset H^s \text{ for any } s < m , \tag{3}$$

where $m \in \mathbb{R}$ is given. The Aubin-Nitsche lemma hinges on the following properties of V_h.

Approximation property AP(m, l). If $\sigma \leqq s < m \leqq l$ and $s \leqq r \leqq l$, then, for any $u \in H^r$ and any $h > 0$, there exists $\varphi_h \in V_h$ such that

$$\|u - \varphi_h\|_t \leqq Ch^{r-t} \|u\|_r \tag{4}$$

for all $t \in [\sigma, s]$, where $C = C(t)$ denotes a constant independent of u and h. (In the following C and c denote generic constants with possibly different meanings and values at different places.)

Inverse property IP(m). For any real numbers t and s satisfying $t \leqq s < m$, there exists a constant C such that

$$\|\varphi\|_s \leqq Ch^{t-s} \|\varphi\|_t \quad \text{for all } \varphi \in V_h. \tag{5}$$

Let $\alpha \in \mathbb{R}$ and let A be a linear operator with the property that

$$A \in \mathscr{L}(H^{s+\alpha}, H^{s-\alpha}) \quad \text{is bijective for all } s \in \mathbb{R}. \tag{6}$$

The number 2α is said to be the *order* of the operator A. Let $u_h \in V_h$ be the Galerkin approximation to the solution $u \in H^\alpha$ of the equation

$$Au = f. \tag{7}$$

That is, u_h satisfies the *Galerkin equations*

$$(Au_h, v)_0 = (f, v)_0 \quad \text{for all } v \in V_h. \tag{8}$$

From the assumption (3) it follows that the Galerkin equations (8) make sense if

$$\sigma \leqq \alpha < m \quad \text{and} \quad f \in H^r, \qquad r > -m. \tag{9}$$

In the sequel we assume (9) to be fulfilled.

It is easy to see that the Galerkin equation (8) is nothing else than the projection equation

$$P^*_{\alpha,h} A u_h = P^*_{\alpha,h} f, \tag{10}$$

where $P^*_{\alpha,h} \in \mathscr{L}(H^{-\alpha})$ is the 0-adjoint operator to the orthogonal projection $P_{\alpha,h}\colon H^\alpha \to V_h$.

The operator A is called *strongly elliptic*, if A satisfies a *Gårding inequality* of the form

$$\operatorname{Re}(Au, u)_0 \geqq \gamma \|u\|_\alpha^2 - |k[u, u]| \qquad \text{for all } u \in H^\alpha \tag{11}$$

with some constant $\gamma > 0$ and a compact bilinear form k.

1.33. The Céa-Polski Lemma. *Let $A \in \mathscr{L}(H^\alpha, H^{-\alpha})$ be an invertible strongly elliptic operator. Then the Galerkin method* 1.32(8) *is stable in H^α, i.e. the Galerkin equations* 1.32(8) *admit a unique solution $u_h \in V_h$ for all h sufficiently small and for each $f \in H^{-\alpha}$. Moreover,*

$$\|u - u_h\|_\alpha \leqq c \inf_{\varphi \in V_h} \|u - \varphi\|_\alpha, \tag{1}$$

where c is independent of h.

This lemma is a simple consequence of Propositions 1.23 and 1.25 and Corollary 1.26. The very important inequality (1) reduces the problem of estimating the error $\|u - u_h\|_\alpha$ to a problem of approximation theory: calculate the distance $d_\alpha(u, V_h) := \inf_{\varphi \in V_h} \|u - \varphi\|_\alpha$. Thus, if the V_h's have the approximation property AP(m, l), then the asymptotic convergence estimate

$$\|u - u_h\|_\alpha \leqq ch^{s-\alpha} \|u\|_s \tag{2}$$

holds for $u \in H^s$ and $\sigma < \alpha \leqq s \leqq l$, where c is independent of u and h.

The inequality (2) gives optimal error estimates in the "energy" norm $\|\cdot\|_\alpha$ under the assumption that $f \in H^{-\alpha}$. To obtain convergence results in spaces H^t for $t < \alpha$ we apply the following Nitsche trick.

1.34. The Aubin-Nitsche Lemma. *Let A satisfy* 1.32(6) *and* 1.33(1). *If the spaces V_h have the approximation property* AP(m, l), *then the Galerkin method* 1.32(8) *has the following super-approximation property:*

$$\|u - u_h\|_t \leqq ch^{s-t} \|u\|_s \tag{1}$$

for $u \in H^s$ with

$$2\alpha - l \leqq t \leqq \alpha \leqq s \leqq l, \tag{2}$$

where c is a constant that does not depend on u and h.

Proof. Let us denote by

$$e := u - u_h$$

the error term from 1.32(7) and 1.32(8). Then

$$(Ae, v)_0 = (e, A^*v)_0 = 0 \quad \text{for all } v \in V_h. \tag{3}$$

From 1.32(6) it follows that the adjoint equation

$$A^*w = g \tag{4}$$

is uniquely solvable for every $g \in H^{-t}$, where $w \in H^{2\alpha-t}$. Moreover, the boundedness of $(A^*)^{-1}$ implies that

$$\|w\|_{2\alpha-t} \leqq c\, \|g\|_{-t}. \tag{5}$$

Thus, we have, with 1.32(1), (4) and (3),

$$\begin{aligned}\|e\|_t &\leqq \sup_{\|g\|_{-t}=1} |(e, g)_0| = \sup |(e, A^*w)_0| \\ &= \sup |(e, A^*(w - v))_0| \\ &\leqq \sup \|e\|_\alpha \|A^*(w - v)\|_{-\alpha} \\ &\leqq c \sup \|e\|_\alpha \|w - v\|_\alpha \quad \text{for every } v \in V_h.\end{aligned}$$

By the approximation property AP(m, l), the estimate

$$\|w - v\|_\alpha \leqq ch^{(2\alpha-t)-\alpha} \|w\|_{2\alpha-t} \tag{6}$$

is valid for $\alpha \leqq 2\alpha - t \leqq l$, $\sigma \leqq \alpha < m$ and, consequently, for all α and t satisfying (2) and 1.32(9). Inserting (6) and 1.33(2) into the above, we find

$$\|e\|_t \leqq c \sup h^{s-\alpha} \|u\|_s \|w\|_{2\alpha-t}\, h^{\alpha-t}.$$

Finally, we use (5) to find the desired estimate

$$\|e\|_t \leqq c \sup_{\|g\|_{-t}=1} h^{s-t} \|u\|_s \|g\|_{-t} = ch^{s-t} \|u\|_s. \ \blacksquare$$

1.35. Corollary. *Let A satisfy* 1.32(6) *and* 1.33(1). *Let V_h have the approximation property* AP(m, l) *and, in addition, satisfy the inverse property* IP(m). *Then, for $u \in H^s$, the Galerkin method* 1.32(8) *converges, and we have*

$$\|u - u_h\|_t \leqq ch^{s-t} \|u\|_s, \tag{1}$$

where

$$2\alpha - l \leqq t \leqq s \leqq l, \qquad t < m, \qquad \alpha \leqq s. \tag{2}$$

Proof. In view of Lemma 1.34, it remains to consider the case $\alpha < t$. From 1.32(4) it follows that

$$\|u - \varphi_h\|_t \leqq c h^{s-t} \|u\|_s . \tag{3}$$

The inverse property 1.32(5) implies that

$$\|\varphi_h - u_h\|_t \leqq C h^{\alpha-t} \|\varphi_h - u_h\|_\alpha . \tag{4}$$

Again using 1.32(4) and 1.33(2), we find

$$\|\varphi_h - u_h\|_\alpha \leqq \|\varphi_h - u\|_\alpha + \|u - u_h\|_\alpha \leqq c h^{s-\alpha} \|u\|_s . \tag{5}$$

Inserting (5) into (4), we get

$$\|\varphi_h - u_h\|_t \leqq c h^{s-t} \|u\|_s . \tag{6}$$

Thus, (1) is an immediate consequence of (3) and (6). ■

Our next concern is the convergence of the Galerkin method for $f \notin H^{-\alpha}$ as well as in spaces H^t for $t > \alpha$.

To this end we assume $\{H^t\}$ to be a Hilbert scale generated by the positive operator D (see Sect. 1.32). Then the orthogonal projection $P_{\alpha,h}\colon H^\alpha \to V_h$ is a bounded operator in H^t for all t satisfying $2\alpha - m < t < m$.

In fact, let $\{\varphi_k\}_{k=1}^n$ $(n = \dim V_h)$ be an orthogonal basis of V_h in H^α. Then we have

$$P_{\alpha,h} u = \sum_{k=1}^n (u, \varphi_k)_\alpha \varphi_k, \qquad u \in H^\alpha .$$

In view of 1.32(3), $\varphi_k \in H^{m'}$, $k = 1, \ldots, n$, for all $m' < m$. Choosing m' such that $2\alpha - m < 2\alpha - m' < t$, it follows from 1.32(1′) that

$$|(u, \varphi_k)_\alpha| \leqq \|u\|_{2\alpha-m'} \|\varphi_k\|_{m'} \leqq \|u\|_t \|\varphi_k\|_{m'}$$

for each $u \in H^t$. Hence,

$$\|P_{\alpha,h} u\|_t \leqq c \|u\|_t, \qquad c = \sum_{k=1}^n \|\varphi_k\|_{m'} \|\varphi_k\|_t, \qquad u \in H^t .$$

Thus, $P_{\alpha,h} \in \mathcal{L}(H^t)$ and, consequently, $P_{\alpha,h}^* \in \mathcal{L}(H^{t-2\alpha})$ for all t satisfying $2\alpha - m < t < m$.

Notice that if V_h has the inverse property IP(m), then $V_h^* := \operatorname{im} P_{\alpha,h}^*$ satisfies the inverse property IP($m - 2\alpha$). Indeed, it follows from the definition of the form $(\cdot, \cdot)_\alpha$ that $P_{\alpha,h}^* = D^{2\alpha} P_{\alpha,h} D^{-2\alpha}$. Using IP($m$), we get that, for $f = P_{\alpha,h}^* g \in V_h^*$ and $t \leqq s < m - 2\alpha$,

$$\begin{aligned} \|f\|_s &= \|D^{2\alpha} P_{\alpha,h} D^{-2\alpha} g\|_s = \|P_{\alpha,h} D^{-2\alpha} g\|_{s+2\alpha} \\ &\leqq c h^{t-s} \|P_{\alpha,h} D^{-2\alpha} g\|_{t+2\alpha} = c h^{t-s} \|f\|_t . \end{aligned}$$

1.36. Corollary. *Let A and V_h satisfy the assumptions of Corollary* 1.35.

(i) *The projections $P_{\alpha,h}$ are uniformly bounded (with respect to h) in H^t for $2\alpha - m < t < m$. Furthermore,*

$$\|u - P_{\alpha,h} u\|_t \leqq c h^{s-t} \|u\|_s, \qquad u \in H^s, \tag{1a}$$

for all $s, t \in \mathbb{R}$ satisfying

$$2\alpha - l \leqq t < m, \qquad \sigma \leqq t \leqq s, \qquad 2\alpha - m < s \leqq l. \tag{2}$$

(ii) *The Galerkin method* 1.32(8) *is stable in H^t for $2\alpha - l \leqq t \leqq m$. Moreover, we have the estimate*

$$\|u - u_h\|_t \leqq ch^{s-t} \|u\|_s \tag{1b}$$

provided $u \in H^s$ and (2) *holds.*

Note that (2) implies that $f = Au \in H^r$, where $-m < r < m$.

Proof. (i) Let $2\alpha - m < t < m$ and $u \in H^t$. We first consider the case $\alpha < t$. In view of the property AP(m, l), there exists an element $\varphi_h \in V_h$ such that

$$\|u - \varphi_h\|_t \leqq c \|u\|_t, \qquad \|u - \varphi_h\|_\alpha \leqq ch^{t-\alpha} \|u\|_t.$$

Applying the inverse property IP(m), we find that

$$\begin{aligned}\|u - P_{\alpha,h}u\|_t &\leqq \|u - \varphi_h\|_t + \|\varphi_h - P_{\alpha,h}u\|_t \\ &\leqq c\big(\|u\|_t + h^{\alpha-t}(\|u - \varphi_h\|_\alpha + \|u - P_{\alpha,h}u\|_\alpha)\big) \\ &\leqq c \|u\|_t. \text{ Thus, } \|P_{\alpha,h}u\|_t \leqq c \|u\|_t.\end{aligned}$$

If $2\alpha - m < t < \alpha$, then we choose $w \in H^{2\alpha-t}$ such that $\|w\|_{2\alpha-t} = 1$ and (cf. 1.32(1'))

$$\begin{aligned}\|P_{\alpha,h}u\|_t &= |(P_{\alpha,h}u, w)_\alpha| = |(u, P_{\alpha,h}w)_\alpha| \\ &\leqq \|u\|_t \|P_{\alpha,h}w\|_{2\alpha-t} \leqq c \|u\|_t.\end{aligned}$$

Hence, the projections $P_{\alpha,h}$ are uniformly bounded in H^t for $2\alpha - m < t < m$. Consequently,

$$\begin{aligned}\|u - P_{\alpha,h}u\|_t &\leqq (1 + \|P_{\alpha,h}\|_t) \|u - P_{t,h}u\|_t \\ &\leqq c \|u - P_{t,h}u\|_t.\end{aligned}$$

This together with (1b), applied to $A = D^{2\alpha}$, yields (1a).
(ii) Let us first prove the stability in H^t for $2\alpha - l \leqq t \leqq \alpha$. Put $A_h := P^*_{\alpha,h}A|_{V_h}$ and fix an arbitrary $v_h \in V_h$. Choosing $f = f_h := A_h v_h$, we obtain from 1.34(1) that

$$\|A^{-1}f_h - v_h\|_t \leqq ch^{\alpha-t} \|A^{-1}f_h\|_\alpha \leqq ch^{\alpha-t} \|f_h\|_{-\alpha}. \tag{3}$$

Applying IP$(m - 2\alpha)$ to $f_h \in V_h^*$, we find

$$\|f_h\|_{-\alpha} \leqq ch^{t-\alpha} \|f_h\|_{t-2\alpha}. \tag{4}$$

From (3) and (4) it follows that

$$\|v_h\|_t \leqq \|A^{-1}f_h\|_t + \|A^{-1}f_h - v_h\|_t \leqq c \|f_h\|_{t-2\alpha},$$

i.e., we have the stability condition

$$\|v_h\|_t \leqq c \|A_h v_h\|_{t-2\alpha} \quad \text{for each} \quad v_h \in V_h. \tag{5}$$

If $\alpha < t < m$, then (5) is an immediate consequence of Corollary 1.35. By Proposition 1.23, (5) implies the quasioptimal error estimate

$$\|u - u_h\|_t \leqq c \inf_{v_h \in V_h} \|u - v_h\|_t \tag{6}$$

for $2\alpha - m < t < m$ provided $P_{\alpha,h} \in \mathcal{L}(H^t)$, i.e. $P^*_{\alpha,h} \in \mathcal{L}(H^{t-2\alpha})$. From (6) and AP$(m, l)$ the estimate (1b) follows in the case $2\alpha - m < t < m$.

It remains to prove (1b) for $2\alpha - l \leqq t \leqq 2\alpha - m$. To this end we again apply Nitsche's trick in analogy to Section 1.34. Choose an element $w \in H^{2\alpha-t}$ and a number $m_1 < m$ satisfying $\|A^*w\|_{-t} = 1$, $\|e\|_t = (e, A^*w)_0$ and $t < 2\alpha - m_1 < s$, $2\alpha - m_1 \leqq \alpha$. Repeating the argument utilized in the proof of the Aubin-Nitsche Lemma we arrive at the estimate

$$\|e\|_t \leqq c\,\|e\|_{2\alpha-m_1}\,\|w - v\|_{m_1} \tag{7}$$

for every $v \in V_h$. By the approximation property there exists $v \in V_h$ such that

$$\|w - v\|_{m_1} \leqq ch^{(2\alpha-t)-m_1}\,\|w\|_{2\alpha-t}\,. \tag{8}$$

Since $2\alpha - m < 2\alpha - m_1 < m$, we have

$$\|e\|_{2\alpha-m_1} = \|u - u_h\|_{2\alpha-m_1} \leqq ch^{s-(2\alpha-m_1)}\,\|u\|_s \tag{9}$$

for arbitrary s satisfying (2). Inserting (8) and (9) in (7), we obtain the desired estimate (1b). ■

The following theorem on "shifting" the stability of a sequence of approximation operators in the Hilbert scale H^s is crucial for the convergence analysis of approximation methods in some of the subsequent chapters.

1.37. Theorem. *Let V_h satisfy the inverse property* IP(m). *Suppose the linear operator A is bijective in H^s and in H^t as well, where $t < s < m$. Moreover, assume*

$$\|Au - A_hP_hu\|_t \leqq Ch^{s-t}\,\|u\|_s\,, \qquad \forall\, u \in H^s, \tag{1}$$

where $C > 0$ is a constant independent of u and h, $A_h \in \mathcal{L}(V_h)$ and $P_h\colon H^t \to V_h$ is a projection satisfying $P_h \in \mathcal{L}(H^t) \cap \mathcal{L}(H^s)$. Then $\{A_h\}$ is stable in H^t if and only if it is stable in H^s.

Proof. Let us first assume $\{A_h\}$ to be stable in H^t. Then, for each $u_h \in V_h$,

$$\|A_h^{-1}u_h\|_s \leqq \|P_h\|_s\,\|A^{-1}\|_s\,\|u_h\|_s + \|A_h^{-1}u_h - P_hA^{-1}u_h\|_s\,. \tag{2}$$

Using IP(m) and the stability of $\{A_h\}$ in H^t, we obtain

$$\|A_h^{-1}u_h - P_hA^{-1}u_h\|_s \leqq Ch^{t-s}\,\|(A - A_hP_h)\,A^{-1}u_h\|_t\,. \tag{3}$$

Thus formulas (1), (2) and (3) yield $\|A_h^{-1}u_h\|_s \leqq C'\,\|u_h\|_s$, i.e., $\{A_h\}$ is stable in H^s.

Conversely, if $\{A_h\}$ is stable in H^s, then, for each $u_h \in V_h$,

$$A_h^{-1}u_h = A^{-1}(A - A_hP_h)\,A_h^{-1}u_h + A^{-1}u_h\,.$$

Hence

$$\|A_h^{-1}u_h\|_t \leqq \|A^{-1}\|_t\,[\|u_h\|_t + \|(A - A_hP_h)\,A_h^{-1}u_h\|_t]\,.$$

We now use (1), the stability of $\{A_h\}$ in H^s and IP(m) to get

$$\begin{aligned}\|(A - A_hP_h)\,A_h^{-1}u_h\|_t &\leqq Ch^{s-t}\,\|A_h^{-1}u_h\|_s\\ &\leqq C'h^{s-t}\,\|u_h\|_s \leqq C''\,\|u_h\|_t\,.\end{aligned}$$

Thus $\|A_h^{-1}u_h\|_t \leqq C'''\,\|u_h\|_t$. ■

Notes and comments

1.1.—1.2. See any book on functional analysis.

1.3. These things are well-known and can be found, for instance, in Mikhlin/Prössdorf [1] or Gohberg/Krupnik [4].

1.4.—1.7. Results of the type of Theorem 1.5 were first obtained by Krupnik (see Kesler/Krupnik [1]). Theorem 1.5 itself was first established by Köhler and Silbermann [1], [2] and independently also by Markus and Feldman [2]. Theorem 1.6 is due to Markus and Feldman [1]. The proofs of these two theorems are also in Chapter 1 of Krupnik [1].

1.8.—1.16. This material, with the exception of Theorem 1.16, is well-known and a major part of it is in almost each book on Banach or C^*-algebras. The evolution of Theorem 1.16 involves many contributors. We here only refer to the original papers of Halmos [1] and Peterson [1]. A proof can also be found in Power [1]. Banach algebras generated by two idempotents are studied in Roch/Silbermann [6].

1.17.—1.21. Here we present the local principles of Gohberg/Krupnik and Douglas. A few historical remarks are in Böttcher/Silbermann [3]. The comparison of several known local principles (theories) should be the subject of further investigations. Steps in this direction have been made by several people (see Böttcher/Silbermann [3]). Also see the paper Böttcher/Krupnik/Silbermann [1] which is closely related to the topic of Sections 1.17.—1.21.

1.22.—1.31. The notion of the convergence manifold was introduced by Gohberg and Levchenko [1]. We here essentially follow Prössdorf/Silbermann [4, 6, 8]. Theorem 1.25 was stated (without proof) in Prössdorf/Silbermann [9] for the first time. Proofs appeared in Junghanns/Silbermann [1] and Prössdorf [4]. Proposition 1.28 is new and due to the authors. The concept of collectively compact operators (including Theorem 1.29) is due to Anselone [1]. The results of these sections also yield the well-known theorems of Kantorovich and Vainikko on the Galerkin method with perturbations for Riesz-Schauder equations (see Krasnoselski/Vainikko et al. [1], Chap. 4, § 17).

1.32.—1.37. Here we follow Prössdorf [10]. These results are generalizations of theorems established by Hsiao and Wendland [2].

Chapter 2
Approximation theory

The theory of the approximation of functions is a cornerstone of numerical analysis and is essential to the development of most algorithms. The aim of approximation theory is to investigate the approximation of functions from certain infinite-dimensional spaces (primarily Sobolev or Hölder-Zygmund spaces) by means of simpler functions, generally coming from a finite-dimensional space. From the point of view of numerical analysis the approximating space should have at least the following three properties. At first, it should be possible to get arbitrarily good approximations to a given (sufficiently smooth) function by making the dimension of the approximating space sufficiently large. Secondly, the elements of the approximating space must be simple, so that they can be easily integrated and differentiated. Finally, there should exist a well-developed theory to facilitate the analysis of the resulting computional procedures. Polynomials and splines (piecewise polynomials) are an ideal choice on all three qualities.

In Part A of this chapter (Sections 2.1 to 2.33) we prove the basic properties of one-dimensional splines (e.g. approximation and inverse properties of periodic splines and the commutator property of (not necessarily periodic) splines). Furthermore, periodic spline interpolation on uniform and arbitrary meshes is studied in detail. Finally, we demonstrate the Arnold-Wendland lemma, which plays a significant role in the analysis of interpolation and collocation procedures. In Part B (Sections 2.34 to 2.47) we investigate the approximation of 2π-periodic functions by Fourier sums, de la Vallée Poussin means, and trigonometric interpolatory polynomials in norms of Hölder-Zygmund type (including Sobolev norms). The constants in all these estimates are calculated explicitly.

Part A. Spline approximation in periodic Sobolev spaces

Periodic Sobolev spaces

2.1. Notation. Let $\mathbf{C}^\infty$ be the set of all 1-periodic infinitely differentiable (complex-valued) functions on the real axis $\mathbb{R}$. By $\mathbf{H}^s$, $s \in \mathbb{R}$, we denote the *periodic Sobolev space* of order s, i.e. the completion of $\mathbf{C}^\infty$ with respect to the norm

$$\|u\|_s := \Big(|\hat{u}_0|^2 + \sum_{0 \neq k \in \mathbb{Z}} |\hat{u}_k|^2 \, |2\pi k|^{2s}\Big)^{1/2}, \tag{1}$$

where $\hat{u}_k = \int_0^1 u(x)\, e^{-2\pi i k x}\, dx$ are the Fourier coefficients of the function u. Obviously, $\mathbf{H}^s$ is a Hilbert space with the scalar product

$$\langle u, v\rangle_s := \hat{u}_0 \bar{\hat{v}}_0 + \sum_{0 \neq k \in \mathbb{Z}} \hat{u}_k \bar{\hat{v}}_k\, |2\pi k|^{2s}. \tag{2}$$

For $\mathbf{L}^2 = \mathbf{H}^0$, we have

$$(u, v) = \langle u, v\rangle_0 = \sum_{k \in \mathbb{Z}} \hat{u}_k \bar{\hat{v}}_k = \int_0^1 u(x)\, \overline{v(x)}\, dx. \tag{3}$$

It follows from the definition that, for $s > t$, $\mathbf{H}^s$ is a dense subset of $\mathbf{H}^t$ and the imbedding $\mathbf{H}^s \subset \mathbf{H}^t$ is continuous and compact. Moreover, $\mathbf{H}^s \subset \mathbf{C}$ if $s > 1/2$, where $\mathbf{C}$ stands for the space of all 1-periodic continuous functions on $\mathbb{R}$. It is well known that, for $s \geqq 0$, the norm (1) is equivalent to the norm $|||u|||_s$, where

$$|||u|||_s^2 = \|u\|_0^2 + \|D^s u\|_0^2 \qquad \text{if } s \in \mathbb{N} \tag{4}$$

and

$$|||u|||_s^2 = \|u\|_0^2 + \int_0^1\int_0^1 \frac{|D^m u(x) - D^m u(y)|^2}{|x - y|^{1+2\delta}}\, dx\, dy \tag{5}$$

if $s = m + \delta$, $m \in \mathbb{N}_0$, $0 < \delta < 1$ (cf. Lions and Magenes [1, Chap. 1.7]). Here $D = D_x$ stands for the differentiation d/dx. For any $s, t \in \mathbb{R}$, $\mathbf{H}^{s-t}$ can be identified with the anti-dual of $\mathbf{H}^{s+t}$ with respect to the Hermitian form (2), extended to $u \in \mathbf{H}^{s-t}$ and $v \in \mathbf{H}^{s+t}$, and one has

$$\|u\|_{s-t} = \sup_{0 \neq v \in \mathbf{H}^{s+t}} |\langle u, v\rangle_s| / \|v\|_{s+t}. \tag{6}$$

Let us introduce the operator

$$\Lambda u(x) := \hat{u}_0 + \sum_{0 \neq k \in \mathbb{Z}} \hat{u}_k\, |2\pi k|\, e^{2\pi i k x}, \qquad u \in \mathbf{C}^\infty.$$

Its closure in $\mathbf{L}^2$, which will be denoted by Λ as well, is a positive definite selfadjoint operator with the domain $\mathcal{D}(\Lambda) = \mathbf{H}^1$. This operator is called the *Bessel potential operator*. The eigenvalues of Λ are $\lambda_0 = 1$, $\lambda_k = |2\pi k|$ ($k \in \mathbb{Z}$, $k \neq 0$) and correspond to the orthonormal system of eigenfunctions $\{e^{2\pi i k x}\}$ ($k \in \mathbb{Z}$). Hence, for all $\alpha \in \mathbb{R}$, the operators Λ^α are defined by

$$\Lambda^\alpha u(x) = \hat{u}_0 + \sum_{0 \neq k \in \mathbb{Z}} \hat{u}_k\, |2\pi k|^\alpha\, e^{2\pi i k x}, \qquad \mathcal{D}(\Lambda^\alpha) = \mathbf{H}^\alpha, \tag{7}$$

and we have $\langle u, v\rangle_s = \langle \Lambda^s u, \Lambda^s v\rangle_0$ for all $s \in \mathbb{R}$ and $u, v \in \mathbf{H}^s$. Moreover, $\Lambda^{s-t}\colon \mathbf{H}^s \to \mathbf{H}^t$ is an isometrical isomorphism, i.e. $\|u\|_s = \|\Lambda^{s-t} u\|_t$ for all $s, t \in \mathbb{R}$ and $u \in \mathbf{H}^s$. Therefore, $\{\mathbf{H}^s : s \in \mathbb{R}\}$ is a Hilbert scale, and for any linear operator A satisfying

$$\|A\|_{\mathcal{L}(\mathbf{H}^{s_j}, \mathbf{H}^{t_j})} \leqq c_j \qquad (j = 1, 2),\ s_1 < s_2,\ t_1 < t_2,$$

one obtains the estimate

$$\|A\|_{\mathcal{L}(\mathbf{H}^s, \mathbf{H}^t)} \leqq c_1^{1-\Theta} c_2^{\Theta}, \tag{8}$$

where $s = (1 - \Theta)\, s_1 + \Theta s_2$, $t = (1 - \Theta)\, t_1 + \Theta t_2$, $0 < \Theta < 1$ (cf. Berezanski [1, Chap. III. 6.9], Triebel [1, Chap. 1. 18.10]).

Let Γ be a simple closed $\mathbf{C}^\infty$ curve in the complex plane $\mathbb{C}$ which is the positively oriented boundary of a bounded domain and has the parametrization $z = \gamma(x)$, $x \in [0, 1]$. Thus $\gamma \in \mathbf{C}^\infty$, $\gamma'(x) \neq 0$, $\gamma(x) \neq \gamma(y)$ for $x \neq y$, $x, y \in [0, 1)$ and $\gamma(0) = \gamma(1)$. In the following we often identify functions $u(z)$ defined on Γ with 1-periodic functions on $\mathbb{R}$:

$$u(x) = (u \circ \gamma)(x) = u(\gamma(x)), \qquad x \in \mathbb{R}. \tag{9}$$

Finally, by $\mathbf{H}^s(\Gamma)$ we denote the closure of $\mathbf{C}^\infty(\Gamma)$ with respect to the norm $\|u \circ \gamma\|_s$, $s \in \mathbb{R}$. The spaces $\mathbf{H}^s(\Gamma)$ and $\mathbf{C}^\infty(\Gamma)$ may be identified with $\mathbf{H}^s$ and $\mathbf{C}^\infty$, respectively, in a natural way.

In what follows we need several discrete Sobolev norms. To this end we consider the n-dimensional space $\mathbb{C}^n$ provided with the usual scalar product

$$(\boldsymbol{x}, \boldsymbol{y}) = \sum_{k=0}^{n-1} x_k \bar{y}_k, \qquad \boldsymbol{x} = (x_k)_{k=0}^{n-1}, \qquad \boldsymbol{y} = (y_k)_{k=0}^{n-1} \in \mathbb{C}^n.$$

Given $\boldsymbol{x} \in \mathbb{C}^n$ and $s \in \mathbb{N}$, let $h = 1/n$ and define

$$|||\boldsymbol{x}|||_{s,h}^2 := h \left(\sum_{k=0}^{n-1} |x_k|^2 + \sum_{k=0}^{n-1} |(\nabla_h^s \boldsymbol{x})_k|^2 \right), \tag{10}$$

where $(\nabla_h^s \boldsymbol{x})_k := h^{-1}\big((\nabla_h^{s-1}\boldsymbol{x})_k - (\nabla_h^{s-1}\boldsymbol{x})_{k-1}\big)$, $\nabla_h^0 \boldsymbol{x} := \boldsymbol{x}$, and $x_{-1} := x_{n-1}$.

2.2. Lemma. *Let $u \in \mathbf{H}^s$, $s \in \mathbb{N}$, and let $\boldsymbol{u} = \big(u(kh)\big)_{k=0}^{n-1}$, $h = 1/n$. There exists a constant c independent of n and u such that*

$$|||\boldsymbol{u}|||_{s,h} \leqq c\, \|u\|_s. \tag{1}$$

Proof. Since $\mathbf{H}^s \subset \mathbf{C}$, we have

$$h \sum_{k=0}^{n-1} |u(kh)|^2 \leqq \max_k |u(kh)|^2 \leqq c\, \|u\|_s^2.$$

For $s = 1$, we obtain

$$h \sum_{k=0}^{n-1} |(\nabla_h \boldsymbol{u})_k|^2 = h^{-1} \sum_{k=0}^{n-1} \left| \int_{(k-1)h}^{kh} u'(x)\, \mathrm{d}x \right|^2 \leqq \sum_{k=0}^{n-1} \int_{(k-1)h}^{kh} |u'(x)|^2\, \mathrm{d}x = \|u'\|_0^2. \tag{2}$$

Hence, (1) holds for $s = 1$.

If $s > 1$, then we set $\partial_h u(x) := h^{-1}\big(u(x) - u(x-h)\big)$ to get

$$\nabla_h^s \boldsymbol{u} = \nabla_h \boldsymbol{v}, \qquad v = \partial_h^{s-1} u. \tag{3}$$

Because $\dfrac{\mathrm{d}}{\mathrm{d}x}\partial_h = \partial_h \dfrac{\mathrm{d}}{\mathrm{d}x}$, it follows from (2) and (3) that

$$h \sum_{k=0}^{n-1} |(\nabla_h^s \boldsymbol{u})_k|^2 \leqq \|v'\|_0^2 = \|\partial_h^{s-1} u'\|_0^2. \tag{4}$$

Furthermore,

$$\int_0^1 |\partial_h u(x)|^2\, \mathrm{d}x = \int_0^1 \left| \int_0^1 u'(x - th)\, \mathrm{d}t \right|^2 \mathrm{d}x$$

$$\leqq \int_0^1 \int_0^1 |u'(x - th)|^2\, \mathrm{d}x\, \mathrm{d}t = \int_0^1 |u'(x)|^2\, \mathrm{d}x.$$

Combining this with (4), we find

$$h \sum_{k=0}^{n-1} |(\nabla_h^s \boldsymbol{u})_k|^2 \leqq \|u^{(s)}\|_0^2 . \blacksquare$$

To define a discrete Sobolev norm for each index $s \in \mathbb{R}$, we proceed as follows. Notice that, for each $\boldsymbol{x} \in \mathbb{C}^n$, we have

$$h^{-1} |||\boldsymbol{x}|||_{1,h}^2 = \sum_{k=0}^{h-1} \{(1 + 2h^{-2}) |x_k|^2 - h^{-2}(x_k \overline{x}_{k-1} + x_k \overline{x}_{k+1})\}$$

$$= (B_n \boldsymbol{x}, \boldsymbol{x}),$$

where the $n \times n$ matrix B_n is a circulant. We recall that a matrix of the form $A_n = (a_{j-k})_{j,k=0}^{n-1}$ is said to be a *circulant*, if

$$a_{-k} = a_{n-k}, \qquad k = 1, \ldots, n-1 .$$

Via simple computations one can verify that the circulant A_n may be written as

$$A_n = U_n M_n U_n^{-1}, \tag{5}$$

where $U_n = h^{1/2}(\mathrm{e}^{2\pi \mathrm{i} jkh})_{j,k=0}^{n-1}$ is a unitary matrix,

$$M_n = (\alpha_j \delta_{jk})_{j,k=0}^{n-1}, \qquad \text{and}$$

$$\alpha_j = \sum_{k=0}^{n-1} a_k \, \mathrm{e}^{-2\pi \mathrm{i} jkh} \tag{6}$$

(see, e.g., Marcus/Minc [1]). Obviously, α_j $(j = 0, \ldots, n-1)$ are just the eigenvalues of A_n.

Hence, the eigenvalues β_j $(j = 0, \ldots, n-1)$ of B_n take the form

$$\beta_j = 1 + 2h^{-2} - h^{-2}(\mathrm{e}^{-2\pi \mathrm{i} jh} + \mathrm{e}^{2\pi \mathrm{i} jh})$$

$$= 1 + 4h^{-2} \sin^2 \pi jh .$$

For each $s \in \mathbb{R}$ and $\boldsymbol{x}, \boldsymbol{y} \in \mathbb{C}^n$, we now define

$$(\boldsymbol{x}, \boldsymbol{y})_{s,h} := h(B_n^s \boldsymbol{x}, \boldsymbol{y}) = h(U_n M_n^s U_n^{-1} \boldsymbol{x}, \boldsymbol{y}), \tag{7}$$

where $M_n^s = \left(\delta_{j,k}(1 + 4h^{-2} \sin^2 \pi jh)^s\right)_{j,k=0}^{n-1}$. Thus,

$$(\boldsymbol{x}, \boldsymbol{y})_{s,h} = h^2 \sum_{j,k,l=0}^{n-1} (1 + 4h^{-2} \sin^2 \pi lh)^s \, \mathrm{e}^{2\pi \mathrm{i} lh(j-k)} \, x_k \overline{y}_j . \tag{8}$$

Setting $\|\boldsymbol{x}\|_{s,h} := \sqrt{(\boldsymbol{x}, \boldsymbol{x})_{s,h}}$, we have

$$\|\boldsymbol{x}\|_{s,h}^2 = h^2 \sum_{j,k,l=0}^{n-1} (1 + 4h^{-2} \sin^2 \pi lh)^s \, \mathrm{e}^{2\pi \mathrm{i} lh(j-k)} \, x_k \overline{x}_j . \tag{9}$$

Obviously, for $s \in \mathbb{N}$, we get

$$|||\boldsymbol{x}|||_{s,h}^2 = h(C_n \boldsymbol{x}, \boldsymbol{x}),$$

where C_n is a circulant with the eigenvalues

$$\gamma_j = 1 + (4h^{-2} \sin^2 \pi jh)^s .$$

Comparing γ_j and the eigenvalues of M_n^s in (7), we see that

$$|||\boldsymbol{x}|||_{s,h} \leqq \|\boldsymbol{x}\|_{s,h} \leqq c\,|||\boldsymbol{x}|||_{s,h}$$

for all $s \in \mathbb{N}$ with a constant c independent of $\boldsymbol{x} \in \mathbb{C}^n$ and $h = 1/n$. Hence, by Lemma 2.2, the restriction operators $u \to \boldsymbol{u} = \big(u(kh)\big)_{k=0}^{n-1}$ are uniformly bounded from $\mathbf{H}^s$ into $\mathbb{C}^n$ endowed with the discrete Sobolev norm $\|\cdot\|_{s,h}$ for all $s \in \mathbb{N}$. By interpolation (see 2.1(8)) we obtain the following theorem.

2.3. Theorem. *For any $u \in \mathbf{H}^s$, $s \in \mathbb{R}$, $s \geqq 1$, one has*

$$\|\boldsymbol{u}\|_{s,h} \leqq c\,\|u\|_s, \tag{1}$$

where $\boldsymbol{u} = \big(u(kh)\big)_{k=0}^{n-1} \in \mathbb{C}^n$ *and c does not depend on u and h.*

In Corollary 2.27 we will prove that (1) holds also for $1/2 < s < 1$.

2.4. Remark. If Z_n is an $n \times n$-circulant with the eigenvalues ζ_k, $k = 0, \ldots, n-1$, then

$$\|Z_n\|_{t,h\to t,h} = \max_k |\zeta_k|, \qquad t \in \mathbb{R},$$

where $\|\cdot\|_{t,h\to t,h}$ stands for the operator norm in $\mathbb{C}^n$, the latter space equipped with the norm $\|\cdot\|_{t,h}$.

This is an immediate consequence of 2.2(7).

Approximation and inverse properties of splines

2.5. Periodic smoothest splines. Let $\Delta = \{x_k\}_{k=-\infty}^{\infty}$ be a partition of $\mathbb{R}$ by mesh points x_k satisfying $x_0 = 0$ and $x_{k+n} = x_k + 1$ for fixed $n \in \mathbb{N}$ and all $k \in \mathbb{Z}$. Let

$$h = \max_{k=1,\ldots,n} (x_k - x_{k-1}), \qquad \underline{h} = \min_{k=1,\ldots,n} (x_k - x_{k-1}).$$

By $\mathscr{S}^\delta(\Delta)$ ($\delta \in \mathbb{N}_0$) we denote the space of all 1-periodic smoothest *splines* of degree δ subordinate to the partition Δ, i.e. each function of $\mathscr{S}^\delta(\Delta)$ together with its derivatives of order $\leqq \delta - 1$ is 1-periodic and continuous on $\mathbb{R}$, and its restriction to any interval (x_k, x_{k+1}) ($k \in \mathbb{Z}$) is a polynomial of degree $\leqq \delta$; $\mathscr{S}^0(\Delta)$ stands for the set of all 1-periodic step functions subordinate to Δ. For every $\delta \in \mathbb{N}_0$, $\mathscr{S}^\delta(\Delta)$ is an n-dimensional space and it has a basis $\{B_{j,\delta}\}_{j=0}^{n-1}$ of *B-splines* which are defined as follows (cf. DE BOOR [3, Chap. 10]).

Let Θ_j be the characteristic function of the interval $[x_j, x_{j+1})$ and define $B_{j,0}$ ($j \in \mathbb{Z}$) by $B_{j,0}(x) = \Theta_j(x)$ for $x \in [x_j, x_{j+n})$ and 1-periodic extension for all $x \in \mathbb{R}$. For $\delta \geqq 1$ and $j \in \mathbb{Z}$, set recurrently

$$B_{j,\delta}(x) = \frac{x - x_j}{x_{j+\delta} - x_j}\, B_{j,\delta-1}(x) + \frac{x_{j+\delta+1} - x}{x_{j+\delta+1} - x_{j+1}}\, B_{j+1,\delta-1}(x)$$

for $x \in [x_j, x_{j+n})$ and take the 1-periodic extension for $x \in \mathbb{R}$.

We have $\mathscr{S}^\delta(\Delta) \subset \mathbf{H}^s$ if and only if $s < \delta + 1/2$. The following properties of the spline spaces $\mathscr{S}^\delta(\Delta)$ play a significant role in the error analysis of spline approximation methods.

2.6. Theorem. (Approximation property AP($\delta + 1/2$)). *If $s \leqq r \leqq \delta + 1$, $s < \delta + 1/2$ and $\sigma < s$, then for any $u \in \mathbf{H}^r$ and any partition Δ there exists $u_\Delta \in \mathcal{S}^\delta(\Delta)$ such that*

$$\|u - u_\Delta\|_t \leqq c(t)\, h^{r-t}\, \|u\|_r \tag{1}$$

for all $t \in [\sigma, s]$, where $c(t)$ denotes a constant independent of u and Δ.

The main ingredient of the proof of Theorem 2.6 is the following theorem.

2.7. Theorem. *If $s \leqq r \leqq \delta + 1$ and $s < \delta + 1/2$, then there exists a constant c independent of u and Δ such that*

$$\|u - P_{s,\Delta} u\|_s \leqq c h^{r-s}\, \|u\|_r, \qquad u \in \mathbf{H}^r, \tag{1}$$

where $P_{s,\Delta}$ is the orthogonal projection onto $\mathcal{S}^\delta(\Delta)$ in $\mathbf{H}^s$.

Proof. *Step* 1. First, we shall verify inequality (1) for $\delta = 0$, $r = 1$ and $s \in [0, 1/2)$. We use the equivalent norms 2.1 (4)–2.1 (5). Let $u \in \mathbf{H}^1$. We have

$$\varphi(x) := P_{0,\Delta} u = h_j^{-1} \int_{I_j} u(y)\, dy, \qquad x \in I_j \; (j = 0, \ldots, n-1),$$

where $I_j = (x_j, x_{j+1})$ and $h_j = x_{j+1} - x_j$. Let first $s = 0$. Using the Cauchy-Schwarz inequality, we then obtain

$$\begin{aligned} |u(x) - \varphi(x)|^2 &= \sum_{0 \leqq j < n} \Theta_j(x)\, \Big|h_j^{-1} \int_{I_j} \big(u(x) - u(y)\big)\, dy\Big|^2 \\ &\leqq \sum_{0 \leqq j < n} \Theta_j(x)\, h_j^{-1} \int_{I_j} |u(x) - u(y)|^2\, dy, \qquad x \in (0, 1). \end{aligned} \tag{2}$$

Furthermore, for any $x, y \in I_j$,

$$|u(x) - u(y)|^2 = \left| \int_x^y (\mathrm{D}u)(t)\, dt \right|^2 \leqq |y - x| \int_{I_j} |\mathrm{D}u|^2\, dx \leqq h_j \int_{I_j} |\mathrm{D}u|^2\, dx. \tag{3}$$

Integrating this inequality on I_j, we get

$$h_j^{-1} \int_{I_j} |u(x) - u(y)|^2\, dy \leqq h_j \int_{I_j} |\mathrm{D}u|^2\, dx.$$

Integrating the last estimate on I_j and using (2), we find that

$$\|u - \varphi\|_0^2 \leqq \sum_{0 \leqq j < n} h_j^2 \int_{I_j} |\mathrm{D}u|^2\, dx \leqq h^2\, \|\mathrm{D}u\|_0^2 \tag{4}$$

which implies (1) for $\delta = s = 0$ and $r = 1$.

Let now $s \in (0, 1/2)$. Then in the light of (4), inequality (1) is a consequence of the estimate

$$J := \int_0^1 \int_0^1 |u(x) - \varphi(x) - u(y) + \varphi(y)|^2\, |x - y|^{-1-2s}\, dx\, dy \leqq c h^{2-2s}\, \|\mathrm{D}u\|_0^2, \tag{5}$$

which will be shown now. Henceforth the letter c will denote various positive constants independent of u and Δ. We have

$$J = \sum_{0 \leqq j,k < n} J_{jk},$$

where J_{jk} denotes the corresponding integral on the rectangle $I_{jk} = I_j \times I_k$. Using (3), we obtain

$$J_{jj} = \int\limits_{I_{jj}} |u(x) - u(y)|^2 \, |x - y|^{-1-2s} \, \mathrm{d}x \, \mathrm{d}y \leqq \int\limits_{I_j} |\mathrm{D}u|^2 \, \mathrm{d}x \int\limits_{I_{jj}} |x - y|^{-2s} \, \mathrm{d}x \, \mathrm{d}y$$

$$\leqq ch^{2-2s} \int\limits_{I_j} |\mathrm{D}u|^2 \, \mathrm{d}x, \qquad j = 0, \ldots, n-1. \tag{6}$$

Applying (2) and (3), for $j \neq k$ we get

$$J_{jk} \leqq 2 \left\{ \int\limits_{I_{jk}} |u(x) - \varphi(x)|^2 \, |x - y|^{-1-2s} \, \mathrm{d}x \, \mathrm{d}y + \int\limits_{I_{jk}} |u(y) - \varphi(y)|^2 \, |x - y|^{-1-2s} \, \mathrm{d}x \, \mathrm{d}y \right\}$$

$$\leqq 2h \left\{ \int\limits_{I_j} |\mathrm{D}u|^2 \, \mathrm{d}x + \int\limits_{I_k} |\mathrm{D}u|^2 \, \mathrm{d}x \right\} \int\limits_{I_{jk}} |x - y|^{-1-2s} \, \mathrm{d}x \, \mathrm{d}y .$$

This implies that

$$\sum_{j \neq k} J_{jk} \leqq 8h \sum_{0 \leqq j < n} \left(\sum_{j<k<n} c_{jk} \right) \int\limits_{I_j} |\mathrm{D}u|^2 \, \mathrm{d}x, \tag{7}$$

where

$$c_{jk} = \int\limits_{I_k} \int\limits_{I_j} (y - x)^{-1-2s} \, \mathrm{d}x \, \mathrm{d}y = \big(2s(1 - 2s)\big)^{-1} \{(x_{k+1} - x_{j+1})^{1-2s} + (x_k - x_j)^{1-2s} - (x_{k+1} - x_j)^{1-2s} - (x_k - x_{j+1})^{1-2s}\} .$$

Finally, combining the estimate

$$\sum_{j<k<n} c_{jk} = \big(2s(1 - 2s)\big)^{-1} \{(x_{j+1} - x_j)^{1-2s} + (x_n - x_{j+1})^{1-2s} - (x_n - x_j)^{1-2s}\}$$

$$\leqq \big(2s(1 - 2s)\big)^{-1} (x_{j+1} - x_j)^{1-2s}$$

with (6) and (7), one obtains (5).

Step 2. Next, we prove (1) for $r = \delta + 1$, $s \in [\delta, \delta + 1/2)$. Introduce the spaces

$$\mathring{\mathbf{H}}^s := \left\{ u \in \mathbf{H}^s : \int\limits_0^1 u \, \mathrm{d}x = 0 \right\}, \quad s \geqq 0, \quad \mathring{\mathcal{S}}^\delta(\Delta) := \mathcal{S}^\delta(\Delta) \cap \mathring{\mathbf{H}}^0 .$$

Then $\mathbf{H}^s = \mathring{\mathbf{H}}^s \dotplus \mathbb{C}$, $\mathcal{S}^\delta(\Delta) = \mathring{\mathcal{S}}^\delta(\Delta) \dotplus \mathbb{C}$, and it is easy to check that the operator D^δ is an isomorphism of $\mathring{\mathbf{H}}^{s+\delta}$ and $\mathring{\mathcal{S}}^\delta(\Delta)$ onto $\mathring{\mathbf{H}}^s$ and $\mathring{\mathcal{S}}^0(\Delta)$, respectively. Therefore it is sufficient to verify that, for any $u \in \mathring{\mathbf{H}}^{\delta+1}$, there exists $\psi \in \mathring{\mathcal{S}}^0(\Delta)$ such that

$$\|\mathrm{D}^\delta u - \psi\|_{s-\delta} \leqq ch^{\delta+1-s} \, \|\mathrm{D}^\delta u\|_1 .$$

This, however, follows immediately from the first part of the proof.

Step 3. We now prove that (1) holds for all integers s and r satisfying $s \leqq r \leqq \delta + 1$ and $s \leqq \delta$. This follows from step 2 when $r = \delta + 1$ and $s = \delta$ and is trivial when $r = s$. Suppose the assertion is true for all $s \geqq j$ and $r > s$, where j is some integer $\leqq \delta$.

Using 2.1 (6) and the assumption, we obtain

$$\|u - P_{j-1,\Delta}u\|_{j-1} \leqq \|u - P_{j,\Delta}u\|_{j-1} = \sup |\langle u - P_{j,\Delta}u, \psi\rangle_j|$$
$$\leqq \sup \|\psi - P_{j,\Delta}\psi\|_j \, \|u - P_{j,\Delta}u\|_j \leqq ch^{r+1-j} \, \|u\|_r, \tag{8}$$

where the supremum is taken over all $\psi \in \mathbf{H}^{j+1}$ with norm $\leqq 1$. Thus, by induction on $j = \delta, \delta - 1, \ldots$, the assertion is proved.

Step 4. We complete the proof of Theorem 2.7 by interpolation. In view of (8), $I - P_{j,\Delta}$ ($j = r - 1, r - 2, \ldots; r = \delta + 1, \delta, \ldots$) is a continuous map of $\mathbf{H}^r$ into $\mathbf{H}^{j-1}$ and of $\mathbf{H}^r$ into $\mathbf{H}^j$ with norm $\leqq ch^{r+1-j}$ and $\leqq ch^{r-j}$, respectively. Therefore 2.1(8) implies (1) for $s \in (j - 1, j)$. Moreover, interpolation for the operators

$$I - P_{s,\Delta}: \mathbf{H}^{\delta+1} \,(\text{resp } \mathbf{H}^r) \to \mathbf{H}^s,$$

yields (1) for all $s \leqq \delta$ and $s \leqq r \leqq \delta + 1$. Finally, by interpolation applied to the maps

$$I - P_{s,\Delta}: \mathbf{H}^{\delta+1} \,(\text{resp } \mathbf{H}^s) \to \mathbf{H}^s,$$

we obtain Theorem 2.7 in the case where $\delta < s < \delta + 1/2$, $s \leqq r \leqq \delta + 1$. ■

2.8. Corollary. *If* $0 \leqq s < 1/2$ *and* $u \in \mathbf{H}^1$, *then*

$$\|u - P_{0,\Delta}u\|_s \leqq ch^{1-s} \, \|u\|_1,$$

where $P_{0,\Delta}: \mathbf{L}^2 = \mathbf{H}^0 \to \mathscr{S}^0(\Delta)$ *is the orthogonal projection.*

This follows from the first step of the preceding proof.

Remark. Let u be a piecewise continuous function possessing a square integrable derivative on each interval $I_j = (x_j, x_{j+1})$. Then 2.7(4)–2.7(7) gives

$$\|u - P_{0,\Delta}u\|_s^2 \leqq ch^{2-2s} \sum_{j=0}^{n-1} \int_{I_j} |\mathrm{D}u|^2 \, \mathrm{d}x$$

for all s, $0 \leqq s < 1/2$.

2.9. Corollary. *For every integer* $l \leqq \delta$, *we have*

$$\mathscr{S}^{\delta-l}(\Delta) = \mathsf{D}^l\big(\mathscr{S}^\delta(\Delta)\big), \tag{1}$$

where D *is the operator defined by*

$$\mathsf{D} := \frac{1}{\mathrm{i}} \mathrm{D} + J, \qquad Ju = \int_0^1 u(x) \, \mathrm{d}x, \qquad u \in \mathbf{H}^1, \tag{2}$$

or by

$$\mathsf{D}u(x) := \hat{u}_0 + \sum_{0 \neq k \in \mathbb{Z}} \hat{u}_k 2\pi k \, \mathrm{e}^{2\pi i k x}, \qquad u \in \mathbf{H}^t \ (t \in \mathbb{R}). \tag{3}$$

Proof. It follows from the second step of the preceding proof that (1) holds for $l = \delta$. This together with the invertibility of the operator $\mathsf{D} \in \mathscr{L}(\mathbf{H}^t, \mathbf{H}^{t-1})$, $t \in \mathbb{R}$, implies (1) for every integer $l \leqq \delta$. ■

2.10. Proof of Theorem 2.6. Let $u \in \mathbf{H}^s$ and

$$E(u) := \inf_{\varphi \in \mathscr{S}^\delta(\Delta)} \{\|u - \varphi\|_\sigma + h^{s-\sigma} \|u - \varphi\|_s\}, \qquad \beta = \sup_{\|u\|_s \leqq 1} E(u).$$

From the definition of β and Theorem 2.7 it follows that

$$E(v) = \inf_{\psi \in \mathcal{S}^\delta(\Delta)} E(v - \psi) \leqq \beta \inf_{\psi \in \mathcal{S}^\delta(\Delta)} \|v - \psi\|_s \leqq ch^{r-s}\beta \|v\|_r, \qquad v \in \mathbf{H}^r. \tag{1}$$

Hence, for any $v \in \mathbf{H}^r$,

$$E(u) \leqq E(u - v) + E(v) \leqq \|u - v\|_\sigma + h^{s-\sigma} \|u - v\|_s + ch^{r-s}\beta \|v\|_r, \tag{2}$$

where we have used (1), the triangle inequality and the fact that $0 \in \mathcal{S}^\delta(\Delta)$. Furthermore, for any $u \in \mathbf{H}^s$ and $\varepsilon > 0$, there exists $v \in \mathbf{H}^r$ such that

$$\|u - v\|_s \leqq \|u\|_s, \qquad \|u - v\|_\sigma \leqq \varepsilon^{s-\sigma} \|u\|_s, \qquad \|v\|_r \leqq \varepsilon^{s-r} \|u\|_s. \tag{3}$$

Indeed, setting

$$v = \sum_{|m| \leqq \bar{m}} \hat{u}_m \, \mathrm{e}^{\mathrm{i}m2\pi x},$$

where $\bar{m}$ is the greatest natural number $\leqq 1/\varepsilon$ and $\hat{u}_m$ are the Fourier coefficients of u, one obtains (3) immediately from the definition 2.1(1) of the norm. With the choice of v according to (3), estimate (2) gives

$$E(u) \leqq \{h^{s-\sigma} + \varepsilon^{s-\sigma} + c(h/\varepsilon)^{r-s} \beta\} \|u\|_s.$$

Choosing $\varepsilon = kh$, where k is large enough, we obtain $c(h/\varepsilon)^{r-s} \leqq 1/2$. Taking the supremum over all $\|u\|_s \leqq 1$, we get $\beta \leqq ch^{s-\sigma}$. This together with (1) implies that

$$E(u) \leqq ch^{r-\sigma} \|u\|_r, \qquad u \in \mathbf{H}^r.$$

Hence, for any $u \in \mathbf{H}^r$ and any partition Δ, there exists $u_\Delta \in \mathcal{S}^\delta(\Delta)$ such that estimate 2.6(1) holds for $t = \sigma$ and $t = s$. Finally, using the inequality

$$\|u\|_t \leqq \|u\|_\sigma^{1-\Theta} \|u\|_s^{\Theta}, \qquad u \in \mathbf{H}^s, \qquad t = (1 - \Theta)\,\sigma + \Theta s, \qquad 0 < \Theta < 1, \tag{4}$$

which is an immediate consequence of 2.1(1) and Hölder's inequality, we obtain 2.6(1) for $t \in (\sigma, s)$. ■

Remark. The bounds for r and s indicated in Theorem 2.6 are sharp.

This follows immediately from a result of Gaier [1]. A mesh Δ is said to be γ-*quasiuniform* ($\gamma \geqq 1$) if $h/\underline{h} \leqq \gamma$. The set of all γ-quasiuniform meshes is denoted by Part γ.

2.11. Theorem (Inverse property IP($\delta + 1/2$)). *For any numbers t, s and γ satisfying $t \leqq s < \delta + 1/2$ and $\gamma \geqq 1$, there exists a constant c such that*

$$\|\varphi\|_s \leqq ch^{t-s} \|\varphi\|_t \tag{1}$$

for all $\varphi \in \mathcal{S}^\delta(\Delta)$ and $\Delta \in$ Part γ.

Proof. *Step* 1. We first verify (1) for $\delta < t \leqq s < \delta + 1/2$. We use the equivalent norms 2.1(4)–2.1(5). Since $\mathrm{D}^\delta\varphi \in \mathcal{S}^\circ(\Delta)$ for $\varphi \in \mathcal{S}^\delta(\Delta)$, it remains to prove the estimate

$$J_s(\varphi) \leqq ch^{2(t-s)} J_t(\varphi), \qquad \varphi \in \mathcal{S}^0(\Delta), \qquad \Delta \in \text{Part}\,\gamma, \tag{2}$$

where $0 < t \leqq s < 1/2$ and

$$J_s(\varphi) = \int_0^1\!\!\int_0^1 |\varphi(x) - \varphi(y)|^2 \, |x - y|^{-1-2s} \, \mathrm{d}x \, \mathrm{d}y.$$

Let $\varphi = c_0\Theta_0 + \dots + c_{n-1}\Theta_{n-1}$, $c_j \in \mathbb{C}$. Recall that Θ_j is the characteristic function of the interval $I_j = [x_j, x_{j+1})$. We have

$$J_s(\varphi) = \sum_{0 \leqq j,k < n} J_{jk,s}, \qquad J_{jk,s} = \int_{I_j} \int_{I_k} |c_j - c_k|^2 \, |x - y|^{-1-2s} \, dx \, dy \tag{3}$$

and for $|j - k| \geqq 2$,

$$J_{jk,s} = \int_{I_j}\int_{I_k} \frac{|c_j - c_k|^2}{|x - y|^{1+2t}} \frac{dx \, dy}{|x - y|^{2(s-t)}} \leqq \underline{h}^{2(t-s)} J_{jk,t} \tag{4}$$

while for $k = j + 1$,

$$J_{jj+1,s} = |c_{j+1} - c_j|^2 \left(2s(1 - s)\right)^{-1} \left(h_j^{1-2s} + h_{j+1}^{1-2s} - (h_j + h_{j+1})^{1-2s}\right). \tag{5}$$

Therefore

$$J_{jj+1,s}/J_{jj+1,t} \leqq c h_j^{2(t-s)} f_s(\bar{x})/f_t(\bar{x}),$$

where $\bar{x} = h_{j+1}/h_j$ and $f_s(x) = 1 + x^{1-2s} - (1 + x)^{1-2s}$. For any $s \in (0, 1/2)$, f_s is increasing on $(0, \infty)$. Moreover, since $\Delta \in \operatorname{Part} \gamma$, we have $\bar{x} \in [1/\gamma, \gamma]$. Thus we get

$$J_{jj+1,s}/J_{jj+1,t} \leqq c\left(f_s(\gamma)/f_t(1/\gamma)\right) \underline{h}^{2(t-s)}$$

which implies an estimate of the form (4) again. By virtue of (3) and since $\Delta \in \operatorname{Part} \gamma$, the proof of inequality (2) is complete.

Step 2. We now extend (1) to the general case. For $\delta < t < s < \delta + 1/2$, we obtain

$$\|\varphi\|_t^2 \leqq \|\varphi\|_s \, \|\varphi\|_{2t-s} \leqq c h^{t-s} \, \|\varphi\|_t \, \|\varphi\|_{2t-s}$$

by 2.10(4) and the first step of the proof. Using the estimate $\|\varphi\|_s \leqq c h^{t-s} \|\varphi\|_t$, this gives $\|\varphi\|_s \leqq c h^{2t-2s} \|\varphi\|_{2t-s}$. By induction,

$$\|\varphi\|_s \leqq c h^{k(t-s)} \, \|\varphi\|_{t+k(t-s)}, \qquad k = 1, 2, \dots.$$

Hence, (1) holds for $s \in (\delta, \delta + 1/2)$ and any $t \leqq s$. If $s \leqq \delta$, then we choose some $r \in (\delta, \delta + 1/2)$ and use 2.10(4) again to obtain

$$\begin{aligned} \|\varphi\|_s &\leqq \|\varphi\|_t^{(r-s)/(r-t)} \, \|\varphi\|_r^{(s-t)/(r-t)} \\ &\leqq c \, \|\varphi\|_t^{(r-s)/(r-t)} \, h^{(t-r)(s-t)/(r-t)} \, \|\varphi\|_t^{(s-t)/(r-t)} \leqq c h^{t-s} \, \|\varphi\|_t . \ \blacksquare \end{aligned}$$

Commutator property of splines

2.12. Theorem. (i) *If* $2s - (\delta + 1) \leqq t \leqq s \leqq r \leqq \delta + 1$, *then there exists a constant* c *independent of* u *and* Δ *such that*

$$\|u - P_{s,\Delta} u\|_t \leqq c h^{r-t} \, \|u\|_r, \qquad u \in \mathbf{H}^r. \tag{1}$$

(ii) *If* $\Delta \in \operatorname{Part} \gamma$ $(\gamma \geqq 1)$, *then the projections* $P_{s,\Delta}$ *are uniformly bounded in* $\mathbf{H}^t$ *for* $2s - \delta - 1/2 < t < \delta + 1/2$. *Moreover, estimate* (1) *holds for any real numbers* t *and* r *satisfying* $2s - (\delta + 1) \leqq t \leqq r \leqq \delta + 1$, $t < \delta + 1/2$, $r > 2s - \delta - 1/2$.

Proof. Assertion (i) is a consequence of Theorem 2.6 and the Aubin-Nitsche lemma (see Sect. 1.34) applied to the operator $A = \Lambda^{2s}$. Assertion (ii) follows from Theorems 2.6 and 2.11 and Corollary 1.36 (i). ■

2.13. Theorem (Commutator property). *Let $\Delta \in \text{Part}\,\gamma$ ($\gamma \geqq 1$) and $f \in \mathbf{C}^{\delta+1}$ ($\delta \in \mathbb{N}_0$). For every real number $r < \delta + 1/2$ there exist projections P_h satisfying* $\operatorname{im} P_h = \mathcal{S}^\delta(\Delta)$ *such that*

$$\|P_h\|_t \leqq C \quad \text{and} \quad \|(I - P_h) f v\|_t \leqq C h^\varrho \|v\|_t \tag{1}$$

for all $v \in \mathcal{S}^\delta(\Delta)$ and $r < t < \delta + 1/2$, where $\varrho = \min\{1, \delta + 1 - t\}$ and the constant C does not depend on h or v.

Proof. For l a positive integer, we define the periodic piecewise Sobolev space $\mathbf{H}^l(\Delta)$ as the space of all $u \in \mathbf{H}^{l-1}$ such that $u^{(l-1)}$ has a square integrable derivative on each interval $I_j = (x_j, x_{j+1})$. For $u \in \mathbf{H}^l(\Delta)$, we denote by $u^{(l)} \in \mathbf{H}^0 = \mathbf{L}^2$ the piecewise (not distributional) l-th derivative.

Let $\delta = 0$ and $u \in \mathbf{H}^1(\Delta)$. Due to Corollary 2.8,

$$\|u - P_{0,\Delta} u\|_s \leqq c h^{1-s} \|u^{(1)}\|_{\mathbf{L}^2}, \qquad 0 \leqq s < 1/2, \tag{2}$$

where $P_{0,\Delta}: \mathbf{L}^2 \to \mathcal{S}^0(\Delta)$ is the orthogonal projection. For $-1 \leqq t < 0$, we choose $w \in \mathbf{H}^{-t}$ such that $\|w\|_{-t} = 1$ and

$$\begin{aligned} \|(I - P_{0,\Delta}) u\|_t &= |\langle (I - P_{0,\Delta}) u, w\rangle_0| \\ &\leqq \|(I - P_{0,\Delta}) u\|_0 \|(I - P_{0,\Delta}) w\|_0 \leqq c h^{1-t} \|u^{(1)}\|_{\mathbf{L}^2} \end{aligned} \tag{3}$$

(recall (2) and Theorem 2.6).

We now consider the orthogonal projection $P_{-1,\Delta}: \mathbf{H}^{-1} \to \mathcal{S}^0(\Delta)$. Since, by Theorem 2.12 (ii), $\|P_{-1,\Delta}\|_t \leqq c$ for $-5/2 < t < 1/2$, it follows from (2) and (3) that, for $-1 \leqq s < 1/2$,

$$\begin{aligned} \|(I - P_{-1,\Delta}) u\|_s &\leqq (1 + \|P_{-1,\Delta}\|_s) \|(I - P_{0,\Delta}) u\|_s \\ &\leqq c h^{1-s} \|u^{(1)}\|_{\mathbf{L}^2}. \end{aligned}$$

Arguing as before we find that

$$\|(I - P_{-1,\Delta}) u\|_t \leqq c h^{1-t} \|u^{(1)}\|_{\mathbf{L}^2}, \qquad -3 \leqq t < -1.$$

By continuing this procedure we get a sequence of projections $P_h^j := P_{-2^j+1,\Delta}$, $j = 0, 1, \ldots$, satisfying $\|P_h^j\|_t \leqq c$, $-2^{j+1} + 3/2 < t < 1/2$, and

$$\|(I - P_h^j) u\|_t \leqq c h^{1-t} \|u^{(1)}\|_{\mathbf{L}^2}, \qquad -2^{j+1} + 1 < t < 1/2. \tag{4}$$

We choose j so that $2^j > \delta + 1 - r$.

Introduce now the mapping

$$\mathsf{D} g(x) = \frac{1}{\mathrm{i}} \frac{\mathrm{d}g}{\mathrm{d}x} + \hat{g}_0, \qquad g \in \mathbf{C}^\infty, \tag{5}$$

which defines an isometry of $\mathbf{H}^t$ onto $\mathbf{H}^{t-1}$ for all $t \in \mathbb{R}$. Set

$$P_h := \mathsf{D}^{-\delta} P_h^j \mathsf{D}^\delta. \tag{6}$$

Obviously, P_h is a projection onto $\mathscr{S}^\delta(\Delta)$ with $\|P_h\|_t \leq c$, $r < t < \delta + 1/2$. Moreover, if $v \in \mathscr{S}^\delta(\Delta)$ then $\mathsf{D}^\delta(fv) \in \mathbf{H}^1(\Delta)$, so (4) implies

$$\begin{aligned}\|(I - P_h) fv\|_t &= \|\mathsf{D}^{-\delta}(I - P_h^j) \mathsf{D}^\delta(fv)\|_t \\ &= \|(I - P_h^j) \mathsf{D}^\delta(fv)\|_{t-\delta} \leq ch^{\delta+1-t} \|[\mathsf{D}^\delta(fv)]^{(1)}\|_{\mathbf{L}^2} \\ &\leq ch^{\delta+1-t} \|(fv)^{(\delta+1)}\|_{\mathbf{L}^2}. \end{aligned} \tag{7}$$

Now since $v^{(\delta+1)} = 0$, the Leibniz rule gives

$$\|(fv)^{(\delta+1)}\|_{\mathbf{L}^2} \leq C \|v\|_\delta,$$

where C depends on f. Applying the inverse property (Theorem 2.11), we obtain

$$\|v\|_\delta \leq Ch^{\min\{t-\delta,0\}} \|v\|_t,$$

and so (1) follows. ■

Our next objective is to extend the commutator property of spline spaces to the case of piecewise continuous f and $\mathbf{L}^2$ norms.

Let Γ be a simple closed or open oriented Lyapunov curve in the complex plane $\mathbb{C}$ given by a regular parametrization $z = \gamma(x)$, $x \in [0, 1]$. By $\mathbf{L}^2(\Gamma)$ we denote the Hilbert space of all Lebesgue-integrable complex-valued functions on Γ with scalar product and norm

$$(f, g) := \int_\Gamma f(t) \overline{g(t)} \, |\mathrm{d}t|, \qquad \|f\| = (f, f)^{1/2}.$$

Let $\mathbf{PC}(\Gamma)$ denote the algebra of all piecewise continuous complex-valued functions on Γ, i.e. for $a \in \mathbf{PC}(\Gamma)$ and each $t \in \Gamma$, there exist the one-sided finite limits $a(t \pm 0)$ (with respect to the orientation of Γ) and the number of discontinuities of a is finite. Let $f \in \mathbf{PC}(\Gamma)$ and let $s_1, \ldots, s_N \in [0, 1]$ be all points of discontinuity of $f \circ \gamma$. A sequence of partitions $\{\Delta_k\}_{k\in\mathbb{N}}$, $\Delta_k = \{\sigma_0^k, \ldots, \sigma_{n_k}^k\}$, $0 = \sigma_0^k < \sigma_1^k < \cdots < \sigma_{n_k}^k = 1$, is said to be *admissible* for f if $\{s_0, \ldots, s_N\} \subset \Delta_k$ $(k \in \mathbb{N})$ and if $h_{\Delta_k} := \max\,(\sigma_{j+1}^k - \sigma_j^k) \to 0$ as $k \to \infty$. We denote by $\mathbf{P}\mathscr{S}^\delta(\Delta_k)$ the space of all $\varphi \in \mathbf{PC}(\Gamma)$ such that $\varphi \circ \gamma$ is $(\delta - 1)$ times continuously differentiable on $[0, 1] \setminus \{s_0, \ldots, s_N\}$ and the restriction of $\varphi \circ \gamma$ to each interval $(\sigma_j^k, \sigma_{j+1}^k)$ is a polynomial of degree $\leq \delta$. Let P_{Δ_k} denote the orthogonal projection of $\mathbf{L}^2(\Gamma)$ onto $\mathbf{P}\mathscr{S}^\delta(\Delta_k)$.

2.14. Theorem. *If $\{\Delta_k\}$ is an admissible sequence of partitions for $f \in \mathbf{PC}(\Gamma)$, then*

$$\|(I - P_{\Delta_k}) fP_{\Delta_k}\| \to 0, \qquad \|P_{\Delta_k} f(I - P_{\Delta_k})\| \to 0$$

as $k \to \infty$, where $\|\cdot\|$ stands for the operator norm in $\mathbf{L}^2(\Gamma)$.

Proof. We only need to prove that $\|(I - P_{\Delta_k}) fP_{\Delta_k}\| \to 0$, since the other assertion follows by taking the operator adjoints.

We introduce a new scalar product in $\mathbf{L}^2(\Gamma)$ by

$$(f, g)_0 := \int_0^1 (f \circ \gamma)(s) \, \overline{(g \circ \gamma)(s)} \, \mathrm{d}s.$$

Clearly, the norms $\|f\| = (f, f)^{1/2}$ and $\|f\|_0 = (f, f)_0^{1/2}$ are equivalent, and thus it suffices to show that $\|(I - P_{\Delta_k}) fP_{\Delta_k}\|_0 \to 0$, where $\|\cdot\|_0$ denotes the operator norm induced by the norm $\|\cdot\|_0$ in $\mathbf{L}^2(\Gamma)$. Let $\tilde{P}_{\Delta_k}$ be the orthogonal projection of $\mathbf{L}^2(\Gamma)$ onto $\mathbf{P}\mathcal{S}^\delta(\Delta_k)$ corresponding to $(\cdot,\cdot)_0$. Since $\sup_k \|P_{\Delta_k}\|_0 < \infty$, we obtain (see e.g. ALEXITS [1])

$$\begin{aligned}\|(I - P_{\Delta_k})\, fP_{\Delta_k}g\|_0 &= \|fP_{\Delta_k}g - P_{\Delta_k}(fP_{\Delta_k}g)\|_0\\ &\leqq C\, \|(fP_{\Delta_k}g) - \tilde{P}_{\Delta_k}(fP_{\Delta_k}g)\|_0\\ &= C\, \|(I - \tilde{P}_{\Delta_k})\, f\tilde{P}_{\Delta_k}P_{\Delta_k}g\|_0,\end{aligned}$$

where C is a positive constant. Consequently, we only have to show that $\|(I - \tilde{P}_{\Delta_k}) \times f\tilde{P}_{\Delta_k}\|_0 \to 0$. Hence, we may assume that $\Gamma = [0, 1]$.

Let f^i $(i = 0, \ldots, N - 1)$ denote the function which vanishes on $[0, 1] \setminus [s_i, s_{i+1}]$ and is equal to f on $[s_i, s_{i+1}]$, and $P^i_{\Delta_k}$ be the orthogonal projection onto the subspace of all functions $f \in \mathbf{P}\mathcal{S}^\delta(\Delta_k)$ vanishing on $[0, 1] \setminus [s_i, s_{i+1}]$. Thus

$$(I - P_{\Delta_k})\, fP_{\Delta_k} = \sum_{i=0}^{N-1} (I - P^i_{\Delta_k})\, f^iP^i_{\Delta_k},$$

and we only need to demonstrate that $\|(I - P^i_{\Delta_k})\, f^iP^i_{\Delta_k}\| \to 0$ $(i = 0, \ldots, N - 1)$. We may therefore assume that $f \in \mathbf{C}[0, 1]$ and that P_{Δ_k} is the orthogonal projection of $\mathbf{L}^2(0, 1)$ onto $\mathcal{S}^\delta(\Delta_k)$, where $\mathcal{S}^\delta(\Delta_k)$ is the subspace of all $f \in \mathbf{L}^2(0, 1)$ which are $(\delta - 1)$ times continuously differentiable and coincide with a polynomial of degree not exceeding δ on each interval $[\sigma_i^k, \sigma_{i+1}^k]$ $(i = 0, \ldots, n_k - 1)$.

We set $\sigma^k_{-r} := -r(\sigma_1^k - \sigma_0^k)/2$ and $\sigma^k_{n_k+r} := 1 + r(\sigma^k_{n_k} - \sigma^k_{n_k-1})/2$ $(r = 1, 2, \ldots, \delta)$. A basis $\{\psi_i : i = -\delta, \ldots, n_k - 1\}$ of $\mathcal{S}^\delta(\Delta_k)$ can now be given as follows (see DE BOOR [3]):

$$g(s, t) := (s - t)_+^\delta,$$

$$\psi_i(t) := \sqrt{\delta + 1}\,\sqrt{\sigma^k_{i+\delta+1} - \sigma_i^k}\; g(\sigma_i^k, \ldots, \sigma^k_{i+\delta+1}; t) \qquad (t \in [0, 1]),$$

wherein $g(\sigma_i^k, \ldots, \sigma^k_{i+\delta+1}; t)$ represents the divided difference of $g(s, t)$ at $\sigma_i^k, \ldots, \sigma^k_{i+\delta+1}$. It is well known that $\psi_i \geqq 0$ and ψ_i has support $[\sigma_i^k, \sigma^k_{i+\delta+1}]$ (see DE BOOR [3]). Moreover, there exists a positive constant c such that

$$c^{-1} \left\|\sum_{i=-\delta}^{n_k-1} \xi_i\psi_i\right\| \leqq \left(\sum_{i=-\delta}^{n_k-1} |\xi_i|^2\right)^{1/2} \leqq c \left\|\sum_{i=-\delta}^{n_k-1} \xi_i\psi_i\right\| \tag{1}$$

(see DE BOOR [1]). We obtain

$$\left\|f\sum_{i=-\delta}^{n_k-1} \xi_i\psi_i - P_{\Delta_k}f\sum_{i=-\delta}^{n_k-1} \xi_i\psi_i\right\| \leqq \left\|f\sum_{i=-\delta}^{n_k-1} \xi_i\psi_i - \sum_{i=-\delta}^{n_k-1} f(\sigma_i^k)\, \xi_i\psi_i\right\|, \tag{2}$$

where $f(\sigma_i^k) := f(0)$ $(i = -\delta, -\delta + 1, \ldots, -1)$. For $t \in [\sigma_j^k, \sigma_{j+1}^k]$,

$$\begin{aligned}&\left|f(t)\sum_{i=-\delta}^{n_k-1} \xi_i\psi_i(t) - \sum_{i=-\delta}^{n_k-1} f(\sigma_i^k)\, \xi_i\psi_i(t)\right| = \left|\sum_{i=j-\delta}^{j+1} \big(f(t) - f(\sigma_i^k)\big)\, \xi_i\psi_i(t)\right|\\ &\leqq \omega(f, \delta h_{\Delta_k}) \sum_{i=-\delta}^{j+1} |\xi_i|\, |\psi_i(t)| \leqq \omega(f, \delta h_{\Delta_k}) \left|\sum_{i=-\delta}^{n_k-1} |\xi_i|\, \psi_i(t)\right|,\end{aligned} \tag{3}$$

with $\omega(f, \delta) := \sup \{|f(t_1) - f(t_2)| : t_1, t_2 \in [0, 1], |t_1 - t_2| < \delta\}$ (modulus of continuity). (1) and (3) yield that

$$\left\| f \sum_{i=-\delta}^{n_k-1} \xi_i \psi_i - \sum_{i=-\delta}^{n_k-1} f(\sigma_i^k) \xi_i \psi_i \right\| \leq \omega(f, \delta h_{\Delta_k}) \left\| \sum_{i=-\delta}^{n_k-1} |\xi_i| \psi_i \right\| \leq \omega(f, \delta h_{\Delta_k}) c^2 \left\| \sum_{i=-\delta}^{n_k-1} \xi_i \psi_i \right\|$$

and (2) implies that

$$\left\| (I - P_{\Delta_k}) f \sum_{i=-\delta}^{n_k-1} \xi_i \psi_i \right\| \leq c^2 \omega(f, \delta h_k) \left\| \sum_{i=-\delta}^{n_k-1} \xi_i \psi_i \right\|.$$

Since $h_{\Delta_k} \to 0$, the proof is complete. ■

2.15. Corollary. *P_{Δ_k} converges strongly to I as $k \to \infty$.*

The Arnold-Wendland lemma

Let $\Delta = \{x_k\}_{k=-\infty}^{\infty}$ be an arbitrary 1-periodic partition. Define the mappings J and J_Δ: $\mathbf{H}^s \to \mathbb{C}$ $(s > 1/2)$ by the integral and its numerical counterpart, the trapezoidal rule, namely

$$Ju := \int_0^1 u(x)\, dx \qquad \text{and} \qquad J_\Delta u := \sum_{k=0}^{n-1} u(x_k)\, (x_{k+1} - x_{k-1})/2.$$

2.16. Lemma. *For $1/2 < s \leqq 2$, we have*

$$|(J - J_\Delta)\, u| \leqq ch^s\, \|u\|_s, \qquad u \in \mathbf{H}^s, \tag{1}$$

where c does not depend on u or Δ.

Proof. *Step 1.* Let $s = 2$. Introduce the piecewise linear function

$$v(x) = u(x_k) + \big(u(x_{k+1}) - u(x_k)\big)\, (x - x_k)/h_k, \qquad x_k \leqq x \leqq x_{k+1}.$$

Obviously, $J_\Delta u = Jv$. Thus we obtain

$$|(J - J_\Delta)\, u|^2 = \left| \int_{x_0}^{x_n} \big(u(x) - v(x)\big)\, dx \right|^2 \leqq \int_{x_0}^{x_n} |u(x) - v(x)|^2\, dx$$

$$= \sum_{k=0}^{n-1} \int_{x_k}^{x_{k+1}} |u(x) - v(x)|^2\, dx.$$

For $x_k \leqq x \leqq x_{k+1}$, we have the representations

$$u(x) = u(x_k) + (x - x_k)\, u'(x_k) + \int_{x_k}^{x} (x - t)\, u''(t)\, dt$$

and

$$w(x) := u(x) - v(x) = \frac{x - x_k}{h_k} \int_{x_k}^{x_{k+1}} \big(u'(x_k) - u'(t)\big)\, dt + \int_{x_k}^{x} (x - t)\, u''(t)\, dt.$$

Applying the Cauchy-Schwarz inequality we find

$$|w(x)|^2 \leqq 2\left(\left|\int_{x_k}^{x_{k+1}} \mathrm{d}t \int_{x_k}^{t} u''(y)\,\mathrm{d}y\right|^2 + \left|\int_{x_k}^{x} (x-t)\,u''(t)\,\mathrm{d}t\right|^2\right) \leqq 2h_k^3 \int_{x_k}^{x_{k+1}} |u''(t)|^2\,\mathrm{d}t.$$

Hence

$$|(J - J_\Delta)\,u|^2 \leqq \sum_{k=0}^{n-1} \int_{x_k}^{x_{k+1}} |w(x)|^2\,\mathrm{d}x$$

$$\leqq 2 \sum_{k=0}^{n-1} h_k^4 \int_{x_k}^{x_{k+1}} |u''(t)|^2\,\mathrm{d}t \leqq 2h^4 \int_0^1 |u''(t)|^2\,\mathrm{d}t.$$

Step 2. If $s = 1$, then set $v(x) = \big(u(x_k) + u(x_{k+1})\big)/2$ for $x_k \leqq x \leqq x_{k+1}$, $k = 0, \ldots, n-1$, to get

$$|(J - J_\Delta)\,u|^2 = \left|\int_{x_0}^{x_n} \big(u(x) - v(x)\big)\,\mathrm{d}x\right|^2$$

$$\leqq \frac{1}{2} \sum_{k=0}^{n-1} \left(\int_{x_k}^{x_{k+1}} |u(x) - u(x_k)|^2\,\mathrm{d}x + \int_{x_k}^{x_{k+1}} |u(x) - u(x_{k+1})|^2\,\mathrm{d}x\right). \tag{2}$$

Using the Cauchy-Schwarz inequality once more, we see that, for $x_k \leqq x \leqq x_{k+1}$,

$$|u(x) - u(x_k)|^2 = \left|\int_{x_k}^{x} u'(t)\,\mathrm{d}t\right|^2 \leqq h_k \int_{x_k}^{x_{k+1}} |u'(t)|^2\,\mathrm{d}t.$$

Hence

$$|(J - J_\Delta)\,u|^2 \leqq \sum_{k=0}^{n-1} h_k^2 \int_{x_k}^{x_{k+1}} |u'(t)|^2\,\mathrm{d}t \leqq h^2 \int_0^1 |u'(t)|^2\,\mathrm{d}t.$$

Thus we obtain (1) for $1 \leqq s \leqq 2$ by interpolation.

Step 3. Let $1/2 < s < 1$. Applying the well-known estimate

$$\sup_{0 \leqq x,y \leqq 1} |f(x) - f(y)|^2 \leqq c \int_0^1 \int_0^1 \frac{|f(x) - f(y)|^2}{|x - y|^{1+2s}}\,\mathrm{d}x\,\mathrm{d}y, \tag{3}$$

where $f \in \mathbf{H}^s(0, 1)$ (nonperiodic Sobolev space), to the function $f(x) = u(h_k x + x_k)$, $y = 0 < x < 1$, we find that

$$|u(h_k x + x_k) - u(x_k)|^2 \leqq c \int_0^1 \int_0^1 \frac{|f(x) - f(y)|^2}{|x - y|^{1+2s}}\,\mathrm{d}x\,\mathrm{d}y$$

$$= c h_k^{2s-1} \int_{x_k}^{x_{k+1}} \int_{x_k}^{x_{k+1}} \frac{|u(x) - u(y)|^2}{|x - y|^{1+2s}}\,\mathrm{d}x\,\mathrm{d}y.$$

Combining this estimate with (2) and 2.1(5), we obtain

$$|(J - J_\Delta)\,u|^2 \leqq c \sum_{k=0}^{n-1} h_k^{2s} \int_{x_k}^{x_{k+1}} \int_{x_k}^{x_{k+1}} \frac{|u(x) - u(y)|^2}{|x - y|^{1+2s}}\,\mathrm{d}x\,\mathrm{d}y \leqq c h^{2s}\,\|u\|_s^2. \ \blacksquare$$

Notice that (3) is an immediate consequence of the compactness of the embedding $\mathbf{H}^s(0, 1) \subset \mathbf{C}[0, 1]$, $1/2 < s < 1$.

Remark. For arbitrary partitions Δ, estimate (1) is sharp. However, in the case of equidistant partitions $\Delta_h = \{kh\}_{k=-\infty}^{\infty}$, $h^{-1} = n \in \mathbb{N}$, (1) holds for all $s > 1/2$ and $u \in \mathbf{H}^s$.

In fact, since $J\, e^{2\pi i k x} = \delta_{k0}$ and

$$J_{\Delta_h} e^{2\pi i k x} = h \sum_{j=0}^{n-1} e^{2\pi i k j h} = \begin{cases} 1 & \text{for } k = rn,\ r \in \mathbb{Z}, \\ 0 & \text{otherwise}, \end{cases}$$

we see that

$$\begin{aligned} |(J - J_{\Delta_h})\, u| &= \Big|\hat{u}_0 - \sum_{r \in \mathbb{Z}} \hat{u}_{rn}\Big| = \Big|\sum_{0 \neq r \in \mathbb{Z}} \hat{u}_{rn}\Big| \\ &\leqq \Big(\sum_{0 \neq r \in \mathbb{Z}} |\hat{u}_{rn}|^2\, |2\pi r n|^{2s}\Big)^{1/2} \Big(\sum_{0 \neq r \in \mathbb{Z}} |2\pi r n|^{-2s}\Big)^{1/2} \\ &\leqq n^{-s} \Big(\sum_{0 \neq r \in \mathbb{Z}} |2\pi r|^{-2s}\Big)^{1/2} \|u\|_s. \ \blacksquare \end{aligned}$$

We now introduce the operator

$$I_\Delta := I - J + J_\Delta \in \mathcal{L}(\mathbf{H}^s), \qquad s > 1/2.$$

Notice that I_Δ is invertible and $I_\Delta^{-1} = I + J - J_\Delta$, since $J^2 = J_\Delta J = J$ and $J_\Delta^2 = J J_\Delta = J_\Delta$. We also remark that the continuity of the embedding $\mathbf{H}^s \subset \mathbf{C}$ $(s > 1/2)$ implies that the operators I_Δ and I_Δ^{-1} are uniformly bounded in $\mathbf{H}^s$, $s > 1/2$, for all meshes Δ.

The following lemma plays an important part in the analysis of interpolation and collocation procedures.

2.17. Arnold-Wendland lemma. *Let $u \in \mathbf{H}^s$, $s > 1/2$. Then the equations*

$$u(x_k) = 0, \qquad k = 0, \ldots, n-1, \tag{1}$$

hold if and only if

$$\langle \mathsf{D} I_\Delta u, v\rangle_0 = 0 \qquad \text{for all } v \in \mathcal{S}^0(\Delta), \tag{2}$$

where $\Delta = \{x_k\}_{k=0}^{n-1}$ and D is the operator defined by 2.13(5).

Proof. Let $u \in \mathbf{C}^\infty$. Observe that

$$\mathsf{D} I_\Delta u = \frac{1}{i} \frac{du}{dx} + J_\Delta u$$

and

$$\langle \mathsf{D} I_\Delta u, v\rangle_0 = \frac{1}{i} \int_{x_0}^{x_n} u'(x)\, \overline{v(x)}\, dx + J_\Delta u \int_{x_0}^{x_n} \overline{v(x)}\, dx.$$

Define $v_j \in \mathcal{S}^0(\Delta)$, $j = 1, \ldots, n-1$, by

$$v_j(x) = \begin{cases} h_j^{-1} & \text{for } x \in [x_j, x_{j+1}), \\ -h_{j-1}^{-1} & \text{for } x \in [x_{j-1}, x_j), \\ 0 & \text{otherwise, } x \in [x_0, x_n), \end{cases}$$

where $h_j = x_{j+1} - x_j$, $x_0 = x_n - 1$. Then

$$\langle \mathrm{D}I_\Delta u, v_j\rangle_0 = \frac{1}{i}\left[h_j^{-1}\big(u(x_{j+1}) - u(x_j)\big) - h_{j-1}^{-1}\big(u(x_j) - u(x_{j-1})\big)\right].$$

Hence, $\langle \mathrm{D}I_\Delta u, v_j\rangle_0 = 0$ for $j = 1, \ldots, n-1$ if and only if there exists a constant $\mu \in \mathbb{C}$ such that

$$h_j^{-1}\big(u(x_{j+1}) - u(x_j)\big) = \mu, \qquad j = 0, \ldots, n-1. \tag{3}$$

Applying (3) repeatedly with $j = 0, \ldots, n-1$, we find that

$$u(x_n) = u(x_0) + \mu(h_0 + h_1 + \cdots + h_{n-1}) = u(x_0) + \mu.$$

Since u is 1-periodic we deduce that $\mu = 0$. Thus (3) implies

$$u(x_j) = u(x_0) = \text{const}, \qquad j = 0, \ldots, n-1. \tag{4}$$

In fact, since $\mathcal{S}^0(\Delta)$ is spanned by the v_j and the constant function $v_0 \equiv 1$, we see that, for $u \in \mathbf{C}^\infty$, (2) holds if and only if (4) is in force and, in addition,

$$0 = \langle \mathrm{D}I_\Delta u, v_0\rangle_0 = \frac{1}{i}\int_{x_0}^{x_n} u'(x)\,\mathrm{d}x + J_\Delta u = u(x_0)\sum_{k=0}^{n-1}(x_{k+1} - x_{k-1})/2 = u(x_0).$$

So the proof of the assertion is complete in the case $u \in \mathbf{C}^\infty$. For $u \in \mathbf{H}^s$, $s > 1/2$, the lemma follows from the continuity of the embedding $\mathbf{H}^s \subset \mathbf{C}$ and the estimate

$$|\langle \mathrm{D}I_\Delta u, v\rangle_0| \leqq c\,\|u\|_s\,\|v\|_{1-s}. \ ■$$

2.18. Corollary. *Let δ be odd and $u \in \mathbf{H}^s$, $s > 1/2$. Then* 2.17 (1) *holds if and only if*

$$\langle I_\Delta u, w\rangle_j = 0 \qquad \textit{for all } w \in \mathcal{S}^\delta(\Delta),$$

where $j = (\delta + 1)/2$.

Proof. It follows from the definition 2.1(2) that $\langle I_\Delta u, w\rangle_j = \langle I_\Delta u, \mathrm{D}^{2j} w\rangle_0 = \langle \mathrm{D}I_\Delta u, \mathrm{D}^\delta w\rangle_0$. So the assertion is a consequence of Lemma 2.17 and Corollary 2.9. ■

Given an arbitrary 1-periodic partition $\Delta = \{x_k\}_{k=-\infty}^{\infty}$, we now consider the space

$$\mathcal{S}^{-1}(\Delta) := \operatorname{span}\{\delta_{x_k} : k = 0, \ldots, n-1\},$$

where $\delta_{x_k} \in \mathbf{H}^t$, $t < -1/2$, denotes the 1-periodic Dirac function, i.e. $\langle u, \delta_{x_k}\rangle_0 = u(x_k)$ for all $u \in \mathbf{H}^s$, $s > 1/2$.

2.19. Corollary. *$\mathcal{S}^{-1}(\Delta)$ has the approximation property* AP$(-1/2)$. *If $\Delta \in$ Part γ ($\gamma \geqq 1$), then $\mathcal{S}^{-1}(\Delta)$ possesses the inverse property* IP$(-1/2)$.

Proof. Let $u \in \mathbf{H}^s$, $s > 1/2$. In view of the Arnold-Wendland lemma, u is orthogonal to $\mathcal{S}^{-1}(\Delta)$ (with respect to the scalar product $\langle\cdot,\cdot\rangle_0$) if and only if it is orthogonal to $I_\Delta^*\mathrm{D}\big(\mathcal{S}^0(\Delta)\big)$, where $I_\Delta^* = I - J + J_\Delta^*$ is the adjoint operator to I_Δ (with respect to $\langle\cdot,\cdot\rangle_0$). Since $I_\Delta^*\mathrm{D}\big(\mathcal{S}^0(\Delta)\big) \subset \mathbf{H}^t$ for $t < -1/2$, we see that $\mathcal{S}^{-1}(\Delta) = I_\Delta^*\mathrm{D}\big(\mathcal{S}^0(\Delta)\big)$. It follows from Lemma 2.16 that

$$\|I_\Delta^*\mathrm{D}\|_{t+1\to t} \leqq c, \qquad \|(I_\Delta^*\mathrm{D})^{-1}\|_{t\to t+1} \leqq c, \qquad t < -1/2.$$

This together with Theorem 2.11 implies IP$(-1/2)$ for $\mathcal{S}^{-1}(\Delta)$ if $\Delta \in$ Part γ.

To prove AP$(-1/2)$ we first consider the case where $s \leqq r < -1/2$ and $u \in \mathbf{H}^r$. Since $\|(I_\Delta^* \mathsf{D})^{-1} u\|_{r+1} < c \|u\|_r$, Theorem 2.6 yields that there exists a spline $u_\Delta \in \mathcal{S}^0(\Delta)$ such that

$$\|(I_\Delta^* \mathsf{D})^{-1} u - u_\Delta\|_{t+1} \leqq ch^{r-t} \|u\|_r$$

for all $t \leqq s$. Hence, for $v_\Delta = I_\Delta^* \mathsf{D} u_\Delta \in \mathcal{S}^{-1}(\Delta)$,

$$\|u - v_\Delta\|_t = \|I_\Delta^* \mathsf{D}\big((I_\Delta^* \mathsf{D})^{-1} u - u_\Delta\big)\|_t \leqq ch^{r-t} \|u\|_r .$$

Let $-2 \leqq s < -1/2 \leqq r \leqq 0$ and $u \in \mathbf{H}^r$. Then we can find $u_\Delta \in \mathcal{S}^0(\Delta)$ such that

$$\|\mathsf{D}^{-1} u - u_\Delta\|_{s+1} \leqq ch^{r-s} \|\mathsf{D}^{-1} u\|_{r+1} = ch^{r-s} \|u\|_r .$$

Setting $w_\Delta = I_\Delta^* \mathsf{D} u_\Delta$, we get

$$\|u - w_\Delta\|_s = \|I_\Delta^* \mathsf{D}\big((I_\Delta^* \mathsf{D})^{-1} u - u_\Delta\big)\|_s \leqq c\big(\|\mathsf{D}^{-1} u - u_\Delta\|_{s+1} + \|\mathsf{D}^{-1}\big(I - (I_\Delta^*)^{-1}\big) u\|_{s+1}\big) \leqq ch^{r-s} \|u\|_r + \|(J - J_\Delta^*) u\|_s .$$

Finally, Lemma 2.16 in conjunction with 2.1(6) implies

$$\|(J - J_\Delta^*) u\|_s = \sup_{\|v\|_{-s} \leqq 1} |\langle u, (J - J_\Delta) v\rangle_0| \leqq ch^{-s} \|u\|_r .$$

Hence, $\|u - w_\Delta\|_s \leqq ch^{r-s} \|u\|_r$. ■

Periodic spline interpolation

2.20. Notation. Besides the definitions 2.1(7) and 2.9(3), we introduce the operators D^l $(l \in \mathbb{Z})$ and S by

$$\mathsf{D}^l u(x) := \hat{u}_0 + \sum_{0 \neq k \in \mathbb{Z}} \hat{u}_k (2\pi k)^l \, \mathrm{e}^{2\pi i k x}, \qquad l \in \mathbb{Z}, \tag{1}$$

$$Su(x) := \sum_{0 \leqq k \in \mathbb{Z}} \hat{u}_k \, \mathrm{e}^{2\pi i k x} - \sum_{0 > k \in \mathbb{Z}} \hat{u}_k \, \mathrm{e}^{2\pi i k x}. \tag{2}$$

Notice that the operators Λ^α $(\alpha \in \mathbb{R})$, D^l $(l \in \mathbb{Z})$ and S are examples of so-called pseudo-differential operators (see Chapter 6). Obviously, these operators commute and possess the following properties:

$$\|\Lambda^\alpha u\|_s = \|u\|_{s+\alpha}, \qquad \langle \Lambda^\alpha u, v\rangle_s = \langle u, \Lambda^\alpha v\rangle_s, \qquad \Lambda^\alpha \Lambda^\beta = \Lambda^{\alpha+\beta};$$

$$\|\mathsf{D}^l u\|_s = \|u\|_{s+l}, \qquad \langle \mathsf{D}^l u, v\rangle_s = \langle u, \mathsf{D}^l v\rangle_s, \qquad \mathsf{D}^l \mathsf{D}^k = \mathsf{D}^{l+k};$$

$$\|Su\|_s = \|u\|_s, \qquad \langle Su, v\rangle_s = \langle u, Sv\rangle_s, \qquad S^2 = I;$$

$$\Lambda^l = (\mathsf{D}S)^l .$$

Here $s, \alpha, \beta \in \mathbb{R}$ and $k, l \in \mathbb{Z}$. For $l = 1$, $\mathsf{D} = \mathsf{D}^1$ coincides with the operator defined by 2.9(3) or 2.13(5).

In what follows we consider uniform meshes $\Delta_h := \{kh\}_{h=-\infty}^{\infty}$, where $h = n^{-1}$, $n \in \mathbb{N}$, and denote by $\mathcal{S}_h^\delta$ the 1-periodic spline space $\mathcal{S}^\delta(\Delta_h)$. In view of Corollary 2.9, we have $\mathcal{S}_h^{\delta-l} = \mathsf{D}^l(\mathcal{S}_h^\delta)$ for all integers $\delta \geqq 0$ and $l \leqq \delta$. Hence

$$\mathcal{S}_h^\delta = \begin{cases} \Lambda^{-\delta}(\mathcal{S}_h^0) & \text{if } \delta \text{ is even}, \\ S\Lambda^{-\delta}(\mathcal{S}_h^0) & \text{if } \delta \text{ is odd}. \end{cases} \tag{4}$$

This leads to the definition

$$\mathcal{M}_h^\alpha := \Lambda^{-\alpha}(\mathcal{S}_h^0), \qquad \mathcal{N}_h^\alpha := S\Lambda^{-\alpha}(\mathcal{S}_h^0), \qquad \alpha \in \mathbb{R}. \tag{5}$$

Since $\mathcal{S}_h^0 \subset \mathbf{H}^s$ if and only if $s < 1/2$, we see that $\mathcal{M}_h^\alpha \subset \mathbf{H}^s$ (resp. $\mathcal{N}_h^\alpha \subset \mathbf{H}^s$) if and only if $s < \alpha + 1/2$. Furthermore, it follows from Theorems 2.6 and 2.11 that $\mathcal{M}_h^\alpha$ and $\mathcal{N}_h^\alpha$ have the properties AP($\alpha + 1/2$) and IP($\alpha + 1/2$).

In connection with ε-collocation of pseudodifferential equations (see Chapters 10 and 13) the following interpolation problem for $\mathcal{M}_h^\alpha$ and $\mathcal{N}_h^\alpha$ is of particular interest. Given any $h = n^{-1}$ and $\boldsymbol{y} = (y_0, \ldots, y_{n-1}) \in \mathbb{C}^n$, we ask for $\varepsilon \in [0, 1)$ and $\delta \in \mathbb{N}_0$ such that there exists $u_h \in \mathcal{M}_h^\alpha$ ($u_h \in \mathcal{N}_h^\alpha$) satisfying

$$u_h(\xi_r) = y_r, \qquad r = 0, \ldots, n-1; \qquad \xi_r = (r + \varepsilon)\, h. \tag{6}$$

To this end we define the bases $\{\Lambda^{-\alpha}\Theta_j\}_{j=0}^{n-1} \subset \mathcal{M}_h^\alpha$ and $\{S\Lambda^{-\alpha}\Theta_j\}_{j=0}^{n-1} \subset \mathcal{N}_h^\alpha$, where Θ_j is the 1-periodic prolongation of the characteristic function of the interval $[jh, (j+1)\,h)$. Since Θ_j has the Fourier expansion

$$\Theta_j(x) = h + \sum_{0 \neq k \in \mathbb{Z}} \frac{\sin \pi k h}{\pi k}\, \mathrm{e}^{-\pi i k h(2j+1)}\, \mathrm{e}^{2\pi i k x}, \tag{7}$$

we deduce from 2.1(7) and (2) that

$$\Lambda^{-\alpha}\Theta_j(x) = h + \sum_{0 \neq k \in \mathbb{Z}} \frac{\sin \pi k h}{2^\alpha \pi^{\alpha+1} k\, |k|^\alpha}\, \mathrm{e}^{-\pi i k h(2j+1)}\, \mathrm{e}^{2\pi i k x}, \tag{8}$$

$$S\Lambda^{-\alpha}\Theta_j(x) = h + \sum_{0 \neq k \in \mathbb{Z}} \frac{\sin \pi k h}{2^\alpha \pi^{\alpha+1}\, |k|^{1+\alpha}}\, \mathrm{e}^{-\pi i k h(2j+1)}\, \mathrm{e}^{2\pi i k x}. \tag{9}$$

From the theory of Fourier series we infer that the series (8) and (9) converge uniformly for $\alpha > 0$; if $-1 < \alpha \leqq 0$, then they converge uniformly outside some arbitrarily small neighborhoods of the points jh and $(j+1)\,h$ (see e.g. ZYGMUND [1]). In particular, equalities (8) and (9) hold for $x = \xi_r = (r + \varepsilon)\,h$, $r, j \in \mathbb{Z}$, if $\varepsilon \in [0, 1)$ and $\alpha > 0$ or $\varepsilon \in (0, 1)$ and $-1 < \alpha \leqq 0$. Hence, the interpolation problem (6) makes sense if and only if either

$$0 \leqq \varepsilon < 1 \qquad \text{and} \qquad \alpha > 0$$

or

$$0 < \varepsilon < 1 \qquad \text{and} \qquad -1 < \alpha \leqq 0.$$

Obviously, (6) has a unique solution $u_h \in \mathcal{M}_h^\alpha$ or $u_h \in \mathcal{N}_h^\alpha$, respectively, exactly if the matrices

$$\left(\Lambda^{-\alpha}\Theta_j(\xi_r)\right)_{r,j=0}^{n-1} \qquad \text{or} \qquad \left(S\Lambda^{-\alpha}\Theta_j(\xi_r)\right)_{r,j=0}^{n-1}, \tag{10}$$

respectively, are invertible. Since these matrices are circulants, their eigenvalues can be calculated by formula 2.2(6). A simple computation shows that the eigenvalues μ_l^n, $l = 0, \ldots, n-1$, of $\left(\Lambda^{-\alpha}\Theta_j(\xi_r)\right)_{r,j=0}^{n-1}$ are

$$\mu_l^n = \sum_{r=0}^{n-1} \mathrm{e}^{-2\pi i r l h} \left(h + \sum_{k \neq 0} \frac{\sin \pi k h}{2^\alpha \pi^{\alpha+1} k\, |k|^\alpha}\, \mathrm{e}^{2\pi i k h r}\, \mathrm{e}^{\pi i k h(2\varepsilon - 1)}\right).$$

Hence, $\mu_0^n = 1$ and

$$\mu_l^n = \frac{n}{2^\alpha \pi^{\alpha+1}} \sum_{j\in\mathbb{Z}} \frac{\sin \pi h(jn+l)}{(jn+l)\,|jn+l|^\alpha}\, e^{\pi i(j+lh)(2\varepsilon-1)}$$
$$= \frac{h^\alpha \sin \pi lh\, e^{\pi i lh(2\varepsilon-1)}}{2^\alpha \pi^{\alpha+1}} \sum_{j\in\mathbb{Z}} \frac{e^{2\pi i j\varepsilon}}{(j+lh)\,|j+lh|^\alpha}$$

for $l = 1, \ldots, n-1$. Calculating the eigenvalues ν_l^n, $l = 0, \ldots, n-1$, of the second matrix in (10) in a similar manner, we arrive at the following lemma.

2.21. Lemma. *Let $0 < \varepsilon < 1$ and $\alpha > -1$ or $\varepsilon = 0$ and $\alpha > 0$. The eigenvalues μ_l^n and ν_l^n, $l = 0, \ldots, n-1$, of the matrices 2.20(10) are*

$$\mu_0^n = \nu_0^n = 1,$$

$$\mu_l^n = h^\alpha \varphi_{\varepsilon,\alpha}(lh)\, \sigma_{\varepsilon,\alpha}^-(lh), \qquad \nu_l^n = h^\alpha \varphi_{\varepsilon,\alpha}(lh)\, \sigma_{\varepsilon,\alpha}^+(lh), \qquad l = 1, \ldots, n,$$

where

$$\varphi_{\varepsilon,\alpha}(\Theta) = 2^{-\alpha}\pi^{-\alpha-1} \sin \pi\Theta\, e^{\pi i\Theta(2\varepsilon-1)} \qquad (0 < \Theta < 1), \tag{1}$$

$$\sigma_{\varepsilon,\alpha}^\pm(\Theta) = \sum_{k=0}^\infty (k+\Theta)^{-\alpha-1} e^{2\pi i k\varepsilon} \pm \sum_{k=1}^\infty (k-\Theta)^{-\alpha-1} e^{-2\pi i k\varepsilon}. \tag{2}$$

We now determine the zeros of the functions $\sigma_{\varepsilon,\alpha}^\pm(\Theta)$, $0 < \Theta < 1$. Note that these functions can be represented in the form

$$\sigma_{\varepsilon,\alpha}^+(\Theta) = \frac{a_0}{2} + \sum_{k=1}^\infty (a_k \cos 2\pi k\varepsilon - i b_k \sin 2\pi k\varepsilon),$$

$$\sigma_{\varepsilon,\alpha}^-(\Theta) = \frac{a_0}{2} + \sum_{k=1}^\infty (-b_k \cos 2\pi k\varepsilon + i a_k \sin 2\pi k\varepsilon),$$

where $a_0 = 2\Theta^{-\alpha-1}$,

$$a_k = (k+\Theta)^{-\alpha-1} + (k-\Theta)^{-\alpha-1}, \qquad b_k = (k-\Theta)^{-\alpha-1} - (k+\Theta)^{-\alpha-1}.$$

Obviously, the sequences $\{a_k\}_{k=1}^\infty$ and $\{b_k\}_{k=1}^\infty$ are positive, strictly decreasing and convex. More precisely, $a_1 \geqq a_2 \geqq \cdots$, $a_k \to 0$, $\sum_{k=1}^\infty \frac{a_k}{k} < \infty$ and $a_{k+2} - 2a_{k+1} + a_k \geqq 0$. Consequently (see ZYGMUND [1, V. 1.14])

$$\sum_{k=1}^\infty a_k \sin 2\pi k\varepsilon > 0, \qquad \sum_{k=1}^\infty b_k \sin 2\pi k\varepsilon > 0 \qquad \text{for} \qquad 0 < \varepsilon < \frac{1}{2}.$$

Hence, the functions $\sigma_{\varepsilon,\alpha}^\pm$ may have zeros only for $\varepsilon = 0$ and $\varepsilon = 1/2$. Setting $\varepsilon = 0$ and $\varepsilon = 1/2$, we conclude that, for admissible α,

$$\sigma_{\varepsilon,\alpha}^+(\Theta) = 0 \quad \text{if and only if} \quad \varepsilon = 1/2,\ \Theta = 1/2,$$

$$\sigma_{\varepsilon,\alpha}^-(\Theta) = 0 \quad \text{if and only if} \quad \varepsilon = 0,\ \Theta = 1/2.$$

This along with Lemma 2.21 implies the following theorem.

2.22. Theorem. *Let* $0 < \varepsilon < 1$ *and* $\alpha > -1$ *or* $\varepsilon = 0$ *and* $\alpha > 0$. *The interpolation problem* 2.20(6) *has a unique solution* $u_h \in \mathcal{M}_h^\alpha$ (*resp.* $u_h \in \mathcal{N}_h^\alpha$) *except for the case where* $\varepsilon = 0$ (*resp.* $\varepsilon = 1/2$) *and* n *is even.*

In what follows we estimate the error $\|u - u_h\|_t$, where u is a given bounded 1-periodic function and $u_h \in \mathcal{M}_h^\alpha$ (resp. $u_h \in \mathcal{N}_h^\alpha$) is the solution of the interpolation problem

$$u_h(\xi_r) = u(\xi_r), \qquad r = 0, \dots, n-1; \qquad \xi_r = (r + \varepsilon)\, h. \tag{1}$$

Of course, the conditions of Theorem 2.22 are assumed to be satisfied. Then each of the spaces $\mathcal{M}_h^\alpha$ and $\mathcal{N}_h^\alpha$ possesses an interpolation basis $\{\chi_k\}_{k=0}^{n-1}$ such that $\chi_k(\xi_r) = \delta_{kr}$, $0 \leqq k, r \leqq n-1$. Hence, u_h has the representation

$$u_h(x) = \sum_{k=0}^{n-1} u(\xi_k)\, \chi_k(x).$$

Note that if $\{\chi_k\}_{k=0}^{n-1} \subset \mathcal{M}_h^\alpha$ ($\mathcal{N}_h^\alpha$), then

$$\chi_k(x) = \sum_{j=0}^{n-1} a_{kj} \Lambda^{-\alpha} \Theta_j(x) \qquad \left(\chi_k(x) = \sum_{j=0}^{n-1} a_{kj} S \Lambda^{-\alpha} \Theta_j(x)\right), \tag{2}$$

where

$$(a_{kj})_{k,j=0}^{n-1} = \left[\left(\Lambda^{-\alpha}\Theta_j(\xi_r)\right)_{r,j=0}^{n-1}\right]^{-1} \qquad \left((a_{kj})_{k,j=0}^{n-1} = \left[\left(S\Lambda^{-\alpha}\Theta_j(\xi_r)\right)_{r,j=0}^{n-1}\right]^{-1}\right).$$

2.23. Theorem. *Let* $\varepsilon \neq 0$, $\alpha > -1$ ($\varepsilon \neq 0$, $\varepsilon \neq 1/2$, $-1 < \alpha \leqq 0$ *or* $\varepsilon \neq 1/2$, $\alpha > 0$). *Let* $\chi_k \in \mathcal{M}_h^\alpha$ ($\mathcal{N}_h^\alpha$), $k = 0, \dots, n-1$, *be the interpolation basis* 2.22(2). *For every* $\boldsymbol{y} = (y_0, \dots, y_{n-1}) \in \mathbb{C}^n$ *one has*

$$c_1 \|\boldsymbol{y}\|_{t,h} \leqq \left\| \sum_{k=0}^{n-1} y_k \chi_k \right\|_t \leqq c_2 \|\boldsymbol{y}\|_{t,h},$$

where $t < \alpha + 1/2$ *and the constants* c_1 *and* c_2 *do not depend on* $\boldsymbol{y}$ *or* n. *Here* $\|\cdot\|_{t,h}$ *is the norm defined by* 2.2(9).

Proof. Let $\chi_k \in \mathcal{M}_h^\alpha$. It follows from 2.22(2) that

$$\left\| \sum_{k=0}^{n-1} y_k \chi_k \right\|_t^2 = \left\| \sum_{k,j=0}^{n-1} y_k a_{kj} \Theta_j \right\|_{t-\alpha}^2 = \sum_{k,j,s,r=0}^{n-1} \bar{a}_{sr} \langle \Theta_j, \Theta_r \rangle_{t-\alpha}\, a_{kj} y_k \bar{y}_s .$$

Since

$$\langle \Theta_j, \Theta_r \rangle_{t-\alpha} = h^2 + \sum_{0 \neq k \in \mathbb{Z}} \frac{\sin^2 \pi k h}{\pi^2 k^2}\, \mathrm{e}^{2\pi i k h (r-j)}\, |2\pi k|^{2(t-\alpha)}$$
$$= h^2 + 2^{2(t-\alpha)} \pi^{2(t-\alpha-1)} \sum_{k=0}^{n-1} \sin^2 \pi k h\, \mathrm{e}^{2\pi i k h (r-j)} \sum_{s \in \mathbb{Z}} |sn + k|^{2(t-\alpha-1)}$$

with $\sum_{s \in \mathbb{Z}}$ interpreted as $\sum_{0 \neq s \in \mathbb{Z}}$ for $k = 0$, the matrix $(\langle \Theta_j, \Theta_k \rangle_{t-\alpha})_{j,k=0}^{n-1}$ is a circulant with the eigenvalues (cf. 2.2(6))

$$\vartheta_l^n = h\delta_{l0} + 2^{2(t-\alpha)} \pi^{2(t-\alpha-1)} (\sin^2 \pi l k)\, h^{1-2(t-\alpha)} \sum_{j \in \mathbb{Z}} |j + lh|^{2(t-\alpha-1)},$$

$l = 0, \dots, n-1$. Hence, in view of Lemma 2.21 and 2.22(2), the matrix

$$\left(\sum_{j,r=0}^{h-1} \bar{a}_{sr} \langle \Theta_j, \Theta_r \rangle_{t-\alpha}\, a_{kj} \right)_{k,s=0}^{n-1}$$

has the eigenvalues

$$\gamma_l^n = |\mu_l^n|^{-2}\,\vartheta_l^n = \begin{cases} h^{1-2t}(2\pi)^{2t}\,|\sigma_{\varepsilon,\alpha}^-(lh)|^{-2}\sum_{j\in\mathbb{Z}} |j+lh|^{2(t-\alpha-1)}, & l = 1,\dots,n-1,\\ h, & l = 0.\end{cases}$$

In the case $\chi_k \in \mathcal{N}_h^\alpha$ we obtain a similar formula with $\sigma_{\varepsilon,\alpha}^+$ instead of $\sigma_{\varepsilon,\alpha}^-$. Notice that the function $\sigma_{\varepsilon,\alpha}^-$ (resp. $\sigma_{\varepsilon,\alpha}^+$) does not vanish because of the assumptions of Theorem 2.23.

Since $\|\boldsymbol{y}\|_{t,h}^2 = h(B_n^t\boldsymbol{y}, \boldsymbol{y})$, hB_n^t being a circulant with the eigenvalues

$$\beta_l^n = h^{1-2t}(h^2 + 4\sin^2\pi lh)^t, \qquad l = 0,\dots,n-1$$

(see formulas 2.2(7) and 2.2(8)), it remains to prove that there exist constants c_1 and c_2 such that

$$c_1\beta_l^n \leqq \gamma_l^n \leqq c_2\beta_l^n \qquad \text{for all} \qquad n\in\mathbb{N},\ 0\leqq l\leqq n-1.$$

For $l \neq 0$, we have to compare the functions

$$f_1(\Theta) := |\sigma_{\varepsilon,\alpha}^-(\Theta)|^2\,|\sin\pi\Theta|^{2t}, \qquad f_2(\Theta) := \sum_{j\in\mathbb{Z}} |j+\Theta|^{2(t-\alpha-1)}$$

$0 < \Theta < 1$. Setting

$$f(\Theta) = \begin{cases} \Theta^{2(\alpha+1-t)} & \text{for} \quad 0<\Theta\leqq 1/2,\\ (1-\Theta)^{2(\alpha+1-t)} & \text{for} \quad 1/2\leqq\Theta<1,\end{cases}$$

we observe that the functions $f_1(\Theta)\,f(\Theta)$ and $f_2(\Theta)\,f(\Theta)$ admit continuous prolongations to the interval $[0, 1]$ which do not vanish. Thus, there exist positive constants c_1 and c_2 such that

$$c_1 f_1(\Theta) \leqq f_2(\Theta) \leqq c_2 f_1(\Theta) \qquad \text{for all} \qquad \Theta\in(0,1). \;\blacksquare$$

2.24. Corollary. *For any $t \leqq s$ and any $\boldsymbol{y} \in \mathbb{C}^n$, one has*

$$c_1 h^{s-t}\,\|\boldsymbol{y}\|_{s,h} \leqq \|\boldsymbol{y}\|_{t,h} \leqq c_2\,\|\boldsymbol{y}\|_{s,h},$$

where the constants c_1 and c_2 do not depend on $\boldsymbol{y}$ or n.

This is an immediate consequence of Theorem 2.23 and the inverse property IP($\alpha + 1/2$) for $\mathcal{M}_h^\alpha$ and $\mathcal{N}_h^\alpha$.

In the degenerate case where the interpolation problem 2.20(6) is solvable but the function $\sigma_{\varepsilon,\alpha}^-$ resp. $\sigma_{\varepsilon,\alpha}^+$ vanishes, one can proceed as follows. Let $\varepsilon = 0$, n be odd and $\chi_k \in \mathcal{M}_h^\alpha$, $\alpha > 0$. Since, for a neighborhood of $\Theta_0 = 1/2$,

$$\sigma_{0,\alpha}^-(\Theta) = 2(\alpha+1)\left(\frac{1}{2} - \Theta\right)\sum_{k=0}^{\infty}(k+1/2)^{-\alpha-2} + O\big((1/2-\Theta)^3\big),$$

in analogy to the preceding reasoning one can show that

$$c_1\left|\frac{1}{2}-\Theta\right|^{-2}|\sin\pi\Theta|^{2t} \leqq |\sigma_{0,\alpha}^-(\Theta)|^{-2}\sum_{j\in\mathbb{Z}}|j+\Theta|^{2(t-\alpha-1)}$$

$$\leqq c_2\left|\frac{1}{2}-\Theta\right|^{-2}|\sin\pi\Theta|^{2t}, \qquad \Theta\in(0,1/2)\cup(1/2,1).$$

Taking into account that n is odd, we find

$$|\sin \pi lh|^2 \, |1/2 - lh|^2 \geqq ch^2 .$$

Hence, there exist constants c_3 and c_4 such that, for $1 \leqq l \leqq n - 1$,

$$c_3 \beta_l^n \leqq \gamma_l^n \leqq c_4 \beta_l^n h^{-2}(h^2 + 4 \sin^2 \pi lh) .$$

This implies the estimate

$$c_1 \, \|\boldsymbol{y}\|_{t,h} \leqq \left\| \sum_{k=0}^{n-1} y_k \chi_k \right\|_t \leqq c_2 \, \|\boldsymbol{y}\|_{t+1,h} . \tag{1}$$

Since an analogous representation holds for the function $\sigma_{1/2,\alpha}^+$, the following theorem is established.

2.25. Theorem. *Let $n = 1/h$ be odd and let $\varepsilon = 0$, $\alpha > 0$ ($\varepsilon = 1/2$, $\alpha > -1$). If $\chi_k \in \mathcal{M}_h^\alpha$ ($\mathcal{N}_h^\alpha$), $k = 0, \ldots, n - 1$, is the interpolation basis 2.22(2), then for every $\boldsymbol{y} = (y_0, \ldots, y_{n-1}) \in \mathbb{C}^n$ and $t < \alpha + 1/2$ the estimate 2.24(1) holds with constants c_1 and c_2 not depending on $\boldsymbol{y}$ or n.*

Note that estimate 2.24(1) is no longer true if the norm on the right-hand side is replaced by $\|\cdot\|_{t+1-\varrho,h}$, $\varrho > 0$.

Let ε and α satisfy the conditions of Theorem 2.22. Define Q_h as the interpolation projection satisfying

$$\operatorname{im} Q_h = \mathcal{M}_h^\alpha \qquad (\text{resp. } \operatorname{im} Q_h = \mathcal{N}_h^\alpha) \tag{1}$$

and

$$Q_h u(\xi_r) = u(\xi_r), \qquad r = 0, \ldots, n - 1, \qquad \xi_r = (r + \varepsilon) \, h . \tag{2}$$

2.26. Theorem. *Let $\alpha > 0$ and $\varepsilon \neq 0$ (resp. $\varepsilon \neq 1/2$). Then for any $u \in \mathbf{H}^s$, $1/2 < s \leqq \alpha + 1$, the interpolation projection Q_h defined by 2.25(1)–2.25(2) satisfies the estimate*

$$\|u - Q_h u\|_t \leqq c h^{s-t} \, \|u\|_s , \tag{1}$$

where $0 \leqq t < \alpha + 1/2$, $t \leqq s$, and the constant c does not depend on u or h.

Proof. Let $\alpha > 1/2$. Since

$$Q_h u = \sum_{k=0}^{n-1} u(\xi_k) \, \chi_k ,$$

where $\{\chi_k\}_{k=0}^{n-1}$ is the corresponding interpolation basis, Theorem 2.23 implies that $\|Q_h u\|_t \leqq c_1 \, \|\boldsymbol{u}\|_{t,h}$ for $t < \alpha + 1/2$ and $\boldsymbol{u} = \big(u(\xi_k)\big)_{k=0}^{n-1} \in \mathbb{C}^n$. By Theorem 2.3,

$$\|Q_h u\|_t \leqq c_1 \, \|\boldsymbol{u}\|_{t,h} \leqq c \, \|u\|_t \qquad \text{for} \qquad 1 \leqq t < \alpha + 1/2 .$$

Hence, the operators Q_h are uniformly bounded in $\mathbf{H}^t$. So it follows from the approximation property AP($\alpha + 1/2$) for $\mathcal{M}_h^\alpha$ (resp. $\mathcal{N}_h^\alpha$) that (1) holds for $1 \leqq t \leqq s \leqq \alpha + 1$, $t < \alpha + 1/2$.

We now denote by Q_h^* the 1-adjoint operator of Q_h defined by

$$\langle Q_h u, v \rangle_1 = \langle u, Q_h^* v \rangle_1 , \qquad u, v \in \mathbf{H}^1 .$$

Because

$$\langle Q_h u, v\rangle_1 = \sum_{k=0}^{n-1} u(\xi_k)\,\langle \chi_k, v\rangle_1 = \sum_{k=0}^{n-1} \langle u, \delta_{\xi_k}\rangle_0\,\langle \chi_k, v\rangle_1$$
$$= \left\langle u, \sum_{k=0}^{n-1} \langle v, \chi_k\rangle_1\, \Lambda^{-2}\delta_{\xi_k}\right\rangle_1,$$

we obtain $\operatorname{im} Q_h^* = \Lambda^{-2}(\mathcal{S}^{-1}(\tilde{\Delta}))$, where $\tilde{\Delta} = \{(k+\varepsilon)\,h\}_{k=-\infty}^{\infty}$ (see Sect. 2.18). In view of Corollary 2.19, $\Lambda^{-2}(\mathcal{S}^{-1}(\tilde{\Delta}))$ has the properties AP(3/2) and IP(3/2). Since the projections Q_h^* are uniformly bounded, we get

$$\|u - Q_h^* u\|_1 \leqq c h^{s-1}\,\|u\|_s \qquad \text{for} \qquad 1 \leqq s < 3/2.$$

Hence,

$$\|u - Q_h^* u\|_s \leqq \|u - z_h\|_s + \|z_h - Q_h^* u\|_s$$
$$\leqq c_1\,\|u\|_s + c_2 h^{1-s}(\|u - Q_h^* u\|_1 + \|u - z_h\|_1)$$
$$\leqq c\,\|u\|_s,$$

where $z_h \in \Lambda^{-2}(\mathcal{S}^{-1}(\tilde{\Delta}))$ stands for the approximating element defined, for all $t \leqq s$, by the property AP(3/2). Taking into account that

$$\|Q_h u\|_t = \sup_{\|v\|_{2-t}=1} |\langle Q_h u, v\rangle_1| \leqq \sup_{\|v\|_{2-t}=1} \|u\|_t\,\|Q_h^* v\|_{2-t} \leqq c\,\|u\|_t \tag{2}$$

for $1/2 < t < 1$, we can derive (1) for $1/2 < t \leqq s \leqq \alpha + 1$, $t < \alpha + 1/2$.

We now establish (1) when $0 \leqq t \leqq 1/2$ and $u \in \mathbf{H}^s$, $1/2 < s \leqq \alpha + 1$. There exists $v \in \mathbf{H}^{2s-t}$ such that $\|v\|_{2s-t} = 1$ and $\|u - Q_h u\|_t = |\langle u - Q_h u, v\rangle_s|$ (see 2.1(6)). Denoting the s-adjoint operator of Q_h by Q_h^*, we have

$$\|u - Q_h u\|_t = |\langle u, v - Q_h^* v\rangle_s| \leqq \|u\|_s\,\|v - Q_h^* v\|_s.$$

The operators Q_h^* are uniformly bounded in $\mathbf{H}^s$, and $\operatorname{im} Q_h^* = \Lambda^{-2s}(\mathcal{S}^{-1}(\tilde{\Delta}))$. From the property AP($2s - 1/2$) (see Corollary 2.19) we conclude that $\|v - Q_h^* v\|_s \leqq c h^{s-t}\,\|v\|_{2s-t}$. Hence, (1) is proved for $0 \leqq t \leqq s \leqq \alpha + 1$, $t < \alpha + 1/2$, $s > 1/2$ under the additional assumption that $\alpha > 1/2$.

In the case $0 < \alpha \leqq 1/2$ the proof runs analogously; one has merely to apply the estimate

$$\|\boldsymbol{u}\|_{t,h} \leqq c\,\|u\|_t \qquad \text{for all} \qquad u \in \mathbf{H}^t,\ 1/2 < t < 1,$$

which is a simple consequence of (2). ■

2.27. Corollary. *For any $u \in \mathbf{H}^t$, $1/2 < t < 1$, one has*

$$\|\boldsymbol{u}\|_{t,h} \leqq c\,\|u\|_t,$$

where $\boldsymbol{u} = (u(\xi_k))_{k=0}^{n-1} \in \mathbb{C}^n$ and c does not depend on u and h.

Proof. If follows from Theorem 2.23 that $\|\boldsymbol{u}\|_{t,h} \leqq c_1\,\|Q_h u\|_t$ for $t < \alpha + 1/2$. If $\alpha > 1/2$, then 2.26(2) implies

$$\|\boldsymbol{u}\|_{t,h} \leqq c_1\,\|Q_h u\|_t \leqq \|u\|_t$$

for $1/2 < t < 1$. ■

2.28. Lemma. *Let $\alpha > 0$ and $\varepsilon \neq 0$ (resp. $\varepsilon \neq 1/2$). Then, for any $f \in \mathbf{C}^\infty$ and $-1/2 < t < \alpha + 1/2$, the interpolation projection Q_h defined by 2.25(1)–2.25(2), satisfies*

$$\|(I - Q_h) f|_{\mathrm{im}\,Q_h}\|_t \to 0 \quad as \quad h \to 0.$$

Proof. Consider, for example, the case $\mathrm{im}\, Q_h = \mathcal{M}_h^\alpha$. By Theorem 2.13 there exist projections P_h^0 satisfying $\mathrm{im}\, P_h^0 = \mathcal{S}_h^0$, $\|P_h^0\|_{t-\alpha} \leqq c$ and

$$\|(I - P_h^0)\, u\|_{t-\alpha} \leqq ch^{1-t+\alpha}\, \|u\|_1, \qquad \|(I - P_h^0)\, f|_{\mathcal{S}_h^0}\|_{t-\alpha} \leqq ch^r,$$
$$-1/2 < t < \alpha + 1/2, \tag{1}$$

where $r = \min\{1, 1 - t + \alpha\}$. Setting $P_h := \Lambda^{-\alpha} P_h^0 \Lambda^\alpha$, we find that $\mathrm{im}\, P_h = \mathcal{M}_h^\alpha$ and, for $v_h \in \mathcal{M}_h^\alpha$,

$$\|(I - P_h)\, f v_h\|_t = \|(I - P_h^0)\, \Lambda^\alpha f v_h\|_{t-\alpha}$$
$$\leqq \|(I - P_h^0)\, f \Lambda^\alpha v_h\|_{t-\alpha} + \|(I - P_h^0)\, (\Lambda^\alpha f - f \Lambda^\alpha)\, v_h\|_{t-\alpha}. \tag{2}$$

In view of the well known commutator property of pseudodifferential operators (see Chapter 6), we have

$$\|(\Lambda^\alpha f - f\Lambda^\alpha)\, v\|_1 \leqq c\, \|v\|_\alpha, \qquad v \in \mathbf{H}^\alpha.$$

Hence, using property $\mathrm{IP}(\alpha + 1/2)$ for $\mathcal{M}_h^\alpha$ and (1), we arrive at

$$\|(I - P_h^0)\, (\Lambda^\alpha f - f\Lambda^\alpha)\, v_h\|_{t-\alpha} \leqq ch^r\, \|v_h\|_t. \tag{3}$$

Combining (1), (2) and (3), we get

$$\|(I - P_h)\, f v_h\|_t \leqq ch^r\, \|v_h\|_t.$$

This implies that

$$\|(I - Q_h)\, f v_h\|_t \leqq (1 + \|Q_h\|_t)\, \|(I - P_h)\, f v_h\|_t \leqq ch^r\, \|v_h\|_t,$$

since the interpolation projections Q_h are uniformly bounded in $\mathbf{H}^t$ for $1/2 < t < \alpha + 1/2$ (see the proof of Theorem 2.26).

For $-1/2 < t \leqq 1/2$, $1/2 < \beta < \alpha + 1/2$ and $v_h \in \mathrm{im}\, Q_h$, we choose $w \in \mathbf{H}^{2\beta - t}$, $\|w\|_{2\beta - t} = 1$, such that

$$\|(I - Q_h)\, f v_h\|_t = |\langle (I - Q_h)\, f v_h, w\rangle_\beta| \leqq \|(I - Q_h)\, f v_h\|_\beta\, \|(I - Q_h^*)\, w\|_\beta.$$

Taking into account that $\mathrm{im}\, Q_h^* = \Lambda^{-2\beta}\big(\mathcal{S}^{-1}(\tilde{\Delta})\big)$ and proceeding as in the proof of Theorem 2.26, we obtain that $\|(I - Q_h^*)\, w\|_\beta \leqq ch^{\min\{\beta - t, \beta\}}\, \|w\|_{2\beta - t}$. Consequently,

$$\|(I - Q_h)\, f v_h\|_t \leqq ch^{\min\{\beta - t, \beta\} + \min\{1, 1 - \beta + \alpha\}}\, \|v_h\|_\beta$$
$$\leqq ch^{\min\{1, 1 + t\} + \min\{0, \alpha - \beta\}}\, \|v_h\|_t.$$

Thus the lemma follows.

In particular, for $\alpha > 1/2$ and $1/2 < \beta < \alpha$, we get that

$$\|(I - Q_h)\, f v_h\|_t \leqq ch^{\min\{1, 1 + t\}}\, \|v_h\|_t:$$

Hence, for $-1/2 < t < \alpha + 1/2$, and $1/2 < \alpha$ if $t \leqq 1/2$,

$$\|(I - Q_h) f|_{\mathrm{im}\, Q_h}\|_t \leqq ch^\varrho, \qquad \varrho = \min\{1, 1 + t, \alpha + 1 - t\}. \quad \blacksquare \tag{4}$$

In comparison with Theorem 2.26, a worse estimate holds in the degenerate case.

2.29. Theorem. *Let $n = 1/h$ be odd and let $\varepsilon = 0$, $\alpha > 0$ (resp. $\varepsilon = 1/2$, $\alpha > -1$). Then, for any $u \in \mathbf{H}^s$, $-1/2 < s \leqq \alpha$, the interpolation projection Q_h defined by 2.25(1)–2.25(2), satisfies the estimate*

$$\|u - Q_h u\|_t \leqq ch^{s-t} \|u\|_{s+1}, \tag{1}$$

where $-1/2 \leqq t < \alpha - 1/2$, $t \leqq s$, and the constant c does not depend on u or h. The order of estimate (1) *is sharp.*

Proof. Let $P_{t,h}: \mathbf{H}^t \to \mathrm{im}\, Q_h$ be the orthogonal projection. These operators are uniformly bounded in $\mathbf{H}^{t+1}$ (see the proof of Theorem 2.12). It follows from Theorems 2.25 and 2.3 and from Corollary 2.27 that

$$\|Q_h u\|_t \leqq c\, \|\boldsymbol{u}\|_{t+1,h} \leqq c\, \|u\|_{t+1}, \qquad -1/2 < t < \alpha - 1/2.$$

Utilize the property $\mathrm{AP}(\alpha + 1/2)$ of $\mathrm{im}\, Q_h$ to get

$$\begin{aligned}\|u - Q_h u\|_t &\leqq \|u - P_{t,h} u\|_t + \|Q_h\|_{t+1\to t}\, \|u - P_{t,h} u\|_{t+1} \\ &\leqq c(h^{s+1-t}\, \|u\|_{s+1} + h^{s-t}\, \|u\|_{s+1}) \leqq ch^{s-t}\, \|u\|_{s+1}.\end{aligned}$$

We now suppose that

$$\|Q_h\|_{t+\varrho\to t} \leqq c$$

for certain $t < \alpha - 1/2$ and $\varrho \in [0, 1)$. Let $\{\tilde{\chi}_k\}_{k=0}^{n-1}$ denote the corresponding interpolation basis of $S(\mathrm{im}\, Q_h)$ (cf. 2.22(2)). In view of Theorem 2.23, we have that, for any $u = \sum_{k=0}^{n-1} u(\xi_k)\, \tilde{\chi}_k \in S(\mathrm{im}\, Q_h)$,

$$\|u\|_{t+\varrho} \leqq c\, \|\boldsymbol{u}\|_{t+\varrho,h}.$$

Hence,

$$\|Q_h u\|_t = \left\| \sum_{k=0}^{n-1} u(\xi_k)\, \tilde{\chi}_k \right\|_t \leqq c\, \|\boldsymbol{u}\|_{t+\varrho,h},$$

which is a contradiction to the remark after Theorem 2.25. $\blacksquare$

We are now in a position to formulate consequences of Theorems 2.26 and 2.29 concerning spline interpolation. Given a bounded 1-periodic function u and a number $\varepsilon \in [0, 1)$, we seek a spline $u_h \in \mathcal{S}_h^\delta$ such that

$$u_h(\xi_r) = u(\xi_r), \qquad r = 0, \ldots, n - 1, \qquad \xi_r = (r + \varepsilon)\, h. \tag{2}$$

2.30. Theorem. *For any $h = n^{-1}$ the uniquely determined interpolation spline u_h exists if and only if either δ is odd and $\varepsilon \neq 1/2$ or δ is even and $\varepsilon \neq 0$. Moreover, if $u \in \mathbf{H}^s$, $1/2 < s \leqq \delta + 1$, then*

$$\|u - u_h\|_t \leqq ch^{s-t}\, \|u\|_s, \tag{1}$$

where $0 \leqq t < \delta + 1/2$, $t \leqq s$, and the constant c does not depend on u or h.

In the cases where δ is odd and $\varepsilon = 1/2$ resp. δ is even and $\varepsilon = 0$, the interpolation spline u_h exists only for odd n and $\delta \geqq 1$. In these cases, if $u \in \mathbf{H}^s$, $1/2 < s \leqq \delta + 1$, then

$$\|u - u_h\|_t \leqq ch^{s-t-1} \|u\|_s ,$$

where $-1/2 < t < \delta - 1/2$, $t \leqq s - 1$.

Proof. Since $\mathcal{S}_h^\delta = \mathcal{M}_h^\delta$ for even δ and $\mathcal{S}_h^\delta = \mathcal{N}_h^\delta$ for odd δ (see Corollary 2.9), it remains to prove estimate (1) for $\delta = 0$, i.e. $u_h \in \mathcal{S}_h^0$. Obviously, $u(x) - u_h(x) = \sum_{k=0}^{n-1} \big(u(x) - u(\xi_k)\big) \times \Theta_k(x)$. Applying 2.16(3) to the function $f(x) = u(hx + kh)$, $0 \leqq x \leqq 1$, we find that

$$|u(hx + kh) - u(\xi_k)|^2 \leqq ch^{2s-1} \int_{kh}^{(k+1)h} \int_{kh}^{(k+1)h} \frac{|u(x) - u(y)|^2}{|x - y|^{1+2s}} \, dx \, dy .$$

Hence,

$$\|u - u_h\|_0^2 = \sum_{k=0}^{n-1} \int_{kh}^{(k+1)h} |u(x) - u(\xi_k)|^2 \, dx$$

$$\leqq ch^{2s} \int_0^1 \int_0^1 \frac{|u(x) - u(y)|^2}{|x - y|^{1+2s}} \, dx \, dy .$$

If $s = 1$, then

$$|u(x) - u(\xi_k)|^2 = \left| \int_x^{\xi_k} u' \, dx \right|^2 \leqq h \int_{kh}^{(k+1)h} |u'|^2 \, dx$$

and

$$\|u - u_h\|_0^2 \leqq h^2 \int_0^1 |u'|^2 \, dx .$$

Let now $0 < t < 1/2$. Choosing $v_h \in \mathcal{S}_h^0$ from AP(1/2) and applying IP(1/2), we conclude that

$$\|u - u_h\|_t \leqq \|u - v_h\|_t + \|v_h - u_h\|_t \leqq ch^{s-t} \|u\|_s + ch^{-t}(\|v_h - u\|_0 + \|u - u_h\|_0)$$

$$\leqq ch^{s-t} \|u\|_s . \ \blacksquare$$

Remark. If we choose

$$Q_h u(x) = \sum_{k=0}^{n-1} u(kh) \, \Theta_k(x) \qquad \text{for} \quad \varepsilon = 0,$$

where Θ_k is the characteristic function of the interval $[kh, (k + 1) h)$, then estimate (1), Theorem 2.23 and Lemma 2.28 remain valid.

2.31. Properties of the function $\Phi_{\varepsilon,\alpha}$. Finally, we exhibit some properties of the function $\Phi_{\varepsilon,\alpha}$ defined on $[0, 1]$ by

$$\Phi_{\varepsilon,\alpha}(\Theta) = \begin{cases} 1 & \text{if} \quad \Theta = 0, \\ \sigma^+_{\varepsilon,\alpha}(\Theta)/\sigma^-_{\varepsilon,\alpha}(\Theta) & \text{if} \quad 0 < \Theta < 1, \\ -1 & \text{if} \quad \Theta = 1, \end{cases} \tag{1}$$

where $\sigma^{\pm}_{\varepsilon,\alpha}$ are the functions introduced in Lemma 2.21. Some of the following properties are important in the analysis of spline approximation methods for singular integral and pseudodifferential equations on closed curves (see Chapters 10 and 13).

(i) $\Phi_{\varepsilon,\alpha} \in \mathbf{C}^\infty[0, 1]$ for fixed $\alpha > -1$, $\varepsilon \in (0, 1)$.

$\Phi^{-1}_{\varepsilon,\alpha} \in \mathbf{C}^\infty[0, 1]$ for $\alpha > 0$, $\varepsilon \neq 1/2$ and for $-1 < \alpha < 0$, $\varepsilon \neq 0$, $\varepsilon \neq 1/2$.

(ii) range $\Phi_{1/2,\alpha}$ = range $\Phi^{-1}_{0,\alpha} = [-1, 1]$.

(iii) $\Phi_{\varepsilon,\alpha}(\Theta) = 0$ if and only if $\varepsilon = 1/2$, $\Theta = 1/2$, $\alpha > -1$.

$\Phi^{-1}_{\varepsilon,\alpha}(\Theta) = 0$ if and only if $\varepsilon = 0$, $\Theta = 1/2$, $\alpha > 0$.

(iv) $\Phi_{\varepsilon,\alpha}(\Theta) = -\Phi_{1-\varepsilon,\alpha}(1 - \Theta) = -\overline{\Phi_{\varepsilon,\alpha}(1 - \Theta)}$.

(v) $\operatorname{Im} \Phi_{\varepsilon,\alpha}(\Theta) < 0$ for $0 < \varepsilon < 1/2$, $0 < \Theta < 1$.

(vi) $\forall z \in \mathbb{C},\ \alpha > 0\ \exists \varepsilon \in [0, 1),\ \Theta \in [0, 1]\colon \Phi_{\varepsilon,\alpha}(\Theta) = z$.

$\forall z \in [-1, 1] \cup \mathbb{C} \setminus \mathbb{R},\ \alpha > -1\ \exists \varepsilon \in (0, 1),\ \Theta \in [0, 1]\colon \Phi_{\varepsilon,\alpha}(\Theta) = z$.

(vii) The relations

$$\sigma^{\pm}_{\varepsilon,\alpha+1}(\Theta) = -(\alpha + 1)^{-1} \frac{\mathrm{d}}{\mathrm{d}\Theta}\, \sigma^{\mp}_{\varepsilon,\alpha}(\Theta),$$

$$\sigma^{\pm}_{\varepsilon,\alpha}(\Theta) = 2 \int_0^1 \frac{\sigma^{\mp}_{y,\alpha}(\Theta)}{1 - \mathrm{e}^{2\pi\mathrm{i}(\varepsilon-y)}}\, \mathrm{d}y \qquad \text{and} \qquad \sigma^{-}_{\varepsilon,0}(\Theta) = \frac{\pi\, \mathrm{e}^{\pi\mathrm{i}\Theta}\mathrm{e}^{-2\pi\mathrm{i}\varepsilon\theta}}{\sin \pi\Theta}$$

imply representations of $\Phi_{\varepsilon,\alpha}(\Theta)$ via singular integrals (see Chap. 6) if $\alpha \in \mathbb{N}_0$. For example, $\Phi_{\varepsilon,0}(\Theta) = 2 \int_0^1 \frac{\mathrm{e}^{2\pi\mathrm{i}\Theta(\varepsilon-y)}}{1 - \mathrm{e}^{2\pi\mathrm{i}(\varepsilon-y)}}\, \mathrm{d}y$.

(viii) $\frac{\mathrm{d}}{\mathrm{d}\Theta} \Phi^{\pm}_{\varepsilon,\alpha}(\Theta) \neq 0$, $\Theta \in (0, 1)$, provided the derivative exists.

Properties (i) and (iii) follow from the results obtained in Section 2.21. The other properties can be proved straightforwardly.

The subsequent two theorems on spline interpolation are particular cases of general results on the collocation of pseudodifferential equations on arbitrary meshes (see Chapter 13). Their proofs are essentially based on the Arnold-Wendland lemma.

2.32. Theorem. *Let $\Delta = \{x_k\}_{k=-\infty}^{\infty}$ be an arbitrary 1-periodic partition of $\mathbb{R}$. If $\delta \in \mathbb{N}$ is odd, then for any bounded 1-periodic function u there exists a uniquely determined interpolation spline $u_\Delta \in \mathcal{S}^\delta(\Delta)$ satisfying*

$$u_\Delta(x_k) = u(x_k) \quad \text{for all} \quad k \in \mathbb{Z}.$$

Moreover, if $u \in \mathbf{H}^s$, $(\delta + 1)/2 \leqq s \leqq \delta + 1$, then

$$\|u - u_\Delta\|_t \leqq c h^{s-t}\, \|u\|_s, \tag{1}$$

where $0 \leqq t \leqq (\delta + 1)/2$ and the constant c does not depend on u or Δ.

If $\Delta \in$ Part γ ($\gamma \geqq 1$) then estimate (1) *holds for all $0 \leqq t \leqq s \leqq \delta + 1$, $t < \delta + 1/2$ and $s > 1/2$.*

This is a consequence of Corollary 13.20.

2.33. Theorem. *Let $\Delta = \{x_k\}_{k=-\infty}^{\infty}$ be an arbitrary 1-periodic partition of $\mathbb{R}$. If $\delta \geqq 0$ is even and $u \in \mathbf{H}^s$, $s > -1/2$, then there exists a uniquely determined spline $u_\Delta \in \mathcal{S}^\delta(\Delta)$ such that*

$$\int_{x_k}^{x_{k+1}} u_\Delta(x)\,\mathrm{d}x = \int_{x_k}^{x_{k+1}} u(x)\,\mathrm{d}x \qquad \textit{for all } k \in \mathbb{Z}. \tag{1}$$

For $-1 \leqq t \leqq \delta/2 \leqq s \leqq \delta + 1$ one has

$$\|u - u_\Delta\|_t \leqq ch^{s-t}\,\|u\|_s,$$

where the constant c does not depend on u or Δ.

If $\Delta \in$ Part γ ($\gamma \geqq 1$) then the estimate holds for all $-1 \leqq t \leqq s \leqq \delta + 1$, $t < \delta + 1/2$ and $s > -1/2$.

This is a consequence of Theorem 13.29.

Part B. Trigonometric approximation

Generalized Hölder-Zygmund spaces

2.34. Definitions. Let $X = X_p$ be one of the usual spaces $\mathbf{L}^p$ (if $1 \leqq p < \infty$) or $\mathbf{C}$ (if $p = \infty$) of 2π-periodic complex-valued functions, the norm in X being

$$\|f\|_X := \begin{cases} \|f\|_p := \left\{\dfrac{1}{2\pi}\displaystyle\int_0^{2\pi} |f(x)|^p\,\mathrm{d}x\right\}^{1/p} & \text{if } 1 \leqq p < \infty, \\ \|f\|_\infty := \max\limits_{x\in\mathbb{R}} |f(x)| & \text{if } p = \infty. \end{cases}$$

Write $\mathrm{D} = \mathrm{d}/\mathrm{d}x$ and define

$$\mathbf{C}^s := \{f \in \mathbf{C} : \mathrm{D}^j f \in \mathbf{C} \text{ for } 0 \leqq j \leqq s\}, \qquad s \in \mathbb{N}_0.$$

In the usual way, $\mathbf{C}^s$ is made into a Banach space by putting

$$\|f\|_{\mathbf{C}^s} := \sum_{j=0}^{s} \|\mathrm{D}^j f\|_\infty, \qquad s \in \mathbb{N}_0.$$

For $k \in \mathbb{N}$, we introduce the set of *modulus type functions* Ω^k as the set of all functions ω satisfying the following conditions:

(a) $\omega : \mathbb{R}^+ \to \mathbb{R}^+$ (here $\mathbb{R}^+ := [0, \infty)$);

(b) ω is monotonically increasing;

(c) $\lim\limits_{h\to 0+} \omega(h) = \omega(0) = 0$;

(d) $\omega(h)\,h^{-k}$ is monotonically decreasing.

Given $\omega \in \Omega^k$, define the seminorm $\|f\|_{X^\omega}$ by

$$\|f\|_{X^\omega} := \sup_{h>0} \frac{\|\Delta_h^k f\|_X}{\omega(h)},$$

where $\Delta_h^1 f(x) = f(x+h) - f(x)$ and $\Delta_h^k f = \Delta_h^1(\Delta_h^{k-1} f)$, $k = 2, 3, \ldots$. In what follows, we will often use the well-known inequalities (see e.g. TIMAN [1])

$$\|\Delta_h^k f\|_X \leqq 2^{k-l} \|\Delta_h^l f\|_X, \qquad k \geqq l, \qquad f \in X$$

and

$$\|\Delta_h^k f\|_X \leqq h^{k-l} \|\Delta_h^l f^{(k-l)}\|_X, \qquad k \geqq l, \qquad f^{(k-l)} \in X, \tag{1}$$

where $f^{(k-l)}$ stands for the (distributional) $(k-l)$-th derivative of f.

We denote by $X^{r,\omega}$, $r \in \mathbb{N}_0$, the set $X^{r,\omega} := \{f \in X : f^{(j)} \in X,\ 1 \leqq j \leqq r, \text{ and } \|f^{(r)}\|_{X^\omega} < \infty\}$. Note that $X^{r,\omega}$ provided with the norm

$$\|f\|_{X^{r,\omega}} := \sum_{j=0}^{r} \|f^{(j)}\|_X + \|f^{(r)}\|_{X^\omega}$$

is a Banach space. Furthermore, let $\tilde{X}^{r,\omega}$ be the subspace of $X^{r,\omega}$ defined by

$$\tilde{X}^{r,\omega} := \left\{ f \in X^{r,\omega} : \lim_{h \to 0+} \frac{\|\Delta_h^k f^{(r)}\|_X}{\omega(h)} = 0 \right\}.$$

Notice that this subspace is only of interest in the case where $\omega(h)\, h^{-k} \to \infty$ as $h \to 0+$, because

$$\lim_{h \to 0+} \|\Delta_h^k f\|_X\, h^{-k} = 0$$

implies that $f \equiv$ const.

The spaces $X^{r,\omega}$ generalize the classical Hölder-Zygmund spaces, in which case $\omega(h) = h^\beta$ with $0 \leqq \beta \leqq k$. In fact, given $s > 0$, write

$$s = m + \alpha, \quad \text{where} \quad m \in \mathbb{N}_0 \quad \text{and} \quad 0 < \alpha \leqq 1. \tag{2}$$

Then the *Hölder-Zygmund space* is defined by

$$\mathcal{H}_p^s := \{f \in X_p : f^{(j)} \in X_p,\ 1 \leqq j \leqq m, \text{ and } [f^{(m)}]^\alpha < \infty\}$$

with the norm

$$\|f\|_{\mathcal{H}_p^s} := \sum_{j=0}^{m} \|f^{(j)}\|_p + [f^{(m)}]^\alpha,$$

where

$$[g]^\alpha := \begin{cases} \sup\limits_{h>0} \dfrac{\|\Delta_h^1 g\|_p}{h^\alpha}, & 0 < \alpha < 1, \\[2ex] \sup\limits_{h>0} \dfrac{\|\Delta_h^2 g\|_p}{h^\alpha}, & \alpha = 1 \end{cases}$$

(see TRIEBEL [1]). If $p = \infty$, we write $\mathcal{H}^s$ instead of $\mathcal{H}_\infty^s$.

We shall now list a few basic properties of $\mathcal{H}^s$, and refer the reader to the book of TRIEBEL [1]. The space $\mathcal{H}^s$ is a Banach space, but fails to be either reflexive or separable. It is possible to construct $\mathcal{H}^s$ by real interpolation between $\mathbf{C}^{s_0}$, $\mathbf{C}^{s_1}$ if $s_0, s_1 \in \mathbb{N}_0$ and $s = (1-\Theta)\, s_0 + \Theta s_1$ with $0 < \Theta < 1$, $s_0 \neq s_1$ and between $\mathcal{H}^{s_0}$, $\mathcal{H}^{s_1}$ as well.

There is a close connection with the more familiar (periodic) *Hölder space* $\mathbf{C}^{s,\alpha}$, defined for $s \in \mathbb{N}_0$ and $0 < \alpha \leqq 1$ by

$$\|f\|_{\mathbf{C}^{s,\alpha}} := \|f\|_{\mathbf{C}^s} + \sup_{h>0} \frac{\|\Delta_h^1 f\|_\infty}{h^\alpha}.$$

In fact, we have

$$\mathbf{C}^{s,\alpha} = \mathscr{H}^{s+\alpha}, \qquad s \in \mathbb{N}_0, \qquad 0 < \alpha < 1,$$

however, when $\alpha = 1$, the inclusions

$$\mathbf{C}^{s+1} \subsetneqq \mathbf{C}^{s,1} \subsetneqq \mathscr{H}^{s+1}, \qquad s \in \mathbb{N}_0$$

are strict (and continuous). Thus, if $f \in \mathscr{H}^1$, then f need not be Lipschitz continuous, although (see Zygmund [1, Vol. 1, p. 44]) its modulus of continuity can be at worst $O(\delta\,|\log \delta|)$.

The space $\mathscr{H}^s$ can be characterized in terms of best uniform approximation by trigonometric polynomials. Indeed, let

$$E_n(f, Y) := \inf_{v \in T_n} \|f - v\|_Y,$$

where

$$T_n := \operatorname{span}\{\mathrm{e}^{\mathrm{i}kt} : |k| \leqq n\}, \qquad n \in \mathbb{N}_0.$$

If $s > 0$ and $f \in \mathbf{C}$, then $f \in \mathscr{H}^s$ if and only if

$$E_n(f, \mathbf{C}) = O(n^{-s}) \quad \text{as} \quad n \to \infty$$

(see Timan [1, pp. 260, 333]).

Finally, if P_n denotes the orthogonal projection of $\mathbf{L}^2$ onto T_n, then

$$P_n f(x) = S_n f(x) := \sum_{k=-n}^{n} \hat{f}_k \,\mathrm{e}^{\mathrm{i}kx}, \qquad n \in \mathbb{N}_0,$$

where

$$\hat{f}_k := \frac{1}{2\pi} \int_{-\pi}^{\pi} f(t)\, \mathrm{e}^{-\mathrm{i}kt}\, \mathrm{d}t, \qquad k \in \mathbb{Z}.$$

In other words, $P_n f$ is the n-th partial sum of the Fourier series of $f \in X$. The n-th *de la Vallée Poussin mean* of f is defined by

$$\sigma_{n,l} f := \frac{1}{l+1} \sum_{k=n-l}^{n} S_k f, \qquad 0 \leqq l \leqq n.$$

To prove estimates in Hölder-Zygmund norms we need the following well-known results.

2.35. Lemma (Jackson's theorem; see Akhiezer [2], Timan [1], Shuk [1]). *If* $f^{(m)} \in X$ *then*

$$E_n(f, X) \leqq \frac{\pi}{2}\, (n+1)^{-m}\, \|f^{(m)}\|_X$$

and, for all $l \in \mathbb{N}$,

$$E_n(f, X) \leqq C_l (n+1)^{-m} \sup_{0 < h < 1/(n+1)} \|\Delta_h^l f^{(m)}\|_X.$$

The constant C_l can be estimated by

$$C_l \leqq \begin{cases} 3 & \text{if } l = 1, 2;\ m \geqq 0, \\ 18 & \text{if } l = 3 \\ 53 & \text{if } l = 4 \\ 2^l & \text{if } l \geqq 5 \end{cases} \quad m \geqq 1.$$

2.36. Lemma. (Zygmund [1], Dahmen [1]). *The well-known estimate*

$$\|f - S_n f\|_X \leqq B_p E_n(f, X)$$

holds with

$$B_p \leqq \begin{cases} 2.5 + 4\pi^{-2} \ln n & \text{if } 1 \leqq p \leqq \infty, \\ 4\big(p/(p-1)\big)^{1/p} + 2 & \text{if } 1 < p < 2, \\ 1 & \text{if } p = 2, \\ 4p^{1-1/p} + 2 & \text{if } 2 < p < \infty. \end{cases}$$

Furthermore,

$$\|f - \sigma_{n,l} f\|_X \leqq 36 \sum_{i=0}^{n+l} \frac{E_{i+n-l}(f, X)}{l + i + 1}. \tag{1}$$

Approximation by Fourier sums

2.37. Theorem. *Let $r, m \in \mathbb{N}_0$, $k, l, n \in \mathbb{N}$, $\omega_\alpha \in \Omega^l$ and $\omega_\beta \in \Omega^k$. Assume one of the following two conditions to be satisfied.*

(i) $0 \leqq m - r \leqq k - l$ *and* $q(h) := h^{m-r} \omega_\alpha(h)/\omega_\beta(h)$

is monotonically increasing or

(ii) $k \leqq m - r$ *(implying that $q(h)$ is monotonically increasing).*
Then for $f \in X^{m,\omega_\alpha}$ we have the estimate

$$\|f - S_n f\|_{X^{r,\omega_\beta}} \leqq D B_p q\left(\frac{1}{n+1}\right) \sup_{0<h<1/(n+1)} \|\Delta_h^l f^{(m)}\|_X / \omega_\alpha(h), \tag{1}$$

where

$$D := \begin{cases} C_l 2^{k+1} & \text{if } k \leqq m - r, \\ \max\left(C_l 2^{k+1}, 2C_l + \pi 2^{k-m+r-l-1}\right) & \text{if } m - r \leqq k - l, \end{cases}$$

or shortly

$$\|f - S_n f\|_{X^{r,\omega_\beta}} = \begin{cases} O\left(q\left(\frac{1}{n}\right)\right) & \text{if } 1 < p < \infty, \\ O\left(q\left(\frac{1}{n}\right) \log n\right) & \text{if } p = 1 \text{ or } p = \infty \end{cases} \tag{2}$$

as $n \to \infty$.

Proof. Put $H := \{h: 0 < h \leqq 1/(n+1)\}$, $G := \{h: h > 1/(n+1)\}$. In view of Lemmas 2.35 and 2.36 we obtain for $0 \leqq j \leqq r$ that

$$\|(f - S_n f)^{(j)}\|_X \leqq B_p E_n(f^{(j)}, X) \leqq B_p C_l (n+1)^{j-m}\, \omega_\alpha\left(\frac{1}{n+1}\right) \sup_{h\in H} \frac{\|\Delta_h^l f^{(m)}\|_X}{\omega_\alpha(h)}, \tag{3}$$

where we have used that $S_n f^{(j)} = (S_n f)^{(j)}$. Hence, by summation, we find that

$$\sum_{j=0}^{r} \|(f - S_n f)^{(j)}\|_X \leqq 2(n+1)^{r-m}\, B_p C_l \omega_\alpha\left(\frac{1}{n+1}\right) \sup_{h\in H} \frac{\|\Delta_h^l f^{(m)}\|_X}{\omega_\alpha(h)}. \tag{4}$$

To estimate the second term of the norm we split the supremum into two parts. First write

$$A := \sup_{h\in H} \frac{\|\Delta_h^k (f - S_n f)^{(r)}\|_X}{\omega_\beta(h)} \leqq B_p \sup_{h\in H} \frac{E_n(\Delta_h^k f^{(r)}, X)}{\omega_\beta(h)}.$$

If condition (i) is fulfilled then

$$\begin{aligned} E_n(\Delta_h^k f^{(r)}, X) &\leqq \frac{\pi}{2} \|\Delta_h^k f^{(r)}\|_X \leqq \frac{\pi}{2} \|\Delta_h^{k-m+r} f^{(m)}\|_X\, h^{m-r} \\ &\leqq \frac{\pi}{2}\, 2^{k-m+r-l} h^{m-r}\, \|\Delta_h^l f^{(m)}\|_X \end{aligned} \tag{5}$$

which gives that

$$A \leqq B_p \pi 2^{k-m+r-l-1} \sup_{h\in H} q(h) \sup_{h\in H} \frac{\|\Delta_h^l f^{(m)}\|_X}{\omega_\alpha(h)}. \tag{6}$$

Let now $k \leqq m - r$. Using Lemma 2.35 once more and taking into account that $\Delta_\delta^l \Delta_h^k f = \Delta_h^k \Delta_\delta^l f$, we get

$$\begin{aligned} E_n(\Delta_n^k f^{(r)}, X) &\leqq C_l \sup_{\delta\in H} \|\Delta_h^k \Delta_\delta^l f^{(m-k)}\|_X\, (n+1)^{r+k-m} \\ &\leqq C_l (n+1)^{r+k-m}\, h^k \sup_{\delta\in H} \frac{\|\Delta_\delta^l f^{(m)}\|_X}{\omega_\alpha(\delta)} \sup_{\delta\in H} \omega_\alpha(\delta). \end{aligned} \tag{7}$$

Hence,

$$A \leqq B_p C_l \frac{(n+1)^{r-m}}{\omega_\beta\left(\frac{1}{n+1}\right)}\, \omega_\alpha\left(\frac{1}{n+1}\right) \sup_{\delta\in H} \frac{\|\Delta_\delta^l f^{(m)}\|_X}{\omega_\alpha(\delta)}. \tag{8}$$

Combining (6) and (8) we obtain

$$A \leqq B_p q\left(\frac{1}{n+1}\right) \sup_{h\in H} \frac{\|\Delta_h^l f^{(m)}\|_X}{\omega_\alpha(h)} \begin{cases} \pi 2^{k-m+r-l-1} & \text{if } m - r \leqq k - l, \\ C_l & \text{if } k \leqq m - r. \end{cases} \tag{9}$$

The case $h \in G$ can be handled in a similar way. This gives

$$\begin{aligned} \sup_{h\in G} \frac{\|\Delta_h^k (f - S_n f)^{(r)}\|_X}{\omega_\beta(h)} &\leqq B_p 2^k \sup_{h\in G} \frac{E_n(f^{(r)}, X)}{\omega_\beta(h)} \\ &\leqq B_p 2^k C_l \frac{(n+1)^{r-m}}{\omega_\beta\left(\frac{1}{n+1}\right)}\, \omega_\alpha\left(\frac{1}{n+1}\right) \sup_{h\in H} \frac{\|\Delta_h^l f^{(m)}\|_X}{\omega_\alpha(h)}. \end{aligned} \tag{10}$$

Putting (4), (9) and (10) together, we arrive at the assertion. ■

2.38. Remarks. 1°. There is a gap $m - r < k < m - r + l$ in the conditions (i) and (ii) of Theorem 2.37. To make this clear we quote the following example. Let $m = 1$, $r = 0$, $k = l = 2$, $\omega_\alpha(h) = h$, $\omega_\beta(h) = h^2$. Then

$$X^{r,\omega_\beta} = \{f : \|\Delta_h^2 f\|_X = O(h^2), h \to 0+\} \subsetneqq$$
$$X^{m,\omega_\alpha} = \{f : E_n(f, X) = O(n^{-2}), n \to \infty\},$$

so that Theorem 2.37 does not make sense.

2°. Assertion 2.37(2) remains true, without any restrictions, for all $k, l \in \mathbb{N}$, $m, r \in \mathbb{N}_0$, $m \geqq r + 1$, $\omega_\alpha \in \Omega^l$, $\omega_\beta \in \Omega^{m-r} \cap \Omega^k$ and q monotonically increasing.

This is an immediate consequence of Theorem 2.37 and the inequality

$$\sup_{h>0} \frac{\|\Delta_h^k f^{(r)}\|_X}{\omega_\beta(h)} \leqq 2^{k-k'} \sup_{h>0} \frac{\|\Delta_h^{k'} f^{(r)}\|_X}{\omega_\beta(h)}$$

if $k' < k$ and $\omega_\beta \in \Omega^{k'} \subset \Omega^k$.

However, in the case $m = r$ we must assume $k \geqq l$, in general, because

$$\begin{aligned} &\{f \in X : \|\Delta_h^1 f\|_X = O(h), h \to 0+\} \subsetneqq \\ &\{f \in X : \|\Delta_h^2 f\|_X = O(h), h \to 0+\}. \end{aligned} \tag{1}$$

3°. Note that if the conditions formulated in Theorem 2.37 or in Remark 2° are fulfilled, then $X^{r,\omega_\beta} \supset X^{m,\omega_\alpha}$.

2.39. Corollary. *Under the assumptions of Theorem* 2.37 (*see also Remark* 2.38.2°) *we have, for* $f \in \tilde{X}^{m,\omega_\alpha}$,

$$\|f - S_n f\|_{X^{r,\omega_\beta}} = \begin{cases} o\left(q\left(\dfrac{1}{n}\right)\right) & \text{if } 1 < p < \infty, \\ o\left(q\left(\dfrac{1}{n}\right) \cdot \log n\right) & \textit{if } p = 1 \text{ or } p = \infty \end{cases}$$

as $n \to \infty$.

The following two corollaries illustrate what happens for particular choices of the parameters.

2.40. Corollary. *Let* $\omega_\alpha \in \Omega^2$, $\omega_\beta \in \Omega^2$, $m - r \geqq 2$, $2 \leqq p < \infty$ *and* $f \in X^{m,\omega_\alpha}$. *Then*

$$\|f - S_n f\|_{X^{r,\omega_\beta}} \leqq 120p\, q\left(\frac{1}{n+1}\right) \|f^{(m)}\|_{X^{\omega_\alpha}}.$$

We now are interested in estimates of $\|f - S_n f\|_{\mathscr{H}_p^s}$, where $f \in \mathscr{H}_p^t$, $t > s$. Theorem 2.37 gives good constants in these estimates, except for the case $t \in \mathbb{N}$, $t - 1 < s < t$ (where $k = 1$, $l = 2$ and $m = r$). In view of 2.38(1) Theorem 2.37 does not include this particular case. However, in the special situation of spaces $\mathscr{H}_p^s$ we can compute the constant by standard methods. It is natural that this constant tends to infinity as $t - s \to 0$.

2.41. Corollary. *Let* $f \in \mathscr{H}_p^t$. *Then, for* $t > s$,

$$\|f - S_n f\|_{\mathscr{H}_p^s} \leqq \varkappa(s, t)\, B_p n^{s-t} \|f\|_{\mathscr{H}_p^t}, \tag{1}$$

where

$$\varkappa(s, t) = \begin{cases} 6\left(3 + \dfrac{2}{1 - 2^{s-t}}\right) & \text{if } t \in \mathbb{N},\ t - 1 < s < t, \\ 24 & \text{otherwise.} \end{cases}$$

Proof. *Step* 1. If $t \notin \mathbb{N}$ or $t \in \mathbb{N}$ and $s < t - 1$, then the assertion is an immediate consequence of Theorem 2.37.

Step 2. Let $t \in \mathbb{N}$ and $s = t - 1$. Obviously,

$$\|f - S_n f\|_{\mathscr{H}_p^s} \leqq \sum_{j=0}^{t-2} \|(f - S_n f)^{(j)}\|_p + 2 \sup_{h>0} h^{-1} \|\Delta_h^1 (f - S_n f)^{(t-2)}\|_p .$$

Applying Theorem 2.37 (with $k = 1$ and $m - r = 1$) to each term of the right-hand side, we obtain (1).

Step 3. It remains to consider the case, where $t \in \mathbb{N}$ and $t - 1 < s < t$. This case can not be reduced to Theorem 2.37, since $r = m = t - 1$, $\omega_\alpha(h) = h$, $\omega_\beta(h) = h^{s-t+1}$, $k = 1$ and $l = 2$.

We have

$$\|f - S_n f\|_{\mathscr{H}_p^s} = \sum_{j=0}^{t-1} \|(f - S_n f)^{(j)}\|_p + \sup_{h>0} h^{t-s-1} \|\Delta_h^1 (f - S_n f)^{(t-1)}\|_p =: S_1 + S_2 .$$

Using 2.37(4), we find that

$$S_1 \leqq 6B_p (n+1)^{-1} \sup_{h \in H} \frac{\|\Delta_h^2 f^{(t-1)}\|_p}{h} \leqq 6B_p (n+1)^{s-t} \|f\|_{\mathscr{H}_p^t} . \tag{2}$$

We now estimate the term S_2. Applying Lemmas 2.35 and 2.36, we get

$$\begin{aligned} &\sup_{h \in G} h^{t-s-1} \|\Delta_h^1 (f - S_n f)^{(t-1)}\|_p \\ &\leqq 2(n+1)^{s+1-t}\, 3B_p (n+1)^{-1} \sup_{h \in H} \frac{\|\Delta_h^2 f^{(t-1)}\|_p}{h} \leqq 6B_p (n+1)^{s-t} \|f\|_{\mathscr{H}^t} . \end{aligned} \tag{3}$$

In view of Lemma 2.36, we obtain that

$$\sup_{h \in H} h^{t-s-1} \|\Delta_h^1 (f - S_n f)^{(t-1)}\|_p \leqq B_p \sup_{h \in H} h^{t-s-1} E_n(\Delta_h^1 f^{(t-1)}, X_p) . \tag{4}$$

Let $p_n \in T_n$ be the polynomial of best approximation for $f^{(t-1)}$, i.e. $E_n(f^{(t-1)}) = \|f^{(t-1)} - p_n\|_p$. Choose $a \in \mathbb{N}$ with $2^{a-1} < n \leqq 2^a$ to get

$$f^{(t-1)} - p_n = -p_n + p_{2^a} - \sum_{j=a}^{\infty} (p_{2^j} - p_{2^{j+1}}) ,$$

the series being convergent in X_p. Hence,

$$\begin{aligned} S_3 := \sup_{h \in H} \frac{E_n(\Delta_h^1 f^{(t-1)}, X_p)}{h^{s+1-t}} &\leqq \sup_{h \in H} \frac{\|\Delta_h^1 (p_n - p_{2^a})\|_p}{h^{s+1-t}} \\ &+ \sum_{j=a}^{\infty} \sup_{h \in H} \frac{\|\Delta_h^1 (p_{2^j} - p_{2^{j+1}})\|_p}{h^{s+1-t}} . \end{aligned}$$

Using Bernstein's inequality (see ZYGMUND [1, Volume II, p. 11]), we arrive at

$$\sup_{h\in H} \frac{\|\Delta_h^1 p_n\|_p}{h^{s+1-t}} \leqq \sup_{h\leqq 1/n} h^{t-s} \|p_n'\|_p \leqq n^{s-t+1} \|p_n\|_p .$$

This implies

$$S_3 \leqq 2^{(s-t+1)a} \|p_n - p_{2^a}\|_p + \sum_{j=a}^{\infty} 2^{(j+1)(s-t+1)} \|p_{2^j} - p_{2^{j+1}}\|_p$$
$$\leqq 2(2n)^{s-t+1} E_n(f^{(t-1)}, X_p) + 2\sum_{j=a}^{\infty} 2^{(j+1)(s-t+1)} E_{2^j}(f^{(t-1)}, X_p) .$$

Applying Lemma 2.35 once more, we obtain that

$$E_k(f^{(t-1)}, X_p) \leqq \frac{3}{k} \sup_{h<1/(k+1)} \frac{\|\Delta_h^2 f^{(t-1)}\|_p}{h} \leqq \frac{3}{k} \|f\|_{\mathscr{H}_p^t}$$

for $k \in \mathbb{N}$. Consequently,

$$S_3 \leqq \|f\|_{\mathscr{H}_p^t} \left(2\,(2n)^{s-t+1} \frac{3}{n} + 2\sum_{j=a}^{\infty} 2^{(j+1)(s-t+1)} \frac{3}{2^j}\right)$$
$$\leqq 12 \|f\|_{\mathscr{H}_p^t} \left(n^{s-t} + \frac{2^{(a+1)(s-t)}}{1-2^{s-t}}\right) \leqq 12 \|f\|_{\mathscr{H}_p^t} n^{s-t} \left(1 + \frac{1}{1-2^{s-t}}\right). \tag{5}$$

Combining (2), (3), (4) and (5), we arrive at the assertion. ■

Best approximation and operator norms

By a well-known result of BARI/STECHKIN [1], the properties

$$E_n(f, X) = O\left(\omega\left(\frac{1}{n}\right)\right), \qquad n \to \infty,$$

and

$$\|\Delta_h^k f\|_X = O(\omega(h)), \qquad h \to 0+,$$

are equivalent if and only if $\omega \in \Omega^k$ satisfies the estimate

$$1 < \varliminf_{h\to 0+} \frac{\omega(ch)}{\omega(h)} \leqq \varlimsup_{h\to 0+} \frac{\omega(ch)}{\omega(h)} < c^k$$

for some c. For this reason we do not try to characterize the best approximation in X^{r,ω_β} for the general choice of ω_α and ω_β. However, with the help of the de la Vallée Poussin means of f we can remove the $\log n$ term in the estimates. Then we obtain an order of convergence which is best possible at least for ω_α and ω_β from some subclasses of Ω^l and Ω^k containing $\omega_\alpha(h) = h^\alpha$, $\omega_\beta(h) = h^\beta$, $0 < \beta < k$, $0 < \alpha < l$.

2.42. Lemma. *Under the assumptions of Theorem* 2.37 *we have*

$$\|f - \sigma_{n,v} f\|_{X^{r,\omega_\beta}} \leqq 36D \sum_{i=0}^{n+v} \frac{1}{v+i+1} q\left(\frac{1}{i+n-v+1}\right) \sup_{0<h<1/(n-v+1)} \frac{\|\Delta_h^l f^{(m)}\|_X}{\omega_\alpha(h)} .$$

Remark. For $q(h)\, h^{-\gamma} < c$, a simple calculation yields that, as $n \to \infty$,

$$\|f - \sigma_{n,v}f\|_{X^{r,\omega_\beta}} = \begin{cases} O\left(n^{-\gamma} \log \dfrac{n}{v+1}\right) & \text{if } 0 \leqq \gamma < 1, \\ O\left(\dfrac{1}{n} \log \dfrac{n^2}{(v+1)(n-v+1)}\right) & \text{if } \gamma = 1, \\ O\left(\dfrac{1}{n} (n-v+1)^{1-\gamma} \log \dfrac{n}{v+1}\right) & \text{if } \gamma > 1. \end{cases}$$

Proof of the Lemma. By 2.36(1),

$$\|f - \sigma_{n,v}f\|_{X^{r,\omega_\beta}} \leqq 36 \sum_{i=0}^{n+v} \frac{1}{v+i+1} \left(\sum_{j=0}^{r} E_{i+n-v}(f^{(j)}, X) + \sup_{h>0} \frac{E_{i+n-v}(\Delta_h^k f^{(r)}, X)}{\omega_\beta(h)} \right).$$

Estimating the terms in the parantheses in the same way as in 2.37(4), 2.37(5), 2.37(7) and 2.37(10) gives the assertion immediately. ■

To have the best possible order of convergence let $v = [n/2]$. Using that $q(2/n) \leqq 2^{m-r+l} q(1/n)$, we obtain the following theorem.

2.43. Theorem. *Assume the conditions of Theorem* 2.37 *are fulfilled. Then*

$$E_n(f, X^{r,\omega_\beta}) \leqq 108 \cdot 2^{r-m+k} D q\left(\frac{1}{n}\right) \|f^{(m)}\|_{X^{\omega_\alpha}}.$$

In particular, if $f \in \tilde{X}^{m,\omega_\alpha}$, then

$$E_n(f, X^{r,\omega_\beta}) = o\left(q\left(\frac{1}{n}\right)\right), \qquad n \to \infty.$$

From the last result it follows immediately that the spaces $\tilde{X}^{m,\omega_\alpha}$ are separable for any choice of $\omega_\alpha \in \Omega^l$.

Recall that the n-th Fourier sum of a function $f \in X$ has the representation

$$S_n f(x) = \frac{1}{2\pi} \int_{-\pi}^{\pi} D_n(x-t)\, f(t)\, \mathrm{d}t \tag{1}$$

with the *Dirichlet kernel*

$$D_n(t) := \sum_{|k| \leqq n} \mathrm{e}^{ikt} = \frac{\sin\,[(n+1/2)\,t]}{\sin\,(t/2)} = 1 + 2 \sum_{k=1}^{n} \cos kx.$$

For $f \in \mathbf{C}$ and $n \in \mathbb{N}_0$, let $L_n f$ be the trigonometric interpolatory polynomial of degree n which interpolates f at the points

$$t_k = t_k^{(n)} := 2\pi k/(2n+1), \qquad |k| \leqq n, \tag{2}$$

so that $L_n f \in T_n$ and

$$(L_n f)\,(t_k^{(n)}) = f(t_k^{(n)}) \quad \text{for} \quad |k| \leqq n. \tag{3}$$

The Dirichlet kernel satisfies

$$D_n(t_k^{(n)}) = \begin{cases} 2n+1 & \text{if } k = 0 \pmod{2n+1}, \\ 0 & \text{if } k \neq 0 \pmod{2n+1}. \end{cases} \tag{4}$$

Therefore

$$(L_n f)(x) = \frac{1}{2n+1} \sum_{|k| \leqq n} f(t_k^{(n)})\, D_n(x - t_k^{(n)}) \tag{5}$$

and

$$(L_n f)_l^\wedge = \frac{1}{2n+1} \sum_{|k| \leqq n} \mathrm{e}^{-\mathrm{i} l t_k^{(n)}} f(t_k^{(n)}), \qquad |l| \leqq n.$$

(Of course, $(L_n f)_l^\wedge = 0$ for $|l| \geqq n+1$.)

Finally, let Λ_n denote the *Lebesgue constants*

$$\Lambda_n := \frac{1}{2\pi} \int_{-\pi}^{\pi} |D_n(x)|\, \mathrm{d}x, \qquad n \in \mathbb{N}_0.$$

The norms of the operators $P_n = S_n$ and L_n can be estimated as follows.

2.44. Lemma. *For $n \geqq 1$ we have*

$$\|S_n\|_{X_p \to X_p} \leqq \begin{cases} \Lambda_n & \text{if } 1 \leqq p \leqq \infty, \\ A_p & \text{if } 1 < p < \infty, \end{cases} \tag{1}$$

where

$$A_p := \begin{cases} 4\big(p/(p-1)\big)^{1/p} + 1 & \text{if } 1 < p < 2, \\ 1 & \text{if } p = 2, \\ 4p^{1-1/p} + 1 & \text{if } 2 < p < \infty \end{cases}$$

and

$$\Lambda_n \leqq \begin{cases} 4\pi^{-2} \log n + 1.436 & \text{if } n = 1, \\ 4\pi^{-2} \log n + 1.362 & \text{if } n > 1. \end{cases} \tag{2}$$

Proof. To prove the first part of (1) we use the well-known convolution theorem $\|S_n f\|_p \leqq \Lambda_n \|f\|_p$ for $f \in X_p$ (see also Chapter 4). Estimate (2) follows from the representation of the remainder by Hardy [1] (cf. also Zygmund [1, Chap. II.12]). In the case $1 < p < \infty$ inequality (1) is an immediate consequence of the estimate $\|S_n f\|_p \leqq A_p \|f\|_p$ for $f \in \mathbf{L}^p$, $1 < p < \infty$ (see Zygmund [1, Chap. VII]). ∎

2.45. Lemma. *For $n \geqq 1$ one has*

$$\|L_n\|_{\mathbf{C} \to X_p} \leqq \begin{cases} 1 & \text{if } 1 \leqq p \leqq 2, \\ 3A_p & \text{if } 2 < p < \infty, \\ \dfrac{2}{\pi} \log n + F_n & \text{if } 2 < p \leqq \infty \end{cases}$$

with

$$F_n = \begin{cases} 5/3 & \text{for } n = 1, \\ 1.548 & \text{for } n > 1. \end{cases}$$

Proof. The Parseval equation (see ZYGMUND [1, Chap. X.2])

$$\|L_n f\|_2^2 = \frac{1}{2n+1} \sum_{|k| \leq n} |(L_n f)(t_k)|^2$$

yields

$$\sup_{\|f\|_\infty = 1} \|L_n f\|_p \leq \sup_{\|f\|_\infty = 1} \|L_n f\|_2 = 1 \quad \text{if} \quad 1 \leq p \leq 2.$$

For $1 < p < \infty$ the assertion follows from (see ZYGMUND [1, Chap. X])

$$\|L_n f\|_p \leq 3A_p \left\{\frac{1}{2n+1} \sum_{|k| \leq n} |(L_n f)(t_k)|^p\right\}^{1/p}.$$

If $p = \infty$ then we have (EHLICH/ZELLER [1])

$$\|L_n\|_{\mathbf{C}\to\mathbf{C}} = \frac{1}{2n+1} \left\{1 + 2\sum_{k=0}^{n-1} \left(\sin \frac{2k+1}{4n+2}\pi\right)^{-1}\right\}. \tag{1}$$

Since

$$|(\sin x)^{-1} - x^{-1}| \leq 1 - 2/\pi \quad \text{for} \quad |x| \leq \pi/2, \qquad x \neq 0,$$

and

$$|(\sin x)^{-1} - x^{-1}| \leq \sqrt{2} - 4/\pi \quad \text{for} \quad |x| \leq \pi/4, \qquad x \neq 0,$$

in view of the monotonicity of the left-hand side for $0 < x \leq \pi/2$ it follows that

$$\|L_n\|_{\mathbf{C}\to\mathbf{C}} \leq \frac{1}{2n+1}\left[1 + n\left(1 + \sqrt{2} - \frac{6}{\pi}\right) + 2\sum_{k=0}^{n-1} \frac{(2n+1)\,2}{(2k+1)\,\pi}\right].$$

For $n \geq 20$ we can estimate

$$\|L_n\|_{\mathbf{C}\to\mathbf{C}} \leq \frac{1}{39} + \frac{1}{2} + \frac{\sqrt{2}}{2} - \frac{3}{\pi} + \frac{4}{\pi}\sum_{k=0}^{19} \frac{1}{2k+1}$$
$$+ \frac{2}{\pi}\log(2n-1) - \frac{2}{\pi}\log 39 < 1.545 + \frac{2}{\pi}\log n.$$

For $1 \leq n \leq 19$ we get the assertion by simple calculations in (1). ■

Remark. Using the same methods but sharper estimates, one can show that $F_n < 1.5$ for $n \geq 4$ and $\Lambda_n - 4\pi^{-2}\log n < 1.3$ for $n \geq 7$.

Trigonometric interpolation

We first estimate the difference between the Fourier sum $S_n f$ and the interpolatory polynomial $L_n f$ for functions $f \in \mathcal{H}^t$.

2.46. Theorem. *If $0 < s < t < \infty$ and $f \in \mathcal{H}^t$, then*

$$\|S_n f - L_n f\|_{\mathcal{H}_p^s} \leq \gamma_n(\|S_n\|_{\mathbf{C}\to X_p} + \|L_n\|_{\mathbf{C}\to X_p})\, E_n(f, \mathbf{C}), \tag{1}$$

where

$$\gamma_n \leq \begin{cases} s+3 & \text{if} \quad n = 1, \\ 4n^s & \text{if} \quad n > 1. \end{cases}$$

Proof. Let $s = m + \alpha$ as in 2.34(2) and put $g_n := S_n f - L_n f$. Applying Bernstein's inequality, we get

$$\|g_n^{(k)}\|_p \leqq n^k \|g_n\|_p \leqq n^k E_n(f, \mathbf{C}) \left(\|S_n\|_{\mathbf{C}\to X_p} + \|L_n\|_{\mathbf{C}\to X_p}\right). \tag{2}$$

We claim that if $0 < \alpha \leqq 1$, then

$$[g]^\alpha \leqq 2n^\alpha \|g\|_p \qquad \text{for all } g \in T_n. \tag{3}$$

Indeed, suppose first that $0 < \alpha < 1$. On the one hand, if $h \geqq n^{-1}$, then

$$\frac{\|\Delta_h g\|_p}{h^\alpha} \leqq \frac{2\,\|g\|_p}{h^\alpha} \leqq 2n^\alpha \|g\|_p,$$

and on the other hand, if $h < n^{-1}$, then

$$\frac{\|\Delta_h g\|_p}{h^\alpha} \leqq h^{1-\alpha} \|g^{(1)}\|_p \leqq n^\alpha \|g\|_p$$

because of 2.34(1). Now suppose $\alpha = 1$. If $h \geqq 2/n$, then

$$\frac{\|\Delta_h^2 g\|_p}{h} \leqq \frac{4\,\|g\|_p}{h} \leqq 2n \|g\|_p.$$

For $h < 2/n$ use 2.34(1) and Bernstein's inequality once more to get

$$\frac{\|\Delta_h^2 g\|_p}{h} \leqq h \|g^{(2)}\|_p \leqq 2n \|g\|_p.$$

This completes the proof of (3).

Hence,

$$[g_n^{(m)}]^\alpha \leqq 2n^\alpha \|g_n^{(m)}\|_p \leqq 2n^s \|g_n\|_p. \tag{4}$$

Collecting (2) for $0 \leqq k \leqq m$ and (4), we obtain that

$$\|g_n\|_{\mathcal{H}_p^s} = \sum_{k=0}^m \|g_n^{(k)}\|_p + [g_n^{(m)}]^\alpha \leqq \left(\sum_{k=0}^m n^k + 2n^s\right) E_n(f, \mathbf{C}) \left(\|S_n\|_{\mathbf{C}\to X_p} + \|L_n\|_{\mathbf{C}\to X_p}\right),$$

which implies (1). ■

2.47. Corollary. *If $0 < s < t < \infty$ and $f \in \mathcal{H}^t$, then*

$$\|f - L_n f\|_{\mathcal{H}_p^s} \leqq D(n, p, s, t)\, n^{s-t} \|f\|_{\mathcal{H}^t},$$

where

$$D(n, p, s, t) \leqq \begin{cases} 24 + \varkappa & \text{if } 1 \leqq p \leqq 2, \\ (48 + \varkappa)\, A_p + \varkappa & \text{if } 2 < p < \infty, \\ 4\pi^{-2}(12 + 6\pi + \varkappa) \\ \quad \times \log n + 34.92 + 2.5\varkappa & \text{if } 2 < p \leqq \infty \end{cases}$$

with $\varkappa = \varkappa(s, t)$ defined as in Corollary 2.41.

This is an immediate consequence of Theorem 2.46, Corollary 2.41, Lemmas 2.44 and 2.45, and the estimates $\|\cdot\|_{\mathcal{H}_p^s} \leqq \|\cdot\|_{\mathcal{H}_2^s}$ for $1 \leqq p \leqq 2$ and

$$E_n(f, \mathbf{C}) \leqq 3(n + 1)^{-t} \|f\|_{\mathcal{H}^t},$$

the latter following from Lemma 2.35. ■

Notes and comments

2.1.—2.4. Discrete Sobolev norms of the form 2.1(10) are used, for instance, in the theory of difference methods. In the proof of Lemma 2.2 we follow VAINIKKO [1]. Formula 2.2(9) and Theorem 2.3 are due to SCHMIDT [3].

2.6.—2.12. Theorems 2.6 and 2.11 were stated in ARNOLD/WENDLAND [1]. For uniform meshes and $s \leqq \delta$, Theorem 2.6 can be deduced from the results of AUBIN [1, Chap. 4] and HELFRICH [1] or BRAMBLE/SCOTT [1]. A proof of Theorem 2.11 in the case $s \leqq \delta$ is sketched in ARNOLD/WENDLAND [1]. Complete proofs of these theorems were given in ELSCHNER/SCHMIDT [1]. In our proofs of Theorems 2.6, 2.7 and 2.11 we follow ELSCHNER [2, Chap. 6.1]. The proof of Theorem 2.6 is based on an argument of HELFRICH [1], which may be extended to derive 2.6(1) simultaneously for all $t \in (-\infty, s]$. In the second part of the proof to Theorem 2.7 we used an argument of BABUSKA/AZIZ [1]. Corollary 2.9 is a well-known property of periodic splines (see e.g. SCHOENBERG [1], ARNOLD/WENDLAND [1]). Theorem 2.12 was established by SCHMIDT [2]; his proof is based upon an idea of NITSCHE [2].

2.13.—2.15. Approximation properties analogous to 2.13(1) have longly been used in finite element analysis, cf. NITSCHE/SCHATZ [1, Proposition 5] and NITSCHE/SCHATZ [2, Hypothesis A.3]. Theorem 2.13 was proved independently by PRÖSSDORF [5, 6] and by ARNOLD/WENDLAND [2] and is a cornerstone of the localization techniques developed in these papers (see also SCHMIDT [4, 5] and Chapter 13). Theorem 2.14 is due to PRÖSSDORF/RATHSFELD [3].

2.16.—2.19. Estimate 2.16(1) is a well-known property of the trapezoidal rule. Our proof follows SCHMIDT [4]. Lemma 2.17 and Corollary 2.18 are due to ARNOLD/WENDLAND [1]. Corollary 2.19 goes back to SCHMIDT [3, 4].

2.20.—2.32. The material of these sections is taken from SCHMIDT [4, Sect. 3.2—3.4] (see also SCHMIDT [3] for the case of integer α). The first part of Theorem 2.30 is due to SUBBOTIN [1], who studied ε-interpolation with non-periodic spline functions. All other results are SCHMIDT's (see also DUBEAU/SAVOIE [1] for estimates of the periodic quadrature spline interpolation).

2.34.—2.47. The asymptotic convergence order in Hölder-Zygmund norms for the Fourier sum and the interpolatory polynomials on equidistant nodes has been studied in a series of papers in the last 15 years. In this connection, we mention CHANDRA [1—3], GAZEWSKI/LENSKI [1], KROTOV [1], MC LEAN/PRÖSSDORF/WENDLAND [1], MC LEAN/WENDLAND [1], LEINDLER [1], MOHAPATRA/CHANDRA [1—2], MOHAPATRA/RODRIGUEZ [1], PRESTIN [1—3], PRESTIN/PRÖSSDORF [1], PRÖSSDORF [1, 2], PRÖSSDORF/SILBERMANN [8], SICKEL [1], STYPINSKI [1], ZOLOTAREVSKI [1].

In the case $0 < s < t < 1$ and without indicating explicit constants, Corollary 2.41 and Lemma 2.42 with $v = n$ were first proved by PRÖSSDORF [1, 2]; Corollaries 2.41 and 2.47 can be found with different proofs in the book PRÖSSDORF/SILBERMANN [8]. For arbitrary real numbers s and t and without calculating the constants, proofs of Corollaries 2.41 and 2.47 are given in MC LEAN/WENDLAND [1] and MC LEAN/PRÖSSDORF/WENDLAND [1], respectively. The constants in the estimates of Sections 2.41 and 2.44—2.47 were first determined by PRESTIN [3] for non-integer values of s and t. In Sect. 2.34—2.43 we follow PRESTIN/PRÖSSDORF [1]. In a sense, these results generalize theorems obtained by LEINDLER [1]. The idea of the proof of Theorem 2.46 is due to the authors [1, Chap. 2.3].

Chapter 3
Approximation methods for Fredholm integral equations

This chapter primarily concentrates on simple applications of the abstract operator approximation theory developed in Chapter 1 to Fredholm integral equations. Moreover, in the setting of Fredholm integral equations of the second kind, we anticipate a major part of the approximation methods which are applied in the next chapters in order to handle more difficult types of problems, such as integral equations with singular kernels or with discontinuous coefficients.

There are, of course, many other methods for the approximate solution of a wide variety of Fredholm and Volterra integral equations. Among the numerous references on the subject we mention the books by ANSELONE [1], ATKINSON [1], BAKER [1], BÜCKNER [1], FENYÖ/STOLLE [1], KANTOROVICH/AKILOV [1], KANTOROVICH/KRYLOV [1], KRASNOSELSKI et al. [1], MIKHLIN [4–5], MIKHLIN/SMOLITSKI [1], NEDELEC [1], NOBLE [2], VAINIKKO [1] and the recent monographs by BRUNNER/VAN DER HOUWEN [1], HACKBUSCH [1] and KRESS [1].

Polynomial approximation of Fredholm integral equations

3.1. Polynomial collocation method. Consider the Fredholm integral equation

$$x(t) + \int_{-1}^{1} k(t, s)\, x(s)\, \mathrm{d}s = y(t), \qquad -1 \leqq t \leqq 1. \tag{1}$$

Let ϱ be a nonnegative integrable function defined on $(-1, 1)$ such that ϱ^{-1} is integrable, too. Let $\mathbf{L}^p_\varrho(-1, 1)$ $(1 \leqq p < \infty)$ be the Banach space of all complex-valued functions on $(-1, 1)$ which are p-integrable with respect to the weight ϱ. The norm in $\mathbf{L}^p_\varrho(-1, 1)$ is defined by

$$\|f\|_{\mathbf{L}^p_\varrho} = \left(\int_{-1}^{1} \varrho(t)\, |f(t)|^p\, \mathrm{d}t \right)^{1/p}.$$

Assume the kernel function k to satisfy the following two conditions:

$$\sup_{-1 \leqq t \leqq 1} \int_{-1}^{1} \varrho^{-1}(s)\, |k(t, s)|^2\, \mathrm{d}s < \infty, \tag{A}$$

$$\sup_{|t_1 - t_2| \leqq \delta} \int_{-1}^{1} \varrho^{-1}(s)\, |k(t_1, s) - k(t_2, s)|^2\, \mathrm{d}s \underset{\delta \to 0}{\to} 0. \tag{B}$$

In view of (A) and (B), the integral operator T defined by

$$Tx(t) = \int_{-1}^{1} k(t, s)\, x(s)\, ds$$

is a compact operator from $X = \mathbf{L}^2_\varrho(-1, 1)$ to the space $Y = \mathbf{C}[-1, 1]$ of all continuous functions (see e.g. KANTOROVICH/AKILOV [1]).

Let ω_n, $n \in \mathbb{N}_0$, be polynomials of degree n which are orthogonal with respect to the weight ϱ, i.e.

$$\int_{-1}^{1} \omega_n(t)\, \omega_m(t)\, \varrho(t)\, dt = \delta_{nm}.$$

It is well known that all roots t_{jn} $(j = 0, \ldots, n)$ of ω_{n+1} are simple and $-1 < t_{0n} < t_{1n} < \cdots < t_{nn} < 1$ (see e.g. FREUD [1]). The collocation method determines an approximate solution to equation (1),

$$x_n(t) = \sum_{k=0}^{n} \alpha_{kn} t^k, \tag{2}$$

by the collocation equations

$$x_n(t_{jn}) + \int_{-1}^{1} k(t_{jn}, s)\, x_n(s)\, ds = y(t_{jn}), \qquad j = 0, \ldots, n. \tag{3}$$

Let $\mathcal{R}[-1, 1]$ denote the Banach space of (properly) Riemann integrable functions on $[-1, 1]$ equipped with the supremum norm.

3.2. Theorem. *Assume* 3.1(A), 3.1(B), *and* $\dim \ker (I + T) = 0$ *in* X. *Then there exist* $n_0 \in \mathbb{N}$ *and* $c > 0$ *such that for* $n \in \mathbb{N}$, $n \geqq n_0$, *and any function* $y \in \mathcal{R}[-1, 1]$ *the equations* 3.1(3) *are uniquely solvable and* $\|x - x_n\|_X \to 0$ *as* $n \to \infty$. *If* $y \in \mathbf{C}[-1, 1]$, *then*

$$\|x - x_n\|_X \leqq c \min_{\gamma_0, \ldots, \gamma_n} \max_{-1 \leqq t \leqq 1} \left| x(t) - \sum_{k=0}^{n} \gamma_k t^k \right|. \tag{1}$$

Proof. It is well known that for any $f \in \mathcal{R}[-1, 1]$ there exists a uniquely determined interpolatory polynomial $Q_n f$ of degree n satisfying

$$Q_n f(t_{jn}) = f(t_{jn}), \qquad j = 0, \ldots, n.$$

By a theorem of Erdös and Turan (cf. FREUD [1]), $\|Q_n f - f\|_X \to 0$ as $n \to \infty$. Hence,

$$\mathcal{R}[-1, 1] \subset \mathfrak{K}(I; \{Q_n, Q_n\})$$

(see Definition 1.22). By virtue of Theorem 1.25 (see also Remark 1.27, 1°), we have

$$\mathcal{R}[-1, 1] \subset \mathfrak{K}(I + T; \{Q_n(I + T)\, Q_n, Q_n\}) = \mathfrak{K}(I; \{Q_n, Q_n\}).$$

If $y \in \mathbf{C}[-1, 1]$, then $x \in \mathbf{C}[-1, 1]$. Applying the estimate 1.25(1), we get

$$\|x - x_n\|_X \leqq c_1 \|x - Q_n x\|_X \leqq c_2 (1 + \|Q_n\|_{\mathbf{C} \to X}) \|x - p_n\|_{\mathbf{C}}, \tag{2}$$

where p_n is an arbitrary polynomial of degree n. Hence, (1) follows from (2) because of the Banach-Steinhaus theorem. ■

In many applications the weight $\varrho(t) = (1 - t^2)^{-1/2}$ plays an important role. In this case, ω_n are the Chebyshev polynomials and

$$t_{jn} = \cos \frac{2j+1}{2(n+1)} \pi, \qquad j = 0, \ldots, n, \tag{3}$$

are the Chebyshev nodes. Moreover, $\|L_n f - f\|_{L^p_\varrho} \to 0$ as $n \to \infty$ for any function $f \in C[-1, 1]$. Consequently, in the case of the Chebyshev polynomials, estimate (1) remains true with $X = L^p_\varrho(-1, 1)$ $(1 \leqq p < \infty)$ provided $y \in C[-1, 1]$ and $k \in C([-1, 1]^2)$.

3.3. Trigonometric collocation method. Consider the Fredholm integral equation

$$x(s) + \int_0^\pi K(s, \sigma)\, x(\sigma)\, d\sigma = y(s), \qquad 0 \leqq s \leqq \pi, \tag{1}$$

in the space $L^p(0, \pi)$ $(1 \leqq p < \infty)$ or $C[0, \pi]$, where the kernel function K is assumed to be continuous on $[0, \pi]^2$. Let

$$s_j = s_j^{(n)} = 2\pi j/(2n+1), \qquad j = 0, \ldots, n, \tag{2}$$

and determine an approximate solution to equation (1) in the form of a trigonometric polynomial

$$x_n(s) = \sum_{k=0}^{n} \gamma_k \cos ks \tag{3}$$

such that

$$x_n(s_j) + \int_0^\pi K(s_j, \sigma)\, x_n(\sigma)\, d\sigma = y(s_j), \qquad j = 0, \ldots, n. \tag{4}$$

Note that the integral operator

$$Tx(s) = \int_0^\pi K(s, \sigma)\, x(\sigma)\, d\sigma$$

is a compact operator from $L^p(0, \pi)$ into $C[0, \pi]$. Let $\mathcal{H}^\mu[0, \pi]$ $(0 < \mu < 1)$ be the space of all functions $f \in C[0, \pi]$ satisfying a Hölder condition with exponent μ, i.e.

$$|f(s) - f(t)| \leqq c\, |s - t|^\mu, \qquad 0 \leqq s, \quad t \leqq \pi.$$

3.4. Theorem. *Assume* $\dim \ker (I + T) = 0$ *in* $L^p(0, \pi)$. *For each function* $y \in \mathcal{R}[0, \pi]$ *and for all sufficiently large* n, *the system* 3.3(4) *has exactly one solution* x_n *and, as* $n \to \infty$, *the functions* x_n *converge in the norm of* $L^p(0, \pi)$ *to the solution* $x \in L^p(0, \pi)$ *of equation* 3.3(1). *If* $y \in C[0, \pi]$, *then the error estimate*

$$\|x - x_n\|_{L^p} \leqq c e_n(x) \tag{1}$$

holds, where $c = \text{const} > 0$ *and*

$$e_n(x) = \min_{\gamma_0, \ldots, \gamma_n} \max_{0 \leqq s \leqq \pi} \left| x(s) - \sum_{k=0}^{n} \gamma_k \cos ks \right|.$$

Moreover, if $y \in \mathcal{H}^\mu[0, \pi]$, *and* $K \in \mathcal{H}^\mu([0, \pi]^2)$ $(0 < \mu < 1)$, *then*

$$\|x - x_n\|_{L^p} = O(n^{-\mu}), \qquad \|x - x_n\|_C = O\left(\frac{\log n}{n^\mu}\right). \tag{2}$$

Proof. For $f \in \mathcal{R}[0, \pi]$, let

$$Q_n f(s) = \sum_{k=0}^{n} c_k^{(n)} \cos ks, \qquad 0 \leqq s \leqq \pi,$$

be the trigonometric interpolatory polynomial to f with nodes $s_j = 2\pi j/(2n + 1)$, $j = 0, \ldots, n$. It is well-known (see e.g. NATANSON [1]) that the coefficients $c_k^{(n)}$ are defined by

$$c_0^{(n)} = (2n + 1)^{-1} \left[f(0) + 2 \sum_{j=1}^{n} f(s_j)\right],$$

$$c_k^{(n)} = 2(2n + 1)^{-1} \left[f(0) + 2 \sum_{j=1}^{n} f(s_j) \cos ks_j\right], \qquad k = 1, \ldots, n.$$

Let $\tilde{f} \in \mathcal{R}[-\pi, \pi]$ be the 2π-periodic even continuation of the function f, and let $L_n\tilde{f}$ denote the trigonometric interpolatory polynomial defined by 2.43(3). It is easily seen that the restriction of $L_n\tilde{f}$ to the interval $[0, \pi]$ equals $Q_n f$. Hence, using Lemma 2.45 and the estimate

$$\|\tilde{f} - L_n\tilde{f}\|_{\mathbf{L}^p} \leqq c_1(1 + \|L_n\|_{\mathbf{C}\to\mathbf{L}^p})\, E_n(\tilde{f}, \mathbf{C})$$

we obtain that, for $f \in \mathcal{H}^\mu[0, \pi]$,

$$\|f - Q_n f\|_{\mathbf{L}^p} = O(n^{-\mu}), \qquad 1 \leqq p < \infty, \tag{3}$$

and

$$\|f - Q_n f\|_{\mathbf{C}} = O\left(\frac{\log n}{n^\mu}\right). \tag{4}$$

Moreover, $\|f - Q_n f\|_{\mathbf{L}^p} \to 0$ as $n \to \infty$ for any function $f \in \mathcal{R}[0, \pi]$. Consequently, the first part of Theorem 3.4 can be derived in the same manner as in the proof of Theorem 3.2.

It remains to prove (2) provided $y \in \mathcal{H}^\mu[0, \pi]$ and $K \in \mathcal{H}^\mu([0, \pi]^2)$. In this case we have $Tx \in \mathcal{H}^\mu[0, \pi]$. Hence, $x \in \mathcal{H}^\mu[0, \pi]$ and the first estimate in (2) is an immediate consequence of (1). To deduce the second estimate in (2), we consider equation 3.3(1) in the space $\mathbf{C}[0, \pi]$. Applying Theorem 1.25 and taking into account that T is a compact operator from $\mathbf{C}[0, \pi]$ into $\mathcal{H}^\nu[0, \pi]$, $0 < \nu < \mu$, we get

$$\|x - x_n\|_{\mathbf{C}} \leqq c_2 \|x - Q_n x\|_{\mathbf{C}}. \tag{5}$$

Thus (2) follows from (4) and (5). ■

3.5. Method of mechanical quadratures. Consider again the integral equation 3.3(1) with a continuous kernel function K. The method of mechanical quadratures consists in finding an approximate solution in the form 3.3(3), but the coefficients $\gamma_k = \gamma_k^{(n)}$ $(k = 0, \ldots, n)$ are now determined through the system

$$\sum_{k=0}^{n} (\alpha_{jk} + \beta_{jk})\, \gamma_k = y(s_j), \qquad s_j = \frac{2\pi j}{2n + 1} \qquad (j = 0, \ldots, n), \tag{1}$$

where

$$\alpha_{jk} = \cos ks_j,$$

$$\beta_{jk} = \frac{2\pi}{2n + 1} \left[\sum_{l=1}^{n} K(s_j, \sigma_l) \cos k\sigma_l\right] + \frac{\pi}{2n + 1} K(s_j, 0) \tag{2}$$

$$(j, k = 0, 1, \ldots, n)$$

and

$$\sigma_l = \frac{2\pi l}{2n+1}, \qquad l = 0, 1, \ldots, n.$$

3.6. Theorem. *Assume* $\dim \ker (I + T) = 0$. *Then, for each function* $y \in \mathcal{R}[0, \pi]$ *and for all sufficiently large* n, *the system* 3.5(1) *has precisely one solution* x_n *and, as* $n \to \infty$, *the functions* x_n *converge in the norm of* $\mathbf{L}^p(0, \pi)$ $(1 \leqq p < \infty)$ *to the solution* $x \in \mathbf{L}^p(0, \pi)$ *of equation* 3.3(1).

Moreover, if $y \in \mathcal{H}^\mu[0, \pi]$ *and* $K \in \mathcal{H}^\mu([0, \pi]^2)$ $(0 < \mu < 1)$, *then the estimates* 3.4(2) *hold.*

Proof. *Step 1.* Let Q_n be the operator introduced in the proof of Theorem 3.4. For any function $f \in \mathcal{R}[0, \pi]$ satisfying $f(0) = 0$, define

$$P_n f(s) = \sum_{k=1}^{n} d_k^{(n)} \sin ks, \qquad 0 \leqq s \leqq \pi,$$

as the trigonometric interpolatory polynomial to f with nodes $s_j = 2\pi j/(2n+1)$, $(j = 1, \ldots, n)$, i.e. (cf. Natanson [1])

$$d_k^{(n)} = 4(2n+1)^{-1} \sum_{j=1}^{n} f(s_j) \sin ks_j, \qquad j = 1, \ldots, n.$$

A simple computation shows that $P_n f = L_n \hat{f} \mid [0, \pi]$, where $\hat{f}$ is the odd continuation of f to $[-\pi, \pi]$.

We first prove that the system 3.5(1) coincides with the operator equation

$$x_n + W_n x_n = Q_n y,$$

where

$$W_n x_n := Q_n \int_0^\pi \cos\sigma \, Q_{n\sigma}[K(s, \sigma) \cos\sigma \, x_n(\sigma)] \, d\sigma + Q_n \int_0^\pi \sin\sigma \, P_{n\sigma}[K(s, \sigma) \sin\sigma \, x_n(\sigma)] \, d\sigma.$$

By a simple computation we obtain that

$$\int_0^\pi \cos\sigma \, Q_{n\sigma}[K(s, \sigma) \cos\sigma \, x_n(\sigma)] \, d\sigma = \int_0^\pi \cos\sigma \sum_{k=0}^{n} c_k(s) \cos k\sigma \, d\sigma = \frac{\pi}{2} c_1(s),$$

$$c_1(s) := \frac{2}{2n+1} \left[K(s, 0) \sum_{k=0}^{n} \gamma_k + 2 \sum_{l=1}^{n} K(s, \sigma_l) \cos^2 \sigma_l \, x_n(\sigma_l) \right]$$

$$= \frac{2}{2n+1} \left[K(s, 0) \sum_{k=0}^{n} \gamma_k + 2 \sum_{k,l=1}^{n} \gamma_k K(s, \sigma_l) \cos^2 \sigma_l \cos k\sigma_l \right],$$

and

$$\int_0^\pi \sin\sigma \, P_{n\sigma}[K(s, \sigma) \sin\sigma \, x_n(\sigma)] \, d\sigma = \int_0^\pi \sin\sigma \sum_{k=1}^{n} d_k(s) \sin k\sigma \, d\sigma = \frac{\pi}{2} d_1(s),$$

$$d_1(s) := \frac{4}{2n+1} \sum_{k,l=1}^{n} \gamma_k K(s, \sigma_l) \sin^2 \sigma_l \cos k\sigma_l.$$

Hence,

$$\frac{\pi}{2}\left(c_1(s)+d_1(s)\right)=\frac{\pi}{2n+1}\,K(s,0)\sum_{k=0}^{n}\gamma_k+\frac{2\pi}{2n+1}\sum_{k,l=1}^{n}\gamma_k K(s,\sigma_l)\cos k\sigma_l\,.$$

Now it is easy to see that the equation

$$x_n+W_nx_n=x_n+2^{-1}\pi Q_n[c_1(s)+d_1(s)]=Q_nf(s)$$

coincides with the system 3.5(1).

Step 2. Setting

$$T_nx_n:=W_nx_n-Q_nTx_n$$

we obviously get

$$x_n+W_nx_n=x_n+Q_nTx_n+T_nx_n\,, \tag{1}$$

where $x_n+Q_nTx_n$ agrees with the left-hand side of the collocation system 3.3(4).

Let $h(s,\sigma)$ be the even continuation (with respect to each variable) of $K(s,\sigma)$ to $[-\pi,\pi]^2$, and let $\tilde f$ be the even continuation of f to $[-\pi,\pi]$. Putting

$$G_n\tilde f(s)=-\int_{-\pi}^{\pi}h(s,\sigma)\,\tilde f(\sigma)\,d\sigma+\int_{-\pi}^{\pi}\tilde f(\sigma)\,L_{n\sigma}[h(s,\sigma)]\,d\sigma$$

we can easily see that

$$T_nx_n=\frac{1}{2}\,L_nG_n\tilde x_n\big|_{[0,\pi]}\,. \tag{2}$$

Hölder's inequality gives

$$\|G_n\tilde x_n\|_{\mathbf{C}}\leqq\varepsilon_n\,\|\tilde x_n\|_{\mathbf{L}^p}\,, \tag{3}$$

where

$$\varepsilon_n:=c\max_{z\in[-\pi,\pi]}\left\{\int_{-\pi}^{\pi}|h(z,\varrho)-L_{n\varrho}[h(z,\varrho)]|^q\,d\varrho\right\}^{1/q}$$

with $q=p/(p-1)$ and $p>1$. It is well known that $\varepsilon_n=o(1)$ if $h\in\mathbf{C}([-\pi,\pi]^2)$ and $\varepsilon_n=o(n^{-\mu})$ if $h\in\mathcal{H}^\mu([-\pi,\pi]^2)$ (see Corollary 2.47).

We also know that the solution x of equation 3.3(1) belongs to $\mathbf{L}^p(0,\pi)$ for every p, $1\leqq p<\infty$. Thus, it is enough to consider the case where p is sufficiently large. Now Theorem 3.4, the relations (1) to (3) and Theorem 1.31 (with $X=Y=\mathbf{L}^p(0,\pi)$, $Z=\mathbf{C}[0,\pi]$ and $y_n=0$) give all assertions of Theorem 3.6, except for the second estimate in 3.4(2).

Step 3. In order to prove the latter assertion, we apply Theorem 1.31 with $X=Y=\mathbf{C}[0,\pi]$ and $Z=\mathcal{H}^\nu[0,\pi]$ $(0<\nu<\mu)$. Since the second estimate in 3.4(2) holds for the collocation solution (see Theorem 3.4), it only remains to estimate the norm $\|A_n^{-1}T_n\|_{\mathbf{C}}$, where $A_nx_n=x_n+Q_nTx_n$. From (2), (3) and the stability inequality

$$\|A_n^{-1}y_n\|_{\mathbf{C}}\leqq c_1\,\|y_n\|_{\mathbf{C}}\qquad (y_n\in\operatorname{im}Q_n)$$

(cf. Theorem 1.25 and Corollary 1.26) we deduce that

$$\|A_n^{-1}T_nx_n\|_{\mathbf{C}}\leqq c_1\,\|T_nx_n\|_{\mathbf{C}}\leqq c_1\,\|L_nG_n\tilde x_n\|_{\mathbf{C}}\leqq c_2\,(\log n)\,\varepsilon_n\,\|x_n\|_{\mathbf{C}}\,.$$

Hence, $\|A_n^{-1}T_n\|_{\mathbf{C}} = O\left(\frac{\log n}{n^\mu}\right)$ and the assertion follows from Theorem 1.31. ■

3.7. Remarks. 1°. Assume $K \in \mathbf{C}([0, \pi]^2)$ and $\dim \ker (I + T) = 0$. If the solution x of equation 3.3(1) is of bounded variation, then

$$\|x - x_n\|_{\mathbf{L}^p} = O(n^{-1/p}), \qquad 1 < p < \infty,$$

if only x_n is the solution of one of the systems 3.3(4) or 3.5(1).

This is an immediate consequence of the proofs of Theorems 3.4 and 3.6 and of the estimate

$$\|f - L_n f\|_{\mathbf{L}^p} = O(n^{-1/p}),$$

which is true for any function f of bounded variation on $[-\pi, \pi]$ (see Prestin [4], Zacharias [1]; cf. also Prössdorf/Silbermann [8, Chap. 2, Sect. 4.2]).

2°. Identifying each function on $[0, \pi]$ with its even continuation to $[-\pi, \pi]$, we obtain from the preceding proofs that

$$\|x - x_n\|_{\mathbf{H}^\nu} = O(n^{-\mu+\nu}) \qquad \text{if} \quad y \in \mathbf{H}^\mu, \qquad 0 \leqq \nu < \mu < \infty,$$

$$\|x - x_n\|_{\mathcal{H}^\nu} = O(n^{-\mu+\nu} \log n) \quad \text{if} \quad y \in \mathcal{H}^\mu, \qquad 0 < \nu < \mu < \infty,$$

provided K is sufficiently smooth.

The numerical solution of first-kind integral equations with logarithmic kernels

3.8. Notation. We now consider the equation

$$\int_\Gamma \log |t - \tau| \, g(\tau) \, |d\tau| = h(t), \qquad t \in \Gamma, \tag{1}$$

with Γ a simple smooth open curve in the complex plane. This equation arises in a variety of contexts, one being the study of plane elasticity crack problems.

Let Γ has a parametrization

$$\gamma(s) = \big(\xi(s), \eta(s)\big), \qquad -1 \leqq s \leqq 1, \tag{2}$$

with $|\gamma'(s)| \neq 0$, $-1 \leqq s \leqq 1$. For simplicity, assume γ is C^∞. Changing the variable,

$$s = \cos x, \qquad 0 \leqq x \leqq \pi, \tag{3}$$

the equation (1) can be written as

$$-\frac{1}{\pi} \int_0^\pi \log |a(x) - a(y)| \, u(y) \, dy = f(x), \qquad 0 \leqq x \leqq \pi, \tag{4}$$

where

$$a(x) = \gamma(\cos x), \qquad u(x) = g\big(a(x)\big) \, |\gamma'(\cos x)| \sin x,$$

$$f(x) = -\frac{1}{\pi} h\big(a(x)\big).$$

Note that $a \in \mathbf{C}^\infty(\mathbb{R})$.

The equation (4) is split as

$$(A_e + B_e)\, u = f, \tag{5}$$

where

$$A_e u(x) = -\frac{1}{\pi} \int_0^{\pi} u(y) \log\,[2e^{-1}\,|\cos x - \cos y|]\, dy, \tag{6}$$

$$B_e u(x) = \int_0^{\pi} b(x, y)\, u(y)\, dy, \tag{7}$$

$$b(x, y) = \begin{cases} -\dfrac{1}{\pi} \log \left| \dfrac{e}{2} \dfrac{a(x) - a(y)}{\cos x - \cos y} \right|, & x \pm y \neq 2\pi m, \\[2ex] -\dfrac{1}{\pi} \log \left| \dfrac{e}{2}\, \gamma'(\cos x) \right|, & x - y \quad \text{or} \quad x + y = 2\pi m, \end{cases} \tag{8}$$

with m an arbitrary integer. With the assumptions on γ, $b(x, y)$ is a C^∞ function of x and y. It is also 2π-periodic and even with respect to each variable.

For real t, let $\mathbf{H}^t$ denote the Sobolev space of 2π-periodic functions (distributions)

$$u(x) = \sum_{k=-\infty}^{\infty} \hat{u}_k\, e^{ikx}, \tag{9}$$

$$\hat{u}_k = \frac{1}{2\pi} \int_0^{2\pi} u(x)\, e^{-ikx}\, dx,$$

whose Fourier coefficients $\hat{u}_k$ satisfy

$$\|u\|_t := \Big[|\hat{u}_0|^2 + \sum_{k \neq 0} |k|^{2t}\, |\hat{u}_k|^2 \Big]^{1/2} < \infty.$$

We will be especially interested in the subspace $\mathbf{H}_e^t$ of even functions,

$$\mathbf{H}_e^t = \{f \in \mathbf{H}^t : f(-x) = f(x)\}.$$

Obviously,

$$A_e u = A u, \qquad u \in \mathbf{H}_e^t, \tag{10}$$

where

$$A u(x) = -\frac{1}{\pi} \int_0^{2\pi} u(y) \log \left| 2e^{-1/2} \sin \frac{x - y}{2} \right| dy. \tag{11}$$

From formulas 6.57(3) and 6.57(5), it follows that the operator A has the representation

$$A u(x) = \hat{u}_0 + \sum_{k \neq 0} \frac{\hat{u}_k}{|k|}\, e^{ikx}, \tag{12}$$

where $u \in \mathbf{H}^t$ is given by (9). Hence, for $u \in \mathbf{H}^t_{\mathrm{e}}$, (9), (10) and (12) yield

$$\begin{aligned} u(x) &= \hat{u}_0 + \sum_{k=1}^{\infty} \hat{u}_k \cos kx, \\ A_{\mathrm{e}} u(x) &= \hat{u}_0 + \sum_{k=1}^{\infty} \frac{\hat{u}_k}{k} \cos kx. \end{aligned} \tag{13}$$

This shows that the operator $A_{\mathrm{e}} : \mathbf{H}^t_{\mathrm{e}} \to \mathbf{H}^{t+1}_{\mathrm{e}}$ is invertible, with $\|A_{\mathrm{e}} u\|_{t+1} = \|u\|_t$ and

$$A_{\mathrm{e}}^{-1} u(x) = \hat{u}_0 + \sum_{k=1}^{\infty} k \hat{u}_k \cos kx. \tag{14}$$

Since the kernel function (8) is C^∞, the operator $B_{\mathrm{e}} : \mathbf{H}^t_{\mathrm{e}} \to \mathbf{H}^{t+1}_{\mathrm{e}}$ is compact for each real t. Thus the operator $A_{\mathrm{e}} + B_{\mathrm{e}} : \mathbf{H}^t_{\mathrm{e}} \to \mathbf{H}^{t+1}_{\mathrm{e}}$ is Fredholm with index zero. In what follows, we assume

$$\dim \ker (A_{\mathrm{e}} + B_{\mathrm{e}}) = 0. \tag{15}$$

Remark. It is known that (15) is satisfied if and only if the transfinite diameter of the curve Γ is different from 1 (see Hille [1]). For example, (15) is valid for each interval of length $l \neq 4$. Note that $A^{-1} = \Lambda$, where Λ is the Bessel potential operator defined in Sect. 2.1.

3.9. Trigonometric collocation method. This method for solving the equation 3.8(5) consists in determining a trigonometric polynomial

$$u_n(x) = \sum_{k=0}^{n} \gamma_k \cos kx \tag{1}$$

such that

$$(A_{\mathrm{e}} + B_{\mathrm{e}})\, u_n(s_j) = f(s_j), \qquad j = 0, \dots, n, \tag{2}$$

where s_j $(j = 0, \dots, n)$ is defined by 3.3(2). Since

$$A_{\mathrm{e}} \cos kx = \frac{1}{\max\{1, k\}} \cos kx, \qquad k = 0, 1, \dots,$$

(see 3.8(13)), we obtain the linear system

$$\sum_{k=0}^{n} \gamma_k \left(\frac{\cos ks_j}{\max\{1, k\}} + \int_0^{\pi} b(s_j, y) \cos ky \, \mathrm{d}y \right) = f(s_j), \qquad j = 0, \dots, n. \tag{3}$$

Similarly to the case of the equation 3.3(1), the integrals in (3) will be approximated by using a trapezoidal rule for even periodic functions. This leads to the following discrete collocation method.

3.10. Method of mechanical quadratures. This method consists in finding an approximate solution of the form 3.9(1), where $\{\gamma_k\}$ is determined by requiring

$$\sum_{k=0}^{n} \left(\frac{\cos ks_j}{\max\{1, k\}} + \beta_{jk} \right) \gamma_k = f(s_j), \qquad j = 0, \dots, n, \tag{1}$$

where

$$\beta_{jk} = \frac{2\pi}{2n+1}\left[\sum_{l=1}^{n} b(s_j, s_l)\cos ks_l\right] + \frac{\pi}{2n+1}\, b(s_j, 0),$$

$$s_j = 2\pi j/(2n+1), \qquad j, k = 0, \ldots, n.$$

3.11. Theorem. *Assume* 3.8(15). *Let* $f \in \mathbf{H}_e^{t+1}$, $t > -\frac{1}{2}$ (*resp.* $f \in \mathcal{H}_e^{t+1}$, $t > -1$). *Then for all sufficiently large* n, *both systems* 3.9(3) *and* 3.10(1) *have exactly one solution* u_n *and*

$$\|u - u_n\|_s = O\left(\frac{1}{n^{t-s}}\right), \qquad -1 \leqq s \leqq t$$

$$(\textit{resp. } \|u - u_n\|_{\mathcal{H}^s} = O\left(\frac{\log n}{n^{t-s}}\right), -1 < s \leqq t).$$

Proof. Substituting $\tilde{u}_n = A_e u_n$, we obtain that 3.9(2) and 3.10(1) take the form

$$\tilde{u}_n(s_j) + B_e A_e^{-1}\tilde{u}_n(s_j) = f(s_j), \qquad j = 0, \ldots, n,$$

and

$$\tilde{u}_n + W_n A_e^{-1}\tilde{u}_n = Q_n f,$$

respectively, where W_n and Q_n are the operators introduced in the proof of Theorem 3.6 with replacing $K(s, \sigma)$ by $b(x, y)$. Repeat the argumentations of the proofs to Theorems 3.4 and 3.6 with replacing $\mathbf{L}^p$ and $\mathbf{C}$ by $\mathbf{H}^{s+1}$ to find that

$$\|\tilde{u} - \tilde{u}_n\|_{s+1} = O\left(\frac{1}{n^{t-s}}\right), \qquad -1 \leqq s \leqq t$$

(resp. $\|\tilde{u} - \tilde{u}_n\|_{\mathcal{H}^{s+1}} = O\left(\frac{\log n}{n^{t-s}}\right)$, $-1 < s \leqq t$), where $\tilde{u} = A_e u$. It remains to note that A_e is an isometric isomorphism from $\mathbf{H}_e^s$ onto $\mathbf{H}_e^{s+1}$ (resp. from $\mathcal{H}_e^s$ onto $\mathcal{H}_e^{s+1}$). ■

3.12. Corollary. *Assume* 3.8(15). *Then for all sufficiently large* n,

$$\|u - u_n\|_{\mathbf{C}} = O\left(\frac{1}{n^{t-s}}\right), \qquad f \in \mathbf{H}_e^{t+1}, \qquad \frac{1}{2} < s \leqq t, \tag{1}$$

$$\|u - u_n\|_{\mathbf{C}} = O\left(\frac{\log n}{n^{t}}\right), \qquad f \in \mathcal{H}_e^{t+1}, \qquad 0 < t. \tag{2}$$

Spline approximation of Fredholm integral equations

3.13. Notation. Let $\mathbf{W}^{p,l}$ ($l \in \mathbb{N}_0$, $1 \leqq p \leqq \infty$) refer to the usual Sobolev space of functions defined on $[0, 1]$ and let $\mathbf{W}^{p,0} = \mathbf{L}^p = \mathbf{L}^p(0, 1)$ denote the Lebesgue space with norm $\|\cdot\|_p$. Put $\mathbf{C} = \mathbf{C}[0, 1]$.

Consider the integral equation

$$Au(x) := u(x) + Ku(x) = f(x), \qquad x \in [0, 1], \tag{1}$$

where $Ku(x) = \int_0^1 k(x, y)\, u(y)\, dy$. For simplicity, the following two conditions are assumed to be in force: (i) $k \in \mathbf{C}^\infty([0, 1]^2)$ and (ii) $u \in \mathbf{C}^\infty[0, 1]$ and $Au = 0$ imply $u = 0$. Hence, the operator A is invertible in $\mathbf{C}$ and in each space $\mathbf{W}^{p,l}$ as well.

Let $\mathcal{S}_n^d$ denote the space of piecewise polynomials of degree d on the uniform mesh $\{x_k = k/n : 0 \leqq k \leqq n\}$, i.e. $u \in \mathcal{S}_n^d$ if and only if $u|_{[x_k, x_{k+1}]}$ is a polynomial of degree $\leqq d$ for all k, $0 \leqq k \leqq n - 1$.

3.14. Galerkin's method with piecewise polynomials. This method determines an approximate solution $u_n \in \mathcal{S}_n^d$ to equation 3.13(1) such that

$$(u_n + Ku_n, v_n) = (f, v_n) \quad \text{for all} \quad v_n \in \mathcal{S}_n^d, \tag{1}$$

where $(.,.)$ stands for the scalar product in $\mathbf{L}^2$,

$$(u, v) = \int_0^1 u(x)\, \overline{v(x)}\, dx, \qquad u, v \in \mathbf{L}^2.$$

Note that (1) can be written in the form

$$P_n A u_n = u_n + P_n K u_n = P_n f, \tag{2}$$

P_n denoting the orthogonal projection $P_n : \mathbf{L}^2 \to \mathcal{S}_n^d$.

3.15. Theorem. *The Galerkin method* 3.14(2) *is stable in* $\mathbf{L}^p$ $(1 \leqq p \leqq \infty)$, *i.e. there are a constant* $c > 0$ *and an integer* n_0 *such that*

$$\|P_n A u_n\|_p \geqq c\, \|u_n\|_p, \qquad u_n \in \mathcal{S}_n^d, \qquad n \geqq n_0. \tag{1}$$

Proof. This is trivial if $K = 0$. For $K \neq 0$ our claim follows from Corollary 1.26 and Remark 1.27, 1°, since $K \in \mathcal{K}(\mathbf{L}^p) \cap \mathcal{K}(\mathbf{L}^\infty, \mathbf{C})$ $(1 \leqq p \leqq \infty)$, and $P_n \to I$ as $n \to \infty$ on $\mathbf{L}^p$ $(p < \infty)$ and on $\mathbf{C} \subset \mathbf{L}^\infty$ as well.

Consequently, if n is sufficiently large then the Galerkin equations 3.14(2) are uniquely solvable for any $f \in \mathbf{L}^p$ $(1 \leqq p \leqq \infty)$ and the error estimate

$$\|u - u_n\|_p \leqq c\, \|u - P_n u\|_p \tag{2}$$

holds, where $u \in \mathbf{L}^p$ is the exact solution of equation 3.13(1). ■

3.16. Theorem. *Let* $1 \leqq p \leqq \infty$ *and* $f \in \mathbf{W}^{p,d+1}$. *Then the Galerkin method* 3.14(2) *converges with optimal rate:*

$$\|u - u_n\|_p = O(n^{-d-1}) \quad as \quad n \to \infty. \tag{1}$$

Proof. Since $u \in \mathbf{W}^{p,d+1}$, (1) follows from 3.15(2) and the approximation property of the spline spaces $\mathcal{S}_n^d$ (see Lemma 5.21). ■

The *iterated Galerkin solution* is defined by $u_n^* = f - Ku_n$, where $u_n \in \mathcal{S}_n^d$ is the solution of 3.14(2).

3.17. Theorem. *Let* $1 \leqq p \leqq \infty$ *and* $f \in \mathbf{W}^{p,d+1}$. *Then* $\|u - u_n^*\|_p = O(n^{-2d-2})$ *as* $n \to \infty$.

Proof. It is easy to verify the equality

$$(I + KP_n)\,(u - u_n^*) = K(P_n - I)\, u. \tag{1}$$

Since $\|K(P_n - I)\|_{\mathbf{L}^p} \to 0$ as $n \to \infty$ (see Section 1.1(h)), the operators $I + KP_n$ are invertible in $\mathbf{L}^p$ and have uniformly bounded inverses as $n \to \infty$. Thus (1) gives the inequality

$$\|u - u_n^*\|_p \leqq c\,\|K(I - P_n)\,u\|_p. \tag{2}$$

Now the assertion follows from (2) and the estimate $\|K(I - P_n)\,u\|_p = O(n^{-2d-2})$, which will be proved in Section 5.54 for the more general case of weighted Sobolev spaces (and with K replaced by a Mellin integral operator). ■

3.18. Collocation with piecewise polynomials. Let $0 \leqq \xi_0 < \xi_1 < \cdots < \xi_d \leqq 1$ be fixed points and put $x_{ij} = (i-1)/n + \xi_j/n$, $i = 1, \ldots, n$, $j = 0, \ldots, d$. (In what follows x_{id} and $x_{i+1,0}$, $1 \leqq i \leqq n-1$, are viewed as different points if $\xi_0 = 0$ and $\xi_d = 1$.) Let $Q_n : \mathbf{L}^\infty \to \mathscr{S}_n^d$ denote the (uniformly bounded) interpolatory projections such that

$$Q_n u(x_{ij}) = u(x_{ij}), \qquad 1 \leqq i \leqq n, \qquad 0 \leqq j \leqq d,$$

for all $u \in \mathbf{C}[0,1] + \mathscr{S}_n^d$ (cf. 5.27(2) and Lemma 5.28 with $i_0 = 1$).

For the approximate solution of equation 3.13(1), we now consider the following collocation method: find $u_n \in \mathscr{S}_n^d$ such that

$$Q_n A u_n = u_n + Q_n K u_n = Q_n f. \tag{1}$$

3.19. Theorem. *The collocation method* 3.18(1) *is stable in* $\mathbf{L}^\infty$, *i.e.*

$$\|Q_n A u_n\|_\infty \geqq c\,\|u_n\|_\infty, \qquad u_n \in \mathscr{S}_n^d, \qquad n \geqq n_0, \tag{1}$$

when n_0 is large enough. Moreover, if $f \in \mathbf{W}^{\infty,d+1}$, then $\|u - u_n\|_\infty = O(n^{-d-1})$ as $n \to \infty$.

Proof. Using the corresponding properties of the projections Q_n (cf. Chap. 5), we may deduce (1) in the same manner as 3.15(1). As a consequence of (1) and Theorem 1.25, we have the estimate $\|u - u_n\|_\infty \leqq c\,\|u - Q_n u\|_\infty$, which implies the second assertion (cf. 5.59(2) and Lemma 5.23). ■

3.20. Iterated collocation (or product integration) method. This method determines u_n^* by

$$u_n^* = f - K u_n, \tag{1}$$

where u_n is defined by 3.18(1). In contrast to the collocation method 3.18(1), the order of convergence of the iterated collocation depends on the choice of the points $\{\xi_j\}$. In order to characterize this dependence we introduce the $(d+1)$-point interpolatory quadrature rule on $[0, 1]$

$$\int_0^1 v(x)\,dx \approx \sum_{j=0}^d w_j v(\xi_j), \qquad v \in \mathbf{C}[0,1], \tag{2}$$

of order $R = d + 1 + \nu$ $(\nu \geqq 0)$ with weights w_j and nodes ξ_j, i.e. (2) is exact for any polynomial v of degree at most $d + \nu$. For example, in the case of Simpson's rule, which corresponds to the choice $\xi_0 = 0$, $\xi_1 = 1/2$ and $\xi_2 = 1$, we have $\nu = 1$. Notice that $0 \leqq \nu \leqq d + 1$, and that $\nu = d + 1$ if and only if the nodes $\{\xi_j\}$ are the Gauss-Legendre points on $[0, 1]$ (see e.g. Brunner/van der Houwen [1, Sect. 2.3]).

3.21. Theorem. *Assume* $f \in \mathbf{W}^{\infty,R}$. *Then* $\|u - u_n^*\|_\infty = O(n^{-R})$ *as* $n \to \infty$.

Proof. From 3.18(1) and 3.20(1) it follows that $Q_n u_n^* = u_n$ and

$$(I + KQ_n)(u_n^* - u) = K(I - Q_n)\,u. \tag{1}$$

Using 3.19(1) and the uniform boundedness of Q_n, as in the proof of Theorem 3.17 we derive that the operators $I + KQ_n$ are invertible in $\mathbf{L}^\infty$ and have uniformly bounded inverses as $n \to \infty$ (cf. also the proof of Theorem 5.43). Thus (1) gives the inequality

$$\|u - u_n^*\|_\infty \leqq c\,\|K(I - Q_n)\,u\|_\infty. \tag{2}$$

Hence, the assertion is a consequence of (2) and the estimate $\|K(I - Q_n)\,u\|_\infty = O(n^{-R})$, which will be demonstrated in Section 5.62 for the more general case of weighted Sobolev spaces (and with K replaced by a Mellin integral operator). ■

3.22. Nyström's method and discrete collocation. Retaining the notations of the preceding sections, we introduce the composite quadrature rule corresponding to 3.20(2),

$$\int_0^1 u(x)\,\mathrm{d}x \approx \int_0^1 Q_n u(x)\,\mathrm{d}x = \sum_{i=1}^n \sum_{j=0}^d w_j u(x_{ij})/n. \tag{1}$$

Obviously, (1) is the result of shifting 3.20(2) to each interval $[(i-1)/n, i/n]$ and of summing over $i = 1, \ldots, n$. Using (1) we approximate the integral operator K in 3.13(1) by

$$K_n u(x) = \int_0^1 Q_{ny}\{k(x, y)\,u(y)\}\,\mathrm{d}y = \sum_{i=1}^n \sum_{j=0}^d w_j k(x, x_{ij})\,u(x_{ij})/n.$$

The *discrete collocation method* determines an approximate solution $u_n \in \mathcal{S}_n^d$ to equation 3.13(1) by the equation

$$u_n + Q_n K_n u_n = Q_n f. \tag{2}$$

The *Nyström method* (or the *discrete iterated collocation*) defines u_n^* by

$$u_n^* = f - K_n u_n. \tag{3}$$

Note that (2) and (3) imply that $Q_n u_n^* = u_n$. Thus (3) can be written as

$$(I + K_n Q_n)\,u_n^* = (I + K_n)\,u_n^* = f. \tag{4}$$

3.23. Theorem. *The Nyström method* 3.22(4) *is stable in* $\mathbf{L}^\infty$, *i.e.*

$$\|(I + K_n)\,u\|_\infty \geqq c\,\|u\|_\infty, \qquad u \in \mathbf{L}^\infty, \qquad n \geqq n_0, \tag{1}$$

when n_0 *is large enough. Moreover, the error estimate* $\|u - u_n^*\|_\infty = O(n^{-R})$, $n \to \infty$, *holds provided* $f \in \mathbf{W}^{\infty,R}$.

Proof. The estimate (1) follows from 3.19(1) and the operator norm convergence $\|(K - K_n)\,Q_n\|_\infty \to 0$ (see Lemma 5.81(ii)). By 3.22(4) we have $u_n^* - u = Ku - K_n u_n^*$. Hence $(I + K_n)(u_n^* - u) = (K - K_n)\,u$, and so (1) implies the estimate

$$\|u - u_n^*\|_\infty \leqq c\,\|(K - K_n)\,u\|_\infty. \tag{2}$$

Since $u \in \mathbf{W}^{\infty,R}$, the right-hand side of (2) has the order $O(n^{-R})$ (cf. the proof of Lemma 5.83). ■

Using the theory of collectively compact operators (cf. Chap. 1), one can establish the stability of Nyström's method for much more general quadrature rules and discontinuous kernel functions K (see ANSELONE [1] and ATKINSON [1]).

3.24. Theorem. *If $f \in \mathbf{W}^{\infty,d+1}$, then the discrete collocation method* 3.22(2) *converges with the optimal rate $\|u - u_n\|_\infty = O(n^{-d-1})$ as $n \to \infty$.*

Proof. Obviously,

$$\|u - u_n\|_\infty = \|u - Q_n u_n^*\|_\infty \leqq \|(I - Q_n)\, u\|_\infty + \|Q_n(u - u_n^*)\|_\infty$$
$$\leqq c\, \|(I - Q_n)\, u\|_\infty + c\, \|u - u_n^*\|_\infty .$$

Using 3.23(2) and taking into account that $u \in \mathbf{W}^{\infty,d+1}$, we get the assertion (cf. the proof of Theorem 3.19). ■

3.25. Remark. Theorems 3.15 to 3.24 remain true with $\mathcal{S}_n^d$ replaced by the spline spaces $\mathcal{S}^d(\Delta_n)$ of all piecewise polynomials on graded meshes $\Delta_n \in \mathcal{M}_q$ (see Definition 5.17).

Notes and comments

The approximate solution of integral equations goes back to NEUMANN [1], VOLTERRA [1] and FREDHOLM [1]. Convergence questions were studied by HILBERT [1] and E. SCHMIDT [1]. Apparently, NYSTRÖM [1] was the first to employ numerical integration with a quadrature formula for the approximate solution of a Fredholm integral equation in particular cases. Convergence theorems and error estimates for approximate solutions of Fredholm equations have been obtained under various assumptions by many investigators, including BRAKHAGE [1—2], BÜCKNER [1—3], FOX/GOODWIN [1], KANTOROVICH/KRYLOV [1], and MYSOVSKIKH [1—3]. The contributions of BRAKHAGE, KANTOROVICH/KRYLOV, and MYSOVSKIKH were direct historical antecedents of the collectively compact operator approximation theory, which has subsequently been developed and applied to a wide variety of Fredholm integral equations (with continuous and with discontinuous kernel functions as well) by ANSELONE [1], ATKINSON [1], VAINIKKO [1], and other mathematicians. For further details, more general results, and examples, the reader is referred to the books quoted at the beginning of this chapter.

3.1.—3.2. These results are due to VAINIKKO [1] (see also KRASNOSELSKI et al. [1]).

3.3.—3.7. The material of these sections is adapted from the book by PRÖSSDORF/SILBERMANN [8, Chap. 2, 3] (cf. also SCHLEIFF [3] for collocation).

3.8.—3.12. In Section 3.8 we essentially follow ATKINSON [2] and ATKINSON/SLOAN [1] (see also SLOAN/YAN [1]). Estimate 3.12(1) was established in ATKINSON/SLOAN [1] by different methods. The other results of these sections are published here for the first time.

3.13.—3.25. Here we essentially follow ELSCHNER [6], who studied the corresponding approximation methods for Mellin convolution equations and for Wiener-Hopf integral equations (see Chapter 5). Theorems 3.15 and 3.16 are standard (cf. ANSELONE [1], ATKINSON [1], BAKER [1]). The iterated Galerkin method goes back to SLOAN [1], see also SLOAN/THOMÉE [1]. For recent results on discrete Galerkin and collocation methods in the case of Fredholm integral equations with smooth kernel functions, we refer to ATKINSON/BOGOMOLNY [1], and JOE [1—2]. For further comments, e.g. concerning the use of graded meshes in the numerical solution of Fredholm and Volterra integral equations, see also the "Notes and comments" to Chapter 5.

Chapter 4
Functions of shifts on Banach spaces and their finite sections

Several classes of convolution operators can be interpreted as concrete realizations of functions of shifts given on the underlying Banach spaces. In this chapter we discuss this idea and study the finite section method for concrete classes of convolution operators from a unique point of view.

Continuous functions of shifts

4.1. Definitions. Throughout this chapter let X denote a Banach space and V a shift on X, i.e. V is a linear bounded operator on X which is only invertible from the left. Fix one of its left inverses, say $V^{(-1)}$, and write for brevity

$$V_n = \begin{cases} V^n, & \text{if } n \geq 0, \\ (V^{(-1)})^{-n}, & \text{if } n < 0. \end{cases}$$

In this and the next sections we present some results on the invertibility of continuous functions of shifts. Since some of these facts are well-known we omit their proofs and refer to the monographs Gohberg/Feldman [1] and Prössdorf [1]. For a given polynomial $R(t) = \sum_{j=-n}^{n} a_j t^j$ on the complex unit circle $\mathbb{T}$, we define an operator $R(V)$ by $\sum_{j=-n}^{n} a_j V_j$ and call $R(V)$ a *polynomial* of V. Let $\mathfrak{L}_0(V)$ stand for the set of all polynomials of V and $\mathfrak{L}_0^+(V)$ $\left(\mathfrak{L}_0^-(V)\right)$ for the set

$$\left\{\sum_{i=0}^{n} a_i V_i : n \in \mathbb{Z}^+\right\} \quad \left(\left\{\sum_{i=-n}^{0} a_i V_i : n \in \mathbb{Z}^+\right\}\right).$$

It is worth mentioning that for $R_1 \in \mathfrak{L}_0^-(V)$, $R_2 \in \mathfrak{L}_0(V)$, and $R_3 \in \mathfrak{L}_0^+(V)$ the equality

$$(R_1 R_2 R_3)(V) = R_1(V)\, R_2(V)\, R_3(V) \tag{1}$$

holds. This property is important in what follows and will be frequently used without comment.

Note that there is a one-to-one correspondence between the operators $R(V)$ in $\mathfrak{L}_0(V)$ and the polynomials on $\mathbb{T}$.

Indeed, if $0 = \sum_{j=-n}^{n} a_j V_j$ and $a_{-n} \neq 0$ then $0 = \sum_{j=-n}^{n} a_j V_j V_n = \sum_{j=0}^{2n} b_j V_j$ with $b_j = a_{j-n}$.

Since $0 \neq a_{-n}I = V\left(-\sum_{j=1}^{2n} b_j V_{j-1}\right)$, this yields the right invertibility of V, which contradicts our hypothesis.

The polynomial $R(t)$ is called the *symbol* of $R(V)$. The following assumption (H1) will play an exceptionally important role in deriving invertibility criteria for polynomials and functions of shifts.

(H1) The spectra sp V and sp V_{-1} are contained in $\{z \in \mathbb{C}: |z| \leqq 1\}$.

4.2. Theorem. *Let* (H1) *be fulfilled. Then*

(i) $\operatorname{sp} V = \operatorname{sp} V_{-1} = \{z: |z| \leqq 1\}$.

(ii) *An operator $R \in \mathfrak{L}_0(V)$ is at least one-sided invertible if its symbol $R(t)$ has no zeros on $\mathbb{T}$. If $R(t) \neq 0$ on $\mathbb{T}$ then the invertibility of R corresponds with the index of $R(t)$, i.e. R is invertible, invertible only from the left or only from the right, respectively, if the winding number of $R(t)$,*

$$\operatorname{wind} R(t) := \frac{1}{2\pi} [\arg R(e^{iz})]_{z=0}^{2\pi},$$

is zero, positive, or negative, respectively.

(iii) *If $R(t_0) = 0$ for some $t_0 \in \mathbb{T}$ then R is neither a Φ_+- nor a Φ_--operator.*

(iv) *If $R(V)$ is one-sided invertible then there exists a one-sided inverse of $R(V)$ in the algebra generated by V and V_{-1}.*

(v) *The spectral radius of $R(V)$ equals* $\max_{t \in \mathbb{T}} |R(t)|$.

Proof.

(i) See Gohberg/Feldman [1], Lemma 1.1, Chap. 1.

(ii) If $R \in \mathfrak{L}_0(V)$ and $R(t) \neq 0$ for all $t \in \mathbb{T}$ then

$$R(t) = c \prod_{j=1}^{k} (1 - t_j^+ t^{-1})\, t^{\varkappa} \prod_{j=1}^{s} (t - t_j^-),$$

where $|t_j^+| < 1$ $(j = 1, \ldots, k)$, $|t_j^-| > 1$ $(j = 1, \ldots, s)$, and $\varkappa := \operatorname{wind} R(t)$. It is easily seen that

$$R = \prod_{j=1}^{k} (I - t_j^+ V_{-1})\, V_{\varkappa} \prod_{j=1}^{s} (V - t_j^- I). \tag{1}$$

Because of (i) the operators $R_- := c \prod_{j=1}^{k} (I - t_j^+ V_{-1})$ and $R_+ := \prod_{j=1}^{s} (V - t_j^- I)$ are invertible. Thus, the operator R is invertible from the left or the right if the operator $V_{\varkappa}$ is invertible from the left or the right, respectively. Consequently, in the case $\varkappa = 0$ the operator R is invertible, R is only left invertible for $\varkappa > 0$ and only right invertible for $\varkappa < 0$.

(iii) See Gohberg/Feldman [1], Chapter 1, § 11.

(iv) This assertion is a consequence of (iii), of the representation (1), and of Lemma 2.2, Chap. 1 in Gohberg/Feldman [1].

(v) See Gohberg/Feldman [1], Lemma 2.1, Chap. 1. ∎

4.3. The closure of $\mathfrak{L}_0(V)$. Let $\mathfrak{L}(V)$ be the closure of $\mathfrak{L}_0(V)$ in the Banach space $\mathcal{L}(X)$. The operators in $\mathfrak{L}(V)$ will be called *continuous functions* of V. Our next concern is to construct a symbol for $R \in \mathfrak{L}(V)$. To start with we remember of assertion (*v*) of Theorem 4.2 (established in GOHBERG/FELDMAN [1]): Let (H1) be fulfilled, let R be in $\mathfrak{L}_0(V)$ and $R(t)$ its symbol. Then the spectral radius $\varrho(R)$ equals $\max_{t\in\mathbb{T}} |R(t)|$.

By this assertion, each operator R in $\mathfrak{L}(V)$ defines a uniquely determined continuous function $R(t)$ on $\mathbb{T}$, which will be called the *symbol* of R: If R_n is in $\mathfrak{L}_0(V)$ and R_n converges to R then $\{R_n(t)\}$ is a Cauchy sequence (in $\mathbf{C}(\mathbb{T})$) whose uniform limit is $R(t)$. Furthermore one can observe that

$$\max_{t\in\mathbb{T}} |R(t)| \leqq \|R\|$$

for all R in $\mathfrak{L}(V)$. A natural and instructive question is whether an operator in $\mathfrak{L}(V)$ is uniquely determined by its symbol or not. The following example shows that the correspondence between the operators from $\mathfrak{L}(V)$ and their symbols is not necessarily one-to-one.

Example. Let $\mathbf{m}$ be the space of all bounded sequences of complex numbers and define

$$E := \left\{\{\xi_k = ak + \eta_k\}_{k=1}^{\infty} : a \in \mathbb{C},\ \{\eta_k\}_{k=1}^{\infty} \in \mathbf{m}\right\}.$$

Obviously,

$$E \ni \xi = \{\xi_k\}_{k=1}^{\infty} \mapsto \|\xi\|_E := \left|\lim_{k\to\infty} \frac{\xi_k}{k}\right| + \sup_n \left|\xi_n - n \lim_{k\to\infty} \frac{\xi_k}{k}\right|$$

defines a norm on E. Moreover, E equipped with this norm actually forms a Banach space. Let V and $V^{(-1)}$ be the operators defined by

$$V\{\xi_k\}_{k=1}^{\infty} = \{\xi_{k-1}\}_{k=1}^{\infty} \qquad \text{with } \xi_0 = 0,$$

$$V^{-1}\{\xi_k\}_{k=1}^{\infty} = \{\xi_{k+1}\}_{k=1}^{\infty}.$$

We claim that the operators V^n and $V^{(-n)} := (V^{(-1)})^n$ are bounded on E and

$$\|V^n\| = \|V^{(-n)}\| = n + 1$$

for all $n = 0, 1, \ldots$. Indeed, if $\{\eta_k\} = V^n\xi$, $\xi = \{ak + \zeta_k\}_{k=1}^{\infty} \in E$, then

$$\eta_k = \begin{cases} 0, & k \leqq n, \\ (k-n)\,a + \zeta_{k-n} = ka + \zeta_{k-n} - na, & k > n. \end{cases}$$

Hence, $\{\eta_k\}_{k=1}^{\infty} = \{ak + \nu_k\}_{k=1}^{\infty}$ with

$$\nu_k = \begin{cases} -ak, & k \leqq n, \\ \zeta_{k-n} - na, & k > n \end{cases}$$

and $\{\eta_k\}_{k=1}^{\infty} \in E$. Moreover,

$$\|V^n\xi\|_E = \|\{\eta_k\}\|_E = |a| + \sup_k |\nu_k| \leqq |a| + |a|\,n + \sup_k |\zeta_k| \leqq (n+1)\|\xi\|_E.$$

This gives the boundedness of V^n and because of

$$\|V^n\{k\}_{k=1}^{\infty}\| = n + 1$$

we get $\|V^n\| = n + 1$. Analogously it can be shown that $V^{(-n)}$ is bounded on E and $\|V^{(-n)}\| = n + 1$. Consequently, the spectral radii of V and $V^{(-1)}$ are equal to one (note that $\lim \sqrt[n]{n+1} = 1$) and (H1) is fulfilled. Of course, V is left invertible but not right invertible.

Now, let A be the operator defined on E by

$$A\xi = \mu\{1, 1, \ldots\},$$

where $\mu = \lim\limits_{k\to\infty} \frac{\xi_k}{k}$, $\xi = \{\xi_k\}_{k=1}^{\infty} \in E$. Obviously, A is bounded on E and $\|A\| = 1$. We prove that $\left\|\frac{1}{n} V^{(-n)} - A\right\| \to 0$ as $n \to \infty$:

For $\xi = \{ak + \zeta_k\}_{k=1}^{\infty} \in E$ we obtain

$$\left(\frac{1}{n} V^{(-n)} - A\right) \xi = \left\{\frac{k+n}{n} a + \frac{\zeta_{k+n}}{n} - a\right\}_{k=1}^{\infty} = \frac{1}{n} \{ka + \zeta_{k+n}\}_{k=1}^{\infty}$$

and

$$\left\|\left(\frac{1}{n} V^{(-n)} - A\right) \xi\right\|_E = \frac{1}{n} \left(|a| + \sup_k |\zeta_{k+n}|\right) \leqq \frac{1}{n} \|\xi\|_E .$$

Thus $A \in \mathfrak{L}(V)$ and $A \neq 0$, but the symbol of A is zero because the symbol of $\frac{1}{n} V^{(-n)}$ is $\frac{1}{n} t^{-n}$ and the sequence $\left\{\frac{1}{n} t^{-n}\right\}$ converges uniformly to zero as $n \to \infty$. ■

Now we ask what conditions are sufficient to guarantee that an operator in $\mathfrak{L}(V)$ is uniquely determined by its symbol. To give a (partial) answer we define a matrix representation for certain operators which is of its own interest too.

Put $Q_n = V_n V_{-n}$ and $P_n = I - Q_n$. Obviously the operators P_n and Q_n are projections, and $Q_n = I$ for $n \leqq 0$. In what follows we assume the hypothesis (H2) to be fulfilled:

$$\text{(H2)} \qquad \bigcap_{n\geqq 0} \operatorname{im} Q_n = \{0\}.$$

Now define $\tilde{X}$ to be the direct product of a countable number of spaces im P_1 and let $X' \subset \tilde{X}$ be the set of those elements $(x_0, x_1, \ldots)$ in $\tilde{X}$ ($x_i \in \operatorname{im} P_1$), for which one can find an x in X so that

$$\sum_{j=0}^{m} V_j x_j = P_{m+1} x \tag{1}$$

for all $m \in \mathbb{Z}^+$.

4.4. Proposition. (i) *Under the hypothesis* (H2), *there exists for every* $(x_0, x_1, \ldots)$ *in* X' *exactly one* x *in* X *such that* (1) *holds.*

(ii) *By defining on* X' *the element-wise addition and scalar multiplication, and a norm by*

$$\|(x_0, x_1, \ldots)\|_{X'} = \|x\|_X, \tag{2}$$

where x *is the uniquely determined element from* X *which corresponds with* $(x_0, x_1, \ldots)$ *by* (i), *the set* X' *becomes a Banach space.*

(iii) The mapping $J: X \to X'$ given by $Jx = (x_0, x_1, \ldots)$ with $x_i = V_{-i}(P_{i+1} - P_i)\,x = P_1 V_{-i} x$ is an isometrical isomorphism of X onto X'.

Proof. (i): By (1) it suffices to prove that $P_m x = 0$ for all $m \geqq 0$ implies $x = 0$. Since $P_m x = 0$ is equivalent to the inclusion $x \in \operatorname{im} Q_m$, this is an immediate consequence of (H2).

(ii): Obviously, X' is a linear space. To prove that (2) defines a norm we have only to show that $(x_0, x_1, \ldots) = 0$ if $\|(x_0, x_1, \ldots)\|_{X'} = 0$. Let x be the element in X corresponding with $(x_0, x_1, \ldots)$. Then $x = 0$. Hence, by (1) we get $\sum_{j=0}^{m} V_j x_j = P_{m+1} x = 0$ for all $m \in \mathbb{Z}^+$. For $m = 0$ this gives $x_0 = 0$.

For $m = 1$ we obtain $V_1 x_1 = 0$, i.e. $V_{-1} V_1 x_1 = x_1 = 0$. Via induction we deduce that $(x_0, x_1, \ldots) = 0$. Finally, that X' is complete can be checked easily using (2) and the continuity of the projections P_m $(m \geqq 0)$.

(iii): Let $x \in X$ and $Jx = (x_0, x_1, \ldots) \in \tilde{X}$. From $V_j x_j = V_j V_{-j}(P_{j+1} - P_j)\,x = V_j V_{-j}(V_j V_{-j} - V_{j+1} V_{-j-1})\,x = (P_{j+1} - P_j)\,x$ it follows that $P_{m+1} x = \sum_{j=0}^{m} V_j x_j$ and, hence, Jx is in X' with $\|Jx\| = \|x\|$. That J is onto becomes clear from (2).

4.5. Remark. It is obvious that all finite sequences $(x_0, x_1, \ldots, x_k, 0, \ldots)$ with $x_i \in \operatorname{im} P_1$ for $i = 0, \ldots, k$ lie in X'. For these elements we have $x := \sum_{i=0}^{m} V_i x_i \in X$ and

$$\|(x_0, \ldots, x_k, 0, \ldots)\|_{X'} = \|x\|_X .$$

4.6. A matrix representation. We associate with each operator A on X' an infinite matrix

$$\mu(A) := (A_{ij})_{i,j=1}^{\infty},$$

where $A_{ij} = P_1 V_{-i} J^{-1} A J V_j P_1$.

The matrix representations for the operators JVJ^{-1}, $JV_{-1}J^{-1}$ and JP_1J^{-1} look as follows:

$$\mu(JVJ^{-1}) = \begin{bmatrix} 0 & 0 & 0 & 0 & \ldots \\ P_1 & 0 & 0 & 0 & \ldots \\ 0 & P_1 & 0 & 0 & \ldots \\ 0 & 0 & P_1 & 0 & \ldots \\ \vdots & \vdots & \vdots & \vdots & \end{bmatrix}, \quad \mu(JV_{-1}J^{-1}) = \begin{pmatrix} 0 & P_1 & 0 & 0 & \ldots \\ 0 & 0 & P_1 & 0 & \ldots \\ 0 & 0 & 0 & P_1 & \ldots \\ \vdots & \vdots & \vdots & \vdots & \end{pmatrix},$$

$$\mu(JP_1J^{-1}) = \begin{pmatrix} P_1 & 0 & \ldots \\ 0 & 0 & \ldots \\ \vdots & \vdots & \end{pmatrix}.$$

Moreover, the matrix representation for an operator $J^{-1}RJ$, $R \in \mathfrak{L}(V)$, is given by

$$\mu(JRJ^{-1}) = (r_{i-j} P_1)_{i,j=0}^{\infty}, \tag{1}$$

where r_i is the ith Fourier coefficient of the symbol $R(t)$ of R. For R in $\mathfrak{L}_0(V)$ this is obvious, and for R in $\mathfrak{L}(V)$ the matrix representation can be obtained by a simple limit passage.

Note that matrices the elements of which depend only on the difference of the indices are called *Toeplitz matrices*. Therefore, (1) is a Toeplitz matrix.

4.7. Theorem. *Let* (H2) *be fulfilled. If at least one of the conditions* (H3) *or* (H4),

(H3) $\operatorname{clos} \bigcup_{n \geq 0} \operatorname{im} P_n = X$,

(H4) $\sup_{n \geq 0} \|Q_n\| < \infty$,

is in force then every operator R in $\mathfrak{L}(V)$ is uniquely determined by its symbol $R(t)$.

Proof. We have to prove that $R = 0$ if $R(t) = 0$ for all $t \in \mathbb{T}$. Clearly, this is equivalent to the statement that $JRJ^{-1} = 0$ if the matrix representation $\mu(JRJ^{-1})$ is the zero matrix. First, it is an easy observation that if $R(t) = 0$ for all $t \in \mathbb{T}$ then the operator JRJ^{-1} acts on finite sequences $(x_0, \ldots, x_k, 0, \ldots)$ as the zero operator. Under hypothesis (H3) it follows immediately that JRJ^{-1} is the zero operator.

Now suppose that $M := \sup \|P_n\| < \infty$. For any element $y \in X'$ we have $JRJ^{-1}JP_m \times J^{-1}y = 0$ or, equivalently, $JRJ^{-1}y = JRJ^{-1}JQ_mJ^{-1}y =: z$. If $z \neq 0$ then there is (by Proposition 4.4) an $n \in \mathbb{Z}^+$ such that $JP_nJ^{-1}z \neq 0$. Therefore we have $JP_nRQ_mJ^{-1}y = JP_nJ^{-1}z \neq 0$ for all m in $\mathbb{Z}^+$. We claim that $\|P_nRQ_m\| \to 0$ as $m \to \infty$, which contradicts the fact that $JP_nJ^{-1}z \neq 0$. Indeed, for arbitrarily given $\varepsilon > 0$ we find an R_ε in $\mathfrak{L}_0(V)$ so that $\|R - R_\varepsilon\| < \varepsilon$. The operator R_ε has the unique representation $R_\varepsilon = \sum_{i=-k}^{k} a_i V_i$. Consider the expression $P_n V_i Q_m$. Since $V_i Q_m = Q_{m+i} V_i$ we have $P_n V_i Q_m = 0$ for $m + i > n$. Thus, $P_n(R - R_\varepsilon) Q_m = P_nRQ_m$ for $m > n + k$, and this shows that $\|P_nRQ_m\| < M(M+1)\,\varepsilon$ for $m > n + k$. Hence, JRJ^{-1} acts on X' as the zero operator. ■

So far we have discussed the correspondence between operators in $\mathfrak{L}(V)$ and their symbols. Now we return to the study of the invertibility of operators in $\mathfrak{L}(V)$. Unfortunately, there is no analogue to Theorem 4.2 without additional assumptions. So our next concern is to extend the assertions of Theorem 4.2 to a subset of $\mathfrak{L}(V)$ which is as large as possible. To that end we shall explain the role of so-called *V-dominating* Banach algebras.

Again, let V be an operator on X which is only invertible from the left and let V_{-1} be one of its left inverses. We do not assume the hypothesis (H1) to be fulfilled.

A commutative Banach algebra $\mathfrak{D}$ with unit element e is called *(V_1, V_{-1})-dominating* (or, shortly, *V-dominating*) if

(a) there is an invertible element d in $\mathfrak{D}$ which together with its inverse d^{-1} spans a dense subalgebra of $\mathfrak{D}$.

(b) the spectra $\operatorname{sp}_{\mathfrak{D}} d$ and $\operatorname{sp}_{\mathfrak{D}} d^{-1}$ of d and d^{-1} in $\mathfrak{D}$ are contained in $\{z \in \mathbb{C} : |z| \leq 1\}$.

(c) there exists a constant $M > 0$ such that

$$\|p(V)\|_{\mathfrak{L}(V)} \leq M\, \|p(d)\|_{\mathfrak{D}} \tag{1}$$

for any polynomial $p(t) = \sum_{j=-n}^{n} a_j t^j$, $|t| = 1$.

4.8. Proposition: (a) *If a (V, V_{-1})-dominating algebra $\mathfrak{D}$ exists, then V and V_{-1} are subject to the hypothesis* (H1). *More generally, if p is a polynomial on $\mathbb{T}$, then the spectral radii $\varrho_{\mathfrak{L}(X)}(p(V))$ and $\varrho_{\mathfrak{D}}(p(d))$ satisfy the inequality*

$$\varrho_{\mathfrak{L}(X)}\big(p(V)\big) \leq \varrho_{\mathfrak{D}}\big(p(d)\big). \tag{1}$$

(b) *The maximal ideal space* $\mathbf{M}(\mathfrak{D})$ *of* $\mathfrak{D}$ *is homeomorphic to the unit circle* $\mathbb{T}$, *and the Gelfand transform maps* d *into the function* $t \mapsto t$ $(t \in \mathbb{T})$.

Proof. (a): It is sufficient to verify the estimate (1). By 4.7(1), $\|p(V)^n\|_{\mathfrak{L}(X)} \leq M \|p(d)^n\|_{\mathfrak{D}}$ $(n \in \mathbb{Z}^+)$. Thus, $\|p(V)^n\|^{1/n} \leq M^{1/n} \|p(d)^n\|_{\mathfrak{D}}^{1/n}$, and passage to the limit yields the assertion.

(b): We shall show that the spectrum of d equals $\mathbb{T}$. The remaining assertions follow immediately from the general theory of commutative Banach algebras.

By 4.7(b) the spectrum $\mathrm{sp}_{\mathfrak{D}}\, d$ of d in $\mathfrak{D}$ is contained in $\mathbb{T}$. Assume that $\mathrm{sp}_{\mathfrak{D}}\, d \neq \mathbb{T}$. Then there is an inner point z of $\mathbb{T} \setminus \mathrm{sp}_{\mathfrak{D}}\, d$, and we can find a continuous function f on $\mathbb{T}$ such that $0 \leq f(t) \leq 1$, $(t \in \mathbb{T})$, $f(z) = 1$ and $f(t) = 0$ for $t \in \mathrm{sp}_{\mathfrak{D}}\, d$. Given $\varepsilon > 0$ we choose a polynomial $p \in \mathbf{C}(\mathbb{T})$ so that $\max_{t \in \mathbb{T}} |f(t) - p(t)| < \varepsilon$. Since $|p(t)| < \varepsilon$ for all $t \in \mathrm{sp}_{\mathfrak{D}}\, d$ we obtain $\varrho_{\mathfrak{D}}(p(d)) \leq \varepsilon$. On the other hand, by Theorem 4.2(v) we have $\varrho_{\mathfrak{L}(X)}(p(V)) > 1 - \varepsilon$. These two inequalities contradict the inequality (1) for ε sufficiently small. ■

Since $p(d)$ is uniquely determined by its Gelfand transform $p(t)$ (even in the case where $\mathfrak{D}$ has a nontrivial radical) the mapping $p(d) \mapsto p(V)$ is well-defined, linear, and by 4.7(1) bounded. Hence, this mapping can be continuously extended to the whole algebra $\mathfrak{D}$, and the image, denoted by $\mathfrak{L}_{\mathfrak{D}}(V)$, is contained in $\mathfrak{L}(V)$. For an element $a \in \mathfrak{D}$, let $a(V)$ be the image element under this mapping. Note that

$$\|a(V)\|_{\mathfrak{L}(X)} \leq M \|a\|_{\mathfrak{D}} \tag{2}$$

for all $a \in \mathfrak{D}$ and that the symbol of $a(V)$ coincides with the Gelfand transform of $a \in \mathfrak{D}$. If the radical of $\mathfrak{D}$ is trivial then an element of $\mathfrak{L}_{\mathfrak{D}}(V)$ is uniquely determined by its symbol (even in the case where this is unknown for elements from $\mathfrak{L}(V)$).

4.9. Theorem. *Let* (H1) *be fulfilled.*

(i) *An operator* $R \in \mathfrak{L}_{\mathfrak{D}}(V)$ ($\mathfrak{D}$ *being a* V*-dominating Banach algebra) is at least one-sided invertible if* $R(t) \neq 0$ *for all* $t \in \mathbb{T}$. *If the symbol* $R(t)$ *does not vanish on* $\mathbb{T}$, *then the invertibility of* R *corresponds with the winding number* $\varkappa = \mathrm{wind}\, R(t)$, *i.e.* R *is invertible, invertible only from the left or only from the right if* $\varkappa$ *is zero, positive or negative, respectively.*

(ii) *If* $R \in \mathfrak{L}(V)$ *and* $R(t_0) = 0$ *for some* $t_0 \in \mathbb{T}$ *then* R *is neither a* Φ_+*- nor a* Φ_-*-operator.*

Proof. (i): Since $R \in \mathfrak{L}_{\mathfrak{D}}(V)$, there is an element $a \in \mathfrak{D}$ such that $R = a(V)$. The element a is invertible since $R(t) \neq 0$ for all $t \in \mathbb{T}$ and the Gelfand transform of a coincides with $R(t)$. From the general theory of Banach algebras we deduce the existence of a polynomial $p(t) = \sum_{j=-k}^{k} a_j t^j$, $|t| = 1$, such that $r = p(d)$ is invertible and

$$r^{-1}a = e + c, \qquad \|c\|_{\mathfrak{D}} < 1/M.$$

Similarly as in the proof of Theorem 4.2 we have a representation of r in the form

$$r = r_- d^{\varkappa} r_+,$$

where $r_\pm(V) \in \mathfrak{L}_0^\pm(V)$ are invertible. Now write

$$a = r_- d^{\varkappa}(e + c)\, r_+, \qquad \varkappa \leq 0,$$

$$a = r_-(e + c)\, d^{\varkappa} r_+, \qquad \varkappa \geq 0.$$

Then

$$a(V) = r_-(V)\, V_\varkappa\big(I + c(V)\big)\, r_+(V), \qquad \varkappa \leqq 0,$$

$$a(V) = r_-(V)\big(I + c(V)\big)\, V_\varkappa r_+(V), \qquad \varkappa \geqq 0.$$

Since $\|c(V)\|_{\mathcal{L}(X)} < 1$, the element $I + c(V)$ is invertible, and we are done.

(ii): Assume $R(t_0) = 0$ and $R \in \Phi_+(X)$ or $R \in \Phi_-(X)$. If $R \in \Phi_+(X)$ then there exists a $\delta > 0$ such that

$$\|R - r\| < \delta \quad \text{implies} \quad r \in \Phi_+(X).$$

Now take $r \in \mathfrak{L}_0(V)$ such that $\|R - r\| < \delta/2$. Obviously, $|r(t_0)| < \delta/2$ and $\big\|R - r + r(t_0)\, I\big\| < \delta$. Hence, $r - r(t_0)\, I \in \mathfrak{L}_0(V)$ is a Φ_+-operator. Since $r(t) - r(t_0)$ vanishes at t_0 this contradicts Theorem 4.2(iii). The case $R \in \Phi_-(X)$ can be treated analogously. ■

Examples of V-dominating algebras

4.10. First example. Let X^0 be a Banach space and $X \subset X^0$ be a complementable subspace of X^0. Thus, there exists a bounded projection P^+ from X^0 onto X. An operator $U \in \mathcal{L}(X^0)$ is called a *dilation* (with respect to the choice of P^+) of the operator $V \in \mathcal{L}(X)$ if $P^+U^m x = V_m x$ for all $x \in X$ and $m \in \mathbb{Z}^+$. Now suppose that V and $V^{(-1)}$ fulfil (H1). Let U be a bounded and invertible dilation of V acting on a Banach space $X^0 \supset X$ (with respect to a projection P^+ from X^0 onto X) such that

(i) $\varrho(U) \leqq 1, \quad \varrho(U^{-1}) \leqq 1;$

(ii) $P^+UP^+ = UP^+, P^+U^{-1}P^+ = P^+U^{-1}$ and $P^+U^{-1}x = V^{(-1)}x$ for all $x \in X$. (1)

It follows immediately that

$$P^+U \neq UP^+.$$

Note that such a dilation always exists: take $X^0 = X \dotplus X$ and define operators P^+, U, and U^{-1} by

$$P^+ = \begin{pmatrix} I & 0 \\ 0 & 0 \end{pmatrix}, \quad U = \begin{pmatrix} V & P_1 \\ 0 & V_{-1} \end{pmatrix}, \quad U^{-1} = \begin{pmatrix} V_{-1} & 0 \\ P_1 & V \end{pmatrix}.$$

Define $\mathfrak{L}_0(U)$ and $\mathfrak{L}(U)$ analogously to $\mathfrak{L}_0(V)$ and $\mathfrak{L}(V)$. Now assign to every operator $R = \sum_{i=-k}^{k} a_i U^i \in \mathfrak{L}_0(U)$ the polynomial $R(t) = \sum_{i=-k}^{k} a_i t^i$, $|t| = 1$. Due to (1) the coefficients of $R(t)$ are uniquely determined, and we get

$$\max_{t \in \mathbb{T}} |R(t)| \leqq \|R\|.$$

This inequality allows us to associate with each operator R in $\mathfrak{L}(U)$ a unique continuous function $R(t)$ on $\mathbb{T}$, which we shall call the *symbol* of R.

Now we summarize some properties of the algebra $\mathfrak{L}(U)$. Let (H1) be fulfilled. Then

1°. The spectra sp U and sp U^{-1} in $\mathcal{L}(X^0)$ and in $\mathfrak{L}(U)$, respectively, coincide both with the unit circle $\mathbb{T}$.

2°. The maximal ideal space of $\mathfrak{L}(U)$ coincides with the unit circle. The symbol of an operator in $\mathfrak{L}(U)$ is nothing else than its Gelfand transform. Hence, $R \in \mathfrak{L}(U)$ is invertible in $\mathfrak{L}(U)$ (or in $\mathcal{L}(X^0)$) if and only if $R(t) \neq 0$ for all $t \in \mathbb{T}$.

3°. Now suppose $\bigcap_{n \geq 0} \operatorname{im} (U^{-n}P^-U^n + U^nP^+U^{-n}) = \{0\}$ and assume that at least one of the following conditions is fulfilled:

$$\operatorname{clos} \bigcup_{n \geq 0} \operatorname{im} (I - U^{-n}P^-U^n - U^nP^+U^{-n}) = X^0$$

or

$$\sup_{n \geq 0} \|U^{-n}P^-U^n + U^nP^+U^{-n}\| < \infty,$$

where $P^- := I - P^+$. Then the radical of $\mathfrak{L}(U)$ is trivial. The proof is left to the reader; note that the proof of 3° runs parallely to that of Theorem 4.7.

The algebra $\mathfrak{L}(U)$ is a V-dominating Banach algebra which is of considerable interest in the theory of paired operators (see Section 4.65). Note that $\mathfrak{L}_{\mathfrak{L}(U)}(V)$ may be a proper subset of $\mathfrak{L}(V)$. An example can be obtained if one uses the operators and spaces introduced in Section 4.46.

4.11. Second example. First recall the notion of *Wiener algebras with weights*. We call a sequence of positive real numbers $\{v_n\}_{n=-\infty}^{\infty}$ a weight if

$$\lim_{n \to \infty} (v_n)^{1/n} = \lim_{n \to \infty} (v_{-n})^{1/n} = 1, \tag{1}$$

$$v^* := \sup_{k,n \in \mathbb{Z}} \frac{v_{n+k}}{v_n v_k} < \infty. \tag{2}$$

Note that

$$\inf_{n \in \mathbb{Z}} v_n \geq (v^*)^{-1}. \tag{3}$$

Indeed, $\frac{v_{n+0}}{v_n v_0} \leq v^*$ yields $v_0 \geq (v^*)^{-1}$. Now assume there is some $j \neq 0$ such that $\vartheta := v_j v^* < 1$. Then $v_{n+j} \leq v_n \vartheta$ for all $n \in \mathbb{Z}$, and, in particular, $v_{k \cdot j} \leq v_0 \vartheta^k$ for $k = 1, 2, \ldots$ Consequently, we have $\lim_{k \to \infty} (v_{k \cdot j})^{1/(k|j|)} \leq \lim_{k \to \infty} (v_0)^{1/(k|j|)} (\vartheta)^{1/|j|} < 1$, which contradicts (1).

Denote by $\mathbf{W}(v)$ the collection of all functions a on the unit circle $\mathbb{T}$ the Fourier coefficients of which satisfy

$$\sum_{k=-\infty}^{\infty} |a_k| \, v_k < \infty,$$

and put

$$\|a\|_{\mathbf{W}(v)} := v^* \sum_{k=-\infty}^{\infty} |a_k| \, v_k. \tag{4}$$

The set $\mathbf{W}(v)$ actually forms a commutative Banach algebra under the norm (4) whose maximal ideal space is $\mathbb{T}$.

Indeed, it is easy to see that for $a, b \in \mathbf{W}(v)$ the equality

$$(ab)(t) = \sum_{n=-\infty}^{\infty} \left(\sum_{k=-\infty}^{\infty} a_k b_{n-k} \right) t^n$$

holds, where the series converges absolutly. Since $v_n \leqq v^* v_{n-k} v_k$ for all $n, k \in \mathbb{Z}$, one gets

$$\|ab\|_{\mathbf{W}(v)} = v^* \sum_{n=-\infty}^{\infty} \sum_{k=-\infty}^{\infty} |a_k b_{n-k}| \, v_n \leqq (v^*)^2 \left(\sum_{k=-\infty}^{\infty} |a_k| \, v_k \right) \left(\sum_{n=-\infty}^{\infty} |b_n| \, v_n \right)$$

$$= \|a\|_{\mathbf{W}(v)} \, \|b\|_{\mathbf{W}(v)} .$$

Now we are going to identify the maximal ideal space of $\mathbf{W}(v)$. Define the operators U, U^{-1} and P^+ on $\mathbf{W}(v)$ as follows:

$$(Ua)(t) = ta(t), \quad (U^{-1}a)(t) = t^{-1}a(t) \quad \text{and} \quad P^+ \left(\sum_{n=-\infty}^{\infty} a_n t^n \right) = \sum_{n=0}^{\infty} a_n t^n .$$

We have $\varrho(U) = \varrho(U^{-1}) = 1$. Indeed, for $n = 0, \pm 1, \ldots$, we get

$$\|U^n a\| = \|t^n a\| = v^* \sum_{k=-\infty}^{\infty} |a_k| \, v_{k+n} \leqq (v^*)^2 \, v_n \sum_{k=-\infty}^{\infty} |a_k| \, v_k .$$

Thus, $\|U^n\| \leqq v^* v_n$. By (1) this gives the assertion. Now it is clear that U, U^{-1}, P^+ satisfy conditions (i) and (ii) of Section 4.10 with respect to the operators $V := P^+UP^+|_{\operatorname{im} P^+}$ and $V^{(-1)} := P^+U^{-1}P^+|_{\operatorname{im} P^+}$.

Assertion 2° of Section 4.10 shows that the maximal ideal space of $\mathfrak{L}(U)$ coincides with $\mathbb{T}$. Finally, note that $\mathfrak{L}(U)$ and $\mathbf{W}(v)$ are isometrically isomorphic to one another, which gives our claim.

Suppose (H1) is fulfilled and put $v_n = \|V_n\|$, $n \in \mathbb{Z}$. Then (1) and (2) are fulfilled. The Wiener algebra associated with this weight will also be denoted by $\mathbf{W}(V)$. The algebra $\mathbf{W}(V)$ is an algebra with trivial radical and is, obviously, V-dominating.

In the sequel we shall frequently be concerned with algebras whose weights are of the special form

$$v_n = \begin{cases} (1+n)^\beta, & n \geqq 0 \\ (1+|n|)^\alpha, & n < 0, \end{cases}$$

where $\alpha, \beta \geqq 0$ are fixed. In that case we write $\mathbf{W}^{\alpha,\beta}$ instead of $\mathbf{W}(v)$. A further important example of a V-dominating algebra occurs in connection with the notion of decomposing algebras (see Section 4.25, 1° and 3°).

The algebra alg (V, V_{-1})

4.12. Definitions. Our next objective is to study the smallest closed subalgebra alg (V, V_{-1}) of $\mathscr{L}(X)$ containing V and V_{-1}. We are mainly interested in the algebraic structure of this algebra and in whether an invertible operator is invertible within this algebra.

By $\mathbf{QC}(V)$ we denote the smallest closed two-sided ideal in alg (V, V_{-1}) containing all operators of the form

$$(p_1 p_2)(V) - p_1(V)\, p_2(V),$$

where p_1, p_2 are arbitrary polynomials. The ideal $\mathbf{QC}(V)$ is called the *quasicommutator ideal* of the algebra alg (V, V_{-1}).

4.13. Lemma. *The following two conditions are equivalent for* $K \in$ alg (V, V_{-1}):

(a) $K \in \mathbf{QC}(V)$;

(b) $K \in$ id (P_1), *where* id (P_1) *refers to the smallest closed two-sided ideal of* alg (V, V_{-1}) *containing* P_1.

If $\|Q_n\| \leqq M < \infty$ *for all* n, *then* (a) *is equivalent to each of the following two conditions:*

(c) $\|Q_nK\| \to 0$ *as* $n \to \infty$;

(d) $\|KQ_n\| \to 0$ *as* $n \to \infty$.

Proof. (a) $\Rightarrow$ (b): The quasicommutator ideal is generated by all operators $(P_1P_2)(V) - P_1(V)P_2(V)$, where P_1, P_2 are polynomials. Since $V_iV_j = V_{i+j} - P_iV_{i+j}$ $(i, j \in \mathbb{Z})$ and $P_i = \sum_{j=0}^{i-1} V_jP_1V_{-j}$, the inclusion $\mathbf{QC}(V) \subset$ id (P_1) holds.

(b) $\Rightarrow$ (a): Trivial, since P_1 is the quasicommutator of V and V_{-1}: $I - VV_{-1} = P_1$.

(b) $\Rightarrow$ (c): We only prove that (b) implies $\|Q_nK\| \to 0$. First we show that $\|Q_nAP_1\| \to 0$ as $n \to \infty$ if $A \in$ alg (V, V_{-1}). For given $\varepsilon > 0$ write $A = A_\varepsilon + A - A_\varepsilon$ with $\|A - A_\varepsilon\| < \varepsilon$ and A_ε a finite sum of products of shifts. We have

$$Q_nV_{i_1}V_{i_2} \dots V_{i_k}P_1 = Q_nP_{i_1+\dots+i_k+1}V_{i_1} \dots V_{i_k} \qquad (i_1, \dots, i_k \in \mathbb{Z})$$

because $V_sP_r = P_{r+s}V_s$ for all $r, s \in \mathbb{Z}$. Hence, $Q_nV_{i_1} \dots V_{i_k}P_1 = 0$ if $i_1 + \dots + i_k + 1 < n$. This yields that, for given ε, we can find an n_0 such that $\|Q_nAP_1\| < M(M+1)\,\varepsilon$ if $n \geqq n_0$. Now let $K \in$ id (P_1). Then we write $K = K_\varepsilon + (K - K_\varepsilon)$ with $\|K - K_\varepsilon\| < \varepsilon$ and $K_\varepsilon = \sum_{j=1}^{r} B_jP_1C_j$ (here $B_j, C_j \in$ alg (V, V_{-1})). By what has already been shown, each item of Q_nK converges to zero, and we are done with the implication (b) $\Rightarrow$ (c).

(c) $\Rightarrow$ (a): If $\|Q_nK\| \to 0$ then K is the uniform limit of the operators P_nK, which are in $\mathbf{QC}(V)$ since P_n is the quasicommutator of V_n, V_{-n}.

(d) $\Rightarrow$ (a): Proceed as in the proof of the implication (c) $\Rightarrow$ (a). ■

4.14. Lemma. *If* $K \in \mathbf{QC}(V)$ *and if* $I + K$ *is at least one-sided invertible then* $I + K$ *is two-sided invertible. The inverse operator is in* alg (V, V_{-1}), *and* $(I + K)^{-1} = I + K'$ *with* $K' \in \mathbf{QC}(V)$.

Proof. By Lemma 4.13 we can approximate K as closely as desired by operators K_n of the form

$$K_n = \sum_{i=1}^{n} M_i^{(n)}P_1N_i^{(n)},$$

where $M_i^{(n)}, N_i^{(n)}$ are finite sums of products of the operators V_j, $j \in \mathbb{Z}$. Again as in the proof of Lemma 4.13 we see that there must exist a sequence $\{a_n\} \subset \mathbb{Z}^+$ which tends to infinity as $n \to \infty$ such that

$$Q_{a_n}K_n = K_nQ_{a_n} = 0 \qquad \text{for all } n \in \mathbb{Z}^+.$$

Hence, $P_{a_n}K_nP_{a_n} = K_n$, and the one-sided invertibility of $I + K$ yields that $I + K_n = I + P_{a_n}K_nP_{a_n}$ is invertible from the same side as $I + K$ if only n is large enough. Hence, $P_{a_n} + P_{a_n}K_nP_{a_n}$ is one-sided invertible in im P_{a_n} for sufficiently large n.

Now a little thought shows that the algebra of all operators $P_{a_n}RP_{a_n}|_{\operatorname{im} P_{a_n}}$ with arbitrary $R \in \operatorname{alg}(V, V_{-1})$ is isomorphic to the algebra $\mathbb{C}_{a_n \times a_n}$ (recall the matrix representation defined in 4.6, which can be proved to be valid for operators acting from im P_{a_n} to im P_{a_n} without hypothesis (H2)). Thus, the one-sided invertibility of the operators $P_{a_n} + P_{a_n}K_nP_{a_n}$ (in im P_{a_n}) gives their two-sided invertibility, and their inverses are in alg (V, V_{-1}). So it becomes clear that the operators $I + K_n = Q_{a_n} + P_{a_n} + P_{a_n}K_nP_{a_n}$ must be two-sided invertible in alg (V, V_{-1}).

For definiteness assume that $A = I + K$ is left invertible and that A' is a left inverse for A. Then $A'(I + K_n)$ tends in the norm to the identity operator I. Thus, for n large enough, we have the invertibility of $A'(I + K_n)$ and that of $I + K_n$. Therefore, the operator A' must be invertible, and, consequently, the operator A is also invertible.

Finally we have $(I + K_n)^{-1} = I + R_n$ with $R_n \in \mathbf{QC}(V)$, and so the closedness of $\mathbf{QC}(V)$ gives $A' - I = (I + K)^{-1} - I \in \mathbf{QC}(V)$. ■

4.15. Corollary. *If* codim (im V) $< \infty$, *then* $\mathbf{QC}(V)$ *consists only of compact operators. If, moreover,* codim (im V) $= 1$ *and* $\{P_n\}$, $\{P_n^*\}$ *converge strongly to the identity operator, then* $\mathbf{QC}(V)$ *equals* $\mathcal{K}(X)$.

Proof. The first assertion is immediate from Lemma 4.13(b). In order to prove the second assertion we remark that the algebra of all operators P_nRP_n with arbitrary $R \in \operatorname{alg}(V, V_{-1})$ is isomorphic to the algebra $\mathbb{C}_{n \times n}$. Since codim (im V) $= 1$, any operator P_nKP_n, $K \in \mathcal{K}(X)$, belongs to this algebra, i.e. $P_nKP_n \in \mathbf{QC}(V)$. It remains to apply 1.1(h). ■

4.16. Definition. Let $R = \sum_{i=1}^{n} \prod_{j=1}^{m} R_{ij}$, where the operators R_{ij} belong to $\mathfrak{L}_0(V)$. The operator $R_0 \in \mathfrak{L}_0(V)$ defined uniquely by $R_0(t) = \sum_{i=1}^{n} \prod_{j=1}^{m} R_{ij}(t)$ is called the *essential part* of R. It can easily be checked that $R - R_0 \in \mathbf{QC}(V)$. If A is an element belonging to alg (V, V_{-1}), then any operator $B \in \mathfrak{L}(V)$ such that $A - B \in \mathbf{QC}(V)$ is called a *main part* of A. There are examples which show that a main part does not always exist. If the main part is uniquely determined we call it also *essential* part. There are important cases in which the main part can be proved to be unique provided it exists. See, for instance, the Sections 4.18 and 4.25, 2°. We shall see that the notion of the essential part plays an important role in numerical analysis. Now denote by $\operatorname{alg}^\pi(V, V_{-1})$ the quotient algebra $\operatorname{alg}(V, V_{-1})/\mathbf{QC}(V)$ and by π the canonical homomorphism from alg (V, V_{-1}) onto $\operatorname{alg}^\pi(V, V_{-1})$. Obviously, $\operatorname{alg}^\pi(V, V_{-1})$ is a commutative Banach algebra generated by $\pi(V)$ and by its inverse $\pi(V^{-1})$.

4.17. Proposition. *Assume that* (H1) *is fulfilled. Then the spectrum* $\operatorname{sp} \pi(V)$ *of* $\pi(V)$ *coincides with the unit circle* $\mathbb{T}$, *and for each polynomial* $p(t) = \sum_{i=-k}^{k} a_it^i$, $|t| = 1$, *we have*

$$\max_{t \in \mathbb{T}} |p(t)| \leq \|p(\pi(V))\| = \|\pi(p(V))\| \leq \|p(V)\|. \tag{1}$$

Hence, the maximal ideal space of $\operatorname{alg}^\pi(V, V_{-1})$ *may be identified with* $\mathbb{T}$, *and by* (1) *the symbol of an operator* $A \in \mathfrak{L}(V)$ *coincides with the Gelfand transform of* $\pi(A)$.

Proof. We have only to verify that $\operatorname{sp} \pi(V) = \mathbb{T}$. Then the other assertions follow immediately from the general theory of Banach algebras. By (H1) the spectra of $\pi(V)$ and $(\pi(V))^{-1} = \pi(V_{-1})$ are contained in $\mathbb{T}$. Now assume $\operatorname{sp} \pi(V) \neq \mathbb{T}$ and choose

$\lambda_0 \in \mathbb{T} \setminus \operatorname{sp} \pi(V)$. Then there exist operators B, K with $B \in \operatorname{alg}(V, V_{-1})$ and $K \in \mathbf{QC}(V)$ such that $B(V - \lambda_0 I) = I + K$. Approximate B by $R = \sum_{i=1}^{n} \prod_{j=1}^{m} R_{ij}$, $R_{ij} \in \mathfrak{L}_0(V)$, so that $\|(B - R)(V - \lambda_0 I)\| < 1/2$. Then $R(V - \lambda_0 I) = I + C_1 + K$ with $\|C_1\| < 1/2$. Further, approximate K by $K_0 = \sum_{i=1}^{n_1} \prod_{j=1}^{m_1} K_{ij} \in \mathbf{QC}(V)$, $K_{ij} \in \mathfrak{L}_0(V)$, so that $\|K - K_0\| < 1/2$. What results is

$$R(V - \lambda_0 I) = I + C + K_0 \quad \text{with} \quad \|C\| < 1.$$

Let R_0 denote the essential part of R and write

$$R_0(V - \lambda_0 I) = I + C + K_0 - (R - R_0)(V - \lambda_0 I).$$

Now it is easy to see that there is an $n \in \mathbb{N}$ such that

$$[K_0 - (R - R_0)(V - \lambda_0 I)] V_n = 0.$$

Thus,

$$R_0(V - \lambda_0 I) V_n = (I + C) V_n,$$

and the right-hand side is invertible from the left. On the other hand, $R_0(V - \lambda_0 I) V_n \in \mathfrak{L}_0(V)$ and $R_0(t)(t - \lambda_0) t^n$ vanishes at $t = \lambda_0$. It results a contradiction to Theorem 4.2(iii) and we are done. ■

4.18. Definition. Let (H1) be fulfilled. The previous theorem enables us to assign to every element $A \in \operatorname{alg}(V, V_{-1})$ a well-defined continuous function $\operatorname{smb} A$ on $\mathbb{T}$, called the *symbol* of A, namely the Gelfand transform of $\pi(A)$. Obviously, the mapping $A \mapsto \operatorname{smb} A$ is a homomorphism and every element $B \in \mathbf{QC}(V)$ belongs to the kernel of this homomorphism. The kernel equals $\mathbf{QC}(V)$ if and only if $\operatorname{alg}^{\pi}(V, V_{-1})$ is an algebra with trivial radical. Proposition 4.17 implies that an element $A \in \operatorname{alg}(V, V_{-1})$ is regularizable with respect to the ideal $\mathbf{QC}(V)$ (that is, there exists an element $B \in \operatorname{alg}(V, V_{-1})$ such that both $AB - I$, $BA - I$ belong to $\mathbf{QC}(V)$) if and only if $(\operatorname{smb} A)(t) \neq 0$ for all $t \in \mathbb{T}$. The notion of the symbol enables us to give a sufficient condition for the main part to be uniquely determined provided that it exists. Assume $A \in \operatorname{alg}(V, V^{(-1)})$ can be represented in the form

$$A = D + K = D_1 + K_1, \qquad D, D_1 \in \mathfrak{L}(V) \quad \text{and} \quad K, K_1 \in \mathbf{QC}(V).$$

Then $\operatorname{smb} A = \operatorname{smb} D = \operatorname{smb} D_1$. If an operator in $\mathfrak{L}(V)$ is uniquely determined by its symbol (see Section 4.7), then $D = D_1$.

4.19. Proposition. *Let* (H1) *be fulfilled and let* $A \in \operatorname{alg}(V, V_{-1})$ *be a* Φ_+*- or a* Φ_-*-operator. Then*

$$(\operatorname{smb} A)(t) \neq 0 \quad \textit{for all } t \in \mathbb{T}.$$

Moreover, if A is a Φ-operator and $\operatorname{codim}(\operatorname{im} V) =: \varkappa < \infty$ *then*

$$\operatorname{ind} A = -\varkappa \operatorname{wind}(\operatorname{smb} A). \tag{1}$$

Proof. First we prove the assertion for elements $R = \sum_{i=1}^{n} \prod_{j=1}^{m} R_{ij}$, $R_{ij} \in \mathfrak{L}_0(V)$. Suppose, for instance, that R is a Φ_+-operator. Denote by R_0 the essential part of R and take

$n \in \mathbb{N}$ so that $(R - R_0) V_n = 0$. Hence,

$$RV_n = R_0 V_n, \tag{2}$$

and RV_n is a Φ_+-operator by 1.3(d). Since $R_0 V_n \in \mathfrak{L}_0(V)$ and smb $R_0 V_n = (\text{smb } R_0)\, t^n$, Theorem 4.2(iii) yields $(\text{smb } R)(t) = (\text{smb } R_0)(t) \neq 0$ for all $t \in \mathbb{T}$.

If R is a Φ_--operator the assertion may be shown analogously using that $V_{-n}(R - R_0) = 0$ for n large enough. Now assume $R \in \Phi(X)$. The identity (2) gives

$$\text{ind } R - n\varkappa = \text{ind } R_0 V_n .$$

Since $R_0 V_n \in \mathfrak{L}_0(V)$, one has

$$\text{ind } R_0 V_n = -\varkappa \text{ wind smb } R_0 V_n = -(\text{wind smb } R_0 + \text{wind smb } V^n)$$
$$= -\varkappa(\text{wind smb } R_0 + n).$$

Thus ind $R = -\varkappa$ wind smb R_0, and because smb R = smb R_0 we get the assertion. Now assume $A \in \text{alg}(V, V_{-1})$ to be a Φ_+-operator with $(\text{smb } A)(t_0) = 0$ for a certain $t_0 \in \mathbb{T}$. First of all there is a $\delta > 0$ such that $B \in \mathscr{L}(X)$ and $\|A - B\| < \delta$ implies B in $\Phi(X)$. Approximate A by $R = \sum_{i=1}^{n} \prod_{j=1}^{m} R_{ij}$, $R_{ij} \in \mathfrak{L}_0(V)$, so that $\|A - R\| < \delta/2$. Since $(\text{smb } A)(t_0) = 0$ we have $|(\text{smb } R)(t_0)| < \delta/2$ and

$$\|A - R + (\text{smb } R)(t_0)\, I\| < \delta .$$

Thus, $k := R - (\text{smb } R)(t_0)\, I$ is a Φ_+-operator with $(\text{smb } k)(t_0) = 0$, which contradicts the part already proved. Hence,

$$(\text{smb } A)(t) \neq 0 \quad \text{for all} \quad t \in \mathbb{T} .$$

Analogously one shows the assertion if A is a Φ_--operator. It remains to prove the index formula (1). This formula, however, is immediate from what was already proved and from the stability of wind and ind under small perturbations. ■

4.20. Theorem. *Assume* (H1) *is satisfied and* $\varkappa :=$ codim (im V) $< \infty$. *Then* $A \in \text{alg}(V, V_{-1})$ *is a* Φ*-operator if and only if*

$$(\text{smb } A)(t) \neq 0 \quad \textit{for all} \quad t \in \mathbb{T} . \tag{1}$$

If this condition is fulfilled then

$$\text{ind } A = -\varkappa \text{ wind } (\text{smb } A) .$$

Proof. Assume (1) to be fulfilled. By Section 4.18 the element A is regularizable with respect to $\mathbf{QC}(V)$. Under the assumption of the theorem, $\mathbf{QC}(V)$ consists only of compact operators (Corollary 4.15). Thus, A is a Φ-operator by 1.3(a). The other assertions follow from Proposition 4.19. ■

4.21. Remark. If codim (im V) $= \infty$, it may happen that $(\text{smb } A)(t) \neq 0$ for all $t \in \mathbb{T}$ while A is neither a Φ_+- nor a Φ_--operator. A simple example is the operator VV_{-1}. Its symbol is the constant function 1, but VV_{-1} is a projection with dim ker $VV_{-1} = \infty$.

4.22. Proposition. *Let* (H1) *be fulfilled.*

(i) *Assume that* $A \in \text{alg}(V, V_{-1})$ *is invertible in* $\mathscr{L}(X)$. *Then* $A^{-1} \in \text{alg}(V, V_{-1})$ *and, moreover,* smb $A^{-1} = (\text{smb } A)^{-1}$ *and* wind smb $A = 0$.

(ii) *Assume there is a V-dominating algebra $\mathfrak{D}$. If $A = B + K$ (with $B \in \mathfrak{L}_{\mathfrak{D}}(V)$ and $K \in \mathbf{QC}(V)$) is an invertible operator then A and B are both invertible in* alg (V, V_{-1}). *Moreover, there exist operators $B' \in \mathfrak{L}_{\mathfrak{D}}(V)$, $K' \in \mathbf{QC}(V)$ so that $A^{-1} = (B + K)^{-1} = B' + K'$, and for the symbols we have*

$$\operatorname{smb} A^{-1} = \operatorname{smb} B^{-1} = (\operatorname{smb} B)^{-1} = (\operatorname{smb} A)^{-1} = \operatorname{smb} B'.$$

Proof. (i): Approximate A by operators A_n, where A_n is of the form $A_n = \sum_{i=1}^{k_n} \prod_{j=1}^{m_n} R_{ij}^{(n)}$, $R_{ij}^{(n)} \in \mathfrak{L}_0(V)$. If $\|A_n - A\| < \|A^{-1}\|^{-1}/2$ then A_n is invertible and $\|A_n^{-1}\| < 2\,\|A^{-1}\|$. Thus, for large n,

$$\|A^{-1} - A_n^{-1}\| \leqq \|A_n^{-1}\|\,\|A^{-1}\|\,\|A_n - A\| < 2\,\|A^{-1}\|^2\,\|A_n - A\|,$$

and this estimate shows that it suffices to verify the assertion for operators $A = \sum_{i=1}^{k} \times \prod_{j=1}^{m} R_{ij}$, $R_{ij} \in \mathfrak{L}_0(V)$. Given such an operator A, let $B \in \mathfrak{L}_0(V)$ be the essential part of A. Now take $l \in \mathbb{N}$ so that $(A - B)\,V_l = 0$. Thus, $AV_l = BV_l$, and the invertibility of A implies the invertibility of BV_l from the left. Theorem 4.2 shows that then B must be one-sided invertible and that its one-sided inverse can be assumed to belong to alg (V, V_{-1}). Let, e.g., B be invertible from the left and take $C \in \operatorname{alg}(V, V_{-1})$ so that $CB = I$. Then

$$A = B + K = B + KCB = (I + KC)\,B \tag{1}$$

with $KC \in \mathbf{QC}(V)$. Since A is invertible we conclude from (1) that $I + KC$ must be invertible from the right, and this leads by Lemma 4.14 to the two-sided invertibility of $I + KC$ in alg (V, V_{-1}). Put $T := (I + KC)^{-1} - I \in \mathbf{QC}(V)$. Then $C(I + T)\,A = C(I + T)\,(I + KC)\,B = I$, i.e. $C(I + T)$ is a left inverse for A. Hence, $C(I + T)$ is an inverse for A and we obtain that $A^{-1} = C(I + T) \in \operatorname{alg}(V, V_{-1})$. Finally, the equation $\operatorname{smb} A^{-1} = (\operatorname{smb} A)^{-1}$ is obvious and the index formula follows from (1) by invoking Theorem 4.2.

(ii): If A is invertible then, by assertion (i), $A^{-1} \in \operatorname{alg}(V, V_{-1})$ and wind smb $A = 0$. Because $\operatorname{smb} A = \operatorname{smb} B$ and $B \in \mathfrak{L}_{\mathfrak{D}}(V)$ we conclude via Theorem 4.9 that B is invertible and via (i) that $B^{-1} \in \operatorname{alg}(V, V_{-1})$. Let $B = a(V)$, $a \in \mathfrak{D}$. The invertibility of B implies the invertibility of a in $\mathfrak{D}$. Put $B' = a^{-1}(V) \in \mathfrak{L}_{\mathfrak{D}}(V)$. Then $B'(B + K) = a^{-1}(V) \times \big(a(V) + K\big) = I + K_1$ with $K_1 \in \mathbf{QC}(V)$. To prove this we note that there are polynomials a_n and b_n such that $a_n(d)$ and $b_n(d)$ converge in the norm of $\mathfrak{D}$ to a and $b = a^{-1}$, respectively. Obviously, $\|b_n(V)\,a_n(V) - b(V)\,a(V)\| \to 0$ as $n \to \infty$. The essential part of $b_n(V)\,a_n(V)$ is given by $(b_na_n)\,(V)$, and since

$$\|(b_na_n)\,(V) - I\|_{\mathscr{L}(X)} \leqq M\,\|b_n(d)\,a_n(d) - e\|_{\mathfrak{D}},$$

we get that $(b_na_n)\,(V) \to I$ in the norm of $\mathscr{L}(X)$ as $n \to \infty$. Thus the sequence $\|b_n(V) \times a_n(V) - (b_na_n)\,(V)\|$ also converges to a certain element of $\mathbf{QC}(V)$, which gives the assertion.

Again by Theorem 4.9, the operator B' is invertible. So the operator $I + K_1$ must be invertible, too, and Lemma 4.14 entails that $(I + K_1)^{-1} = I + K_2$ with $K_2 \in \mathbf{QC}(V)$. Finally, the operator $(I + K_2)\,B' =: B' + K'$ is an inverse for A. The symbol identity is obvious. ■

Decomposing algebras alg (V, V_{-1})

4.23. Definition. Generally, all what we know about the relations between $\mathfrak{L}(V)$, $\mathbf{QC}(V)$ and alg (V, V_{-1}) is that the algebraical sum $\mathfrak{L}(V) + \mathbf{QC}(V)$ is dense in the algebra alg (V, V_{-1}). Under additional conditions we can get essentially more information about the structure of alg (V, V_{-1}). In the sequel the algebra alg (V, V_{-1}) will be said to *decompose* into the direct sum $\mathfrak{L}(V) \dotplus \mathbf{QC}(V)$ if there is a continuous projection S mapping alg (V, V_{-1}) onto $\mathfrak{L}(V)$ parallel to $\mathbf{QC}(V)$.

4.24. Theorem. *Assume that* (H1) *is fulfilled and that, for all* $R_{ij} \in \mathfrak{L}_0(V)$,

$$\|R_0\| \leqq M \left\| \sum_{i=1}^{n} \prod_{j=1}^{m} R_{ij} \right\|, \tag{1}$$

where R_0 *is the essential part of* $\sum_{i=1}^{n} \prod_{j=1}^{m} R_{ij}$ *(see Section* 4.16). *Then the algebra* alg (V, V_{-1}) *decomposes into the direct sum*

$$\text{alg}\,(V, V_{-1}) = \mathfrak{L}(V) \dotplus \mathbf{QC}(V).$$

Proof. Let N denote the set

$$N := \left\{ \sum_{i=1}^{n} \prod_{j=1}^{m} R_{ij},\ m, n \in \mathbb{N} \text{ and } R_{ij} \in \mathfrak{L}_0(V) \right\},$$

which is dense in alg (V, V_{-1}). Define a linear mapping S on N by

$$S\left(\sum_{i=1}^{n} \prod_{j=1}^{m} R_{ij} \right) = R_0,$$

where R_0 is the essential part of $\sum_{i=1}^{n} \prod_{j=1}^{m} R_{ij}$. Note that S is well-defined and continuous on N by (1).

Since $S(R) = R$, $R \in \mathfrak{L}_0(V)$, we conclude that S is a continuous projection on N. Its continuous extension to all of alg (V, V_{-1}) is denoted by S again. Thus,

$$\text{alg}\,(V, V_{-1}) = \operatorname{im} S \dotplus \ker S.$$

It is immediately clear that $\operatorname{im} S = \mathfrak{L}(V)$. To finish the proof it remains to show that $\ker S = \mathbf{QC}(V)$. So let $A \in \ker S$. Then there exists a sequence $\{A_n\}$ in N with $\|A_n - A\| \to 0$ and $\|S(A_n)\| \to 0$. Since $S(A_n) - A_n$ is in $\mathbf{QC}(V)$ we get that A is in $\mathbf{QC}(V)$. Finally note that the inclusion $\mathbf{QC}(V) \subset \ker S$ is obvious, and the proof is complete. ■

4.25. Remarks. 1°. Since $\text{alg}^\pi\,(V, V_{-1})$ is isomorphic to $\mathfrak{L}(V)$ for a decomposing algebra alg (V, V_{-1}), we can endow $\mathfrak{L}(V)$ with an equivalent norm and the structure of a commutative Banach algebra, which we denote by $(\mathfrak{L}(V), \circ)$. By Proposition 4.17 the maximal ideal space of $(\mathfrak{L}(V), \circ)$ can be identified with $\mathbb{T}$ and the Gelfand transform for elements A in $(\mathfrak{L}(V), \circ)$ coincides with the symbol of A as defined in Section 4.3. Obviously, $(\mathfrak{L}(V), \circ)$ is also a V-dominating Banach algebra and Theorem 4.9 is true with $\mathfrak{L}(V)$ in place of $\mathfrak{L}_{\mathfrak{D}}(V)$.

2°. Assume alg (V, V_{-1}) is decomposing. Then for any element the essential part exists.

3°. A consequence of 4.24(1) is the inequality

$$\|R_0\| \leqq M \|R_1\| \|R_2\|, \tag{1}$$

where R_0 is the essential part of $R_1R_2(R_1, R_2 \in \mathfrak{L}_0(V))$. Assume that (1) is fulfilled (without any knowledge about the validity of 1.24(1)). It turns out that in this case the space $\mathfrak{L}(V)$ can be endowed with an equivalent norm and with a multiplication $\circ$ such that $(\mathfrak{L}(V), \circ)$ becomes a Banach algebra. The Banach algebra $(\mathfrak{L}(V), \circ)$ defined in 1° is a special kind of this type of algebras. Indeed, put $R_0 = R_1 \circ R_2$. It is easily seen that the commutative multiplication $\circ$ can be extended to a continuous multiplication on all of $\mathfrak{L}(V)$, which will also be denoted by $\circ$. From the general theory of Banach algebras it follows that $\mathfrak{L}(V)$ can be endowed with an equivalent norm such that $(\mathfrak{L}(V), \circ)$ becomes a Banach algebra. We claim that the maximal ideal space of this algebra can be identified with $\mathbb{T}$. Indeed, the spectrum of V, viewed as an element of alg (V), coincides with $\{z \in \mathbb{C}: |z| \leqq 1\}$ by Theorem 4.2(i) and 1.8(c). Obviously, alg (V) is also a subalgebra of $(\mathfrak{L}(V), \circ)$ and hence, the spectrum of V in $(\mathfrak{L}(V), \circ)$ must be $\mathbb{T}$ (the inclusion sp $V \subset \mathbb{T}$ is clear and the equality sp $V = \mathbb{T}$ then results from 1.8(c)). From the general theory of Banach algebras it follows that the maximal ideal space can be identified with $\mathbb{T}$. Note that the Gelfand transform of $A \in (\mathfrak{L}(V), \circ)$ coincides with the symbol of A defined as in Section 4.3. Now it is immediate that $(\mathfrak{L}(V), \circ)$ is also a V-dominating Banach algebra.

4°. Suppose X is a Hilbert space, V is an isometry and $V^{(-1)} := V^*$. Then alg (V, V_{-1}) is decomposing and a C^*-algebra. Hence $(\mathfrak{L}(V), \circ)$ is also a C^*-algebra. Moreover, $(\mathfrak{L}(V), \circ)$ is isometrically *-isomorphic to $\mathbf{C}(\mathbb{T})$ by 1.13(a).

4.26. Proposition. *Let* (H1) *be fulfilled and assume there is a family* $\{W_\omega\}_{\omega\in\Omega}$ ($\Omega \subset \mathbb{R}^+$ *unbounded*) *of left-invertible operators with left-inverses* $W_\omega^{(-1)}$ *such that*

(i) $VW_\omega = W_\omega V$, $V_{-1}W_\omega^{(-1)} = W_\omega^{(-1)}V_{-1}$ *for all* $\omega \in \Omega$;

(ii) $\sup\limits_\omega \|W_\omega\| < \infty$, $\sup\limits_\omega \|W_\omega^{(-1)}\| < \infty$;

(iii) $\|W_\omega^{(-1)}kW_\omega\| \to 0$ *as* $\omega \to \infty$ *for all* $K \in \mathbf{QC}(V)$.

Then the algebra alg (V, V_{-1}) *decomposes:*

$$\text{alg}\,(V, V_{-1}) = \mathfrak{L}(V) \dotplus \mathbf{QC}(V).$$

In particular, this is true if $\sup\limits_{n\in\mathbb{Z}} \|V_n\| < \infty$ *(in which case we may take* $\{V_n\}_{n\in\mathbb{Z}}$ *as the family* $\{W_\omega\}_{\omega\in\Omega}$*).*

Proof. Notice that (i) yields $W_\omega^{(-1)}R_0W_\omega = R_0$ for any $R_0 \in \mathfrak{L}_0(V)$ and any $\omega \in \Omega$. Consider $R = \sum\limits_{i=1}^{n} \prod\limits_{j=1}^{m} R_{ij}$, $R_{ij} \in \mathfrak{L}_0(V)$, and let R_0 be the essential part of R. Then

$$\lim_{\omega\to\infty} W_\omega^{(-1)}RW_\omega = R_0 \qquad \text{(by (i) and (iii))}.$$

Using (ii) one gets

$$\|R_0\| \leqq M \left\| \sum_{i=1}^{n} \prod_{j=1}^{m} R_{ij} \right\|$$

and Theorem 4.24 applies. Now suppose $\sup\limits_{n\in\mathbb{Z}} \|V_n\| < \infty$.

It remains only to show that $\|V_{-n}KV_n\| \to 0$ as $n \to \infty$ for any $k \in \mathbf{QC}(V)$. Due to Corollary 4.13(c) one has

$$\|V_{-n}KV_n\| \leqq \|V_{-n}\| \, \|KQ_n\| \, \|V_n\| \to 0$$

as $n \to \infty$ and we are done. ■

Finite sections

4.27. Notation. Let X be a Banach space, Ω be an unbounded set of nonnegative numbers and let $\{P_\omega\}_{\omega \in \Omega}$ be a family of bounded projections with $\sup_\omega \|P_\omega\| < \infty$. The operators $P_\omega A P_\omega$ are called *finite sections* of the operator $A \in \mathcal{L}(X)$ with respect to the family $\{P_\omega\}$. We are mainly interested in whether for a given operator A the (generalized) sequence $\{P_\omega A P_\omega\}_{\omega \in \Omega}$ is stable, i.e. whether $P_\omega A P_\omega$ is invertible in im P_ω for all ω large enough, say for $\omega \geqq \omega_0$, and

$$\sup_{\omega \geqq \omega_0} \|(P_\omega A P_\omega)^{-1} P_\omega\| < \infty .$$

In the sequel we consider this question for certain invertible operators from alg (V, V_{-1}) without the extra assumption that s-lim P_ω exists. We shall see that this enables us also to prove the convergence of the finite section method in a somewhat weaker sense. It turns out that the following observation is very useful.

(a) Let $A: X \to X$ be a linear and invertible operator, $P: X \to X$ a linear projection and put $Q = I - P$. Then PAP: im $P \to$ im P is invertible if and only if $QA^{-1}Q$: im $Q \to$ im Q is so. In that case

$$(PAP)^{-1} P = PA^{-1}P - PA^{-1}Q(QA^{-1}Q)^{-1} QA^{-1}P. \tag{1}$$

Indeed, we have

$$\begin{aligned} &PAP\big(PA^{-1}P - PA^{-1}Q(QA^{-1}Q)^{-1} QA^{-1}P\big) \\ &= PAPA^{-1}P - PAA^{-1}Q(QA^{-1}Q)^{-1} QA^{-1}P + PAQA^{-1}Q(QA^{-1}Q)^{-1} QA^{-1}P \\ &= PAPA^{-1}P + PAQA^{-1}P = P. \end{aligned}$$

Hence, PAP is right-invertible in im P if $QA^{-1}Q$ is invertible in im Q. In a similar manner one shows the left-invertibility of PAP in im P. To finish the proof one only needs to change the roles of PAP and $QA^{-1}Q$.

(b) If $Q = WW^{(-1)}$, where W is only left-invertible and $W^{(-1)}$ is one of its left inverses, then the preceding assertion can be reformulated: PAP is invertible in im P if and only if $W^{(-1)}A^{-1}W$ is invertible in X. The proof is very simple: if $W^{(-1)}A^{-1}W$ is invertible and if C is its inverse, then $QWCW^{(-1)}Q$ is the inverse of $QA^{-1}Q$ in im Q, since

$$QA^{-1}QWCW^{(-1)}Q = WW^{(-1)}A^{-1}WCW^{(-1)} = WW^{(-1)} = Q,$$

$$QWCW^{(-1)}QA^{-1}Q = W(CW^{(-1)}A^{-1}W)\, W^{(-1)} = WW^{(-1)} = Q;$$

conversely, if $QA^{-1}Q$ is invertible in im Q and C is its inverse, then $W^{(-1)}QCQW$ is the inverse of $W^{(-1)}A^{-1}W$:

$$W^{(-1)}QCQWW^{(-1)}A^{-1}WW^{(-1)}W = W^{(-1)}QCQA^{-1}QW = W^{(-1)}W = I,$$

$$W^{(-1)}A^{-1}WW^{(-1)}QCQW = W^{(-1)}QA^{-1}QCQW = W^{(-1)}W = I.$$

4.28. Proposition. *Let X be a Banach space, let $\{P_\omega\}_{\omega\in\Omega}$ be a family of projections with $\sup_\omega \|P_\omega\| < \infty$ and let $A \in \mathcal{L}(X)$ be invertible.*

(i) *The sequence $\{P_\omega A P_\omega\}$ is stable if and only if $\{Q_\omega A^{-1} Q_\omega\}$ is stable ($Q_\omega := I - P_\omega$, $\omega \in \Omega$).*

(ii) *Suppose in addition that there is a family $\{W_\omega\}_{\omega\in\Omega}$ of left-invertible operators with left-inverses $W_\omega^{(-1)}$ such that $W_\omega W_\omega^{(-1)} = Q_\omega$ (i.e. $P_\omega = I - W_\omega W_\omega^{(-1)}$). Then the sequence $\{P_\omega A P_\omega\}$ is stable if and only if the operators $W_\omega^{(-1)} A^{-1} W_\omega$ are invertible for ω large enough, say for $\omega \geqq \omega_0$, and*

$$\sup_{\omega\geqq\omega_0} \|W_\omega (W_\omega^{(-1)} A^{-1} W_\omega)^{-1} W_\omega^{(-1)}\| < \infty. \tag{1}$$

Proof. Immediate from 4.27(a) and (b). ■

Notice that the families $\{W_\omega\}_{\omega\in\Omega}$ defined in the Sections 4.26 and 4.28 have, at least formally, nothing to do with each other. However, in concrete applications it will frequently occur that these two families in fact coincide.

4.29. Definition. In this section we connect with the family $\{P_\omega\}_{\omega\in\Omega}$ $\left(\sup_\omega \|P_\omega\| = M < \infty\right)$ a Banach algebra which is of interest in what follows. Denote by $\mathcal{A}$ the collection of all operators $B \in \mathcal{L}(X)$ such that $\|P_{\omega_1} B Q_\omega\| \to 0$ as $\omega \to \infty$ for each $\omega_1 \in \Omega$. It is easily seen that $\mathcal{A}$ actually forms a (closed) subalgebra of $\mathcal{L}(X)$. Hence, $\mathcal{A}$ itself is a Banach algebra. Obviously, the collection $\mathcal{J} := \{B \in \mathcal{L}(X): \|BQ_\omega\| \to 0 \text{ as } \omega \to \infty\}$ is contained in $\mathcal{A}$. Moreover, $\mathcal{J}$ is a closed two-sided ideal in $\mathcal{A}$. Of course, there is a second Banach algebra $\mathcal{B}$ which can be obtained by changing the roles of P_{ω_1} and Q_ω in the previous definition. The finite section method will be studied in more detail in what follows. In the meanwhile we confine ourselves to stating the following result of technical nature.

4.30. Proposition. *Let $A \in \mathfrak{L}(X)$ be an invertible operator such that $A^{-1} \in \mathcal{A}$. If $A + K$ is invertible, $K \in \mathcal{J}$, and the sequence $\{P_\omega A P_\omega\}$ is stable then the sequence $\{P_\omega (A + K) P_\omega\}$ is also stable. This also holds if one replaces $\mathcal{A}$ by $\mathcal{B}$.*

Proof. By Proposition 4.28 it is clear that $Q_\omega A^{-1} Q_\omega$ is stable and so it suffices to show that $\{Q_\omega (A + K)^{-1} Q_\omega\}$ is also stable. Write $A + K = (I + KA^{-1}) A$, where $I + KA^{-1}$ is invertible. Put $B := (I + KA^{-1})^{-1}$. Then $B(I + KA^{-1}) = I$ and $B = I - BKA^{-1}$. Therefore, $(A + K)^{-1} = A^{-1}(I - BKA^{-1}) = A^{-1} - A^{-1}BKA^{-1}$. Since $KA^{-1} \in \mathcal{J}$ we get $\|Q_\omega A^{-1} BKA^{-1} Q_\omega\| \to 0$ as $\omega \to \infty$. This shows that $\{Q_\omega (A + K)^{-1} Q_\omega\}$ must be stable. ■

4.31. Remark. The proof of the previous proposition shows that under its assumptions the operator $(A + K)^{-1}$ also belongs to $\mathcal{A}$.

Convergence of the finite section method

4.32. Notation. If the sequence $\{P_\omega A P_\omega\}$ is stable, if the operator A is invertible and if s-lim $P_\omega = I$ then s-lim $(P_\omega A P_\omega)^{-1} P_\omega = A^{-1}$. This means that the finite section method converges in the usual sense. If the family $\mathcal{P} = \{P_\omega\}$ is only uniformly bounded this is no longer true. However, in this case a weaker convergence can be considered. More precisely, we say that a (generalized) sequence $\{x_\omega\} \subset X$ is *$\mathcal{P}$-convergent* to $x \in X$ (and write $x_\omega \xrightarrow{\mathcal{P}} x$) if there exists an ω^* such that

(i) the sequence $\{x_\omega\}_{\omega \geqq \omega^*}$ is bounded;
(ii) $P_{\omega_0} x_\omega \to P_{\omega_0} x$ in the norm of X for each fixed $\omega_0 \in \Omega$.

If

$$\bigcap_{\omega} \operatorname{im} Q_\omega = \{0\} \tag{1}$$

then the $\mathcal{P}$-limit x is seen to be uniquely determined. In what follows we always assume that (1) is fulfilled. An operator $A \in \mathcal{L}(X)$ is said to be *$\mathcal{P}$-continuous* if it maps $\mathcal{P}$-convergent sequences into $\mathcal{P}$-convergent sequences. Obviously, the sum and the product of $\mathcal{P}$-continuous operators is also $\mathcal{P}$-continuous.

4.33. Proposition. *Each operator $A \in \mathcal{A} \subset \mathcal{L}(X)$ is $\mathcal{P}$-continuous.*

Proof. Let $x_\omega \xrightarrow{\mathcal{P}} x$. We have to show that $P_{\omega_1} A x_\omega$ converges in the norm to $P_{\omega_1} A x$ for each $\omega_1 \in \Omega$. Put $M = \sup\limits_{\omega \geqq \omega^*} \|x_\omega - x\|$. For given $\varepsilon > 0$, choose $\tau \in \Omega$ such that $\|P_{\omega_1} A Q_\tau\| < \varepsilon/(2M)$. Now take $\mu \in \Omega$, $\mu \geqq \omega^*$, such that $\|P_{\omega_1} A P_\tau x_\omega - P_{\omega_1} A P_\tau x\| < \varepsilon/2$ for all $\omega > \mu$. Consequently,

$$\begin{aligned}\|P_{\omega_1} A(x_\omega - x)\| &\leqq \|P_{\omega_1} A Q_\tau (x_\omega - x)\| + \|P_{\omega_1} A P_\tau (x_\omega - x)\| \\ &< \varepsilon/2 + \varepsilon/2 = \varepsilon. \quad \blacksquare\end{aligned}$$

4.34. Theorem. *Let $\sup\limits_{\omega} \|P_\omega\| < \infty$, suppose $P_{\omega_1} P_{\omega_2} = P_{\omega_1}$ for $\omega_2 \geqq \omega_1$, and let A be invertible with $A^{-1} \in \mathcal{A}$. If $\{P_\omega A P_\omega\}$ is stable then*

$$(P_\omega A P_\omega)^{-1} P_\omega x_\omega \xrightarrow{\mathcal{P}} A^{-1} x$$

for every sequence $\{x_\omega\}$ such that $x_\omega \xrightarrow{\mathcal{P}} x$.

Proof. First note that if $\{x_\omega\}_{\omega \geqq \omega^*}$ is any bounded sequence, then $Q_\omega x_\omega \xrightarrow{\mathcal{P}} 0$ and, consequently, $P_\omega x_\omega \xrightarrow{\mathcal{P}} x$ if in addition $x_\omega \to x$. Further, for ω large enough we have

$$(P_\omega A P_\omega)^{-1} P_\omega x_\omega = P_\omega A^{-1} P_\omega x_\omega - P_\omega A^{-1} Q_\omega (Q_\omega A^{-1} Q_\omega)^{-1} Q_\omega A^{-1} P_\omega x_\omega,$$

which results from 4.27(1). Since $\{Q_\omega A^{-1} Q_\omega\}$ is stable, the second item on the right-hand side $\mathcal{P}$-converges to zero. Obviously, $P_\omega A^{-1} P_\omega x_\omega$ $\mathcal{P}$-converges to $A^{-1}x$. Putting these things together we get the assertion. $\blacksquare$

4.35. Remark. This theorem can be viewed as an assertion about the $\mathcal{P}$-convergence of a special operator sequence. Our next concern is to consider this notion more deeper. There are at least two reasons for doing so. In order to prove Theorem 4.34 we have assumed that A^{-1} belongs to $\mathcal{A}$. Unfortunately we have not been able to prove that A^{-1} belongs to $\mathcal{A}$ whenever A is in $\mathcal{A}$ and A is invertible in $\mathcal{L}(X)$. To overcome this difficulty we introduce a Banach algebra $\mathfrak{A}$ having the following property: If A belongs to $\mathfrak{A}$ and is invertible in $\mathcal{L}(X)$, then $A^{-1} \in \mathfrak{A}$ and $A^{-1} \in \mathcal{A}$. The second reason can be described as follows. Suppose $P_\omega \to I$ strongly as $\omega \to \infty$ and suppose that $\{P_\omega A P_\omega\}$ is stable. Then it is wellknown that A is an operator of regular type. We are interested in whether there are conditions on A which guarantee the invertibility of A. We shall see that this question is closely connected with the notion of the $\mathcal{P}$-convergence of operator sequences.

$\mathcal{P}$-convergent operator sequences

4.36. Notations. Let X be a Banach space. By $\mathcal{M}$ we denote the set of all finite unions of (open, closed or half-open) intervals of $\mathbb{R}^+$. Assume that with each $U \in \mathcal{M}$ a projection operator $P_U \in \mathcal{L}(X)$ is associated so that

(i) $P_U + P_{\mathbb{R}^+ \setminus U} = I$,

$P_U P_V = P_V P_U = P_{U \cap V}$,

$P_U + P_V = P_{U \cup V}$ if $U \cap V = \emptyset$;

(ii) $\sup\limits_{U \in \mathcal{M}} \|P_U\| = C < \infty$;

(iii) $\bigcap\limits_{w \in \mathbb{R}^+} \ker P_{[0,w]} = \{0\}$.

Put $\mathcal{P}_{\mathcal{M}} = \{P_U : U \in \mathcal{M}\}$ and $\mathcal{P} = \{P_{[0,w]} : w \in \mathbb{R}^+\}$. Note that $\mathcal{P}^*_{\mathcal{M}} := \{P^*_U : U \in \mathcal{M}\}$ fulfils (i) and (ii). Let Ω stand for a certain unbounded subset of $\mathbb{R}^+$ and let $\{x_\omega\}_{\omega \in \Omega}$ be a sequence of elements belonging to X. Notice that condition (iii) ensures that the $\mathcal{P}$-limit of $\{x_\omega\}$ is unique if it exists (see Section 4.32).

A (generalized) sequence $\{A_\omega\}_{\omega \in \Omega} \subset \mathcal{L}(X)$ is said to be *convergent with respect to* $\mathcal{P}$ (or simply to be $\mathcal{P}$-convergent) to the operator $A \in \mathcal{L}(X)$ if for each $w \in \mathbb{R}^+$ and each $x \in X$ we have

$$\|P_{[0,w]} A_\omega x - P_{[0,w]} A x\| \to 0, \qquad \|A_\omega P_{[0,w]} x - A P_{[0,w]} x\| \to 0$$

as $\omega \to \infty$ and if there is an ω^* such that $\sup\limits_{\omega \geqq \omega^*} \|A_\omega\| < \infty$.

This will be abbreviated to $A_\omega \xrightarrow{\mathcal{P}} A$ or $A = \mathcal{P}\text{-lim}\, A_\omega$. It is worth noticing that condition (iii) also ensures that the $\mathcal{P}$-limit of a given operator sequence is unique provided that it exists. We mention that strong convergence is included in the notion of $\mathcal{P}$-convergence. Indeed, choose the family $\mathcal{P}$ as follows: if $U \in \mathcal{M}$ contains an interval of the form $[0, w]$, then put $P_U := I$, and if $U \in \mathcal{M}$ is not of this form put, $P_U = 0$.

Now consider the set $\mathfrak{A}$ of all (generalized) sequences $\{A_\omega\}_{\omega \in \Omega}$ of bounded linear operators on X which are subject to the following conditions.

(a) $\sup\limits_{\omega \in \Omega} \|A_\omega\| < \infty$.

(b) The $\mathcal{P}$-limit of $\{A_\omega\}$ exists.

(c) Given two subsets $U, V \subset \mathbb{R}^+$ put $\varrho(U, V) = \inf\{|u - v| : u \in U, v \in V\}$. For $h \in \mathbb{R}^+$ set

$$\varphi_{\{A_\omega\}}(h) = \sup\{\|P_U A_\omega P_V\| : \omega \in \Omega,\ U, V \in \mathcal{M} \text{ such that } \varrho(U, V) > h\}.$$

Assume $\lim_{h\to\infty} \varphi_{\{A_\omega\}}(h) = 0$.
If the constant sequence $\{A_\omega\} = \{A\}$ belongs to $\mathfrak{A}$ we write $A \in \mathfrak{A}$ for brevity. Such an operator is also called an *operator of local type*. Condition (i) shows that $I \in \mathfrak{A}$ and $P_U \in \mathfrak{A}$ for any $U \in \mathcal{M}$.

4.37. Proposition. (i) *$\mathfrak{A}$ is a Banach algebra when provided with the norm $\|\{A_\omega\}\| := \sup_\omega \|A_\omega\|$ and with elementwise operations. Furthermore, $\mathcal{P}$-lim $(\{A_\omega\} + \{B_\omega\}) = \mathcal{P}$-lim $\{A_\omega\} + \mathcal{P}$-lim $\{B_\omega\}$ and $\mathcal{P}$-lim $(\{A_\omega\}\{B_\omega\}) = (\mathcal{P}$-lim $\{A_\omega\})\,(\mathcal{P}$-lim $\{B_\omega\})$.*
(ii) *If $A \in \mathfrak{A}$, then $A \in \mathcal{A}$ (with respect to the family $\mathcal{P}$).*
(iii) *Let $\mathcal{N}$ denote the collection of all sequences $\{A_\omega\} \in \mathfrak{A}$ such that $\|P_{(w,\infty)}A_\omega\| \to 0$ and $\|A_\omega P_{(w,\infty)}\| \to 0$ uniformly with respect to ω as $w \to \infty$. Then $\mathcal{N}$ is a closed two-sided ideal in $\mathfrak{A}$. In particular, if $A \in \mathcal{L}(X)$ is such that $\|P_{(w,\infty)}A\| \to 0$ and $\|AP_{(w,\infty)}\| \to 0$ as $w \to \infty$, then A is an operator of local type and $A \in \mathcal{N}$.*

Proof. It suffices to prove assertion (i). Let $\{A_\omega\}, \{B_\omega\} \in \mathfrak{A}$. Then, obviously, $\{A_\omega\} + \{B_\omega\} = \{A_\omega + B_\omega\} \in \mathfrak{A}$. Consider $\{A_\omega\}\{B_\omega\} = \{A_\omega B_\omega\}$. Put $A = \mathcal{P}$-lim A_ω, $B = \mathcal{P}$-lim B_ω. The identity

$$\begin{aligned} P_{[0,w]}(A_\omega B_\omega x - ABx) &= P_{[0,w]}(A_\omega B_\omega x - A_\omega Bx + A_\omega Bx - ABx) \\ &= P_{[0,w]}A_\omega P_{[0,v]}(B_\omega - B)\,x \\ &\quad + P_{[0,w]}A_\omega P_{(v,\infty)}(B_\omega - B)\,x + P_{[0,w]}(A_\omega - A)\,Bx \end{aligned}$$

shows (choose v large enough) that $P_{[0,w]}A_\omega B_\omega \to P_{[0,w]}AB$ strongly. The existence of s-lim $A_\omega B_\omega P_{[0,w]} = ABP_{[0,w]}$ can be proved analogously. Hence, $A_\omega B_\omega \xrightarrow{\mathcal{P}} AB$. Further 4.36(c) is a consequence of

$$\|P_U A_\omega B_\omega P_V\| \leqq \|P_U A_\omega\|\,\|P_W B_\omega P_V\| + \|P_U A_\omega P_{\mathbb{R}^+\setminus W}\|\,\|B_\omega P_V\|,$$

and standard arguments show that $\mathfrak{A}$ is a Banach algebra. ■

4.38. Corollary. *If $\{A_\omega\} \in \mathfrak{A}$, $A = \mathcal{P}$-lim A_ω, then $A \in \mathfrak{A}$ and $A_\omega \in \mathfrak{A}$ for each $\omega \in \Omega$.*

Proof: Obviously, $\{A_\omega\} \in \mathfrak{A}$ implies that $A_\omega \in \mathfrak{A}$ for each $\omega \in \Omega$. Now consider $P_U A P_V$. A little thought shows that we can assume $U = [0, w]$, $V = (v, \infty)$ or $U = (v, \infty)$, $V = [0, w]$ without loss of generality. In the first case we obtain

$$P_U A_\omega P_V = P_{[0,w]}A_\omega P_V \to P_{[0,w]}AP_V = P_U A P_V \quad \text{strongly},$$

hence, $\|P_U A P_V\| \leqq \sup \|P_U A_\omega P_V\|$. If U is the unbounded interval (v, ∞) and V is the interval $[0, w]$, the proof is analogous. Thus, $\varphi_A(h) := \varphi_{\{A\}}(h) = \sup \{\|P_U A P_V\| : U, V \in \mathcal{M}, \varrho(U, V) > h\} \leqq \sup \{\|P_U A_\omega P_V\| : \omega \in \Omega,\ U, V \in \mathcal{M}, \varrho(U, V) > h\} = \varphi_{\{A_\omega\}}(h)$, and we are done. ■

4.39. Theorem. *An element $\{A_\omega\} \in \mathfrak{A}$ is invertible in $\mathfrak{A}$ if and only if* (i) *the operators A_ω $(\omega \in \Omega)$ and $A = \mathcal{P}$-lim A_ω are invertible,* (ii) $\sup_\omega \|A_\omega^{-1}\| < \infty$.

If $\{A_\omega\}$ is invertible in $\mathfrak{A}$, then $A_\omega^{-1} \xrightarrow{\mathcal{P}} A^{-1}$

Proof. If $\{A_\omega\}$ is invertible in $\mathfrak{A}$, then, obviously, (ii) is fulfilled. Put $B = \mathcal{P}$-lim A_ω^{-1}. Proposition 4.37 shows that $AB = BA = I$. Conversely, assume that (i) and (ii) are fulfilled. First we show that

$$\varphi_{\{A_\omega^{-1}\}}(t) \leqq \frac{C^4 \sup_\omega \|A_\omega\| \sup_\omega \|A_\omega^{-1}\|^2}{n} + 4C^2 \sup_\omega \|A_\omega^{-1}\|^2 \varphi_{\{A_\omega\}}\left(\frac{t}{4n-1}\right) \tag{1}$$

for any positive integer n, not providing A to be invertible. This implies that

$$\lim_{t\to\infty} \varphi_{\{A_\omega^{-1}\}}(t) = 0 .$$

Let $U, V \in \mathcal{M}$ and $\varrho(U, V) = r > t$. Let h, r_1, r_2, r_3, r_4 be any real numbers satisfying $0 \leqq r_1 < r_2 < r_3 < r_4$, $0 < h < r_2 - r_1$, $h < r_3 - r_2$, $h < r_4 - r_3$ and put $U_1 := \{w \in \mathbb{R}^+ : r_1 \leqq \varrho(w, U) \leqq r_3\}$, $V_1 := \{w \in \mathbb{R}^+ : r_2 \leqq \varrho(w, U) \leqq r_4\}$. We claim that

$$P_U A_\omega^{-1} P_V = -P_U A_\omega^{-1} P_{U_1} A_\omega P_{V_1} A_\omega^{-1} P_V + E,$$

where $\|E\| \leqq 3C^2 \sup_\omega \|A_\omega^{-1}\|^2 \varphi_{\{A_\omega\}}(h)$. Put $V' := \{w \in \mathbb{R}^+ : \varrho(w, U) < r_2\}$, $U' := \{w \in \mathbb{R}^+ : \varrho(w, U) \leqq r_3\}$. Then we have $P_U A_\omega^{-1} P_V = P_U P_{V'} A_\omega^{-1} P_V = P_U A_\omega^{-1} P_{U'} A_\omega P_{V'} A_\omega^{-1} P_V + e_1$, where $\|e_1\| = \|P_U A_\omega^{-1} P_{\mathbb{R}^+ \setminus U'} A_\omega P_{V'} A_\omega^{-1} P_V\| \leqq C^2 \sup_\omega \|A_\omega^{-1}\|^2 \varphi_{\{A_\omega\}}(h)$, because $\varrho(\mathbb{R}^+ \setminus U', V') > h$. Further

$$P_U A_\omega^{-1} P_{U'} A_\omega P_{V'} A_\omega^{-1} P_V = -P_U A_\omega^{-1} P_{U'} A_\omega P_{\mathbb{R}^+ \setminus V'} A_\omega^{-1} P_V,$$

since $P_U A_\omega^{-1} P_{U'} A_\omega A_\omega^{-1} P_V = 0$. Finally,

$$P_U A_\omega^{-1} P_{U'} A_\omega P_{\mathbb{R}^+ \setminus V'} A_\omega^{-1} P_V = P_U A_\omega^{-1} P_{U_1} A_\omega P_{V_1} A_\omega^{-1} P_V + e_2,$$

where

$$\begin{aligned} \|e_2\| &\leqq \|P_U A_\omega^{-1} P_{U' \setminus U_1} A_\omega P_{V_1} A_\omega^{-1} P_V\| + \|P_U A_\omega^{-1} P_{U'} A_\omega P_{(\mathbb{R}^+ \setminus V') \setminus V_1} A_\omega^{-1} P_V\| \\ &\leqq 2C^2 \sup_\omega \|A_\omega^{-1}\|^2 \varphi_{\{A_\omega\}}(h), \end{aligned}$$

because $\varrho(U' \setminus U_1, V_1) > h$, $\varrho\big(U', (\mathbb{R}^+ \setminus V') \setminus V_1\big) > h$. Putting these things together we get our claim. Now let n be any positive integer. Put $h := t/(4n-1)$, $l = r/(4n-1)$,

$$U_i := \{w \in \mathbb{R}^+ : (4i-4)\, l \leqq \varrho(w, U) \leqq (4i-2)\, l\},$$

$$V_i := \{w \in \mathbb{R}^+ : (4i-3)\, l \leqq \varrho(w, U) \leqq (4i-1)\, l\},$$

where $i = 1, \ldots, n$. From what has just been proved (with $r_1 = (4i-4)\, l$, $r_2 = (4i-3)\, l$, $r_3 = (4i-2)\, l$, $r_4 = (4i-1)\, l$) we obtain

$$P_U A_\omega^{-1} P_V = -P_U A_\omega^{-1} P_{U_i} A_\omega P_{V_i} A_\omega^{-1} P_V + E_i \tag{2}$$

with $\|E_i\| \leqq 3C^2 \sup_\omega \|A_\omega\|^{-2} \varphi_{\{A_\omega\}}(h)$ for $i = 1, \ldots, n$. Adding the n equalities (2) we arrive at the equality

$$\begin{aligned} nP_U A_\omega^{-1} P_V &= -\sum_{i=1}^n P_U A_\omega^{-1} P_{U_i} A_\omega P_{V_i} A_\omega^{-1} P_V + \sum_{i=1}^n E_i \\ &= -P_U A_\omega^{-1} P_{(U_1 \cup \cdots \cup U_n)} A P_{(V_1 \cup \cdots \cup V_n)} A_\omega^{-1} P_V + E + \sum_{i=1}^n E_i, \end{aligned}$$

where $E := \sum_{i=1}^n P_U A_\omega^{-1} P_{U_i} A_\omega P_{(V_1 \cup \cdots \cup V_{i-1} \cup V_{i+1} \cup \cdots \cup V_n)} A_\omega^{-1} P_V$.

Since $\varrho(U_i, V_j) > h$ for $i \neq j$, it follows that

$$\|E\| < nC^2 \sup_\omega \|A_\omega^{-1}\|^2 \varphi_{\{A_\omega\}}(h),$$

and thus

$$\|P_U A_\omega^{-1} P_V\| \leqq C^4 \sup_\omega \|A_\omega\| \sup_\omega \|A_\omega^{-1}\|^2/n + 4C^2 \sup_\omega \|A_\omega^{-1}\|^2 \varphi_{\{A_\omega\}}(h),$$

which gives 4.36(c). We finally prove that $A_\omega^{-1} \xrightarrow{\mathcal{P}} A^{-1}$. We have

$$P_{[0,w]}(A_\omega^{-1}x - A^{-1}x) = P_{[0,w]}A_\omega^{-1}(Ay - A_\omega y)$$
$$= P_{[0,w]}A_\omega^{-1}P_{[0,v]}(Ay - A_\omega y) + P_{[0,w]}A^{-1}P_{(v,\infty)}(Ay - A_\omega y)$$

with $y = A^{-1}x$. Hence, if $v > w$,

$$\|P_{[0,w]}(A_\omega^{-1}x - A^{-1}x)\|$$
$$\leqq C \sup_\omega \|A_\omega^{-1}\| \, \|P_{[0,v]}(Ay - A_\omega y)\| + \varphi_{\{A_\omega^{-1}\}}(v - w) \sup_\omega \|A - A_\omega\| \, \|y\|.$$

Choose $v_0 > w$ so that $\varphi_{\{A_\omega^{-1}\}}(v_0 - w) < \varepsilon \Big/ \Big(2 \sup_\omega \|A - A_\omega\| \, \|y\|\Big)$. Then we can find an $\omega_0 \in \Omega$ so that for $\omega > \omega_0$

$$\|P_{[0,v_0]}(Ay - A_\omega y)\| \leqq \varepsilon \Big/ \Big(2C \sup_\omega \|A_\omega^{-1}\|\Big),$$

which yields $P_{[0,w]}(A_\omega^{-1} - A) \to 0$ strongly. Similarly one can prove that $(A_\omega^{-1} - A) P_{[0,w}$ $\to 0$ strongly, and this completes the proof. ■

4.40. Corollary. (i) *Let* $\{A_\omega\} \in \mathfrak{A}$, $A = \mathcal{P}\text{-lim}\, A_\omega$. *If* A_ω *is invertible for all* $\omega \in \Omega$ *and if* $\sup \|A_\omega^{-1}\| < \infty$ *then* $A_\omega^{-1}A \xrightarrow{\mathcal{P}} I$, *and* A *is one-to-one.*
(ii) *Assume* $A \in \mathcal{L}(X)$ *is of local type, i.e.* $A \in \mathfrak{A}$, $P_{[0,w]} \to I$ *strongly as* $w \to \infty$, *and suppose* $\{P_{[0,w]}AP_{[0,w]}\}$ *to be stable. Then* A *is invertible.*

Proof. (i): The proof of Theorem 4.39 shows that the hypotheses of the corollary imply that $\varphi_{\{A_\omega^{-1}\}}(h) \to 0$ as $h \to \infty$. Hence,

$$P_{[0,w]}(A_\omega^{-1}Ax - x) = P_{[0,w]}(A_\omega^{-1}Ax - A_\omega^{-1}A_\omega x) = P_{[0,w]}A_\omega^{-1}P_{[0,v]}(Ax - A_\omega x)$$
$$+ P_{[0,w]}A_\omega^{-1}P_{[v,\infty)}(Ax - A_\omega x)$$

becomes as small as desired if $v - w$ and ω are large enough. On the other hand,

$$A_\omega^{-1}AP_{[0,w]}x - P_{[0,w]}x = A_\omega^{-1}(A - A_\omega) P_{[0,w]}x,$$

and this becomes small if ω is large enough. Thus, $A_\omega^{-1}A \xrightarrow{\mathcal{P}} I$, and this shows, moreover, that A must be one-to-one.

(ii): First we prove that $\mathcal{P}^*$ also satisfies 4.36(iii). Indeed if $y \in \bigcap_{w \in \mathbb{R}^+} \ker P_{[0,w]}^*$ then $\langle P_{[0,w]}x, y\rangle = \langle x, P_{[0,w]}^*y\rangle = 0$ for all $w \in \mathbb{R}^+$ and all $x \in X$. Since the elements of the form $P_{[0,w]}x$ are dense in X, the element $y \in X^*$ must be the zero functional. Now choose w_0 so that $P_{[0,w]}AP_{[0,w]}$ is invertible for $w \geqq w_0$ and

$$\sup_{w \geqq w_0} \|(P_{[0,w]}AP_{[0,w]})^{-1} P_{[0,w]}\| < \infty.$$

We consider the sequence $\{A_w\}$, A_w defined by

$$A_w := \begin{cases} P_{[0,w]}AP_{[0,w]} + P_{(w,\infty)}, & w \geqq w_0, \\ I, & w < w_0. \end{cases}$$

It is easily seen that $\{A_w\} \in \mathfrak{A}$ and $A_w \to A$ strongly. Now we claim that the $\mathcal{P}^*$-limit of $\{A_w^*\}$ exists and equals A^*. First note that $\varphi_{\{A^*\}}(h) \to 0$ as $h \to \infty$. Hence, $A^* \in \mathcal{A}$ (with respect to $\mathcal{P}^*$) and this yields the $\mathcal{P}^*$-continuity of A^* by Proposition 4.33. Thus,

$$P_{[0,w]}^*A^*P_{[0,w]}^*y \xrightarrow{\mathcal{P}^*} A^*y$$

for all $y \in X^*$. Using this result it can be easily deduced that

$$A_w^*y \xrightarrow{\mathcal{P}^*} A^*y.$$

Finally, for fixed w_1 we have ($w > \max\{w_0, w_1\}$)

$$(A^* - A_w^*)\,P_{[0,w_1]}^* = P_{(w,\infty)}^*A^*P_{[0,w_1]}^*$$

which converges in the norm of $\mathcal{L}(X^*)$ to zero as $w \to \infty$ because $\varphi_{\{A^*\}}(h) \to 0$ as $h \to \infty$. Now it is clear that $\{A_w^*\}$ belongs to $\mathfrak{A}$ (with respect to $\mathcal{P}_{\mathcal{M}}^*$). Part (i) applies and, consequently, $\ker A^* = \{0\}$. This together with the fact that $\|Ax\| > M\,\|x\|$ ($M > 0$) for all x gives the assertion (compare Proposition 1.23). ■

Non-strongly converging approximation methods

4.41. Notation. Besides the family $\mathcal{P}$ of projections which defines the convergence we consider another family $\mathcal{R} = \{R_\omega\}_{\omega\in\Omega}$ of projection operators which is related to $\mathcal{P}$ by the condition

$$\{R_\omega\}_{\omega\in\Omega} \in \mathfrak{A} \quad \text{with} \quad R_\omega \xrightarrow{\mathcal{P}} I \quad \text{as} \quad \omega \to \infty. \tag{1}$$

Let A be an operator on X and consider the equation $Ax = y$. Let $\{A_\omega\}_{\omega\in\Omega}$, $A_\omega\colon \operatorname{im} R_\omega \to \operatorname{im} R_\omega$, be a sequence of operators for which $\{A_\omega R_\omega\}$ converges with respect to $\mathcal{P}$ to A as $\omega \to \infty$, and assume that there is an $\omega_0 \in \Omega$ so that the approximate equation $A_\omega x_\omega = R_\omega y$ has a unique solution x_ω for all right-hand sides y and for all $\omega \in \Omega$ with $\omega \geqq \omega_0$.

We shall say that the approximation method $\Pi_{\mathcal{P}}\{A_\omega\}$ converges for the operator A (and write $A \in \Pi_{\mathcal{P}}\{A_\omega\}$ in that case) if and only if the sequence $\{x_\omega\}_{\omega\geqq\omega_0}$ converges with respect to $\mathcal{P}$ to a solution x of the equation $Ax = y$. In the case where $A_\omega = R_\omega A R_\omega$ we write $\Pi_{\mathcal{P}}\{R_\omega\}$ instead of $\Pi_{\mathcal{P}}\{R_\omega A R_\omega\}$; this special approximation method is also called the *finite section method*. Put $Q_\omega := I - R_\omega$.

4.42. Proposition. *Let $A_\omega\colon \operatorname{im} R_\omega \to \operatorname{im} R_\omega$ and $\{A_\omega R_\omega\} \in \mathfrak{A}$ with $A = \mathcal{P}\text{-}\lim A_\omega R_\omega$. Then $A \in \Pi_{\mathcal{P}}\{A_\omega\}$ if and only if A is invertible and if there exists an $\omega_0 \in \Omega$ such that A_ω: $\operatorname{im} P_\omega \to \operatorname{im} P_\omega$ is invertible for $\omega \in \Omega$, $\omega \geqq \omega_0$, and $\sup\limits_{\omega\geqq\omega_0} \|A_\omega^{-1}R_\omega\| < \infty$.*

Proof. Let $A \in \Pi_{\mathcal{P}}\{A_\omega\}$. Then, by definition, the operators $A_\omega\colon \operatorname{im} R_\omega \to \operatorname{im} R_\omega$ are invertible for ω large enough. Denote by $A_\omega^{-1}\colon \operatorname{im} R_\omega \to \operatorname{im} R_\omega$ the inverse of A_ω. Since $A_\omega^{-1}R_\omega y \xrightarrow{\mathcal{P}} x$, there is an ω_0 such that $\sup\limits_{\omega\geqq\omega_0} \|A_\omega^{-1}R_\omega\| < \infty$. Indeed, if this were not the case, then there would be a sequence $\{\omega_n\}_{n\in\mathbf{N}}$ such that $\omega_n \to \infty$ as $n \to \infty$, $\|A_{\omega_n}^{-1}R_{\omega_n}\| \to \infty$ as $n \to \infty$, and $A_{\omega_n}^{-1}R_{\omega_n}y \xrightarrow{\mathcal{P}} x$. Then, by definition, $A_{\omega_n}^{-1}R_{\omega_n}y$ would be uniformly bounded for each $y \in X$, and the uniform boundedness principle shows that $\sup \|A_{\omega_n}^{-1}R_{\omega_n}\| < \infty$. This is a contradiction which proves our assertion. Next we verify the invertibility of A. By the definition of the approximation method, A is onto. Put $\tilde{A}_\omega := A_\omega R_\omega + Q_\omega$ for $\omega \geqq \omega_0$ and $\tilde{A}_\omega = I$ for $\omega < \omega_0$. By 4.41(1) the sequence $\{\tilde{A}_\omega\}$ is in $\mathfrak{A}$ and $\tilde{A}_\omega \xrightarrow{\mathcal{P}} A$. It is immediate from what has been proved above, that $\tilde{A}_\omega$ is invertible for all $\omega \in \Omega$ and that $\sup\limits_\omega \|\tilde{A}_\omega^{-1}\| < \infty$. Hence, by Corollary 4.40, A is one-to-one. The other direction is an evident consequence of Theorem 4.39 applied to $\{\tilde{A}_\omega\}$. ■

The previous proof also yields the following result.

4.43. Corollary. *Let $\{A_\omega\} \in \mathfrak{A}$, $A = \mathcal{P}$-lim A_ω, and denote by $\mathfrak{C}$ the two-sided closed idea in $\mathfrak{A}$ defined by*

$$\mathfrak{C} := \big\{\{A_\omega\} \in \mathfrak{A} \colon \|A_\omega\| \to 0 \text{ as } \omega \to \infty\big\}.$$

The following statements are equivalent.

(i) *The coset $\{A_\omega\}^0 := \{A_\omega\} + \mathfrak{C}$ is invertible in the quotient algebra $\mathfrak{A}/\mathfrak{C}$.*

(ii) *A is invertible, the operators A_ω are invertible for ω large enough (say $\omega \geqq \omega_0$), and $\sup\limits_{\omega \geqq \omega_0} \|A_\omega^{-1}\| < \infty$.*

(iii) $A \in \Pi_{\mathcal{P}}\{A_\omega\}$.

Stability of finite sections

4.44. Notation. Suppose we are given a left invertible operator V and a left inverse $V^{(-1)}$ satisfying (H1). In what follows we present an approach to the finite section method which slightly differs from the one presented in Gohberg/Feldman [1]. The reason for doing so is the following. We are interested in a theory of the finite section method which works equally good in the scalar and system case. So we shall overcome some difficulties arising in the treatment of Gohberg/Feldman [1] for the system case. However, it should be noted that, from the theoretical point of view, our considerations are less general. Nevertheless, all cases of interest can be reduced to them.

Assume there is a family $\{W_\omega\}_{\omega\in\Omega}$ of left invertible operators with left inverses W_ω^{-1} such that

(i) $VW_\omega = W_\omega V$, $V^{(-1)}W_\omega^{(-1)} = W_\omega^{(-1)}V^{(-1)}$ for all $\omega \in \Omega$;

(ii) $\sup\limits_\omega \|W_\omega\| < \infty$, $\sup\limits_\omega \|W_\omega^{(-1)}\| < \infty$;

(iii) with $Q_\omega := W_\omega W_\omega^{(-1)}$ and $P_\omega := I - Q_\omega$, we have alg $(V, V^{(-1)}) \subset \mathcal{A}$ and $\mathbf{QC}(V) \subset \mathcal{J}$ (or that these conditions hold with respect to the algebra $\mathcal{B}$, where $\mathcal{A}$, $\mathcal{J}$, and $\mathcal{B}$ are defined in Section 4.29).

Conditions (i) – (iii) imply that

$$\text{alg}\,(V, V^{(-1)}) = \mathfrak{L}(V) \dotplus \mathbf{QC}(V). \tag{1}$$

Indeed, since $\mathbf{QC}(V) \subset \mathcal{J}$, we have for $B \in \mathbf{QC}(V)$

$$W_\omega^{(-1)}BW_\omega = W_\omega^{(-1)}(BQ_\omega)\, W_\omega \to 0$$

and Proposition 4.26 applies.

4.45. Theorem. *Suppose* 4.44 (i)–(iii) *are fulfilled. Let* $A \in \mathrm{alg}\,(V, V_{-1})$ *be an invertible operator. Then* $\{P_\omega A P_\omega\}$ *is stable if and only if the essential part* $S(A^{-1})$ *is invertible.*

Proof. (a) Suppose $\{P_\omega A P_\omega\}$ is stable. According to Proposition 4.28 (ii), $W_\omega^{(-1)} A^{-1} W_\omega$ is invertible for $\omega \geqq \omega_0$. From Remark 4.25, 2° we conclude that $A^{-1} \in \mathrm{alg}\,(V, V_{-1})$ and $A^{-1} = S(A^{-1}) + B$, $B \in \mathbf{QC}(V)$. Because $\|BQ_\omega\| \to 0$ as $\omega \to \infty$ we obtain

$$W_\omega^{(-1)} A^{-1} W_\omega = S(A^{-1}) + W_\omega^{(-1)} B W_\omega$$

with $\|W_\omega^{(-1)} B W_\omega\| \to 0$ as $\omega \to \infty$. Now it follows that $S(A^{-1})$ is invertible.

(b) Let $S(A^{-1})$ be invertible. Consider $\{P_\omega S^{-1}(A^{-1}) P_\omega\}$. By Proposition 4.28 this sequence is stable if and only if $W_\omega^{(-1)} S(A^{-1}) W_\omega = S(A^{-1})$ is invertible and $\sup_\omega \|W_\omega S^{-1}(A^{-1}) W_\omega^{(-1)}\| < \infty$. Hence $\{P_\omega S^{-1}(A^{-1}) P_\omega\}$ is stable. Since $A = S(A) + B_1$, $B_1 \in \mathbf{QC}(V)$, we get $AS(A^{-1}) = I + C$, $C \in \mathbf{QC}(V)$, and $A = S^{-1}(A^{-1}) + D$, $D \in \mathbf{QC}(V)$. Proposition 4.30 applies and this gives the stability of $\{P_\omega A P_\omega\}$. ■

Notice that, as a matter of fact, the invertibility of $A \in \mathrm{alg}\,(V, V_{-1})$ already implies the invertibility of $S(A^{-1})$. Indeed, this a consequence of Proposition 4.22 and Theorem 4.9. However, the preceding proof applies to the system case without any changes (see Section 4.102).

Toeplitz operators

4.46. Spaces and notation. Let G be the linear space of all (one-sided) sequences of com plex numbers $\xi = \{\xi_n\}_{n=0}^\infty$.

In the sequel E stands for any of the following linear spaces:

(a) $\mathbf{l}^p := \left\{\xi = \{\xi_n\} \in G : \|\xi\| = \left(\sum_{n=0}^\infty |\xi_n|^p\right)^{1/p} < \infty\right\}, \quad 1 \leqq p < \infty;$

(b) $\mathbf{m} := \left\{\xi = \{\xi_n\} \in G : \|\xi\| = \sup_n |\xi_n| < \infty\right\};$

(c) $\mathbf{m}_c := \left\{\xi = \{\xi_n\} \in \mathbf{m} : \lim_{n\to\infty} \xi_n \text{ exists}\right\};$

(d) $\mathbf{m}_0 := \left\{\xi = \{\xi_n\} \in \mathbf{m}_c : \lim_{n\to\infty} \xi_n = 0\right\}.$

The spaces $\mathbf{l}^p$ and $\mathbf{m}$ actually form Banach spaces under the norms defined above. Moreover, $\mathbf{m}_c$ and $\mathbf{m}_0$ are (closed) subspaces of $\mathbf{m}$. Further, for $\mu \in \mathbb{R}$, let E^μ denote the space of all sequences $\eta = \{(1+n)^{-\mu} \xi_n\}$ with $\xi = \{\xi_n\} \in E$. Obviously, E^μ equipped with the norm

$$\|\eta\|_{E^\mu} = \|\xi\|_E$$

becomes a Banach space. The linear operator $\Lambda^\mu : E \to E^\mu$, $\xi \mapsto \eta$ is clearly an isometry between E and E^μ. The inverse operator to Λ^μ will be denoted by $\Lambda^{-\mu}$. Define operators V and $V^{(-1)}$ on E^μ by

$$V\{\xi_n\}_{n=0}^\infty = \{0, \xi_0, \xi_1, \ldots\}, \qquad V^{(-1)}\{\xi_n\}_{n=0}^\infty = \{\xi_{n+1}\}_{n=0}^\infty.$$

The operators V and $V^{(-1)}$ are usually called forward and backward shift, respectively. Obviously, $V^{(-1)}V = I$ and V is not invertible.

(i) *If* $\mu \geqq 0$ *then* $\|V_{-k}\| = 1$ *and* $\|V_k\| = (1 + k)^\mu$.
If $\mu < 0$ *then* $\|V_{-k}\| = (1 + k)^{|\mu|}$ *and* $\|V_k\| = 1, \quad k \in \mathbb{N}$.
(ii) V *and* V_{-1} *fulfil* (H1), (H2) *and* (H4).

Proof. Let $\mu \geqq 0$ and $\eta = V_{-k}\{\xi_n\}$. Then ($E = \mathbf{l}^p$)

$$\|\eta\|_{E^\mu}^p = \sum_{n=0}^{\infty} [(1 + n)^\mu |\xi_{n+k}|]^p \leqq \sum_{n=0}^{\infty} [(1 + n + k)^\mu |\xi_{n+k}|]^p \leqq \|\xi\|_{E^\mu}^p.$$

Hence, $\|V_{-k}\| \leqq 1$. Moreover it is easy to see that $\|V_{-k}\| = 1$. Assume $\eta = V_k\{\xi_n\}$. Then

$$\|\eta\|_{E^\mu}^p = \sum_{n=k}^{\infty} [(1 + n)^\mu |\xi_{n-k}|]^p.$$

The inequality $(1 + n) \leqq (1 + n - k)(1 + k)$, which holds for $n \geqq k$, shows that $\|\eta\|_{E^\mu} \leqq (1 + k)^\mu \|\xi\|_{E^\mu}$ and $\|V_k\| = (1 + k)^\mu$. Now let $\mu < 0$ and $E = \mathbf{l}^p$. By means of the inequality $1/(1 + n) \leqq (1 + k)/(1 + n + k)$ one immediately obtains

$$\|\eta\|_{E^\mu} \leqq (1 + k)^{|\mu|} \|\xi\|_{E^\mu} \quad \text{and} \quad \|V_{-k}\| = (1 + k)^{|\mu|}.$$

The equality $\|V_k\| = 1$ is almost obvious. The cases $E = \mathbf{m}, \mathbf{m}_c, \mathbf{m}_0$ can be treated analogously. Consequently, (H1) is fulfilled. The conditions (H2) and (H4) are obvious. ∎

The operators belonging to $\mathfrak{L}(V)$ are called *Toeplitz operators* or *discrete Wiener-Hopf operators*.

The action of an operator $A \in \mathfrak{L}(V) \subset \mathcal{L}(E^\mu)$ can be easily described. According to (H2) and (H4) an element of $\mathfrak{L}(V)$ is uniquely determined by its symbol. Let a_n be the Fourier coefficients of the symbol $A(t)$, $A \in \mathfrak{L}(V)$. Put $\eta := A\xi$, $\xi = \{\xi_n\}_0^\infty$, $\eta = \{\eta_n\}_0^\infty$. Then

$$\sum_{k=0}^{\infty} a_{j-k}\xi_k = \eta_j \qquad (j = 0, 1, \ldots). \tag{1}$$

Indeed, for $A \in \mathfrak{L}_0(V)$ the assertion is obvious and by passing to the limit one obtains the assertion in full generality.

A more general look at Toeplitz operators can be obtained in the following manner. Take a function $a \in \mathbf{L}^\infty(\mathbb{T})$ and consider the operator $T(a)$ given on finite sequences by the rule (1), where the numbers a_j are the Fourier coefficients of the function a. If this operator can be continuously extended to the whole space E^μ it is called the Toeplitz operator generated by the function a. This notion includes, for instance, Toeplitz operators on $\mathbf{l}^p$ $(1 \leqq p < \infty)$ generated by classes of discontinuous functions. The theory of these operators can be found in Böttcher/Silbermann [3].

In order to formulate the next theorem we recall the definition of a special Wiener algebra with weight (see Section 4.11): For $\mu \in \mathbb{R}$ define $\mu_- := \max\{0, -\mu\}$, $\mu_+ := \max\{0, \mu\}$ and let $\mathbf{W}^{\mu_-,\mu_+}$ be the Wiener algebra $\mathbf{W}(v)$ with the weight $\{v_n\}_{-\infty}^\infty$, where

$$v_n := (1 + |n|)^{\mu_-}, \quad n < 0 \quad \text{and} \quad v_n := (1 + n)^{\mu_+}, \quad n \geqq 0.$$

4.47. Theorem. (a) *Let* E *be any of the spaces defined in Section* 4.46. *Then an operator* $A \in \mathfrak{L}(V) \subset \mathcal{L}(E^\mu)$ *is at least one-sided invertible if*

$$A(t) \neq 0 \quad \textit{for all} \quad t \in \mathbb{T}.$$

If the symbol $A(t)$ does not vanish on $\mathbb{T}$, the invertibility of A corresponds with the winding number wind $A(t)$, *i.e. A is invertible, invertible only from the left or only from the right, respectively, if the winding number of $A(t)$ is zero, positive or negative, respectively. If, on the other hand, $A(t_0) = 0$ for some $t_0 \in \mathbb{T}$ then A is neither a Φ_+- nor a Φ_--operator.*

(b) *For any $a(t) \in W^{\mu_-,\mu_+}$ there exists an operator $A \in \mathfrak{L}(V) \subset \mathcal{L}(E^\mu)$ such that the symbol $A(t)$ of A coincides with $a(t)$.*

(c) *For $\mu = 0$ the algebra* alg (V, V_{-1}) *decomposes and for $E = \mathbf{l}^2$ the algebra $(\mathfrak{L}(V), \circ)$ is isometrically isomorphic to the algebra $\mathbf{C}(\mathbb{T})$ of all continuous functions defined on $\mathbb{T}$.*

(d) *Let $E = \mathbf{l}^1, \mathbf{m}, \mathbf{m}_c, \mathbf{m}_0$. Then $A \in \mathfrak{L}(V) \subset \mathcal{L}(E^\mu)$ implies that $A(t) \in \mathbf{W}$.*

Proof. (a): For $\mu = 0$ the assertion can be deduced from assertion (c) and Remark 4.25, 1°. If $\mu \in \mathbb{R}$, Theorem 4.20 applies, that is, $A \in \mathfrak{L}(V)$ is a Φ-operator if and only if

$$A(t) \neq 0 \quad \text{for all} \quad t \in \mathbb{T}.$$

Moreover, if this condition is fulfilled, one has

$$\operatorname{ind} A = -\operatorname{wind} A(t).$$

Now assume $E = \mathbf{l}^p, \mathbf{m}_0$. Our next concern is to show that if $A \in \mathfrak{L}(V) \subset \mathcal{L}(E^\mu)$, then there is an operator $B \in \mathfrak{L}(V) \subset \mathcal{L}(E)$ such that

$$A|_E = B \quad \text{for} \quad \mu < 0, \qquad B|_{E^\mu} = A \quad \text{for} \quad \mu > 0. \tag{1}$$

It is easily seen that this assertion is proved if one can obtain the following inequality for any polynomial $p(t) = \sum_{i=-n}^{n} p_i t^i$, $|t| = 1$:

$$\|p(V)\|_{\mathcal{L}(E)} \leqq \|p(V)\|_{\mathcal{L}(E^\mu)}. \tag{2}$$

Let Λ^μ be the isometry defined in Section 4.46 and consider $\Lambda^{-\mu} p(V) \Lambda^\mu \in \mathcal{L}(E)$ $(p(V) \in \mathcal{L}(E^\mu))$. We prove that

$$\operatorname*{s-lim}_{n\to\infty} V_{-n}\Lambda^{-\mu} p(V) \Lambda^\mu V_n = p(V) \qquad (\in \mathcal{L}(E)). \tag{3}$$

Put $e_n := V_n\{1, 0, \ldots\}$ $(n = 0, 1, \ldots)$ and note that span $\{e_n\}_{n=0}^\infty$ is dense in E^μ. A straightforward computation gives

$$\lim_{n\to\infty} V_{-n}\Lambda^{-\mu} p(V) \Lambda^\mu V_n e_k = p(V)\, e_k.$$

Using 1.1(d) and (e) we arrive at (3) and (2). It remains to prove assertion (a) for $E = \mathbf{m}, \mathbf{m}_c$. Obviously, this will be a consequence of (b) and (d).

(b): Immediate from 4.46(i) and Section 4.11.

(c): According to Proposition 4.26, the estimate $\sup_{n\in\mathbb{Z}} \|V_n\| < \infty$ yields that alg $(V, V_{-1}) = \mathfrak{L}(V) \dotplus \mathbf{QC}(V)$. Now, let $E = \mathbf{l}^2$. Then $V_{-1} = V^*$ and due to Remark 4.25, 4°, $(\mathfrak{L}(V), \circ)$ is isometrically isomorphic to $\mathbf{C}(\mathbb{T})$.

(d): Assume $A \in \mathfrak{L}(V) \subset \mathcal{L}(\mathbf{l}^1)$ and consider $Ae_k = \{a_{-k}, a_{-k+1}, \ldots, a_0, a_1 \ldots\}$, where a_j are the Fourier coefficients of $A(t)$. Since $\|Ae_k\| \leqq \|A\|$ we get

$$\sum_{i=-k}^{\infty} |a_i| \leqq \|A\|$$

for all $k \in \mathbb{N}$. Consequently, $\sum_{-\infty}^{\infty} |a_i| \leqq \|A\|$ and $A(t) \in \mathbf{W}^{0,0} =: \mathbf{W}$. Now we use that $(\mathbf{m}_0)^* = \mathbf{l}^1$ (see KANTOROVICH/AKILOV [1], Chap. VI) in order to obtain the assertion for $\mathbf{m}_0$. Further, note that $A \in \mathfrak{L}(V) \subset \mathcal{L}(E)$, $E = \mathbf{m}, \mathbf{m}_c$, implies $A \in \mathcal{L}(\mathbf{m}_0)$ and, hence, $A(t) \in \mathbf{W}$. To finish the proof it remains to apply (2). ■

For any $U \in \mathcal{M}$ (see Section 4.36) we define a projection P_U on E^μ by

$$P_U\{\xi_n\}_0^\infty = \{\eta_n\}_0^\infty,$$

where

$$\eta_n = \begin{cases} 0, & n \notin U, \\ \xi_n, & n \in U. \end{cases}$$

The collection $\mathcal{P} := \{P_U : U \in \mathcal{M}\}$ satisfies (i)–(iii) of Section 4.36. Obviously, $P_{[0,n]} = P_n := I - V_{n+1}V_{-n-1}$ and $P_{(n,\infty)} = Q_n := V_{n+1}V_{-n-1}$.

4.48. Corollary. *For any E^μ the following assertions are true.*

(a) alg $(V, V_{-1}) \subset \mathfrak{A}$ *with respect to $\mathcal{P}$, that is, any element $A \in$ alg (V, V_{-1}) is of local type.*

(b) alg $(V, V_{-1}) \subset \mathcal{A}$, $\mathbf{QC}(V) \subset \mathcal{J}$ (*where $\mathcal{A}$ and $\mathcal{J}$ are defined in Section* 4.29).

(c) $\mathbf{QC}(V) = \{K \in \mathcal{K}(E^\mu) : \|Q_nK\| \to 0$ *and* $\|KQ_n\| \to 0$ *as* $n \to \infty\}$.

(d) *For* $E = \mathbf{l}^p, \mathbf{m}_0$ $(1 < p < \infty)$,

$$\mathbf{QC}(V) = \mathcal{K}(E^\mu).$$

Proof. (a): It is easily seen that $V, V_{-1} \in \mathfrak{A}$ and since $\mathfrak{A}$ forms an algebra, it follows that alg $(V, V_{-1}) \subset \mathfrak{A}$.

(b): The inclusion alg $(V, V_{-1}) \subset \mathcal{A}$ is immediate from (a), and the inclusion $\mathbf{QC}(V) \subset \mathcal{J}$ is a consequence of Lemma 4.13(d).

(c): Since codim(im V) = 1, we have $\mathbf{QC}(V) \subset \mathcal{K}(E^\mu)$ (by Corollary 4.15), and $\|Q_nK\| \to 0$, $\|KQ_n\| \to 0$ as $n \to \infty$ for any $K \in \mathbf{QC}(V)$ (by Lemma 4.13(c), (d)). Now, let $K \in \mathcal{K}(E^\mu)$ be such that $\|Q_nK\| \to 0$ and $\|KQ_n\| \to 0$ as $n \to \infty$. Since $P_nKP_n \in \mathbf{QC}(V)$ and (H4) is fulfilled, the conditions yield that $\|P_nKP_n - K\| \to 0$ as $n \to \infty$, and that $K \in \mathbf{QC}(V)$.

(d): Immediate from Corollary 4.15 and the equalities $(\mathbf{l}^{p,\mu})^* = \mathbf{l}^{q,-\mu}$, $(\mathbf{m}_0^\mu)^* = \mathbf{l}^{1,-\mu}$. ■

In order to prepare Theorem 4.50 we prove the following Lemma 4.49.

4.49. Lemma. *Put $Q_n = V_nV_{-n}$ and $P_n = I - Q_n$ for all integers $n \geqq 0$ and let $\lambda > \mu$. Assume there is an invertible linear operator A which is bounded on both spaces E^λ and E^μ together with its inverse. Then $\{P_nAP_n\}$ is stable on E^λ if and only if this sequence is stable on E^μ.*

Proof. *Step 1*: We prove that for $x_n \in$ im P_n

$$\|x_n\|_{E^\lambda} \leqq C_1 n^{\lambda-\mu} \|x_n\|_{E^\mu}. \tag{1}$$

First, let $E = \mathbf{l}^p$, $1 \leqq p < \infty$, and $x_n = \{\xi_0, \ldots, \xi_n, 0, \ldots\}$. Then

$$\|x_n\|_{E^\mu}^p = \sum_{i=0}^{n} (1+i)^{\lambda p} |\xi_i|^p = (1+n)^{(\lambda-\mu)p} \sum_{i=0}^{n} (1+i)^{\lambda p} (1+n)^{(\mu-\lambda)p} |\xi_i|^p$$
$$\leqq (1+n)^{(\lambda-\mu)p} \sum_{i=0}^{n} (1+i)^{\mu p} |\xi_i|^p$$

and (1) follows. The cases $E = \mathbf{m}, \mathbf{m}_c, \mathbf{m}_0$ can be treated analogously.

Step 2: We claim that for any $x \in E^\lambda$

$$\|(I - P_n)\, x\|_{E^\mu} \leqq C_2 n^{-(\lambda-\mu)} \|x\|_{E^\lambda}. \tag{2}$$

First, let $E = \mathbf{l}^p$, $1 \leqq p < \infty$, and $x = \{\xi_0, \xi_1, \ldots\}$. We have

$$\|\{0, \ldots, 0, \xi_{n+1}, \ldots\}\|_{E^\mu}^p = \sum_{i=n+1}^{\infty} (1+i)^{\mu p} |\xi_i|^p = (1+n)^{-(\lambda-\mu)p}$$
$$\times \sum_{i=n+1}^{\infty} (1+i)^{\mu p} (1+n)^{(\lambda-\mu)p} |\xi_i|^p$$
$$\leqq (1+n)^{-(\lambda-\mu)p} \sum_{i=n+1}^{\infty} (1+i)^{\lambda p} |\xi_i|^p$$

and (2) follows. The case $E = \mathbf{m}, \mathbf{m}_c, \mathbf{m}_0$ can be treated similarly.

Step 3: We show that for any $x \in E^\lambda$

$$\|(P_n A P_n - A)\, x\|_{E^\mu} \leqq C_3 n^{-(\lambda-\mu)} \|x\|_{E^\lambda}. \tag{3}$$

Indeed,

$$\|(P_n A P_n - A)\, x\|_{E^\mu} \leqq \|(P_n A P_n - P_n A)\, x\|_{E^\mu} + \|(P_n A - A)\, x\|_{E^\mu}$$
$$\leqq \|P_n A\| \, \|(I - P_n)\, x\|_{E^\mu} + \|(I - P_n)\, Ax\|_{E^\mu}$$
$$\leqq C_3 n^{-(\lambda-\mu)} \|x\|_{E^\lambda}$$

due to (2) and since $Ax \in E^\lambda$.

Step 4: Let $\{P_n A P_n\}$ be stable in E^μ. For $x_n \in \operatorname{im} P_n$ and n large enough we have (with $A_n^{-1} := (P_n A P_n)^{-1} P_n$)

$$\|A_n^{-1} x_n\|_{E^\lambda} \leqq \|A_n^{-1} x_n - A^{-1} x_n\|_{E^\lambda} + \|A^{-1}\| \, \|x_n\|_{E^\lambda}. \tag{4}$$

Further,

$$\|A_n^{-1} x_n - A^{-1} x_n\|_{E^\lambda} = \|A_n^{-1} x_n - Q_n A^{-1} x_n - P_n A^{-1} x_n\|_{E^\lambda}$$
$$\leqq \|A^{-1}\| \, \|x_n\|_{E^\lambda} + \|A_n^{-1} x_n - P_n A^{-1} x_n\|_{E^\lambda}$$
$$\leqq \|A^{-1}\| \, \|x_n\|_{E^\lambda} + C_1 n^{\lambda-\mu} \|A_n^{-1} x_n - P_n A^{-1} x_n\|_{E^\mu} \quad \text{(by (1))}$$
$$\leqq \|A^{-1}\| \, \|x_n\|_{E^\lambda} + C_1 n^{\lambda-\mu} \big(\|A_n^{-1}\|_{\mathcal{L}(E^\mu)} \, \|(A - A_n P_n)\, A^{-1} x_n\|_{E^\mu}\big)$$
$$\leqq \|A^{-1}\| \, \|x_n\|_{E^\lambda} + C_1 n^{\lambda-\mu} (\|A_n^{-1}\|_{\mathcal{L}(E^\mu)} C_3 n^{-(\lambda-\mu)} \|A^{-1} x_n\|_{E^\lambda}) \quad \text{(by (3))}$$
$$\leqq C_4 \|x_n\|_{E^\lambda} \quad \Big(\text{since } \sup_{n \geqq n_0} \|A_n^{-1}\|_{\mathcal{L}(E^\mu)} < \infty\Big).$$

This in conjunction with (4) yields the stability of $\{P_n A P_n\}$ in E^λ.

Step 5: Let $\{P_nAP_n\}$ be stable in E^λ. Then we have for any $x_n \in \operatorname{im} P_n$ (n large enough)

$$\begin{aligned}\|(A^{-1} - A_n^{-1})\, x_n\|_{E^\mu} &= \|A^{-1}(A - A_nP_n)\, A_n^{-1}P_nx_n\|_{E^\mu} \\ &\leqq \|A^{-1}\|\, \|(A - A_nP_n)\, A_n^{-1}P_nx_n\|_{E^\mu} \\ &\leqq \|A^{-1}\|\, C_3n^{-(\lambda-\mu)}\, \|A_n^{-1}P_nx_n\|_{E^\lambda} \quad \text{(by (3))} \\ &\leqq C_3n^{-(\lambda-\mu)}\, \|A_n^{-1}\|_{\mathcal{L}(E^\lambda)}\, \|x_n\|_{E^\lambda} \\ &\leqq C_3n^{-(\lambda-\mu)}\, \|A_n^{-1}\|_{\mathcal{L}(E^\lambda)}\, C_1n^{\lambda-\mu}\, \|x_n\|_{E^\mu} \quad \text{(by (1))} \\ &\leqq C_4\, \|x_n\|_{E^\mu} \quad \Big(\text{due to } \sup_{n \geqq n_0} \|A_n^{-1}\|_{\mathcal{L}(E^\lambda)} < \infty\Big);\end{aligned}$$

Hence, we have proved that $\|(A^{-1} - A_n^{-1})\, x_n\|_{E^\mu} \leqq C_4\, \|x_n\|_{E^\mu}$ and, thus,

$$\|A_n^{-1}x_n\|_{E^\mu} \leqq C_5\, \|x_n\|_{E^\mu} \qquad (n \text{ large enough}). \ \blacksquare$$

4.50. Theorem. *Put* $Q_n = V_nV_{-n}$ *and* $P_n = I - Q_n$ *for all integers* $n \geqq 0$. *Then for* $A \in \mathfrak{L}(V)$ *and* $K \in \mathbf{QC}(V)$ $\big(\subset \mathcal{L}(E^\mu)\big)$ *the sequence*

$$\{P_n(A + K)\, P_n\}_{n=0}^\infty$$

is stable on E^μ *if and only if*

(a) $A + K$ *is invertible on* E^μ,

(b) *the operator* $A' \in \mathfrak{L}(V) \subset \mathcal{L}(E)$ *defined by the symbol* $A(t)^{-1}$ *is invertible on* E.

Proof. First of all note that A' is well-defined on E if $A + K$ is a Φ-operator on E^μ. Indeed, if $A + K \in \Phi(E^\mu)$ then also $A \in \Phi(E^\mu)$ and $A(t) \neq 0$ for all $t \in \mathbb{T}$ by Theorem 4.20. The proof of Theorem 4.47 shows that there exists an operator in $\mathfrak{L}(V) \subset \mathcal{L}(E)$ with symbol $A(t)^{-1}$.

1. Assume that $\{P_n(A + K)\, P_n\}$ is stable on E^μ. If $E = \mathbf{l}^p, \mathbf{m}_0$ $(1 \leqq p < \infty)$, then $A + K$ is invertible by Corollary 4.40(ii) and Corollary 4.48(a). Now, let $E = \mathbf{m}, \mathbf{m}_c$. Since for $K \in \mathbf{QC}(V) \subset \mathcal{L}(E^\mu)$ one has $\|Q_nK\| \to 0$ as $n \to \infty$, the inclusion $K \in \mathcal{K}(E^\mu, \mathbf{m}_0^\mu)$ follows. Therefore, the stability of $\{P_n(A + K)\, P_n\}$ on E^μ implies the stability of this sequence on $\mathbf{m}_0^\mu$.

Hence, by the part already proved, the operator $A + K$ is invertible on $\mathbf{m}_0^\mu$. Using Theorem 4.20 it is easy to deduce that $A + K$ is a Φ-operator on E_μ with index zero. We claim that $A + K$ is even invertible on E^μ. To prove this we need only to show that $\ker(A + K) = \{0\}$. For $C \in \mathfrak{L}(V)$, define operators H_C and $\tilde{H}_C$ by

$$\begin{aligned} H_C\{\xi_j\}_{j=0}^\infty &= \left\{\sum_{k=0}^\infty c_{k+j+1}\xi_k\right\}_{j=0}^\infty, \\ \tilde{H}_C\{\xi_j\}_{j=0}^\infty &= \left\{\sum_{k=0}^\infty c_{-(k+j+1)}\xi_k\right\}_{j=0}^\infty, \end{aligned} \tag{1}$$

respectively, where c_j are the Fourier coefficients of $C(t)$. Assume $C(t) \in \mathbf{W}^{\mu_-,\mu_+}$. Then it is readily seen that $\tilde{H}_C \in \mathcal{K}(E^\mu, E)$, $H_C \in \mathcal{K}(E, E^\mu)$ for any space E defined in Section 4.46, and that

$$\|H_C\| \leqq \sum_{k=0}^\infty (1 + k)^{\mu_+}\, |c_k|, \qquad \|\tilde{H}_C\| \leqq \sum_{k=0}^\infty (1 + k)^{\mu_-}\, |c_{-k}| \tag{2}$$

(take $C = V_n$, $n \in \mathbb{Z}$, and compute the norm). The operators (1) are examples of so-called *Hankel operators*. If $A \in \mathfrak{L}(V)$ with $A(t) \in \mathbf{W}^{\mu_-,\mu_+}$ and $A(t) \neq 0$ for all $t \in \mathbb{T}$ then $A^{-1}(t) \in \mathbf{W}^{\mu_-,\mu_+}$ and there exists an operator $B \in \mathfrak{L}(V)$ with $B(t) = A^{-1}(t)$. Moreover, we have

$$AB = I + H_A\tilde{H}_B, \qquad BA = I + H_B\tilde{H}_A. \tag{3}$$

Note that this is true for any space defined in Section 4.46. Now let us return to the proof that $\ker(A + K) = \{0\}$. Consider $(A + K)\varphi = 0 \big(A + K \in \mathcal{L}(E^\mu)\big)$. Then

$$0 = B(A + K)\varphi = \varphi + (H_B\tilde{H}_A + BK)\varphi.$$

Since $H_B\tilde{H}_A + BK \in \mathbf{QC}(V)$ we have $H_B\tilde{H}_A + BK \in \mathcal{K}(E^\mu, \mathbf{m}_0^\mu)$, which yields $\varphi \in \mathbf{m}_0^\mu$. Since $A + K$ is invertible on $\mathbf{m}_0^\mu$ we get $\varphi = 0$ and the invertibility of $A + K$ on E^μ.

In the case under consideration assertion (b) is obvious. Nevertheless we prefer to give a proof which also works in the system case. Since $K \in \mathbf{QC}(V)$ we can K approximate by $K' = P_{n_0}KP_{n_0}$ close enough, so that $A + K'$ is invertible and $\{P_n(A + K')P_n\}$ is stable (by Proposition 4.30 in conjunction with Theorem 4.39 and $\mathfrak{A} \subset \mathcal{A}$). Note that $K' \in \mathbf{QC}(V) \subset \mathcal{L}(E)$. Hence, $A + K'$ is defined on both spaces E^μ and E (see the proof of Theorem 4.47). Since $A + K'$ is invertible on E^μ we conclude by Theorem 4.20 that $A + K'$ is a Φ-operator on E with index zero. We prove the invertibility of this operator, i.e. that $\ker(A + K') = \{0\}$. If $\mu < 0$ this is obvious. Assume $\mu > 0$. For $E = \mathbf{l}^p$, $\mathbf{m}_0$ $(1 \leqq p < \infty)$, 1.3(g) applies, which yields the invertibility of $A + K'$ on E. Assume $E = \mathbf{m}, \mathbf{m}_c$ and consider $(A + K')\varphi = 0$, $A + K' \in \mathcal{L}(E)$. Then $0 = B(A + K')\varphi = \varphi + (H_B\tilde{H}_A + BK')\varphi$.

Since $H_B\tilde{H}_A \in \mathcal{K}(E, E^\mu)$ and $BK'\varphi \in E^\mu$ we get $\varphi \in E^\mu$. The invertibility of $A + K'$ on E^μ gives $\varphi = 0$. Therefore, $A + K'$ is invertible on both E and E^μ, and $\{P_n(A + K') \times P_n\}$ is stable on E^μ (E anyone of the spaces $\mathbf{l}^p$, $\mathbf{m}$, $\mathbf{m}_c$, $\mathbf{m}_0$). Of course, Lemma 4.49 applies, that is, the sequence $\{P_n(A + K)P_n\}$ is stable on E. Now it remains to apply Theorem 4.45 and we are done.

2. Suppose the conditions of the theorem are fulfilled. Let K' be the operator defined in the previous part, so that $A + K'$ is invertible on E^μ (and on E). By Theorem 4.45 the sequence $\{P_n(A + K')P_n\}$ is stable on E, and by Lemma 4.49 it is so on E^μ. Using Proposition 4.30 (in conjunction with Theorem 4.39 and $\mathfrak{A} \subset \mathcal{A}$) we get the stability of $\{P_n(A + K)P_n\}$ on E^μ. ■

4.51. Remarks. 1°. Condition (a) of Theorem 4.50 implies condition (b). This is no longer true if one considers systems of discrete Wiener-Hopf equations. The formulation of Theorem 4.50 and its proof is completely adapted to the system case, which will be considered later.

2°. Theorem 4.50 can be proved under weaker conditions. Namely, for $A \in \operatorname{alg}(V, V_{-1}) \subset \mathcal{L}(E^\mu)$ the sequence $\{P_nAP_n\}$ is stable on E^μ if and only if the operator A is invertible on E^μ.

The proof rests on some ideas which are of its own interest. Let Λ^μ be the isometry defined in Section 4.46 and denote $\operatorname{alg}(V, V_{-1}) \subset \mathcal{L}(E^\mu)$ for $\mu \neq 0$ by $\operatorname{alg}_\mu(V, V_{-1})$. Then we have

$$\operatorname{alg}(V, V_{-1}) = \Lambda^{-\mu} \operatorname{alg}_\mu(V, V_{-1})\,\Lambda^\mu. \tag{1}$$

Indeed, $\Lambda^{-\mu}V\Lambda^\mu = V + K$, where $K \in \mathcal{K}(E)$. Moreover, using Proposition 4.48(c) it is readily seen that $K \in \mathbf{QC}(V)$. This is also true for V_{-1}. Therefore,

$$\Lambda^{-\mu} \operatorname{alg}_\mu (V, V_{-1}) \Lambda^\mu \subset \operatorname{alg} (V, V_{-1}).$$

The reverse inclusion can be proved analogously. For $V_{\pm 1} \in \mathcal{L}(E)$ define $\hat{V}_{\pm 1} := \Lambda^\mu V_{\pm 1} \times \Lambda^{-\mu} \in \mathcal{L}(E^\mu)$. Obviously, $\hat{V}_{\pm 1}$ fulfil (H1) and $\|\hat{V}_{\pm n}\| = 1$ for all $n \in \mathbb{Z}^+$. Moreover, $P_n := I - V_n V_{-n} = I - \hat{V}_n \hat{V}_{-n}$, $n \in \mathbb{Z}^+$. Hence, Theorem 4.45 is applicable to get the stability of $\{P_n A P_n\}$ for every invertible $A \in \operatorname{alg} (\hat{V}, \hat{V}_{-1}) = \operatorname{alg}_\mu (V, V_{-1})$. Notice further that the decomposition $\operatorname{alg} (\hat{V}, \hat{V}_{-1}) = \mathfrak{L}(\hat{V}) \dotplus \mathbf{QC}(\hat{V})$ shows that the algebras $\operatorname{alg}_\mu^\pi (V, V_{-1})$ and $\operatorname{alg}^\pi (V, V_{-1})$ are naturally isometrically isomorphic. In particular, smb: $\operatorname{alg}_\mu^\pi (V, V_{-1}) \to \mathbf{C}(\mathbb{T})$ is an isometric isomorphism for $E = \mathbf{l}^2$.

4.52. Theorem. *Let*

$$\sum_{k=0}^{\infty} a_{j-k} \xi_k = \eta_j \qquad (j = 0, 1, \ldots) \tag{1}$$

be an equation in E^μ given by an operator $A \in \mathfrak{L}(V) \subset \mathcal{L}(E^\mu)$ and consider the truncated equations

$$\sum_{k=0}^{n} a_{j-k} \xi_k = \eta_j \qquad (j = 0, 1, \ldots, n). \tag{2}$$

(a) *Let $E = \mathbf{l}^p, \mathbf{m}_0$ $(1 \leq p < \infty)$ and the operator A be invertible. Then for each $\eta = \{\eta_j\}_{j=0}^\infty \in E^\mu$ and for all sufficiently large n the system* (2) *has exactly one solution $\{\xi_k^{(n)}\}_{k=0}^n$ and, as $n \to \infty$, the elements $\xi^{(n)} := \{\xi_0^{(n)}, \ldots, \xi_n^{(n)}, 0, \ldots\}$ converge in the norm of E^μ to the solution $\xi = \{\xi_j\}_{j=0}^\infty$ of the equation* (1). *This convergence is quasioptimal, that means,*

$$\|\xi - \xi^{(n)}\|_{E^\mu} \leq C \inf_{x_n \in \operatorname{im} P_n} \|\xi - x_n\|. \tag{3}$$

If, in addition, $\eta \in E^\lambda$, $A \in \mathfrak{L}(V) \subset \mathcal{L}(E^\lambda)$ with $\lambda > \mu$, then

$$\|\xi - \xi^{(n)}\|_{E^\mu} \leq C n^{-(\lambda-\mu)} \|\eta\|_{E^\lambda}. \tag{4}$$

(b) *Let E be any of the spaces $\mathbf{l}^p$, $\mathbf{m}$, $\mathbf{m}_c$, $\mathbf{m}_0$ and the operator $A \in \mathfrak{L}(V) \subset \mathcal{L}(E^\mu)$ be invertible. Then for each $\eta = \{\eta_j\}_{j=0}^\infty \in E^\mu$ and for all sufficiently large n the system* (2) *has exactly one solution $\{\xi_k^{(n)}\}_{k=0}^n$ and, as $n \to \infty$,*

$$\|P_k(\xi^{(n)} - \xi)\|_{E^\mu} \to 0$$

for any $k \in \mathbb{N}$. Moreover, if, in addition, the symbol $A(t)$ belongs to $\mathbf{W}^{\mu_-,\mu_+}$, then for any $k \in \mathbb{N}$ and all n large enough

$$\|P_k(\xi^{(n)} - \xi)\|_{E^\mu} \leq C \sum_{i=n+1-k}^{\infty} (1 + i)^{\mu_-} |b_{-i}|, \tag{5}$$

where b_i are the Fourier coefficients of $B(t) := A(t)^{-1} \in \mathbf{W}^{\mu_-,\mu_+}$.

Proof. Note that the conditions ensure the stability of $\{P_n A P_n\}$ (Theorem 4.50 and Remark 4.51). Equation (2) can be rewritten in the form $P_n A \xi^{(n)} = P_n \eta$, $\xi^{(n)} \in \operatorname{im} P_n$.

(a): Remark 1.27.3°(a) applies and gives the assertions without (4). The estimate (4) is a consequence of (3) and 4.49(2).

(b): Using Corollary 4.48 and Proposition 4.22 it becomes clear that Theorem 4.34 proves the assertions without (5). Let us prove (5). According to 4.27(1) and Proposition 4.28 we have for n large enough

$$\begin{aligned} P_k(\xi - \xi^{(n)}) &= P_k\big(A^{-1} - P_nA^{-1}P_n + P_nA^{-1}(Q_nA^{-1}Q_n)^{-1}\,Q_nA^{-1}P_n\big)\,\eta \\ &= \big(P_kA^{-1}Q_n + P_kA^{-1}Q_n(Q_nA^{-1}Q_n)^{-1}\,A^{-1}P_n\big)\,\eta \\ &= P_kA^{-1}Q_n\big(I + (Q_nA^{-1}Q_n)^{-1}\,Q_nA^{-1}P_n\big)\,\eta, \end{aligned}$$

where $\|I + (Q_nA^{-1}Q_n)^{-1}Q_nA^{-1}P_n\eta\| \leqq M$ independently of n. An estimate for $\|P_kA^{-1}Q_n\|$ can be obtained in the following manner. Let B be the element of $\mathfrak{L}(V)$ for which $B(t) = A(t)^{-1} \in \mathbf{W}^{\mu_-,\mu_+}$. Then

$$AB = I + H_A\tilde{H}_B$$

with $\tilde{H}_B \in \mathcal{K}(E^\mu, E)$, $H_A \in \mathcal{K}(E, E^\mu)$, and

$$\|H_A\| \leqq \sum_{k=0}^{\infty} (1 + k)^{\mu_+}\,|a_k|, \qquad \|\tilde{H}_B\| \leqq \sum_{k=0}^{\infty} (1 + k)^{\mu_-}\,|b_{-k}|, \tag{6}$$

where a_k, b_k are the Fourier coefficients of $A(t)$ and $B(t)$, respectively (see 4.50(1), (2), and (3)). The operator A^{-1} can be represented in the form

$$A^{-1} = B - A^{-1}H_A\tilde{H}_B.$$

Let $B_n \in \mathfrak{L}_0(V)$ be the operator with symbol $B_n(t) = \sum_{k=0}^{n} b_{-k}t^{-k}$. Then $\tilde{H}_{B_n}Q_n = 0$ and $\tilde{H}_{B-B_n}Q_n = \tilde{H}_BQ_n$. By (6) we get

$$\|\tilde{H}_BQ_n\| \leqq \sum_{k=n+1}^{\infty} (1 + k)^{\mu_-}\,|b_{-k}|. \tag{7}$$

Since $P_kB_{n-k}Q_n = 0$, $P_kDQ_n = 0$ for $k < n$ and for $D \in \mathfrak{L}(V)$ with $d_l = 0$, $l < 0$, we get

$$\|P_kBQ_n\| = \|P_k(B - B_{n-k})\,Q_n\| \leqq \sum_{i=n+1-k}^{\infty} (1 + i)^{\mu_-}\,|b_{-i}|. \tag{8}$$

Taking into account (7) and (8) we obtain

$$\|P_kA^{-1}Q_n\| = \|P_k(B - A^{-1}H_A\tilde{H}_B)\,Q_n\| \leqq C_1 \sum_{i=n+1-k}^{\infty} (1 + i)^{\mu_-}\,|b_{-i}|,$$

and this yields the assertion. ■

Wiener-Hopf integral operators

4.53. Notation and definitions. Let G be the linear space of all measurable complex-valued functions defined on $\mathbb{R}^+$. In the following E will stand for any of the linear spaces

(a) $\mathbf{L}^p := \left\{f \in G \colon \|f\| := \left(\int_0^\infty |f|^p \,dt\right)^{1/p} < \infty\right\}$, $1 \leqq p < \infty$,

(b) $\mathbf{M} := \left\{f \in G \colon \|f\| := \operatorname{ess\,sup}_{t \geqq 0} |f(t)| < \infty\right\}$,

(c) $\mathbf{M}_l := \left\{ f \in M : \text{there exists a constant function } f_0 \text{ such that } \operatorname*{ess\,sup}_{t \geq \alpha} |f(t) - f_0| \to 0 \right.$

$\left. \text{as} \quad \alpha \to \infty \right\}$,

(d) $\mathbf{M}_0 := \left\{ f \in M : \operatorname*{ess\,sup}_{t \geq \alpha} |f(t)| \to 0 \text{ as } \alpha \to \infty \right\}$,

(e) $\mathbf{C}$, $\mathbf{C}_l$, $\mathbf{C}_0$, which are the spaces of continuous functions belonging to $\mathbf{M}$, $\mathbf{M}_l$, $\mathbf{M}_0$, respectively.

The spaces $\mathbf{L}^p$ and $\mathbf{M}$ actually form Banach spaces under the above norms. Moreover, $\mathbf{M}_l$, $\mathbf{M}_0$, $\mathbf{C}$, $\mathbf{C}_l$, $\mathbf{C}_0$ are (closed) subspaces of $\mathbf{M}$. If E is anyone of the above spaces and if $\mu \in \mathbb{R}$, then E^μ will denote the linear space of all functions f in G for which $(1 + t)^\mu f(t)$ is in E. Provided with the norm

$$\|f\|_{E^\mu} := \|(1 + t)^\mu f(t)\|_E$$

E^μ becomes a Banach space.

Let $\mathbf{L}^1(\mathbb{R})$ be the space of Lebesgue integrable functions on $\mathbb{R}$ with the usual norm $\|f\| = \int_{-\infty}^{\infty} |f| \, \mathrm{d}t$. Note that $\mathbf{L}^1(\mathbb{R})$ is a commutative Banach algebra (without identity element) if the product of two functions $f_1, f_2 \in \mathbf{L}^1(\mathbb{R})$ is defined as their convolution,

$$(f_1 * f_2)(t) = \int_{-\infty}^{\infty} f_1(t - s) f_2(s) \, \mathrm{d}s .$$

Let $\tilde{\mathbf{L}}$ denote the Banach algebra resulting from $\mathbf{L}^1(\mathbb{R})$ by adjoining an identity element (the Dirac δ-function) in the usual way. For $c\delta + f \in \tilde{\mathbf{L}}$ $(c \in \mathbb{C}, f \in \mathbf{L}^1(\mathbb{R}))$ the Fourier transform is defined by

$$F(\lambda) = c + \int_{-\infty}^{\infty} \mathrm{e}^{\mathrm{i}\lambda t} f(t) \, \mathrm{d}t \qquad (\lambda \in \mathbb{R}).$$

For $\mu \in \mathbb{R}$, we define $\mathbf{L}^{1,\mu}(\mathbb{R})$ analogously to E^μ. If $\mu \geqq 0$, we let $\mathcal{F}^\mu$ denote the Banach algebra (with identity element) consisting of all functions of the form

$$F(\lambda) = c + \int_{-\infty}^{\infty} \mathrm{e}^{\mathrm{i}\lambda t} f(t) \, \mathrm{d}t,$$

where $f \in \mathbf{L}^{1,\mu}(\mathbb{R})$ and $c \in \mathbb{C}$. The norm of $F(\lambda) \in \mathcal{F}^\mu$ is defined by

$$\|F(\lambda)\|_{\mathcal{F}^\mu} := |c| + \|f\|_{\mathbf{L}^{1,\mu}(\mathbb{R})} .$$

Now let $k \in \mathbf{L}^{1,|\mu|}(\mathbb{R})$ and $\varphi(\lambda) = c - \int_{-\infty}^{\infty} k(t) \, \mathrm{e}^{\mathrm{i}\lambda t} \, \mathrm{d}t$, $c \in \mathbb{C}$. The *Wiener-Hopf integral operator* with symbol φ is the operator W_φ given by

$$(W_\varphi f)(t) = cf(t) - \int_0^\infty k(t - s) f(s) \, \mathrm{d}s, \qquad t > 0. \tag{1}$$

If $k \in \mathbf{L}^{1,|\mu|}(\mathbb{R})$, then W_φ is a bounded operator on each of the above spaces E^μ (see GOHBERG/FELDMAN [1] and PRÖSSDORF [3]), moreover,

$$\|W_\varphi\|_{E^\mu} \leqq \|\varphi\|_{\mathcal{F}^{|\mu|}}. \tag{2}$$

It turns out that the theory of such operators can be developed analogously to that of the Toeplitz operators. Indeed, define

$$(Vf)(t) = f(t) - \int_0^\infty l_1(t-s)\, f(s)\, ds, \qquad (0 < t < \infty)$$

$$(V^{(-1)}f)(t) = f(t) - \int_0^\infty l_{-1}(t-s)\, f(s)\, ds,$$

where

$$l_1(t) = \begin{cases} 2e^{-t}, t > 0, \\ 0, \quad t < 0 \end{cases} \quad l_{-1}(t) = l_1(-t).$$

Obviously, the operator $V^{(-1)}$ is a left inverse for V. Note that

$$1 - \int_{-\infty}^{\infty} l_{\pm 1}(t)\, e^{i\lambda t}\, dt = \left(\frac{\lambda - i}{\lambda + i}\right)^{\pm 1}. \tag{3}$$

Since an element $\varphi \in \mathcal{F}^{|\mu|}$ is invertible in $\mathcal{F}^{|\mu|}$ if and only if $\lim_{\lambda \to \infty} \varphi(\lambda) \neq 0$ and $\varphi(\lambda) \neq 0$ for all $\lambda \in \mathbb{R}$, it results that the maximal ideal space of $\mathcal{F}^{|\mu|}$ can be identified with $\mathbb{T}$ and the Gelfand transform of $\varphi \in \mathcal{F}^{|\mu|}$ is given on $\mathbb{T}$ by $\varphi\left(i\frac{1+t}{1-t}\right)$. Now we show that $\mathcal{F}^{|\mu|}$ is generated by $\left(\frac{\lambda - i}{\lambda + i}\right)^{\pm 1}$. The space $\mathfrak{L}_0(V)$ consists of all operators

$$(Rf)(t) = cf(t) - \int_0^\infty r(t-s)\, f(s)\, ds,$$

where

$$r(t) = \begin{cases} e^{-t}P_1(t), & 0 < t, \\ e^{t}P_2(t), & t < 0 \end{cases} \tag{4}$$

$c \in \mathbb{C}$ and $P_1(t)$, $P_2(t)$ are any polynomials in t (containing, however, only nonnegative powers of t). The collection of all functions (4) is dense in $\mathbf{L}^{1,|\mu|}(\mathbb{R})$ and hence, $\mathcal{F}^{|\mu|}$ is generated by $\left(\frac{\lambda - i}{\lambda + i}\right)^{\pm 1}$. Therefore, $\mathcal{F}^{|\mu|}$ is V-dominating and Theorem 4.9 applies. Notice that the symbol $W_\varphi(t)$ (in the sense of Section 4.3) is connected with the symbol φ by

$$\operatorname{smb} W_\varphi(t) = \varphi\left(i\frac{1+t}{1-t}\right).$$

Thus, the following assertion is almost obvious.

4.54. Theorem. (a) *The operator W_φ with $\varphi \in \mathcal{F}^{|\mu|}$ is in $\mathcal{L}(E^{|\mu|})$ at least one-sided invertible if $\varphi(\lambda) \neq 0$ for all $-\infty \leqq \lambda \leqq \infty$. If this condition is fulfilled, then the invertibility of W_φ corresponds with the winding number* $\operatorname{wind} \varphi(\lambda) := 1/2\pi[\arg \varphi(\lambda)]_{-\infty}^{\infty}$, *i.e. W_φ is* *invertible, invertible only from the left or only from the right, respectively, if the winding number of $\varphi(\lambda)$ is zero, positive or negative, respectively.*

(b) *If* $\varphi(\lambda) \neq 0$ *for all* $-\infty \leqq \lambda \leqq \infty$, *then*

$$\dim \ker W_\varphi = \max\{-\text{wind}\,\varphi, 0\},$$

$$\text{codim}\,(\text{im}\, W_\varphi) = \max\{\text{wind}\,\varphi, 0\}.$$

(c) *If* $\varphi(\lambda_0) = 0$ *for some* λ_0, $-\infty \leqq \lambda_0 \leqq \infty$, *then* W_φ *is neither a* Φ_+- *nor a* Φ_--*operator.*

Proof. We have only to prove (b). Any solution of the equation

$$V_{-n} g = 0 \qquad (n \in \mathbb{N})$$

is of the form

$$g(t) = \sum_{j=1}^{n} c_j t^{j-1} \mathrm{e}^{-t} \qquad (0 < t < \infty),$$

where c_j are arbitrary complex numbers. Hence, $\dim \ker V_{-1} = 1$ and $\text{codim}\,(\text{im}\, V) = 1$. It remains to apply Theorem 4.20. ■

Let $E = \mathbf{L}^p, \mathbf{M}, \mathbf{M}_1, \mathbf{M}_0$ and define for $U \in \mathcal{M}$ (see Section 4.36) a projection P_U on E^μ by

$$P_U f = \chi_U f,$$

where χ_U is the characteristic function of U. The collection $\mathcal{P} := \{P_U : U \in \mathcal{M}\}$ satisfies (i)–(iii) of Section 4.36. For $\tau \geqq 0$ put $P_\tau := P_{[0,\tau]}$, $Q_\tau := P_{(\tau,\infty)}$. Obviously, $P_\tau = I - Q_\tau$.

4.55. Proposition. *Let* $E = \mathbf{L}^p, \mathbf{M}, \mathbf{M}_1, \mathbf{M}_0$. *Then, with respect to* E^μ *and* $\mathcal{P}$,

(a) $\text{alg}\,(V, V_{-1}) \subset \mathfrak{A}$, *that is, any element of* $\text{alg}\,(V, V_{-1})$ *is of local type,*

(b) $\text{alg}\,(V, V_{-1}) \subset \mathcal{A}$ *and* $\mathbf{QC}(V) \subset \mathcal{J}$ *and if E is anyone of the spaces defined in Section* 4.53 *then*

$$\mathbf{QC}(V) \subset \mathcal{K}(E^\mu).$$

(c) *If* $E = \mathbf{M}$ *then every element* $K \in \mathbf{QC}(V) \subset \mathcal{K}(\mathbf{M}^\mu)$ *belongs to* $\mathcal{K}(\mathbf{M}^\mu, \mathbf{C}_0^\mu)$.

Proof. (a): It is easily seen that $V, V_{-1} \in \mathfrak{A}$, and because $\mathfrak{A}$ is an algebra, we get $\text{alg}\,(V, V_{-1}) \subset \mathfrak{A}$.

(b): The inclusion $\text{alg}\,(V, V_{-1}) \subset \mathcal{A}$ is immediate from (a). Let us prove that $\mathbf{QC}(V) \subset \mathcal{J}$. For $k \in \mathbf{L}^{1,|\mu|}(\mathbb{R})$ and $\varphi(\lambda) := c - \int_{-\infty}^{\infty} k(t)\, \mathrm{e}^{\mathrm{i}\lambda t}\, \mathrm{d}t$ ($c \in \mathbb{C}$), the so called *Hankel integral operator* is defined on E^μ by

$$(H_\varphi f)(t) = \int_0^\infty k(t+s)\, f(s)\, \mathrm{d}s, \qquad t > 0. \tag{1}$$

Moreover,

$$\|H_\varphi\|_{\mathcal{L}(E^\mu)} \leqq \|k\|_{\mathbf{L}^{1,|\mu|}(\mathbb{R}^+)}. \tag{2}$$

The following identity can be verified without difficulty:

$$W_{\varphi_1\varphi_2} = W_{\varphi_1} W_{\varphi_2} + H_{\varphi_1} \tilde{H}_{\varphi_2}, \tag{3}$$

where $\tilde{H}_{\varphi_2} := H_{\tilde{\varphi}_2}$, $\tilde{\varphi}_2(\lambda) := \varphi_2(-\lambda)$. The quasicommutator ideal is, by definition, the smallest closed two-sided ideal containing all operators $H_{\varphi_1}\tilde{H}_{\varphi_2}$, where $\varphi_i(\lambda) = c_i - \int_{-\infty}^{\infty} r_i(t)\, e^{i\lambda t}\, dt$ $(i = 1, 2)$ and r_i is of the form 4.53(4). Now it is easy to see that

$$\|H_{\varphi_1}\tilde{H}_{\varphi_2}Q_\tau\| \to 0 \quad \text{as} \quad \tau \to \infty,$$

hence $H_{\varphi_1}H_{\tilde{\varphi}_2} \in \mathcal{J}$. Since $\mathcal{J}$ is an ideal we get $\mathbf{QC}(V) \subset \mathcal{J}$. Since codim im $V_{-1} = 1$, Corollary 4.15 gives $\mathbf{QC}(V) \subset \mathcal{K}(E^\mu)$ at once.

(c): By Lemma 4.13 we have $\mathbf{QC}(\mathbf{M}^\mu) = \operatorname{clos\,id}(I - VV_{-1})$. On the other hand, $I - VV_{-1}$ is a continuous projection onto $\ker V_{-1} = \{ce^{-t} : c \in \mathbb{C}\} \subset \mathbf{C}_0^\mu$. Since any Wiener-Hopf operator continuous on $\mathbf{M}^\mu$ is also continuous on $\mathbf{C}_0^\mu$ (recall that $\mathbf{C}_0^\mu$ is a closed subspace of $\mathbf{M}^\mu$) we get the assertion. ■

4.56. Corollary. *Let* $E = \mathbf{L}^p, \mathbf{M}, \mathbf{M}_1, \mathbf{M}_0$. *Then* $\operatorname{alg}(V, V_{-1}) \subset \mathcal{L}(E)$ *is decomposing.*

Proof. Define a family of operators $\{\tilde{W}_\omega\}_{\omega \in (0,\infty)}$ on E by

$$(\tilde{W}_\omega f)(t) := \begin{cases} 0, & 0 < t < \omega, \\ f(t - \omega), & t > \omega, \end{cases}$$

and put $(\tilde{W}_{-\omega} f)(t) := f(t + \omega)$. Obviously,

$$\sup_{\omega > 0} \|\tilde{W}_\omega\| < \infty, \qquad \sup_{\omega > 0} \|\tilde{W}_{-\omega}\| < \infty, \qquad \tilde{W}_{-\omega}\tilde{W}_\omega = I \quad \text{and} \quad \tilde{W}_\omega\tilde{W}_{-\omega} = Q_\omega$$

and $\tilde{W}_{-\omega}W_\varphi\tilde{W}_\omega = W_\varphi$ for any Wiener-Hopf operator W_φ. Since

$$\|\tilde{W}_{-\omega}K\tilde{W}_\omega\| \leqq \|\tilde{W}_{-\omega}\|\, \|KQ_\omega\|\, \|\tilde{W}_\omega\| \to 0$$

(by Proposition 4.55(b)) all conditions of Proposition 4.26 are fulfilled and we are done. ■

In order to prepare Theorem 4.58 we mention the following lemma.

4.57. Lemma. *Let E denote any of the spaces* $\mathbf{L}^p, \mathbf{M}, \mathbf{M}_1, \mathbf{M}_0$ *and put* $P_\tau = I - \tilde{W}_\tau\tilde{W}_{-\tau}$ $(\tau \in (0, \infty))$. *Assume* $\lambda > \mu$ *and that there is an invertible linear operator A which is bounded on both spaces* E^λ *and* E^μ *together with its inverse. Then* $\{P_\tau AP_\tau\}$ *is stable on* E^λ *if and only if this sequence is stable on* E^μ.

Proof. The proof is essentially the same as that of Lemma 4.49 and is based on the following two assertions:

1°. For $x_\tau \in \operatorname{im} P_\tau$,

$$\|x_\tau\|_{E^\lambda} \leqq C_1\tau^{\lambda-\mu}\, \|x_\tau\|_{E^\mu}. \tag{1}$$

2°. For any $x \in E^\lambda$,

$$\|(I - P_\tau)\, x\|_{E^\mu} \leqq C_2\tau^{\mu-\lambda}\, \|x\|_{E^\lambda}. \tag{2}$$

The inequalities (1) and (2) can be proved similar to 4.49(1) and (2). ■

4.58. Theorem. *Let* $E = \mathbf{L}^p, \mathbf{M}, \mathbf{M}_1, \mathbf{M}_0$. *Then, for* W_φ, $\varphi \in \mathcal{F}^{|\mu|}$, *and* $K \in \mathcal{K}(E^\mu)$ *with* $\|KQ_\tau\| \to 0$, $\|Q_\tau K\| \to 0$ *as* $\tau \to \infty$ *the sequence*

$$\{P_\tau(W_\varphi + K)\, P_\tau\}$$

is stable if and only if

(a) $W_\varphi + K$ *is invertible on* E^μ,

(b) *the operator* $W_{\varphi^{-1}}$ *is invertible on* E.

Proof. 1. Let $\{P_\tau(W_\varphi + K)\,P_\tau\}$ be stable and consider $E = \mathbf{L}^p, \mathbf{M}_0$. We have $P_\tau \to I$ strongly as $\tau \to \infty$. Since $\|KQ_\tau\| \to 0$, $\|Q_\tau K\| \to 0$ it is clear that $K \in \mathfrak{A}$, whence $W_\varphi + K \in \mathfrak{A}$ by Proposition 4.55(a). Corollary 4.40 gives the invertibility of $W_\varphi + K$. Further, let $E = \mathbf{M}, \mathbf{M}_1$. Obviously, the conditions of the theorem ensure that $\{P_\tau(W_\varphi + K)\,P_\tau\}$ is also stable on $\mathbf{M}_0^\mu$. By the part already proved the operator $W_\varphi + K$ is invertible on $\mathbf{M}_0^\mu$. Similarly as in the proof of Theorem 4.50 one shows that $W_\varphi + K$ is invertible on $\mathbf{M}_1^\mu$ or $\mathbf{M}^\mu$, respectively. Let us prove the invertibility of $W_{\varphi^{-1}}$ on E. Since $W_\varphi + K \in \mathfrak{A}$ and $W_\varphi + K$ is invertible we get $(W_\varphi + K)^{-1} \in \mathfrak{A}$ by Theorem 4.39. Hence, $(W_\varphi + K)^{-1} \in \mathcal{A}$. Approximate K by $K' = P_{\tau_0} K P_{\tau_0}$ close enough so that $W_\varphi + K'$ is also invertible. Proposition 4.30 gives the stability of

$$\{P_\tau(W_\varphi + K')\,P_\tau\},$$

where the compact operator K' is defined on both E^μ and E. The operator $W_\varphi + K'$ is invertible on E^μ by what has already been proved. Since $W_\varphi + K'$ is also a Φ-operator on E with index 0 the invertibility of $W_\varphi + K'$ on E can be proved without difficulty. Since Lemma 4.57 holds in the case under consideration we get the stability of $\{P_\tau(W_\varphi + K')\,P_\tau\}$ in the space E.

Theorem 4.45 does not apply directly but an analysis of the proof shows that this theorem applies if one only knows that $(W_\varphi + K')^{-1} = W_{\varphi^{-1}} + K_1$, $K_1 \in \mathcal{J}$. Using that $W_\varphi + K'$ is an operator of local type this can easily be proved. Thus, $W_{\varphi^{-1}}$ is invertible on E.

2. Let the conditions be fulfilled. Then $W_{\varphi^{-1}}$ is invertible on both E and E^μ, and $\{P_\tau W_{\varphi^{-1}}^{-1} P_\tau\}$ is stable on E (see the proof of Theorem 4.45, part (b)). Then $\{P_\tau W_{\varphi^{-1}}^{-1} P_\tau\}$ is also stable on E^μ by Lemma 4.57. Since $W_\varphi - W_{\varphi^{-1}}^{-1} \in \mathcal{J}$, Proposition 4.30 applies and we get the stability of $\{P_\tau(W_\varphi + K)\,P_\tau\}$ on E^μ. The proof is complete. ■

4.59. Remarks. 1°. Condition (a) implies condition (b). However, these conditions are independent of each other if one considers systems of Wiener-Hopf integral equations.

2°. For $E = \mathbf{L}^p$ $(1 < p < \infty)$ and for $K \in \mathcal{K}(E^\mu)$ the conditions $\|KQ_\tau\| \to 0$, $\|Q_\tau K\| \to 0$ as $\tau \to \infty$ are automatically fulfilled. If $E = \mathbf{L}^1, \mathbf{M}_0$, then automatically $\|Q_\tau K\| \to 0$ as $\tau \to \infty$.

3°. If $E = \mathbf{L}^1, \mathbf{M}_0$ the sufficiency can be proved under the sole assumption that $K \in \mathcal{K}(E^\mu)$.

4.60. Theorem. *Let*

$$(W_\varphi f)\,(t) := cf(t) - \int_0^\infty k(t-s)\,f(s)\,\mathrm{d}s = g(t), \qquad t > 0 \tag{1}$$

be a Wiener-Hopf equation on E^μ $(\varphi \in \mathcal{F}^{|\mu|})$, *where* E *is anyone of the spaces* $\mathbf{L}^p, \mathbf{M}, \mathbf{M}_1, \mathbf{M}_0$ *and consider the truncated equations*

$$(W_\varphi^\tau f_\tau)\,(t) := cf_\tau(t) - \int_0^\tau k(t-s)\,f_\tau(s)\,\mathrm{d}s = (P_\tau g)\,(t), \qquad 0 < t < \tau. \tag{2}$$

(a) *Let $E = \mathbf{L}^p, \mathbf{M}_0$ and the operator W_φ be invertible. Then for each $g \in E^\mu$ and for all sufficiently large τ the equation* (2) *has exactly one solution f_τ in* $\operatorname{im} P_\tau$ *and, as $\tau \to \infty$, the elements f_τ (assumed to be zero on (τ, ∞)) converge in the norm of E^μ to the solution f of equation* (1). *This convergence is quasioptimal, that means,*

$$\|f - f_\tau\|_{E^\mu} \leqq c \inf_{x_\tau \in \operatorname{im} P_\tau} \|f - x_\tau\|_{E^\mu}. \tag{3}$$

If, in addition, $g \in E^\lambda$, $\varphi \in \mathcal{F}^{|\lambda|}$ with $\lambda > \mu$, then

$$\|f - f_\tau\|_{E^\mu} \leqq C\tau^{-(\lambda-\mu)} \|g\|_{E^\lambda}. \tag{4}$$

(b) *Let E be any of the spaces $\mathbf{L}^p, \mathbf{M}, \mathbf{M}_1, \mathbf{M}_0$ and the operator W_φ be invertible. Then for each $g \in E^\mu$ and for all sufficiently large τ the equation* (2) *has exactly one solution f_τ in* $\operatorname{im} P_\tau$ *and, as $\tau \to \infty$,*

$$\|P_\omega(f - f_\tau)\|_{E^\mu} \to 0$$

for any $\omega \in (0, \infty)$. Moreover, for any $\omega \in (0, \infty)$ and all τ large enough

$$\|P_\omega(f - f_\tau)\|_{E^\mu} \leqq C \int_{-\infty}^{-\tau+\omega} (1 + |t|)^{|\mu|} |k_1(t)| \,dt, \tag{5}$$

where $\varphi^{-1}(\lambda) = c - \int_{-\infty}^{\infty} k_1(t)\, e^{i\lambda t} \,dt.$

Proof. The proof is that of Theorem 4.52 with minor modifications concerning (5). In order to prove (5) we have only to estimate $\|P_\omega W_\varphi^{-1} Q_\tau\|$, where $Q_\tau := I - P_\tau$ (see Section 4.52.). Due to 4.55(3) we have

$$I = W_\varphi W_{\varphi^{-1}} + H_\varphi \tilde{H}_{\varphi^{-1}},$$

where the Hankel integral operators H_φ and $\tilde{H}_{\varphi^{-1}}$ are bounded on E^μ and

$$\|\tilde{H}_{\varphi^{-1}}\|_{\mathcal{L}(E^\mu)} \leqq \|\tilde{k}\|_{\mathbf{L}^{1,|\mu|}(\mathbb{R}^+)}, \tag{6}$$

$\varphi^{-1}(\lambda) := c - \int_{-\infty}^{\infty} k_1(t)\, e^{i\lambda t} dt$, $\tilde{k}(t) := k_1(-t)$. The operator W_φ^{-1} can be represented in the form

$$W_\varphi^{-1} = W_{\varphi^{-1}} - W_\varphi^{-1} H_\varphi \tilde{H}_{\varphi^{-1}}.$$

Let W_{φ_τ} be the Wiener-Hopf integral operator with symbol

$$\varphi_\tau(\lambda) := c - \int_{-\infty}^{\infty} \chi_{(-\tau,0)}\, k_1(t)\, e^{i\lambda t} \,dt,$$

where χ_F denotes the characteristic function of a measurable subset F of $\mathbb{R}$. Then $\tilde{H}_{\varphi_\tau} Q_\tau = 0$ and $\tilde{H}_{\varphi^{-1}-\varphi_\tau} Q_\tau = \tilde{H}_{\varphi^{-1}} Q_\tau$. By (6) we get

$$\|\tilde{H}_{\varphi^{-1}} Q_\tau\| \leqq \|\chi_{(-\infty,-\tau)} k_1\|_{\mathbf{L}^{1,|\mu|}(\mathbb{R})}. \tag{7}$$

Since $P_\omega W_{\varphi_{\tau-\omega}} Q_\tau = 0$, $P_\omega W_\psi Q_\tau = 0$ for $\omega < \tau$ and $\psi := c - \int_{-\infty}^{\infty} l(t)\, e^{i\lambda t} \,dt$, $l(t) = 0$ for $t < 0$, we obtain

$$\|P_\omega W_{\varphi^{-1}} Q_\tau\| \leqq \|\chi_{(-\infty,-\tau+\omega)}\, k\|_{\mathbf{L}^{1,|\mu|}(\mathbb{R})} \tag{8}$$

and this yields the assertion. ∎

4.61. Remark. Theorem 4.60 can also be proved for equations of the form

$$(W_\varphi + K) f = g,$$

where $W_\varphi + K$ is invertible and K is a compact operator satisfying the conditions of Theorem 4.58. The estimate (5) must be replaced by the following. Let $\beta(\tau)$ be any increasing function on $(0, \infty)$ so that $\tau - \beta(\tau) \to \infty$ as $\tau \to \infty$. Then

$$\|P_\omega(f - f_\tau)\|_{E^\mu} \leqq C \left(\int_{-\infty}^{-\tau+\omega+\beta(\tau)} (1 + |t|)^\mu |k(t)| \, \mathrm{d}t + \|KQ_{\beta(\tau)}\| \right). \tag{1}$$

Again we need only to estimate $\|P_\omega(W_\varphi + K)^{-1} Q_\tau\|$ (see Section 4.52). Using the representation

$$(W_\varphi + K)^{-1} = W_{\varphi^{-1}} + (W_\varphi + K)^{-1} H_\varphi \tilde{H}_{\varphi^{-1}} - (W_\varphi + K)^{-1} (K W_{\varphi^{-1}}),$$

the inequalities 4.60(7) and (8), and the obvious estimate

$$\|K W_{\varphi^{-1}} Q_\tau\| \leqq \|KQ_{\beta(\tau)} W_{\varphi^{-1}} Q_\tau\| + \|KP_{\beta(\tau)} W_{\varphi^{-1}} Q_\tau\|$$
$$\leqq C_1 \left(\|KQ_{\beta(\tau)}\| + \int_{-\infty}^{-\tau+\beta(\tau)} (1 + |t|)^{|\mu|} |k(t)| \, \mathrm{d}t \right),$$

we arrive at (1).

Modified finite sections

4.62. Notation. The finite sections 4.60(2) are not defined on spaces of continuous functions, that means, on the spaces E^μ, where $E = \mathbf{C}, \mathbf{C}_{\mathrm{l}}, \mathbf{C}_0$. However, it turns out that the *modified finite sections*

$$(W_{\varphi,\tau} f)(t) := c f(t) - \int_0^\tau k(t - s) f(s) \, \mathrm{d}s, \qquad 0 < t < \infty, \tag{1}$$

are well-defined on E^μ (if only $\varphi \in \mathcal{F}^{|\mu|}$). To give a sufficiently general look at this circle of ideas assume that we are given an integral operator $K\colon E^\mu \to E^\mu$ by

$$(Kf)(t) = \int_0^\infty k(t, s) f(s) \, \mathrm{d}s, \qquad t \in \mathbb{R}^+, \quad f \in E^\mu$$

and put $k_t(s) := k(t, s)$. We require that $k_t \in \mathbf{L}^{1,-\mu}(\mathbb{R}^+)$ for all $t \in \mathbb{R}^+$. Furthermore, throughout the following suppose that

(a) $\sup\limits_t \|(1 + t)^\mu k_t\|_{1,-\mu} := \sup\limits_t (1 + t)^\mu \int_0^\infty (1 + s)^{-\mu} |k(t, s)| \, \mathrm{d}s < \infty$,

(b) $\|k_{t'} - k_t\|_{1,-\mu} \to 0$ as $t' \to t$.

It can be easily seen that K belongs to $\mathcal{L}(E^\mu)$. Moreover, we have $K \in \mathcal{L}(\mathbf{M}^\mu, \mathbf{C}^\mu)$. Suppose the following condition (which is stronger than (b)) is satisfied:

(b′) $\|(1 + t')^\mu k_{t'} - (1 + t)^\mu k_t\|_{1,-\mu} \to 0$ as $t' \to t$ uniformly with respect to $t \in \mathbb{R}^+$.

Consequently, $\{(1 + t)^\mu (Kf)(t) : \|f\|_{E^\mu} \leqq 1\}$ is bounded and equicontinuous on $\mathbb{R}^+$. This combined with

(c) $\|(1+t)^\mu k_t\|_{1,-\mu} \to 0$ as $t \to \infty$

implies that $K \in \mathcal{K}(\mathbf{M}^\mu, \mathbf{C}_0^\mu)$. For $\tau \in \mathbb{R}^+$ define K_τ by

$$(K_\tau f)(t) = \int_0^\tau k(t, s) f(s)\, ds = \int_0^\infty k_\tau(t, s) f(s)\, ds,$$

where, for $t \in \mathbb{R}^+$,

$$k_\tau(t, s) := \begin{cases} k(t, s), & 0 \leqq s \leqq \tau \\ 0, & s > \tau \end{cases}$$

(and k is assumed to satisfy the conditions (a) and (b)). Let

$$v_\tau(t) := (1+t)^\mu \int_\tau^\infty |k(t, s)|\, (1+s)^{-\mu}\, ds, \qquad \tau \in \mathbb{R}^+, \quad t \in \mathbb{R}^+.$$

Obviously,

$$v_\tau(t) \leqq \|(1+t)^\mu k_t\|_{1,-\mu}, \qquad v_\tau \in \mathbf{C},$$

and $\{v_\tau\}_{\tau \geq 0}$ is bounded, (pointwise) equicontinuous and converges uniformly to zero on each compact interval from $\mathbb{R}^+$. The last assertion results from the following facts.

(i) For fixed t, we have $v_\tau(t) \to 0$ as $\tau \to \infty$.

(ii) Convergence of equicontinuous functions to a continuous function is uniform on compact intervals.

4.63. Proposition. (i) *If k is subject to the conditions* (a) *and* (b) *of the previous section, then $K, K_\tau \in \mathcal{L}(\mathbf{M}^\mu, \mathbf{C}^\mu)$ and*

$$\|(K - K_\tau) f\| \leqq \sup_t v_\tau(t) \sup_{t \geqq \tau} |(1+t)^\mu f(t)|. \tag{1}$$

(ii) *If k satisfies the conditions* (a), (b′), *and* (c) *then*

$$K, K_\tau \in \mathcal{K}(\mathbf{M}^\mu, \mathbf{C}_0^\mu)$$

and

$$\|K - K_\tau\|_{\mathcal{L}(\mathbf{M}^\mu)} \to 0 \quad as \quad \tau \to \infty. \tag{2}$$

(iii) *If $k \in \mathbf{L}^{1,|\mu|}(\mathbb{R})$ and $(Kf)(t) := \int_0^\infty k(t-s) f(s)\, ds$ $(t \geqq 0)$, then*

$$K \in \mathcal{L}(\mathbf{M}^\mu, \mathbf{C}^\mu), \qquad K \in \mathcal{L}(\mathbf{M}_0^\mu, \mathbf{C}_0^\mu), \qquad K_\tau \in \mathcal{K}(\mathbf{M}^\mu, \mathbf{C}_0^\mu).$$

Proof. (i): Obvious. (ii): It remains to show (2). First of all, (c) implies that $v_\tau(t) \to 0$ as $t \to \infty$, uniformly with respect to τ. Let $\varepsilon > 0$ be arbitrarily given. Then there is a t_0 such that $\sup_{t \geqq t_0} v_\tau(t) < \varepsilon$ whenever $\tau \geqq 0$. Because $\{v_\tau\}$ converges uniformly to zero on each interval, there exists a τ_0 such that $\sup_{0 \leqq t \leqq t_0} v_\tau < \varepsilon$ whenever $\tau \geqq \tau_0$. Thus, by (1), we then have

$$\|K - K_\tau\| \leqq \sup v_\tau(t) < \varepsilon.$$

(iii): We claim that the convolution operator satisfies 4.62(a). First let $\mu \geqq 0$ and note that $1 + t \leqq (1 + |t - s|)(1 + s)$, so that

$$\sup_t \|(1 + t)^\mu k(t - s)\|_{1,-\mu} = \sup_t \int_0^\infty (1 + t)^\mu |k(t - s)| (1 + s)^{-\mu} \, ds$$

$$\leqq \sup_t \int_0^\infty |k(t - s)| (1 + |t - s|)^\mu \, ds = \|k\|_{\mathbf{L}^{1,\mu}(\mathbb{R})}.$$

Using the inequality $1 + s \leqq (1 + |t - s|)(1 + t)$, the assertion can be proved analogously in the case $\mu < 0$. Hence, $K \in \mathcal{L}(\mathbf{M}^\mu, \mathbf{C}^\mu)$. We claim $K \in \mathcal{L}(\mathbf{M}_0^\mu, \mathbf{C}_0^\mu)$. We have

$$|(1 + t)^\mu (Kf)(t)| \leqq \int_0^\infty (1 + t)^\mu (1 + s)^{-\mu} |k(t - s)| (1 + s)^\mu |f(s)| \, ds$$

$$\leqq \int_0^\infty |k(t - s)| (1 + |t - s|)^{|\mu|} |f(s)| (1 + s)^\mu \, ds$$

$$\leqq \left(\int_0^\alpha |k(t - s)| (1 + |t - s|)^{|\mu|} \, ds \right) \left(\sup_{0 \leqq t \leqq \alpha} |(1 + t)^\mu f(t)| \right)$$

$$+ \|k\|_{\mathbf{L}^{1,\mu}(\mathbb{R})} \sup_{t \geqq \alpha} |(1 + t)^\mu f(t)|$$

which converges to zero as $t \to \infty$ since

$$\int_0^\alpha |k(t - s)| (1 + |t - s|)^\mu \, ds \to 0 \quad \text{as} \quad t \to \infty.$$

To see that $K_\tau \in \mathcal{K}(\mathbf{M}^\mu, \mathbf{C}_0^\mu)$ note that $K_\tau = KP_\tau$ where P_τ can be viewed as an operator acting from $\mathbf{M}^\mu$ into $\mathbf{M}_0^\mu$. Therefore, $K_\tau \in \mathcal{L}(\mathbf{M}^\mu, \mathbf{C}_0^\mu)$ and, moreover, $K_\tau \in \mathcal{K}(\mathbf{M}^\mu, \mathbf{C}_0^\mu)$ because k satisfies (b′). ■

4.64. Theorem. *Let $E = \mathbf{C}, \mathbf{C}_1, \mathbf{C}_0$ and $W_\varphi + K$ be invertible on E^μ, where $\varphi \in \mathcal{F}^{|\mu|}$ and k satisfies the conditions* (a), (b′), (c) *of Section* 4.62. *Then the sequence $\{W_{\varphi,\tau} + K_\tau\}$ is stable on E^μ. Moreover, the uniquely determined solutions f_τ of*

$$(W_{\varphi,\tau} + K_\tau) f_\tau = g$$

(τ large enough, say $\tau \geqq \tau_0$) converge uniformly on any compact interval of $\mathbb{R}^+$ to the unique solution f of the equation

$$(W_\varphi + K) f = g.$$

The speed of convergence can be estimated by

$$\sup_{0 \leqq t \leqq \omega} \left|(1 + t)^\mu \big(f(t) - f_\tau(t)\big)\right| \leqq C \left(\int_{-\infty}^{-\tau+\omega+\beta(\tau)} (1 + |t|)^{|\mu|} |k(t)| \, dt + \sup_t v_{\beta(\tau)}(t) \right), \quad (1)$$

where the notations of Theorem 4.60 *and Remark* 4.61 *are used.*

Proof. Due to Theorem 4.58, Remark 4.59, 1°, and Proposition 4.63(ii) the conditions of the theorem imply the stability of $\{P_\tau(W_\varphi + K) P_\tau\}$ on $\mathbf{M}^\mu$. Since (on $\mathbf{M}^\mu$)

$$W_{\varphi,\tau} + K_\tau = \big(P_\tau(W_\varphi + K) P_\tau + cQ_\tau\big)\big(I + c^{-1}Q_\tau(W_\varphi + K) P_\tau\big),$$

the stability of $\{P_\tau(W_\varphi + K) P_\tau\}$ implies the stability of $\{W_{\varphi,\tau} + K_\tau\}$ on $\mathbf{M}^\mu$ (use $\big(I + c^{-1} \times Q_\tau(W_\varphi + K) P_\tau\big)^{-1} = I - c^{-1}Q_\tau(W_\varphi + K) P_\tau$).

According to Proposition 4.63(ii) and (iii) we also obtain the stability of $\{W_{\varphi,\tau} + K_\tau\}$ on E^μ. Finally, using that $P_\tau(W_{\varphi,\tau} + K_\tau) = P_\tau(W_\varphi + K) P_\tau$ (on $\mathbf{M}^\mu$) we get (1) by means of Remark 4.61. ■

Abstract paired operators

4.65. Notation. Let $\mathfrak{M}$ be a closed subalgebra of the algebra $\mathscr{L}(X)$ of all linear bounded operators on a Banach space X, P^+ a continuous projection on X and $P^- = I - P^+$. Assume $A, B \in \mathfrak{M}$. Every operator of the form $AP^+ + BP^-$ or $P^+A + P^-B$ is called a *paired operator*. The paired operators $AP^+ + BP^-$ or $P^+B + P^-A$ are called transposed to one another. Let U be an invertible operator on X such that

(a) the spectral radii of the operators U and U^{-1} are both equal to one: $\varrho(U) = \varrho(U^{-1}) = 1$;

(b) $UP^+ = P^+UP^+$, $UP^+ \neq P^+U$, $P^+U^{-1} = P^+U^{-1}P^+$.

For a given polynomial $R(t) = \sum_{j=-n}^{n} a_j t^j$ on the complex unit circle $\mathbb{T}$ we define an operator $R(U)$ by $\sum_{j=-n}^{n} a_j U^j$ and call $R(U)$ a polynomial of U. Let $\mathfrak{L}_0(U)$ stand for the set of all polynomials of U and let $\mathfrak{L}(U)$ be the closure of $\mathfrak{L}_0(U)$ in $\mathscr{L}(X)$.

By V and $V^{(-1)}$ we denote the restrictions of the operators P^+UP^+ and $P^+U^{-1}P^+$, respectively, to the closed subspace $X_+ = \operatorname{im} P^+$. As is easily seen, the operator V is only left invertible, and $V^{(-1)}$ is a left inverse for V. Moreover, the spectral radii of V and $V^{(-1)}$ are both equal to one. Hence, V and $V^{(-1)}$ satisfy (H1) and U is a dilation of V in the sense of Section 4.10 satisfying 4.10(i) and (ii). Hence, $\mathfrak{L}(U)$ is a V-dominating Banach algebra. It is worth noticing that the last observation is also valid if one replaces U, P^+, X_+ by U^{-1}, P^-, $X_- = \operatorname{im} P^-$, respectively. Ofcourse, we have $\mathfrak{L}(U) = \mathfrak{L}(U^{-1})$.

In the following we consider paired operators with coefficients belonging to $\mathfrak{L}(U)$, i.e., we put $\mathfrak{M} = \mathfrak{L}(U)$. With the operator $AP^+ + BP^-(P^+A + P^-B)$, $A \in \mathfrak{L}(U)$ and $B \in \mathfrak{L}(U)$, we associate the following function $C(t, \theta)$ of the two variables $t \in \mathbb{T}$ and $\theta = \pm 1$:

$$C(t, \theta) = A(t)\,\frac{1 + \theta}{2} + B(t)\,\frac{1 - \theta}{2},$$

where $A(t)\,\big(B(t)\big)$ is the Gelfand transform of $A(B)$ with respect to the algebra $\mathfrak{L}(U)$. The function $C(t, \theta)$ is called the *symbol* of $AP^+ + BP^-(P^+A + P^-B)$. If $C(t, \theta) \neq 0$ for all $t \in \mathbb{T}$ and $\theta = \pm 1$, then the *index* of $C(t, \theta)$ is defined as the winding number of the quotient $A(t)/B(t)$:

$$\operatorname{ind} C(t, \theta) := \operatorname{wind}\big(A(t)/B(t)\big) = \operatorname{wind} A(t) - \operatorname{wind} B(t).$$

4.66. Definitions. Our next objective is to study the smallest closed subalgebra $\operatorname{alg}(U, U^{-1}, P^+) \subset \mathcal{L}(X)$ containing U, U^{-1}, and P^+. By $\mathbf{QC}(U)$ we denote the smallest closed two-sided ideal in $\operatorname{alg}(U, U^{-1}, P^+)$ containing

$$\{[R, P^+] : R \in \mathfrak{L}_0(U)\} \qquad ([M_1, M_2] := M_1M_2 - M_2M_1).$$

The ideal $\mathbf{QC}(U)$ is called the *quasicommutator ideal* of the algebra $\operatorname{alg}(U, U^{-1}, P^+)$. Note that if $C = A_1P^+ + B_1P^-$, $D = A_2P^+ + B_2P^-$, then

$$CD = A_1A_2P^+ + B_1B_2P^- + K,$$

where

$$K = A_1[P^+, A_2]\,P^+ + B_1[P^-, B_2]\,P^- + A_1[P^+, B_2]\,P^- + B_1[P^-, A_2]\,P^+.$$

Using that $[M_1, M_2] = -[M_2, M_1]$ we get $K \in \mathbf{QC}(U)$. Put $Q_n := U^{-n}P^-U^n + U^nP^+U^{-n}$ and $P_n := I - Q_n$ $(n \in \mathbb{Z}^+)$. Obviously, P_n and Q_n are continuous projections.

4.67. Proposition. *For* $K \in \operatorname{alg}(U, U^{-1}, P^+)$ *the following conditions are equivalent.*

(a) $K \in \mathbf{QC}(U)$.

(b) $K \in \operatorname{id}(P^+P_1, P^-P_1)$ ($\operatorname{id}(P^+P_1, P^-P_1)$ *refers to the smallest closed two-sided ideal of* $\operatorname{alg}(U, U^{-1}, P^+)$ *containing* P^+P_1 *and* P^-P_1).

If $\|Q_n\| \leqq M < \infty$ *for all* n, *then* (a) *is equivalent to each of the following two conditions:*

(c) $\|Q_nK\| \to 0$ *as* $n \to \infty$;

(d) $\|KQ_n\| \to 0$ *as* $n \to \infty$.

Proof. (a) $\Rightarrow$ (b): Since $\mathbf{QC}(U)$ is generated by all operators of the form $[R, P^+]$ with $R \in \mathcal{L}_0(U)$ it remains to consider $[U^i, P^+]$, $i \in \mathbb{Z}$. If $i > 0$ then

$$U^iP^+ - P^+U^i = (U^iP^+U^{-i} - P^+)\,U^i = -(P^+P_i)\,U^i.$$

If $i < 0$ then $U^iP^+ - P^+U^i = U^i(P^+ - U^{-i}P^+U^i) = U^i(P^+P_{|i|})$. Using $P^+P_k = \sum_{j=0}^{k-1} P^+ \times U^jP^+P_1P^+U^{-j}P^+$ we obtain $[U^i, P^+] \in \operatorname{id}(P^+P_1, P^-P_1)$. Thus, $\mathbf{QC}(U) \subset \operatorname{id}(P^+P_1, P^-P_1)$.

(b) $\Rightarrow$ (a): Trivial, since $-P^+P_1 = (UP^+ - P^+U)\,U^{-1}$, $P^-P_1 = U^{-1}(UP^- - P^-U)$ are elements of $\mathbf{QC}(U)$.

(a) $\Rightarrow$ (c), (d): We only prove that (a) implies $\|Q_nK\| \to 0$. For $i \in \mathbb{Z}$ consider

$$Q_n[U^i, P^+] = U^{-n}P^-U^{n+i}P^+ + U^nP^+U^{-n+i}P^+ - U^nP^+U^{-n+i} = 0$$

if only $n > |i|$. Thus, $[U^i, P^+]$ belongs to the ideal $\mathcal{J}$ defined in Section 4.29 with respect to the algebra $\mathcal{B}$. Let us prove that $\operatorname{alg}(U, U^{-1}, P^+) \subset \mathcal{B}$. To this end it suffices to consider U, U^{-1}, P^+. Obviously, $P^+ \in \mathcal{B}$. A straigthforward computation gives $Q_nU^{\pm 1}P_{n_0} = 0$ for $n > n_0$. Now it is clear that $\mathbf{QC}(U) \subset \mathcal{J}$ and we are done.

(c) $\Rightarrow$ (a): If $\|Q_nK\| \to 0$ then K is the uniform limit of the operators P_nK which are in $\mathbf{QC}(U)$ since

$$\begin{aligned} P_n &= I - U^{-n}P^-U^{-n} - U^nP^+U^n = (P^+ - U^nP^+U^{-n}) + (P^- - U^{-n}P^-U^n) \\ &= U^n(U^{-n}P^+ - P^+U^{-n}) + U^{-n}(U^nP^- - P^-U^n). \end{aligned}$$

(d) $\Rightarrow$ (a): Proceed as in the proof of the implication (c) $\Rightarrow$ (a). ■

4.68. Corollary. *If* codim $(\operatorname{im} P^+UP^+|_{\operatorname{im} P^+}) < \infty$, *then* $\mathbf{QC}(U)$ *consists only of compact operators. If, in particular,* codim $(\operatorname{im} P^+UP^+|_{\operatorname{im} P^+}) = 1$ *and* $\{P_n\}$, $\{P_n^*\}$ *converge strongly to the identity operator, then* $\mathbf{QC}(U)$ *equals* $\mathcal{K}(X)$.

Proof. First of all,

$$\operatorname{codim}\,(\operatorname{im} P^+UP^+|_{\operatorname{im} P^+}) = \operatorname{codim}\,(\operatorname{im} P^-U^{-1}P^-|_{\operatorname{im} P^-}),$$

due to the equality $P^- + P^+UP^+ = U(P^-U^{-1}P^- + P^+)$ and the invertibility of U. Hence, P^+P_1 and P^-P_1 are finite-dimensional. Proposition 4.67(b) shows our first assertion. The second one can be proved as in Corollary 4.15. ■

4.69. Definitions. For $A \in \operatorname{alg}(U, U^{-1}, P^+)$, we call an operator $B \in \mathbf{D}(U) := \operatorname{clos} \mathbf{D}_0(U)$, $\mathbf{D}_0(U) := \{P^+A_1P^+ + P^-B_1P^- : A_1, B_1 \in \mathfrak{L}(U)\}$, a *main part* if $A - B \in \mathbf{QC}(U)$. There are examples which show that a main part does not always exist.

Let $R = \sum_{i=1}^{n} \prod_{j=1}^{m} (A_{ij}P^+ + B_{ij}P^-)$, where $A_{ij}, B_{ij} \in \mathfrak{L}_0(U)$. Let $A_0 \in \mathfrak{L}_0(U)$ and $B_0 \in \mathfrak{L}_0(U)$ be the elements uniquely defined by $\sum_{i=1}^{n} \prod_{j=1}^{m} A_{ij}$ and $\sum_{i=1}^{n} \prod_{j=1}^{m} B_{ij}$, respectively. The operator $R_0 := P^+A_0P^+ + P^-B_0P^-$ is called the *essential part* of the operator R. Clearly, we have $R - R_0 \in \mathbf{QC}(U)$. If the main part is uniquely determined we call it also essential part.

Now denote by $\operatorname{alg}^\pi(U, U^{-1}, P^+)$ the quotient algebra $\operatorname{alg}(U, U^{-1}, P^+)/\mathbf{QC}(U)$ and by π the canonical homomorphism from $\operatorname{alg}(U, U^{-1}, P^+)$ onto $\operatorname{alg}^\pi(U, U^{-1}, P^+)$. Obviously, $\operatorname{alg}^\pi(U, U^{-1}, P^+)$ is a commutative Banach algebra generated by $\pi(U)$, $\pi(U^{-1})$, and $\pi(P^+)$.

4.70. Proposition. (i) *The maximal ideal space of* $\operatorname{alg}^\pi(U, U^{-1}, P^+)$ *can be identified with the disjoint union of two copies of* $\mathbb{T}$.

(ii) *The Gelfand transform of* $\pi(C)$, $C = AP^+ + BP^-$ *or* $C = P^+A + P^-B$ $(A, B \in \mathfrak{L}(U))$, *is given by*

$$C(t, \theta) = A(t)\,\frac{1 + \theta}{2} + B(t)\,\frac{1 - \theta}{2}, \qquad t \in \mathbb{T} \quad \text{and} \quad \theta = \pm 1.$$

Proof. Since $(P^+)^2 = P^+$ and $(P^-)^2 = P^-$ the algebra $\operatorname{alg}^\pi(U, U^{-1}, P^+)$ contains the idempotents $\pi(P^+)$ and $\pi(P^-)$. A well-known theorem of Shilov says that $\operatorname{alg}^\pi(U, U^{-1}, P^+)$ is the direct sum of the ideals $\mathcal{J}_1 = \pi(P^+) \operatorname{alg}^\pi(U, U^{-1}, P^+)\,\pi(P^+)$ and $\mathcal{J}_2 = \pi(P^-) \times \operatorname{alg}^\pi(U, U^{-1}, P^+)\,\pi(P^-)$, that $\pi(P^+)$ is the unit element in $\mathcal{J}_1$ and $\pi(P^-)$ is the unit element in $\mathcal{J}_2$. Moreover, the maximal ideal space of $\operatorname{alg}^\pi(U, U^{-1}, P^+)$ is the disjoint union of the maximal ideal spaces of $\mathcal{J}_1$ and $\mathcal{J}_2$.

First we verify that $\operatorname{sp} \pi(P^+UP^+) = \mathbb{T} \cup \{0\}$. By (H1) the spectra of $\pi(P^+UP^+)$ and $\pi(P^+UP^+)^{-1}$ (regarded as elements of $\mathcal{J}_1$) are contained in $\mathbb{T}$. Now assume $\operatorname{sp} \pi(P^+UP^+) \neq \mathbb{T}$ and choose $\lambda_0 \in \mathbb{T} \setminus \operatorname{sp} \pi(P^+UP^+)$. Then there exist operators B, K with $P^+DP^+ = D$ and $K \in \mathbf{QC}(U)$ such that

$$D(P^+UP^+ - \lambda_0 P^+) = P^+ + K$$

or, equivalently,

$$(D + P^-)\big(P^+(U - \lambda_0 I)\,P^+ + P^-\big) = I + K.$$

Exactly as in the proof of Proposition 4.17 we get a contradiction. Hence, $\operatorname{sp} \pi(P^+UP^+) \supset \mathbb{T}$. Analogously, $\operatorname{sp} \pi(P^-U^{-1}P^-) \supset \mathbb{T}$. Thereby we have proved that neither $\pi(P^+)$ nor $\pi(P^-)$ equal zero. Hence,

$$\operatorname{sp} \pi(P^+UP^+) = \mathbb{T} \cup \{0\}, \qquad \operatorname{sp} \pi(P^-U^{-1}P^-) = \mathbb{T} \cup \{0\}.$$

The other assertions follow immediately from the general theory of Banach algebras. ■

4.71. Definition. The previous theorem enables us to assign to every element $A \in \operatorname{alg}(U, U^{-1}, P^+)$ a well-defined continuous function $\operatorname{smb} A$ on $\mathbb{T} \times \{-1, 1\}$, called the *symbol* of A, namely the Gelfand transform of $\pi(A)$. Obviously, the mapping $A \mapsto \operatorname{smb} A$ is a homomorphism and any element $B \in \mathbf{QC}(U)$ belongs to the kernel of this homomorphism. The kernel equals $\mathbf{QC}(U)$ if and only if $\operatorname{alg}^\pi(U, U^{-1}, P^+)$ is an algebra with trivial radical. It is easy to see that an element $A \in \operatorname{alg}(U, U^{-1}, P^+)$ is regularizable with respect to the ideal $\mathbf{QC}(U)$ if and only if $(\operatorname{smb} A)(t, \theta) \neq 0$ for all $t \in \mathbb{T}$ and $\theta = \pm 1$. It should also be noted that this notion of the symbol is related to the uniqueness problem of the main part (compare Section 4.18).

4.72. Proposition. *Let* $A \in \operatorname{alg}(U, U^{-1}, P^+)$ *be a* Φ_+- *or a* Φ_--*operator. Then*

$$(\operatorname{smb} A)(t, \theta) \neq 0 \quad \textit{for all} \quad t \in \mathbb{T}, \theta = \pm 1.$$

Moreover, if A *is a* Φ-*operator and* $\operatorname{codim}(P^+UP^+|_{\operatorname{im} P^+}) =: \varkappa < \infty$ *then*

$$\operatorname{ind} A = -\varkappa \operatorname{ind}(\operatorname{smb} A), \tag{1}$$

where $\operatorname{ind} \operatorname{smb} A := \operatorname{wind}(\operatorname{smb} A)(t, 1)/(\operatorname{smb} A)(t, -1)$.

Proof. Note that each of the sets

$$\left\{\sum_{i=1}^{n} \prod_{j=1}^{m} (A_{ij}P^+ + B_{ij}P^-) : A_{ij}, B_{ij} \in \mathfrak{L}_0(U), n, m = 1, 2, \ldots\right\},$$

$$\left\{\sum_{i=1}^{n} \prod_{j=1}^{m} (P^+A_{ij} + P^-B_{ij}) : A_{ij}, B_{ij} \in \mathfrak{L}_0(U), n, m = 1, 2, \ldots\right\}$$

is dense in $\operatorname{alg}(U, U^{-1}, P^+)$.

Assume $A \in \operatorname{alg}(U, U^{-1}, P^+)$ to be a Φ_+-operator with $(\operatorname{smb} A)(t_0, \theta_0) = 0$ for certain $t_0 \in \mathbb{T}$ and $\theta_0 \in \{-1, 1\}$. First of all there is a $\delta > 0$ such that $B \in \mathcal{L}(X)$ and $\|A - B\| < \delta$ implies $B \in \Phi_+(X)$. Approximate A by $R = \sum_{i=1}^{n} \prod_{j=1}^{m} (A_{ij}P^+ + B_{ij}P^-)$, $A_{ij}, B_{ij} \in \mathfrak{L}_0(U)$ so that $\|A - R\| < \delta/2$. Since $(\operatorname{smb} A)(t_0, \theta_0) = 0$ we have $|(\operatorname{smb} R)(t_0, \theta_0)| < \delta/2$ and

$$\|A - R + (\operatorname{smb} R)(t_0, \theta_0) I\| < \delta.$$

Thus, $K := R - (\operatorname{smb} R)(t_0, \theta_0) I$ is a Φ_+-operator with $(\operatorname{smb} K)(t_0, \theta_0) = 0$. Denote by K_0 the essential part of K. It is easy to see that there is an $n \in \mathbb{N}$ so that $(K - K_0) D = 0$, $D := U^nP^+ + U^{-n}P^-$. Hence,

$$KD = K_0D = P^+K_0DP^+ + P^-K_0DP^-$$

and KD is a Φ_+-operator by 1.3(d) (note that D is a Φ_+-operator since $(U^{-n}P^+ + U^nP^-) \times (U^nP^+ + U^{-n}P^-) = I$). Moreover, $(\operatorname{smb} KD)(t_0, \theta_0) = (\operatorname{smb} K_0D)(t_0, \theta_0) = 0$. Since $P^+K_0DP^+$ and $P^-K_0DP^-$ are Φ_+-operators on $\operatorname{im} P^+$ and $\operatorname{im} P^-$, respectively, we get a contradiction to Theorem 4.2(iii). Hence, if A is a Φ_+-operator then $(\operatorname{smb} A)(t, \theta) \neq 0$ for all $t \in \mathbb{T}$ and $\theta \in \{-1, 1\}$. Analogously one shows the assertion if A is a Φ_--operator. The index formula (1) can be proved as in 4.19. ■

4.73. Theorem. *Assume* $\operatorname{codim}(\operatorname{im} P^+UP^+|_{\operatorname{im} P^+}) =: \varkappa < \infty$. *Then* $A \in \operatorname{alg}(U, U^{-1}, P^+)$ *is a* Φ*-operator if and only if*

$$(\operatorname{smb} A)(t, \theta) \neq \text{ for all } t \in \mathbb{T}, \theta = \pm 1.$$

If this condition is fulfilled,

$$\operatorname{ind} A = -\varkappa \operatorname{ind}(\operatorname{smb} A).$$

Proof. See the proof of Theorem 4.20. ■

4.74. Theorem. *Let* $A, B \in \mathfrak{L}(U)$. *For the paired operator* $C = AP^+ + BP^-$ $(C = P^+A + P^-B)$ *to be at least one-sided invertible it is necessary and sufficient that its symbol does not degenerate, that is*

$$C(t, \theta) \neq 0 \quad \text{for all} \quad t \in \mathbb{T} \quad \text{and} \quad \theta = \pm 1. \tag{1}$$

If this condition is fulfilled, then the invertibility of the operator C corresponds to the index of the symbol.

If (1) is not satisfied then the operator C is neither a Φ_+- nor a Φ_--operator.

Proof. If (1) is fulfilled then A and B are invertible in $\mathfrak{L}(U)$. For $C = AP^+ + BP^-$ we have

$$C = B(P^+B^{-1}AP^+ + P^-)(I + P^-B^{-1}AP^+).$$

Since $I + P^-B^{-1}AP^+$ is invertible (the inverse is given by $I - P^-B^{-1}AP^+$), the invertibility properties of C coincide with those of $P^+B^{-1}AP^+|_{\operatorname{im} P^+}$ and it remains to apply Theorem 4.9(i) in conjunction with Example 4.10. The other assertions follow from Proposition 4.72. ■

4.75. Definition. In the sequel the algebra $\operatorname{alg}(U, U^{-1}, P^+)$ will be said to decompose into the direct sum $\mathbf{D}(U) \dotplus \mathbf{QC}(U)$ if there is a continuous projection S mapping $\operatorname{alg}(U, U^{-1}, P^+)$ onto $\mathbf{D}(U)$ parallel to $\mathbf{QC}(U)$. It is worth noticing that if $\operatorname{alg}(U, U^{-1}, P^+)$ decomposes, the essential part is well-defined for any element $A \in \operatorname{alg}(U, U^{-1}, P^+)$.

4.76. Corollary. *Assume that for all* $A_{ij} \in \mathfrak{L}_0(U)$ *and all* $B_{ij} \in \mathfrak{L}_0(U)$

$$\|C_0\| \leq M \left\| \sum_{i=1}^{n} \prod_{j=1}^{m} (A_{ij}P^+ + B_{ij}P^-) \right\|,$$

where C_0 is the essential part of $\sum_{i=1}^{n} \prod_{j=1}^{m} (A_{ij}P^+ + B_{ij}P^-)$. *Then the algebra* $\operatorname{alg}(U, U^{-1}, P^+)$ *decomposes into the direct sum*

$$\operatorname{alg}(U, U^{-1}, P^+) = \mathbf{D}(U) \dotplus \mathbf{QC}(U).$$

Proof. See the proof of Theorem 4.24. ■

Assume we are given a family $\{\widetilde{W}_\omega\}_{\omega \in \Omega}$ of invertible operators on X ($\Omega \subset \mathbb{R}^+$ an unbounded set) such that

(i) $U\widetilde{W}_\omega = \widetilde{W}_\omega U$ for all $\omega \in \Omega$ and $\sup\limits_\omega \|\widetilde{W}_\omega\| < \infty$, $\sup\limits_\omega \|\widetilde{W}_\omega^{-1}\| < \infty$,

(ii) $\widetilde{W}_\omega P^+ = P^+\widetilde{W}_\omega P^+, \widetilde{W}_\omega P^+ \neq P^+\widetilde{W}_\omega, P^+\widetilde{W}_\omega^{-1} = P^+\widetilde{W}_\omega^{-1}P^+$ for all $\omega \in \Omega$.
Condition (ii) implies

$$\widetilde{W}_\omega^{-1}P^- = P^-\widetilde{W}_\omega^{-1}P^-, \qquad \widetilde{W}_\omega^{-1}P^- \neq P^-\widetilde{W}_\omega^{-1}, \qquad P^-\widetilde{W}_\omega = P^-\widetilde{W}_\omega P^-.$$

Put $G_\omega := \widetilde{W}_\omega^{-1}P^- + \widetilde{W}_\omega P^+$ and $G_\omega^{(-1)} := P^-\widetilde{W}_\omega + P^+\widetilde{W}_\omega^{-1}$. A straigthforward computation gives $G_\omega^{(-1)}G_\omega = I$ and hence, $Q_\omega := G_\omega G_\omega^{(-1)} = \widetilde{W}_\omega^{-1}P^-\widetilde{W}_\omega + \widetilde{W}_\omega P^+\widetilde{W}_\omega^{-1}$ is a continuous projection $\neq I$. Put $P_\omega := I - Q_\omega$.

4.77. Proposition. *Assume there is a family* $\{\widetilde{W}_\omega\}_{\omega\in\Omega}$ *of invertible operators on* X *satisfying* 4.76(i), (ii) *and the following condition:*

$$\|G_\omega^{(-1)}KG_\omega\| \to 0 \quad \textit{as} \quad \omega \to \infty \quad \textit{for all} \quad K \in \mathbf{QC}(U).$$

Then the algebra alg (U, U^{-1}, P^+) *decomposes,*

$$\text{alg}\,(U, U^{-1}, P^+) = \mathbf{D}(U) \dotplus \mathbf{QC}(U),$$

and the essential part is well-defined for every element $A \in$ alg (U, U^{-1}, P^+). *In particular, this is true if* $\sup_{n\in\mathbb{Z}} \|U^n\| < \infty$.

Proof. See the proof of Proposition 4.26. ■

4.78. Remark. Assume that, in addition to the conditions of Proposition 4.77, $Q_\omega \to 0$ strongly as $\omega \to \infty$. Then $\mathbf{D}_0(U) = \mathbf{D}(U)$.

Indeed, $Q_\omega P^- = \widetilde{W}_\omega^{-1}P^-\widetilde{W}_\omega \to 0$ strongly as $\omega \to \infty$. Thus, $\widetilde{W}_\omega^{-1}P^+\widetilde{W}_\omega = I - \widetilde{W}_\omega^{-1}P^-\widetilde{W}_\omega \to I$ strongly as $\omega \to \infty$.

Let $M \in \mathfrak{L}(U)$ and consider

$$\widetilde{W}_\omega^{-1}P^+MP^+\widetilde{W}_\omega = \widetilde{W}_\omega^{-1}P^+\widetilde{W}_\omega M\widetilde{W}_\omega^{-1}P^+\widetilde{W}_\omega.$$

The latter operator sequence converges strongly to M. Due to the Banach-Steinhaus theorem there exists a constant $m > 0$ such that $\|M\| \leqq m\,\|P^+MP^+\|$ for all $M \in \mathfrak{L}(U)$. Analogously, $\|M\| \leqq m\,\|P^-MP^-\|$ for all $M \in \mathfrak{L}(U)$ which proves our claim. ■

4.79. Proposition. *Assume* $A \in$ alg (U, U^{-1}, P^+) *is invertible in* $\mathcal{L}(X)$. *Then* A *is also invertible in* alg (U, U^{-1}, P^+). *If* A *has a main part* $B \in \mathbf{D}_0(U)$ *then the inverse of* A *has a main part* $C \in \mathbf{D}_0(U)$.

Proof. First assume that A has a main part $B \in \mathbf{D}_0(U)$. Then

$$B = P^+M_1P^+ + P^-M_2P^-, \qquad M_1, M_2 \in \mathfrak{L}(U).$$

Since A is invertible we have

$$M_1(t) \neq 0, \qquad M_2(t) \neq 0 \quad \text{for all} \quad t \in \mathbb{T}$$

(by Proposition 4.72). Hence, M_1 and M_2 are invertible and $M_1^{-1}P^+ + M_2^{-1}P^-$ is at least one-sided invertible (by Theorem 4.74). Since

$$(M_1^{-1}P^+ + M_2^{-1}P^-)\,A = I + K, \qquad K \in \mathbf{QC}(U),$$

we obtain the one-sided invertibility of $I + K$. Similarly to Lemma 4.14 one proves the two-sided invertibility of $I + K$ and that $(I + K)^{-1} \in$ alg (U, U^{-1}, P^+). This yields

$A^{-1} \in \operatorname{alg}(U, U^{-1}, P^+)$. Now it is easy to see that $P^+M_1^{-1}P^+ + P^-M_2^{-1}P^-$ is a main part of A^{-1}.

If A is invertible in $\mathcal{L}(X)$, we approximate A in the norm by elements A_n whose main parts B_n exist and belong to $\mathbf{D}_0(U)$. Since

$$\left\{\sum_{i=1}^{n} \prod_{j=1}^{m} (A_{ij}P^+ + B_{ij}P^-) : A_{ij}, B_{ij} \in \mathfrak{L}_0(U),\, n, m \in \mathbb{N}\right\}$$

is dense in $\operatorname{alg}(U, U^{-1}, P^+)$ this can be made. Thus, A_n is invertible for n large enough and, by the part already proved, $A_n^{-1} \in \operatorname{alg}(U, U^{-1}, P^+)$. Because $\|A_n^{-1} - A^{-1}\| \to 0$ as $n \to \infty$ we get $A^{-1} \in \operatorname{alg}(U, U^{-1}, P^+)$. ■

4.80. Proposition. *Assume we are given a family* $\{\widetilde{W}_\omega\}_{\omega\in\Omega}$ *of invertible operators on* X *satisfying* 4.76(i) *and* (ii). *Define* $Q_\omega = G_\omega G_\omega^{(-1)}$ *and put* $P_\omega = I - Q_\omega$. *Assume further that*

$$\operatorname{alg}(U, U^{-1}, P^+) \subset \mathcal{A}, \qquad \mathbf{QC}(U) \subset \mathcal{J}$$

($\mathcal{A}$ and $\mathcal{J}$ being defined with respect to $\{P_\omega\}_{\omega\in\Omega}$).

Let $A \in \operatorname{alg}(U, U^{-1}, P)$ *be an invertible operator. Then* $\{P_\omega A P_\omega\}$ *is stable if and only if the essential part* $S(A^{-1})$ *is invertible.*

Proof. Using that $G_\omega^{(-1)} B G_\omega = B$ for $B \in \mathbf{D}(U)$ the proof can be carried out similarly to that of Theorem 4.45. ■

Paired discrete Wiener-Hopf operators

4.81. Spaces and notation. Let G be the linear space of all (two-sided) sequences of complex numbers $\xi = \{\xi_n\}_{n=-\infty}^{\infty}$. In the following E stands for anyone of the linear spaces

(a) $$\tilde{\mathrm{l}}^p := \left\{\xi = \{\xi_n\} \in G : \|\xi\| = \left(\sum_{n=-\infty}^{\infty} |\xi_n|^p\right)^{1/p} < \infty\right\},\quad 1 \leq p < \infty,$$

(b) $$\tilde{\mathrm{m}} := \left\{\xi = \{\xi_n\} \in G : \|\xi\| = \sup_n |\xi_n| < \infty\right\},$$

(c) $$\tilde{\mathrm{m}}_c := \left\{\xi = \{\xi_n\} \in \mathrm{m} : \lim_{n\to\infty} \xi_n \text{ and } \lim_{n\to-\infty} \xi_n \text{ exist}\right\},$$

(d) $$\tilde{\mathrm{m}}_0 := \left\{\xi = \{\xi_n\} \in \mathrm{m}_c : \lim_{n\to\infty} \xi_n = 0 \text{ and } \lim_{n\to-\infty} \xi_n = 0\right\}.$$

It is well-known that the spaces $\tilde{\mathrm{l}}^p$ and $\tilde{\mathrm{m}}$ are Banach spaces under the norms defined above. Moreover, $\tilde{\mathrm{m}}_c$ and $\tilde{\mathrm{m}}_0$ are (closed) subspaces of $\tilde{\mathrm{m}}$. Let further $\mu \in \mathbb{R}$ and let E^μ denote the space of all sequences $\eta = \{(1 + |n|)^{-\mu}\, \xi_n\}_{n=-\infty}^{\infty}$ with $\xi = \{\xi_n\}_{n=-\infty}^{\infty} \in E$. Obviously, E^μ equipped with norm

$$\|\eta\|_{E^\mu} = \|\xi\|_E$$

becomes a Banach space.

Let U be the operator defined on E^μ by

$$U\{\xi_n\} = \{\xi_{n-1}\}.$$

Obviously, U is invertible, and the corresponding inverse operator is given by $U^{-1}\{\xi_n\} = \{\xi_{n+1}\}$. As in Section 4.46 it can be proved that for $k \in \mathbb{Z}$

$$\|U^k\| \leqq (1 + |k|)^{|\mu|}.$$

We denote by P^+ the projection on E^μ defined by

$$P^+\{\xi_n\} = \{\eta_n\}, \qquad \eta_n = \begin{cases} \xi_n, & n \in \mathbb{Z}^+ \\ 0, & -n \in \mathbb{N}. \end{cases}$$

Then the operators U, U^{-1} and P^+ obviously satisfy the conditions (a) and (b) of Section 4.65. Put $Q_n := U^{-n}P^-U^n + U^nP^+U^{-n}$ ($n \in \mathbb{Z}^+$). Since $\sup\limits_n \|Q_n\| < \infty$ and $\bigcap\limits_n \ker P_n = \{0\}$ ($P_n := I - Q_n$), Section 4.10, 3° applies and we obtain that $\mathfrak{L}(U)$ is a Banach algebra without radical. Note that for any function $f \in \mathrm{W}^{|\mu|} := \mathrm{W}^{|\mu|,|\mu|}$ there is a uniquely determined element $A \in \mathfrak{L}(U) \subset \mathcal{L}(E^\mu)$ the Gelfand transform of which coincides with f.

Let A and B be elements from $\mathfrak{L}(U)$ and denote by a_n, b_n ($n \in \mathbb{Z}$) the Fourier coefficients of the symbols of A and B, respectively. It is not hard to see that the action of $AP^+ + BP^-$ and $P^+A + P^-B$ is given by

$$\sum_{k=0}^{\infty} a_{j-k}\xi_k + \sum_{k=-\infty}^{-1} b_{j-k}\xi_k = \eta_j \qquad (j \in \mathbb{Z}), \tag{1}$$

and

$$\sum_{k=-\infty}^{\infty} a_{j-k}\xi_k = \eta_j \quad (j \in \mathbb{Z}^+), \qquad \sum_{k=-\infty}^{\infty} b_{j-k}\xi_k = \eta_j \quad (-j \in \mathbb{N}), \tag{2}$$

respectively.

4.82. Theorem. *For the operator C defined on the space E^μ by* 4.81(1) *or* 4.81(2) *to be at least one-sided invertible it is necessary and sufficient that*

$$A(t) \neq 0, \qquad B(t) \neq 0 \quad \text{for all} \quad t \in \mathbb{T}. \tag{1}$$

If (1) *is satisfied, then the invertibility of the operator C corresponds with the index of the symbol,*

$$\varkappa = \operatorname{wind} A(t)/B(t),$$

and $\dim\ker C = \max\{-\varkappa, 0\}$, $\operatorname{codim}(\operatorname{im} C) = \max\{\varkappa, 0\}$. *If at least one of the conditions* (1) *does not hold, then C is neither a Φ_+- nor a Φ_--operator on the space E.*

Proof. Apply Theorem 4.74. ■

For $V \in \mathcal{M}$ (see Section 4.36) define the projection P_V on E^μ by

$$P_V\{\xi_n\} = \{\eta_n\}, \qquad \eta_n = \begin{cases} 0, & n \notin V \\ \xi_n, & n \in V \end{cases} \quad (n \geqq 0),$$

$$\eta_n = \begin{cases} 0, & -n+1 \notin V \\ \xi_n, & -n+1 \in V \end{cases} \qquad (n < 0).$$

The collection $\mathcal{P} := \{P_V : V \in \mathcal{M}\}$ satisfies 4.36(i)–(iii). Obviously, $P_{[0,n]} = P_{n+1} := I - U^{-(n+1)}P^-U^{n+1} - U^{n+1}P^+U^{-(n+1)}$ and $P_{(n,\infty)} = Q_{n+1}$.

4.83. Corollary. *For any E^μ the following assertions are true.*

(a) $\mathrm{alg}\,(U, U^{-1}, P^+) \subset \mathfrak{A}$ *with respect to* $\mathcal{P}$, *that is, any element* $A \in \mathrm{alg}\,(U, U^{-1}, P^+)$ *is of local type.*

(b) $\mathrm{alg}\,(U, U^{-1}, P^+) \subset \mathcal{A}$, $\mathbf{QC}(U) \subset \mathcal{J}$ *(where* $\mathcal{A}$ *and* $\mathcal{J}$ *are defined in Section* 4.29).

(c) $\mathbf{QC}(U) = \{K \in \mathcal{K}(E^\mu): \|Q_n K\| \to 0$ *and* $\|KQ_n\| \to 0$ *as* $n \to \infty\}$.

(d) *For* $E = \tilde{\mathbf{l}}^p, \tilde{\mathbf{m}}_0$ $(1 < p < \infty)$, *we have* $\mathbf{QC}(U) = \mathcal{K}(E^\mu)$.

Proof. The proof is essentially the same as that of Corollary 4.48. ■

4.84. Theorem. *For* $C = AP^+ + BP^-$ $(C = P^+A + P^-B)$ *and* $K \in \mathbf{QC}(U)$ $\big(A, B \in \mathfrak{L}(U)\big)$ *the sequence* $\{P_n(C + K) P_n\}$ *is stable on* E^μ *if and only if*

(a) $C + K$ *is invertible,*

(b) *the operator* $C' = P^+A^{-1}P^+ + P^-B^{-1}P^-$ *is invertible on* E.

Proof. Using Proposition 4.80 the proof can be carried out (with minor modifications) as that of Theorem 4.50. ■

4.85. Theorem. *Let* $A, B \in \mathfrak{L}(U) \subset \mathcal{L}(E^\mu)$ *and* a_i, b_i *be the Fourier coefficients of the symbols* $A(t), B(t)$, *respectively. Consider the equations*

$$\sum_{k=0}^{\infty} a_{j-k}\xi_k + \sum_{k=-\infty}^{-1} b_{j-k}\xi_k = \eta_j \qquad (j \in \mathbb{Z}), \tag{1}$$

$$\left(\sum_{k=-\infty}^{\infty} a_{j-k}\xi_k = \eta_j \quad (j \in \mathbb{Z}^+), \qquad \sum_{k=-\infty}^{\infty} b_{j-k}\xi_k = \eta_j \quad (-j \in \mathbb{N})\right)$$

and the truncated equations

$$\sum_{k=0}^{n} a_{j-k}\xi_k + \sum_{k=-n-1}^{-1} b_{j-k}\xi_k = \eta_j \qquad (-n-1 \leqq j \leqq n) \tag{2}$$

$$\left(\sum_{k=-n-1}^{n} a_{j-k}\xi_k = \eta_j \quad (0 \leqq j \leqq n), \qquad \sum_{k=-n-1}^{n} b_{j-k}\xi_k = \eta_j \quad (-n-1 \leqq j \leqq -1)\right)$$

(a) *Let* $E = \tilde{\mathbf{l}}^p, \tilde{\mathbf{m}}_0$ $(1 \leqq p < \infty)$ *and the operators* $C = AP^+ + BP^-$ $(C = P^+A + P^-B)$ *and* $C' = P^+A^{-1}P^+ + P^-B^{-1}P^-$ *be invertible. Then for each* $\eta = \{\eta_j\} \in E^\mu$ *and for all sufficiently large* n *the system* (2) *has exactly one solution* $\{\xi_k^{(n)}\}_{k=-n-1}^{n}$ *and, as* $n \to \infty$, *the elements* $\xi^{(n)} := \{\ldots, 0, \xi_{-n-1}^{(n)}, \ldots, \xi_n^{(n)}, 0, \ldots\}$ *converge in the norm of* E^μ *to the solution* $\xi = \{\xi_j\}$ *of equation* (1). *This convergence is quasioptimal, that means,*

$$\|\xi - \xi^{(n)}\|_{E^\mu} \leqq C \inf_{x_n \in \mathrm{im}\, P_n} \|\xi - x_n\|_{E^\mu}. \tag{3}$$

If, in addition, $\eta \in E^\lambda$ *and* $A, B \in \mathfrak{L}(U) \subset \mathcal{L}(E^\lambda)$ *with* $\lambda > \mu$, *then*

$$\|\xi - \xi^{(n)}\|_{E^\mu} \leqq Cn^{-(\lambda-\mu)}\, \|\eta\|_{E^\lambda}. \tag{4}$$

(b) *Let* E *be anyone of the spaces* $\tilde{\mathbf{l}}^p, \tilde{\mathbf{m}}, \tilde{\mathbf{m}}_c, \tilde{\mathbf{m}}_0$ *and the conditions of* (a) *be fulfilled. Then for each* $\eta = \{\eta_j\}$ *and for all sufficiently large* n *the system* (2) *has exactly one solution* $\{\xi_k^{(n)}\}_{k=-n-1}^{n}$ *and, as* $n \to \infty$,

$$\|P_k(\xi^{(n)} - \xi)\|_{E^\mu} \to 0$$

for every $k \in \mathbb{N}$.

Moreover if, in addition, the symbols $A(t)$, $B(t)$ belong to $\mathbf{W}^{|\mu|}$ then for each $k \in \mathbb{N}$ and all n large enough

$$\|P_k(\xi^{(n)} - \xi)\|_{E^\mu} \leqq C\left(\sum_{i=n+1-k}^{\infty} (1+i)^{|\mu|}\,|c_{-i}| + \sum_{i=n+1-k}^{\infty} (1+i)^{|\mu|}\,|d_i|\right), \tag{5}$$

$$\left(\|P_k(\xi^{(n)} - \xi)\|_{E^\mu} \leqq C\left(\sum_{i=[(n+1)/2]}^{\infty} (1+i)^{\mu}\,(|c_{-i}| + |b_{-i}|) + \sum_{i=[(n+1)/2]}^{\infty} (1+i)^{\mu}\,(|d_i| + |a_i|)\right)\right),$$

where c_j, d_j are the Fourier coefficients of $A^{-1}(t)$ and $B^{-1}(t)$, respectively, and $[(n+1)/2]$ denotes the largest integer equal to or less than $(n+1)/2$.

Proof. The proof is essentially the same as that of Theorem 4.52. However, some remarks concerning the proof of (5) are in order. Again, it suffices to estimate $P_kC^{-1}Q_n$ (see Section 4.52). For $C = AP^+ + BP^-$ we have

$$C^{-1} = P^+A^{-1}P^+ + P^-B^{-1}P^- + C^{-1}(P^+AP^- + P^-AP^-)\,P^-A^{-1}P^+$$
$$+ C^{-1}(P^-BP^+ + P^+BP^+)\,P^+B^{-1}P^-$$

and this immediately yields the first inequality in (5). If $C = P^+A + P^-B$ we have

$$C^{-1} = P^+A^{-1}P^+ + P^-B^{-1}P^- + C^{-1}P^+AP^-(P^-A^{-1}P^+ - P^-B^{-1}P^-)$$
$$+ C^{-1}P^-BP^+(P^+B^{-1}P^- - P^+A^{-1}P^+).$$

It is not hard to see that this representation of C^{-1} leads to the second inequality in (5). ■

Paired Wiener-Hopf integral operators

4.86. Spaces and notation. Let G be the linear space of all measurable complex-valued functions defined on $\mathbb{R}$. In the following let E stand for any of the linear spaces

(a) $\tilde{\mathbf{L}}^p := \left\{f \in G\colon \|f\| := \left(\int_{-\infty}^{\infty} |f|^p \, dt\right)^{1/p} < \infty\right\}$, $1 \leqq p < \infty$,

(b) $\tilde{\mathbf{M}} := \left\{f \in G\colon \|f\| := \operatorname{ess\,sup}_{t\in\mathbf{R}} |f(t)| < \infty\right\}$,

(c) $\tilde{\mathbf{M}}_1 := \{f \in \tilde{\mathbf{M}}\colon$ there exist constant functions f_1, f_2 such that $\operatorname{ess\,sup}_{t\geqq\alpha} |f(t) - f_1| \to 0$, $\operatorname{ess\,sup}_{t\leqq-\alpha} |f(t) - f_2| \to 0$ as $\alpha \to \infty\}$,

(d) $\tilde{\mathbf{M}}_0 := \{f \in \tilde{\mathbf{M}}_1\colon f_1 = f_2 = 0\}$.

(e) $\tilde{\mathbf{C}}, \tilde{\mathbf{C}}_1, \tilde{\mathbf{C}}_0$, the subspaces of $\tilde{\mathbf{M}}, \tilde{\mathbf{M}}_1, \tilde{\mathbf{M}}_0$, respectively, consisting of all functions which are continuous at each point $t \neq 0$ and for which each of the one-sided limits exist at $t = 0$.

Clearly, the spaces $\tilde{\mathbf{L}}^p$ and $\tilde{\mathbf{M}}$ are Banach spaces under the above norms. Moreover, $\tilde{\mathbf{M}}_1, \tilde{\mathbf{M}}_0, \tilde{\mathbf{C}}, \tilde{\mathbf{C}}_1, \tilde{\mathbf{C}}_0$ are (closed) subspaces of $\tilde{\mathbf{M}}$. If E is anyone of the spaces we have just introduced and if $\mu \in \mathbb{R}$, then E^μ will denote the linear space of all functions in G for which $(1 + |t|)^\mu f(t)$ is in E. Provided with the norm

$$\|f\|_{E^\mu} := \|(1 + |t|)^\mu f(t)\|_E ,$$

E^μ becomes a Banach space.

Let P^+ be the projection acting on E^μ by the rule

$$(P^+f)(t) = \begin{cases} f(t), & 0 < t < \infty, \\ 0, & -\infty < t < 0. \end{cases}$$

Now we consider operators U and U^{-1} defined on the space E^μ by the following equations:

$$\left.\begin{aligned} (Uf)(t) &= f(t) - 2\int_{-\infty}^{t} e^{s-t} f(s)\, ds , \\ (U^{-1}f)(t) &= f(t) - 2\int_{t}^{\infty} e^{t-s} f(s)\, ds . \end{aligned}\right\} \quad (-\infty < t < \infty)$$

Obviously, U and U^{-1} are the operators of convolution by the function $\delta - l_1$, $\delta - l_{-1}$, respectively, where l_1 and l_{-1} are as in Section 4.53. The operators U, U^{-1} and P^+ clearly satisfy the conditions (a) and (b) of Section 4.65. Note that $\mathfrak{L}(U)$ contains every operator of the form

$$(Mf)(t) = cf(t) - \int_{-\infty}^{\infty} k(t - s) f(s)\, ds , \qquad (-\infty < t < \infty)$$

where $k \in \tilde{\mathbf{L}}^{1,|\mu|}$, $c \in \mathbb{C}$. This operator is invertible on E^μ if and only if $c \neq 0$ and

$$F(\lambda) = c - \int_{-\infty}^{\infty} e^{i\lambda t} f(t)\, dt \neq 0 \quad \text{for all} \quad \lambda \in \mathbb{R} .$$

If the operators $A \in \mathfrak{L}(U)$ and $B \in \mathfrak{L}(U)$ are of the form

$$Af = f(t) - \int_{-\infty}^{\infty} k_1(t - s) f(s)\, ds ,$$

$$Bf = f(t) - \int_{-\infty}^{\infty} k_2(t - s) f(s)\, ds ,$$

where $k_j \in \tilde{\mathbf{L}}^{1,|\mu|}$, then $AP^+ + BP^-$ and $P^+A + P^-B$ are defined on E^μ by

$$(AP^+ + BP^-) f = f(t) - \int_{0}^{\infty} k_1(t - s) f(s)\, ds - \int_{-\infty}^{0} k_2(t - s) f(s)\, ds , \tag{1}$$

$$(P^+A + P^-B) f = \begin{cases} f(t) - \int_{-\infty}^{\infty} k_1(t - s) f(s)\, ds , & 0 < t < \infty , \\ f(t) - \int_{-\infty}^{\infty} k_2(t - s) f(s)\, ds , & -\infty < t < 0 , \end{cases} \tag{2}$$

respectively.

According to the statements made in 4.65 the symbol of the operators defined by the equations (1) and (2) is the function

$$C(\lambda, \theta) = A(\lambda)\frac{1+\theta}{2} + B(\lambda)\frac{1-\theta}{2} \qquad (-\infty \leqq \lambda \leqq \infty, \theta = \pm 1)$$

with

$$A(\lambda) = 1 - \int_{-\infty}^{\infty} e^{i\lambda t} k_1(t)\,dt, \qquad B(\lambda) = 1 - \int_{-\infty}^{\infty} e^{i\lambda t} k_2(t)\,dt.$$

Theorem 4.74 immediately yields the following theorem.

4.87. Theorem. *For the operator* $AP^+ + BP^-(P^+A + P^-B)$ *defined on the space* E^μ *by* 4.86(1) (4.86(2)) *to be at least one-sided invertible it is necessary and sufficient that*

$$A(\lambda) \neq 0, \qquad B(\lambda) \neq 0 \quad \text{for all} \quad \lambda, \quad -\infty \leqq \lambda \leqq \infty. \tag{1}$$

If the conditions (1) *are satisfied then the invertibility of the operator* $AP^+ + BP^-(P^+A + P^-B)$ *corresponds with the index of the symbol,*

$$\varkappa = \operatorname{wind} A(\lambda)/B(\lambda),$$

and $\dim\ker(AP^+ + BP^-) = \max\{-\varkappa, 0\}$, $\operatorname{codim}\operatorname{im}(AP^+ + BP^-) = \max\{\varkappa, 0\}$ $\big(\dim\ker(P^+A + P^-B) = \max\{-\varkappa, 0\}$, $\operatorname{codim}\operatorname{im}(P^+A + P^-B) = \max\{\varkappa, 0\}\big)$. *If at least one of the conditions* (1) *does not hold, then* $AP^+ + BP^-$ $(P^+A_+P^-B)$ *is neither a* Φ_+- *nor a* Φ_-*-operator on the space* E^μ.

Let $E = \tilde{\mathbf{L}}^p, \tilde{\mathbf{M}}, \tilde{\mathbf{M}}_1, \tilde{\mathbf{M}}_0$. For $V \in \mathcal{M}$ (see Section 4.36) define the projection P_V on E^μ by

$$P_V f = \chi_W f,$$

where χ_W is the characteristic function of the set

$$W := V \cup \{t \in \mathbb{R} : -t \in V\}.$$

The collection $\mathcal{P} := \{P_V : V \in \mathcal{M}\}$ satisfies 4.36(i)–(iii). Define operators $\tilde{W}_\tau$ $(\tau \in \mathbb{R}^+)$ on E^μ by

$$(\tilde{W}_\tau f)(t) = f(t - \tau).$$

Obviously, the operators $\tilde{W}_\tau$ are invertible and $(\tilde{W}_\tau^{-1} f)(t) = f(t + \tau)$. Put $G_\tau := \tilde{W}_\tau^{-1}P^- + \tilde{W}_\tau P^+$, $G_\tau^{(-1)} := P^-\tilde{W}_\tau + P^+\tilde{W}_\tau^{(-1)}$. A straigthforward computation gives $G_\tau^{(-1)}G_\tau = I$ and, hence, $Q_\tau := G_\tau G_\tau^{(-1)}$ is a continuous projection $\neq I$ $(\tau > 0)$. Notice that $P_{[0,\tau]} = P_\tau := I - Q_\tau$ and that the family $\{\tilde{W}_\tau\}_{\tau \in \mathbb{R}^+}$ satisfies the conditions 4.76(i) and (ii) if $\mu = 0$. The following corollary is easy to prove.

4.88. Corollary. *For any* E^μ $(E = \tilde{\mathbf{L}}^p, \tilde{\mathbf{M}}, \tilde{\mathbf{M}}_1, \tilde{\mathbf{M}}_0)$ *the following assertions are true.*

(a) $\operatorname{alg}(U, U^{-1}, P^+) \subset \mathfrak{A}$ *with respect to* $\mathcal{P}$, *that is, every element* $A \in \operatorname{alg}(U, U^{-1}, P^+)$ *is of local type.*

(b) $\operatorname{alg}(U, U^{-1}, P^+) \subset \mathcal{A}$, $\mathbf{QC}(U) \subset \mathcal{J}$ (*where* $\mathcal{A}$ *and* $\mathcal{J}$ *are defined in Section* 4.29). *If* E *is anyone of the spaces defined in Section* 4.86, *then* $\mathbf{QC}(U) \subset \mathcal{K}(E^\mu)$.

4.89. Theorem. *Let $E = \tilde{\mathbf{L}}^p, \tilde{\mathbf{M}}, \tilde{\mathbf{M}}_1, \tilde{\mathbf{M}}_0$ and C be one of the operators defined by 4.86(1) or 4.86(2). Let further $K \in \mathcal{K}(E^\mu)$ be an operator such that $\|KQ_\tau\| \to 0$, $\|Q_\tau K\| \to 0$ as $\tau \to \infty$. Then the sequence $\{P_\tau(C + K) P_\tau\}$ is stable on E^μ if and only if*

(a) *$C + K$ is invertible,*

(b) *the operator $C' = P^+A^{-1}P^+ + P^-B^{-1}P^-$ is invertible.*

Proof. Using Proposition 4.80, the proof can be carried out (with minor modifications) as that of Theorem 4.58. ■

4.90. Theorem. *Let $E = \tilde{\mathbf{L}}^p, \tilde{\mathbf{M}}, \tilde{\mathbf{M}}_1, \tilde{\mathbf{M}}_0$. Consider the equation*

$$(AP^+ + BP^- + K) f = g \tag{1}$$

on E^μ and the truncated equations

$$P_\tau(AP^+ + BP^- + K) f_\tau = P_\tau g \qquad (f_\tau \in \operatorname{im} P_\tau), \tag{2}$$

where $AP^+ + BP^-$ is defined by 4.86(1) and $K \in \mathcal{K}(E^\mu)$ is an operator such that $\|KQ_\tau\| \to 0$, $\|Q_\tau K\| \to 0$ as $\tau \to \infty$.

(a) *Let $E = \tilde{\mathbf{L}}^p, \tilde{\mathbf{M}}_0$ and the operators $AP^+ + BP^- + K$, $P^+A^{-1}P^+ + P^-B^{-1}P^-$ be invertible. Then for each $g \in E^\mu$ and for all sufficiently large τ the equation (2) has exactly one solution f_τ in $\operatorname{im} P_\tau$ and, as $\tau \to \infty$, the elements f_τ converge in the norm of E^μ to the solution f of equation (1). This convergence is quasioptimal, that is*

$$\|f - f_\tau\|_{E^\mu} \leqq C \inf_{x_\tau \in \operatorname{im} P_\tau} \|f - x_\tau\|_{E^\mu}. \tag{3}$$

If, in addition, $g \in E^\lambda$ and $A, B \in \mathfrak{L}(U) \subset \mathcal{L}(E^\lambda)$ with $\lambda > \mu$, then

$$\|f - f_\tau\|_{E^\mu} \leqq C\tau^{-(\lambda-\mu)} \|g\|_{E^\lambda}. \tag{4}$$

(b) *Let E be any of the spaces $\tilde{\mathbf{L}}^p, \tilde{\mathbf{M}}, \tilde{\mathbf{M}}_1, \tilde{\mathbf{M}}_0$ and the operators $AP^+ + BP^- + K$, $P^+A^{-1}P^+ + P^-B^{-1}P^-$ be invertible. Then for each $g \in E^\mu$ and for all sufficiently large τ the equation (2) has exactly one solution f_τ in $\operatorname{im} P_\tau$ and, as $\tau \to \infty$,*

$$\|P_\omega(f - f_\tau)\|_{E^\mu} \to 0$$

for any $\omega \in (0, \infty)$. Moreover, if $K = 0$, then for each $\omega \in (0, \infty)$ and all τ large enough we have

$$\|P_\omega(f - f_\tau)\|_{E^\mu} \leqq C \left(\int_{-\infty}^{-\tau+\omega} (1 + |t|)^\mu |k_3(t)| \, dt + \int_{\tau-\omega}^{\infty} (1 + t)^\mu |k_4(t)| \, dt \right), \tag{5}$$

where k_3, k_4 are defined by

$$A(\lambda)^{-1} = 1 - \int_{-\infty}^{\infty} e^{i\lambda t} k_3(t) \, dt; \qquad B(\lambda)^{-1} = 1 - \int_{-\infty}^{\infty} e^{i\lambda t} k_4(t) \, dt.$$

Proof. See the proof of the Theorems 4.60 and 4.85. ■

4.91. Remark. 1°. If $K \neq 0$, the inequality (5) can be replaced as it was done in Remark 4.61.

2°. The theorem can also be proved for the operator $PA + QB + K$, however, (5) must be replaced by an estimate analogous to the second one in 4.85(5).

Singular integral operators on spaces $\mathbf{L}^p(\mathbb{T}, \varrho)$

4.92. Notation. Let $\tau_1, \ldots, \tau_m$ be distinct points on the unit circle $\mathbb{T}$ and $\beta_1, \ldots, \beta_m$ be real numbers. Put

$$\varrho(t) = \prod_{k=1}^{m} |t - \tau_k|^{\beta_k}.$$

The space $\mathbf{L}^p(\mathbb{T}, \varrho)$ $(1 < p < \infty)$ is (by definition) the space of all measurable functions φ on $\mathbb{T}$ such that

$$\|\varphi\|_{\mathbf{L}^p(\mathbb{T},\varrho)} := \left(\int_{\mathbb{T}} |\varphi(t)|^p \, \varrho(t) \, |\mathrm{d}t|\right)^{1/p} < \infty.$$

(a) $\mathbf{L}^p(\mathbb{T}, \varrho)$ is a Banach space.

(b) The linear hull of $\{t^n\}_{n\in\mathbb{Z}}$ is dense in $\mathbf{L}^p(\mathbb{T}, \varrho)$ if

$$\beta_k > -1 \qquad (k = 1, \ldots, m).$$

(c) It is well-known that the singular integral operator defined by

$$(S\varphi)(t) := \frac{1}{\pi \mathrm{i}} \int_{\mathbb{T}} \frac{\varphi(\tau)}{\tau - t} \, \mathrm{d}\tau$$

is bounded on $\mathbf{L}^p(\mathbb{T}, \varrho)$ if only

$$1 < p < \infty, \qquad -1 < \beta_k < p - 1 \qquad (k = 1, \ldots, m). \tag{1}$$

Moreover one has

$$S^2 = I.$$

(d) The projection $P^+ := 1/2(I + S)$ maps $\mathbf{L}^p(\mathbb{T}, \varrho)$ onto the subspace $H^p(\varrho)$ consisting of all functions the Fourier coefficients of which vanish for negative indices. More precisely, $P^+t^n = t^n$ for $n \geqq 0$ and $P^+t^n = 0$ for $n < 0$ (see Gohberg/Krupnik [4]).

Let U be the bounded operator defined on $\mathbf{L}^p(\mathbb{T}, \varrho)$ $(1 < p < \infty, -1 < \beta_k < p - 1$ for $k = 1, \ldots, m)$ by

$$(U\varphi)(t) = t\varphi(t).$$

Obviously, the operator U is invertible and its inverse is given by

$$(U^{-1}\varphi)(t) = t^{-1}\varphi(t).$$

The operators U, U^{-1} and P^+ fulfil conditions 4.65(a) and (b). The underlying algebra $\mathfrak{L}(U)$ is an algebra without radical since the projections $Q_n := U^{-n}P^-U^n + U^nP^+U^{-n}$ $(P^- := I - P^+)$ are uniformly bounded and satisfy $\bigcap_{n=0}^{\infty} \operatorname{im} Q_n = \{0\}$ (see Section 4.10, 3°). Moreover the algebra $\mathfrak{L}(U)$ is precisely the algebra of all multiplication operators with continuous functions. Note that $\operatorname{codim} \operatorname{im} (P^+UP^+|_{\operatorname{im} P^+}) = 1$. Let a and b be continuous functions defined on $\mathbb{T}$. The singular integral operator C with the coefficients a and b is defined on $\mathbf{L}^p(\mathbb{T}, \varrho)$ $(1 < p < \infty, -1 < \beta_k < p - 1)$ by

$$(C\varphi)(t) := a(t)\,\varphi(t) + \frac{b(t)}{\pi \mathrm{i}} \int_{\mathbb{T}} \frac{\varphi(\tau)}{\tau - t} \, \mathrm{d}\tau. \tag{2}$$

It is easy to verify that C can be written in the form $C = cP^+ + dP^-$, where $c = a + b$ and $d = a - b$. Hence, C is a paired operator and Theorem 4.74 immediately leads to the following theorem.

4.93. Theorem. *The operator C defined by 4.92(2) is at least one-sided invertible on $\mathbf{L}^p(\mathbb{T}, \varrho)$ if and only if*

$$(a^2 - b^2)(t) \neq 0 \quad \text{for all} \quad t \in \mathbb{T}. \tag{1}$$

If condition (1) is fulfilled, then the invertibility of the operator C corresponds with the winding number

$$\varkappa = \operatorname{wind} \frac{(a+b)(t)}{(a-b)(t)},$$

and $\dim \ker C = \max\{-\varkappa, 0\}$, $\dim \operatorname{coker} C = \max\{\varkappa, 0\}$. *If condition (1) does not hold, then C is neither a Φ_+- nor a Φ_--operator on the space $\mathbf{L}^p(\mathbb{T}, \varrho)$.*

4.94. Corollary. (a) $\operatorname{alg}(U, U^{-1}, P^+) \subset \mathcal{A}$, $\mathbf{QC}(U) \subset \mathcal{J}$, *where* $P_n := I - Q_n$ *and* $Q_n := U^{-n}P^-U^n + U^nP^+U^{-n}$.
(b) $\mathbf{QC}(U) = \mathcal{K}(\mathbf{L}^p(\mathbb{T}, \varrho))$.

Proof. (a): Obvious.
(b): Immediate from Corollary 4.68 since $\operatorname{codim}(\operatorname{im} P^+UP^+|_{\operatorname{im} P^+}) = 1$, $(\mathbf{L}^p(\mathbb{T}, \varrho))^* = \mathbf{L}^q(\mathbb{T}, \varrho^{1-q})$ $(1/p + 1/q = 1)$ and $-1 < (1-q)\,\beta_k < q - 1$. ■

4.95. Theorem. *Let $P_n := I - Q_n$ and $T \in \mathcal{K}(\mathbf{L}^p(\mathbb{T}, \varrho))$. Then the sequence $\{P_n(C + T)\,P_n\}$ (C defined by 4.92(2)) is stable on $\mathbf{L}^p(\mathbb{T}, \varrho)$ if and only if*

(a) *$C + T$ is invertible on $\mathbf{L}^p(\mathbb{T}, \varrho)$*
(b) *the operator $C' = P^+(a+b)^{-1}P^+ + P^-(a-b)^{-1}P^-$ is invertible on $\mathbf{L}^p(\mathbb{T}, \varrho)$.*

Proof. Since $P_n \to I$ and $P_n^* \to I$ strongly as $n \to \infty$ the invertibility of $C + T$ is a consequence of Proposition 1.23(a) provided that $\{P_n(C + T)\,P_n\}$ is stable. The other assertions are immediate from Proposition 4.80. ■

Obviously, the projections P_n defined above act by the rule

$$(P_nf)(t) = \sum_{i=-n}^{n-1} f_i t^i,$$

where f_i $(i \in \mathbb{Z})$ are the Fourier coefficients of the function $f \in \mathbf{L}^p(\mathbb{T}, \varrho)$. Suppose we are given the singular integral equation

$$(Cf)(t) = a(t)\,f(t) + \frac{b(t)}{\pi i} \int_{\mathbb{T}} \frac{f(\tau)}{\tau - t}\,d\tau = g(t) \tag{1}$$

in $\mathbf{L}^p(\mathbb{T}, \varrho)$, where a and b are arbitrary continuous functions on $\mathbb{T}$. Put $c := a + b$, $d := a - b$ and let c_i, d_i, f_i denote the Fourier coefficients of the functions c, d, f, respectively $(i \in \mathbb{Z})$.

4.96. Theorem. *Assume the conditions of Theorem 4.95 are fulfilled. Then, for all n large enough, the system*

$$\sum_{k=0}^{n-1} c_{j-k}\xi_k + \sum_{k=-n}^{-1} d_{j-k}\xi_k = g_j \qquad (-n \leqq j \leqq n - 1)$$

has exactly one solution $\{\xi_j^{(n)}\}_{j=-n}^{n-1}$, *and the functions*

$$f_n(t) = \sum_{j=-n}^{n-1} \xi_j^{(n)} t^j$$

converge in the norm of $\mathbf{L}^p(\mathbb{T}, \varrho)$ *to the solution* f *of* 4.95(1). *This convergence is quasioptimal, that is*

$$\|f - f_n\|_{\mathbf{L}^p(\mathbf{T},\varrho)} \leqq M \inf_{x_n \in \operatorname{im} P_n} \|f - x_n\|_{\mathbf{L}^p(\mathbf{T},\varrho)} .$$

Proof. Use Remark 1.27, 3°(a). ∎

Systems

4.97. Notation. If X is an arbitrary vector space, then by X_N (N being a natural number) we denote the set of all N-dimensional vectors with components from X and by $X_{N\times N}$ the set of all quadratic matrices of order N with elements from X.

If X is a Banach space, then X_N equipped with the norm defined by

$$|||x||| = \sum_{j=1}^{N} \|x_j\|$$

$\big(x = (x_1, \ldots, x_N) \in X_N\big)$ is also a Banach space. The norm of a matrix $A = (A_{jk})_{j,k=1}^N \in X_{N\times N}$ can be defined by

$$|||A||| = N \max_{j,k} \|A_{jk}\| . \tag{1}$$

If X is a Banach algebra, then $X_{N\times N}$ with this norm is also a Banach algebra.

If X is an arbitrary Banach space and if $\mathcal{L}(X)$ is the set of all linear bounded operators on X, then $\mathcal{L}_{N\times N}(X)$ can be identified with $\mathcal{L}(X_N)$. In other words, each operator $A \in \mathcal{L}(X_N)$ can be written as a matrix $A = (A_{jk})_{j,k=1}^N$, where $A_{jk} \in \mathcal{L}(X)$. The operator A is compact if and only if all the operators A_{jk} are compact. Obviously, the operator norm of $A = (A_{jk})_{j,k=1}^N \in \mathcal{L}(X_N)$ is less than or equal to the norm defined by (1). Given an operator $A \in \mathcal{L}(X)$, we denote by the same symbol A the operator $\operatorname{diag}(A, \ldots, A)$ acting on X_N. As a rule, this practice will not cause any confusion.

Assume we are given operators V, $V^{(-1)} \in \mathcal{L}(X)$ such that $V^{(-1)}V = I$, $VV^{(-1)} \neq I$ and (H1) is fulfilled. According to Section 4.18, with each operator $A = (A_{jk})_{j,k=1}^N \in \operatorname{alg}_{N\times N}(V, V_{-1}) \big(:= (\operatorname{alg}(V, V_{-1}))_{N\times N}\big)$ we associate the continuous matrix function on the unit circle

$$\operatorname{smb} A := (\operatorname{smb} A_{ij})_{j,k=1}^N ,$$

which will be called *symbol* of A.

Further, we introduce the notions of the quasicommutator ideal, of the main part, and of the essential part similar to their counterparts in the case $N = 1$.

It is not hard to see that

$$\mathbf{QC}_{N\times N}(V) = \big(\mathbf{QC}(V)\big)_{N\times N} .$$

Finally, the algebra $\mathrm{alg}_{N\times N}(V, V_{-1})$ is called decomposing if $\mathrm{alg}\,(V, V_{-1})$ is decomposing. Obviously, for a decomposing algebra we have

$$\mathrm{alg}_{N\times N}(V, V_{-1}) = \mathfrak{L}_{N\times N}(V) \dotplus \mathbf{QC}_{N\times N}(V).$$

4.98. Theorem. *If* $A \in \mathrm{alg}_{N\times N}(V, V_{-1})$ *is a* Φ_+*- or a* Φ_-*-operator, then*

$$(\det \mathrm{smb}\, A)\,(t) \neq 0 \quad \textit{for all} \quad t \in \mathbb{T}.$$

Proof. Assume A is a Φ_+-operator. Then there is a $\delta > 0$ such that any operator $B \in \mathcal{L}(X_N)$ satisfying $\|A - B\| < \delta$ is also a Φ_+-operator. Assume further that there is a $t_0 \in \mathbb{T}$ such that $(\det \mathrm{smb}\, A)\,(t_0) = 0$. We choose an element $R = \sum_{i=1}^{n} \prod_{j=1}^{m} R_{ij}$, R_{ij} in $\big(\mathfrak{L}_0(V)\big)_{N\times N}$, such that the norm $|||A - R|||$ defined by 4.97(1) is less then $\delta/(2N)$. Since $|(\mathrm{smb}\, R)\,(t_0)| < \delta/(2N)$, we have

$$|||A - R + (\mathrm{smb}\, R)\,(t_0)\, I||| < \delta$$

and, hence, $S := R - (\mathrm{smb}\, R)\,(t_0)\, I$ is a Φ_+-operator with

$$(\det \mathrm{smb}\, S)\,(t_0) = 0. \tag{1}$$

If n is large enough then SV^n belongs to $\big(\mathfrak{L}_0^+(V)\big)_{N\times N}$ (and is, of course, a Φ_+-operator). Because $\mathfrak{L}_0^+(V)$ is commutative, this implies that $(\det SV^n) \in \mathfrak{L}_0^+(V)$ is a Φ_+-operator (by Theorem 1.5) and that $\mathrm{smb} \det SV^n = \det \mathrm{smb}\, SV^n = (\det \mathrm{smb}\, S)\,(\det \mathrm{smb}\, V^n)$. Because of (1) this contradicts Theorem 4.2(iii). If A is a Φ_--operator the assertion can analogously be proved. ■

4.99. Theorem. *Suppose* (H1) *is fulfilled and* $\dim \ker V_{-1} =: \varkappa < \infty$ *(in* X*). Then* $A \in \mathrm{alg}_{N\times N}(V, V_{-1})$ *is a* Φ_+*- or a* Φ_-*-operator if and only if*

$$(\det \mathrm{smb}\, A)\,(t) \neq 0 \quad \textit{for all} \quad t \in \mathbb{T}. \tag{1}$$

If (1) *is satisfied, then* A *is a* Φ*-operator and*

$$\mathrm{ind}\, A = -\varkappa\, \mathrm{wind}\, (\det \mathrm{smb}\, A). \tag{2}$$

Proof. Since $\varkappa < \infty$, the quasicommutator ideal $\mathbf{QC}(V)$ consists only of compact operators. Hence, for any $A = (A_{ij})_{i,j=1}^N \in \mathrm{alg}_{N\times N}\,(V, V_{-1})$ the operators A_{ij} $\big(\in \mathrm{alg}\,(V, V_{-1})\big)$ commute pairwise up to a compact operator. Theorem 1.5 yields that A is a Φ_+- or a Φ_--operator if and only if $\det\,(A_{ij})_{i,j=1}^N \in \mathrm{alg}\,(V, V_{-1})$ is a Φ_+- or a Φ_--operator, respectively. Using Theorem 4.20 we get (1). It remains to show (2). The pertubation argument 1.3(c) shows that it is sufficient to prove (2) for $A = (A_{ij})_{i,j=1}^N$ with $A_{ij} = \sum_{l=1}^{n} \prod_{k=1}^{m} R_{lk}^{(ij)}$, $R_{lk}^{(ij)} \in \mathfrak{L}_0(V)$. Since the commutator of any two entries of $(A_{ij})_{i,j=1}^N$ is finite-dimensional Theorem 1.6 along with Theorem 4.20 gives the index formula (2). ■

4.100. Remark. It is worth noticing that there is a significant difference between the cases $N = 1$ and $N > 1$. For $N = 1$ the conditions $R \in \mathfrak{L}_0(V)$ and $(\mathrm{smb}\, R)\,(t) \neq 0$ for all $t \in \mathbb{T}$ imply the one-sided invertibility of R. If $N > 1$ and $R \in \big(\mathfrak{L}_0(V)\big)_{N\times N}$ and $(\det \mathrm{smb}\, R)\,(t) \neq 0$ for all $t \in \mathbb{T}$, it may happen that R is neither a Φ_+- nor a Φ_--operator. A simple example is given by $\begin{pmatrix} V & 0 \\ 0 & V_{-1} \end{pmatrix}$, where $\dim \ker V_{-1} = \infty$. If we know that

$R \in (\mathfrak{L}_0(V))_{N\times N}$ is a Φ-operator ($N > 1$) we cannot conclude that R is one-sided invertible (take the above example with $\dim\ker V_{-1} < \infty$). A series of other complications arising in the case $N > 1$ rest at long last upon these facts. For instance, assume we are given an invertible operator $A \in \mathrm{alg}_{N\times N}(V, V_{-1})$. In case $N = 1$ we are able to prove that $A^{-1} \in \mathrm{alg}_{N\times N}(V, V_{-1})$. However, this proof does not work for $N > 1$. On the other hand, $A^{-1} \in \mathrm{alg}_{N\times N}(V, V_{-1})$ was needed to prove that there is an element $D \in \mathfrak{L}_{N\times N}(V)$ such that $A^{-1} - D \in \mathbf{QC}_{N\times N}(V)$ provided that $\mathrm{alg}_{N\times N}(V, V_{-1})$ is decomposing. In order to overcome this difficulty we note the following.

1°. If $\mathbf{QC}_{N\times N}(V) = \mathcal{K}(X_N)$ then the invertibility of $A \in \mathrm{alg}(V, V_{-1})$ implies that $A^{-1} \in \mathrm{alg}(V, V_{-1})$. Indeed, if A has a main part $B \in (\mathfrak{L}_{\mathfrak{D}}(V))_{N\times N}$ then there is an element $D \in (\mathfrak{L}_{\mathfrak{D}}(V))_{N\times N}$ such that

$$BD - I, \quad DB - I \in \mathbf{QC}_{N\times N}(V) \tag{1}$$

(for $N = 1$ see Section 4.22). Hence,

$$DA = I + K, \qquad K \in \mathcal{K}(X_N),$$

and $A^{-1} = D - KA^{-1} \in \mathrm{alg}_{N\times N}(V, V_{-1})$. The general case can be proved by means of the approximation argument used in the proof of Proposition 4.22.

2°. Suppose that $A \in \mathrm{alg}_{N\times N}(V, V_{-1})$ is an invertible operator of local type. Assume A has a main part $B \in (\mathfrak{L}_{\mathfrak{D}}(V))_{N\times N}$ and $\mathbf{QC}_{N\times N}(V) \subset \mathcal{J}$ (see Section 4.29). Then there is an element $D \in (\mathfrak{L}_{\mathfrak{D}}(V))_{N\times N}$ satisfying (1) such that

$$A^{-1} - D \in \mathcal{J}.$$

This, in fact, is enough for our purposes. If the conditions of 2° are satisfied, the operator D will also be called a main part of A^{-1}. If the algebra $\mathrm{alg}_{N\times N}(V, V_{-1})$ decomposes then the operator D is uniquely defined and will be said to be the essential part of A^{-1}.

4.101. Definition. Let $\Omega \subset \mathbb{R}^+$ be an unbounded set and $\{P_\omega\}_{\omega\in\Omega}$ be a family of projections on X. An operator $A \in \mathcal{L}(X)$ is called an *operator of local type* with respect to $\{P_\omega\}$ if there is a system $\mathcal{P}_{\mathcal{M}} \subset \mathcal{L}(X)$ of projections satisfying (i)–(iii) of Section 4.36 such that $P_\omega = P_{[0,\omega]}$ and $A \in \mathfrak{A}$, where $\mathfrak{A}$ corresponds to $\mathcal{P}_{\mathcal{M}}$.

4.102. Theorem. *Assume that* V, V_{-1} *fulfil* (H1) *and that we are given a family* $\{\widetilde{W}_\omega\}_{\omega\in\Omega}$ *of left-invertible operators with left-inverses* $\widetilde{W}_\omega^{-1}$ *such that*

(a) $V\widetilde{W}_\omega = \widetilde{W}_\omega V$, $V_{-1}\widetilde{W}_\omega^{(-1)} = \widetilde{W}_\omega^{(-1)}V_{-1}$ *for all* $\omega \in \Omega$,

(b) $\sup_\omega \|\widetilde{W}_\omega\| < \infty$, $\sup_\omega \|\widetilde{W}_\omega^{(-1)}\| < \infty$,

(c) $\mathrm{alg}_{N\times N}(V, V_{-1}) \subset \mathfrak{A}$, $\mathbf{QC}_{N\times N}(V) \subset \mathcal{J}$, *where* $Q_\omega := \widetilde{W}_\omega\widetilde{W}_\omega^{(-1)}$ *and* $P_\omega = I - Q_\omega$.

Consider $\{P_\omega AP_\omega\}$, $A \in \mathrm{alg}_{N\times N}(V, V_{-1})$. *Then:*

(i) *If* $\{P_\omega AP_\omega\}$ *is stable and if* A *is invertible, we get the invertibility (in* $\mathcal{L}(X_N)$*) of the essential part* $S(A^{-1})$ *(defined in* Section 4.100, 2°*).*

(ii) *The invertibility of* A *and* $S(A^{-1})$ *on* X *implies the stability of* $\{P_\omega AP_\omega\}$.

Proof. First of all note that the conditions of the theorem imply the decomposition $\mathrm{alg}_{N\times N}(V, V_{-1}) = (\mathfrak{L}(V))_{N\times N} \dotplus \mathbf{QC}_{N\times N}(V)$, so that the essential part $S(A)$ of any element A from $\mathrm{alg}_{N\times N}(V, V_{-1})$ exists. Put $\mathfrak{D} = (\mathfrak{L}(V), \circ)$. Remark 4.100, 2° shows that the essential part $S(A^{-1})$ exists.

The proof of (i): We have $A^{-1} - S(A^{-1}) \in \mathcal{J}$ and it remains to apply the argumentation of the proof to Theorem 4.45.

The proof of (ii): A can be written in the form

$$A = S(A^{-1})^{-1} + K, \qquad K \in \mathcal{J},$$

and it remains to apply the reasoning of the proof to Theorem 4.45. ■

4.103. Remark. If $N > 1$, the invertibility of $A + K$ does not imply the invertibility of $S(A^{-1})$.

4.104. Notation. Assume we are given an invertible operator U and a projection P^+ satisfying the conditions (a) and (b) of Section 4.65. Put $P^- = I - P^+$. According to Section 4.71, with each operator $A = (A_{jk})_{j,k=1}^N \in \mathrm{alg}_{N\times N}(U, U^{-1}, P^+) := (\mathrm{alg}\,(U, U^{-1}, P^+))_{N\times N}$ we associate the continuous matrix function on $\mathbb{T}\times\{-1, 1\}$ given by

$$\mathrm{smb}\, A := (\mathrm{smb}\, A_{jk})_{j,k=1}^N$$

and called *symbol* of A.

The notions of the quasicommutator ideal, of the main and essential parts, and of a decomposing algebra can be introduced as for $N = 1$ (see also Section 4.100, 2°).

4.105. Theorem. *If $A \in \mathrm{alg}_{N\times N}(U, U^{-1}, P)$ is a Φ_+- or a Φ_--operator, then*

$$(\det \mathrm{smb}\, A)\,(t, \theta) \neq 0 \quad \textit{for all} \quad t \in \mathbb{T}, \quad \theta \in \{-1, 1\}.$$

Proof. Combine the proofs of Proposition 4.72 and Theorem 4.98. ■

4.106. Theorem. *Let* $\mathrm{codim}\,(\mathrm{im}\, P^+UP^+|_{\mathrm{im}\,P^+}) =: \varkappa < \infty$. *Then $A \in \mathrm{alg}_{N\times N}(U, U^{-1}, P^+)$ is a Φ_+- or a Φ_--operator if and only if*

$$(\det \mathrm{smb}\, A)\,(t, \theta) \neq 0 \quad \textit{for all} \quad t \in \mathbb{T}, \theta \in \{-1, 1\}. \tag{1}$$

If (1) *is fulfilled, then A is a Φ-operator and*

$$\mathrm{ind}\, A = -\varkappa\, \mathrm{ind}\, (\det \mathrm{smb}\, A).$$

Proof. See the proof of Theorem 4.99. ■

Suppose we are given a family $\{\widetilde{W}_\omega\}_{\omega\in\Omega}$ of invertible operators on X ($\Omega \subset \mathbb{R}^+$ an unbounded set) satisfying 4.76(i), (ii). Put $Q_\omega := \widetilde{W}_\omega^{-1}P^-\widetilde{W}_\omega + \widetilde{W}_\omega P^+\widetilde{W}_\omega^{-1}$ and $P_\omega := I - Q_\omega$ (see Section 4.76). Suppose in addition that

$$\|G_\omega^{(-1)}KG_\omega\| \to 0$$

as $\omega \to \infty$ for any $K \in \mathbf{QC}_{N\times N}(U)$, where $G_\omega^{(-1)} := P^-\widetilde{W}_\omega + P^+\widetilde{W}_\omega^{-1}$, $G_\omega := \widetilde{W}_\omega^{-1}P^- + \widetilde{W}_\omega P^+$. Using Proposition 4.77 we get that $\mathrm{alg}_{N\times N}(U, U^{-1}, P)$ decomposes:

$$\mathrm{alg}_{N\times N}(U, U^{-1}, P^+) = \mathbf{D}_{N\times N}(U) \dotplus \mathbf{QC}_{N\times N}(U).$$

Note that the essential part is well-defined in this case.

4.107. Theorem. *Assume we are given a family $\{\widetilde{W}_\omega\}_{\omega\in\Omega}$ of invertible operators on X satisfying* 4.76(i) *and* (ii). *Define $Q_\omega = G_\omega G_\omega^{(-1)}$ and put $P_\omega = I - Q_\omega$. Assume further*

$$\mathrm{alg}_{N\times N}(U, U^{-1}, P^+) \subset \mathfrak{A}, \qquad \mathbf{QC}_{N\times N}(U) \subset \mathcal{J} \tag{1}$$

or

$$\mathrm{alg}_{N\times N}(U, U^{-1}, P^+) \subset \mathcal{A}, \qquad \mathbf{QC}_{N\times N}(U) = \mathcal{K}(X_N) \subset \mathcal{J}. \tag{2}$$

Let $A \in \operatorname{alg}(U, U^{-1}, P^+)$ *be an invertible operator. Then* $\{P_\omega A P_\omega\}$ *is stable if and only if the essential part* $S(A^{-1})$ *is invertible. (If only condition* (1) *is fulfilled, the existence of* $S(A^{-1})$ *is understood in the sense of Remark* 4.100, 2°).

Proof. See the proof of Theorem 4.45. ■

Block Toeplitz operators

4.108. Theorem. (a) *Let* E^μ *be anyone of the spaces defined in Section* 4.46, *let* V *and* $V^{(-1)}$ *be the operators defined on* E^μ *by*

$$V\{\xi_n\}_{n=0}^{\infty} = \{0, \xi_0, \xi_1, \ldots\}, \qquad V^{(-1)}\{\xi_n\}_{n=0}^{\infty} = \{\xi_{n+1}\}_{n=0}^{\infty}.$$

Then an operator $A \in \mathfrak{L}_{N\times N}(V) \subset \mathcal{L}(E_N^\mu)$ *is a* Φ_+*- or a* Φ_-*-operator if and only if*

$$(\det \operatorname{smb} A)(t) \neq 0 \quad \textit{for all} \quad t \in \mathbb{T}. \tag{1}$$

If condition (1) *is fulfilled then* A *is a* Φ*-operator the index of which equals*

$$\operatorname{ind} A = -\operatorname{wind}(\det \operatorname{smb} A).$$

(b) *For any* $a \in \mathbf{W}_{N\times N}^{\mu_-,\mu_+}$ *there exists an operator* $A \in \mathcal{L}_{N\times N}(V) \subset \mathcal{L}(E_N^\mu)$ *such that the symbol* $A(t)$ *of* A *coincides with* $a(t)$.

(c) *For* $\mu = 0$ *the algebra* $\operatorname{alg}_{N\times N}(V, V^{(-1)})$ *decomposes and for* $E = \mathbf{l}^2$ *the algebra* $\left(\mathfrak{L}_{N\times N}(V), \circ\right) := \left(\mathfrak{L}(V), \circ\right)_{N\times N}$ *is isometrically isomorphic to the algebra* $\mathbf{C}_{N\times N}(\mathbb{T})$ *of all continuous matrix valued functions on* $\mathbb{T}$ *(equipped with the natural* C^**-norm).*

(d) *Let* $E = \mathbf{l}^1, \mathbf{m}, \mathbf{m}_c, \mathbf{m}_0$. *Then* $A \in \mathcal{L}_{N\times N}(V) \subset \mathcal{L}(E_N^\mu)$ *implies that* $\operatorname{smb} A \in \mathbf{W}_{N\times N}$.

Proof. The proof is essential the same as that of Theorem 4.47. However, Theorem 4.99 has to be taken into account. ■

The operators belonging to $\mathcal{L}_{N\times N}(V)$ are called *block Toeplitz operators*.

4.109. Theorem. *Put* $Q_n := V_n V_{-n}$ *and* $P_n := I - Q_n$ *for all integers* $n \geqq 0$. *Then for* $A \in \mathcal{L}_{N\times N}(V)$ *and* $K \in \mathbf{QC}_{N\times N}(V)$ $\left(\subset \mathcal{L}(E_N^\mu)\right)$ *the sequence*

$$\{P_n(A + K) P_n\}_{n=1}^{\infty}$$

is stable on E_N^μ *if and only if*

(a) $A + K$ *is invertible on* E_N^μ,

(b) *the block Toeplitz operator* $A' \in \mathfrak{L}_{N\times N}(V) \subset \mathcal{L}(E_N)$ *defined by the symbol* $(\operatorname{smb} A)^{-1}$ *is invertible on* E_N.

Proof. Due to Theorem 4.102 and Corollary 4.48, the proof can be carried out exactly as that of Theorem 4.50 (see also Remark 4.51). ■

The action of an operator $A \in \mathfrak{L}_{N\times N}(V) \subset \mathcal{L}(E_N^\mu)$ can be easily described. According to (H2) and (H4) an element of $\mathfrak{L}_{N\times N}(V)$ is uniquely determined by its symbol. Let a_n be the block Fourier coefficients of the symbol smb A. Put

$$\eta := A\xi, \qquad \eta = \{\eta_n\}_{n=0}^{\infty}, \qquad \xi = \{\xi_n\}_{n=0}^{\infty} \qquad (\xi_n, \eta_n \in \mathbb{C}_N).$$

Then

$$\sum_{k=0}^{\infty} a_{j-k}\xi_k = \eta_j \qquad (j = 0, 1, \ldots).$$

Indeed, for $A \in \left(\mathfrak{L}_0(V)\right)_{N\times N}$ the assertion is obvious and by passing to the limit one obtains the assertion in full generality. ■

4.110. Theorem. *Let*

$$\sum_{k=0}^{\infty} a_{j-k}\xi_k = \eta_j \qquad (j = 0, 1, \ldots) \tag{1}$$

be an equation in E_N^μ *given by an operator* $A \in \mathfrak{L}_{N\times N}(V) \subset \mathcal{L}(E_N^\mu)$ *and consider the truncated equations*

$$\sum_{k=0}^{n} a_{j-k}\xi_k = \eta_j \qquad (j = 0, 1, \ldots, n), \tag{2}$$

$n \in \mathbb{N}$. *Let the operators* A *and* A' (*see Section* 4.109(b)) *be invertible on the spaces* E_N^μ *and* E_N, *respectively.*

(a) *Let* $E = \mathbf{l}^p, \mathbf{m}_0$ $(1 \leqq p < \infty)$. *Then for each* $\eta = \{\eta_j\}_{j=0}^\infty \in E_N^\mu$ *and for all sufficiently large* n *the system* (2) *has exactly one solution* $\{\xi_k^{(n)}\}_{k=0}^n$ *and, as* $n \to \infty$, *the elements* $\xi^{(n)} := \{\xi_0^{(n)}, \ldots, \xi_n^{(n)}, 0, \ldots\}$ *converge in the norm of* E_N^μ *to the solution* $\xi = \{\xi_j\}_{j=0}^\infty$ *of equation* (1). *This convergence is quasioptimal, that means,*

$$\|\xi - \xi^{(n)}\|_{E_N^\mu} \leqq C \inf_{x_n \in \operatorname{im} P_n} \|\xi - x_n\|_{E_N^\mu}. \tag{3}$$

If, in addition, $\eta \in E_N^\lambda$, $A \in \mathfrak{L}_{N\times N}(V) \subset \mathcal{L}(E_N^\lambda)$ *with* $\lambda > \mu$, *then*

$$\|\xi - \xi^{(n)}\|_{E_N^\mu} \leqq C n^{-(\lambda-\mu)} \|\eta\|_{E_N^\lambda}. \tag{4}$$

(b) *Let* E *be any of the spaces* $\mathbf{l}^p, \mathbf{m}, \mathbf{m}_c, \mathbf{m}_0$. *Then for each* $\eta = \{\eta_j\}_{j=0}^\infty \in E_N^\mu$ *and for all sufficiently large* n *the system* (2) *has exactly one solution* $\{\xi_k^{(n)}\}_{k=0}^n$ *and, as* $n \to \infty$,

$$\|P_k(\xi^{(n)} - \xi)\|_{E_N^\mu} \to 0$$

for each $k \in \mathbb{N}$. *Moreover if, in addition, the symbol* smb A *belongs to* $\mathbf{W}_{N\times N}^{\mu_-,\mu_+}$, *then for each* $k \in \mathbb{N}$ *and all* n *large enough*

$$\|P_k(\xi^{(n)} - \xi)\|_{E_N^\mu} \leqq C \sum_{i=n+1-k}^{\infty} (1+i)^{\mu_-} |b_{-i}|,$$

where b_{-i} *are the block Fourier coefficients of* $B = (\operatorname{smb} A)^{-1} \in \mathbf{W}_{N\times N}^{\mu_-,\mu_+}$ *and* $|g|$, $g = (g_{ij})_{i,j=1}^N$ $(g_{ij} \in \mathbb{C})$, *refers to* $N \max_{1 \leqq i,j \leqq N} |g_{ij}|$.

Proof. See the proof of Theorem 4.52. ■

Systems of Wiener-Hopf integral equations

4.111. Notation. Let E be anyone of the spaces $\mathbf{L}^p$ $(1 \leq p < \infty)$, $\mathbf{M}$, $\mathbf{M}_1$, $\mathbf{M}_0$, $\mathbf{C}$, $\mathbf{C}_1$, $\mathbf{C}_0$ introduced in Section 4.53 and let V and $V^{(-1)}$ be the operators defined on E^μ by

$$(Vf)(t) = f(t) - \int_0^\infty l_1(t-s)\, f(s)\, ds,$$

$$(0 < t < \infty)$$

$$(V^{(-1)}f)(t) = f(t) - \int_0^\infty l_{-1}(t-s)\, f(s)\, ds,$$

where

$$l_1(t) = \begin{cases} 2e^{-t}, & t > 0 \\ 0, & t < 0 \end{cases}, \qquad l_{-1}(t) = l_1(-t).$$

Assume $\varphi = (\varphi_{ij})_{i,j=1}^N \in \mathcal{F}^{|\mu|}_{N\times N}$ and put $W_\varphi := \{W_{\varphi_{ij}}\}_{i,j=1}^N$. Due to Section 4.53 it is clear that

$$W_\varphi \in \mathfrak{L}_{N\times N}(V) \subset \mathcal{L}(E_N^\mu).$$

4.112. Theorem. *The operator $W_\varphi \subset \mathcal{L}(E_N^\mu)$, $\varphi \in \mathcal{F}^{|\mu|}_{N\times N}$, is a Φ_+- or a Φ_--operator if and only if*

$$(\det \varphi)(\lambda) \neq 0 \text{ for all } -\infty \leqq \lambda \leqq \infty. \tag{1}$$

If condition (1) *is fulfilled then W_φ is a Φ-operator the index of which equals*

$$\operatorname{ind} W_\varphi = -\operatorname{wind} \det \varphi.$$

Proof. Combine the proof of Theorem 4.54 with Theorem 4.99. ■

Let P_τ, Q_τ $(\tau > 0)$ be the projections defined in Section 4.54.

4.113. Theorem. *Let $E = \mathbf{L}^p, \mathbf{M}, \mathbf{M}_1, \mathbf{M}_0$. Then, for $W_\varphi \in \mathcal{L}(E_N^\mu)$, $\varphi \in \mathcal{F}^{|\mu|}_{N\times N}$, and $K \in \mathcal{K}(E_N^\mu)$ with $\|KQ_\tau\| \to 0$, $\|Q_\tau K\| \to 0$ as $\tau \to \infty$, the sequence*

$$\{P_\tau(W_\varphi + K)\, P_\tau\}$$

is stable if and only if

(a) *$W_\varphi + K$ is invertible on E_N^μ,*
(b) *the operator $W_{\varphi^{-1}}$ is invertible on E_N.*

Proof. See the proof of Theorem 4.58. ■

4.114. Theorem. *Let*

$$(W_\varphi f)(t) := cf(t) - \int_0^\infty k(t-s)\, f(s)\, ds = g(t), \qquad s > 0 \tag{1}$$

be a system of Wiener-Hopf equations on E_N^μ ($\varphi \in \mathcal{F}^{|\mu|}_{N\times N}$), where E is anyone of the spaces $\mathbf{L}^p$, $\mathbf{M}$, $\mathbf{M}_1$, $\mathbf{M}_0$ and consider the truncated equations

$$(W_\varphi^\tau f_\tau)(t) := cf_\tau(t) - \int_0^\tau k(t-s)\, f_\tau(s)\, ds = (P_\tau g)(t), \qquad 0 < t < \tau. \tag{2}$$

Let the operators W_φ and $W_{\varphi^{-1}}$ be invertible on the spaces E_N^μ and E_N, respectively.

(a) *Let $E = \mathbf{L}^p, \mathbf{M}_0$. Then for each $g \in E_N^\mu$ and for all sufficiently large τ the equation* (2) *has exactly one solution f_τ in* $\operatorname{im} P_\tau$ *and, as $\tau \to \infty$, the elements f_τ (assumed to be zero on (τ, ∞)) converge in the norm of E_N^μ to the solution f of equation* (1). *This convergence is quasioptimal, that is,*

$$\|f - f_\tau\|_{E_N^\mu} \leqq C \inf_{x_\tau \in \operatorname{im} P_\tau} \|f - x_\tau\|_{E_N^\mu}. \tag{3}$$

If, in addition, $g \in E_N^\lambda$, $\varphi \in \mathcal{F}_{N\times N}^\nu$ with $\lambda > \mu$, and $\nu = \max\{|\mu|, |\lambda|\}$, then

$$\|f - f_\tau\|_{E_N^\mu} \leqq C\tau^{-(\lambda-\mu)} \|g\|_{E_N^\lambda}. \tag{4}$$

(b) *Let E be any of the spaces $\mathbf{L}^p, \mathbf{M}, \mathbf{M}_1, \mathbf{M}_0$. Then for each $g \in E_N^\mu$ and for all sufficiently large τ the equation* (2) *has exactly one solution f_τ in* $\operatorname{im} P_\tau$ *and, as $\tau \to \infty$,*

$$\|P_\omega(f - f_\tau)\|_{E_N^\mu} \to 0$$

for each $\omega \in (0, \infty)$. Moreover, for each $\omega \in (0, \infty)$ and all τ large enough

$$\|P_\omega(f - f_\tau)\|_{E_N^\mu} \leqq C \max_{1 \leqq i,j \leqq N} \|\chi_{(-\infty, -\tau+\omega)} k_{ij}\|_{\mathbf{L}^{1,|\mu|}(\mathbb{R})},$$

where $k_1 = (k_{ij})_{i,j=1}^N \in \mathbf{L}_{N\times N}^{1,|\mu|}$ is defined by

$$\varphi^{-1}(\lambda) = c - \int_{-\infty}^{\infty} k_1(t)\, e^{i\lambda t}\, dt.$$

Proof. See the proof of Theorem 4.60. ∎

4.115. Notation. Our next concern is to study the modified finite sections for systems of Wiener-Hopf integral equations (see Section 4.62) in the spaces E^μ, $E = \mathbf{C}, \mathbf{C}_1, \mathbf{C}_0$. First we recall and extend some definitions from Section 4.62. Assume we are given a matrix-valued function $k(t, s)$ on $\mathbb{R}^+ \times \mathbb{R}^+$ such that $k_t \in \mathbf{L}^{1,-\mu}(\mathbb{R}^+)$ for all $t \in \mathbb{R}^+$, where $k_t(s) := k(t, s)$. Throughout the following suppose

(a) $\sup_t \|(1 + t)^\mu k_t\|_{1,-\mu} := N \max_{1 \leqq i,j \leqq N} \sup_t (1 + t)^\mu \int_0^\infty (1 + s)^{-\mu} |k_{ij}(t, s)|\, ds < \infty$, where $k = (k_{ij})_{i,j=1}^N$.

(b) $\|(1 + t')^\mu k_{t'} - (1 + t)^\mu k_t\|_{1,-\mu} \to 0$ as $t' \to t$.

Consider $(Kf)(t) := \int_0^\infty k(t, s) f(s)\, ds$, $t \in \mathbb{R}^+$, $f \in E_N^\mu$. It can be easily seen that K belongs to $\mathcal{L}(E_N^\mu)$. Moreover, we have $K \in \mathcal{L}(\mathbf{M}_N^\mu, \mathbf{C}_N^\mu)$. Suppose the following condition is satisfied:

(b') $\|(1 + t')^\mu k_{t'} - (1 + t)^\mu k_t\|_{1,-\mu} \to 0$ as $t' \to t$ uniformly with respect to $t \in \mathbb{R}^+$.

Conditions (a), (b') together with

(c) $\|(1 + t)^\mu k_t\|_{1,-\mu} \to 0$ as $t \to \infty$

yield that

$$K \in \mathcal{K}(\mathbf{M}_N^\mu, \mathbf{C}_0^\mu)_N.$$

Put

$$v_\tau(t) := N \max_{1\leq i,j\leq N} (1+t)^\mu \int_\tau^\infty |k_{ij}(t,s)|\,(1+s)^{-\mu}\,ds, \qquad \tau\in\mathbb{R}^+, t\in\mathbb{R}^+,$$

$$(K_\tau f)(t) = \int_0^\tau k(t,s)\,f(s)\,ds = \int_0^\infty k_\tau(t,s)\,f(s)\,ds,$$

where

$$k_\tau(t,s) = \begin{cases} k(t,s), & 0 \leqq s \leqq \tau\ (\tau\in\mathbb{R}^+),\\ 0, & s > \tau \end{cases} \qquad t\in\mathbb{R}^+.$$

4.116. Proposition. (i) *If K is subject to the conditions* (a) *and* (b) *of the previous section then $K, K_\tau \in \mathcal{L}(\mathbf{M}_N^\mu, \mathbf{C}_N^\mu)$ and*

$$\|(K-K_\tau)f\| \leqq \sup_t |v_\tau(t)| \max_{1\leqq i\leqq N} \sup_t |(1+t)^\mu f_i(t)| \quad \big(f = (f_1,\dots,f_N)\big).$$

(ii) *If K satisfies the conditions* (a), (b'), *and* (c) *then*

$$K, K_\tau \in \mathcal{K}(\mathbf{M}_N^\mu, (\mathbf{C}_0^\mu)_N)$$

and

$$\|K-K_\tau\|_{\mathcal{L}(\mathbf{M}_N^\mu)} \to 0 \quad as \quad \tau\to\infty. \tag{2}$$

(iii) If $k \in \mathbf{L}^{1,|\mu|}_{N\times N}(\mathbb{R})$ *then for* $(Kf)(t) := \int_0^\infty k(t-s)\,f(s)\,ds\ (t\in\mathbb{R}^+)$ *we have*

$$K\in\mathcal{L}(\mathbf{M}_N^\mu, \mathbf{C}_N^\mu), \qquad K\in\mathcal{L}\big((\mathbf{M}_0^\mu)_N, (\mathbf{C}_0^\mu)_N\big), \qquad K_\tau\in\mathcal{K}(\mathbf{M}_N^\mu, (\mathbf{C}_0^\mu)_N).$$

Proof. See the proof of Proposition 4.63. ■

Define $(W_{\varphi,\tau}f)(t) := cf(t) - \int_0^\tau k(t-s)\,f(s)\,ds,\ t\in\mathbb{R}^+$.

4.117. Theorem. *Let $E = \mathbf{C}, \mathbf{C}_1, \mathbf{C}_0$ and $\varphi \in \mathcal{F}^{|\mu|}_{N\times N}$. Assume $W_\varphi + K$ and $W_{\varphi^{-1}}$ are invertible on E_N^μ and E_N, respectively, where K satisfies the conditions* (a), (b'), (c) *of Section* 4.115. *Then the sequence $\{W_{\varphi,\tau} + K_\tau\}$ is stable on E^μ. Moreover, the uniquely determined solutions f_τ of*

$$(W_{\varphi,\tau} + K_\tau)\,f_\tau = g$$

(τ large enough, say $\tau \geqq \tau_0$) converge uniformly on every compact interval of $\mathbb{R}^+$ to the unique solution f of the equation

$$(W_\varphi + K)\,f = g.$$

The speed of convergence can be estimated by

$$\max_{1\leqq i\leqq N}\ \sup_{0\leqq t\leqq\omega} \big|\big((1+t)^\mu\,(f_i(t) - f_{\tau,i}(t)\big)\big|$$

$$\leqq C \max_{1\leqq i,j\leqq N} \int_{-\infty}^{-\tau+\omega+\beta(\tau)} (1+|t|)^\mu\,|k_{ij}(t)|\,dt + \sup_t v_{\beta(\tau)}(t),$$

where $\beta(\tau)$ *is as in Section* 4.61 *and* $k_1 = (k_{ij})_{i,j=1}^N$ *is given by*

$$\varphi^{-1}(\lambda) = c - \int_{-\infty}^{\infty} k_1(t)\, e^{i\lambda t}\, dt.$$

Proof. See the proof of Theorem 4.64. ■

4.118. Remark. Systems of paired discrete Wiener-Hopf equations or paired Wiener-Hopf integral equations can be considered in a quite similar manner.

Systems of singular integral equations

4.119. Notation. According to Section 4.92 we consider singular integral operators with continuous matrix-valued coefficients on $\mathbf{L}_N^p(\mathbb{T}, \varrho)$ $(1 < p < \infty,\ -1 < \beta_k < p - 1)$:

$$(C\varphi)(t) := a(t)\,\varphi(t) + \frac{b(t)}{\pi i} \int_{\mathbb{T}} \frac{\varphi(\tau)}{\tau - t}\, d\tau. \tag{1}$$

Let U, U^{-1}, P^+ be the operators defined on $\mathbf{L}^p(\mathbb{T}, \varrho)$ (cf. Section 4.92). Then $C \in \mathrm{alg}_{N\times N}(U, U^{-1}, P^+)$ and C can be represented in the form

$$C = (a + b)\, P^+ + (a - b)\, P^-.$$

4.120. Theorem. *The singular integral operator* (1) *is a* Φ_+- *or a* Φ_--*operator if and only if*

$$\det\,(a + b)\,(a - b)^{-1}\,(t) \neq 0 \quad \textit{for all} \quad t \in \mathbb{T}. \tag{1}$$

If (1) *is fulfilled, then* C *is a* Φ-*operator the index of which equals*

$$\mathrm{ind}\, C = -\mathrm{wind}\, \det\,(a + b)\,(a - b)^{-1}.$$

Proof. Apply Theorem 4.106 in conjunction with $\mathrm{codim}\,(\mathrm{im}\, P^+UP^+|_{\mathrm{im}\,P^+}) = 1$. ■

4.121. Theorem. *Let* $P_n := I - Q_n$, $Q_n := U^{-n}P^-U^n + U^nP^+U^{-n}$ *and* $T \in \mathcal{K}(\mathbf{L}_N^p(\mathbb{T}, \varrho))$. *Then the sequence* $\{P_n(C + T)\, P_n\}$ *(*C *defined by* 4.119(1)*) is stable on* $\mathbf{L}_N^p(\mathbb{T}, \varrho)$ *if and only if*

(a) $C + T$ *is invertible on* $\mathbf{L}_N^p(\mathbb{T}, \varrho)$,

(b) *the operator* $C' := P^+(a + b)^{-1}\, P^+ + P^-(a - b)^{-1}\, P^-$ *is invertible on* $\mathbf{L}_N^p(\mathbb{T}, \varrho)$.

Proof. Apply Theorem 4.107, Corollary 4.94, and see the proof of Theorem 4.95. ■

Clearly, Theorem 4.96 can also be proved in the system case. Since no new ideas are needed we omit the details.

Singular integral operators on the interval (−1, 1)

Our next concern is to demonstrate how the theory pointed out in this chapter is connected with the theory of singular integral operators acting on the space $\mathbf{L}^2(w)$.

4.122. Definitions. Let $\mathbf{L}^2(w)$ be the space of all measurable functions on $(-1, 1)$ such that

$$\|\varphi\|_{\mathbf{L}^2(w)} := \left(\int_{-1}^{1} |\varphi(z)|^2\, w(z)\, dz \right)^{1/2} < \infty,$$

where $w(z) := (1 - z)^{\alpha_1}\,(z + 1)^{\alpha_2}$ with $-1 < \alpha_1, \alpha_2 < 1$.

Write $\mathbf{L}^2$ instead $\mathbf{L}^2(w)$ if $w(z) \equiv 1$. It is well-known (see GOHBERG/KRUPNIK [4]) that the singular integral operator defined by

$$(S\varphi)(x) := \frac{1}{\pi \mathrm{i}} \int_{-1}^{1} \frac{\varphi(z)}{z - x} \, \mathrm{d}z$$

is bounded on $\mathbf{L}^2(w)$ (the integral being understood in the sense of Cauchy's principal value).

We introduce the Chebychev polynomials $\{T_n(x)\}_{n=0}^{\infty}$ and $\{U_n(x)\}_{n=0}^{\infty}$. For each $n \geqq 0$,

$$T_n(x) = \frac{\left(x + \mathrm{i}\sqrt{1 - x^2}\right)^n + \left(x - \mathrm{i}\sqrt{1 - x^2}\right)^n}{2} \quad \left(T_n(\cos\theta) = \cos n\theta\right),$$

$$U_n(x) = \frac{\left(x + \mathrm{i}\sqrt{1 - x^2}\right)^{n+1} - \left(x - \mathrm{i}\sqrt{1 - x^2}\right)^{n+1}}{2\mathrm{i}\sqrt{1 - x^2}}$$

$$\left(U_n(\cos\theta) = \frac{\sin(n + 1)\theta}{\sin\theta}\right).$$

These polynomials are special kinds of Jacobi polynomials. They are (in the terminology of Sect. 9.15, 2°) associated to the coefficients $a = 0$, $b = 1$ and the parameters $\alpha = \beta = -1/2$ and $\alpha = \beta = 1/2$, respectively.

Notice the identities

$$U_{n+2}(x) = 2xU_{n+1}(x) - U_n(x), \qquad n \geqq 0, \tag{1}$$

$$T_{n+1}(x) = \frac{1}{2}\left(U_{n+1}(x) - U_{n-1}(x)\right), \qquad n \geqq 1. \tag{2}$$

4.123. Theorem (ROSENBLUM/ROVNYAK [1]). *There is a unique shift operator V on $\mathbf{L}^2$ such that $V\left((1 - x^2)^{1/4} U_n(x)\right) = (1 - x^2)^{1/4} U_{n+1}(x)$, $n \geqq 0$. For each $f \in \mathbf{L}^2$, the action of V on f is given by*

$$f(x) \mapsto xf(x) - \frac{1}{\pi} \int_{-1}^{1} \frac{(1 - x^2)^{1/4} (1 - z^2)^{1/4}}{z - x} f(z) \, \mathrm{d}z. \tag{1}$$

Proof. The existence of V follows from the fact that the functions $\{(2/\pi)^{1/2} (1 - x^2)^{1/4} \times U_n(x)\}_{n=0}^{\infty}$ form an orthonormal basis for $\mathbf{L}^2$. Using 4.122(1), (2), and the identity

$$1/\pi \int_{-1}^{1} \frac{(1 - z^2)^{1/2} U_n(z)}{z - x} \, \mathrm{d}z = -T_{n+1}(x), \qquad |x| < 1$$

(see Section 9.15, 2°), one can check (1) for the basis elements and then obtain the general case by linearity and approximation. ■

Put $V_{-1} := V^*$. Since codim im $V = 1$ and V is an isometry, Corollary 4.15, Theorem 4.20, and Remark 4.25, 4° are in force. So we get

4.124. Theorem. (i) $\mathcal{K}(\mathbf{L}^2) \subset \operatorname{alg}(V, V_{-1})$.

(ii) *$A \in \operatorname{alg}(V, V_{-1})$ is a Φ-operator if and only if* $(\operatorname{smb} A)(t) \neq 0$ *for all* $t \in \mathbb{T}$. *If this condition is fulfilled then*

$$\operatorname{ind} A = -\operatorname{wind}(\operatorname{smb} A).$$

(iii) $\operatorname{alg}(V, V_{-1})/\mathcal{K}(\mathbf{L}^2)$ *is isometrically isomorphic to* $\mathbf{C}(\mathbb{T})$.

4.125. Proposition. (i) *For each continuous function* $a \in \mathbf{C}[-1, 1]$, *the operator of multiplication* aI *belongs to* $\operatorname{alg}(V, V_{-1})$, *and*

$$\operatorname{smb}(xI) = 1/2(t + t^{-1}) = x \qquad (t = x + iy \in \mathbb{T}).$$

(ii) *Let* $b \in \mathbf{C}[-1, 1]$ *be a continuous function such that* $b(-1) = b(1) = 0$. *Then* $bw^{-1}Sw \in \operatorname{alg}(V, V_{-1})$, *$w$ being any weight as in Sec. 4.122. In particular,* $bS \in \operatorname{alg}(V, V_{-1})$, *and* $\operatorname{smb} bS = \operatorname{smb} bw^{-1}Sw$.

Proof. (i): We compute the matrix representation of $J(xI)J^{-1}$ (see Sec. 4.6). Using 4.122(1) we immediately obtain the matrix

$$\begin{pmatrix} a_0 & 1/2 & 0 & \cdot & \cdot & \cdot & \cdot \\ a_1 & 0 & 1/2 & 0 & \cdot & \cdot & \cdot \\ a_2 & 1/2 & 0 & 1/2 & 0 & \cdot & \cdot \\ a_3 & 0 & 1/2 & 0 & 1/2 & 0 & \cdot \\ \cdot & \cdot & \cdot & \cdot & \cdot & \cdot & \cdot \end{pmatrix}.$$

Hence, $xI = 1/2(V + V_{-1}) + T$, $T \in \mathcal{K}(\mathbf{L}^2)$, and it remains to apply Weierstrass' approximation theorem.

(ii): Put $w_0(x) = (1 - x)^{1/4}$. Then $xI - V = i(1 - x^2)^{1/2} w_0^{-1}Sw_0$. According to Sec. 1.14 we have $bw_0^{-1}Sw_0 \in \operatorname{alg}(V, V_{-1})$ for each $b \in \mathbf{C}[-1, 1]$ with $b(-1) = b(1) = 0$. Finally, if $b \in \mathbf{C}[-1, 1]$ is such that $b(-1) = b(1) = 0$, then $b(w_0^{-1}Sw_0 - w^{-1}Sw) \in \mathcal{K}(\mathbf{L}^2)$, since, for $a \in \mathbf{C}[-1, 1]$, $aS - SaI \in \mathcal{K}(\mathbf{L}^2)$ (see GOHBERG/KRUPNIK [1]). ■

Obviously, $\operatorname{alg}(V, V_{-1}) \subset \operatorname{alg}\{S, \mathbf{C}[-1, 1]\}$. Moreover, $\operatorname{alg}(V, V_{-1})$ and $\operatorname{alg}\{S\}$ generate $\operatorname{alg}\{S, \mathbf{C}[-1, 1]\}$. Let A_1 and A_2 denote the image of $\operatorname{alg}(V, V_{-1})$ and $\operatorname{alg}\{S\}$ under the canonical homomorphism $\pi\colon \operatorname{alg}\{S, \mathbf{C}[-1, 1]\} \to \operatorname{alg}\{S, \mathbf{C}[-1, 1]\}/\mathcal{K}(\mathbf{L}^2) =: B$, respectively. The algebras A_1 and A_2 are C^*-subalgebras of the commutative C^*-algebra B (recall that $aS - SaI \in \mathcal{K}(\mathbf{L}^2)$ for every $a \in \mathbf{C}[-1, 1]$). By Theorem 4.124(iii), the algebra A_1 is isometrically isomorphic to $\mathbf{C}(\mathbb{T})$. Further, A_2 is isometrically isomorphic to $\mathbf{C}[-1, 1]$. This can be proved using the following argument. Since $S^* = S$, the algebra A_2 turns out to be a C^*-algebra and by Sec. 1.11 it remains to prove that $\operatorname{sp}(\pi(S)) = [-1, 1]$. This can be done be means of the Hartman-Wintner theorem (however, with respect to the real line, see BÖTTCHER/SILBERMANN [3], Sec. 2.36, and GOHBERG/KRUPNIK [4], Chap. 1, § 5) and a simple algebraic argument (see RUDIN [1], Exerc. 10.2). Since B is generated by A_1 and A_2, the maximal ideal space $\mathbf{M}(B)$ can be identified with a compact subset of $\mathbf{M}(A_1) \times \mathbf{M}(A_2)$. Indeed, consider the mapping

$$\tau_i\colon \mathbf{M}(B) \to \mathbf{M}(A_i), \qquad \alpha \mapsto \alpha|_{A_i},$$

which assigns to each functional $\alpha \in \mathbf{M}(B)$ its restriction to A_i $(i = 1, 2)$. It is not hard to see that τ_i is continuous $(i = 1, 2)$. Define

$$\tau\colon \mathbf{M}(B) \to \mathbf{M}(A_1) \times \mathbf{M}(A_2), \qquad \tau(\alpha) = (\tau_1(\alpha), \tau_2(\alpha)). \tag{1}$$

Since τ is continuous and one-to-one, it follows that τ is a homeomorphism of $\mathbf{M}(B)$ onto a compact subset M_1 of $\mathbf{M}(A_1) \times \mathbf{M}(A_2)$. Our next aim is to describe M_1.

Let $\varphi \in \mathbf{M}(B)$ be any functional the restriction of which on A_1 coincides with the functional $\varphi_s \in \mathbf{M}(A_1)$ given by $f \mapsto f(s)$, $f \in \mathbf{C}(\mathbb{T})$. Assume $s \notin \{-1, 1\}$. Then $\varphi\big(\pi(\mathrm{i}(1 - x^2)^{1/2} S)\big) = \varphi_s\big(\pi(\mathrm{i}(1 - x^2)^{1/2})\big)\, \varphi\big(\pi(S)\big)$. Since $\pi\big(\mathrm{i}(1 - x^2)^{1/2} S\big) \in A_1$ (by Proposition 4.125(ii)), we get (again by Proposition 4.125(ii)),

$$\begin{aligned}\varphi\big(\pi(\mathrm{i}(1 - x^2)^{1/2} S)\big) &= \varphi\big(\pi(\mathrm{i}(1 - x^2)^{1/2} w_0^{-1} S w_0)\big)\\ &= \varphi\big(\pi(xI - V)\big) = 1/2(s^{-1} - s).\end{aligned}$$

On the other hand,

$$\begin{aligned}\pi\big(\mathrm{i}(1 - x^2)^{1/2} I\big) &= \operatorname{smb}\big(\mathrm{i}(1 - x^2)^{1/2} I\big)\\ &= \mathrm{i}\big(1 - 1/4(t + t^{-1})^2\big)^{1/2} = \big((1/2t - 1/2t^{-1})^2\big)^{1/2}.\end{aligned}$$

Putting $s = x + \mathrm{i}y$, we therefore get

$$\varphi\big(\pi(S)\big) = \frac{\varphi\big(\pi(\mathrm{i}(1 - x^2)^{1/2} w_0^{-1} S w_0)\big)}{\varphi\big(\pi(\mathrm{i}(1 - x^2)^{1/2} I)\big)} = \frac{y}{|y|}$$

and

$$(\{s\} \times [-1, 1]) \cap M_1 = \begin{cases} (s, 1), & \text{if } \operatorname{Im} s > 0,\\ (s, -1), & \text{if } \operatorname{Im} s < 0.\end{cases} \tag{2}$$

Consider $s \in \{-1, 1\}$. We claim that $\{s\} \times [-1, 1] \subset M_1$. For this aim introduce the isomorphism ψ of $\operatorname{alg}\{S, \mathbf{C}[-1, 1]\}$ given by $A \mapsto U^{-1}AU$, $(Uf)(x) = f(-x)$. Notice that $U^{-1}SU = -S$ and $U^{-1}aIU = a(-x)I$. This isomorphism induces an isomorphism ψ^π on B and a homeomorphism on $\mathbf{M}(B)$. This shows that the sets $N_{-1} := (\{-1\} \times [-1, 1]) \cap M_1$ and $N_1 := (\{1\} \times [-1, 1]) \cap M_1$ are homeomorphic to each other. Moreover, if $N_{-1} \cap M_1 \neq N_{-1}$, then also $N_1 \cap M_1 \neq N_1$, and M_1 is not connected by (2). If M_1 is not connected then there exists an element $\chi \in B$ such that $\chi \neq 0, 1$, and $\chi^2 = 1$. Using (2) it is easy to see that $\chi \in A_2$. This is impossible since $\mathbf{M}(A_2) = [-1, 1]$, hence $N_{-1}, N_1 \subset M_1$. Thus, $M_1 \subset \mathbb{T} \times [-1, 1]$ can be identified with the curve $\Lambda \subset \mathbb{R}^3(x, y, z)$ consisting of $\{\mathbb{T}^+\} \times 1$, $\mathbb{T}^- \times \{-1\}$, $\{(-1, 0)\} \times [-1, 1]$, $\{(1, 0)\} \times [-1, 1]$, where $\mathbb{T}^+ = \{x + \mathrm{i}y \in \mathbb{T} : y \geqq 0\}$, $\mathbb{T}^- = \{x + \mathrm{i}y \in \mathbb{T} : y \leqq 0\}$. The orientation of Λ is chosen so that it coincides with the orientation of $\mathbb{T}^+$ in the plane $z = 1$.

Now put

$$\operatorname{smb} xI = x, \qquad (x, y, z) \in \Lambda,$$

$$\operatorname{smb} S = z, \qquad (x, y, z) \in \Lambda.$$

4.126. Theorem. (i) *The algebra* $\operatorname{alg}\{S, \mathbf{C}[-1, 1]\}/\mathcal{K}(\mathbf{L}^2)$ *is isometrically isomorphic to* $\mathbf{C}(\Lambda)$. *The isomorphism can be chosen so that it coincides on the generating elements* $\pi(xI)$ *and* $\pi(S)$ *with* $\operatorname{smb} xI$ *and* $\operatorname{smb} S$, *respectively. Let us denote the composition of this isomorphism and* π *by* smb.

(ii) *An element* $A \in \operatorname{alg}\{S, \mathbf{C}[-1, 1]\}$ *is Fredholm if and only if*

$$(\operatorname{smb} A)(s) \neq 0 \quad \textit{for all} \quad s \in \Lambda. \tag{1}$$

If (1) *is fulfilled then*

$$\text{ind } A = -\text{wind smb } A. \tag{2}$$

Proof. It remains to prove (2). By Theorem 4.124 the assertion is true for $A \in \text{alg}(V, V_{-1})$. Since Λ is homeomorphic to $\mathbb{T}$ and two invertible functions $f, g \in \mathbf{C}(\mathbb{T})$ are homotopic if and only if wind f = wind g, we get the analogous assertion for invertible elements in $\mathbf{C}(\Lambda)$. Now it is easy to obtain (2). ■

4.127. Remarks. 1°. Similar results can by proved for the spaces $\mathbf{L}^2(w)$ and in the system case.

2°. For $A \in \text{alg}(V, V_{-1})$ the projection method $\{P_nAP_n\}$ can be studied, where $P_n = I - V_nV_{-n}$. In particular, for $a \in \mathbf{C}[-1, 1]$, the sequence $\{P_naIP_n\}$ is stable if and only if $a(x) \neq 0$ for all $x \in [-1, 1]$ (use the results of Sec. 4.45).

Notes and comments

4.1.—4.26. Functions of shifts (in our terminology) were investigated by Gohberg [1] and Gohberg/Feldman [1]. We mainly follow the papers Roch/Silbermann [2], [9], here further contributions are made. The example in Section 4.3 is due to A. Pomp (private communication).

4.27.—4.28. It is a delicate problem to say who made such observations for the first time. The essence of 4.27, except for formula 4.27(1) and 4.27(b), is already contained in Devinatz/Shinbrot [1]. We learned formula 4.27(1) from A. V. Kozak (private communication). Other applications of 4.27(1) can be found in Böttcher/Silbermann [3, Chapt. 10].

4.29.—4.45. Roch/Silbermann [10]. The proof of Theorem 4.39 is borrowed from Kozak/Simonenko [1] (see also Böttcher/Silbermann [3], Prop. 8.56).

4.46.—4.52. A main part of these results is well-known (Gohberg/Feldman [1], Douglas [1], Böttcher/Silbermann [3]), the presentation of the material, however, is in the spirit of Roch/Silbermann [2], [9].

4.53.—4.61. Wiener-Hopf integral operators with continuous symbols have been the subject of deep investigations by many people. We only refer to the fundamental work of M. G. Krein [1] and Gohberg/Feldman [1]. The finite section method was studied in Gohberg/Feldman [1], Böttcher/Silbermann [3].

4.62.—4.64. Anselone/Sloan [1], de Hoog/Sloan [1], Silbermann [3].

4.65.—4.80. Abstract paired operators were studied by Gohberg and Feldman [1]. We here present a new variant of their approach, which is based on Roch/Silbermann [9] and is published here for the first time.

4.81.—4.121. All concrete classes of operators considered in these sections have been subject of numerous investigations (see Gohberg/Feldman [1]). In comparision with Gohberg/Feldman [1], our presentation of the matter involves some simplifications and also a few supplements.

4.122.—4.127. The algebra alg $\{S, \mathbf{C}[-1, 1]\}/\mathcal{K}(\mathbf{L}^2)$ was studied by Gohberg and Krupnik (see Gohberg/Krupnik [3]). The approach developed here is due to Silbermann (unpublished).

Chapter 5
Spline approximation methods for classes of convolution equations

5.0. Introduction. We consider the solution of the Mellin convolution equation

$$u(x) - \int_0^1 k(x/y)\, u(y)\, y^{-1}\, \mathrm{d}y = f(x), \qquad x \in [0, 1] \tag{1}$$

and the Wiener-Hopf integral equation

$$u(x) - \int_0^\infty \varkappa(x - y)\, u(y)\, \mathrm{d}y = f(x), \qquad x \in [0, \infty) \tag{2}$$

by spline approximation methods. Here f, k and $\varkappa$ are given functions, and u is the unknown function. Note that equations (1) and (2) are related by a simple transformation of the interval $[0, 1]$ to the half-axis (cf. Section 5.1). Such equations arise in a variety of applications; for example, (2) occurs when boundary integral equation methods are applied to potential problems in a plane region with corners. In particular, when the double layer potential is used to solve the Dirichlet problem and the domain contains a corner with an interior angle $(1 - \lambda)\,\pi$, $\lambda \in (-1, 1)$, the kernel function

$$k(x) = \frac{\sin \lambda\pi}{\pi}\, \frac{x}{1 + 2x \cos \lambda\pi + x^2} \tag{3}$$

appears.

In this chapter we present a unified approach to convergence proofs for Galerkin and collocation methods based on piecewise polynomials together with their iterated and discrete versions. This approach, which relies heavily on the Wiener-Hopf factorization, yields sharp stability results for the numerical methods under consideration. It also allows us to prove optimal orders of convergence which are standard for second kind equations with smooth kernels or weakly singular kernels on finite intervals. Note that the integral operators in (1) and (2) are not compact, in general, so that standard theories for the numerical analysis of second kind Fredholm integral equations cannot be applied.

Finally, we remark that the results of the present chapter extend easily to the situation when an integral operator is added in (1) or (2) whose kernel possesses suitable smoothness and growth properties (see Chapter 3).

A class of non-compact integral operators

5.1. Spaces and operators. To derive some analytical results about equations 5.0(1) and 5.0(2), we first introduce weighted Sobolev spaces. For any interval $J \subset \mathbb{R}$ and $1 \leq p \leq \infty$, $\mathbf{L}^p(J)$ will denote the usual Lebesgue space on J with norm

$$\begin{aligned} \|u\|_{p,J} &= \Big\{ \int_J |u|^p \, \mathrm{d}x \Big\}^{1/p}, \qquad p < \infty; \\ \|u\|_{\infty,J} &= \operatorname*{ess\,sup}_{x \in J} |u(x)| . \end{aligned} \tag{1}$$

Let $\mathbb{R}^+ = [0, \infty)$, and let $X^p(\mathbb{R}^+)$ denote the space $\mathbf{L}^p(0, \infty)$ if $p < \infty$ and anyone of the spaces $\mathbf{L}^\infty(\mathbb{R}^+)$, $\mathbf{C}(\mathbb{R}^+)$, $\mathbf{C}(\overline{\mathbb{R}}^+)$, $\mathring{\mathbf{C}}(\overline{\mathbb{R}}^+)$ if $p = \infty$. Here $\mathbf{C}(\mathbb{R}^+)$ is the space of bounded continuous functions on $\mathbb{R}^+$ provided with the uniform norm, $\mathbf{C}(\overline{\mathbb{R}}^+)$ refers to its subspace of functions which have well-defined limits at infinity, and $\mathring{\mathbf{C}}(\overline{\mathbb{R}}^+) \subset \mathbf{C}(\overline{\mathbb{R}}^+)$ stands for the subspace of functions for which those limits are zero. For $l \in \mathbb{N}$, $\varrho \in \mathbb{R}$, we now define the weighted spaces

$$X^{p,l}_\varrho(\mathbb{R}^+) = \{u \in \mathrm{D}'(0, \infty) \colon \mathrm{e}^{\varrho x} \mathrm{D}^j u \in X^p(\mathbb{R}^+),\ j = 0, \ldots, l\}$$

equipped with the canonical norm

$$\|u\|_{p,l,\varrho,\mathbb{R}^+} = \sum_{0 \leq j \leq l} \|\mathrm{e}^{\varrho x} \mathrm{D}^j u\|_{p,\mathbb{R}^+}, \qquad \mathrm{D} = \mathrm{d}/\mathrm{d}x. \tag{2}$$

For a finite interval $J = (\alpha, \beta)$, let $X^p(J)$ denote the space $\mathbf{L}^p(J)$ if $p < \infty$ and anyone of the spaces $\mathbf{L}^\infty(J)$, $\mathbf{C}(\alpha, \beta]$, $\mathbf{C}[\alpha, \beta]$ and $\mathring{\mathbf{C}}[\alpha, \beta]$ provided with the uniform norm if $p = \infty$, where $\mathbf{C}(\alpha, \beta]$ ($\mathbf{C}[\alpha, \beta]$) refers to the space of functions which are bounded and continuous on $(\alpha, \beta]$ (continuous on $[\alpha, \beta]$) and $\mathring{\mathbf{C}}[\alpha, \beta] = \{u \in \mathbf{C}[\alpha, \beta] \colon u(\alpha) = 0\}$. Then we define the weighted Sobolev spaces

$$X^{p,l}_\varrho(J) = \{u \in \mathrm{D}'(J) \colon x^{j-\varrho} \mathrm{D}^j u \in X^p(J), j = 0, \ldots, l\}$$

endowed with the norm

$$\|u\|_{p,l,\varrho,J} = \sum_{0 \leq j \leq l} \|x^{j-\varrho} \mathrm{D}^j u\|_{p,J} . \tag{3}$$

In the sequel we use the convention that $X^{p,l}_0 = X^{p,l}$, $X^{p,0}_\varrho = X^p_\varrho$, and we drop the ϱ from the notation in (2), (3) if $\varrho = 0$. The notation $\mathbf{C}^l_\varrho$ refers to the case $X^\infty = \mathbf{C}$ etc. Note that $\mathbf{L}^{p,l}_\varrho(J)$ coincides with the usual Sobolev space $\mathbf{W}^{p,l}(J)$ when $J = \mathbb{R}^+$ and $\varrho = 0$ or $J = (\alpha, \beta)$, $0 < \alpha < \beta < \infty$ and $\varrho \in \mathbb{R}$. For the Wiener-Hopf integral operator A defined by

$$A = I - W, \qquad Wu(x) = \int_0^\infty \varkappa(x - y)\, u(y)\, \mathrm{d}y, \tag{4}$$

where I is the identity and $\varkappa$ a measurable function on $\mathbb{R}$, we make the following assumptions throughout this chapter:

$$\int_{-\infty}^{\infty} |\varkappa(x)| \, \mathrm{d}x < \infty, \tag{A1}$$

$$1 - a(\xi) \neq 0, \qquad \xi \in \mathbb{R}; \qquad \{\arg(1 - a)\}_{-\infty}^{\infty} = 0 . \tag{A2}$$

Here a is the Fourier transform of the kernel of W, i.e.

$$a(\xi) = \int_{-\infty}^{\infty} \mathrm{e}^{\mathrm{i}x\xi} \varkappa(x)\,\mathrm{d}x, \tag{5}$$

and $\{\arg\}_{-\infty}^{\infty}$ denotes the variation of the argument when ξ runs from $-\infty$ to ∞. The function $1 - a$ (resp. a) is called the symbol of A (resp. W). Note that (A1) implies that $a(\xi)$ is a continuous function on $\mathbb{R}$ vanishing at infinity. Furthermore, (A1) ensures that W is bounded on each of the spaces $X^p(\mathbb{R}^+)$, and (A2) is then equivalent to the invertibility of A in $X^p(\mathbb{R}^+)$ (cf. Section 4.54).

For the Mellin (convolution) operator B defined by

$$B = I - K, \qquad Ku(x) = \int_0^1 k(x/y)\, y^{-1} u(y)\,\mathrm{d}y \tag{6}$$

and fixed p, $1 \leqq p \leqq \infty$, we make the set of assumptions

$$\int_0^{\infty} x^{1/p-1}\, |k(x)|\,\mathrm{d}x < \infty, \tag{H1^p}$$

$$1 - b(z) \neq 0, \qquad z \in \Gamma_{1/p}; \qquad \{\arg\,(1 - b(1/p + \mathrm{i}\xi))\}_{-\infty}^{\infty} = 0 \tag{H2^p}$$

where $\Gamma_\alpha := \{z \in \mathbb{C}\colon \operatorname{Re} z = \alpha\}$, $\alpha \in \mathbb{R}$, $1/p := 0$ if $p = \infty$, and $b(z)$ is the Mellin transform of the kernel of K:

$$b(z) = \tilde{k}(z) := \int_0^{\infty} x^{z-1} k(x)\,\mathrm{d}x.$$

Note that $b(z)$ is a continuous function on $\Gamma_{1/p}$ vanishing at infinity if (H1^p) is satisfied. Furthermore, (H1^p) guarantees the boundedness of K on $X^p(0, 1)$, and (H2^p) is then equivalent to the invertibility of B in $X^p(0, 1)$. These assertions are easily checked via the results on Wiener-Hopf operators mentioned above.

Indeed, using the map

$$R_p u(x) := \mathrm{e}^{-x/p} u(\mathrm{e}^{-x}), \tag{7}$$

which is an isomorphism of $X^p(0, 1)$ onto $X^p(\mathbb{R}^+)$, we observe that $R_p B R_p^{-1}$ is a Wiener-Hopf operator with kernel $\varkappa(x) = \mathrm{e}^{-x/p}\, k(\mathrm{e}^{-x})$ and symbol $1 - a(\xi)$ given by

$$\begin{aligned} a(\xi) &= \int_{-\infty}^{\infty} \mathrm{e}^{\mathrm{i}x\xi - x/p} k(\mathrm{e}^{-x})\,\mathrm{d}x = \int_0^{\infty} x^{1/p - \mathrm{i}\xi - 1} k(x)\,\mathrm{d}x \\ &= b(1/p - \mathrm{i}\xi), \qquad \xi \in \mathbb{R}. \end{aligned} \tag{8}$$

For this operator, (A1) and (A2) are obviously equivalent to (H1^p) and (H2^p), respectively.

The function $1 - b(z)$ resp. $b(z)$, $z \in \Gamma_{1/p}$, is called the symbol of the Mellin operator B resp. K (on $X^p(0, 1)$). Note that for $1 < p < \infty$ the operator B can be regarded as the continuous extension of the map

$$C_0^{\infty}(\mathbb{R}^+) \ni u \mapsto \frac{1}{2\pi\mathrm{i}} \int_{\Gamma_{1/p}} x^{-z} [1 - b(z)]\, (\widetilde{\chi u})(z)\,\mathrm{d}z, \tag{9}$$

which is a pseudodifferential operator of Mellin type on the interval [0, 1] (cf. LEWIS/PARENTI [1], where more general operators of this kind are treated). Here χ denotes the characteristic function of the unit interval.

5.2. Remark. In ELSCHNER [8], [5] and LEWIS/PARENTI [1] the following class of symbols was used. Let $m \in \mathbb{R}$ and $\alpha < 1/p < \beta$. The function $b(z)$ belongs to the class $\Sigma^m_{\alpha,\beta}$ if it is analytic in the strip $\Gamma_{\alpha,\beta} = \{z \in \mathbb{C} : \alpha < \operatorname{Re} z < \beta\}$ and satisfies the estimates

$$(\mathrm{d}/\mathrm{d}z)^k\, b(z) = O\big((1 + |z|)^{m-k}\big), \qquad z \in \Gamma_{\gamma,\delta} \tag{1}$$

for all $\alpha < \gamma < \delta < \beta$ and $k \in \mathbb{N}$. Now we observe that $b(z) = \tilde{k}(z) \in \Sigma^m_{\alpha,\beta}$ and $m < -1/2$ imply (H1^p). This is a consequence of the inequalities

$$\begin{aligned}\int_0^\infty x^{1/p-1}\, |k(x)|\, \mathrm{d}x &\leqq \int_0^1 x^{\delta-1/2}\, |x^{1/p-1/2-\delta} k(x)|\, \mathrm{d}x + \int_1^\infty x^{-\delta-1/2}\, |x^{1/p-1/2+\delta} k(x)|\, \mathrm{d}x \\ &\leqq c_\delta \{\|x^{1/p-1/2-\delta} k(x)\|_{2,\mathbb{R}^+} + \|x^{1/p-1/2+\delta} k(x)\|_{2,\mathbb{R}^+}\} \\ &\leqq c_\delta \{\|b\|_{\mathbf{L}^2(\Gamma_{1/p+\delta})} + \|b\|_{\mathbf{L}^2(\Gamma_{1/p-\delta})}\} < \infty,\end{aligned}$$

which hold for sufficiently small $\delta > 0$, since the last two expressions are finite by estimate (1) (for $k = 0$ and $m < -1/2$) and the Mellin transform is an isomorphism of $\mathbf{L}^2_\varrho(\mathbb{R}^+)$ onto $\mathbf{L}^2(\Gamma_{1/2-\varrho})$ for any $\varrho \in \mathbb{R}$.

5.3. Factorization. We first recall some results on the Wiener-Hopf factorization which are crucial in proving the stability of our spline approximation methods. Following Section 4.53, we introduce the operators

$$\begin{aligned}&V_+u(x) = \int_0^x \mathrm{e}^{y-x}\, u(y)\, \mathrm{d}y, \qquad V_-u(x) = \int_x^\infty \mathrm{e}^{x-y}\, u(y)\, \mathrm{d}y, \\ &V = I - 2V_+, \qquad V^{(-1)} = I - 2V_-.\end{aligned} \tag{1}$$

Note that V and $V^{(-1)}$ are Wiener-Hopf operators with kernels $\mathrm{e}^{-x}\lambda_+$ and $\mathrm{e}^{x}(1 - \lambda_+)$ and symbols $(\mathrm{i}\xi + 1)/(\mathrm{i}\xi - 1)$ and $(\mathrm{i}\xi - 1)/(\mathrm{i}\xi + 1)$, respectively, where λ_+ denotes the characteristic function of $\mathbb{R}^+$. Moreover, $V^{(-1)}V = I$, and the kernel $\ker V^{(-1)}$ of $V^{(-1)}$ coincides with the one-dimensional space $\{c\, \mathrm{e}^{-x} : c \in \mathbb{C}\}$ in each of the spaces $X^p(\mathbb{R}^+)$. Let $\mathfrak{L}^p(V)$ be the closure of all polynomials in $V^{(-1)}$ and V with respect to the operator norm on $\mathbf{L}^p(\mathbb{R}^+)$, which is denoted by $\|\cdot\|_p$. Then each Wiener-Hopf operator satisfying (A1) belongs to $\mathfrak{L}^p(V)$ for any p, $1 \leqq p \leqq \infty$.

5.4. Theorem. (cf. Section 4.9). *Assume* (A1), (A2), *and let* $\delta > 0$, $1 \leqq p \leqq \infty$.

(i) *Then the operator A defined in* 5.1(4) *admits the factorization*

$$A = (I - W_-)\,(I - W_0)\,(I - W_+), \tag{1}$$

where $W_0 \in \mathfrak{L}^p(V)$, $\|W_0\|_p < \delta$, $W_\pm$ are finite linear combinations of the operators $V^j_\pm$ $(j = 1, 2, \ldots)$, and the operators $I - W_\pm$ are invertible in $\mathbf{L}^p(\mathbb{R}^+)$.

(ii) *The factorization* (1) *can also be written in the form*

$$A = (I - W_+)\,(I - W_0)\,(I - W_-) + T, \tag{2}$$

where T is compact on $\mathbf{L}^p(\mathbb{R}^+)$. *For* $p = \infty$, *T is even a compact map of* $\mathbf{L}^\infty(\mathbb{R}^+)$ *into* $\mathring{\mathbf{C}}(\overline{\mathbb{R}}^+)$ (*see Section* 4.55).

In the case of Mellin operators, for fixed p we set

$$F_{p,+}u(x) = \int_x^1 (x/y)^{1-1/p}\, y^{-1}u(y)\, dy,$$

$$F_{p,-}u(x) = \int_0^x (x/y)^{-1-1/p}\, y^{-1}u(y)\, dy. \tag{3}$$

Note that $F_{p,\pm} = R_p^{-1} V_\pm R_p$, where $V_\pm$ are the operators defined in 5.3(1) and R_p denotes the isomorphism 5.1(7). Therefore, the following theorem is an immediate consequence of the preceding one. The operator norm on $\mathbf{L}^p(0, 1)$ is denoted again by $\|\cdot\|_p$.

5.5. Theorem. *Let* $\delta > 0$, $1 \leqq p \leqq \infty$, *and assume* (H1^p), (H2^p). *Then the operator B defined by* 5.1(6) *admits the representations*

$$B = (I - K_-)(I - K_0)(I - K_+), \tag{1}$$

$$B = (I - K_+)(I - K_0)(I - K_-) + T, \tag{2}$$

where $\|K_0\|_p < \delta$, $K_\pm$ *are finite linear combinations in* $F_{p,\pm}^j$ ($j = 1, 2, \ldots$), *the operators* $I - K_\pm$ *are invertible in* $\mathbf{L}^p(0, 1)$, *and T is compact on* $\mathbf{L}^p(J)$. *For* $p = \infty$, *T is even a compact map of* $\mathbf{L}^\infty(J)$ *into* $\mathring{\mathbf{C}}[0, 1]$.

We finally collect some simple properties of the operators $W_\pm$ and $K_\pm$ occurring in 5.4(1) and 5.5(1), respectively, which are needed in the sequel.

5.6. Remark. (i) *Let* $h \in (0, \infty)$ *and* $u \in \mathbf{L}^p(\mathbb{R}^+)$. *If* $u = 0$ *on* (h, ∞), *then* $W_-u = 0$ *on* (h, ∞); $u = 0$ *on* $(0, h)$ *implies that* $W_+u = 0$ *on* $(0, h)$. *Moreover,* $W_- \mathrm{e}^{-x} = c\, \mathrm{e}^{-x}$ *with some* $c \in \mathbb{C}$.

(ii) *For any p and l,* $W_\pm$ *are continuous operators of* $\mathbf{L}^{p,l}(\mathbb{R}^+)$ *into* $\mathbf{L}^{p,l+1}(\mathbb{R}^+)$.

Proof. It is sufficient to prove this for the elementary operators $V_\pm$ (cf. 5.3(1)) instead of $W_\pm$. Then assertion (i) is obvious, and (ii) follows by successive application of the relations

$$\mathrm{D}V_\pm u = \pm(I - V_\pm)\, u, \qquad u \in \mathbf{L}^p(\mathbb{R}^+). \blacksquare$$

5.7. Remark. (i) *Let* $h \in (0, 1)$, $1 \leqq p \leqq \infty$ *and* $u \in \mathbf{L}^p(0, 1)$. *If* $u = 0$ *on* $(0, h)$, *then* $K_-u = 0$ *on* $(0, h)$; *for* $u = 0$ *on* $(h, 1)$, *we have* $K_+u = 0$ *on* $(h, 1)$. *Furthermore,* $u = x^\mu$, $\mu \geqq 0$, *implies that* $K_-u = cx^\mu$ *with some constant c.*

(ii) *For any l, the operators* $K_\pm$ *map* $\mathbf{L}^{p,l}(0, 1)$ *continuously into* $\mathbf{L}^{p,l+1}(0, 1)$.

Proof. It suffices again to consider the elementary operators defined in 5.4(3). Then (i) is obvious, and (ii) is a consequence of the relations

$$x\mathrm{D}F_{p,\pm}u = \pm(F_{p,\pm} - I)\, u - p^{-1}F_{p,\pm}u, \qquad u \in \mathbf{L}^p(0, 1). \blacksquare$$

Smoothness of solutions

5.8. Assumptions. Let W be the operator defined in 5.1(4) with kernel $\varkappa$ and symbol a, and let $A = I - W$. To prove a regularity result for solutions to equation 0(2), we need stronger assumptions on $\varkappa$ and a. For $l \in \mathbb{N}$ and $\varrho \geqq 0$, we introduce the conditions

$$\int_{-\infty}^{\infty} e^{\varrho x} |D^j \varkappa| \, dx < \infty, \qquad j = 0, \dots, l; \tag{A1$^l_\varrho$}$$

$$1 - a(\xi - i\varrho) \neq 0, \qquad \xi \in \mathbb{R}; \qquad \{\arg(1 - a(\xi - i\varrho))\}_{-\infty}^{\infty} = 0, \tag{A2$_\varrho$}$$

where the derivatives involved in (A1$^l_\varrho$) are assumed to exist in the sense of distributions on $\mathbb{R}$ and the l or ϱ is dropped from the notation if $l = 0$ or $\varrho = 0$.

Remark. Let $\varrho > 0$. Then (A1) and (A1$_\varrho$) imply that $a(\xi)$ is an analytic function in the strip $\gamma_\varrho = \{\xi \in \mathbb{C}: -\varrho < \operatorname{Im} \xi < 0\}$ which is continuous on $\bar{\gamma}_\varrho$ and vanishes at infinity. This follows from 5.1(5) and well known properties of the Fourier transform. Moreover, if the assumptions (A1), (A1$_\varrho$) and (A2) are in force, then (A2$_\varrho$) is satisfied if and only if the symbol $1 - a$ has no zero in $\bar{\gamma}_\varrho$.

5.9. Theorem. *Assume* $l \geqq 1$, $\varrho \geqq 0$, (A1), (A1$_\varrho$) (A2), (A2$_\varrho$) *and* $\varkappa \in X^{p,l-1}_\varrho(\mathbb{R}^+)$. *Then* $u \in L^p(\mathbb{R}^+)$ *and* $Au \in X^{p,l}_\varrho(\mathbb{R}^+)$ *imply* $u \in X^{p,l}_\varrho(\mathbb{R}^+)$. *Furthermore, if* $u \in L^\infty(\mathbb{R}^+)$ *and* $Au \in X^{\infty,l}_\varrho(\mathbb{R}^+) \dotplus \mathbb{C}$, $\varrho > 0$, *then* $u \in X^{\infty,l}_\varrho(\mathbb{R}^+) \dotplus \mathbb{C}$, $\dotplus$ *denoting the direct topological sum.*

Proof. Consider the Wiener-Hopf operator $A_\varrho = e^{\varrho x} A \, e^{-\varrho x}$ with kernel $e^{\varrho x}\varkappa(x)$ and symbol $1 - a(\xi - i\varrho)$, $\xi \in \mathbb{R}$. It follows from (A1$_\varrho$) and (A2$_\varrho$) that A_ϱ is an invertible operator on $X^p(\mathbb{R}^+)$. Note that the multiplication operator $e^{-\varrho x}$ is an isomorphism of $X^{p,m}(\mathbb{R}^+)$ onto $X^{p,m}_\varrho(\mathbb{R}^+)$ for any $m \in \mathbb{N}$. Thus A is invertible on $X^p_\varrho(\mathbb{R}^+)$.

Next we prove that A is also an isomorphism of $X^{p,l}_\varrho(\mathbb{R}^+)$ onto itself. In order to do so, we first verify the relation

$$DWu(x) = WDu(x) + \varkappa(x)\, u(0), \qquad u \in X^{p,1}(\mathbb{R}^+). \tag{1}$$

This follows easily by differentiating the integral

$$Wu(x) = \int_0^\infty \varkappa(x - y)\, u(y)\, dy = \int_{-\infty}^x \varkappa(z)\, u(x - z)\, dz$$

with respect to x. Applying successively (1) and the last assumption, one obtains that W is a continuous map of $X^{p,l}_\varrho(\mathbb{R}^+)$ onto itself and that, for any $c \in \mathbb{C}$, the equality

$$(D + c)^l Au = A(D + c)^l u + T_l u, \qquad u \in X^{p,l}(\mathbb{R}^+) \tag{2}$$

holds with some finite dimensional operator T_l of $X^{p,l}(\mathbb{R}^+)$ into $X^p_\varrho(\mathbb{R}^+)$. We now observe that, for $c < \varrho$, $(D + c)^l: X^{p,l}_\varrho(\mathbb{R}^+) \to X^p_\varrho(\mathbb{R}^+)$ is an isomorphism since $e^{\varrho x}(D + c)^l \times e^{-\varrho x} = (D - \varrho + c)^l$ is an isomorphic map of $X^{p,l}(\mathbb{R}^+)$ onto $X^p(\mathbb{R}^+)$ if $\varrho - c > 0$ (cf. e.g. Elschner [2, Chap. 3]). Therefore, it follows from (2) and the invertibility of A in $X^p_\varrho(\mathbb{R}^+)$ that A is a Fredholm operator with index 0 in $X^{p,l}_\varrho(\mathbb{R}^+)$ and hence it is invertible there. Together with the invertibility of A on $L^p(\mathbb{R}^+)$, which is a consequence of (A1) and (A2), we obtain the first assertion of the theorem. The second one is derived analogously: By using (2) for $c = 0$ and the fact that D^l is a surjective operator of $X_\varrho^{\infty,l}(\mathbb{R}^+) \dotplus \mathbb{C}$ onto $X_\varrho^{\infty}(\mathbb{R}^+)$ with kernel $\mathbb{C}$, it follows that $A(\mathbb{C}) \subset X_\varrho^{\infty,l}(\mathbb{R}^+) \dotplus \mathbb{C}$ and that A is a Fredholm operator with index 0 on $X_\varrho^{\infty,l}(\mathbb{R}^+) \dotplus \mathbb{C}$, and as $X_\varrho^{\infty,l}(\mathbb{R}^+) \dotplus \mathbb{C} \subset \mathbf{L}^\infty(\mathbb{R}^+)$, its kernel is trivial. ∎

5.10. Remark. If assumption $(\mathrm{A}1_\varrho^l)$ is fulfilled, then W is a continuous operator of $X_\varrho^p(\mathbb{R}^+)$ into $X_\varrho^{p,l}(\mathbb{R}^+)$. For $l = 0$, this follows by passing to the operator A_ϱ defined in the above proof. If $l \geqq 1$, we observe that $\mathrm{D}^l W$ is a convolution operator (on $\mathbb{R}^+$) with the kernel $\mathrm{D}^l \varkappa$, hence the assertion is reduced to the case $l = 0$.

Furthermore, $(\mathrm{A}1_\varrho^l)$, $l \geqq 1$, implies that $\varkappa \in \mathbf{L}_\varrho^{p,l-1}(\mathbb{R}^+)$ for any $p \in [1, \infty)$ and any $\varkappa \in \mathring{\mathbf{C}}^{l-1}(\mathbb{R}^+)$. This is a consequence of 5.9(1) if we choose an element $u \in \mathbf{C}_0^\infty(\mathbb{R}^+)$ such that $u(0) \neq 0$.

We now study smoothness of solutions to equation 5.0(1). Let K be the operator defined in 5.1(6) with kernel function $k(x)$ and symbol b, and let $B = I - K$. We introduce the hypotheses

$$\int_0^\infty x^{1/p-1-\varrho} \, |x^j \mathrm{D}^j k(x)| \, \mathrm{d}x < \infty, \qquad j = 0, \ldots, l; \tag{$\mathrm{H}1_\varrho^{p,l}$}$$

$$\left.\begin{aligned} &1 - b(z) \neq 0, \qquad z \in \Gamma_{1/p-\varrho};\\ &\{\arg\left(1 - b(1/p - \varrho + \mathrm{i}\xi)\right)\}_{-\infty}^{\infty} = 0, \end{aligned}\right. \tag{$\mathrm{H}2_\varrho^p$}$$

where $1 \leqq p \leqq \infty$, $\varrho \geqq 0$, $l \in \mathbb{N}$, the derivatives in $(\mathrm{H}1_\varrho^{p,l})$ are supposed to exist in the sense of distributions on $(0, 1)$ and the notation is abbreviated by dropping the l or ϱ when $l = 0$ or $\varrho = 0$.

5.11. Remark. Let $\varrho > 0$. Then $(\mathrm{H}1^p)$ and $(\mathrm{H}1_\varrho^p)$ imply that $b(z)$ is an analytic function in the strip $\Gamma_{1/p-\varrho,1/p}$ which is continuous on $\overline{\Gamma}_{1/p-\varrho,1/p}$ and vanishes at infinity. Furthermore, under the assumptions $(\mathrm{H}1^p)$, $(\mathrm{H}2^p)$, $(\mathrm{H}1_\varrho^p)$, the assumption $(\mathrm{H}2_\varrho^p)$ is equivalent to $1 - b(z) \neq 0$, $z \in \overline{\Gamma}_{1/p-\varrho,1/p}$. This follows immediately from Remark 5.8 and 5.1(8).

Let $\mathbb{P}_d$ denote the space of polynomials of degree $\leqq d$, and set $\mathbb{P}_d = \{0\}$ if $d < 0$. The following assertion is the analogue of Theorem 5.9 for equation 5.0(1).

5.12. Theorem. *Let* $l \geqq 1$, $\varrho \geqq 0$, *and let* k *be an integer such that* $k < \min\{l, \varrho - 1/p\}$. *Furthermore, assume* $(\mathrm{H}1^p)$, $(\mathrm{H}1_\varrho^p)$, $(\mathrm{H}2^p)$, $(\mathrm{H}2_\varrho^p)$, *and* $k(x) \in X_\varrho^{p,l-1}(0, 1)$. *Then* $u \in \mathbf{L}^p(0, 1)$ *and* $Bu \in X_\varrho^{p,l}(0, 1) \dotplus \mathbb{P}_k$ *imply that* $u \in X_\varrho^{p,l}(0, 1) \dotplus \mathbb{P}_k$.

Proof. First we observe that the Mellin operator $B_\varrho = x^{-\varrho} B x^\varrho$ has the kernel function $x^{-\varrho} k(x)$ and the symbol $1 - b(z - \varrho)$, $z \in \Gamma_{1/p}$ (on $X^p(0, 1)$). Moreover, the multiplication operator x^ϱ is an isomorphism of $X^{p,m}(0, 1)$ onto $X_\varrho^{p,m}(0, 1)$ for any $m \in \mathbb{N}$. The operator B_ϱ is invertible on $X^p(0, 1)$ by $(\mathrm{H}1_\varrho^p)$ and $(\mathrm{H}2_\varrho^p)$, hence B is invertible on $X_\varrho^p(0, 1)$.

Next we verify that B is an isomorphic map of $X_\varrho^{p,l}(0, 1)$ onto itself. Together with the invertibility of B on $\mathbf{L}^p(0, 1)$ (by $(\mathrm{H}1^p)$ and $(\mathrm{H}2^p)$), this completes the proof of the theorem in the case $\mathbb{P}_k = \{0\}$. Applying the differential operator $x\mathrm{D}$ to the integral

$$Ku(x) = \int_0^1 k(x/y) \, y^{-1} u(y) \, \mathrm{d}y = \int_x^\infty k(z) \, u(x/z) \, z^{-1} \, \mathrm{d}z,$$

one obtains the relation

$$x\mathrm{D}Ku(x) = Kx\mathrm{D}u(x) - k(x) \, u(1), \qquad u \in X^{p,1}(0, 1). \tag{1}$$

Using successively (1) and the last assumption, we get the continuity of B on $X^{p,l}_\varrho(0, 1)$ and the identity

$$(xD + 2)^l Bu = B(xD + 2)^l u + T_l u, \qquad u \in X^{p,l}(0, 1) \tag{2}$$

with some finite dimensional operator T_l of $X^{p,l}(0, 1)$ into $X^p_\varrho(0, 1)$. Since $(xD + 2)^l$ is an isomorphism of $X^{p,l}_\varrho(0, 1)$ onto $X^p_\varrho(0, 1)$ for $1 \leqq p \leqq \infty$ and $\varrho \geqq 0$ (cf. ELSCHNER [2, Chap. 1]), (2) and the invertibility of B on $X^p_\varrho(0, 1)$ imply that B is a Fredholm operator with index 0 in $X^{p,l}_\varrho(0, 1)$, hence it is invertible there.

Finally, let $k \geqq 0$. The verification of the relation

$$x^{k+1}D^{k+1}Bu = Bx^{k+1}D^{k+1}u + T'_k u, \qquad u \in X^{p,l}(0, 1), \tag{3}$$

T'_k being a compact map of $X^{p,l}(0, 1)$ into $X^{p,l-k-1}_\varrho(0, 1)$, is analogous to that of (2). Note that $x^{k+1}D^{k+1}$ is a surjective operator of $X^{p,l}_\varrho(0, 1) \dotplus \mathbb{P}_k$ onto $X^{p,l-k-1}_\varrho(0, 1)$ with kernel $\mathbb{P}_k$ since $k < \varrho - 1/p$ (cf. ELSCHNER [2, Chap. 1]). Now it follows from (3) that $B(\mathbb{P}_k) \subset X^{p,l}_\varrho(0, 1) \dotplus \mathbb{P}_k$ and that B is a Fredholm operator with index 0 on $X^{p,l}_\varrho(0, 1) \dotplus \mathbb{P}_k$. Because $X^{p,l}_\varrho(0, 1) \dotplus \mathbb{P}_k \subset L^p(0, 1)$, the kernel of this operator is trivial, which finishes the proof of the theorem. ■

5.13. Remark. 1°. If assumption $(H1^{p,l}_\varrho)$ holds, then K is a continuous operator of $X^p_\varrho(0, 1)$ into $X^{p,l}_\varrho(0, 1)$. For $l = 0$, this is easily checked by considering the operator B_ϱ defined in the proof of the preceding theorem. For $l \geqq 1$, one uses the fact that $(xD)^l K$ is a Mellin convolution operator with the kernel $(xD)^l k$.

Moreover, if $(H1^{p,l}_\varrho)$ is satisfied, then $k \in L^{p,l-1}_\varrho(0, 1)$ for $p < \infty$ and $k \in \mathring{C}^{l-1}[0, 1]$ for $p = \infty$ (choose $u \in C_0^\infty(0, 1]$, $u(1) \neq 0$, in 5.12(1)).

2°. Let $b \in \Sigma^m_{\alpha,\beta}$, $\alpha < 1/p - \varrho < 1/p < \beta$ and $m < -l - 1/2$. Then condition $(H1^{p,l}_\varrho)$ is satisfied. This may be proved by similar arguments as in Remark 5.2 (cf. also ELSCHNER [2]). Moreover, if in addition $(H1^p)$, $(H2^p)$ and $b(z) \neq 0$, $z \in \overline{\Gamma}_{1/p-\varrho,1/p}$, hold, then $(H2^p_\varrho)$ is valid (cf. Remark 5.11).

5.14. Example. Consider the operator K with kernel defined by 5.0(3). The symbol of K is given by $b(z) = \sin \lambda\pi z/\sin \pi z$ (cf. COSTABEL/STEPHAN [1]), and we have $b \in \Sigma^m_{-1,1}$ for any $m \in \mathbb{R}$, and $\max_{\Gamma_\nu} |b(z)| < 1$ for any $\nu \in \left(-(1 + |\lambda|)^{-1}, (1 + |\lambda|)^{-1}\right)$. Therefore, by Remark 5.13, 2°, condition $(H1^{p,l}_\varrho)$ is satisfied for all $l \in \mathbb{N}$, $1 < p \leqq \infty$, $0 \leqq \varrho < 1 + 1/p$, and $(H2^p_\varrho)$ holds for all $1 + |\lambda| < p \leqq \infty$, $0 \leqq \varrho < 1/p + 1/(1 + |\lambda|)$; note that $1 - |b(z)|$ vanishes at the points $z = \pm 1/(1 + |\lambda|)$.

The symbol of the L^2-adjoint of K which arises when the single layer potential is used to solve the exterior Neumann problem is given by $\overline{b(1 - \overline{z})}$ and belongs to $\Sigma^m_{0,2}$ for any m. For this symbol, assumption $(H1^{p,l}_\varrho)$ is valid with $l \in \mathbb{N}$, $1 \leqq p < \infty$, $0 \leqq \varrho < 1/p$, and $(H2^p_\varrho)$ holds for all $1 \leqq p < 1 + 1/|\lambda|$, $0 \leqq \varrho < -|\lambda|/(1 + |\lambda|) + 1/p$.

Meshes

Let Δ_n (resp. Δ^n) be a sequence of meshes $\{x_i^{(n)}: 0 \leqq i \leqq n\}$ on $[0, 1]$ (resp. $\mathbb{R}^+$), where $0 = x_0^{(n)} < x_1^{(n)} < \cdots < x_n^{(n)} = 1$ (resp. $0 = x_0^{(n)} < x_1^{(n)} < \cdots < x_n^{(n)} = \infty$). For convenience write x_i for $x_i^{(n)}$ whenever possible, and let $h_i^{(n)} = h_i = x_i - x_{i-1}$ and $J_i = (x_{i-1}, x_i)$.

5.15. Definition. The sequence of meshes Δ_n belongs to the class $\mathcal{M}$ ($\Delta_n \in \mathcal{M}$) if $x_1^{(n)} \to 0$ and $\max\limits_{i=2,\ldots,n} h_i^{(n)}/x_i^{(n)} \to 0$ as $n \to \infty$. Furthermore, $\Delta^n \in \mathcal{M}$ if $x_{n-1}^{(n)} \to \infty$ and $\max\limits_{i=1,\ldots,n-1} h_i^{(n)} \to 0$ as $n \to \infty$.

Note that $\Delta_n = \{x_i^{(n)}\} \in \mathcal{M}$ if and only if the partitions $\Delta^n = \{y_i^{(n)}\}$ on $\mathbb{R}^+$ belong to the class $\mathcal{M}$, where $y_{n-i} = -\log x_i$ (i.e. $x_i = \exp(-y_{n-i})$). This is an easy consequence of the definition and the obvious relations

$$(x_i^{(n)} - x_{i-1}^{(n)})/x_i^{(n)} = 1 - \exp(y_{n-i}^{(n)} - y_{n-i+1}^{(n)}). \tag{1}$$

5.16. Example. 1°. (cf. Sloan/Spence [1]). Let $x_i^{(n)} = q_n \log\big(n/(n-i)\big)$, where $q_n = (1 + \mu \log n)^{-1/2}$ and μ is a positive constant. Then $\Delta^n = \{x_i^{(n)}\} \in \mathcal{M}$.

2°. For any $n = 1, 2, \ldots$, let N be the largest natural number $\leqq qn \log n$, where $q \geqq 1$ is fixed. Then $N \to \infty$ if and only if $n \to \infty$, and we define the sequences of partitions $\Delta_N = \{x_i^{(N)}\}$ and $\Delta^N = \{y_i^{(N)}\}$, where $x_i^{(N)} = n^{-q} \exp(i/n)$ and $y_i^{(N)} = -\log x_{N-1-i}^{(N)} = -(N-1-i)/n + q \log n$ ($i = 1, \ldots, N-1$). Then both sequences of meshes are of class $\mathcal{M}$ since $y_{N-1}^{(N)} = q \log n$ and $y_i^{(N)} - y_{i-1}^{(N)} = 1/n$.

We now define classes of graded meshes which do not belong to $\mathcal{M}$, in general.

5.17. Definition. Let $q \geqq 1$. The sequence of partitions Δ_n belongs to the class $\mathcal{M}_q$ if $h_i^{(n)} \sim i^{-1}(i/n)^q$ ($i = 1, \ldots, n$), i.e.

$$c^{-1} i^{-1} (i/n)^q \leqq h_i^{(n)} \leqq c i^{-1} (i/n)^q, \qquad i = 1, \ldots, n, \tag{1}$$

with some constant $c > 0$ independent of i and n. The sequence of meshes Δ^n is of class $\mathcal{M}_q$ if

$$\exp(-x_i^{(n)}) - \exp(-x_{i+1}^{(n)}) \sim n^{-1}\big((n-i)/n\big)^{q-1}, \qquad i = 0, \ldots, n-1. \tag{2}$$

Note that $\Delta_n \in \mathcal{M}_1$ is equivalent to the quasi-uniformity of the mesh sequence.

5.18. Remark. Using 5.15(1) we observe that $\Delta_n = \{x_i^{(n)}\} \in \mathcal{M}_q$ if and only if $\Delta^n = \{-\log x_{n-i}^{(n)}\} \in \mathcal{M}_q$. Moreover, 5.17(1) (resp. 5.17(2)) implies that $x_i^{(n)} \sim (i/n)^q$ and $h_i^{(n)}/x_i^{(n)} \sim 1/i$ (resp. $x_i^{(n)} \sim q \log(n/(n-i))$ and $h_i^{(n)} \sim 1/(n-i)\big)$, $i = 1, \ldots, n-1$. It suffices to prove this in the case of meshes on $[0, 1]$, and then the assertion follows from 5.17(1) and the estimates

$$x_i^{(n)} = h_1^{(n)} + \cdots + h_i^{(n)} \sim \sum_{j=1}^{i} j^{-1}(j/n)^q \sim (i/n)^q.$$

5.19. Example. Let $x_i^{(n)} = (i/n)^q$ and $y_i^{(n)} = q \log\big(n/(n-i)\big)$. Then we have

$$\Delta_n = \{x_i^{(n)}\} \in \mathcal{M}_q \quad \text{and} \quad \Delta^n = \{y_i^{(n)}\} \in \mathcal{M}_q, \quad \text{but} \quad \Delta_n, \Delta^n \notin \mathcal{M}.$$

For the convergence analysis of Nyström methods in Sect. 5.79–5.95 we still need the following subclass of partitions from $\mathcal{M}$.

5.20. Definition. The sequence of meshes Δ_n belongs to the class $\tilde{\mathcal{M}}$ if

$$x_1^{(n)} \to 0, \qquad \sum_{i=2}^{n} (h_i^{(n)}/x_i^{(n)})^2 \to 0 \quad \text{as} \quad n \to \infty.$$

Moreover, $\Delta^n \in \tilde{\mathcal{M}}$ if

$$x_{n-1}^{(n)} \to \infty, \qquad \sum_{i=1}^{n-1} (h_i^{(n)})^2 \to 0 \quad \text{as} \quad n \to \infty.$$

It can easily be checked that the meshes introduced in Examples 5.16, 1° and 2° are of class $\tilde{\mathcal{M}}$.

Spline spaces and projections

We now introduce certain spaces of polynomial spline functions and study their approximation properties.

Let $\Delta_n = \{x_i : 0 \leq i \leq n\}$ be a sequence of meshes on $[0, 1]$, and let $\mathcal{S}^d(\Delta_n)$ denote the space of piecewise polynomials of degree d on the grid Δ_n, i.e. $u \in \mathcal{S}^d(\Delta_n)$ iff $u|_{J_i} \in \mathbb{P}_d$ for all i. The notation $\mathcal{S}_0^d(\Delta_n)$ refers to the subspace $\left\{u \in \mathcal{S}^d(\Delta_n) : u|_{J_1} = 0\right\}$. For the sequence Δ_n, we further introduce the modifications

$$\tilde{\mathcal{S}}^d(\Delta_n) = \left\{u \in \mathcal{S}^d(\Delta_n) : u\big|_{(0,x_{i_0})} \in \mathbb{P}_d\right\}, \tag{1}$$

$$\tilde{\mathcal{S}}_0^d(\Delta_n) = \left\{u \in \mathcal{S}^d(\Delta_n) : u\big|_{(0,x_{i_0})} = 0\right\} \tag{2}$$

of $\mathcal{S}^d(\Delta_n)$ and $\mathcal{S}_0^d(\Delta_n)$, respectively, where the index i_0 is independent of n. Let P_n (resp. $\mathring{P}_n$) be the orthogonal projection of $\mathbf{L}^2(0, 1)$ onto $\tilde{\mathcal{S}}^d(\Delta_n)$ (resp. $\tilde{\mathcal{S}}_0^d(\Delta_n)$). In the following c denotes a generic constant independent of u, n and i which may vary at each occurrence.

5.21. Lemma. *Let* $1 \leq p \leq \infty$. *Then*

$$\|(I - P_n)\, u\|_{p,J_i} \leq c h_i^l \, \|\mathrm{D}^l u\|_{p,J_i}, \qquad u \in \mathbf{L}^{p,l}(0, 1)$$

or $l = 0, \ldots, d + 1$; $i = i_0 + 1, \ldots, n$.

Proof. With the orthogonal projection $P : \mathbf{L}^2(0, 1) \to \mathbb{P}_d$, the estimates

$$\|Pv\|_{p,(0,1)} \leq c\, \|v\|_{p,(0,1)}, \qquad v \in \mathbf{L}^p(0, 1);$$

$$\begin{aligned} \|(I - P)\, v\|_{p,(0,1)} &\leq \left\|(I - P)\left(v - \sum_{j<l} x^j \mathrm{D}^j v(0)/j!\right)\right\|_{p,(0,1)} \\ &\leq c\, \|\mathrm{D}^l v\|_{p,(0,1)}, \qquad v \in \mathbf{W}^{p,l}(0, 1),\ l = 1, \ldots, d + 1, \end{aligned}$$

are obviously valid. Applying these inequalities to $v(x) = u(x_{i-1} + h_i x)$ and using the fact that

$$(P_n u)\,(x) = (Pv)\left(h_i^{-1}(x - x_{i-1})\right), \qquad x \in J_i,$$

one gets the result. ■

5.22. Corollary. *For all* n *and* i_0, *the projections* P_n *are uniformly bounded on* $\mathbf{L}^p(0, 1)$.

Proof. It is sufficient to check the estimate

$$\|P_n u\|_{p,(0,h)} \leq c\, \|u\|_{p,(0,h)}, \qquad u \in \mathbf{L}^p(0, 1), \tag{1}$$

where $h := x_{i_0}$. This follows by the same arguments as in the above lemma (for $l = 0$). ■

Using graded meshes, we now derive results on the optimal order of approximation.

5.23. Lemma. *Let* $1 \leqq p \leqq \infty$, $\mu < \varrho$, $\Delta_n \in \mathcal{M}_q$, $q \geqq (d+1)/(\varrho - \mu)$ *and* $u \in \mathbf{L}^{p,d+1}_\varrho(0, 1)$. *Then*

$$\|(I - \mathring{P}_n)\, u\|_{p,0,\mu,(0,1)} \leqq cn^{-d-1}\, \|u\|_{p,d+1,\varrho,(0,1)}.$$

Proof. For each $i > i_0$, Lemma 5.21 (with $l = d + 1$) implies

$$\begin{aligned} \|x^{-\mu}(I - \mathring{P}_n)\, u\|_{p,J_i} &\leqq cx_i^{-\mu}\, \|(I - \mathring{P}_n)\, u\|_{p,J_i} \\ &\leqq cx_i^{-\mu} h_i^{d+1}\, \|\mathrm{D}^{d+1} u\|_{p,J_i} \\ &\leqq cx_i^{\varrho-\mu-d-1} h_i^{d+1}\, \|x^{d+1-\varrho} \mathrm{D}^{d+1} u\|_{p,J_i} \end{aligned} \tag{1}$$

since $x_{i-1}/x_i \sim 1$ (cp. Remark 5.18). By 5.17(1), and since $x_i \sim (i/n)^q$ and $(\varrho - \mu)\, q \geqq d + 1$, we have

$$x_i^{\varrho-\mu-d-1} h_i^{d+1} \leqq cn^{-d-1} (i/n)^{(\varrho-\mu)q-d-1} \leqq cn^{-d-1}.$$

Thus the right-hand side of (1) can be estimated by $cn^{-d-1}\, \|u\|_{p,d+1,\varrho,J_i}$.

Together with the inequality

$$\begin{aligned} \|x^{-\mu} u\|_{p,(0,h)} &\leqq h^{\varrho-\mu}\, \|x^{-\varrho} u\|_{p,(0,h)} \\ &\leqq c(i_0/n)^{q(\varrho-\mu)}\, \|x^{-\varrho} u\|_{p,(0,h)} \\ &\leqq cn^{-d-1}\, \|u\|_{p,d+1,\varrho,(0,h)}, \end{aligned} \tag{2}$$

this yields the result. ■

5.24. Remark. Using partitions of class $\mathcal{M}$, one can at least obtain an "almost" optimal order of approximation. Consider the meshes Δ_N defined in Example 5.16, 2°, and let $\mathring{P}_N$ be the corresponding projection onto $\mathcal{S}^d_0(\Delta_N)$. Then, under the assumptions of Lemma 5.23, the estimate

$$\|(I - \mathring{P}_N)\, u\|_{p,0,\mu,(0,1)} \leqq c(\log N/N)^{d+1}\, \|u\|_{p,d+1,\varrho,(0,1)} \tag{1}$$

holds.

Proof. For each $i > i_0$, Lemma 5.21 yields the inequality

$$\begin{aligned} \left\|x^{-\mu}\left(I - \mathring{P}_N\right) u\right\|_{p,J_i} &\leqq cx_i^{\varrho-\mu-d-1} h_i^{d+1}\, \|u\|_{p,d+1,\varrho,J_i} \\ &\leqq c(h_i/x_i)^{d+1}\, \|u\|_{p,d+1,\varrho,J_i} \leqq cn^{-d-1}\, \|u\|_{p,d+1,\varrho,J_i} \end{aligned}$$

since $x_{i-1}/x_i \sim 1$ and $h_i/x_i \sim 1/n$. Moreover, estimate 5.23(2) holds again. Finally, $N \sim qn \log n$ implies that $\log N \sim \log n + \log(\log n) \sim \log n$ as $N \to \infty$, hence $n \sim N/\log N$ which completes the proof. ■

Next, we consider analogous spline spaces subordinate to partitions on the half-axis. Let $\Delta^n = \{x_i : 0 \leqq i \leqq n\}$ be a sequence of meshes on $\mathbb{R}^+$, and let $\mathcal{S}^d(\Delta^n)$ denote the space

$$\mathcal{S}^d(\Delta^n) = \left\{u \in \mathbf{L}^\infty(\mathbb{R}^+) : u\big|_{J_i} \in \mathbb{P}_d,\ i = 1, \ldots, n-1,\ u\big|_{J_n} = 0\right\}.$$

To the sequence Δ^n we further associate the modifications

$$\bar{\mathcal{S}}^d(\Delta^n) = \left\{u \in \mathcal{S}^d(\Delta^n) : u\big|_{(x_{n-i_0},\infty)} = 0\right\} \tag{2}$$

of $\mathcal{S}^d(\Delta^n)$, where the index i_0 does not depend on n. The orthogonal projection of $\mathbf{L}^2(\mathbb{R}^+)$ onto $\bar{\mathcal{S}}^d(\Delta^n)$ will be denoted by P^n.

5.25. Lemma. *Let* $1 \leqq p \leqq \infty$. *Then*

$$\|(I - P^n)\, u\|_{p,J_i} \leqq c h_i^l \, \|D^l u\|_{p,J_i}, \qquad u \in \mathbf{L}^{p,l}(\mathbb{R}^+)$$

for $l = 0, \ldots, d + 1$, $i = 1, \ldots, n - i_0$.
The proof is completely analogous to that of Lemma 5.21.

5.26. Lemma. *Let* $1 \leqq p \leqq \infty$, $\mu < \varrho$, $\Delta^n \in \mathcal{M}_q$, $q \geqq (d+1)/(\varrho - \mu)$ *and* $u \in \mathbf{L}_\varrho^{p,d+1}(\mathbb{R}^+)$. *Then*

$$\|(I - P^n)\, u\|_{p,0,\mu,\mathbb{R}^+} \leqq c n^{-d-1} \, \|u\|_{p,d+1,\varrho,\mathbb{R}^+}.$$

Proof. For each $i \leqq n - i_0$, it follows from Lemma 5.25 (with $l = d + 1$) that

$$\begin{aligned} \|e^{\mu x}(I - P^n)\, u\|_{p,J_i} &\leqq c\, e^{\mu x_i} \, \|(I - P^n)\, u\|_{p,J_i} \\ &\leqq c\, e^{(\mu-\varrho)x_i} \, h_i^{d+1} \, \|e^{\varrho x} D^{d+1} u\|_{p,J_i} \end{aligned} \tag{1}$$

since $e^{x_i}/e^{x_{i-1}} \sim 1$. According to Remark 5.18 we further have $e^{-x_i} \sim \big((n - i)/n\big)^q$ and $h_i \sim 1/(n - i)$ which yields the estimate

$$e^{(\mu-\varrho)x_i} \, h_i^{d+1} \leqq c n^{-d-1} \big((n - i)/n\big)^{(\varrho-\mu)q-d-1} \leqq c n^{-d-1}.$$

Therefore, the right-hand side of (1) can be dominated by $c n^{-d-1} \, \|u\|_{p,d+1,\varrho,J_i}$. Setting $h = x_{n-i_0}$, we finally obtain

$$\begin{aligned} \|u\|_{p,0,\mu,(h,\infty)} &\leqq e^{(\mu-\varrho)h} \, \|e^{\varrho x} u\|_{p,(h,\infty)} \\ &\leqq c (i_0/n)^{(\varrho-\mu)q} \, \|u\|_{p,0,\varrho,(h,\infty)} \\ &\leqq c n^{-d-1} \, \|u\|_{p,d+1,\varrho,(h,\infty)}, \end{aligned} \tag{2}$$

which finishes the proof. ■

5.27. Remark. Consider the meshes Δ^N defined in Example 5.16, 2°, and let P^N be the corresponding projections onto $\mathcal{S}^d(\Delta^N)$. Then, under the assumptions of Lemma 5.26, we have the estimate

$$\|(I - P^N)\, u\|_{p,0,\mu,\mathbb{R}^+} \leqq c (\log N/N)^{d+1} \, \|u\|_{p,d+1,\varrho,\mathbb{R}^+}. \tag{1}$$

The proof of (1) is similar to that of 5.24(1) (note that the inequalities 5.26(1) and 5.26(2) are valid again).

In order to introduce interpolatory projections onto the spline spaces $\bar{\mathcal{S}}_0^d(\Delta_n)$, set $x_{ij} = x_{i-1} + \xi_j h_i$, where ξ_j are fixed points with $0 \leqq \xi_0 < \xi_1 < \cdots < \xi_d \leqq 1$. For any function $u \in \mathbf{C}(0, 1]$, define $Q_n u \in \bar{\mathcal{S}}_0^d(\Delta_n)$ by

$$(Q_n u)(x_{ij}) = u(x_{ij}), \qquad j = 0, \ldots, d;\ i = i_0 + 1, \ldots, n. \tag{2}$$

Here x_{id} and $x_{i+1,0}$, $i_0 \leqq i < n$, have to be considered as distinct nodes if $\xi_0 = 0$ and $\xi_d = 1$.

5.28. Lemma. *The operators* Q_n *extend to bounded projections of* $\mathbf{L}^\infty(0, 1)$ *onto* $\bar{\mathcal{S}}_0^d(\Delta_n)$ *(denoted again by* Q_n*) such that*

$$\|Q_n u\|_{\infty,J_i} \leqq c \, \|u\|_{\infty,J_i}, \qquad u \in \mathbf{L}^\infty(0, 1), \qquad i = i_0 + 1, \ldots, n. \tag{1}$$

Proof. Consider the operator Q defined by

$$Qu(x) = \sum_{i=0}^{d} w(x)\, u(\xi_i)/(x - \xi_i)\, (Dw)(\xi_i), \qquad w(x) = \prod_{j=0}^{d} (x - \xi_j),$$

$u \in \mathbf{C}[0, 1]$.

The point evaluation functionals $u \mapsto u(\xi_i)$ admit bounded extensions to the whole of $\mathbf{L}^\infty(0, 1)$ with norm 1 (according to the Hahn-Banach theorem). Thus one obtains a bounded extension of Q to $\mathbf{L}^\infty(0, 1)$. Transforming each interval $\bar{J}_i$ ($i > i_0$) to $[0, 1]$ as in the proof of Lemma 5.21, we get an extension of Q_n having the desired properties (cf. ATKINSON/GRAHAM/SLOAN [1] for similar constructions). ■

5.29. Lemma. *Let* $\varrho \in \mathbb{R}$ *and* $\Delta_n \in \mathcal{M}_q$ *or* $\Delta_n \in \mathcal{M}$. *Then*

$$\|x^{-\varrho}Q_n u\|_{\infty,J_i} \leqq c\,\|x^{-\varrho}u\|_{\infty,J_i}, \qquad u \in \mathbf{L}_\varrho^\infty(0, 1)$$

for $i = i_0 + 1, \ldots, n$.

Proof. By Definition 5.15 and Remark 5.18, we have $x_{i-1}/x_i \sim 1$, hence

$$\|x^{-\varrho}Q_n u\|_{\infty,J_i} \leqq c x_i^{-\varrho}\,\|Q_n u\|_{\infty,J_i} \leqq c x_i^{-\varrho}\,\|u\|_{\infty,J_i} \leqq c\,\|x^{-\varrho}u\|_{\infty,J_i}$$

for each $i > i_0$, where we have used 5.28(1). ■

We now consider the interpolation projections $Q^n\colon \mathbf{C}(\mathbb{R}^+) \to \bar{\mathcal{S}}^d(\Delta^n)$ given by

$$(Q^n u)\,(x_{ij}) = u(x_{ij}), \qquad j = 0, \ldots, d,\ i = 1, \ldots, n - i_0, \tag{1}$$

where the points x_{ij} are defined as above and x_{id}, $x_{i+1,0}$, $1 \leqq i < n - i_0$, are considered as distinct nodes if $\xi_0 = 0$ and $\xi_d = 1$. The following assertion is the analogue of Lemmas 5.28 and 5.29; its proof runs in an analogous manner.

5.30. Lemma. *The operators* Q^n *extend to bounded projections of* $\mathbf{L}^\infty(\mathbb{R}^+)$ *onto* $\bar{\mathcal{S}}^d(\Delta^n)$ *satisfying the estimates*

$$\|e^{\varrho x}Q^n u\|_{\infty,J_i} \leqq c\,\|e^{\varrho x}u\|_{\infty,J_i}, \qquad u \in \mathbf{L}_\varrho^\infty(\mathbb{R}^+)$$

for $i = 1, \ldots, n - i_0$, *where for* $\varrho \neq 0$ *it is, in addition, assumed that* $\Delta^n \in \mathcal{M}_q$ *or* $\Delta^n \in \mathcal{M}$.

Finally, we introduce spaces of continuous piecewise polynomials and appropriate interpolating projections. To a sequence of partitions $\Delta_n = \{x_i\colon 0 \leqq i \leqq n\}$ we associate the spaces $\mathcal{S}^{d,1}(\Delta_n)$ of piecewise polynomials of degree $d \geqq 1$ on Δ_n, which are continuous on $[0, 1]$, and their modifications

$$\bar{\mathcal{S}}^{d,1}(\Delta_n) = \left\{u \in \mathcal{S}^{d,1}(\Delta_n)\colon u\big|_{(0,x_{i_0})} \in \mathbb{C}\right\}, \tag{1}$$

where the index i_0 is independent of n. Let ξ_j be fixed points with $0 = \xi_0 < \xi_1 < \cdots < \xi_d = 1$, and define $x_{ij} = x_{i-1} + \xi_j h_i$ as above. Consider the interpolating projections $\Pi_n\colon \mathbf{L}^{p,1}(0, 1) \to \bar{\mathcal{S}}^{d,1}(\Delta_n)$ given by

$$(\Pi_n u)\,(x_{ij}) = u(x_{ij}), \qquad j = 0, i = i_0 + 1, j = 1, \ldots, d, i = i_0 + 1, \ldots, n. \tag{2}$$

5.31. Lemma. *Let* $1 \leqq p \leqq \infty$. *Then*

(i) $\|D^k(I - \Pi_n)\,u\|_{p,J_i} \leqq c h_i^{l-k}\,\|D^l u\|_{p,J_i}$
for $u \in \mathbf{L}^{p,l}(0, 1)$, $i = i_0 + 1, \ldots, n$,
$0 \leqq k \leqq d$, $1 \leqq l \leqq d + 1$, $k \leqq l$;

(ii) $\|\Pi_n u\|_{p,1,(0,h)} \leqq c\,\|u\|_{p,1,(0,h)}$
for $u \in \mathbf{L}^{p,1}(0, 1)$, $h := x_{i_0}$.

Proof. Let $W^{p,l}$ denote the usual Sobolev space of order l on $(0, 1)$ with the norm $\|u\|_{W^{p,l}} = \sum_{j \leq l} \|D^j u\|_{p,(0,1)}$.

For $v \in C[0, 1]$, let Πv be the polynomial of degree d interpolating v at $\xi_0, \ldots, \xi_d$. We show the inequalities

$$\|D^k(I - \Pi) v\|_{p,(0,1)} \leq c \|D^l v\|_{p,(0,1)}, \qquad v \in W^{p,l}, \tag{1}$$

where $0 \leq k \leq d$, $1 \leq l \leq d + 1$, $k \leq l$. Indeed, with the Taylor polynomial v_0 of degree $l - 1$ for v about $x = 0$, we have the estimates

$$\begin{aligned}\|D^k(I - \Pi) v\|_{p,(0,1)} &\leq \|D^k(v - v_0)\|_{p,(0,1)} + \|D^k \Pi(v - v_0)\|_{p,(0,1)} \\ &\leq c \|v - v_0\|_{W^{p,l}} \leq c \|D^l v\|_{p,(0,1)}.\end{aligned}$$

Using the scaling argument from the proof of Lemma 5.21, we then obtain (i) from (1).

Let $u \in L^{p,1}(0, 1)$. To prove (ii), we note that $\Pi_n u = u(h)$ on $[0, h]$ and

$$\begin{aligned}\|\Pi_n u\|_{p,(0,h)} + \|x D \Pi_n u\|_{p,(0,h)} &= |u(h)|\, h^{1/p} = h^{1/p-1} \left| \int_0^h D(xu)\, dx \right| \\ &\leq \|D(xu)\|_{p,(0,h)} \leq c \|u\|_{p,1,(0,h)}. \quad \blacksquare\end{aligned} \tag{2}$$

5.32. Corollary. *If $\Delta_n \in \mathcal{M}_q$ or $\Delta_n \in \mathcal{M}$, then for all n and i_0 the interpolatory projections Π_n are uniformly bounded on $L^{p,1}(0.1)$.*

Proof. Applying Lemma 5.31 (i) (with $k = 0$, $l = 1$ and $k = l = 1$), we get for $i > i_0$

$$\begin{aligned}\|\Pi_n u\|_{p,J_i} + \|x D \Pi_n u\|_{p,J_i} &\leq c(h_i/x_{i-1}) \|x D u\|_{p,J_i} + c(x_i/x_{i-1}) \|x D u\|_{p,J_i} \\ &\leq c \|u\|_{p,1,J_i}\end{aligned}$$

since h_i/x_{i-1} and x_i/x_{i-1} are uniformly bounded from above according to Definition 5.15 and Remark 5.18. Together with these estimates, Lemma 5.31(ii) implies the result. ■

We now examine the order of approximation in weighted Sobolev norms.

5.33. Lemma. *Let $1 \leq p \leq \infty$, $\mu < \varrho$, $\Delta_n \in \mathcal{M}_q$, $q \geq l/(\varrho - \mu)$, $1 \leq l \leq d + 1$. Then*

(i) $\|(I - \Pi_n) u\|_{p,0,\mu,(h,1)} \leq c n^{-l} \|u\|_{p,l,\varrho,(h,1)}$ *for $u \in L^{p,l}_\varrho(0, 1)$;*

(ii) $\|(I - \Pi_n) u\|_{p,1,\mu,(h,1)} \leq c n^{-l} \|u\|_{p,l+1,\varrho,(h,1)}$ *for $u \in L^{p,l+1}_\varrho(0, 1)$, if $l \leq d$.*

Proof. Applying Lemma 5.31(i) (with $k = 0$) and arguing as in the proof of Lemma 5.23, for each $i > i_0$ one obtains the estimate

$$\begin{aligned}\|x^{-\mu}(I - \Pi_n) u\|_{p,J_i} &\leq c x_i^{\varrho-\mu-l} h_i^l \|x^{l-\varrho} D^l u\|_{p,J_i} \\ &\leq c n^{-l} (i/n)^{(\varrho-\mu)q-l} \|u\|_{p,l,\varrho,J_i} \leq c n^{-l} \|u\|_{p,l,\varrho,J_i}\end{aligned}$$

and hence (i). Analogously, using Lemma 5.31(i) (with $k = 1$), one gets

$$\begin{aligned}\|x^{1-\mu} D(I - \Pi_n) u\|_{p,J_i} &\leq c x_i^{\varrho-\mu-l} h_i^l \|x^{l+1-\varrho} D^{l+1} u\|_{p,J_i} \\ &\leq c n^{-l} \|u\|_{p,l+1,\varrho,J_i},\end{aligned}$$

which yields (ii). ■

On the interval $(0, h)$, the following approximation result holds.

5.34. Lemma. *Under the assumptions of Lemma 5.33 with $\mu = 0$, we have*

$$\|(I - \Pi_n)\, u\|_{p,1,(0,h)} \leqq cn^{-l}\, \|u\|_{p,1,\varrho,(0,h)} \quad \textit{for} \quad u \in \mathbf{L}_\varrho^{p,1}(0, 1).$$

Proof. Using 5.31(2) we get

$$\begin{aligned}\|(I - \Pi_n)\, u\|_{p,1,(0,h)} &\leqq c\, \|u\|_{p,1,(0,h)} \\ &\leqq ch^\varrho\, \|x^{-\varrho}u\|_{p,1,(0,h)} \leqq cn^{-\varrho q}\, \|u\|_{p,1,\varrho,(0,h)} \\ &\leqq cn^{-l}\, \|u\|_{p,1,\varrho,(0,h)}. \quad \blacksquare\end{aligned}$$

Let $\Delta^n = \{x_i : 0 \leqq i \leqq n\}$ be a sequence of meshes on $\mathbb{R}^+$, and let $\mathcal{S}^{d,1}(\Delta^n)$, $d \geqq 1$, denote the space of piecewise polynomials of degree d on the grid Δ^n, which are continuous on $\mathbb{R}^+$. To a sequence Δ^n we further associate the modifications

$$\bar{\mathcal{S}}^{d,1}(\Delta^n) = \left\{u \in \mathbf{C}(\mathbb{R}^+) : u|_{J_i} \in \mathbb{P}_d,\ i = 1, \ldots, n - i_0,\ \mathrm{e}^x u\big|_{(x_{n-i_0}, \infty)} \in \mathbb{C}\right\} \tag{1}$$

of $\mathcal{S}^{d,1}(\Delta^n)$, where the index i_0 does not depend on n. Defining the points x_{ij} as above, we introduce the interpolatory projections $\Pi^n : \mathbf{L}^{p,1}(\mathbb{R}^+) \to \bar{\mathcal{S}}^{d,1}(\Delta^n)$ by

$$\begin{aligned}(\Pi^n u)\,(x_{ij}) = u(x_{ij}), \qquad & j = 0, \ldots, d - 1; \qquad i = 1, \ldots, n - i_0; \\ & j = d,\ i = n - i_0.\end{aligned} \tag{2}$$

5.35. Lemma. *Let $1 \leqq p \leqq \infty$. Then*

(i) $\|\mathrm{D}^k(I - \Pi^n)\, u\|_{p,J_i} \leqq ch_i^{l-k}\, \|\mathrm{D}^l u\|_{p,J_i}$
for $u \in \mathbf{L}^{p,l}(\mathbb{R}^+)$, $i = 1, \ldots, n - i_0$, $0 \leqq k \leqq d$, $1 \leqq l \leqq d + 1$, $k \leqq l$;

(ii) $\|\Pi^n u\|_{p,1,(h,\infty)} \leqq c\, \|u\|_{p,1,(h,\infty)}$,
for $u \in \mathbf{L}^{p,1}(\mathbb{R}^+)$, $h := x_{n-i_0}$.

Proof. To verify (ii), we observe that $\Pi^n u = \mathrm{e}^{h-x} u(h)$ on (h, ∞) and

$$\begin{aligned}\|\Pi^n u\|_{p,1,(h,\infty)} &\leqq c\, \mathrm{e}^h\, |u(h)|\, \|\mathrm{e}^{-x}\|_{p,(h,\infty)} \leqq c\, |u(h)| \\ &\leqq c\, \|u\|_{p,1,(h,\infty)}.\end{aligned}$$

The last estimate is trivial if $p = \infty$; for $p = 1$ it follows from the identity $u(h) = -\int_h^\infty \mathrm{D}u(t)\, \mathrm{d}t$ (which holds at least for the dense subset of smooth functions vanishing at infinity), hence it is valid for all p by interpolation. The proof of (i) is completely analogous to that of Lemma 5.31(i). $\blacksquare$

The following lemma contains the analogues of Corollary 5.32 and Lemmas 5.33, 5.34; the proof follows the same lines as there, using of course Lemma 5.35 instead of Lemma 5.31.

5.36. Lemma. *Let $1 \leqq p \leqq \infty$.*

(i) *If $\Delta^n \in \mathcal{M}_q$ or $\Delta^n \in \mathcal{M}$, then for all n and i_0 the projections Π^n are uniformly bounded on $\mathbf{L}^{p,1}(\mathbb{R}^+)$.*

(ii) *Assume $\mu < \varrho$, $\Delta^n \in \mathcal{M}_q$, $q \geqq l/(\varrho - \mu)$ and $1 \leqq l \leqq d + 1$. Then*

$$\|(I - \Pi^n)\, u\|_{p,0,\mu,(0,h)} \leqq cn^{-l}\, \|u\|_{p,l,\varrho,(0,h)} \ \textit{for}\ u \in \mathbf{L}_\varrho^{p,l}(\mathbb{R}^+);$$

$$\|(I - \Pi^n)\, u\|_{p,1,\mu,(0,h)} \leqq cn^{-l}\, \|u\|_{p,l+1,\varrho,(0,h)} \ \textit{for}\ u \in \mathbf{L}_\varrho^{p,l+1}(\mathbb{R}^+),\ \textit{if}\ l \leqq d.$$

(iii) *Suppose* $\varrho > 0$, $\Delta^n \in \mathcal{M}_q$, $q \geqq l/\varrho$ *and* $1 \leqq l \leqq d+1$. *Then*

$$\|(I - \Pi^n)\, u\|_{p,1,(h,\infty)} \leqq c n^{-l}\, \|u\|_{p,1,\varrho,(h,\infty)} \quad \text{for } u \in \mathbf{L}^{p,1}_{\varrho}(\mathbb{R}^+).$$

Galerkin methods with piecewise polynomials: Wiener-Hopf equations

Let $A = I - W$ be the Wiener-Hopf operator defined by 5.1(4). For the approximate solution of equation 5.0(2), we first consider the classical Galerkin method with splines from $\bar{\mathcal{S}}^d(\Delta^n)$ (cf. 5.24(2)).

5.37. Theorem. *Consider the Galerkin method*

$$P^n A u_n = u_n - P^n W u_n = P^n f \qquad \big(u_n \in \bar{\mathcal{S}}^d(\Delta^n)\big), \tag{1}$$

where P^n *denotes the orthogonal projection of* $\mathbf{L}^2(\mathbb{R}^+)$ *onto* $\bar{\mathcal{S}}^d(\Delta^n)$. *Assume* $1 \leqq p \leqq \infty$, (A1), (A2), *and let* $\bar{\mathcal{S}}^d(\Delta^n)$ *be a modification of* $\mathcal{S}^d(\Delta^n)$.

(i) *If* $\Delta^n \in \mathcal{M}_q$ *and* i_0 *is sufficiently large, then the Galerkin method* 5.37(1) *is stable in* $\mathbf{L}^p(\mathbb{R}^+)$, *i.e.*

$$\|P^n A u_n\|_{p,\mathbb{R}^+} \geqq c\, \|u_n\|_{p,\mathbb{R}^+}, \qquad u_n \in \bar{\mathcal{S}}^d(\Delta^n), \qquad n \geqq n_0 \tag{2}$$

when n_0 *is large enough.*

(ii) *If* $\Delta^n \in \mathcal{M}$, *then* (2) *is valid for* $i_0 = 1$ (*i.e.* $\bar{\mathcal{S}}^d(\Delta^n) = \mathcal{S}^d(\Delta^n)$).

To prove (2), we consider the representations 5.4(2). The following lemma is the key to the stability proof.

5.38. Lemma. *Let* W_- *be the operator occurring in* 5.4(2), *where* δ *is fixed, and let* $\bar{\mathcal{S}}^d(\Delta^n)$ *be a modification of* $\mathcal{S}^d(\Delta^n)$. *Then, given* $\varepsilon > 0$, *there exists an integer* n_0 *such that*

$$\|(I - P^n)\, W_- u_n\|_{p,\mathbb{R}^+} \leqq \varepsilon\, \|u_n\|_{p,\mathbb{R}^+}, \qquad u_n \in \bar{\mathcal{S}}^d(\Delta^n), \qquad n \geqq n_0, \tag{1}$$

if either $\Delta^n \in \mathcal{M}$ *and* $i_0 = 1$, *or* $\Delta^n \in \mathcal{M}_q$ *and the index* i_0 *is sufficiently large.*

Proof. We first estimate the $\mathbf{L}^p$ norm on $(0, h)$, where $h = x_{n-i_0}$. Let $u \in \mathbf{L}^p(\mathbb{R}^+)$. By Lemma 5.25 (for $l = 1$), we obtain

$$\|(I - P^n)\, W_- u\|_{p,J_i} \leqq c h_i\, \|\mathrm{D} W_- u\|_{p,J_i}, \qquad i \leqq n - i_0. \tag{2}$$

Furthermore, Definition 5.15 (resp. Remark 5.18) implies that $h_i \leqq \varepsilon$, $i \leqq n - i_0$, when $\Delta^n \in \mathcal{M}$ (resp. $\Delta^n \in \mathcal{M}_q$) and n (resp. n, i_0) are large enough. Thus, for those indices i_0 and n, (2) and the continuity of the operator $W_-\colon \mathbf{L}^p(\mathbb{R}^+) \to \mathbf{L}^{p,1}(\mathbb{R}^+)$ (cf. Remark 5.6(ii)) yield the estimate

$$\|(I - P^n)\, W_- u\|_{p,(0,h)} \leqq c\varepsilon\, \|D W_- u\|_{p,\mathbb{R}^+} \leqq c\varepsilon\, \|u\|_{p,\mathbb{R}^+}.$$

To complete the proof of (1), we note that $(I - P^n)\, W_- u_n = 0$ on (h, ∞) for all $u_n \in \bar{\mathcal{S}}^d(\Delta^n)$ (cf. Remark 5.6(i)). ■

Proof of Theorem 5.37. *Step 1:* By Theorem 5.4(ii), for any $\delta > 0$ we have the representation

$$P^n A P^n = P^n (A_1 + T)\, P^n, \qquad A_1 := (I - W_+)\,(I - W_0)\,(I - W_-), \tag{3}$$

where $\|W_0\|_p < \delta$. It follows from Lemma 5.25 (with $l = 0$) that the projections P^n are uniformly bounded on $\mathbf{L}^p(\mathbb{R}^+)$ (for arbitrary modifications of the spline spaces $\mathcal{S}^d(\Delta^n)$). Choosing now δ such that $\sup_n \|P^n\|_p \delta \leqq 1$, we obtain that the operators $P^n(I - W_0) P^n$ are stable, i.e. 5.37(2) holds with A replaced by $I - W_0$ (where c does not depend on i_0).

Step 2: Next we verify the stability of the operators $P^n(I - W_\pm) P^n$ in $\mathbf{L}^p(\mathbb{R}^+)$. Choosing ε sufficiently small in Lemma 5.38 and using the invertibility of $I - W_-$, we get the result for W_-. On the other hand, one can apply the following duality argument. Associating to the operator W (with kernel $\varkappa(x)$) the convolution operator W^* with kernel $\overline{\varkappa(-x)}$, we observe that $W^*: \mathbf{L}^{p'}(\mathbb{R}^+) \to \mathbf{L}^{p'}(\mathbb{R}^+)$ is the adjoint of $W: \mathbf{L}^p(\mathbb{R}^+) \to \mathbf{L}^p(\mathbb{R}^+)$, where $p < \infty$ and $p' = p/(p-1)$. Since W_+ is a polynomial in the elementary operator V_+ and $V_+^* = V_-$ (cf. 5.3(1)), the operators $P^n(I - W_+^*) P^n$ are stable in $\mathbf{L}^{p'}(\mathbb{R}^+)$ for any p' in view of the things proved above. Passing to adjoint operators, one obtains the stability of the sequence $P^n(I - W_+) P^n$ in $\mathbf{L}^p(\mathbb{R}^+)$ when $p > 1$. For $p = 1$, the assertion follows from the stability of the adjoint operators $P^n(I - W_+^*) P^n$ on $\mathbf{L}^\infty(\mathbb{R}^+)$.

Step 3: We now show the stability of the operators $P^n A_1 P^n$. Using the above duality argument and applying Lemma 5.32 to the operator W_+^*, we see that $\|P^n W_+(I - P^n)\|_p$ can be made as small as desired if $\Delta^n \in \mathcal{M}$ (resp. $\Delta^n \in \mathcal{M}_q$) and n (resp. i_0, n) are sufficiently large. Together with (3), this implies that the norm of the operators

$$P^n A_1 P^n - P^n(I - W_+) P^n(I - W_0) P^n(I - W_-) P^n$$

becomes arbitrarily small when $n \to \infty$ and i_0 is chosen large enough. Therefore, by the considerations in the first and second steps, Theorem 5.37 holds with A replaced by A_1.

Step 4: Let $p < \infty$. Then $\mathbf{C}_0^\infty(\mathbb{R}^+)$ is dense in $\mathbf{L}^p(\mathbb{R}^+)$, and the operators P^n converge strongly to the identity on $\mathbf{L}^p(\mathbb{R}^+)$ (according to their uniform boundedness and Lemma 5.25). 5.38(3) and a standard perturbation theorem (Remark 1.27, 3°(6)) now yields the stability of $P^n A P^n$ since the operators A and A_1 are invertible and T is compact on $\mathbf{L}^p(\mathbb{R}^+)$.

Finally, let $p = \infty$. Arguing by contradiction, we assume that there is a sequence $u_n \in \overline{\mathcal{S}}^d(\Delta^n)$ such that $\|u_n\|_{\infty,\mathbb{R}^+} = 1$, $\|P^n A u_n\|_{\infty,\mathbb{R}^+} \to 0$ as $n \to \infty$. Since T is a compact map of $\mathbf{L}^\infty(\mathbb{R}^+)$ into $\mathring{\mathbf{C}}(\overline{\mathbb{R}}^+)$, Tu_n converges to some element $v \in \mathring{\mathbf{C}}(\overline{\mathbb{R}}^+)$ in the uniform norm (at least for a certain subsequence). This implies $\|P^n T u_n - v\|_{\infty,\mathbb{R}^+} \to 0$ and, by the second assumption on u_n and the stability of $P^n A_1 P^n$, $\|u - u_n\|_{\infty,\mathbb{R}^+} \to 0$ for some $u \in \mathbf{L}^\infty(\mathbb{R}^+)$; hence Wu_n converges to Wu in $\mathbf{C}(\mathbb{R}^+)$ and $\|P^n W u_n - Wu\|_{\infty,J} \to 0$ for each compact subinterval $J \subset \mathbb{R}^+$. (Here we have used Lemma 5.25 and the fact that $\mathbf{C}_0^\infty(\mathbb{R}^+)$ is dense in $\mathring{\mathbf{C}}(\overline{\mathbb{R}}^+)$.) Thus we obtain $Au = (I - W) u = 0$, hence $u = 0$, and this contradiction proves 5.37(2) when $p = \infty$. ■

5.39. Corollary. *Suppose that the assumptions of Theorem* 5.37 *are satisfied.*

(i) *For all i_0 and n sufficiently large, the Galerkin equations* 5.37(1) *are uniquely solvable for any $f \in \mathbf{L}^p(\mathbb{R}^+)$ and the error estimate*

$$\|u - u_n\|_{p,\mathbb{R}^+} \leqq c \, \|(I - P^n) u\|_{p,\mathbb{R}^+} \tag{1}$$

holds, where $u \in \mathbf{L}^p(\mathbb{R}^+)$ denotes the exact solution of 5.0(2). *Moreover, for $p < \infty$, the left-hand side of* (1) *converges to* 0 *as $n \to \infty$.*

(ii) *If $f \in \mathring{\mathbf{C}}(\overline{\mathbb{R}}^+)$, then u_n converges uniformly to u on the half-axis.*

(iii) *If $f \in \mathbf{C}(\mathbb{R}^+)$, then u_n converges uniformly to u on each compact subinterval of* $\mathbb{R}^+$.

Proof. Inequality (1) follows in a standard way from Theorem 5.37 (see (Rem. 1.27, 3°(a)). The second assertion of (i) and (ii) hold according to the strong convergence of P^n to the identity on $\mathbf{L}^p(\mathbb{R}^+)$, $p < \infty$, and $\mathring{\mathbf{C}}(\overline{\mathbb{R}}^+)$.

Let $f \in \mathbf{C}(\mathbb{R}^+)$. Then $P^n f \to f$ (which means uniform convergence on each compact subset of $\mathbb{R}^+$ in the sequel), and the sequence u_n is bounded in $\mathbf{L}^\infty(\mathbb{R}^+)$ by virtue of 5.37(2). We now use the fact that the sequence Wu_n is bounded in $\mathbf{C}(\mathbb{R}^+)$ and (uniformly) equicontinuous on $\mathbb{R}^+$. By the Arzela-Ascoli theorem, one obtains (passing to a certain subsequence) $Wu_n \to w$ for some $w \in \mathbf{C}(\mathbb{R}^+)$, hence $P^n Wu_n \to w$. Therefore, $u_n \to v$ for some element $v \in \mathbf{C}(\mathbb{R}^+)$. Furthermore, $Wu_n \to Wv$ (cf. Secs. 4.33, 4.55) so that $w = Wv$ and $u_n - P^n W u_n \to Av = f$, hence $v = u$. Thus any subsequence of $\{u_n\}$ contains a further subsequence converging uniformly to u on each compact interval of the half-axis, and the proof of (iii) is complete. ■

5.40. Remark. Corollary 5.39 (ii) can also be deduced from PRÖSSDORF/SILBERMANN [1, Chap. 1.3, Th. 2] since the convergence manifold (with respect to the Galerkin method in $\mathbf{L}^\infty$) of the operator A_1 defined in 5.38(3) contains $\mathring{\mathbf{C}}(\overline{\mathbb{R}}^+)$. Moreover, the considerations in PRÖSSDORF/SILBERMANN [1, Chap. 1.3, Th. 1] yield another proof of the stability estimate 5.37(2) for $p = \infty$. Finally, note that the same kind of convergence as in Corollary 5.39(iii) has already been studied in the case of the finite section method for Wiener-Hopf integral equations; see Sects. 4.60, 4.64.

The next result shows that one obtains the optimal order of convergence when suitably graded meshes are used.

5.41. Theorem. *Assume* $1 \leqq p \leqq \infty$, $\varrho > 0$, (A1), (A1$_\varrho$), (A2), (A2$_\varrho$) *and* $\varkappa \in \mathbf{L}^{p,d}_\varrho(\mathbb{R}^+)$. *Suppose further that* $f \in \mathbf{L}^{p,d+1}_\varrho(\mathbb{R}^+)$, $\Delta^n \in \mathcal{M}_q$ *with* $q \geqq (d+1)/\varrho$, *and let* $\tilde{\mathcal{S}}^d(\Delta^n)$ *be a modification of* $\mathcal{S}^d(\Delta^n)$ *such that the number* i_0 *in* 5.24(2) *is sufficiently large. Then the Galerkin method* 5.37(1) *converges with the error bound*

$$\|u - u_n\|_{p,\mathbb{R}^+} \leqq cn^{-d-1} \|u\|_{p,d+1,\varrho,\mathbb{R}^+}. \tag{1}$$

Proof. Note that $u \in \mathbf{L}^{p,d+1}_\varrho(\mathbb{R}^+)$ in view of Theorem 5.9. Estimate (1) is now a consequence of 5.39(1) and Lemma 5.26. ■

5.42. Remark. Suppose that A satisfies the assumptions of the preceding theorem. Let Δ^N be the meshes defined in Example 5.16, 2° with $q \geqq (d+1)/\varrho$, and let $f \in \mathbf{L}^{p,d+1}_\varrho(\mathbb{R}^+)$ and $i_0 = 1$. Then the Galerkin method 5.37(1) (with n replaced by N) converges with the error estimate

$$\|u - u_N\|_{p,\mathbb{R}^+} \leqq c(\log N/N)^{d+1} \|u\|_{p,d+1,\varrho,\mathbb{R}^+}. \tag{1}$$

This follows from 5.39(1) and Remark 5.27. Inequality (1) shows that, in a certain sense, one can obtain "almost" optimal rates of convergence when partitions of class $\mathcal{M}$ and unmodified spline spaces are used.

Finally, we study the rate of convergence for the iterated Galerkin method. The iterated Galerkin solution is defined by

$$u_n^* := f + Wu_n,$$

where u_n is given by 5.37(1), or equivalently,

$$u - u_n - P^n W(u - u_n) = (I - P^n)\, u,$$

where u is the exact solution of 5.0(2). Multiplying the last equality by W from the left and using the relation $u - u_n^* = W(u - u_n)$, one obtains

$$(I - WP^n)(u - u_n^*) = W(I - P^n)\,u. \tag{2}$$

Now we are in a position to prove

5.43. Theorem. *Assume* $1 \leqq p \leqq \infty$, $\varrho > 0$, $(\mathrm{A1}^{d+1})$, $(\mathrm{A1}_\varrho^{d+1})$, $(\mathrm{A2})$ *and* $(\mathrm{A2}_\varrho)$. *If* $f \in \mathbf{L}_\varrho^{p,d+1}(\mathbb{R}^+)$, $\Delta^n \in \mathcal{M}_q$ *with* $q \geqq 2(d+1)/\varrho$, *and* $\bar{\mathcal{S}}^d(\Delta^n)$ *is an appropriate modification of* $\mathcal{S}^d(\Delta^n)$, *then*

$$\|u - u_n^*\|_{p,\mathbb{R}^+} \leqq cn^{-2(d+1)}\,\|u\|_{p,d+1,\varrho,\mathbb{R}^+} \quad as \quad n \to \infty. \tag{1}$$

Proof. First we verify that the operators $I - WP^n$ are invertible in $\mathbf{L}^p(\mathbb{R}^+)$ and have uniformly bounded inverses as $n \to \infty$. Indeed, regarding $I - WP^n$ as the matrix operator

$$\begin{pmatrix} P^n(I - W)\,P^n & 0 \\ (I - P^n)\,WP^n & I \end{pmatrix}$$

with respect to the direct topological sum $\mathbf{L}^p(\mathbb{R}^+) = \bar{\mathcal{S}}^d(\Delta^n) \dotplus (I - P^n)\,\mathbf{L}^p(\mathbb{R}^+)$, one easily obtains the assertion from the stability of P^nAP^n and the uniform boundedness of P^n.

Thus 5.42(2) gives the estimate $\|u - u_n^*\|_{p,\mathbb{R}^+} \leqq c\,\|W(I - P^n)\,u\|_{p,\mathbb{R}^+}$, and we have to show that the last expression can be bounded by the right-hand side of (1). Note that $u \in \mathbf{L}_\varrho^{p,d+1}(\mathbb{R}^+)$ (see Remark 5.10 and Theorem 5.9) and $W(I - P^n)\,u = W(I - P^n)^2\,u$. Since $q \geqq (d+1)/(\varrho/2)$, Lemma 5.26 with $\mu = \varrho/2$ implies that the norm of the operators $I - P^n\colon \mathbf{L}_\varrho^{p,d+1}(\mathbb{R}^+) \to \mathbf{L}_{\varrho/2}^p(\mathbb{R}^+)$ can be dominated by cn^{-d-1}. Thus it remains to check that the norm of the operators $W(I - P^n)\colon \mathbf{L}_{\varrho/2}^p(\mathbb{R}^+) \to \mathbf{L}^p(\mathbb{R}^+)$, or equivalently, $(I - P^n) \times W^*\colon \mathbf{L}^{p'}(\mathbb{R}^+) \to \mathbf{L}_{-\varrho/2}^{p'}(\mathbb{R}^+)$, $p' = p/(p-1)$, can be estimated in the same way. Here W^* denotes the convolution operator with kernel $\overline{\varkappa(-x)}$ (cf. Step 2 in the proof of Theorem 5.37) which also satisfies the condition $(\mathrm{A1}^{d+1})$. By Remark 5.10, W^* is a continuous map of $\mathbf{L}^{p'}(\mathbb{R}^+)$ into $\mathbf{L}^{p',d+1}(\mathbb{R}^+)$. To complete the proof of (1), we observe that the norm of the operators $I - P^n\colon \mathbf{L}^{p',d+1}(\mathbb{R}^+) \to \mathbf{L}_{-\varrho/2}^{p'}(\mathbb{R}^+)$ can be estimated by cn^{-d-1} again according to Lemma 5.26. ■

Remark. Under the assumptions of the preceding theorem and the stronger hypothesis $(\mathrm{A1}^{d+1+l})$ for some $l \geqq 1$, we have

$$\|u - u_n^*\|_{p,l,\mathbb{R}^+} = O(n^{-2d-2}), \qquad n \to \infty. \tag{2}$$

Indeed, using the relations

$$\begin{aligned} &\mathrm{D}^j(u - u_n^*) = W_j(u - u_n) = W_j(I - P^n)\,u + W_jP^n(u - u_n^*), \\ &W_j := \mathrm{D}^jW, \qquad j \leqq l \end{aligned} \tag{3}$$

and the fact that W_j is a convolution operator with kernel $\mathrm{D}^j\varkappa$, one obtains the estimates $\|W_j(I - P^n)\,u\|_{p,\mathbb{R}^+} = O(n^{-2d-2})$ by applying the above proof to the operators W_j. Moreover,

$$\|W_jP^n(u - u_n^*)\|_{p,\mathbb{R}^+} \leqq c\,\|u - u_n^*\|_{p,\mathbb{R}^+} = O(n^{-2d-2}), \qquad j \leqq l,$$

which completes the proof of (2).

Note that such error estimates in Sobolev norms are already known in the case of a Fredholm integral equation with smooth kernel on a finite interval; see SLOAN/THOMÉE [1].

Galerkin methods with piecewise polynomials: Mellin convolution equations

5.44. Notation. Consider the Mellin operator $B = I - K$ defined by 5.1(6), and let P_n denote the orthogonal projection of $\mathbf{L}^2(0, 1)$ onto the modified spline space $\bar{\mathcal{S}}^d(\Delta_n)$ (cf. 5.20(1)). Then the Galerkin method for the approximate solution of eq. 5.0(1) is the following: find $u_n \in \bar{\mathcal{S}}^d(\Delta_n)$ such that

$$P_n B u_n = u_n - P_n K u_n = P_n f. \tag{1}$$

We have the following stability result.

5.45. Theorem. *Assume* $1 \leqq p \leqq \infty$, ($\mathrm{H1}^p$), ($\mathrm{H2}^p$), *and let* $\bar{\mathcal{S}}^d(\Delta_n)$ *be a modification of* $\mathcal{S}^d(\Delta_n)$.

(i) *If* $\Delta_n \in \mathcal{M}_q$ *and* i_0 *is sufficiently large, then the estimate*

$$\|P_n B u_n\|_{p,(0,1)} \geqq c\,\|u_n\|_{p,(0,1)}, \qquad u_n \in \bar{\mathcal{S}}^d(\Delta_n), \qquad n \geqq n_0 \tag{1}$$

holds when n_0 *is large enough.*

(ii) *If* $\Delta_n \in \mathcal{M}$, *then* (1) *is valid for* $i_0 = 1$ (*i.e.* $\bar{\mathcal{S}}^d(\Delta_n) = \mathcal{S}^d(\Delta_n)$).

The proof is essentially parallel to that of Theorem 5.37. Consider the representations 5.5(2) for fixed p. The analogue of Lemma 5.38 is

5.46. Lemma. *Let* K_- *be the operator occurring in Theorem* 5.5, *where* δ *is fixed. Then, given* $\varepsilon > 0$, *there is an integer* n_0 *such that*

$$\|(I - P_n)\,K_- u_n\|_{p,(0,1)} \leqq \varepsilon\,\|u_n\|_{p,(0,1)}, \qquad u_n \in \bar{\mathcal{S}}^d(\Delta_n), \qquad n \geqq n_0$$

if either $\Delta_n \in \mathcal{M}$ *and* $i_0 = 1$, *or* $\Delta_n \in \mathcal{M}_q$ *and the index* i_0 *in* 5.20(1) *is sufficiently large.*

Proof. Let $u \in \mathbf{L}^p(0, 1)$ and $h = x_{i_0}$. By Lemma 5.21 (for $l = 1$), one has

$$\begin{aligned}\|(I - P_n)\,K_- u\|_{p,J_i} &\leqq c h_i\,\|\mathrm{D}K_- u\|_{p,J_i}\\ &\leqq c(h_i/x_{i-1})\,\|x\mathrm{D}K_- u\|_{p,J_i}, \qquad i > i_0.\end{aligned}$$

Further, Definition 5.15 (resp. Remark 5.18) implies that $h_i/x_{i-1} \leqq \varepsilon$, $i > i_0$, when $\Delta_n \in \mathcal{M}$ (resp. $\Delta_n \in \mathcal{M}_q$) and n (resp. i_0, n) are large enoguh. Consequently, for those indices i_0 and n, the above estimates and the continuity of the operator $K_-\colon \mathbf{L}^p(0, 1) \to \mathbf{L}^{p,1}(0, 1)$ (cf. Remark 5.7(ii)) imply

$$\|(I - P_n)\,K_- u\|_{p,(h,1)} \leqq c\varepsilon\,\|x\mathrm{D}K_- u\|_{p,(0,1)} \leqq c\varepsilon\,\|u\|_{p,(0,1)}.$$

Together with the relation (cp. Remark 5.7(i))

$$(I - P_n)\,K_- u_n|_{(0,h)} = 0, \qquad u_n \in \bar{\mathcal{S}}^d(\Delta_n),$$

this yields the results. ∎

Proof of Theorem 5.45. *Step 1:* By Collary 5.22, the projections P_n are uniformly bounded on $\mathbf{L}^p(0, 1)$ (for arbitrary modifications of the spline spaces $\mathcal{S}^d(\Delta_n)$). 5.5(2) implies

$$P_nBP_n = P_n(B_1 + T)\,P_n, \qquad B_1 := (I - K_+)\,(I - K_0)\,(I - K_-),$$
$$\|K_0\|_p < \delta, \tag{1}$$

and choosing δ such that $\sup_n \|P_n\|_p\,\delta \leqq 1$, we observe that the operators $P_n(I - K_0)\,P_n$ are stable in $\mathbf{L}^p(0, 1)$ (even uniformly with respect to i_0).

Step 2: Using Lemma 5.46 and the invertibility of $I - K_-$, we obtain the stability of the operators $P_n(I - K_-)\,P_n$. To prove the stability of $P_n(I - K_+)\,P_n$, one uses a duality argument which is analogous to that in Theorem 5.37. Note that the Mellin convolution operator K^* with kernel $\overline{k(x^{-1})}\,x^{-1}$ acting on $\mathbf{L}^{p'}(0, 1)$ is the adjoint of $K: \mathbf{L}^p(0, 1) \to \mathbf{L}^p(0, 1)$, where $p' = p/(p - 1)$ and $p < \infty$. Moreover, $F^*_{p,+} = F_{p',-}$, where $F_{p,\pm}$ are the elementary operators defined in 5.4(3).

Step 3: Since Lemma 5.46 also applies to the operator K^*_+ (acting on $\mathbf{L}^{p'}(0, 1)$), the norm of the operators

$$P_nB_1P_n - P_n(I - K_+)\,P_n(I - K_0)\,P_n(I - K_-)\,P_n$$

becomes as small as desired when $n \to \infty$ and i_0 is chosen large enough. Consequently, by the considerations in the first and second steps, Theorem 5.45 is valid with B replaced by B_1.

Step 4: If $p < \infty$, then $\mathbf{C}_0^\infty(0, 1)$ is dense in $\mathbf{L}^p(0, 1)$, and the operators P_n converge strongly to the identity on $\mathbf{L}^p(0, 1)$ (according to their uniform boundedness and Lemma 5.21). Now (1) and Remark 1.27, 3°(b) imply the stability of P_nBP_n since the operator T is compact on $\mathbf{L}^p(0, 1)$.

Let $p = \infty$. Using the compactness of the map $T: \mathbf{L}^\infty(0, 1) \to \mathring{\mathbf{C}}[0, 1]$ and the fact that P_n converges strongly to the identity on $\mathring{\mathbf{C}}[0, 1]$ (cf. Lemma 5.21 again), we may obtain 5.45(1) by the same arguments as in Theorem 5.37. We only have to check that if the sequence $\{u_n\} \subset \mathbf{L}^\infty(0, 1)$ converges uniformly on each compact subinterval of $(0, 1]$ then so does the sequence $\{Ku_n\}$. But this can be reduced to the corresponding assertion for Wiener-Hopf operators, with the help of the transformation R_∞ given by 5.1(7). ■

5.47. Remark. Suppose that the Mellin operator 5.1(6) with symbol $1 - b$ fulfils condition (H1²). B is called strongly elliptic if there exist constants $\lambda \in \mathbb{C}$, $|\lambda| = 1$, and $c > 0$ such that

$$\operatorname{Re}\{\lambda(1 - b(z))\} \geqq c, \qquad z \in \Gamma_{1/2}. \tag{1}$$

If B is strongly elliptic, then the method 5.44(1) with $i_0 = 1$ is stable in $\mathbf{L}^2(0, 1)$ for arbitrary sequences of meshes. Indeed, using Parseval's relation for the Mellin transform, from (1) and 5.1(9) (for $p = 2$) we obtain the estimate

$$\begin{aligned}\operatorname{Re}\left\{\lambda\int_0^1 Bu\,\bar{u}\,\mathrm{d}x\right\} &= \operatorname{Re}\,(2\pi)^{-1}\left\{\lambda\int_{\Gamma_{1/2}}\big(1 - b(z)\big)\,|\widetilde{\chi u}(z)|^2\,|\mathrm{d}z|\right\}\\ &\geqq (2\pi)^{-1}\int_{\Gamma_{1/2}}|\widetilde{\chi u}(z)|^2\,|\mathrm{d}z| = c\,\|u\|^2_{2,(0,1)}, \qquad u \in \mathbf{L}^2(0, 1)\end{aligned}$$

which obviously implies

$$\|P_nBu_n\|_{2,(0,1)} \geqq c\,\|u_n\|_{2,(0,1)}, \qquad u_n \in \mathcal{S}^d(\Delta_n), \qquad n \in \mathbb{N}.$$

Note that by Proposition 6.17 the operator B is strongly elliptic if and only if it takes the form $B = \lambda_1(I + C)$, where $\lambda_1 \in \mathbb{C} \setminus \{0\}$ and the $\mathbf{L}^2$ operator norm of C is smaller than one.

5.48. Remark. We now give an example showing that Theorem 5.45 does not hold, in general, when $\Delta_n \in \mathcal{M}_q$ and $i_0 = 1$. Consider the convolution operator

$$Ku(x) = x^{-1} \int_0^x u(y)\,\mathrm{d}y \quad \text{on} \quad \mathbf{L}^\infty(0,1)$$

with kernel $k(x) = x^{-1}(1 - \chi(x))$ and symbol $\tilde{k}(z) = -(z-1)^{-1}$, $z \in \Gamma_0$. Then $K^2u(x) = x^{-1} \int_0^x \log(x/y)\,u(y)\,\mathrm{d}y$, and the symbol of K^2 is given by $(z-1)^{-2}$, $z \in \Gamma_0$. For $\lambda \in \mathbb{R}$, we define the Mellin operator

$$B_\lambda = I - K_\lambda, \qquad K_\lambda = 2\lambda K - \lambda^2 K^2,$$

having the symbol $\{1 + \lambda(z-1)^{-1}\}^2$. Note that the kernel function $2\lambda k(x) - \lambda^2 \log x\, k(x)$ of K_λ fulfils condition (H1$^\infty$). Furthermore, for any $\lambda < 1$, B_λ satisfies (H2$^\infty$) since then

$$\{\arg\,[1 + \lambda/(\mathrm{i}\xi - 1)]\}_{\xi=-\infty}^{\infty} = 0.$$

Let Δ_n be partitions of the unit interval. A basis of the spline space $\mathcal{S}^0(\Delta_n)$ is given by the characteristic functions θ_i of the intervals J_i, $i = 1, \ldots, n$. With respect to this basis, the operator $P_nB_\lambda P_n$ can be represented as the matrix $(\alpha_{ij})_{i,j=1}^n$, $\alpha_{ij} = (B_\lambda\theta_j, \theta_i)$. Here P_n denotes the orthogonal projection of $\mathbf{L}^2(0,1)$ onto $\mathcal{S}^0(\Delta_n)$ and $(\cdot,\cdot)$ stands for the scalar product in $\mathbf{L}^2(0,1)$. Since the kernel of K_λ vanishes on $(0,1)$, we observe that the matrix (α_{ij}) is in lower triangular form.

We now compute the diagonal element α_{22}. For $x \in J_2$, one has

$$K\theta_2(x) = 1 - x_1/x, \qquad K^2\theta_2(x) = 1 - x_1/x - (x_1/x)\log(x/x_1),$$

hence

$$B_\lambda\theta_2(x) = (1-\lambda)^2 + 2\lambda(x_1/x) - \lambda^2\big(x_1/x + (x_1/x)\log(x/x_1)\big).$$

Setting $x_2 = \eta x_1$, $\eta > 1$ and $\varrho(\lambda, \eta) := \alpha_{22}/x_1$, we thus obtain

$$\varrho(\lambda,\eta) = (B_\lambda\theta_2, \theta_2)/x_1 = (1-\lambda)^2(\eta - 1) + (2\lambda - \lambda^2)\log\eta - \lambda^2\log^2\eta/2.$$

Note that $\varrho(1,\eta) = \log\eta - \log^2\eta/2$ is strictly positive for $\eta \in (1, \mathrm{e}^2)$ and strictly negative for $\eta > \mathrm{e}^2$. Therefore, choosing a number $\lambda^0 < 1$ sufficiently close to 1, we have $\varrho(\lambda^0, 5) > 0$ and $\varrho(\lambda^0, 10) < 0$, hence there exists a number $\eta^0 \in (5, 10)$ satisfying $\varrho(\lambda^0, \eta^0) = 0$.

Choosing the sequence of partitions

$$\Delta_n = \{0, 1/n, \eta^0/n; \eta^0/n + i(1 - \eta^0/n)/(n-1),\ i = 1, \ldots, n-1\}, \qquad n \geqq 10,$$

which belongs to the class $\mathcal{M}_1$, we see that the operators $P_nB_{\lambda^0}P_n$ cannot be stable in $\mathbf{L}^\infty(0,1)$ since the corresponding matrix operator has a nontrivial kernel for any n.

The next theorem extends the stability result of Remark 5.47 to the spaces $\mathbf{L}^p$ when graded meshes are used. For its proof, we apply an idea of CHANDLER and GRAHAM [1], who studied the special case when $p = \infty$ and the $\mathbf{L}^2$ and $\mathbf{L}^\infty$ operator norms of K are smaller than one.

5.49. Theorem. *Assume* (H1²), (H1ᵖ), (H2ᵖ), $p \neq 2$, *and suppose that* B *is strongly elliptic. Then, for* $\Delta_n \in \mathcal{M}_q$ *and* $i_0 = 1$, *the Galerkin method* 5.44(1) *is stable in* $\mathbf{L}^p(0, 1)$.

Proof. By Remark 5.47, the inequality

$$\|P_n B u_n\|_{2,(0,1)} \geqq c\, \|u_n\|_{2,(0,1)}, \qquad u_n \in \mathcal{S}^d(\Delta_n), \qquad n \in \mathbb{N} \tag{1}$$

holds. We have to show that this estimate is valid if the $\mathbf{L}^2$ norm is replaced by the $\mathbf{L}^p$ norm and n is large enough ($n \geqq n_0$, say).

Let $h = x_{j_0}$, where the index j_0 will be chosen sufficiently large, but independent of n. Let χ_h be the characteristic function of $[0, h]$, and define the operators $\pi_0 u = \chi_h u$ and $\pi_1 = I - \pi_0$, the norm of which in $\mathbf{L}^p(0, 1)$ is bounded by one for any p, n and j_0. We further define the operators $L_n = P_n - \pi_1 P_n K P_n$ which admit the matrix representations

$$\begin{pmatrix} I & 0 \\ -\pi_1 P_n K \pi_0 P_n & \pi_1 P_n B \pi_1 P_n \end{pmatrix} \tag{2}$$

with respect to the direct sums $\mathcal{S}^d(\Delta_n) = \pi_0 \mathcal{S}^d(\Delta_n) \dotplus \pi_1 \mathcal{S}^d(\Delta_n)$. By Remark 5.53 below, we can choose j_0 so large that the operators $\pi_1 P_n B \pi_1 P_n$ are stable in $\mathbf{L}^p(0, 1)$ and $\mathbf{L}^2(0, 1)$, which implies the stability of the operators L_n in the same spaces according to representation (2). Note that $P_n B P_n = M_n L_n$, where $M_n = I - \pi_0 P_n K L_n^{-1}$. Thus it remains to verify the stability of the operators M_n, and in order to do so, it is enough to prove the inequalities

$$\|M_n u\|_{p,(0,1)} \geqq c\, \|u\|_{p,(0,1)}, \qquad u \in Y_n^p, \qquad n \geqq n_0, \tag{3}$$

where Y_n^p is the finite dimensional space $\pi_0 \mathcal{S}^d(\Delta_n)$ equipped with the $\mathbf{L}^p$ norm. We see that (3) holds for $p = 2$ since the operators $P_n B P_n$ and L_n are stable with respect to the $\mathbf{L}^2$ norm. Furthermore, because $h_i \sim i^{q-1}/n^q$, $i = 1, \ldots, j_0$ (cf. Definition 5.17), a simple scaling argument can be used to construct isomorphisms $J_n \colon Y_n^p \to Y_{n_0}^p$ which are independent of p and satisfy

$$\|J_n u\|_{p,(0,1)} \sim n^{q/p}\, \|u\|_{p,(0,1)}, \qquad u \in Y_n^p, \qquad n \geqq n_0 \tag{4}$$

for any p ($\sim$ means uniform equivalence with respect to n). Since all norms on $Y_{n_0}^p$ are equivalent, (4) yields

$$\|u\|_{p,(0,1)} \sim \|u\|_{2,(0,1)}\, n^{q/2 - q/p}, \qquad u \in Y_n^p, \qquad n \geqq n_0,$$

and since (3) is already valid for $p = 2$, this implies the desired result. ∎

We now study the convergence of the Galerkin method 5.44(1). First, Theorem 5.45 yields the following

5.50. Corollary. *Suppose that the assumptions of Theorem* 5.45 *are satisfied.*

(i) *For all* i_0 *and* n *sufficiently large, the Galerkin equations* 5.44(1) *are uniquely solvable for any* $f \in \mathbf{L}^p(0, 1)$ *and the error bound*

$$\|u - u_n\|_{p,(0,1)} \leqq c\, \|(I - P_n)\, u\|_{p,(0,1)} \tag{1}$$

holds, where $u \in \mathbf{L}^p(0,1)$ *denotes the exact solution of* 5.0(1). *Moreover, for* $p < \infty$, *the left-hand side of* (1) *converges to* 0 *as* $n \to \infty$.
(ii) *If* $f \in \mathbf{C}[0,1]$, *then* u_n *converges uniformly to* u *on* $[0,1]$.
(iii) *If* $f \in \mathbf{C}(0,1]$, *then* u_n *converges uniformly to* u *on each compact subinterval of* $(0,1]$.

Proof. (i) Inequality (1) is a standard consequence of 5.45(1). The second assertion follows from the strong convergence of the projections P_n to the identity on $\mathbf{L}^p(0,1)$ (cf. Step 4 in the proof of Theorem 5.45).

(ii) Since $\mathbf{C}[0,1] = \mathring{\mathbf{C}}[0,1] \dotplus \mathbb{C}$ and the operators P_n converge strongly to the identity on $\mathring{\mathbf{C}}[0,1]$, estimate (1) implies the result.

(iii) Arguing as in Corollary 5.39(iii), we only have to check that if the sequence u_n is bounded in $\mathbf{L}^\infty(0,1)$ then Ku_n is (uniformly) equicontinuous on $(0,1]$ and that if u_n converges uniformly on each compact subset of $(0,1]$ then so does the sequence Ku_n. This follows, however, by using the transformation R_∞ (see 5.1(7)) and the corresponding assertions for Wiener-Hopf equations. ■

Secondly, we show that one obtains the optimal rate of convergence for Galerkin's method when suitably graded meshes are used.

5.51. Theorem. *Assume* $1 \leqq p \leqq \infty$, $\varrho > 0$, $(\mathrm{H1}^p)$, $(\mathrm{H1}^p_\varrho)$, $(\mathrm{H2}^p)$, $(\mathrm{H2}^p_\varrho)$ *and* $k(x) \in \mathbf{L}^{p,d}_\varrho(0,1)$. *Suppose further that* $f \in \mathbf{L}^{p,d+1}_\varrho(0,1) \dotplus \mathbb{P}_k$, $k < \min\{\varrho - 1/p,\ d+1\}$, $\Delta_n \in \mathcal{M}_q$ *with* $q \geqq (d+1)/\varrho$, *and let* $\bar{\mathcal{F}}^d(\Delta_n)$ *be a modification of* $\mathcal{S}^d(\Delta_n)$ *such that the index* i_0 *in* 5.20(1) *is large enough. Then the Galerkin method* 5.44(1) *converges with the error bound*

$$\|u - u_n\|_{p,(0,1)} = O(n^{-d-1}) \quad as \quad n \to \infty. \tag{1}$$

Proof. By Theorem 5.12, the solution of 5.0(1) admits the representation $u = u_0 + \varphi$, where $u_0 \in \mathbf{L}^{p,d+1}_\varrho(0,1)$ and $\varphi \in \mathbb{P}_k$. Now 5.50(1) implies that

$$\|u - u_n\|_{p,(0,1)} \leqq c\,\|(I - P_n)\,u_0\|_{p,(0,1)}. \tag{2}$$

Consider the projections $\mathring{P}_n$ onto the subspaces $\bar{\mathcal{F}}^d_0(\Delta_n)$ of $\bar{\mathcal{F}}^d(\Delta_n)$ defined in Sect. 5.20. Since the operators P_n are uniformly bounded on $\mathbf{L}^p(0,1)$ and $P_n\mathring{P}_n = \mathring{P}_n$, one obtains the estimate

$$\begin{aligned}\|(I - P_n)\,u_0\|_{p,(0,1)} &\leqq \|(I - \mathring{P}_n)\,u_0\|_{p,(0,1)} + \|P_n u_0 - \mathring{P}_n u_0\|_{p,(0,1)}\\ &\leqq (1 + \|P_n\|_p)\,\|(I - \mathring{P}_n)\,u_0\|_{p,(0,1)}\\ &\leqq c\,\|(I - \mathring{P}_n)\,u_0\|_{p,(0,1)}.\end{aligned} \tag{3}$$

Since the last term in (3) can be dominated by cn^{-d-1} according to Lemma 5.23, the estimate (1) is proven. ■

5.52. Remark. Suppose that B satisfies the assumptions of the preceding theorem. Let Δ_N be the partitions defined in Example 5.16, 2° with $q \geqq (d+1)/\varrho$, and let $f \in \mathbf{L}^{p,d+1}_\varrho(0,1) \dotplus \mathbb{P}_k$, $k < \min\{\varrho - 1/p, d+1\}$ and $i_0 = 1$. Then the Galerkin method 5.44(1) (with n replaced by N) converges with the error estimate

$$\|u - u_N\|_{p,(0,1)} = O\big((\log N/N)^{d+1}\big) \quad \text{as} \quad N \to \infty. \tag{1}$$

This is a consequence of 5.51(2), 5.51(3) and Remark 5.24.

5.53. Remark. For the convergence analysis of the collocation method with piecewise polynomials, which will be studied in Sect. 5.56–5.68, we need the stability of the Galerkin method

$$\mathring{P}_n B u_n = \mathring{P}_n f, \qquad u_n \in \bar{\mathcal{S}}_0^d(\Delta_n). \tag{1}$$

By an inspection of the corresponding proofs, we observe that, with obvious modifications, the assertions in 5.45 to 5.52 continue to hold for the method (1). In Corollary 5.50(ii), the stronger assumption $f \in \check{C}[0, 1]$ is needed, and one has to assume $\mathbb{P}_k = \{0\}$ in Theorem 5.51 and Remark 5.52.

Finally, we study the rate of convergence for the iterated Galerkin method. Set $u_n^* = f + Ku_n$, where u_n is given by 5.44(1).

5.54. Theorem. *Assume* $1 \leqq p \leqq \infty$, $\varrho > 0$, $(\mathrm{H1}^{p,d+1})$, $(\mathrm{H1}_\varrho^{p,d+1})$ $(\mathrm{H2}^p)$ *and* $(\mathrm{H2}_\varrho^p)$. *If* $f \in \mathbf{L}_\varrho^{p,d+1}(0, 1) \dotplus \mathbb{P}_k$, $k < \min\{\varrho - 1/p\, d + 1\}$, $\Delta_n \in \mathcal{M}_q$ *with* $q \geqq 2(d + 1)/\varrho$, *and* $\bar{\mathcal{S}}^d(\Delta_n)$ *is a suitable modification of* $\mathcal{S}^d(\Delta_n)$, *then*

$$\|u - u_n^*\|_{p,(0,1)} = O(n^{-2d-2}) \quad as \quad n \to \infty. \tag{1}$$

Proof. The arguments are essentially parallel to that in Theorem 5.43.

We have the estimate $\|u - u_n^*\|_{p,(0,1)} \leqq c\, \|K(I - P_n)\, u\|_{p,(0,1)}$, where $u = u_0 + \varphi$, $u_0 \in \mathbf{L}_\varrho^{p,d+1}(0, 1)$, $\varphi \in \mathbb{P}_k$ in view of Remark 5.13, 1° and Theorem 5.12. Moreover,

$$K(I - P_n)\, u = K(I - P_n)\, u_0 = K(I - P_n)\, (I - \mathring{P}_n)\, u_0 .$$

To verify (1), we thus have to show that the norms of the operators

$$I - \mathring{P}_n : \mathbf{L}_\varrho^{p,d+1}(0, 1) \to \mathbf{L}_{\varrho/2}^p(0, 1), \qquad (I - P_n)\, K^* : \mathbf{L}^{p'}(0, 1) \to \mathbf{L}_{-\varrho/2}^{p'}(0, 1)$$

can be estimated by cn^{-d-1}. Here $p' = p/(p - 1)$, and K^* denotes the Mellin convolution operator with kernel $\overline{k(x^{-1})}\, x^{-1}$ (see Step 2 in the proof of Theorem 5.45). The first assertion follows immediately from Lemma 5.23 (for $\mu = \varrho/2$). To prove the second, we observe that K^* satisfies the condition $(\mathrm{H1}^{p',d+1})$, hence K^* is a continuous map of $\mathbf{L}^{p'}(0, 1)$ into $\mathbf{L}^{p',d+1}(0, 1)$ (cf. Remark 5.13, 1°). Therefore it is sufficient to show that the norm of the operators $I - P_n : \mathbf{L}^{p',d+1}(0, 1) \to \mathbf{L}_{-\varrho/2}^{p'}(0, 1)$ can be estimated by cn^{-d-1}. But this is a consequence of Lemma 5.23 and the inequalities

$$\begin{aligned} \|x^{\varrho/2}\, (I - P_n)\, u\|_{p',(0,h)} &\leqq ch^{\varrho/2}\, \{\|u\|_{p',(0,h)} + \|P_n u\|_{p',(0,h)}\} \\ &\leqq cn^{-q\varrho/2}\, \|u\|_{p',(0,h)} \\ &\leqq cn^{-d-1}\, \|u\|_{p',(0,h)}, \end{aligned}$$

where we have used 5.22(1) and $q\varrho/2 \geqq d + 1$. ■

5.55. Remarks. 1°. Under the assumptions of the preceding theorem and the stronger hypothesis $(\mathrm{H1}^{p,d+1+l})$ for some $l \geqq 1$, the error estimate

$$\|u - u_n^*\|_{p,l,(0,1)} = O(n^{-2d-2}), \qquad n \to \infty,$$

holds. As in Remark 5.44, one may reduce the assertion to the case $l = 0$ by using the relation

$$u - u_n^* = K(u - u_n) = K(I - P_n)\, u + KP_n(u - u_n^*)$$

and the fact that $(x\mathrm{D})^j K$, $j \leqq l$, is a Mellin convolution operator with kernel function $(x\mathrm{D})^j\, k(x)$.

2°. It is possible to extend the results of this section to Galerkin's method with continuous piecewise polynomials. To prove this, one has to use Lemma 5.31(i), Lemma 5.33(i) and the (non-trivial) fact that the orthogonal projections onto $\bar{\mathscr{S}}^{d,1}(\Delta_n)$ are uniformly bounded on $\mathbf{L}^p$, $1 \leqq p \leqq \infty$; see DE BOOR [2], CROUZEIX/THOMÉE [1]. Furthermore, most of the above results may be generalized to smoothest polynomial splines, at least in the case $p = 2$ (cf. ELSCHNER [10]).

Collocation with piecewise polynomials: Mellin convolution equations

5.56. Notation. Using the notation of Section 5.20, we consider the following collocation method for the approximate solution of equation 5.0(1): find $u_n \in \bar{\mathscr{S}}_0^d(\Delta_n)$ such that (cf. 5.27(2))

$$Q_n B u_n = Q_n f. \tag{1}$$

First, we investigate the stability of the method (1) in $\mathbf{L}^\infty(0, 1)$.

5.57. Theorem. *Assume* (H1$^\infty$), (H2$^\infty$), *and let* $\bar{\mathscr{S}}_0^d(\Delta_n)$ *be a modification of* $\mathscr{S}_0^d(\Delta_n)$.

(i) *If* $\Delta_n \in \mathscr{M}_q$ *and the index* i_0 *in* 5.20(2) *is sufficiently large, then*

$$\|Q_n B u_n\|_{\infty,(0,1)} \geqq c\, \|u_n\|_{\infty,(0,1)}, \qquad u_n \in \bar{\mathscr{S}}_0^d(\Delta_n), \qquad n \geqq n_0 \tag{1}$$

when n_0 *is large enough.*

(ii) *If* $\Delta_n \in \mathscr{M}$, *then* (1) *holds for* $i_0 = 1$ (*i.e.* $\bar{\mathscr{S}}_0^d(\Delta_n) = \mathscr{S}_0^d(\Delta_n)$).

Proof. We regard $Q_n B Q_n$ as a small perturbation of $\mathring{P}_n B \mathring{P}_n$ and apply the stability of the Galerkin method 5.53(1) (cf. Remark 5.53).

For any $\delta > 0$, B admits the representation $B = I - K_\delta - K_\delta'$, where $\|K_\delta'\|_\infty < \delta$ and K_δ is a polynomial in the elementary operators $F_{\infty,\pm}$ (cf. Sect. 5.5). According to the stability of 5.53(1) in $\mathbf{L}^\infty(0, 1)$ and the uniform boundedness of $\mathring{P}_n$, there exists a positive constant c which does not depend on u, n, i_0 and δ such that, for all sufficiently small $\delta > 0$,

$$\|\mathring{P}_n(I - K_\delta)\, u_n\|_{\infty,(0,1)} \geqq c\, \|u_n\|_{\infty,(0,1)}, \qquad u_n \in \bar{\mathscr{S}}_0^d(\Delta_n) \tag{2}$$

when $\Delta_n \in \mathscr{M}$ (resp. $\Delta_n \in \mathscr{M}_q$) and n (resp. i_0, n) are large enough. Moreover, applying the continuity of the operators $K_\delta\colon \mathbf{L}^\infty(0, 1) \to \mathbf{L}^{\infty,1}(0, 1)$ and arguing as in the proof of Lemma 5.46, for any fixed $\varepsilon, \delta > 0$ one obtains the estimate

$$\|(I - \mathring{P}_n)\, K_\delta u_n\|_{\infty,(h,1)} \leqq \varepsilon\, \|u_n\|_{\infty,(0,1)}, \qquad u_n \in \bar{\mathscr{S}}_0^d(\Delta_n) \tag{3}$$

whenever i_0 and n are sufficiently large. Choosing ε sufficiently small and using the uniform boundedness of the projections Q_n (cf. Lemma 5.28), we deduce from (3) that

$$\begin{aligned} \|(\mathring{P}_n - Q_n)\,(I - K_\delta)\, u_n\|_{\infty,(0,1)} &= \|Q_n(I - \mathring{P}_n)\, K_\delta u_n\|_{\infty,(0,1)} \\ &\leqq (c/2)\, \|u_n\|_{\infty,(0,1)}, \qquad u_n \in \bar{\mathscr{S}}_0^d(\Delta_n), \end{aligned} \tag{4}$$

where i_0 and n are large enough. Selecting δ sufficiently small, (2) and (4) imply the estimate

$$\|Q_n(I - K_\delta)\, u_n\|_{\infty,(0,1)} \geqq (c/2)\, \|u_n\|_{\infty,(0,1)}, \qquad u_n \in \bar{\mathscr{S}}_0^d(\Delta_n)$$

for those indices i_0 and n, hence the result. ∎

5.58. Remark. We now give a counter-example to the preceding theorem when $\Delta_n \in \mathcal{M}_q$ and $i_0 = 1$. Let B_λ, $\lambda < 1$, be the operator introduced in Remark 5.48 and consider the partitions

$$\Delta_n = \{x_i : i = 0, \ldots, n\}, \qquad x_1 = 1/n, \qquad x_2 = \eta/n,$$
$$x_{2+i} = \eta/n + i(1 - \eta/n)/(n - 1), \qquad i \geqq 1$$

again, where $\eta > 1$ will be suitably chosen later on. Consider the projections Q_n onto $\mathcal{S}_0^0(\Delta_n)$ which correspond to the choice $\xi_0 = 1$ (cf. 5.27(2)). A basis of $\mathcal{S}_0^0(\Delta_n)$ is given by the characteristic functions θ_i, $i = 2, \ldots, n$, and with respect to this basis, the operator $Q_n B_\lambda Q_n$ can be represented as the matrix $(\beta_{ij})_{i,j=2}^n$, $\beta_{ij} = (B_\lambda \theta_j)(x_i)$, which is in lower triangular form. We have (cf. Remark 5.48)

$$\sigma(\lambda, \eta) := \beta_{22} = (1 - \lambda)^2 + (2\lambda - \lambda^2)\,\eta^{-1} - \lambda^2 \eta^{-1} \log \eta\,.$$

Note that $\sigma(1, \eta) \gtreqless 0$ if $\eta \lesseqgtr e$, hence one may choose a number $\lambda^0 < 1$ sufficiently close to 1 such that $\sigma(\lambda^0, 2) > 0$ and $\sigma(\lambda^0, 3) < 0$. Consequently, there exists a number $\eta = \eta^0 \in (2, 3)$ satisfying $\sigma(\lambda^0, \eta^0) = 0$. With this choice of η, the operators $Q_n B_{\lambda^0} Q_n$ are obviously not stable in $\mathbf{L}^\infty(0, 1)$.

Note that Chandler and Graham [2] detected numerically a counter-example showing that Theorem 5.57(i) does not hold for $i_0 = 1$, in general, even if $\|K\|_\infty < 1$.

Next we study the convergence of our collocation method.

5.59. Corollary. *Suppose that the assumptions of Theorem* 5.57 *are fulfilled.*

(i) *For all i_0 and n sufficiently large, the collocation equations* 5.56(1) *are uniquely solvable for any $f \in \mathbf{L}^\infty(0, 1)$ and the error estimate*

$$\|u - u_n\|_{\infty,(0,1)} \leqq c\,\|(I - \mathring{P}_n)\,u\|_{\infty,(0,1)} \tag{1}$$

holds, where $u \in \mathbf{L}^\infty(0, 1)$ denotes the exact solution of equation 5.0(1).

(ii) *If $f \in \mathring{\mathbf{C}}[0, 1]$, then the left-hand side of* (1) *converges to* 0 *as $n \to \infty$.*

(iii) *If $f \in \mathbf{C}(0, 1]$, then u_n converges uniformly to u on each compact subinterval of* (0, 1].

Proof. As a consequence of 5.57(1), we have the estimate $\|u - u_n\|_{\infty,(0,1)} \leqq c\,\|(I - Q_n) \times u\|_{\infty,(0,1)}$. Furthermore, by the uniform boundedness of the projections Q_n,

$$\|(I - Q_n)\,u\|_{\infty,(0,1)} \leqq c\,\|(I - \mathring{P}_n)\,u\|_{\infty,(0,1)}, \qquad u \in \mathbf{L}^\infty(0, 1) \tag{2}$$

so that Q_n converges strongly to the identity on $\mathring{\mathbf{C}}[0, 1]$. With these observations, the proof runs analogously to that of Corollary 5.50 (for $p = \infty$). ■

With suitably graded meshes, one now obtains the optimal order of convergence for the method 5.56(1).

5.60. Theorem. *Assume $\varrho > 0$,* (H1$^\infty$), (H1$_\varrho^\infty$), (H2$^\infty$), (H2$_\varrho^\infty$) *and $k(x) \in \mathbf{L}_\varrho^{\infty,d}(0, 1)$. Suppose further that $f \in \mathbf{L}_\varrho^{\infty,d+1}(0, 1)$, $\Delta_n \in \mathcal{M}_q$ with $q \geqq (d + 1)/\varrho$, and let $\tilde{\mathcal{S}}_0^d(\Delta_n)$ be a modification of $\mathcal{S}_0^d(\Delta_n)$ such that the number i_0 in* 5.20(2) *is large enough. Then*

$$\|u - u_n\|_{\infty,(0,1)} = O(n^{-d-1}) \quad as \quad n \to \infty\,.$$

Proof. Notice that Theorem 5.12 yields $u \in \mathbf{L}_\varrho^{\infty,d+1}(0, 1)$. It remains to apply estimate (1) and Lemma 5.23. ■

Further, we note that the "almost" optimal error estimate 5.52(1) in the $\mathbf{L}^\infty$ norm is also valid for the collocation solutions u_N if the assumptions of Remark 5.52 are satisfied with $p = \infty$ and $\mathbb{P}_k = \{0\}$.

5.61. Remark. The preceding theorem extends to the case $f \in \mathbf{L}_\varrho^{\infty,d+1}(0, 1) \dot{+} \mathbb{C}$ when $u_n = u_n^0 + c$, $u_n^0 \in \mathcal{S}_0^d(\Delta_n)$, $c \in \mathbb{C}$, is sought as the solution of the following modified collocation equations:

$$c\big(1 - b(0)\big) = f(0), \qquad Q_n B u_n^0 = Q_n g, \tag{1}$$

where b is the symbol of K and $g = f - c + cK1$. Note that $c = u(0)$ and $g \in \mathbf{L}_\varrho^{\infty,d+1}(0, 1)$ since

$$(K1)(x) = \int_0^1 k(x/y)\, y^{-1}\, dy = \int_x^\infty k(y)\, y^{-1}\, dy$$

implies that $(K1)(0) = b(0)$ and $(Bu)(0) = \big(1 - b(0)\big)\, u(0) = f(0)$.

Finally, we investigate the convergence of the iterated collocation (or product integration) method. The iterated collocation solution is defined by

$$u_n^* := f + K u_n, \tag{2}$$

where u_n is given by 5.56(1). As in 5.42(2), it follows from (2) and 5.56(1) that

$$(I - KQ_n)(u - u_n^*) = K(I - Q_n)\, u. \tag{3}$$

Note that $Q_n u_n^* = u_n$. Thus the values $u_n^*(x_{ij})$ can be obtained by solving the collocation equations 5.56(1), whereas from these values the right-hand side of (2) may be computed giving $u_n^*(x)$ for all $x \in [0, 1]$. Here $x_{ij} = x_{i-1} + \xi_j h_i$ ($j = 0, \ldots, d$; $i = i_0 + 1, \ldots, n$) denotes the collocation points (cf. 5.27(2)). In contrast to the method 5.56(1), the order of convergence of (2) depends on the choice of the points ξ_j, $0 \leqq \xi_0 < \cdots < \xi_d \leqq 1$.

Let $v \in \mathbf{C}[0, 1]$ and introduce the $(d + 1)$-point interpolatory quadrature rule on $[0, 1]$

$$\int_0^1 v\, dx \approx \int_0^1 Qv\, dx =: \sum_{j=0}^{d} \omega_j v(\xi_j) \tag{4}$$

with weights ω_j, where $Qv \in \mathbb{P}_d$ interpolates v at $\xi_0, \ldots, \xi_d$. Let $R = d + 1 + r$ be the order of this rule, i.e.

$$\int_0^1 v\, dx = \int_0^1 Qv\, dx, \qquad v \in \mathbb{P}_{d+r}. \tag{5}$$

(5) is of course equivalent to each of the following two conditions:

$$\int_0^1 \left\{ \prod_{j=0}^{d} (\xi - \xi_j) \right\} \varphi(\xi)\, d\xi = 0, \qquad \varphi \in \mathbb{P}_{r-1}; \tag{6}$$

$$\int_0^1 \psi\varphi\, dx = \int_0^1 \psi Q\varphi\, dx, \qquad \varphi \in \mathbb{P}_{d+r-\nu}, \qquad \psi \in \mathbb{P}_\nu, \qquad \nu = 0, 1, \ldots, r. \tag{7}$$

Simpson's rule, for example, corresponds to the choice $\xi_0 = 0$, $\xi_1 = 1/2$, $\xi_2 = 1$, and in this case we have $r = 1$. Note that $0 \leqq r \leqq d + 1$ and $r = d + 1$ if and only if ξ_j are the Gauss-Legendre points on [0, 1], i.e. $\xi_j = (z_j + 1)/2$, where the z_j are the $d + 1$ zeros of the corresponding Legendre polynomial (cf. BRUNNER/VAN DER HOUWEN [1], Sect. 2.3).

Now we are in a position to prove

5.62. Theorem. *Assume* $\varrho > 0$, 5.61(5), $(\mathrm{H}1^{\infty,r})$, $(\mathrm{H}1_\varrho^{\infty,R})$, $(\mathrm{H}2^\infty)$ *and* $(\mathrm{H}2_\varrho^\infty)$. *If* $f \in \mathbf{L}_\varrho^{\infty,R}(0, 1)$, $\Delta_n \in \mathcal{M}_q$ *with* $q \geqq R/\varrho$, *and* $\bar{\mathcal{S}}_0^d(\Delta_n)$ *is an appropriate modification of* $\mathcal{S}_0^d(\Delta_n)$, *then*

$$\|u - u_n^*\|_{\infty,(0,1)} = O(n^{-R}) \quad as \quad n \to \infty. \tag{1}$$

Proof. Using Theorem 5.57(i) and the uniform boundedness of Q_n, as in Theorem 5.43 one derives that the operators $I - KQ_n$ are invertible in $\mathbf{L}^\infty(0, 1)$ and have uniformly bounded inverses as $n \to \infty$. Thus 5.61(3) gives the estimate

$$\|u - u_n^*\|_{\infty,(0,1)} \leqq c\, \|K(I - Q_n)\, u\|_{\infty,(0,1)}. \tag{2}$$

Further, we have

$$\|K(I - Q_n)\, u\|_{\infty,(0,1)} = \sup \left|\left((I - Q_n)\, u,\, K^*v\right)\right|, \tag{3}$$

where the supremum is taken over all $v \in \mathbf{L}^1(0, 1)$, $\|v\|_{1,(0,1)} \leqq 1$, K^* is the Mellin convolution operator with kernel $\overline{k(x^{-1})}\, x^{-1}$ and $(.,.)$ stands for the scalar product in $\mathbf{L}^2(0, 1)$. Using the notation P_n^d for the projection $\mathring{P}_n$ onto $\bar{\mathcal{S}}_0^d(\Delta_n)$, we can write

$$\begin{aligned}\left((I - Q_n)\, u,\, K^*v\right) &= \left((I - Q_n)\,(I - P_n^{d+r})\, u,\, K^*v\right)\\ &\quad + \sum_{j=0}^{r-1} \left((I - Q_n)\,(P_n^{d+r-j} - P_n^{d+r-j-1})\, u,\, (I - P_n^j)\, K^*v\right)\end{aligned} \tag{4}$$

because of 5.61(7). Note that $u \in \mathbf{L}_\varrho^{\infty,R}(0, 1)$ (according to Theorem 5.12 and Remark 5.13, 1°), and since K^* satisfies $(\mathrm{H}1^{1,r})$, K^* is a continuous map of $\mathbf{L}^1(0, 1)$ into $\mathbf{L}^{1,r}(0, 1)$. Applying Lemma 5.23, the first term on the right-hand side of (4) can be estimated by

$$\begin{aligned}\|K^*v\|_{1,(0,1)}\, \|(I - Q_n)\,(I - P_n^{d+r})\, u\|_{\infty,(0,1)} &\leqq c\, \|v\|_{1,(0,1)}\, \|(I - P_n^{d+r})\, u\|_{\infty,(0,1)}\\ &\leqq cn^{-R}\, \|u\|_{\infty,R,\varrho,(0,1)}.\end{aligned}$$

Moreover, the j-th term in the sum in (4) can be dominated by

$$\begin{aligned}&\|x^{-\varrho(j+1)/R}(I - Q_n)\, [(I - P_n^{d+r-j-1}) - (I - P_n^{d+r-j})]\, u\|_{\infty,(0,1)}\\ &\times \|x^{\varrho(j+1)/R}(I - P_n^j)\, K^*v\|_{1,(0,1)}.\end{aligned}$$

According to Lemmas 5.29 and 5.23, the first factor of this expression is bounded by $cn^{-d-r+j}\, \|u\|_{\infty,d+r-j,\varrho,(0,1)}$ since $q \geqq (d + r - j)/\{\varrho - \varrho(j + 1)/R\}$. Analogously, by Lemma 5.23 the second factor may be estimated by

$$cn^{-j-1}\, \|K^*v\|_{1,j+1,(0,1)} \leqq cn^{-j-1}\, \|v\|_{1,(0,1)}.$$

Combining the above estimates with (2) to (4), we obtain (1). ■

Remark. Under the assumptions of the preceding theorem and the stronger hypothesis $(\mathrm{H}1^{\infty,r+l})$ for some $l \geqq 1$, we have the error bound

$$\|u - u_n^*\|_{\infty,l,(0,1)} = O(n^{-R}), \qquad n \to \infty.$$

This may be proved as in Remark 5.55, 1°.

Collocation with piecewise polynomials: Wiener-Hopf equations

5.63. Notation. Retaining the notation of Section 5.20, the analysis of the last sections will now be carried over to the collocation method

$$Q^n A u_n = Q^n f, \qquad u_n \in \tilde{\mathcal{S}}^d(\Delta^n) \tag{1}$$

for the approximate solution of equation 5.0(2), where the projections Q^n are defined as in 5.29(1). The following stability result is the analogue of Theorem 5.57.

.64. Theorem. *Suppose* (A1), (A2), *and let* $\tilde{\mathcal{S}}^d(\Delta^n)$ *be a modification of* $\mathcal{S}^d(\Delta^n)$.

(i) *If* $\Delta^n \in \mathcal{M}_q$ *and the index* i_0 *in* 5.24(2) *is sufficiently large, then*

$$\|Q^n A u_n\|_{\infty,\mathbb{R}^+} \geqq c\,\|u_n\|_{\infty,\mathbb{R}^+}, \qquad u_n \in \tilde{\mathcal{S}}^d(\Delta^n),\ n \geqq n_0 \tag{1}$$

when n_0 *is large enough.*

(ii) *If* $\Delta^n \in \mathcal{M}$, *then* (1) *is valid with* $i_0 = 1$ (*i.e.* $\tilde{\mathcal{S}}^d(\Delta^n) = \mathcal{S}^d(\Delta^n)$).

Proof. For any $\delta > 0$, A takes the form $I - W_\delta - W'_\delta$, where $\|W'_\delta\|_\infty < \delta$ and W_δ is a polynomial in the elementary operators $V_\pm$ (cf. Sect. 5.4). Thus W_δ is a continuous map of $\mathbf{L}^\infty(\mathbb{R}^+)$ into $\mathbf{L}^{\infty,1}(\mathbb{R}^+)$. Furthermore, the projections Q^n are uniformly bounded by virtue of Lemma 5.30. Arguments similar to those employed in Lemma 5.38 and Theorem 5.57 now show that $Q^n A Q^n$ is a small perturbation of $P^n A P^n$, hence Theorem 5.37 implies the result. ■

5.65. Corollary. *Suppose that the conditions of Theorem* 5.64 *are satisfied.*

(i) *For all* i_0 *and* n *sufficiently large, the collocation equations* 5.63(1) *are uniquely solvable for any* $f \in \mathbf{L}^\infty(\mathbb{R}^+)$ *and the error estimate*

$$\|u - u_n\|_{\infty,\mathbb{R}^+} \leqq c\,\|(I - P^n)\,u\|_{\infty,\mathbb{R}^+} \tag{1}$$

holds, where $u \in \mathbf{L}^\infty(\mathbb{R}^+)$ *is the exact solution of eq.* 5.0(2).

(ii) *If* $f \in \mathring{\mathbf{C}}(\overline{\mathbb{R}}^+)$, *then the left-hand side of* (1) *converges to* 0 *as* $n \to \infty$.

(iii) *If* $f \in \mathbf{C}(\mathbb{R}^+)$, *then* u_n *converges uniformly to* u *on each compact subset of* $\mathbb{R}^+$.

The proof of these assertions is analogous to that of Corollary 5.39. Note that

$$\|(I - Q^n)\,u\|_{\infty,\mathbb{R}^+} \leqq c\,\|(I - P^n)\,u\|_{\infty,\mathbb{R}^+},\ u \in \mathbf{L}^\infty(\mathbb{R}^+) \tag{2}$$

because of the uniform boundedness of Q^n. With the help of (1), Theorem 5.9 and Lemma 5.26, one obtains the following result on the optimal order of convergence.

5.66. Theorem. *Under the assumptions of Theorem* 5.41 *for* $p = \infty$, *the collocation method* 5.63(1) *converges with the error bound*

$$\|u - u_n\|_{\infty,\mathbb{R}^+} = O(n^{-d-1}) \quad as \quad n \to \infty.$$

Notice that the "almost" optimal error estimate 5.42(1) in the uniform norm continues to hold for the collocation solutions u_N if the conditions of Remark 5.42 are satisfied with $p = \infty$.

Remark. The analogue of 5.61(1) for Wiener-Hopf equations is the system

$$c\big(1 - a(0)\big) = f(\infty), \qquad Q^n A u_n^0 = Q^n g, \tag{1}$$

where $c \in \mathbb{C}$, $u_n^0 \in \bar{\mathcal{S}}^d(\Delta^n)$, $f \in \mathbf{L}_\varrho^{\infty,d+1}(\mathbb{R}^+) \dotplus \mathbb{C}$, $g = f - c + cW1$, and a is the symbol of W. If we set $u_n = c + u_n^0$, then Theorem 5.66 extends to this case, since

$$(W1)(\infty) = \int_{-\infty}^{\infty} \varkappa(x)\,dx = a(0)$$

implies that $(Au)(\infty) = \big(1 - a(0)\big)\,u(\infty) = f(\infty)$, whence $c = u(\infty)$ and $g \in \mathbf{L}_\varrho^{\infty,d+1}(\mathbb{R}^+)$. Note that $u \in \mathbf{L}_\varrho^{\infty,d+1}(\mathbb{R}^+) \dotplus \mathbb{C}$ by Theorem 5.9. (1) coincides with the so-called collocation at infinity introduced by SLOAN and SPENCE [1], who studied the case of piecewise constant basis functions and partitions of class $\mathcal{M}$.

Finally, we state a superconvergence result for the iterated collocation solution $u_n^* = f + Wu_n$, the proof of which is parallel to that of Theorem 5.62 and uses Theorem 5.64(i) and Lemmas 5.26 and 5.30.

5.67. Theorem. *Assume* $\varrho > 0$, 5.61(5), (A1^r), (A1$_\varrho^R$), (A2) *and* (A2$_\varrho$). *If* $f \in \mathbf{L}_\varrho^{\infty,R}(\mathbb{R}^+)$, $\Delta^n \in \mathcal{M}_q$ *with* $q \geqq R/\varrho$, *and* $\bar{\mathcal{S}}^d(\Delta^n)$ *is a suitable modification of* $\mathcal{S}^d(\Delta^n)$, *then*

$$\|u - u_n^*\|_{\infty,\mathbb{R}^+} = O(n^{-R}) \quad as \quad n \to \infty. \tag{1}$$

Note that the maximal order $O(n^{-2d-2})$ in (1) is achieved by selecting the Gauss-Legendre points $\xi_0, \ldots, \xi_d$ (on $[0, 1]$) for the definition of the collocation points x_{ij}.

Collocation with continuous piecewise polynomials: Mellin convolution equations

5.68. Notation. In what follows we give a convergence analysis in the scale of weighted Sobolev spaces $\mathbf{L}^{p,1}$, $1 \leqq p \leqq \infty$, for collocation methods based on continuous piecewise polynomials.

Let $B = I - K$ be the Mellin operator defined in 5.1(6) with kernel k, and let Π_n be the interpolating projections onto the modified spline spaces $\bar{\mathcal{S}}^{d,1}(\Delta_n)$ (cf. 5.30(2)). Then the collocation method for the approximate solution of eq. 5.0(1) is the following: find $u_n \in \bar{\mathcal{S}}^{d,1}(\Delta_n)$ such that

$$\Pi_n B u_n = \Pi_n f. \tag{1}$$

We first establish a stability result.

5.69. Theorem. *Assume* $1 \leqq p \leqq \infty$, (H1^p), (H2^p), *and let* $\bar{\mathcal{S}}^{d,1}(\Delta_n)$ *be a modification of* $\mathcal{S}^{d,1}(\Delta_n)$. *Suppose further that* $k \in \mathbf{L}^p(0, 1)$ *if* $p < \infty$ *and that* $k \in \mathring{\mathrm{C}}[0, 1]$ *if* $p = \infty$.

(i) *If* $\Delta_n \in \mathcal{M}_q$ *and the number* i_0 *in* 5.30(1) *is sufficiently large, then*

$$\|\Pi_n B u_n\|_{p,1,(0,1)} \geqq c\,\|u_n\|_{p,1,(0,1)}, \qquad u_n \in \bar{\mathcal{S}}^{d,1}(\Delta_n), \qquad n \geqq n_0 \tag{1}$$

when n_0 *is large enough.*

(ii) *If* $\Delta_n \in \mathcal{M}$, *then* (1) *holds with* $i_0 = 1$ *(i.e.* $(\bar{\mathcal{S}}^{d,1}\Delta_n) = \mathcal{S}^{d,1}(\Delta_n)$*).*

In the sequel we use the notation $X^p(0, 1) = \mathbf{L}^p(0, 1)$ for $p < \infty$, $X^\infty(0, 1) = \mathring{\mathrm{C}}[0, 1]$ (cf. Sect. 5.1), and $\|\cdot\|_{p,1}$ for the operator norm on $\mathbf{L}^{p,1}(0, 1)$. The following lemmas are essential for the stability proof.

5.70. Lemma. *Under the assumptions of the preceding theorem on the operator B, for any $\delta > 0$ the representation*

$$B = (I - K_+)(I - K_0)(I - K_-) + T_1 \tag{1}$$

holds, where $\|K_0\|_{p,1} < \delta$, T_1 is a compact map of $\mathbf{L}^{p,1}(0, 1)$ *into* $X^{p,1}(0, 1)$ *and* $K_\pm$ *are as in* 5.5(1).

Proof. Consider the differential operator $\Lambda = x\mathrm{D} + 2$, which is an isomorphism of $\mathbf{L}^{p,1}(0, 1)$ onto $\mathbf{L}^p(0, 1)$, $1 \leqq p \leqq \infty$, and of $X^{\infty,1}(0, 1)$ onto $X^\infty(0, 1)$ (cf. Sect. 5.12), and let $T_2, T_3, \ldots$ denote compact operators of $\mathbf{L}^{p,1}(0, 1)$ into $X^p(0, 1)$. To prove the lemma, it suffices to show the identities

$$\Lambda K_+ = K_+\Lambda + T_2, \qquad \Lambda K_- = K_-\Lambda, \tag{2}$$

$$\Lambda K_0 = K_0\Lambda + T_3, \tag{3}$$

$$\Lambda T = T\Lambda + T_4 \tag{4}$$

on $\mathbf{L}^{p,1}(0, 1)$, with K_0 and T defined as in 5.5(2).

To verify (2), it is enough to consider the elementary operators $F_{p,\pm}$ (cf. 5.4(3)), and in this case the assertion is an easy consequence of the relation 5.12(1). Since the operators $I - K_\pm$ are invertible on $\mathbf{L}^p(0, 1)$, (2) and Remark 5.7(ii) imply that they are invertible on $\mathbf{L}^{p,1}(0, 1)$, too, and that

$$\Lambda(I - K_+)^{-1} = (I - K_+)^{-1}\Lambda + T_5, \qquad \Lambda(I - K_-)^{-1} = (I - K_-)^{-1}\Lambda \tag{5}$$

on $\mathbf{L}^{p,1}(0, 1)$. Furthermore, applying the relation 5.12(1) to the operator B and using the fact that $k(x) \in X^p(0, 1)$, we obtain

$$\Lambda B = B\Lambda + T_6 \quad (\text{on } \mathbf{L}^{p,1}(0, 1)). \tag{6}$$

The representation 5.5(1) now gives $I - K_0 = (I - K_-)^{-1} B(I - K_+)^{-1}$, so that (3) follows from (5) and (6). Finally, (4) can be deduced from 5.5(2), (2) and (3). ■

5.71. Lemma. *Let $K_\pm$ be the operators occurring in the preceding lemma, where δ is fixed. Then, given $\varepsilon > 0$, there exists an integer n_0 such that for all $u \in \mathbf{L}^{p,1}(0, 1)$ and $n \geqq n_0$*

$$\|(I - \Pi_n) K_-\Pi_n u\|_{p,1,(0,1)} \leqq \varepsilon \|\Pi_n u\|_{p,1,(0,1)}, \tag{1}$$

$$\|\Pi_n K_+(I - \Pi_n) u\|_{p,1,(0,1)} \leqq \varepsilon \|u\|_{p,1,(0,1)} \tag{2}$$

if either $\Delta_n \in \mathcal{M}$ and $i_0 = 1$, or $\Delta_n \in \mathcal{M}_q$ and the index i_0 in 5.30(1) *is sufficiently large.*

Proof. First we verify (1). Let $u \in \mathbf{L}^{p,1}(0, 1)$ and $h = x_{i_0}$. Note that $K_-: \mathbf{L}^{p,1}(0, 1) \to \mathbf{L}^{p,2}(0,1)$ is continuous according to Remark 5.7(ii). By Lemma 5.31, we then have

$$\begin{aligned}\|(I - \Pi_n) K_-u\|_{p,1,J_i} &\leqq ch_i \|\mathrm{D}K_-u\|_{p,J_i} + cx_ih_i \|\mathrm{D}^2K_-u\|_{p,J_i}\\ &\leqq c(h_i/x_{i-1}) \|x\mathrm{D}K_-u\|_{p,J_i} + c(x_ih_i/x_{i-1}^2) \|x^2\mathrm{D}^2K_-u\|_{p,J_i}, \quad i > i_0.\end{aligned}$$

Furthermore, Definition 5.15 (resp. Remark 5.18) implies that $h_i/x_{i-1}, x_ih_i/x_{i-1}^2 \leqq \varepsilon$, $i > i_0$, when $\Delta_n \in \mathcal{M}$ (resp. $\Delta_n \in \mathcal{M}_q$) and n (resp. i_0, n) are large enough. Therefore, for those indices i_0 and n, one obtains

$$\|(I - \Pi_n) K_-u\|_{p,1,(h,1)} \leqq c\varepsilon \|K_-u\|_{p,2,(0,1)} \leqq c\varepsilon \|u\|_{p,1,(0,1)}.$$

Together with the relation

$$(I - \Pi_n) K_- u_n|_{(0,h)} = 0, \qquad u_n \in \bar{\mathcal{S}}^{d,1}(\Delta_n)$$

(cf. Remark 5.7(i)), this implies the result.

To derive (2), let $u \in \mathbf{L}^{p,1}(0, 1)$, $v_1 = \chi_h(I - \Pi_n) u$ and $v_2 = (I - \Pi_n) u - v_1$, where χ_h denotes the characteristic function of $[0, h]$. Then we obtain

$$\begin{aligned}\|\Pi_n K_+ v_2\|_{p,1,(0,1)} &\leqq c \|K_+ v_2\|_{p,1,(0,1)} \leqq c \|v_2\|_{p,(0,1)} \\ &= c \|(I - \Pi_n) u\|_{p,(h,1)} \leqq c\varepsilon \|u\|_{p,1,(0,1)}\end{aligned}$$

provided that $\Delta_n \in \mathcal{M}$ (resp. $\Delta_n \in \mathcal{M}_q$) and n (resp. i_0, n) are large enough. Here we have used Corollary 5.32 to get the first inequality, whereas the verification of the last inequality relies on Lemma 5.31(i), Definition 5.15 and Remark 5.18. To complete the proof of (2), observe that $K_+ v_1 = 0$ on $(h, 1)$ (cf. Remark 5.7(i)), hence $\Pi_n K_+ v_1 = 0$ on $[0, 1]$. ■

Proof of Theorem 5.69. *Step 1:* Lemma 5.70 implies that

$$\Pi_n B \Pi_n = \Pi_n (B_1 + T_1) \Pi_n, \qquad B_1 := (I - K_+)(I - K_0)(I - K_-), \tag{3}$$

and choosing δ such that $\sup_n \|\Pi_n\|_{p,1} \delta \leqq 1$, we obtain the stability of the operators $\Pi_n(I - K_0) \Pi_n$, i.e. inequality 5.69(1) with $I - K_0$ instead of B. Note that the choice of δ does not depend on i_0 by Corollary 5.32.

Step 2: Next we examine the stability of the operators $\Pi_n(I - K_\pm) \Pi_n$. Using (1) and the invertibility of $I - K_-$, one obtains that the operators $\Pi_n(I - K_-) \Pi_n$ are stable in $\mathbf{L}^{p,1}(0, 1)$ when i_0 and n are large enough. Applying (2) and taking adjoints in $\mathbf{L}^{p,1}(0, 1)$, we get the stability of $(\Pi_n(I - K_+) \Pi_n)^*$ in $(\mathbf{L}^{p,1}(0, 1))^*$, hence the operators $\Pi_n(I - K_+) \times \Pi_n$ are stable in $\mathbf{L}^{p,1}(0, 1)$, too.

Step 3: By Lemma 5.71, the $\mathbf{L}^{p,1}$ operator norm of

$$\Pi_n B_1 \Pi_n - \Pi_n(I - K_+) \Pi_n(I - K_0) \Pi_n(I - K_-) \Pi_n$$

can be made as small as desired when $n \to \infty$ and i_0 is chosen large enough. Therefore, by the considerations in the first and second steps, estimate 5.69(1) holds with B replaced by B_1.

Step 4: If $p < \infty$, then $\mathbf{C}_0^\infty((0, 1])$ is dense in $\mathbf{L}^{p,1}(0, 1)$, and the projections Π_n converge strongly to the identity on $\mathbf{L}^{p,1}(0, 1)$ (due to their uniform boundedness and Lemma 5.31). From Remark 1.27, 3°(b), (3) and the compactness of T_1 we now deduce the stability of $\Pi_n B \Pi_n$.

Let $p = \infty$. Since $T_1 : \mathbf{L}^{\infty,1}(0, 1) \to \mathring{\mathbf{C}}^1[0, 1]$ is compact and Π_n converges strongly to the identity on $\mathring{\mathbf{C}}^1[0, 1]$, we may complete the proof of 5.69(1) by the same arguments as in Theorem 5.37. In order to do so, we only have to verify that if the sequence $\{u_n\} \subset \mathbf{L}^{\infty,1}(0, 1)$ converges in the norm of $\mathbf{L}^{\infty,1}(\alpha, 1)$ for all $\alpha \in (0, 1)$ then so does the sequence $\{K u_n\}$. Applying the relation 5.70(6), this assertion reduces to the case which was already considered in the proof of Theorem 5.45. ■

We now investigate the convergence of our collocation method. Note that, under the assumptions of the preceding theorem, B is invertible in $\mathbf{L}^{p,1}(0, 1)$ (according to Theorem 5.12 for $l = 1$, $\varrho = 0$).

5.72. Corollary. *Assume that the conditions of Theorem* 5.69 *are satisfied.*

(i) *For all i_0 and n sufficiently large, the collocation equations* 5.69(1) *are uniquely solv-*

able for any $f \in \mathbf{L}^{p,1}(0, 1)$ *and the error estimate*

$$\|u - u_n\|_{p,1,(0,1)} \leq c \,\|(I - \Pi_n)\, u\|_{p,1,(0,1)} \tag{1}$$

holds, where $u \in \mathbf{L}^{p,1}(0, 1)$ *denotes the exact solution of* 5.0(1). *Moreover, for* $p < \infty$, *the left-hand side of* (1) *converges to* 0 *as* $n \to \infty$.

(ii) *If* $f \in \mathbf{C}^1[0, 1]$, *then* u_n *converges to* u *in the norm of* $L^{\infty,1}(0, 1)$.

(iii) *If* $f \in \mathbf{C}^1(0, 1]$, *then* u_n *converges to* u *in the norm of the Sobolev space* $\mathbf{W}^{\infty,1}(\alpha, 1)$ *for all* $\alpha \in (0,1)$.

Proof. (i) follows in a standard way from 5.69(1) and the strong convergence of Π_n to the identity on $\mathbf{L}^{p,1}$, $p < \infty$. To deduce (ii) from (1), we note that $u = c + u_0$, $c \in \mathbb{C}$, $u_0 \in \mathring{\mathbf{C}}^1[0, 1]$, whence

$$\|(I - \Pi_n)\, u\|_{\infty,1,(0,1)} = \|(I - \Pi_n)\, u_0\|_{\infty,1,(0,1)} \to 0 \quad \text{as} \quad n \to \infty.$$

The proof of (iii) is analogous to that of Corollary 5.39 (iii). We only have to check that if the sequence u_n is bounded in $\mathbf{L}^{\infty,1}(0, 1)$ then Ku_n and $x\mathrm{D}Ku_n$ are (uniformly) equicontinuous on (0, 1] and that if u_n converges in the norm of $\mathbf{L}^{\infty,1}(\alpha, 1)$ for all $\alpha \in (0, 1)$ then so does the sequence Ku_n. But this follows by the identity 5.70(6), which reduces the assertion to the situation considered in the proof of Corollary 5.50(iii). ■

The next result establishes the optimal rate of convergence for the collocation method in $\mathbf{L}^{p,1}$.

5.73. Theorem. *Assume* $1 \leq p \leq \infty$, $\varrho > 0$, $(\mathrm{H1}^p)$, $(\mathrm{H1}^p_\varrho)$, $(\mathrm{H2}^p)$, $(\mathrm{H2}^p_\varrho)$ *and* $k \in \mathbf{L}^{p,d}_\varrho(0, 1)$. *Suppose further that* $f \in \mathbf{L}^{p,d+1}_\varrho(0, 1) \dotplus \mathbb{P}_k$, $k < \min\{\varrho - 1/p, 1\}$, $\Delta_n \in \mathcal{M}_q$ *with* $q \geq d/\varrho$, *and let* $\tilde{\mathcal{S}}^{d,1}(\Delta_n)$ *be a modification of* $\mathcal{S}^{d,1}(\Delta_n)$ *such that the index* i_0 *in* 5.30(1) *is large enough. Then the collocation method* 5.68(1) *converges with the error bound*

$$\|u - u_n\|_{p,1,(0,1)} = O(n^{-d}) \quad as \quad n \to \infty.$$

Proof. By Theorem 5.12, the solution of 5.0(1) takes the form $u = u_0 + c$, where $c \in \mathbb{C}$ and $u_0 \in \mathbf{L}^{p,d+1}_\varrho(0, 1)$. Since 5.72(1) gives

$$\|u - u_n\|_{p,1,(0,1)} \leq c \,\|(I - \Pi_n)\, u\|_{p,1(0,1)}$$

and the last term can be bounded by cn^{-d} according to Lemmas 5.33 and 5.34, we get the result. ■

Finally, we consider the iterated collocation (or product integration) method. The product integration solution is defined by $u_n^* = f + Ku_n$, where u_n is given by 5.68(1). As in Sect. 5.61, 5.62, the convergence rate of this method depends on the choice of the points ξ_j, $0 = \xi_0 < \xi_1 < \cdots < \xi_d = 1$, occuring in the definition of the collocation points x_{ij}. Let $R = d + r + 1$ be the order of the quadrature rule 5.61(4). Then $0 \leq r \leq d - 1$ and $r = d - 1$ if and only if $\xi_j = (\bar{z}_j + 1)/2$, $j = 1, \ldots, d - 1$, where the $\bar{z}_j$ are the $d - 1$ zeros of $\mathrm{D}P_d$, P_d being the Legendre polynomial of degree d (cf. Brunner/van der Houwen [1, Sect. 2.4]). In the case $r = d - 1$, 5.61(4) is called the $(d + 1)$-point Lobatto formula.

5.74. Theorem. *Assume* $\varrho > 0$, $1 \leq p \leq \infty$, 5.61(5), $(\mathrm{H1}^{p,r+1})$, $(\mathrm{H1}^{p,R}_\varrho)$, $(\mathrm{H2}^p)$ *and* $(\mathrm{H2}^p_\varrho)$. *If* $f \in \mathbf{L}^{p,R}_\varrho(0, 1) \dotplus \mathbb{P}_k$, $k < \min\{\varrho - 1/p, 1\}$, $\Delta_n \in \mathcal{M}_q$ *with* $q \geq R/\varrho$, *and* $\tilde{\mathcal{S}}^{d,1}(\Delta_n)$ *is a suitable modification of* $\mathcal{S}^{d,1}(\Delta_n)$, *then*

$$\|u - u_n^*\|_{p,1,(0,1)} = O(n^{-R}) \quad as \quad n \to \infty. \tag{1}$$

Proof. Arguing as in Theorem 5.43, one obtains the estimate

$$\|u - u_n^*\|_{p,1,(0,1)} \leqq c \, \|K(I - \Pi_n) \, u\|_{p,1,(0,1)} \tag{2}$$

from Theorem 5.69(i) and the uniform boundedness of Π_n in $\mathbf{L}^{p,1}(0, 1)$. Using the operator Λ defined in the proof of Lemma 5.70, we have (cf. 5.62(3))

$$\begin{aligned}\|K(I - \Pi_n) \, u\|_{p,1,(0,1)} &\leqq c \, \|\Lambda K(I - \Pi_n) \, u\|_{p,(0,1)} \\ &= \sup \left|\left((I - \Pi_n) \, u, K_1 v\right)\right|,\end{aligned} \tag{3}$$

where the supremum is taken over all $v \in \mathbf{L}^{p'}(0, 1)$, $\|v\|_{p',(0,1)} \leqq 1$, $p' = p/(p-1)$, and K_1 is the Mellin convolution operator with kernel $x^{-1}\overline{(\Lambda k)} \, (x^{-1})$. By virtue of Theorem 5.12 and Remark 5.13, 1°, $u = u_0 + c$ with $u_0 \in \mathbf{L}_\varrho^{p,R}(0, 1)$, $c \in \mathbb{C}$, and K_1 is a continuous operator of $\mathbf{L}^{p'}(0, 1)$ into $\mathbf{L}^{p',r}(0, 1)$ since its kernel statisfies (H1$^{p',r}$). In analogy to 5.62(4), we can write

$$\begin{aligned}\left((I - \Pi_n) \, u_0, K_1 v\right) &= \left((I - \Pi_n^d) \, (I - \Pi_n^{d+r}) \, u_0, K_1 v\right) \\ &\quad + \sum_{j=0}^{r-1} \left((I - \Pi_n^d) \, (\Pi_n^{d+r-j} - \Pi_n^{d+r-j-1}) \, u_0, (I - P_n^j) \, K_1 v\right),\end{aligned} \tag{4}$$

where Π_n^j stands for the interpolatory projection onto $\mathcal{S}^{j,1}(\Delta_n)$. Applying Lemmas 5.33 and 5.34, we observe that the first term on the right-hand side of (4) can be estimated by

$$\begin{aligned}&\|K_1 v\|_{p',(0,1)} \, \|(I - \Pi_n^d) \, (I - \Pi_n^{d+r}) \, u_0\|_{p,(0,1)} \\ &\leqq c \, \|v\|_{p',(0,1)} \, n^{-1} \, \|(I - \Pi_n^{d+r}) \, u_0\|_{p,1,\varrho/R,(0,1)} \leqq cn^{-R} \, \|u_0\|_{p,R,\varrho,(0,1)}.\end{aligned}$$

Furthermore, the j-th term of the sum in (4) may be bounded by

$$\begin{aligned}&\|(I - \Pi_n^d) \, (\Pi_n^{d+r-j} - \Pi_n^{d+r-j-1}) \, u_0\|_{p,0,\varrho(j+1)/R,(0,1)} \\ &\times \|(I - P_n^j) \, K_1 v\|_{p',0,-\varrho(j+1)/R,(0,1)}.\end{aligned}$$

As in Theorem 5.62, the second factor may be estimated by cn^{-j-1}. The first can be dominated by

$$cn^{-1} \, \|(\Pi_n^{d+r-j} - \Pi_n^{d+r-j-1}) \, u_0\|_{p,1,\varrho(j+2)/R,(0,1)} \leqq cn^{-1-(d+r-j-1)} \, \|u_0\|_{p,d+r-j,\varrho,(0,1)},$$

according to Lemmas 5.33 and 5.34. Combining the above inequalities with (2) to (4), one obtains (1). ∎

Remark. The results of this section extend easily to the case when the spline spaces 5.30(1) are modified by requiring $u = cx^k$ on $(0, x_{i_0})$ for some $k > 0$. Moreover, it is possible to generalize the stability and convergence theorems to the weighted spaces $\mathbf{L}_\mu^{p,1}$, $\mu \in \mathbb{R}$, and to collocation methods based on weighted splines; see Elschner [4] for some results in this direction.

Collocation with continuous piecewise polynomials: Wiener-Hopf equations

5.75. Notation. Let $A = I - W$ be the operator 5.1(4) with kernel $\varkappa$, and let Π^n be the interpolatory projections onto the modified spline spaces $\tilde{\mathcal{S}}^{d,1}(\Delta^n)$ introduced in 5.34(2). We now give the corresponding convergence analysis of the collocation method

$$\Pi^n A u_n = \Pi^n f, \qquad u_n \in \tilde{\mathcal{S}}^{d,1}(\Delta^n) \tag{1}$$

for the approximate solution of equation 5.0(2). The following stability result is the analogue of Theorem 5.69. In the sequel we use the notation $X^p(\mathbb{R}^+) = \mathbf{L}^p(\mathbb{R}^+)$ for $p < \infty$, $X^\infty(\mathbb{R}^+) = \mathring{\mathbf{C}}(\overline{\mathbb{R}}^+)$ (cf. Sect. 5.1), and $\|\cdot\|_{p,1}$ for the operator norm on $\mathbf{L}^{p,1}(\mathbb{R}^+)$.

5.76. Theorem. *Assume* $1 \leqq p \leqq \infty$, (A1), (A2), $\varkappa \in X^p(\mathbb{R}^+)$, *and let* $\tilde{\mathcal{S}}^{d,1}(\Delta^n)$ *be a modification of* $\mathcal{S}^{d,1}(\Delta^n)$.

(i) *If* $\Delta^n \in \mathcal{M}_q$ *and the number* i_0 *in* 5.34(1) *is sufficiently large, then*

$$\|\Pi^n A u_n\|_{p,1,\mathbb{R}^+} \geqq c\,\|u_n\|_{p,1,\mathbb{R}^+}, \qquad u_n \in \tilde{\mathcal{S}}^{d,1}(\Delta^n), \qquad n \geqq n_0 \tag{1}$$

when n_0 *is large enough.*

(ii) *If* $\Delta^n \in \mathcal{M}$, *then* (1) *is valid in the case* $i_0 = 1$.

Proof. The proof is parallel to that of Theorem 5.69 and is based on the representation

$$A = (I - W_+)(I - W_1)(I - W_-) + T_1, \tag{2}$$

where $\|W_1\|_{p,1} < \delta$, $W_\pm$ are as in 5.4(1) and T_1 is a compact operator of $\mathbf{L}^{p,1}(\mathbb{R}^+)$ into $X^{p,1}(\mathbb{R}^+)$. This may be proved as in Lemma 5.70, using Theorem 5.4 and the isomorphisms $\Lambda := \mathrm{D} - 1\colon \mathbf{L}^{p,1}(\mathbb{R}^+) \to \mathbf{L}^p(\mathbb{R}^+)$, $1 \leqq p \leqq \infty$, and $\Lambda\colon X^{\infty,1}(\mathbb{R}^+) \to X^\infty(\mathbb{R}^+)$ (cf. Sect. 5.9). Note that, by the assumption $\varkappa \in X^p(\mathbb{R}^+)$, A is a continuous map of $\mathbf{L}^{p,1}(\mathbb{R}^+)$ into itself and

$$\Lambda A = A\Lambda + T_2, \tag{3}$$

where $T_2\colon \mathbf{L}^{p,1}(\mathbb{R}^+) \to X^p(\mathbb{R}^+)$ is compact (cf. 5.9(1)). Furthermore, applying the continuity of the operators (cf. Remark 5.6(ii))

$$W_-\colon \mathbf{L}^{p,1}(\mathbb{R}^+) \to \mathbf{L}^{p,2}(\mathbb{R}^+), \qquad W_+\colon \mathbf{L}^p(\mathbb{R}^+) \to \mathbf{L}^{p,1}(\mathbb{R}^+),$$

Lemmas 5.35 and 5.36 and Remark 5.6(i), it follows that

$$\|(I - \Pi^n)\,W_-\Pi^n\|_{p,1} \to 0, \qquad \|\Pi^n W_+(I - \Pi^n)\|_{p,1} \to 0$$

when $\Delta^n \in \mathcal{M}$ and $n \to \infty$ resp. $\Delta^n \in \mathcal{M}_q$ and $n, i_0 \to \infty$. This gives the stability of the operators $\Pi^n(I - W_+)(I - W_1)(I - W_-)\,\Pi^n$ if we only choose δ sufficiently small in (2). Since $\mathbf{C}_0^\infty(\mathbb{R}^+)$ is dense in $X^{p,1}(\mathbb{R}^+)$, Lemmas 5.35(i) and 5.36(i) imply that the projections Π^n converge strongly to the identity on $X^{p,1}(\mathbb{R}^+)$. Using the equalities (2) and (3), the rest of the proof is now analogous to that of Theorem 5.69. ■

5.77. Corollary. *Suppose that the assumptions of the preceding theorem are satisfied.*

(i) *For all* i_0 *and* n *sufficiently large, the collocation equations* 5.75(1) *are uniquely solvable for any* $f \in \mathbf{L}^{p,1}(\mathbb{R}^+)$ *and the error bound*

$$\|u - u_n\|_{p,1,\mathbb{R}^+} \leqq c\,\|(I - \Pi^n)\,u\|_{p,1,\mathbb{R}^+} \tag{1}$$

holds, where $u \in \mathbf{L}^{p,1}(\mathbb{R}^+)$ *is the exact solution of eq.* 5.0(2). *Moreover, for* $p < \infty$, *the left-hand side of* (1) *converges to* 0 *as* $n \to \infty$.

(ii) *If* $f \in \overset{\circ}{\mathbf{C}}{}^1(\mathbb{R}^+)$, *then* u_n *converges to* u *in the norm of* $\mathbf{L}^{\infty,1}(\mathbb{R}^+)$.

(iii) *If* $f \in \mathbf{C}^1(\overline{\mathbb{R}}^+)$, *then* u_n *converges to* u *in the norm of the Sobolev space* $\mathbf{W}^{\infty,1}(0, \alpha)$ *for all* $\alpha > 0$.

The proof of these assertions is analogous to that of Corollary 5.72. Note that, under the conditions of Theorem 5.76, A is invertible in $\mathbf{L}^{p,1}(\mathbb{R}^+)$ according to Theorem 5.9 (for $l = 1$, $\varrho = 0$).

The next result, which is a consequence of Theorem 5.9 (1) and Lemma 5.36, gives the optimal order of convergence in $\mathbf{L}^{p,1}$.

5.78. Theorem. *Assume* $1 \leqq p \leqq \infty$, $\varrho > 0$, (A1), (A2), $(\mathrm{A1}_\varrho)$, $(\mathrm{A2}_\varrho)$ *and* $\varkappa \in \mathbf{L}^{p,d}_\varrho(\mathbb{R}^+)$. *Suppose further that* $f \in \mathbf{L}^{p,d+1}_\varrho(\mathbb{R}^+)$, $\Delta^n \in \mathcal{M}_q$ *with* $q \geqq d/\varrho$, *and let* $\tilde{\mathcal{S}}^{d,1}(\Delta^n)$ *be a modification of* $\mathcal{S}^{d,1}(\Delta^n)$ *such that the number* i_0 *in* 5.34(1) *is large enough. Then the collocation mehod* 5.75(1) *converges with the error estimate*

$$\|u - u_n\|_{p,1,\mathbb{R}^+} = O(n^{-d}) \quad as \quad n \to \infty .$$

Finally, we establish a superconvergence result for the iterated collocation solution $u_n^* = f + Wu_n$. Its proof parallels that of Theorem 5.74 and relies on Theorem 5.76(i) and Lemma 5.36.

5.79. Theorem. *Assume* $1 \leqq p \leqq \infty$, $\varrho > 0$, 5.61(5), $(\mathrm{A1}^{r+1})$, $(\mathrm{A1}^R_\varrho)$, (A2) *and* $(\mathrm{A2}_\varrho)$. *If* $f \in \mathbf{L}^{p,R}_\varrho(\mathbb{R}^+)$, $\Delta^n \in \mathcal{M}_q$ *with* $q \geqq R/\varrho$, *and* $\tilde{\mathcal{S}}^{d,1}(\Delta^n)$ *is a suitable modification of* $\mathcal{S}^{d,1}(\Delta^n)$, *then*

$$\|u - u_n^*\|_{p,1,\mathbb{R}^+} = O(n^{-R}) \quad as \quad n \to \infty . \tag{1}$$

Note that one obtains the maximal convergence rate $O(n^{-2d})$ in (1) by using the $(d + 1)$-point Lobatto formula in 5.61(4).

Discrete methods: Nyström methods and discrete collocation for Mellin convolution equations

5.80. Notation. We now study stability and convergence of certain quadrature methods based on discontinuous piecewise polynomials. To define our numerical methods for eq. 5.0(1), consider the $(d + 1)$-point interpolatory quadrature rule 5.61(4) with weights ω_j and points $0 \leqq \xi_0 < \xi_1 < \cdots < \xi_d \leqq 1$, and let Q_n be the interpolating projections onto $\mathcal{S}^d_0(\Delta_n)$ (cf. 5.27(2)) which correspond to the collocation points

$$x_{ij} = x_{i-1} + \xi_j h_i , \quad (i, j) \in \mathcal{J} , \quad \mathcal{J} := \{(i, j) : 0 \leqq j \leqq d , \, i_0 + 1 \leqq i \leqq n\}. \tag{1}$$

Let $R = d + r + 1$ be the order of the quadrature rule 5.61(4), where $0 \leqq r \leqq d + 1$ (cf. Sect. 5.61). Then the composite quadrature rule obtained by shifting 5.61(4) to each J_i, and summing over $i > i_0$, is

$$\int_0^1 u \, dx \approx \sum_{(i,j)\in\mathcal{J}} \omega_j u(x_{ij}) \, h_i = \int_0^1 Q_n u \, dx . \tag{2}$$

Using (2) we approximate the integral operator K in 5.1(6) by

$$K_n u(x) = \sum_{(i,j)\in\mathcal{J}} \omega_j \mathsf{k}(x, x_{ij})\, u(x_{ij})\, h_i = \int_0^1 Q_n\{\mathsf{k}(x, .)\, u(.)\}\,, \tag{3}$$

where $\mathsf{k}(x, y) = k(x/y)\, y^{-1}$. Note that, under appropriate assumptions on the kernel k and the meshes Δ_n, the operators K_n are well-defined on $\mathbf{C}(0, 1]$ and $\mathcal{S}_0^d(\Delta_n)$ and admit bounded extensions to the whole of $\mathbf{L}^\infty(0, 1)$ (cf. Lemma 5.81 below).

For the integral equation 5.0(1), the discrete collocation method is the following: find $u_n \in \mathcal{S}_0^d(\Delta_n)$ such that

$$u_n - Q_n K_n u_n = Q_n f\,. \tag{4}$$

The Nyström solution u_n^* to 5.0(1) is then defined by

$$u_n^* = f + K_n u_n\,. \tag{5}$$

Note that $Q_n u_n^* = u_n$; thus (5) can be written in the form

$$(I - K_n Q_n)\, u_n^* = (I - K_n)\, u_n^* = f\,. \tag{6}$$

(4) is a linear system in the values $u_n^*(x_{ij})$, and then the right-hand side of (5) may be computed giving $u_n^*(x)$ for all $x \in [0, 1]$. The Nyström method (5) represents a discrete iterated collocation method.

Remark. (4) and (5) may also be interpreted as discrete Galerkin and discrete iterated Galerkin methods, respectively. Choose the basis $\{\varphi_{ij}: (i, j) \in \mathcal{J}\}$ of $\mathcal{S}_0^d(\Delta_n)$, where $\operatorname{supp} \varphi_{ij} \subset \bar{J}_i$, $\varphi_{ij}|_{J_i} \in \mathbb{P}_d$, $\varphi_{ij}(x_{ij}) = 1$ and $\varphi_{ij}(x_{il}) = 0$ if $l \neq j$. A realisation of the Galerkin method 5.53(1) is

$$(u_n, \varphi_{kl}) = (f, \varphi_{kl}) + (K u_n, \varphi_{kl}), \qquad (k, l) \in \mathcal{J}\,.$$

If K is replaced by K_n and the inner products in this system are calculated using (3), the resulting scheme is

$$\omega_l h_k u_n(x_{kl}) = \omega_l h_k f(x_{kl}) + \omega_l h_k \Big\{ \sum_{(i,j)\in\mathcal{J}} \omega_j \mathsf{k}(x_{kl}, x_{ij})\, u_n(x_{ij})\, h_i \Big\}, \qquad (k, l) \in \mathcal{J}\,,$$

which is equivalent to (4) (if $\omega_i \neq 0$ for all j). For related observations, see Atkinson/Bogomolny [1], Chandler/Graham [4], Joe [1].

For our convergence analysis, the following technical lemmas are needed. As in the preceding sections, c denotes various constants which do not depend on u, n and i.

5.81. Lemma. *Assume* $(\mathrm{H1}^{\infty,2})$.

(i) *If* $\Delta_n \in \tilde{\mathcal{M}}$ *or* $\Delta_n \in \mathcal{M}_q$, *then for all* n *and* i_0, 5.80(3) *extends to a bounded operator on* $\mathbf{L}^\infty(0, 1)$ *such that* $\|K_n\|_\infty \leq c$.

(ii) *If* $\Delta_n \in \tilde{\mathcal{M}}$ *and* $n \to \infty$ *resp.* $\Delta_n \in \mathcal{M}_q$ *and* $n, i_0 \to \infty$, *then* $\|(K - K_n)\, Q_n\|_\infty \to 0$.

Proof. Using the basis functions φ_{ij} introduced above, for any $u \in \mathcal{S}_0^d(\Delta_n)$ we have

$$(K - K_n)\, u(x) = \sum_{(i,j)\in\mathcal{J}} \int_{J_i} [\mathsf{k}(x, y) - \mathsf{k}(x, x_{ij})]\, \varphi_{ij}(y)\, u(x_{ij})\, \mathrm{d}y\,, \qquad x \in [0, 1], \tag{1}$$

since

$$u = \sum_{(i,j)\in\mathcal{J}} u(x_{ij})\, \varphi_{ij}\,, \qquad h_i \omega_j = \int_{J_i} \varphi_{ij}\, \mathrm{d}x\,.$$

Each term of the sum in (1) can be estimated by

$$\int_{J_i} |\mathsf{k}(x, y) - \mathsf{k}(x, x_{ij})|\, |u(x_{ij})|\, |\varphi_{ij}(y)|\, \mathrm{d}y$$
$$\leqq c\, \|u\|_{\infty,(0,1)} \int_{J_i} |\mathsf{k}(x, y) - \mathsf{k}(x, x_{ij})|\, \mathrm{d}y, \tag{2}$$

since $\|\varphi_{ij}\|_{\infty,(0,1)} \leqq c$ for all i, j and n (by a scaling argument). Moreover, it is easily seen that $(\mathrm{H1}^{\infty,2})$ implies that $k(x)$ and $x\mathrm{D}k(x)$ are bounded on $\mathbb{R}^+$. Therefore, for any $x \in [0, 1]$,

$$|\mathsf{k}(x, y) - \mathsf{k}(x, x_{ij})| \leqq h_i\, \|\mathrm{D}_y \mathsf{k}(x, y)\|_{\infty, J_i}$$
$$\leqq c h_i x_{i-1}^{-2} \{\|k(x)\|_{\infty,\mathbb{R}^+} + \|x\mathrm{D}k(x)\|_{\infty,\mathbb{R}^+}\} \leqq c h_i x_{i-1}^{-2}.$$

Combining the last estimate with (1) and (2), we obtain

$$|(K - K_n)\, u(x)| \leqq c\, \|u\|_{\infty,(0,1)} \sum_{i>i_0} h_i^2/x_{i-1}^2, \qquad u \in \mathcal{F}_0^d(\Delta_n), \qquad x \in [0, 1]. \tag{3}$$

By Definition 5.20, the sum in (3) tends to zero when $\Delta_n \in \tilde{\mathcal{M}}$, $i_0 = 1$ and $n \to \infty$. If $\Delta_n \in \mathcal{M}_q$, then $\sum_{i>i_0} h_i^2/x_{i-1}^2 \leqq c \sum_{i>i_0} i^{-2}$ by virtue of Remark 5.18, and the last expression tends to zero when $i_0 \to \infty$. Therefore, together with the uniform boundedness of the projections Q_n (cf. Lemma 5.28), (3) implies (ii). Similarly, it follows from (3) that the operators $(K - K_n)\, Q_n$ are uniformly bounded on $\mathbf{L}^\infty(0, 1)$ with respect to n and i_0 if $\Delta_n \in \tilde{\mathcal{M}}$ or $\Delta_n \in \mathcal{M}_q$. Setting $K_n = 0$ on $(I - Q_n)\, \mathbf{L}^\infty(0, 1)$, we get (i). ■

5.82. Lemma. *If* $u \in \mathbf{L}^{1,l}(0, 1)$, $1 \leqq l \leqq R$, $i \geqq i_0 + 1$, *then*

$$\left| \int_{J_i} u \,\mathrm{d}x - \sum_{0\leqq j\leqq d} \omega_j u(x_{ij})\, h_i \right| \leqq c h_i^l\, \|\mathrm{D}^l u\|_{1,J_i}. \tag{1}$$

Proof. If u_0 is the Taylor polynomial of degree $l - 1$ for u about x_i,

$$\left| \int_{J_i} u \,\mathrm{d}x - \sum_j \omega_j u(x_{ij})\, h_i \right| = \left| \int_{J_i} (u - u_0) \,\mathrm{d}x - \sum_j \omega_j (u - u_0)(x_{ij})\, h_i \right|$$
$$\leqq \|u - u_0\|_{1,J_i} + c h_i\, \|u - u_0\|_{\infty,J_i}.$$

The last expressions can be bounded by $c h_i^l\, \|\mathrm{D}^l u\|_{1,J_i}$, using Taylor's formula with integral remainder. ■

5.83. Lemma. *Assume* $(\mathrm{H1}^{\infty,l})$, *where* $1 \leqq l \leqq R$. *If* $u \in \mathbf{L}_\varrho^{\infty,l}(0, 1)$, $\varrho > 0$, $q \geqq l/\varrho$ *and* $\Delta_n \in \mathcal{M}_q$, *then*

$$\|(K - K_n)\, u\|_{\infty,(0,1)} \leqq c n^{-l}\, \|u\|_{\infty,l,\varrho,(0,1)}.$$

Proof. With $h = x_{i_0}$, we have

$$|(K - K_n)\, u(x)| \leqq \left| \int_0^h \mathsf{k}(x, y)\, u(y)\, \mathrm{d}y \right|$$
$$+ \sum_{i>i_0} \left| \int_{J_i} \mathsf{k}(x, y)\, u(y)\, \mathrm{d}y - \sum \omega_j \mathsf{k}(x, x_{ij})\, u(x_{ij})\, h_i \right|. \tag{1}$$

The $\mathbf{L}^\infty$ norm of the first term on the right-hand side of (1) may be bounded by

$$\|K\|_\infty \|u\|_{\infty,(0,h)} \leqq ch^\varrho \|x^{-\varrho}u\|_{\infty,(0,h)} \leqq cn^{-l} \|u\|_{\infty,l,\varrho,(0,1)}, \tag{2}$$

because $h \sim (i_0/n)^q \sim n^{-q}$ (cf. Remark 5.18) and $q\varrho \geqq l$. Furthermore, by Lemma 5.82,

$$\begin{aligned}&\Big|\int\limits_{J_i} \mathsf{k}(x, y)\, u(y)\, \mathrm{d}y - \sum_j \omega_j \mathsf{k}(x, x_{ij})\, u(x_{ij})\, h_i\Big| \\ &\leqq ch_i^l \|\mathrm{D}_y \mathsf{k}(x, y)\, u(y)\|_{1,J_i} \\ &\leqq c \sum_{k=0}^{l} h_i^l \|y^{-k}\mathrm{D}^{l-k}u\|_{\infty,J_i} \|y^k \mathrm{D}_y^k \mathsf{k}(x, y)\|_{1,J_i}, \qquad x \in [0, 1].\end{aligned} \tag{3}$$

Since $x_i \sim (i/n)^q$, $h_i \sim n^{-1}(i/n)^{q-1}$ (according to Remark 5.18) and $\varrho q \geqq l$, we obtain the estimates

$$\begin{aligned} h_i^l \|y^{-k}\mathrm{D}^{l-k}u\|_{\infty,J_i} &\leqq h_i^l x_i^{\varrho-l} \|y^{l-k-\varrho}\mathrm{D}^{l-k}u\|_{\infty,J_i} \\ &\leqq cn^{-l}(i/n)^{(\varrho-l)q+l(q-1)} \|u\|_{\infty,l,\varrho,J_i} \\ &\leqq cn^{-l} \|u\|_{\infty,l,\varrho,(0,1)}, \qquad 0 \leqq k \leqq l. \end{aligned} \tag{4}$$

Moreover, since $y^k\mathrm{D}_y^k\mathsf{k}(x, y)$ is a linear combination of the expressions $y^{-1}k_i(x/y), i \leqq k$, $k_i(x) := x^i\mathrm{D}^i k(x)$, condition $(\mathrm{H}1^{\infty,l})$ implies that

$$\begin{aligned}\sum_{i>i_0} \sum_{j\leqq l} \|y^j\mathrm{D}_y^j\mathsf{k}(x, y)\|_{1,J_i} &= \sum_{i>i_0} \sum_{j\leqq l} \int\limits_{x_{i-1}}^{x_i} |k_j(x/y)|\, y^{-1}\, \mathrm{d}y \leqq \sum_{i>i_0} \sum_{j\leqq l} \int\limits_{x/x_i}^{x/x_{i-1}} |k_j(y)|\, y^{-1}\, \mathrm{d}y \\ &\leqq \sum_{j\leqq l} \int\limits_0^\infty |k_j(y)|\, y^{-1}\, \mathrm{d}y \leqq c, \qquad x \in [0, 1].\end{aligned}$$

Combining the last inequality with (1) to (4), one gets the result. ■

We are now in a position to prove stability and error estimates for the Nyström methods.

5.84. Theorem. *Assume* $(\mathrm{H}1^{\infty,2})$ *and* $(\mathrm{H}2^\infty)$.

(i) *If* $\Delta_n \in \mathcal{M}_q$ *and the index* i_0 *is sufficiently large, then*

$$\|(I - K_n)\, u\|_{\infty,(0,1)} \geqq c\, \|u\|_{\infty,(0,1)}, \qquad u \in \mathbf{L}^\infty(0, 1), \qquad n \geqq n_0 \tag{1}$$

when n_0 *is large enough.*

(ii) *If* $\Delta_n \in \tilde{\mathcal{M}}$, *then* (1) *holds in the case* $i_0 = 1$.

Proof. Theorem 5.57 and the uniform boundedness of Q_n imply that the operators $I - KQ_n$ are stable in $\mathbf{L}^\infty(0, 1)$ when $\Delta_n \in \tilde{\mathcal{M}} \subset \mathcal{M}$ or $\Delta_n \in \mathcal{M}_q$ and i_0 is sufficiently large (cf. the proof of Theorem 5.43). Since $K_n(I - Q_n) = 0$, Lemma 5.81(ii) gives the stability of $I - K_n$. ■

Assertion (i) is due to Chandler and Graham [4] who gave another proof, using a stability result of Anselone and Sloan [1] on the finite section method for Wiener-Hopf equations.

5.85. Theorem. *Assume* $\varrho > 0$, $(\mathrm{H1}^{\infty,k})$ *with* $k = \max\{R, 2\}$, $(\mathrm{H1}_\varrho^{\infty,R})$, $(\mathrm{H2}^\infty)$ *and* $(\mathrm{H2}_\varrho^\infty)$. *Suppose further that* $f \in \mathbf{L}_\varrho^{\infty,R}(0,1)$, $\Delta_n \in \mathcal{M}_q$ *with* $q \geqq R/\varrho$, *and let* i_0 *be sufficiently large. Then the Nyström method* 5.80(5) *converges with the error bound*

$$\|u - u_n^*\|_{\infty,(0,1)} = O(n^{-R}) \quad \text{as} \quad n \to \infty. \tag{1}$$

Proof. By 5.80(6) we have $u - u_n^* = Ku - K_n u_n^*$, hence $(I - K_n)(u - u_n^*) = (K - K_n)u$, so that Theorem 5.84(i) implies the estimate

$$\|u - u_n^*\|_{\infty,(0,1)} \leqq c\,\|(K - K_n)\,u\|_{\infty,(0,1)}$$

when n and i_0 are large enough. Since $u \in \mathbf{L}_\varrho^{\infty,R}(0,1)$ according to Theorem 5.12 and Remark 5.13, 1°, the estimate (1) now follows from Lemma 5.83. ■

5.86. Corollary. *Under the conditions of the preceding theorem with* R *replaced by* $d + 1$, *the discrete collocation method* 5.80(4) *converges with the error estimate*

$$\|u - u_n\|_{\infty,(0,1)} = O(n^{-d-1}) \quad \text{as} \quad n \to \infty.$$

Proof. Note that

$$\begin{aligned}\|u - u_n\|_{\infty,(0,1)} &= \|u - Q_n u_n^*\|_{\infty,(0,1)}\\ &\leqq \|(I - Q_n)\,u\|_{\infty,(0,1)} + \|Q_n(u - u_n^*)\|_{\infty,(0,1)}\\ &\leqq \|(I - Q_n)\,u\|_{\infty,(0,1)} + c\,\|u - u_n^*\|_{\infty,(0,1)}.\end{aligned} \tag{1}$$

Since $u \in \mathbf{L}_\varrho^{\infty,d+1}(0,1)$ by the assumptions and Theorem 5.12, the second last term in (1) can be estimated by $cn^{-d-1}\,\|u\|_{\infty,d+1,\varrho,(0,1)}$ (cf. 5.59(2) and Lemma 5.23). Furthermore, the proof of Theorem 5.85 and Lemma 5.83 show that the last term may be bounded in the same way. ■

5.87. Remark. Suppose that K satisfies the assumptions of Theorem 5.85. Let Δ_N be the meshes defined in Example 5.16 with $q \geqq R/\varrho$, and let $f \in \mathbf{L}_\varrho^{\infty,R}(0,1)$ and $i_0 = 1$. Then the Nyström method 5.80(5) (with n replaced by N) converges with the "almost" optimal error bound

$$\|u - u_N^*\|_{\infty,(0,1)} = O\big((\log N/N)^R\big) \quad \text{as} \quad N \to \infty. \tag{1}$$

Proof. Note that $\Delta_N \in \tilde{\mathcal{M}}$. Since $h_i/x_i \sim 1/n \sim \log N/N$ (see Remark 5.24), the proof of Lemma 5.83 can easily be modified to give the estimate $\|(K - K_N)\,u\|_{\infty,(0,1)} = O\big((\log N/N)^R\big)$. Now (1) may be proved as in Theorem 5.85. ■

5.88. Remark. 1°. Theorem 5.85 extends to the case $f \in \mathbf{L}_\varrho^{\infty,R}(0,1) \dotplus \mathbb{C}$ when $u_n^* = v_n^* + c$, $c \in \mathbb{C}$, is determined by the following modified Nyström method:

$$\big(1 - b(0)\big)\,c = f(0), \qquad (I - K_n)\,v_n^* = g,$$

where $b(z)$ is the symbol of K and $g \in \mathbf{L}_\varrho^{\infty,R}(0,1)$ is defined as in Remark 5.61.

2°. Under the assumptions of Theorem 5.85 and the stronger hypothesis $(\mathrm{H1}^{\infty,R+l})$ for some $l \geqq 1$, the error estimate

$$\|u - u_n^*\|_{\infty,l,(0,1)} = O(n^{-R})$$

holds.

Proof. Applying the operators $(x\mathrm{D})^i$, $i \leq l$, to the relation

$$u - u_n^* = Ku - K_n u_n^* = (K - K_n)\, u + K_n(u - u_n^*),$$

we obtain

$$(x\mathrm{D})^i\, (u - u_n^*) = (K^{(i)} - K_n^{(i)})\, u + K_n^{(i)}(u - u_n^*), \tag{1}$$

where $K^{(i)} = (x\mathrm{D})^i\, K$ is the Mellin convolution operator with kernel function $(x\mathrm{D})^i\, k(x)$ and $K_n^{(i)}$ is the approximate operator 5.80(3) corresponding to $K^{(i)}$. Applying Lemma 5.83 to $K^{(i)}$, we get $\|(K^{(i)} - K_n^{(i)})\, u\|_{\infty,(0,1)} = O(n^{-R})$, and by the uniform boundedness of $K_n^{(i)}$ (resulting from Lemma 5.81(i)) and 5.85(1), the second term in (1) can be estimated in the same manner. ■

Discrete methods: Nyström methods and discrete collocation for Wiener-Hopf equations

5.89. Notation. The analysis of quadrature methods based on piecewise polynomials will now be carried out for the Wiener-Hopf equation 5.0(2). With the quadrature rule 5.61(4) of order R we associate the composite rule

$$\int_0^\infty u \,\mathrm{d}x \approx \sum_{(i,j)\in\mathcal{J}} \omega_j u(x_{ij})\, h_i = \int_0^\infty Q^n\, u \,\mathrm{d}x, \tag{1}$$

where Q^n are the interpolatory projections onto $\bar{\mathcal{S}}^d(\Delta^n)$ (cf. 5.29(1)) corresponding to the collocation points

$$x_{ij} = x_{i-1} + \xi_j h_i \qquad (i, j) \in \mathcal{J},$$
$$\mathcal{J} := \{(i, j) : j = 0, \ldots, d,\ i = 1, \ldots, n - i_0\}.$$

Using (1) we approximate the integral operator W in 5.1(4) by

$$W_n u(x) = \sum_{(i,j)\in\mathcal{J}} \omega_j w(x, x_{ij})\, u(x_{ij})\, h_i = \int_0^\infty Q^n\, \{w(x, y)\, u(y)\}\, \mathrm{d}y, \tag{2}$$

where $w(x, y) = \varkappa(x - y)$. The discrete collocation solution $u_n \in \bar{\mathcal{S}}^d(\Delta^n)$ to 5.0(2) is then defined by

$$u_n - Q^n W_n u_n = Q^n f, \tag{3}$$

and the Nyström solution to 5.0(2) is given by

$$u_n^* = f + W_n u_n, \tag{4}$$

or equivalently, $(I - W_n)\, u_n^* = f$. The following assertions are the analogues of Lemmas 5.81 to 5.83.

5.90. Lemma. *Assume* (A1²).

(i) *If $\Delta^n \in \tilde{\mathcal{M}}$ or $\Delta^n \in \mathcal{M}_q$, then for all n and i_0,* 5.89(2) *extends to a continuous operator on $\mathrm{L}^\infty(\mathbb{R}^+)$ such that $\|W_n\|_\infty \leq c$.*

(ii) *If $\Delta^n \in \tilde{\mathcal{M}}$ and $n \to \infty$ resp. $\Delta^n \in \mathcal{M}_q$ and n, $i_0 \to \infty$, then*

$$\|(W - W_n)\, Q^n\|_\infty \to 0.$$

Proof. Defining the functions $\varphi_{ij} \in \bar{\mathcal{F}}^d(\Delta^n)$, $(i, j) \in \mathcal{J}$, as in the Remark at the end of section 5.80, in analogy to the proof of Lemma 5.81 we obtain

$$\begin{aligned}|(W - W_n)\,u(x)| &= \Big|\sum_{(i,j)\in\mathcal{J}} \int_{J_i} [w(x, y) - w(x, x_{ij})]\,\varphi_{ij}(y)\,u(x_{ij})\,\mathrm{d}y\Big| \\ &\leqq c\,\|u\|_{\infty,\mathbb{R}^+} \sum_{(i,j)\in\mathcal{J}} \int_{J_i} |w(x, y) - w(x, x_{ij})|\,\mathrm{d}y \\ &\leqq c\,\|u\|_{\infty,\mathbb{R}^+} \sum_{i\leqq n-i_0} h_i^2\,\|\mathrm{D}_y w(x, y)\|_{\infty,J_i} \\ &\leqq c\,\|u\|_{\infty,\mathbb{R}^+} \sum_{i\leqq n-i_0} h_i^2, \qquad u \in \bar{\mathcal{F}}^d(\Delta^n), \qquad x \in \mathbb{R}^+, \end{aligned} \tag{1}$$

where we have used that $(\mathrm{A1}^2)$ obviously implies the boundedness of $\mathrm{D}\varkappa$ on $\mathbb{R}^+$. By Definition 5.20 and Remark 5.18, the last sum in (1) tends to zero when $\Delta^n \in \tilde{\mathcal{M}}$ (resp. $\Delta^n \in \mathcal{M}_q$) and $n \to \infty$ (resp. $n, i_0 \to \infty$). This proves (ii) since the projections Q^n are uniformly bounded (cf. Lemma 5.30). Similary, (1) implies that the operators $(W - W_n)\,Q^n$ are uniformly bounded in n and i_0 if $\Delta^n \in \tilde{\mathcal{M}}$ or $\Delta^n \in \mathcal{M}_q$. Thus we get (i), setting $W_n = 0$ on $(I - Q^n)\,\mathrm{L}^\infty(\mathbb{R}^+)$. ■

5.91. Lemma. *If* $u \in \mathrm{L}^{1,l}(\mathbb{R}^+)$, $1 \leqq l \leqq R$ *and* $i \leqq n - i_0$, *then the estimate* 5.82(1) *holds.*

The proof is completely analogous to that of Lemma 5.82.

5.92. Lemma. *Assume* $(\mathrm{A1}^l)$, *where* $1 \leqq l \leqq R$. *If* $u \in \mathrm{L}_\varrho^{\infty,l}(\mathbb{R}^+)$, $\varrho > 0$, $q \geqq l/\varrho$ *and* $\Delta^n \in \mathcal{M}_q$, *then*

$$\|(W - W_n)\,u\|_{\infty,\mathbb{R}^+} \leqq cn^{-l}\,\|u\|_{\infty,l,\varrho,\mathbb{R}^+}.$$

Proof. Setting $h = x_{n-i_0}$, we have

$$\begin{aligned}|(W - W_n)\,u(x)| \leqq &\Big|\int_h^\infty w(x, y)\,u(y)\,\mathrm{d}y\Big| \\ &+ \sum_{i\leqq n-i_0} \Big|\int_{J_i} w(x, y)\,u(y)\,\mathrm{d}y - \sum_j \omega_j w(x, x_{ij})\,u(x_{ij})\,h_i\Big|. \end{aligned} \tag{1}$$

The norm of the first term may be estimated by

$$\|W\|_\infty\,\|u\|_{\infty,(h,\infty)} \leqq c\,\mathrm{e}^{-\varrho h}\,\|\mathrm{e}^{\varrho x}u\|_{\infty,(h,\infty)} \leqq cn^{-l}\,\|u\|_{\infty,l,\varrho,\mathbb{R}^+}, \tag{2}$$

because $h \sim q \log(n/i_0) \sim q \log n$ (see Remark 5.18) and $\varrho q \geqq l$. By Lemma 5.91,

$$\begin{aligned}&\Big|\int_{J_i} w(x, y)\,u(y)\,\mathrm{d}y - \sum_j \omega_j w(x, x_{ij})\,u(x_{ij})\,h_i\Big| \\ &\leqq ch_i^l\,\|\mathrm{D}_y w(x, y)\,u(y)\|_{1,J_i} \leqq c \sum_{k\leqq l} h_i^l\,\|\mathrm{D}^{l-k}u\|_{\infty,J_i}\,\|\mathrm{D}_y^k w(x, y)\|_{1,J_i}, \qquad x \in \mathbb{R}^+. \end{aligned} \tag{3}$$

Since $h_i \sim (n - i)^{-1}$ and $\mathrm{e}^{-x_i} \sim \big((n - i)/n\big)^q$ (due to Remark 5.18) and $\varrho q \geqq l$, we further obtain the estimates

$$\begin{aligned}h_i^l\,\|\mathrm{D}^{l-k}u\|_{\infty,J_i} &\leqq h_i^l\,\mathrm{e}^{-\varrho x_i}\,\|\mathrm{e}^{\varrho x}\mathrm{D}^{l-k}u\|_{\infty,J_i} \\ &\leqq cn^{-l}\big((n - i)/n\big)^{\varrho q-l}\,\|u\|_{\infty,l,\varrho,\mathbb{R}^+} \leqq cn^{-l}\,\|u\|_{\infty,l,\varrho,\mathbb{R}^+}, \\ 0 \leqq k \leqq l.& \end{aligned} \tag{4}$$

Moreover, $(\mathrm{H1}^{\infty,l})$ implies that

$$\sum_{i \leq n-i_0} \sum_{k \leq l} \|D_y^k w(x, y)\|_{1,J_i} \leq \sum_{k \leq l} \int_{-\infty}^{\infty} |D^k \varkappa| \, dy \leq c, \qquad x \in \mathbb{R}^+,$$

which completes the proof, taking into account the relations (1) to (4). ■

We now establish the stability of the method 5.89(4).

5.93. Theorem. *Assume* $(\mathrm{A1}^2)$ *and* (A2).

(i) *If* $\Delta^n \in \mathcal{M}_q$ *and* i_0 *is sufficiently large, then*

$$\|(I - W_n) u\|_{\infty,\mathbb{R}^+} \geq c \|u\|_{\infty,\mathbb{R}^+}, \qquad u \in \mathbf{L}^\infty(\mathbb{R}^+), \qquad n \geq n_0 \tag{1}$$

when n_0 *is large enough.*

(ii) *If* $\Delta_n \in \tilde{\mathcal{M}}$, *then* (1) *is valid in the case* $i_0 = 1$.

Proof. It follows from Theorem 5.64 and the uniform boundedness of Q^n that the operators $I - WQ^n$ are stable in $\mathbf{L}^\infty(\mathbb{R}^+)$ if $\Delta^n \in \tilde{\mathcal{M}} \subset \mathcal{M}$ or $\Delta^n \in \mathcal{M}_q$ and i_0 is sufficiently large (cf. the proof of Theorem 5.43). Because $W_n(I - Q^n) = 0$, Lemma 5.90(ii) then yields the stability of $I - W_n$. ■

Note that Theorem 5.93(i) was proved by Chandler and Graham [3], applying the result of Anselone and Sloan [1] on the stability of the finite section method.

Next, we present results on the optimal order of convergence.

5.94. Theorem. *Assume* $\varrho > 0$, $(\mathrm{A1}^k)$ *with* $k = \max\{R, 2\}$, $(\mathrm{A1}_\varrho^R)$, (A2) *and* $(\mathrm{A2}_\varrho)$. *Suppose further that* $f \in \mathbf{L}_\varrho^{\infty,R}(\mathbb{R}^+)$, $\Delta^n \in \mathcal{M}_q$ *with* $q \geq R/\varrho$, *and let* i_0 *be sufficiently large. Then the Nyström method* 5.89(4) *converges with the error bound*

$$\|u - u_n^*\|_{\infty,\mathbb{R}^+} = O(n^{-R}) \quad as \quad n \to \infty. \tag{1}$$

Proof. By 5.89(4), $(I - W_n)(u - u_n^*) = (W - W_n) u$, and hence Theorem 5.93(i) implies the estimate $\|u - u_n^*\|_{\infty,\mathbb{R}^+} \leq c \|(W - W_n) u\|_{\infty,\mathbb{R}^+}$ for n and i_0 sufficiently large. Because $u \in \mathbf{L}_\varrho^{\infty,R}(\mathbb{R}^+)$ in view of Theorem 5.9 and Remark 5.10, the estimate (1) is now a consequence of Lemma 5.92. ■

5.95. Corollary. *Under the conditions of the preceding theorem with* R *replaced by* $d + 1$, *the discrete collocation method* 5.89(3) *converges with the error estimate*

$$\|u - u_n\|_{\infty,\mathbb{R}^+} = O(n^{-d-1}) \quad as \quad n \to \infty.$$

Proof. In analogy to 5.86(1), we have

$$\|u - u_n\|_{\infty,\mathbb{R}^+} \leq \|(I - Q^n) u\|_{\infty,\mathbb{R}^+} + c \|u - u_n^*\|_{\infty,\mathbb{R}^+}, \tag{1}$$

where $u \in \mathbf{L}_\varrho^{\infty,d+1}(\mathbb{R}^+)$. By Lemma 5.26 and 5.65(2), the first term in (1) can be bounded by $cn^{-d-1} \|u\|_{\infty,d+1,\varrho,\mathbb{R}^+}$, and the proof of Theorem 5.94 and Lemma 5.92 show that the second term may be estimated in the same way. ■

Note that the results presented in Remarks 5.87 and 5.88 carry over, with obvious modifications, to Wiener-Hopf equations, but we refrain from giving them explicitly.

Systems

5.96. Notations. In what follows we use the agreements formulated in Section 4.97. If the equations 5.0(1) and 5.0(2) are replaced by systems, the analysis of the present chapter can be carried out under suitable assumptions. The key of such an investigation are again stability results. However, to prove the stability of some of the methods described in this chapter, more care is in place if we consider the system case. First of all the invertibility of the corresponding operator 5.1(4) or 5.1(6) does not imply the factorization 5.4(2) or 5.5(2). To overcome this difficulty we present factorization-free proofs. To exemplify this philosophy, we consider the method 5.37(1) for equation 5.0(2). For the sake of further simplification we restrict ourselves to the case of $\mathbf{L}^p_N(\mathbb{R}^+)$, $1 < p < \infty$. Consider the operator $A = W_\varphi + K$ on $\mathbf{L}^p_N(\mathbb{R}^+)$, where W_φ, $\varphi \in \mathcal{F}_{N\times N}$, is a system of Wiener-Hopf operators and K is compact (see Section 4.111).

5.97. Theorem. *Assume $\Delta^n \in \mathcal{M}$ or $\Delta^n \in \mathcal{M}_q$ and suppose i_0 to be sufficiently large* (*see* 5.24(2)). *The sequence $\{P^nAP^n\}$ is stable if and only if the operators A and $W_{\varphi^{-1}}$ are invertible.*

Proof. (a) First of all notice that the stability of $\{P^nAP^n\}$ is equivalent to the stability of $\{Q^nA^{-1}Q^n\}$, $Q^n := I - P^n$, (Section 4.27(a)) provided that A is invertible. Of course, if $\{P^nAP^n\}$ is stable, then standard arguments show the invertibility of A. Therefore, it is sufficient to consider the stability of $\{Q^nA^{-1}Q^n\}$.

(b) Let $\{P_\tau\}_{\tau\in\mathbb{R}^+}$ be the collection of projections defined in Section 4.54. With the projections P^n we connect the projections $P_{\tau(n,i)}$, where $\tau(n, i) := x_{n-i}$. If no confusion will occur we write τ instead of $\tau(n, i)$.

The proof of the convergence of the usual finite section method (see Sections 4.58 and 4.113) was mainly supported upon the following conclusions. If W_ψ is a Wiener-Hopf operator then (with $Q_\tau := I - P_\tau$, $\tau \in \mathbb{R}^+$)

$$Q_\tau W_\psi Q_\tau \text{ is invertible in } \operatorname{im} Q_\tau \text{ for a certain } \tau \Rightarrow W_\psi \text{ is invertible,} \tag{1}$$

$$W_\psi \text{ is invertible} \Rightarrow Q_\tau W_\psi Q_\tau \text{ is invertible in } \operatorname{im} Q_\tau \text{ for all } \tau > 0 \text{ and } \sup_{\tau>0} \|(Q_\tau W_\psi Q_\tau)^{-1} Q_\tau\| < \infty. \tag{2}$$

(c) Obviously, we have

$$P^nP_\tau = P_\tau P^n = P^n,$$
$$Q_\tau Q^n = Q^nQ_\tau = Q_\tau \qquad \big(\tau = \tau(n, i_0)\big),$$

and $\operatorname{im} P_\tau Q^n$, $\operatorname{im} Q_\tau$ are (complementary) subspaces of $\operatorname{im} Q^n$. Further we need the following assertion (which can be proved analogously to Lemma 5.38 after approximating by smooth kernels). Let W_ψ be a Wiener-Hopf operator, $\psi \in \mathcal{F}_{N\times N}$. Then, given $\delta > 0$, there exists an integer n_0 such that

$$\big\|P_\tau Q^n\big(\psi(\infty)\, I - W_\psi\big)\, u\big\|_{p,\mathbb{R}^+} \leqq \delta\, \|u\|_{p,\mathbb{R}^+}, \tag{3}$$

$$\big\|\big(\psi(\infty)\, I - W_\psi\big)\, P_\tau Q^n u\big\|_{p,\mathbb{R}^+} \leqq \delta\, \|u\|_{p,\mathbb{R}^+} \quad \text{for} \quad n \geqq n_0, \tag{4}$$

if either $\Delta^n \in \mathcal{M}$ and $i_0 = 1$, or $\Delta^n \in \mathcal{M}_q$ and the index i_0 is sufficiently large.

Consider $Q^n W_\psi Q^n$. We have

$$Q^n W_\psi Q^n = P_\tau Q^n W_\psi P_\tau Q^n + Q_\tau W_\psi Q_\tau + P_\tau Q^n W_\psi Q_\tau + Q_\tau W_\psi P_\tau Q^n .$$

Given $\varepsilon > 0$ we get (by (3) and (4))

$$\|Q^n W_\psi Q^n - (\psi(\infty) P_\tau Q^n + Q_\tau W_\psi Q_\tau)\| < \varepsilon \qquad (n \geqq n_1) \tag{5}$$

if either $\Delta^n \in \mathcal{M}$ and $i_0 = 1$, or $\Delta^n \in \mathcal{M}_q$ and the index i_0 is large enough.

Now we are ready to prove the theorem.

(d) *if-part*: Let A and $W_{\varphi^{-1}}$ be invertible. Since $A^{-1} = W_{\varphi^{-1}} + T$, $T \in \mathcal{K}(\mathbf{L}_N^p(\mathbb{R}^+))$ and $\|Q^n T Q^n\| \to 0$ as $n \to \infty$ (Section 1.1(h)), we obtain from (5)

$$\|Q^n A^{-1} Q^n - (\varphi^{-1}(\infty) P_\tau Q^n + Q_\tau W_{\varphi^{-1}} Q_\tau)\| < 2\varepsilon \qquad (n \geqq n_2) \tag{6}$$

for the i_0 ensuring (5). Now use (2) in order to get the stability of $\{Q^n A^{-1} Q^n\}$.

(e) *only if-part*: (6) shows that the invertibility of $\{Q^n A^{-1} Q^n\}$ in im Q^n implies the invertibility of $\{Q_\tau W_{\varphi^{-1}} Q_\tau\}$ in im Q^n (n sufficiently large). Apply (1) to get the assertion. ■

5.98. Example. The problem of determining the distribution of stress in a thin elastic plate in the vicinity of a cruciform crack, when the crack faces of equal length are subject to the same constant normal pressure, has been studied by several authors (see e.g. Panasjuk et al. [1], Rooke/Sneddon [1], Stallybrass [1]). Most of these authors start from a boundary-value formulation of the problem and reduce this to a linear integral equation using Mellin or Fourier transforms. In particular, Rooke/Sneddon [1] and Stallybrass [1] obtain the equation

$$Bu(x) := u(x) + \frac{4}{\pi} \int_0^1 \frac{xy^2}{(x^2 + y^2)^2} u(y) \, dy = 1, \qquad 0 < x \leqq 1. \tag{1}$$

Obviously, this is a Mellin convolution equation of the form 5.0(1), where

$$k(x) = -\frac{4}{\pi} \frac{x}{(1 + x^2)^2}$$

and $f(x) \equiv 1$. It can be easily seen that $k(x)$ satisfies the assumptions ($\mathrm{H1}^p$), ($\mathrm{H1}_\varrho^{p,1}$) and that $k(x) \in X_\varrho^{p,l-1}(0, 1)$ ($l \geqq 1$) for $1 \leqq p \leqq \infty$, $l \in \mathbb{N}_0$ and $0 \leqq \varrho < 1 + 1/p$. Furthermore, taking into account that the integral

$$I(z) := \frac{4}{\pi} \int_0^\infty \frac{t^z}{(1 + t^2)^2} \, dt = \frac{1 - z}{\cos(\pi z/2)} \qquad (z \in \mathbb{C})$$

admits the estimate $\operatorname{Re} I(z) > -2\sqrt{2}/\pi > -1$ if $-1 < \operatorname{Re} z < 1$ (see Monegato/Prössdorf [1]) we observe that conditions ($\mathrm{H2}^p$) and ($\mathrm{H2}_\varrho^p$) are fulfilled for all ϱ such that $0 \leqq \varrho < 1/p + 1$. Thus, the operator B is invertible in $\mathcal{L}(\mathbf{L}^p(0, 1))$ for each p, $1 \leqq p \leqq \infty$. Moreover, it follows from Theorem 5.12 that the unique solution $u \in \mathbf{L}^\infty(0,1)$ of equation (1) has the form $u = v + c$, where $c \in \mathbb{C}$ and $v \in \mathbf{C}_\varrho^l[0, 1]$ for all $l \geqq 1$. In particular, $x^{-\varepsilon} v(x) \in \mathbf{C}[0, 1]$ for each $\varepsilon \in [0, 1)$. It can be easily checked that $c = 1/2$.

Numerical methods for solving (1) and related equations have been proposed by several authors (see e.g. GERASOULIS [1], JEN/SRIVASTAV [1], ROOKE/SNEDDON [1], STALLYBRASS [1]), however they usually do not take into account the behavior of the solution at the critical point $x = 0$; furthermore, no stability results have been derived for these methods. The convergence results of the present chapter applied to equation (1), in particular, for Galerkin and Nyström methods, have been confirmed with high accuracy by numerical computations in MONEGATO/PRÖSSDORF [1].

Notes and comments

Except for Sect. 5.96—5.98, the material of this chapter is taken from ELSCHNER [6]. Unless stated otherwise, the results were published there for the first time.

5.8.—5.14. The supremum of the numbers $\varrho > 0$ for which conditions ($H1^p$) and ($H2^p$) in Theorem 5.12 and Remark 5.13 are satisfied reflects the principal term of the asymptotics of solutions to equation 5.0(1) at the point $x = 0$; cf. ELSCHNER [5] concerning more precise results for pseudodifferential operators of Mellin type in weighted $\mathbf{L}^2$ spaces. Analogously, the number ϱ in Theorem 5.9 is related to the exponential decay at infinity of solutions to the Wiener-Hopf equation 5.0(2).

5.15.—5.36. The usage of graded meshes in the approximation by polynomial splines goes back to RICE [1]. Graded meshes were then used in the numerical solution of second kind Fredholm integral equations and Volterra integral equations with weakly singular kernels by CHANDLER [1], SCHNEIDER [1], GRAHAM [1], VAINIKKO, PEDAS and UBA [1], and BRUNNER and VAN DER HOUWEN [1]. Recently CHANDLER and GRAHAM [1]—[4] applied meshes of class $\mathcal{M}_q$ to the approximate solution of Mellin convolution and Wiener-Hopf equations.

The results of Sects. 5.21—5.36 on the approximation properties of polynomial spline functions are mostly well-known (cf. DE BOOR [3], SCHUMAKER [1]), but for the reader's convenience we gave a self-contained exposition.

5.37.—5.56. The idea of applying the Wiener-Hopf factorization in order to obtain necessary and sufficient conditions for the stability of spline approximation methods applied to non-compact integral equations is due to ELSCHNER [8], who studied Galerkin methods with piecewise polynomials for equation 5.0(1) on $\mathbf{L}^2$ using graded meshes of class $\mathcal{M}_q$. In Sect. 5.37, 5.38, 5.45 and 5.46 this approach was extended to prove the stability of such Galerkin methods in the case of equations 5.0(1) and 5.0(2) on $\mathbf{L}^p$, $1 \leqq p \leqq \infty$, and meshes of class $\mathcal{M}$.

The optimal convergence rate obtained in Theorems 5.41 and 5.51 is standard for second kind integral equations on finite intervals, see ATKINSON [1], BAKER [1]. We do not know whether the optimal order of convergence can be achieved by using meshes of class $\mathcal{M}$. The iterated Galerkin method goes back to SLOAN [1], see also SLOAN and THOMÉE [1]. CHANDLER and GRAHAM [2] obtained Theorems 5.51 and 5.54 if $p = 2$ and the $\mathbf{L}^2$ operator norm of K is smaller than one (cf. also CHANDLER [2] concerning the double layer potential on curves with corners). In this case, however, the stability of the Galerkin method is obvious and holds for arbitrary meshes and spline spaces.

5.57.—5.68. CHANDLER and GRAHAM [2] established the stability in the uniform norm of collocation methods with piecewise polynomials applied to equation 5.0(1) if the $\mathbf{L}^\infty$ operator norm of K is smaller than one and the spline spaces are modified in that the high order polynomial approximations are replaced by piecewise constant functions on some of the closest intervals to the point $x = 0$. Moreover, under these assumptions, they proved the error estimates of Theorems 5.60 and 5.62. A result on the stability of piecewise constant collocation for Wiener-Hopf equations (a special case of Theorem 6.54(ii)) was first shown by SLOAN and SPENCE [1].

Using another approach, which is directly based on the Wiener-Hopf factorization, the results of Sects. 5.57—5.68 may be extended to the spaces $\mathbf{L}^p$, $1 \leqq p \leqq \infty$; cf. ELSCHNER [9].

5.69.—5.79. For Mellin convolution equations, $p = 2$ and meshes of class $\mathcal{M}_q$, Theorems 5.69, 5.73 and 5.74 were first obtained in ELSCHNER [8].

5.80.–5.89. Lemmas 5.82, 5.83 and Theorems 5.84(i), 5.85 are due to CHANDLER and GRAHAM [4]. Generalizations concerning stability and error estimates in the spaces $\mathbf{L}^p$, $1 \leqq p \leqq \infty$, can be found in ELSCHNER [9]. For recent results on discrete Galerkin and collocation methods in the case of Fredholm integral equations with smooth kernels, we refer to ATKINSON and BOGOMOLNY [1], and JOE [1], [2].

5.90.–5.95. Lemmas 5.91, 5.92 and Theorems 5.93(i), 5.94 are due to CHANDLER and GRAHAM [3]. For results on stability and error estimates in $\mathbf{L}^p$, $1 \leqq p \leqq \infty$, for discrete collocation and Nyström methods applied to equation 5.0(1), we refer to ELSCHNER [9]. Other approaches to quadrature methods for second kind integral equations on the half-axis were given by ANSELONE and SLOAN [2], S. GÄHLER and W. GÄHLER [1], and ROCH and SILBERMANN [10]. These papers, however, do not contain results on the rate of convergence.

5.96.–5.97. The extension of Theorem 5.37 to systems of Wiener-Hopf integral equations is due to the authors. The recent paper SILBERMANN [7] contains a Banach algebra approach to the theory of the approximation method considered in these sections (see also Chapters 7 and 10).

Chapter 6
Singular integral equations and pseudodifferential equations on curves

We here focus our attention on those aspects of the general theory of singular integral operators, Toeplitz operators, and pseudodifferential operators which are of importance for the following chapters. Our main concern will be symbol constructions, strong ellipticity and asymptotics of solutions.

Curves and singular integral operators in L^2-spaces

6.1. The curve zoo. A *bounded Lyapunov* arc Γ is an oriented bounded curve in the complex plane $\mathbb{C}$ which is homeomorphic to the closed interval $[0, 1]$ and which fulfils the Lyapunov condition, i.e. the first derivative of the parametric representation t: $[0, L] \to \Gamma$ of Γ by its arc length must be Hölder continuous.

Let $\Gamma_1, \dots, \Gamma_m$ be bounded Lyapunov arcs possessing a finite number of common points. The union $\Gamma = \bigcup_{i=1}^{m} \Gamma_i$ is called a (bounded) *admissible curve* if whenever Γ_i and Γ_j have a common point, z, the curve $\Gamma_i \cup \Gamma_j$ is either a Lyapunov arc or the tangents of Γ_i and Γ_j at the point z do not coincide.

Let $k \in \mathbb{Z}^+$. A point $z_0 \in \Gamma$ is said to have the *order k* if for each sufficiently small closed neighborhood U of z_0 the set $U \cap \Gamma$ is the union of k Lyapunov arcs $\Gamma_1, \dots, \Gamma_k$ such that z_0 is one of the end points of each of these arcs $\Gamma_1, \dots, \Gamma_k$ and the arcs have no other points in common. The order of a point $z \in \Gamma$ will be denoted by $k(z)$. A curve Γ is called *simple* if the order of each point of Γ is at most 2. A simple curve Γ is said to be *closed* if it divides the closed complex plane into two non-empty open sets D_Γ^+ and D_Γ^- such that Γ is the boundary of either of these sets. The *positive direction* on a closed curve will be chosen so that if Γ is traced out in this direction, D_Γ^+ is always on the left. For the sake of simplicity we shall henceforth assume that D_Γ^+ contains the origin $z = 0$ and that the point at infinity, $z = \infty$, belongs to D_Γ^-.

In what follows we shall only consider curves of the following two types: (a) bounded, admissible and simple curves, or (b) unbounded curves of the form $\Gamma = \mathbb{R}^+ \cup e^{i\beta}\mathbb{R}^+$ $(0 < \beta < 2\pi)$. These and only these curves will henceforth be called admissible.

Let Γ be an admissible curve and let $z_0 \in \Gamma$ be a point of order 2. Choose a neighborhood U of z_0 such that $U \cap \Gamma$ is the union of two Lyapunov arcs $\Gamma_1(z_0)$, $\Gamma_2(z_0)$ having z_0 as one of their end points. We denote by $\beta = \beta(z_0)$ the angle between Γ_1 and Γ_2 at the point z_0. More precisely, if $t_i: [0, L_i] \to \Gamma_i$ is the parametric representation of Γ_i by its arc length and if we put

$$\arg t_i'(z_0) := \lim_{\substack{z \to z_0 \\ z \in \Gamma_i}} \arg t_i'(t_i^{-1}(z))$$

then $\beta := \arg t_2'(z_0) - \arg t_1'(z_0)$ in case Γ_1 and Γ_2 are both directed to z_0 or away from z_0, and $\beta := \arg t_2'(z_0) - \arg t_1'(z_0) + \pi$ in case one of the arcs Γ_1, Γ_2 is directed to z_0 and the other one away from z_0. In both cases we assume the branches of the argument function to be chosen so that $0 \leqq \beta_i < 2\pi$. An admissible curve Γ is called *Lyapunov* if $\beta(z) = \pi$ for all $z \in \Gamma$ with $k(z) = 2$.

Let Γ be an admissible curve. The *singular integral operator* S on Γ is defined by

$$(S_\Gamma f)\ (t) := \frac{1}{\pi \mathrm{i}} \int_\Gamma \frac{f(z)}{z - t}\, \mathrm{d}z \ (t \in \Gamma), \tag{1}$$

the integral understood as the Cauchy principal value. Further, let $t_1, \ldots, t_n \in \Gamma$, $\alpha_1, \ldots, \alpha_n \in \mathbb{R}$, and put

$$w(z) := \prod_{i=1}^{n} |z - t_i|^{\alpha_i}.$$

By $\mathbf{L}^2(\Gamma, w)$ we denote the weighted Lebesgue space on Γ with the norm

$$\|f\|_{2,w} := \Big(\int_\Gamma |f(z)|^2\, w^2(z)\, |\mathrm{d}z|\Big)^{1/2}.$$

Notice that the norm used here differs from the norm introduced in Section 4.92.

Now we list some properties of the operator S_Γ, which can be found in Gohberg/Krupnik [4] or Mikhlin/Prössdorf [1], for example.

(i) *Let Γ be an admissible bounded curve. Then $S_\Gamma \in \mathcal{L}(\mathbf{L}^2(\Gamma, w))$ if and only if $-1/2 < \alpha_i < 1/2$ for $i = 1, \ldots, n$.*

(ii) *Let Γ be an admissible unbounded curve. Then $S_\Gamma \in \mathcal{L}(\mathbf{L}^2(\Gamma, w))$ if and only if $-1/2 < \alpha_i < 1/2$ for $i = 1, \ldots, n$, and $-1/2 < \sum_{i-1}^{n} \alpha_i < 1/2$.*

(iii) *If Γ is an admissible closed curve then $S_\Gamma^2 = I$.*

(iv) *If Γ is an admissible bounded curve, then $aS - SaI \in \mathcal{K}(\mathbf{L}^2(\Gamma, w))$ for all $a \in \mathbf{C}(\Gamma)$, where $\mathbf{C}(\Gamma)$ stands for the algebra of all continuous (complex-valued) functions on Γ.*

Since the dual space to $\mathbf{L}^2(\Gamma, w)$ can be identified with $\mathbf{L}^2(\Gamma, w^{-1})$, the singular integral operator S_Γ is defined on both spaces if condition (i) is satisfied. Let $z \in \Gamma$. It is easily seen that

$$\mathrm{d}z = h_\Gamma(z)\, |\mathrm{d}z|, \qquad h_\Gamma(z) = \mathrm{e}^{\mathrm{i}\Theta_\Gamma(z)},$$

where $\Theta_\Gamma(z)$ denotes the angle between the tangent to the admissible curve Γ at the point z and the positive direction of the x-axis. Obviously, the function $h_\Gamma(z)$ is defined at all points for which $\beta(z) = \pi$, and it is a bounded piecewise continuous function.

(v) *The adjoint operator S_Γ^*, acting on the space $\mathbf{L}^2(\Gamma, w^{-1})$, is given by*

$$S_\Gamma^* = -H_\Gamma S_\Gamma H_\Gamma,$$

where H_Γ is the operator defined on $\mathbf{L}^2(\Gamma, w^{-1})$ by the rule

$$(H_\Gamma \varphi)\ (z) = \overline{h_\Gamma(z)\ \varphi(z)}.$$

(vi) *Let Γ be a Lyapunov curve. If condition* (i) *is in force, then*

$$S_\Gamma^* - S_\Gamma \in \mathcal{K}(\mathbf{L}^2(\Gamma, w^{-1}))$$

(vii) *Let Γ be a circle, a circular arc, or an interval. Then, provided condition* (i) *holds,* $S_\Gamma^* = S_\Gamma$.

(viii) *If $S_{\mathbb{T}}$ is bounded on $\mathbf{L}^2(\mathbb{T}, w)$ (which occurs, for instance, if it is subject to the condition* (i)*), then the action of $S_{\mathbb{T}}$ on $\mathbf{L}^2(\mathbb{T}, w)$ can be described in terms of Fourier coefficients as follows:*

$$S_{\mathbb{T}}\colon \sum_{k=-\infty}^{\infty} c_k t^k \mapsto \sum_{k=0}^{\infty} c_k t^k - \sum_{k=-\infty}^{-1} c_k t^k \qquad (t \in \mathbb{T}).$$

The latter property can be used to define the singular integral operator on spaces of periodic functions (given on $\mathbb{R}$).

Let Γ be an admissible bounded curve. A *piecewise continuous function* on Γ is a function which, at each point $z_0 \in \Gamma$, has finite limits as $z \to z_0$ along each Lyapunov arc Γ_i $(i = 1, \dots, k)$ terminating in z_0. By $\mathbf{PC}(\Gamma)$ we denote the Banach algebra of all piecewise continuous functions a on Γ provided with the norm

$$\|a\|_\infty = \sup_{z\in\Gamma} |a(z)| . \tag{2}$$

Obviously, for each $a \in \mathbf{PC}(\Gamma)$ the operator of multiplication by a is bounded on $\mathbf{L}^2(\Gamma, w)$ and its norm coincides with (2). So it makes sense to consider the Banach algebra $\operatorname{alg}\{S_\Gamma, \mathbf{PC}(\Gamma)\}$ which, by definition, is the smallest closed subalgebra of the algebra $\mathcal{L}(\mathbf{L}^2(\Gamma, w))$ containing S_Γ and all multiplication operators generated by functions in $\mathbf{PC}(\Gamma)$.

(ix) *Let $a, b \in \mathbf{PC}(\Gamma)$ satisfy* $\inf_{t\in\Gamma} |a(t)| > 0$, $\inf_{t\in\Gamma} |b(t)| > 0$ *and let A stand for any of the operators* $aP_\Gamma + bQ_\Gamma$, $P_\Gamma a + Q_\Gamma b$, $P_\Gamma a P_\Gamma + Q_\Gamma$, $P_\Gamma + Q_\Gamma b Q_\Gamma$ $(P_\Gamma := 1/2(I + S_\Gamma)$, $Q_\Gamma := 1/2(I - S_\Gamma))$. *Then at least one of the integers* $\dim\ker A$ *and* $\dim\ker A^*$ *is equal to zero.*

Note, that $\mathbf{L}^2(\Gamma, w)$ is a Hilbert space with respect to the scalar product

$$(f, g)_w := \int_\Gamma f(z)\, \overline{g(z)}\, w^2(z)\, |dz| .$$

Therefore, $\mathcal{L}(\mathbf{L}^2(\Gamma, w))$ actually forms a C^*-algebra.

6.2. Theorem. *Let Γ be admissible and bounded.*

(a) $\operatorname{alg}\{S_\Gamma, \mathbf{PC}(\Gamma)\}$ *is a C^*-subalgebra of* $\mathcal{L}(\mathbf{L}^2(\Gamma, w))$.

(b) $\mathcal{K}(\mathbf{L}^2(\Gamma, w)) \subset \operatorname{alg}\{S_\Gamma, \mathbf{PC}(\Gamma)\}$.

(c) *The quotient algebra* $\operatorname{alg}^\pi\{S_\Gamma, \mathbf{PC}(\Gamma)\} := \operatorname{alg}\{S_\Gamma, \mathbf{PC}(\Gamma)\}/\mathcal{K}(\mathbf{L}^2(\Gamma, w))$ *contains a copy of the C^*-algebra* $\mathbf{C}(\Gamma)$.

We shall not present a proof of this theorem here, but a few remarks seem to be in order.

Assertion (a) is perhaps well-known to specialists. A proof is given in ROCH/SILBERMANN [7]. Since $\operatorname{alg}\{S_\Gamma, \mathbf{PC}(\Gamma)\}$ is irreducible and contains at least one compact operator (by 6.1(iv)), we have

$$\mathcal{K}(\mathbf{L}^2(\Gamma, w)) \subset \text{alg}\,\{S_\Gamma, \mathbf{PC}(\Gamma)\}$$

(by Theorem 5.39 in Douglas [1]). Finally, (c) is a consequence of 6.1(iv).

Our next (and main) aim is the description of $\text{alg}^\pi\,\{S_\Gamma, \mathbf{PC}(\Gamma)\}$ in a local language. The key ingredient for doing this is the algebra $\dot{\Sigma}^k(\alpha)$, we are going to define now.

The algebra $\dot{\Sigma}^k(\alpha)$

6.3. Notations. Consider the Hilbert space $\mathbf{L}^2(\mathbb{R}^+, w)$ with the special weight $w(t) = t^\alpha$, $-1/2 < \alpha < 1/2$. The singular integral operator (or the *Hilbert transform*)

$$(Sf)\,(t) = \frac{1}{\pi \mathrm{i}} \int_0^\infty \frac{f(s)}{s - t}\, \mathrm{d}s, \qquad t \in \mathbb{R}^+$$

plays an outstanding role in the local theory of singular integral operators. Besides the operator S, we shall use the *weighted Hilbert transform*

$$(S_\gamma f)\,(t) = \frac{1}{\pi \mathrm{i}} \int_0^\infty \left(\frac{s}{t}\right)^\gamma \frac{f(s)}{s - t}\, \mathrm{d}s, \qquad t \in \mathbb{R}^+, \ \gamma \in \mathbb{C}$$

and the *generalized Hankel operator*

$$(N_\beta f)\,(t) = \frac{1}{\pi \mathrm{i}} \int_0^\infty \frac{f(s)}{s - \mathrm{e}^{\mathrm{i}\beta} t}\, \mathrm{d}s, \qquad t \in \mathbb{R}^+, \ \ \mathrm{Re}\,\beta \in (0, 2\pi).$$

6.4. Proposition. (a) *If* $0 < 1/2 + \alpha - \mathrm{Re}\,\gamma < 1$ *then the operator* S_γ *is bounded on* $\mathbf{L}^2(\mathbb{R}^+, t^\alpha)$

(b) *Let* $0 < 1/2 + \alpha < 1$. *Then the Hankel operator* N_β *is bounded on* $\mathbf{L}^2(\mathbb{R}^+, t^\alpha)$ *for each* β *with* $\mathrm{Re}\,\beta \in (0, 2\pi)$.

Proof. See Gohberg/Krupnik [4], Krupnik [1], Costabel [2], or Roch/Silbermann [7], for example. ■

Let $\Sigma(\alpha)$ stand for the smallest closed subalgebra of $\mathcal{L}(\mathbf{L}^2(\mathbb{R}^+, t^\alpha))$ that contains the singular integral operator S and the identity operator. We shortly write Σ for $\Sigma(0)$.

6.5. Theorem. *Let* $0 < 1/2 + \alpha < 1$.

(a) *For* $0 < 1/2 + \alpha - \mathrm{Re}\,\gamma < 1$ *the weighted Hilbert transform* S_γ *belongs to* $\Sigma(\alpha)$.

(b) *For* $0 < \mathrm{Re}\,\beta < 2\pi$ *the generalized Hankel operator* N_β *belongs to* $\Sigma(\alpha)$.

(c) *The algebras* $\Sigma(\alpha)$ *and* Σ *are isometrically isomorphic. The image of the operator* $S \in \Sigma(\alpha)$ *under this isomorphism is the operator* $S_{-\alpha} \in \Sigma$.

(d) *Let* $0 < 1/2 + \alpha - \mathrm{Re}\,\gamma < 1$. *Then the smallest closed subalgebra of* $\mathcal{L}(\mathbf{L}^2(\mathbb{R}^+, t^\alpha))$ *which contains the operators* I *and* S_α *coincides with* $\Sigma(\alpha)$.

(e) $\Sigma(\alpha)$ *is a* C^**-subalgebra of* $\mathcal{L}(\mathbf{L}^2(\mathbb{R}^+, t^\alpha))$.

A proof of (a) and (b) can be found in COSTABEL [2]. The assertions (c) and (d) can easily be deduced from (a). The last assertion is a consequence of the following facts: the mapping $A \mapsto t^\alpha A t^{-\alpha}$ is a *-isomorphism from $\mathcal{L}(\mathbf{L}^2(\mathbb{R}^+, t^\alpha))$ onto $\mathcal{L}(\mathbf{L}^2(\mathbb{R}^+))$ and Σ is a C^*-algebra, by 6.1(vii).

6.6. Theorem. *The maximal ideal space of $\Sigma(\alpha)$ can be identified with $\tilde{\mathbb{R}}$, i.e. the two-point compactification of $\mathbb{R}$. Under this identification, the Gelfand transforms of S_γ* $(0 < 1/2 + \alpha - \operatorname{Re}\gamma =: \delta < 1)$ *and N_β* $(0 < \operatorname{Re}\beta < 2\pi)$ *are given for $z \in \mathbb{R}$ by*

$$s_\delta(z) = \coth(z + i\delta)\,\pi \quad \text{and} \quad n_\beta(tz) = \frac{e^{(z+i(1/2+\alpha)(\pi-\beta))}}{\sinh\big(z + i(1/2 + \alpha)\big)\,\pi},$$

respectively (e^{a+ib} *being defined as* $e^a(\cos b + i\sin b)$).

*Moreover $\Sigma(\alpha)$ is *-isomorphic to $\mathbf{C}(\tilde{\mathbb{R}})$, the isomorphism being nothing else than the Gelfand transform.*

We omit the proof of this theorem here. We merely remark that the Gelfand transforms of S_γ and N_β can be computed be means of the Mellin transformation.

Let $N(\alpha)$ denote the smallest closed two-sided ideal of the Banach algebra $\Sigma(\alpha)$ which contains the operator N_π.

6.7. Theorem. (a) *Let $A \in \Sigma(\alpha)$. Then A lies in the ideal $N(\alpha)$ if and only if $a(\pm\infty) = 0$, where a stands for the Gelfand transform of A*

(b) *Let* $0 < \operatorname{Re}\beta < 2\pi$. *Then the smallest closed two-sided ideal of $\Sigma(\alpha)$ containing the operator N_β coincides with $N(\alpha)$.*

Proof. By Theorem 6.6, the Gelfand transforms of N_π and N_β vanish only at the points $-\infty$ and $+\infty$. Taking into account Sec. 1.14 we get our claim. ■

Given $k \in \mathbb{Z}^+$, denote by $\mathbf{L}_k^2(\mathbb{R}^+, t^\alpha)$ the Banach space of all vectors $(f_1, \ldots, f_k)^T$, $f_j \in L^2(\mathbb{R}^+, t^\alpha)$ for $i = 1, \ldots, k$, endowed with the norm

$$\|(f_1, \ldots, f_k)^T\|_{\mathbf{L}_k^2(\mathbb{R}^+, t^\alpha)} := \left(\sum_{j=1}^{k} \|f_j\|^2_{\mathbf{L}^2(\mathbb{R}^+, t^\alpha)}\right)^{1/2}.$$

Further, let $\dot{\Sigma}^k(\alpha) \subset \mathcal{L}(\mathbf{L}_k^2(\mathbb{R}^+, t^\alpha))$ refer to the Banach algebra of all $k \times k$ matrices (A_{ij}) such that $A_{ii} \in \Sigma(\alpha)$ $(i = 1, \ldots, k)$ and $A_{ij} \in N(\alpha)$ for $i \neq j$ $(i, j = 1, \ldots, k)$.

The following result is a simple consequence of the commutativity of the algebra $\Sigma(\alpha)$ and of the fact that all underlying algebras are C^*-algebras.

6.8. Corollary. *Let $A \in \dot{\Sigma}^k(\alpha)$. If A is invertible in $\mathcal{L}(\mathbf{L}_k^2(\mathbb{R}^+, t^\alpha))$ then $A^{-1} \in \dot{\Sigma}^k(\alpha)$. A is invertible if and only if* $\det A$ *is invertible in $\mathcal{L}(\mathbf{L}^2(\mathbb{R}^+, t^\alpha))$.*

The algebra alg $\{S_\Gamma, \mathbf{PC}(\Gamma)\}$ and symbols

6.9. Notations. Let Γ be an admissible and bounded curve and

$$w(z) = \prod_{i=1}^{n} |z - t_i|^{\alpha_i} \tag{1}$$

a weight (on Γ) satisfying $-1/2 < \alpha_i < 1/2$, $i = 1, \ldots, n$.

Put

$$\alpha(z_0) = \begin{cases} \alpha_i & \text{if } z_0 \in \{t_1, \ldots, t_n\} \text{ and } z_0 = t_i, \\ 1 & \text{if } z_0 \notin \{t_1, \ldots, t_n\}. \end{cases}$$

Then define a new weight function depending on z_0 by $w_{z_0}(z) = |z|^{\alpha(z_0)}$ $(z \in \mathbb{R}^+)$ and, finally, set $n_i(z_0) = 1$ if Γ_i is directed away from z_0 and $n_i(z_0) = -1$ if Γ_i is directed to z_0 (see Section 6.1).

Now we are in a position to state our main result.

6.10. Theorem. *Let Γ be an admissible and bounded curve. Consider* alg $\{S_\Gamma, \mathbf{PC}(\Gamma)\} \subset \mathcal{L}(\mathbf{L}^2(\Gamma, w))$, *where w is given by* 6.9(1).

(a) *For each $z_0 \in \Gamma$, there exists a homomorphism $A \mapsto \widetilde{\text{smb}}\,(A, z_0)$ from* alg $\{S_\Gamma, \mathbf{PC}(\Gamma)\}$ *onto the algebra $\mathring{\Sigma}^{k(z_0)}(\alpha(z_0))$ such that*

$$\widetilde{\text{smb}}\,(S_\Gamma, z_0) = \begin{cases} \begin{pmatrix} S & N_{2\pi-\beta} \\ N & S \end{pmatrix} \begin{pmatrix} n_1(z_0) & 0 \\ 0 & n_2(z_0) \end{pmatrix} & \text{for } k(z_0) = 2 \\ S n_1(z_0) & \text{for } k(z_0) = 1, \end{cases}$$

and, if we let $a_i(z_0) := \lim\limits_{\substack{z \to z_0 \\ z \in \Gamma_i(z_0)}} a(z)$ *for piecewise continuous* a,

then

$$\widetilde{\text{smb}}\,(a, z_0) = \begin{cases} \operatorname{diag}\big(a_1(z_0)\, I, a_2(z_0)\, I\big) & \text{for } k(z_0) = 2 \\ a_1(z_0)\, I & \text{for } k(z_0) = 1 \end{cases}$$

(b) *$A \in$ alg $\{S_\Gamma, \mathbf{PC}(\Gamma)\}$ is Fredholm if and only if $\widetilde{\text{smb}}\,(A, z)$ is invertible for all $z \in \Gamma$.*

(c) *If $\pi(A)$ stands for the image of A in the Calkin algebra, then, for $A \in$* alg $\{S_\Gamma, \mathbf{PC}(\Gamma)\}$,

$$\|\pi(A)\| = \sup_{z \in \Gamma} \|\widetilde{\text{smb}}\,(A, z)\|.$$

(d) *$A \in$ alg $\{S_\Gamma, \mathbf{PC}(\Gamma)\}$ is compact if and only if $\widetilde{\text{smb}}\,(A, z) = 0$ for all $z \in \Gamma$.*

(e) *If $A = aP_\Gamma + bQ_\Gamma$ is Fredholm $(a, b \in \mathbf{PC}(\Gamma))$, then it is at least one-sided invertible. The invertibility properties of A correspond with the index* ind A, *i.e. A is invertible, invertible only from the left or only from the right in dependence on whether* ind A *is zero, negative or positive, respectively.*

The proof is not trivial and we limit ourselves to the main steps.

1. Since $\mathbf{C}(\Gamma) \subset$ Cen alg$^\pi$ $\{S_\Gamma, \mathbf{PC}(\Gamma)\}$ and Γ can be identified with the maximal ideal space of $\mathbf{C}(\Gamma)$, we can, for each $z_0 \in \Gamma$, form the corresponding local ideal J_{z_0} in alg$^\pi$ $\{S_\Gamma, \mathbf{PC}(\Gamma)\}$ (see Sect. 1.20). Put

$$\text{alg}^\pi_{\{z_0\}}\, \{S_\Gamma, \mathbf{PC}(\Gamma)\} := \text{alg}^\pi\, \{S_\Gamma, \mathbf{PC}(\Gamma)\}/J_{z_0}.$$

2. It can be proved that alg$^\pi_{\{z_0\}}$ $\{S_\Gamma, \mathbf{PC}(\Gamma)\}$ is isometrically isomorphic to $\Sigma(\alpha(z_0))$ if $k(z_0) = 1$ and to alg $\{S_{\mathbb{R}^+ \cup e^{i\beta}\mathbb{R}^+}, \Lambda\} \subset \mathcal{L}(\mathbf{L}^2(\mathbb{R}^+ \cup e^{i\beta}\mathbb{R}^+, x^{\alpha(z_0)}))$ if $k(z_0) = 2$. Here Λ denotes the algebra of all functions which are constant on both $\mathbb{R}^+$ and $e^{i\beta}\mathbb{R}^+$. Moreover the local direction of Γ at the point z_0 determines the direction of $\mathbb{R}^+$ or $\mathbb{R}^+ \cup e^{i\beta}\mathbb{R}^+$, respectively. The isomorphism can be chosen so that it takes S_Γ into S or $S_{\mathbb{R}^+ \cup e^{i\beta}\mathbb{R}^+}$ and $a \in \mathbf{PC}(\Gamma)$ into $a_1\chi_1$ or $a_1\chi_1 + a_2\chi_2$, respectively, where χ_1 is the characteristic function of $\mathbb{R}^+$ and χ_2 that of $e^{i\beta}\mathbb{R}^+$.

3. Define the mapping $\eta: \mathbf{L}^2(\mathbb{R}^+ \cup e^{i\beta}\mathbb{R}^+, x^{\alpha(z_0)}) \to \mathbf{L}_2^2(\mathbb{R}^+, t^{\alpha(z_0)})$ by

$$(\eta f)(t) = \begin{pmatrix} f(t) \\ f(e^{i\beta}t) \end{pmatrix}.$$

It is easy to see that the mapping $A \mapsto \eta A \eta^{-1}$ is an isometrical isomorphism between the algebras

$$\mathcal{L}\left(\mathbf{L}^2(\mathbb{R}^+ \cup e^{i\beta}\mathbb{R}^+, x^{\alpha(z_0)})\right) \quad \text{and} \quad \mathcal{L}\left(\mathbf{L}_2^2(\mathbb{R}^+, t^{\alpha(z_0)})\right).$$

The restriction of this isomorphism to alg $\{S_{\mathbb{R}^+ \cup e^{i\beta}\mathbb{R}^+}, \Lambda\}$ has as its image precisely $\mathring{\Sigma}^2(\alpha(z_0))$, where, for $a \in \Lambda$

$$\eta a \eta^{-1} = \operatorname{diag}(a_1 I, a_2 I),$$

and

$$\eta S_{\mathbb{R}^+ \cup e^{i\beta}\mathbb{R}^+} \eta^{-1} = \begin{pmatrix} S & N_{2\pi-\beta} \\ N_\beta & S \end{pmatrix} \begin{pmatrix} n_1(z_0) & 0 \\ 0 & n_2(z_0) \end{pmatrix}.$$

4. It remains to apply Theorem 1.21. ■

We shall see in Chapter 11 that some pieces of the above approach play an important role in the numerical analysis of singular integral operators on Lyapunov curves Γ having corner points, that is, having points $z \in \Gamma$ of order 2 with $\beta(z) \neq \pi$.

Let $\mathring{\mathbf{C}}^2$ be the algebra of all 2×2 matrices (f_{ij}) with entries $f_{ij} \in \mathbf{C}(\tilde{\mathbb{R}})$ $(i, j = 1, 2)$ such that $f_{ij}(\pm\infty)$ vanishes for $i \neq j$. Note that $\mathring{\mathbf{C}}^2$ actually forms a C^*-algebra under the norm

$$\|f\| = \sup_{t \in \tilde{\mathbb{R}}} \left(\lambda(t)\right)^{1/2},$$

where $\lambda(t)$ is the largest eigenvalue of the matrix $(ff^*)(t)$. Using Theorem 6.6 we see that the algebras $\mathring{\Sigma}^2(\alpha)$ and $\mathring{\mathbf{C}}^2$ are *-isomorphic. The *-isomorphism can be chosen so that the image of $\begin{pmatrix} S & N_{2\pi-\beta} \\ N_\beta & S \end{pmatrix}$ and diag (a_1, a_2) (where $a_1, a_2 \in \mathbb{C}$) is

$$\begin{pmatrix} \coth(z + i\delta)\pi & \dfrac{e^{(z+i\delta)(\pi-\beta)}}{\sinh(z+i\delta)\pi} \\ \dfrac{e^{(z+i\delta)(\pi-\beta)}}{\sinh(z+i\delta)\pi} & \coth(z+i\delta)\pi \end{pmatrix} \quad \left(\delta = \frac{1}{2} + \alpha\right) \tag{1}$$

and

$$\operatorname{diag}(a_1, a_2), \tag{2}$$

respectively.

As a first consequence we find that the quotient algebra $\mathrm{alg}^\pi\,\{S_\Gamma, \mathbf{PC}(\Gamma)\}$ is *-isomorphic to an algebra of matrix-valued functions on Γ whose values at the point z are in $\mathring{\mathbf{C}}^2(\tilde{\mathbb{R}})$ if $k(z) = 2$ and in $\mathbf{C}(\tilde{\mathbb{R}})$ if $k(z) = 1$, i.e. we have a matrix-valued symbol for the algebra $\mathrm{alg}^\pi\,\{S_\Gamma, \mathbf{PC}(\Gamma)\}$.

Another remarkable consequence is reflected in the following fact: the property of the operator $A = aP + bQ$ $\left(a, b \in \mathbf{PC}(\Gamma)\right)$ to be Fredholm does not depend on the values of the angles made at the corner points of the curve Γ. Indeed, put

$$D = \operatorname{diag}(1, e^{-(z+i\delta)\beta})$$

and let A_1, A_2 stand for the matrices in 6.10(1) and 6.10(2), respectively. Then

$$D^{-1}A_1D = \begin{pmatrix} \coth(z+i\delta)\pi & \dfrac{e^{(z+i\delta)\pi}}{\sinh(z+i\delta)\pi} \\ \dfrac{e^{-(z+i\delta)\pi}}{\sinh(z+i\delta)\pi} & \coth(z+i\delta)\pi \end{pmatrix}$$

and $D^{-1}A_2D = A_2$, implying our claim.

The aim of the next sections is to show how Theorem 6.10 can be used to obtain the well-known Gohberg-Krupnik symbol for the algebra $\operatorname{alg}^\pi\{S_\Gamma, \mathbf{PC}(\Gamma)\}$.

6.11. The Gohberg/Krupnik symbol. Continuous coefficients. Let Γ be a (not necessarily closed) Lyapunov curve. Let $z_1, \ldots, z_{2n}$ be all points of order one and suppose that the curve is directed away from z_m $(m = 1, \ldots, n)$ and directed to z_m $(m = n+1, \ldots, 2n)$. Consider

$$\operatorname{alg}^\pi\{S_\Gamma, \mathbf{C}(\Gamma)\} := \operatorname{alg}\{S_\Gamma, \mathbf{C}(\Gamma)\}/\mathcal{K}\big(\mathbf{L}^2(\Gamma, w)\big),$$

where $w(z) := \prod_{i=1}^{2n} |z - z_i|^{\alpha_i}$, $-1/2 < \alpha_i < 1/2$.

Given Γ, we define the curve $\Lambda(\Gamma) \subset \mathbb{R}^3$ as the curve consisting of all the points $(x, y, t) \in \mathbb{R}^3$ such that

$$x + iy \in \Gamma, \qquad -1 \leqq t \leqq 1, \qquad (1 - t^2)\prod_{k=1}^{2n}(x + iy - z_k) = 0.$$

The orientation of $\Lambda(\Gamma)$ is chosen so that it coincides with the orientation of Γ in the plane $t = 1$ and is opposite to the orientation of Γ in the plane $t = -1$.

Let $\Omega_{2,w}$ be the function given on $\Lambda(\Gamma)$ by

$$\Omega_{2,w}(z, t) = \begin{cases} \dfrac{t(1 + a_k^2) - i(1 - t^2)\,a_k}{1 + t^2a_k^2} & \text{for } z = z_k \ (k \leqq n\}, \\ t & \text{for } z \in \Gamma \setminus \{z_1, \ldots, z_{2n}\}, \\ \dfrac{t(1 + a_k^2) + i(1 - t^2)\,a_k}{1 + t^2a_k^2} & \text{for } z = z_k \ (k > n + 1) \end{cases}$$

where $a_k = \cot(1/2 + \alpha_k)\pi$. Following Gohberg/Krupnik [4] we associate with

$$A = cI + dS_\Gamma \qquad \big(c, d \in \mathbf{C}(\Gamma)\big) \tag{1}$$

the symbol

$$(\operatorname{smb} A)(z, t) = c(z) + d(z)\,\Omega_{2,w}(z, t), \qquad (z, t) \in \Lambda(\Gamma). \tag{2}$$

According to Theorems 6.10 and 6.6, the spectrum of $\widetilde{\operatorname{smb}}(A, z_0)$ coincides with $\{c(z_0) + b(z_0)\,\Omega_{2,w}(z_0, t) \colon (z_0, t) \in \Lambda(\Gamma)\}$.

Moreover, again by Theorem 6.10, the algebra $\operatorname{alg}^\pi\{S_\Gamma, \mathbf{C}(\Gamma)\}$ turns out to be commutative and the maximal ideal space of this C^*-algebra can be identified with $\Lambda(\Gamma)$.

Thereby, the Gelfand transform of $cI + dS + \mathcal{K}(\mathbf{L}^2(\Gamma, w))$ $(c, d \in \mathbf{C}(\Gamma))$ is given by (2). This observation gives the following result.

Let Γ be a Lyapunov curve and let $c, d \in \mathbf{C}(\Gamma)$. The singular integral operator (1) *is Fredholm if and only if*

$$(\operatorname{smb} A)(z, t) \neq 0 \quad \textit{for all} \quad (z, t) \in \Lambda(\Gamma). \tag{3}$$

If (3) *is fulfilled, then A is one-sided invertible and*

$$\operatorname{ind} A = -\operatorname{wind} \operatorname{smb} A.$$

For details see GOHBERG/KRUPNIK [4].

It is frequently useful to reformulate the latter result. To do so, consider the following special but important case. Let $J := [a, b] \subset \mathbb{R}$ $(a < b)$ and put $w(z) = (z - a)^{\alpha_1} \times (b - z)^{\alpha_2}$, $-1/2 < \alpha_i < 1/2$. *Assume c, d are continuous functions on J such that* $(c^2 - d^2)(z) \neq 0$ *for all $z \in J$ and consider the operator* (1) *in the space* $\mathbf{L}^2(J, w)$. *Let $\varkappa$ denote the continuous function*

$$\varkappa(z) := 1/(2\pi \mathrm{i}) \log \frac{c(z) + d(z)}{c(z) - d(z)}$$

satisfying $-1/2 - \alpha_1 < \operatorname{Re} \varkappa(a) < 1/2 - \alpha_1$, *where* log *denotes that branch of the logarithm which is continous in* $\mathbb{C} \setminus (-\infty, 0]$ *and takes real values on the positive real line. If there exists an integer $m_0 \in \mathbb{Z}$ such that the inequality*

$$\alpha_2 - \frac{1}{2} < \operatorname{Re} \varkappa(b) + m_0 < \frac{1}{2} + \alpha_2$$

is fulfilled, then the operator (1) *is Fredholm and one-sided invertible with*

$$\operatorname{ind} A = m_0$$

(see GOHBERG/KRUPNIK [4]).

6.12. The Gohberg/Krupnik symbol. Piecewise continuous coefficients. In this section we restrict ourselves to the construction of the Gohberg-Krupnik symbol for the algebra alg $\{S_\Gamma, \mathbf{PC}(\Gamma)\}$ in the special case in which Γ is a closed Lyapunov curve and $w(z) = 1$.

Let $z_0 \in \Gamma$ be an arbitrary point and let Γ_1 and Γ_2 be Lyapunov arcs connected with z_0 as in Section 6.1. We assume that Γ_1 is directed to z_0 and Γ_2 away from z_0. Then, in accordance with Theorem 6.10, the homomorphism $A \mapsto \widetilde{\operatorname{smb}}(A, z_0)$ takes S into $\begin{pmatrix} S & -N_\pi \\ N_\pi & -S \end{pmatrix}$. Since $S^2 = I$, the operator $1/2(I + S)$ is obviously a projection. Hence, the elements

$$p = \begin{pmatrix} \frac{1}{2}(I + S) & -\frac{1}{2} N_\pi \\ \frac{1}{2} N_\pi & \frac{1}{2}(I - S) \end{pmatrix} \quad \text{and} \quad q = \begin{pmatrix} I & 0 \\ 0 & 0 \end{pmatrix}$$

are idempotents belonging to $\mathring{\Sigma}^2$. Moreover, the algebra $\mathring{\Sigma}^2$ is generated by p, q and $e = \begin{pmatrix} I & 0 \\ 0 & I \end{pmatrix}$. It is not hard to see that p and q are self-adjoint. Since $qpq = \begin{pmatrix} (1/2)/(I + S) & 0 \\ 0 & 0 \end{pmatrix}$, the spectrum of qpq coincides with $\mathrm{sp}\,\big((1/2)(I+S)\big) = [0, 1]$. Using that $qpq = (qp)\,(pq)$ and $(pq)\,(qp) = pqp$ we get by a well-known theorem (see RUDIN [1], Chap. 10, Exerc. 2) $\mathrm{sp}\,(pqp) = \mathrm{sp}\,(qpq) = [0, 1]$. Hence, Theorem 1.16 is applicable and we conclude that $\mathring{\Sigma}^2$ is *-isomorphic to a C^*-algebra, $\mathring{\mathbf{C}}^2[0, 1]$, of continuous matrix functions on $[0, 1]$ with values in $\mathbb{C}^{2\times 2}$. The algebra $\mathring{\mathbf{C}}^2[0, 1]$ is generated by

$$\hat{p} = \begin{pmatrix} 1 & 0 \\ 0 & 0 \end{pmatrix}, \qquad \hat{q} = \begin{pmatrix} x & \sqrt{x(1-x)} \\ \sqrt{x(1-x)} & 1-x \end{pmatrix}, \qquad \hat{e} = \begin{pmatrix} 1 & 0 \\ 0 & 1 \end{pmatrix}.$$

The isomorphism $\Phi\colon \mathring{\Sigma}^2 \to \mathring{\mathbf{C}}^2[0, 1]$ can be taken so that $\Phi(p) = \hat{p}$, $\Phi(q) = \hat{q}$.

Thus, we have a homomorphism from alg $\{S_\Gamma, \mathbf{PC}(\Gamma)\}$ onto $\mathring{\mathbf{C}}^2[0, 1]$ acting by the rule

$$A \mapsto \Phi\big(\widetilde{\mathrm{smb}}\,(A, z_0)\big) =: \mathrm{smb}\,(A, z_0).$$

Let $a \in \mathbf{PC}(\Gamma)$ and put

$$a(z_0 - 0) = \lim_{\substack{z\to z_0 \\ z\in\Gamma_1}} a(z), \qquad a(z_0 + 0) = \lim_{\substack{z\to z_0 \\ z\in\Gamma_2}} a(z).$$

Then we get, for $P_\Gamma := (1/2)(I + S_\Gamma)$ and a,

$$\mathrm{smb}\,(P_\Gamma, z_0) = \begin{pmatrix} 1 & 0 \\ 0 & 0 \end{pmatrix}, \tag{1}$$

$$\mathrm{smb}\,(aI, z_0) = a(z+0) \begin{pmatrix} x & \sqrt{x(1-x)} \\ \sqrt{x(1-x)} & 1-x \end{pmatrix} + a(z-0) \begin{pmatrix} 1-x & -\sqrt{x(1-x)} \\ -\sqrt{x(1-x)} & x \end{pmatrix} \tag{2}$$

To compute $\mathrm{smb}\,(aP_\Gamma + bQ_\Gamma, z_0)$, where $Q_\Gamma = (1/2)\,(I - S_\Gamma) = I - P_\Gamma$ and $a, b \in \mathbf{PC}(\Gamma)$ we can make use of (1) and (2), which together with a straightforward calculation leads to

$$\mathrm{smb}\,(aP_\Gamma + bQ_\Gamma, z_0) = \begin{pmatrix} a(z_0+0)\,x + a(z_0-0)\,(1-x) & \big(b(z_0+0) - b(z_0-0)\big)\sqrt{x(1-x)} \\ \big(a(z_0+0) - a(z_0-0)\big)\sqrt{x(1-x)} & b(z_0+0)\,(1-x) + b(z_0-0)\,x \end{pmatrix}$$

This, however, is nothing else but the Gohberg-Krupnik symbol for the operator $aP_\Gamma + bQ_\Gamma$. So we are offered a possibility of reformulating Theorem 6.10 in the case at hand. This rephrasement is the content of the next proposition. We also give a fairly simple (and independent) proof of this proposition. Moreover, this proof is intended to serve as a preparation of a certain approach which is of great importance in the numerical analysis of integral operators (see Sections 10.31–10.42).

6.13. Proposition. *Let Γ be a closed Lyapunov curve and consider* alg $\{S_\Gamma, \mathbf{PC}(\Gamma)\}$ $\subset \mathcal{L}\big(\mathbf{L}^2(\Gamma)\big)$.

(a) *For each $z_0 \in \Gamma$, there exists a homomorphism $A \mapsto \mathrm{smb}\,(A, z_0)$ from* alg $\{S_\Gamma, \mathbf{PC}(\Gamma)\}$ *onto $\mathring{\mathbf{C}}^2[0, 1]$ such that*

$$\mathrm{smb}\,(P_\Gamma, z_0) = \begin{pmatrix} 1 & 0 \\ 0 & 0 \end{pmatrix},$$

and for $a \in \mathbf{PC}(\Gamma)$,

$$\operatorname{smb}(aI, z_0) = a(z_0 + 0) \begin{pmatrix} x & \sqrt{x(1-x)} \\ \sqrt{x(1-x)} & 1-x \end{pmatrix} + a(z_0 - 0) \begin{pmatrix} 1-x & -\sqrt{x(1-x)} \\ -\sqrt{x(1-x)} & x \end{pmatrix}.$$

(b) $A \in \operatorname{alg}\{S_\Gamma, \mathbf{PC}(\Gamma)\}$ *is Fredholm if and only if* smb (A, z) *is invertible for all* $z \in \Gamma$.

(c) *If* $\pi(A)$ *stands for the image of* A *in the Calkin algebra, then, for* $A \in \operatorname{alg}\{S_\Gamma, \mathbf{PC}(\Gamma)\}$,

$$\|\pi(A)\| = \sup_{z\in\Gamma} \|\operatorname{smb}(A, z)\|.$$

(d) $A \in \operatorname{alg}\{S_\Gamma, \mathbf{PC}(\Gamma)\}$ *is compact if and only if* smb $(A, z) = 0$ *for all* $z \in \Gamma$.

Proof. 1. Since $\mathbf{C}(\Gamma) \subset \operatorname{Cen} \operatorname{alg}^\pi \{S_\Gamma, \mathbf{PC}(\Gamma)\}$ and Γ can be identified with the space of maximal ideals of $\mathbf{C}(\Gamma)$, we can construct the corresponding local algebra $\mathfrak{A}(z) := \operatorname{alg}^\pi_{\{z\}} \{S_\Gamma, \mathbf{PC}(\Gamma)\}$ for each $z \in \Gamma$ (we use the notations of Section 6.10). Note that all these algebras are C^*-algebras. We claim that $\mathfrak{A}(z)$ is generated by two selfadjoint idempotents p_z, q_z and the unit element e_z such that sp $(p_z q_z p_z) = [0, 1]$. By 6.1(vi), the element P_Γ^π is selfadjoint (recall that $P_\Gamma := (1/2)(I + S_\Gamma)$). Hence, the image of P_Γ in $\mathfrak{A}(z)$ is a selfadjoint idempotent, which will be denoted by p_z. Fix z and let Γ_1 be a Lyapunov arc which is a proper subset of Γ and has z as one of its endpoints. The other endpoint is denoted by z_0. We assume further that when tracing out Γ_1 in the direction induced by the orientation of Γ, z will follow after z_0. Let χ_1, χ_2 be the characteristic functions of Γ_1 and $\Gamma \setminus \Gamma_1$, respectively. Given a function $a \in \mathbf{PC}(\Gamma)$, we clearly may represent a in the form

$$a = a(z - 0)\,\chi_1 + a(z + 0)\,\chi_2 + b,$$

where $b \in \mathbf{PC}(\Gamma)$, $b(z - 0) = b(z + 0) = 0$. In particular, b is continuous at z. Let q_z be the image of χ_2 in $\mathfrak{A}(z)$. Obviously, q_z is a self-adjoint idempotent and the image of a in $\mathfrak{A}(z)$ is $a(z - 0)\,(1 - q_z) + a(z + 0)\,q_z$. Thus, the algebra $\mathfrak{A}(z)$ is generated by the selfadjoint idempotents p_z, q_z and by the unit element. What we are left with is to prove that sp $(p_z q_z p_z) = [0, 1]$, which, in point of fact, is the main difficulty in the proof.

2. In order to determine sp $(p_z q_z p_z)$ we may assume that $\Gamma = \mathbb{T}$, where $\mathbb{T}$ refers to the unit circle in the complex plane. (This can be seen by resorting to an idea of the proof of Theorem 4.4 in Gohberg/Krupnik [4]).

The well-known Hartman-Wintner Theorem (see Böttcher/Silbermann [3]) claims that the spectra of both $P\chi_1P$, $P\chi_2P$ coincide with $[0, 1]$. Since $P\chi_2P$ is selfadjoint (cf. 6.1(vii)), we have ind $(P\chi_2P - \lambda I) = 0$ if only $B := P\chi_2P - \lambda I$ is Fredholm. From 6.1(ix) we know that at least one of the numbers dim ker B, dim coker B is equal to zero. Therefore, B is Fredholm if and only if it is invertible, or, what is the same, sp $\left((P\chi_2P)^\pi\right) = [0, 1]$. By Theorem 1.21 we have

$$\operatorname{sp}\left((P\chi_2P)^\pi\right) = \bigcup_{s\in\mathbb{T}} \operatorname{sp}\left(p_s \chi_2^{(s)} p_s\right), \tag{1}$$

where $\chi_2^{(s)}$ denotes the coset in $\mathfrak{A}(s)$ containing χ_2. Obviously for $s \neq z, z_0$ the spectrum of $p_s\chi_2^{(s)}p_s$ consists of at most two points, namely 0 and 1. Suppose that $z = 1$ and that there is a $\mu \in (0, 1)$ such that $\mu \notin \operatorname{sp}(p_1 q_1 p_1)$. Then, by (1),

$$\mu \in \operatorname{sp}(p_{z_0} \chi_2^{(z_0)} p_{z_0}) = \operatorname{sp}\big(p_{z_0}(e_{z_0} - q_{z_0})\, p_{z_0}\big). \tag{2}$$

We claim that then also $\mu \notin \operatorname{sp}\big(p_1(e_1 - q_1)\, p_1\big)$. To show this we introduce the operators $(Vf)(t) := f(1/t)$ and $(Cf)(t) := \overline{f(t)}$ $\big(f \in \mathbf{L}^2(\mathbb{T})\big)$. An easy computation gives that

$$CVSVC = S, \qquad CVaVC = \overline{a(1/t)}\, I\,.$$

for all $a \in \mathbf{PC}(\mathbb{T})$. Thus, the mapping $A \mapsto CVAVC$ is actually an automorphism Φ of $\operatorname{alg}\{S_\mathbb{T}, \mathbf{PC}(\mathbb{T})\}$. Since $\Phi\big(\mathcal{K}(\mathbf{L}^2(\mathbb{T}))\big) \subset \mathcal{K}\big(\mathbf{L}^2(\mathbb{T})\big)$ and, for $a \in \mathbf{C}(\mathbb{T})$, $\Phi(a)$ and a simultanuously vanish at the point 1, the automorphism Φ induces an automorphism Φ_1 on $\mathfrak{A}(1)$. Moreover, $\Phi_1(p_1) = p_1$ and $\Phi_1(q_1) = e_1 - q_1$. This reveals that $\operatorname{sp}(p_1 q_1 p_1) = \operatorname{sp}\big(p_1 \times (e_1 - q_1)\, p_1\big)$ and therefore proves our claim.

Now choose $\varphi \in (0, 2\pi)$ and define an operator W_φ by $(W_\varphi f)(t) = f(\mathrm{e}^{\mathrm{i}\varphi} t)$ for $f \in \mathbf{L}^2(\mathbb{T})$. Then

$$W_\varphi^{-1} S W_\varphi = S, \qquad W_\varphi^{-1} a W_\varphi = a(\mathrm{e}^{-\mathrm{i}\varphi} t)\, I\,,$$

and the mapping $A \mapsto W_\varphi^{-1} A W_\varphi$ induces an isomorphism between $\mathfrak{A}(s)$ and $\mathfrak{A}(s\, \mathrm{e}^{\mathrm{i}\varphi})$. This isomorphism takes p_s and q_s into $p_{s\mathrm{e}^{\mathrm{i}\varphi}}$ and $q_{s\mathrm{e}^{\mathrm{i}\varphi}}$, respectively. So we have

$$\operatorname{sp}(p_s q_s p_s) = \operatorname{sp}(p_{s\mathrm{e}^{\mathrm{i}\varphi}} q_{s\mathrm{e}^{\mathrm{i}\varphi}} p_{s\mathrm{e}^{\mathrm{i}\varphi}})\,, \tag{3}$$

$$\operatorname{sp}\big(p_s(e_s - q_s)\, p_s\big) = \operatorname{sp}\big(p_{s\mathrm{e}^{\mathrm{i}\varphi}}(e_{s\mathrm{e}^{\mathrm{i}\varphi}} - q_{s\mathrm{e}^{\mathrm{i}\varphi}})\, p_{s\mathrm{e}^{\mathrm{i}\varphi}}\big). \tag{4}$$

Now put $s = z_0$ and choose φ so that $z_0\, \mathrm{e}^{i\varphi} = 1$. From (2) and (4), it follows that $\mu \in \operatorname{sp}\big(p_1 \times (e_1 - q_1)\, p_1\big)$, which is impossible since $\mu \notin \operatorname{sp}(p_1 q_1 p_1) = \operatorname{sp}\big(p_1(e_1 - q_1)\, p_1\big)$. Thus $\operatorname{sp}(p_1\, q_1\, p_1) = [0, 1]$. Once more applying (3) we see that

$$\operatorname{sp}(p_z q_z p_z) = [0, 1].$$

3. In order to finish the proof apply Theorems 1.16 and 1.21. ■

6.14. Remarks. 1°. Proposition 6.11 can be also established for $\operatorname{alg}\big\{S_\Gamma, \mathbf{PC}(\Gamma)\big\} \subset \mathcal{L}\big(\mathbf{L}^2(\Gamma, w)\big)$, where Γ is admissible, bounded and closed. The symbol mapping $\operatorname{smb}(A, z)$ at the point $z \in \Gamma$ is defined via

$$\operatorname{alg}^\pi\{S_\Gamma, \mathbf{PC}(\Gamma)\} \xrightarrow{\widetilde{\operatorname{smb}}} \mathring{\Sigma}^2(\alpha) \xrightarrow{\xi} \mathring{\Sigma}^2 \xrightarrow{\zeta} \mathring{\mathbf{C}}^2[0, 1],$$

where ξ and ζ are the *-isomorphisms between $\mathring{\Sigma}^2(\alpha)$ and $\mathring{\Sigma}^2$ and between $\mathring{\Sigma}^2$ and $\mathring{\mathbf{C}}^2$ defined above.

2°. Let χ^+ denote the characteristic function of $\{t \in \mathbb{T} : \operatorname{im} t > 0\}$. Then $\operatorname{sp}(P_\mathbb{T} \chi^+ P_\mathbb{T}) = [0, 1]$ (by the Hartman-Wintner Theorem), and $P_\mathbb{T}$, $\chi^+ I$ are selfadjoint idempotents. Thus, $\operatorname{alg}\{P_\mathbb{T}, \chi^+\}$ is isometrically isomorphic to $\mathring{\mathbf{C}}^2$. Therefore the symbol map can be slightly modified so as to take values in $\operatorname{alg}\{P_\mathbb{T}, \chi^+\}$. This point of view facilitates the understanding of some phenomena which will arise in Chapter 11.

3°. Theorem 6.10 and Proposition 6.11 remain true for singular integral operators with matrix-valued coefficients in all parts which do not concern the one-sided invertibility. This can be proved by the following tensor product argument: if $0 \to A_1 \to A_2 \to A_3 \to 0$ is an exact sequence of C^*-algebras then

$$0 \to A_1 \otimes \mathbb{C}^{n \times n} \to A_2 \otimes \mathbb{C}^{n \times n} \to A_3 \otimes \mathbb{C}^{n \times n} \to 0$$

is also exact. Note that $A_i \otimes \mathbb{C}^{n\times n}$ can be identified with the C^*-algebra of all $n \times n$ matrices with entries in A_i ($i = 1, 2, 3$).

4°. Finally, we mention an index formula for Fredholm operators from alg $\{S_\Gamma, \mathbf{PC}(\Gamma)\}$ (under the conditions of Proposition 6.11). Let $A \in$ alg $\{S_\Gamma, \mathbf{PC}(\Gamma)\}$ be Fredholm. Then, for each $z \in \Gamma$ and each $x \in [0, 1]$, we have

$$\det \operatorname{smb}(A, z)(x) \neq 0.$$

Put $\begin{pmatrix} A_{11}(z, x) & A_{12}(z, x) \\ A_{21}(z, x) & A_{22}(z, x) \end{pmatrix} := \operatorname{smb}(A, z)(x)$.

Obviously, $A_{22}(z, 1)\, A_{22}(z, 0) \neq 0$ (because $A_{12}(z, 1) = A_{12}(z, 0) = 0$). Hence, $\psi(z, x) := \det \operatorname{smb}(A, z)(x)/\big(A_{22}(z, 1)\, A_{22}(z, 0)\big)$ is well-defined. It turns out that ind A equals $(1/2\pi)[\arg \psi(z, x)]_{(x,z)\in\Gamma_0}$ where Γ_0 is a curve defined as follows. First suppose that smb (A, z) is diagonal with exception of finitely many points $z_1, \ldots, z_e$. Let Γ_0 denote a curve obtained from Γ by adding certain loops in the following way: think of each point z_k ($k = 1, \ldots, e$) as split up into two points z_k^- and z_k^+, and join z_k^- and z_k^+ by a loop; along each added loop let the parameter x run from 0 to 1. Clearly, $\psi(z, x)$ is a continuous matrix function on Γ_0. The general case can be settled by an approximation argument.

Toeplitz operators

6.15. Notations. Let $\mathbf{L}^{2,\alpha}(\mathbb{T}) := \mathbf{H}^\alpha$ denote the Sobolev-Slobodetski space defined on the unit circle $\mathbb{T}$ ($\alpha \in \mathbb{R}$) (see Chap. 2). It is well-known that $\mathbf{L}^{2,\alpha}(\mathbb{T})$ can be identified with $\tilde{\mathbf{l}}^{2,\alpha}$ (for the definition of $\tilde{\mathbf{l}}^{2,\alpha}$ see Section 4.81) via the correspondence of $f \in \mathbf{L}^{2,\alpha}(\mathbb{T})$ with the sequence $\{f_n\}_{-\infty}^{\infty}$ of the Fourier coefficients of f. The singular integral operator S is defined and bounded on $\mathbf{L}^{2,\alpha}(\mathbb{T})$ if $\alpha \geqq 0$. For $\alpha \in \mathbb{Z}^+$ this is obvious, and for arbitrary $\alpha \in \mathbb{R}^+$ it can be obtained by interpolation. Note that the spaces $\mathbf{L}^{2,\alpha}(\mathbb{T})$ and $\mathbf{L}^{2,-\alpha}(\mathbb{T})$ are dual to each other. Therefore, for $\alpha < 0$ the operator S can be introduced as the adjoint to $S \in \mathcal{L}\big(\mathbf{L}^{2,-\alpha}(\mathbb{T})\big)$. The action of S in $\tilde{\mathbf{l}}^{2,\alpha}$ can be described very easily:

$$S\{f_n\}_{n=-\infty}^{\infty} = \{\eta_n\}_{n=-\infty}^{\infty},$$

where $\eta_n = f_n$ if $n \geqq 0$ and $\eta_n = -f_n$ if $n < 0$.

Since $S^2 = I$, the operator $P := (1/2)(I + S)$ turns out to be a projection. Now let a be a multiplier on $\mathbf{L}^{2,\alpha}(\mathbb{T})$. The *Toeplitz operator* $T(a)$ with symbol a is defined on $\mathbf{L}_+^{2,\alpha}(\mathbb{T}) := P\big(\mathbf{L}^{2,\alpha}(\mathbb{T})\big)$ by the rule

$$T(a)\, f = Paf.$$

Equivalently, $T(a)$ is defined on $\mathbf{l}^{2,\alpha}$ by the rule

$$\{\xi_n\}_{n=0}^{\infty} \mapsto \{\eta_n\}_{n=0}^{\infty}, \qquad \eta_n := \sum_{i=0}^{\infty} a_{n-i}\xi_i, \tag{1}$$

where a_i is the i-th Fourier coefficient of a. Here are some properties of Toeplitz operators on $\mathbf{l}^2 := \mathbf{l}^{2,0}$ (see Böttcher/Silbermann [3]).

(i) *An operator defined by* (1) *is bounded on* $\mathbf{l}^2$ *if and only if the sequence* $\{a_i\}_{i=-\infty}^{\infty}$ *is the sequence of Fourier coefficients of a function* $a \in \mathbf{L}^\infty(\mathbb{T})$.
(ii) *The mapping* $a \mapsto T(a)$, $a \in \mathbf{L}^\infty(\mathbb{T})$, *is norm-preserving*: $\|T(a)\| = \|a\|_\infty$.
(iii) $T(a)^* = T(\bar{a})$.

Let $\mathcal{T}(\mathbf{PC})$ be the collection of all Toeplitz operators on $\mathbf{l}^2$ with symbols belonging to **PC** and let alg $\mathcal{T}(\mathbf{PC})$ denote the smallest closed subalgebra of $\mathcal{L}(\mathbf{l}^2)$ containing $\mathcal{T}(\mathbf{PC})$. Obviously, alg $\mathcal{T}(\mathbf{PC})$ forms a C^*-algebra.

6.16. Theorem. (i) $\mathcal{K}(\mathbf{l}^2) \subset \text{alg}\, \mathcal{T}(\mathbf{PC})$.
(ii) $\text{alg}^\pi\, \mathcal{T}(\mathbf{PC}) := \text{alg}\, \mathcal{T}(\mathbf{PC})/\mathcal{K}(\mathbf{l}^2)$ *is isometrically isomorphic to the algebra of all continuous functions defined on* $\mathbb{T} \times [0, 1]$, *the latter space being equipped with a special topology: the neighbourhoods of a point* (ζ, μ), $\zeta = \exp i\varphi$, $0 \leqq \mu \leqq 1$, *are given by*

$$U_{\varepsilon,\delta}(\zeta_0, 0) = \{(\zeta, \mu) : \varphi_0 - \delta < \arg \zeta < \varphi_0 ; 0 \leqq \mu \leqq 1\} \cup \{(\zeta_0, \mu) : 0 \leqq \mu < \varepsilon\},$$
$$U_{\varepsilon,\delta}(\zeta_0, 1) = \{(\zeta, \mu) : \varphi_0 < \arg \zeta < \varphi_0 + \delta, 0 \leqq \mu \leqq 1\} \cup \{(\zeta_0, \mu) : \varepsilon < \mu \leqq 1\},$$
$$U_{\delta_1,\delta_2}(\zeta_0, \mu_0) = \{(\zeta_0, \mu) : \mu_0 - \delta_1 < \mu < \mu_0 + \delta_2\} \; (\mu_0 \neq 0, 1),$$

where $0 < \delta_1 < \mu_0$, $0 < \delta_2 < 1 - \mu_0$, $0 < \varepsilon < 1$.

(iii) *This isomorphism takes the coset* $T(a) + \mathcal{K}(\mathbf{l}^2)$ *into the function*

$$a(t, \mu) = a(t + 0)\, \mu + a(t - 0)\, (1 - \mu) \quad \big((t, \mu) \in \mathbb{T} \times [0, 1]\big).$$

Proof. The theorem is an easy consequence of Proposition 6.11: alg $\mathcal{T}(\mathbf{PC})$ is the compression of alg $\{S_{\mathbb{T}}, \mathbf{PC}\}$ to im P. ■

It is worth noticing that the topology defined above is Hausdorff. Moreover it coincides with the Gelfand topology on $\mathbb{T} \times [0, 1]$ ($\mathbb{T} \times [0, 1]$ is the maximal ideal space of the commutative algebra $\text{alg}^\pi \mathcal{T}(\mathbf{PC})$).

Now we are going to study Toeplitz operators with piecewise continuous symbols on the space $\mathbf{l}^{2,\alpha}$, $-1/2 < \alpha < 1/2$. Let **PK** be the set of all piecewise constant functions on $\mathbb{T}$ having only a finite number of jumps, and let alg $\mathcal{T}_\alpha(\mathbf{PK})$ stand for the smallest closed subalgebra of $\mathcal{L}(\mathbf{l}^{2,\alpha})$ containing all Toeplitz operators $T(a)$ with $a \in \mathbf{PK}$. Note that for $a \in \mathbf{PK}$ the operator $T(a)$ is bounded on $\mathbf{l}^{2,\alpha}$, $-1/2 < \alpha < 1/2$, since a has finite total variation (see Böttcher/Silbermann [3], Section 6.2(f)). Put $V := T(t)$, $V^{(-1)} := T(t^{-1})$. Then V and $V^{(-1)}$ are the operators defined in Sect. 4.46, and it is not hard to see that

$$\mathcal{K}(\mathbf{l}^{2,\alpha}) \subset \text{alg}\,(V, V^{(-1)}) \subset \text{alg}\, \mathcal{T}_\alpha(\mathbf{PK}).$$

The first inclusion is a consequence of Corollary 4.48, and the second inclusion can be easily deduced using some arguments from Böttcher/Silbermann [3], Proposition 6.26. Note that for $\alpha = 0$ the algebra alg $\mathcal{T}_\alpha(\mathbf{PK})$ coincides with alg $\mathcal{T}(\mathbf{PC})$. Define

$$\text{alg}^\pi\, \mathcal{T}_\alpha(\mathbf{PK}) := \text{alg}\, \mathcal{T}_\alpha(\mathbf{PK})/\mathcal{K}(\mathbf{l}^{2,\alpha}).$$

It turns out that this algebra is commutative (Böttcher/Silbermann [3], Proposition 6.30). Moreover, by Remark 4.51, 2° the algebra $\text{alg}^\pi\, \mathcal{T}_\alpha(\mathbf{PK})$ contains a copy of $\mathbf{C}(\mathbb{T})$. This circumstance may be used to localize the algebra $\text{alg}^\pi\, \mathcal{T}_\alpha(\mathbf{PK})$ over $\mathbb{T}$ in order to describe the maximal ideal space and the Gelfand transform. The details of such an investigation can be found in BÖTTCHER/SILBERMANN [3], Chapter 6. We here wish to add the following fact. Namely, the algebra alg $\mathcal{T}_\alpha(\mathbf{PK})$ is actually a C^*-subalgebra of $\mathcal{L}(\mathbf{l}^{2,\alpha})$.

The proof is essentially based on a formula of Duduchava (see BÖTTCHER/SILBERMANN [3], Theorem 6.20). Using this formula one can show that

$$\operatorname{alg} \mathcal{T}_\alpha(\mathbf{PK}) = \Lambda^\alpha \operatorname{alg} \mathcal{T}(\mathbf{PC}) \Lambda^{-\alpha} \qquad \left(-\frac{1}{2} < \alpha < \frac{1}{2}\right),$$

where Λ^α is the isomorphism between $\mathbf{l}^2$ and $\mathbf{l}^{2,\alpha}$ defined in Section 4.46. This, as a matter of fact, gives the assertion since the map $\psi: \mathcal{L}(\mathbf{l}^2) \to \mathcal{L}(\mathbf{l}^{2,\alpha})$, $\psi(A) := \Lambda^\alpha A \Lambda^{-\alpha}$, is actually a C^*-isomorphism. A generalization of Theorem 6.16 can be stated as follows:

6.17. Theorem. *Let* $-1/2 < \alpha < 1/2$. *Then*

(i) $\mathcal{K}(\mathbf{l}^{2,\alpha}) \subset \operatorname{alg} \mathcal{T}_\alpha(\mathbf{PK})$,

(ii) $\operatorname{alg}^\pi \mathcal{T}_\alpha(\mathbf{PK}) := \operatorname{alg} \mathcal{T}_\alpha(\mathbf{PK})/\mathcal{K}(\mathbf{l}^{2,\alpha})$ *is isometrically isomorphic to the algebra of all continuous functions defined on* $\mathbb{T} \times [0, 1]$, *the cylinder endowed with the topology described in Theorem* 6.16,

(iii) *this isomorphism takes the coset* $T(a) + \mathcal{K}(\mathbf{l}^{2,\alpha})$ $\big(T(a) \in \operatorname{alg} \mathcal{T}_\alpha(\mathbf{PK})\big)$ *into the function*

$$a(t+0)\,\sigma_r(\mu) + a(t-0)\,\big(1 - \sigma_r(\mu)\big) \qquad (t \in \mathbb{T},\ \mu \in [0, 1])$$

where $r := (1/2 - \alpha)^{-1}$ *and*

$$\sigma_r(\mu) = \begin{cases} \mu & \text{for } r = 2 \\ \dfrac{\sin \vartheta\mu \exp(i\vartheta\mu)}{\sin \vartheta \exp i\vartheta} & \text{for } r \neq 2,\ \vartheta := \pi\left(1 - \dfrac{2}{r}\right). \end{cases}$$

Strong ellipticity

6.18. Definitions. Let A be a linear bounded operator acting on a Hilbert space. As usual, A is said to be *strongly elliptic* if it admits a representation

$$A = \alpha D + K, \tag{1}$$

where $\alpha \in \mathbb{C}$, $\alpha \neq 0$, K is a compact operator, and $\operatorname{Re} D := (1/2)(D + D^*) > 0$. When considering certain approximation methods for solving the equation $Ax = y$ it is often useful to have a more general concept of strong ellipticity. In particular, the representations

$$A = BDC + K, \tag{2}$$

B and C being, in a sense, simple operators, will play an important role in the subsequent chapters.

Notice that (2) is equivalent to the representation

$$A = \lambda B(I + T)\,C + K, \tag{3}$$

where λ is a positive real number and $\|T\| < 1$. This is a consequence of the next result.

6.19. Proposition. *Let A be an arbitrary element of a C^*-algebra (with unit element e). Then the following are equivalent:*

(i) $\operatorname{Re} A \geqq C > 0$, *where C is a positive number,*

(ii) *there exists a positive number λ such that $A = \lambda(e + T)$, $\|T\| < 1$.*

Proof. (ii) $\Rightarrow$ (i) is obvious. (i) $\Rightarrow$ (ii): It follows from the general theory of C^*-algebras that we can assume A to be an operator given on a Hilbert space H. Let λ be a positive number and consider $B = (1/\lambda) A - I$. For $x \in H$ with $\|x\| = 1$ we have

$$(Bx, Bx) = \frac{1}{\lambda^2}(Ax, Ax) - \frac{1}{\lambda}(Ax, x) - \frac{1}{\lambda}\overline{(Ax, x)} + 1$$
$$\leqq \frac{1}{\lambda^2}\|A\|^2 - \frac{1}{\lambda}C + 1 \leqq 1 - \delta$$

if λ is sufficiently large and $\delta > 0$ is sufficiently small. Hence, $\|B\| < 1$. ∎

To decide whether a given operator admits a representation of the form 6.18(3) is by no means trivial. In what follows we consider this problem for singular integral operators with piecewise continuous coefficients. Our starting point is a rather abstract result. Let $\mathfrak{B}$ be a C^*-algebra with unit e generated by two selfadjoint idempotents p, q such that $\operatorname{sp}(pqp) = [0, 1]$. Let Λ be the C^*-subalgebra of $\mathfrak{B}$ defined by

$$\Lambda := \{a = (a_+ - a_-) q + a_- e : a_+, a_- \in \mathbb{C}\}.$$

6.20. Theorem. *Let $A = ap + b(e - p) \in \mathfrak{B}$ be an element with $a = (a_+ - a_-) q + a_- e$, $b = (b_+ - b_-) q + b_- e \in \Lambda$.*

(i) *A admits a representation*

$$A = (e + T) B,$$

where $\|T\| < 1$ and $B = \alpha p + \beta(e - p)$ is an invertible element with $\alpha, \beta \in \mathbb{C} \setminus \{0\}$, if and only if

$$\frac{a_+}{a_-}\mu + \frac{b_+}{b_-}(1 - \mu) \notin (-\infty, 0], \qquad 0 \leqq \mu \leqq 1. \tag{1}$$

(ii) *The representation*

$$A = C(e + T),$$

where $\|T\| < 1$ and $C \in \Lambda$ is invertible, holds if and only if

$$\frac{a_+}{b_+}\mu + \frac{a_-}{b_-}(1 - \mu) \notin (-\infty, 0], \qquad 0 \leqq \mu \leqq 1. \tag{2}$$

6.21. Proof. (i) In view of Proposition 6.19 we have to show that 6.20(1) holds if and only if $\operatorname{Re}(AB^{-1}) \geqq C > 0$, where $B = \alpha p + \beta(e - p)$ is an invertible element with $\alpha, \beta \in \mathbb{C}$. By Theorem 1.16, $\mathfrak{B}$, p and q can be replaced by $\mathring{C}^2[0, 1]$, $\hat{p}$ and $\hat{q}$, respectively (cf. Sect. 6.12). Consequently, it suffices to show that 6.20(1) is satisfied if and only if there exist complex numbers $\gamma, \delta \neq 0$ ($\gamma = \alpha^{-1}$, $\delta = \beta^{-1}$) such that

$$\operatorname{Re}\begin{pmatrix}\{x\gamma a_+ + (1-x)\gamma a_-\} & \left\{\sqrt{x(1-x)}\,(\delta b_+ - \delta b_-)\right\} \\ \left\{\sqrt{x(1-x)}\,(\gamma a_+ - \gamma a_-)\right\} & \{(1-x)\,\delta b_+ + x\delta b_-\}\end{pmatrix} \geqq C\begin{pmatrix}1 & 0\\ 0 & 1\end{pmatrix}, \tag{1}$$

$$0 \leqq x \leqq 1,\ C > 0.$$

Without loss of generality, we may suppose $a_- = b_- = 1$, i.e., $(a_+/a_-) = a_+$, $(b_+/b_-) = b_+$. Then the left-hand side of (1) takes the form

$$\operatorname{Re}\begin{pmatrix}\sqrt{x} & -\sqrt{1-x}\\ \sqrt{1-x} & \sqrt{x}\end{pmatrix}\begin{pmatrix}\{x\gamma a_+ + (1-x)\,\delta b_+\} & \left\{\sqrt{x(1-x)}\,(\delta b_+ - \gamma a_+)\right\}\\ \left\{\sqrt{x(1-x)}\,(\delta - \gamma)\right\} & \{(1-x)\,\gamma + x\delta\}\end{pmatrix}$$

$$\times\begin{pmatrix}\sqrt{x} & \sqrt{1-x}\\ -\sqrt{1-x} & \sqrt{x}\end{pmatrix}.$$

Since $\begin{pmatrix}\sqrt{x} & -\sqrt{1-x}\\ \sqrt{1-x} & \sqrt{x}\end{pmatrix}$ is unitary, we have to show that 6.20(1) is equivalent to the existence of numbers $\gamma, \delta \in \mathbb{C}$ satisfying

$$\operatorname{Re}\begin{pmatrix}\{x\gamma a_+ + (1-x)\,\delta b_+\} & \left\{\sqrt{x(1-x)}\,(\delta b_+ - \gamma a_+)\right\}\\ \left\{\sqrt{x(1-x)}\,(\delta - \gamma)\right\} & \{(1-x)\,\gamma + x\delta\}\end{pmatrix}$$

$$= \begin{pmatrix}\{x\operatorname{Re}\gamma a_+ + (1-x)\operatorname{Re}\delta b_+\} & \left\{\sqrt{x(1-x)}\,\frac{1}{2}\,(\delta b_+ + \bar{\delta} - \gamma a_+ - \bar{\gamma})\right\}\\ \left\{\sqrt{x(1-x)}\,\frac{1}{2}\,(\overline{\delta b_+} + \delta - \overline{\gamma a_+} - \gamma)\right\} & \{(1-x)\operatorname{Re}\gamma + x\operatorname{Re}\delta\}\end{pmatrix}$$

$$\geqq C\begin{pmatrix}1 & 0\\ 0 & 1\end{pmatrix},\quad 0 \leqq x \leqq 1,\quad C > 0. \tag{2}$$

The relation (2) holds if and only if

$$\begin{aligned}&\{x\operatorname{Re}\gamma a_+ + (1-x)\operatorname{Re}\delta b_+\} > 0,\\ &\{(1-x)\operatorname{Re}\gamma + x\operatorname{Re}\delta\} > 0,\\ &\left|\frac{1}{2}\sqrt{x(1-x)}\,(\delta b_+ + \bar{\delta} - \gamma a_+ - \bar{\gamma})\right|^2 < \{x\operatorname{Re}\gamma a_+ + (1-x)\operatorname{Re}\delta b_+\}\\ &\times\{(1-x)\operatorname{Re}\gamma + x\operatorname{Re}\delta\}.\end{aligned} \tag{3}$$

Dividing the last inequality by $x(1-x)$ and taking the infimum over $[0, 1]$ with respect to x, we arrive at the inequality

$$|(\delta b_+ + \bar{\delta}) - (\gamma a_+ + \bar{\gamma})| < 2\sqrt{\operatorname{Re}\gamma a_+}\sqrt{\operatorname{Re}\gamma} + 2\sqrt{\operatorname{Re}\delta b_+}\sqrt{\operatorname{Re}\delta}.$$

The first two inequalities in (3) are equivalent to requiring $\operatorname{Re}\gamma a_+ > 0$, $\operatorname{Re}\gamma > 0$, $\operatorname{Re}\delta b_+ > 0$, $\operatorname{Re}\delta > 0$. Consequently, there exist numbers $\gamma, \delta \in \mathbb{C}$ satisfying (1) if and only if there exist $\gamma, \delta, \varkappa \in \mathbb{C}$ such that

$$\operatorname{Re}\gamma a_+ > 0,\qquad \operatorname{Re}\gamma > 0,\qquad |\gamma a_+ + \bar{\gamma} - \bar{\varkappa}| < 2\sqrt{\operatorname{Re}\gamma a_+ \operatorname{Re}\gamma},$$

$$\operatorname{Re}\delta b_+ > 0,\qquad \operatorname{Re}\delta > 0,\qquad |\delta b_+ + \bar{\delta} - \bar{\varkappa}| < 2\sqrt{\operatorname{Re}\delta b_+ \operatorname{Re}\delta},$$

i.e., if and only if $\mathfrak{M}(a_+) \cap \mathfrak{M}(b_+) \neq 0$, where

$$\mathfrak{M}(c) := \left\{\varkappa \in \mathbb{C}: \exists \gamma \in \mathbb{C} \text{ with } \operatorname{Re} \gamma c > 0, \operatorname{Re} \gamma > 0, |\gamma c + \bar{\gamma} - \bar{\varkappa}| < 2\sqrt{\operatorname{Re} \gamma c \operatorname{Re} \gamma}\right\}.$$

We shall prove that

$$\mathfrak{M}(c) = \{\varkappa \in \mathbb{C}: \operatorname{Re} \varkappa > 0, \operatorname{Re} \varkappa c > 0\}. \tag{4}$$

Once this has been proved, the existence of numbers $\gamma, \delta \in \mathbb{C}$ satisfying (1) is equivalent to the existence of $\varkappa \in \mathbb{C}$ such that $\operatorname{Re} \varkappa > 0$, $\operatorname{Re} \varkappa a_+ > 0$ and $\operatorname{Re} \varkappa b_+ > 0$, i.e., is equivalent to 6.20(1).

Let $\mathfrak{N}(c)$ denote the right-hand side of (4). The inequality

$$|\gamma c + \bar{\gamma} - \bar{\varkappa}|^2 < 4 \operatorname{Re} \gamma c \operatorname{Re} \gamma \tag{5}$$

is equivalent to

$$|\gamma(c-1) - \bar{\varkappa} + 2 \operatorname{Re} \gamma|^2 = |\gamma(c-1) - \bar{\varkappa}|^2 + 4(\operatorname{Re} \gamma)^2 + 4 \operatorname{Re} \gamma \times \operatorname{Re} \left(\gamma(c-1) - \varkappa\right) < 4 \operatorname{Re} \gamma c \operatorname{Re} \gamma,$$

$$|\gamma(c-1) - \bar{\varkappa}|^2 < 4 \operatorname{Re} \gamma \operatorname{Re} \varkappa \tag{6}$$

and to

$$\begin{aligned} &|\bar{\varkappa} + \bar{\gamma}(c-1) - (c-1)\, 2 \operatorname{Re} \gamma|^2 < 4 \operatorname{Re} \gamma \operatorname{Re} \varkappa, \\ &|\bar{\varkappa} + \bar{\gamma}(c-1)|^2 < 4 \operatorname{Re} \gamma \operatorname{Re} c\varkappa. \end{aligned} \tag{7}$$

Hence the inclusion $\mathfrak{M}(c) \subset \mathfrak{N}(c)$ follows easily from (6) and (7). Vice versa, for $\varkappa$ belonging to $\mathfrak{N}(c)$, we choose $\gamma := \{\varkappa^{-1} + c\bar{\varkappa}^{-1}\}^{-1}$. Since $\operatorname{Re} \varkappa > 0$ and $\operatorname{Re} \varkappa c > 0$, we obtain $\operatorname{Re} \varkappa^{-1} > 0$ as well as $\operatorname{Re} c\varkappa\, |\varkappa|^{-2} = \operatorname{Re} c\bar{\varkappa}^{-1} > 0$. Hence, $\operatorname{Re} \{\varkappa^{-1} + c\bar{\varkappa}^{-1}\} > 0$ and $\operatorname{Re} \gamma > 0$. Moreover, $\operatorname{Re} \bar{\gamma}^{-1} c = \operatorname{Re} c\bar{\varkappa}^{-1} + |c|^2 \operatorname{Re} \varkappa^{-1} > 0$ and $\operatorname{Re} \gamma c > 0$. Furthermore, since $\operatorname{Re} \gamma^{-1} = \operatorname{Re} \varkappa^{-1} + \operatorname{Re} c\bar{\varkappa}^{-1} > \operatorname{Re} \varkappa^{-1}$, we get

$$\operatorname{Re} \gamma^{-1} > \operatorname{Re} \varkappa^{-1}. \tag{8}$$

By definition, $\gamma^{-1} = \varkappa^{-1} + c\bar{\varkappa}^{-1}$, i.e., $\gamma^{-1} - (c-1)\,\bar{\varkappa}^{-1} = 2 \operatorname{Re} \bar{\varkappa}^{-1}$. Hence, $\bar{\varkappa} - \gamma(c-1) = 2 \operatorname{Re} \varkappa^{-1} \gamma\bar{\varkappa}$ and $|\bar{\varkappa} - (c-1)\,\gamma|^2 = 4(\operatorname{Re} \varkappa^{-1})^2\, |\varkappa|^2\, |\gamma|^2$. Inequality (8) implies that

$$|\bar{\varkappa} - (c-1)\,\gamma|^2 < 4 \operatorname{Re} \varkappa^{-1}\, |\varkappa|^2 \operatorname{Re} \gamma^{-1}\, |\gamma|^2 = 4 \operatorname{Re} \varkappa \operatorname{Re} \gamma.$$

Using the equivalence of (5) and (6), we deduce that (5) is valid. Hence, $\mathfrak{N}(c) \subset \mathfrak{M}(c)$.

(ii) We have to show that 6.20(2) is equivalent to the existence of an invertible $C = \alpha q + \beta(e - q)$ satisfying $\operatorname{Re}(C^{-1}A) > 0$, i.e., to the existence of complex numbers $\gamma, \delta \neq 0$ ($\gamma = \alpha^{-1}, \delta = \beta^{-1}$) such that

$$\operatorname{Re} \begin{pmatrix} \{x\gamma a_+ + (1-x)\,\delta a_-\} & \left\{\sqrt{x(1-x)}\,(\gamma b_+ - \delta b_-)\right\} \\ \left\{\sqrt{x(1-x)}\,(\gamma a_+ - \delta a_-)\right\} & \{(1-x)\,\gamma b_+ + x\delta b_-\} \end{pmatrix} > 0, \quad 0 \leqq x \leqq 1. \tag{9}$$

Without loss of generality we may suppose $b_+ = b_- = 1$, i.e., $a_+/b_+ = a_+$ and $a_-/b_- = a_-$. Repeating the arguments presented after (2), we conclude that (9) holds if and only if 6.20(2) is satisfied. ■

We now consider a closed and bounded Lyapunov curve Γ in $\mathbb{C}$. Let S_Γ be given by 6.1(1) and put $P_\Gamma := 1/2(I + S_\Gamma)$, $Q_\Gamma := 1/2(I - S_\Gamma)$. Let A denote the singular integral operator $aP_\Gamma + bQ_\Gamma \in \mathcal{L}(\mathbf{L}^2(\Gamma))$, where $a, b \in \mathbf{PC}(\Gamma)$. We call A *locally strongly elliptic* if it admits a representation of the form

$$A = gD + K, \tag{10}$$

where $g \in \mathbf{PC}(\Gamma)$ is an invertible function, $K \in \mathcal{L}(\mathbf{L}^2(\Gamma))$ is compact and $\operatorname{Re} D > 0$. We have seen in Sect. 6.19 (cf. the equivalence of 6.18(2) and 6.18(3)) that the condition $\operatorname{Re} D > 0$ in (10) may be replaced by inequality $\|I - D\| < 1$.

6.22. Theorem. *Suppose Γ is a closed and bounded Lyapunov curve, $a, b \in \mathbf{PC}(\Gamma)$ and $A = aP_\Gamma + bQ_\Gamma$.*

(i) *The operator A admits a representation $A = D(cP_\Gamma + dQ_\Gamma) + K$, where $K \in \mathcal{L}(\mathbf{L}^2(\Gamma))$ is compact, $\|I - D\| < 1$, and the functions $c, d \in \mathbf{C}(\Gamma)$ do not vanish, if and only if*

$$\mu \frac{a(z+0)}{a(z-0)} + (1-\mu) \frac{b(z+0)}{b(z-0)} \notin (-\infty, 0], \quad 0 \leqq \mu \leqq 1, \quad z \in \Gamma. \tag{1}$$

(ii) *The operator A is locally strongly elliptic if and only if*

$$\mu \frac{a(z+0)}{b(z+0)} + (1-\mu) \frac{a(z-0)}{b(z-0)} \notin (-\infty, 0], \quad 0 \leqq \mu \leqq 1, \quad z \in \Gamma. \tag{2}$$

Proof. Let us start with proving (ii). If A admits a representation 6.21(10) with $\|D - I\| < 1$ and if $z \in \Gamma$, then Proposition 6.13 yields that $\operatorname{smb}(A, z) = \operatorname{smb}(g, z) \operatorname{smb}(D, z)$, where

$$\begin{aligned} &\|\operatorname{smb}(D, z) - \hat{e}\| < 1, \\ &\operatorname{smb}(g, z) = g(z+0)\,\hat{q} + g(z-0)\,(\hat{e} - \hat{q}), \\ &\operatorname{smb}(A, z) = [a(z+0)\,\hat{q} + a(z-0)\,(\hat{e} - \hat{q})]\,\hat{p} \\ &\qquad\qquad + [b(z+0)\,\hat{q} + b(z-0)\,(\hat{e} - \hat{q})]\,(\hat{e} - \hat{p}). \end{aligned}$$

The elements $\hat{p}, \hat{q}, \hat{e} \in \dot{\mathbf{C}}^2[0, 1]$ are defined in Sect. 6.12. From Theorem 6.20(ii) we conclude that (2) is satisfied.

Vice versa, suppose (2) holds. Assertion (ii) of Theorem 6.20 shows that, for arbitrary $z \in \Gamma$, there exist $g_+, g_- \in \mathbb{C} \setminus \{0\}$ satisfying

$$\|\{g_+\hat{q} + g_-(\hat{e} - \hat{q})\}^{-1} \operatorname{smb}(A, z) - \hat{e}\| < 1. \tag{3}$$

We shall put these local values g_+, g_- together to one function $g \in \mathbf{PC}(\Gamma)$ such that

$$\begin{aligned} &\|\{\operatorname{smb}(g, z)\}^{-1} \operatorname{smb}(A, z) - \hat{e}\| \\ &= \|\{g(z+0)\,\hat{q} + g(z-0)\,(\hat{e} - \hat{q})\}^{-1} \operatorname{smb}(A, z) - \hat{e}\| < 1 \end{aligned} \tag{4}$$

for any $z \in \Gamma$. Once this is done, Proposition 6.13 guarantees the existence of a compact operator $K \in \mathcal{L}(\mathbf{L}^2(\Gamma))$ such that $\|g^{-1}A - I - K\| < 1$. Setting $D := g^{-1}A - K$, we obtain $\|D - I\| < 1$ as well as $A = gD + gK$. Hence A is locally strongly elliptic. Thus it remains to prove (4). If we choose the argument x of the functions $\operatorname{smb}(A, z)$,

$\hat{e}, \hat{q} \in \dot{C}^2[0, 1]$ to be 0 or 1, we arrive at

$$\|\{g_+\hat{e}\}^{-1} \operatorname{smb}\big(a(z+0)\, P_\Gamma + b(z+0)\, Q_\Gamma, z\big) - \hat{e}\| < 1,$$

$$\|\{g_-\hat{e}\}^{-1} \operatorname{smb}\big(a(z-0)\, P_\Gamma + b(z-0)\, Q_\Gamma, z\big) - \hat{e}\| < 1.$$

Let ε denote a sufficiently small positive number. By Proposition 6.13(a), there exists a neighborhood U_z of z on Γ such that

$$\big\|\operatorname{smb}(A, \zeta) - \operatorname{smb}\big(a(z+0)\, P_\Gamma + b(z+0)\, Q_\Gamma, \zeta\big)\big\| < \varepsilon$$

if $\zeta \in U_z$ is on the right-hand side of z (with respect to the orientation on Γ) and

$$\big\|\operatorname{smb}(A, \zeta) - \operatorname{smb}\big(a(z-0)\, P_\Gamma + b(z-0)\, Q_\Gamma, \zeta\big)\big\| < \varepsilon$$

if $\zeta \in U_z$ is on the left-hand side of z. Consequently,

$$\|\{g_+\hat{e}\}^{-1} \operatorname{smb}(A, \zeta) - \hat{e}\| < 1 \tag{5}$$

if $\zeta \in U_z$, $\zeta > z$ and

$$\|\{g_-\hat{e}\}^{-1} \operatorname{smb}(A, \zeta) - \hat{e}\| < 1 \tag{6}$$

if $\zeta \in U_z$, $\zeta < z$. Now we choose a $g^z \in \mathbf{PC}(\Gamma)$ satisfying

$$g^z(\zeta) = \begin{cases} g_+ & \text{if} \quad \zeta \in U_z, \quad \zeta > z, \\ g_- & \text{if} \quad \zeta \in U_z, \quad \zeta < z, \end{cases}$$

and so we arrive at the inequality

$$\|\{\operatorname{smb}(g^z, \zeta)\}^{-1} \operatorname{smb}(A, \zeta) - \hat{e}\| < 1 \tag{7}$$

for any $\zeta \in U_z$ (cf. (3), (5) and (6)). Since Γ is compact and $\Gamma \subset \bigcup_{z \in \Gamma} U_z$, we can find $z_i \in \Gamma$, $i = 1, \dots, N$ satisfying $\Gamma \subset \bigcup_{i=1}^{N} U_{z_i}$. Furthermore, there exist functions $f^i \in \mathbf{C}(\Gamma)$, $i = 1, \dots, N$ such that $\sum_{i=1}^{N} f^i \equiv 1$, $0 \leqq f^i \leqq 1$ and $\operatorname{supp} f^i \subset U_{z_i}$. We now define $g \in \mathbf{PC}(\Gamma)$ by setting

$$g := \left\{\sum_{i=1}^{N} f^i [g^{z_i}]^{-1}\right\}^{-1}.$$

For this function g, we get

$$\{\operatorname{smb}(g, z)\}^{-1} = \operatorname{smb}(g^{-1}, z) = \sum_{i=1}^{N} \operatorname{smb}(f^i, z) \operatorname{smb}([g^{z_i}]^{-1}, z)$$

$$= \sum_{i=1}^{N} f^i(z) \{\operatorname{smb}(g^{z_i}, z)\}^{-1}.$$

Since $\operatorname{supp} f^i \subset U_{z_i}$, the last sum may be taken only over the i for which $z \in U_{z_i}$. If we replace ζ by z and z by z_i in (7), we obtain

$$\|\{\operatorname{smb}(g, z)\}^{-1} \operatorname{smb}(A, z) - \hat{e}\|$$

$$= \left\|\sum_{i=1}^{N} f^i(z) \left[\{\operatorname{smb}(g^{z_i}, z)\}^{-1} \operatorname{smb}(A, z) - \hat{e}\right]\right\|$$

$$\leqq \sum_{i=1}^{N} f^i(z) \,\|\{\operatorname{smb}(g^{z_i}, z)\}^{-1} \operatorname{smb}(A, z) - \hat{e}\| < 1.$$

Hence (4) is satisfied. This finishes the proof of assertion (ii) in Theorem 6.22.

Assertion (i) can be shown in a completely analogous manner, and hence we omit the details. For a different proof we refer to Sec. 7.15. ■

6.23. Corollary. 1°. *Suppose* $A = aP_\Gamma + bQ_\Gamma \in \mathcal{L}(\mathbf{L}^2(\Gamma))$, *where* $a, b \in \mathbf{C}(\Gamma)$ *and* Γ *is a closed Lyapunov curve. Then the following assertions are equivalent.*

(i) *The operator* A *is locally strongly elliptic.*

(ii) *The operator* A *admits the representation* 6.21(10) *with continuous* g.

(iii) $\lambda a(z) + (1 - \lambda)\, b(z) \neq 0$ *for all* $\lambda \in [0, 1]$ *and* $z \in \Gamma$.

(iv) *There exists a continuous function* θ *on* Γ *satisfying* $\operatorname{Re}(\theta a)(z) > 0$, $\operatorname{Re}(\theta b)(z) > 0$ *for all* $z \in \Gamma$.

2°. *Suppose* $A = cI + dS_J \in \mathcal{L}(\mathbf{L}^2(J, \omega))$, *where* $c, d \in \mathbf{C}(J)$ (*see Sec.* 6.11). *Then the following assertions are equivalent.*

(i) *The operator* A *admits a representation in the form*

$$A = gD + K, \tag{1}$$

where $g \in \mathbf{C}(J)$ *is invertible,* $K \in \mathcal{L}(\mathbf{L}^2(J, w))$ *is compact and* $\operatorname{Re} D > 0$. (*In that case* A *is called locally strongly elliptic*).

(ii) $c(z) + \lambda d(z) \neq 0$ *for all* $\lambda \in [-1, 1]$ *and* $z \in J$, *and*

$$|\operatorname{Re} \varkappa_j + \alpha_j| < 1/2, \qquad j = 1, 2,$$

where

$$\varkappa_1 = \frac{1}{2\pi i} \log \frac{c(a) + d(a)}{c(a) - d(a)}, \qquad \varkappa_2 = \frac{1}{2\pi i} \log \frac{c(b) - d(b)}{c(b) + d(b)} \qquad (J = [a, b])$$

and log *denotes that branch of the logarithm which is continuous in* $\mathbb{C} \setminus (-\infty, 0]$ *and takes real values on the positive real line.*

If $w \equiv 1$, *then* (i) *and* (ii) *are equivalent to the following condition.*

(iii) *There exists a function* $\theta \in \mathbf{C}(J)$ *satisfying*

$$\operatorname{Re} \theta(z)\, [c(z) \pm d(z)] > 0 \quad \text{for all} \quad z \in J.$$

Proof. 1°(i) ⇒ 1°(ii): Obviously, if $a_+ = a_-$ and $b_+ = b_-$ in Theorem 6.20 and $0 \notin [a_+, b_+]$, then $a_+p + b_+(e - p) = c(e + T)$, where $c \in \mathbb{C} \setminus \{0\}$ and $\|T\| < 1$. Using this and following the proof of Theorem 6.22, we get the representation 6.21(10) with $g \in \mathbf{C}(\Gamma)$ (provided 6.22(2) holds).

1°(ii) ⇒ 1°(i): Obvious.

1°(i) ⇔ 1°(iii): Immediate from 6.22(2).

1°(ii) ⇒ 1°(iv): Use the Gohberg-Krupnik symbol.

1°(iv) ⇒ 1°(iii): Obvious.

2°(i) ⇒ 2°(ii): Using 6.1(ix) it is easily seen that A is invertible. Further, the representation (1) shows that

$$c(z) + d(z)\, \lambda \neq 0, \qquad \lambda \in \operatorname{conv}\{\Omega_{2,\omega}(z, t) : -1 \leq t \leq 1\}, \tag{2}$$

($\Omega_{2,\omega}$ being the function defined in Sec. 6.11 and conv denoting the convex hull), whence the assertion follows.

2°(ii) ⇒ 2°(i): First of all, 2°(ii) implies (2). Therefore at each fixed point z the function

$(\operatorname{smb} A)(z, t)$ can be represented in the form

$$\lambda(1 + m)$$

where $\lambda \in \mathbb{C} \setminus \{0\}$ and $\|m\|_\infty < 1$. By repeating the arguments of the proof to Theorem 6.22 we get the representation (1).

The equivalence of 2(ii) and 2(iii) is obvious. ■

Consider the Sobolev-Slobodetski space $\mathbf{H}^\alpha = \mathbf{L}^{2,\alpha}(\mathbb{T})$ and let P be the projection defined in Sec. 6.15, put $tI = U$, $t^{-1}I = U^{-1}$. Then the algebra $\operatorname{alg}_\alpha(U, U^{-1}, P)$ $:= \operatorname{alg}(U, U^{-1}, P) \subset \mathcal{L}(\mathbf{H}^\alpha)$ is well-defined (see Sec. 4.65, 4.66). We call an operator $A \in \operatorname{alg}_\alpha\{U, U^{-1}, P\}$ *locally strongly elliptic* if it admits a representation of the form

$$A = gD + K, \tag{3}$$

where $g \in \mathbf{C}^\infty(\mathbb{T})$ is invertible, $\operatorname{Re} D > 0$ and $K \in \mathcal{K}(\mathbf{H}^\alpha)$. In what follows we identify the space $\mathbf{H}^\alpha$ with $\tilde{\mathbf{l}}^{2,\alpha}$. Thus, it makes sense to define the operator

$$\tilde{\Lambda}^\alpha\colon \tilde{\mathbf{l}}^{2,0} \to \tilde{\mathbf{l}}^{2,\alpha}, \qquad \{\xi_i\}_{-\infty}^{\infty} \mapsto \{(1 + |i|)^{-\alpha}\, \xi_i\}_{-\infty}^{\infty}.$$

6.24. Theorem. (i) $\operatorname{alg}_0(U, U^{-1}, P) = \tilde{\Lambda}^{-\alpha} \operatorname{alg}_\alpha(U, U^{-1}, P)\, \tilde{\Lambda}^\alpha$, *and, for* $A \in \operatorname{alg}_\alpha(U, U^{-1}, P)$,

$$\operatorname{smb} A = \operatorname{smb} \tilde{\Lambda}^{-\alpha} A \tilde{\Lambda}^\alpha.$$

(ii) $A \in \operatorname{alg}_\alpha(U, U^{-1}, P)$ *is locally strongly elliptic if and only if*

$$\lambda(\operatorname{smb} A)(t, 1) + (1 - \lambda)(\operatorname{smb} A)(t, -1) \neq 0 \tag{1}$$

for all $t \in \mathbb{T}$ *and* $\lambda \in [0, 1]$.

Proof. (i) Repeat the reasoning of the proof of equality 4.51(1).

(ii) Let (1) be fulfilled. Using (i) and Proposition 4.77, we obtain $B := \tilde{\Lambda}^\alpha A \tilde{\Lambda}^{-\alpha}$ $= (\operatorname{smb} A)(t, 1)\, P + (\operatorname{smb} A)(t, -1)\, Q + T$, where $T \in \mathcal{K}(\tilde{\mathbf{l}}^{2,0})$. Further, by Corollary 6.23(1°) the operator B can be represented in the form

$$B = gD + T_1$$

where $g \in \mathbf{C}(\mathbb{T})$ is invertible, $\operatorname{Re} D > 0$, and $T_1 \in \mathcal{K}(\tilde{\mathbf{l}}^{2,0})$. It is not hard to see that g can be chosen to belong to $\mathbf{C}^\infty(\mathbb{T})$. Since $C \mapsto \tilde{\Lambda}^{-\alpha} C \tilde{\Lambda}^\alpha$ is a *-homomorphism, we get

$$A = \tilde{\Lambda}^{-\alpha} B \tilde{\Lambda}^\alpha = gD_1 + K,$$

where $\operatorname{Re} D_1 > 0$ and $K \in \mathcal{K}(\mathbf{l}^{2,\alpha})$, and 6.23(3) is proved. The converse assertion can be proved analogously. ■

Singular integral operators on Hölder-Zygmund spaces

6.25. Basic facts. Let Γ be a plane Jordan curve given by a regular parameter representation

$$z = \big(z_1(t), z_2(t)\big) = z_1(t) + \mathrm{i} z_2(t)$$

where z is a 2π-periodic function of a real variable t and $\mathrm{d}z/\mathrm{d}t \neq 0$ for all t. Due to this representation we have a one-to-one correspondence between functions on Γ and 2π-periodic functions. Denote by $\mathbf{C}$ the set of continuous functions $f\colon \mathbb{R} \to \mathbb{C}$ which are 2π-periodic, that is,

$$f(x + 2\pi) = f(x) \quad \text{for all} \quad x \in \mathbb{R},$$

and equip **C** with the maximum norm

$$\|f\|_\infty := \max |f(x)| .$$

We write $\mathrm{D} := \mathrm{d}/\mathrm{d}x$, define

$$\mathbf{C}^s := \{f \in \mathbf{C} : \mathrm{D}^j f \in \mathbf{C} \text{ for } 0 \leqq j \leqq s\}, \qquad s \in \mathbb{Z}_+,$$

and introduce the norm

$$\|f\|_{\mathbf{C}^s} := \sum_{j=0}^{s} \|\mathrm{D}^j f\|_\infty .$$

The Hölder-Zygmund seminorm is defined by

$$[f]^\alpha := \begin{cases} \sup\limits_{h>0} \dfrac{\|\Delta_h f\|_\infty}{h^\alpha}, & 0 < \alpha < 1 \\ \sup\limits_{h>0} \dfrac{\|\Delta_h^2 f\|_\infty}{h}, & \alpha = 1, \end{cases}$$

where Δ_h is the forward difference operator

$$(\Delta_h f)(x) := f(x+h) - f(x),$$

and Δ_h^2 is the second forward difference operator

$$(\Delta_h^2 f)(x) := f(x+2h) - 2f(x+h) + f(x).$$

Given $s > 0$, write

$$s = m + \alpha, \qquad m \in \mathbb{Z}_+ \text{ and } 0 < \alpha \leqq 1,$$

and define the periodic Hölder-Zygmund space

$$\mathscr{H}^s := \{f \in \mathbf{C}^m : [\mathrm{D}^m f]^\alpha < \infty\},$$

the norm being given by

$$\|f\|_{\mathscr{H}^s} := \|f\|_{\mathbf{C}^m} + [\mathrm{D}^m f]^\alpha$$

(see Triebel [1] and Section 2.34).

The space $\mathscr{H}^s$ is complete but neither reflexive nor separable. Further, for $s_1 < s_2$, the space $\mathscr{H}^{s_2}$ is compactly embedded in $\mathscr{H}^{s_1}$. Notice that $\mathscr{H}^s$ actually forms an algebra on which the operator aI of multiplication (this operator will also be denoted by M_a) is continuous (use Mc Lean/Wendland [1], Lemma 4.7).

Given $f \in \mathscr{H}^s$ define the complex Fourier coefficients by

$$f_n := \frac{1}{2\pi} \int_{-\pi}^{\pi} f(x)\, \mathrm{e}^{-\mathrm{i}nx}\, \mathrm{d}x, \qquad n \in \mathbb{Z},$$

and the singular integral operator S by

$$S\left(\sum_{-\infty}^{\infty} f_n\, \mathrm{e}^{\mathrm{i}nx}\right) = \sum_{-\infty}^{\infty} (\operatorname{sgn} n)\, f_n\, \mathrm{e}^{\mathrm{i}nx}, \tag{0}$$

where $\operatorname{sgn} n = -1$ for $n \leqq -1$ and $\operatorname{sgn} n = 1$ for $n \geqq 0$. More precisely, S assigns to each function f the function Sf whose Fourier coefficients are $(\operatorname{sgn} n) f_n$, $n \in \mathbb{Z}$. It is well-known (cf. MIKHLIN/PRÖSSDORF [1], Chapt. 2, § 6) that, for $s \notin \mathbb{N}$, $S: \mathcal{H}^s \to \mathcal{H}^s$ is a bounded linear operator. Since one can interpolate between Hölder-Zygmund spaces (real interpolation, cf. TRIEBEL [1]) S also proves to be a bounded operator on $\mathcal{H}^s$, $s \in \mathbb{N}$. Put $P := (1/2)(I + S)$, $Q := (1/2)(I - S)$. The operators P and Q are projections since $S^2 = I$ (by (0)).

In what follows we shall use the Kuratowski measure of non-compactness

$$\|A\|_d \tag{1}$$

instead of the essential norm

$$|||A||| := \inf \{\|A - K\| : K \in \mathcal{K}(\mathcal{H}^s)\}. \tag{2}$$

We recall that $\|A\|_d$ is the infimum of all real numbers $r > 0$ such that the set $\{Ax : \|x\|_{\mathcal{H}^s} \leqq 1\}$ can be covered by a finite number of sets the diameter of which does not exceed $2r$ (the use of the norm (1) instead of (2) is advantageous when studying the Fredholm properties of singular integral equations given on $\mathcal{H}^s$). Note that

(i) *the essential norm and the Kuratowski measure of non-compactness are equivalent norms on* $\mathcal{L}(\mathcal{H}^s)/\mathcal{K}(\mathcal{H}^s)$;

(ii) $\|M_a\|_d = \|a\|_\infty$ *for all* $a \in \mathcal{H}^s$.

Property (i) follows from SCHECHTER [1], since $\mathcal{H}^s$ is isomorphic to $\mathbf{L}^\infty$ (cf. PEETRE [1]) and $\mathbf{L}^\infty$ has the metric approximation property (see PIETSCH [1]). To prove (ii) we only consider the most difficult case, namely the case where $s \in \mathbb{N}$. Since the unit ball $F := \{f \in \mathcal{H}^s : \|f\|_{\mathcal{H}^s} \leqq 1\}$ is compact in $\mathcal{H}^{s-(1/2)}$, for any $\varepsilon > 0$ there is a partition $F = \bigcup_{m=1}^{n} U_m$ such that $\operatorname{diam}_{\mathcal{H}^{s-(1/2)}} U_m < \varepsilon$ $(m = 1, \ldots, n)$. Fix such a partition and estimate $\|a(f - g)\|_{\mathcal{H}^s}$ for arbitrarily given $f, g \in U_m \cap \mathcal{H}^s$, $m = 1, \ldots, n$:

$$\|a(f-g)\|_{\mathcal{H}^s} = \|a(f-g)\|_{\mathbf{C}^{s-1}} + \sup_{x,h} \frac{\left|\Delta_h^2 \mathrm{D}^{s-1}\big(a(f-g)\big)(x)\right|}{h}$$

$$\leqq C_1 \varepsilon \|a\|_{\mathbf{C}^{s-1}} + \sup_{x,h} \frac{\left|\Delta_h^2 \mathrm{D}^{s-1}\big(a(f-g)\big)(x)\right|}{h} =: C_1\varepsilon \|a\|_{\mathbf{C}^{s-1}} + \sigma. \tag{3}$$

Further,

$$\sigma \leqq \sup_{x,h} \frac{\left|\Delta_h^2\big(a\mathrm{D}^{s-1}(f-g)\big)(x)\right|}{h} + C_2 \sum_{j=1}^{s-1} \sup_{x,h} \frac{\left|\Delta_h^2\big(\mathrm{D}^j a \mathrm{D}^{s-1-j}(f-g)\big)(x)\right|}{h}$$

$$=: \sigma_1 + \sigma_2.$$

Obviously,

$$\sigma_1 \leqq \|a\|_\infty \sup_{x,h} \frac{|\Delta_h^2 \mathrm{D}^{s-1}(f-g)(x)|}{h} + \|D^{s-1}(f-g)\|_\infty \sup_{x,h} \frac{|\Delta_h^2 a(x)|}{h}$$

$$+ 2 \sup_{x,h} \frac{|\Delta_h a(x)|}{h^{1/2}} \frac{|\Delta_h \mathrm{D}^{s-1}(f-g)(x+h)|}{h^{1/2}}$$

$$\leqq \|a\|_\infty \left(\sup_{x,h} \frac{|\Delta_h^2 \mathrm{D}^{s-1} f(x)|}{h} + \sup_{x,h} \frac{|\Delta_h^2 \mathrm{D}^{s-1} g(x)|}{h}\right) + \varepsilon \sup_{x,h} \frac{|\Delta_h^2 a(x)|}{h} + C_3 \varepsilon$$

$$\leqq 2\|a\|_\infty + \varepsilon(\|a\|_{\mathcal{H}^s} + C_3),$$

where C_3 is independent of f, g, a, and ε.

The term σ_2 can be estimated analogously. Hence, $\|M_a\|_d \leqq \|a\|_\infty + \varepsilon c$ for every given ε. Consequently,

$$\|M_a\|_d \leqq \|a\|_\infty . \tag{4}$$

That actually equality holds is an immediate consequence of the following fact.

(iii) *Let* $a = \sum_{-N}^{N} a_n \mathrm{e}^{inx}$, $b = \sum_{-N}^{N} b_n \mathrm{e}^{inx}$. *Then the singular integral operator*

$$A := aP + bQ \left(\in \mathcal{L}(\mathcal{H}^s),\, s > 0\right)$$

is a Φ_+- *or* Φ_--*operator if and only if*

$$a(x) \neq 0, \qquad b(x) \neq 0 \quad \textit{for all} \quad x \in [-\pi, \pi]. \tag{5}$$

If (5) *is satisfied, then* A *is a* Φ-*operator with* $\dim \ker A = \max\{\varkappa, 0\}$, $\operatorname{codim}(\operatorname{im} A) = \max\{-\varkappa, 0\}$, *where*

$$\varkappa = -\frac{1}{2\pi}\,[\arg (a/b)\,(x)]_{x=-\pi}^{\pi} .$$

Indeed, putting $U = M_{\mathrm{e}^{ix}}$, all assumptions of Sec. 4.65 are fulfilled and Theorems 4.73, 4.74 apply. In particular, if $b = \sum_{-N}^{N} a_n \mathrm{e}^{inx}$, then $M_b \in \Phi(\mathcal{H}^s)$ if and only if

$$b(x) \neq 0 \quad \text{for all} \quad x \in [-\pi, \pi].$$

Now it is easy to see that in (4) equality holds. Indeed, the essential spectrum of M_b coincides with $\{b(x) : x \in [-\pi, \pi]\}$. Therefore, $\|b\|_\infty \leqq \|M_b\|_d$ and so for trigonometric polynomials b one gets $\|M_b\|_d = \|b\|_\infty$. Choose a sequence of trigonometric polynomials b_n such that $\|a - b_n\|_\infty \to 0$ as $n \to \infty$. By (4) we have $\|M_a - M_{b_n}\|_d \leqq \|a - b_n\|_\infty$. Therefore $\|b_n\|_\infty = \|M_{b_n}\|_d \to \|a\|_\infty$ and hence $\|M_a\|_d = \|a\|_\infty$.

Now we are in a position to prove the following properties (iv) and (v).

(iv) $a \in \mathcal{H}^s$ *is invertible in* $\mathcal{H}^s$ *if and only if* $a(x) \neq 0$ *for all* $x \in [-\pi, \pi]$.

(v) $aS - Sa \in \mathcal{K}(\mathcal{H}^s)$ *for all* $a \in \mathcal{H}^s$.

Proof of (iv): Obviously, if $a \in \mathcal{H}^s$ is invertible in $\mathcal{H}^s$ then $a(x) \neq 0$ for all $x \in [-\pi, \pi]$. Conversely, let a have no zeros on $[-\pi, \pi]$. Then there is a sequence of trigonometric polynomials a_n such that $\|M_a - M_{a_n}\|_d \to 0$, $M_{a_n} + \mathcal{K}(\mathcal{H}^s)$ is invertible in the Calkin algebra for all n large enough ($n \geqq n_0$, say), and $\left\|M_{a_n^{-1}}\right\|_d \leqq C$ for all $n \geqq n_0$. Consequently, $M_a + \mathcal{K}(\mathcal{H}^s)$ is invertible in the Calkin algebra. So we have $M_a \in \Phi(\mathcal{H}^s)$, and for given $\varepsilon > 0$ there is an n and a $T \in \mathcal{K}(\mathcal{H}^s)$ such that

$$\|M_a - (M_{a_n} + T)\| < \varepsilon \tag{6}$$

From this we deduce that $\operatorname{ind} M_a = 0$, which implies the invertibility of M_a. Therefore, the equation $M_a b = 1$ has a unique solution $b \in \mathcal{H}^s$ and we are done.

Proof of (v): The assertion is obvious if a is a trigonometric polynomial. The general case can be proved by the help of (6) (which is, of course, also true without the requirement that a does not vanish). ■

We finish this section by considering the Wiener-Hopf factorization of functions belonging to $\mathscr{H}^s$. To this end we introduce the closed subalgebras

$$\mathscr{H}^s_+ = \{f \in \mathscr{H}^s : f_n = 0 \text{ for } n < 0\},$$
$$\mathscr{H}^s_- = \{f \in \mathscr{H}^s : f_n = 0 \text{ for } n > 0\}.$$

Since S is bounded on $\mathscr{H}^s$ and the assertions (iv) and (v) are in force, $\mathscr{H}^s$ is seen to possess the factorization property (see Clancey/Gohberg [1] or Litvinchuk/Spitkovski [1]):
(vi) *Let* $a \in \mathscr{H}^s$ *satisfy* $a(x) \neq 0$ *for all* $x \in [-\pi, \pi]$ *and put* $\varkappa = (1/2\pi)[\arg a(x)]_{x=-\pi}^{\pi}$. *Then there exist* $a_\pm \in \mathscr{H}^s_\pm$ *such that* $a(x) = a_+(x)\, e^{i\varkappa x}\, a_-(x)$ *for all* $x \in \mathbb{R}$ *and* $1/a_\pm \in \mathscr{H}^s_\pm$.

6.26. Theorem. *Let* $a, b \in \mathscr{H}^s$. *Then the singular integral operator*

$$A := aP + bQ\ \left(\in \mathscr{L}(\mathscr{H}^s),\ s > 0\right)$$

is a Φ_+- *or* Φ_-*-operator if and only if*

$$a(x) \neq 0, \qquad b(x) \neq 0 \quad \textit{for all} \quad x \in [-\pi, \pi]. \tag{1}$$

If (1) *is fulfilled, then* A *is a* Φ*-operator with*

$$\dim \ker A = \max\{\varkappa, 0\}, \qquad \operatorname{codim}(\operatorname{im} A) = \max\{-\varkappa, 0\}, \tag{2}$$

where $\varkappa = (-1/2\pi)[\arg (a/b)(x)]_{x=-\pi}^{\pi}$.

Proof. If (1) is in force then, by (iv) and (v), the operator $a^{-1}P + b^{-1}Q$ is a two-sided regularizer of A and by 1.3(a) we see that $A \in \Phi(\mathscr{H}^s)$. In order to prove (2) use that $c := b^{-1}a$ is factorizable:

$$c(x) = c_+(x)\, e^{-i\varkappa x} c_-(x) \quad \text{for all} \quad x \in \mathbb{R} \quad \text{and} \quad c_\pm \in \mathscr{H}^s_\pm .$$

Because $Pc_+^{\pm 1}P = c_+^{\pm 1}P$, $Qc_-^{\pm 1}Q = c_-^{\pm 1}Q$, we obtain

$$\begin{aligned} A &= b(cP + Q) = b(c_+ e^{-i\varkappa x} c_- P + Q) \\ &= \begin{cases} bc_-(P + c_-^{-1}Q)\,(e^{-i\varkappa x}P + Q)\,(c_+P + Q) & \text{for} \quad \varkappa \leqq 0, \\ bc_-(e^{-i\varkappa x}P + Q)\,(P + c_-^{-1}Q)\,(c_+P + Q) & \text{for} \quad \varkappa > 0. \end{cases} \end{aligned}$$

Due to 6.25 (iii) we have

$$\dim \ker (e^{-i\varkappa x}P + Q) = \max\{\varkappa, 0\},$$
$$\operatorname{codim} \operatorname{im} (e^{-i\varkappa x}P + Q) = \max\{-\varkappa, 0\}.$$

Obviously, the operators M_{bc_-}, $P + c_-^{-1}Q$, and $c_+P + Q$ are invertible, and so (2) follows.

Now suppose that A is a Φ_+-operator but (1) fails. For instance, let $x_0 \in [-\pi, \pi]$ be a point such that $a(x_0) = 0$. Then, first of all, there is a $\delta > 0$ such that $B \in \mathscr{L}(\mathscr{H}^s)$ and $\|A - B\|_d < \delta$ imply that $B \in \Phi_+(\mathscr{H}^s)$ (recall 1.3(c) and 6.25(i)).

Approximate a by a trigonometric polynomial $a_0 = \sum_{-n}^{n} c_k e^{ikx}$ so that $\|a - a_0\|_\infty < (\delta/2)\,\|P\|_d$ and put $A_0 = a_0P + bQ$. Obviously,

$$\|A - A_0\|_d < \delta/2, \qquad \|a_0(x_0)\, P\|_d < \delta/2.$$

This shows that $\|A - A_1\|_d < \delta$ where $A_1 = (a_0 - a_0(x_0))\,P + bQ$. Therefore, A_1 is in $\Phi_+(\mathcal{H}^s)$. But this is a contradiction to 6.25(iii) (recall that $A_1 \in \Phi_+(\mathcal{H}^s) \Leftrightarrow P(a_0 - a_0(x_0)) \times P + QbQ \in \Phi_+(\mathcal{H}^s) \Rightarrow (a_0 - a_0(x_0))\,P + Q \in \Phi_+(\mathcal{H}^s)$). The case $A \in \Phi_-(\mathcal{H}^s)$ can be treated analogously. ■

The analysis of this section can easily be extended to systems of singular integral equations. In particular, Theorem 6.26 remains valid with condition (1) replaced by the condition

$$\det a(x) \neq 0, \qquad \det b(x) \neq 0 \quad \text{for all} \quad x \in [-\pi, \pi]. \tag{3}$$

If (3) is fulfilled, then A is a Φ-operator with

$$\operatorname{ind} A = -\frac{1}{2\pi}\,[\arg \det (a/b)\,(x)]_{x=-\pi}^{\pi}.$$

Smoothness and asymptotics of solutions

6.27. Notation. In this section we study the smoothness of the solutions to the equation

$$(Au)\,(x) := a(x)\,u(x) + b(x)\,(Su)\,(x) = f(x) \tag{1}$$

and their asymptotic behavior at the end-points 0 and 1, where a and b are continuous functions on the interval $\mathbb{I} = [0, 1]$ and

$$(Su)\,(x) = \frac{1}{\pi \mathrm{i}} \int_0^1 \frac{u(y)}{y - x}\,\mathrm{d}y.$$

For analytic a, b and f, the complete asymptotics has been known for a long time (see MacCamy [1], Muskhelishvili [1]). In what follows we prove a more precise version of the complete asymptotics for a scale of Sobolev spaces which plays an important role in the error analysis of spline approximation methods for equation (1) (see Chapter 12).

Assume the operator $A \in \mathcal{L}(\mathbf{L}^2(\mathbb{I}))$ to be invertible. By virtue of Section 6.11 and the assertion 6.1(ix), this hypothesis is equivalent to the requirement that

$$a(x) \pm b(x) \neq 0, \qquad 0 \leqq x \leqq 1, \qquad \text{and wind } \Gamma = 0, \tag{2}$$

where Γ is the curve defined by

$$\Gamma = \left\{\frac{a(0) + b(0)}{a(0) - b(0)}\,\mu + (1 - \mu): 0 \leqq \mu \leqq 1\right\} \cup \left\{\frac{a(x) + b(x)}{a(x) - b(x)} : 0 \leqq x \leqq 1\right\}$$
$$\cup \left\{\mu + \frac{a(1) + b(1)}{a(1) - b(1)}\,(1 - \mu): 0 \leqq \mu \leqq 1\right\}.$$

Because of (2), we have $-1/2 < \operatorname{Re} \varkappa_j < 1/2$ ($j = 0, 1$), where

$$\varkappa_j := \frac{(-1)^j}{2\pi \mathrm{i}} \log \frac{a(j) + b(j)}{a(j) - b(j)}, \qquad j = 0, 1, \tag{3}$$

and log denotes that branch of the logarithm which is continuous in $\mathbb{C} \setminus (-\infty, 0]$ and takes real values on the positive real axis.

For a real number $k \geqq 0$, let $\mathbf{H}^k$ denote the usual Sobolev space of order k on $(0, 1)$ and $\|\cdot\|_k$ the norm in $\mathbf{H}^k$. Furthermore, let $\tilde{\mathbf{H}}^k$ $(k \geqq 0)$ be the closed subspace $\{u \in \mathbf{H}^k(\mathbb{R})$: $\operatorname{supp} u \subset [0, 1]\}$ of $\mathbf{H}^k(\mathbb{R})$. We introduce by duality (with respect to the $\mathbf{L}^2$ scalar product) $\mathbf{H}^{-k} = (\tilde{\mathbf{H}}^k)'$ and $\tilde{\mathbf{H}}^{-k} = (\mathbf{H}^k)'$, $k \geqq 0$.

The following properties can be found in LIONS/MAGENES [1] and ESKIN [1].

(i) $\mathbf{H}^k$ *and* $\tilde{\mathbf{H}}^k$ $(k \in \mathbb{R})$ *are Hilbert scales.*

(ii) $\tilde{\mathbf{H}}^k = \mathbf{H}^k$ *when* $|k| < 1/2$, *and* $\tilde{\mathbf{H}}^k$ *is a subset of* $\mathbf{H}^k$ *when* $k \geqq 1/2$. *The topology of* $\tilde{\mathbf{H}}^k$, *however, is stronger than that of* $\mathbf{H}^k$ *if* $k = m + 1/2$ *and* m *is an integer* $\geqq 0$.

(iii) *The operator* S *is a continuous linear map of* $\tilde{\mathbf{H}}^k$ *into* $\mathbf{H}^k$ *when* $k \geqq 0$: $S \in \mathcal{L}(\tilde{\mathbf{H}}^k, \mathbf{H}^k)$. *For* $k < 0$, *we have* $S \in \mathcal{L}(\mathbf{H}^k, \tilde{\mathbf{H}}^k)$.

We also need the following result.

6.28. Lemma. *For any* $\varkappa \in \mathbb{C}$ *and* $k \in \mathbb{R}$, $x^\varkappa I$, $(1 - x)^\varkappa I$ *and* $x^\varkappa(1 - x)^\varkappa I$ *are continuous maps of* $\tilde{\mathbf{H}}^k$ *into* $\tilde{\mathbf{H}}^{k+\min\{0,\mathrm{Re}\,\varkappa\}}$.

Proof. By Theorems 11.2, 11.3 and 11.8 in LIONS/MAGENES [1, Chap. 1], the lemma holds for $k = -\varkappa$ and $k \in [-1, 0)$ and for $k \in \mathbb{N}$ and $\varkappa = -1, \ldots, -k$. Consider the operator $x^\varkappa I$, for example; in the other cases the rest of the proof is analogous.

Let first $\varkappa < 0$. For $k = m - \varkappa$ and $m \in \mathbb{N}$, we may assume that $\varkappa \in [-1, 0)$ and by means of the relation

$$\mathrm{D}x^\varkappa = x^\varkappa \mathrm{D} + \varkappa x^{\varkappa-1} I, \qquad \mathrm{D} = \mathrm{d}/\mathrm{d}x, \tag{1}$$

we can successively reduce the assertion to the case $m = 0$ which is already clear. By interpolation we now obtain the lemma for all $k \geqq -\varkappa$ and by duality for all $k \leqq 0$. Finally, by interpolation we get the assertion for $k \in (0, -\varkappa)$, which completes the proof for $\varkappa < 0$. It remains to consider the case $\mathrm{Re}\,\varkappa \geqq 0$. Using the relation (1) again, we deduce by induction on k that $x^\varkappa I$ is a continuous map of $\tilde{\mathbf{H}}^k$ into itself for $k \in \mathbb{N}_0 = \mathbb{N} \cup \{0\}$. By interpolation one obtains the assertion for all $k \geqq 0$ and by duality for $k < 0$. ■

We now study the singular integral operator 6.27(1) in a scale of Sobolev spaces. Let $\varkappa_j$ $(j = 0, 1)$ be the numbers defined by 6.27(3).

6.29. Theorem. *Under the hypotheses* 6.27(2), *if* $a, b \in \mathbf{H}^1$ *and*

$$\mathrm{Re}\,\varkappa_j - 1/2 < s < \mathrm{Re}\,\varkappa_j + 1/2 \qquad (j = 0, 1), \tag{1}$$

then $A = aI + bS \in \mathcal{L}(\tilde{\mathbf{H}}^s, \mathbf{H}^s)$ *is invertible when* $s \geqq 0$, *while* $A \in \mathcal{L}(\mathbf{H}^s, \tilde{\mathbf{H}}^s)$ *is invertible when* $s < 0$.

6.30. Corollary. *Under the hypotheses of Theorem* 6.29, $u \in \mathbf{H}^0$ *and* $Au \in \mathbf{H}^s$ $(s > 0)$ *imply that* $u \in \tilde{\mathbf{H}}^s$.

6.31. Proof of Theorem 6.29. *Step 1.* Let $s > 0$. Choosing a function $\chi \in \mathrm{C}_0^\infty(\mathbb{R})$ such that $\chi(x) = 1$ for $x \in [0, 1/2]$ and $\chi(x) = 0$ for $x \geqq 2/3$, we set

$$\varphi_0(x) = \chi(x), \qquad \varphi_1(x) = \chi(1 - x). \tag{1}$$

We shall show that, under hypothesis 6.29(1) for $s > 0$, the unique $\mathbf{L}^2$ solution of the equation $Au = f \in \mathbf{H}^s$ satisfies $\varphi_j u \in \tilde{\mathbf{H}}^s$ $(j = 0, 1)$, which proves the theorem. For $\delta > 0$, set

$$\chi_\delta(x) = \chi(x/\delta) \quad \text{and} \quad u_\delta(x) = \chi_\delta(x)\, \tilde{u}(x),$$

where $\tilde{u}$ denotes the extension of u by 0 to $(1, \infty)$. Moreover, let

$$a_\delta(x) = \begin{cases} a(x), & x \in (0, \delta), \\ a(x)\,(2\delta - x)/\delta + a(0)\,(x - \delta)/\delta, & x \in [\delta, 2\delta], \\ a(0), & x \geqq 2\delta, \end{cases}$$

and define $b_\delta(x)$ in the same way by means of $b(x)$. Multiplying the equation $Au = f$ by χ_δ, we obtain

$$a_\delta(x)\, u_\delta(x) + b_\delta(x)\, S_{\mathbb{R}^+} u_\delta(x) = g_\delta(x), \qquad x \in (0, \infty), \;\; \delta \in (0, 1),$$

where $g_\delta(x) = \chi_\delta(x)\, \tilde{f}(x) + b_\delta(x) \int_0^1 [\chi_\delta(x) - \chi_\delta(y)]\,(x - y)^{-1} u(y)\, \mathrm{d}y$ and $S_{\mathbb{R}^+}$ is the Cauchy integral operator on the positive real axis. Since for $m = 0, 1$

$$\mathrm{D}^m \left\{ \int_0^1 [\chi_\delta(x) - \chi_\delta(y)]\,(x - y)^{-1}\, u(y)\, \mathrm{d}y \right\} = O(x^{-1}) \quad \text{as} \quad x \to +\infty,$$

we have $g_\delta \in \mathbf{H}^s(\mathbb{R}^+)$. By assumption 6.29(1), $a(0)\, I + b(0)\, S_{\mathbb{R}^+}$ is an isomorphic mapping of $\overline{\mathbf{H}}^s(\mathbb{R}^+) = \{u \in \mathbf{H}^s(\mathbb{R})\colon \operatorname{supp} u \subseteqq \overline{\mathbb{R}}^+\}$ onto $\mathbf{H}^s(\mathbb{R}^+)$ (see e.g. Eskin [1, Theorem 7.1]). We shall show that

$$\|c_\delta I + d_\delta S_{\mathbb{R}^+}\|_{\mathscr{L}(\overline{\mathbf{H}}^s(\mathbb{R}^+), \mathbf{H}^s(\mathbb{R}^+))} \to 0 \quad \text{as} \quad \delta \to 0, \tag{2}$$

where $c_\delta(x) = a_\delta(x) - a(0)$ and $d_\delta(x) = b_\delta(x) - b(0)$, which will imply that $u_\delta \in \overline{\mathbf{H}}^s(\mathbb{R}^+)$ for some sufficiently small δ, whence $\varphi_0 u \in \overline{\mathbf{H}}^s$. In order to check (2) we note that

$$\max_{x \in \mathbb{R}^+} |c_\delta(x)| \to 0, \qquad \int_{\mathbb{R}^+} |\mathrm{D} c_\delta|^2\, \mathrm{d}x \leqq c \quad (\delta \to 0),$$

which yields

$$\|c_\delta I\|_{\mathscr{L}(\mathbf{L}^2(\mathbb{R}^+))} \to 0, \qquad \|c_\delta I\|_{\mathscr{L}(\mathbf{H}^1(\mathbb{R}^+))} \leqq c \quad (\delta \to 0).$$

By interpolation, it follows from the last two inequalities that $\|c_\delta I\|_{\mathscr{L}(\mathbf{H}^s(\mathbb{R}^+))} \to 0$ as $\delta \to 0$. Since this holds with c_δ replaced by d_δ and since $I, S_{\mathbb{R}^+} \in \mathscr{L}\big(\overline{\mathbf{H}}^s(\mathbb{R}^+), \mathbf{H}^s(\mathbb{R}^+)\big)$, we finally obtain (2). The proof that $\varphi_1 u \in \overline{\mathbf{H}}^s$ is analogous.

Step 2. Let $s < 0$. We consider the adjoint $A^* = \bar{a} I + S \bar{b} I$ of A as a mapping of $\overline{\mathbf{H}}^{-s}$ into $\mathbf{H}^{-s}$. An easy modification of the proof in step 1 gives that A^* is invertible. Consequently, $A \in \mathscr{L}(\mathbf{H}^s, \overline{\mathbf{H}}^s)$ is invertible. ∎

Our aim now is to generalize Theorem 6.29 to the case where

$$\operatorname{Re} \varkappa_j + 1/2 + k - 1 < s < \operatorname{Re} \varkappa_j + 1/2 + k \quad (j = 0, 1), \;\; k \in \mathbb{N}, \tag{3}$$

which also gives us the complete asymptotics of solutions at the end-points. For this we need a sequence of preliminary lemmas.

Denote by $\log z$ the continuous branch of the logarithm in $\mathbb{C} \setminus \overline{\mathbb{R}}^+$ which takes real values $\log x$ as $z \to x \in \mathbb{R}^+$, $\operatorname{Im} z > 0$. For $\varkappa \notin \mathbb{Z}$, set $z^\varkappa = \mathrm{e}^{\varkappa \log z}$ $(z \in \mathbb{C} \setminus \overline{\mathbb{R}}^+)$ and $x^\varkappa = \mathrm{e}^{\varkappa \log x}$ $(x \in \mathbb{R}^+)$. Furthermore, with some $\varepsilon \in (0, 1)$, we set $U = (0, \varepsilon)$ and $V = \{z \in \mathbb{C}\colon 0 < |z| < \varepsilon\}$. Let us define the functions

$$\Omega_m^\varkappa(z) = (\pi \mathrm{i})^{-1} \int_0^1 y^\varkappa (\log y)^m\, (y - z)^{-1}\, \mathrm{d}y, \quad z \in \mathbb{C} \setminus \{0\}, \; \operatorname{Re} \varkappa > -1, \; m \in \mathbb{N}_0, \tag{4}$$

where the integral is to be interpreted as a Cauchy principal value when $z \in (0, 1]$. We first study the asymptotics of the functions (4) at the origin.

6.32. Lemma. *If $\varkappa \notin \mathbb{N}_0$, then there exist complex numbers $c^{\varkappa}_{mj}$ $(j = 0, \ldots, m)$ and a real-analytic function ψ in $\overline{U}$ such that*

$$\Omega^{\varkappa}_m(x) = \sum_{j=0}^{m} c^{\varkappa}_{mj} x^{\varkappa} \log^j x + \psi(x), \qquad x \in U. \tag{1}$$

In particular, $c^{\varkappa}_{mm} = \mathrm{i} \cot \pi\varkappa$ and $c^{\varkappa}_{mm-1} = -\pi\, \mathrm{i}m/\sin^2 \pi\varkappa$.

Proof. We proceed by induction on m. For $m = 0$, the assertion is proved in Muskhelishvili [1, § 23]. Assume that the representation (1) holds for all $m \leqq k - 1$, $k \in \mathbb{N}$. Setting

$$\Omega(z) = \Omega^{\varkappa}_k(z) + c_{k-1}\Omega^{\varkappa}_{k-1}(z) + \cdots + c_0\Omega^{\varkappa}_0(z)$$

with complex numbers $c_0, \ldots, c_{k-1}$, we obtain by the Sokhotski-Plemelj formulas (see e.g. Muskhelishvili [1, § 16])

$$\begin{aligned} \Omega^+(x) + \Omega^-(x) &= 2\Omega(x), \\ \Omega^+(x) - \Omega^-(x) &= 2x^{\varkappa}(\log^k x + c_{k-1} \log^{k-1} x + \cdots + c_0), \qquad x \in U, \end{aligned} \tag{2}$$

where Ω^+ and Ω^- denote the limits of $\Omega(z)$ as $z \to x$ such that $\operatorname{Im} z > 0$ and $\operatorname{Im} z < 0$, respectively. Moreover, we have

$$(x^{\varkappa} \log^k x)^+ - (x^{\varkappa} \log^k x)^- = x^{\varkappa}\{\log^k x - \mathrm{e}^{2\pi\mathrm{i}\varkappa}(\log x + 2\pi\mathrm{i})^k\}. \tag{3}$$

Choosing the numbers $c_0, \ldots, c_{k-1}$ such that

$$\mathrm{e}^{2\pi\mathrm{i}\varkappa}(1 - \mathrm{e}^{2\pi\mathrm{i}\varkappa})^{-1} [\log^k z - (\log z + 2\pi\mathrm{i})^k] = c_{k-1} \log^{k-1} z + \cdots + c_0$$

and setting $w(z) = (1 - \mathrm{e}^{2\pi\mathrm{i}\varkappa})^{-1}\, 2z^{\varkappa} \log^k z$, we deduce from (2) and (3) that

$$[\Omega - w]^+ = [\Omega - w]^- \quad \text{on} \quad U. \tag{4}$$

Note that $c_{k-1} = -2\pi\mathrm{i}k\, \mathrm{e}^{2\pi\mathrm{i}\varkappa}(1 - \mathrm{e}^{2\pi\mathrm{i}\varkappa})^{-1} = \pi k\, \mathrm{e}^{\pi\mathrm{i}\varkappa}/\sin \pi\varkappa$. Since $\operatorname{Re} \varkappa > -1$, one obtains the estimate

$$|\Omega(z) - w(z)| \leqq c\, |z|^{-r}, \qquad z \in V$$

with some $r < 1$ (see Muskhelishvili [1, § 23]), hence the origin is a removable singularity of $\Omega - w$ and so (4) implies that $\Omega(z) = w(z) + \psi_1(z)$, $z \in V$, with some analytic function ψ_1 in $\overline{V}$. Consequently,

$$\begin{aligned} \Omega(x) = (\Omega^+ + \Omega^-)/2 &= \mathrm{i} \cot \pi\varkappa \cdot x^{\varkappa} \log^k x \\ &+ (\mathrm{i}\, \mathrm{e}^{\pi\mathrm{i}\varkappa}/2 \sin \pi\varkappa)\, x^{\varkappa}[(\log x + 2\pi\mathrm{i})^k - \log^k x] + \psi_1(x), \qquad x \in U \end{aligned} \tag{5}$$

by virtue of the definition of w, (2) and the equality

$$(1 - \mathrm{e}^{2\pi\mathrm{i}\varkappa})^{-1}\, (1 + \mathrm{e}^{2\pi\mathrm{i}\varkappa}) = \mathrm{i} \cot \pi\varkappa .$$

Inserting the representations (1) for $\Omega^{\varkappa}_0, \ldots, \Omega^{\varkappa}_{k-1}$ and the relation

$$\Omega^{\varkappa}_k = \Omega - c_{k-1}\Omega^{\varkappa}_{k-1} - \cdots - c_0\Omega^{\varkappa}_0$$

into (5), we finally obtain (1) for $m = k$, where

$$c^{\varkappa}_{kk-1} = (-1)\, c_{k-1}(1 + \mathrm{i}\cot \pi\varkappa) = -\mathrm{i}\pi k/\sin^2 \pi\varkappa\,. \blacksquare$$

An inspection of the above proof shows that $c^{\varkappa+k}_{mj} = c^{\varkappa}_{mj}$ for all $\varkappa$, j, m and $k \in \mathbb{N}$. Similarly one can prove the following result.

6.33. Lemma. *If $\varkappa \in \mathbb{N}_0$, then there exist complex numbers c_{mj} ($j = 0, \ldots, m$) and a real-analytic function ψ in $\overline{U}$ such that*

$$\Omega^{\varkappa}_m(x) = \sum_{j=0}^{m} c_{mj} x^{\varkappa} \log^{j+1} x + \psi(x), \qquad x \in U, \quad \textit{where} \quad c_{mm} = -1/\pi\mathrm{i}(m+1)\,.$$

Let now $A \in \mathcal{L}(\mathbf{L}^2(\mathit{\Pi}))$ be an invertible operator of the form 6.27(1). Then $\varkappa_0 = 0$ if and only if $b(0) = 0$, and for $\varkappa_0 \neq 0$ we have

$$a(0) + b(0)\,\mathrm{i}\cot \pi\varkappa_0 = a(0) - b(0)\,(\mathrm{e}^{2\pi\mathrm{i}\varkappa_0} + 1)/(\mathrm{e}^{2\pi\mathrm{i}\varkappa_0} - 1) = 0\,. \tag{1}$$

Let further φ_0 and φ_1 be the cut-off-functions defined by 6.31(1).

6.34. Lemma. *Assume* 6.27(2) *and $a, b \in \mathbf{H}^{k+1}$, $k \in \mathbb{N}$. Then there exist functions $\psi^0_{k,l}(x) = x^{k-l}\eta_l(x)$ ($l = 1, \ldots, k$), where η_l is of the form*

$$\varphi_0(x)\, x^{\varkappa_0} \left(1 + \sum_{m=1}^{l-1} \sum_{j=1}^{m} \alpha_{mj} x^m \log^j x\right), \qquad \alpha_{mj} \in \mathbb{C}\,, \tag{1}$$

such that $A\psi^0_{k,l} \in \mathbf{H}^s$ for all $s < \operatorname{Re}\varkappa_0 + 1/2 + k$.

P r o o f. We shall write $u \sim v$ if $u - v \in \mathbf{C}^\infty[0, 1]$. Set

$$A_l = (a_0 + \cdots + a_l x^l)\, I + (b_0 + \cdots + b_l x^l)\, S\,,$$

where $a_l = a^{(l)}(0)/l!$ and $b_l = b^{(l)}(0)/l!$. Let us show by induction on l that there exist functions η_l of the form (1) satisfying

$$A_l\eta_l(x) \sim x^{\varkappa_0 + l} \sum_{m,j=0}^{l-1} \alpha_{lmj} x^m \log^j x \quad \text{if} \quad \varkappa_0 \neq 0\,, \tag{2}$$

$$A_l\eta_l(x) \sim x^l \sum_{m,j=0}^{l-1} \alpha_{lmj} x^m \log^{j+1} x \quad \text{if} \quad \varkappa_0 = 0\,, \tag{3}$$

with certain complex numbers α_{lmj}. Let first $\varkappa_0 \neq 0$. By Lemma 6.32 and 6.33(1), we obtain

$$A_1\varphi_0 x^{\varkappa_0} \sim a_1 x^{\varkappa_0 + 1} + b_1\,\mathrm{i}\cot \pi\varkappa_0 \cdot x^{\varkappa_0 + 1}\,,$$

which proves (2) for $l = 1$. Let $l \geqq 2$. By the induction hypothesis, we have

$$A_{l-1}\eta_{l-1} \sim x^{\varkappa_0 + l - 1} \sum_{m,j=0}^{l-2} \alpha_{l-1mj} x^m \log^j x\,,$$

which together with Lemma 6.32 implies that

$$A_l\eta_{l-1} \sim x^{\varkappa_0 + l - 1} \sum_{j=0}^{l-2} \alpha_{l-10j} \log^j x + x^{\varkappa_0 + l} \sum_{m,j=0}^{l-2} \alpha'_{mj} x^m \log^j x\,, \qquad \alpha'_{mj} \in \mathbb{C}\,.$$

Now we choose numbers $\alpha_{l-11}, \ldots, \alpha_{l-1l-1}$ such that

$$A_0\varphi_0\left(\sum_{j=1}^{l-1}\alpha_{l-1j}x^{\varkappa_0+l-1}\log^j x\right) \sim x^{\varkappa_0+l-1}\sum_{j=0}^{l-2}\alpha_{l-10j}\log^j x. \tag{4}$$

Since by 6.33(1), Lemma 6.32 and the remark following it

$$A_0\varphi_0 x^{\varkappa_0+l-1}\log^j x \sim \sum_{r=0}^{j-1} b_0 c_{jr}^{\varkappa_0} x^{\varkappa_0+l-1}\log^r x, \qquad j = 1, \ldots, l-1,$$

(4) leads to a linear system in the unknowns α_{l-1j} whose matrix is in lower triangular form with the diagonal elements $b_0c_{jj-1}^{\varkappa_0} \neq 0$. Therefore this system is uniquely solvable, and setting

$$\eta_l = \eta_{l-1} - \varphi_0\sum_{j=1}^{l-1}\alpha_{l-1j}x^{\varkappa_0+l-1}\log^j x,$$

we obtain (2).

Let now $\varkappa_0 = 0$. Then Lemma 6.33 implies that

$$A_1\varphi_0 \sim b_1c_{00}x\log x,$$

and hence (3) holds for $l = 1$. Using Lemma 6.33 and taking into account that $b_0 = 0$ and $a_0c_{jj} \neq 0$, one can carry out the induction step similarly to the case $\varkappa_0 \neq 0$.

Now it follows from (2) and (3) that $A_l\psi_{k,l}^0 \in \mathbf{H}^s$ for all $s < \operatorname{Re}\varkappa_0 + k + 1/2$. Indeed, if $\varkappa_0 \neq 0$ for example, then (2) implies that

$$A_l\psi_{k,l}^0 \sim x^{k-l}A_l\eta_l \sim x^{\varkappa_0+k}\sum_{m,j=0}^{l-1}\alpha_{lmj}x^m\log^j x,$$

and the right-hand side belongs to $\mathbf{H}^s$ whenever $s < \operatorname{Re}\varkappa_0 + k + 1/2$. Since by assumption and Lemma 6.28

$$x^{-j}[a(x) - a_0 - \cdots - a_lx^l] \in \mathbf{H}^{k+1-j}, \qquad j = 1, \ldots, l,$$

and the same is true with a replaced by b, we finally obtain with the help of Lemma 6.32 that $(A - A_l)\,\psi_{k,l}^0 \in \mathbf{H}^s$ for all $s < \operatorname{Re}\varkappa_0 + k + 1/2$, which completes the proof of the lemma. ■

In an analogous manner one can prove the next result.

6.35. Lemma. *Under the assumptions of Lemma* 6.34, *there exist functions* $\psi_{k,l}^1(x) = (1-x)^{k-l}\,\eta_l(1-x)$ $(l = 1, \ldots, k)$ *such that* $A\psi_{k,l}^1 \in \mathbf{H}^s$ *for all* $s < \operatorname{Re}\varkappa_1 + 1/2 + k$, *where* η_l *is of the form* 6.34(1) *with* $\varkappa_0$ *replaced by* $\varkappa_1$.

The functions $\psi_{k,l}^0$ *and* $\psi_{k,l}^1$ $(l = 1, \ldots, k)$ *are linearly independent and for their span,* L_k, *we have* $\overline{\mathbf{H}}^s \cap L_k = \{0\}$ *for all* s *satisfying* 6.31(3).

Let $\mathbf{H}_k^s = \overline{\mathbf{H}}^s + L_k$ for those s. We are now ready to prove the main theorem of this section.

6.36. Theorem. *Assume* 6.27(2), 6.31(3) *are in force and let* $a, b \in \mathbf{H}^{k+1}$, $k \in \mathbb{N}$. *Then* A *is an invertible mapping of* $\mathbf{H}_k^s$ *onto* $\mathbf{H}^s$.

Proof. Since $\dim \mathbf{H}_k^s/\overline{\mathbf{H}}^s = 2k$, it is sufficient to verify that $A \in \mathscr{L}(\overline{\mathbf{H}}^s, \mathbf{H}^s)$ is a Fredholm operator with index $-2k$. Let $s \geq k$ and consider $A \in \mathscr{L}(\overline{\mathbf{H}}^s, \mathbf{H}^s)$, $\mathrm{D}^k \in \mathscr{L}(\mathbf{H}^s, \mathbf{H}^{s-k})$.

Then we have

$$AD^k + T = D^kA \tag{1}$$

with the compact operator $T = [D^k, A] = [D^k, aI] + [D^k, bI] S$, the last equality following from the facts that $[D, S] u = 0$ for all $u \in C_0^\infty(0, 1)$ (see PRÖSSDORF [3], Chap. 6, Lemma 1.5]) and that C_0^∞ is dense in $\overline{\mathbf{H}}^s$.

Furthermore, $D^k \in \mathscr{L}(\mathbf{H}^s, \mathbf{H}^{s-k})$ and $D^k \in \mathscr{L}(\overline{\mathbf{H}}^s, \overline{\mathbf{H}}^{s-k})$ are Fredholm operators with index k and $-k$, respectively. Indeed, this is easily seen for an integer $s \geqq k$, and by interpolation one gets the result for any $s \geqq k$. Finally, by Theorem 6.29, $A : \overline{\mathbf{H}}^{s-k} \to \mathbf{H}^{s-k}$ is invertible, so $A : \overline{\mathbf{H}}^s \to \mathbf{H}^s$ has index $-2k$ in view of (1).

For $s < k$, the assertion can be reduced to Theorem 6.29 in an analogous manner. ■

6.37. Corollary. *Let $k \geqq 2$ be an integer. Under the assumptions 6.27(2), if $a, b \in \mathbf{H}^{k+1}$ and $f \in \mathbf{H}^k$, then the unique $\mathbf{L}^2$ solution of the equation $Au = f$ has the representation*

$$u(x) = \varphi_0(x)\, x^{\varkappa_0} \sum_{m=0}^{k-1} \sum_{j=0}^{m} c_{mj} x^m \log^j x + \varphi_1(x)\,(1 - x)^{\varkappa_1}$$
$$\times \sum_{m=0}^{k-2} \sum_{j=0}^{m} d_{mj} (1 - x)^m \log^j (1 - x) + x^{\varkappa_0}(1 - x)^{\varkappa_1}\, u_{k-1}(x), \tag{1}$$

where $c_{mj}, d_{mj} \in \mathbb{C}$ and $u_{k-1} \in \overline{\mathbf{H}}^s$ for all $s < k - 1/2$.

Proof. By Theorem 6.36 for $s = k$, the solution u admits the representation (1) with

$$u_{k-1}(x) = \tilde{\varphi}_0(x)\, x^{k-1} \sum_{j=0}^{k-1} c_{k-1j} \log^j x + \tilde{\varphi}_1(x)\,(1 - x)^{k-1} \sum_{j=0}^{k-1} d_{k-1j} \log^j (1 - x)$$
$$+ x^{-\varkappa_0}(1 - x)^{-\varkappa_1}\, v(x),$$

where $c_{k-1,j}, d_{k-1,j} \in \mathbb{C}$, $\tilde{\varphi}_0 = (1 - x)^{-\varkappa_1} \varphi_0$, $\tilde{\varphi}_1 = x^{-\varkappa_0}\varphi_1$ and $v \in \overline{\mathbf{H}}^k$. Using Lemma 6.28 and the fact that $\tilde{\varphi}_0 x^{k-1} \log^j x \in \overline{\mathbf{H}}^s$ $(s < k - 1/2)$, we see that $u_{k-1} \in \overline{\mathbf{H}}^s$ for all $s < k - 1/2$. ■

Remark. Similar formulas can be established in the case of piecewise continuous functions a, b and of weighted $\mathbf{L}^2$ spaces (cf. ELSCHNER [5]). It is not true, in general, that $u_{k-1} \in \mathbf{H}^s$ for some $s \geqq k - 1/2$ in (1). In the case of constant coefficients, however, an inspection of the proof of Lemma 6.34 shows that the logarithmic terms do not occur in (1). Setting $\varkappa = (2\pi i)^{-1} \log (a + b)/(a - b)$, we have $\varkappa_0 = \varkappa$, $\varkappa_1 = -\varkappa$ and thus obtain by Corollary 6.37 and Lemma 6.28 the following result.

6.38. Corollary. *Let $a, b \in \mathbb{C}$, $a + \lambda b \neq 0$, $-1 \leqq \lambda \leqq 1$, and let $k \geqq 2$ be an integer. If $f \in \mathbf{H}^k$, then the solution of equation 6.27(1) has the representation $u = x^\varkappa(1 - x)^{-\varkappa} u_{k-1}$, where $u_{k-1} \in \mathbf{H}^{k-|\mathrm{Re}\,\varkappa|}$.*

Pseudodifferential operators on a closed curve

This section is devoted to the study of some properties of pseudodifferential operators (ψdo's) on a closed curve. One of the main results is the Fourier series representation of ψdo's on the unit circle, which generalizes formula 6.25(0) for the Cauchy singular integral operator. This representation is due to AGRANOVITCH [2, 3] and has become extremely useful in applications. In particular, it has been exploited succesfully for the analysis of spline approximation methods for pseudodifferential equations on closed curves (see ARNOLD/WENDLAND [1, 2], SARANEN/WENDLAND [1], and Chapter 13 of this book). To begin with, for a ψdo A on the circle we define and study the global complete symbol, which is periodic in x (see 6.43(1)). The correspondence "ψdo $\leftrightarrow$ symbol" is not one-to-one, but the "degree of noninjectiveness" is described and turns out to be unessential. In what follows we extend this correspondence by considering a ψdo modulo the addition of any operator of order $-\infty$, and the symbol modulo the addition of functions from the class $\mathbf{CS}^{-\infty}(\mathbb{T})$ (defined in Sect. 6.41). Thus a symbol, periodic in x, is associated with every (classical) ψdo on $\mathbb{T}$, and we give a method of recovering this symbol from the ψdo. We then derive the representation of the ψdo A in terms of its symbol in the form of a Fourier type series instead of the usual Fourier type integral (see 6.51(1) and 6.61(1)). Note that this representation allows elementary proofs of theorems on the calculus of ψdo's. Further we consider a variety of examples. In particular, the class of ψdo's on a closed curve contains Fredholm integral equations of the second kind, certain first kind Fredholm equations (e.g. equations with logarithmic kernels), singular integral equations, involving Cauchy kernels and Hilbert kernels, integro-differential equations, and two-point boundary value problems for ordinary differential equations.

6.39. Notation. In order to take account of the peculiarities of a ψdo on the unit circle $\mathbb{T}$, we begin by recalling some facts and using some obvious arguments.

We shall use an atlas on $\mathbb{T}$ comprising local charts $U_{\alpha\beta}$ with the polar angle x as the local coordinate varying over an interval (α, β) with $\beta - \alpha \leqq 2\pi$. If $U_{\alpha\beta} \cap U_{\alpha'\beta'}$ is nonempty, then it consists of at most two connected components; in each of them the local coordinates are related by a translation.

For convenience, a function u on $\mathbb{T}$ will be identified with the corresponding 2π-periodic function $u(x)$ on $\mathbb{R}$. The Sobolev space $\mathbf{H}^t = \mathbf{H}^t(\mathbb{T})$, $t \in \mathbb{R}$, is defined as the completion of $\mathbf{C}^\infty(\mathbb{T})$ in the norm

$$\|u\|_t = \Big[\sum_{k\in\mathbb{Z}} (1 + |k|)^{2t}\, |\hat{u}_k|^2\Big]^{1/2}, \tag{1}$$

where

$$\hat{u}_k = \frac{1}{2\pi} \int_{-\pi}^{\pi} u(x)\, \mathrm{e}^{-\mathrm{i}kx}\, \mathrm{d}x, \qquad k \in \mathbb{Z}, \tag{2}$$

are the Fourier coefficients of the function u. Note that norm (1) is equivalent to norm 2.1(1). We remark that $\mathbf{C}^\infty = \bigcap_{t\in\mathbb{R}} \mathbf{H}^t$ and $\mathscr{D}' = \bigcup_{t\in\mathbb{R}} \mathbf{H}^t$. Let $r \in \mathbb{R}$. An operator A acting in $\mathbf{C}^\infty(\mathbb{T})$ is said to be an operator of *order* (at most) r (in the scale $\mathbf{H}^t$, $t \in \mathbb{R}$) if

$$\|Au\|_{t-r} \leqq C_t\, \|u\|_t \tag{3}$$

for all $t \in \mathbb{R}$, where C_t does not depend on $u \in \mathbf{C}^\infty(\mathbb{T})$. In this case we also say that A belongs to the class $\mathbf{OP}^r$. Any operator $A \in \mathbf{OP}^r$ can be extended to a bounded operator from $\mathbf{H}^t$ into $\mathbf{H}^{t-r}$ for all t, with (3) being preserved. If an operator A belongs to $\mathbf{OP}^r$ for all (negative) values of r, we call A an operator of *order* $-\infty$ and write $A \in \mathbf{OP}^{-\infty}$. We denote by a_{mn} the entries of the matrix of the operator $A \in \mathbf{OP}^r$ corresponding to the basis $\{e^{inx}\}_{n=-\infty}^{\infty}$, i.e.

$$a_{mn} = a_{mn}(A) = [A(e^{inx})]_m^\wedge. \tag{4}$$

6.40. Proposition. *Let $r < -1$ and $A \in \mathbf{OP}^r$. Then the function*

$$K(x, y) := (2\pi)^{-1} \sum_m e^{imx} \sum_n a_{mn} e^{-iny} \tag{1}$$

is continuous, and A is an integral operator with the kernel (1):

$$(Au)(x) = \int_{\alpha-\pi}^{\alpha+\pi} K(x, y)\, u(y)\, dy, \qquad u \in \mathbf{H}^0, \tag{2}$$

where α is arbitrary.

Proof. Inequality 6.39(3) can be rewritten in the form

$$\sum_m (1 + |m|)^{2(t+s)} \Big| \sum_k a_{mk} c_k \Big|^2 \leqq C_t^2 \sum_k (1 + |k|)^{2t} |c_k|^2, \tag{3}$$

where c_k is a rapidly decreasing sequence of reals, and $s = |r| > 1$. We prove first that (1) is a continuous function. In (3) we set $t = -s$ and $c_k = \delta_n^k$, multiply both sides by $(1 + |n|)^s$ and sum over n, to obtain

$$\sum_m \sum_n (1 + |n|)^s |a_{mn}|^2 \leqq C^{(1)}.$$

By Schwarz' inequality we have

$$\sum_n |a_{mn}| \leqq \Big[\sum_n (1 + |n|)^s |a_{mn}|^2 \cdot \sum_n (1 + |n|)^{-s}\Big]^{1/2} \leqq C^{(2)}, \tag{4}$$

so that the function

$$h_m(y) = \sum_n a_{mn} e^{-iny}$$

is continuous for all y. Setting $t = -s/2$ and $c_k = e^{-iky}$ (with y fixed) in (3), we obtain $\sum_m (1 + |m|)^s |h_m(y)|^2 \leqq C^{(3)}$. Hence again by Schwarz' inequality

$$\sum_m |h_m(y)| \leqq C^{(4)}, \tag{5}$$

so that (1) defines for all (x, y) a function $K(x, y)$ which is continuous in x for each y. We now estimate $|K(x, y) - K(x, y')|$. By setting in (3) $t = -s/2$ and $c_k = e^{-iky} - e^{-iky'}$ and taking $\varepsilon = \min\{1, (s - 1)/4\}$, we obtain

$$\sum_m (1 + |m|)^s |h_m(y) - h_m(y')|^2 \leqq C^{(5)} \sum_k \frac{|e^{-iky} - e^{-iky'}|^2}{(1 + |k|)^s}$$

$$\leqq C^{(5)} 2^{2-2\varepsilon} \sum_k \frac{|e^{-iky} - e^{-iky'}|^{2\varepsilon}}{(1 + |k|)^s} \leqq C^{(5)} 2^{2-\varepsilon} \sum_k \frac{|k|^{2\varepsilon}}{(1 + |k|)^s} |y - y'|^{2\varepsilon}$$

by Lagrange's mean value theorem. The right-hand side is at most $C^{(6)}\,|y-y'|^{2\varepsilon}$. Using Schwarz' inequality once more, we get

$$|K(x,y)-K(x,y')| \leqq \sum |h_m(y)-h_m(y')| \leqq C^{(7)}\,|y-y'|^{\varepsilon}.$$

Thus we see that $K(x, y)$ is continuous.

For $u \in \mathbf{C}^\infty(\mathbb{T})$ we have, using (4),

$$(Au)^\wedge_m = \sum_n a_{mn}\hat{u}_n = (2\pi)^{-1}\int_{\alpha-\pi}^{\alpha+\pi} h_m(y)\,u(y)\,\mathrm{d}y;$$

we now use (5) and Lebesgue's convergence theorem to obtain (2). It remains to carry over (2) to the case where $u \in \mathbf{H}^0$ by passing to the limit. ■

If $r < -2$, then by taking in (3) $t = r/2$ and $c_k = \delta_n^k$ we obtain, after applying Schwarz' inequality and summing over n,

$$\sum_{m,n} |a_{mn}| < \infty; \tag{6}$$

in this case the order of summation in (1) makes no difference. It is not difficult to verify also that A has order $-\infty$ if and only if the a_{mn} decrease rapidly:

$$|a_{mn}| \leqq C_k(1+|m|)^{-k}\,(1+|n|)^{-k} \tag{7}$$

for every k, where the C_k do not depend on m or n. This is equivalent to the requirement that $K(x, y) \in \mathbf{C}^\infty$.

6.41. Definition. Let $r \in \mathbb{R}$. We define the class $\mathbf{C}_0\mathbf{S}^r(\alpha, \beta)$ as the set of all functions $a(x, \xi) \in \mathbf{C}^\infty[(\alpha, \beta) \times \mathbb{R}]$ such that a is equal to 0 for x sufficiently close to α or β,

$$|\partial_x^p\,\partial_\xi^q a(x,\xi)| \leqq C_{pq}(1+|\xi|)^{r-q} \tag{1}$$

and there exist functions $a_j(x, \xi) \in \mathbf{C}^\infty[(\alpha, \beta) \times (\mathbb{R} \setminus \{0\})]$, $j \in \mathbb{N}_0$, which are positively homogeneous of degree $r - j$ in ξ, vanish when x is sufficiently close to α or β, and satisfy the estimates

$$|\partial_x^p\partial_\xi^q[a(x,\xi) - \sum_{j=0}^{N} a_j(x,\xi)]| \leqq C_{pqN}\,|\xi|^{r-q-N-1} \tag{2}$$

for every $p, q, N \in \mathbb{N}_0$ and $|\xi| \geqq 1$, the constants C_{pq} and C_{pqN} being independent of (x, ξ). Here and in the following we use the abbreviations $\partial_x = \partial/\partial x$, $\partial_\xi = \partial/\partial\xi$.

The fact that (2) holds will be simply expressed by writing the asymptotic expansion

$$a(x,\xi) \approx a_0(x,\xi) + a_1(x,\xi) + \cdots \tag{3}$$

as $|\xi| \to \infty$.

We denote the intersection of all the classes $\mathbf{C}_0\mathbf{S}^r(\alpha, \beta)$ by $\mathbf{C}_0\mathbf{S}^{-\infty}(\alpha, \beta)$. The functions $a \in \mathbf{C}_0\mathbf{S}^{-\infty}(\alpha, \beta)$ are characterized by the inequalities

$$|\partial_x^p\partial_\xi^q a(x,\xi)| \leqq C_{pqN}(1+|\xi|)^{-N}\ (p, q, N \in \mathbb{N}_0). \tag{4}$$

In this case, all the terms of an expansion of the form (3) vanish. In analogy to the class $\mathbf{C}_0\mathbf{S}^r(\alpha, \beta)$, we now define $\mathbf{CS}^r(\mathbb{T})$ to be the class of functions $a(x, \xi) \in \mathbf{C}^\infty(\mathbb{R} \times \mathbb{R})$

which are 2π-periodic in x, satisfy (1), and admit an asymptotic expansion (3) in the sense of (2) for $x \in \mathbb{R}$, where the $a_j(x, \xi) \in \mathbf{C}^\infty[\mathbb{R} \times (\mathbb{R} \setminus \{0\})]$, $j \in \mathbb{N}_0$, are 2π-periodic in x and positively homogeneous of degree $r - j$ in ξ. The intersection of all spaces $\mathbf{CS}^r(\mathbb{T})$ will be denoted by $\mathbf{CS}^{-\infty}(\mathbb{T})$.

6.42. Definition of a ψdo on $\mathbb{T}$. A is called a (classical) ψdo *of order* $r \in \mathbb{R}$ *on* $\mathbb{T}$ if, for any functions $\varphi, \psi \in \mathbf{C}^\infty(\mathbb{T})$ the union of whose supports is not all of $\mathbb{T}$ and for any neighborhood $U = U_{\alpha\beta}$ containing these supports, we have

$$[\varphi A(\psi u)](x) = (2\pi)^{-1} \int e^{i\xi x} a_{U\varphi\psi}(x, \xi)\, \hat{u}_0(\xi)\, d\xi, \qquad u \in \mathbf{C}^\infty(\mathbb{T}), \quad x \in (\alpha, \beta), \tag{1}$$

where

$$a_{U\varphi\psi} \in \mathbf{C}_0\mathbf{S}^r(\alpha, \beta),$$

and $u_0 \in \mathbf{C}_0^\infty(\mathbb{R})$ is any function which is equal to u on $\operatorname{supp} \psi \cap (\alpha, \beta)$, and $\hat{u}_0(\xi)$ is its Fourier transform,

$$\hat{u}_0(\xi) = \int e^{-iy\xi} u_0(y)\, dy. \tag{2}$$

The function $a_{U\varphi\psi}$ is called the *symbol of the operator* $\varphi A(\psi\cdot)$ in U. If the supports of φ and ψ are disjoint, then $a_{U\varphi\psi} \in \mathbf{C}_0\mathbf{S}^{-\infty}(\alpha, \beta)$ and all the terms of an expansion of the form 6.41(3) for $a_{U\varphi\psi}$ vanish. In this case, the operator $\varphi A(\psi\cdot)$ has an infinitely smooth kernel and is of order $-\infty$.

If we take a finite partition of unity

$$\sum \varphi_\mu \equiv 1 \tag{3}$$

on $\mathbb{T}$, where $\varphi_\mu \in \mathbf{C}^\infty(\mathbb{T})$ and the union of the supports of any two terms is not the whole of $\mathbb{T}$, then we can write

$$A = \sum \varphi_\mu A(\varphi_\nu\cdot); \tag{4}$$

here, each term of the right-hand side in (4) can be represented in the form (1).

6.43. Definition of the global symbol. We call a function $a(x, \xi)$ a *symbol* or, more precisely, an *ε-symbol* of the ψdo A of order r on $\mathbb{T}$ ($\varepsilon > 0$), if $a \in \mathbf{CS}^r(\mathbb{T})$ and if for all $u \in \mathbf{C}^\infty(\mathbb{T})$ with support in any coordinate neighborhood $U = U_{\alpha\beta}$ of length $\beta - \alpha = 2\pi - \varepsilon$,

$$(Au)(x) = (2\pi)^{-1} \int e^{ix\xi} a(x, \xi)\, \hat{u}_0(\xi)\, d\xi, \qquad x \in (\alpha, \beta), \tag{1}$$

where u_0 is a function on $\mathbb{R}$ equal to u on (α, β) and 0 outside (α, β). The term a_0 in the expansion 6.41(3) is called the *principal symbol* of A.

6.44. Construction of the symbol. The case $r < -1$. We now discuss a method o constructing an ε-symbol.

Let $K(x, y)$ be the kernel of a ψdo A of order r. Since $r < -1$, it is continuous (Proposition 6.40), and since A is a ψdo it is $\mathbf{C}^\infty$ for $x \neq y \bmod 2\pi$ (see e.g. SHUBIN [1] or TREVES [1]). Let $L(x, t)$ be any function on $\mathbb{R} \times \mathbb{R}$ with the following properties:

(a) $L(x, t) = K(x, x - t)$ for $|t| \leqq 2\pi - \varepsilon$.

(b) $L(x, t) \in \mathbf{C}^\infty\big(\mathbb{R} \times (\mathbb{R} \setminus \{0\})\big)$.

(c) For any non negative integers p and q the function $\partial_x^p \partial_t^q L(x, t)$ is 2π-period ic in$x$ and decreases rapidly as $|t| \to \infty$, uniformly in x:

$$|\partial_x^p\, \partial_t^q\, L(x, t)| \leqq C_{pqN}\, |t|^{-N} \qquad (|t| \geqq 1) \tag{1}$$

for every natural number N, where C_{pqN} does not depend on (x, t). We define $a(x, \xi)$ to be the Fourier transform of $L(x, t)$ with respect to t:

$$a(x, \xi) = \int e^{-i\xi t} L(x, t) \, dt. \tag{2}$$

For instance, we can take $L(x, t) = K(x, x - t) \, \varrho(t)$, where $\varrho(t)$ is a function from $C_0^\infty(\mathbb{R})$ which is equal to 1 when $|t| \leqq 2\pi - \varepsilon$ and to 0 when $|t| \geqq 2\pi - \varepsilon/2$; then

$$a(x, \xi) = \int e^{i(y-x)\xi} \, K(x, y) \, \varrho(x - y) \, dy.$$

Notice that this formula is analogous to the definition of the symbol of the nonsingular ψdo in $\mathbb{R}^n$ obtained from a singular ψdo by "correcting" the kernel of the latter outside a neighborhood of the diagonal in $\mathbb{R}^n \times \mathbb{R}^n$ (see Shubin [1], Treves [1]).

6.45. Theorem. (i) *Any ε-symbol of the* ψdo A *has the form* 6.44(2).
(ii) *Any function of the form* 6.44(2) *is an ε-symbol of the operator* A.

Proof. (i) Let $a(x, \xi)$ be an ε-symbol of the operator A and let $L(x, t)$ be its inverse Fourier transform:

$$L(x, t) = (2\pi)^{-1} \int e^{i\xi t} a(x, \xi) \, d\xi. \tag{1}$$

The integral converges absolutely since $r < -1$. When $t \neq 0$,

$$L(x, t) = (2\pi)^{-1} (-it)^{-k} \int e^{i\xi t} \partial_\xi^k a(x, \xi) \, d\xi \qquad (k = 1, 2, \ldots); \tag{2}$$

whence, clearly, L has properties (b) and (c). Let $U = U_{\alpha\beta}$, $u(x)$ and $u_0(x)$ be as in the definition of an ε-symbol. Substituting 6.42(2) into 6.43(1) and changing the order of integration with respect to y and ξ, we obtain

$$(Au)(x) = \int_\alpha^\beta L(x, x - y) \, u(y) \, dy \quad \text{for} \quad x \in (\alpha, \beta),$$

so that $K(x, y) = L(x, x - y)$ for $x, y \in (\alpha, \beta)$; that is, condition (a) holds.

(ii) Suppose that $a(x, \xi)$ is defined by 6.44(2); let U, u and u_0 be as in the definition of an ε-symbol. For $x \in (\alpha, \beta)$,

$$(Au)(x) = \int K(x, y) \, u_0(y) \, dy = \int L(x, x - y) \, u_0(y) \, dy.$$

We substitute

$$u_0(y) = (2\pi)^{-1} \int e^{i\xi y} \hat{u}_0(\xi) \, d\xi \tag{3}$$

in the right-hand side, change the order of integration with respect to ξ and y, and substitute $x - y = t$ in the inner integral. We then obtain 6.43(1).

It remains to verify that $a(x, \xi) \in \mathbf{CS}^r(\mathbb{T})$, using the fact that A is a ψ do of order r on $\mathbb{T}$. It suffices to do this for the operator $\varphi_\mu A(\varphi_\nu \cdot)$ rather than A. We retain for this case the notation $K(x, y)$, $L(x, t)$, and 6.44(2). The latter two functions as well as the first are 2π-periodic in x. If the supports of φ_μ and φ_ν are disjoint then $K \in \mathbf{C}^\infty(\mathbb{R} \times \mathbb{R})$, $L(x, t) \in \mathbf{CS}^{-\infty}(\mathbb{T})$, and $a \in \mathbf{CS}^{-\infty}(\mathbb{T})$. Next suppose that the supports of φ_μ and φ_ν have a nonempty intersection; we can assume that they lie in a neighborhood $U = U_{\alpha\beta}$ of length $2\pi - \varepsilon$. The functions $a(x, \xi)$, $K(x, y)$, and $L(x, t)$ all have $\varphi_\mu(x)$ as a factor, and therefore vanish when $x \notin \operatorname{supp} \varphi(x)$. Hence it suffices to show that $a \in \mathbf{C}_0\mathbf{S}^r(\alpha, \beta)$. For this, it is enough to verify that

$$a(x, \xi) - a^{(1)}(x, \xi) \in \mathbf{C}_0\mathbf{S}^{-\infty}(\alpha, \beta), \tag{4}$$

where $a^{(1)}$ is the symbol of $\varphi_\mu A(\varphi_\nu \cdot)$ in U. We go over from symbols to kernels. If in the formula of type 6.42(1) for the operator $\varphi_\mu A(\varphi_\nu \cdot)$ we substitute 6.42(2) and change the order of integration with respect to y and ξ, we obtain

$$\varphi_\mu A(\varphi_\nu u)(x) = \int K^{(1)}(x, y)\, u_0(y)\, \mathrm{d}y, \qquad x \in (\alpha, \beta), \tag{5}$$

where $u_0(y)$ is any function in $\mathbf{C}_0^\infty(\mathbb{R})$ equal to $u(y)$ on $Y = (\alpha, \beta) \cap \operatorname{supp} \varphi_\nu(y)$, and

$$K^{(1)}(x, y) = (2\pi)^{-1} \int \mathrm{e}^{\mathrm{i}(x-y)\xi}\, a^{(1)}(x, \xi)\, \mathrm{d}\xi.$$

Because of this arbitrariness of choice, $K^{(1)}(x, y)$ vanishes when $y \notin Y$ and $x \in (\alpha, \beta)$, and we can replace $u_0(y)$ in (5) by $u(y)$. Hence $K(x, y) = K^{(1)}(x, y)$ when $x \in (\alpha, \beta)$ and $y \in (\alpha - \varepsilon, \beta + \varepsilon)$. Then

$$L(x, t) = K^{(1)}(x, x - t), \tag{6}$$

at least for $x \in (\alpha, \beta)$ and $|t| < \varepsilon$. Both sides of (6) belong to $\mathbf{C}^\infty$ when $t \neq 0$ (cf. (2)); the right-hand side vanishes when $|t|$ is large and when $x \in (\alpha, \beta) \setminus \operatorname{supp} \varphi_\mu(x)$. It is now easily verified that the difference between the right and left sides of (6) belongs to $\mathbf{C}_0\mathbf{S}^{-\infty}(\alpha, \beta)$. By going over to the Fourier transform we obtain (4). ■

Remark. We explain the role of the positive number ε in the definition of the symbol for $r < -1$. If 6.43(1) were to hold in neighborhoods of length 2π, then we would have $L(x, t) = K(x, x - t)$ for $|t| \leqq 2\pi$. In this case the function $L(x, t)$ would not be infinitely smooth at $t = \pm 2\pi$, which is inconsistent with the fact that $a(x, \xi)$ belongs to $\mathbf{CS}^r(\mathbb{T})$.

6.46. Corollary. *Let $a^{(k)}(x, \xi) \in \mathbf{CS}^r(\mathbb{T})$ $(k = 1, 2)$ and let $a^{(1)}$ be the symbol of the* ψdo *A of order r on $\mathbb{T}$. Then $a^{(2)}$ is also a symbol of this* ψdo *if and only if the inverse Fourier transforms $L^{(k)}(x, t)$ $(k = 1, 2)$ of these functions coincide for $|t| \leqq \delta$ with some $\delta > 0$. In this case, in particular, $a^{(1)} - a^{(2)} \in \mathbf{CS}^{-\infty}(\mathbb{T})$.*

6.47. Corollary. *Let A_1 and A_2 be two* ψdo's *of order r on $\mathbb{T}$ and let $a(x, \xi)$ be the symbol of A_1. Then $a(x, \xi)$ is also a symbol of A_2 if and only if the kernels $K_j(x, y)$ $(j = 1, 2)$ of these* ψdo's *coincide for $|x - y| < \delta$ with some $\delta > 0$. In this case, in particular, $A_1 - A_2$ is an operator of order $-\infty$.*

6.48. The case $r \geqq -1$. Let $r = s + 2l$, where $s < -1$ and l is a natural number, for definiteness the least possible such number. Then $A = \hat{A}(1 + \mathbf{D}^2)^l$, where $\hat{A}$ is a classical ψdo of order s and $\mathbf{D} = -i\partial/\partial x$. The general method for constructing its ε-symbol $\hat{a}(x, \xi)$ has already been described. We obtain the following general formula for the ε-symbol of A:

$$a(x, \xi) = \hat{a}(x, \xi)\,(1 + \xi^2)^l. \tag{1}$$

Corollaries 6.46 and 6.47 remain valid for the case $r \geqq -1$ provided we make the following corrections: the $L^{(k)}$ have to be the inverse Fourier transforms of the $\hat{a}_k$, and the K_j the kernels of the operators $\hat{A}_j$.

6.49. Theorem. *Any function $a(x, \xi)$ from $\mathbf{CS}^r(\mathbb{T})$ is an ε-symbol of some* ψdo *of order r on $\mathbb{T}$, at least when $\pi < \varepsilon < 2\pi$.*

Proof. It suffices to verify this for $r < -1$. Let $a(x, \xi) \in \mathbf{CS}^r(\mathbb{T})$ and $\varepsilon \in (\pi, 2\pi)$. Defining $L(x, t)$ by 6.45(1), we set $L_1(x, t) = L(x, t)\,\alpha(t)$ for $|t| \leqq \pi$, where $\alpha(t)$ is a function from $C^\infty[-\pi, \pi]$ equal to 1 when $|t| \leqq 2\pi - \varepsilon$ $(< \pi)$ and to 0 when t is close to π or $-\pi$. We extend $L_1(x, t)$ to a function on $\mathbb{R} \times \mathbb{R}$ which is 2π-periodic in t, and set $K(x, y) = L_1(x, x - y)$. Let A be an operator on $\mathbb{T}$ with the kernel $K(x, y)$. It is not difficult to verify that A is a classical ψdo with ε-symbol $a(x, \xi)$. ■

6.50. Remarks. 1°. When $\varepsilon < \pi$, the ψdo A can be recovered uniquely from its ε-symbol: its kernel $K(x, y)$, if $r < -1$, and the kernel of $\hat{A}$, if $r \geqq -1$, turns out to be defined for all x and $|y - x| \leqq \pi$.

2°. It is clear that any operator of order $-\infty$ can be considered as a ψdo with a symbol of class $\mathbf{CS}^{-\infty}(\mathbb{T})$, and any function from this class can be considered to be a symbol of some ψdo of order $-\infty$. Since such operators and functions are not discerned by symbolic calculus, we now extend the above correspondence between ψdo's and their symbols. Namely, let A be a ψdo of order r with a as its ε-symbol, T any operator of order $-\infty$, and a_∞ any function from $\mathbf{CS}^{-\infty}(\mathbb{T})$. From now on, we shall call $a + a_\infty$ the symbol of the ψdo $A + T$ and we shall write smb $(A + T) = a + a_\infty$ and $A + T \in \mathbf{OPCS}^r$.

6.51. Theorem. (Representation of a ψdo on the circle by a Fourier type series). *Let A be a ψdo of order r on $\mathbb{T}$ with symbol $a(x, \xi)$. Then there exists an operator T of order $-\infty$ such that*

$$(A - T)\, u(x) = \sum_{n=-\infty}^{\infty} e^{inx} a(x, n)\, \hat{u}_n \quad \textit{for} \quad u \in C^\infty(\mathbb{T}). \tag{1}$$

Proof. Assume $r < -1$. Let $K(x, y)$ be the kernel of A and $\varrho(t)$ a function from $C_0^\infty(\mathbb{R})$ equal to 1 when $|t| < \delta_1$ and to 0 when $|t| \geqq \delta_2$, where $0 < \delta_1 < \delta_2 < \pi$. We set

$$(A_1 u)(x) = \int K(x, y)\, \varrho(x - y)\, u(y)\, dy, \qquad T_1 = A - A_1.$$

Choosing α in 6.40(2) close to x, it is fairly easy to check that T_1 is an operator of order $-\infty$. The function

$$a^{(1)}(x, \xi) = \int e^{i(y-x)\xi} K(x, y)\, \varrho(x - y)\, dy$$

is a $(2\pi - \delta_1)$-symbol of A. Also $a^{(1)}(x, n) = e^{-inx} A_1(e^{inx})$; by applying the operator A_1 term by term to the Fourier series of the function $u(x) \in C^\infty(\mathbb{T})$ we obtain

$$(A_1 u)(x) = \sum e^{inx} a^{(1)}(x, n)\, \hat{u}_n.$$

The operator

$$T_2 u(x) = \sum e^{inx} [a(x, n) - a^{(1)}(x, n)]\, \hat{u}_n$$

has the kernel

$$(2\pi)^{-1} \sum e^{in(x-y)} [a(x, n) - a^{(1)}(x, n)],$$

belonging to the class C^∞, since $a(x, \xi) - a^{(1)}(x, \xi) \in \mathbf{CS}^{-\infty}(\mathbb{T})$. Hence T_2 has order $-\infty$, and it remains to set $T = T_1 - T_2$.

The passage to the case $r \geqq -1$ is dealt with as before. ■

6.52. Corollary. *Let $a(x, \xi)$ belong to $\mathbf{CS}^r(\mathbb{T})$. Then the formula*

$$(Au)(x) = \sum e^{inx} a(x, n)\, \hat{u}_n \tag{1}$$

defines (one of the) ψdo's with symbol $a(x, \xi)$.

6.53. Definition of a ψdo on a closed curve. Let Γ be a simple closed $\mathbf{C}^\infty$ curve in the complex plane given by a regular parametric representation $z = \gamma(x)$, $x \in [0, 2\pi]$, i.e. $\gamma \in \mathbf{C}^\infty$, $\gamma'(x) \neq 0$, $\gamma(x) \neq \gamma(y)$ for $x \neq y$, $x, y \in [0, 2\pi)$ and $\gamma(0) = \gamma(2\pi)$. If $v: \Gamma \to \mathbb{C}$ is a function on Γ, then the function $u = K_\gamma v$ defined on $\mathbb{R}$ by $K_\gamma v(x) = v(\gamma(x))$ is 2π-periodic. By the definition of the norm in $\mathbf{H}^t(\Gamma)$ $(t \in \mathbb{R})$ (see Sec. 2.1), K_γ is an isometric isomorphism from $\mathbf{H}^t(\Gamma)$ onto $\mathbf{H}^t = \mathbf{H}^t(\mathbb{T})$.

A is called a ψdo *of order* $r \in \mathbb{R}$ *on* Γ if $K_\gamma A K_\gamma^{-1}$ is a ψdo of order r on $\mathbb{T}$ (see Definition 6.42). If $a(x, \xi)$ is the symbol of $K_\gamma A K_\gamma^{-1}$, then $a(z, \xi) := a(x, \xi)$ is said to be the *symbol* of the ψdo A on Γ.

6.54. Smoothing operators. An integral operator of the form

$$(Kv)\,(t) = \int_\Gamma k(t, \tau)\, v(\tau)\, d\tau, \qquad k \in \mathbf{C}^\infty(\Gamma \times \Gamma),$$

is called a *smoothing operator* on Γ. This is an operator of order $-\infty$ (in the scale $\mathbf{H}^t(\Gamma)$, $t \in \mathbb{R}$).

Indeed, $\mathsf{K} := K_\gamma K K_\gamma^{-1}$ is an integral operator with kernel $\mathsf{k}(x, y) := k(\gamma(x), \gamma(y)) \times \gamma'(y)$, i.e.

$$(\mathsf{K}u)\,(x) = \int_{-\pi}^{\pi} \mathsf{k}(x, y)\, u(y)\, dy. \tag{1}$$

Since $\mathrm{D}^l\mathsf{K}$ is an integral operator with kernel $\mathrm{D}_x^l\mathsf{k}$, we obviously have $\mathsf{K} \in \mathscr{L}(\mathbf{H}^0, \mathbf{H}^l)$ for all $l \in \mathbb{N}_0$. Let K^* be the integral operator with kernel $\overline{\mathsf{k}(y, x)}$. Using that

$$\langle \mathsf{K}u, w\rangle_0 = \langle u, \mathsf{K}^* w\rangle_0, \qquad u, w \in \mathbf{C}^\infty,$$

and taking into account 2.1(6), we obtain that $\mathsf{K} \in \mathscr{L}(\mathbf{H}^{-l}, \mathbf{H}^0)$ for all $l \in \mathbb{N}_0$. Thus $\mathsf{K} \in \mathscr{L}(\mathbf{H}^l, \mathbf{H}^{l-r})$ for all $l, r \in \mathbb{Z}$, and by interpolation, this is valid for any $l, r \in \mathbb{R}$. ∎

Remark. If a linear mapping K: $\mathbf{C}^\infty(\Gamma) \to \mathbf{C}^\infty(\Gamma)$ has continuous extensions $K \in \mathscr{L}(\mathbf{H}^l(\Gamma), \mathbf{H}^{l-r}(\Gamma))$ for all $l, r \in \mathbb{R}$, then K is a smoothing operator. This is a consequence of the Schwartz kernel theorem (see e.g. Gelfand/Vilenkin [1] and Sec. 6.40).

6.55. Example. The Cauchy singular integral operator on Γ (see Sec. 6.1),

$$(S_\Gamma v)\,(t) = \frac{1}{\pi \mathrm{i}} \int_\Gamma \frac{v(\tau)}{\tau - t}\, d\tau, \qquad t \in \Gamma,$$

is a ψdo of order 0 on Γ with symbol $\xi/|\xi|$. In the case of the unit circle $\mathbb{T} = \{z = \mathrm{e}^{\mathrm{i}x}: x \in [0, 2\pi]\}$ this is an immediate consequence of the formula

$$(S_{\mathbb{T}} u)\,(z) = \sum_{n \geq 0} \hat{u}_n z^n - \sum_{n<0} \hat{u}_n z^n, \qquad u \in \mathbf{C}^\infty(\mathbb{T}) \tag{1}$$

(see Sec. 6.1).

In the general case $K_\gamma S_\Gamma K_\gamma^{-1}$ is an integral operator of the form 6.54(1) with the singular kernel

$$k(x, y) = \frac{1}{\pi \mathrm{i}} \frac{\gamma'(y)}{\gamma(y) - \gamma(x)}$$

(see e.g. MIKHLIN/PRÖSSDORF [1, Theorem II.1.2]). Hence

$$K_\gamma S_\Gamma K_\gamma^{-1} = S_{\mathbb{T}} + K_1\,, \tag{2}$$

where K_1 is a smoothing operator on $\mathbb{T}$ whose kernel is

$$k_1(x, y) = \frac{1}{\pi \mathrm{i}} \left\{ \frac{\gamma'(y)}{\gamma(y) - \gamma(x)} - \frac{\mathrm{i}\, \mathrm{e}^{\mathrm{i}y}}{\mathrm{e}^{\mathrm{i}y} - \mathrm{e}^{\mathrm{i}x}} \right\} \in \mathbf{C}^\infty(\mathbb{T} \times \mathbb{T})\,.$$

Thus the assertion follows from (2) and Sec. 6.54. ■

6.56. The Hilbert singular integral operator. This is the operator defined by

$$(Hu)\,(x) = \frac{1}{2\pi} \int_0^{2\pi} \cot \frac{y - x}{2}\, u(y)\, \mathrm{d}y\,, \qquad u \in \mathbf{C}^\infty(\mathbb{T})\,, \tag{1}$$

where the integral is of course to be understood in the sense of the Cauchy principal value. The operator H is a ψdo of order 0 on $\mathbb{T}$ with symbol $\mathrm{i}\xi/|\xi|$.

Indeed, the difference between the Cauchy and Hilbert kernel is

$$\frac{\mathrm{d}\zeta}{\zeta - z} - \frac{1}{2} \cot \frac{y - x}{2}\, \mathrm{d}y = \frac{1}{2} \frac{\mathrm{d}\zeta}{\zeta}\,,$$

where $z = \mathrm{e}^{\mathrm{i}x}$, $\zeta = \mathrm{e}^{\mathrm{i}y} \in \mathbb{T}$. Consequently, by setting $u(x) = u(\mathrm{e}^{\mathrm{i}x}) = u(z)$ we arrive at the following relation between the Hilbert and Cauchy singular integral operators:

$$(Hu)\,(x) = \mathrm{i}(S_{\mathbb{T}} u)\,(z) - \frac{1}{2\pi} \int_{\mathbb{T}} \frac{u(\zeta)}{\zeta}\, \mathrm{d}\zeta\,. \tag{2}$$

Thus the assertion is a consequence of (2) and Sec. 6.54. ■

Notice that H has the following Fourier series representation:

$$(Hu)\,(z) = \mathrm{i} \sum_{n=1}^{\infty} \hat{u}_n z^n - \mathrm{i} \sum_{n=-\infty}^{-1} \hat{u}_n z^n\,, \qquad z = \mathrm{e}^{\mathrm{i}x}\,. \tag{3}$$

6.57. The single layer potential operator. This operator, which is defined by

$$(V_\Gamma u)\,(x) = -\frac{1}{\pi} \int_0^{2\pi} \log |\gamma(x) - \gamma(y)|\, u(y)\, \mathrm{d}y\,, \qquad u \in \mathbf{C}^\infty(\mathbb{T})\,, \tag{1}$$

is a ψ do of order -1 on $\mathbb{T}$ with symbol $1/|\xi|$. (V_Γ is also called *Symm's integral operator*.) Obviously, V_Γ can be written in the form

$$V_\Gamma = V_{\mathbb{T}} + L\,, \tag{2}$$

where

$$(V_{\mathbb{T}} u)\,(x) = -\frac{1}{\pi} \int_0^{2\pi} \log |\mathrm{e}^{\mathrm{i}x} - \mathrm{e}^{\mathrm{i}y}|\, u(y)\, \mathrm{d}y \tag{3}$$

and L is a smoothing operator with kernel $L(x, y) = -\pi^{-1} \log |[\gamma(x) - \gamma(y)]/[e^{ix} - e^{iy}]| \in C^\infty(\mathbb{T} \times \mathbb{T})$. The operator $V_{\mathbb{T}}$ is closely related to the Hilbert singular operator by the well-known formula

$$Hu = V_{\mathbb{T}} u', \qquad u'(y) = du/dy \tag{4}$$

(see Mikhlin [1] or Mikhlin/Prössdorf [1, Sec. II.6.2]). Using (4) and (56.3) we obtain that, for $n \neq 0$,

$$V_{\mathbb{T}} e_n = (in)^{-1} He_n = |n|^{-1} e_n, \qquad e_n = e^{inx}.$$

Since ker $V_{\mathbb{T}} = \mathbb{C}$ (see e.g. Muskhelishvili [1, § 65.4]), we have

$$(V_{\mathbb{T}} u)(x) = \sum_{n \neq 0} e^{inx} \frac{\hat{u}_n}{|n|}, \qquad u \in C^\infty(\mathbb{T}). \tag{5}$$

Hence, the assertion follows from (2), (5) and Theorem 51. ■

6.58. Example. The *differential operator*

$$A = \sum_{j=0}^{m} a_j(x) D^j, \qquad a_j \in C^\infty(\Gamma), \qquad D = -i(d/dx),$$

is a ψdo of order $m (\in \mathbb{N}_0)$ on Γ with symbol $a_0(x) + a_1(x)\xi + \cdots + a_m(x)\xi^m$.

6.59. Example. The singular integral operator

$$A = cI + dS_\Gamma + K, \qquad c, d \in C^\infty(\Gamma),$$

K being a smoothing operator, is a ψdo of order 0 on Γ with symbol $c + d\xi/|\xi|$.

6.60. Example. The *Bessel potential operator* Λ^α, $\alpha \in \mathbb{R}$, defined by (cf. 2.1(7))

$$(\Lambda^\alpha u)(x) = \hat{u}_0 + \sum_{0 \neq n \in \mathbb{Z}} \hat{u}_n |n|^\alpha e^{inx}, \qquad u \in C^\infty(\mathbb{T}), \tag{1}$$

is a ψdo of order α on $\mathbb{T}$ with symbol $|\xi|^\alpha$.

This is an immediate consequence of (1) and Theorem 6.51. The following property plays an important part in the numerical analysis of pseudodifferential equations (see Chapter 13). It turns out that the operators 6.54(1), $S_{\mathbb{T}}$ and Λ^α (or, what is the same, $V_{\mathbb{T}}$) already generate all of our pseudodifferential operators. In order to show this property we introduce the projections

$$P_{\mathbb{T}} u = 2^{-1}(I + S_{\mathbb{T}}) u = \sum_{n \geq 0} \hat{u}_n e^{inx}, \tag{2}$$

$$Q_{\mathbb{T}} u = 2^{-1}(I - S_{\mathbb{T}}) u = \sum_{n < 0} \hat{u}_n e^{inx}, \qquad u \in C^\infty(\mathbb{T}),$$

which are ψdo's of order 0 on $\mathbb{T}$ with symbols

$$h(\xi) = 2^{-1}(1 + \xi/|\xi|), \qquad h(-\xi) = 2^{-1}(1 - \xi/|\xi|). \tag{3}$$

6.61. Theorem. *Let A be a ψdo of order $r \in \mathbb{R}$ on $\mathbb{T}$ with symbol $a(x, \xi)$. For any $N \in \mathbb{N}_0$, A can be written as*

$$A = \sum_{j=0}^{N} [a_j^+ P_{\mathbb{T}} + a_j^- Q_{\mathbb{T}}] \Lambda^{r-j} + A_N + T, \tag{1}$$

where $T \in \mathbf{OP}^{-\infty}$, $A_N \in \mathbf{OPCS}^{r-N-1}$, and $a_j^\pm(x) = a_j(x, \pm 1)$.

Proof. Since the functions $a_j(x, \xi)$ ($j \in \mathbb{N}_0$) in the asymptotic expansion 6.41(3) of the symbol $a(x, \xi)$ are positively homogeneous of degree $r - j$ in $\xi \in \mathbb{R} \setminus \{0\}$, we obviously have the representation

$$a_j(x, \xi) = [a_j^+(x) \, h(\xi) + a_j^-(x) \, h(-\xi)] \, |\xi|^{r-j}. \tag{2}$$

Let $\varphi \in \mathbf{C}^\infty(\mathbb{R})$ be some excision function satisfying $\varphi(\xi) = 0$ for $|\xi| \leqq 1/2$ and $\varphi(\xi) = 1$ for $|\xi| \geqq 1$. Setting

$$b_N(x, \xi) = \varphi(\xi) \sum_{j=0}^{N} a_j(x, \xi)$$

and using (2), by definitions 6.60(1) and 6.60(2) we obtain that

$$\sum_{j=0}^{N} [a_j^+ P_{\mathbb{T}} + a_j^- Q_{\mathbb{T}}] \, \Lambda^{r-j} u = \sum_{n \in \mathbb{Z}} \mathrm{e}^{\mathrm{i}nx} b_N(x, n) \, \hat{u}_n + Ku,$$

where $u \in \mathbf{C}^\infty(\mathbb{T})$ and K is the smoothing operator with kernel $k(x, y) = \sum\limits_{0 \leqq j \leqq N} a_j^+(x)$. Since $a - b_N \in \mathbf{CS}^{r-N-1}$ by Definition 6.41, (1) follows from Theorem 6.51 and Section 6.54. ∎

6.62. Corollary. *Every* ψdo A *of order* $r \in \mathbb{R}$ *on* Γ *can be extended to an operator* $A \in \mathcal{L}(\mathbf{H}^k(\Gamma), \mathbf{H}^{k-r}(\Gamma))$ *for all* $k \in \mathbb{R}$.

Proof. Obviously, we only need to consider the case $\Gamma = \mathbb{T}$. In view of 6.61(1), it is sufficient to prove that $A_N \in \mathcal{L}(\mathbf{H}^k, \mathbf{H}^{k-r})$ for some $N \in \mathbb{N}_0$, since $P_{\mathbb{T}}, Q_{\mathbb{T}} \in \mathcal{L}(\mathbf{H}^k)$, $\Lambda^{r-j} \in \mathcal{L}(\mathbf{H}^k, \mathbf{H}^{k-r+j})$ and $aI \in \mathcal{L}(\mathbf{H}^k)$ for any $k \in \mathbb{R}$ and $a \in \mathbf{C}^\infty(\mathbb{T})$. Further, we may assume that

$$(A_N u)(x) = \sum_{n \neq 0} \mathrm{e}^{\mathrm{i}nx} a_N(x, n) \, \hat{u}_n, \qquad a_N \in \mathbf{CS}^{r-N-1}.$$

Choose a natural number $m \geqq k - r$. Using 6.41(2) and the Cauchy-Schwarz inequality, we get

$$\begin{aligned} |\mathrm{D}^l(A_N u)(x)|^2 &\leqq c \left(\sum_n |\hat{u}_n| \, (1 + |n|)^{r-N-1+l} \right)^2 \\ &\leqq c_1 \sum_n |\hat{u}_n|^2 \, (1 + |n|)^{2k}, \end{aligned}$$

where $u \in \mathbf{C}^\infty(\mathbb{T})$, $x \in \mathbb{R}$, $l = 0, 1, \ldots, m$, and N is sufficiently large. Consequently, $\|A_N u\|_{k-r}^2 \leqq \|A_N u\|_m^2 \leqq c_1 \|u\|_k^2$, $u \in \mathbf{C}^\infty(\mathbb{T})$, so that $A_N \in \mathcal{L}(\mathbf{H}^k, \mathbf{H}^{k-r})$ for suitable N. ∎

6.63. Corollary. *Let A be a* ψdo *of order* $r \in \mathbb{R}$ *on* Γ *with principal symbol* a_0. *Then A can be represented as*

$$A = (a_0^+ P_{\mathbb{T}} + a_0^- Q_{\mathbb{T}}) \, \Lambda^r + T_1, \tag{1}$$

where $T_1 \in \mathbf{OP}^{r-1}$ *and* $a_0^\pm(x) = a_0(x, \pm 1)$.

In particular, it follows from Corollary 6.63 that the principal symbol a_0 is uniquely defined by the ψdo A. Indeed, suppose $A = 0$ in (1) to get $a_0^+ P_{\mathbb{T}} + a_0^- Q_{\mathbb{T}} \in \mathcal{K}(\mathbf{L}^2(\mathbb{T}))$ because of the compactness of the embedding $\mathbf{H}^1 \subset \mathbf{L}^2(\mathbb{T})$. Hence, $a_0^\pm \equiv 0$ in view of Theorem 6.13(d).

6.64. Corollary. *If two ψdo's A_1 and A_2 have the same symbol, then $A_1 - A_2$ is a smoothing operator.*

Proof. By our assumption, the operator $A = A_1 - A_2$ has the representation 6.51(1), where $a \in \mathbf{CS}^r(\mathbb{T})$ for all $r \in \mathbb{R}$. From Corollary 6.62 we obtain that $A \in \mathcal{L}(\mathbf{H}^k(\Gamma), \mathbf{H}^{k-r}(\Gamma))$ for any $k, r \in \mathbb{R}$. Hence, A is a smoothing operator in view of the remark in Sec. 6.54. ■

It is clear from 6.43(1) that the rules for calculating the symbol of the composition of two ψdo's on $\mathbb{R}$ and the symbol of a formally adjoint ψdo on $\mathbb{R}$ (see e.g. Treves [1] or Shubin [1]) are preserved for ψdo's on $\mathbb{T}$.

6.65. Theorem. *Let $A \in \mathbf{OPCS}^\alpha$ and $B \in \mathbf{OPCS}^\beta$ with symbols a and b, respectively. Then AB belongs to $\mathbf{OPCS}^{\alpha+\beta}$ and has the symbol*

$$\operatorname{smb} AB = \sum_{l \geq 0} (\partial_\xi^l a)\,(-\mathrm{i}\partial_x)^l\, b/l!\,.$$

Proofs of Theorem 6.65 and 6.67 which are based on the Fourier series representation 6.51(1) of ψdo's on $\mathbb{T}$ can be found in Elschner [1]. ■

6.66. Corollary. *Let $A \in \mathbf{OPCS}^\alpha$ and $B \in \mathbf{OPCS}^\beta$ with principal symbols a_0 and b_0, respectively. Then*

(i) *AB has the principal symbol $a_0 b_0$,*

(ii) *$[A, B] := AB - BA \in \mathbf{OP}^{\alpha+\beta-1}$.*

We define the *formal adjoint* A^* of a ψdo A on $\mathbb{T}$ by

$$(Au, w) = (u, A^*w), \qquad u, w \in \mathbf{C}^\infty(\mathbb{T}),$$

where (., .) stands for the scalar product in $\mathbf{L}^2(\mathbb{T})$,

$$(u, w) = \int_{-\pi}^{\pi} u(x)\,\overline{w(x)}\,\mathrm{d}x\,.$$

6.67. Theorem. *Let $A \in \mathbf{OPCS}^\alpha$ with symbol a. Then $A^* \in \mathbf{OPCS}^\alpha$ and we have*

$$\operatorname{smb} A^* = \sum_{l \geq 0} \partial_\xi^l (-\mathrm{i}\partial_x)^l\, \bar{a}/l!\,.$$

6.68. Definition. $A \in \mathbf{OPCS}^\alpha$ is called *elliptic* if its principal symbol satisfies $a_0(x, \xi) \neq 0$ for all $x \in \mathbb{R}$ and $\xi = \pm 1$. (1)

6.69. Theorem. *If $A \in \mathbf{OPCS}^\alpha$ is elliptic, then, for any $t \in \mathbb{R}$, $A \in \mathcal{L}(\mathbf{H}^t, \mathbf{H}^{t-\alpha})$ is a Fredholm operator with index*

$$\operatorname{ind} A = (2\pi)^{-1} \{\arg a_0(x, -1)/a_0(x, +1)\}_{x=0}^{2\pi}. \tag{1}$$

If A is not elliptic, then $A \in \mathcal{L}(\mathbf{H}^t, \mathbf{H}^{t-\alpha})$ is not Fredholm.

Proof. Since $\Lambda^t : \mathbf{H}^t \to \mathbf{L}^2$ is an isomorphism for any $t \in \mathbb{R}$, $A \in \mathcal{L}(\mathbf{H}^t, \mathbf{H}^{t-\alpha})$ is Fredholm if and only if $B := \Lambda^{t-\alpha} A \Lambda^{-t} \in \mathcal{L}(\mathbf{L}^2)$ is so. By Theorem 6.61 and Corollary 6.66, we have $B = a_0^+ P_{\mathbb{T}} + a_0^- Q_{\mathbb{T}} + T$, where $T \in \mathbf{OP}^{-1}$. Since the embedding $\mathbf{L}^2 \subset \mathbf{H}^{-1}$ is compact, we obtain that $T \in \mathcal{K}(\mathbf{L}^2)$. Hence, B is Fredholm if and only if 6.68 (1) holds (see Section 6.11 and Theorem 6.13 (b)). Moreover, if 6.68 (1) is fulfilled, then B has index (1) (cf. Remark 6.14.4°).

6.70. Corollary (Elliptic regularity). *Let $A \in \mathbf{OPCS}^\alpha$ be elliptic. Then for any $s < t$, $u \in \mathbf{H}^{s+\alpha}$ and $Au \in \mathbf{H}^t$ implies $u \in \mathbf{H}^{t+\alpha}$. Moreover, if $u \in \mathcal{D}'$ and $Au \in \mathbf{C}^\infty$, then $u \in \mathbf{C}^\infty$.*

Proof. The first assertion is a simple consequence of Theorem 6.69, since $\mathbf{H}^t$ is dense in $\mathbf{H}^s$. The second claim is then obvious, because $\mathbf{C}^\infty = \bigcap_{k\in\mathbf{R}} \mathbf{H}^k$ and $\mathcal{D}' = \bigcup_{k\in\mathbf{R}} \mathbf{H}^k$. ∎

6.71. Example. The *singular integro-differential operator*

$$A = \sum_{k=0}^{m} (c_k I + d_k S_\Gamma)\, \mathsf{D}^k, \qquad c_k, d_k \in \mathbf{C}^\infty(\Gamma), \qquad \mathsf{D} = -\mathrm{i}(\mathrm{d}/\mathrm{d}x)$$

is a ψ do of order m on Γ with symbol

$$\operatorname{smb} A = \sum_{k=0}^{m} (c_k + d_k \xi/|\xi|)\, \xi^k.$$

This follows from Examples 6.55 and 6.58 and Theorem 6.65.

Notes and comments

6.3.—6.8. Costabel [1], Simonenko/Chin Ngok Minh [1].

6.9.—6.14. In a series of papers, Gohberg and Krupnik established a theory for the image $A^\pi := \operatorname{alg}^\pi \{S_\Gamma, \mathbf{PC}(\Gamma)\}$ in the Calkin algebra to algebras of singular integral operators with piecewise continuous coefficients, which can be viewed as an essential generalization of the Gelfand theory for commutative Banach algebras (see Gohberg/Krupnik [3] and the literature cited there). Note that they considered operators on the Lebesgue space $\mathbf{L}^p(\Gamma, w)$, with Γ a system of Lyapunov arcs and w a Khvedelidze weight. Gohberg and Krupnik's approach gives full information about the Fredholm theory in terms of a certain matrix-valued symbol which is associated to each operator, and for $p = 2$ and $w \equiv 1$ their theory yields a precise description of the quotient algebra A^π up to isomorphism and isometry. The symbols described in Sec. 6.11 and 6.12 are particular cases of the Gohberg-Krupnik symbol. The proof of Proposition 6.13 essentially differs from the original one given by Gohberg/Krupnik [2] and is published here for the first time. During the last 10—15 years several authors have aimed at making progress into the following directions:

— Curves with corners and intersections.

For $p = 2$ all irreducible representations of A^π were described in Plamenevski/Senichkin [1], [2]. In the case where $p \neq 2$ and Γ is a simple curve (with corners but without intersections) a symbol calculus was created by Costabel [3]. The general case was covered by Roch/Silbermann [8], [7].

— The description of A^π up to isomorphism.

Steps into this direction were made in Costabel [1], and Simonenko/Chin Ngok Minh [1] for simple closed Lyapunov curves.

The general case (corners and intersections) was treated by Roch/Silbermann [8], [7]. We here essentially follow the paper of Roch/Silbermann [8].

6.15. This material is well-known, see e.g. Böttcher/Silbermann [3].

6.16. Gohberg/Krupnik [1].

6.17. Maybe, this theorem is published here for the first time in this generality.

6.18.—6.22. In our opinion, one has to distinguish between strong ellipticity and locally strong ellipticity. First of all, the property of an operator to be locally strongly elliptic is of local nature. On the other hand, this terminology is in accordance with the notion of sectoriality and local sectoriality already used in the theory of Toeplitz operators for a long time (see Böttcher/Silbermann [3]).

Assertion 6.20(i) was established by ROCH/SILBERMANN [3] (see also Section 7.11), whereas 6.20(ii) is implicitely contained in PRÖSSDORF/RATHSFELD [6]. The first part of Theorem 6.22 is taken from ROCH [1] and the second one from PRÖSSDORF/RATHSFELD [6].

6.23. These results are widely scattered in the literature. See PRÖSSDORF/SCHMIDT [2], [3], ELSCHNER/PRÖSSDORF [1].

6.24. This theorem is published for the first time and is due to the authors.

6.25. The importance of the Kuratowski measure of noncompactness in this connection was recognized by PÖLTZ [1]. He also proved assertion (ii) for $s > 0$ and $s \notin \mathbb{N}$. The case $s \in \mathbb{N}$ was considered in SCHULZE/SILBERMANN [1]. Assertions (iv) and (v) are also taken from SCHULZE/SILBERMANN [1]. In the case where $s \notin \mathbb{N}$ they are well-known (see GOHBERG/KRUPNIK [4], PRÖSSDORF [1]).

6.26. For $s \notin \mathbb{N}$ all statements concerning Fredholmness and the index are well-known (e.g. GOHBERG/KRUPNIK [4], MUSKHELISHVILI [1], PRÖSSDORF [1]). The remaining assertions were established in SCHULZE/SILBERMANN [1].

6.27.—6.38. For analytic functions a, b and f, the complete asymptotics of the solution of equation 6.27(1) has long been known (see MAC CAMY [1] and MUSKHELISHVILI [1]). For the general case, the results of these sections are due to ELSCHNER [1]. Here we follow his paper [1], Sec. 2.

6.39.—6.71. Pseudodifferential operators (ψdo's) are natural generalizations of both singular integral operators and differential operators. MIKHLIN [2] (see also [1, 3]) was the first to introduce the concept of the symbol $a(x, \xi)$ of a singular integral operator A on $\mathbb{R}^m$ (in the case $m = 2$). It was CALDERON and ZYGMUND [1, 2] who first applied Fourier transform techniques to singular integrals. Calderon and Zygmund as well as Mikhlin set up the extremely important formulas

$$a(x, \xi) = c(x) + \hat{k}(x, \xi), \qquad A = F^{-1}_{\xi\to x} a(x, \xi)\, F_{x\to\xi},$$

where F refers to the Fourier transform and $\hat{k}(x, \xi)$ denotes the Fourier transform of the kernel $k(x, y)$ of A with respect to the variable y. Moreover, the systematic use of the apparatus of the Fourier transform promoted a further synthesis of multidimensional singular integral equations and partial differential equations, which eventually culminated in the theory of pseudodifferential operators. The foundation of the theory of ψdo's was laid in 1965 by the work of AGRANOVICH [1], CORDES [1], HÖRMANDER [1], KOHN and NIRENBERG [1], and SEELEY [1] (for more about the history of integral operators and ψdo's, we refer the reader to the survey PRÖSSDORF [8]). A detailed account of all the different aspects of the modern theory of ψdo's as well as their applications can be found in the monographs ESKIN [1], HÖRMANDER [1], SHUBIN [1], TREVES [1], TAYLOR [1], CORDES [2], GRUBB [1], JOURNÉ [1], KUMANO-GO [1], MIKHLIN/PRÖSSDORF [1], PLAMENEVSKI [1], SEELEY [2]. ψdo's have turned out to be a powerful tool in many fields of applied mathematics, for example, in boundary element methods for approximately solving exterior and interior boundary value problems in engineering (see Chapters 9—13 and the references there).

The Fourier series representation for the basic boundary integral operators such as Cauchy and Hilbert singular integral operators or the single layer potential operator has long been known (see MIKHLIN [1]). The general results (Theorems 6.51 and 6.61, Corollaries 6.52 and 6.63) are due to AGRANOVICH [2, 3]. The material of Sec. 6.39—6.52 is taken from AGRANOVICH [3, § 1]. Another elementary proof of Corollary 6.63, based on composing the ψdo A as a product of Symm's integral operator, tangential differentiation and a Cauchy type integral with a kernel of fractional order has been given by SARANEN and WENDLAND [2]. For a detailed study of (generally nonelliptic) ψdo's defined by the representation 6.51(1) we refer the reader to ELSCHNER's book [2, Chapters 5—7]. The basic properties of systems of integral equations of the first kind with the principal part V_Γ (see Sec. 6.57), which can be found in many applications, were given in MAC CAMY [1], HSIAO/MAC CAMY [1], HSIAO/WENDLAND [1] and LE ROUX [1] (see also WENDLAND [1] and HSIAO/KOPP/WENDLAND [1]).

Chapter 7
Polynomial approximation methods for the solution of singular integral equations with piecewise continuous coefficients on the unit circle

The approximation methods

7.1. Preliminaries. We now consider singular integral equations whose coefficients are merely supposed to be piecewise continuous. When studying approximation methods for these equations there arise substantial difficulties, which have been overcome only to a certain degree at the present time. This chapter is devoted to the finite section method and the collocation method for such equations. We shall demonstrate the advantage of the Banach algebra approach (including localization techniques) to the treatment of these questions.

Let $\mathbb{T}$ denote the counter-clockwise oriented unit circle on the complex plane. Let $\mathbf{PC}$ be the algebra of all functions $a \in \mathbf{L}^\infty(\mathbb{T})$ that are continuous with the exception of finitely many points $t_1, \ldots, t_R \in \mathbb{T}$, where, however, the finite limits $a(t_r+) := a(t_r + 0)$ and $a(t_r-) := a(t_r - 0)$ exist ($r = 1, \ldots, R$). We denote by $\overline{\mathbf{PC}}$ the closure of $\mathbf{PC}$ in $\mathbf{L}^\infty(\mathbb{T})$. Note that a function in $\overline{\mathbf{PC}}$ has at most countably many discontinuities of the first kind (jumps). Without loss of generality we shall assume that the value of a function $a \in \overline{\mathbf{PC}}$ coincides at any point $t \in \mathbb{T}$ with one of the limits $a(t+)$ or $a(t-)$. Let us recall some basic facts from the theory of singular integral operators

$$(A\varphi)(t) := (c\varphi)(t) + \frac{d(t)}{\pi \mathrm{i}} \int_{\mathbb{T}} \frac{\varphi(\tau)}{\tau - t}\, d\tau + (K\varphi)(t) \tag{1}$$

with coefficients $c, d \in \overline{\mathbf{PC}}$ and $K \in \mathcal{K}(\mathbf{L}^p(\mathbb{T}))$ (we assume that the operator (1) is given on the space $\mathbf{L}^p(\mathbb{T})$, $1 < p < \infty$). Because $S^2 = I$, where

$$(S\varphi)(t) := \frac{1}{\pi \mathrm{i}} \int_{\mathbb{T}} \frac{\varphi(\tau)}{\tau - t}\, d\tau,$$

the operators $P := (1/2)(I + S)$ and $Q := (1/2)(I - S)$ are linear bounded projections and so the operator (1) can be rewritten as follows:

$$A = aP + bQ + K, \tag{2}$$

where $a(t) := c(t) + d(t)$, $b(t) := c(t) - d(t)$.

Now we set up some statements and notations needed in what follows.

(i) $aP + bQ \in \Phi(\mathbf{L}^p(\mathbb{T}))$ *implies that* $a, b \in \mathbf{GL}^\infty(\mathbb{T})$ (*cf.* MIKHLIN/PRÖSSDORF [1]).

(ii) *Let* $p \in (1, \infty)$. *Then the following assertions are equivalent:*

(a) $a^{-1}P + b^{-1}Q$ *is invertible;*

(b) $\tilde{a}P + \tilde{b}Q$ *is invertible, where* $\tilde{c}(t) := c(t^{-1})$ *for* $c \in \mathbf{L}^\infty(\mathbb{T})$.

If, in addition, $p = 2$, *then each of* (a), (b) *is equivalent to*

(c) $aP + bQ$ *is invertible.*

Proof. (a) $\Rightarrow$ (b): Define the linear isometry W on $\mathbf{L}^2(\mathbb{T})$ by

$$W\left(\sum_{k=-\infty}^{\infty} \xi_k t^k\right) := \sum_{k=-\infty}^{\infty} \xi_{-k} t^k. \tag{3}$$

The equalities $(a \in \mathbf{L}^\infty(\mathbb{T}))$

$$P\left(\frac{1}{t} W\right) = \left(\frac{1}{t} W\right) Q, \quad \left(\frac{1}{t} W\right)^2 = I, \quad \left(\frac{1}{t} W\right) a \left(\frac{1}{t} W\right) = \tilde{a} I$$

can be easily checked. Therefore, we get the invertibility of

$$\left(\frac{1}{t} W\right)(a^{-1}P + b^{-1}Q)\left(\frac{1}{t} W\right) = \tilde{a}^{-1}Q + \tilde{b}^{-1}P = \tilde{a}^{-1}\tilde{b}^{-1}(\tilde{a}P + \tilde{b}Q).$$

(b) $\Rightarrow$ (a): Write $\tilde{a}P + \tilde{b}Q = (\tilde{a}^{-1})^{-1} P + (\tilde{b}^{-1})^{-1} Q = (\widetilde{a^{-1}})^{-1} P + (\widetilde{b^{-1}})^{-1} Q$. By the part already proved the invertibility of the last operator implies that of

$$(\widetilde{\widetilde{a^{-1}}})\, P + (\widetilde{\widetilde{b^{-1}}})\, Q = a^{-1}P + b^{-1}Q.$$

(c) $\Rightarrow$ (a): Define the antilinear isometry $\overline{W}$ on $\mathbf{L}^2(\mathbb{T})$ by

$$\overline{W}\left(\sum_{k=-\infty}^{\infty} \xi_k t^k\right) := \sum_{k=-\infty}^{\infty} \bar{\xi}_{-k} t^k.$$

It is easily seen that for $a \in \mathbf{L}^\infty(\mathbb{T})$

$$P\left(\frac{1}{t} \overline{W}\right) = \left(\frac{1}{t} \overline{W}\right) Q, \quad \left(\frac{1}{t} \overline{W}\right)^2 = I, \quad \left(\frac{1}{t} \overline{W}\right) a \left(\frac{1}{t} \overline{W}\right) = \bar{a} I.$$

Therefore, we get the invertibility of

$$\left(\frac{1}{t} \overline{W}\right)(aP + bQ)\left(\frac{1}{t} \overline{W}\right) = \bar{b}P + \bar{a}Q.$$

Passage to adjoints yields the invertibility of $Pb + Qa$ and then that of $Pba^{-1} + Q$. Further, $Pba^{-1} + Q = (I + Pba^{-1}Q)(Pba^{-1}P + Q)$, $ba^{-1}P + Q = (Pa^{-1}bP + Q) \times (I + Qa^{-1}bP)$, where the operators $I + Pba^{-1}Q$, $I + Qa^{-1}bP$ are invertible. Thus the operator $a^{-1}P + b^{-1}Q$ must be invertible.

(a) $\Rightarrow$ (c): By what has already been proved the operator

$$(a^{-1})^{-1} P + (b^{-1})^{-1} Q = aP + bQ$$

is invertible. ∎

(iii) *Let $p = 2$ and $a, b \in \overline{\mathbf{PC}}$. Then the operator $A = aP + bQ + K$, K being compact, is Fredholm if and only if*

$$a(t+)\, b(t-)\, \mu + a(t-)\, b(t+)\, (1 - \mu) \neq 0 \qquad (t \in \mathbb{T}, 0 \leqq \mu \leqq 1).$$

If this condition is fulfilled then the index ind A is given by

$$\operatorname{ind} A = -\operatorname{ind}_2 (a/b),$$

where $\operatorname{ind}_2 (a/b)$ is defined as follows. Let $V_2(a/b)$ be the curve obtained from the range of a/b by filling in the segments $(a/b)(t+)\, \mu + (a/b)(t-)\, (1 - \mu)$ $(0 \leqq \mu \leqq 1)$. The curve $V_2(a/b)$ is oriented in a natural way: on the intervals of continuity of the function a/b the motion of the point t along $\mathbb{T}$ in the positive direction induces a direction on $V_2(a/b)$, and the intervals $(a/b)(t+)\, \mu + (a/b)(t-)\, (1 - \mu)$ $(0 \leqq \mu \leqq 1)$ are oriented from $(a/b)(t-)$ to $(a/b)(t+)$. The winding number of the curve $V_2(a/b)$ about the origin is referred to as the 2-index of the function a/b and is denoted by $\operatorname{ind}_2 (a/b)$ (cf. Mikhlin/Prössdorf [1]).

In our following considerations we need the notion of the symbol. For our aim it is favourable to use that notion in a somewhat restrictive form. Let $E \subset \mathbb{T}$ be a measurable subset such that the measures of E and $\mathbb{T} \setminus E$ are not equal to 0. Let χ_E denote the characteristic function of E. Obviously, χ_E is a selfadjoint projection. Since P is also selfadjoint one can try to apply Theorem 1.16. For this end, consider the spectrum of $P\chi_E P$ in $\mathbf{L}^2(\mathbb{T})$. The well-known Hartman-Wintner theorem (cf. Böttcher/Silbermann [3]) claims that for every real-valued function a from $\mathbf{L}^\infty(\mathbb{T})$ the spectrum of $PaP|_{\operatorname{im} P}$ coincides with [ess inf a, ess sup a]. Now it is easy to see that $\sigma(P\chi_E P) = [0, 1]$ and so Theorem 1.16 applies to get

7.2. Theorem. *Let* A *denote the smallest C^*-subalgebra of $\mathcal{L}(\mathbf{L}^2(\mathbb{T}))$ containing the elements χ_E, P, I. Further, let $\mathbb{C}_{2\times 2}[0, 1]$ be the C^*-algebra of all continuous $\mathbb{C}_{2\times 2}$-valued functions on $[0, 1]$ and let $\hat{\mathsf{A}} \subset \mathbb{C}_{2\times 2}[0, 1]$ be the C^*-subalgebra which is generated by $\hat{P}$, $\hat{\chi}_E$, $\hat{I}$, where*

$$\hat{P}(\mu) = \begin{pmatrix} 1 & 0 \\ 0 & 0 \end{pmatrix}, \quad \hat{\chi}_E(\mu) = \begin{pmatrix} \mu & \sqrt{\mu(1 - \mu)} \\ \sqrt{\mu(1 - \mu)} & 1 - \mu \end{pmatrix},$$

$$\hat{I}(\mu) = \begin{pmatrix} 1 & 0 \\ 0 & 1 \end{pmatrix} \quad (\mu \in [0, 1]).$$

Then A *is isometrically star-isomorphic to* $\hat{\mathsf{A}}$. *This isomorphism takes P, χ_E, I into $\hat{P}$, $\hat{\chi}_E$, $\hat{I}$, respectively.*

The image of an element $A \in \mathsf{A}$ under this isomorphism is called the *symbol* of A. Therefore, A is invertible in A (or in $\mathcal{L}(\mathbf{L}^2(\mathbb{T}))$) if and only if the symbol of A is invertible. In particular, the operator

$$\{(a_+ - a_-)\, \chi_E + a_-\}\, P + \{(b_+ - b_-)\, \chi_E + b_-\}\, Q,$$

$a_\pm, b_\pm \in \mathbb{C}$, is invertible if and only if its symbol

$$\begin{pmatrix} a_+\mu + a_-(1 - \mu) & (b_+ - b_-)\sqrt{\mu(1 - \mu)} \\ (a_+ - a_-)\sqrt{\mu(1 - \mu)} & b_+(1 - \mu) + b_-\mu \end{pmatrix} \quad (\mu \in [0, 1]) \tag{1}$$

is invertible.

7.3. Description of the methods. Now we are going to describe two approximation methods for the solution of singular integral equations with piecewise continuous coefficients. The first one is the finite section method with respect to the projections

$$P_n\left(\sum_{k=-\infty}^{\infty} f_k t^k\right) = \sum_{k=-n}^{n} f_k t^k$$

($t \in \mathbb{T}$, $n = 0, 1, \ldots$) in the space $\mathbf{L}^2(\mathbb{T})$, previously studied for such equations with continuous coefficients in Chapter 4.

The second method is the polynomial collocation method. In order to solve the equation

$$(aP + bQ)\,\varphi = f$$

approximately, one considers the collocation equations

$$a(t_j) \sum_{k=0}^{n} t_j^k \xi_k + b(t_j) \sum_{k=-n}^{-1} t_j^k \xi_k = f(t_j) \qquad (j = 0, \pm 1, \ldots, \pm n), \tag{2}$$

where $t_j = \exp\,(2\pi \mathrm{i}\, j)/(2n+1)$. We shall later be concerned with this method in spaces of continuous functions (Chapter 9), where the value of a function at any point $t \in \mathbb{T}$ is well-defined. Since the space $\mathbf{L}^2(\mathbb{T})$ is a natural candidate for the underlying space in which we shall carry out the investigation in the case of piecewise continuous coefficients, we must restrict ourselves to right-hand sides f from suitably chosen subclasses. Such a subclass is the set $\mathcal{R}(\mathbb{T})$ of all Riemann-integrable functions f. The reason for this is the following.

Let $L_n f \in \operatorname{im} P_n$ be the interpolation polynomial of a bounded function f, uniquely determined by the conditions

$$(L_n f)\,(t_j) = f(t_j), \qquad t_j = \exp \frac{2\pi \mathrm{i} j}{2n+1} \qquad (j = 0, \pm 1, \ldots, \pm n).$$

(a) *One has* $\|f - L_n f\|_2 \to 0$ *in* $\mathbf{L}^2(\mathbb{T})$ *for* $f \in \mathcal{R}(\mathbb{T})$ *as* $n \to \infty$. (*cf.* Freud [1]).

Obviously, the equations (2) are equivalent to the operator equation

$$A_n \varphi_n := L_n(aP + bQ)\,\varphi_n = L_n f, \tag{3}$$

considered as an equation in $\operatorname{im} P_n$.

(b) *For any bounded and measurable function* a, *and any polynomial* $u_n \in \operatorname{im} P_n$, *we have*

$$\|L_n a u_n\|_2 \leqq \|a\|^\infty \|u_n\|_2,$$

where $\|a\|^\infty$ *is understood to be* $\sup_{t \in \mathbb{T}} |a(t)|$.

The proof is based on the following assertions: For any integer m with $|m| \leqq 2n$, one has $\sum_{k=-n}^{n} t_k^m = (2n+1)\,\delta_{0m}$ and

$$(L_n a)\,(t) = \sum_{k=-n}^{n} a_k t^k, \qquad a_k := (2n+1)^{-1} \sum_{j=-n}^{n} a(t_j)\, t_j^{-k}. \tag{4}$$

Indeed, the norm of $L_n f$, f bounded and measurable, can be computed in the space $\mathbf{L}^2(\mathbb{T})$ by

$$\|L_n f\|_2^2 = \frac{1}{2\pi} \int_0^{2\pi} |(L_n f)(e^{i\vartheta})|^2 \, d\vartheta = \sum_{k=-n}^{n} |f_k|^2$$

$$= (2n+1)^{-2} \sum_{k=-n}^{n} \left(\sum_{j=-n}^{n} f(t_j)\, t_j^{-k} \right) \left(\sum_{l=-n}^{n} \overline{f(t_l)}\, t_l^{k} \right)$$

$$= (2n+1)^{-2} \sum_{j=-n}^{n} f(t_j) \left(\sum_{l=-n}^{n} \overline{f(t_l)} \left(\sum_{k=-n}^{n} t_k^{l-j} \right) \right)$$

$$= (2n+1)^{-1} \sum_{j=-n}^{n} |f(t_j)|^2 .$$

Hence, we obtain

$$\|L_n a u_n\|_2^2 = (2n+1)^{-1} \sum_{j=-n}^{n} |a(t_j)|^2 \, |u_n(t_j)|^2 \leqq (\|a\|^\infty)^2 \, \|L_n u_n\|_2^2 = (\|a\|^\infty)^2 \, \|u_n\|_2^2 .$$

(c) $A_n P_n := L_n(aP + bQ) P_n$ *converges strongly to* $A = aP + bQ$ $(a, b \in \overline{\mathbf{PC}})$. Indeed, we have, for $u \in \mathbf{L}^2(\mathbb{T})$,

$$\|A_n P_n u\|_2 = \|L_n(aP + bQ) P_n u\|_2 \leqq \|L_n a P P_n u\|_2 + \|L_n b Q P_n u\|_2$$

$$\leqq \|a\|^\infty \, \|P P_n u\|_2 + \|b\|^\infty \, \|Q P_n u\|_2 . \qquad \text{(by (b))}$$

Thus the sequence $\{A_n P_n\}$ is uniformly bounded. Let $u \in \operatorname{im} P_m$. Then $P_n u = u$ for $n \geqq m$ and we get $(n \geqq m)$

$$\|L_n(aP + bQ) P_n u - (aP + bQ) u\|_2$$

$$= \|L_n(aP + bQ) u - (aP + bQ) u\|_2 \to 0 \qquad \text{(by (a))}$$

as $n \to \infty$ since $(aP + bQ) u \in \mathcal{R}(\mathbb{T})$. The Banach-Steinhaus theorem gives the assertion at once. In the light of Proposition 1.23 the following question arises. What conditions are needed to guarantee the stability of $\{A_n\}$? Of course, the same question emerges for the sequence $\{B_n\}$ of finite sections, where $B_n := P_n(aP + bQ) P_n|_{\operatorname{im} P_n}$.

Algebraization

7.4. Notations. Let X be a Banach space and let $\{P_n\} \subset \mathcal{L}(X)$ be a sequence of projections such that $P_n \to I$ and $P_n^* \to I^*$ strongly as $n \to \infty$. We denote by

$\mathcal{S}$ the collection of all sequences $\{A_n\}_{n=0}^\infty$ of operators $A_n : \operatorname{im} P_n \to \operatorname{im} P_n$ such that $\sup_n \|A_n P_n\| < \infty$; on defining $\{A_n\} + \{B_n\} := \{A_n + B_n\}$, $\{A_n\}\{B_n\} := \{A_n B_n\}$ and $\|\{A_n\}\|_{\mathcal{S}} := \sup_n \|A_n P_n\|$ the set $\mathcal{S}$ becomes a Banach algebra;

$\mathcal{S}_c$ the closed subalgebra of $\mathcal{S}$ consisting of all sequences $\{A_n\}$ which possess a strong limit $A \in \mathcal{L}(X)$;

$\mathcal{N}$ the collection of all sequences $\{C_n\} \in \mathcal{S}$ such that $\|C_nP_n\| \to 0$ as $n \to \infty$; it is easy to see that $\mathcal{N}$ forms a closed two-sided ideal in both $\mathcal{S}$ and $\mathcal{S}_c$;

π the canonical projection of $\mathcal{S}$ onto $\mathcal{S}/\mathcal{N}$;

π_c the canonical projection of $\mathcal{S}_c$ onto $\mathcal{S}_c/\mathcal{N}$.

Frequently we think of A_n as being identified with A_nP_n and we may therefore regard A_n as an element of $\mathcal{L}(X)$.

7.5. Proposition. *Suppose* $P_n \to I$, $P_n^* \to I^*$, $A_nP_n \to A$, $A_n^*P_n^* \to A^*$ *strongly as* $n \to \infty$. *Then the following statements are equivalent:*

(a) *the sequence* $\{A_n\}$ *is stable;*

(b) $A \in \mathfrak{G}\mathcal{L}(X)$ *and* $\pi_c\{A_n\} \in \mathfrak{G}(\mathcal{S}_c/\mathcal{N})$;

(c) $A \in \mathfrak{G}\mathcal{L}(X)$ *and* $\pi\{A_n\} \in \mathfrak{G}(\mathcal{S}/\mathcal{N})$.

Proof. (a) $\Rightarrow$ (b): For any $x \in X$ we have (n large enough, say $n \geqq n_0$)

$$\|P_nx\| \leqq C\,\|A_nP_nx\|, \tag{1}$$

where C is independent of n and x. Passage to the limit $n \to \infty$ in (1) leads to $\|x\| \leqq C\,\|Ax\|$, that is, $\ker A = \{0\}$ and $\operatorname{im} A$ is closed. Passage to adjoints gives the same for A^*. Therefore $A \in \mathfrak{G}\mathcal{L}(X)$. Put $B_n := P_n$ for $n < n_0$ and $B_n := A_n^{-1}P_n$ for $n \geqq n_0$. Then

$$\begin{aligned}\|B_nx - P_nA^{-1}x\| &\leqq C\,\|A_n(B_nx - P_nA^{-1}x)\|\\ &= C\,\|P_nx - A_nP_nA^{-1}x\| \quad \text{for} \quad n \geqq n_0.\end{aligned}$$

Since $P_n \to I$ and $A_nP_n \to A$ strongly, it follows that $B_n \to A^{-1}$ strongly. Hence $\{B_n\} \in \mathcal{S}_c$. The assertion is now a consequence of the inclusions

$$\{B_n\}\,\{A_n\} - \{P_n\}, \quad \{A_n\}\,\{B_n\} - \{P_n\} \in \mathcal{N}.$$

(b) $\Rightarrow$ (c): Obvious.

(c) $\Rightarrow$ (a): Suppose there is a sequence $\{B_n\} \in \mathcal{S}$ such that

$$B_nA_n = P_n + C_n, \qquad A_nB_n = P_n + D_n, \qquad \{C_n\}, \{D_n\} \in \mathcal{N}.$$

Then there is an n_0 such that $\|C_n\| < 1/2$ and $\|D_n\| < 1/2$ for all $n \geqq n_0$. Thus $P_n + C_n$ and $P_n + D_n$ regarded as acting on $\operatorname{im} P_n$ are invertible for $n \geqq n_0$, and the norms of their inverses are uniformly bounded by $\dfrac{1}{1 - (1/2)} = 2$. This implies the invertibility of $A_n : \operatorname{im} P_n \to \operatorname{im} P_n$ for $n \geqq n_0$ and the inequality

$$\sup_{n \geqq n_0} \|A_n^{-1}P_n\| \leqq 2 \sup_{n \geqq n_0} \|B_nP_n\| =: C < \infty. \quad \blacksquare$$

We shall see that in the theory of the finite section method as well as in the theory of the collocation method certain operators W_n ($n = 0, 1, \ldots$) are of great importance. Therefore, we have to insert these operators into our consideration. This can be done in the following manner.

Let $\{W_n\} \subset \mathcal{L}(X)$ be a uniformly bounded sequence such that

$$W_n^2 = P_n, \qquad W_nP_n = P_nW_n = W_n,$$

and W_n as well as W_n^* tend weakly to zero as $n \to \infty$.

Now define $\mathcal{A}$ to be the subset of $\mathcal{S}$ consisting of all sequences $\{A_n\}$ such that there exist $A \in \mathcal{L}(X)$ and $\tilde{A} \in \mathcal{L}(X)$ satisfying $A_nP_n \to A$, $A_n^*P_n^* \to A^*$, $\tilde{A}_nP_n \to \tilde{A}$, $\tilde{A}_n^*P_n^* \to \tilde{A}^*$ strongly (as $n \to \infty$), where $\tilde{A}_n := W_nA_nW_n|_{\operatorname{im} P_n}$.

It is easy to see that $\mathcal{A}$ forms a closed subalgebra of $\mathcal{S}$ (even of $\mathcal{S}_c$), that is, $\mathcal{A}$ itself is a Banach algebra.

Obviously, $\mathcal{N} \subset \mathcal{A}$. Moreover, $\{P_nKP_n\}$ and $\{W_nKW_n\}$ are elements of $\mathcal{A}$ for any $K \in \mathcal{K}(X)$ due to Section 1.1, (f) and (h). Now, denote by $\mathcal{J}$ the collection of all sequences $\{A_n\} \in \mathcal{S}$ of the form

$$A_n = P_nGP_n + W_nKW_n + C_n,$$

where $G, K \in \mathcal{K}(X)$ and $\{C_n\} \in \mathcal{N}$.

7.6. Proposition. *$\mathcal{J}$ is a two-sided closed ideal in $\mathcal{A}$.*

Proof. Obviously, $\mathcal{J}$ is a linear subspace of $\mathcal{A}$. Let us prove that $\mathcal{J}$ is closed. If

$$\{A_n\} = \{P_nGP_n + W_nKW_n + C_n\} \in \mathcal{J},$$

then $A_n \to G$ and $\tilde{A}_n \to K$, whence

$$\|G\| \leqq \liminf \|A_n\|, \qquad \|K\| \leqq \liminf \|\tilde{A}_n\|.$$

Thus if $\{\{A_n^{(j)}\}\}_{j=0}^{\infty} \subset \mathcal{J}$ is a Cauchy sequence, then so are $\{G^{(j)}\}$ and $\{K^{(j)}\}$ in $\mathcal{K}(X)$. Consequently, there exist compact operators G and K such that $\|G^{(j)} - G\| \to 0$, $\|K^{(j)} - K\| \to 0$ as $j \to \infty$. Now it follows almost at once that there is a sequence $\{A_n\} \in \mathcal{J}$ such that $\|\{A_n^{(j)}\} - \{A_n\}\| \to 0$ as $j \to \infty$.

Finally, let $\{A_n\} \in \mathcal{J}$ and $\{B_n\} \in \mathcal{A}$. Then

$$\begin{aligned} B_nA_n &= B_nP_nGP_n + B_nW_nKW_n + B_nC_n \\ &= P_nB_nP_nGP_n + W_n\tilde{B}_nP_nKW_n + B_nC_n \\ &= P_nBGP_n + W_n\tilde{B}KW_n + C_n', \end{aligned}$$

where $\{C_n'\} \in \mathcal{N}$. Analogously it can be proved that $\{A_nB_n\} \in \mathcal{J}$. ∎

We put $\mathcal{A}^\circ = \mathcal{A}/\mathcal{J}$ and denote by $\{A_n\}^\circ \in \mathcal{A}^\circ$ the coset containing $\{A_n\} \in \mathcal{A}$.

7.7. Theorem. *Let $\{A_n\} \in \mathcal{A}$ and s-lim $A_n = A$. Then $\{A_n\}$ is stable if and only if*

$$A \in \mathbf{G}\mathcal{L}(X), \qquad \tilde{A} \in \mathbf{G}\mathcal{L}(X) \quad \text{and} \quad \{A_n\}^\circ \in \mathbf{G}\mathcal{A}^\circ.$$

Moreover, if $\{R_n\}^\circ$ is the inverse of $\{A_n\}^\circ$ in $\mathcal{A}^\circ$, then $\pi_c\{R_n + P_n(A^{-1} - R)P_n + W_n(\tilde{A}^{-1} - \tilde{R})W_n\}$ is the inverse of $\pi_c\{A_n\}$ in $\mathcal{S}_c/\mathcal{N}$ (here $R :=$ s-lim R_nP_n, $\tilde{R} =$ s-lim $W_nR_nW_n$).

Proof. Let $\{A_n\} \in \mathcal{A}$ be stable. Then $A \in \mathbf{G}\mathcal{L}(X)$ by Proposition 7.5. Let us prove that $\tilde{A} \in \mathbf{G}\mathcal{L}(X)$. Obviously, $\tilde{A}_n = W_nA_nW_n$ is invertible on $\operatorname{im} P_n$ for all sufficiently large n and

$$\sup_{n \geqq n_0} \|\tilde{A}_n^{-1}P_n\| = \sup_{n \geqq n_0} \|W_nA_n^{-1}W_n\| \leqq C_1 \sup_{n \geqq n_0} \|A_n^{-1}P_n\| := C < \infty.$$

Again from Proposition 7.5 we may conclude that $\tilde{A} \in \mathbf{G}\mathcal{L}(X)$. Let $\{B_n\} \in \mathcal{S}_c$ be the inverse of $\{A_n\} \in \mathcal{S}_c$ modulo $\mathcal{N}$, that is $B_nA_n = P_n + C_n$, $\{C_n\} \in \mathcal{N}$. Then for n sufficiently large

$$\tilde{B}_n = (P_n + W_nC_nW_n)\,\tilde{A}_n^{-1}P_n \to \tilde{A}^{-1},$$

$$B_n^* = (A_n^*)^{-1}\,(P_n^* + C_n^*)\,P_n^* \to (A^*)^{-1},$$

$$\tilde{B}_n^* = (\tilde{A}_n^*)^{-1}\,(P_n^* + W_n^*C_n^*W_n^*)\,P_n^* \to (\tilde{A}^*)^{-1}.$$

Since $(D^{-1})^* = (D^*)^{-1}$ for any operator $D \in \mathfrak{G}\mathcal{L}(X)$, we consequently get that $\{B_n\}^\circ$ is the inverse of $\{A_n\}^\circ$ in $\mathcal{A}^\circ$, i.e. $\{A_n\}^\circ \in \mathfrak{G}\mathcal{A}^\circ$. Conversely, suppose now A and $\tilde{A}$ are invertible in $\mathcal{L}(X)$ and $\{A_n\}^\circ$ is so in $\mathcal{A}^\circ$. Then there is a sequence $\{R_n\} \in \mathcal{A}$ such that

$$A_nR_n = P_n + P_nGP_n + W_nKW_n + C_n,$$

where G and K are compact and $\{C_n\} \in \mathcal{N}$. Passage to the limit $n \to \infty$ gives

$$AR = I + G, \qquad \tilde{A}\tilde{R} = I + K.$$

Thus, $A^{-1} - R =: G_1 \in \mathcal{K}(X)$, $\tilde{A}^{-1} - \tilde{R} =: K_1 \in \mathcal{K}(X)$.

Put $R_n' = R_n + P_nG_1P_n + W_nK_1W_n$. Then, obviously, $\{R_n'\} \in \mathcal{A}$ and,

$$A_nR_n' = P_n + P_n(G + A_nP_nG_1)\,P_n + W_n(K + \tilde{A}_nP_nK_1)\,W_n + C_n$$

$$= P_n + P_n(G + AG_1)\,P_n + W_n(K + \tilde{A}K_1)\,W_n + C_n' = P_n + C_n',$$

where $\{C_n'\} \in \mathcal{N}$. In the same way we may show that $R_n'A_n = P_n + C_n''$, $\{C_n''\} \in \mathcal{N}$. This gives the remaining assertion of the theorem. ■

The last theorem is supplemented by the following observations.

(a) *$\{A_n\}^\circ \in \mathfrak{G}\mathcal{A}^\circ$ implies that $A, \tilde{A} \in \Phi(X)$.*

This is immediate from the previous proof.

(b) *Let $\{A_n\}^\circ \in \mathcal{A}^\circ$ be left invertible (right invertible) and let* $\ker A = \ker \tilde{A} = \{0\}$ ($\ker A^* = \ker \tilde{A}^* = \{0\}$). *Then, if $A_n \in \Phi(\operatorname{im} P_n)$ with* $\operatorname{ind} A_n = 0$ *for all n large enough, the sequence $\{A_n\}$ is stable.*

Proof. Let $\{A_n\}^\circ \in \mathcal{A}^\circ$ be left invertible. Then there exists an element $\{B_n\}^\circ \in \mathcal{A}^\circ$ such that

$$B_nA_n = P_n + P_nGP_n + W_nKW_n + C_n,$$

where $G, K \in \mathcal{K}(X)$, and $\|C_n\| \to 0$ as $n \to \infty$. Passing to the limit, this yields $BA = I + G$ and $\tilde{B}\tilde{A} = I + K$. So A and $\tilde{A}$ are regularizable from the left. Therefore $\operatorname{im} A$ and $\operatorname{im} \tilde{A}$ are complementable subspaces of X (cf. Mikhlin/Prössdorf [1]). The hypothesis that $\ker A = \ker \tilde{A} = \{0\}$ shows that one can find operators C and D so that

$$CA = I, \qquad D\tilde{A} = I.$$

Put $B_n' = B_n - P_nGCP_n - W_nKDW_n$. Then

$$B_n'A_n = P_n + C_n', \qquad \|C_n'\| \to 0 \quad \text{as} \quad n \to \infty.$$

Thus, the operators A_n are left invertible for n large enough, and since $\operatorname{ind} A_n = 0$, we get the invertibility of A_n (on $\operatorname{im} P_n$). Now it follows that $\{A_n\}$ is stable. ■

7.8. A realization. Let X be the Banach space $\mathbf{L}^p(\mathbb{T})$ $(1 < p < \infty)$ and define the projections P_n and the flip operators W_n by

$$P_n\left(\sum_{k=-\infty}^{\infty} f_k t^k\right) = \sum_{k=-n}^{n} f_k t^k, \tag{1}$$

$$W_n\left(\sum_{k=-\infty}^{\infty} f_k t^k\right) = f_{-1}t^{-n} + \dots + f_{-n}t^{-1} + f_n + f_{n-1}t + \dots + f_0 t^n, \tag{2}$$

respectively. Obviously,

(a) $W_n^2 = P_n$, $W_n P_n = P_n W_n = W_n$.

Since the projections P_n converge strongly to I on $\mathbf{L}^p(\mathbb{T})$ $(1 < p < \infty)$ and W_n can be expressed in the form

$$(W_n f)(t) = t^{-n-1}(QP_n f)\left(\frac{1}{t}\right) + t^n(PP_n f)\left(\frac{1}{t}\right)$$

(P and Q being the projections introduced in Section 7.1) one obtains that

(b) *the sequence W_n is uniformly bounded.*

Notice that the operators P_n^* and W_n^* (acting on $\mathbf{L}^q(\mathbb{T})$, $1/p + 1/q = 1$) are also defined by (1) and (2), respectively. So it is easy to see that

(c) *the sequences $\{W_n\}$, $\{W_n^*\}$ converge weakly to zero.*

Therefore, in the given case the algebra $\mathcal{A}$, the ideal $\mathcal{J} \subset \mathcal{A}$ and the quotient algebra $\mathcal{A}^\circ$ are well-defined.

The finite section method

7.9. Proposition. (a) *For any $a, b \in \mathbf{L}^\infty(\mathbb{T})$ and any $T \in \mathcal{K}(\mathbf{L}^p(\mathbb{T}))$ $(1 < p < \infty)$ the sequence $\{A_n\} := \{P_n A P_n\}$, $A = aP + bQ + T$, belongs to the algebra $\mathcal{A}$ (associated with the space $\mathbf{L}^p(\mathbb{T})$).*

In particular, s-lim $W_n A W_n = P\tilde{a}P + Q\tilde{b}Q$, *where*

$$\tilde{a}(t) := a(1/t), \qquad \tilde{b}(t) := b(1/t).$$

(b) *If $\{P_n(aP + bQ)P_n\}^\circ$ is invertible, then A, $\tilde{A}$ are Fredholm and*

$$\operatorname{ind} A = -\operatorname{ind} \tilde{A}.$$

Proof. (a) In view of Section 7.6 it is sufficient to carry out the proof for $T = 0$. Now write

$$W_n A_n W_n = W_n P_n PaPP_n W_n + W_n P_n QbQP_n W_n$$
$$+ W_n P_n QaPP_n W_n + W_n P_n PbQP_n W_n.$$

An easy computation gives $W_n P_n PaPP_n W_n = P_n P\tilde{a}PP_n$, $W_n P_n QbQP_n W_n = P_n Q\tilde{b}QP_n$. Since the sequences $QaPW_n$, $PbQW_n$ converge strongly to zero, s-lim $W_n A W_n$ exists and equals $P\tilde{a}P + Q\tilde{b}Q$. Because $A_n^* = P_n(P\bar{a} + Q\bar{b})P_n$ $(\in \mathcal{L}(\mathbf{L}^q(\mathbb{T}))$, $1/p + 1/q = 1)$, both s-lim $A_n^* P_n$ and s-lim $\tilde{A}_n^* P_n$ exist and are equal to A^* and $P\tilde{\bar{a}}P + Q\tilde{\bar{b}}Q = \widetilde{A^*} = (\tilde{A})^*$, respectively.

(b) The first assertion follows from 7.7(a). Now put $A_{m,r} := at^{-m}P + bt^{-r}Q$ (m, r integers). Evidently,

$$\tilde{A}_{m,r} = P\tilde{a}t^{m}P + Q\tilde{b}t^{r}Q = (P\tilde{a}t^{m}P + Q)(P + Q\tilde{b}t^{r}Q).$$

Now the integers m, r may be chosen so that

$$\operatorname{ind}(P\tilde{a}t^{m}P + Q) = \operatorname{ind}(P + Q\tilde{b}t^{r}Q) = 0.$$

Coburn's theorem (see 6.1(ix)) states that the operators $P\tilde{a}t^{m}P + Q$, $P + Q\tilde{b}t^{r}Q$ are invertible and, hence, so also is $\tilde{A}_{m,r}$. Again by Coburn's theorem the kernel or the cokernel of $A_{m,r}$ is trivial. Note that $\{P_nA_{m,r}P_n\}^\circ$ is invertible. Indeed, this is a consequence of the following identity:

$$\{P_nA_{m,r}P_n\}^\circ = \{P_nAP_n\}^\circ \{P_n(t^{-m}P + t^{-r}Q) P_n\}^\circ.$$

The first factor is invertible by the hypothesis; the inverse of the second is $\{P_n(t^{m}P + t^{r}Q) P_n\}^\circ$ (see the proof of the following Theorem 7.10). Therefore 7.7(b) applies, that means $\{P_nA_{m,r}P_n\}$ is stable and, hence, $A_{m,r}$ and $\tilde{A}_{m,r}$ are invertible. Now we have

$$0 = \operatorname{ind} A_{m,r} = \operatorname{ind}(aP + bQ) + \operatorname{ind}(Pt^{-m}P + Qt^{-r}Q),$$
$$0 = \operatorname{ind} \tilde{A}_{m,r} = \operatorname{ind}(P\tilde{a}P + Q\tilde{b}Q) + \operatorname{ind}(Pt^{m}P + Qt^{r}Q).$$

Since $\operatorname{ind}(Pt^{-m}P + Qt^{-r}Q) = -\operatorname{ind}(Pt^{m}P + Qt^{r}Q)$ we get

$$\operatorname{ind} A = -\operatorname{ind} \tilde{A}. \ ■$$

7.10. Theorem. (a) *For any $f \in \mathbf{L}^\infty(\mathbb{T})$ we have*

$$\begin{aligned} Q_nPfPP_n &= t^{n}WQ\tilde{f}PW_n, \\ P_nPfPQ_n &= W_nP\tilde{f}QWt^{-n}, \\ Q_nQfQP_n &= t^{-n-1}WP\tilde{f}QW_n, \\ P_nQfQQ_n &= W_nQ\tilde{f}PWt^{n+1}, \end{aligned} \tag{1}$$

where $Q_n := I - P_n$ and W is defined by 7.1(3).

(b) *Consider $A = aP + bQ$, $B = PcP + QcQ$ in $\mathbf{L}^p(\mathbb{T})$ with $a, b \in \mathbf{L}^\infty(\mathbb{T})$ and $c \in \mathbf{C}(\mathbb{T})$. Then*

$$c_1 := \{P_nAP_n\}\{P_nBP_n\} - \{P_n(acP + bcQ) P_n\} \in \mathcal{J},$$
$$c_2 := \{P_nBP_n\}\{P_nAP_n\} - \{P_n(acP + bcQ) P_n\} \in \mathcal{J}.$$

A similar assertion holds if one replaces A by $Pa + Qb$ or $PaP + QbQ$.

Proof. (a) Compute.
(b) We have $c_1 = I_1 + I_2$,

$$I_1 = \{P_nPaPP_nPcPP_n\} + \{P_nQbQP_nQcQP_n\} - \{P_nPacPP_n + P_nQbcQP_n\},$$
$$I_2 = (\{P_nQaPP_nPcPP_n\} - \{P_nQacPP_n\}) + (\{P_nPbQP_nQcQP_n\} - \{P_nPbcQP_n\}).$$

Obviously,

$$\begin{aligned} P_nPacPP_n &= P_nPaPcPP_n + P_nPaQcPP_n \\ &= P_nPaPP_nPcPP_n + P_nPaQcPP_n + P_nPaPQ_nPcPP_n, \end{aligned}$$

where $Q_n = I - P_n$. Since $P_nPaPQ_nPcPP_n = W_nP\tilde{a}Q\tilde{c}PW_n$ (by (a)), QcP, $Q\tilde{c}P \in \mathcal{K}(\mathbf{L}^p(\mathbb{T}))$, we obtain

$$P_nPacPP_n = P_nPaPP_nPcPP_n + P_nG_1P_n + W_nG_2W_n, \qquad G_1, G_2 \in \mathcal{K}(\mathbf{L}^p(\mathbb{T})).$$

Analogously,

$$P_nQbcQP_n = P_nQbQP_nQcQP_n + P_nK_1P_n + W_nK_2W_n, \qquad K_1, K_2 \in \mathcal{K}(\mathbf{L}^p(\mathbb{T})).$$

Putting these things together we obtain that $I_1 \in \mathcal{J}$. Let us consider the first item I_2' in I_2:

$$\begin{aligned} I_2' &= \{P_nQaPP_nPcPP_n\} - \{P_nQacPP_n\} \\ &= \{P_nQaP_ncPP_n\} - \{P_nQacPP_n\} - \{P_nQaQP_nQcPP_n\}. \end{aligned}$$

Since QcP is compact it remains to show that $\{P_nQacPP_n\} - \{P_nQaP_ncPP_n\}$ belongs to $\mathcal{J}$. For any elements $d, e \in \mathbf{L}^\infty$ we have

$$\begin{aligned} P_nQdePP_n &= QP_ndeP_nP = Q(t^{-n}P_{2n}PdePP_{2n}t^n)\,P \\ &= Q(t^{-n}P_{2n}PdPP_{2n}t^nt^{-n}P_{2n}PePP_{2n}t^n)\,P \\ &\quad + Q(t^{-n}P_{2n}PdQePP_{2n}t^n)\,P + Q(t^{-n}W_{2n}P\tilde{d}Q\tilde{e}PW_{2n}t^n)\,P. \end{aligned}$$

This leads to

$$\begin{aligned} P_nQdePP_n &= P_nQdP_nePP_n + P_nQt^{-n}P_{2n}PdQePP_{2n}t^nPP_n \\ &\quad + P_nQt^{-n}W_{2n}P\tilde{d}Q\tilde{e}PW_{2n}t^nPP_n. \end{aligned} \tag{2}$$

If one of the functions d, e is continuous, then the last two items in (2) converge in the norm to zero. This follows immediately from 1.1(h) and the fact that

$$(t^nP)^* = (Pt^nP)^* = Pt^{-n}P \to 0, \qquad Qt^{-n}W_{2n}P = Qt^{-n}PW_{2n}P \to 0$$

strongly as $n \to \infty$. Thus, $I_2' \in \mathcal{J}$. Analogously one can show that $I_2 - I_2' \in \mathcal{J}$. ■

This theorem allows us to apply localization techniques to the finite section method. Our first concern is to consider the finite section method in a more simple case. The underlying operator (respectively sequence) is called a local representative, because locally this operator will have the same behavior as the general operator under consideration.

7.11. An abstract local representative. Let $\mathcal{B}$ be a C^*-algebra with unit element e and $p, q \in \mathcal{B}$ be selfadjoint idempotents such that $\sigma(pqp) = [0, 1]$. Denote by Λ the C^*-subalgebra of $\mathcal{B}$ generated by q and e. It is readily seen that every element $a \in \Lambda$ has the form $a = (a_+ - a_-)\,q + a_-e$, where a_+ and a_- are complex numbers. Moreover, for such an element $a = (a_+ - a_-)\,q + a_-e$ we have the equality $\|a\| = \max\{|a_+|, |a_-|\}$ since a is normal (i.e. $aa^* = a^*a$) and the spectrum of a consists of the points a_+ and a_- only.

Further, let

$$a = (a_+ - a_-)\,q + a_-e, \qquad b = (b_+ - b_-)\,q + b_-e \tag{1}$$

be elements from Λ such that the complex numbers $a_\pm$, $b_\pm$ satisfy

$$\left|\arg \frac{a_+}{a_-}\right| < \pi, \qquad \left|\arg \frac{b_+}{b_-}\right| < \pi, \qquad \left|\arg \frac{a_+}{a_-} - \arg \frac{b_+}{b_-}\right| < \pi, \tag{2}$$

or equivalently

$$\mu \frac{a_+}{a_-} + (1 - \mu) \frac{b_+}{b_-} \notin (-\infty, 0]$$

for all $\mu \in [0, 1]$.

For definiteness we choose $-\pi \leqq \arg z < \pi$ for any complex number $z \neq 0$. Now let $A = ap + b(e - p)$ be an element the coefficients a and b of which satisfy (1) and (2). Then this element can be represented as

$$A = (e + T)\, B, \tag{3}$$

where $\|T\| < 1$ and $B = \alpha p + \beta(e - p)$ is an invertible element with $\alpha, \beta \in \mathbb{C}$. Indeed, this is an immediate consequence of Theorem 6.20. However, there is a second proof which we present now. First we show that there are complex numbers λ, μ such that

$$\|\lambda a - e\| < 1, \qquad \|\mu b - e\| < 1, \tag{4}$$

$$(\lambda a_+ - 1)\,(\mu b_- - 1) = (\lambda a_- - 1)\,(\mu b_+ - 1). \tag{5}$$

Indeed if $a_+ = a_-$ $(= a)$ then put $\lambda = a^{-1}$. Obviously, $\lambda a - 1 = 0$ and (5) holds. Since $|\arg b_+/b_-| < \pi$ there is a $\mu_0 \in \mathbb{C}$ so that $\operatorname{Re} \mu_0 b_\pm > 0$. Clearly, for sufficiently small $t > 0$ we have $\|(t\mu_0)\, b - e\| < 1$. The case $b_+ = b_-$ can be treated analogously. Now let $a_+ \neq a_-$, $b_+ \neq b_-$.

For $t \in \mathbb{R}$ put

$$\lambda^{-1} = \frac{it(a_+ - a_-) + a_+ + a_-}{2}, \qquad \mu^{-1} = \frac{it(b_+ - b_-) + b_+ + b_-}{2}.$$

Then (5) is fulfilled. An easy computation shows that the inequalities (4) will follow once we have proved that

$$|t - \mathrm{i}| < \left|t - \mathrm{i}\,\frac{a_+ + a_-}{a_+ - a_-}\right|, \qquad |t - \mathrm{i}| < \left|t - \mathrm{i}\,\frac{b_+ + b_-}{b_+ - b_-}\right|, \tag{6}$$

or, what is essentially the same,

$$2t \operatorname{Re} z_i < |z_i|^2 - 1 \qquad (i = 1, 2), \tag{7}$$

where

$$z_1 = \mathrm{i}\,\frac{a_+ + a_-}{a_+ - a_-}, \qquad z_2 = \mathrm{i}\,\frac{b_+ + b_-}{b_+ - b_-}.$$

Because of (2) we have $a_+/a_- = \varrho_1\, e^{2\pi i\alpha_1}$, $b_+/b_- = \varrho_2\, e^{2\pi i \alpha_2}$ with $\varrho_i > 0$, $-1/2 < \alpha_i < 1/2$ $(i = 1, 2)$, $|\alpha_1 - \alpha_2| < 1/2$. Using

$$2 \operatorname{Re} z_i = \frac{4\varrho_i \sin 2\pi\alpha_i}{\varrho_i^2 - 2\varrho_i \cos 2\pi\alpha_i + 1}, \qquad |z_i|^2 - 1 = \frac{4\varrho_i \cos 2\pi\alpha_i}{\varrho_i^2 - 2\varrho_i \cos 2\pi\alpha_i + 1}$$

we get the inequalities

$$t \sin 2\pi\alpha_i < \cos 2\pi\alpha_i \qquad (i = 1, 2) \tag{8}$$

which are equivalent to those in (7). Clearly, if $\alpha_1\alpha_2 \geqq 0$ then (8) has a solution. Let $\alpha_1\alpha_2 < 0$ and without loss of generality assume $0 < \alpha_1 < 1/2$, $-1/2 < \alpha_2 < 0$. Then $0 < \alpha_1 - \alpha_2 < 1/2$ and

$$\frac{\cos 2\pi\alpha_1}{\sin 2\pi\alpha_1} - \frac{\cos 2\pi\alpha_2}{\sin 2\pi\alpha_2} = \frac{\sin 2\pi(\alpha_2 - \alpha_1)}{\sin 2\pi\alpha_1 \sin 2\pi\alpha_2} > 0.$$

Thus, in this case there exists also a value $t \in \mathbb{R}$ satisfying (8) (take t strongly between $(\cos 2\pi\alpha_1)/(\sin 2\pi\alpha_1)$ and $(\cos 2\pi\alpha_2)/(\sin 2\pi\alpha_2)$). Hence, the relations (4) and (5) are proved.

Our next concern is to prove the estimate

$$\|cp + d(e - p)\| \leqq \max\{\|c\|, \|d\|\} + (2\,\|c\|\,\|d\|\,|c_+d_- - c_-d_+|)^{1/4} \tag{9}$$

for coefficients of the form (1).

According to Theorem 7.2 and Section 1.15 the norm of $cp + d(e - p)$ can be computed as follows:

$$\|cp + d(e - p)\|^2 = \max_{\mu \in [0,1]} \left(\max\{\lambda_1(\mu), \lambda_2(\mu)\}\right),$$

where $\lambda_1(\mu)$ and $\lambda_2(\mu)$ are the roots of the polynomial

$$\det\left(B(\mu)\,B^*(\mu) - \lambda I\right),$$

I is the identity matrix and

$$B(\mu) = \begin{pmatrix} c_+\mu + c_-(1 - \mu) & (d_+ - d_-)\sqrt{\mu(1 - \mu)} \\ (c_+ - c_-)\sqrt{\mu(1 - \mu)} & d_+(1 - \mu) + d_-\mu \end{pmatrix}.$$

Put $\lambda(\mu) = \max\{\lambda_1(\mu), \lambda_2(\mu)\}$. By computation it can be checked immediately that

$$\lambda(\mu) = \frac{1}{2}\left(c(\mu) + \sqrt{c^2(\mu) - 4d(\mu)}\right),$$

where $c(\mu) = (|c_+|^2 + |d_-|^2)\,\mu + (|c_-|^2 + |d_+|^2)\,(1 - \mu)$,

$$d(\mu) = |c_+d_-\mu + c_-d_+(1 - \mu)|^2.$$

Suppose, for instance, that $|c_-|^2 + |d_+|^2 \geqq |c_+|^2 + |d_-|^2$. Then

$$c(\mu) \leqq |c_-|^2 + |d_+|^2,$$

$d(\mu) = |c_-d_+ + \mu(c_+d_- - c_-d_+)|^2 \geqq (|c_-d_+| - \mu\,|c_+d_- - c_-d_+|)^2$. Now we get

$$\begin{aligned} c^2(\mu) - 4d(\mu) &\leqq (|c_-|^2 + |d_+|^2)^2 - 4(|c_-d_+| - \mu\,|c_+d_- - c_-d_+|)^2 \\ &\leqq (|c_-|^2 - |d_+|^2)^2 + 8\,\|c\|\,\|d\|\,|c_+d_- - c_-d_+|. \end{aligned}$$

Thus,

$$\begin{aligned} |\lambda(\mu)| &\leqq \frac{1}{2}\left[|c_-|^2 + |d_+|^2 + \left||c_-|^2 - |d_+|^2\right| + \sqrt{8\,\|c\|\,\|d\|\,|c_+d_- - c_-d_+|}\right] \\ &\leqq \max\{\|c\|^2, \|d\|^2\} + \sqrt{2\,\|c\|\,\|d\|\,|c_+d_- - c_-d_+|}, \end{aligned}$$

which proves (9).

The element $A = ap + b(e - p)$ with coefficients a, b satisfying (1) and (2) can be written as

$$ap + b(e - p) = \big(\lambda ap + \mu b(e - p)\big)\big(\lambda^{-1}p + \mu^{-1}(e - p)\big)$$
$$= (I + T)\big(\lambda^{-1}p + \mu^{-1}(e - p)\big), \qquad \text{(by (4))}$$

where $T = (\lambda a - e)\,p + (\mu b - e)\,(e - p)$. The estimate $\|T\| < 1$ is an immediate consequence of (4), (5) and (9), and the representation (3) is proved. Now it is clear that the following corollary holds true.

7.12. Corollary. *Let χ be the characteristic function of any measurable subset E of $\mathbb{T}$ such that the measures of both E and $\mathbb{T} \setminus E$ are nonzero and let $a = (a_+ - a_-)\,\chi + a_-$, $b = (b_+ - b_-)\,\chi + b_-$, where $a_\pm$, $b_\pm$ are complex numbers satisfying*

$$\left|\arg \frac{a_+}{a_-}\right| < \pi, \qquad \left|\arg \frac{b_+}{b_-}\right| < \pi, \qquad \left|\arg \frac{a_+}{a_-} - \arg \frac{b_+}{b_-}\right| < \pi.$$

Then the sequence $\{P_n(aP + bQ)\,P_n\}$ is stable and moreover, the operators $P_n(aP + bQ)\,P_n$ are invertible on $\operatorname{im} P_n$ *for all $n \in \mathbb{N}_0$.*

7.13. Proposition. *Let $A = aP + bQ$, $a, b \in \overline{\mathbf{PC}}$, and suppose that $\{P_nAP_n\}^\circ$ is invertible in $\mathcal{A}^\circ$. Then*

$$\left|\arg \frac{a(t+)}{a(t-)}\right| < \pi, \qquad \left|\arg \frac{b(t+)}{b(t-)}\right| < \pi,$$
$$\left|\arg \frac{a(t+)}{a(t-)} - \arg \frac{b(t+)}{b(t-)}\right| < \pi, \tag{1}$$

or, equivalently,

$$\mu\,\frac{a(t+)}{a(t-)} + (1 - \mu)\,\frac{b(t+)}{b(t-)} \notin (-\infty, 0] \tag{2}$$

for all $t \in \mathbb{T}$, $\mu \in [0, 1]$.

Proof. We shall prove the assertion by means of local techniques. For any $\tau \in \mathbb{T}$, let N_τ denote the collection of all continuous functions f on $\mathbb{T}$ such that

(i) $0 \leqq f(t) \leqq 1$ for all $t \in \mathbb{T}$,

(ii) there exists a neighborhood U of τ so that the restriction of f to U is the constant function 1.

Define $M_\tau \in \mathcal{A}^\circ$ to be the set

$$M_\tau = \big\{\{P_nPfPP_n + P_nQfQP_n\}^\circ : f \in N_\tau\big\}.$$

Theorem 7.10 implies that M_τ actually forms a localizing class in the sense of Chapter 1. Moreover, the system $\{M_\tau\}_{\tau\in\mathbb{T}}$ is covering. Indeed, for any family $\{f_\tau\}_{\tau\in\mathbb{T}}$, $f_\tau \in N_\tau$, there exists a finite number of functions $f_{\tau_1}, \ldots, f_{\tau_n}$ such that

$$f(t) := \sum_{i=1}^{n} f_{\tau_i}(t) \geqq 1 \quad \text{for all} \quad t \in \mathbb{T}.$$

Consequently, the sequence $\{P_n(PfP + QfQ) P_n\}$ is stable by Theorem 4.84. Theorem 7.10 also gives that $\{P_n(aP + bQ) P_n\}^\circ$ $(a, b \in \mathbf{L}^\infty(\mathbb{T}))$ commutes with any element from $\bigcup_{\tau\in\mathbb{T}} M_\tau$. Let us prove that the conditions (1) are fulfilled. For simplicity take $t = 1$ and let χ_1 (resp. χ_2) be the characteristic function of the arc $(1, -1)$ (resp. $(1, \mathrm{i})$). Put $B := \{(a(1+) - a(1-))\chi_1 + a(1-)\} P + \{(b(1+) - b(1-))\chi_2 + b(1-)\} Q$. First we show the invertibility of $\{P_nBP_n\}^\circ$. For this end it suffices to show that $\{P_nBP_n\}^\circ$ is M_τ-equivalent to an invertible element $\{P_nB_\tau P_n\}^\circ$ for each $\tau \in \mathbb{T}$ (cf. Theorem 1.19). For the point $\tau = 1$ this is clear, since $\{P_nAP_n\}^\circ$ is invertible and $\{P_nBP_n\}^\circ$ is M_1-equivalent to this element. For $\tau = \mathrm{i}$ one has the M_i-equivalence of $\{P_nBP_n\}^\circ$ and $\{P_nB_\mathrm{i}P_n\}^\circ$, where

$$B_\mathrm{i} = a(1+) P + \{(b(1+) - b(1-))\chi_2 + b(1-)\} Q.$$

Since $P\tilde{a}P + Q\tilde{b}Q \in \Phi(\mathbf{L}^2(\mathbb{T}))$, we get by 7.1(iv)

$$b(1+)\mu + b(1-)(1-\mu) \neq 0, \qquad \mu \in [0, 1],$$

$$a(1+)\mu + a(1-)(1-\mu) \neq 0, \qquad \mu \in [0, 1].$$

This is equivalent to

$$\left|\arg \frac{a(1-)}{a(1+)}\right| < \pi, \qquad \left|\arg \frac{b(1+)}{b(1-)}\right| < \pi.$$

Now it is easy to see that B_i fulfils the conditions of Corollary 7.12, hence $\{P_nB_\mathrm{i}P_n\}^\circ$ is invertible. The case $\tau = -1$ can be treated analogously. Finally, for $\tau \neq \pm 1$, i there exist complex numbers $c = c(\tau)$, $d = d(\tau)$, $c \neq 0$, $d \neq 0$, such that $\{P_nBP_n\}^\circ$ is M_τ-equivalent to $\{P_n(cP + dQ) P_n\}^\circ$, which is, of course, invertible too. Therefore by virtue of Proposition 7.9(b), $\operatorname{ind} B = -\operatorname{ind} \tilde{B}$. Now we are going to compute these indices. Without loss of generality one can assume that $a(1-) = b(1-) = 1$, and for brevity put $a(1+) = \alpha$, $b(1+) = \beta$. The operator $\tilde{B}$ is clearly given by $\tilde{B} = B_1B_2 = B_2B_1$, where

$$B_1 = P\{(\alpha - 1)\tilde{\chi}_1 + 1\} P + Q, \qquad B_2 = P + Q\{(\beta - 1)\tilde{\chi}_2 + 1\} Q.$$

Since $\tilde{B}$ is Fredholm, each factor on the right-hand side must also be Fredholm (by 1.3(d)). Much more is true: since the coefficients of the right-hand side operators take only two values, their Fredholmness implies their invertibility (use 7.1(iv) and Theorem 7.2). Consequently, $\operatorname{ind} B = \operatorname{ind} \tilde{B} = 0$ and

$$\operatorname{ind}\left[\left(\frac{(\alpha - 1)\chi_1 + 1}{(\beta - 1)\chi_2 + 1}\right) P + Q\right] = 0.$$

By virtue 7.1(iv) this can only happen if the origin does not belong to the closed convex hull of the points α/β, α and 1. Now it is easily seen that the last assertion is equivalent to

$$|\arg \alpha - \arg \beta| < \pi.$$

The equivalence of (1) and (2) can be easily checked using that the first two inequalities of (1) are equivalent to $\dfrac{a(t+)}{a(t-)}, \dfrac{b(t+)}{b(t-)} \notin (-\infty, 0]$. ■

7.14. Theorem. *Let $a, b \in \overline{\mathbf{PC}}$ and $A = aP + bQ + T$, where $T \in \mathcal{K}(\mathbf{L}^2(\mathbb{T}))$. Put $A_n = P_nAP_n$.*

(a) *The element $\{A_n\}^\circ$ is invertible in $\mathcal{A}^\circ$ if and only if the operators $\tilde{a}P + Q$, $P + \tilde{b}Q$ are both Fredholm and one of the following conditions holds:*

(i) $$\left|\arg\frac{a(t+)}{a(t-)} - \arg\frac{b(t+)}{b(t-)}\right| < \pi, \quad t \in \mathbb{T};$$

(ii) $$\frac{a(t+)}{a(t-)}\mu + \frac{b(t+)}{b(t-)}(1-\mu) \notin (-\infty, 0], \quad t \in \mathbb{T}, \quad \mu \in [0, 1].$$

(b) *The finite section method converges for A if and only if the operators A, $\tilde{a}P + Q$, $P + \tilde{b}Q$ are invertible and at least one of the conditions* (i), (ii) *is fulfilled.*

Proof. (a) If $\{P_n(aP + bQ)P_n\}^\circ$ is invertible, then the conditions are fulfilled by Proposition 7.13. Suppose now that the conditions are satisfied. We shall prove the invertibility of $\{P_n(aP + bQ)P_n\}^\circ$ again by means of local techniques. Let $\{M_\tau\}_{\tau\in\mathbb{T}}$ be the covering system of localizing classes introduced in the proof of Proposition 7.13. For a given point $\tau \in \mathbb{T}$, let the functions $a_\tau(t)$, $b_\tau(t)$ be defined as follows:

$$a_\tau(t) = \big(a(\tau+) - a(\tau-)\big)\chi_\tau + a(\tau-),$$

$$b_\tau(t) = \big(b(\tau+) - b(\tau-)\big)\chi_\tau + b(\tau-),$$

where $\chi_\tau(t)$ is the characteristic function of the arc $(\tau, -\tau)$. Theorem 7.10 gives almost at once that the elements $\{P_n(aP + bQ)P_n\}^\circ$ and $\{P_n(a_\tau P + b_\tau Q)P_n\}^\circ$ are M_τ-equivalent for each $\tau \in \mathbb{T}$. The Fredholmness of the operators $\tilde{a}P + Q$, $P + \tilde{b}Q$ together with (i) shows that the conditions of Corollary 7.12 are fulfilled for the operator $a_\tau P + b_\tau Q$. Therefore $\{P_n(a_\tau P + b_\tau Q)P_n\}^\circ$ is invertible in $\mathcal{A}^\circ$. Theorem 1.19 applies and we get the invertibility of $\{P_n(aP + bQ)P_n\}^\circ$.
(b) Immediate from (a) and Theorem 7.7. ■

7.15. Theorem. *Let $A = aP + bQ$ be a singular integral operator on $\mathbf{L}^2(\mathbb{T})$ with coefficients $a, b \in \overline{\mathbf{PC}}$. This operator can be represented in the form*

$$A = (I + T)(cP + dQ) + K \tag{1}$$

with $\|T\| < 1$, $K \in \mathcal{K}(\mathbf{L}^2(\mathbb{T}))$, $c, d \in \mathbf{GC}(\mathbb{T})$ if and only if

$$\left|\arg\frac{a(t+)}{a(t-)}\right| < \pi, \quad \left|\arg\frac{b(t+)}{b(t-)}\right| < \pi, \quad \left|\arg\frac{a(t+)}{a(t-)} - \arg\frac{b(t+)}{b(t-)}\right| < \pi \tag{2}$$

for all $t \in \mathbb{T}$.

Proof. If A has the form (1), then $\{P_nAP_n\}^\circ$ proves to be in $\mathbf{G}\mathcal{A}^\circ$. Indeed, $ac^{-1}P + bd^{-1}Q = I + T + K_1$, with $K_1 \in \mathcal{K}(\mathbf{L}^2(\mathbb{T}))$. Thus, $\{P_nTP_n\} \in \mathcal{A}$ and $\{P_nTP_n\}^\circ$ commutes with every element from $\bigcup_{\tau\in\mathbb{T}} M_\tau$, M_τ being the localizing classes introduced in the proof of Proposition 7.13. By using local techniques (similarly as in the previous proof) we get the invertibility of $\{P_nAP_n\}^\circ$ in $\mathcal{A}^\circ$. So Proposition 7.13 applies, which leads to (2).

Conversely, suppose (2) is true. Fix $\tau \in \mathbb{T}$ and let χ_τ be the characteristic function of the arc $(\tau, -\tau)$. From Section 7.11 we know that there are complex numbers $\lambda_\tau \neq 0$, $\mu_\tau \neq 0$ such that

$$\|(\lambda_\tau a_\tau - 1) P + (\mu_\tau b_\tau - 1) Q\| < 1,$$

where

$$a_\tau(t) = \big(a(\tau+) - a(\tau-)\big) \chi_\tau + a(\tau-), \qquad b_\tau(t) = \big(b(\tau+) - b(\tau-)\big) \chi_\tau + b(\tau-).$$

We claim that the function

$$\mathbb{T} \ni \omega \mapsto \|(\lambda_\tau a_\omega - 1) P + (\mu_\tau b_\omega - 1) Q\| \tag{3}$$

is upper-semicontinuous. If $\omega_0 \in \mathbb{T}$ is a point such that a and b are continuous in a neighborhood of ω_0, then, obviously, the function (3) is continuous at ω_0. If the functions a, b are continuous at ω_0, but in every neighborhood of ω_0 there are points of discontinuity, the continuity of the function (3) follows from 7.11(9). Now let ω_0 be a point at which at least one of the functions a, b has a jump. Using the inequality

$$\max \{\|c\|_\infty, \|d\|_\infty\} \leqq \|cP + dQ\|, \qquad c, d \in \mathbf{L}^\infty(\mathbb{T})$$

one can easily obtain that the function (3) is upper-semicontinuous at ω_0. Hence, there is an open subarc U_τ containing τ such that

$$\|(\lambda_\tau a_\omega - 1) P + (\mu_\tau b_\omega - 1) Q\| < 1 \tag{4}$$

for all $\omega \in U_\tau$. The compactness of $\mathbb{T}$ implies that we can select a finite number of sets U_{τ_i} such that $\bigcup_{i=1}^{k} U_{\tau_i} = \mathbb{T}$. Without loss of generality we can assume that $U_{\tau_i} \cap U_{\tau_{i+1}} \neq \emptyset$, $i = 1, \ldots, k$, where $\tau_{k+1} = \tau_1$, and $U_{\tau_i} \cap U_{\tau_j} = \phi$, $\tau_j \notin \{\tau_{i-1}, \tau_i, \tau_{i+1}\}$. Now we are going to construct continuous functions $\lambda(t)$ and $\mu(t)$ as in Fig. 1.

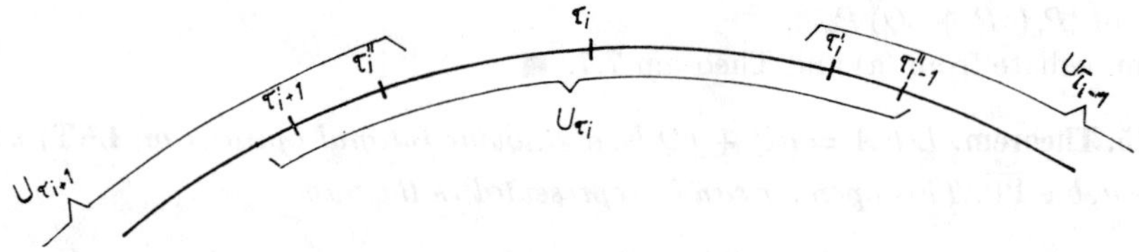

Fig. 1

Take continuous f_i on (τ_i'', τ_{i+1}') such that $0 \leqq f_i(t) \leqq 1$ for all $t \in (\tau_i'', \tau_{i+1}')$ and $f_i(\tau_i'') = 1$, $f_i(\tau_{i+1}') = 0$ $(i = 1, \ldots, k)$ and put

$$\lambda(t) = \begin{cases} \lambda_{\tau_i}, & t \in (\tau_i', \tau_i'') \\ \big(1 - f_i(t)\big) \lambda_{\tau_{i+1}} + f_i(t) \lambda_{\tau_i}, & t \in (\tau_i'', \tau_{i+1}') \end{cases}$$

Define $\mu(t)$ analogously. These functions are obviously continuous and do not vanish on $\mathbb{T}$. Consider the essential norm, $\|\cdot\|_{\text{ess}}$, of $A_0 = (\lambda a - 1) P + (\mu b - 1) Q$, that is the norm of the coset

$$A_0 + \mathcal{K}\big(\mathbf{L}^2(\mathbb{T})\big)$$

in the Calkin algebra $\mathcal{L}(\mathbf{L}^2(\mathbb{T}))/\mathcal{K}(\mathbf{L}^2(\mathbb{T}))$. We claim that this norm is less than one. Indeed, by virtue of the Theorems 7.2 and 1.21 we have

$$\|A_0\|_{\mathrm{ess}} = \max_{\tau\in\mathbb{T}} \|A_\tau\|_{\mathrm{ess}},$$

$A_\tau = (\lambda(\tau)\, a_\tau - 1)\, P + (\mu(\tau)\, b_\tau - 1)\, Q$. Therefore, it suffices to check that $\|A_\tau\| < 1$ for each $\tau \in \mathbb{T}$. If $\tau \in (\tau_i', \tau_i'')$ for a certain i or $\tau = \tau_i', \tau_i''$ then obviously $\|A_\tau\| < 1$. If $\tau \in (\tau_i'', \tau_{i+1}')$ for a certain i, then (because of (4))

$$\|A_\tau\| = \Big\|(1 - f_i(\tau))\, \{(\lambda_{\tau_{i+1}} a_\tau - 1)\, P + (\mu_{\tau_{i+1}} b_\tau - 1)\, Q\} + f_i(\tau)\, \{(\lambda_{\tau_i} a_\tau - 1)\, P + (\mu_{\tau_i'} b_\tau - 1)\, Q\}\Big\| < 1.$$

Hence $\|A_0\|_{\mathrm{ess}} < 1$ and there is a $K_1 \in \mathcal{K}(\mathbf{L}^2(\mathbb{T}))$ such that

$$\|A_0 + K_1\| < 1.$$

Now $aP + bQ$ can be represented in the form

$$aP + bQ = (I + (A_0 + K_1))\, (\lambda^{-1}P + \mu^{-1}Q) + K,$$

$K \in \mathcal{K}(\mathbf{L}^2(\mathbb{T}))$. ■

7.16. Theorem. *Let $A = Pa + Qb$ be a singular integral operator on $\mathbf{L}^2(\mathbb{T})$ with coefficients $a, b \in \overline{\mathbf{PC}}$.*

(a) *A can be represented in the form*

$$A = (Pc + Qd)\, (I + T) + K$$

with $\|T\| < 1$, $K \in \mathcal{K}(\mathbf{L}^2(\mathbb{T}))$, $c, d \in \mathbf{GC}(\mathbb{T})$ if and only if

$$\left|\arg \frac{a(t+)}{a(t-)}\right| < \pi, \qquad \left|\arg \frac{b(t+)}{b(t-)}\right| < \pi, \qquad \left|\arg \frac{a(t+)}{a(t-)} - \arg \frac{b(t+)}{b(t-)}\right| < \pi$$

for all $t \in \mathbb{T}$.

(b) *The finite section method converges for the operator $Pa + Qb + T$, $T \in \mathcal{K}(\mathbf{L}^2(\mathbb{T}))$, if and only if the operators $Pa + Qb + T$, $P\tilde{a} + Q$, $P + Q\tilde{b}$ are invertible and*

$$\left|\arg \frac{a(t+)}{a(t-)} - \arg \frac{b(t+)}{b(t-)}\right| < \pi$$

for all $t \in \mathbb{T}$.

Proof. Take adjoints and use Theorems 7.14 and 7.15. ■

The collocation method

7.17. Theorem. *Let $a, b \in \overline{\mathbf{PC}}$ and $A_n := L_n(aP + bQ)P_n(\in \mathbf{L}^2(\mathbb{T}))$. Then $\{A_n\} \in \mathcal{A}$ and $\tilde{A} = \tilde{a}P + \tilde{b}Q$.*

Proof. (i) The existence of s-lim $L_n(aP + bQ)\, P_n = aP + bQ$ was already proved in 3(c).

(ii) We prove that $A_n^*P_n^* = (PL_n\bar{a}I + QL_n\bar{b}I)\, P_n$ and $A_n^*P_n^*$ converge strongly to $P\bar{a}I + Q\bar{b}I$. Denote by (x, y) the value of

$$\frac{1}{2\pi} \int_0^{2\pi} x(\mathrm{e}^{\mathrm{i}\varphi})\, \overline{y(\mathrm{e}^{\mathrm{i}\varphi})}\, d\varphi,$$

where $x, y \in \mathbf{L}^2(\mathbb{T})$. From 7.3(4) we obtain

$$\left(A_nP_n\left(\sum_{k=-\infty}^{\infty} x_kt^k\right), \quad \sum_{k=-\infty}^{\infty} y_kt^k\right)$$

$$= \sum_{k=-n}^{n} (2n+1)^{-1} \sum_{j=-n}^{n} \left(a(t_j) \sum_{m=0}^{n} x_mt_j^{m-k} + b(t_j) \sum_{m=-n}^{-1} x_mt_j^{m-k}\right) \overline{y_k}$$

$$= \sum_{m=0}^{n} x_m(2n+1)^{-1} \overline{\sum_{j=-n}^{n} \overline{a(t_j)} \sum_{k=-n}^{n} y_kt_j^{k-m}} + \sum_{m=-n}^{-1} x_m(2n+1)^{-1} \overline{\sum_{j=-n}^{n} \overline{b(t_j)} \sum_{k=-n}^{n} y_kt_j^{k-m}}$$

$$= \left(x, (PL_n\bar{a}I + QL_n\bar{b}I)\, P_ny\right).$$

The strong convergence of the operators $A_n^*P_n^* = A_n^*P_n$ follows analogously to (i).

(iii) Consider $W_nA_nW_n = \tilde{A}_nP_n$. Taking into account that $t_j^{2n+1} = 1$ ($j = 0, \pm 1, \ldots, \pm n$) we get

$$W_nL_naW_nx = W_n \sum_{m=-n}^{n} (2n+1)^{-1} \sum_{j=-n}^{n} a(t_j) \left(\sum_{k=0}^{n} x_{n-k}t_j^{k-m} + \sum_{k=-n}^{-1} x_{-n-k-1}t_j^{k-m}\right) t^m$$

$$= W_n \sum_{m=-n}^{n} (2n+1)^{-1} \sum_{j=-n}^{n} a(t_j) \sum_{k=-n}^{n} x_kt_j^{n-k-m}t^m$$

$$= \sum_{m=-n}^{n} (2n+1)^{-1} \sum_{j=-n}^{n} a(t_j) \sum_{k=-n}^{n} x_kt_j^{m-k}t^m$$

$$= \sum_{m=-n}^{n} (2n+1)^{-1} \sum_{j=-n}^{n} a(t_j^{-1}) \sum_{k=-n}^{n} x_kt_j^{k-m}t^m = L_n\tilde{a}P_nx,$$

where $\tilde{a}(t) := a(1/t)$. Consequently

$$W_nA_nW_n = L_n(\tilde{a}P + \tilde{b}Q)\, P_n \to \tilde{a}P + \tilde{b}Q =: \tilde{A} \quad \text{strongly by (i).}$$

(iv) Using the results of (ii) and (iii) we obtain that $\tilde{A}_n^*P_n^* = (PL_n\bar{\tilde{a}}I + QL_n\bar{\tilde{b}}I)\, P_n \to P\bar{\tilde{a}}I + Q\bar{\tilde{b}}I = \tilde{A}^*$ strongly. ■

The following result will prove to be of great importance for the application of local techniques.

7.18. Corollary. *For all $f \in \mathbf{C}(\mathbb{T})$ the sequences $\{QL_nfPP_n\}$ and $\{PL_nfQP_n\}$ are elements of the ideal $\mathcal{J}$.*

Proof. To begin with, we remark that

$$Q_nPfPW_n = Q_nPfPP_nW_n = t^nWQ\tilde{f}PP_n$$

(by Theorem 7.10(a)). Thus,

$$QL_nfPP_n = QP_nfPP_n + QL_nQ_nfPP_n$$

$$= P_nQfPP_n + Q(L_n - P_n)\, QfPP_n + W_nK_nW_n,$$

where $K_n = W_nQL_nt^nWQ\tilde{f}PP_n$. This is equivalent to

$$QL_nfPP_n = P_nT_1P_n + W_nT_2W_n + C_n$$

with $T_1 = QfP$, $T_2 = Q\tilde{f}P$ and

$$C_n = Q(L_n - P_n)\, QfPP_n - W_n(Q\tilde{f}P - K_n)\, W_n.$$

If the function f is a polynomial (in t and $1/t$), and, consequently, QfP, $Q\tilde{f}P$ are elements from $\mathcal{K}(\mathbf{L}^2(\mathbb{T}), \mathbf{C}(\mathbb{T}))$, we obtain with the help of 1.1(h) that

$$\|Q(L_n - P_n)\, QfPP_n\| \to 0 \quad \text{as} \quad n \to \infty.$$

We have

$$\begin{aligned}(QL_nt^nWg)(t) &= \sum_{k=-n}^{-1} (2n+1)^{-1} \sum_{j=-n}^{n} t_j^n g(t_{-j})\, t_j^{-k} t^k \\ &= \sum_{k=-n}^{-1} (2n+1)^{-1} \sum_{j=-n}^{n} g(t_j)\, t_j^{k-n} t^k \\ &= \sum_{m=-1}^{-n} (2n+1)^{-1} \sum_{j=-n}^{n} g(t_j)\, t_j^{-m} t^{-(m+n+1)} \\ &= (W_nQL_ng)(t).\end{aligned}$$

This gives $K_n = QL_nQ\tilde{f}PP_n$ and $\|K_n - Q\tilde{f}P\| \to 0$ as $n \to \infty$. Therefore, $\{QL_nfPP_n\} \in \mathcal{J}$ for any polynomial f. Now the assertion follows for all $f \in \mathbf{C}(\mathbb{T})$ from the estimate

$$\|\{QL_nfPP_n\}\| \leqq \|f\|_\infty,$$

which is a consequence of 7.3(b). ■

7.19. Theorem. *Let $a, b \in \overline{\mathbf{PC}}$ and let $A = aP + bQ + T \in \mathcal{L}(\mathbf{L}^2(\mathbb{T}))$, $T \in \mathcal{K}(\mathbf{L}^2(\mathbb{T}), \mathcal{R}(\mathbb{T}))$, $A_n := L_nAP_n$.*

(a) *The element $\{L_n(aP + bQ)\, P_n\}^\circ$ is invertible in $\mathcal{A}^\circ$ if and only if $aP + bQ$ is Fredholm.*

(b) *The sequence $\{A_n\}$ is stable if and only if the operator A is invertible in $\mathbf{L}^2(\mathbb{T})$.*

(c) *The collocation method converges for every right-hand side $f \in \mathcal{R}(\mathbb{T})$ if the operator A is invertible.*

Proof. (a): Because of (a) in Section 7.7 it suffices to show that $\{L_n(aP + bQ)\, P_n\}^\circ$ is invertible in $\mathcal{A}^\circ$ if $aP + bQ$ is Fredholm. If $aP + bQ$ is Fredholm, then $a, b \in \mathbf{G}(\overline{\mathbf{PC}})$ and we have $L_n(aP + bQ)\, P_n = (L_nbP_n)\, \big(L_n(cP + Q)\, P_n\big)$ with $c = b^{-1}a$, and $L_nbP_nL_nb^{-1} \times P_n = L_nb^{-1}P_nL_nbP_n = P_n$. Thus, we have only to prove the invertibility of $\{L_n(cP + Q)\, P_n\}^\circ$, where the operator $cP + Q$ is, of course, Fredholm. For any point $\tau \in \mathbb{T}$ define the function

$$c_\tau(t) := \big(c(\tau+) - c(\tau-)\big)\, \chi_\tau + c(\tau-),$$

where χ_τ is the characteristic function of the arc $(\tau, -\tau)$. By 7.1(iv),

$$c(\tau+)\, \mu + c(\tau-)\, (1 - \mu) \neq 0 \qquad (0 \leqq \mu \leqq 1)$$

and, consequently, there is a number $\lambda \in \mathbb{C} \setminus \{0\}$ such that

$$\|1 - \lambda c_\tau\| < 1.$$

This implies $L_n(c_\tau P + Q) P_n = \lambda^{-1}\big(P_n - L_n(1 - \lambda c_\tau) PP_n\big)(P_nP + \lambda P_nQ)$, and by virtue of 7.3(b) we obtain the invertibility of $L_n(c_\tau P + Q) P_n$. Thus, the element $\{L_n(c_\tau P + Q) P_n\}^\circ$ is invertible in $\mathcal{A}^\circ$. Now define M_τ to be the set

$$M_\tau = \big\{\{L_n f P_n\}^\circ : f \in N_\tau\big\},$$

where N_τ is defined as in the proof of Proposition 7.13. Obviously, the system $\{M_\tau\}_{\tau\in\mathbb{T}}$ forms a covering system of localizing classes in $\mathcal{A}^\circ$. Moreover, by a simple computation we get

$$\{L_n(dP + eQ) P_n\} \{L_n f P_n\} - \{L_n f P_n\} \{L_n(dP + eQ) P_n\}$$
$$= \{L_n(d - e) P_n\} \{PL_n f QP_n - QL_n f PP_n\}$$

which, by Corollary 7.18, means that $\{L_n(dP + eQ) P_n\}^\circ$ ($d, e \in \overline{\mathbf{PC}}$) commutes with any element of $\bigcup_{\tau\in\mathbb{T}} M_\tau$. Now it is clear that $\{L_n(cP + Q) P_n\}^\circ$ is M_τ-equivalent to the invertible element $\{L_n(c_\tau P + Q) P_n\}^\circ$ for any $\tau \in \mathbb{T}$. So Theorem 1.19 can be applied to deduce that $\{L_n(cP + Q) P_n\}^\circ$ is invertible.

(b): Since L_n converges strongly on $\mathcal{R}(\mathbb{T})$ to the continuous embedding operator of $\mathcal{R}(\mathbb{T})$ into $\mathbf{L}^2(\mathbb{T})$ and because $T \in \mathcal{K}\big(\mathbf{L}^2(\mathbb{T}), \mathcal{R}(\mathbb{T})\big)$, we see that $\|L_nTP_n - T\| \to 0$ as $n \to \infty$ (by 1.1(h)). Thus, $\{L_nTP_n\} \in \mathcal{J}$ and $\{A_n\} \in \mathcal{A}$. Assume now the stability of $\{A_n\}$. By virtue of Theorem 7.7 we get the invertibility of A.

Vice versa, let A be invertible. By 1.3(c) the operator $aP + bQ$ is Fredholm with index zero and Coburn's theorem shows the invertibility of $aP + bQ$. Thus, by virtue of 7.1(ii), the operator $\tilde{a}P + \tilde{b}Q$ is invertible too. Part (a) and Theorem 7.7 gives the stability of $\{A_n\}$ at once.

(c): Immediate from (b) and 7.3(a). ■

Systems

7.20. Definitions. For any linear operator A given on a Banach space X or on a linear subset of X, the operator

$$(\delta_{ij}A)_{i,j=1}^N$$

defined on X_N or on a linear subset of X_N, respectively, will also be denoted by A.

In particular this definition applies to the operators $P, Q, P_n, W_n, L_n, \ldots$ in the case $X = \mathbf{L}^2(\mathbb{T})$. The definitions of the algebra $\mathcal{A}$ and the ideal $\mathcal{J}$ (cf. Sections 7.5 and 7.6) extend to the system case in the natural manner. We shall denote them by $\mathcal{A}_N$ and $\mathcal{J}_N$. In the following section we need the space $\mathbf{L}^2_N(-1, 1)$ and the operators $P_{[-1,1]}$, $Q_{[-1,1]}$, uI, vI, defined on $\mathbf{L}^2_N(-1, 1)$ by

$$P_{[-1,1]} = \frac{1}{2}\,(S_{[-1,1]} + I), \qquad Q_{[-1,1]} = \frac{1}{2}\,(I - S_{[-1,1]}),$$

where

$$(S_{[-1,1]}\varphi)(t) = \frac{1}{\pi \mathrm{i}} \int_{-1}^{1} \frac{\varphi(\tau)}{\tau - t}\,\mathrm{d}\tau,$$

and

$$v(t) = \begin{cases} 0, & \text{if } t \in (-1, 0), \\ 1, & \text{if } t \in (0, 1) \end{cases} \qquad uI = I - vI.$$

7.21. Theorem. *Let a und b be matrix valued functions from* $(\overline{\mathbf{PC}})_{N\times N}$ *and consider the operator* $A = aP + bQ + T$ *on* $\mathbf{L}^2_N(\mathbb{T})$, $T \in \mathcal{K}(\mathbf{L}^2_N(\mathbb{T}))$. *The finite section method (with respect to the system* $\{P_n\}_{n=0}^\infty$*) converges for the operator A if and only if the following conditions are fulfilled.*

(i) *The operators A and* $\tilde{a}P + Q$, $P + \tilde{b}Q$ *are invertible;*

(ii) *The operator-valued function*

$$\begin{aligned}&\operatorname{smb}_{\{P_n\}}(aP + bQ)\colon \mathbb{T} \to \mathcal{L}(\mathbf{L}^2_N(-1, 1)),\\ &\tau \mapsto P_{[-1,1]}\{a(\tau+)\, vI + b(\tau+)\, uI\} + Q_{[-1,1]}\{a(\tau-)\, vI + b(\tau-)\, uI\}\end{aligned} \tag{1}$$

takes invertible values only.

The proof of this theorem is based upon local techniques; we omit the (nontrivial) details. We merely mention that in general there is no obvious procedure for deciding whether the operator $\operatorname{smb}_{\{P_n\}}(aP + bQ)(\tau)$ is invertible or not. Without doubt further investigations are needed. Recently, however, this question was completely solved in Spitkowski/Tashbaev [1] for $N = 2$.

7.22. Theorem. *Let* $a, b \in (\overline{\mathbf{PC}})_{N\times N}$ *and let* $A = aP + bQ + T$, $T \in \mathcal{K}(\mathbf{L}^2_N(\mathbb{T}), \mathcal{R}_N(\mathbb{T}))$, $A_n = L_n A P_n$.

(a) *The element* $\{A_n\}^\circ$ *is invertible in* $\mathcal{A}^\circ_N$ *if and only if* $aP + bQ$ *is Fredholm.*

(b) *The sequence* $\{A_n\}$ *is stable if and only if the operators A and* $\tilde{a}P + \tilde{b}Q$ *are invertible on* $\mathbf{L}^2_N(\mathbb{T})$.

(c) *The collocation method converges for every right-hand side* $f \in \mathcal{R}_N(\mathbb{T})$ *if the operators A and* $\tilde{a}P + \tilde{b}Q$ *are invertible.*

Proof. The proof is a copy of the proof of Theorem 7.19, with only minor modifications concerning part (a). More precisely, some complications arise in the proof that $\{L_n(aP + bQ)\, P_n\}^\circ$ is invertible in $\mathcal{A}^\circ_N$ if $aP + bQ$ is Fredholm. Exactly as in the proof of Theorem 7.19, it is sufficient to assume b is the identity matrix. Thus, let $aP + Q$ be Fredholm. Then, for any $\tau \in \mathbb{T}$, the segment

$$a(\tau+)\, \mu + a(\tau-)\, (1 - \mu), \qquad \mu \in [0, 1]$$

consists only of invertible matrices (cf. Mikhlin/Prössdorf [1]). Now, fix $\tau \in \mathbb{T}$ and assume (without loss of generality) that $a(\tau-)$ is the identity matrix. Define the operator A_τ to be

$$\{(a(\tau+) - I)\, \chi_\tau + I\}\, P + Q.$$

Of course, $\{L_n(aP + Q)\, P_n\}^\circ$ and $\{L_n A_\tau P_n\}^\circ$ are M_τ-equivalent (compare with the proof of Theorem 7.19). We have to show the invertibility of $\{L_n A_\tau P_n\}^\circ$. Let $a(\tau+) = DFD^{-1}$, where F is the Jordan canonical form of $a(\tau+)$

$$\begin{pmatrix} \lambda_1 & 1 & & & & & & \\ & \ddots & \ddots & & & & & \\ & & \lambda_1 & 1 & & & & \\ & & & \ddots & \ddots & & & \\ & & & & \lambda_k & 1 & & \\ & & & & & \ddots & \ddots & \\ & & & & & & \lambda_k & 1 \\ & & & & & & & \lambda_k \end{pmatrix}.$$

Obviously, the segment $\{\mu F + (1-\mu) I : \mu \in [0, 1]\}$ consists only of invertible matrices and $\{L_n A_\tau P_n\}^\circ$ is invertible in $\mathcal{A}^\circ$ if and only if $\{L_n(\{(F - I)\chi_\tau + I\} P + Q) P_n\}^\circ$ is so. Let $F_1 = F - \operatorname{diag}(\lambda_1, \ldots, \lambda_k)$ and write

$$\begin{aligned} L_n(\{(F-I)\chi_\tau + I\} P + Q) P_n &= L_n(\{(E-I)\chi_\tau + I\} P + Q) P_n + L_n F_1 \chi_\tau P P_n \\ &= L_n C P_n + L_n F_1 \chi_\tau P P_n \qquad (E = F - F_1). \end{aligned}$$

Since $\lambda_i \mu + (1-\mu) \neq 0$ for $\mu \in [0, 1]$, the element $L_n\{(\{(\lambda_i - 1)\chi_\tau + 1\} P + Q) P_n\}^\circ$ is invertible in $\mathcal{A}_1^\circ$ by Theorem 7.19. Thus, $\{L_n F_1 P_n\}^\circ$ must be invertible in $\mathcal{A}_N^\circ$. Write

$$\begin{aligned} \{L_n(\{(F-I)\chi_\tau + I\} P + Q) P_n\}^\circ = \{L_n C P_n\}^\circ \big(\{P_n\}^\circ + (\{L_n C P_n\}^\circ)^{-1} \\ \times \{L_n F_1 \chi_\tau P P_n\}^\circ\big), \end{aligned}$$

where $\big((\{L_n C P\}^\circ)^{-1} \{L_n F_1 \chi_\tau P P_n\}^\circ\big)^N = 0$. This shows the invertibility of $\{L_n(\{(F - I)\chi_\tau + I\} P + Q) P_n\}^\circ$ and, thus, that of $\{L_n A_\tau P_n\}^\circ$. Now the proof can be finished as the proof of Theorem 7.19. ∎

Subalgebras

7.23. Definitions. Define $\operatorname{alg}_{\{P_n\}} S(\overline{PC})$ $\big(\operatorname{alg}_{\{L_n\}} S(\overline{PC})\big)$ to be the smallest C^*-subalgebra of $\mathcal{A}_N$ containing $\{\{P_n(aP + bQ) P_n\} : a, b \in (\overline{PC})_{N\times N}\}$ and $\mathcal{J}_N\big(\{\{L_n(aP + bQ) P_n\} : a, b \in (\overline{PC})_{N\times N}\}$ and $\mathcal{J}_N\big)$ (recall that we consider the case $\mathbf{L}_N^2(\mathbb{T})$). Thus, an element c from $\operatorname{alg}_{\{P_n\}} S(\overline{PC})$ $\big(\operatorname{alg}_{\{L_n\}} S(\overline{PC})\big)$ is invertible in this algebra if and only if c is invertible in $\mathcal{A}_N$. Put

$$\operatorname{alg}^\circ_{\{P_n\}} S(\overline{PC}) := \operatorname{alg}_{\{P_n\}} S(\overline{PC})/\mathcal{J}_N, \qquad \operatorname{alg}^\circ_{\{L_n\}} S(\overline{PC}) := \operatorname{alg}_{\{L_n\}} S(\overline{PC})/\mathcal{J}_N.$$

Of course, these algebras turn out to be C^*-subalgebras of $\mathcal{A}_N^\circ$. If an element c belongs to one of these subalgebras, then it is invertible in this subalgebra if and only if it is invertible in $\mathcal{A}_N^\circ$. There are some situations in which the knowledge of the structure of these subalgebras is of interest. We mention here only one of them. If $\{A_n\}$ is stable, then $\sup_{n \geq n_0} \|A_n^{-1} P_n\|$ is a quantity on which some constants arising in convergence speed estimates depend (cf. Section 1.23). If these constants are large then, as a rule, difficulties emerge in practical numerical computations.

Obviously, we have

$$\|(\{A_n\}^\circ)^{-1}\| \leq \sup_{n \geq n_0} \|A_n^{-1} P_n\|$$

in the case under consideration. Therefore, norm estimates for $(\{A_n\}^\circ)^{-1}$ are of interest.

Introduce the C^*-algebra $\mathcal{B}$ of all bounded functions $f : \mathbb{T} \to \mathcal{L}\big(\mathbf{L}_N^2(-1, 1)\big)$ with the norm $\|f\| = \sup_{\tau \in \mathbb{T}} \|f(\tau)\|_{\mathcal{L}(\mathbf{L}_N^2(-1,1))}$ and denote by $\operatorname{smb}_{\{P_n\}} (\overline{PC})$ the smallest C^*-subalgebra of $\mathcal{B}$ containing all functions of the form 7.21(1).

7.24. Theorem. *The algebra* $\operatorname{alg}^\circ_{\{P_n\}} S(\overline{PC})$ *is isometrically isomorphic to the algebra* $\operatorname{smb}_{\{P_n\}} (\overline{PC})$. *The isomorphism can be chosen so that it takes* $\{P_n(aP + bQ) P_n\}^\circ$ *into* $\operatorname{smb}_{\{P_n\}} (aP + bQ)$.

We shall not prove this theorem, but a few remarks are in order.

(i) For $N = 1$ and fixed $\tau \in \mathbb{T}$ the spectrum of the operator $\operatorname{smb}_{\{P_n\}}(aP + bQ)(\tau)$ coincides with the convex hull $\Lambda(\tau)$ of the points $\frac{a(\tau+)}{a(\tau-)}, \frac{b(\tau+)}{b(\tau-)}$, 1 (use Proposition 7.13). If the distance of the origin to $\Lambda(\tau)$ is small for a certain $\tau \in \mathbb{T}$, then $\left\|\left(\{P_n(aP + bQ) P_n\}^\circ\right)^{-1}\right\|$ must be large.

(ii) The function $\{P_n(aP + bQ + T) P_n\} \mapsto \operatorname{smb}_{\{P_n\}}(aP + bQ)$ plays the same role in the study of the finite section method as the usual symbol of $aP + bQ + T$ does in the Fredholm theory $\left(a, b \in (\overline{\mathbf{PC}})_{N\times N},\ T \in \mathcal{K}(\mathbf{L}^2_N(\mathbb{T}))\right)$. Thus, the symbol of $aP + bQ + T$ with respect to the finite section method is operator-valued.

7.25. Theorem. *The algebra* $\operatorname{alg}^\circ_{\{L_n\}} \mathbf{S}(\overline{\mathbf{PC}})$ *is isometrically isomorphic to the algebra* $\operatorname{alg} \mathbf{S}(\overline{\mathbf{PC}})/\mathcal{K}\left(\mathbf{L}^2_N(\mathbb{T})\right)$. *The isomorphism can be chosen so that it takes*

$$\{L_n(aP + bQ) P_n\}^\circ \quad \text{into the coset} \quad aP + bQ + \mathcal{K}\left(\mathbf{L}^2_N(\mathbb{T})\right).$$

We again omit the proof and restrict ourselves to some remarks.

(i) $\operatorname{alg} \mathbf{S}(\overline{\mathbf{PC}})$ denotes the smallest C^*-subalgebra containing the set $\{aP + bQ\colon a, b \in (\overline{\mathbf{PC}})_{N\times N}\}$. It is well-known that $\mathcal{K}\left(\mathbf{L}^2_N(\mathbb{T})\right) \subset \operatorname{alg} \mathbf{S}(\overline{\mathbf{PC}})$ (cf. Mikhlin/Prössdorf [1, Chapt. V, § 8]).

(ii) A proof of this theorem can be given by means of Theorem 1.15 and Theorem 1.21.

(iii) The symbol of $aP + bQ$ with respect to the collocation method coincides with the usual symbol of $aP + bQ$ with respect to Fredholmness.

A generalization

7.26. Preliminaries and notations. We have seen that Banach algebra techniques are a powerful tool for investigating the finite section method and the collocation method. It should be noted that the approach discussed so far is to restrictive in numerous cases. This is mainly due to the following two facts. First, it may happen, that the sequence $\{W_n\}$ is not uniformly bounded and secondly, it may occur that the ideal $\mathcal{J}$ is to small. In order to overcome these difficulties we shall have a general look at these things and explain some examples. The generalization given below will play an important role in what follows (especially in Chapter 10).

Suppose we are given a Banach space X and a uniformly bounded sequence $\{P_n\}_{n=0}^\infty$ of projections on X.

Let Ω be a certain index set, which may be infinite. Let X_ω $(\omega \in \Omega)$ be Banach spaces and $\{L_n^{(\omega)}\}_{n=0}^\infty$ be sequences of projections on X_ω. Suppose that $L_n^{(\omega)} \to I$, $I \in \mathcal{L}(X_\omega)$ and $(L_n^{(\omega)})^* \to I^*$ strongly and assume there exist invertible operators $E_n^{(\omega)}\colon \operatorname{im} P_n \to \operatorname{im} L_n^{(\omega)}$ such that

$$\sup_{n,\omega} \|(E_n^{(\omega)})^{\pm 1}\| = M < \infty.$$

Now define $\mathcal{A}$ to be the collection of all sequences $\{A_n\}_{n=0}^\infty$ with $A_n\colon \operatorname{im} P_n \to \operatorname{im} P_n$ such that

(a) $\sup\limits_n \|A_nP_n\| < \infty$.

(b) there exist operators $W_\omega\{A_n\} \in \mathcal{L}(X_\omega)$ so that

$$E_n^{(\omega)} A_n (E_n^{(\omega)})^{-1} \to W_\omega\{A_n\} \quad \text{strongly},$$
$$(E_n^{(\omega)} A_n (E_n^{(\omega)})^{-1})^* \to W_\omega^*\{A_n\} \quad \text{strongly}$$

for all $\omega \in \Omega$.

On defining $\{A_n\} + \{B_n\} = \{A_n + B_n\}$, $\{A_n\}\{B_n\} = \{A_n B_n\}$, $\|\{A_n\}\|_{\mathcal{A}} = \sup_n \|A_n P_n\|$ the set $\mathcal{A}$ becomes a Banach algebra with unit element $\{P_n\}$.

Obviously, a sequence $\{A_n\}$, $A_n : \operatorname{im} P_n \to \operatorname{im} P_n$, with $\|A_n P_n\| \to 0$ belongs to $\mathcal{A}$. Clearly, the collection $\mathcal{N}$ of all such sequences forms a two-sided closed ideal in $\mathcal{A}$.

Finally, assume

(c) for any $\omega \in \Omega$ and any $T \in \mathcal{K}(X_\omega)$ and $\omega' \neq \omega$ we have

$$E_n^{(\omega')} (E_n^{(\omega)})^{-1} L_n^{(\omega)} T L_n^{(\omega)} E_n^{(\omega)} (E_n^{(\omega')})^{-1} \to 0 \quad \text{strongly},$$
$$\left(E_n^{(\omega')} (E_n^{(\omega)})^{-1} L_n^{(\omega)} T L_n^{(\omega)} E_n^{(\omega)} (E_n^{(\omega')})^{-1}\right)^* \to 0 \quad \text{strongly}.$$

The last condition means that for any ω and any $T \in \mathcal{K}(X_\omega)$ the sequence $\{A_n\}$, $A_n = (E_n^{(\omega)})^{-1} L_n^{(\omega)} T L_n^{(\omega)} E_n^{(\omega)}$, belongs to $\mathcal{A}$ and

$$W_\tau\{A_n\} = \begin{cases} T, & \tau = \omega \\ 0, & \tau \neq \omega. \end{cases}$$

Define

$$\mathcal{J}_0 := \left\{ \left\{ \sum_{k=1}^{r} (E_n^{(\omega_k)})^{-1} L_n^{(\omega_k)} T_k L_n^{(\omega_k)} E_n^{(\omega_k)} \right\} + \{C_n\} : \right.$$
$$\left. \omega_k \in \Omega, \qquad T_k \in \mathcal{K}(X_{\omega_k}), \qquad \{C_n\} \in \mathcal{N}, r \in \mathbb{N} \right\}.$$

One can easily check that $\mathcal{J}_0$ forms a two-sided ideal in $\mathcal{A}$, which, however, need not to be closed. Denote by $\mathcal{J}$ the closure of $\mathcal{J}_0$ and note that $\mathcal{J}$ forms already a closed two-sided ideal in $\mathcal{A}$.

7.27. Theorem. *A sequence $\{A_n\} \in \mathcal{A}$ is stable if and only if the operators $W_\omega\{A_n\}$ are invertible for any $\omega \in \Omega$ and the coset $\{A_n\}^\circ := \{A_n\} + \mathcal{J}$ is invertible in $\mathcal{A}^\circ := \mathcal{A}/\mathcal{J}$.*

Proof. The proof runs widely parallel to the proof of Theorem 7.7. A slight modification is needed for the proof of the sufficiency. Let $\{A_n\}^\circ$ be invertible. Then there is a sequence $\{R_n\} \in \mathcal{A}$ such that

$$A_n R_n = P_n + D_n, \qquad \{D_n\} \in \mathcal{J}. \tag{1}$$

Since $\mathcal{J}_0$ is dense in $\mathcal{J}$, we can find an element $\{D_n'\} \in \mathcal{J}_0$ such that $\|\{D_n\} - \{D_n'\}\| < 1/2$,

$$D_n' = \sum_{k=1}^{r} (E_n^{(\omega_k)})^{-1} L_n^{(\omega_k)} T_k L_n^{(\omega_k)} E_n^{(\omega_k)} + C_n,$$

$\|C_n\| \to 0$ and $T_k \in \mathcal{K}(X_{\omega_k})$. Put

$$R_n' = R_n - \sum_{k=1}^{r} (E_n^{(\omega_k)})^{-1} L_n^{(\omega_k)} W_{\omega_k}^{-1}\{A_n\} T_k L_n^{(\omega_k)} E_n^{(\omega_k)}.$$

Then, obviously, $\{R'_n\} \in \mathcal{A}$ and

$$A_n R'_n = P_n + \sum_{k=1}^{r} (E_n^{(\omega_k)})^{-1} \left(L_n^{(\omega_k)} T_k L_n^{(\omega_k)} - E_n^{(\omega_k)} A_n (E_n^{(\omega_k)})^{-1} L_n W_{\omega_k}^{-1}\{A_n\} T_k L_n^{(\omega_k)}\right)$$
$$\times E_n^{(\omega_k)} + (D_n - D'_n) + C'_n = P_n + (D_n - D'_n) + C''_n,$$

$\|C''_n\| \to 0$ as $n \to \infty$.

In the same way we may show $R'_n A_n = P_n + (F_n - F'_n) + K_n$, $\|F_n - F'_n\| < 1/2$, $\|K_n\| \to 0$ as $n \to \infty$. This gives the stability of $\{A_n\}$. ■

7.28. Examples.

Example 1. The theory of the Sections 7.4 to 7.7 is included in this general case.

Example 2. Let $\mathbf{L}^p(\mathbb{T}, \varrho)$ $\big(p \in (1, \infty)\big)$ be spaces with a weight ϱ satisfying 4.92(1). In these spaces the operators P_n and W_n (cf. 7.8) are well-defined, but, in general, $\{W_n\}_{n=0}^{\infty}$ is not uniformly bounded. Let $\overline{\mathbf{L}}^p(\mathbb{T}, \varrho)$ be the Banach space defined as follows: $\overline{\mathbf{L}}^p(\mathbb{T}, \varrho)$ is the set $\mathbf{L}^p(\mathbb{T}, \varrho)$ with the usual addition and norm, but the multiplication by a scalar $\lambda \in \mathbb{C}$ is given by $\lambda * f := \bar{\lambda} \cdot f$, where $\bar{\lambda} \cdot f$ refers to the usual multiplication by the scalar $\bar{\lambda}$ in $\mathbf{L}^p(\mathbb{T}, \varrho)$. Define

$$\overline{W}_n \left(\sum_{-\infty}^{\infty} f_i t^i\right) = \bar{f}_{-1} t^{-n} + \cdots + \bar{f}_{-n} t^{-1} + \bar{f}_n + \bar{f}_{n-1} t + \cdots + \bar{f}_0 t^n.$$

The operator $\overline{W}_n$ is a linear bounded operator from $\mathbf{L}^p(\mathbb{T}, \varrho)$ into $\overline{\mathbf{L}}^p(\mathbb{T}, \varrho)$ and moreover, the set $\{\overline{W}_n\}_{n=0}^{\infty}$ is uniformly bounded; this is a consequence of the equality

$$\overline{W}_n = \overline{W}(t^{n+1} Q P_n + t^{-n} P P_n),$$

where $\overline{W}$ is as in Section 7.1.

ow put $X_1 = \mathbf{L}^p(\mathbb{T}, \varrho)$, $X_2 = \overline{\mathbf{L}}^p(\mathbb{T}, \varrho)$, $L_n^{(i)} = P_n$ $(i = 1, 2)$, $E_n^{(1)} = P_n$, $E_n^{(2)} = \overline{W}_n$. All conditions of 7.26 are fulfilled. We mention that the finite section method for singular integral operators with respect to the system $\{P_n\}_{n=0}^{\infty}$ can be studied with the help of this modification.

Example 3. Let $X = \mathbf{l}^{p,\mu}$ and $X_1 = \mathbf{l}^{p,\mu}$, $X_2 = \mathbf{l}^{p,0}$ ($\mathbf{l}^{p,\gamma}$ being the space of all one-sided sequences $x = \{x_i\}_0^{\infty}$ with the norm

$$\|x\| = \left(\sum_{i=0}^{\infty} |x_i|^p \, (i+1)^{\gamma p}\right)^{1/p}.$$

Choose $L_n^{(i)} = P_n$ $(i = 1, 2)$, where

$$P_n\{x_0, x_1, \ldots\} = \{x_0, \ldots, x_n, 0, \ldots\},$$
$$E_n^{(1)} = P_n, \; E_n^{(2)} = W_n \operatorname*{diag}_{0 \le j \le n} \{(j+1)^{-\gamma}\},$$
$$W_n\{x_0, \ldots\} = \{x_n, \ldots, x_0, 0, \ldots\}.$$

Again all conditions of Section 7.26 are satisfied (cf. Roch/Silbermann [3]).

Example 4. Let $\Omega = \mathbb{T} \cup \{2, -2\}$ and put $X = \mathbf{L}^2(\mathbb{T})$, $X_{\pm 2} = \mathbf{L}^2(\mathbb{T})$, $X_\tau = \mathbf{L}^2(-1, 1)$ $(\tau \in \mathbb{T})$. Let $L_n^{(\pm 2)} = P_n$, $P_n \left(\sum_i f_i t^i\right) = \sum_{i=-n}^{n-1} f_i t^i$. Define functions $\varphi_i^{(n)}$ $(i = -n, \ldots, n-1)$ by

$$\varphi_i^{(n)} = \sqrt{n}\, \chi_{\left[\frac{i}{n}, \frac{i+1}{n}\right]}$$

and denote by $L_n = L_n^{(\tau)}$ the orthoprojection of $\mathbf{L}^2(-1, 1)$ onto span $\{\varphi_{-n}^{(n)}, \ldots, \varphi_{n-1}^{(n)}\}$ (χ_E always denotes the characteristic function of the measurable set E).

Finally, put $E_n^{(2)} = P_n$, $E_n^{(-2)} = W_n$, where

$$W_n\left(\sum_i f_i t^i\right) = f_{-1}t^{-n} + \cdots + f_{-n}t^{-1} + f_{n-1} + \cdots + f_0 t^{n-1},$$

$$E_n^{(\tau)}: \operatorname{im} P_n \to \operatorname{im} L_n, \qquad \sum_{i=-n}^{n-1} f_i t^i \mapsto \sum_{i=-n}^{n-1} \tau^i f_i \varphi_i^{(n)}.$$

Again all conditions of Section 7.26 are fulfilled. Notice that the examples 3 and 4 are not trivial and that checking the conditions is very expensive. In Chapter 10 we shall carry out the details in another (but similar) situation.

Notes and comments

7.5.—7.10. SILBERMANN [1], [2]. Proposition 7.9(b) goes back to ROCH/SILBERMANN [3].

7.11.—7.13. ROCH/SILBERMANN [3]. Some technical details are taken from VERBITSKI [1].

7.14. Assertion (a) was proved by ROCH and SILBERMANN [3]. Assertion (b) is the main ingredient of the theorem. The sufficiency part of (b) for $a, b \in \mathbf{PC}$ with finitely many jumps was proved by VERBITSKI [1]. Verbitski's original proof is hard and rests on a series of subtle function theoretical constructions. It has been considerably simplified by SILBERMANN (unpublished).

Moreover, with the help of localization arguments SILBERMANN [2] proved the sufficiency part of (b). The necessity part of (b) was established by VERBITSKI in a special case and independently by RATHSFELD [1] and ROCH/SILBERMANN [3] in the general case. The proofs given by RATHSFELD and ROCH/SILBERMANN are quite different, and each of them is of its own interest. Results for other operator equations are contained in DIDENKO [1], MEYER [1] (see also BÖTTCHER/SILBERMANN [3]).

7.15.—7.16. ROCH [1].

7.17.—7.19. JUNGHANNS/SILBERMANN [3]. For continuous coefficients the convergence of the collocation method was investigated by BOIKOV [1] for $p = 2$ and by JUNGHANNS [7] for general p. For certain classes of piecewise continuous coefficients the collocation method in the spaces $\mathbf{L}^p(\mathbb{T})$ was studied by JUNGHANNS [7] and PRÖSSDORF [2]. On the basis of a further development of the ideas of JUNGHANNS and SILBERMANN [3], RATHSFELD [1] generalized Theorem 7.19 to coefficients $a, b \in \mathbf{PC}(\mathbb{T})$ and the spaces $\mathbf{L}^p(\mathbb{T})$.

7.21. This theorem was proved by SILBERMANN [2] in the case where the coefficients have no common jumps. The general case was disposed of by RATHSFELD [1]. It is by no means easy to decide whether condition 7.21(i) is fulfilled or not. For $N = 2$ this question is completely solved in SPITKOVSKI/TASHBAEV [1].

7.22. JUNGHANNS/SILBERMANN [3].

7.24. ROCH/SILBERMANN [5]. The reader is also referred to ROCH [2], [3], [4], [5], [6].

7.25. HAGEN/SILBERMANN [2].

7.27. This theorem is a generalization of Theorem 7.7 and is taken from ROCH/SILBERMANN [4].

Chapter 8
Polynomial approximation methods for singular integral equations in Zygmund-Hölder spaces

Factorization

8.1. Preliminaries. Let $\mathcal{H}^s$ $(s > 0)$ denote the classical Zygmund-Hölder space on $\mathbb{T}$ and $\mathbf{C}^m$ the space of all m-times continuously differentiable functions on $\mathbb{T}$ (see Section 2.34). As pointed out in Chapter 6, the Cauchy singular integral operator S,

$$(Sf)(t) := \frac{1}{\pi i} \int_{\mathbb{T}} \frac{f(\tau)}{\tau - t} \, d\tau \qquad (t \in \mathbb{T})$$

is a bounded linear operator on the space $\mathcal{H}^s$.

We mention the following important properties of S (see Chapt. 6).

1°. $S^2 = I$

2°. The operators $P := (I + S)/2$ and $Q := (I - S)/2$ are complementary projections on $\mathcal{H}^s$.

3°. The projection $P(Q)$ maps the space $\mathcal{H}^s$ onto the subspace $\mathcal{H}^s_+$ $(\dot{\mathcal{H}}^s_-)$ consisting of all functions f the Fourier coefficients f_n of which vanish for $n < 0$ $(n \geqq 0)$.

4°. If $a \in \mathcal{H}^s$ and $0 < r < s$, then the commutator $[a, S] := aS - Sa$ is a compact operator from $\mathcal{H}^v$ $(v > 0)$ into $\mathcal{H}^r$. For $s < 1$ much more is true: $aS - Sa \in \mathcal{K}(\mathcal{H}^r, \mathcal{H}^s)$ (see Prössdorf [3], Chap. 3).

5°. For $a \in \mathcal{H}^s$ one has $aS - Sa \in \mathcal{K}(\mathcal{H}^s)$.

Let P_n be the projection defined on $\mathbf{C}^s$ $(s \in \mathbb{Z}^+)$ or on $\mathcal{H}^t$ $(t > 0)$ by

$$P_n \sum_{k=-\infty}^{\infty} f_k t^k = \sum_{k=-n}^{n} f_k t^k .$$

Define further $L_n f$ for $f \in \mathbf{C}$ by

$$(L_n f)(t) = \sum_{k=-n}^{n} a_k t^k, \qquad a_k := \frac{1}{2n+1} \sum_{j=-n}^{n} f(t_j) \, t_j^{-k},$$

where $t_j = t_j^{(n)} = e^{2\pi i j/(2n+1)}$ $(j = 0, \pm 1, \ldots, \pm n)$. Thus $L_n f$ is the interpolation polynomial for the function $f \in \mathbf{C}$ with respect to the knots t_j. In what follows we need the following estimates.

6°. *Let $f \in \mathcal{H}^t$. Suppose $0 < s < t$ and denote by S_n one of the operators P_n or L_n. Then*

$$\|(I - S_n) f\|_{\mathcal{H}^s} \leqq M \frac{\log n}{n^{t-s}} \|f\|_{\mathcal{H}^t} \qquad (n \geqq 2). \tag{1}$$

If $s \in \mathbb{Z}^+$ and $s < t$, then

$$\|(I - S_n) f\|_{\mathbf{C}^s} \leqq M \frac{\log n}{n^{t-s}} \|f\|_{\mathscr{H}^t} \quad (n \geqq 2). \tag{2}$$

Proof of 6°. The estimate (1) follows from Corollary 2.41 for $S_n = P_n$ and from Corollary 2.47 for $S_n = L_n$. Now let $s = 0$. Then $\|(I - P_n) f\|_{\mathbf{C}} \leqq \|f - h_n\|_{\mathbf{C}} + \|P_n(f - h_n)\|_{\mathbf{C}}$, where h_n is any trigonometric polynomial of degree less than or equal to n. Hence,

$$\|(I - P_n) f\|_{\mathbf{C}} \leqq (I + \|P_n\|_{\mathbf{C}}) E_n(f, \mathbf{C})$$

and it remains to use the Lemmas 2.35 and 2.44 in order to get (2) for $S_n = P_n$ and $s = 0$. For $s \in \mathbb{N}$, the estimate (2) is almost obvious in case $S_n = P_n$, since P_n commutes with the operator of differentiation. Now suppose $S_n = L_n$. Consider

$$\|(I - L_n) f\|_{\mathbf{C}^s} \leqq \|(I - P_n) f\|_{\mathbf{C}^s} + \|(P_n - L_n) f\|_{\mathbf{C}^s}.$$

Put $g_n := (P_n - L_n) f$. Applying Bernstein's inequality we get for $k \leqq s$

$$\|g_n^{(k)}\|_{\mathbf{C}} \leqq n^k \|g_n\|_{\mathbf{C}} \leqq n^k E_n(f, \mathbf{C}) (\|L_n\|_{\mathbf{C}} + \|P_n\|_{\mathbf{C}}),$$

and using the Lemmas 2.35 and 2.45 this leads to the estimate

$$\|g_n^{(k)}\|_{\mathbf{C}} \leqq M n^k \frac{\log n}{n^t} \|f\|_{\mathscr{H}^t}.$$

We so obtain (2) also for $S_n = L_n$. ∎

Frequently we also make use of the Sections 6.25–6.26 without comment.

8.2. Definition. Let $A \in \mathscr{H}^s_{N\times N}$ be a non-singular matrix function, that is $(\det A)(t) \neq 0$ for all $t \in \mathbb{T}$. A representation of the form

$$A(t) = A_-(t)\, D(t)\, A_+(t) \qquad (t \in \mathbb{T}) \tag{1}$$

where D is a diagonal matrix function of the form

$$D(t) = \operatorname{diag}(t^{\varkappa_1}, \ldots, t^{\varkappa_N})$$

with certain integers $\varkappa_1 \geqq \varkappa_2 \geqq \cdots \geqq \varkappa_N$ and $A_-^{\pm 1} \in (\mathscr{H}^s_-)_{N\times N}$ $(\mathscr{H}^s_- := \mathscr{H}^s_- \dotplus \mathbb{C})$, $A_+^{\pm 1} \in (\mathscr{H}^s_+)_{N\times N}$, is called a right canonical factorization. A factorization arising from (1) by changing the roles of A_+ and A_- is called *a left canonical factorization.*

The following facts are well known.

1°. If the function $A \in \mathscr{H}^s_{N\times N}$ admits a right (left) factorization, then the integers $\varkappa_j$ $(j = 1, \ldots, N)$ are uniquely determined by the matrix function A.

2°. Every non-singular matrix function $A \in \mathscr{H}^s_{N\times N}$ admits a right (left) canonical factorization (see Clancey/Gohberg [1] or Mikhlin/Prössdorf [1])

Finite sections

8.3. Notations. We are going to consider the finite section method for the approximate solution of singular integral equations over the unit circle $\mathbb{T}$ of the form

$$(A\varphi)(t) := c(t)\,\varphi(t) + \frac{d(t)}{\pi i} \int_{\mathbb{T}} \frac{\varphi(\tau)}{\tau - t}\, d\tau + (T\varphi)(t) = f(t) \qquad (t \in \mathbb{T}) \tag{1}$$

in the spaces $\mathscr{H}^{\mu}_{N\times N}$. We suppose that $c, d \in \mathscr{H}^{\lambda+\delta}_{N\times N}$ $(\delta > 0)$ and that T is a compact linear operator from $\mathscr{H}^{\mu}_{N\times N}$ into $\mathscr{H}^{\lambda}_{N\times N}$ $(0 < \mu < \lambda)$. Henceforth let

$$a(t) = c(t) + d(t), \qquad b(t) = c(t) - d(t)$$

and denote by f_j, a_j, b_j and c_{jk} $(j, k = 0, \pm 1, \ldots)$ the (matrix) Fourier coefficients of the functions $f \in \mathscr{H}^{\lambda}_{N}$, $a, b \in \mathscr{H}^{\lambda+\delta}_{N\times N}$ and Tt^k, respectively. In addition to (1) consider the system

$$\sum_{k=0}^{n} a_{j-k}\xi_k + \sum_{k=-n}^{-1} b_{j-k}\xi_k + \sum_{k=-n}^{n} c_{jk}\xi_k = f_j \qquad (j = 0, \pm 1, \ldots, \pm n). \tag{2}$$

In the following M denotes a generic constant independent of n which may vary at each occurrence.

8.4. Theorem. *Assume the singular integral operator* 8.3(1) *fulfils the conditions of Section* 8.3 *and*

(i) *A is invertible on $\mathscr{H}^{\mu}_{N}$,*
(ii) *$Pa^{-1}P + Qb^{-1}Q$ is invertible on $\mathscr{H}^{\mu}_{N}$ (recall that the invertibility of A implies the existence of $a^{-1}, b^{-1} \in \mathscr{H}^{\lambda+\delta}_{N\times N}$).*

Then for all sufficiently large n and for each $f \in \mathscr{H}^{\lambda}_{N}$ the system 8.3(2) *has precisely one solution $\{\xi_k^{(n)}\}_{k=-n}^{n}$ and, as $n \to \infty$, the functions*

$$\varphi_n(t) = \sum_{k=-n}^{n} \xi_k^{(n)} t^k$$

converge in the norm of $\mathscr{H}^{\mu}_{N}$ to the solution $\varphi \in \mathscr{H}^{\mu}_{N}$ of equation 8.3(1). *Furthermore, one has the estimate*

$$\|\varphi - \varphi_n\|_{\mathscr{H}^{\mu}_{N}} = O\left(\frac{\log n}{n^{\lambda-\mu}}\right), \tag{1}$$

Moreover, one can put $\delta = 0$ in the conditions of the theorem if $\lambda < 1$.

Proof. First of all the operator A can be written in the form

$$A = aP + bQ + T.$$

Let P_n $(n = 1, 2, \ldots)$ denote the projection on the space $\mathscr{H}^{\mu}_{N}$ which sends a function $\varphi(t) = \sum_{j=-\infty}^{\infty} \varphi_j t^j \in \mathscr{H}^{\mu}_{N}$ to $(P_n\varphi)(t) := \sum_{j=-n}^{n} \varphi_j t^j$ $(t \in \mathbb{T})$. It is easy to see that the system 8.3(2) is nothing else than the projection equation $A_n\varphi_n := P_nA\varphi_n = P_nf$, $\varphi_n \in \operatorname{im} P_n$. It is well known that the projections P_n are not uniformly bounded on $\mathscr{H}^{\mu}_{N}$. However, there is a constant M independent of n such that for $g \in \mathscr{H}^{\lambda}_{N}$ $(\lambda > \mu)$ we have

$$\|g - P_ng\|_{\mathscr{H}^{\mu}_{N}} \le M \frac{\log n}{n^{\lambda-\mu}} \|g\|_{\mathscr{H}^{\lambda}_{N}} \tag{2}$$

(Corollary 2.41).

The conditions of our theorem imply that the functions a and b are non-singular. Hence, $a(b)$ admits a left (right) canonical factorization and condition (ii) tells us that the middle factors must be equal to the identity matrix. Hence

$$a = a_+a_-, \qquad b = b_-b_+,$$

and

$$Pa^{-1}P + Qb^{-1}Q = (Pa_-^{-1}P + Qb_+^{-1}Q)\,(Pa_+^{-1}P + Qb_-^{-1}Q).$$

Let B denote the inverse of $Pa^{-1}P + Qb^{-1}Q$. It follows that $B = (Pa_+P + Qb_-Q)\,(Pa_-P + Qb_+Q)$. Put $B_n := P_nBP_n|_{\operatorname{im} P_n}$. Since $P_n(Pa_-P + Qb_+Q)\,P_n = (Pa_-P + Qb_+Q)\,P_n$ and $P_n(Pa_+P + Qb_-Q)\,P_n = P_n(Pa_+P + Qb_-Q)$, we get the invertibility of B_n on im P_n and

$$B_n^{-1}P_n = (Pa_-^{-1}P + Qb_+^{-1}Q)\,P_n(Pa_+^{-1}P + Qb_-^{-1}Q).$$

Using (2), a straightforward computation shows that for $g \in \mathcal{H}_N^\lambda$

$$\|B^{-1}g - B_n^{-1}P_ng\|_{\mathcal{H}_N^\mu} \leq M\,\frac{\log n}{n^{\lambda-\mu}}\,\|g\|_{\mathcal{H}_N^\lambda}. \tag{3}$$

Hence, $\mathcal{H}_N^\lambda \subset \mathfrak{R}(B, \{B_n, P_n\})$. Our next aim is to show that $A = B + T_1$, where T_1 is compact from $\mathcal{H}_N^\mu$ into $\mathcal{H}_N^\lambda$. To this end consider

$$A(Pa^{-1}P + Qb^{-1}Q) = I - aQa^{-1}P - bPb^{-1}Q + T(Pa^{-1}P + Qb^{-1}Q).$$

Hence,

$$A = B - (aQa^{-1}P + bPb^{-1}Q)\,B + T = B + T_1.$$

According to assertion 8.1.4° the operator $aQa^{-1}P + bPb^{-1}Q$ is compact from $\mathcal{H}_N^\mu$ into $\mathcal{H}_N^\lambda$. Thus, Remark 1.27.2° applies and we obtain that

$$\mathcal{H}_N^\lambda \subset \mathfrak{R}(A, \{A_n, P_n\}).$$

It remains to prove the estimate (1). Obviously, since $C = Pa_+P + Qb_-Q$ is invertible and $P_nC^{\pm 1} = P_nC^{\pm 1}P_n$, the equations

$$\begin{aligned} &(aP + bQ + T)\,\varphi = f, \\ &C^{-1}(aP + bQ + T)\,\varphi = (Pa_-P + Qb_+Q + C^{-1}T_1)\,\varphi = C^{-1}f \end{aligned} \tag{4}$$

are equivalent to one another. Clearly, so also are the equations

$$\begin{aligned} &P_n(aP + bQ + T)\,\varphi_n = P_nf, \\ &P_n(Pa_-P + Qb_+Q + C^{-1}T_1)\,\varphi_n = P_nC^{-1}f, \end{aligned} \tag{5}$$

$\varphi_n \in \operatorname{im} P_n$. Now it is easy to see that Remark 1.27.2° can be applied, which, in view of the inclusion $(Pa_-P + Qb_+Q)\,\varphi \in \mathcal{H}_N^\lambda$ (note that $f \in \mathcal{H}_N^\lambda$) gives the estimate (1). ■

8.5. Remark. For $N = 1$ condition (ii) of Theorem 8.4 is equivalent to the equalities wind a = wind b = 0.

If one is interested in results on pointwise rates of convergence, the best what can be expected is the estimate

$$\|\varphi - \varphi_n\|_{\mathbf{C}} \leq M\left(\frac{\log n}{n^s}\right)\|f\|_{\mathcal{H}_N^s}, \tag{1}$$

since the same rate of convergence holds for the partial sums of the Fourier series of a function in $\mathcal{H}^s$. In order to prove (1) we have to set up some notations. For $m \in \mathbb{Z}^+$, define

$$\mathcal{A}^{(m)} = \{f \in \mathbf{C}^m(\mathbb{T}) : f_n = 0 \text{ for all } n < 0\}.$$

Here f_n denotes the n-th Fourier coefficient of the function f. Put $\mathcal{A}^{(0)} = \mathcal{A}$. The set $\mathcal{A}$ is nothing else than the familiar disk algebra. All sets $\mathcal{A}^{(m)}$ ($m \in \mathbb{Z}^+$) are closed subalgebras of $\mathbf{C}^m(\mathbb{T})$. We need the following interesting assertion.

8.6. Theorem. *For $a \in \mathcal{H}^s$ ($s > m$) the Toeplitz operators*

$$Pa : \mathcal{A}^{(m)} \to \mathcal{A}^{(m)}, \qquad \varphi \mapsto Pa\varphi, \tag{1}$$

$$Qa : \bar{\mathcal{A}}_0^{(m)} \to \bar{\mathcal{A}}_0^{(m)}, \qquad \psi \mapsto Qa\psi, \tag{2}$$

are well-defined and bounded, where $\bar{\mathcal{A}}_0^{(m)} = \{f \in \mathbf{C}^m(\mathbb{T}) : f_n = 0$ for all $n \geqq 0\}$.

Proof. Let $0 < \gamma < 1$. Given $a \in \mathcal{H}^{m+\gamma}$ consider $Pa\varphi$ for $\varphi \in \mathcal{A}^{(m)}$. Since $aS - SaI : \mathbf{C} \to \mathcal{H}^{m+\gamma/2}$ is compact (by a well-known theorem on the compactness of integral operators with kernels of potential type, see e.g. Kantorovich/Akilov [1]) the operators $aP - PaI$, $aQ - QaI$ turn out to be also compact from $\mathbf{C}$ into $\mathcal{H}^{m+\gamma/2}$. Using $aP\varphi = a\varphi$ for $\varphi \in \mathcal{A}^{(m)}$ ($aQ\varphi = a\varphi$ for $\varphi \in \bar{\mathcal{A}}_0^{(m)}$) we get $Pa\varphi = a\varphi + Pa\varphi - aP\varphi \in \mathcal{A}^{(m)}$ ($Qa\varphi = a\varphi + Qa\varphi - aQ\varphi \in \bar{\mathcal{A}}_0^{(m)}$) and we are done. ∎

8.7. Theorem. *Let $r \in \mathbb{Z}^+$ and $r < s$. Suppose $a, b \in \mathcal{H}_N^{s+\delta}$, $T \in \mathcal{K}(\mathcal{H}_N^r, \mathcal{H}_N^{s+\delta})$ if $r > 0$, and $T \in \mathcal{K}(\mathbf{C}_N, \mathcal{H}_N^{s+\delta})$ if $r = 0$, where $\delta > 0$ is fixed. Then, under the conditions of Theorem 8.4 ($\lambda = s$), we have the estimate*

$$\|\varphi - \varphi_n\|_{\mathbf{C}_N^r} \leqq M \left(\frac{\log n}{n^{s-r}}\right) \|f\|_{\mathcal{H}_N^s}. \tag{1}$$

In particular, for $s \in \mathbb{N}$,

$$\|\varphi - \varphi_n\|_{\mathbf{C}_N^r} \leqq M \left(\frac{\log n}{n^{s-r}}\right) \|f\|_{\mathbf{C}_N^s}. \tag{2}$$

Proof. Putting $K := C^{-1}T_1$ and using $P_nDP_n = DP_n$, we get from 8.4(4) and 8.4(5)

$$(D + P_nK)(\varphi_n - \varphi) = (P_n - I) D\varphi, \tag{3}$$

where $D := Pa_-P + Qb_+Q$. Rewrite (3) as

$$(I + D^{-1}P_nK)(\varphi - \varphi_n) = D^{-1}(P_n - I) D\varphi. \tag{4}$$

Since $K : \mathbf{C}_N^r \to \mathcal{H}_N^{r+\delta/2}$ is compact (note that $\mathbf{C}_N^r$ is continuously embedded in $\mathcal{H}_N^r$ for $r \in \mathbb{N}$, and, for $d \in \mathcal{H}_N^{r+\delta}$, $dS - Sd \in \mathcal{K}(\mathbf{C}_N, \mathcal{H}_N^{r+\delta/2})$; the validity of the last assertion is a consequence of a well-known theorem on the compactness of integral operators with kernels of potential type, see e.g. Kantorovich/Akilov [1]) and $D^{-1} = Pa_-^{-1}P + Qb_+^{-1}Q : \mathcal{H}_N^{r+\delta/2} \to \mathcal{H}_N^{r+\delta/2}$ is bounded, it follows that $D^{-1}K : \mathbf{C}_N^r \to \mathbf{C}_N^r$ is compact. This implies that $I + D^{-1}K$ is invertible on $\mathbf{C}_N^r$, since it is easily seen to be one-to-one. The estimate

$$\begin{aligned}\|(D^{-1}P_nK - D^{-1}K)\varphi\|_{\mathbf{C}_N^r} &\leqq M\, \|(P_n - I) K\varphi\|_{\mathcal{H}_N^{r+\delta/4}} \\ &\leqq M(1/n)^{\delta/4} \log n\, \|K\varphi\|_{\mathcal{H}_N^{r+\delta/2}} \leqq M(1/n)^{\delta/4} \log n\, \|\varphi\|_{\mathbf{C}_N^r}\end{aligned}$$

shows that $D^{-1}P_nK$ converges in the operator norm to $D^{-1}K$, which implies the bound

$$\|(I + D^{-1}P_nK)^{-1}\|_{\mathbf{C}^r_N \to \mathbf{C}^r_N} \leq M,$$

and therefore it follows from (4) that

$$\|\varphi_n - \varphi\|_{\mathbf{C}^r_N} \leq M \|D^{-1}(P_n - I) D\varphi\|_{\mathbf{C}^r_N}. \tag{5}$$

So we are left with the estimation of (5). Let $u \in \mathcal{H}^s_N$. Since $Du \in \mathcal{H}^s_N$ we get

$$\|(P_n - I) Du\|_{\mathbf{C}^r_N} \leq M \frac{\log n}{n^{s-r}} \|u\|_{\mathcal{H}^s_N} \qquad (n \geq 2).$$

Note further that

$$(P_n - I) D = P(P_n - I) Pa_-P + Q(P_n - I) Qb_+Q = B_1 + B_2.$$

Hence, $B_1u \in \mathcal{A}^{(r)}_N$, $B_2u \in (\bar{\mathcal{A}}^{(r)}_0)_N$, and so Theorem 8.6 applies. The estimate (1) is proved. Since, for $s \in \mathbb{N}$, $\mathbf{C}^s_N$ is continuously embedded in $\mathcal{H}^s_N$, we immediately get (2). ■

The collocation method

8.8. Notations. Our next subject is the collocation method for the approximate solution of equation 8.3(1) with the following collocation points on the unit circle.

$$t_j = t_j^{(n)} = e^{2\pi i j/(2n+1)} \qquad (j = 0, \pm 1, \ldots, \pm n). \tag{1}$$

This method consists in finding an approximate solution of equation 8.3(1) in the form

$$\varphi_n(t) = \sum_{k=-n}^{n} \xi_k^{(n)} t^k,$$

where the (vector-valued) coefficients $\xi_k = \xi_k^{(n)}$ are to be determined so that equation $A\varphi_n = f$ is satisfied at the collocation points (1), that is, so that

$$(A\varphi_n)(t_j) = f(t_j) \qquad (j = 0, \pm 1, \ldots, \pm n). \tag{2}$$

It is easily seen that equation (2) written down in full takes the form

$$a(t_j) \sum_{k=0}^{n} t_j^k \xi_k + b(t_j) \sum_{k=-n}^{-1} t_j^k \xi_k + \sum_{k=-n}^{n} d_{jk} \xi_k = f(t_j) \qquad (j = 0, \pm 1, \ldots, \pm n) \tag{3}$$

where $d_{jk} = (Tt^k)(t_j)$ $(j, k = 0, \pm 1, \ldots, \pm n)$.

8.9. Theorem. *Assume the singular integral operator* 8.3(1) *fulfils the conditions of Section* 8.3 *and*

(i) *A is invertible on $\mathcal{H}^\mu_N$,*
(ii) *$a^{-1}P + b^{-1}Q$ is invertible on $\mathcal{H}^\mu_N$.*

Then for all sufficiently large n and for each $f \in \mathcal{H}^\lambda_N$ ($\lambda > \mu$) the system 8.8(2) *has exactly one solution $(\xi_k^{(n)})_{k=-n}^{n}$ and, as $n \to \infty$, the functions*

$$\varphi_n(t) = \sum_{k=-n}^{n} \xi_k^{(n)} t^k$$

converge in the norm of $\mathscr{H}_N^\mu$ *to the solution* $\varphi \in \mathscr{H}_N^\mu$ *of equation* 8.3(1). *Furthermore, one has the estimate*

$$\|\varphi - \varphi_n\|_{\mathscr{H}_N^\mu} \leqq M \frac{\log n}{n^{\lambda-\mu}} \|f\|_{\mathscr{H}_N^\lambda} \tag{1}$$

Moreover, one can take $\delta = 0$ *in the assumptions of the theorem if* $\lambda < 1$.

Proof. First of all the operator A can be written in the form

$$A = aP + bQ + T.$$

It is easy to see that the system 8.8(3) is equivalent to the projection equation

$$A_n\varphi_n = L_nA\varphi_n = L_nf, \qquad \varphi_n \in \operatorname{im} P_n \tag{2}$$

(L_n and P_n being as in Section 8.1). A straigthforward computation shows that for any $g, h \in \mathscr{H}_N^\lambda$

$$L_n(gh) = L_n(gL_nh). \tag{3}$$

Thus, without loss of generality we can assume that the coefficient b of the operator A is equal to the identity matrix. The conditions of the theorem imply that the function a is non-singular. Hence, a admits a left canonical factorization and condition (ii) of the theorem implies that the middle factor must be equal to the identity matrix. Thus, $a = a_+a_-$. Consider the operators $B := a_+(Pa_-P + Qa_+^{-1}Q)$ and $B_n := L_nB|_{\operatorname{im}P_n}$. For $f \in \mathscr{H}_N^\lambda$ we obtain

$$B^{-1}f = (Pa_-^{-1}P + Qa_+Q)\, a_+^{-1}f,$$

$$B_n^{-1}L_nf = (Pa_-^{-1}P + Qa_+Q)\, L_na_+^{-1}f$$

(using (3) and $P_n(Pa_-^{\pm1}P + Qa_+^{\mp1}Q)\, P_n = (Pa_-^{\pm1}P + Qa_+^{\mp1}Q)\, P_n$). Since for $g \in \mathscr{H}_N^\lambda$ the estimate

$$\|g - L_ng\|_{\mathscr{H}_N^\mu} \leqq M \frac{\log n}{n^{\lambda-\mu}} \|g\|_{\mathscr{H}_N^\lambda}$$

is valid, we get

$$\|B^{-1}f - B_n^{-1}L_nf\|_{\mathscr{H}_N^\mu} \leqq M \frac{\log n}{n^{\lambda-\mu}} \|f\|_{\mathscr{H}_N^\lambda}.$$

Hence, $\mathscr{H}_N^\lambda \subset \mathfrak{R}(B, \{B_n, L_n\})$. Since $A = B + T_1$, where T_1 is compact from $\mathscr{H}_N^\mu$ into $\mathscr{H}_N^\lambda$ (by Section 8.1, 4°), Remark 1.27.2° applies and it follows that

$$\mathscr{H}_N^\lambda \subset \mathfrak{R}(A, \{A_n, L_n\}).$$

Now the proof can be finished similar to that of Theorem 8.4. ■

8.10. Remark. If $N = 1$ then condition (i) implies (ii).

Our next aim is to study pointwise rates of convergence.

8.11. Theorem. *Let* $r \in \mathbb{Z}^+$ *and* $r < s$. *Suppose* $a, b \in \mathscr{H}_N^{s+\delta}$ $(\delta > 0)$, $T \in \mathscr{K}(\mathscr{H}_N^r, \mathscr{H}_N^{s+\delta})$ *if* $r > 0$, *and* $T \in \mathscr{K}(\mathbf{C}_N, \mathscr{H}_N^{s+\delta})$ *if* $r = 0$. *Then, under the conditions of Theorem* 8.9 ($\lambda = s$),

$$\|\varphi - \varphi_n\|_{\mathbf{C}_N^r} \leqq M \left(\frac{\log n}{n^{s-r}}\right) \|f\|_{\mathscr{H}_N^s} \qquad (n > 1). \tag{1}$$

In particular, for $s \in \mathbb{N}$,

$$\|\varphi - \varphi_n\|_{\mathbf{C}_N^r} \leqq M \left(\frac{\log n}{n^{s-r}}\right) \|f\|_{\mathbf{C}_N^s}. \tag{2}$$

Proof. Put $D := Pa_-P + Qa_+^{-1}Q$. One can prove in a similar fashion as in Section 8.7 that

$$\|\varphi - \varphi_n\|_{\mathbf{C}_N^r} \leqq M \, \|D^{-1}(L_n - I) \, D\varphi\|_{\mathbf{C}_N^r}. \tag{3}$$

Consider $D^{-1}(L_n - I) \, D\varphi = D^{-1}(P_n - I) \, D\varphi + D^{-1}(L_n - P_n) \, D\varphi =: K_n\varphi + D^{-1}(L_n - P_n) \, D\varphi$. The estimate

$$\|K_n\varphi\|_{\mathbf{C}_N^r} \leqq M \left(\frac{\log n}{n^{s-r}}\right) \|f\|_{\mathbf{C}_N^s}$$

was obtained in the proof of Theorem 8.7. Write $D^{-1}(L_n - P_n) \, D\varphi = D^{-1}P(L_n - P_n) \, D\varphi + D^{-1}Q(L_n - P_n) \, D\varphi$ and estimate both terms on the right-hand side. Let $\sigma_n(a)$ denote the n-th Fejer mean of a. Since $P_nDP_n = DP_n$, we get

$$D^{-1}P(L_n - P_n) \, D\varphi = D^{-1}P(L_n - P_n) \, D\big(\varphi - \sigma_n(P\varphi) - \sigma_n(Q\varphi)\big).$$

Notice that

$$\|P(L_n - P_n)\|_{\mathcal{A}_N^{(r)} \to \mathcal{A}_N^{(r)}} \leqq M \log n, \qquad \|P(L_n - P_n)\|_{(\bar{\mathcal{A}}_0^{(r)})_N \to \mathcal{A}_N^{(r)}} \leqq M \log n,$$

$$\|Q(L_n - P_n)\|_{\mathcal{A}_N^{(r)} \to (\bar{\mathcal{A}}_0^{(r)})_N} \leqq M \log n, \qquad \|Q(L_n - P_n)\|_{(\bar{\mathcal{A}}_0^{(r)})_N \to (\bar{\mathcal{A}}_0^{(r)})_N} \leqq M \log n.$$

For $r = 0$, this follows from Ivanov [1], Theorem 10.1. If $r \in \mathbb{N}$ consider, for instance, $PL_n - PP_n$. Using Bernstein's inequality and proceeding as in the proof of Theorem 2.46 we get our claim.

Since $\varphi \in \mathcal{H}_N^s$ implies $P\varphi, Q\varphi \in \mathcal{H}_N^s$, we deduce from Lemma 2.42 that

$$\|P\varphi - \sigma_n(P\varphi)\|_{\mathbf{C}_N^r} \leqq M \frac{\|\varphi\|}{n^{s-r}} \mathcal{H}_N^s, \qquad \|Q\varphi - \sigma_n(Q\varphi)\|_{\mathbf{C}_N^r} \leqq M \frac{\|\varphi\|}{n^{s-r}} \mathcal{H}_N^s.$$

In view of Theorem 8.6 it remains to put these things together in order to obtain

$$\|D^{-1}P(L_n - P_n) \, D\varphi\|_{\mathbf{C}_N^r} \leqq M \frac{\log n}{n^{s-r}} \|f\|_{\mathcal{H}_N^s}.$$

The remaining assertions can be shown in a similar way. ■

Mechanical quadratures

8.12. Notations. We now consider the singular integral equation 8.3(1) with a compact integral operator T:

$$(A\varphi)(t) = c(t) \, \varphi(t) + \frac{d(t)}{\pi i} \int_{\mathbf{T}} \frac{\varphi(\tau)}{\tau - t} \, d\tau + \int_{\mathbf{T}} k(t, \tau) \, \varphi(\tau) \, d\tau = f(t), \tag{1}$$

where $k(t, \tau) \in \mathcal{H}_N^\lambda(\mathbb{T} \times \mathbb{T})$ and the conditions of Section 8.3 are satisfied. The method of mechanical quadratures again consists in finding an approximate solution of equation (1) in the form

$$\varphi_n(t) = \sum_{k=-n}^{n} \xi_k^{(n)} t^k,$$

but the coefficients $\xi_k = \xi_k^{(n)}$ $(k = 0, \ldots, \pm n)$ are now determined through the system

$$a(t_j) \sum_{k=0}^{n} t_j^k \xi_k + b(t_j) \sum_{k=-n}^{-1} t_j^k \xi_k + \frac{2\pi \mathrm{i}}{2n+1} \sum_{k=-n}^{n} \gamma_{jk} \xi_k = f(t_j), \qquad (2)$$

$$(j = 0, \pm 1, \ldots, \pm n)$$

where

$$\gamma_{jk} = \sum_{r=-n}^{n} k(t_j, t_r)\, t_r^{k+1}, \qquad t_j = \mathrm{e}^{2\pi \mathrm{i} j/(2n+1)}.$$

It is not difficult to check that the system (2) is equivalent to the equation

$$A_n \varphi_n + L_n G_n \varphi_n = L_n f, \qquad (3)$$

where

$$A_n \varphi_n = L_n(aP + bQ)\, \varphi_n + L_n \int_{\mathbb{T}} k(t, \tau)\, \varphi_n(\tau)\, \mathrm{d}\tau,$$

$$G_n \varphi_n = -\int_{\mathbb{T}} k(t, \tau)\, \varphi_n(\tau)\, \mathrm{d}\tau + \int_{\mathbb{T}} \tau^{-1} L_{n\tau}[\tau k(t, \tau)]\, \varphi_n(\tau)\, \mathrm{d}\tau,$$

$\varphi_n \in \operatorname{im} P_n$. Indeed, it suffices to verify the identity

$$\int_{\mathbb{T}} L_n\big(k(t)\, \varphi_n(t)\big)\, \mathrm{d}t = \int_{\mathbb{T}} t^{-1} L_n[tk(t)]\, \varphi_n\, \mathrm{d}t.$$

We have

$$L_n\big(k(t)\, \varphi_n(t)\big) = \sum_{l=-n}^{n} f_l t^l,\ f_l = 1/(2n+1) \sum_{j=-n}^{n} k(t_j)\, \varphi_n(t_j)\, t_j^{-l}$$

which implies that $1/(2\pi \mathrm{i}) \int_{\mathbb{T}} L_n\big(k(t)\, \varphi_n(t)\big)\, \mathrm{d}t = f_{-1}$. On the other hand,

$$t^{-1} L_n\big(tk(t)\big)\, \varphi_n(t) = t^{-1} \sum_{l=-n}^{n} g_l t^l \sum_{l=-n}^{n} \varphi_{ln} t^l.$$

Therefore,

$$\begin{aligned} 1/2\pi \mathrm{i} \int_{\mathbb{T}} t^{-1} L_n[tk(t)]\, \varphi_n(t)\, \mathrm{d}t &= \sum_{l=-n}^{n} g_{-l} \varphi_{ln} \\ &= \sum_{l=-n}^{n} \frac{1}{(2n+1)^2} \sum_{j=-n}^{n} t_j k(t_j)\, t_j^l \sum_{m=-n}^{n} \varphi_n(t_m)\, t_m^{-l} \\ &= 1/(2n+1) \sum_{j=-n}^{n} t_j k(t_j)\, \varphi_n(t_j), \end{aligned}$$

and we are done.

Of course, $A_n\varphi_n = L_n f$ is the collocation equation considered in Section 8.9. Therefore, (3) can be interpreted as a perturbed collocation equation and our next aim is to study the operator G_n on im P_n.

Notice that here and in what follows $L_{n\tau}$ denotes the operator L_n applied with respect to the variable τ.

8.13. Proposition. *If $k \in \mathcal{H}^\lambda_{N\times N}(\mathbb{T}\times\mathbb{T})$ then, for any $\varphi_n \in \operatorname{im} P_n$,*

$$\|G_n\varphi_n\|_{\mathbf{C}_N} \leqq \eta_n \|\varphi_n\|_{\mathbf{L}^2_N},$$

where $\eta_n = O(n^{-\lambda})$.

Proof. First of all note that $\{L_n\} \subset \mathcal{L}(\mathbf{C}, \mathbf{L}^2)$ is uniformly bounded (Lemma 2.45). We have

$$\begin{aligned}\|G_n\varphi_n\|_{\mathbf{C}} &= \max_{t\in\mathbb{T}} \left|\int_{\mathbb{T}} \big(k(t,\tau) - \tau^{-1}L_{n\tau}[\tau k(t,\tau)]\big)\,\varphi_n(\tau)\,d\tau\right| \\ &\leqq \max_{t\in\mathbb{T}} \int_{\mathbb{T}} |k(t,\tau) - \tau^{-1}L_{n\tau}[\tau k(t,\tau)]|\,|\varphi_n(\tau)|\,|d\tau| \\ &\leqq M \max_{t\in\mathbb{T}} \|k(t,\tau)\,\tau - L_{n\tau}[k(t,\tau)\,\tau]\|_{\mathbf{L}^2} \|\varphi_n\|_{\mathbf{L}^2}\end{aligned}$$

(by Hölder's inequality). Since $t \mapsto \|k(t,\tau)\,\tau - L_{n\tau}[k(t,\tau)\,\tau]\|_{\mathbf{L}^2}$ is a continuous function on $\mathbb{T}$, there is a $t_n \in \mathbb{T}$ such that

$$\max_{t\in\mathbb{T}} \|k(t,\tau)\,\tau - L_{n\tau}[k(t,\tau)\,\tau]\|_{\mathbf{L}^2}$$

$$= \|k(t_n,\tau)\,\tau - L_{n\tau}[k(t_n,\tau)\,\tau]\|_{\mathbf{L}^2} = O(n^{-\lambda}) \qquad \text{(by Corollary 2.47).} \ \blacksquare$$

8.14. Theorem. *Assume that the operator* 8.12(1) *fulfils the hypotheses of Theorem* 8.9 *and assume $k(t,\tau) \in \mathcal{H}^\lambda_{N\times N}(\mathbb{T}\times\mathbb{T})$, $f \in \mathcal{H}^\lambda_N$. Then for all sufficiently large n the system* 8.12(2) *has precisely one solution $\{\xi_k^{(n)}\}_{k=-n}^n$ and, as $n \to \infty$, the functions*

$$\varphi_n(t) := \sum_{k=-n}^{n} \xi_k^{(n)} t^k$$

converge in the norm of the space $\mathcal{H}^\mu_N$ to the solution $\varphi \in \mathcal{H}^\mu_N$ of the equation 8.12(1). *One has*

$$\|\varphi - \varphi_n\|_{\mathcal{H}^\mu_N} = O\left(\frac{\log n}{n^{\lambda-\mu}}\right).$$

Proof. The system 8.12(2) is equivalent to the system

$$A_n\varphi_n + L_nG_n\varphi_n = L_n f,$$

where we have used the notations of Section 8.12. Without loss of generality we can assume that the coefficient b of the operator A is equal to the identity matrix. The operator A_n can be written in the form $B_n + L_nT_1$, where $B_n := L_nB|_{\operatorname{im} P_n}$, $B := a_+(Pa_-P + Qa_+^{-1}Q)$ and the compact operator $T_1 \in \mathcal{K}(\mathcal{H}^\mu_N, \mathcal{H}^\lambda_N)$ is defined in Section 8.9. In that section we have also proved that $\mathcal{H}^\lambda_N \subset \mathfrak{R}(B, \{B_nL_n\})$.

Put $\varepsilon_n := \|L_nT_1 - (L_nT_1 - L_nG_n)\| = \|L_nG_n\|$. Then,

$$\|L_nG_n\varphi_n\|_{\mathcal{H}^\mu_N} \leqq Mn^\mu \|L_nG_n\varphi_n\|_{\mathbf{C}} \leqq Mn^\mu(\log n)\, \eta_n\, \|\varphi_n\|_{\mathbf{L}^2}$$

$$\leqq M \frac{\log n}{n^{\lambda-\mu}} \|\varphi_n\|_{\mathcal{H}^\mu_N}.$$

Hence, $\varepsilon_n = O(n^{-\lambda+\mu} \log n)$. So remark 1.27.2° applies and we get all assertions of the theorem. ■

8.14. Remark. The assertions of Theorem 8.11 remain valid for mechanical quadratures.

Notes and comments

8.4. The first part of Theorem 8.4 was independently of each other and by different methods established by Zolotarevski [2], [3] and Prössdorf [2]. Zolotarevski gave error estimates which were somewhat rougher than 8.4(1) ($\log^4 n$ in place of $\log n$). The estimate 8.4(1) is from Prössdorf and Silbermann [3], [5].

8.6. This result is due to Böttcher and the second author (see also Silbermann [8]). The simple proof given here was suggested by Mc Lean (private communication).

8.7. Mc Lean/Wendland [1]. The proof presented here is taken from Silbermann [8].

8.9. Under the hypothesis $N = 1$, the convergence of the collocation method in a somewhat weaker norm was proved by Ivanov [1] and subsequently, in the norm of $\mathcal{H}^\lambda$ $(0 < \lambda < 1)$, by Gabdulkhaev [1]. For $N > 1$, the collocation method was studied by Zolotarevski [1], but his error estimate was not optimal ($\log^3 n$ in place of $\log n$). The estimate 8.9(1) goes back to Prössdorf and Silbermann [3], [5].

8.11. Theorem 8.11 is due to Mc Lean, Prössdorf and Wendland [1], however with $\log^2 n$ in place of $\log n$. The estimate 8.11(1) was first obtained by Silbermann [8].

8.14. This theorem was obtained with slightly worse error estimates by Gabdulkhaev [2] ($N = 1$), Zolotarevski [4], and Prössdorf/Silbermann [8] ($N > 1$). The error estimate given here was proved by Junghanns in his diploma paper (Konvergenzmannigfaltigkeit und reguläre Konvergenz von Folgen linearer Operatoren, Karl-Marx-Stadt, 1979) for the first time. See also Prössdorf/Silbermann [9]. Finally we mention Amosov's paper [1], in which further contributions are made (in the set-up of Sobolev spaces).

Chapter 9
Polynomial approximation methods for singular integral equations on intervals

This chapter is devoted to the numerical analysis of polynomial approximation methods for the singular integral equation

$$a(x)\,u(x) + \frac{b(x)}{\pi}\int_{-1}^{1}\frac{u(t)}{t-x}\,\mathrm{d}t + \frac{1}{\pi}\int_{-1}^{1} h(x,t)\,u(t)\,\mathrm{d}t = f(x), \qquad x \in (-1, 1).$$

We shall consider Galerkin, collocation, and quadrature methods. In order to carry out this program we need some facts from the general theory of these equations, which can not be assumed to be well-known. Thus, we start with a brief survey of some important properties of singular integrals.

Singular integrals and piecewise holomorphic functions

9.1. Preliminaries. Given an integer $m \geqq 0$ and a number $0 < \mu < 1$, we denote by $\mathbf{C}^{m,\mu}[d_1, d_2]$ $(d_1, d_2 \in \mathbb{R}, d_2 > d_1)$ the collection of all functions f on $[d_1, d_2]$ which are m-times continuously differentiable on $[d_1, d_2]$ and whose m-th derivative $f^{(m)}$ satisfies a Hölder condition with the exponent μ, that is,

$$\sup_{\substack{t_1 \neq t_2 \\ t_1, t_2 \in [d_1, d_2]}} \frac{|f^{(m)}(t_2) - f^{(m)}(t_1)|}{|t_2 - t_1|^{\mu}} < \infty.$$

To a function $f \in \mathbf{C}^{m,\mu}[d_1, d_2]$ we assign the norm

$$\|f\|_{\mathbf{C}^{m,\mu}} = \sum_{i=0}^{m-1} \frac{\max\limits_{t \in [d_1, d_2]} |f^{(i)}(t)|}{i!} + \sup_{\substack{t_1 \neq t_2 \\ t_1, t_2 \in [d_1, d_2]}} \frac{|f^{(m)}(t_2) - f^{(m)}(t_1)|}{|t_2 - t_1|^{\mu}}.$$

Define $\mathcal{H} := \bigcup\limits_{0<\mu<1} \mathcal{H}^{\mu}$, where $\mathcal{H}^{\mu} := \mathbf{C}^{0,\mu}$. The function class $\mathcal{H}^*$ is, by definition, the collection of all functions $u(t)$ such that

$$u(t) = \prod_{r=1}^{s} |t - \tilde{d}_r|^{\gamma_r}\, u^*(t),$$

where $u^* \in \mathcal{H}[d_1, d_2]$, $\gamma_r > -1$, $\tilde{d}_1 = d_1 < \cdots < \tilde{d}_s = d_2$, $s \geqq 2$.

Finally, let Δ be the set of all piecewise holomorphic functions $\phi(z)$ the discontinuity line of which is $[d_1, d_2]$. More precisely, $\phi \in \Delta$ if and only if ϕ is holomorphic in $\mathbb{C} \setminus [d_1, d_2]$, if there exist $\varepsilon > 0$, $\beta < 1$ and $K > 0$ such that

$$|\phi(z)| \leqq \frac{K}{|z - c|^\beta}$$

for $|z - c| < \varepsilon$, $z \in \mathbb{C} \setminus [d_1, d_2]$, $c = \tilde{d}_j$, $j = 1, \ldots, s$, and if the (non-tangential) limits

$$\phi^\pm(t) := \lim_{z \to t,\, \operatorname{Im} z \gtrless 0} \phi(z)$$

exist for all $t \in [d_1, d_2] \setminus \{\tilde{d}_1, \ldots, \tilde{d}_s\}$.

9.2. Theorem. *If $u \in \mathcal{H}^*[d_1, d_2]$ and*

$$\phi(z) := \frac{1}{2\pi i} \int_{d_1}^{d_2} \frac{u(t)}{t - z} \, dt, \qquad z \in \mathbb{C} \setminus [d_1, d_2], \tag{1}$$

then

(a) $\phi \in \Delta$ *and* $\phi(\infty) = 0$

(b) $\phi^+, \phi^- \in \mathcal{H}^*[d_1, d_2]$ *and*

$$\phi^\pm(x) = \pm \frac{1}{2} u(x) + \frac{1}{2\pi i} \int_{d_1}^{d_2} \frac{u(t)}{t - x} \, dt, \; x \in [d_1, d_2] \setminus \{\tilde{d}_1, \ldots, \tilde{d}_s\} \tag{2}$$

or, equivalently,

$$u(x) = \phi^+(x) - \phi^-(x), \quad \text{and} \quad (Su)(x) := \frac{1}{\pi i} \int_{d_1}^{d_2} \frac{u(t)}{t - x} \, dt = \phi^+(x) + \phi^-(x), \tag{3}$$

for $x \in [d_1, d_2] \setminus \{\tilde{d}_1, \ldots, \tilde{d}_2\}$. (The latter integrals are understood in the sense of the Cauchy principal value.)

Proof. Muskhelishvili [1], § 26. ■

9.3. Corollary. *Assume $u \in \mathcal{H}^\mu[d_1, d_2]$. Then the functions ϕ^+, ϕ^- belong to $\mathcal{H}^\mu$ on $[d_1, d_2] \setminus (\{|z - d_1| < \varepsilon\} \cup \{|z - d_2| < \varepsilon\})$ for every $\varepsilon > 0$.*

Proof. Muskhelishvili [1], § 18. ■

9.4. Remark. *If $u \in \mathcal{H}^\mu[d_1, d_2]$ and $u(d_1) = u(d_2) = 0$, then the function defined by*

$$\frac{1}{\pi i} \int_{d_1}^{d_2} \frac{u(x)}{t - x} \, dt \tag{1}$$

belongs to $\mathcal{H}^\mu[d_1, d_2]$.

Proof. Choose $\delta > 0$ and extend the function u by zero onto the interval $[d_1 - \delta, d_2 + \delta]$. Obviously, this function belongs to $\mathcal{H}^\mu[d_1 - \delta, d_2 + \delta]$. By Corollary 9.3 and the last equality in 9.2(3) the function

$$\frac{1}{\pi i} \int_{d_1}^{d_2} \frac{u(t)}{t - x} \, dt, \qquad x \in [d_1 - \delta, d_2 + \delta]$$

belongs to $\mathcal{H}^\mu$ on $[d_1, d_2]$. ■

In the sequel we specify $[d_1, d_2]$ to $[-1, 1]$.

9.5. Conventions. In what follows assume that $a, b \in \mathcal{H}$ are real-valued and

$$a^2(x) + b^2(x) > 0 \quad \text{for all} \quad x \in [-1, 1]. \tag{1}$$

We then can define

$$g_\pm(x) := \log [a(x) \pm ib(x)], \qquad g(x) := \frac{1}{2\pi i} [g_-(x) - g_+(x)]$$

in such a manner that these functions are continuous. We have

$$r(x) := e^{(1/2)[g_+(x) + g_-(x)]} = \sqrt{a^2(x) + b^2(x)} > 0.$$

Moreover, $g(x)$ is real-valued. Indeed, $g(x) = \frac{1}{2\pi i} \log \frac{a(x) - ib(x)}{a(x) + ib(x)}$ and $\left| \frac{a(x) - ib(x)}{a(x) + ib(x)} \right| = 1$ for every $x \in [-1, 1]$. Hence,

$$\frac{a(x) - ib(x)}{a(x) + ib(x)} = e^{i2\pi\alpha(x)},$$

where $\alpha(x)$ is real-valued, implying that $g(x)$ is also real-valued. Further, choose integers λ and ν such that

$$-1 < \alpha_0, \qquad \beta_0 < 1,$$

where $\alpha_0 := \lambda + g(1)$, $\beta_0 = \nu - g(-1)$. Define $\varkappa := -(\lambda + \nu)$ and, for $z \in \mathbb{C} \setminus [-1, 1]$, let

$$X(z) := (1 - z)^\lambda (1 + z)^\nu \exp \int_{-1}^{1} \frac{g(t)}{t - z} \, dt. \tag{2}$$

From Theorem 9.2 it follows that $X \in \Delta$ and that

$$X^\pm(t) = e^{\pm \pi i g(t)} X(t), \qquad t \in (-1, 1), \tag{3}$$

where

$$X(t) = (1 - t)^\lambda (1 + t)^\nu \exp \int_{-1}^{1} \frac{g(x)}{x - t} \, dx. \tag{4}$$

Because $e^{\pm \pi i g(t)} = \frac{e^{g_\mp(t)}}{r(t)} = \frac{a(t) \mp ib(t)}{r(t)}$, equality (3) yields that

$$X^\pm(t) = [a(t) \mp ib(t)] \, \sigma(t), \qquad t \in (-1, 1), \tag{5}$$

where

$$\sigma(t) = X(t)/r(t).$$

Note that in the important case $b \in \mathbb{R} \setminus \{0\}$ the integer $\varkappa$ takes only the values 1, 0, −1. Indeed, it is easily seen that the continuous function $g(t)$ defined above can be chosen so that $0 \leqq g(x) \leqq 1$. Therefore, λ and ν assume only the values 0, −1 and 0, 1, respectively. This gives that $-1 \leqq \varkappa \leqq 1$.

9.6. Proposition. *For $t \in (-1, 1)$ we have (in the sense of the Cauchy principal value)*

(a) $$\int_{-1}^{1} \frac{1}{x - t} \, dx = \log \frac{1 - t}{1 + t},$$

(b) $$\int_{-1}^{1} \frac{x}{x - t} \, dx = 2 + t \log \frac{1 - t}{1 + t}.$$

Proof. (a) Clearly, $\log |x - t|$ is a primitive of $(x - t)^{-1}$ on $\mathbb{R} \setminus \{t\}$. Thus, for sufficiently small $\varepsilon > 0$,

$$\int_{-1}^{t-\varepsilon} \frac{1}{x - t} \, dx = \log |x - t| \Big|_{x=-1}^{x=t-\varepsilon} = \log \varepsilon - \log (1 + t),$$

$$\int_{t+\varepsilon}^{1} \frac{1}{x - t} \, dx = \log |x - t| \Big|_{x=t+\varepsilon}^{x=1} = \log (1 - t) - \log \varepsilon,$$

and (a) follows.

(b) Immediate from the decomposition

$$\int_{-1}^{1} \frac{x}{x - t} \, dx = \int_{-1}^{1} \frac{x - t}{x - t} \, dx + t \int_{-1}^{1} \frac{1}{x - t} \, dt$$

and (a). ■

9.7. Proposition. *Assume $a, b \in \mathcal{H}^\mu$. Then*

$$\sigma(t) = (1 - t)^{\alpha_0} (1 + t)^{\beta_0} w(t),$$

where $w \in \mathcal{H}^\mu$ and $w(t) \neq 0$ for all $t \in [-1, 1]$.

Proof. According to 9.5(4) we have

$$X(t) = (1 - t)^\lambda (1 + t)^\nu \exp \int_{-1}^{1} \frac{g(x) - \frac{1 - x}{2} g(-1) - \frac{1 + x}{2} g(1)}{x - t} \, dx$$

$$\times \exp \int_{-1}^{1} \frac{\frac{1 - x}{2} g(-1) + \frac{1 + x}{2} g(1)}{x - t} \, dx.$$

Using Proposition 9.6 we obtain that

$$\exp \int_{-1}^{1} \frac{\frac{1-x}{2}\, g(-1) + \frac{1+x}{2}\, g(1)}{x-t}\, dt$$

$$= \exp \left[\left(\frac{1}{2}\, g(-1) + \frac{1}{2}\, g(1)\right) \log \frac{1-t}{1+t} + \left(\frac{1}{2}\, g(1) - \frac{1}{2}\, g(-1)\right)\left(2 + t \log \frac{1-t}{1+t}\right)\right]$$

$$= \left(\frac{1-t}{1+t}\right)^{(1/2)(g(-1)+g(1))} \exp\big(g(1) - g(-1)\big) \cdot \left(\frac{1-t}{1+t}\right)^{(1/2)(g(1)-g(-1))t}.$$

Since $\alpha_0 = \lambda + g(1)$ and $\beta_0 = \nu - g(-1)$, an easy computation shows that

$$X(t) = (1-t)^{\alpha_0}\,(1+t)^{\beta_0}\left((1-t)^{1-t}\,(1+t)^{1+t}\right)^{(1/2)(g(-1)-g(1))} \exp\big(g(1) - g(-1)\big)$$

$$\times \exp \int_{-1}^{1} \frac{g(x) - \frac{1-x}{2}\, g(-1) - \frac{1+x}{2}\, g(1)}{x-t}\, dx.$$

Taking into account that $\tau^\tau = e^{\tau \ln \tau} \in \mathcal{H}^{1-\varepsilon}[0, 2]$ for all $\varepsilon > 0$ and that $1/r \in \mathcal{H}^\mu$ (since $a^2(x) + b^2(x) > 0$), we get the assertion from Remark 9.4. ■

9.8. Corollary. *Assume* $u \in H^*$, $\phi \in \Delta$, $\phi(\infty) = 0$, *and*

$$\phi^+(t) - \phi^-(t) = u(t), \qquad t \in (-1, 1) \setminus \{\tilde{d}_1, \ldots, \tilde{d}_s\}.$$

Then

$$\phi(z) = \frac{1}{2\pi i} \int_{-1}^{1} \frac{u(t)}{t-z}\, dt, \qquad z \in \mathbb{C} \setminus [-1, 1].$$

Proof. MUSKHELISHVILI [1], § 31. ■

Special mapping properties

9.9. Theorem. *Let* $A = aI + SbI$. *If* $p_m = \sum_{j=0}^{m} p_{mj} t^j$ *is a polynomial of degree* m, $p_{mm} \neq 0$, *then*

$$q_{m-\varkappa} = A\sigma p_m \tag{1}$$

is a polynomial of degree $m - \varkappa$,

$$q_{m-\varkappa}(t) = \sum_{k=0}^{m-\varkappa} q_{m-\varkappa,k} t^k,$$

($q_{m-\varkappa} \equiv 0$, if $m - \varkappa < 0$) *with* $q_{m-\varkappa,m-\varkappa} = (-1)^\lambda p_{mm}$, *if* $m - \varkappa \geqq 0$. *Moreover, the polynomial* $q_{m-\varkappa}$ *is defined by*

$$\lim_{z \to \infty} [X(z)\, p_m(z) - q_{m-\varkappa}(z)] = 0. \tag{2}$$

Proof. Due to definition 9.5(2) we have

$$X(z) = \sum_{k=\varkappa}^{\infty} x_k z^{-k}$$

at infinity, where $x_\varkappa = (-1)^\lambda$. If one defines $q_{m-\varkappa}$ by (2) then obviously $q_{m-\varkappa,m-\varkappa} = (-1)^\lambda p_{mm}$. Let us prove that $(aI + SbI)\,\sigma p_m$ is the polynomial $q_{m-\varkappa}$. Put $\phi(z) = X(z)\,p_m(z) - q_{m-\varkappa}(z)$. In view of 9.5(5) we get

$$\phi^+(t) - \phi^-(t) = -2ib(t)\,\sigma(t)\,p_m(t), \qquad t \in (-1, 1).$$

Because $\phi(\infty) = 0$ and $b\sigma p_m \in \mathcal{H}^*$, it follows by Corollary 9.8 that

$$\phi(z) = -\frac{1}{\pi}\int_{-1}^{1} \frac{b(t)\,p_m(t)\,\sigma(t)}{t-z}\,dt, \qquad z \in \mathbb{C} \setminus [-1, 1].$$

Theorem 9.2(b) in conjunction with 9.5(5) yields that

$$-\frac{2}{\pi}\int_{-1}^{1} \frac{b(t)\,p_m(t)\,\sigma(t)}{t-x}\,dt = \phi^+(x) + \phi^-(x)$$

$$= 2a(x)\,\sigma(x)\,p_m(x) - 2q_{m-\varkappa}(x),$$

and so our claim is proved. ■

9.10. Remark. *Let $p_m(t)$ be a polynomial of degree m and let $\mu(t) := 1/(X(t)\,r(t))$ (this weight is said to be dual to $\sigma(t)$). Then*

$$q_{m+\varkappa} = B\mu p_m := (aI - Sb)\,\mu p_m$$

is a polynomial of degree $m + \varkappa$ ($q_{m+\varkappa} \equiv 0$, if $m + \varkappa < 0$) satisfying

$$\lim_{z\to\infty} [p_m(z)/X(z) - q_{m+\varkappa}(z)] = 0.$$

Moreover, for each polynomial $p(t)$ we have

$$B\mu(A\sigma p) = p, \quad \text{if} \quad \varkappa \leqq 0,$$
$$A\sigma(B\mu p) = p, \quad \text{if} \quad \varkappa \geqq 0.$$

Proof. The first assertion is an immediate consequence of Theorem 9.9, with the coefficient b replaced by $-b$. Let $\varkappa \leqq 0$ and $X(z) = \sum_{k=\varkappa}^{\infty} x_k z^{-k}$. Then, in virtue of Theorem 9.9,

$$q_{m-\varkappa}(t) = (A\sigma p)\,(t) = \sum_{l=0}^{m-\varkappa} \left(\sum_{k=\varkappa}^{m-l} x_k p_{l+k}\right) t^l,$$

where $p(t) = \sum_{j=0}^{m} p_j t^j$ and $p_{l+k} = 0$ if $l + k < 0$. Because

$$\frac{1}{X(z)} = \sum_{j=-\varkappa}^{\infty} y_j z^{-j},$$

Theorem 9.9 again yields

$$q_{m+\varkappa}(t) = (B\mu p)(t) = \sum_{l=0}^{m+\varkappa} \left(\sum_{j=-\varkappa}^{m-l} y_j p_{l+j} \right) t^l.$$

Thus,

$$\begin{aligned}(B\mu A\sigma p)(t) &= \sum_{l=0}^{m} \left(\sum_{j=-\varkappa}^{m-\varkappa-l} y_j \left(\sum_{k=\varkappa}^{m-l-j} x_k p_{l+j+k} \right) \right) t^l \\ &= \sum_{l=0}^{m} \left(\sum_{j=0}^{m-l} y_{-\varkappa+j} \left(\sum_{k=\varkappa}^{m+\varkappa-l-j} x_k p_{l-\varkappa+j+k} \right) \right) t^l \\ &= \sum_{l=0}^{m} \left(\sum_{j=0}^{m-l} y_{-\varkappa+j} \left(\sum_{k=0}^{m-l-j} x_{\varkappa+k} p_{l+j+k} \right) \right) t^l \\ &= \sum_{l=0}^{m} \left(\sum_{j=0}^{m-l} y_{-\varkappa+j} \left(\sum_{k=j}^{m-l} x_{\varkappa+k-j} p_{l+k} \right) \right) t^l \\ &= \sum_{l=0}^{m} \left(\sum_{k=0}^{m-l} \left(\sum_{j=0}^{k} y_{-\varkappa+j} x_{\varkappa+k-j} \right) p_{l+k} \right) t^l = \sum_{l=0}^{m} p_l t^l,\end{aligned}$$

since

$$x_\varkappa y_{-\varkappa} = 1 \quad \text{and} \quad \sum_{l=0}^{k} x_{\varkappa+k-l} y_{-\varkappa+l} = 0 \qquad (k = 1, 2, \ldots).$$

The case $\varkappa \geqq 0$ can be treated by changing the roles of A and B. ■

9.11. Proposition. *If $u, v \in \mathscr{H}^*$ and $uv \in \mathscr{H}^*$ then*

$$\int_{-1}^{1} v(x) \left(\int_{-1}^{1} \frac{u(t)}{t-x} \, \mathrm{d}t \right) \mathrm{d}x = -\int_{-1}^{1} u(x) \left(\int_{-1}^{1} \frac{v(t)}{t-x} \, \mathrm{d}t \right) \mathrm{d}x$$

Proof. Mushkelishvili [1], § 28.

9.12. Theorem. *Assume $p_m(t)$ to be a polynomial of degree m such that*

$$\int_{-1}^{1} t^j p_m(t)\, b(t)\, \sigma(t) \mathrm{d}t = 0 \qquad (j = 0, 1, \ldots, m-1),$$

and let $q_{m-\varkappa}$ be defined by 9.9(1). Then

$$\int_{-1}^{1} t^k q_{m-\varkappa}(t)\, b(t)\, \mu(t)\, \mathrm{d}t = 0 \qquad (k = 0, 1, \ldots, m-\varkappa-1),$$

where $\mu(t)$ is as in Remark 9.10.

Proof. Due to Remark 9.10, the function $(aI - SbI)\,\mu t^j$ is a polynomial of degree $j + \varkappa$. For $j = 0, 1, \ldots, m - \varkappa - 1$, it follows with the help of Proposition 9.11

that

$$0 = \int_{-1}^{1} b(x)\,\sigma(x)\,p_m(x)\left[a(x)\,\mu(x)\,x^j - \frac{1}{\pi}\int_{-1}^{1}\frac{b(t)\,t^j}{t-x}\,\mu(t)\,\mathrm{d}t\right]\mathrm{d}x$$

$$= \int_{-1}^{1} b(x)\,\mu(x)\,x^j\left[a(x)\,\sigma(x)\,p_m(x) + \frac{1}{\pi}\int_{-1}^{1}\frac{b(t)\,p_m(t)}{t-x}\,\sigma(t)\,\mathrm{d}t\right]\mathrm{d}x$$

$$= \int_{-1}^{1} b(x)\,\mu(x)\,x^j q_{m-\varkappa}(x)\,\mathrm{d}x$$

and so we are done. ■

9.13. Remark. *If the polynomial $p_m(t)$ satisfies the conditions of Theorem 9.12, then*

$$(-1)^\lambda \int_{-1}^{1} t^m p_m(t)\,b(t)\,\sigma(t)\,\mathrm{d}t = \int_{-1}^{1} t^{m-\varkappa} q_{m-\varkappa}(t)\,b(t)\,\mu(t)\,\mathrm{d}t$$

for $m \geqq \varkappa$, where $q_{m-\varkappa}$ is again defined by 9.9(1).

Proof. In view of Theorem 9.9 we have

$$(aI - SbI)\,\mu t^{m-\varkappa} = (-1)^\lambda\, t^m + \delta_{m-1} t^{m-1} + \cdots + \delta_0 .$$

Now in the same way as in the proof of Theorem 9.12 we obtain that

$$(-1)^\lambda \int_{-1}^{1} x^m p_m(x)\,b(x)\,\sigma(x)\,\mathrm{d}x = \int_{-1}^{1}\left((-1)^\lambda x^m + \delta_{m-1}\,x^{m-1} + \cdots + \delta_0\right) \times p_m(x)\,b(x)\,\sigma(x)\,\mathrm{d}x$$

$$= \int_{-1}^{1} x^{m-\varkappa} q_{m-\varkappa}(x)\,b(x)\,\mu(x)\,\mathrm{d}x . \ ■$$

In the following we shall consider the important case in which there exists a function $c(x)$ on $[-1, 1]$ such that the following properties 1°–3° are fulfilled:

1°. $B(x) := c(x)\,b(x)$ is a polynomial, say of degree r:

$$B(x) = \sum_{k=0}^{r} \beta_k x^k .$$

2°. $c(x) \geqq 0$; $\sigma_0 := c^{-1}\sigma \in \mathcal{H}^*$, $\mu_0 := c^{-1}\mu \in \mathcal{H}^*$.
3°. If $B(x_0) = 0$ for $x_0 \in [-1, 1]$, then $b(x_0) = 0$.

9.14. Theorem. *Let $p_m(t)$ be a polynomial of degree m. Then the function $q_m(t)$ defined by $q_m = (a\sigma I + BS\sigma_0 I)\,p_m$ is a polynomial of degree at most* $\max\{m - \varkappa, r - 1\}$ *and of degree $m - \varkappa$ if $m - \varkappa > r - 1$. If $p_{m,\sigma_0}(t)$ is an orthogonal polynomial of degree m with respect to the weight $\sigma_0(t)$ and if $m - \varkappa > r - 1$, then $q_{m,\mu_0} := (a\sigma I + BS\sigma_0 I)\,p_{m,\sigma_0}$ is an orthogonal polynomial of degree $m - \varkappa$ with respect to the weight $\mu_0(t)$. Moreover, if $m - \varkappa > r - 1$, then*

$$\|p_{m,\sigma_0}\|_{\sigma_0} = \|q_{m,\mu_0}\|_{\mu_0}, \tag{1}$$

where

$$\|p_{m,\sigma_0}\|_{\sigma_0} = \left(\frac{1}{\pi}\int_{-1}^{1} |p_{m,\sigma_0}|^2\,\sigma_0\,dx\right)^{1/2},$$

$$\|q_{m,\mu_0}\|_{\mu_0} = \left(\frac{1}{\pi}\int_{-1}^{1} |q_{m,\mu_0}|^2\,\mu_0\,dx\right)^{1/2}.$$

Proof. Since $\dfrac{B(x) - B(t)}{t - x}$ is a polynomial of degree $r - 1$ with respect to either variable, the function q_m defined by

$$q_m = (aI + SbI)\,\sigma p_m + (BS - SBI)\,\sigma_0 p_m$$

is a polynomial of degree at most $\max\{m - \varkappa, r - 1\}$ and of degree $m - \varkappa$ if $m - \varkappa > r - 1$ (recall Theorem 9.9). Because $(SbI\sigma - SBI\sigma_0) = 0$, we get the first assertion of the theorem. From Remark 9.10 we conclude that

$$\tilde{q}_j := (a\mu I - Sb\mu I)\,t^j$$

is a polynomial of degree $j + \varkappa$. Thus, the proof of the orthogonality properties of $q_{m,\mu}$ is analogous to the proof of Theorem 9.12. Indeed, for $j + \varkappa < m$ we have

$$0 = \int_{-1}^{1} \sigma_0(x)\,p_{m,\sigma_0}(x)\left[a(x)\,\mu(x)\,x^j - \frac{1}{\pi}\int_{-1}^{1}\frac{b(t)\,t^j\mu(t)}{t - x}\,dt\right]dx$$

$$= \int_{-1}^{1} \mu_0(x)\,x^j\left[a(x)\,\sigma(x)\,p_{m,\sigma_0}(x) + \frac{B(x)}{\pi}\int_{-1}^{1}\frac{p_{m,\sigma_0}(t)}{t - x}\,\sigma_0(t)\,dt\right]dx.$$

Here we used that $\sigma_0 \in \mathcal{H}^*$ and we also applied Proposition 9.11. It remains to prove (1). For this aim we remark that

$$\tilde{q}_m(t) = (aI - SbI)\,\mu t^{m-\varkappa} \qquad (m - \varkappa \geqq 0)$$

is of the form

$$\tilde{q}_m(t) = (-1)^\lambda\,t^m + \varepsilon_{m-1}\,t^{m-1} + \cdots + \varepsilon_0.$$

Hence (cf. the proof of Remark 9.13),

$$(-1)^\lambda \int_{-1}^{1} t^m p_{m,\sigma_0}(t)\,\sigma_0(t)\,dt = \int_{-1}^{1} t^{m-\varkappa} q_{m,\mu_0}(t)\,\mu_0(t)\,dt.$$

On the other hand, if $m - \varkappa > r - 1$, then, by Theorem 9.9 and the orthogonality of $\{p_{m,\sigma_0}\}$,

$$q_{m,\mu_0}(t) = (-1)^\lambda\,p_{mm}t^{m-\varkappa} + \cdots,$$

where $p_{m,\sigma_0}(t) = p_{mm}t^m + \cdots$. Now it is easily seen that

$$\|p_{m,\sigma_0}\|_{\sigma_0}^2 = p_{mm}\int_{-1}^{1} t^m p_{m,\sigma_0}\sigma_0\,dt$$

$$= (-1)^\lambda\,p_{mm}\int_{-1}^{1} t^{m-\varkappa} q_{m,\mu_0}\mu_0\,dt = \|q_{m,\mu_0}\|_{\mu_0}^2. \quad \blacksquare$$

9.15. Remarks. In this section we specify some of the previous assertions to the case where a and b are real numbers such that $b \neq 0$.

1°. We have $a + ib = (a^2 + b^2)^{1/2} \cdot e^{i\pi\gamma_0}$ ($\gamma_0 \in \mathbb{R}$, $0 < |\gamma_0| < 1$). Thus, according to Section 9.5,

$$g_\pm(t) = \ln (a \pm ib) = \pm i\pi\gamma_0 + \ln \sqrt{a^2 + b^2}$$

and

$$g(t) = -\gamma_0 .$$

Choose integers λ and ν so that

$$-1 < \alpha = \lambda - \gamma_0 < 1, \qquad -1 < \beta = \nu + \gamma_0 < 1 .$$

Then

$$\sigma(t) = \frac{1}{(a^2 + b^2)^{1/2}} (1 - t)^\lambda (1 + t)^\nu \exp \int_{-1}^{1} \frac{-\gamma_0}{x - t} \, dx .$$

Since

$$\int_{-1}^{1} \frac{1}{x - t} \, dx = \ln \frac{1 - t}{1 + t} \qquad \text{(Proposition 9.6)},$$

we get

$$\sigma(t) = \frac{1}{(a^2 + b^2)^{1/2}} (1 - t)^\alpha (1 + t)^\beta . \tag{1}$$

In particular, if $a = 0$ and $b = 1$ then $\gamma_0 = 1/2$ and the following cases are possible:

(a) $\lambda = \nu = 0$, $\sigma(t) = \left(\frac{1 + t}{1 - t}\right)^{1/2}$, $\varkappa = 0$.

(b) $\lambda = 1$, $\nu = 0$, $\sigma(t) = \sqrt{1 - t^2}$, $\varkappa = -1$.

(c) $\lambda = 0$, $\nu = -1$, $\sigma(t) = (1 - t^2)^{-1/2}$, $\varkappa = 1$.

(d) $\lambda = 1$, $\nu = -1$, $\sigma(t) = \left(\frac{1 - t}{1 + t}\right)^{1/2}$, $\varkappa = 0$.

Note that the $\mu(t)$ defined in Remark 9.10 takes the form

$$\mu(t) = \frac{1}{\sqrt{a^2 + b^2}} (1 - t)^{-\alpha} (1 + t)^{-\beta} .$$

2°. It is well-known that the Jacobi polynomials

$$P_n^{(\alpha,\beta)}(t) = \frac{(-1)^n}{2^n n!} (1 - t)^{-\alpha} (1 + t)^{-\beta} \frac{d^n}{dt^n} [(1 - t)^{\alpha+n} (1 + t)^{\beta+n}]$$

form a system of orthogonal polynomials with respect to the weight $\sigma(t)$ defined by (1), the coefficient of t^n being equal to

$$\frac{1}{2^n} \binom{2n + \alpha + \beta}{n} = \frac{1}{2^n} \binom{2n - \varkappa}{n} .$$

From Theorems 9.12 and 9.9 we infer that

$$(aI + bS)\,\sigma P_n^{(\alpha,\beta)} = c_n P_{n-\varkappa}^{(-\alpha,-\beta)}$$

with

$$c_n = \frac{1}{2^n}\binom{2n-\varkappa}{n}(-1)^\lambda \Big/ \frac{1}{2^{n-\varkappa}}\binom{2(n-\varkappa)+\varkappa}{n-\varkappa} = \frac{(-1)^\lambda}{2^\varkappa}.$$

Using the relation

$$\sin \pi\alpha = \sin \pi(\lambda - \gamma_0) = (-1)^{\lambda+1} \sin \pi\gamma_0 = \frac{(-1)^{\lambda+1}\,b}{(a^2+b^2)^{1/2}}$$

we obtain the important relation

$$a(1-t)^\alpha (1+t)^\beta P_n^{(\alpha,\beta)}(t) + \frac{b}{\pi}\int_{-1}^{1} \frac{P_n^{(\alpha,\beta)}(x)}{x-t}(1-x)^\alpha (1+x)^\beta\,dx$$

$$= -\frac{2^{-\varkappa}b}{\sin(\pi\alpha)} P_{n-\varkappa}^{(-\alpha,-\beta)}(t), \qquad -1 < t < 1,\ n = 0, 1, 2, \ldots, \tag{2}$$

where the parameters α, β and $\varkappa$ are defined by 1° and $P_{n-\varkappa}^{(-\alpha,-\beta)} \equiv 0$ if $n - \varkappa < 0$.

Fredholm properties

9.16. Notations. Let $\sigma(t)$ be a nonnegative integrable function defined on $(-1, 1)$. Let $\mathbf{L}_\sigma^2$ be the space of all complex-valued functions on $(-1,1)$ square-summable with respect to the weight σ. The norm is given by

$$\|u\|_\sigma = \left(\frac{1}{\pi}\int_{-1}^{1} |u(t)|^2\,\sigma(t)\,dt\right)^{1/2}.$$

Note that $\mathbf{L}_\sigma^2$ provided with scalar product

$$(u, v)_\sigma := \frac{1}{\pi}\int_{-1}^{1} u(t)\,\overline{v(t)}\,\sigma(t)\,dt$$

actually forms a Hilbert space. Assume we are given a singular integral operator $a\sigma I + bS\sigma I$, where σ is defined as in Section 9.5. Suppose further there is a function c satisfying the conditions 1°–3° of Section 9.13. Recall the notations $B(t) = c(t)\,b(t)$, $\sigma_0 = c^{-1}\sigma$ and $\mu_0 = c^{-1}\mu$. Let A be the operator defined by

$$A = a\sigma I + BS\sigma_0 I. \tag{1}$$

Our next concern is to study the operator A in the pair of Hilbert spaces $(\mathbf{L}_{\sigma_0}^2, \mathbf{L}_{\mu_0}^2)$.

9.17. Theorem. *The operator A can be extended to a bounded linear operator acting in the pair of Hilbert spaces $(\mathbf{L}_{\sigma_0}^2, \mathbf{L}_{\mu_0}^2)$. Moreover,*

(i) $\dim \ker A = \max\{0, \varkappa\}$,

(ii) $f \in \operatorname{im} A$ *if and only if* $(t^j, f)_{\mu_0} = 0$ $(j = 0, 1, \ldots, -\varkappa - 1)$,

(iii) $A^* = a\mu I - Sb\mu I$,

(iv) $\dim \ker A^* = \max\{0, -\varkappa\}$,
(v) *if* $A^{(-1)} = a\mu I - BS\mu_0 I$, *then* $A^{(-1)} A = I$ *for* $\varkappa \leqq 0$, *and* $AA^{(-1)} = I$ *for* $\varkappa \geqq 0$,
(vi) $\ker A = \operatorname{span}\{B, Bt, \ldots, Bt^{\varkappa-1}\}$ *if* $\varkappa > 0$.

Proof. Let Π_n be the linear space of all polynomials the degree of which is less than n. Decompose $\mathbf{L}^2_{\sigma_0}$ and $\mathbf{L}^2_{\mu_0}$ into the orthogonal sums

$$\mathbf{L}^2_{\sigma_0} = \Pi_n \oplus \overline{\operatorname{span}\{\tilde{p}_{n+k,\sigma_0}\}_{k=0}^{\infty}}$$
$$\mathbf{L}^2_{\mu_0} = \Pi_{n-\varkappa} \oplus \overline{\operatorname{span}\{\tilde{q}_{n+k-\varkappa,\mu_0}\}_{k=0}^{\infty}}$$

where $\tilde{p}_{j,\sigma_0}$, $\tilde{q}_{j,\mu_0}$ are normalized orthogonal polynomials such that

$$A\tilde{p}_{n+k,\sigma_0} = \tilde{q}_{n+k-\varkappa,\mu_0}$$

and n is some integer greater than $r + |\varkappa|$, with r the degree of the polynomial B (cf. Theorem 9.14). Hence, Au is defined for any $u \in \mathbf{L}^2_{\sigma_0}$ by

$$Au_n + \sum_{k=0}^{\infty} (u, \tilde{p}_{n+k,\sigma_0})_{\sigma_0} \tilde{q}_{n+k-\varkappa,\mu_0}, \tag{1}$$

where u_n is the orthogonal projection of u onto Π_n in $\mathbf{L}^2_{\sigma_0}$. In view of Proposition 9.11 we have, for any $u, v \in \bigcup_{k=1}^{\infty} \Pi_k$,

$$\begin{aligned}(Au, v)_{\mu_0} &= (a\sigma u, v)_{\mu_0} + (S\sigma_0 u, Bv)_{\mu_0} \\ &= (u, a\mu v)_{\sigma_0} - (u, SB\mu_0 v)_{\sigma_0} \\ &= (u, a\mu v - Sb\mu v)_{\sigma_0}\end{aligned}$$

which proves (iii). Now we are going to prove (v). First of all, the operator $a\sigma I + Sb\sigma I$ is also bounded in the pair $(\mathbf{L}^2_{\sigma_0}, \mathbf{L}^2_{\mu_0})$. Indeed,

$$a\sigma I + Sb\sigma I = a\sigma I + SB\sigma_0 I = A + (SB - BS)\,\sigma_0 I.$$

Since B is a polynomial, the operator $(SB - BS)\,\sigma_0 I$ turns out to be a compact operator from $\mathbf{L}^2_{\sigma_0}$ into $\mathbf{L}^2_{\mu_0}$. Assume $\varkappa \leqq 0$. Then

$$A^*(a\sigma I + Sb\sigma I) = I$$

due to Remark 9.10. Hence $(a\sigma I + Sb\sigma I)^* A = I$. Taking into account that

$$(a\sigma I + Sb\sigma I)^* = a\mu I - BS\mu_0 I,$$

we obtain (v). The validity of (v) in the case where $\varkappa \geqq 0$ follows in a similar fashion. By what has already been proved,

$$\dim \ker A = 0 \quad \text{if} \quad \varkappa \leqq 0, \qquad \dim \ker A^* = 0 \quad \text{if} \quad \varkappa \geqq 0.$$

Now assume $\varkappa > 0$. Since the action of A is given by (1), we have

$$\ker A = \ker (A|_{\Pi_n}), \qquad A(\Pi_n) \subset \Pi_{n-\varkappa}.$$

As $A^* = a\mu I - Sb\mu I$ and because of Remark 9.10, we obtain that

$$1, t, \ldots, t^{\varkappa-1} \notin \operatorname{im} A^*,$$

and that

$$t^{\varkappa}, \ldots, t^n \in \operatorname{im} A^*.$$

Thus, dim ker A = codim im A^* = $\varkappa$. If $\varkappa < 0$ then span $\{1, t, \ldots, t^{-\varkappa-1}\}$ equals ker A^*, which implies (ii) and (iv). From Theorem 9.9 it follows that, for $\varkappa > 0$,

$$ABt^j = B(a\sigma t^j + Sb\sigma t^j) = 0 \qquad (j = 0, 1, \ldots, \varkappa - 1).$$

This together with (i) implies (vi). ■

Quadrature rules

9.18. Notations. Let $\sigma(x)$ be a given non-negative summable function on $(-1, 1)$ such that

$$\int_{-1}^{1} \sigma(x)\, dx > 0.$$

In order to compute the integral

$$\frac{1}{\pi} \int_{-1}^{1} f(x)\, \sigma(x)\, dx \tag{1}$$

for a given continuous function f, it is frequently useful to apply a quadrature rule

$$\frac{1}{\pi} \int_{-1}^{1} f(x)\, \sigma(x)\, dx \approx \sum_{k=1}^{n} A_k^{(n)} f(x_k^{(n)}),$$

where $-1 \leqq x_n^{(n)} < x_{n-1}^{(n)} < \cdots < x_1^{(n)} \leqq 1$. One of the possibilities of obtaining quadrature rules can be described as follows. Assume we are given n points $x_i^{(n)} \in [-1, 1]$, $1 \leqq i \leqq n$, such that

$$-1 \leqq x_n^{(n)} < x_{n-1}^{(n)} < \cdots < x_1^{(n)} \leqq 1,$$

and let $L_n(f)$ be the interpolation polynomial of least degree such that

$$[L_n(f)]\,(x_i^{(n)}) = f(x_i^{(n)}), \qquad i = 1, \ldots, n.$$

It is well-known that such an interpolation polynomial always exists and that it is uniquely determined. The degree of $L_n(f)$ is equal to or less than $n - 1$. Moreover,

$$[L_n(f)]\,(x) = \sum_{k=1}^{n} l_k^{(n)}(x)\, f(x_k^{(n)}),$$

where

$$l_k^{(n)}(x) = \frac{w_n(x)}{(x - x_k^{(n)})\, w_n'(x_k^{(n)})}, \qquad w_n(x) = (x - x_1^{(n)})\,(x - x_2^{(n)}) \cdots (x - x_n^{(n)}),$$

$k = 1, 2, \ldots, n$. The mapping $f \mapsto L_n(f)$ is linear and is referred to as the interpolation operator L_n. If in (1) we replace f by $L_n(f)$ we get

$$\frac{1}{\pi} \int_{-1}^{1} L_n f(x)\, \sigma(x)\, dx = \sum_{k=1}^{n} A_k^{(n)} f(x_k^{(n)}) =: Q_n(f) \tag{2}$$

with

$$A_k^{(n)} = \frac{1}{\pi} \int_{-1}^{1} l_k^{(n)}(x)\, \sigma(x)\, \mathrm{d}x .$$

Quadrature rules obtained in such a manner are also called interpolatory quadrature rules. If f is a polynomial of degree at most $n - 1$ then $L_n f = f$ and, hence, we have equality in (2). The largest integer m such that in relation (2) equality holds for all polynomials of degree at most m is referred to as the algebraic accuracy of the rule (2).

Assume we are given the operator $A = a\sigma I + BS\sigma_0 I$ defined by 9.16(1), let p_n be a sequence of orthogonal polynomials with respect to the weight σ_0, and let the degree of p_n be n. Choose sequences of real numbers $\{a_n\}_{n=1}^{\infty}$, $\{b_n\}_{n=1}^{\infty}$ with $b_n \leqq 0$ for all n and put

$$g_n(t) = p_n(t) + a_n p_{n-1}(t) + b_n p_{n-2}(t) \; \big(p_{-1}(t) = p_{-2}(t) := 0\big). \tag{3}$$

It is well-known (cf. Freud [1]) that all roots of $g(t)$ are simple and real. Take the functions g_n defined by (3) and consider

$$k_{n-\varkappa} := A g_n .$$

By Theorem 9.14 we have, for $n - \varkappa > r + 1$,

$$k_{n-\varkappa} = q_{n-\varkappa} + a_n q_{n-\varkappa-1} + b_n q_{n-\varkappa-2}, \tag{4}$$

where $q_{n-\varkappa} := Ap_n$ are orthogonal polynomials of degree $n - \varkappa$ with respect to the weight μ_0. We now make the following convention: all roots of both g_n and $k_{n-\varkappa}$ are assumed to be contained in $[-1, 1]$ $(n - \varkappa > r + 1)$.

Let $L_n^{\sigma_0}$ and $L_{n-\varkappa}^{\mu_0}$ denote the corresponding interpolation operators, which under our assumptions can be written in the form

$$(L_n^{\sigma_0} v)(t) = \sum_{k=1}^{n} v(t_k^{(n)}) \frac{g_n(t)}{(t - t_k^{(n)})\, g_n'(t_k^{(n)})},$$

$$(L_{n-\varkappa}^{\mu_0} f)(x) = \sum_{j=1}^{n-\varkappa} f(x_j^{(n-\varkappa)}) \frac{k_{n-\varkappa}(t)}{(x - x_j^{(n-\varkappa)})\, k_{n-\varkappa}'(x_j^{(n-\varkappa)})},$$

where $t_n^{(n)} < t_{n-1}^{(n)} < \cdots < t_1^{(n)}$, $x_{n-\varkappa}^{(n-\varkappa)} < x_{n-\varkappa-1}^{(n-\varkappa)} < \cdots < x_1^{(n-\varkappa)}$ are the roots of g_n, $k_{n-\varkappa}$, respectively. The weights in the corresponding quadrature rules (2) (which are said to be of Gauss type) we denote by $A_{kn}^{\sigma_0}$ $(k = 1, \ldots, n)$ and $A_{j,n-\varkappa}^{\mu_0}$ $(j = 1, \ldots, n - \varkappa)$, respectively. We remark that

$$A_{kn}^{\sigma_0} > 0 \; (k = 1, \ldots, n) \quad \text{and} \quad A_{j,n-\varkappa}^{\mu_0} > 0 \; (j = 1, \ldots, n - \varkappa),$$

and that the algebraic accuracy is given by

$$\begin{aligned} &2n - 1 \quad \text{if} \quad a_n = b_n = 0, \\ &2n - 2 \quad \text{if} \quad a_n \neq 0,\ b_n = 0, \\ &2n - 3 \quad \text{if} \quad b_n \neq 0 \quad (\text{cf. Freud } [1]). \end{aligned}$$

Later on we specify the sequences $\{a_n\}$, $\{b_n\}$ (see Sec. 9.35).

9.19. Proposition. *We have* $B(t_{k_0}^{(n)}) = 0$ *for some* $k_0 \in \{1, \ldots, n\}$ *if and only if there is a* $j_0 \in \{1, \ldots, n - \varkappa\}$ *such that* $x_{j_0}^{(n-\varkappa)} = t_{k_0}^{(n)}$ $(n - \varkappa > r + 1)$.

Proof. First notice that

$$k_{n-\varkappa}(x) = a(x)\,\sigma(x)\,g_n(x) + \frac{B(x)}{\pi}\int_{-1}^{1}\frac{g_n(t)-g_n(x)}{t-x}\,\sigma_0(t)\,dt$$

$$+\, g_n(x)\,\frac{B(x)}{\pi}\int_{-1}^{1}\frac{\sigma_0(t)}{t-x}\,dt.$$

For fixed x, the function $w(t) = \dfrac{g_n(t)-g_n(x)}{t-x}$ is a polynomial of degree $n-1$. Hence if $x \neq t_k^{(n)}$ $(k = 1, \ldots, n)$, then

$$\frac{1}{\pi}\int_{-1}^{1}\frac{g_n(t)-g_n(x)}{t-x}\,\sigma_0(t)\,dt = \frac{1}{\pi}\int_{-1}^{1} w(t)\,\sigma_0(t)\,dt = \sum_{k=1}^{n} A_{kn}^{\sigma_0} w(t_k^{(n)})$$

$$= -g_n(x)\sum_{k=1}^{n}\frac{A_{kn}^{\sigma_0}}{t_k^{(n)}-x},$$

where we have used that $g_n(t_k^{(n)}) = 0$ for $k = 1, \ldots, n$.

A straightforward computation leads to the equality

$$k_{n-\varkappa}(x) = g_n(x)\left(A1 - B(x)\sum_{k=1}^{n} A_{kn}^{\sigma_0}\frac{1}{t_k^{(n)}-x}\right), \tag{1}$$

where $1(t) \equiv 1$. Now suppose $x_{j_0}^{(n-\varkappa)} = t_{k_0}^{(n)}$ and let $x \to x_{j_0}^{(n-\varkappa)}$ in (1). This yields that

$$B(x_{j_0}^{(n-\varkappa)})\,A_{k_0 n}^{\sigma_0} g_n'(t_{k_0}^{(n)}) = 0.$$

Taking into consideration that the roots of $g_n(t)$ are all simple, it follows that $B(x_{j_0}^{(n-\varkappa)}) = 0$. Now let $B(t_{k_0}^{(n)}) = 0$. Then $\lim_{x\to t_{k_0}^{(n)}} k_{n-\varkappa}(x) = 0$ and hence, there is an $x_{j_0}^{(n-\varkappa)}$ such that $t_{k_0}^{(n)} = x_{j_0}^{(n-\varkappa)}$. ■

9.20. Corollary. *Let Π_n be the collection of polynomials whose degree is equal to or less than $n-1$, where $n - \varkappa > r + 1$. Then, for $u \in \Pi_n$,*

$$(Au_n)\,(x_j^{(n-\varkappa)}) = \begin{cases} B(x_j^{(n-\varkappa)})\displaystyle\sum_{k=1}^{n} A_{kn}^{\sigma_0}\frac{u_n(t_k^{(n)})}{t_k^{(n)}-x_j^{(n-\varkappa)}} & \text{if } x_j^{(n-\varkappa)} \neq t_k^{(n)} \\ & (k = 1, \ldots, n) \\ a(x_j^{(n-\varkappa)})\,\sigma(x_j^{(n-\varkappa)})\,u_n(t_{k_0}^{(n)}) & \text{if } x_j^{(n-\varkappa)} = t_{k_0}^{(n)}. \end{cases}$$

Proof. It is easily seen that, for $x \neq t_k^{(n)}$ $(k = 1, \ldots, n)$,

$$(Au_n)\,(x) = u_n(x)\,(A1)\,(x) + B(x)\sum_{k=1}^{n} A_{kn}^{\sigma_0}\frac{u_n(t_k^{(n)})-u_n(x)}{t_k^{(n)}-x}$$

(recall the preceding proof). Expressing $A1$ by the help of 9.19(1) gives

$$(Au_n)\,(x) = u_n(x)\,\frac{k_{n-\varkappa}(x)}{g_n(x)} + B(x)\sum_{k=1}^{n} A_{kn}^{\sigma_0}\frac{u_n(t_k^{(n)})}{t_k^{(n)}-x}.$$

This immediately implies the assertion in the case $x_j^{(n-\varkappa)} \neq t_k^{(n)}$ $(k = 1, \ldots, n)$. Using that

$$\frac{1}{\pi} \int_{-1}^{1} \frac{u_n(t)}{t - x} \sigma_0(t) \, dt \in \mathcal{H}^*,$$

and taking into account Proposition 9.19 we get, for $x_j^{(n-\varkappa)} = t_{k_0}^{(n)}$, $k_0 \in \{1, \ldots, n\}$,

$$B(x_j^{(n-\varkappa)}) \frac{1}{\pi} \int_{-1}^{1} \frac{u_n(t)}{t - x_j^{(n-\varkappa)}} \sigma_0(t) \, dt = 0,$$

and the proof is complete. ■

9.21. Remark. If $f_n \in \Pi_{n-\varkappa}$, then (cf. Theorem 9.17)

$$(A^{(-1)} f_n)(t_k^{(n)}) = \begin{cases} B(t_k^{(n)}) \sum_{j=1}^{n-\varkappa} A_{j,n-\varkappa}^{\mu_0} \dfrac{f_n(x_j^{(n-\varkappa)})}{t_k^{(n)} - x_j^{(n-\varkappa)}} & \text{if } t_k^{(n)} \neq x_j^{(n-\varkappa)} \quad (j = 1, \ldots, n - \varkappa) \\ a(t_k^{(n)}) \, \mu(t_k^{(n)}) \, f_n(x_{j_0}^{(n-\varkappa)}) & \text{if } t_k^{(n)} = x_{j_0}^{(n-\varkappa)}. \end{cases}$$

9.22. Remark. The conclusions of Corollary 9.20 and Remark 9.21 remain valid if $u_n \in \Pi_{2n-s}$ and $f_n \in \Pi_{2(n-\varkappa)-s}$, where

$$s = \begin{cases} 0 & \text{if } a_n = b_n = 0, \\ 1 & \text{if } a_n \neq 0, \, b_n = 0, \\ 2 & \text{if } b_n \neq 0. \end{cases}$$

This is an immediate consequence of the algebraic accuracy of the corresponding quadrature rule in the regular case.

9.23. Remark. Denote by $\boldsymbol{A}_n = (a_{jk})_{j=1,k=1}^{n-\varkappa,n}$ the matrix whose entries are

$$a_{jk} = B(x_j^{(n-\varkappa)}) \frac{A_{kn}^{\sigma_0}}{t_k^{(n)} - x_j^{(n-\varkappa)}} \qquad (j = 1, \ldots, n - \varkappa; \, k = 1, \ldots, n)$$

if $x_j^{(n-\varkappa)} \neq t_k^{(n)}$ for all j and k. If there exists a $k_0 \in \{1, \ldots, n\}$ such that $x_j^{(n-\varkappa)} = t_{k_0}^{(n)}$, then the j-th row of $\boldsymbol{A}_n$ is given by

$$a_{jk} = \begin{cases} a(x_j^{(n-\varkappa)}) \, \sigma(x_j^{(n-\varkappa)}) & \text{if } k = k_0 \\ 0 & \text{if } k \neq k_0. \end{cases}$$

Analogously, define $\boldsymbol{A}_n^{(-1)} = (\hat{a}_{kj})_{k=1,j=1}^{n,n-\varkappa}$ by

$$\hat{a}_{kj} = \begin{cases} B(t_k^{(n)}) \dfrac{A_{j,n-\varkappa}^{\mu_0}}{t_k^{(n)} - x_j^{(n-\varkappa)}} & \text{if } t_k^{(n)} \neq x_j^{(n-\varkappa)} \quad (j = 1, \ldots, n - \varkappa) \\ a(t_k^{(n)}) \, \mu(t_k^{(n)}) \, \delta_{jj_0} & \text{if } t_k^{(n)} = x_{j_0}^{(n-\varkappa)} \end{cases}$$

where δ_{kj} is the Kronecker symbol. Due to Theorem 9.17(v) and Remark 9.21 we obtain

$$\boldsymbol{A}_n^{(-1)} \boldsymbol{A}_n = \boldsymbol{I} \quad \text{if} \quad \varkappa \leqq 0, \qquad \boldsymbol{A}_n \boldsymbol{A}_n^{(-1)} = \boldsymbol{I} \quad \text{if} \quad \varkappa \geqq 0,$$

where $\boldsymbol{I}$ denotes the identity matrix of appropriate order.

Convergence analysis

9.24. Notations. Consider the singular integral equation

$$a(x)\, u(x) + \frac{b(x)}{\pi} \int_{-1}^{1} \frac{u(t)}{t - x}\, dt + \frac{1}{\pi} \int_{-1}^{1} h_1(x, t)\, u(t)\, dt = f_1(x). \tag{1}$$

We assume that there is a function $c(x)$ on $[-1, 1]$ such that $B(x) = c(x)\, b(x)$ fulfils the conditions $1°-3°$ of Section 9.13 and look for a solution $u(x)$ of (1) in the form $u(x) = \sigma_0(x)\, v(x)$. Write (1) in the equivalent form

$$a(x)\, \sigma(x)\, v(x) + \frac{B(x)}{\pi} \int_{-1}^{1} \frac{\sigma_0(t)}{t - x}\, v(t)\, dt + \frac{1}{\pi} \int_{-1}^{1} h(x, t)\, v(t)\, \sigma_0(t)\, dt = f(x), \tag{2}$$

where $f(x) = c(x)\, f_1(x)$, $h(x, t) = c(x)\, h_1(x, t)$. If $\varkappa > 0$, we agree upon requiring that the solution $v(t)$ of (2) satisfies the additional conditions

$$\int_{-1}^{1} v(x)\, x^j \sigma_0(x)\, dx = 0 \qquad (j = 0, 1, \ldots, \varkappa - 1). \tag{3}$$

Henceforth we write (2), (3) as

$$Av + Hv = f, \tag{4}$$

the operator A defined by 9.16(1) as well as the operator H defined by

$$(Hv)\,(x) = \frac{1}{\pi} \int_{-1}^{1} h(x, t)\, v(t)\, \sigma_0(t)\, dt$$

being considered as operators in the pair of Hilbert spaces $(\mathbf{L}^2_{\sigma_0,s}, \mathbf{L}^2_{\mu_0})$, where $s = \max\,\{0, \varkappa\}$ and

$$\mathbf{L}^2_{\sigma_0,s} := \{v \in \mathbf{L}^2_{\sigma_0}, \text{ with } v \text{ satisfying (3)}\}.$$

Due to Theorem 9.17 the operator A is invertible in the pair $(\mathbf{L}^2_{\sigma_0,\varkappa}, L^2_{\mu_0})$ if only $\varkappa \geqq 0$, and its inverse A^{-1} equals

$$A^{-1} = a\mu I - BS\mu_0 I.$$

In what follows we deal with the case $\varkappa \geqq 0$. A few words about the case $\varkappa < 0$ will be said later on (cf. Sections 9.41–9.43). We seek an approximate solution of equation (4) in the form

$$v_n(t) = \sum_{k=1}^{n} \xi_{kn} \prod_{\substack{j=1\\ j\neq k}}^{n} \frac{t - t_j^{(n)}}{t_k^{(n)} - t_j^{(n)}},$$

where the coefficients ξ_{kn} are determined from the equations

$$\sum_{k=1}^{n} [a_{jk} + A_{kn}^{\sigma_0} h(x_j^{(n-\varkappa)}, t_k^{(n)})]\, \xi_{kn} = f(x_j^{(n-\varkappa)}), \qquad j = 1, \dots, n - \varkappa, \tag{6}$$

$$\sum_{k=1}^{n} A_{kn}^{\sigma_0} (t_k^{(n)})^j\, \xi_{kn} = 0, \qquad j = 0, 1, \dots, \varkappa - 1. \tag{7}$$

In view of Theorem 9.14 and Corollary 9.20 the last two equations are equivalent to

$$(A + H_{n-\varkappa,n})\, v_n = L_{n-\varkappa}^{\mu_0} f, \qquad v_n \in \Pi_n \cap \mathbf{L}^2_{\sigma_0,\varkappa}. \tag{8}$$

Here and in what follows $L_{n-\varkappa}^{\mu_0}$ and $L_n^{\sigma_0}$ denote the interpolation operators associated with the roots of $k_{n-\varkappa}$ and g_n, respectively. The operator $H_{n-\varkappa,n}$ is defined by

$$H_{n-\varkappa,n} v_n = L_{n-\varkappa}^{\mu_0} \int_{-1}^{1} L_{n,t}^{\sigma_0}[h(x, t)\, v_n(t)]\, \sigma_0(t)\, dt.$$

In order to study this approximation method we need some facts from approximation theory. In the following we use the notations and assumptions of Section 9.18. In particular, we assume that all roots of g_n defined by 9.18(3) are located in $[-1, 1]$. The corresponding quadrature rule defined by 9.18(2) is denoted by $Q_n(f)$. Finally, define

$$\nu(b_n) = \begin{cases} 1 & \text{if} \quad b_n = 0, \\ 2 & \text{if} \quad b_n < 0. \end{cases}$$

9.25. Proposition. *For a bounded function $f(t)$, $t \in [-1, 1]$, we have*

$$\|L_n^\sigma f\|_\sigma^2 \leqq \nu(b_n)\, Q_n(|f|^2).$$

Proof. See Freud [1], Chapter III, Lemma 2.4. ∎

9.26. Proposition. *If $f \in \mathcal{R}([-1, 1], \sigma(t)\, dt)$, i.e. if $f(t)$ is bounded and Riemann-Stieltjes integrable with respect to $\sigma(t)\, dt$ on $[-1, 1]$, then*

$$\lim_{n\to\infty} \|f - L_n^\sigma f\|_\sigma = 0.$$

Proof. See Freud [1], Chapter III, Prop. 2.5. ∎

9.27. Proposition. *If $f \in \mathbf{C}^{m,\mu}$ (see Section 9.1), then*

$$\|f - L_{n+1}^\sigma f\|_\sigma \leqq (1 + [\nu(b_{n+1})]^{1/2})\, \|1\|_\sigma \cdot c_{m,\mu}(f)\, n^{-m-\mu},$$

where $c_{0,\mu} := 12\, \|f\|_{\mathscr{H}^\mu}$ if $m = 0$ and

$$c_{m,\mu} = 12\, \frac{(6m)^m}{m!} \left(\frac{m+1}{2}\right)^\mu 2^{m+\mu}\, \|f^{(m)}\|_{\mathscr{H}^\mu}.$$

Proof. For $u_n \in \Pi_{n+1}$, we have $L_{n+1}^\sigma u_n = u_n$. Thus,

$$\|f - L_{n+1}^\sigma f\|_\sigma \leqq \|f - u_n\|_\sigma + \|L_{n+1}^\sigma (f - u_n)\|_\sigma.$$

Proposition 9.25 implies that

$$\|f - L^\sigma_{n+1}f\|_\sigma \leqq \left(\frac{1}{\pi}\int_{-1}^{1} \sigma(t)\,\mathrm{d}t\right)^{1/2} \|f - u_n\|_\infty + [\nu(b_{n+1})\, Q_{n+1}(|f - u_n|^2)]^{1/2}$$

$$\leqq \|1\|_\sigma \cdot (1 + [\nu(b_{n+1})]^{1/2})\, \|f - u_n\|_\infty .$$

Indeed, because $A^\sigma_{k,n+1} > 0$ (see FREUD [1]) and the rule 9.18(2) is exact for the function 1 we get for any continuous function

$$Q_{n+1}(|g|^2) = \left(\sum_{k=1}^{n+1} A^\sigma_{k,n+1}\, |g(x_k^{(n+1)})|^2\right) \leqq \left(\sum_{k=1}^{n+1} A_{k,n+1}\right) \|g\|^2_\infty = \|1\|^2_\sigma\, \|g\|^2_\infty .$$

Since u_n is an arbitrarily chosen polynomial from Π_{n+1}, the assertion follows from the estimate

$$E_n(f) = c_{m,\mu}(f)\, n^{-m-\mu},$$

where $E_n(f) = \inf_{u\in\Pi_{n+1}} \|f - u_n\|_\infty$ (NATANSON [1], Part I, Chap. VI, § 2).

In the following we have often to handle with functions $h(x, t)$ given on $[-1, 1] \times [-1, 1]$. For fixed t or x put $h_t(x) := h(x, t)$ and $h_x(t) := h(x, t)$. The space of all continuous (resp. bounded) functions on $[-1, 1]$ with values in a given Banach space B will be denoted by $\mathbf{C}([-1, 1], B)$ (resp. $\boldsymbol{M}([-1, 1], B)$).

9.28. Corollary. *Let $h(x, t)$ be a function on $[-1, 1]\times[-1, 1]$ such that $\{h_t\}_{t\in[-1,1]} \in \mathbf{C}([-1, 1], \mathbf{C}^{m,\mu})$, i.e. the function $t \mapsto h_t$ belongs to $\mathbf{C}([-1, 1], \mathbf{C}^{m,\mu})$. Then*

$$\|H - L^{\mu_0}_k H\|_{\mathscr{L}(\mathbf{L}^2_{\sigma_0}, \mathbf{L}^2_{\mu_0})} \leqq \|1\|_{\sigma_0}\, \|1\|_{\mu_0} (1 + [\nu(b_k)]^{1/2}) \left(\sup_{t\in[-1,1]} c_{m,\mu}(h_t)\right) \cdot (k-1)^{-m-\mu}.$$

Proof. Since, obviously,

$$\|H - L^{\mu_0}_k H\|_{\mathscr{L}(\mathbf{L}^2_{\sigma_0}, \mathbf{L}^2_{\mu_0})} \leqq \left(\frac{1}{\pi}\int_{-1}^{1} \|h_t - L^{\mu_0}_k h_t\|^2_{\mu_0}\, \sigma_0(t)\,\mathrm{d}t\right)^{1/2},$$

the assertion follows immediately from Proposition 9.27. ■

9.29. Corollary. *Let $h(x, t)$ be a function on $[-1, 1]\times[-1, 1]$ for which $\{h_x\}_{x\in[-1,1]} \in \boldsymbol{M}([-1, 1], \mathbf{C}^{m,\mu})$. Assume further that $b_l = 0$ for all l. Then, for all $v_n \in \Pi_n$,*

$$\|(L^{\mu_0}_k H - H_{kn})\, v_n\|_{\mu_0} \leqq 2\, \|1\|_{\sigma_0}\, \|1\|_{\mu_0} \left(\sup_{x\in[-1,1]} c_{m,\mu}(h_x)\right) \cdot (n-1)^{-m-\mu}\, \|v_n\|_{\sigma_0}.$$

Proof. In case $b_n = 0$ the quadrature rule 9.18(2) is exact for polynomials $v(t)$ of degree less than or equal to $2n - 2$. Consequently, if $v_n \in \Pi_n$ then

$$(H_{kn} v_n)\,(x) = \frac{1}{\pi}\, L^{\mu_0}_k \int_{-1}^{1} [L^{\sigma_0}_n h_x(t)]\, v_n(t)\, \sigma_0(t)\,\mathrm{d}t .$$

In conjunction with Proposition 9.25 this yields that

$$\|(L_k^{\mu_0}H - H_{kn})\, v_n\|_{\mu_0} \leqq \|1\|_{\mu_0} \sup_{x\in[-1,1]} \left| \int_{-1}^{1} [h_x(t) - L_n^{\sigma_0}h_x(t)]\, v_n(t)\, \sigma_0(t)\, dt \right|$$

$$\leqq \|1\|_{\mu_0} \|v_n\|_{\sigma_0} \sup_{x\in[-1,1]} \|h_x - L_n^{\sigma_0}h_x\|_{\sigma_0}.$$

Applying Proposition 9.27 we get the assertion. ■

9.30. Remark. Let $h(x, t)$ be a function on $[-1, 1] \times [-1, 1]$ such that $\{h_t\} \in \mathbf{C}([-1, 1], \mathbf{C}^{m,\gamma})$ and $\{h_x\} \in \boldsymbol{M}([-1, 1], \mathbf{C}^{r,\delta})$. Also suppose that $b_l = 0$ for all l. Then, for arbitrary $v_n \in \Pi_n$,

$$\|(H - H_{kn})\, v_n\|_{\mu_0} \leqq \varepsilon_{kn} \|v_n\|_{\sigma_0}, \tag{1}$$

where

$$\varepsilon_{kn} = 2\, \|1\|_{\mu_0} \|1\|_{\sigma} \left\{ \left[\sup_{t\in[-1,1]} c_{m,\gamma}(h_t)\right] (k-1)^{-m-\gamma} + \left[\sup_{x\in[-1,1]} c_{r,\delta}(h_x)\right] (n-1)^{-r-\delta} \right\}. \tag{2}$$

This is immediate from Corollaries 9.28 and 9.29. ■

If $b_n < 0$, an analogous result is true in the case of a Jacobi weight

$$\sigma_0 = (t+1)^\alpha (1-t)^\beta,$$

where $\alpha, \beta > -1$ and $\max\{\alpha, \beta\} \geqq -1/2$. We denote by $\{p_n\}_n^\infty$ the corresponding sequence of orthogonal polynomials, the degree of p_n being $n - 1$.

9.31. Proposition. *Assume* $\sigma_0(t) = (t+1)^\alpha (1-t)^\beta$ *with* $\alpha, \beta > -1$ *and* $\tau := \max\{\alpha, \beta\} \geqq -1/2$. *Then there exists a constant* c_0 *depending only on* α *and* β *such that*

(a) $\|\tilde{p}_n\|_\infty \leqq c_0 n^{\tau+1/2}$ $(n = 1, 2, \dots)$ *where* $\tilde{p}_n = p_n/\|p_n\|_{\sigma_0}$,
(b) $\|v_n\|_\infty \leqq c_0 n^{\tau+1} \|v_n\|_{\sigma_0}$ *for all* $v_n \in \Pi_n$.

This assertion (with a perhaps other c_0) remains valid in the slightly more general situation where $\tilde{\sigma}_0(t) = (t+1)^\alpha (1-t)^\beta\, w(t)$ and $w(t)$ is a continuous function such that $w(t) \geqq c > 0$.

Proof. (a): See Natanson [1], Chapter VI, § 3, Prop. 2.

(b): Represent $v_n(t)$ in the form

$$v_n(t) = \sum_{j=1}^{n} \alpha_j \tilde{p}_n.$$

Cauchy's inequality in conjunction with part (a) gives

$$|v_n(t)| \leqq c_0 \left(\sum_{j=1}^{n} |\alpha_j|^2\right)^{1/2} \left(\sum_{j=1}^{n} j^{2\tau+1}\right)^{1/2}$$

$$\leqq c_0 \|v_n\|_{\sigma_0} \cdot n^\tau \left(\frac{n(n+1)}{2}\right)^{1/2} \leqq c_0 n^{\tau+1} \|v_n\|_{\sigma_0}.$$

Finally, note that the norms $\|\cdot\|_{\sigma_0}$ and $\|\cdot\|_{\tilde{\sigma}_0}$ are equivalent. ■

9.32. Proposition. *Assume* $\sigma_0(t) = (t+1)^\alpha (1-t)^\beta w_1(t)$, *where* $w_1(t) \geqq c > 0$ *is continuous and* $\alpha, \beta > -1$, $\tau = \max\{\alpha, \beta\} \geqq -1/2$. *Let* $b_l < 0$ *for all* l *and let* $h(x, t)$ *be a function on* $[-1, 1] \times [-1, 1]$ *such that* $\{h_t\} \in \mathbf{C}([-1, 1], \mathbf{C}^{m,\gamma})$, $\{h_x\} \in \boldsymbol{M}([-1, 1], \mathbf{C}^{r,\delta})$ *and* $r + \delta > \tau + 1$. *Then the estimate* 9.30(1) *is valid with*

$$\varepsilon_{kn} = \|1\|_{\sigma_0} \|1\|_{\mu_0} \left[\left(1+\sqrt{2}\right)\left[\sup_{t\in[-1,1]} c_{m,\gamma}(h_t)\right](k-1)^{-m-\gamma} + \sqrt{2}\,\|1\|_{\sigma_0}\left(1+\sqrt{2}\right) c_0 \left[\sup_{x\in[-1,1]} c_{r,\delta}(h_x)\right](n-2)^{\tau+1-r-\delta}\right]. \tag{1}$$

Proof. Given $f \in \mathbf{C}[-1, 1]$, put

$$E_n(f) := \inf\{\|f - p_n\|_\infty : p_n \in \Pi_{n+1}\}.$$

It is well-known that there exists a uniquely determined polynomial p_n^* such that $E_n(f) = \|f - p_n^*\|$. Define $h_{n-1,x} \in \Pi_{n-1}$ such that $E_{n-2}(h_x) = \|h_x - h_{n-1,x}\|_\infty$ for all $x \in [-1, 1]$. Since $E_n(f) \leqq c_{m,\gamma}(f)\, n^{-m-\gamma}$ for $f \in \mathbf{C}^{m,\gamma}$ (Natanson [1], Part I, Chap. VI, § 2), we obtain

$$\sup_{x\in[-1,1]} \|h_x - h_{n-1,x}\| \leqq \left[\sup_{x\in[-1,1]} c_{r,\delta}(h_x)\right](n-2)^{-r-\delta} := d_1(n-2)^{-r-\delta}.$$

In the case considered here the quadrature rule 9.18(2) is exact for polynomials $v(t)$ of degree less than or equal to $2n - 3$. Thus,

$$\begin{aligned}
\|(L_k^{\mu_0}H - H_{kn})\, v_n\|_{\mu_0} &\leqq \sqrt{2}\,\|1\|_{\mu_0} \times \sup_{x\in[-1,1]} \left|\frac{1}{\pi}\int_{-1}^{1} \left(h_x(t)\, v_n(t) - L_n^{\sigma_0}[h_x(t)\, v_n(t)]\right)\sigma_0(t)\, dt\right| \\
&\leqq \sqrt{2}\,\|1\|_{\mu_0} \sup_{x\in[-1,1]} \left\{\left|\frac{1}{\pi}\int_{-1}^{1} [h_x(t) - h_{n-1,x}(t)]\, v_n(t)\, \sigma_0(t)\, dt\right| \right. \\
&\quad \left. + \left|\frac{1}{\pi}\int_{-1}^{1} L_n^{\sigma_0}\{[h_x(t) - h_{n-1,x}(t)]\, v_n(t)\}\, \sigma_0(t)\, dt\right|\right\} \\
&\leqq \sqrt{2}\,\|1\|_{\mu_0}\|1\|_{\sigma_0}\left(1+\sqrt{2}\right) d_1(n-2)^{-r-\delta}\,\|v_n\|_\infty \\
&\leqq \sqrt{2}\,\|1\|_{\mu_0}\|1\|_{\sigma_0}\left(1+\sqrt{2}\right) d_1 c_0 n^{\tau+1-r-\delta}\,\|v_n\|_{\sigma_0}
\end{aligned}$$

due to Proposition 9.25 and Proposition 9.31(b). Using Corollary 9.28 we get the assertion. ■

9.33. Corollary. *Let* $\sigma(t) = (t+1)^\alpha (1-t)^\beta w_1(t)$, *the function* $w_1(t)$ *being continuous and satisfying* $w_1(t) \geqq c > 0$, *and let* $\alpha, \beta > -1$ *and* $\max\{\alpha, \beta\} \geqq -1/2$. *Assume* $v \in \mathbf{L}^2_\sigma$, $v_n \in \Pi_n$ *and* $\|v - v_n\|_\sigma \leqq c_1 n^{-\varrho}$ $(n = 1, 2, \ldots)$ *with* $\varrho > \tau + 1$. *Then the function* $v(t)$ *is continuous (that is,* $v \in \mathbf{C}[-1, 1]$*) and*

$$\|v - v_n\|_\infty \leqq \frac{(1 + 2^\varrho)\, c_1 c_0}{1 - 2^{\tau-\varrho+1}} \cdot n^{\tau+1-\varrho}.$$

Proof. Choose $j \in \mathbb{N}$ so that $2^{j-1} \leqq n < 2^j$. By virtue of Proposition 9.31(b) we have

$$\|v_{2^j} - v_n\|_\infty \leqq c_0 2^{j(\tau+1)}\, \|v_{2^j} - v_n\|_\sigma.$$

Further,

$$\|v_{2^j} - v_n\|_\sigma \leqq \|v_{2^j} - v\|_\sigma + \|v - v_n\|_\sigma \leqq (1 + 2^\varrho)\, c_1 2^{-j\varrho},$$

and consequently

$$\|v_{2^j} - v_n\|_\infty \leqq (1 + 2^\varrho)\, c_1 c_0 2^{j(\tau+1-\varrho)}. \tag{1}$$

Thus, the series

$$v_1(t) + \sum_{j=1}^{\infty} [v_{2^j}(t) - v_{2^{j-1}}(t)]$$

converges uniformly with respect to $t \in [-1, 1]$ to $v(t)$ and hence, $v(t)$ is continuous. In view of (1) it follows that

$$\|v - v_n\|_\infty \leqq \|v - v_{2^j}\|_\infty + \|v_{2^j} - v_n\|_\infty \to 0$$

for $2^{j-1} \leqq n < 2^j$ as $n \to \infty$. If $m = 1, 2, \ldots, j = 1, 2, \ldots$, and $2^{j-1} \leqq n < 2^j$, then from (1) we obtain

$$\begin{aligned} \|v_{2^{j+m}} - v_n\|_\infty &\leqq \|v_{2^{j+m}} - v_{2^{j+m-1}}\|_\infty + \cdots + \|v_{2^j} - v_n\|_\infty \\ &\leqq (1 + 2^\varrho) \cdot c_1 c_0 2^{j(\tau+1-\varrho)} [2^{(\tau+1-\varrho)m} + \cdots + 1] \\ &\leqq (1 + 2^\varrho) \cdot c_1 c_0 \frac{1}{1 - 2^{\tau+1-\varrho}} n^{\tau+1-\varrho}. \end{aligned}$$

Passage to the limit $m \to \infty$ gives the assertion. ∎

9.34. Theorem. *Consider equation*

$$Av + Hv = f, \tag{1}$$

introduced in 9.24(4), *in the pair* $(\mathbf{L}^2_{\sigma_0,\varkappa}, \mathbf{L}^2_{\mu_0})$ *with* $\varkappa \geqq 0$. *Write the equations* 9.24(6) *and* 9.24(7) *in the equivalent form*

$$(A + H_{n-\varkappa,n})\, v_n = L^{\mu_0}_{n-\varkappa} f, \qquad v_n \in \Pi_n \cap \mathbf{L}^2_{\sigma_0,\varkappa}. \tag{2}$$

Assume $f \in \mathbf{C}^{m,\gamma}$, $\{h_t\} \in \mathbf{C}([-1, 1], \mathbf{C}^{m,\gamma})$ *and* $\{h_x\} \in \boldsymbol{M}([-1, 1], \mathbf{C}^{r,\delta})$. *If equation* (1) *has a unique solution* $v \in \mathbf{L}^2_{\sigma_0,\varkappa}$, *then the equations* (2) *are uniquely solvable for all sufficiently large* n *and*

$$\|v - v_n\|_{\sigma_0} = O(n^{-\eta}), \tag{3}$$

where

$$\eta = \begin{cases} \min\{m + \gamma, r + \delta\} & \text{if} \quad b_n = 0, \\ \min\{m + \gamma, r + \delta - \tau - 1\} & \text{if} \quad b_n < 0, \end{cases}$$

and in the case $b_n < 0$ *it is also assumed that*

$$\sigma_0(t) = (t + 1)^\alpha (1 - t)^\beta w_1(t) \tag{4}$$

with

$$w_1(t) \geqq c > 0 \quad \text{and} \quad \tau := \max\{\alpha, \beta\} \geqq -1/2. \tag{5}$$

If (5) *is fulfilled and if* $\eta > \tau + 1$, *then* $v(t)$ *is continuous and*

$$\|v - v_n\|_\infty = O(n^{-(\eta-\tau-1)}). \tag{6}$$

Proof. We have, for $v_n \in \Pi_n \cap \mathbf{L}^2_{\sigma_0,\varkappa}$,

$$\|v_n\|_{\sigma_0} = \|(A+H)^{-1}(A+H)\,v_n\|_{\sigma_0}$$

(recall that A is invertible in the pair $(\mathbf{L}^2_{\sigma_0,\varkappa}, \mathbf{L}^2_{\mu_0})$ and that H is compact). Hence

$$\|v_n\|_{\sigma_0} \leqq \|(A+H)^{-1}\| \left(\|(A+H_{n-\varkappa,n})\,v_n\|_{\mu_0} + \varepsilon_{n-\varkappa,n}\,\|v_n\|_{\sigma_0}\right),$$

where $\varepsilon_{n-\varkappa,n}$ is defined by 9.30(2) or 9.32(1), respectively. Furthermore,

$$\|(A+H_{n-\varkappa,n})\,v_n\|_{\mu_0} \geqq \{\|(A+H)^{-1}\|^{-1} - \varepsilon_{n-\varkappa,n}\}\,\|v_n\|_{\sigma_0}. \tag{7}$$

Since $\varepsilon_{n-\varkappa,n} \to 0$ as $n \to \infty$, the last inequality in connection with the relation

$$\operatorname{im}\left(A|_{\Pi_n}\right) \cap \mathbf{L}^2_{\sigma_0,\varkappa} = \Pi_{n-\varkappa}$$

gives the invertibility of

$$A + H_{n-\varkappa,n} : \Pi_n \cap \mathbf{L}^2_{\sigma_0,\varkappa} \to \Pi_{n-\varkappa} \subset \mathbf{L}^2_{\mu_0}$$

for n large enough (say $n \geqq n_0$). Because of (7) we also get that $\|v_n\|_\sigma \leqq$ const, where $v_n(t)$ is the solution of (2) $(n \geqq n_0)$. Therefore, the estimate

$$\begin{aligned}\|v - v_n\|_{\sigma_0} &= \|(A+H)^{-1}(A+H)(v - v_n)\|_{\sigma_0} \\ &\leqq \|(A+H)^{-1}\| \cdot \|(A+H)\,v - (A+H_{n-\varkappa,n})\,v_n - (H - H_{n-\varkappa,n})\,v_n\|_{\mu_0} \\ &\leqq \|(A+H)^{-1}\| \cdot \left(\|f - L^{\mu_0}_{n-\varkappa}f\|_{\mu_0} + \|(H - H_{n-\varkappa,n})\,v_n\|_{\mu_0}\right)\end{aligned}$$

combined with the assertions of the Secs. 9.30, 9.32 and 9.33 implies (3) and (6). ■

Zero distribution

9.35. Notations. Recall that Theorem 9.34 was proved under the assumption that all zeros of the polynomials g_n and $k_{n-\varkappa}$ are contained in the interval $[-1, 1]$. Therefore, we need some knowledge about the zero distribution of the polynomials g_n and $k_{n-\varkappa}$ defined by 9.18(3) and 9.18(4), respectively. We shall restrict ourselves to the consideration of some important special cases which are of interest in numerical analysis. More precisely, we study the zero distribution of

$$g_n(t) = p_n(t) + a_n p_{n-1}(t) + b_n p_{n-2}(t)$$

(and of the associated polynomials $k_{n-\varkappa}$) in the following three cases:

(i) $a_n = b_n = 0$ for all n;

(ii) $a_n = -\dfrac{p_n(c)}{p_{n-1}(c)}$, $b_n = 0$ for all n, and c takes the value 1 or -1;

(iii) $$\begin{aligned} a_n &= -\frac{p_n(-1)\,p_{n-2}(1) - p_n(1)\,p_{n-2}(-1)}{p_{n-1}(-1)\,p_{n-2}(1) - p_{n-1}(1)\,p_{n-2}(-1)}, \\ b_n &= \frac{p_n(-1)\,p_{n-1}(1) - p_n(1)\,p_{n-1}(-1)}{p_{n-1}(-1)\,p_{n-2}(1) - p_{n-1}(1)\,p_{n-2}(-1)} \end{aligned} \tag{1}$$

for all n.

The underlying Gauss type quadrature rules are called Gauss, Radau and Lobatto quadrature rules, respectively. Notice that the numbers a_n and b_n are well-defined. In the case (ii) this is obvious, since all zeros of the orthogonal polynomial p_{n-1} are located in $(-1, 1)$ (see FREUD [1]). In the third case we can proceed as follows. Put $\psi_n(t, \xi) = p_{n-1}(\xi)\, p_n(t) - p_n(\xi)\, p_{n-1}(t)$. Then (see FREUD [1]) all zeros of $\psi_n(t, \xi)$ (thought of as a polynomial in t) are real and simple, and at least $n - 1$ zeros are contained in $(-1, 1)$. Obviously, $\psi_n(1, 1) = 0$ and hence the polynomial $\psi_n(t, 1)$ cannot vanish at the point -1. Thus $\psi_n(-1, 1) \neq 0$ $\big(\psi_{n-1}(-1, 1) \neq 0\big)$. It is also easy to see that

$$b_n < 0.$$

(b_n being defined by (1)).

The preceding argument also shows that in case (ii) all zeros of g_n are simple and are contained in $[-1, 1]$. This is true in case (iii), too. A proof of this is essentially based upon the following result: if A_n, B_n are sequences of real numbers such that $B_n \leqq 0$ for all n, then the polynomials

$$g_n^*(t) := p_n(t) + A_n p_{n-1}(t) + B_n p_{n-2}(t)$$

have at least $(n - 2)$ distinct zeros in $(-1, 1)$.

Indeed, let $\bar{t}_i^{(n)}$ be the zeros of $p_n(t)$:

$$-1 < \bar{t}_n^{(n)} < \bar{t}_{n-1}^{(n)} < \cdots < \bar{t}_1^{(n)} < 1.$$

Then (see FREUD [1])

$$\bar{t}_{j+1}^{(n)} < \bar{t}_j^{(n-1)} < \bar{t}_j^{(n)} \qquad (j = 1, \ldots, n - 1).$$

Since the leading coefficient of $p_n(t)$ may be assumed to be positive, we get

$$\operatorname{sgn} p_n(\bar{t}_j^{(n-1)}) = (-1)^j,$$

and analogously,

$$\operatorname{sgn} p_{n-2}(\bar{t}_j^{(n-1)}) = (-1)^{j+1}.$$

These two relations in connection with the hypothesis that $B_n \leqq 0$ give

$$\operatorname{sgn} g_n^*(\bar{t}_j^{(n-1)}) = (-1)^j.$$

Hence, at least one zero of $g_n^*(t)$ lies in some interval $(\bar{t}_{j+1}^{(n-1)}, \bar{t}_j^{(n-1)})$, $j = 1, \ldots, n - 2$. ■

Using the latter result and the equalities $g_n(-1) = g_n(1) = 0$ (in case (iii)) we see that all zeros of $g_n(t)$ are simple and belong to $[-1, 1]$. The zero distribution of the polynomials $k_{n-\varkappa}$ is of a more intricated nature. Put $n - \varkappa = m$ $(n - \varkappa > r - 1)$.

9.36. Proposition. *If* $B(t_j^{(n)})\, B(t_{j+1}^{(n)}) > 0$, *then there is an odd number of zeros of* $k_m(x)$ *in the open interval* $(t_{j+1}^{(n)}, t_j^{(n)})$.

Proof. Without loss of generality assume $B(t_j^{(n)}) > 0$ and $B(t_{j+1}^{(n)}) > 0$. Define

$$G_n(x) = (A1)(x) - B(x) \sum_{k=1}^{n} A_{kn}^{\sigma_0}/(t_k^{(n)} - x).$$

Because $A_{kn}^{\sigma_0} > 0$ $(k = 1, \ldots, n)$, it follows that

$$G_n(x) \to \infty \quad \text{as} \quad x \to t_{j+k}^{(n)} + 0 \quad \text{and} \quad G_n(x) \to -\infty \quad \text{as} \quad x \to t_j^{(n)} - 0.$$

Since the function $G_n(x)$ is continuous on $(t_{j+1}^{(n)}, t_j^{(n)})$, there is at least one point c such that $G_n(c) = 0$. Because of 9.19(1) this proves the assertion. ■

9.37. Proposition. *Assume* $B(t_{j+l}^{(n)}) = B'(t_{j+l}^{(n)}) = 0$ *for* $l = 1, \ldots, k-1$ *and* $B(t_{j+k}^{(n)}) \times B(t_j^{(n)}) > 0$. *Then at least k zeros of* $k_m(x)$ *are contained in* $(t_{j+k}^{(n)}, t_j^{(n)})$.

Proof. It is again no loss of generality to suppose that $B(t_j^{(n)}) > 0$ and $B(t_{j+k}^{(n)}) > 0$. Then

$$G_n(x) \to \infty \quad \text{as} \quad x \to t_{j+k}^{(n)} + 0 \quad \text{and} \quad G_n(x) \to -\infty \quad \text{as} \quad x \to t_j^{(n)} - 0$$

and

$$G_n(t_{j+l}^{(n)}) = a(t_{j+l}^{(n)})\, \sigma(t_{j+l}^{(n)})$$

for $l = 1, \ldots, k-1$ (see the proof of Corollary 9.20). In view of assumption 3° of Section 9.13, we have $b(t_{j+l}^{(n)}) = 0$. Thus, taking into account Proposition 9.7, it follows that

$$G_n(t_{j+l}^{(n)}) \neq 0 \qquad (l = 1, \ldots, k-1).$$

Consequently, $G_n(x)$ has at least one zero distinct from all $t_{j+l}^{(n)}$ $(l = 1, \ldots, k-1)$ in $(t_{j+k}^{(n)}, t_j^{(n)})$. Hence the assertion follows again from 9.19(1). ■

9.38. Proposition. *If* $B(t) > 0$ *for all* $t \in (-1, 1)$ *then only the cases* $\varkappa = -1, 0, 1$ *are possible, and for* $n - \varkappa > r - 1$ *all zeros* $x_j^{(n)}$ *of* $k_m(x)$ *are contained in* $[-1, 1]$ *whenever one of the following conditions is fulfilled:*

(a) $a_n = b_n = 0$,
(b) *the Radau quadrature is used* $\left(a_n = -\frac{p_n(1)}{p_{n-1}(1)}\right)$ *and* $\varkappa = 1$, *or* $\varkappa = 0$ *and* $(A1)(-1) \geq B(-1) \sum_{k=1}^{n} \frac{A_{kn}^{\sigma_0}}{t_k^{(n)} + 1}$,
(c) *the Lobatto quadrature is used and* $\varkappa = 1$.

Proof. Let $B(t) > 0$ for $t \in (-1, 1)$. Using 1°–3° of Section 9.13, we get $b(t) \geq 0$ for $t \in (-1, 1)$. Then it is easy to see that the continuous function $g(t)$ defined in Section 9.5 can be chosen so that $0 \leq g(x) \leq 1$. Therefore, λ takes only the values $0, -1$ and ν merely assumes the values $0, 1$. Thus, $-1 \leq \varkappa \leq 1$.

(a) Because $k_{n-\varkappa}$ $(n - \varkappa > r - 1)$ are orthogonal polynomials (by Theorem 9.14), all zeros are in $(-1, 1)$. Moreover, in view of Corollary 9.19, we have $t_k^n \neq x_j^{(n-\varkappa)}$ for all $k = 1, \ldots, n$ and all $j = 1, \ldots, n - \varkappa$.

(b) By Proposition 9.36 each interval $(t_{k+1}^{(n)}, t_k^{(n)})$ $(k = 2, \ldots, n-2)$ contains one zero of $k_{n-\varkappa}(x)$. If $B(1) \neq 0$ it follows again from Proposition 9.36 that one zero of $k_{n-\varkappa}$ is contained in $(t_2^{(n)}, 1]$. If $B(1) = 0$, then the point 1 is a zero due to Proposition 9.19. In either case we obtain the assertion for the Radau quadrature and $\varkappa = 1$. If the second part of condition (b) is in force, we have

$$G_n(x) \to -\infty \quad \text{as} \quad x \to t_n^{(n)} - 0$$

and

$$G_n(-1) = (Ae)(-1) - B(-1) \sum_{k=1}^{n} \frac{A_{kn}^{\sigma_0}}{t_k^{(n)} + 1},$$

which gives the desired result.

(c) The proof is the same as for (b). ■

Numerical construction of Gauss type quadrature rules

9.39. Determination of the zeros and weights. Let σ_0 be a weight (such as in Section 9.18) and denote by $\tilde{p}_n$ the normalized orthogonal polynomial of the degree n with respect to the weight σ_0. It is well-known that the polynomials $\tilde{p}_n$ satisfy the three term recurrence relation (see FREUD [1])

$$\beta_{n+1}\tilde{p}_{n+1}(t) = (t - \alpha_n)\,\tilde{p}_n(t) - \beta_n\tilde{p}_{n-1}(t) \tag{1}$$

($n = 0, 1, 2, \ldots$; $\tilde{p}_{-1} := 0$). The relations (1) can be written in the form

$$\boldsymbol{J}_n\boldsymbol{p}_n(t) = t\boldsymbol{p}_n - \beta_{n+1}\begin{pmatrix} 0 \\ \vdots \\ \tilde{p}_n \end{pmatrix} \tag{2}$$

where

$$\boldsymbol{J}_n = \begin{pmatrix} \alpha_0 & \beta_1 & & 0 \\ \beta_1 & \ddots & \ddots & \\ & \ddots & \ddots & \beta_{n-1} \\ 0 & & \beta_{n-1} & \alpha_{n-1} \end{pmatrix}, \quad \boldsymbol{p}_n = \begin{pmatrix} \tilde{p}_0(t) \\ \tilde{p}_1(t) \\ \vdots \\ \tilde{p}_{n-1}(t) \end{pmatrix}.$$

The equality (2) implies that $t_k^{(n)}$ is a zero of $\tilde{p}_n$ if and only if $t_k^{(n)}$ is an eigenvalue of $\boldsymbol{J}_n$. In that case $\boldsymbol{p}_n(t_k^{(n)})$ is the corresponding eigenvector.

The Gauss quadrature rule

$$\sum_{k=1}^{n} A_{kn}^{\sigma_0} v(t_k^{(n)}) \sim \frac{1}{\pi}\int_{-1}^{1} v(t)\,\sigma_0(t)\,dt$$

can be obtained in the following way. Because of the orthonormality of the polynomials $\tilde{p}_n$ and due to the algebraic accuracy of the Gauss quadrature formula (which equals $2n - 1$) we have

$$\sum_{k=1}^{n} A_{kn}^{\sigma_0}\tilde{p}_j(t_k^{(n)})\,\tilde{p}_l(t_k^{(n)}) = \delta_{jl} \qquad (j, l = 0, \ldots, n-1).$$

Thus,

$$\left(A_{kn}^{\sigma_0}\tilde{p}_l(t_k^{(n)})\right)_{k=1,\,l=0}^{n,\,n-1}$$

is the inverse matrix of

$$\left(\tilde{p}_j(t_k^{(n)})\right)_{j=0,\,k=1}^{n-1,\,n}.$$

Consequently,

$$\sum_{j=0}^{n-1} A_{kn}^{\sigma_0}[\tilde{p}_j(t_k^{(n)})]^2 = 1 \qquad (k = 1, \ldots, n),$$

which implies that

$$A_{kn}^{\sigma_0} = \left(\sum_{j=0}^{n-1}[\tilde{p}_j(t_k^{(n)})]^2\right)^{-1} \qquad (k = 1, \ldots, n). \tag{3}$$

If $\mathbf{c}_k := (c_{jk})_{j=1}^n$ is some eigenvector to the eigenvalue $t_k^{(n)}$, then $\mathbf{c}_k = \varepsilon_k \mathbf{p}_n(t_k^{(n)})$ with

$$\varepsilon_k = c_{1k} \left(\frac{1}{\pi} \int_{-1}^{1} \sigma_0(t) \, dt \right)^{1/2}.$$

From (3) it follows that

$$A_{kn}^{\sigma_0} = \tilde{c}_{1k}^2 \frac{1}{\pi} \int_{-1}^{1} \sigma_0(t) \, dt, \tag{4}$$

where

$$\tilde{c}_{1k}^2 = \frac{\varepsilon_k^2}{\sum_{j=0}^{n-1} [\varepsilon_k \tilde{p}_j(t_k^{(n)})]^2} \cdot \left(\frac{1}{\pi} \int_{-1}^{1} \sigma_0(t) \, dt \right)^{-1} = \frac{c_{1k}^2}{\sum_{j=1}^{n} c_{jk}^2}$$

is the square of the first component of the normalized eigenvector corresponding to the eigenvalue $t_k^{(n)}$. So we have to solve the eigenvalue/eigenvector problem for the matrix J_n (called the Jacobi matrix) in order to construct the Gauss quadrature rule.

In the case of a Radau quadrature rule we proceed as follows. Choose, for example,

$$g_n(t) = p_n(t) + a_n p_{n-1}(t), \qquad a_n = -p_n(1)/p_{n-1}(1).$$

Then $t_1^{(n)} = 1$, and therefore we can write

$$g_n(t) = p_{n-1}^{(1)}(t) \, (1 - t),$$

where $p_{n-1}^{(1)}$ is an orthogonal polynomial of degree $n - 1$ with respect to the weight $(1 - t) \, \sigma_0(t)$. According to (1) we have

$$\beta_{n+1}^{(1)} \tilde{p}_{n+1}^{(1)}(t) = (t - \alpha_n^{(1)}) \, \tilde{p}_n^{(1)}(t) - \beta_n^{(1)} \tilde{p}_{n-1}^{(1)}(t),$$

and $t_2^{(n)}, \ldots, t_n^{(n)}$ are the eigenvalues of

$$J_{n-1}^{(1)} = \begin{pmatrix} \alpha_0^{(1)} & \beta_1^{(1)} & & 0 \\ \beta_1^{(1)} & \ddots & \ddots & \\ & \ddots & \ddots & \beta_{n-2}^{(1)} \\ 0 & & \beta_{n-2}^{(1)} & \alpha_{n-2}^{(1)} \end{pmatrix},$$

while

$$\mathbf{p}_{n-1}^{(1)}(t_k^{(n)}) = \begin{pmatrix} \tilde{p}_0^{(1)}(t_k^{(n)}) \\ \vdots \\ \tilde{p}_{n-2}^{(1)}(t_k^{(n)}) \end{pmatrix} \qquad (k = 2, \ldots, n)$$

are the corresponding eigenvectors. To determine the weights $A_{kn}^{\sigma_0}$ of the quadrature rule we remark that

$$\sum_{k=2}^{n} A_{kn}^{\sigma_0} \tilde{p}_j^{(1)}(t_k^{(n)}) \, \tilde{p}_l^{(1)}(t_k^{(n)}) \, (1 - t_k^{(n)}) = \delta_{jl}$$

for $j, l = 0, 1, \ldots, n-2$. Thus, in the same way as above we obtain

$$A_{kn}^{\sigma_0} = \frac{1}{1 - t_k^{(n)}} \left(\sum_{j=0}^{n-2} [\tilde{p}_j^{(1)}(t_k^{(n)})]^2 \right)^{-1} = \frac{\tilde{c}_{1k}^2}{1 - t_k^{(n)}} \frac{1}{\pi} \int_{-1}^{1} (1 - t)\, \sigma_0(t)\, dt,$$

$(k = 2, \ldots, n)$,

where $\tilde{c}_{1k}$ is the first component of the normalized eigenvector of $J_{n-1}^{(n)}$ corresponding to the eigenvalue $t_k^{(n)}$. Finally, it follows that

$$A_{1n}^{\sigma_0} = \frac{1}{\pi} \int_{-1}^{1} \sigma_0(t)\, dt - \sum_{k=2}^{n} A_{kn}^{\sigma_0}.$$

The construction of the Lobatto quadrature rule can be effected analogously. We have

$$g_n(t) = p_{n-2}^{(2)}(t)\,(1 - t^2),$$

where $p_{n-2}^{(2)}(t)$ is an orthogonal polynomial of degree $n-2$ with respect to the weight $(1 - t^2)\,\sigma_0(t)$. This implies that $t_1^{(n)} = 1$, $t_n^{(n)} = -1$, and

$$A_{kn}^{\sigma_0} = \frac{\tilde{c}_{1k}^2}{1 - (t_k^{(n)})^2} \frac{1}{\pi} \int_{-1}^{1} (1 - t^2)\, \sigma_0(t)\, dt,$$

where $\tilde{c}_{1k}$ is the first component of the normalized eigenvector of

$$J_{n-2}^{(2)} = \begin{pmatrix} \alpha_0^{(2)} & \beta_1^{(2)} & & 0 \\ \beta_1^{(2)} & \ddots & \ddots & \\ & \ddots & \ddots & \beta_{n-3}^{(2)} \\ 0 & & \beta_{n-3}^{(2)} & \alpha_{n-3}^{(2)} \end{pmatrix}$$

corresponding to the eigenvalue $t_k^{(n)}$ $(k = 2, \ldots, n-1)$. The remaining weights of the quadrature formula are given by

$$2A_{1n}^{\sigma_0} = \frac{1}{\pi} \int_{-1}^{1} (1 + t)\, \sigma_0(t)\, dt - \sum_{k=2}^{n-1} A_{kn}^{\sigma_0}(1 + t_k^{(n)}),$$

and

$$2A_{nn}^{\sigma_0} = \frac{1}{\pi} \int_{-1}^{1} (1 - t)\, \sigma_0(t)\, dt - \sum_{k=2}^{n-1} A_{kn}^{\sigma_0}(1 - t_k^{(n)}),$$

or

$$A_{nn}^{\sigma_0} = \frac{1}{\pi} \int_{-1}^{1} \sigma_0(t)\, dt - \sum_{k=1}^{n-1} A_{kn}^{\sigma_0}.$$

9.40. Determination of the recurrence parameters. We have seen that the characteristic quantities of the quadrature rules can be computed effectively provided the parameters α_n and β_n (or $\alpha_n^{(i)}, \beta_n^{(i)}$, $i = 1, 2$) in the recurrence relation 9.39(1) are known. The determination of these parameters is a crucial point for the construction of quadrature rules and is equivalent to an orthogonalizing procedure for a given set of polynomials $P_k(t)$ $(k = 0, 1, \ldots)$ of degree k. Such orthogonalizing procedures, however, are known to be numerically instable and so other ways should be found. We here shall describe the so-called method of modified moments (WHEELER [1]). Suppose we are given a set of polynomials $P_k(t)$ of degree k $(k = 0, 1, \ldots)$ which are orthogonal with respect to some weight $\sigma_1(t)$, which is required to be close to the weight $\sigma_0(t)$ in some sense. Assume further we know the parameters A_l, B_l in the recurrence relation

$$B_{k+1}P_{k+1} = (t - A_k)\, P_k(t) - B_k P_{k-1}(t) \tag{1}$$

and the so called modified moments

$$M_k = \frac{1}{\pi} \int_{-1}^{1} P_k(t)\, \sigma_0(t)\, dt \qquad (k = 0, 1, \ldots, 2n - 1). \tag{2}$$

Using the numbers $\{M_k\}_{k=0}^{2n-1}$ we want to realize the transformation

$$[(A_k)_{k=0}^{n-1}, (B_k)_{k=1}^{n-1}] \xrightarrow{\mathscr{F}} [(\alpha_k)_{k=0}^{n-1}, (\beta_k)_{k=0}^{n-1}],$$

numerically. Define the mixed moments

$$m_{jk} = (\tilde{p}_j, P_k)_{\sigma_0} \qquad (j, k = -1, 0, 1, \ldots)$$

$(\tilde{p}_{-1} \equiv 0, P_{-1} \equiv 0)$, where $\tilde{p}_j$ $(j = 0, 1, \ldots)$ are orthogonal polynomials of degree j with respect to the weight σ_0. We have

$$\beta_{j+1} m_{j+1,k} = (\beta_{j+1}\tilde{p}_{j+1}, P_k)_{\sigma_0} = (t\tilde{p}_j, P_k)_{\sigma_0} - \alpha_j m_{jk} - \beta_j m_{j-1,k}$$

and

$$(t\tilde{p}_j, P_k)_{\sigma_0} = (\tilde{p}_j, tP_k)_{\sigma_0} = B_{k+1} m_{j,k+1} + A_k m_{jk} + B_k m_{j,k-1}.$$

Thus,

$$m_{0k} = M_k \qquad (k = 0, 1, \ldots, 2n - 1),$$

$$\beta_{j+1} m_{j+1,k} = B_{k+1} m_{j,k+1} + (A_k - \alpha_j)\, m_{jk} + B_k m_{j,k-1} - \beta_j m_{j-1,k}.$$

In particular, for $k = j - 1$ and $k = j$ we get

$$0 = B_j m_{jj} - \beta_j m_{j-1,j-1}$$

and

$$0 = B_{j+1} m_{j,j+1} + (A_j - \alpha_j)\, m_{jj} - \beta_j m_{j-1,j},$$

respectively. Thus,

$$\beta_j = B_j \frac{m_{jj}}{m_{j-1,j-1}}, \qquad \alpha_j = A_j + B_{j+1} \frac{m_{j,j+1}}{m_{jj}} - B_j \frac{m_{j-1,j}}{m_{j-1,j-1}}.$$

Putting these things together we obtain the following algorithm:

$$m_{-1,k} = 0 \qquad (k = 1, 2, \ldots, 2n - 2),$$

$$m_{0k} = M_k \qquad (k = 0, 1, \ldots, 2n - 1),$$

$$\beta_0 = 0, \qquad \alpha_0 = A_0 + \frac{M_1}{M_0},$$

and for $j = 1, \ldots, n - 1$,

$$m_{jk} = \frac{1}{\beta_j} [B_{k+1} m_{j-1,k+1} + (A_k - \alpha_{j-1}) m_{j-1,k} + B_k m_{j-1,k-1} - \beta_{j-1} m_{j-2,k}],$$

$$k = j, j + 1, \ldots, 2n - j - 1,$$

$$\alpha_j = A_j + B_{j+1} \frac{m_{j,j+1}}{m_{jj}} - B_j \frac{m_{j-1,j}}{m_{j-1,j-1}},$$

$$\beta_j = B_j \frac{m_{jj}}{m_{j-1,j-1}}.$$

Note that in the case of Radau or Lobatto quadratures the starting situation has to be modified. Viz, in (2) the weight $\sigma_0(t)$ must be replaced by $(1 - t)\, \sigma_0(t)$ or $(1 - t^2)\, \sigma_0(t)$, respectively.

If we specify the weight $\sigma_0(t)$ to be the Jacobi weight $(1 - t)^\alpha (1 + t)^\beta$ with $-1 < \alpha < 1$, $-1 < \beta < 1$, the parameters β_{n+1}, β_n and α_n in the three term recurrence formula 9.39(1) for the normalized orthogonal polynomials $\hat{P}_n^{(\alpha,\beta)}$ with positive leading coefficients can be determined precisely. Indeed, it is well-known (Obreshkov [1]) that the usual Jacobi polynomials $P_n^{(\alpha,\beta)}$ (for whose definition see Section 9.15, 2°) fulfil the equation

$$P_{n+1}^{(\alpha,\beta)}(x) = (a_n x + b_n)\, P_n^{(\alpha,\beta)}(x) - d_n P_{n-1}^{(\alpha,\beta)}(x), \qquad (n = 1, 2, \ldots), \tag{3}$$

with

$$a_n = \frac{c_{n+1}}{c_n}, \qquad b_n = a_n \left(\frac{c'_{n+1}}{c_{n+1}} - \frac{c'_n}{c_n}\right), \qquad d_n = \frac{a_n A_n}{a_{n-1} A_{n-1}},$$

where the notations

$$P_n^{(\alpha,\beta)}(x) = c_n x^n + c'_n x^{n-1} + \cdots \quad \text{for} \quad n > 1 \quad \text{and} \quad P_0^{(\alpha,\beta)}(x) = 1,$$

$$P_1^{(\alpha,\beta)}(x) = (1/2)(\alpha + \beta + 2)\, x + (1/2)(\alpha - \beta),$$

$$A_n = \int_{-1}^{1} [P_n^{(\alpha,\beta)}(x)]^2 (1 - x)^\alpha (1 + x)^\beta \, dx$$

are used. Moreover, in Obreshkov [1] it is proved that

$$a_n = \frac{(2n + \alpha + \beta + 1)(2n + \alpha + \beta + 2)}{2(n + 1)(n + \alpha + \beta + 1)},$$

$$b_n = \frac{(\alpha^2 - \beta^2)(2n + \alpha + \beta + 1)}{2(n + 1)(n + \alpha + \beta + 1)(2n + \alpha + \beta)},$$

$$d_n = \frac{(n + \alpha)(n + \beta)(2n + \alpha + \beta + 2)}{(n + 1)(n + \alpha + \beta + 1)(2n + \alpha + \beta)}.$$

The identity (3) is equivalent to the equality

$$\frac{1}{c_{n+1}} P_{n+1}^{(\alpha,\beta)}(x) = \left(x + \frac{c_n b_n}{c_{n+1}}\right) \frac{1}{c_n} P_n^{(\alpha,\beta)}(x) - \frac{c_{n-1} d_n}{c_{n+1}} \frac{1}{c_{n-1}} P_{n-1}^{(\alpha,\beta)}(x).$$

Thus, we arrive at the formula

$$\tilde{P}_{n+1}^{(\alpha,\beta)}(x) = (x - \alpha_n)\, \tilde{P}_n^{(\alpha,\beta)}(x) - \beta_n \tilde{P}_{n-1}^{(\alpha,\beta)}(x), \tag{4}$$

where $\tilde{P}_n^{(\alpha,\beta)}(x) = \frac{1}{c_n} P_n^{(\alpha,\beta)}(x)$ and

$$\alpha_n = \frac{\beta^2 - \alpha^2}{(2n + \alpha + \beta)(2n + \alpha + \beta + 2)}, \qquad \beta_0 = \frac{4(1 + \alpha)(1 + \beta)}{(2 + \alpha + \beta)^2 (3 + \alpha + \beta)},$$
$$\beta_n = \frac{(n + \alpha)(n + \beta)(n + \alpha + \beta)\, 4n}{(2n + \alpha + \beta - 1)(2n + \alpha + \beta)^2 (2n + \alpha + \beta + 1)}. \tag{5}$$

Now one can easily obtain the three term recurrence formula for the normalized orthogonal polynomials $\hat{P}_n^{(\alpha,\beta)}$ with positive leading coefficients from (4). First of all we have

$$\alpha_n = \frac{(x\tilde{P}_n^{(\alpha,\beta)}, \tilde{P}_n^{(\alpha,\beta)})_\sigma}{(\tilde{P}_n^{(\alpha,\beta)}, \tilde{P}_n^{(\alpha,\beta)})_\sigma}, \qquad \beta_n = \frac{(x\tilde{P}_n^{(\alpha,\beta)}, \tilde{P}_{n-1}^{(\alpha,\beta)})_\sigma}{(\tilde{P}_{n-1}^{(\alpha,\beta)}, \tilde{P}_{n-1}^{(\alpha,\beta)})_\sigma}.$$

Using that $\tilde{P}_n^{(\alpha,\beta)} = (\tilde{P}_n^{(\alpha,\beta)}, \tilde{P}_n^{(\alpha,\beta)})^{1/2}\, \hat{P}_n^{(\alpha,\beta)}$ $(n = 0, 1, \ldots)$, we deduce from (4) that

$$(\beta_{n+1})^{1/2}\, \hat{P}_{n+1}^{(\alpha,\beta)} = (x - \alpha_n)\, \hat{P}_n^{(\alpha,\beta)} - (\beta_n)^{1/2}\, \hat{P}_{n-1}^{(\alpha,\beta)}, \tag{6}$$

with α_k, β_k defined by (5).

It seems to be appropriate to choose the Jacobi weight $(1 - t)^\alpha (1 + t)^\beta$ as the starting point in the method of modified moments in order to compute the quadrature data for the weight $\sigma(t) = (1 - t)^\alpha (1 + t)^\beta w(t)$.

An algorithm for the case $\varkappa = -1$

9.41. Preliminaries. Assume we have to solve an equation

$$Bu = f$$

with $B \in \mathscr{L}(\mathbf{L}_\sigma^2, \mathbf{L}_\mu^2)$, the weights σ and μ being as in Theorem 9.12. Assume further $\ker B = \{0\}$, $\operatorname{im} B$ is closed, and $\dim \operatorname{coker} B = 1$. Let us be given a sequence of bounded operators $B_n : \Pi_n \to \Pi_{n+1}$, where Π_i is defined in Section 9.17, and suppose $\|B_n u_n - B u_n\| \leqq \varepsilon_n \|u_n\|_\sigma$ for all $u_n \in \Pi_n$ and $\varepsilon_n \to 0$ as $n \to \infty$. Obviously, the equation

$$B_n u_n = f_n, \qquad u_n \in \Pi_n, \qquad f_n \in \Pi_{n+1} \tag{1}$$

need not be solvable for all $f_n \in \Pi_{n+1}$. This case occurs if we consider singular integral equations with index -1 and try solving them by the methods investigated in the previous sections. To overcome this difficulty we have to modify the equations (1). We start our program with the following result.

9.42. Proposition. *Suppose the conditions of Section 9.41 are fulfilled. Then* $\ker B_n = \{0\}$ *for all sufficiently large* n.

Proof. First notice that there exists a constant $c > 0$ such that $\|Bu\|_\mu \geqq c \|u\|_\sigma$ for all $u \in \mathbf{L}^2_\sigma$. Hence

$$\|B_n u_n\|_\mu + \|Bu_n - B_n u_n\|_\mu \geqq c \|u_n\|_\sigma$$

for all $u_n \in \Pi_n$ and thus,

$$\|B_n u_n\|_\mu \geqq (c - \varepsilon_n) \|u_n\|_\sigma . \ \blacksquare$$

Let $g_n^0 \in \Pi_{n+1}$ satisfy $\|g_n^0\|_\mu = 1$ and

$$(B_n u_n, g_n^0)_\mu = 0 \quad \text{for all} \quad u_n \in \Pi_n .$$

From Proposition 9.42 we obtain

$$\Pi_{n+1} = B_n(\Pi_n) \dotplus \operatorname{span}(g_n^0) .$$

Instead of equation (1) we consider the modified equation

$$B_n u_n + \gamma_n g_n = f_n , \tag{1}$$

where $g_n \in \Pi_{n+1}$ with $\|g_n\|_\mu = 1$, $\alpha_n := (g_n, g_n^0)_\mu \neq 0$, and $\|f_n - f\|_\mu \to 0$ as $n \to \infty$.

9.43. Theorem. (i) *The equations* 9.42(1) *are uniquely solvable for all n large enough.*

(ii) *Assume* $|\alpha_n| \geqq \varepsilon > 0$ *for all* $n \geqq n_0$. *Then* $f \in \operatorname{im} B$ *if and only if* $\gamma_n \to 0$ *as* $n \to \infty$.

(iii) *Assume* $|\alpha_n| \geqq \varepsilon > 0$ *for all* $n \geqq n_0$ *and* $f \in \operatorname{im} B$. *Then the solutions* u_n *of* (1) *converge to the solution* u *of* $Bu = f$ *in the norm of the space* $\mathbf{L}^2_\sigma$ *and one has*

$$\|u - u_n\|_\sigma \leqq C(\|f - f_n\|_\mu + \varepsilon_n \|f_n\|_\mu) .$$

Proof. (i): Put $\gamma_n = \alpha_n^{-1}(f_n, g_n^0)_\mu$. Then $f_n - \gamma_n g_n \in B_n(\Pi_n)$ and there is an element $u_n \in \Pi_n$ such that $B_n u_n = f_n - \gamma_n g_n$. Hence, a solution of 9.42(1) exists. We show that the solution $(u_n, \gamma_n) \in \Pi_n \times \mathbb{R}$ is unique. Of course, if $B_n u_n^* + \gamma_n^* g_n = f_n$, then $f_n - \gamma_n^* g_n \in B_n(\Pi_n)$. Thus, $(f_n - \gamma_n^* g_n, g_n^0)_\mu = 0$ and, consequently, $\gamma_n^* = \alpha_n^{-1}(f_n, g_n^0)_\mu$. By Proposition 9.42 the element u_n^* is determined uniquely.

(ii): We show that

$$\|u_n\|_\sigma \leqq d < \infty . \tag{1}$$

From the proof of Proposition 9.42 we conclude that

$$\|u_n\|_\sigma \leqq (c - \varepsilon_n)^{-1} \|B_n u_n\|_\mu = (c - \varepsilon_n)^{-1} \|f_n - \gamma_n g_n\|_\mu .$$

Thus,

$$\|u_n\|_\sigma \leqq (c - \varepsilon_n)^{-1} \|f_n\|_\mu (1 + |\alpha_n^{-1}|) \tag{2}$$

and (1) is proved. Now assume that $\gamma_n \to 0$ as $n \to \infty$. Then, obviously, $B_n u_n \to f$ as $n \to \infty$. Because $\|B_n u_n - Bu_n\|_\mu < \varepsilon_n \|u_n\|_\sigma$ with $\varepsilon_n \to 0$ as $n \to \infty$ (by assumption) we get that $Bu_n \to f$. Thus $f \in \operatorname{im} B$, since $\operatorname{im} B$ is closed. On the other hand, let $f \in \operatorname{im} B$. Obviously, there exists a $g \in \mathbf{L}^2_\mu$ such that $(\varphi, g)_\mu = 0$ for all $\varphi \in \operatorname{im} B$ and $\|g\|_\mu = 1$. Let $1 > \varepsilon' > 0$. Choose an $h_{m_0} \in \Pi_{m_0+1}$ such that

$$|(g, h_{m_0})_\mu| \geqq 1 - \frac{\varepsilon'}{2} \quad \text{and} \quad \|h_{m_0}\|_\mu = 1 .$$

Let $n \geqq m_0$ and

$$B_n v_n^* + \gamma_n^* g_n = h_{m_0}.$$

Then

$$1 - \frac{\varepsilon'}{2} \leqq |(g, B_n v_n^* + \gamma_n^* g_n)_\mu| = |(g, B_n v_n^* - B v_n^* + \gamma_n^* g_n)_\mu|$$

$$\leqq |\alpha_n|^{-1} |(g, g_n)_\mu| + \varepsilon_n \|v_n^*\| \leqq |\alpha_n|^{-1} |(g, g_n)_\mu| + \frac{\varepsilon_n}{c - \varepsilon_n} (1 + |\alpha_n|^{-1})$$

(by (2)).

Consequently,

$$1 - \varepsilon' \leqq |\alpha_n|^{-1} |(g, g_n)_\mu| \tag{3}$$

for all $n \geqq m_1(\varepsilon')$. Since

$$|(g, \gamma_n g_n)_\mu| = |(g, Bu_n + \gamma_n g_n)_\mu| = |(g, Bu_n - B_n u_n + f_n)_\mu|$$
$$\leqq \varepsilon_n \|u_n\|_\sigma + |(g, f_n)_\mu|$$

it follows that $(g, \gamma_n g_n)_\mu \to 0$. Combining this with (3) we obtain that $\gamma_n \to 0$ as $n \to \infty$.

(iii): Use the estimate

$$\|u - u_n\|_\sigma \leqq c^{-1} \|Bu - Bu_n\|_\mu = c^{-1} \|f - f_n + (B_n - B) u_n + \gamma_n g_n\|_\mu$$
$$\leqq c^{-1}(\|f - f_n\|_\mu + \varepsilon_n \|u_n\|_\sigma + |\gamma_n|) \leqq c^{-1}(\|f - f_n\|_\mu + \varepsilon_n d + |\gamma_n|)$$

resulting from (1) in order to prove convergence. Further, the proof of part (ii) yields that $|\gamma_n| \leqq c_1(\varepsilon_n \|u_n\|_\sigma + |(g, f_n)_\mu|)$. Because $(g, f)_\mu = 0$ we get $|(g, f_n)_\mu| = |(g, f_n - f)_\mu| \leqq \|f - f_n\|_\mu$. Putting the things together we see that

$$\|u - u_n\|_\sigma \leqq C(\|f - f_n\|_\mu + \varepsilon_n)$$

with some constant $C > 0$. ■

Let us consider the singular integral equation

$$Bu(x) = a(x)\, u(x)\, \sigma(x) + \frac{1}{\pi} \int_{-1}^{1} \frac{u(t)}{t - x}\, \sigma(t)\, dt + \frac{1}{\pi} \int_{-1}^{1} h(x, t)\, u(t)\, \sigma(t)\, dt = f(x), \tag{4}$$

where $a \in \mathcal{H}$. Define the weight $\sigma(x) = (1 - x)^{\alpha_0} (1 + x)^{\beta_0} w(t)$ (with $w \in \mathcal{H}$, $w(t) > 0$ for all $t \in [-1, 1]$) so that $\alpha_0 > 0$, $\beta_0 > 0$. This is precisely the case $\varkappa = -1$ (see Section 9.5). The operator $a\sigma I + S\sigma I$ is bounded and invertible in the pair of Hilbert spaces $(\mathbf{L}_\sigma^2, \mathbf{L}_{\mu,1}^2)$, where μ is the dual weight to σ. If we try to solve (4) by means of the quadrature method described above we get exactly the case in which B_n maps Π_n into Π_{n+1}. Thus, we have to modify the method. We indicate one possibility of determining the elements g_n in the modified equations

$$B_n u_n + \gamma_n g_n = f_n.$$

Namely, we write this equation in the form

$$\sum_{k=1}^{n+1} a_{jk} \xi_{kn} = \eta_{jn} \qquad (j = 1, \ldots, n + 1),$$

where a_{jk} $(j = 1, \ldots, n + 1;\ k = 1, \ldots, n)$ and η_{jn} $(j = 1, \ldots, n + 1)$ are given by

$$a_{jk} = A^{\sigma}_{kn} \left[\frac{1}{t^{(n)}_k - x^{(n)}_j} + h(x^{(n)}_j, t^{(n)}_k) \right]$$

and

$$\eta_{jn} = f(x_{jn}) .$$

The $(n + 1)$st column of the system matrix is determined from the conditions

$$\sum_{j=1}^{n+1} A^{\mu}_{jn} a_{jk} a_{j,n+1} = 0 \qquad (k = 1, \ldots, n), \tag{5}$$

and

$$\sum_{j=1}^{n+1} A^{\mu}_{jn} \, |a_{j,n+1}|^2 = 1 . \tag{6}$$

Notice that (5) and (6) are equivalent to the requirement that

$$(B_n u_n, g_n)_\mu = 0 \quad \text{for all} \quad u_n \in \Pi_n \quad \text{and} \quad \|g_n\|_\mu = 1 .$$

Equations (5) and (6) can be solved with the help of a suitable method for solving non-linear equations.

It should be noted that all hypotheses of Theorem 9.43 are fulfilled if the kernel $h(x, t)$ is subject to some smoothness conditions.

The special case of singular integral operators with constant real coefficients

9.44. Preliminaries. We use the notations of Section 9.15 and consider the quadrature method for solving the singular integral equation with constant real coefficients

$$a\sigma(x)\, v(x) + \frac{b}{\pi} \int_{-1}^{1} \frac{\sigma(t)}{t - x} v(t) \, dt + \frac{1}{\pi} \int_{-1}^{1} h(x, t)\, v(t)\, \sigma(t) \, dt = f(x), \tag{1}$$

where

$$\sigma(x) = \frac{1}{\sqrt{a^2 + b^2}} (1 - x)^\alpha (1 + x)^\beta, \qquad -1 < \alpha, \beta < 1, \qquad \alpha = \lambda - \gamma_0,$$

$$\beta = \nu + \gamma_0, \quad \lambda, \nu \in \mathbb{Z}, \quad a + ib = \sqrt{a^2 + b^2}\, e^{i\pi\gamma_0}, \qquad 0 < |\gamma_0| < 1 .$$

Thus, the Jacobi polynomials play the role of the orthogonal polynomials and the known parameters of these polynomials, especially the coefficients 9.40(5) of the three term recurrence relation 9.40(4), can be used for the determination of the nodes and weights of the quadrature formulas on the basis of the results of Section 9.39. In view of Proposition 9.38, besides the Gauss quadrature rule the Radau quadrature and the Lobatto quadrature can be used in the case where $\varkappa = -(\alpha + \beta) = 1$. Further, let us remark that as a consequence of Proposition 9.19 all nodes $t^{(n)}_k$ $(k = 1, \ldots, n)$ are distinct from all collocation points $x^{(n)}_j$ $(j = 1, \ldots, n - \varkappa)$, which simplifies the construction of the matrices $\boldsymbol{A}_n$ (cf. Sec. 9.23). Moreover, note that in all cases the condition $\tau = \max\{\alpha, \beta\} \geqq -1/2$

(cf. Secs. 9.31–9.34) is fulfilled. If $\varkappa = 1$ then the condition

$$\frac{1}{\pi} \int_{-1}^{1} v(t)\, \sigma(t)\, dt = 0 \tag{2}$$

is added to equation (1). So the approximate solution

$$v_n(t) = \sum_{k=1}^{n} \xi_k^{(n)} \prod_{\substack{l=1 \\ l \neq k}}^{n} \frac{t - t_l^{(n)}}{t_k^{(n)} - t_l^{(n)}} \tag{3}$$

has to be determined by solving

$$\sum_{k=1}^{n} A_{kn}^{\sigma} \left[\frac{1}{t_k^{(n)} - x_j^{(n)}} + h(x_j^{(n)}, t_k^{(n)}) \right] \xi_k^{(n)} = f(x_j^{(n)}), \tag{4}$$

$j = 1, \ldots, n - \varkappa$, and for $\varkappa = 1$ by also requiring that

$$\sum_{k=1}^{n} A_{kn}^{\sigma} \xi_k^{(n)} = 0. \tag{5}$$

The equations (3) have to be modified if $\varkappa = -1$ (cf. Sec. 9.43, especially 9.43(5), 9.43(6)).

9.45. Convergence theorem. Applying Theorem 9.34 we obtain the following result. Let $f \in \mathbf{C}^{m,\gamma}$, $\{h_t\} \in \mathbf{C}([-1, 1], \mathbf{C}^{m,\gamma})$, $\{h_x\} \in \boldsymbol{M}([-1, 1], \mathbf{C}^{r,\delta})$, and $\varkappa \geqq 0$. Further, assume that $r + \delta > \tau + 1$ if the Lobatto quadrature rule is used. If equation 9.44(1) (with 9.44(2) if $\varkappa = 1$) has a unique solution $v \in \mathbf{L}_\sigma^2$, then the equations 9.44(4) (with 9.44(5) if $\varkappa = 1$) are uniquely solvable for all sufficiently large n and the approximate solutions 9.44(3) converge to v in the norm of $\mathbf{L}_\sigma^2$. Moreover, the estimate

$$\|v_n - v\|_\sigma = O(n^\eta)$$

is valid, where $\eta = \min\{m + \gamma, r + \delta - \tau - 1\}$ in the case of Lobatto quadrature and $\eta = \min\{m + \gamma, r + \delta\}$ in the other cases.

An analogous result holds in case $\varkappa = -1$, where the results of Secs. 9.41–9.43 have to be applied.

Some further results

9.46. The special case $b = -1, \varkappa = -1$. Here we deal with the quadrature method for the singular integral equation

$$a(x)\, u(x) - \frac{1}{\pi} \int_{-1}^{1} \frac{u(t)}{t - x}\, dt - D = f(x) \tag{1}$$

with the additional condition

$$u(-1) = u(1) = 0. \tag{2}$$

We assume $a \in \mathcal{H}$ to be real valued and seek $u(t)$ and $D \in \mathbb{C}$. Condition (2) implies that we have to choose $\lambda = 0$, $\nu = 1$ in 9.5(2). Hence

$$\sigma(t) = \frac{(1+t)}{\sqrt{1+a^2(t)}} \exp \int_{-1}^{1} \alpha(x) \frac{dx}{x-t},$$

$$\mu(t) = \frac{1}{[1+a^2(t)]\,\sigma(t)}$$

with

$$a(x) + i = \sqrt{1+a^2(x)}\, e^{i\pi\alpha(x)}, \qquad 0 < \alpha(x) < 1.$$

In order to have a possibility of computing the parameters of the corresponding quadrature rule we simplify the weights by

$$\sigma_1(t) = \frac{1+t}{\sqrt{1+a_1^2(t)}} \exp \int_{-1}^{1} \alpha_1(x) \frac{dx}{x-t},$$

$$\mu_1(t) = \frac{1}{[1+a_1^2(t)]\,\sigma_1(t)},$$

where

$$\alpha_1(x) = \frac{1}{2}\,[(1-x)\,\alpha(-1) + (1+x)\,\alpha(1)]$$

and

$$a_1(x) = \cot\,[\pi\alpha_1(x)].$$

The weight functions $\sigma_1(t)$ and $\mu_1(x)$ are associated with the singular integral operator $A_1 = a_1\sigma_1 I - S\sigma_1 I$ in the sense of Section 9.5. Define $w(t) = \sigma_1(t)/\sigma(t)$.

9.47. Corollary. *We have* $w, w^{-1} \in \mathcal{H}^\delta$ *if* $a \in \mathcal{H}^\delta$.

Proof. This is a consequence of the equality

$$w(t) = \sqrt{\frac{1+a^2(t)}{1+a_1^2(t)}} \exp \int_{-1}^{1} \frac{\alpha_1(x) - \alpha(x)}{x-t}\, dx,$$

$\alpha_1(\pm 1) = \alpha(\pm 1)$, and Remark 9.4. ■

9.48. Notations and a modified quadrature method. The operator defined by the left hand side of 9.46(1) with $u = \sigma_1 v$ will be denoted by $\mathsf{A}: \mathbf{L}^2_\sigma \times \mathbb{C} \to \mathbf{L}^2_\mu$, where $\mathsf{A}(v, D) = a\sigma_1 v - S\sigma_1 v - D$. The norm in $\mathbf{L}^2_\sigma \times \mathbb{C}$ is defined by $\|(v, D)\| = (\|v\|^2_\sigma + |D|^2)^{1/2}$. To get an approximate solution to

$$\mathsf{A}(v, D) = f \tag{1}$$

in the form $(v_n, D_n) \in \Pi_n \times \mathbb{C}$, we solve the collocation equations

$$L^{\mu_1}_{n+1}\, \mathsf{A}(v_n, D_n) = L^{\mu_1}_{n+1} f, \tag{2}$$

where $L^{\mu_1}_{n+1}$ is the Lagrange interpolation operator with respect to the nodes $x_j^{(n+1)}$

($j = 1, \ldots, n + 1$) of the Gauss quadrature rule

$$\frac{1}{\pi} \int_{-1}^{1} f(x)\, \mu_1(x)\, \mathrm{d}x \sim \sum_{j=1}^{n+1} A_{jn}^{\mu_1} f(x_j^{(n+1)}).$$

Analogously the nodes $t_k^{(n)}$, $k = 1, \ldots, n$, and the weights $A_{kn}^{\sigma_1}$, $k = 1, \ldots, n$, are defined. With these notations, the collocation equations (2) are equivalent to

$$\sum_{k=1}^{n} \left\{ [a(x_j^{(n+1)}) - a_1(x_j^{(n+1)})]\, \sigma_1(x_j^{(n+1)})\, l_{kn}(x_j^{(n+1)}) - \frac{A_{kn}^{\sigma_1}}{t_k^{(n)} - x_j^{(n+1)}} \right\}$$
$$\times\, v_n(t_k^{(n)}) - D_n = f(x_j^{(n+1)}),$$

$j = 1, \ldots, n + 1$, where $l_{kn}(t)$ denotes the k-th Lagrange basic polynomial of degree $n - 1$ with respect to the nodes $t_l^{(n)}$, $l_{kn}(t) = \prod_{\substack{l=1 \\ l \neq k}}^{n} \frac{t - t_l^{(n)}}{t_k^{(n)} - t_l^{(n)}}$.

9.49. Convergence theorem. *Let $a \in H^\delta$ with $\delta > 1/2$ and let f be bounded and Riemann integrable with respect to $\mu_1(x)\,\mathrm{d}x$, i.e. $f \in \mathcal{R}([-1, 1], \mu_1(x)\,\mathrm{d}x)$. Then the equations* 9.48(2) *are uniquely solvable for all sufficiently large n and the solutions (v_n^*, D_n^*) converge in the norm of $\mathbf{L}_\sigma^2 \times \mathbb{C}$ to the unique solution $(v^*, D^*) \in \mathbf{L}_\sigma^2 \times \mathbb{C}$ of equation* 9.48(1).

Proof. Equation 9.48(1) can be written in the form

$$wAv - D + wTv = f,$$

where $A = a\sigma I - S\sigma I$ and $T = (S - w^{-1}SwI)\,\sigma I$.

Analogously, equation 9.48(2) can be brought into the form

$$L_{n+1}^{\mu_1} wAv_n - D_n + L_{n+1}^{\mu_1} wTv_n = L_{n+1}^{\mu_1} f.$$

These two equations are equivalent to the equations

$$\mathsf{B}(v, D) = Av - w^{-1}D + Tv = w^{-1}f \tag{1}$$

and

$$\mathsf{B}_n(v_n, D_n) = Av_n - L_{n+1}^{\mu_1} w^{-1} D_n + L_{n+1}^{\mu_1} Tv_n = L_{n+1}^{\mu_1} w^{-1} f, \tag{2}$$

respectively, where we have used the property $L_{n+1}^{\mu_1} w L_{n+1}^{\mu_1} = L_{n+1}^{\mu_1} wI$ of the interpolation operator and the fact that $Av_n \in \Pi_{n+1}$ whenever $v_n \in \Pi_n$. The operator $\mathsf{B}\colon \mathbf{L}_\sigma^2 \times \mathbb{C} \to \mathbf{L}_\mu^2$ is a continuous isomorphism. Indeed, because $\mathsf{B}(v, D) = w^{-1}(Awv - D)$, $\|Av\|_\mu = \|v\|_\sigma$ and $(Av, 1)_\mu = 0$ (cf. Theorems 9.14 and 9.17) we have

$$\|\mathsf{B}(v, D)\|_\mu^2 \geq c_1(Awv - D, Awv - D)_\mu = c_1(\|wv\|_\sigma^2 + |D|^2) \geq c_2\, \|(v, D)\|^2$$

with $c_2 > 0$. Here we used that $\|1\|_\mu = 1$, which follows from the evident equality

$$\frac{1}{\pi} \int_{-1}^{1} \mu(t)\, \mathrm{d}t = (a\mu x + Sx\mu)\,(0),$$

Theorem 9.9 and Remark 9.10. Thus, $\dim \ker \mathsf{B} = 0$. If $f \in \mathbf{L}_\mu^2$ and $D = -(wf, 1)_\mu$ then $wf + D \in \operatorname{im} A$ in view of Theorem 9.17(ii). This implies the existence of $v \in \mathbf{L}_\sigma^2$

such that $Awv = wf + D$. Consequently, im $\mathsf{B} = \mathbf{L}^2_\mu$. The operator T can be written in the form

$$(Tv)(x) = \frac{1}{\pi} \int_{-1}^{1} h(x, t)\, v(t)\, \sigma(t)\, dt,$$

where

$$h(x, t) = \frac{w(x) - w(t)}{w(x)\,(t - x)}.$$

Hence $|h(x, t)| \leqq c\,|t - x|^{\delta-1}$ by Corollary 9.47. Because $\delta > 1/2$, we have

$$\int_{-1}^{1} |h(x, t)|^2\, \sigma(t)\, dt \in \mathcal{R}\big([-1, 1], \mu(x)\, dx\big)$$

and

$$\int_{-1}^{1} h(x, t)\, v_n(t)\, \sigma(t)\, dt \in \mathcal{R}\big([-1, 1], \mu(x)\, dx\big)$$

for any $v_n \in \Pi_n$. Thus,

$$\|(T - L^{\mu_1}_{n+1} T)\, v\|_\mu \leqq \delta_n\, \|v\|_\sigma, \qquad v \in \mathbf{L}^2_\sigma,$$

with $\lim_{n\to\infty} \delta_n = 0$ (Junghanns/Silbermann [1]). By the definition of the operators B and B_n,

$$\|(\mathsf{B} - \mathsf{B}_n)\,(v_n, D_n)\|_\mu \leqq |D_n| \cdot \|w^{-1} - L^{\mu_1}_{n+1} w^{-1}\|_\mu + \delta_n\, \|v_n\|_\sigma \leqq \varepsilon_n\, \|(v_n, D_n)\|,$$
$$\varepsilon_n \to 0.$$

This gives the invertibility of B_n for all sufficiently large n and that $\|\mathsf{B}_n^{-1}\| \leqq 2\,\|\mathsf{B}^{-1}\|$ for $n \geqq n_0$. Consequently, if (v^*, D^*) and (v_n^*, D_n^*) are the solutions of (1) and (2), respectively, then

$$\|(v^*, D^*) - (v_n^*, D_n^*)\|$$
$$\leqq \|\mathsf{B}^{-1}\|\,\big(\|w^{-1}f - L^{\mu_1}_{n+1} w^{-1} f\|_\mu + \varepsilon_n\, \|\mathsf{B}_n^{-1}\|\, \|L^{\mu_1}_{n+1} w^{-1} f\|_\mu\big), \qquad n \geqq n_0.$$

Thus, $\lim_{n\to\infty} \|(v^*, D^*) - (v_n^*, D_n^*)\| = 0$. ■

9.50. A class of nonlinear singular integral equations. Let us consider a singular integral equation of the form

$$g\big(x, u(x)\big) - \frac{1}{\pi} \int_{-1}^{1} \frac{u(t)}{t - x}\, dt - D = 0, \qquad x \in (-1, 1), \tag{1}$$

with the additional condition

$$u(-1) = u(1) = 0. \tag{2}$$

The function $g(x, u)$, $x \in [-1, 1]$, $u \in \mathbb{R}$, is assumed to be real valued, the function $u(x)$ and the real number D are sought. Further, we suppose that the partial derivatives $g_x(x, u)$ and $g_u(x, u)$ exist in $[-1, 1] \times \mathbb{R}$, that $g_x(x, u)$ is nonnegative and bounded and that $g_u(x, u)$ is uniformly Hölder continuous with respect to both variables. Under these assumptions there exists a unique solution (u^*, D^*) of (1), (2) with $u^* \in W_p^1$ for some $p > 1$, where W_p^1 denotes the space of all functions $u(t)$ which possess a generalized derivative $u' \in \mathbf{L}^p(-1, 1)$ (cf. v. WOLFERSDORF [1]).

Let $(u^{(0)}, D^{(0)})$ be an approximation to the solution of (1), (2). Then we can determine a new approximation $u^{(1)}(t) = u^{(0)}(t) + \Delta u^{(0)}(t)$, $D^{(1)} = D^{(0)} + \Delta D^{(0)}$ by $\Delta u^{(0)}(-1) = \Delta u^{(0)}(1) = 0$ and

$$g_u\big(x, u^{(0)}(x)\big)\,\Delta u^{(0)}(x) - \frac{1}{\pi}\int_{-1}^{1} \frac{\Delta u^{(0)}(t)}{t-x}\,dt - \Delta D^{(0)} = f_0(x) \tag{3}$$

with

$$f_0(x) = -g\big(x, u^{(0)}(x)\big) + \frac{1}{\pi}\int_{-1}^{1}\frac{u^{(0)}(t)}{t-x}\,dt + D^{(0)}.$$

Equation (3) belongs to the class of singular integral equations considered in Secs. 9.46 to 9.49. Thus, for the approximate solution of (3) we can use the modified quadrature method described in Section 9.48. In this way we obtain a Newton iteration process for solving (1), (2) approximately, where at each step the same quadrature formula can be used, because in view of (2) the s-th (in the s-th step) coefficient $a^{(s)}(x) = g_u\big(x, u_n^{(s)}(x)\big)$ at $x = \pm 1$ is equal to $g_n(\pm 1, 0)$ independently of the number s. For some theoretical aspects of this method see JUNGHANNS/SILBERMANN [4].

9.51. Collocation and Galerkin methods. Let us again consider equation 9.34(1) in the pair $(\mathbf{L}^2_{\sigma_0,\varkappa}, \mathbf{L}^2_{\mu_0})$, where $\varkappa \geq 0$. Instead of 9.34(2) we use the collocation equations

$$(A + L^{\mu_0}_{n-\varkappa} H)\, v_n = L^{\mu_0}_{n-\varkappa} f, \qquad v_n \in \Pi_n \cap \mathbf{L}^2_{\sigma_0,\varkappa} \tag{1}$$

or the Galerkin equations

$$(A + P^{\mu_0}_{n-\varkappa} H)\, v_n = P^{\mu_0}_{n-\varkappa} f, \qquad v_n \in \Pi_n \cap \mathbf{L}^2_{\sigma_0,\varkappa}, \tag{2}$$

where the projections $P^{\mu_0}_{n-\varkappa}$ are defined by

$$(P^{\mu_0}_{n-\varkappa} f)(x) = \sum_{j=0}^{n-\varkappa-1} \frac{(f, q_j)_{\mu_0}}{\|q_j\|^2_{\mu_0}} \cdot q_j(x)$$

and $q_j(x)$ denotes the orthogonal polynomial of degree j with respect to the weight $\mu_0(x)$ (cf. Sec. 9.18). The theoretical investigation of the convergence of these two methods is easier than for the quadrature method and can be carried out in an analogous manner. If $f \in \mathbf{C}^{m,\gamma}$, $\{h_t\} \in \mathbf{C}([-1, 1], \mathbf{C}^{m,\gamma})$, and 9.34(1) has a unique solution $v \in \mathbf{L}^2_{\sigma_0,\varkappa}$, then the equations (1) (resp. (2)) are uniquely solvable for all sufficiently large n and

$$\|v - v_n\|_{\sigma_0} = O(n^{-m-\gamma}).$$

9.52. An iterative algorithm. The system matrix of 9.24(6), 9.24(7) does not possess any structural properties such as symmetry or positive definiteness, which could be used to solve these equations by effective iteration processes. Here we present an iteration algorithm which leans on the approximation properties of $H_{n-\varkappa,n}$ and which is based on the idea of two-grid methods and can be developed to multiple grid methods.

Let $\varkappa \geqq 0$ and $m < n$. We construct a sequence $\{v_n^{(s)}\} \subset \Pi_n \cap \mathbf{L}^2_{\sigma_0,\varkappa}$ with any given $v_n^{(0)} \in \Pi_n \cap \mathbf{L}^2_{\sigma_0,\varkappa}$ by solving

$$Au_n^{(s+1)} = L_{n-\varkappa}^{\mu_0} f - H_{n-\varkappa,n} v_n^{(s)}, \qquad u_n^{(s+1)} \in \Pi_n \cap \mathbf{L}^2_{\sigma_0,\varkappa}, \tag{1}$$

$$(A + H_{m-\varkappa,m})\, w_m^{(s+1)} = H_{m-\varkappa,n}(v_n^{(s)} - u_n^{(s+1)}), \qquad w_m^{(s+1)} \in \Pi_m \cap \mathbf{L}^2_{\sigma_0,\varkappa} \tag{2}$$

$$v_n^{(s+1)} = u_n^{(s+1)} + w_m^{(s+1)}. \tag{3}$$

Note that (1) can easily be solved by the help of the results of Sec. 9.23. The equations (1)–(3) can be summarized to

$$v_n^{(s+1)} = T_{mn} v_n^{(s)} + [I - (A + H_{m-\varkappa,m})^{-1} H_{m-\varkappa,n}]\, A^{-1} L_{n-\varkappa}^{\mu_0} f, \tag{4}$$

where

$$T_{mn} = (A + H_{m-\varkappa,m})^{-1} H_{m-\varkappa,n}(I + A^{-1} H_{n-\varkappa,n}) - A^{-1} H_{n-\varkappa,n}.$$

We obtain

$$\begin{aligned} T_{mn} &= (A + H_{m-\varkappa,m})^{-1} [H_{m-\varkappa,n}(I + A^{-1} H_{n-\varkappa,n}) - (I + H_{m-\varkappa,m} A^{-1})\, H_{m-\varkappa,n}] \\ &\quad + A^{-1}(H_{m-\varkappa,n} - H_{n-\varkappa,n}) = (A + H_{m-\varkappa,m})^{-1} [(H_{m-\varkappa,n} - H_{m-\varkappa,m})\, A^{-1} H_{n-\varkappa,n} \\ &\quad + H_{m-\varkappa,m} A^{-1}(H_{n-\varkappa,n} - H_{m-\varkappa,n})] + A^{-1}(H_{m-\varkappa,n} - H_{n-\varkappa,n}). \end{aligned}$$

Thus, from the results of Secs. 9.28, 9.29 we deduce the existence of an m_0 such that $\|T_{mn}\| < 1$, $m, n \geqq m_0$. Because the solution v_n^* of 9.24(8) is a fixed point of the operator defined by the right-hand side of (4), the estimate

$$\|v_n^{(s)} - v_n^*\|_{\sigma_0} \leqq \|T_{mn}\|^s \, \|v_n^{(0)} - v_n^*\|_{\sigma_0}$$

is valid, which shows the convergence of the iteration process (1)–(3).

9.53. A quadrature method for a class of singular integro-differential equations. If a singular integro-differential equation of Prandtl type

$$-\frac{1}{\pi} \int_{-1}^{1} \frac{u'(t)}{t - x}\, dt + a(x)\, u(x) + \frac{1}{\pi} \int_{-1}^{1} h(x, t)\, u(t)\, dt = f(x) \tag{1}$$

$x \in (-1, 1)$ is to be solved, where the unknown function $u(t)$ has to fulfil the condition

$$u(-1) = u(1) = 0, \tag{2}$$

one can seek an approximate solution in the form

$$u_n(t) = \sqrt{1 - t^2} \sum_{j=1}^{n} \xi_{jn} \frac{U_n(t)}{(t - t_{jn})\, U_n'(t_{jn})}. \tag{3}$$

Here $U_n(t)$ denotes the Chebyschev polynomial of the second kind, $U_n(t) = \dfrac{\sin (n+1)\,\psi}{\sin \psi}$, $t = \cos \psi$, and $t_{jn} = \cos \psi_{jn}$ $\left(\psi_{jn} = \dfrac{j\pi}{n+1}\right)$, $j = 1, \ldots, n$, are the roots of $U_n(t)$. With

$$u_n'(t) = -\frac{2}{n+1} \sum_{j=1}^{n} \xi_{jn} \frac{\sin \psi_{jn}}{\sin \psi} \sum_{m=1}^{n} m \sin m\psi_{jn} \cos m\psi,$$

the notation $(Su)(x) = \frac{1}{\pi} \int_{-1}^{1} \frac{u(t)}{t - x}\, dt$ and the relation

$$U_{n-1}(x) = (S\sigma T_n)(x), \qquad T_n(x) = \cos n\varphi, \qquad x = \cos\varphi,$$
$$\sigma(t) = (1 - t^2)^{-1/2},$$

which is a consequence of the results of Sec. 9.15, we obtain

$$-(Su_n')(t_{kn}) = \frac{2}{n+1} \sum_{j=1}^{n} \xi_{jn} \frac{\sin\psi_{jn}}{\sin\psi_{kn}} \sum_{m=1}^{n} m \sin m\psi_{jn} \sin m\psi_{kn}.$$

Having recourse to the well known formula

$$\sum_{m=1}^{n} \sin m\psi = \frac{\sin\left(\frac{n+1}{2}\right)\psi}{\sin\frac{\psi}{2}} \sin\frac{n\psi}{2}$$

the last equation becomes

$$-(Su_n')(t_{kn}) = \sum_{j=1}^{n} a_{kj}\xi_{jn},$$

where (cf. Kalandija [1])

$$a_{kj} = \begin{cases} \frac{n+1}{2} & \text{if } k = j, \\ 0 & \text{if } |k-j| = 2, 4, 6, \ldots, \\ \frac{\sin\psi_{jn}}{2(n+1)\sin\psi_{kn}} \left(\frac{1}{\tan^2\frac{\psi_{jn}+\psi_{kn}}{2}} - \frac{1}{\tan^2\frac{\psi_{jn}-\psi_{kn}}{2}} \right), & \\ \text{if } |k-j| = 1, 3, \ldots. & \end{cases}$$

Thus, to solve (1), (2) approximately, we arrive at the equations

$$a_k\xi_{kn} + \sum_{j=1}^{n} (a_{kj} + b_{kj})\,\xi_{jn} = f(t_{kn}), \qquad k = 1, \ldots, n, \tag{4}$$

where $a_k = a(t_{kn}) \cdot \sin\psi_{kn}$ and $b_{kj} = \frac{\sin^2\psi_{jn}}{n+1} h(t_{kn}, t_{jn})$. Write equations (1), (2) and (4) with (3) in the equivalent forms

$$(S_1 + H^{(1)} + H^{(2)})\, u = f, \qquad u \in W_{1,\mu}^2 \tag{5}$$

and

$$(S_1 + H_n^{(1)} + H_n^{(2)})\, u_n = L_n f, \tag{6}$$

respectively, where $W_{1,\mu}^2$ denotes the completion of the normed space $(X_0, \|\cdot\|_0)$ with

$$X_0 = \{u \in \mathbf{C}[-1, 1] \cap \mathbf{C}^{(1)}(-1, 1) : u' \in \mathbf{L}_\mu^2,\ u(-1) = u(1) = 0\}$$

and

$$\|u\|_0 = \|u\|_\infty + \|u'\|_\mu.$$

The operators in (5) and (6) are defined by

$$S_1 u = -Su', \qquad (H^{(1)}u)(x) = a(x)\, u(x),$$

$$(H^{(2)}u)(x) = \frac{1}{\pi} \int_{-1}^{1} h(x, t)\, u(t)\, dt, \qquad (H_n^{(1)}u_n)(x) = (L_n a u_n)(x)$$

$$(H_n^{(2)}u_n)(x) = L_{nx} \frac{1}{\pi} \int_{-1}^{1} L_{nt}[h(x, t)\, \sigma(t)\, u_n(t)]\, \mu(t)\, dt,$$

$$\mu(t) = (1 - t^2)^{1/2}, \qquad (L_n f)(x) = \sum_{j=1}^{n} f(t_{jn}) \frac{U_n(t)}{(t - t_{jn})\, U_n'(t_{jn})}.$$

9.54. Theorem. *Assume that a, f and h are continuous functions. If the homogeneous equation* 9.53(5) *possesses only the trivial solution, then* 9.53(6) *is uniquely solvable for all sufficiently large n and the solutions u_n converge to the unique solution u of* 9.53(5) *in the norm of the space $W^2_{1,\mu}$.*

Proof. Since the operator $S\sigma I$ is invertible in the pair $(\mathbf{L}^2_{\sigma,1}, \mathbf{L}^2_\mu)$ and since the operator $\mu\,(d/dt)$ considered in $(W^2_{1,\mu}, \mathbf{L}^2_{\sigma,1})$ is continuous and invertible (the inverse being $\big(\mu(d/dt)\big)^{-1} f(t) = \int_{-1}^{t} f(s)\, \sigma(s)\, ds$), the operator $S_1 = (S\sigma I)\,\big(\mu(d/dt)\big)$ is an invertible element of $\mathcal{L}(W^2_{1,\mu}, \mathbf{L}^2_\mu)$. Because of Corollary 9.28 we have the estimate

$$\|(H^{(2)} - H_n^{(2)})\, u_n\|_\mu \leqq \varepsilon_n\, \|\sigma u_n\|_\mu \leqq \varepsilon_n\, \|u_n\|_0$$

with $\varepsilon_n \to 0$ as $n \to \infty$ (cf. the proof of Corollary 9.29 and Remark 9.30). It is easily seen that $H^{(1)}$ is a compact operator from $W^2_{1,\mu}$ into $\mathbf{C}[-1, 1]$, which implies that $\|H^{(1)} - H_n^{(1)}\|_{\mathcal{L}(W^2_{1,\mu}, \mathbf{L}^2_\mu)} \to 0$ as $n \to \infty$. Notice that $H^{(2)}$ is also compact from $W^2_{1,\mu}$ into $\mathbf{L}^2_\mu$. Now standard arguments give the assertion. ■

Notes and comments

9.9.—9.10. Junghanns/Silbermann [1], [4], Junghanns [3].

9.12.—9.13. Junghanns/Silbermann [1], [5], Junghanns [3].

9.14. Dow/Elliott [1], Elliott [3], [6], Junghanns/Silbermann [5].

9.15. The important equality 9.15(2) has a long history. Likely, a special case was first treated by Akhiezer [1] (see also Tricomi [1]).

We remark that one more particular case was also considered (for instance) in Prössdorf/Silbermann [8, pp. 79—80], where appropriate substitutions were used in order to obtain some kind of singular integral operators transfering cosine functions into sine functions.

The general version occurs in several papers, e.g. in Karpenko [1], Erdogan/Gupta/Cook [1], and Parton/Perlin [1]. The simple proof given in this section is taken from Junghanns/Silbermann [1].

9.17. The theorem is well-known. The proof given here follows Junghanns/Silbermann [5].

9.19.—9.23. See Junghanns/Silbermann [1], [5], Elliott [3], [4].

9.24.—9.34. We essentially follow Junghanns/Silbermann [1], [4], [5]. The approximate solution of singular integral equations over intervals is the subject of numerous investigations, beginning with the classical paper of Multhopp [1]. It cannot be the aim of these comments to give a complete historical survey on this topic. However, we like to emphasize that there is an enormous number of papers dealing with the approximate solution of singular integral equations with constant coefficients. See, for example, LINZ [1], KRENK [1], IOAKIMIDIS [2], [3], GERASOULIS [1], [3], GERASOULIS/SRIVASTAV [1], [2], GOLBERG/FROMME [1], GOLBERG [1]—[4], IOAKIMIDIS/THEOCARIS [1], [2], [3], SRIVASTAV [1], SRIVASTAV/JEN [1], TSAMASHYROS/THEOCARIS [1], JUNGHANNS/SILBERMANN [2], [4], PRÖSSDORF/SILBERMANN [8].

The question of uniform convergence of the approximate solutions (see the estimate 9.34(6)) is of high interest.

Here we follow JUNGHANNS [4] (cf. also HAFTMANN [1], [2], JUNGHANNS/SILBERMANN [4], [5]), where the uniform convergence is derived from $\mathbf{L}_2$-convergence with sufficiently high speed. ELLIOTT [1], [2], [4], [5], [6] studied the question mentioned by the help of regularization and considered the singular equation in spaces of continuous functions (especially in the case of a polynomial $b(x)$). See also JUNGHANNS [7]. A fairly complete answer to this question is given by BERTHOLD [1]. In the recent paper MASTROIANNI/PRÖSSDORF [1] the estimates 9.34(3) and 9.34(6) were generalized to the case of a weakly singular kernel function h.

9.35.—9.38. JUNGHANNS/SILBERMANN [1], [5], JUNGHANNS [3], [5].

9.39.—9.40. JUNGHANNS/SILBERMANN [5].

9.41.—9.43. JUNGHANNS/SILBERMANN [4]. In JUNGHANNS/SILBERMANN [5] the case $\varkappa \leqq -1$ is considered as well. An analogous approach can also be found in MUSAEV [1].

9.46.—9.49. The idea of simplifying the weight function and the modified quadrature method is due to JUNGHANNS [2] (see also JUNGHANNS [3]). Here we follow JUNGHANNS/SILBERMANN [5].

9.50. This nonlinear singular integral equation is of considerable interest for some free boundary value problems (see ÖSTREICH [1], [2], JUNGHANNS/ÖSTREICH [1]).

9.51. JUNGHANNS/SILBERMANN [5].

9.52. JUNGHANNS [6], JUNGHANNS/SILBERMANN [4]. Multiple grid methods being based on the algorithm presented here are studied in JUNGHANNS/BERTHOLD [1] (see also BERTHOLD [1]).

9.53.—9.54. Prandtl's singular integro-differential equation and methods for its numerical solution have been studied by many people. We confine ourselves to referring the reader to the works KALANDIYA [1], SCHLEIFF [1], [3], PRÖSSDORF/SILBERMANN [8]. Our presentation here follows JUNGHANNS/SILBERMANN [4].

Chapter 10
Spline approximation and quadrature methods for the solution of singular integral equations on closed curves

Preliminaries

10.1. Introduction. For the numerical solution of singular integral equations (and, more generally, pseudodifferential equations, cf. Chapter 13) posed on closed plane curves, spline approximation methods are extensively employed. In fact, collocation and quadrature methods are the most widely used numerical procedures for solving the boundary integral equations arising from exterior or interior boundary value problems of elasticity, aerodynamics, fluid mechanics, electromagnetics, acoustics, and other engineering applications with the boundary element technique. (See, e.g., the books and conference proceedings Banerjee/Butterfield [1], Banerjee/Shaw [1], Brebbia [1–3], Brebbia/Futagami/Tanaka [1], Brebbia/Telles/Wrobel [1], Brebbia/Kuhn/Wendland [1], Crouch/Starfield [1], Filippi [1], Mukherjee [1], Belotserkovski/Lifanov [1], and the discussion of the boundary integral equations arising in applications in Arnold/Wendland [1].)

In this chapter we present a general and uniform approach to the stability theory and the error analysis of spline collocation and spline Galerkin methods as well as quadrature procedures and collocation methods with trigonometric trial functions. The heart of our analysis is a localization technique for numerical schemes in which the matrices of the finite systems are paired circulants. It turns out that each of the aforementioned numerical methods is stable if and only if two families of singular integral operators (depending on the numerical procedure under consideration) contain invertible operators only.

In order to state the results of the present chapter more precisely we now introduce some notation. We shall be concerned with the numerical solution of the singular integral equation

$$(Au)\,(t) := (cu)\,(t) + (dS_\Gamma u)\,(t) + (Ku)\,(t) = f(t) \qquad (t \in \Gamma), \tag{1}$$

where Γ is a simple smooth closed curve on the complex plane oriented counterclockwise,

$$(S_\Gamma u)\,(t) := \frac{1}{\pi i} \int_\Gamma \frac{u(\tau)}{\tau - t}\, d\tau$$

is the Cauchy singular operator,

$$(Ku)\,(t) := \int_\Gamma k(t, \tau)\, u(\tau)\, d\tau,$$

$u \in \mathbf{L}^2(\Gamma)$ is the unknown function, c, d, k and f denote given functions. Suppose c and d are merely piecewise continuous on Γ (i.e. $c, d \in \mathbf{PC}(\Gamma)$), k is sufficiently smooth on $\Gamma \times \Gamma$, f is Riemann integrable $(f \in \mathcal{R}(\Gamma))$, and the operator $A \in \mathcal{L}(\mathbf{L}^2(\Gamma))$ is invertible. Recall that the operator (1) can be rewritten in the form

$$A = aP_\Gamma + bQ_\Gamma + K,$$

where $a := c + d$, $b := c - d$, and $P_\Gamma := \frac{1}{2}(I + S_\Gamma)$, $Q_\Gamma := \frac{1}{2}(I - S_\Gamma)$.

Let the curve Γ be given by a regular parametric representation $\Gamma := \{\gamma(s) : 0 \leqq s \leqq 1\}$, where γ is a 1-periodic function of the real variable s and $|d\gamma/ds| \neq 0$. Via the parametrization we have a one-to-one correspondence between functions x on Γ and the 1-periodic functions $\tilde{\tilde{x}}$ on $\mathbb{R}$, where $\tilde{\tilde{x}}(s) = x(\gamma(s))$. Throughout this chapter we shall identify the functions x and $\tilde{\tilde{x}}$ as well as the function $\tilde{x}$ defined on the unit circle $\mathbb{T}$ by $\tilde{x}(e^{i2\pi s}) = \tilde{\tilde{x}}(s)$.

Let δ be a nonnegative integer and n a natural number. Let $\mathcal{S}_n^\delta := \mathcal{S}^\delta(\Delta)$ denote the space of smooth 1-periodic splines of degree δ on the uniform mesh $\Delta := \{k/n : k = 0, \ldots, n-1\}$. Thus $\mathcal{S}_n^\delta$ $(\delta \geqq 1)$ consists of periodic $\mathbf{C}^{\delta-1}$ piecewise polynomials of degree δ and has dimension n; $\mathcal{S}_n^0$ is defined as the corresponding space of piecewise constant functions. Let ε, $0 \leqq \varepsilon < 1$, be a fixed number, where $\varepsilon > 0$ if $\delta = 0$.

The *ε-collocation method* determines an approximate solution $u_n \in \mathcal{S}_n^\delta$ to Eq. (1) by the equations

$$(Au_n)\big((k+\varepsilon)/n\big) = f\big((k+\varepsilon)/n\big), \qquad k = 0, \ldots, n-1. \tag{2}$$

(It is easily seen that, for $\delta = 0$, the left-hand side of (2) makes sense if and only if $0 < \varepsilon < 1$.)

The *Galerkin method* defines $u_n \in \mathcal{S}_n^\delta$ by

$$(Au_n, \varphi_n) = (f, \varphi_n) \qquad \forall \varphi_n \in \mathcal{S}_n^\delta, \tag{3}$$

where $(\cdot, \cdot)$ denotes the usual scalar product on $\mathbf{L}^2(\Gamma)$.

In the case of the unit circle $\Gamma = \mathbb{T}$, the simplest *quadrature method* for Eq. (1) is the so-called "method of discrete whirls", which reads as follows: let $v_k := \exp\big(i2\pi(k + 1/2)/n\big)$, $t_k := \exp(i2\pi k/n)$ $(k = 0, \ldots, n-1)$, and determine the approximate values ξ_k to $u(t_k)$ by the system

$$c(v_k)\,\xi_k + \frac{d(v_k)}{\pi i} \sum_{j=0}^{n-1} \frac{\xi_j}{t_j - v_k} \frac{2\pi i}{n} t_j + \sum_{j=0}^{n-1} k(v_k, t_j) \frac{2\pi i}{n} t_j \xi_j = f(v_k), \qquad k = 0, \ldots, n-1. \tag{4}$$

If there exists a unique solution (ξ_k) to (4) and $\delta > 0$ is an odd integer, then the interpolating spline $u_n \in \mathcal{S}_n^\delta$ satisfying $u_n(t_j) = \xi_j$ $(j = 0, \ldots, n-1)$ gives an approximate solution to Eq. (1).

For other, more general quadrature methods we refer to Sections 10.12 and 10.25. Here we only mention that all quadrature methods we shall consider are so-called modified quadrature methods, i.e., in order to obtain a finite linear system for determining the values $\xi_j = u_n(t_j)$ we replace the singular integral $(S_\Gamma u)(\tau_k)$ by

$$(S_\Gamma u)(\tau_k) = u(\tau_k) + \frac{1}{\pi i} \int_{\mathbf{T}} \frac{u(\tau) - u(\tau_k)}{\tau - \tau_k} \, d\tau$$

$$\approx u(\tau_k) + \frac{1}{\pi i} \sum_{j=0}^{n-1} \frac{u(t_j) - u(\tau_k)}{t_j - \tau_k} \frac{2\pi i}{n} t_j,$$

where $\tau_k := \exp\big(i2\pi(k + \varepsilon)/n\big)$ $(0 < \varepsilon < 1,\ k = 0, \ldots, n - 1)$.

The present chapter is organized as follows. The next two sections contain auxiliary material on circulants. In Sections 10.4–10.15 and 10.25–10.27 we thoroughly analyse the structure of the finite-dimensional operators arising in the discretized equations to the above mentioned approximation methods. Once a suitable basis of $\mathcal{S}_n^\delta$ is chosen, the approximate equation is reduced to an $n \times n$ linear system for the unknown coefficients of u_n whose matrix has a special form (viz, the matrix is a paired circulant). Applying the localization principle (Section 10.3) we establish necessary and sufficient conditions for the aforementioned approximation methods to be stable in $\mathbf{L}^2$. In Sections 10.16 to 10.24 and 10.28–10.30 we extend our stability and error analysis to the scale of Sobolev spaces H^s. Finally, Sections 10.31–10.42 contain a Banach algebra approach to the localization principle.

10.2. Circulants. Let $\mathbf{l}^2(n)$ denote the n-dimensional complex Hilbert space $\mathbb{C}^n$ provided with the scalar product

$$(\xi, \eta) = \sum_{k=0}^{n-1} \xi_k \bar{\eta}_k, \qquad \xi = (\xi_k)_{k=0}^{n-1}, \qquad \eta = (\eta_k)_{k=0}^{n-1} \in \mathbb{C}^n$$

and the norm $|\xi| = \sqrt{(\xi, \xi)}$. In what follows, each operator $A_n \in \mathcal{L}\big(\mathbf{l}^2(n)\big)$ will be identified with the corresponding matrix in $\mathbb{C}^{n \times n}$. Let $\|A_n\|$ denote the operator norm in $\mathcal{L}\big(\mathbf{l}^2(n)\big)$. Introduce the unitary operators $U_n, U_n^{-1} \in \mathcal{L}\big(\mathbf{l}^2(n)\big)$ by

$$(U_n\xi)_k := \frac{1}{\sqrt{n}} \sum_{j=0}^{n-1} e^{i2\pi kj/n} \xi_j, \qquad (U_n^{-1}\xi)_k := \frac{1}{\sqrt{n}} \sum_{j=0}^{n-1} e^{-i2\pi kj/n} \xi_j.$$

A finite Toeplitz matrix $T_n = (a_{j-k})_{j,k=0}^{n-1}$ is said to be a *circulant*, if

$$a_{-k} = a_{n-k}, \qquad k = 1, \ldots, n - 1.$$

It is easily seen that $A_n \in \mathcal{L}\big(\mathbf{l}^2(n)\big)$ is a circulant if and only if there exists a vector $\zeta = (\zeta_k) \in \mathbb{C}^n$ such that $A_n = U_n M_\zeta U_n^{-1}$, where $M_\zeta = (\zeta_k \delta_{jk})_{j,k=0}^{n-1}$. Obviously, ζ_k $(k = 0, \ldots, n - 1)$ are just the eigenvalues of A_n and $(\xi_j^{(k)})_{j=0}^{n-1} = \left(\frac{1}{\sqrt{n}} \exp(i2\pi jk/n)\right)_{j=0}^{n-1}$ the corresponding eigenvectors.

Given a bounded function ϱ on $\mathbb{T}$, let ϱ_n and $\tilde{\varrho}_n$ denote the diagonal matrices

$$\varrho_n := \big(\varrho(t_k)\, \delta_{jk}\big)_{j,k=0}^{n-1}, \qquad \tilde{\varrho}_n := \big(\varrho(\tau_k)\, \delta_{jk}\big)_{j,k=0}^{n-1},$$

where $t_k := \exp(i2\pi k/n)$ and $\tau_k := \exp\big(i2\pi(k + \varepsilon)/n\big)$ $(0 \leqq \varepsilon < 1,\ k = 0, \ldots, n - 1)$, respectively.

Consider the circulant

$$\hat{\varrho}_n := U_n \varrho_n U_n^{-1}.$$

Thus the values $\varrho(t_k)$ $(k = 0, \ldots, n-1)$ are just the eigenvalues of $\hat{\varrho}_n$. Obviously,

$$\hat{\varrho}_n = \left(\frac{1}{n}\sum_{m=0}^{n-1}\varrho(t_m)\exp\left(\mathrm{i}2\pi m(j-k)/n\right)\right)_{j,k=0}^{n-1} = \left(\frac{1}{n}\sum_{m=0}^{n-1}\varrho(t_m)\,t_{j-k}^m\right)_{j,k=0}^{n-1}$$

and $\|\hat{\varrho}_n\| = \|\varrho_n\| \le \sup_{t\in\mathbb{T}}|\varrho(t)|$, $\|\tilde{\varrho}_n\| \le \sup_{t\in\mathbb{T}}|\varrho(t)|$.

10.3. Paired circulants and a stability theorem. Let $C_n \in \mathcal{L}(\mathrm{l}^2(n))$ satisfy $\|C_n\| \to 0$ as $n \to \infty$. For given $c, d, \alpha, \beta \in \mathbf{PC}(\mathbb{T})$ we introduce the *paired circulants* $B_n \in \mathcal{L}(\mathrm{l}^2(n))$ by

$$B_n := \tilde{c}_n\hat{\alpha}_n + \tilde{d}_n\hat{\beta}_n + C_n.$$

We associate two families of singular integral operators B_τ and B^τ $(\tau \in \mathbb{T})$ with the sequence $\{B_n\}_{n=1}^\infty$. These operators act on $\mathbf{L}^2(\mathbb{T})$ and are defined by

$$B_\tau := [\alpha(\tau+0)\,c + \beta(\tau+0)\,d]\,P_{\mathbb{T}} + [\alpha(\tau-0)\,c + \beta(\tau-0)\,d]\,Q_{\mathbb{T}},$$

$$B^\tau := P_{\mathbb{T}}[c(\tau+0)\,\tilde{\alpha} + d(\tau+0)\,\tilde{\beta}] + Q_{\mathbb{T}}[c(\tau-0)\,\tilde{\alpha} + d(\tau-0)\,\tilde{\beta}],$$

where $\tilde{\alpha}(t) := \alpha(t^{-1})$, $\tilde{\beta}(t) := \beta(t^{-1})$ $(t \in \mathbb{T})$. The pair $\{B_\tau, B^\tau\}$ is called the *symbol* of the sequence $\{B_n\}$ (see also Section 10.40).

Theorem. (i) *Let* $c, d, \alpha, \beta \in \mathbf{PC}(\mathbb{T})$. *The sequence* $\{B_n\}$ *is stable if and only if the operators* B_τ *and* B^τ *are invertible in* $\mathbf{L}^2(\mathbb{T})$ *for each* $\tau \in \mathbb{T}$.

(ii) *Let* $c, d \in \mathbf{PC}(\mathbb{T})$, $\alpha, \beta \in \mathbf{PC}(\mathbb{T})$, $\alpha, \beta \in \mathbf{C}(\mathbb{T} \setminus \{1\})$ (*resp.* $\alpha, \beta \in \mathbf{C}(\mathbb{T} \setminus \{1, -1\})$). *The sequence* $\{B_n\}$ *is stable if and only if all the operators* B^τ $(\tau \in \mathbb{T})$ *and the operator* B_1 (*resp. the operators* B_1, B_{-1}) *are invertible in* $\mathbf{L}^2(\mathbb{T})$.

This theorem will be proved in Sections 10.41–42.

Collocation methods

10.4. Approximation operators of the collocation. Let us introduce the spline space $\tilde{\mathcal{S}}_n^\delta$ by

$$\tilde{\mathcal{S}}_n^\delta := \begin{cases} \mathcal{S}_n^{\delta+1} & \text{if } \varepsilon = 0 \text{ and } \delta \text{ is even or if } \varepsilon = \dfrac{1}{2} \text{ and } \delta \text{ is odd,} \\ \mathcal{S}_n^\delta & \text{otherwise.} \end{cases}$$

Let L_n denote the orthogonal projection from $\mathbf{L}^2(\Gamma)$ onto $\mathcal{S}_n^\delta$, i.e.

$$L_n x \in \mathcal{S}_n^\delta, \quad (L_n x - x, \varphi_n) = 0 \quad \forall x \in \mathbf{L}^2(\Gamma) \quad \forall \varphi_n \in \mathcal{S}_n^\delta.$$

For arbitrary $\varepsilon \in [0, 1)$ and $x \in \mathcal{R}(\Gamma)$, the interpolation projection $K_n = K_n^\varepsilon$ is defined by (see Chapter 2)

$$K_n x \in \tilde{\mathcal{S}}_n^\delta, \quad (K_n x - x)\left((k+\varepsilon)/n\right) = 0, \quad k = 0, 1, \ldots, n-1.$$

Obviously, the collocation equations 10.1(2) are equivalent to the operator equation

$$A_n u_n = K_n f, \quad A_n := K_n A \mid \mathcal{S}_n^\delta \in \mathcal{L}(\mathcal{S}_n^\delta, \tilde{\mathcal{S}}_n^\delta).$$

Now suppose $\Gamma = \mathbb{T}$. We shall analyse the structure of the collocation operator A_n,

$$A_n = \left(K_n cI\big|_{\tilde{\mathcal{S}}_n^\delta}\right)\left(K_n\big|_{\mathcal{S}_n^\delta}\right) + \left(K_n d\big|_{\tilde{\mathcal{S}}_n^\delta}\right)\left(K_n S_{\mathbb{T}}\big|_{\mathcal{S}_n^\delta}\right) + K_n K\big|_{\mathcal{S}_n^\delta}.$$

Set $t_k := \exp(\mathrm{i}2\pi k/n)$, $v_k := \exp\big(\mathrm{i}2\pi(k + (1/2))/n\big)$, $\tau_k := \exp\big(\mathrm{i}2\pi(k + \varepsilon)/n\big)$, $(k = 0, 1, \ldots, n - 1;\, \varepsilon \in [0, 1))$ and let $(\chi_k)_{k=0}^{n-1}$, $(\psi_k)_{k=0}^{n-1}$ denote the interpolation bases of $\mathscr{S}_n^\delta$, $\tilde{\mathscr{S}}_n^\delta$, respectively, satisfying the conditions

$$\chi_k(t_j) = \delta_{jk} \quad \text{if } \delta \text{ is odd},$$

$$\chi_k(v_j) = \delta_{jk} \quad \text{if } \delta \text{ is even},$$

$$\psi_k(\tau_j) = \delta_{jk}; \quad j, k = 0, 1, \ldots, n - 1.$$

As shown in Sections 2.20–2.25 and 2.30, the splines χ_n and ψ_n exist and possess the following properties

$$C_1 |\xi| \leqq n^{1/2} \left\| \sum_{k=0}^{n-1} \xi_k \chi_k \right\|_{\mathbf{L}^2(\Gamma)} \leqq C_2 |\xi|, \tag{1}$$

$$C_1 |\xi| \leqq n^{1/2} \left\| \sum_{k=0}^{n-1} \xi_k \psi_k \right\|_{\mathbf{L}^2(\Gamma)} \leqq C_2 |\xi|, \tag{2}$$

where C_1 and C_2 are positive constants independent of $\xi = (\xi_k)_{k=0}^{n-1} \in \mathbf{l}^2(n)$ and $n \in \mathbb{N}$. The operators of $\mathscr{L}(\mathscr{S}_n^\delta, \tilde{\mathscr{S}}_n^\delta)$ can be identified with their matrix representations corresponding to the bases (χ_k) and (ψ_k) of $\mathscr{S}_n^\delta$ and $\tilde{\mathscr{S}}_n^\delta$, respectively. It is obvious that a sequence $\{B_n\}$ of operators $B_n \in \mathscr{L}(\mathscr{S}_n^\delta, \tilde{\mathscr{S}}_n^\delta)$ is uniformly bounded (stable) in $\mathbf{L}^2(\Gamma)$ if and only if the sequence of the corresponding matrices $B_n \in \mathscr{L}\big(\mathbf{l}^2(n)\big)$ is uniformly bounded (stable). Consequently, it suffices to analyse the corresponding matrices $A_n \in \mathscr{L}\big(\mathbf{l}^2(n)\big)$ for our collocation operators.

Now let δ be odd. With respect to the basis (ψ_k), the operator $K_n cI\big|_{\tilde{\mathscr{S}}_n^\delta} \in \mathscr{L}(\tilde{\mathscr{S}}_n^\delta)$ has the matrix

$$\tilde{c}_n := \big(c(\tau_k)\, \delta_{jk}\big)_{j,k=0}^{n-1}.$$

The matrix of $K_n S_{\mathbb{T}}\big|_{\mathscr{S}_n^\delta} \in \mathscr{L}(\mathscr{S}_n^\delta, \tilde{\mathscr{S}}_n^\delta)$ corresponding to the bases (χ_k), (ψ_k) takes the form $\tilde{S}_n := \big((S_{\mathbb{T}}\chi_k)\,(\tau_j)\big)_{j,k=0}^{n-1}$. Since

$$\chi_k(t) = \chi_0(t_{-k}t), \qquad \big(S_{\mathbb{T}} x(t_k\cdot)\big)\,(t) = (S_{\mathbb{T}} x)\,(t_k t) \qquad (t \in \mathbb{T}),$$

we obtain $\tilde{S}_n = \big((S_{\mathbb{T}}\chi_0)\,(\tau_{j-k})\big)_{j,k=0}^{n-1}$. Hence the matrix $\tilde{S}_n$ is a circulant. Moreover, as an immediate consequence of Sections 10.2 and 2.21, we have the formula

$$K_n S_{\mathbb{T}}\big|_{\mathscr{S}_n^\delta} = \hat{\beta}_n,$$

where

$$\beta(\mathrm{e}^{\mathrm{i}2\pi s}) := \begin{cases} 1 & \text{if } s = 0, \\[2mm] \dfrac{\Phi_{\varepsilon,\delta}(s)\, \sigma_{\varepsilon,\delta}^-(s)}{\Phi_{0,\delta}(s)\, \sigma_{0,\delta}^+(s)} & \text{if } \; 0 < s < 1, \end{cases}$$

$$\sigma_{\varepsilon,\delta}^{\pm}(s) := \sum_{k=0}^{\infty} (k + s)^{-\delta-1}\, \mathrm{e}^{\mathrm{i}2\pi k\varepsilon} \pm \sum_{k=1}^{\infty} (k - s)^{-\delta-1}\, \mathrm{e}^{-\mathrm{i}2\pi k\varepsilon},$$

$$\Phi_{\varepsilon,\delta}(s) := 2^{-\delta}\pi^{-\delta-1} \sin(\pi s)\, \mathrm{e}^{\mathrm{i}\pi s(2\varepsilon-1)} \qquad (0 < s < 1).$$

Finally, the matrix of $K_n\big|_{\mathscr{S}_n^\delta} \in \mathscr{L}(\mathscr{S}_n^\delta, \tilde{\mathscr{S}}_n^\delta)$ corresponding to the bases (χ_k), (ψ_k) equals $\big(\chi_k(\tau_j)\big)_{j,k=0}^{n-1} = \big(\chi_0(\tau_{j-k})\big)_{j,k=0}^{n-1}$ and is a circulant, too. Section 2.21 yields that

$$K_n\big|_{\mathscr{S}_n^\delta} = \hat{\alpha}_n,$$

where

$$\alpha(e^{i2\pi s}) := \begin{cases} 1 & \text{if } s = 0, \\ \dfrac{\Phi_{\varepsilon,\delta}(s)\,\sigma^+_{\varepsilon,\delta}(s)}{\Phi_{0,\delta}(s)\,\sigma^+_{0,\delta}(s)} & \text{if } 0 < s < 1. \end{cases}$$

Thus we obtain

$$A_n = \tilde{c}_n\hat{\alpha}_n + \tilde{d}_n\hat{\beta}_n + K_nK\big|_{\mathcal{S}_n^\delta}. \tag{3}$$

If δ is even, then we can proceed analogously to conclude that Eq. (3) holds with

$$\alpha(e^{i2\pi s}) := \begin{cases} 1 & \text{if } s = 0, \\ \dfrac{\Phi_{\varepsilon,\delta}(s)\,\sigma^-_{\varepsilon,\delta}(s)}{\Phi_{1/2,\delta}(s)\,\sigma^-_{1/2,\delta}(s)} & \text{if } 0 < s < 1, \end{cases}$$

$$\beta(e^{i2\pi s}) := \begin{cases} 1 & \text{if } s = 0, \\ \dfrac{\Phi_{\varepsilon,\delta}(s)\,\sigma^+_{\varepsilon,\delta}(s)}{\Phi_{1/2,\delta}(s)\,\sigma^-_{1/2,\delta}(s)} & \text{if } 0 < s < 1. \end{cases}$$

It is easily seen that $\alpha \in \mathbf{C}^\infty(\mathbb{T})$, $\beta \in \mathbf{C}^\infty(\mathbb{T} \setminus \{1\})$ and $\beta(1 + 0) = 1$, $\beta(1 - 0) = -1$. Moreover, $\sigma^+_{\varepsilon,\delta}(s) = 0$ $\big(\sigma^-_{\varepsilon,\delta}(s) = 0\big)$ if and only if $\varepsilon = s = 1/2$ (if and only if $\varepsilon = 0$ and $s = 1/2$) (see Sections 2.21 and 2.30).

10.5. Stability theorem for the collocation method. *Let $c, d \in \mathbf{PC}(\Gamma)$, $K \in \mathcal{K}\big(\mathbf{L}^2(\Gamma), \mathcal{R}(\Gamma)\big)$ and $\varepsilon \in [0, 1)$. Assume the operator $A := cI + dS_\Gamma + K$ to be invertible in $\mathbf{L}^2(\Gamma)$. Then the ε-collocation method* 10.1(2) *is stable in $\mathbf{L}^2(\Gamma)$ if and only if the operators $B^\tau \in \mathcal{L}\big(\mathbf{L}^2(\mathbb{T})\big)$ are invertible for all $\tau \in \Gamma$, where*

$$B^\tau := P_{\mathbb{T}}[c(\tau + 0)\,\tilde{\alpha} + d(\tau + 0)\,\tilde{\beta}] + Q_{\mathbb{T}}[c(\tau - 0)\,\tilde{\alpha} + d(\tau - 0)\,\tilde{\beta}],$$

$$\tilde{\alpha}(t) := \alpha(t^{-1}), \qquad \tilde{\beta}(t) := \beta(t^{-1}) \quad (t \in \mathbb{T}),$$

and the functions α and β are defined as in Sect. 10.4.

Proof. We first consider the case $\Gamma = \mathbb{T}$. By Theorem 10.3 and Corollary 1.26, the sequence of the operators 10.4(3) is stable if and only if the operators B^τ and B_1 defined in 10.3 are invertible in $\mathbf{L}^2(\mathbb{T})$ for each $\tau \in \mathbb{T}$.

If $\tau = 1$, then

$$B_1 = c(1P_{\mathbb{T}} + 1Q_{\mathbb{T}}) + d(1P_{\mathbb{T}} - 1Q_{\mathbb{T}}) = cI + dS_{\mathbb{T}}.$$

Thus $B_1 \in \mathcal{GL}\big(\mathbf{L}^2(\mathbb{T})\big)$, since $A = B_1 + K$ is assumed to be invertible (cf. Chapter 6).

In case $\Gamma \neq \mathbb{T}$ one has $S_\Gamma - S_{\mathbb{T}} \in \mathcal{K}\big(\mathbf{L}^2(\Gamma), \mathbf{C}(\Gamma)\big)$ (see Mikhlin/Prössdorf [1], p. 58; cf. also Sections 6.55 and 10.26). As a consequence of Corollary 1.26, the method 10.1(2) is stable if and only if the sequence of the operators 10.4(3) is stable. This completes the proof. ■

10.6. Corollary. *Suppose the assumptions of Theorem* 10.5 *are satisfied. If, in addition, the coefficients c and d are continuous, then the ε-collocation method* 10.1(2) *is stable if and only if*

$$c(\tau)\,\alpha(t) + d(\tau)\,\beta(t) \neq 0 \quad \forall \tau \in \Gamma \quad \forall t \in \mathbb{T}.$$

10.7. Corollary. *Let the hypotheses of Theorem* 10.5 *be satisfied.*

(a) *If* δ *is* odd *and* $\varepsilon = 0$ *or if* δ *is even and* $\varepsilon = 1/2$ (*if* δ *is* odd *and* $\varepsilon = 1/2$ *or if* δ *is even and* $\varepsilon = 0$), *then the collocation method* 10.1(2) *is stable if and only if the following two conditions are fulfilled:*

(i) $b(\tau) \neq 0 \quad \forall \tau \in \Gamma$;

(ii) $\mu \dfrac{a}{b} (\tau + 0) + (1 - \mu) \dfrac{a}{b} (\tau - 0) \notin (-\infty, 0] \quad \forall \tau \in \Gamma, \mu \in [0, 1]$

((i) $b(\tau) \neq 0 \quad \forall \tau \in \Gamma$;

(iii) $\mu \dfrac{a}{b} (\tau + 0) + (1 - \mu) \dfrac{a}{b} (\tau - 0) \notin [0, \infty) \quad \forall \tau \in \Gamma, \mu \in [0, 1]$).

Here and hereafter we use the notations $a := c + d$, $b := c - d$ *introduced in Sec.* 10.1.

(b) *Let* $0 < \varepsilon < 1$, $\varepsilon \neq 1/2$. *Then the* ε*-collocation method* 10.1(2) *is stable if and only if the operators* $A^\tau \in \mathcal{L}(\mathbf{L}^2(\mathbb{T}))$ *are invertible for all* $\tau \in \Gamma$, *where*

$$A^\tau := \sigma_A(\tau + 0, \cdot)\, P_{\mathbb{T}} + \sigma_A(\tau - 0, \cdot)\, Q_{\mathbb{T}}$$

and

$$\sigma_A(\tau, e^{2\pi i s}) := a(\tau) \sum_{k=0}^{\infty} (s + k)^{-\delta-1} e^{2\pi i k s} + b(\tau) \sum_{k=-1}^{-\infty} (s + k)^{-\delta-1} e^{2\pi i k s},$$
$$0 < s < 1,$$

$$\sigma_A(\tau, 1) := a(\tau).$$

Proof. (a) Taking into account that

$$\operatorname{im} \sigma_{0,\delta}^- / \sigma_{0,\delta}^+ = \operatorname{im} \sigma_{1/2,\delta}^+ / \sigma_{1/2,\delta}^- = [-1, 1] \tag{1}$$

and using the theorem on the invertibility of the singular integral operator (see Sections 6.13, 6.14, 1°, and 6.1(ix)) we obtain that, for fixed $\tau \in \Gamma$, the operator $B^\tau \in \mathcal{L}(\mathbf{L}^2(\mathbb{T}))$ is invertible if and only if

$$\nu + (1 - \nu) \left[\mu \frac{a}{b} (\tau + 0) + (1 - \mu) \frac{a}{b} (\tau - 0) \right] \neq 0 \tag{2}$$

$$\left(-\nu + (1 - \nu) \left[\mu \frac{a}{b} (\tau + 0) + (1 - \mu) \frac{a}{b} (\tau - 0) \right] \right) \neq 0 \tag{3}$$

$\forall \mu, \nu \in [0, 1]$ (cf. the end of Section 10.13).

Obviously, condition (2) (condition (3)) is equivalent to condition (ii) (condition (iii)).

(b) For $0 < \varepsilon < 1$, $\varepsilon \neq 1/2$, it is easily seen that the operator $B^\tau \in \mathcal{L}(\mathbf{L}^2(\mathbb{T}))$ is invertible if and only if the operator $A^\tau \in \mathcal{L}(\mathbf{L}^2(\mathbb{T}))$ is so. ■

Remark. Condition (i) and (ii) (condition (i) and (iii)) is equivalent to the locally strong ellipticity of the singular integral operator $A_0 := aP_\Gamma + bQ_\Gamma$ ($A_0 S_\Gamma = aP_\Gamma - bQ_\Gamma$) (see Theorem 6.22). If a and b are continuous, then, obviously, condition (i) and (ii) (condition (i) and (iii)) is equivalent to the requirement

$$c(\tau) + \lambda d(\tau) \neq 0 \qquad (d(\tau) + \lambda c(\tau) \neq 0) \qquad \forall \tau \in \Gamma \qquad \forall \lambda \in [-1, 1].$$

Galerkin methods

10.8. Approximation operators of the Galerkin method. We first consider the Galerkin operators $A_n := L_n A|_{\mathcal{S}_n^\delta}$ in the case where $\Gamma = \mathbb{T}$ and c and d are continuous coefficients. Since $\|L_n c(I - L_n)\| \to 0$ as $n \to \infty$ (see Theorems 2.13 and 2.14) we find

$$A_n = L_n c|_{\mathcal{S}_n^\delta} + \left(L_n d|_{\mathcal{S}_n^\delta}\right) \left(L_n S_{\mathbb{T}}|_{\mathcal{S}_n^\delta}\right) + L_n K|_{\mathcal{S}_n^\delta} + B_n, \tag{1}$$

where $B_n \in \mathcal{L}(\mathcal{S}_n^\delta)$ and $\|B_n\| \to 0$ as $n \to \infty$.

Let φ_k ($k = 0, \ldots, n-1$) denote the basis splines in $\mathcal{S}_n^\delta$ (see AUBIN [1, Section 4.2]) which are defined by

$$\varphi_k(s) = (2\pi)^{-\delta/2}\, \Phi^\delta(sn - k), \qquad 0 < s < 1,$$

where Φ^δ is the $(\delta + 1)$-times convolution of the characteristic function of the interval $(-1/2, 1/2)$. Then

$$C_1\, |\xi| \le n^{1/2} \left\| \sum_{k=0}^{n-1} \xi_k \varphi_k \right\|_{\mathbf{L}^2(\Gamma)} \le C_2\, |\xi|,$$

where C_1 and C_2 are positive constants independent of $\xi = (\xi_k)_{k=0}^{n-1} \in \mathbf{l}^2(n)$ and n. The operators of $\mathcal{L}(\mathcal{S}_n^\delta)$ can be identified with their matrix representations corresponding to the basis (φ_k).

In PRÖSSDORF/RATHSFELD [4], the matrices $\left((\varphi_j, \varphi_k)\right)_{j,k=0}^{n-1}$ and $\left((S_{\mathbb{T}}\varphi_j, \varphi_k)\right)_{j,k=0}^{n-1}$ have been shown to be circulants given by

$$\left((\varphi_j, \varphi_k)\right)_{j,k=0}^{n-1} = n^{-1}\hat{\varkappa}_n, \qquad \left((S_{\mathbb{T}}\varphi_j, \varphi_k)\right)_{j,k=0}^{n-1} = n^{-1}\hat{\varrho}_n, \tag{2}$$

where

$$\varkappa(e^{2\pi i s}) := 2^{-\delta}\pi^{-3\delta-2} \sin^{2(\delta+1)}(\pi s)\, \sigma_{0,2\delta+1}^{+}(s),$$

$$\varrho(e^{2\pi i s}) := 2^{-\delta}\pi^{-3\delta-2} \sin^{2(\delta+1)}(\pi s)\, \sigma_{0,2\delta+1}^{-}(s), \qquad 0 < s < 1.$$

Thus the matrix of $L_n S_{\mathbb{T}}|_{\mathcal{S}_n^\delta}$ corresponding to the basis (φ_k) takes the form

$$L_n S_{\mathbb{T}}|_{\mathcal{S}_n^\delta} = \hat{\varkappa}_n^{-1}\hat{\varrho}_n = \hat{\beta}_n, \tag{3}$$

where

$$\beta(e^{2\pi i s}) := \begin{cases} 1 & \text{if } s = 0, \\ \sigma_{0,2\delta+1}^{-}(s)/\sigma_{0,2\delta+1}^{+}(s) & \text{if } 0 < s < 1. \end{cases}$$

Obviously,

$$(c\varphi_j, \varphi_k) = c(\tau_k)\,(\varphi_j, \varphi_k) + O\left(n^{-1}\omega(c, n^{-1})\right), \tag{4}$$

where $\omega(c, \cdot)$ stands for the modulus of continuity of the function $c \in \mathbf{C}(\Gamma)$. Taking into account that the matrix $\left((c\varphi_j, \varphi_k)\right)_{j,k=0}^{n-1}$ contains only $4\delta + 1$ diagonals different from zero, it follows from (2) and (4) that

$$\left\|\left((c\varphi_j, \varphi_k)\right)_{j,k=0}^{n-1} - \tilde{c}_n n^{-1}\hat{\varkappa}_n\right\| = O\left(n^{-1}\omega(c, n^{-1})\right). \tag{5}$$

Eqs. (2) and (5) imply

$$L_n c|_{\mathcal{S}_n^\delta} = n\hat{\varkappa}_n^{-1}\left((c\varphi_j, \varphi_k)\right)_{j,k=0}^{n-1}, \qquad \left\|L_n c|_{\mathcal{S}_n^\delta} - \hat{\varkappa}_n^{-1}\tilde{c}_n\hat{\varkappa}_n\right\| \to 0, \qquad n \to \infty. \tag{6}$$

Combining (1), (3) and (6) we get

$$A_n = \mathscr{R}_n^{-1}[\tilde{c}_n + \tilde{d}_n\hat{\beta}_n]\,\mathscr{R}_n + L_nK\big|_{\mathscr{S}_n^\delta} + C_n, \tag{7}$$

where $C_n \in \mathscr{L}(\mathscr{S}_n^\delta)$ and $\|C_n\| \to 0$ as $n \to \infty$.

10.9. Stability theorem. *Let* $c, d \in \mathbf{C}(\Gamma)$ *and* $K \in \mathscr{K}(\mathbf{L}^2(\Gamma))$. *Assume the operator* $A := cI + dS_\Gamma + K$ *to be invertible in* $\mathbf{L}^2(\Gamma)$. *Then the Galerkin method* 10.1(3) *is stable in* $\mathbf{L}^2(\Gamma)$ *if and only if*

$$c(\tau) + \lambda d(\tau) \neq 0 \quad \forall \tau \in \Gamma \quad \forall \lambda \in [-1, 1]. \tag{1}$$

Proof. As in 10.5, it suffices to consider the case $\Gamma = \mathbb{T}$. By Corollary 1.26 the sequence of the operators 10.8(7) is stable if and only if the sequence $\{\tilde{c}_n + \tilde{d}_n\hat{\beta}_n\}$ is so. Using Theorem 10.3 we conclude that the stability of the Galerkin method 10.1(3) is equivalent to the invertibility of the operators

$$B^\tau = [c(\tau) + d(\tau)\,\beta]\,I \quad \text{and} \quad B_1 = [c + \beta(1+0)\,d]\,P_{\mathbb{T}} + [c + \beta(1-0)\,d]Q_{\mathbb{T}}$$

for all $\tau \in \Gamma$, where $\beta(1 \pm 0) = \pm 1$.

Owing to 10.7(1) the invertibility of the operators $B^\tau \in \mathscr{L}(\mathbf{L}^2(\mathbb{T}))$ is equivalent to condition (1). Moreover, the invertibility of A implies the invertibility of the operator $B_1 \in \mathscr{L}(\mathbf{L}^2(\mathbb{T}))$ (cf. the second part of the proof to Theorem 10.5). ■

10.10. Stability theorem. *Let* $\delta = 0$, $m \in \mathbb{N}$ *and* $K \in \mathscr{K}(\mathbf{L}^2(\Gamma))$. *Suppose the coefficients* $c, d \in \mathbf{PC}(\Gamma)$ *are continuous on* $\Gamma \setminus \{\gamma(k/m) : k = 0, \ldots, m-1\}$, *and the operator* $A := cI + dS_\Gamma + K$ *is invertible in* $\mathbf{L}^2(\Gamma)$. *Consider Eqs.* 10.1(3) *only for* $n = lm$ $(l \in \mathbb{N})$. *Then the Galerkin method* 10.1(3) *is stable in* $\mathbf{L}^2(\Gamma)$ *if and only if condition* (i) *and* (ii) *of* 10.7 *is satisfied.*

Proof. For the case in consideration the points of discontinuity of the coefficients c and d are break points of the spline functions from $\mathscr{S}_n^0$. Consequently, the relations 10.8(6) and 10.8(7) can be proved in the same way as in the case of continuous coefficients. It remains to combine the proof to Theorem 10.9 with the proof to Corollary 10.7. ■

Now let $c, d \in \mathbf{PC}_0(\Gamma)$ be arbitrary piecewise continuous functions. Denote by $s_1, \ldots, s_N \in [0, 1]$ all points of discontinuity of the functions c and d, and choose any (not necessarily equidistant) partition $\Delta := \{0 = x_0 < x_1 < \cdots < x_n = 1\}$ containing all points s_j $(j = 1, \ldots, N)$. Let $\mathbf{P}\mathscr{S}^\delta(\Delta)$ denote the space of all 1-periodic splines which are $\delta - 1$ times continuously differentiable on $[0, 1] \setminus \{s_1, \ldots, s_N\}$ and whose restriction to $[x_k, x_{k+1}]$ $(k = 0, \ldots, n-1)$ is a polynomial of degree less than or equal to δ; $\mathbf{P}\mathscr{S}^0(\Delta) = \mathscr{S}^0(\Delta)$ is defined as the corresponding space of piecewise constant functions. Finally, let $h := \max_k (x_{k+1} - x_k)$.

We consider the following Galerkin method for Eq. 10.1(1): find an approximate solution $u_\Delta \in \mathbf{P}\mathscr{S}^\delta(\Delta)$ satisfying

$$(Au_\Delta, v) = (f, v) \qquad \forall v \in \mathbf{P}\mathscr{S}^\delta(\Delta). \tag{1}$$

Let P_Δ denote the orthogonal projection from $\mathbf{L}^2(\Gamma)$ onto $\mathbf{P}\mathscr{S}^\delta(\Delta)$. Then Eq. (1) is equivalent to the equation

$$P_\Delta A P_\Delta u_\Delta = P_\Delta f.$$

The method (1) is said to be *stable* if there exists a constant $h_0 > 0$ such that the operators $P_\Delta A P_\Delta \in \mathcal{L}(\mathbf{P}\mathcal{S}^\delta(\Delta))$ are invertible for all $h < h_0$ and $\sup_{h<h_0} \|(P_\Delta A P_\Delta)^{-1}\| < \infty$.

10.11. Stability theorem. *Let $c, d \in \mathbf{PC}(\Gamma)$ and $K \in \mathcal{K}(\mathbf{L}^2(\Gamma))$. Assume the operator $A := cI + dS_\Gamma + K \in \mathcal{L}(\mathbf{L}^2(\Gamma))$ to be invertible. If conditions* 10.7(i) *and* (ii) *are satisfied, then the Galerkin method* 10.10(1) *is stable in* $\mathbf{L}^2(\Gamma)$.

Proof. Conditions 10.7(i) and (ii) imply the existence of an invertible function $\theta \in \mathbf{PC}(\Gamma) \cap \mathbf{C}(\Gamma \setminus \{\gamma(s_1), \ldots, \gamma(s_N)\})$ such that

$$A = \theta(D + T),$$

where $T \in \mathcal{K}(\mathbf{L}^2(\Gamma))$ and $\operatorname{Re} D > 0$ (see Section 6.22). Moreover,

$$\|(I - P_\Delta)\,\theta P_\Delta\| \to 0, \qquad \|P_\Delta \theta (I - P_\Delta)\| \to 0 \qquad (h \to 0)$$

(see Section 2.14). Consequently

$$P_\Delta A P_\Delta = (P_\Delta \theta P_\Delta)\, P_\Delta (D + T)\, P_\Delta + B_\Delta, \tag{1}$$

$$(P_\Delta \theta^{-1} P_\Delta)\, P_\Delta \theta P_\Delta - C_\Delta = P_\Delta \theta P_\Delta (P_\Delta \theta^{-1} P_\Delta) - D_\Delta = P_\Delta, \tag{2}$$

where $\|B_\Delta\| \to 0$, $\|C_\Delta\| \to 0$, and $\|D_\Delta\| \to 0$ as $h \to 0$. Thus the assertion of the theorem is an immediate consequence of Eqs. (1), (2) and the stability of the sequence $\{P_\Delta(D + T) \times P_\Delta\}$ (see Section 1.33). ■

Remark. By Theorems 10.9 and 10.10, condition 10.7 (i) + (ii) (i.e. the strong ellipticity of the operator A) is even necessary for the Galerkin method (1) to be stable in $\mathbf{L}^2(\Gamma)$ (at least in the case of equidistant partitions Δ).

Quadrature methods in the case of the unit circle

10.12. Description of the methods. In the present section we shall describe five different quadrature methods for the singular integral equation 10.1(1) in the case of the unit circle $\Gamma = \mathbb{T}$.

To this end fix an arbitrary number $\varepsilon \in [0, 1)$ and set again

$$t_k := \exp(\mathrm{i}2\pi k/n), \qquad v_k := \exp\left(\mathrm{i}2\pi(k + (1/2))/n\right),$$
$$\tau_k := \exp\left(\mathrm{i}2\pi(k + \varepsilon)/n\right), \qquad k = 0, 1, \ldots, n - 1,\ n \in \mathbb{N}.$$

In order to evaluate the regular integral Ku in equation 10.1(1) we use the quadrature rule

$$\int_{\mathbb{T}} f(\tau)\,\mathrm{d}\tau = \mathrm{i} \int_{\mathbb{T}} f(\tau)\,\tau\,|\mathrm{d}\tau| \approx \sum_{j=0}^{n-1} f(t_j)\,\frac{2\pi\mathrm{i}}{n}\,t_j. \tag{1}$$

It is well known that the right-hand side of formula (1) approximates the integral on the left-hand side with an arbitrarily high rate of convergence if f is a smooth function. Hence

$$(cu + Ku)(\tau_k) \approx c(\tau_k)\,u(\tau_k) + \frac{2\pi\mathrm{i}}{n} \sum_{j=0}^{n-1} k(\tau_k, t_j)\,t_j u(t_j). \tag{2}$$

To evaluate the Cauchy singular integral Su we rewrite it in the form

$$(Su)(t) = u(t) + \frac{1}{\pi i} \int_{\mathbb{T}} \frac{u(\tau) - u(t)}{\tau - t} \, d\tau. \tag{3}$$

Now besides (1) we shall make use of the quadrature rules

$$\int_{\mathbb{T}} f(\tau) \, d\tau \approx \sum_{\substack{j=0 \\ j \equiv 0 \bmod 2}}^{n-1} f(t_j) \frac{4\pi i}{n} t_j \tag{4_0}$$

or

$$\int_{\mathbb{T}} f(\tau) \, d\tau \approx \sum_{\substack{j=0 \\ j \equiv 1 \bmod 2}}^{n-1} f(t_j) \frac{4\pi i}{n} t_j, \tag{4_1}$$

where $n \in \mathbb{N}$ is assumed to be even. Applying (1) or (4), respectively, to equation (3) we obtain

$$(Su)(\tau_k) \approx \begin{cases} u(\tau_k) + \dfrac{2}{n} \displaystyle\sum_{j=0}^{n-1} \frac{u(t_j) - u(\tau_k)}{t_j - \tau_k} t_j & \text{if } \varepsilon \neq 0, \\ u(t_k) + \dfrac{2}{n} \left[u'(t_k) \, t_k + \displaystyle\sum_{\substack{j=0 \\ j \neq k}}^{n-1} \frac{u(t_j) - u(t_k)}{t_j - t_k} t_j \right] & \text{if } \varepsilon = 0 \end{cases} \tag{5}$$

or

$$(Su)(t_k) \approx u(t_k) + \frac{4}{n} \sum_{\substack{j=0 \\ j \equiv k+1 \bmod 2}}^{n-1} \frac{u(t_j) - u(t_k)}{t_j - t_k} t_j, \tag{6}$$

respectively. Now by means of the formula

$$\sum_{j=0}^{n-1} z^j = \begin{cases} (z^n - 1)/(z - 1) & \text{if } z \in \mathbb{C}, z \neq 1, \\ n & \text{if } z = 1 \end{cases}$$

it can be easily checked that

$$\begin{aligned} \frac{2}{n} \sum_{j=0}^{n-1} \frac{t_j}{t_j - \tau_k} &= \frac{2}{n} \sum_{j=0}^{n-1} \left[1 - \exp\left(-i \frac{2\pi(j - \varepsilon)}{n} \right) \right]^{-1} \\ &= \frac{2}{n} \frac{1}{1 - \exp(i2\pi\varepsilon)} \sum_{j=0}^{n-1} \frac{1 - \exp\left(-i \dfrac{2\pi(j - \varepsilon) \, n}{n} \right)}{1 - \exp\left(-i \dfrac{2\pi(j - \varepsilon)}{n} \right)} \\ &= 1 + i \cot(\pi\varepsilon). \end{aligned} \tag{7}$$

Hence

$$\begin{aligned} \frac{4}{n} \sum_{\substack{j=0 \\ j \equiv k+1 \bmod 2}}^{n-1} \frac{t_j}{t_j - t_k} &= \frac{2}{n/2} \sum_{m=0}^{(n/2)-1} \left[1 - \exp\left(-i \frac{2\pi(m - 1/2)}{n/2} \right) \right]^{-1} \\ &= 1 + i \cot \frac{\pi}{2} = 1. \end{aligned} \tag{8}$$

Moreover,

$$\sum_{\substack{j=0\\ j\neq k}}^{n-1} \frac{t_j}{t_j - t_k} = \sum_{j=1}^{n-1} [1 - \exp(i2\pi j/n)]^{-1} = \frac{\omega'(1)}{\omega(1)},$$

where $\omega(t) := \prod_{k=1}^{n-1} (t - t_k)$. Since $\omega(t)\,(t - 1) = t^n - 1$, it follows that

$$\sum_{\substack{j=0\\ j\neq k}}^{n-1} \frac{t_j}{t_j - t_k} = (n - 1)/2. \tag{9}$$

Substituting formulas (7), (8), (9) into (5) and (6) we arrive at

$$(Su)\,(\tau_k) \approx \begin{cases} -\mathrm{i} \cot(\pi\varepsilon)\, u(\tau_k) + \dfrac{2}{n} \displaystyle\sum_{j=0}^{n-1} \frac{t_j u(t_j)}{t_j - \tau_k} & \text{if } \varepsilon \neq 0, \\[2ex] \dfrac{1}{n}\, u(t_k) + \dfrac{2}{n}\, u'(t_k)\, t_k + \dfrac{2}{n} \displaystyle\sum_{\substack{j=0\\ j\neq k}}^{n-1} \frac{t_j u(t_j)}{t_j - t_k} & \text{if } \varepsilon = 0 \end{cases} \tag{10}$$

or

$$(Su)\,(t_k) \approx \frac{4}{n} \sum_{\substack{j=0\\ j\equiv k+1 \bmod 2}}^{n-1} \frac{t_j u(t_j)}{t_j - t_k}, \tag{11}$$

respectively.

In order to find terms independent of the values $u'(t_k)$ and $u(\tau_k)$ ($\varepsilon \neq 0$) on the right-hand sides of (10) we shall distinguish four cases. First choose $\varepsilon \neq 0$ and replace $u(\tau_k)$ by $u(t_k)$. Since this substitution produces an error of order n^{-1}, the convergence rate of the corresponding quadrature method is also expected to be at most of this order. For the second case, choose $\varepsilon = 1/2$ and replace $u(\tau_k)$ by $1/2\big(u(t_{k+1}) + u(t_k)\big)$. Now an error of order n^{-2} arises. (Obviously, via better interpolation of $u(\tau_k)$, higher orders of n^{-1} can also be attained.) In the third case, choose $\varepsilon = 0$ and neglect the term $(1/n)u(t_k) + (2/n)\,u'(t_k)\,t_k$. This leads to an error of order n^{-1}. At last, the choice $\varepsilon = 0$ and the substitu- of $u'(t_k)$ by $\big(u(t_{k+1}) - u(t_{k-1})\big)/(t_{k+1} - t_{k-1})$ produce an error of order n^{-3}. (In this case any higher order can also be achieved by using a better difference formula.)

Finally, equations (2), (10) and (11) together with the substitutions described above imply expressions for $f(\tau_k) = (Au)\,(\tau_k)$ depending only on the values $u(t_k)$, $k = 0, \ldots, n - 1$. Substitute $u(t_k)$ by ξ_k to obtain the following quadrature methods: determine approximate values ξ_k for $u(t_k)$ by solving one of the systems

$$\begin{aligned} f(\tau_k) &= \big(c(\tau_k) - \mathrm{i} \cot(\pi\varepsilon)\, d(\tau_k)\big)\, \xi_k \\ &\quad + d(\tau_k)\, \frac{2}{n} \sum_{j=0}^{n-1} \frac{t_j}{t_j - \tau_k}\, \xi_j + \frac{2\pi\mathrm{i}}{n} \sum_{j=0}^{n-1} k(\tau_k, t_j)\, t_j \xi_j, \end{aligned} \tag{12}$$

$$\begin{aligned} f(v_k) &= c(v_k)\, \frac{1}{2}\, (\xi_{k+1} + \xi_k) \\ &\quad + d(v_k)\, \frac{2}{n} \sum_{j=0}^{n-1} \frac{t_j}{t_j - v_k}\, \xi_j + \frac{2\pi\mathrm{i}}{n} \sum_{j=0}^{n-1} k(v_k, t_j)\, t_j \xi_j, \end{aligned} \tag{13}$$

$$f(t_k) = c(t_k)\,\xi_k + d(t_k)\,\frac{2}{n}\sum_{\substack{j=0\\ j\neq k}}^{n-1}\frac{t_j}{t_j - t_k}\,\xi_j + \frac{2\pi\mathrm{i}}{n}\sum_{j=0}^{n-1} k(t_k, t_j)\,t_j\xi_j, \tag{14}$$

$$f(t_k) = \left(c(t_k) + \frac{1}{n}\,d(t_k)\right)\xi_k + d(t_k)\,\frac{2}{n}\,t_k\,\frac{\xi_{k+1} - \xi_{k-1}}{t_{k+1} - t_{k-1}}$$
$$+ d(t_k)\,\frac{2}{n}\sum_{\substack{j=0\\ j\neq k}}^{n-1}\frac{t_j}{t_j - t_k}\,\xi_j + \frac{2\pi\mathrm{i}}{n}\sum_{j=0}^{n-1} k(t_k, t_j)\,t_j\xi_j, \tag{15}$$

$$f(t_k) = c(t_k)\,\xi_k + d(t_k)\,\frac{4}{n}\sum_{\substack{j=0\\ j\equiv k+1 \bmod 2}}^{n-1}\frac{t_j}{t_j - t_k}\,\xi_j + \frac{2\pi\mathrm{i}}{n}\sum_{j=0}^{n-1} k(t_k, t_j)\,t_j\xi_j, \tag{16}$$

$$k = 0, \ldots, n-1; \qquad \xi_{-1} := \xi_{n-1},\ \xi_n := \xi_0.$$

If there exists a unique solution $(\xi_k^{(n)})_{k=0}^{n-1}$, then by trigonometric interpolation one finds the approximate solution

$$u_n := \sum_{k=0}^{n-1} \xi_k^{(n)}\psi_k^{(n)} \tag{17}$$

to equation 10.1(1), where

$$\psi_k^{(n)}(t) := \frac{1}{n}\sum_{j=-[n/2]}^{[(n-1)/2]} (t_k^{(n)})^{-j}\,t^j. \tag{18}$$

Note that, for a Riemann integrable function u, the $\mathbf{L}^2$-convergence $\|u - u_n\|_{\mathbf{L}^2} \to 0$ as $n \to \infty$ implies the discrete convergence

$$n^{-1/2}\left(\sum_{k=0}^{n-1} |\xi_k^{(n)} - u(t_k)|^2\right)^{1/2} \to 0 \quad \text{as} \quad n \to \infty.$$

This is an immediate consequence of the fact that $\left\|\sum_k u(t_k)\,\psi_k^{(n)} - u\right\|_{\mathbf{L}^2} \to 0$ (see ZYGMUND [1]). Moreover, if $u \in \mathbf{H}^s$, $s > 1/2$, and $\|u - u_n\|_{\mathbf{H}^s} \to 0$ as $n \to \infty$, then

$$\sup_{k=0,\ldots,n-1} |\xi_k^{(n)} - u(t_k)| \to 0 \quad \text{as} \quad n \to \infty,$$

because of the embedding $\mathbf{H}^s \subset \mathbf{C}$.

Finally, let us remark that in the case of the method of discrete whirls (i.e. Eq. (13) in the case $c \equiv 0$) no value $u(\tau_k)$ has been substituted by $1/2\big(u(t_{k+1}) + u(t_k)\big)$. Thus, for this method as well as in similar special cases, an arbitrarily high order of convergence can be achieved (cf. also Sect. 10.21).

10.13. Stability theorem for the quadrature methods. *Let* $c, d \in \mathbf{PC}(\mathbb{T})$, $k \in \mathbf{C}(\mathbb{T}\times\mathbb{T})$, *and* $\varepsilon \in (0, 1)$. *Assume the operator* $A := cI + dS_{\mathbb{T}} + K$ *is invertible in* $\mathbf{L}^2(\mathbb{T})$.

(a) *The quadrature method* 10.12(12) *is stable in* $\mathbf{L}^2(\mathbb{T})$ *if and only if the condition*

$$\mu\,\frac{a}{b}\,(\tau + 0) + (1 - \mu)\,\frac{a}{b}\,(\tau - 0) \notin \mathrm{e}^{-\mathrm{i}\pi\varepsilon}(-\infty, 0] \quad \forall \tau \in \mathbb{T}, \mu \in [0, 1], \tag{1}$$

is fulfilled.

(b) *The method* 10.12(13) *or the methods* 10.12(14), 10.12(15), *respectively, are stable in* $\mathbf{L}^2(\mathbb{T})$ *if and only if condition* (1) *is satisfied with* ε *replaced by* 1 *or* 0, *respectively.*

(c) *The method* 10.12(16) *is stable in* $\mathbf{L}^2(\mathbb{T})$.

Proof. Introduce the orthogonal projection P_n from $\mathbf{L}^2(\mathbb{T})$ onto $\mathcal{S}_n^\infty := \operatorname{lin}(\psi_k^{(n)})_{k=0}^{n-1}$ and the interpolation projection $M_n = M_n^\varepsilon$ defined by

$$M_n x := \sum_{k=0}^{n-1} x(\tau_k^{(n)})\, \tilde{\psi}_k^{(n)}, \qquad \tilde{\psi}_k^{(n)}(t) := n^{-1} \sum_{j=-[n/2]}^{[(n-1)/2]} (\tau_k^{(n)})^{-j}\, t^j .$$

Then the system of the quadrature method is equivalent to the operator equation $A_n u_n = M_n f$. Herein $A_n \in \mathcal{L}(\mathcal{S}_n^\infty)$ denotes the operator whose matrix representation with respect to the bases $(\psi_k^{(n)})$, $(\tilde{\psi}_k^{(n)})$ is equal to the matrix of the system 10.12(12), 10.12(13), 10.12(14), 10.12(15) or 10.12(16), respectively. In the same manner as in Section 10.4, we shall identify the operator $A_n \in \mathcal{L}(\mathcal{S}_n^\infty)$ with the corresponding matrix.

Our first concern is to show that A_n takes the form

$$A_n = \tilde{c}_n \hat{\alpha}_n + \tilde{d}_n \hat{\beta}_n + C_n + T_n\,, \tag{2}$$

where $\|C_n\| \to 0$ as $n \to \infty$,

$$T_n := \begin{cases} \left(k(\tau_k, t_j)\, \dfrac{2\pi \mathrm{i}}{n}\, t_j\right)_{k,j=0}^{n-1} & \text{for 10.12(12)},\\[2ex] \left(k(v_k, t_j)\, \dfrac{2\pi \mathrm{i}}{n}\, t_j\right)_{k,j=0}^{n-1} & \text{for 10.12(13)},\\[2ex] \left(k(t_k, t_j)\, \dfrac{2\pi \mathrm{i}}{n}\, t_j\right)_{k,j=0}^{n-1} & \text{for 10.12(14), 10.12(15) and 10.12(16)}, \end{cases}$$

$$\alpha(t) := \begin{cases} 1 & \text{in the case 10.12(12), 10.12(14), 10.12(15) or 10.12(16)},\\ \dfrac{1}{2}\,(1+t) & \text{in the case 10.12(13)}, \end{cases}$$

$$\beta(\exp(\mathrm{i}2\pi\theta)) := \begin{cases} -\mathrm{i}\cot(\pi\varepsilon) + \dfrac{2\exp(\mathrm{i}2\pi\varepsilon\theta)}{1-\exp(\mathrm{i}2\pi\varepsilon)} & \text{for 10.12(12)},\\ \exp(\mathrm{i}\pi\theta) & \text{for 10.12(13)},\\ 1-2\theta & \text{for 10.12(14)},\\ 1-2\theta+\pi^{-1}\sin(2\pi\theta) & \text{for 10.12(15)}, \end{cases}$$

$(0 \leqq \theta < 1)$

and, in the case 10.12(16),

$$\beta(\exp(\mathrm{i}2\pi\theta)) := \begin{cases} +1 & \text{if } 0 \leqq \theta < 1/2,\\ -1 & \text{if } 1/2 \leqq \theta < 1. \end{cases}$$

The matrices $\tilde{c}_n$, $\tilde{d}_n$ and $\hat{\alpha}_n$, $\hat{\beta}_n$ were defined in Section 10.2.

Indeed, let us first consider the method 10.12(12). It is evident that for $A = cI + K$ the matrix of the operator A_n is equal to $\tilde{c}_n + T_n$. Moreover, for $A = S$, the matrix of

A_n assumes the form

$$\left(-\mathrm{i}\cot(\pi\varepsilon)\,\delta_{kj} + \frac{2}{n}\,\frac{t_j}{t_j - \tau_k}\right)_{k,j=0}^{n-1}$$

$$= \left(-\mathrm{i}\cot(\pi\varepsilon)\,\delta_{kj} + \frac{2}{n}\,\frac{1}{1 - \tau_{k-j}}\right)_{k,j=0}^{n-1}$$

$$= \left(-\mathrm{i}\cot(\pi\varepsilon)\,\delta_{kj} + \frac{2}{n}\,\frac{1}{1 - \exp(\mathrm{i}2\pi\varepsilon)}\,\frac{1 - \tau_{k-j}^n}{1 - \tau_{k-j}}\right)_{k,j=0}^{n-1}$$

$$= \left(-\mathrm{i}\cot(\pi\varepsilon)\,\frac{1}{n}\sum_{l=0}^{n-1}\exp\big(\mathrm{i}2\pi(k-j)\,l/n\big)\right)_{k,j=0}^{n-1}$$

$$+ \left(\frac{2}{n}\sum_{l=0}^{n-1}\frac{\exp\big(\mathrm{i}2\pi(k+\varepsilon-j)\,l/n\big)}{1 - \exp(\mathrm{i}2\pi\varepsilon)}\right)_{k,j=0}^{n-1}$$

$$= \left(\frac{1}{n}\sum_{l=0}^{n-1}\beta(t_l)\exp\big(\mathrm{i}2\pi(k-j)\,l/n\big)\right)_{k,j=0}^{n-1} = \hat{\beta}_n,$$

where

$$\beta\big(\exp(\mathrm{i}2\pi\theta)\big) := -\mathrm{i}\cot(\pi\varepsilon) + \frac{2\exp(\mathrm{i}2\pi\varepsilon\theta)}{1 - \exp(\mathrm{i}2\pi\varepsilon)}. \tag{3}$$

Hence, formula (2) is proved in the case of the method 10.12(12).

In a similar way formula (2) can be checked for the other methods considered here. Obviously, for $A = cI + K$ the matrix of the operator A_n takes the form $\tilde{c}_n\hat{\alpha}_n + T_n$, where $\alpha(t) = (1+t)/2$ in the case of the method 10.12(13) and $\alpha(t) \equiv 1$ in the case of the methods 10.12(14), 10.12(15) and 10.12(16). For the operator $A = S$, the matrix of A_n is equal to $\hat{\beta}_n$, where

$$\beta\big(\exp(\mathrm{i}2\pi\theta)\big) := \exp(\mathrm{i}\pi\theta)$$

in the case of the method 10.12(13) (cf. (3)).

We now consider the matrix of the operator S_n for the method 10.12(14). Calculating the entries of the circulant $\hat{\beta}_n$, where

$$\beta\big(\exp(\mathrm{i}2\pi\theta)\big) := 1 - 2\theta \qquad (0 \leqq \theta < 1), \tag{4}$$

we obtain, for $k = j$,

$$\frac{1}{n}\sum_{m=0}^{n-1}\beta(t_m) = \frac{1}{n}\sum_{m=0}^{n-1}\left(1 - \frac{2m}{n}\right) = \frac{1}{n}$$

and, for $k \neq j$,

$$\frac{1}{n}\sum_{m=0}^{n-1}\beta(t_m)\,t_{k-j}^m = -\frac{2}{n^2}\sum_{m=0}^{n-1} m t_{k-j}^m = -\frac{2}{n^2}\sum_{m=0}^{n-1}(m+1)\,t_{k-j}^m$$

$$= -\frac{2}{n^2}\,\frac{\mathrm{d}}{\mathrm{d}t}\,\frac{1 - t^{n+1}}{1 - t}\bigg|_{t = t_{k-j}} = \frac{2}{n}\,\frac{1}{1 - t_{k-j}} = \frac{2}{n}\,\frac{t_j}{t_j - t_k}.$$

Therefore, in the case of the method 10.12(14), the matrix of S_n can be rewritten in the form

$$\left((1-\delta_{kj})\frac{2}{n}\frac{t_j}{t_j-t_k}\right)_{k,j=0}^{n-1} = \hat{\beta}_n - \frac{1}{n}(\delta_{kj})_{k,j=0}^{n-1}. \tag{5}$$

Hence, formula (2) is true with $C_n = -(1/n)\,\tilde{d}_n$.

For the method 10.12(15), we put $\beta^0(t) := 1 - 2\theta + \gamma(t)$, $t = e^{2\pi i\theta}$ $(0 \leqq \theta < 1)$, where $\gamma(t) := \pi^{-1}\sin(2\pi\theta)$. Obviously,

$$\hat{\gamma}_n = \frac{1}{2\pi i}(\delta_{k+1,j} - \delta_{k-1,j})_{k,j=0}^{n-1}$$

and

$$\frac{2}{n}\left(\frac{t_k}{t_{k+1}-t_{k-1}}(\delta_{k+1,j} - \delta_{k-1,j})\right)_{k,j=0}^{n-1} = \hat{\gamma}_n + C_n^0, \tag{6}$$

where

$$C_n^0 := \left(\frac{2\pi/n}{\sin(2\pi/n)} - 1\right)\hat{\gamma}_n, \qquad \|C_n^0\| \to 0 \quad \text{as} \quad n \to \infty.$$

Now it follows from (5) and (6) that the matrix of the system 10.12(15) takes the form

$$\tilde{c}_n + \tilde{d}_n\hat{\beta}_n^0 + \tilde{d}_n C_n^0 + T_n,$$

and formula (2) is valid again.

It remains to consider the matrix $(\sigma_{kj})_{k,j=0}^{n-1}$ of S_n in the case of the method 10.12(16). It is evident, that $\sigma_{kj} = 0$ if $k \equiv j \bmod 2$ and $\sigma_{kj} = 2(\hat{\beta}_n)_{kj}$ if $k \equiv j + 1 \bmod 2$, where β is the function defined by (4). Consequently,

$$(\sigma_{kj})_{k,j=0}^{n-1} = \hat{\beta}_n - W_n\hat{\beta}_n W_n \tag{7}$$

with the diagonal matrix $W_n := \left((-1)^k\,\delta_{kj}\right)_{k,j=0}^{n-1}$. It is easily seen that, for arbitrary $\beta \in \mathbf{PC}$,

$$W_n\hat{\beta}_n W_n = (\beta^-)_n^\wedge, \qquad \beta^-(t) = \beta(-t). \tag{8}$$

Formulas (7) and (8) immediately yield

$$(\sigma_{kj})_{k,j=0}^{n-1} = \hat{\gamma}_n, \qquad \gamma := \beta - \beta^-,$$

$$\gamma\left(\exp(i2\pi\theta)\right) = \begin{cases} +1 & \text{if} \quad 0 \leqq \theta < 1/2 \\ -1 & \text{if} \quad 1/2 \leqq \theta < 1. \end{cases}$$

Hence, formula (2) is proved for each of the methods under consideration.

We now proceed with the proof of the Stability Theorem 10.13. By Corollary 1.26, the sequence of the operators (2) is stable if and only if the sequence $B_n := \tilde{c}_n\hat{\alpha}_n + \tilde{d}_n\hat{\beta}_n + C_n$ is so (cf. Secs. 3.6 and 8.13). As a consequence of Theorem 10.3 and the assumptions of Theorem 10.13, it can be easily seen that the sequence $\{B_n\}$ is stable if and only if the operators

$$B^\tau := P_{\mathbb{T}}[c(\tau+0)\,\tilde{\alpha} + d(\tau+0)\,\tilde{\beta}] + Q_{\mathbb{T}}[c(\tau-0)\,\tilde{\alpha} + d(\tau-0)\,\tilde{\beta}]$$

are invertible in $\mathbf{L}^2(\mathbb{T})$ for all $\tau \in \mathbb{T}$. On the other hand, it is evident that the operator B^τ is invertible if and only if

$$c(\tau - 0)\,\tilde{\alpha}(t) + d(\tau - 0)\,\tilde{\beta}(t) \neq 0 \quad \forall t \in \mathbb{T} \tag{9}$$

and if the operator

$$C^\tau := P_{\mathbb{T}} \frac{ab^{-1}(\tau + 0)\,(\tilde{\alpha} + \tilde{\beta}) + (\tilde{\alpha} - \tilde{\beta})}{(\tilde{\alpha} + \tilde{\beta}) + a^{-1}b(\tau - 0)\,(\tilde{\alpha} - \tilde{\beta})} + Q_{\mathbb{T}}$$

$$= (P_{\mathbb{T}}a(\tau - 0)/b(\tau + 0) + Q_{\mathbb{T}})\,B^\tau[c(\tau - 0)\,\tilde{\alpha} + d(\tau - 0)\,\tilde{\beta}]^{-1}$$

is invertible in $\mathbf{L}^2(\mathbb{T})$. Substituting the values of α and β we get

$$\frac{\tilde{\alpha} - \tilde{\beta}}{\tilde{\alpha} + \tilde{\beta}}\,(\mathrm{e}^{\mathrm{i}2\pi\theta}) = \begin{cases} \mathrm{e}^{-\mathrm{i}\pi\varepsilon} \dfrac{\sin \pi\varepsilon(1 - \theta)}{\sin \pi\varepsilon\theta} & \text{for 10.12(12)}, \\ 1 - 2/(1 - \cos \pi\theta) & \text{for 10.12(13)}, \\ (1 - \theta)/\theta & \text{for 10.12(14)}, \\ \dfrac{1 - \theta + (2\pi)^{-1} \sin 2\pi\theta}{\theta - (2\pi)^{-1} \sin 2\pi\theta} & \text{for 10.12(15)}. \end{cases}$$

Thus, the range of the function $(\tilde{\beta} - \tilde{\alpha})/(\tilde{\alpha} + \tilde{\beta})$ is the half line $\mathbb{R}_\varepsilon^- := \mathrm{e}^{-\mathrm{i}\pi\varepsilon}(-\infty, 0]$, where $\varepsilon = 1$ for 10.12(13) and $\varepsilon = 0$ for 10.12(14) or 10.12(15).

By virtue of the invertibility theorem for singular operators (see Sections 6.13, 6.14, 1°, and 6.1(ix)) the operator C^τ is invertible in $\mathbf{L}^2(\mathbb{T})$ if and only if the origin does not belong to the domain enclosed by the arc

$$\left\{\frac{ab^{-1}(\tau + 0) - \nu}{1 - \nu a^{-1}b(\tau - 0)} : \nu \in \mathbb{R}_\varepsilon^-\right\}$$

and the segment

$$\{\mu ab^{-1}(\tau + 0) + (1 - \mu)\,ab^{-1}(\tau - 0) : \mu \in [0, 1]\}.$$

It is evident that the afore-mentioned condition is equivalent to requiring that

$$\mu \frac{ab^{-1}(\tau + 0) - \nu}{1 - \nu a^{-1}b(\tau - 0)} + (1 - \mu)\,ab^{-1}(\tau - 0) \neq 0 \quad \forall \mu \in [0, 1] \quad \forall \nu \in \mathbb{R}_\varepsilon^-.$$

Obviously, the latter condition holds if and only if condition (1) is fulfilled. Moreover, (1) implies (9), and, consequently, the invertibility of the operator B^τ. This completes the proof to the assertions (a) and (b).

In the case of method 10.12(16), the operator C^τ is invertible in $\mathbf{L}^2(\mathbb{T})$ if and only if

$$\mu \frac{a}{b}\,(\tau - 0) + (1 - \mu)\,\frac{a}{b}\,(\tau + 0) \neq 0 \qquad \forall \mu \in [0, 1]. \tag{10}$$

Since (10) is a consequence of the invertibility of the operator $A_0 = cI + dS_\Gamma$, the assertion (c) of Theorem 10.13 is proved, too. ■

10.14. Remarks. 1°. In case $K = 0$, 10.12.(16) is the system of the *trigonometric collocation method*

$$(Ax_n)(t_k) = f(t_k), \qquad k = 0, \ldots, n-1,$$

where $t_k = \exp(i2\pi k/n)$, and the unknown solution x_n is a polynomial of the form

$$x_n(t) = \sum_{k=0}^{n-1} \xi_k \psi_k(t), \qquad \psi_k(t) := \frac{1}{n} \sum_{j=-[n/2]}^{[(n-1)/2]} t_k^{-j} t^j .$$

Obviously, $\psi_k(t_l) = \delta_{kl}$ $(k, l = 0, \ldots, n-1)$.

Thus Theorem 10.13(c) is equivalent to Theorem 7.24(b), if $K = T = 0$.

2°. Theorem 10.13 remains valid for kernels k of the form $k(t, \tau) = g(t)\, h(t, \tau)\, f(\tau)$, where $g, f \in \mathbf{PC}(\Gamma)$, and the function h satisfies a Hölder condition with respect to both variables.

Indeed, as can be easily seen (cf. Chapter 7), in the afore-mentioned case the approximate operator K_n of K takes the form

$$K_n = M_n^\varepsilon g P_n M_n^\varepsilon H P_n M_n^0 f P_n + N_n = M_n^\varepsilon g H M_n^0 f P_n + N_n, \tag{1}$$

where H denotes the integral operator with kernel h and $\|N_n\| \to 0$ as $n \to \infty$ (with M_n^ε and M_n^0 as in Section 10.13). The perturbations (1) of $(A_0)_n$ can be handled by the methods developed in Chapter 7.

10.15. Theorem. *Let $c, d \in \mathbf{PC}(\mathbb{T})$, and assume the kernel k is sufficiently smooth. Let $A_n \in \mathcal{L}(\mathcal{S}_n^\infty)$ denote the approximate operator of the quadrature method defined in Section 10.13. Then $A_n P_n \to A \in \mathcal{L}(\mathbf{L}^2(\mathbb{T}))$.*

Proof. Consider the operator $\tilde{c}_n \in \mathcal{L}(\mathcal{S}_n^\infty)$, i.e. the operator, whose matrix representation with respect to the basis $(\tilde{\psi}_k^{(n)})$ is the matrix $\left(c(\tau_k^{(n)})\, \delta_{kj}\right)_{k,j=0}^{n-1}$. Obviously, $\tilde{c}_n$ can be rewritten as $\tilde{c}_n = M_n c I \mid \mathcal{S}_n^\infty$. Therefore, $\tilde{c}_n P_n = M_n c P_n \to cI$ (cf. Chapter 7).

For $\varrho \in \mathbf{PC}(\mathbb{T})$, let $\hat{\varrho}_n \in \mathcal{L}(\mathcal{S}_n^\infty)$ denote the operator the matrix representation of which is the circulant (cf. Section 10.2)

$$\hat{\varrho}_n := \left(\frac{1}{n} \sum_{m=0}^{n-1} \varrho(t_m)\, t_{k-j}^m\right)_{k,j=0}^{n-1} .$$

It is easy to check that the matrix representation of $\hat{\varrho}_n \in \mathcal{L}(\mathcal{S}_n^\infty)$ corresponding to the basis (t^j) $(-[n/2] \leqq j \leqq [(n-1)/2])$ is the matrix

$$\left(\varrho(\exp(i2\pi j/n))\right) \exp(-i2\pi j\varepsilon/n)\, \delta_{kj})_{k,j=-[n/2]}^{[(n-1)/2]} .$$

Consequently, $\hat{\varrho}_n P_n \to \varrho(1+0)\, P_\mathbb{T} + \varrho(1-0)\, Q_\mathbb{T} \in \mathcal{L}(\mathbf{L}^2(\mathbb{T}))$, and

$$(A_0)_n P_n = (cI + dS_\mathbb{T})_n P_n = \tilde{c}_n \hat{\alpha}_n P_n + \tilde{d}_n \hat{\beta}_n P_n \to cI + dS_\mathbb{T},$$

because $\alpha(1+0) = \alpha(1-0) = 1$, $\beta(1 \pm 0) = \pm 1$.

Finally, let $K_n \in \mathcal{L}(\mathcal{S}_n^\infty)$ denote the operator, whose matrix representation with respect to the bases $(\psi_k^{(n)})$, $(\tilde{\psi}_k^{(n)})$ is the matrix T_n introduced in Section 10.13. Then the sequence $\{K_n P_n\}$ is uniformly bounded in $\mathcal{L}(\mathbf{L}^2(\mathbb{T}))$, and $K_n P_n u \to Ku$ in $\mathbf{L}^2(\mathbb{T})$ for smooth functions u (cf. Section 10.19). Thus by the Banach-Steinhaus theorem, $K_n P_n \to K$. Taking into account 10.13(2), we find

$$A_n P_n = (\tilde{c}_n \hat{\alpha}_n + \tilde{d}_n \hat{\beta}_n + C_n + K_n) P_n \to A . \quad \blacksquare$$

Remark. Theorem 10.15 remains valid for the kernels k mentioned in Remark 10.14, 3°. To see this notice that $\|K_n - M_n gHM_n fP_n\| \to 0$, $M_n \to I \in \mathcal{L}(\mathbf{PC}(\mathbb{T}), \mathbf{L}^2(\mathbb{T}))$, $gH \in \mathcal{K}(\mathbf{L}^2(\mathbb{T}), \mathbf{PC}(\mathbb{T}))$ and $M_n fP_n \to fI$.

Error estimates

10.16. Approximation theorems. Let $\mathbf{H}^s := \mathbf{H}^s(\mathbb{T})$ denote the periodic Sobolev space of arbitrary real order s, i.e., the closure of all smooth 1-periodic functions with respect to the norm

$$\|u\|_s = \|u\|_{\mathbf{H}^s} := \Big(\sum_{k\in\mathbb{Z}} (1 + |k|)^{2s} |\hat{u}(k)|^2\Big)^{1/2}, \qquad s \in \mathbb{R}, \tag{1}$$

where

$$\hat{u}(k) := \int_0^1 u(x)\, e^{-2\pi i k x}\, dx, \qquad k \in \mathbb{Z},$$

are the Fourier coefficients. In order to obtain error estimates for smooth coefficients c and d and smooth right-hand sides f one needs the subsequent approximation results. In what follows, P_n and M_n denote the projections introduced in Section 10.13.

Lemma. (i) (Jackson's theorem). *Let $0 \leqq s \leqq t$, $r < \infty$ and $t > 1/2$. Then there exists a constant $C > 0$ depending only on s, t, and r such that*

$$\|(I - P_n) u\|_s \leqq Cn^{s-r} \|u\|_r \quad \forall u \in \mathbf{H}^r, \tag{2}$$

$$\|(I - M_n) u\|_s \leqq Cn^{s-t} \|u\|_t \quad \forall u \in \mathbf{H}^t. \tag{3}$$

(ii) (Bernstein's inequality). *If $0 \leqq s \leqq t < \infty$, $n \in \mathbb{N}$, and $u_n \in \mathcal{S}_n^\infty = \operatorname{im} P_n$, then*

$$\|u_n\|_t \leqq n^{t-s} \|u_n\|_s . \tag{4}$$

The lemma is an immediate consequence of the norm definition (1) and the formulas

$$(P_n u)(z) = \sum_{k=-[n/2]}^{[(n-1)/2]} \hat{u}(k)\, z^k,$$

$$(M_n u)(z) = \sum_{k=-[n/2]}^{[(n-1)/2]} z^k \Big(\sum_{l\in\mathbb{Z}} \hat{u}(k + nl)\, e^{2\pi i \varepsilon l}\Big),$$

where $n \in \mathbb{N}$, $0 \leqq \varepsilon < 1$, and $z \in \mathbb{T}$. Notice that, for $n = 2m + 1$, P_n (M_n for $\varepsilon = 0$) coincides with the operator $P_m(L_m)$ introduced in Section 7.3.

10.17. Lemma. *Let $0 \leqq s \leqq t < \infty$. Suppose the operator A to be bounded and invertible in $\mathbf{H}^s$ as well as in $\mathbf{H}^t$ and the sequence $\{A_n\}$ of approximate operators $A_n \in \mathcal{L}(\mathcal{S}_n^\infty)$ to be stable in $\mathbf{H}^s$. Moreover, assume*

$$\|Au - A_n P_n u\|_s \leqq Cn^{s-t} \|u\|_t \quad \forall u \in \mathbf{H}^t, \tag{1}$$

where $C > 0$ is a constant independent of u and n. Then $\{A_n\}$ is stable in $\mathbf{H}^t$.

This follows from 10.16(4) and Theorem 1.37. ∎

By virtue of Theorem 10.13 and Lemma 10.17, one needs estimates of the form (1) in order to prove the stability of the approximation method in Sobolev spaces.

Let $0 \leqq s \leqq t < \infty$, $t > 1/2$, and $cI \in \mathcal{L}(\mathbf{H}^t)$. The inequalities 10.16(2) and 10.16(3) yield that

$$\|\tilde{c}_n P_n u - cu\|_s = \|M_n c P_n u - cu\|_s \leqq C n^{s-t} \|u\|_t \tag{2}$$

for each $u \in \mathbf{H}^t$. Taking into account the uniform boundedness of the sequence $\{\tilde{c}_n P_n\}$ in $\mathcal{L}(\mathbf{L}^2(\mathbb{T}))$ we conclude via the interpolation theory in the scale $\mathbf{H}^s$ that (2) is valid for arbitrary s, t such that $0 \leqq s \leqq t < \infty$.

10.18. Lemma. *Let $\varrho \in \mathbf{PC}(\mathbb{T})$ be m times continuously differentiable in a right-sided as well as in a left-sided neighborhood of the point* 1. *Assume $\gamma^{(k)}(1 \pm 0) = 0$, $k = 0, \ldots, m - 1$, where $\gamma := \varrho - \sigma$,*

$$\sigma(\exp(\mathrm{i}2\pi x)) = \begin{cases} \varrho(1 + 0) \exp(\mathrm{i}2\pi x\varepsilon) & \text{if} \quad x \in (0, 1/2), \\ \varrho(1 - 0) \exp(\mathrm{i}2\pi x\varepsilon) & \text{if} \quad x \in (-1/2, 0). \end{cases}$$

Then there exists a constant $C > 0$ such that

$$\|\hat{\varrho}_n P_n u - [\varrho(1 + 0) P_{\mathbb{T}} + \varrho(1 - 0) Q_{\mathbb{T}}] u\|_s \leqq C \frac{\|u\|_t}{n^{\min\{m, t-s\}}} \tag{1}$$

for $0 \leqq s \leqq t < \infty$ and $u \in \mathbf{H}^t$.

Proof. 1. We first consider the case $\varrho(1 + 0) = \varrho(1 - 0) = 0$. Then $\varrho(z) = \varrho_0(z)(z - 1)^m$, where $\varrho_0 \in \mathbf{PC}$, and $\hat{\varrho}_n = (\varrho_0)_n^\wedge ((z - 1)^m)_n^\wedge$. Corresponding to the basis (t^j) $(-[n/2] \leqq j \leqq [(n - 1)/2])$, the operator $(\varrho_0)_n^\wedge \in \mathcal{L}(\mathcal{S}_n^\infty)$ $(\mathcal{S}_n^\infty \subset \mathbf{H}^s)$ has the matrix representation

$$\left(\varrho_0(\exp(\mathrm{i}2\pi j/n)) \exp(-\mathrm{i}2\pi j\varepsilon/n)\, \delta_{kj}\right)_{k,j=-[n/2]}^{[(n-1)/2]}$$

(cf. Section 10.15). Thus $\|(\varrho_0)_n^\wedge\| \leqq \sup |\varrho_0|$. Consequently, without loss of generality, it can be assumed that $\varrho(z) = (z - 1)^m$ and $\varepsilon = 0$.

Let $V_n \in \mathcal{L}(\mathbf{H}^s)$ denote the operator defined by $(V_n u)(z) := u(z \exp(2\pi\mathrm{i}/n))$. Then the matrix representation of the operator $V_n|_{\operatorname{im} P_n}$ corresponding to the basis $(\psi_k^{(n)})$ equals $\hat{v}_n$, where $v(z) := z$. Therefore

$$\hat{\varrho}_n P_n u = (V_n - I)^m P_n u = (V_n - I)^m (P_n - I) u + (V_n - I)^m u$$

and

$$\|\hat{\varrho}_n P_n u\|_s \leqq C \|(P_n - I) u\|_s + \|(V_n - I)^m u\|_s. \tag{2}$$

Now, by 10.16(1),

$$\|(V_n - I)^m u\|_s = \left(\sum_{k \in \mathbb{Z}} |\hat{u}(k) (1 + |k|)^s (e^{\mathrm{i}(2\pi/n)k} - 1)^m|^2\right)^{1/2}$$

$$\leqq 2^m \left(\sum_{k \in \mathbb{Z}} \left|\hat{u}(k) (1 + |k|)^s \sin^m \frac{\pi k}{n}\right|^2\right)^{1/2}$$

$$\leqq 2^m \left(\frac{\pi}{n}\right)^{\min\{m, t-s\}} \|u\|_t.$$

Using 10.16(2) and (2) we obtain

$$\|\hat{\varrho}_n P_n u\|_s \leqq C n^{-\min\{m,t-s\}} \|u\|_t .$$

2. We now turn to the general case, where $\varrho(1 \pm 0)$ are arbitrary finite values. Since, by 10.16(2),

$$\|(B - P_n B P_n) u\|_s \leqq C n^{s-t} \|u\|_t ,$$

where $B := \varrho(1+0) P_{\mathbb{T}} + \varrho(1-0) Q_{\mathbb{T}}$, it suffices to estimate the term $\|\hat{\varrho}_n P_n u - P_n B P_n u\|_s$. The matrix representation of the operator $P_n B P_n$ corresponding to the bases $(\psi_k^{(n)})$, $(\tilde{\psi}_k^{(n)})$ is equal to $\hat{\sigma}_n$. Since $(\varrho - \sigma)(1 \pm 0) = 0$, the assertion of the lemma is a consequence of the first step of this proof. ■

10.19. Lemma. *Suppose* $k \in \mathbf{C}^\infty(\mathbb{T} \times \mathbb{T})$ *and let* $K_n \in \mathcal{L}(\mathcal{S}_n^\infty)$ *denote the operator defined in Section* 10.15. *Then there exists a constant* $C > 0$ *such that*

$$\|Ku - K_n P_n u\|_s \leqq C n^{s-t} \|u\|_t \tag{1}$$

for all $0 \leqq s \leqq t < \infty$ *and* $u \in \mathbf{H}^t$.

Proof. Let $M_n^\varepsilon = M_n$ $(0 \leqq \varepsilon < 1)$ denote the interpolation projection introduced in Section 10.13. We first assume $t > 1/2$. For $u_n \in \mathcal{S}_n^\infty = \operatorname{im} P_n$, it follows from 10.16(3) that

$$\begin{aligned}
&\left| (K u_n)(\tau_k) - \frac{2\pi \mathrm{i}}{n} \sum_{j=0}^{n-1} k(\tau_k, t_j)\, t_j u_n(t_j) \right| \\
&= \left| \int_{\mathbb{T}} \left\{ k(\tau_k, t)\, u_n(t) - \left(M_n^0[k(\tau_k, \cdot)\, u_n(\cdot)]\right)(t) \right\} \mathrm{d}t \right| \\
&\leqq (2\pi)^{1/2} \|(I - M_n^0)[k(\tau_k, \cdot)\, u_n(\cdot)]\|_0 \\
&\leqq C n^{-t} \|k(\tau_k, \cdot)\, u_n(\cdot)\|_t \leqq C' n^{-t} \|u_n\|_t .
\end{aligned}$$

Collecting these estimates together with 10.16(4) gives

$$\begin{aligned}
\|M_n^\varepsilon K u_n - K_n u_n\|_s &\leqq n^s \|M_n^\varepsilon K u_n - K_n u_n\|_0 \\
&\leqq n^s n^{-1/2} \left(\sum_{k=0}^{n-1} \left| (K u_n)(\tau_k) - \frac{2\pi \mathrm{i}}{n} \sum_{j=0}^{n-1} k(\tau_k, t_j)\, t_j u_n(t_j) \right|^2 \right)^{1/2} \\
&\leqq C' n^{s-t} \|u_n\|_t .
\end{aligned} \tag{2}$$

Moreover,

$$Ku - M_n^\varepsilon K P_n u = K(I - P_n) u + (I - M_n^\varepsilon) K P_n u .$$

Hence, by 10.16(2) and 10.16(3),

$$\begin{aligned}
\|Ku - M_n^\varepsilon K P_n u\|_s &\leqq C \|(I - P_n) u\|_s + \|(I - M_n^\varepsilon) K P_n u\|_s \\
&\leqq C n^{s-t} (\|u\|_t + \|K P_n u\|_t) \leqq C' n^{s-t} \|u\|_t .
\end{aligned} \tag{3}$$

From (2) and (3) we obtain (1) if $t > 1/2$.

Since the entries of the matrix K_n have the order n^{-1}, the sequence $\{K_n\}$ $\left(K_n \in \mathcal{L}(\mathbf{l}^2(n))\right)$ is uniformly bounded. Hence (1) is valid for $s = t = 0$. Finally, an interpolation argument shows that (1) holds for all $0 \leqq s \leqq t < \infty$. ■

As a corrollary of Lemmas 10.17, 18, 19, formula 10.17(2) and the equality $A_n = \tilde{c}_n\hat{\alpha}_n + \tilde{d}_n\hat{\beta}_n + C_n + K_n$ (see Section 10.15), as well as Theorem 10.13 we get the following theorem.

10.20. Theorem. *Let* $s \geqq 0$, $c, d \in \mathbf{C}(\mathbb{T})$ *and* $cI, dI \in \mathcal{L}(\mathbf{H}^s)$, $k \in \mathbf{C}^\infty(\mathbb{T} \times \mathbb{T})$. *Assume the operator* $A := cI + dS_{\mathbb{T}} + K$ *to be invertible in* $\mathbf{L}^2(\mathbb{T})$. *Then the quadrature method* 10.12(16) *is stable in* $\mathbf{H}^s$.

If the condition

$$\frac{c+d}{c-d}(\tau) \notin \mathrm{e}^{-\mathrm{i}\pi\varepsilon}(-\infty, 0] \quad \forall \tau \in \mathbb{T}, \tag{1}$$

is fulfilled, then the quadrature method 10.12(12) *is stable in* $\mathbf{H}^s$. *The method* 10.12(13) *or the methods* 10.12(14), 10.12(15), *respectively, are stable in* $\mathbf{H}^s$ *if condition* (1) *is fulfilled for* $\varepsilon = 1$ *or* $\varepsilon = 0$, *respectively.*

Remark. Condition (1) is also necessary for the corresponding quadrature method to be stable in $\mathbf{H}^s$.

Indeed, in the case of constant coefficients $c, d \in \mathbb{C}$ and $K = 0$ this is an immediate consequence of the circulant structure of the approximate operators. In this case condition (1) means that the eigenvalues $\lambda_k^{(n)}$ of the circulants satisfy $|\lambda_k^{(n)}| \geqq \delta > 0$, where δ is a fixed constant. If c, d are variable coefficients and $K = 0$, then the necessity of condition (1) may be verified by using local principles (cf. Chapters 11 and 13).

10.21. Theorem. *Assume the conditions of Theorem* 10.20 *are fulfilled. If* $0 \leqq s \leqq t < \infty$, $cI, dI \in \mathcal{L}(H^t)$, *and* $u \in \mathbf{H}^t$ *is the solution to* 10.1(1), *then there exist* $n_0 \in \mathbb{N}$ *and* $C > 0$ *such that for* $n \in \mathbb{N}$, $n > n_0$, *and any function* $f \in \mathbf{H}_t$ *the quadrature equations are uniquely solvable and the error estimate*

$$\|u - u_n\|_s \leqq Cn^{-\min\{m,t-s\}} \|u\|_t$$

holds. Here $m = 1$ *for the quadrature methods* 10.12(12) *and* 10.12(14), $m = 2$ *for* 10.12(13), $m = 3$ *for* 10.12(15) *and* $m = +\infty$ *for* 10.12(16).

This theorem follows from Lemma 10.18, Lemma 10.19, 10.17(2), 10.16(2), 10.16(3) and the well-known estimate (see Section 1.23)

$$\|u - u_n\|_s \leqq C\{\|f - M_nf\|_s + \|u - P_nu\|_s + \|(A - A_nP_n)\, u\|_s\} \tag{2}$$

(cf. also the proof of Theorems 10.13 and 10.15).

Remark. For the following particular cases

(i) $A = d(\mathrm{i}\cot(\pi\varepsilon)\, I + S_{\mathbb{T}}) + K$ and the method 10.12(12),

(ii) $A = dS_{\mathbb{T}} + K$ and the method 10.12(13),

or

(iii) $A = dI + K$ and the method 10.12(15),

we even obtain the optimal error estimate (1) with $m = \infty$.

Besides the approximate solution u_n defined by formulas 10.12(17)–10.12(18) one can consider the approximate solution $\tilde{u}_n$ to Eq. 10.1(1) determined by spline interpolation.

To this end let δ be a nonnegative integer and n a natural number. Let $\dot{\mathscr{S}}_n^\delta$ denote the space of smooth 1-periodic splines of degree δ on the uniform mesh $\tilde{\Delta}$, where

$$\tilde{\Delta} = \begin{cases} \{j/n : j \in \mathbb{Z}\} & \text{if } \delta \text{ is odd,} \\ \{(j + 1/2)/n : j \in \mathbb{Z}\} & \text{if } \delta \text{ is even.} \end{cases}$$

Choose the interpolation basis $(\chi_k)_{k=0}^{n-1}$ of $\dot{\mathscr{S}}_n^\delta$ satisfying the conditions $\chi_k(j/n) = \delta_{jk}$, $j, k = 0, \ldots, n-1$. If there exists a unique solution $(\xi_k^{(n)})_{k=0}^{n-1}$ of the quadrature system (see 10.12(12)–10.12(16)), then the interpolating spline $\tilde{u}_n := \sum_{k=0}^{n-1} \xi_k^{(n)} \chi_k$ gives an approximate solution to Eq. 10.1(1).

Obviously, $\tilde{u}_n = \dot{K}_n u_n$, where $\dot{K}_n$ denotes the interpolation projection defined by

$$\dot{K}_n u \in \dot{\mathscr{S}}_n^\delta, \qquad (\dot{K}_n u - u)(j/n) = 0, j = 0, \ldots, n-1.$$

Thus

$$\|u_n - \tilde{u}_n\|_s = \|u_n - \dot{K}_n u_n\|_s \leqq C n^{s-t} \|u_n\|_t \leqq C n^{s-t} \|u\|_t, \tag{3}$$

if $0 \leqq s \leqq t \leqq \delta + 1$, $s < \delta + 1/2$, $t > 1/2$ (see Section 2.30).

As a corrolary of Theorem 10.21 and (3) we get the following theorem.

10.22. Theorem. *Let the conditions of Theorem* 10.21 *be fulfilled. Assume* $0 \leqq s \leqq t \leqq \delta + 1$, $s < \delta + 1/2$ *and* $t > 1/2$. *If* $f \in \mathbf{H}^t$, *and if* $u \in \mathbf{H}^t$ *is the solution to Eq.* 10.1(1), *then the quadrature equations are uniquely solvable for* $n > n_0$ *and we have the error estimate*

$$\|u - \tilde{u}_n\|_s \leqq C n^{-\min\{m, t-s\}} \|u\|_t,$$

the value m being as in Theorem 10.21.

10.23. Theorem. *Let* $0 \leqq s \leqq t \leqq \delta + 1$, $s < \delta + 1/2$ *and* $t > 1/2$. *Suppose the singular integral operator A is invertible in* $\mathbf{L}^2(\mathbb{T})$ *and bounded in* $\mathbf{H}^t$. *If the collocation method* 10.1(2) *is stable in* $\mathbf{L}^2(\mathbb{T})$ *and* $f \in \mathbf{H}^t$, *then the optimal error estimate*

$$\|u - u_n\|_s \leqq C n^{s-t} \|f\|_t \tag{1}$$

holds, where $C > 0$ *is a constant independent of n and f.*

This theorem may be deduced in a similar way as Theorem 10.22 (cf. the proof of Lemma 10.18 and Section 2.30). In particular, for the ε-collocation method 10.1(2), we obtain the estimate

$$\|Au - K_n A L_n u\|_s \leqq C n^{s-t} \|u\|_t \quad \forall u \in \mathbf{H}^t, \tag{2}$$

for each s and t satisfying $0 \leqq s \leqq t \leqq \delta + 1$, $s < \delta + 1/2$, where C is a constant independent of u and n.

A different proof of Theorem 10.23 and Theorem 10.24 will be given in Chapter 13 for the more general case of pseudodifferential equations.

10.24. Theorem. *Let* $-\delta - 1 \leqq s \leqq t \leqq \delta + 1$, $s < \delta + 1/2$ *and* $t > -\delta - 1/2$. *Suppose the singular integral operator A is invertible in* $\mathbf{L}^2(\mathbb{T})$ *and bounded in* $\mathbf{H}^s$ *as well as in* $\mathbf{H}^t$. *If the Galerkin method* 10.1(3) *is stable in* $\mathbf{L}^2(\mathbb{T})$ *and* $f \in \mathbf{H}^t$, *then the error estimate* 10.23(1) *holds.*

Quadrature methods in the case of an arbitrary closed curve

10.25. Description of the methods. We now consider the case of a plane Jordan curve Γ given by a regular parametric representation $\gamma: \mathbb{R} \to \Gamma$, where γ is a 1-periodic function of the real variable s. Assume that $\mathrm{d}\gamma/\mathrm{d}s \neq 0$ and that the second derivative $\mathrm{d}^2\gamma/\mathrm{d}s^2$ satisfies a Hölder condition.

If $\Gamma \neq \mathbb{T}$, then there is no simple quadrature formula approximating the integral of smooth functions with arbitrarily high order. In what follows we shall use the quadrature formulas

$$\int_\Gamma f(t)\,\mathrm{d}t \approx \sum_{j=0}^{n-1} f(t_j^{(n)})\,(t_{j+1}^{(n)} - t_j^{(n)}) \tag{1}$$

or

$$\int_\Gamma f(t)\,\mathrm{d}t \approx \sum_{j=0}^{n-1} f(t_j^{(n)})\,(t_{j+1}^{(n)} - t_{j-1}^{(n)})/2, \tag{2}$$

where $t_j^{(n)} := \gamma\big((j + 1/2)/n\big)$, $j = 0, \ldots, n-1$; $n \in \mathbb{N}$. Formulas (1), (2) approximate the integral of smooth functions with order n^{-1}, n^{-2}, respectively.

We shall analyze four quadrature methods based on formulas (1), (2). Set $\tau_j^{(n)} := \gamma\big((j + 1/2 + \varepsilon)/n\big)$ $(j = 0, \ldots, n-1)$, where $\varepsilon \in (-1/2, 1/2]$, and let u denote the exact solution of Eq. 10.1(1). We determine approximate values $\xi_k = \xi_k^{(n)}$ to $u(t_k^{(n)})$ by one of the following systems:

$$[c(\tau_k) - \mathrm{i}\cot(\pi\varepsilon)\,d(\tau_k)]\,\xi_k + \frac{d(\tau_k)}{\pi\mathrm{i}} \sum_{j=0}^{n-1} \frac{t_{j+1} - t_j}{t_j - \tau_k}\,\xi_j$$
$$+ \sum_{j=0}^{n-1} k(\tau_k, t_j)\,(t_{j+1} - t_j)\,\xi_j = f(\tau_k) \tag{3}$$

if $\varepsilon \neq 0$,

$$c(t_k)\,\xi_k + \frac{d(t_k)}{\pi\mathrm{i}} \sum_{\substack{j=0\\ j\neq k}}^{n-1} \frac{t_{j+1} - t_j}{t_j - t_k}\,\xi_j + \sum_{j=0}^{n-1} k(t_k, t_j)\,(t_{j+1} - t_j)\,\xi_j = f(t_k) \tag{4}$$

or

$$c(\tau_k)\,\frac{1}{2}\,(\xi_{k+1} + \xi_k) + \frac{d(\tau_k)}{\pi\mathrm{i}} \sum_{j=0}^{n-1} \frac{1}{2}\,\frac{t_{j+1} - t_{j-1}}{t_j - \tau_k}\,\xi_j$$
$$+ \sum_{j=0}^{n-1} k(\tau_k, t_j)\,\frac{1}{2}\,(t_{j+1} - t_{j-1})\,\xi_j = f(\tau_k) \tag{5}$$

if $\varepsilon = 1/2$, and

$$c(t_k)\,\xi_k + \frac{d(t_k)}{\pi\mathrm{i}} \sum_{\substack{j=0\\ j=k+1 \bmod 2}}^{n-1} \frac{t_{j+1} - t_{j-1}}{t_j - t_k}\,\xi_j + \sum_{j=0}^{n-1} k(t_k, t_j)\,\frac{1}{2}\,(t_{j+1} - t_{j-1})\,\xi_j = f(t_k) \tag{6}$$

if ε is arbitrary; $k = 0, \ldots, n-1$.

The systems (3)–(6) are the result of a formal replacement of the quadrature weight $(2\pi i/n)\, t_j^{(n)}$ by $(t_{j+1}^{(n)} - t_j^{(n)})$ in 10.12(12), 10.12(14) and by $(1/2)(t_{j+1}^{(n)} - t_{j-1}^{(n)})$ in 10.12(13), 10.12(16), respectively. However, formulas (3)–(6) may be deduced similarly to the formulas 10.12(12)–10.12(16) in Section 10.12. To this end one rewrites the operator A in 10.1(1) in the form $A = A_0 + T$, where T is compact and A_0 denotes the "dominant" singular part of A. More precisely, A_0 turns into the singular operator $cI + dS_{\mathbb{T}} \in \mathcal{L}(\mathbf{L}^2(\mathbb{T}))$ by the transformation $u \to \tilde{u} = u \circ \gamma \circ \gamma_0^{-1}$, where $\gamma_0(s) := e^{2\pi i s}\colon \mathbb{R} \to \mathbb{T}$. After that we apply the modified quadrature method (see Section 10.12) to A_0 and the usual quadrature method to T. Finally, using formulas 10.12(7), 10.12(9) and neglecting the terms of order n^{-1} or n^{-2}, we arrive at formulas (3), (4) or (5), (6), respectively (cf. also the proof of the subsequent Theorem 10.26).

If there exists a unique solution $(\xi_k^{(n)})_{k=0}^{n-1}$ of one of the quadrature systems (3)–(6), then we define an approximate solution u_n of Eq. 10.1(1) by piecewise linear interpolation:

$$u_n := \sum_{k=0}^{n-1} \xi_k^{(n)} \varphi_{k,0}^{(n)},$$

where

$$\varphi_{k,\varepsilon}^{(n)}(\gamma(s)) := \begin{cases} ns - (k - 1/2 + \varepsilon), & k - 1/2 + \varepsilon \leq ns \leq k + 1/2 + \varepsilon, \\ k + 3/2 + \varepsilon - ns, & k + 1/2 + \varepsilon \leq ns \leq k + 3/2 + \varepsilon, \\ 0, & \text{otherwise}. \end{cases}$$

Let L_n^ε denote the orthogonal projection from $\mathbf{L}^2(\Gamma)$ onto lin $(\varphi_{k,\varepsilon}^{(n)})_{k=0}^{n-1}$ and N_n^ε the interpolation projection defined by

$$N_n^\varepsilon u := \sum_{k=0}^{n-1} u(\tau_k^{(n)})\, \varphi_{k,\varepsilon}^{(n)}.$$

Moreover, let $A_n \in \mathcal{L}(\operatorname{im} L_n^0, \operatorname{im} L_n^\varepsilon)$ denote the operator whose matrix representation corresponding to the bases $(\varphi_{k,0}^{(n)})$, $(\varphi_{k,\varepsilon}^{(n)})$ is equal to the matrix of the system (3), (4), (5) or (6), respectively. Then each of the systems (3)–(6) can be written in form of the operator equation $A_n u_n = N_n^\varepsilon f$.

10.26. Stability theorem. *Let $c, d \in \mathbf{PC}(\Gamma)$, $k \in \mathbf{C}^\infty(\Gamma \times \Gamma)$, and $\varepsilon \in (-1/2, 1/2]$. Assume the operator $A := cI + dS_\Gamma + K$ to be invertible in $\mathbf{L}^2(\Gamma)$.*

(a) *The quadrature method* 10.25(3) *is stable in $\mathbf{L}^2(\Gamma)$ if and only if condition* 10.13(1) *is fulfilled for any $\tau \in \Gamma$.*

(b) *The method* 10.25(4) ((10.25(5)) *is stable in $\mathbf{L}^2(\Gamma)$ if and only if condition* 10.13(1) *for $\varepsilon = 0$ ($\varepsilon = 1$) is satisfied.*

(c) *The method* 10.25(6) *is stable in $\mathbf{L}^2(\Gamma)$.*

Proof. Let G denote the linear bounded and invertible operator from $\mathbf{L}^2(\Gamma)$ onto $\mathbf{L}^2(\mathbb{T})$ defined by

$$Gu := \tilde{u}, \qquad \tilde{u} := u \circ \alpha, \qquad \alpha := \gamma \circ \gamma_0^{-1}.$$

It is well known (see e.g. MIKHLIN/PRÖSSDORF [1], p. 58) that then $A = G^{-1}A^{(0)}G$, where

$$A^{(0)} := \tilde{c}I + \tilde{d}S_{\mathbf{T}} + K^{(1)} + K^{(2)} \left(\in \mathcal{L}(\mathbf{L}^2(\mathbf{T}))\right),$$

$$(K^{(1)}\tilde{u})(t) := \frac{\tilde{d}(t)}{\pi \mathrm{i}} \int_{\mathbf{T}} \left\{\frac{\alpha'(\tau)}{\alpha(\tau) - \alpha(t)} - \frac{1}{\tau - t}\right\} \tilde{u}(\tau)\, \mathrm{d}\tau,$$

$$(K^{(2)}\tilde{u})(t) := \int_{\mathbf{T}} k\big(\alpha(t), \alpha(\tau)\big)\, \alpha'(\tau)\, \tilde{u}(\tau)\, \mathrm{d}\tau.$$

Hence $A^{(0)}\tilde{u} = \tilde{f}$ if $Au = f$.

For the operator $A^{(0)}$, the conditions of Theorem 10.13 and Remark 10.14.2° are satisfied. Moreover, Theorem 10.13 remains valid after replacing $\tau_k^{(n)} = \exp\big(\mathrm{i}2\pi(k + \varepsilon)/n\big)$ and $t_k^{(n)} = \exp(\mathrm{i}2\pi k/n)$ by $\tilde{\tau}_k^{(n)} := \exp\big(\mathrm{i}2\pi(k + 1/2 + \varepsilon)/n\big)$ and $\tilde{t}_k^{(n)} := \exp\big(\mathrm{i}2\pi(k + 1/2)/n\big)$, respectively. In particular, the sequence $\{A_n^{(0)}\}$ is uniformly bounded in $\mathcal{L}\big(\mathrm{l}^2(n)\big)$, where $A_n^{(0)}$ denotes the matrix to the corresponding quadrature method for $A^{(0)}$. Furthermore, by Theorem 10.13, the sequence $\{A_n^{(0)}\}$ is stable if and only of the corresponding condition of Theorem 10.26 is fulfilled.

To complete the proof we show that

$$\|A_n - A_n^{(0)}\|_{\mathcal{L}(\mathrm{l}^2(n))} \leqq Cn^{-1}, \tag{1}$$

where $C > 0$ is a constant independent of n. Hence the sequence $\{A_n\}$ is uniformly Nounded in $\mathcal{L}\big(\mathrm{l}^2(n)\big)$ and the stability of $\{A_n^{(0)}\}$ is equivalent to the stability of $\{A_n\}$. bow since the spline bases $(\varphi_{k,0}^{(n)})$, $(\varphi_{k,\varepsilon}^{(n)})$ possess the properties (see Section 2.23)

$$\begin{aligned} C_1 |\xi| &\leqq n^{1/2} \left\| \sum_{k=0}^{n-1} \xi_k \varphi_{k,0}^{(n)} \right\|_{\mathbf{L}^2(\Gamma)} \leqq C_2 |\xi|, \\ C_1 |\xi| &\leqq n^{1/2} \left\| \sum_{k=0}^{n-1} \xi_k \varphi_{k,\varepsilon}^{(n)} \right\|_{\mathbf{L}^2(\Gamma)} \leqq C_2 |\xi|, \end{aligned} \tag{2}$$

Theorem 10.26 follows.

It remains to prove the estimate (1). For simplicity, we confine ourselves to the case of the quadrature method 10.25(3). (In the case of the methods 10.25(4)–10.25(6) the argumentation is similar (cf. also Sections 10.27–29).) Then $A_n^{(0)}$ is the matrix to the method 10.12(12) for the operator $A^{(0)}$, where $\tau_k^{(n)}$, $t_k^{(n)}$ in 10.12(12) must be replaced by $\tilde{\tau}_k^{(n)}$, $\tilde{t}_k^{(n)}$. Consequently,

$$\begin{aligned} A_n^{(0)} &= \big([c(\tau_k^{(n)}) - \mathrm{i}\cot(\pi\varepsilon)\, d(\tau_k^{(n)})]\, \delta_{kj}\big)_{k,j=0}^{n-1} \\ &+ \big(d(\tau_k^{(n)})\, \delta_{kj}\big)_{k,j=0}^{n-1} \left(\frac{2}{n} \frac{\tilde{t}_j^{(n)}}{\tilde{t}_j^{(n)} - \tilde{\tau}_k^{(n)}}\right)_{k,j=0}^{n-1} \\ &+ \left(\frac{2}{n}\, d(\tau_k^{(n)}) \left\{\frac{\alpha'(\tilde{t}_j^{(n)})}{t_j^{(n)} - \tau_k^{(n)}} - \frac{1}{\tilde{t}_j^{(n)} - \tilde{\tau}_k^{(n)}}\right\} \tilde{t}_j^{(n)}\right)_{k,j=0}^{n-1} \\ &+ \left(\frac{2\pi \mathrm{i}}{n}\, k(\tau_k^{(n)}, t_j^{(n)})\, \alpha'(\tilde{t}_j^{(n)})\, \tilde{t}_j^{(n)}\right)_{k,j=0}^{n-1}. \end{aligned}$$

Thus

$$A_n^{(0)} = \tilde{c}_n - \mathrm{i} \cot(\pi\varepsilon)\, \tilde{d}_n + \tilde{d}_n \left(\frac{1}{\pi \mathrm{i}} \frac{1}{n} \frac{\gamma'\big((j+1/2)/n\big)}{t_j^{(n)} - \tau_k^{(n)}}\right)_{k,j=0}^{n-1}$$

$$+ \left(k(\tau_k^{(n)}, t_j^{(n)}) \frac{1}{n} \gamma'\big((j+1/2)/n\big)\right)_{k,j=0}^{n-1}.$$

Comparing the matrix A_n of the system 10.25(3) with $A_n^{(0)}$ we obtain

$$A_n = A_n^{(0)} + B_n^{(0)} \left(\frac{(t_{j+1}^{(n)} - t_j^{(n)})\, n - \gamma'\big((j+1/2)/n\big)}{\gamma'\big((j+1/2)/n\big)}\, \delta_{kj}\right)_{k,j=0}^{n-1}, \tag{3}$$

where

$$B_n^{(0)} := \tilde{d}_n \left(\frac{1}{\pi \mathrm{i} n} \frac{\gamma'\big((j+1/2)/n\big)}{t_j^{(n)} - \tau_k^{(n)}}\right)_{k,j=0}^{n-1} + \left(\frac{1}{n} k(\tau_k^{(n)}, t_j^{(n)})\, \gamma'\left(\frac{j+(1/2)}{n}\right)\right)_{k,j=0}^{n-1}.$$

Obviously, $B_n^{(0)}$ is equal to the matrix of the method 10.12(12) for the operator

$$B^{(0)} := \mathrm{i} \cot(\pi\varepsilon)\, \tilde{d} + \tilde{d} S_{\mathbf{T}} + K^{(1)} + K^{(2)}.$$

Hence $\{B_n^{(0)}\}$ is uniformly bounded in $\mathscr{L}\big(\mathbf{l}^2(n)\big)$. Moreover,

$$\frac{(t_{j+1}^{(n)} - t_j^{(n)})\, n - \gamma'\left(\frac{j+(1/2)}{n}\right)}{\gamma'\left(\frac{j+(1/2)}{n}\right)}$$

$$= \frac{\gamma\left(\frac{j+(3/2)}{n}\right) - \gamma\left(\frac{j+(1/2)}{n}\right) - \frac{1}{n}\gamma'\left(\frac{j+(1/2)}{n}\right)}{\frac{1}{n}\gamma'\left(\frac{j+(1/2)}{n}\right)} = o\left(\frac{1}{n}\right).$$

Thus (1) follows from (3). This completes the proof of Theorem 10.26. ∎

10.27. Theorem. *Let* $c, d \in \mathbf{PC}(\Gamma)$ *and* $k \in \mathbf{C}^\infty(\Gamma \times \Gamma)$. *Then* $A_n L_n^0 \to A \in \mathscr{L}\big(\mathbf{L}^2(\Gamma)\big)$, *where* A_n *denotes the approximate operator of the quadrature methods* 10.25(3)–(6) *(see Section* 10.25).

Proof. We proved in Section 10.26 that the sequence $\{A_n L_n^0\}$ is uniformly bounded in $\mathscr{L}\big(\mathbf{L}^2(\Gamma)\big)$. So it remains to verify the $\mathbf{L}^2$-convergence $A_n L_n^0 u \to Au$ for smooth functions u.

Since $L_n^0 u \to u$ and $N_n^\varepsilon u \to u$ (see Sects. 2.26 and 2.30) we see that $A_n(N_n^0 - L_n^0)\, u \to 0$ and $(I - N_n^\varepsilon)\, Au \to 0$. Thus we must prove that

$$\|A_n N_n^0 u - N_n^\varepsilon A u\|_{\mathbf{L}^2(\Gamma)} \to 0. \tag{1}$$

By 10.26(2), the convergence (1) is equivalent to

$$n^{-1/2} \left\| A_n \big(u(t_k^{(n)})\big)_{k=0}^{n-1} - \big((Au)\,(\tau_k^{(n)})\big)_{k=0}^{n-1} \right\|_{\mathbf{l}^2(n)} \to 0. \tag{2}$$

Taking into account 10.26(1) and the relation $GAu = A^{(0)}Gu = A^{(0)}\tilde{u}$ we conclude that (2) holds if and only if

$$\|A_n^{(0)} M_n^0 \tilde{u} - M_n^\varepsilon A^{(0)} \tilde{u}\|_{\mathbf{L}^2(\mathbf{T})} \to 0. \tag{3}$$

Since $(M_n^0 - P_n)\,\tilde{u} \to 0$, $(I - M_n^\varepsilon)\,A^{(0)}\tilde{u} \to 0$ and the sequence $\{A_n^{(0)}\}$ is uniformly bounded, (3) follows from the fact that

$$\|(A_n^{(0)} P_n - A^{(0)})\,\tilde{u}\|_{\mathbf{L}^2(\mathbf{T})} \to 0$$

(see Theorem 10.15). This completes the proof of Theorem 10.27. ■

Error estimates

10.28. Theorem. *Let $c, d \in \mathbf{H}^1(\Gamma)$. If the quadrature method* 10.25(3) *or* 10.25(4) *is stable in $\mathbf{L}^2(\Gamma)$, then there exist $n_0 \in \mathbb{N}$ and $C > 0$ such that for $n \in \mathbb{N}$, $n > n_0$, and any function $f \in \mathbf{H}^1(\Gamma)$ the quadrature equations are uniquely solvable and the error estimate*

$$\|u - u_n\|_{\mathbf{L}^2(\Gamma)} \leqq Cn^{-1}\,\|u\|_{\mathbf{H}^1(\Gamma)} \tag{1}$$

holds.

Proof. Combining the estimates

$$\begin{aligned} &\|(N_n^0 - L_n^0)\,u\|_{\mathbf{L}^2(\Gamma)} \leqq 2\,\|u - N_n^0 u\|_{\mathbf{L}^2(\Gamma)} \leqq \frac{c}{n}\,\|u\|_{\mathbf{H}^1(\Gamma)},\\ &\|(I - N_n^\varepsilon)\,Au\|_{\mathbf{L}^2(\Gamma)} \leqq \frac{c}{n}\,\|Au\|_{\mathbf{H}^1(\Gamma)} \leqq \frac{c'}{n}\,\|u\|_{\mathbf{H}^1(\Gamma)} \end{aligned} \tag{2}$$

(see Sections 2.26 and 2.30) as well as

$$\|(A_n^{(0)} P_n - A^{(0)})\,\tilde{u}\|_{\mathbf{L}^2(\mathbf{T})} \leqq \frac{c}{n}\,\|\tilde{u}\|_{\mathbf{H}^1(\mathbf{T})} \leqq \frac{c}{n}\,\|u\|_{\mathbf{H}^1(\Gamma)}$$

(see Sections 10.18–21) and arguing as in the previous section we get

$$\|A_n L_n^0 u - Au\|_{\mathbf{L}^2(\Gamma)} \leqq Cn^{-1}\,\|u\|_{\mathbf{H}^1(\Gamma)}. \tag{3}$$

Now (1) follows from (2), (3) and 10.21(2). ■

10.29. Theorem. *Assume the derivative $\mathrm{d}^3\gamma/\mathrm{d}s^3$ to exist and to satisfy a Hölder condition. Let $c, d \in \mathbf{H}^2(\Gamma)$ and the quadrature method* 10.25(5) *or* 10.25(6) *be stable in $\mathbf{L}^2(\Gamma)$. Then there exist $n_0 \in \mathbb{N}$ and $C > 0$ such that for $n \in \mathbb{N}$, $n > n_0$, and any function $f \in \mathbf{H}^2(\Gamma)$ the quadrature equations are uniquely solvable and the error estimate*

$$\|u - u_n\|_{\mathbf{L}^2(\Gamma)} \leqq Cn^{-2}\,\|u\|_{\mathbf{H}^2(\Gamma)}$$

holds.

Proof. Since

$$\frac{\frac{1}{2}\,(t_{j+1}^{(n)} - t_{j-1}^{(n)}) - \frac{1}{n}\,\gamma'\left(\frac{j+(1/2)}{n}\right)}{\frac{1}{n}\,\gamma'\left(\frac{j+(1/2)}{n}\right)}$$

$$= \frac{\frac{1}{2}\left(\gamma\left(\frac{j+(3/2)}{n}\right) - \gamma\left(\frac{j-(1/2)}{n}\right)\right) - \frac{1}{n}\,\gamma'\left(\frac{j+(1/2)}{n}\right)}{\frac{1}{n}\,\gamma'\left(\frac{j+(1/2)}{n}\right)} = O(n^{-2})$$

for the methods 10.25(5) and 10.25(6), by repeating the proof of the estimate 10.26(1), it follows that

$$\|A_n - A_n^{(0)}\|_{\mathcal{L}(\mathbf{l}^2(n))} \leqq Cn^{-2}. \tag{1}$$

Combining the estimates

$$\begin{aligned} \|(N_n^0 - L_n^0)\,u\|_{\mathbf{L}^2(\Gamma)} &\leqq 2\,\|u - N_n^0 u\|_{\mathbf{L}^2(\Gamma)} \leqq \frac{c}{n^2}\,\|u\|_{\mathbf{H}^2(\Gamma)}, \\ \|(I - N_n^\varepsilon)\,Au\|_{\mathbf{L}^2(\Gamma)} &\leqq \frac{c}{n^2}\,\|Au\|_{\mathbf{H}^2(\Gamma)} \leqq \frac{c'}{n^2}\,\|u\|_{\mathbf{H}^2(\Gamma)} \end{aligned} \tag{2}$$

(see Sections 2.26 and 2.30) and

$$\|(A_n^{(0)}P_n - A^{(0)})\,\tilde{u}\|_{\mathbf{L}^2(\mathbb{T})} \leqq \frac{c}{n^2}\,\|\tilde{u}\|_{\mathbf{H}^2(\mathbb{T})} \leqq \frac{c}{n^2}\,\|u\|_{\mathbf{H}^2(\Gamma)}$$

(see Sections 10.18–21) and proceeding as in Section 10.27 we obtain that

$$\|A_n L_n^0 u - Au\|_{\mathbf{L}^2(\Gamma)} \leqq Cn^{-2}\,\|u\|_{\mathbf{H}^2(\Gamma)}. \tag{3}$$

Now (1) follows from (2), (3) and 10.21(2). ■

10.30. Remarks. 1°. Using 10.28(3), 10.29(3) and Lemma 10.17, one can prove as in Sections 10.20–22 that if the quadrature method 10.25(3), 10.25(4) or 10.25(5), 10.25(6) is stable in $\mathbf{L}^2(\Gamma)$, then it is stable in $\mathbf{H}^s(\Gamma)$ for $0 \leqq s \leqq 1$ or $0 \leqq s \leqq 2$, respectively. Moreover, the corresponding error estimates (see Theorem 10.21) hold.

However, in case $3/2 \leqq s \leqq 2$ one has to use spline interpolation of higher order instead of the linear interpolation of the $\xi_k^{(n)} \approx u(t_k^{(n)})$.

2°. Obviously, in Sections 10.19–29 the condition $k \in \mathbf{C}^\infty(\Gamma \times \Gamma)$ can be replaced by less restrictive smoothness conditions for the kernel k.

3°. Theorem 10.3 remains true in the *system case* (i.e. in the case when c, d, α, β are matrix functions; see Section 10.41, Remark 2°). Thus, the results of Sections 10.5, 10.9–11, 10.13 and 10.26 can easily be extended to the case where $c, d \in [\mathbf{PC}(\Gamma)]_{N \times N}$ and $\alpha, \beta \in \mathbf{PC}(\mathbb{T})$ are the functions defined in Sections 10.4, 10.8 and 10.13, respectively. Let $K \in \mathcal{K}(\mathbf{L}_N^2(\Gamma), \mathcal{R}_N(\Gamma))$ and let $A := cI + dS_\Gamma + K$ be invertible in $\mathbf{L}_N^2(\Gamma)$.

Then, for $0 \leqq \varepsilon < 1$, the ε-collocation method 10.1(2) is stable in $\mathbf{L}_N^2(\Gamma)$ if and only if the operators $A^\tau \in \mathcal{L}(\mathbf{L}_N^2(\mathbb{T}))$ defined in Corollary 10.7(b) are invertible for all $\tau \in \Gamma$.

Notice that if $c, d \in [\mathbf{C}(\Gamma)]_{N\times N}$, then the latter stability condition is equivalent to the condition

$$\det\big(c(t) + \lambda d(t)\big) \neq 0, \qquad t \in \Gamma, \lambda \in [-1, 1] \tag{1}$$

in case $\varepsilon = 0$ and δ is odd or $\varepsilon = 1/2$ and δ is even, while it is equivalent to the condition

$$\det\big(d(t) + \lambda c(t)\big) \neq 0, \qquad t \in \Gamma, \lambda \in [-1, 1] \tag{2}$$

in case $\varepsilon = 0$ and δ is even or $\varepsilon = 1/2$ and δ is odd. We remark that condition (1) (condition (2)) holds if and only if the operator A (the operator AS_Γ) is locally strongly elliptic (see Theorem 6.22 and Prössdorf/Rathsfeld [2]).

Furthermore, the quadrature methods 10.12(16) and 10.25(6) are stable in $\mathbf{L}^2_N(\Gamma)$ if and only if the operator $cI - dS_\Gamma$ is invertible in $\mathbf{L}^2_N(\Gamma)$ (cf. also Theorem 7.27).

4°. Recently Prössdorf and Sloan [1] analyzed the following quadrature method for solving the equation 10.1(1). Rewrite the singular integral as

$$\frac{1}{\pi \mathrm{i}} \int_0^1 \frac{\gamma'(y)\, v(y)}{\gamma(y) - \gamma(x)}\, \mathrm{d}y$$

$$= v(x) + \frac{1}{\pi \mathrm{i}} \int_0^1 \frac{\gamma'(y)\big(v(y) - v(x)\big)}{\gamma(y) - \gamma(x)}\, \mathrm{d}y, \quad v(x) = u\big(\gamma(x)\big).$$

Then approximate the integrals by using the rectangle rule with step-length $1/n$ (n being an odd natural number) and replace $v(j/n)$ by the approximation $\xi_j^{(n)}$ ($j = 0, \ldots, n-1$) to obtain

$$\left(c(\gamma(x)) + d(\gamma(x)) \left[1 - \frac{1}{\pi \mathrm{i} n} \sum_{j=0}^{n-1} \frac{\gamma'(j/n)}{\gamma(j/n) - \gamma(x)}\right]\right) v(x)$$

$$+ \frac{d\big(\gamma(x)\big)}{\pi \mathrm{i} n} \sum_{j=0}^{n-1} \frac{\gamma'(j/n)}{\gamma(j/n) - \gamma(x)}\, \xi_j^{(n)}$$

$$+ \frac{1}{n} \sum_{j=0}^{n-1} k\big(\gamma(x), \gamma(j/n)\big)\, \gamma'(j/n)\, \xi_j^{(n)} \approx f\big(\gamma(x)\big), \qquad x \in \mathbb{R}.$$

Collocating this equation at the points $(k + \varepsilon)/n$, $k = 0, \ldots, n-1$, $-1/2 < \varepsilon \leqq 1/2$, $\varepsilon \neq 0$, and approximating $v\big((k+\varepsilon)/n\big)$ by the trigonometric polynomial u_n of degree $(n-1)/2$ that takes the value $\xi_j^{(n)}$ at j/n for $j = 0, \ldots, n-1$, our quadrature method becomes: find $\xi_0^{(n)}, \ldots, \xi_{n-1}^{(n)}$ such that

$$\left(c(\tau_k) + d(\tau_k) \left[1 - \frac{1}{\pi \mathrm{i} n} \sum_{j=0}^{n-1} \frac{\gamma'(j/n)}{t_j - \tau_k}\right]\right) \sum_{j=0}^{n-1} \frac{\sin \pi(k - j + \varepsilon)}{n \sin \pi\big((k - j + \varepsilon)/n\big)}\, \xi_j^{(n)}$$

$$+ \frac{d(\tau_k)}{\pi \mathrm{i} n} \sum_{j=0}^{n-1} \frac{\gamma'(j/n)}{t_j - \tau_k}\, \xi_j^{(n)} + \frac{1}{n} \sum_{j=0}^{n-1} k(\tau_k, t_j)\, \gamma'(j/n)\, \xi_j^{(n)} = f(\tau_k), \tag{1}$$

$$k = 0, \ldots, n-1,$$

where

$$t_j = t_j^{(n)} = \gamma(j/n), \qquad \tau_j = \tau_j^{(n)} = \gamma\big((j + \varepsilon)/n\big).$$

From 10.12(7) it follows that, in the case of the unit circle $\mathbb{T}$, the system (1) becomes

$$\big(c(\tau_k) - \mathrm{i} \cot(\pi\varepsilon)\, d(\tau_k)\big) \sum_{j=0}^{n-1} \frac{\sin \pi(k - j + \varepsilon)}{n \sin \pi\big((k - j + \varepsilon)/n\big)}\, \xi_j^{(n)}$$

$$+ \frac{2d(\tau_k)}{n} \sum_{j=0}^{n-1} \frac{1}{1 - \mathrm{e}^{\mathrm{i}2\pi(k-j+\varepsilon)/n}}\, \xi_j^{(n)}$$

$$+ \frac{2\pi\mathrm{i}}{n} \sum_{j=0}^{n-1} k(\tau_k, t_j)\, t_j \xi_j^{(n)} = f(\tau_k), \qquad k = 0, \ldots, n-1. \tag{2}$$

It can be shown (see PRÖSSDORF/SLOAN [1]) that (2) may be rewritten as

$$(aP_{\mathbb{T}}u_n + bQ_{\mathbb{T}}u_n)(\tau_k)$$

$$+ \frac{2\pi\mathrm{i}}{n} \sum_{j=0}^{n-1} k(\tau_k, t_j)\, t_j u_n(t_j) = f(\tau_k), \qquad k = 0, \ldots, n-1,$$

where $a = c + d$ and $b = c - d$. Thus the method (2) is simply the trigonometric ε-collocation method, combined with numerical quadrature for the compact term. Hence, it follows from Theorem 10.13(c) and Remark 10.14.1° (see also Theorem 7.24(b)) that the method (2) (resp. (1)) is stable in $\mathbf{L}^2(\mathbb{T})$ (resp. $\mathbf{L}^2(\Gamma)$) provided the operator A is invertible. Moreover, if c, d and k are $\mathbf{C}^\infty$ functions and A is invertible in $\mathbf{L}^2(\mathbb{T})$ then the error estimate

$$\|u - u_n\|_s \leqq Cn^{s-t} \|u\|_t$$

holds for all $0 \leqq s \leqq t < \infty$ and the exact solution $u \in \mathbf{H}^t$ to 10.1(1).

Banach algebra approach to the stability of paired circulants

10.31. Introduction. In the previous sections we have studied collocation and Galerkin methods with spline functions as well as quadrature methods for singular integral equations over the unit circle $\mathbb{T}$. As it turned out, all the occuring finite-dimensional operators can be represented in the form of paired circulants (cf. Section 10.3) and their stability is a consequence of Theorem 10.3. Now we shall give a proof for this theorem. Moreover, we shall study a certain C^*-algebra $\mathfrak{Z}$ generated by the paired circulants and some other operator sequences. The algebra $\mathfrak{Z}$ will contain a sufficiently large ideal J, so that it will be possible to get an invertibility criterion for elements of the quotient algebra $\mathfrak{Z}/J$. It will be shown that the center of $\mathfrak{Z}/J$ "contains" the algebra of all functions continuous on the torus, and this circumstance permits a very sensitive localization in $\mathfrak{Z}/J$. Each of the local algebras occuring is generated by two selfadjoint idempotents, which yields a precise description of these algebras and thus the solution of the invertibility problem.

The approach presented in the following sections has some interesting applications. One of them consists in the possibility of associating a certain symbol calculus with each numerical method (such constructions are widely used in Fredholm theory, cf. Chapter 6). Other applications concern approximation methods on curves with corners (cf. Chapter 11).

10.32. Definitions and notation. Let $\mathbb{T}$ be the unit circle in $\mathbb{C}$. We define the functions $g_j^{(n)}(t)$ $(j \in \mathbb{Z})$ by $g_j^{(n)}(t) := \sqrt{n}\, \chi_j^{(n)}(t)$, where $\chi_j^{(n)}$ denotes the characteristic function of the arc $[\exp(2\pi i j/n), \exp(2\pi i(j+1)/n)) \subseteq \mathbb{T}$, and we denote by L_n the orthogonal projection of $\mathbf{L}^2(\mathbb{T})$ onto span $\{g_j^{(n)} : j = 0, \dots, n-1\}$. On identifying the operators of $\mathcal{L}(\operatorname{im} L_n)$ with their matrices corresponding to the basis $\{g_j^{(n)} : j = 0, \dots, n-1\}$, we obtain that $\mathcal{L}(\operatorname{im} L_n) = \mathcal{L}(\mathbf{l}^2(n))$. In particular, the operators U_n introduced in Section 10.2 as well as $\tilde{c}_n$, $\tilde{d}_n$, $\hat{\alpha}_n$, $\hat{\beta}_n$, C_n, and

$$B_n := \tilde{c}_n \hat{\alpha}_n + \tilde{d}_n \hat{\beta}_n + C_n, \tag{1}$$

defined in Section 10.3 will be thought of as belonging to $\mathcal{L}(\operatorname{im} L_n)$. Moreover, setting

$$K_n^\varepsilon f := \sum_{j=0}^{n-1} f\big(\exp[2\pi i(j+\varepsilon)/n]\big)\, \chi_j^{(n)}$$

for $0 \leqq \varepsilon < 1$, we get

$$B_n = K_n^\varepsilon c L_n U_n K_n^0 \alpha L_n U_n^{-1} + K_n^\varepsilon d L_n U_n K_n^0 \beta L_n U_n^{-1} + C_n. \tag{2}$$

Now we introduce two families of isomorphisms. The first is a discretized version of the family of translations on $\mathbb{T}$. For $\tau \in \mathbb{T}$, we choose $\tau_E \in \{\exp(2\pi i k/n) : k = 0, \dots, n-1\}$ such that $\tau_E \approx \tau$. More precisely, we define $k_E = k_E(\varepsilon, \tau, n) \in \{0, \dots, n-1\}$ by $\tau \in \big(\exp(2\pi i(k_E + \varepsilon - 1)/n), \exp(2\pi i(k_E + \varepsilon)/n)\big]$ and put $\tau_E = \tau_E(\varepsilon, \tau, n) := \exp[2\pi i k_E/n]$. We set

$$T_n^{(\tau,\varepsilon)} : \operatorname{im} L_n \to \operatorname{im} L_n, \qquad \sum_{j=0}^{n-1} \xi_j g_j^{(n)} \mapsto \sum_{j=0}^{n-1} \xi_j g_{j-k_E}^{(n)}. \tag{3}$$

Since $g_j^{(n)} = g_{j\pm n}^{(n)}$, the matrix of $T_n^{(\tau,\varepsilon)}$ corresponding to $\{g_j^{(n)} : j = 0, \dots, n-1\}$ is a circulant. Consequently, $T_n^{(\tau,\varepsilon)}$ commutes with any other circulant and we have

$$T_n^{(\tau,\varepsilon)} \hat{\alpha}_n (T_n^{(\tau,\varepsilon)})^{-1} = \hat{\alpha}_n \tag{4}$$

for every $\alpha \in \mathbf{PC}(\mathbb{T})$. Furthermore, it is easy to verify that

$$T_n^{(\tau,\varepsilon)} \tilde{c}_n (T_n^{(\tau,\varepsilon)})^{-1} = \tilde{f}_n, \qquad f(t) := c(t \cdot \tau_E). \tag{5}$$

In accordance with Section 10.2, the matrix of $T_n^{(\tau,\varepsilon)}$ can be written as $T_n^{(\tau,\varepsilon)} = U_n^{-1} \times (\tau_E^{-k} \delta_{k,j})_{k,j=0}^{n-1} U_n$.

In order to get a formula analogous to (5), but for $\hat{\alpha}_n$ instead of $\tilde{c}_n$, we set

$$E_n^{(\tau,1)} : \operatorname{im} L_n \to \operatorname{im} L_n, \qquad E_n^{(\tau,1)} := U_n T_n^{(\tau,0)} U_n^{-1}. \tag{6}$$

If we define $t_E = t_E(\tau, n)$ by $t_E := \exp(2\pi i j_E/n)$, where $j_E = j_E(\tau, n) \in \{0, \dots, n-1\}$ satisfies $\tau \in \big(\exp(2\pi i(j_E - 1)/n), \exp(2\pi i j_E/n)\big]$, then

$$E_n^{(\tau,1)} : \sum_{j=0}^{n-1} \xi_j g_j^{(n)} \mapsto \sum_{j=0}^{n-1} \xi_j t_E^{-j} g_j^{(n)}. \tag{7}$$

Since the matrix of $E_n^{(\tau,1)}$ is diagonal, we get

$$E_n^{(\tau,1)}\tilde{c}_n(E_n^{(\tau,1)})^{-1} = \tilde{c}_n \tag{8}$$

for every $c \in \mathbf{PC}(\mathbb{T})$. The definition of $\hat{\alpha}_n$ in Section 10.2 and the formulas (5), (6) yield that

$$E_n^{(\tau,1)}\hat{\alpha}_n(E_n^{(\tau,1)})^{-1} = \hat{\gamma}_n, \qquad \gamma(t) := \alpha(t \cdot t_E). \tag{9}$$

We define $P_n \in \mathscr{L}(\mathbf{L}^2(\mathbb{T}))$ and M_n (cf. Section 10.13) by

$$P_n f := \sum_{k=-[n/2]}^{[(n-1)/2]} f_k t^k, \qquad f_k := \frac{1}{2\pi \mathrm{i}} \int_{\mathbb{T}} f(t)\, t^{-k-1}\, \mathrm{d}t, \tag{10}$$

$$\begin{aligned} M_n f &:= \sum_{k=0}^{n-1} f(\exp(2\pi \mathrm{i}k/n))\, \tilde{\varphi}_k^{(n)}, \\ \tilde{\varphi}_k^{(n)}(t) &:= n^{-1} \sum_{j=-[n/2]}^{[(n-1)/2]} \exp(-2\pi \mathrm{i}kj/n)\, t^j. \end{aligned} \tag{11}$$

Then there is an isometric isomorphism $E_n : \operatorname{im} L_n \to \operatorname{im} P_n$, viz

$$E_n : \sum_{j=0}^{n-1} \xi_j g_j^{(n)} \mapsto \sum_{j=0}^{[(n-1)/2]} \xi_j t^j + \sum_{j=[(n-1)/2]+1}^{n-1} \xi_j t^{j-n}.$$

For $\alpha \in \mathbf{PC}(\mathbb{T})$ put $\tilde{\alpha}(t) := \alpha(t^{-1})$. The operator $M_n\tilde{\alpha}|_{\operatorname{im} P_n}$ has the matrix $(\tilde{\alpha}(\exp(2\pi \mathrm{i}k/n))$ $(\tilde{\alpha}\delta_{kj})_{k,j=0}^{n-1}$ with respect to the basis $\{\tilde{\varphi}_k^{(n)} : k = 0, \ldots, n-1\}$, i.e., the matrix U_n^{-1} $\times$ $(\exp(2\pi \mathrm{i}k/n))\, \delta_{kj})_{k,j=0}^{n-1} U_n$ with respect to the basis $\{t^j : j = 0, \ldots, [(n-1)/2]\} \cup \{t^{j-n} : j = [(n-1)/2] + 1, \ldots, n-1\}$. The last matrix is equal to $\hat{\alpha}_n = U_n(\alpha(\exp(2\pi \mathrm{i}k/n))$ $\times\, \delta_{kj})_{k,j=0}^{n-1} U_n^{-1}$. Thus we may write

$$B_n = K_n^\varepsilon c L_n E_n^{-1} M_n \tilde{\alpha} P_n E_n + K_n^\varepsilon d L_n E_n^{-1} M_n \tilde{\beta} P_n E_n + C_n. \tag{12}$$

Finally, we set $E_n^{(\tau,2)} := E_n T^{(\tau,\varepsilon)} : \operatorname{im} L_n \to \operatorname{im} P_n$.

10.33. The general theorem. Let X be a Banach space and let $\{L_n\}$ be a uniformly bounded sequence of projections on X. By W we denote a certain (possibly infinite) index set. Suppose we are given, for each $w \in W$, a Banach space X_w and a sequence $L_n^{(w)}$ of projections on X_w. Let $L_n^{(w)} \to I_w$, $(L_n^{(w)})^* \to I_w^*$ strongly and suppose that there exist invertible operators $E_n^{(w)} : \operatorname{im} L_n \to \operatorname{im} L_n^{(w)}$ such that $\sup_{n,w} \|(E_n^{(w)})^{\pm 1}\| < \infty$. Furthermore, let $\mathfrak{B}$ denote the collection of all sequences $\{B_n\}_{n=1}^\infty$, $B_n \in \mathscr{L}(\operatorname{im} L_n)$, subject to the following two requirements:

(a) $\sup_n \|B_n L_n\| < \infty$;

(b) there exist operators $W_w\{B_n\} \in \mathscr{L}(X_w)$ such that

$$E_n^{(w)} B_n (E_n^{(w)})^{-1} L_n^{(w)} \to W_w\{B_n\} \quad \text{and}$$

$$(E_n^{(w)} B_n (E_n^{(w)})^{-1} L_n^{(w)})^* \to (W_w\{B_n\})^* \quad \text{strongly as } n \to \infty \text{ for all } w \in W.$$

On defining $\{A_n\} + \{B_n\} := \{A_n + B_n\}$, $\{A_n\} \cdot \{B_n\} = \{A_n \cdot B_n\}$ and $\|\{A_n\}\| := \sup_n \{\|A_n L_n\|\}$, the collection $\mathfrak{B}$ becomes a Banach algebra with identity $\{I_n\}$, where I_n is the identity in the space $\operatorname{im} L_n$. Finally, assume that

(c) for all $T \in \mathscr{K}(X_w)$ ($\mathscr{K}(X_w)$ referring to the set of all compact operators on X_w) ($w \in W$), the sequences $\{B_n\} := \{(E_n^{(w)})^{-1} L_n^{(w)} T E_n^{(w)}\}$ belong to $\mathfrak{B}$ and, for $w' \neq w$,

$E_n^{(w')}B_n(E_n^{(w')})^{-1} L_n^{(w')} \to 0$ strongly as well as $(E_n^{(w')}B_n(E_n^{(w')})^{-1} L_n^{(w')})^* \to 0$ strongly as $n \to \infty$.

Define

$$J_0 := \left\{\sum_{k=1}^{r} \{(E_n^{(w_k)})^{-1} L_n^{(w_k)} T_k E_n^{(w_k)}\} + \{C_n\} : w_k \in W,\, T_k \in \mathcal{K}(X_k),\right.$$

$$\left. \|C_n\| \to 0,\, r \in \mathbb{N}\right\}.$$

Then J_0 forms a two-sided ideal in $\mathfrak{B}$ and the closure J of J_0 forms a closed two-sided ideal in $\mathfrak{B}$ (see Chapter 1).

We shall say that $\{A_n\} \in \mathfrak{B}$ is *stable* if there exists an $n_0 \in \mathbb{N}$ such that for all $n \geqq n_0$ the operators A_n are invertible and $\sup\limits_{n \geqq n_0} \|(A_n)^{-1} L_n\| < \infty$.

Theorem. *The sequence $\{A_n\} \in \mathfrak{B}$ is stable if and only if for all $w \in W$ the operators $W_w\{A_n\}$ are invertible in $\mathcal{L}(X_w)$ and the coset $\{A_n\}^0 := \{A_n\} + J$ is invertible in $\mathfrak{B}^0 := \mathfrak{B}/J$.*

Proof. Suppose the operators $W_w\{A_n\}$ are invertible and $\{B_n\}^0$ is the inverse of $\{A_n\}^0$. By the definition of J, there exist operators $T_k \in \mathcal{K}(X_{w_k})$ $(k = 1, \ldots, r)$ and sequences $\{C_n\}, \{D_n\} \in \mathfrak{B}$ such that $\|C_n\| < 1/2$, $\|D_n\| \to 0$ and, for $n \geqq n_1$,

$$B_nA_n = I_n + \sum_{k=1}^{r} (E_n^{(w_k)})^{-1} L_n^{(w_k)} T_k E_n^{(w_k)} + C_n + D_n .$$

Without loss of generality we assume $r = 1$ and set $w_1 = w$, $T_1 = T$. We put

$$B_n' := B_n - (E_n^{(w)})^{-1} L_n^{(w)} T (W_w\{A_n\})^{-1} E_n^{(w)}$$

and arrive at the equality

$$B_n'A_n = I_n + (E_n^{(w)})^{-1} L_n^{(w)} T\big(L_n^{(w)} - (W_w\{A_n\})^{-1} E_n^{(w)} A_n (E_n^{(w)})^{-1}\big) E_n^{(w)} + C_n + D_n .$$

Since $\big(L_n^{(w)} - (W_w\{A_n\})^{-1} E_n^{(w)} A_n (E_n^{(w)})^{-1} L_n^{(w)}\big)^* \to 0$ and T is compact, there exists a sequence $\{D_n'\} \in \mathfrak{B}$ such that $\|D_n'\| \to 0$ and $B_n'A_n = I_n + C_n + D_n'$. Hence, for sufficiently large n, $(I_n + C_n + D_n')^{-1} B_n'$ is a uniformly bounded left inverse of A_n. Analogously the existence of a right inverse can be shown.

Vice versa, let $\{A_n\}$ be stable. Then, for $w \in W$, there exists a constant $C > 0$ such that

$$\|E_n^{(w)} A_n (E_n^{(w)})^{-1} L_n^{(w)} f\| \geqq C^{-1} \|L_n^{(w)} f\|$$

and

$$\left\|\big(E_n^{(w)} A_n (E_n^{(w)})^{-1} L_n^{(w)}\big)^* (L_n^{(w)})^* g\right\| \geqq C^{-1} \|(L_n^{(w)})^* g\|$$

for all $f \in X_w$, $g \in X_w^*$. Passage to the limit $n \to \infty$ yields that

$$\|W_w\{A_n\} f\| \geqq C^{-1} \|f\| \quad \text{and} \quad \|(W_w\{A_n\})^* g\| \geqq C^{-1} \|g\| .$$

Thus the kernel of $W_w\{A_n\}$ is trivial and the image of $W_w\{A_n\}$ is closed. Furthermore, the kernel of $(W_w\{A_n\})^*$ is also trivial. Hence the image of $W_w\{A_n\}$ is dense and therefore

$W_w\{A_n\}$ is invertible. We set $B_n := A_n^{-1}$ for $n \geqq n_0$ and $B_n := I_n$ for $n < n_0$. The equality

$$E_n^{(w)} B_n (E_n^{(w)})^{-1} L_n^{(w)} f - (W_w\{A_n\})^{-1} f$$
$$= (L_n^{(w)} - I)(W_w\{A_n\})^{-1} f + E_n^{(w)} B_n (E_n^{(w)})^{-1}$$
$$\times \left(L_n^{(w)} - E_n^{(w)} A_n (E_n^{(w)})^{-1} L_n^{(w)} (W_w\{A_n\})^{-1}\right) f$$

implies that $E_n^{(w)} B_n (E_n^{(w)})^{-1} L_n^{(w)} f \to (W_w\{A_n\})^{-1} f$. Analogously, $\left(E_n^{(w)} B_n (E_n^{(w)})^{-1} L_n^{(w)}\right)^* \to (W_w\{A\}_n)^{-1*}$. Thus $\{B_n\} \in \mathfrak{B}$, and $\{B_n\}^\circ$ is the inverse of $\{A_n\}^\circ$. ■

In accordance with the notation in Section 10.32, we have the following situation: $X = \mathbf{L}^2(\mathbb{T})$, $\{L_n\}$ are the orthogonal projections of Section 10.32, $W = \mathbb{T} \times \{1,2\} = (\mathbb{T}_1, \mathbb{T}_2)$,

$$X_{(w,1)} = X_{(w,2)} = X, \qquad L_n^{(w,1)} = L_n, \qquad L_n^{(v,2)} = P_n,$$

$E_n^{(w,1)}$ and $E_n^{(v,2)}$ are defined as in Section 10.32.

10.34. Lemma. *Let* $\{B_n\}$ *be as in* 10.32(1). *Then we have the following.*

(a) $\sup_n \|B_n L_n\| < \infty$.

(b) *There exist operators* $W_{(w,1)}\{B_n\}$, $W_{(v,2)}\{B_n\} \in \mathcal{L}\left(\mathbf{L}^2(\mathbb{T})\right)$ *such that, for all* $w, v \in \mathbb{T}$,

$$E_n^{(w,1)} B_n (E_n^{(w,1)})^{-1} L_n \to W_{(w,1)}\{B_n\},$$
$$\left(E_n^{(w,1)} B_n (E_n^{(w,1)})^{-1} L_n\right)^* \to (W_{(w,1)}\{B_n\})^*,$$
$$E_n^{(v,2)} B_n (E_n^{(v,2)})^{-1} P_n \to W_{(v,2)}\{B_n\},$$
$$\left(E_n^{(v,2)} B_n (E_n^{(v,2)})^{-1} P_n\right)^* \to (W_{(v,2)}\{B_n\})^*.$$

(c) *For* $T \in \mathcal{K}\left(\mathbf{L}^2(\mathbb{T})\right)$, *set* $D_n := (E_n^{(w,1)})^{-1} L_n T E_n^{(w,1)}$ *and* $G_n := (E_n^{(v,2)})^{-1} P_n T E_n^{(v,2)}$.

(c.1) $$E_n^{(v,j)} D_n (E_n^{(v,j)})^{-1} \to \begin{cases} T, & \text{if } v = w \text{ and } j = 1, \\ 0, & \text{if } v \neq w \text{ or } j = 2, \end{cases}$$

$$\left(E_n^{(v,j)} D_n (E_n^{(v,j)})^{-1}\right)^* \to \begin{cases} T^*, & \text{if } v = w \text{ and } j = 1, \\ 0, & \text{if } v \neq w \text{ or } j = 2, \end{cases}$$

(c.2) $$E_n^{(w,j)} G_n (E_n^{(w,j)})^{-1} \to \begin{cases} T, & \text{if } w = v \text{ and } j = 2, \\ 0, & \text{if } w \neq v \text{ or } j = 1 \end{cases}$$

$$\left(E_n^{(w,j)} G_n (E_n^{(w,j)})^{-1}\right)^* \to \begin{cases} T^*, & \text{if } w = v \text{ and } j = 2, \\ 0, & \text{if } w \neq v \text{ or } j = 1. \end{cases}$$

Here "$\to$*" means strong convergence as* $n \to \infty$.

Proof. (a) The well-known inequality $\|\tilde{c}_n\| \leqq \|c\|_\infty$, (cf. Section 10.2), the equality $\|(U_n)^{\pm 1}\| = 1$ and 10.32(1) give the assertion.

(b) Obviously, we may suppose that $C_n = 0$. In virtue of 10.32(8) and 10.32(9) we get

$$E_n^{(w,1)} B_n (E_n^{(w,1)})^{-1} L_n = \left[K_n^\varepsilon c L_n \left(\alpha(tt_E)\right)_n^\wedge + K_\varepsilon^n d L_n \left(\beta(tt_E)\right)_n^\wedge\right] L_n.$$

Therefore, we have to show the strong convergence of $\{K_n^\varepsilon c L_n\}$ and $\left\{\alpha\left((tt_E)\right)_n^\wedge L_n\right\}$. By the boundedness of the sequences (cf. part (a)), it will be enough to show their convergence on a dense subset of $\mathbf{L}^2(\mathbb{T})$. However, for a piecewise continuous function f,

$$K_n^\varepsilon cL_n f - cf = [K_n^\varepsilon cL_n] (L_n - K_n^\varepsilon) f + (K_n^\varepsilon - I) cf,$$

and since $(L_n - I) f \to 0$ and $(K_n^\varepsilon - I) f \to 0$ it follows that $K_n^\varepsilon cL_n f \to cf$. Moreover, for fixed j, $K_n^0 t^j = \sum_{k=0}^{n-1} \exp(\mathrm{i}2\pi kj/n)\, g_k^{(n)}$ is the eigenvector of the circulant $\big(\alpha(tt_E)\big)_n^\wedge$ corresponding to the eigenvalue $\alpha\big(t_E \exp(\mathrm{i}2\pi j/n)\big)$ (cf. Section 10.2).

So

$$\begin{aligned}
&\big(\alpha(tt_E)\big)_n^\wedge L_n t^j - \big(\alpha(w+0) P_{\mathbb{T}} + \alpha(w-0) Q_{\mathbb{T}}\big) t^j \\
&= [\alpha(tt_E)_n^\wedge] (L_n - K_n^0) t^j + \alpha\big(t_E \exp(\mathrm{i}2\pi j/n)\big) (K_n^0 - I) t^j \\
&\quad + \begin{cases} \big[\alpha\big(t_E \exp(\mathrm{i}2\pi j/n)\big) - \alpha(w+0)\big] t^j, & \text{if } j \geqq 0, \\ \big[\alpha\big(t_E \exp(\mathrm{i}2\pi j/n)\big) - \alpha(w-0)\big] t^j, & \text{if } j < 0, \end{cases}
\end{aligned}$$

and the definition of t_E in Section 10.32 shows that $\big(\alpha(tt_E)\big)_n^\wedge L_n t^j \to \big(\alpha(w+0) P_{\mathbb{T}} + \alpha(w-0) Q_{\mathbb{T}}\big) t^j$. All these facts together lead to the equality

$$\begin{aligned}
W_{(w,1)}\{B_n\} &= c\big(\alpha(w+0) P_{\mathbb{T}} + \alpha(w-0) Q_{\mathbb{T}}\big) \\
&\quad + d\big(\beta(w+0) P_{\mathbb{T}} + \beta(w-0) Q_{\mathbb{T}}\big).
\end{aligned} \tag{1}$$

Furthermore, the system $\{g_j^{(n)} : j = 0, \ldots, n-1\}$ forms an orthogonal basis. Consequently, for any $D_n \in \mathcal{L}(\operatorname{im} L_n)$, the matrix of the adjoint operator D_n^* corresponding to this basis is exactly the adjoint matrix of D_n. We conclude that $\tilde{c}_n^* = (\bar{c})_n^\sim$ and $\alpha_n^* = (\bar{\alpha})_n$, where c and α denote the complex conjugates to c and α. The equality $U_n^{-1} = U_n^*$ implies that $\hat{\alpha}_n^* = (\bar{\alpha})_n^\wedge$ and

$$B_n^* = (\bar{\alpha})_n^\wedge (\bar{c})_n^\sim + (\bar{\beta})_n^\wedge (\bar{d})_n^\sim. \tag{2}$$

Using this, the strong limits established above, and formula (1), we see that $\big(E_n^{(w,1)} B_n (E_n^{(w,1)})^{-1} L_n\big)^* \to (W_{(w,1)}\{B_n\})^*$.

In order to compute $W_{(v,2)}\{B_n\}$ we have to study the strong convergence of $E_n^{(v,2)} \hat{\alpha}_n (E_n^{(v,2)})^{-1} P_n$ and $E_n^{(v,2)} \tilde{c}_n (E_n^{(v,2)})^{-1} P_n$. We have (cf. the definition of $E_n^{(v,2)}$, 10.32(4) and 10.32(12))

$$\begin{aligned}
E_n^{(v,2)} \hat{\alpha}_n (E_n^{(v,2)})^{-1} P_n &= E_n\big(T_n^{(v,\varepsilon)} \hat{\alpha}_n (T_n^{(v,\varepsilon)})^{-1}\big) (E_n)^{-1} P_n \\
&= E_n \hat{\alpha}_n (E_n)^{-1} P_n = M_n \tilde{\alpha} P_n.
\end{aligned}$$

Theorem 7.17 yields that $M_n \tilde{\alpha} P_n \to \tilde{\alpha}$. (Notice that, for $n = 2k+1$, the projections M_n and P_n of this section coincide with L_k and P_k of Chapter 7. Thus Theorem 7.1 applies. For $n = 2k$, the assertion of this theorem can be proved in an analogous manner.) Furthermore, from the definition of $E_n^{(v,2)}$, 10.32(5) and 10.32(12) we conclude that

$$\begin{aligned}
E_n^{(v,2)} \tilde{c}_n (E_n^{(v,2)})^{-1} P_n &= E_n\big(T_n^{(v,\varepsilon)} \tilde{c}_n (T_n^{(v,\varepsilon)})^{-1}\big) (E_n)^{-1} P_n \\
&= E_n\big(c(\tau_E t)\big)_n^\sim (E_n)^{-1} P_n \\
&= E_n\Big(c\big(\tau_E \exp(\mathrm{i}2\pi(j+\varepsilon)/n)\big)\, \delta_{jk}\Big)_{j,k=0}^{n-1} (E_n)^{-1} P_n.
\end{aligned}$$

Again, it suffices to show the strong convergence on a dense subset. We fix an integer l and get (cf. the definition of E_n)

$$E_n^{(v,2)}\tilde{c}_n(E_n^{(v,2)})^{-1}P_nt^l = c\big(\tau_E \exp\left(\mathrm{i}2\pi(l+\varepsilon)/n\right)\big)\,t^l.$$

The definition of τ_E implies that

$$E_n^{(v,2)}\tilde{c}_n(E_n^{(v,2)})^{-1}P_nt^l \to \begin{cases} c(v+0)\,t^l & \text{if} \quad l \geqq 0, \\ c(v-0)\,t^l & \text{if} \quad l < 0, \end{cases}$$

$$E_n^{(v,2)}\tilde{c}_n(E_n^{(v,2)})^{-1}P_n \to c(v+0)\,P_{\mathbb{T}} + c(v-0)\,Q_{\mathbb{T}}.$$

Putting the things together leads to the equality

$$W_{(v,2)}\{B_n\} = \big(c(v+0)\,P_{\mathbb{T}} + c(v-0)\,Q_{\mathbb{T}}\big)\,\tilde{\alpha} + \big(d(v+0)\,P_{\mathbb{T}} + d(v-0)\,Q_{\mathbb{T}}\big)\,\tilde{\beta}. \tag{3}$$

By the strong limits just established and by (2) we get

$$\big(E_n^{(v,2)}B_n(E_n^{(v,2)})^{-1}P_n\big)^* \to (W_{(v,2)}\{B_n\})^*.$$

(c) The first assertions of (c.1) and (c.2) (the strong convergence to T) are obvious, because L_n and P_n are selfadjoint and converge strongly to the identity as $n \to \infty$. Furthermore, since the operators $E_n^{(w,1)}$ and $E_n^{(v,2)}$ are unitary, the strong convergence for the adjoint operators follows from the strong convergence of the sequences of the original operators. Now we prove the second property of (c.1). We consider two cases: $j = 1$ and $v \neq w$, $j = 2$ and $v \in \mathbb{T}$ arbitrary. To prove (c.1) in the first case, it is sufficient to verify that $L_nE_n^{(w,1)}(E_n^{(v,1)})^{-1}L_n \to 0$ weakly as $n \to \infty$ ($v \neq w$!). It is enough to prove the following:

$$sp(n) := \left(g_l^{(s)}, \sum_{j=0}^{n-1} \big(t_E(v,n)/t_E(w,n)\big)^j\,(g_i^{(k)}, g_j^{(n)})\,g_j^{(n)}\right) \to 0 \quad \text{as} \quad n \to \infty,$$

for all $l, s, i, k \in \mathbb{N}$, $0 \leqq l \leqq s-1$, $0 \leqq i \leqq k-1$. Computing the scalar product (recall the definition of $g_j^{(n)}$) we get

$$sp(n) = \frac{\sqrt{s}\,\sqrt{k}}{n} \sum_{j=j_0}^{j_1} \big(t_E(v,n)/t_E(w,n)\big)^j + O\left(\frac{1}{n}\right),$$

where $0 \leqq j_0 < j_1 \leqq n-1$. Since $w' := v/w \neq 1$, we obtain that (if only n is large enough)

$$\left|\sum_{j=j_0}^{j_1} \big(t_E(v,n)/t_E(w,n)\big)^j\right| \leqq \frac{4}{|(1-w')|},$$

and it follows that $sp(n)$ converges to zero as $n \to \infty$ for each $s, k \in \mathbb{N}$. Now we consider the case where $j = 2$ and v is arbitrary. We shall deduce that $L_nE_n^{(w,1)}(E_n^{(v,2)})^{-1}P_n \to 0$ weakly as $n \to \infty$. To prove this it suffices to verify that

$$\big(g_l^{(s)}, E_n^{(w,1)}(E_n^{(v,2)})^{-1}t^k\big) \to 0, \qquad n \to \infty,$$

for all $l, s \in \mathbb{N}$, $k \in \mathbb{Z}$, $0 \leqq l \leqq s-1$. Note that $E_n^{(w,1)}(E_n^{(v,2)})^{-1} t^k$ is equal to $\exp\left(-2\pi i j_E(w, n)\,(k + k_E(\varepsilon, v, n))/n\right) g^{(n)}_{k+k_E(\varepsilon,v,n)}$. Hence $\left|(g_l^{(s)}, E_n^{(w,1)}(E_n^{(v,2)})^{-1} t^k)\right| \leqq \frac{\sqrt{s}}{\sqrt{n}} \to 0$. This completes the proof of (c.1).

To obtain (c.2) we also consider two cases: $j = 1$ and $w \in \mathbb{T}$ arbitrary, $j = 2$ and $w \neq v$. In the first case it is enough to prove that $P_nTP_nE_n^{(v,2)}(E_n^{(w,1)})^{-1} L_n \to 0$ strongly as $n \to \infty$. A little thought shows that it is sufficient to establish convergence for an arbitrary finite rank operator $T_1 : f \mapsto (f, t^p)\, t^q$, where $p, q \in \mathbb{Z}$. Consequently, we have to prove that, for $m, s \in \mathbb{N}$, $0 \leqq m \leqq s-1$, $P_nT_1P_nE_n^{(v,2)}(E_n^{(w,1)})^{-1} L_n g_m^{(s)}$ converges to zero as $n \to \infty$. In view of the definition of $E_n^{(w,1)}$ and $E_n^{(v,2)}$ we have

$$P_nT_1P_nE_n^{(v,2)}(E_n^{(w,1)})^{-1} L_n g_m^{(s)} = (g_m^{(s)}, g_l^{(n)})\, t_E(w, n)^l\, t^q,$$

where l denotes the number in $\{-[n/2], \ldots, [(n-1)/2]\}$ such that $p + k_E(\varepsilon, v, n) - l$ is a multiple of n. This term tends to zero as $n \to \infty$ (cf. the proof of (c.1)). Analogously for $j = 2$ and $w \neq v$, it suffices to show that $P_nT_1P_nE_n^{(v,2)}(E_n^{(w,2)})^{-1} t^m \to 0$ $(n \to \infty)$, where $m \in \mathbb{Z}$ and T_1 is the finite rank operator defined above. We have $E_n^{(v,2)}(E_n^{(w,2)})^{-1} t^m = t^l$, where l denotes the number in $\{-[n/2], \ldots, [(n-1)/2]\}$ such that $m + k_E(\varepsilon, w, n) - k_E(\varepsilon, v, n) - l$ is a multiple of n. Consequently,

$$P_nT_1P_nE_n^{(v,2)}(E_n^{(w,2)})^{-1} t^m = \begin{cases} t^q, & \text{if } p = l, \\ 0, & \text{if } p \neq l. \end{cases}$$

By our assumption that $w \neq v$, it follows that $|k_E(\varepsilon, w, n) - k_E(\varepsilon, v, n)| \to \infty$ as $n \to \infty$ (cf. the definition of $k_E(\varepsilon, \tau, n)$ in Section 10.32). Hence $P_nT_1P_nE_n^{(v,2)}(E_n^{(w,2)})^{-1} t^m = 0$ for n large enough. This completes the proof of (c.2). ■

Due to Lemma 10.34 the operator sequences $\{B_n\}$ in 10.32(1) belong to the C^*-algebra $\mathfrak{B}$. We define the C^*-algebra $\mathfrak{Z}_1$ as the smallest closed C^*-algebra containing the set $\left\{\sum_{i=1}^{m} \prod_{j=1}^{l} \{B_n^{ij}\}\right\}$, where $l, m, n \in \mathbb{N}$, and the $\{B_n^{ij}\}$ are sequences of the form 10.32(1). In view of Lemma 10.34(c) it follows that J forms a two-sided closed ideal in $\mathfrak{B}$, where J is the closure in $\mathfrak{B}$ of

$$J_0 := \left\{\sum_{k=1}^{r} \{(E_n^{(w_k,1)})^{-1} L_nT_kE_n^{(w_k,1)}\} + \sum_{j=1}^{1} \{(E_n^{(v_j,2)})^{-1} P_nT_jE_n^{(v_j,2)}\} + \{C_n\} :\right.$$
$$\left. w_k, v_j \in \mathbb{T};\ T_k, T_j \in \mathcal{K}(\mathbf{L}^2(\mathbb{T}));\ \|C_n\| \to 0;\ r, l \in \mathbb{N}\right\}$$

(cf. Section 10.33). Standard C^*-algebra arguments show that $\mathfrak{Z}_1 + J$ is a C^*-subalgebra of $\mathfrak{B}$. We remark that J is actually contained in $\mathfrak{Z}_1$. This can be proved with the help of the arguments that will be presented in the proof of Lemma 10.36. However, this fact is not needed for what follows and does not influence our further consideration. We put $\mathfrak{Z} := \mathfrak{Z}_1 + J$ and $\mathfrak{Z}^\circ := \mathfrak{Z}/J$.

10.35. Corollary. *The sequence $\{A_n\} \in \mathfrak{Z}$ is stable if and only if the operators $W_{(w,1)}\{A_n\}$, $W_{(v,2)}\{A_n\}$ are invertible for all $w, v \in \mathbb{T}$ and the coset $\{A_n\}^\circ := \{A_n\} + J$ is invertible in $\mathfrak{Z}^\circ$.*

Proof. Since $\mathfrak{Z}^\circ$ is a C^*-subalgebra of $\mathfrak{B}^\circ := \mathfrak{B}/J$, the assertion follows from Theorem 10.33 and standard C^*-algebra arguments. ■

Now we shall study some properties of the algebra $\mathfrak{Z}^\circ$. Note that Corollary 10.35 necessitates the consideration of the invertibility of $\{A_n\}^\circ$ in $\mathfrak{Z}^\circ$. To this end it will be helpful to know something about the center of $\mathfrak{Z}^\circ$.

10.36. Lemma. *The cosets* $\{\tilde{c}_n\}^0$ *and* $\{\hat{\alpha}_n\}^0$ $(c, \alpha \in \mathbf{C}(\mathbb{T}))$ *belong to the center of* $\mathfrak{Z}^0$.

Proof. (a) First of all, we show that $\{\tilde{c}_n\hat{\alpha}_n - \hat{\alpha}_n\tilde{c}_n\} \in J$ for $c \in \mathbf{C}(\mathbb{T})$ and $\alpha \in \mathbf{PC}(\mathbb{T})$. To prove this, we fix $v \in \mathbb{T}$ and set (cf. Section 2.21)

$$\lambda(e^{i2\pi s}) := \begin{cases} 1, & \text{if } s = 0 \\ \dfrac{\sigma_{0,1}^-(s)}{\sigma_{0,1}^+(s)}, & \text{if } 0 < s < 1. \end{cases}$$

Since the only point of discontinuity of λ is 1, we get (cf. Section 10.32) that

$$\left\|\left(\lambda(v^{-1}\exp(i2\pi j/n))\,\delta_{jk}\right)_{j,k=0}^{n-1} - \left(\lambda(t_E(v,n)^{-1}\exp(i2\pi j/n))\,\delta_{jk}\right)_{j,k=0}^{n-1}\right\|$$

goes to zero as n approaches infinity. Hence $\left\|\left(\lambda(v^{-1}t)\right)_n^\wedge - \left(\lambda(t_E(v,n)^{-1}t)\right)_n^\wedge\right\| \to 0$ and 10.32(9) yields that $\left\|\left(\lambda(v^{-1}t)\right)_n^\wedge - (E_n^{(v,1)})^{-1}\hat{\lambda}_n E_n^{(v,1)}\right\| \to 0$. What we arrive at is that

$$\tilde{c}_n\left(\lambda(v^{-1}t)\right)_n^\wedge - \left(\lambda(v^{-1}t)\right)_n^\wedge \tilde{c}_n = C_n + (E_n^{(v,1)})^{-1}(\tilde{c}_n\hat{\lambda}_n - \hat{\lambda}_n\tilde{c}_n)\,E_n^{(v,1)},$$

where $\|C_n\| \to 0$.

It is easy to see that $\|\tilde{c}_n - L_ncL_n\| \to 0$. Because of this and 10.8(3) and since $\|L_nc(I - L_n)\| \to 0$ and $\|(I - L_n)\,cL_n\| \to 0$ (cf. Sections 2.13 and 2.14), we obtain that

$$\tilde{c}_n\left(\lambda(v^{-1}t)\right)_n^\wedge - \left(\lambda(v^{-1}t)\right)_n^\wedge \tilde{c}_n = C_n' + (E_n^{(v,1)})^{-1}L_n(cS_{\mathbb{T}} - S_{\mathbb{T}}c)\,E_n^{(v,1)},$$

where $\|C_n'\| \to 0$. Thus $\tilde{c}_n\left(\lambda(v^{-1}t)\right)_n^\wedge - \left(\lambda(v^{-1}t)\right)_n^\wedge \tilde{c}_n \in J_0$ and so the function $t \mapsto \lambda(v^{-1}t)$ belongs to the set of functions α such that $\{\tilde{c}_n\hat{\alpha}_n - \hat{\alpha}_n\tilde{c}_n\} \in J$. However, this set is a closed subalgebra of $\mathbf{PC}(\mathbb{T})$. Since the smallest closed subalgebra containing all the functions $t \mapsto \lambda(v^{-1}t)$ $(v \in \mathbb{T})$ is $\mathbf{PC}(\mathbb{T})$ we conclude that $\{\tilde{c}_n\hat{\alpha}_n - \hat{\alpha}_n\tilde{c}_n\} \in J$ for any $c \in \mathbf{C}(\mathbb{T})$, $\alpha \in \mathbf{PC}(\mathbb{T})$.

(b) We now show that $\{\tilde{c}_n\hat{\alpha}_n - \hat{\alpha}_n\tilde{c}_n\} \in J$ for $c \in \mathbf{PC}(\mathbb{T})$ and $\alpha \in \mathbf{C}(\mathbb{T})$. We fix $w \in \mathbb{T}$ and set $\alpha(t) := t^{-1}$ as well as $f(\exp(i2\pi s)) = 1 - 2s$, $0 \leq s < 1$. We have (cf. Section 10.32)

$$\left\|\left(f(w^{-1}\exp(i2\pi j/n))\,\delta_{jk}\right)_{j,k=0}^{n-1} - \left(f(\tau_E(\varepsilon, w, n)^{-1}\exp(i2\pi j/n))\,\delta_{jk}\right)_{j,k=0}^{n-1}\right\| \to 0,$$

and (cf. 10.32(5)) $\left\|\left(f(w^{-1}t)\right)_n^\sim - (T_n^{(w,\varepsilon)})^{-1}\tilde{f}_nT_n^{(w,\varepsilon)}\right\| \to 0$. By 10.32(4) it follows that

$$\left(f(w^{-1}t)\right)_n^\sim \hat{\alpha}_n - \hat{\alpha}_n\left(f(w^{-1}t)\right)_n^\sim = C_n + (T_n^{(w,\varepsilon)})^{-1}(\tilde{f}_n\hat{\alpha}_n - \hat{\alpha}_n\tilde{f}_n)\,T_n^{(w,\varepsilon)}, \quad \|C_n\| \to 0. \quad (1)$$

The element in the j-th row and k-th column of $\hat{\alpha}_n$ equals $\tilde{\delta}_{j,k+1}$, where $\tilde{\delta}_{j,k+1}$ is 1 for $j = k + 1$, $k = 0, \ldots, n - 2$ or $j = 0$, $k = n - 1$ and is 0 in any other case. Hence

$$\begin{aligned}\tilde{f}_n\hat{\alpha}_n - \hat{\alpha}_n\tilde{f}_n &= \left(\tilde{\delta}_{j,k+1}\left[f(\exp(i2\pi(k + 1 + \varepsilon)/n)) - f(\exp(i2\pi(k + \varepsilon)/n))\right]\right)_{j,k=0}^{n-1} \\ &= O\left(\frac{1}{n}\right) + (2\delta_{j0}\delta_{k,n-1})_{j,k=0}^{n-1}. \end{aligned} \quad (2)$$

Since $(S_{\mathbb{T}}t - tS_{\mathbb{T}})$ maps $\sum \xi_kt^k$ to $2\xi_{-1}$, we get that $(2\delta_{j0}\delta_{k,n-1})_{j,k=0}^{n-1}$ is the matrix of $E_n^{-1} \times P_n(S_{\mathbb{T}}\tilde{\alpha} - \tilde{\alpha}S_{\mathbb{T}})\,E_n$. Consequently, (1) and (2) yield that

$$\left(f(w^{-1}t)\right)_n^\sim \hat{\alpha}_n - \hat{\alpha}_n\left(f(w^{-1}t)\right)_n^\sim = C_n' + (E_n^{(w,2)})^{-1}P_n(S_{\mathbb{T}}\tilde{\alpha} - \tilde{\alpha}S_{\mathbb{T}})\,E_n^{(w,2)}, \quad \|C_n'\| \to 0,$$

i.e., the function $t \mapsto f(w^{-1}t)$ belongs to the set of all $c \in \mathbf{PC}(\mathbb{T})$ such that $\{\tilde{c}_n\hat{\alpha}_n - \hat{\alpha}_n\tilde{c}_n\} \in J$. Analogously to part (a) of this proof, we conclude that $\{\tilde{c}_n\hat{\alpha}_n - \hat{\alpha}_n\tilde{c}_n\} \in J$ for any $c \in \mathbf{PC}(\mathbb{T})$, where $\alpha(t) = t^{-1}$. The formula

$$\tilde{c}_n(\alpha^{-1})\hat{}_n - (\alpha^{-1})\hat{}_n\, \tilde{c}_n = (\alpha^{-1})\hat{}_n\, (\hat{\alpha}_n\tilde{c}_n - \tilde{c}_n\hat{\alpha}_n)\, (\alpha^{-1})\hat{}_n$$

implies that $\alpha(t) := t$ belongs to the set of all functions such that $\{\tilde{c}_n\hat{\alpha}_n - \hat{\alpha}_n\tilde{c}_n\} \in J$. However, a closed subalgebra of $\mathbf{C}(\mathbb{T})$ containing $\alpha(t) = t$ and $\alpha(t) = t^{-1}$ coincides with $\mathbf{C}(\mathbb{T})$. Thus $\{\tilde{c}_n\hat{\alpha}_n - \hat{\alpha}_n\tilde{c}_n\} \in J$ for any $\alpha \in \mathbf{C}(\mathbb{T})$, $c \in \mathbf{PC}(\mathbb{T})$.

(c) The lemma follows from (a), (b) and the fact that $\tilde{c}_n\tilde{d}_n = \tilde{d}_n\tilde{c}_n = (cd)\tilde{}_n$, $\hat{\alpha}_n\hat{\beta}_n = \hat{\beta}_n\hat{\alpha}_n = (\alpha\beta)\hat{}_n$ for any $c, d, \alpha, \beta \in \mathbf{PC}(\mathbb{T})$. ■

In view of Lemma 10.36 the subalgebra of $\mathfrak{Z}^\circ$ generated by the sequences $\{\tilde{c}_n\}^0$, $\{\hat{\alpha}_n\}^0$ $(c, \alpha \in \mathbf{C}(\mathbb{T}))$ lies in the center of $\mathfrak{Z}^\circ$. We shall identify the maximal ideal space of the above algebra with $\mathbb{T} \times \mathbb{T}$.

Remark. In accordance with 10.34(1), 10.34(3) we have, for $c, \alpha \in \mathbf{C}(\mathbb{T})$ and $v, w \in \mathbb{T}$,

$$W_{(w,1)}\{\tilde{c}_n\hat{\alpha}_n\} = c\alpha(w), \qquad W_{(v,2)}\{\tilde{c}_n\hat{\alpha}_n\} = c(v)\,\tilde{\alpha}.$$

Therefore it is possible to define, for every $(v, w) \in \mathbb{T} \times \mathbb{T}$, a multiplicative linear functional. This shows that the maximal ideal space of the algebra generated by $\{\tilde{c}_n\}^\circ$, $\{\hat{\alpha}_n\}^\circ$ $(c, \alpha \in \mathbf{C}(\mathbb{T}))$ is $\mathbb{T} \times \mathbb{T}$.

Now let us apply the localization principle of Allan-Douglas. For a given maximal ideal $(v, w) \in \mathbb{T} \times \mathbb{T}$, let $J^\circ_{v,w}$ denote the closed ideal in $\mathfrak{Z}^\circ$ generated by (v, w), i.e., the closed ideal generated by all $\{\hat{\alpha}_n\}^\circ$, $\{\tilde{c}_n\}^\circ$ such that $c, \alpha \in \mathbf{C}(\mathbb{T})$, $\alpha(w) = 0$ and $c(v) = 0$. We put $\mathfrak{Z}^\circ_{v,w} := \mathfrak{Z}^\circ/J^\circ_{v,w}$ and $\{A_n\}^\circ_{v,w} := \{A_n\}^\circ + J^\circ_{v,w}$.

10.37. Lemma. *Let p_{v_0} be the characteristic function of the arc $(v_0, -v_0) \subset \mathbb{T}$ and q_{w_0} that of the* arc $(w_0, -w_0) \subset \mathbb{T}$, *put $p := (p_{v_0})\tilde{}_n$ and $q := (q_{w_0})\hat{}_n$. Then, for an arbitrary coset $\{B_n\}^\circ_{v_0,w_0} = \{\tilde{c}_n\hat{\alpha}_n\}^\circ_{v_0,w_0} \in \mathfrak{Z}^\circ_{v_0,w_0}$, we have the following representation*

$$\{B_n\}^\circ_{v_0,w_0} = \big(c(v_0 + 0)\, p^\circ_{v_0,w_0} + c(v_0 - 0)\, (\{I_n\} - p)^\circ_{v_0,w_0}\big)$$
$$\times \big(\alpha(w_0 + 0)\, q^\circ_{v_0,w_0} + \alpha(w_0 - 0)\, (\{I_n\} - q)^\circ_{v_0,w_0}\big).$$

Therefore, the algebra $\mathfrak{Z}^\circ_{v_0,w_0}$ is generated by the cosets $\{I_n\}^\circ_{v_0,w_0}$, $p^\circ_{v_0,w_0}$ and $q^\circ_{v_0,w_0}$.

Proof. This is immediate from the definition of $J^\circ_{v_0,w_0}$. ■

10.38. Lemma. *With the notations of Lemma* 10.37, *the following properties hold:*

(i) $(p^\circ_{v_0,w_0})\,(p^\circ_{v_0,w_0}) = p^\circ_{v_0,w_0}$,

(ii) $(q^\circ_{v_0,w_0})\,(q^\circ_{v_0,w_0}) = q^\circ_{v_0,w_0}$

(iii) *the spectrum* $\mathrm{sp}_{\mathfrak{Z}^\circ_{v_0,w_0}}\big((qpq)^\circ_{v_0,w_0}\big)$ *of* $(qpq)^\circ_{v_0,w_0}$ *in* $\mathfrak{Z}^\circ_{v_0,w_0}$ *is* $[0, 1]$.

Proof. The property (i) follows from the identity $(p_{v_0})\tilde{}_n\, (p_{v_0})\tilde{}_n = (p^2_{v_0})\tilde{}_n = (p_{v_0})\tilde{}_n$. The proof of (ii) is analogous. So we are left with the proof of (iii). Consider $qpq \in \mathfrak{Z}$. This element is selfadjoint, has norm less than or equal to one and is positive (the last assertion can be proved by a straightforward computation). What follows is that the spectrum $\mathrm{sp}_{\mathfrak{Z}}\,(qpq)$ is contained in the segment $[0, 1]$. Consequently,

$$\mathrm{sp}_{\mathfrak{Z}^\circ_{v_0,w_0}}\big((qpq)^\circ_{v_0,w_0}\big) \subset [0, 1]. \tag{1}$$

Let $r_{v_0}(s_{w_0}) \in \mathbf{PC}(\mathbb{T})$ be a function which coincides with $p_{v_0}(q_{w_0})$ on a sufficiently small open arc $U_{v_0} \ni v_0(U_{w_0} \ni w_0)$, is continuous on $\mathbb{T} \setminus \{v_0\}$ $(\mathbb{T} \setminus \{w_0\})$ and takes values in $V := \{z: |z - 1/2| = 1/2, \operatorname{Im} z \geqq 0\}$. We put $r := \{(r_{v_0})_n^{\sim}\}$, $s = \{(s_{w_0})_n^{\wedge}\}$. It is easily seen (cf. Lemma 10.37) that $(srs)^\circ_{v_0,w_0} = (qpq)^\circ_{v_0,w_0}$, i.e.,

$$\operatorname{sp}_{\mathfrak{Z}^\circ_{v_0,w_0}}\left((srs)^\circ_{v_0,w_0}\right) = \operatorname{sp}_{\mathfrak{Z}^\circ_{v_0,w_0}}\left((qpq)^\circ_{v_0,w_0}\right). \tag{2}$$

Furthermore, we have

$$\operatorname{sp}_{\mathfrak{Z}^\circ_{v,w}}\left((srs)^\circ_{v,w}\right) = \begin{cases} \{r_{v_0}(v)\, s_{w_0}(w)^2\} & \text{if } v \neq v_0,\ w \neq w_0, \\ \{0, r_{v_0}(v)\} & \text{if } v \neq v_0,\ w = w_0, \\ \{0, s_{w_0}(w)^2\} & \text{if } v = v_0,\ w \neq w_0 \end{cases} \tag{3}$$

(cf. Lemma 10.37 and the definition of r_{v_0}, s_{w_0}). Formula 10.34(1) yields that

$$\psi_{(w_0,1)}(srs) := W_{(w_0,1)}(srs) = P_{\mathbb{T}} r_{v_0} P_{\mathbb{T}}. \tag{4}$$

The mapping $\psi_{(w_0,1)}$ is a star-homomorphism from the algebra $\mathfrak{Z}$ onto the algebra $\operatorname{alg}\{P_{\mathbb{T}}, \mathbf{PC}\}$ (here $\operatorname{alg}\{P_{\mathbb{T}}, \mathbf{PC}\}$ refers to the subalgebra of $\mathcal{L}(\mathbf{L}^2(\mathbb{T}))$ generated by $\mathbf{PC}$-functions and the singular integral operator $S_{\mathbb{T}}$). From Lemma 10.34 we obtain that $\psi_{(w_0,1)}(J) \subset \mathfrak{K}$, where $\mathfrak{K}$ denotes the two-sided ideal of the compact operators in $\operatorname{alg}\{P_{\mathbb{T}}, \mathbf{PC}\}$. Consequently, $\psi_{(w_0,1)}$ generates a star-homomorphism $\psi^\circ_{(w_0,1)}$ from $\mathfrak{Z}^\circ$ onto $\operatorname{alg}\{P_{\mathbb{T}}, \mathbf{PC}\}^\circ = \operatorname{alg}\{P_{\mathbb{T}}, \mathbf{PC}\}/\mathfrak{K}$. We put $(P_{\mathbb{T}} r_{v_0} P_{\mathbb{T}})^\circ := P_{\mathbb{T}} r_{v_0} P_{\mathbb{T}} + \mathfrak{K}$. Hence, by (4), we deduce that

$$[0, 1] = \operatorname{sp}_{\operatorname{alg}\{P_{\mathbb{T}},\mathbf{PC}\}^\circ}\left((P_{\mathbb{T}} r_{v_0} P_{\mathbb{T}})^\circ\right) \subset \operatorname{sp}_{\mathfrak{Z}^\circ}\left((srs)^\circ\right). \tag{5}$$

According to the localization principle of Allan-Douglas (Theorem 1.21)

$$\operatorname{sp}_{\mathfrak{Z}^\circ}\left((srs)^\circ\right) = \bigcup_{(v,w)\in\mathbb{T}\times\mathbb{T}} \operatorname{sp}_{\mathfrak{Z}^\circ_{v,w}}\left((srs)^\circ_{v,w}\right). \tag{6}$$

Now (5), (6) and (3) (recall the definition of V) give the inclusion $[0, 1] \subset \operatorname{sp}_{\mathfrak{Z}^\circ_{v_0,w_0}}\left((srs)^\circ_{v_0,w_0}\right)$, and (1), (2) yield the equality

$$\operatorname{sp}_{\mathfrak{Z}^\circ_{v_0,w_0}}\left((qpq)^\circ_{v_0,w_0}\right) = [0, 1]. \ \blacksquare$$

10.39. Theorem. *Let $\bar{e}$, $\bar{p}$, $\bar{q} \in \mathbf{C}([0, 1], \mathcal{L}(\mathbb{C}^2))$ be defined by*

$$\begin{pmatrix} 1 & 0 \\ 0 & 1 \end{pmatrix}, \quad \begin{pmatrix} x & \sqrt{x(1-x)} \\ \sqrt{x(1-x)} & 1-x \end{pmatrix}, \quad \begin{pmatrix} 1 & 0 \\ 0 & 0 \end{pmatrix},$$

respectively.

The C^*-algebra $\mathfrak{Z}^\circ_{v,w}(v, w \in \mathbb{T})$ and the subalgebra $\operatorname{alg}(\bar{e}, \bar{p}, \bar{q})$ *of $\mathbf{C}([0, 1], \mathcal{L}(\mathbb{C}^2))$ generated by $\bar{e}$, $\bar{p}$ and $\bar{q}$ are isometrically star-isomorphic. The isomorphism will be denoted by $\varphi^\circ_{v,w}$. It takes $\{I_n\}^\circ_{v,w}$, $p^\circ_{v,w}$, $q^\circ_{v,w}$ into $\bar{e}$, $\bar{p}$, $\bar{q}$, respectively.*

Proof. This is a consequence of the Lemmas 10.37, 10.38 and the "Two-Projections-Theorem" by Halmos (Theorem 1.16) $\blacksquare$

10.40. The local symbol of an operator sequence. For a given $\{A_n\} \in \mathfrak{Z}$ the matrix $\varphi^\circ_{v,w}(\{A_n\}^\circ_{v,w})$ is called the local symbol of $\{A_n\}$ at the point $(v, w) \in \mathbb{T} \times \mathbb{T}$.

Theorem. *The coset* $\{A_n\}^\circ \in \mathfrak{Z}^\circ$ *is invertible in* $\mathfrak{Z}^0$ *if and only if the local symbol* $\varphi^\circ_{v,w}(\{A_n\}^\circ_{v,w}) \in \operatorname{alg}(\bar{e}, \bar{p}, \bar{q})$ *is invertible for all* $(v, w) \in \mathbb{T} \times \mathbb{T}$.

Proof. This is immediate from Theorem 10.39 and the localization principle of Allan-Douglas (Theorem 1.21). ■

Let us, for example, compute the local symbol of the operator sequence $\{B_n\}$ in 10.32(1) at the point (v, w). In accordance to the definition of $\varphi^\circ_{v,w}$ we obtain that (cf. Lemma 10.37)

$$\varphi^\circ_{v,w}(\{B_n\}^\circ_{v,w}) = \big(a_{ij}(v, w, x)\big)^2_{i,j=1}, \tag{1}$$

$$a_{11}(v, w, x) = x[c(v+0)\,\alpha(w+0) + d(v+0)\,\beta(w+0)]$$
$$+ (1-x)\,[c(v-0)\,\alpha(w+0) + d(v-0)\,\beta(w+0)],$$

$$a_{12}(v, w, x) = \sqrt{x(1-x)}\,\big[\big(c(v+0) - c(v-0)\big)\,\alpha(w-0)$$
$$+ \big(d(v+0) - d(v-0)\big)\,\beta(w-0)\big],$$

$$a_{21}(v, w, x) = \sqrt{x(1-x)}\,\big[\big(c(v+0) - c(v-0)\big)\,\alpha(w+0)$$
$$+ \big(d(v+0) - d(v-0)\big)\,\beta(w+0)\big],$$

$$a_{22}(v, w, x) = x[c(v-0)\,\alpha(w-0) + d(v-0)\,\beta(w-0)]$$
$$+ (1-x)\,[c(v+0)\,\alpha(w-0) + d(v+0)\,\beta(w-0)].$$

10.41. Theorem. *The operator sequence* $\{A_n\} \in \mathfrak{Z}$ *is stable if and only if the operators* $W_{(w,1)}\{A_n\}$ *and* $W_{(v,2)}\{A_n\}$ *are invertible for all* $v, w \in \mathbb{T}$.

Proof. In view of the Theorems 10.40 and 10.33 the sequence $\{A_n\}$ is stable if and only if the operators $W_{(w,1)}\{A_n\}$ and $W_{(v,2)}\{A_n\}$ as well as the matrix functions $\varphi^\circ_{v,w}(\{A_n\}^\circ_{v,w})$ are invertible for all $v, w \in \mathbb{T}$. So Theorem 10.41 will follow once we have shown that the invertibility of $W_{(w,1)}(\{A_n\})$ implies $\varphi^\circ_{v,w}(\{A_n\}^\circ_{v,w})$ to be invertible.

Let $\operatorname{alg}\{P_{\mathbb{T}}, \mathbf{PC}\}$, $\mathfrak{K}$ and $\operatorname{alg}\{P_{\mathbb{T}}, \mathbf{PC}\}^\circ$ be defined as in the proof of Lemma 10.38. It is well known that $\mathbf{C}(\mathbb{T}) \subset \operatorname{alg}\{P_{\mathbb{T}}, \mathbf{PC}\}$ and that $\mathbf{C}(\mathbb{T})^\circ$ is contained in the center of $\operatorname{alg}\{P_{\mathbb{T}}, \mathbf{PC}\}^\circ$. Furthermore, for $v \in \mathbb{T}$, let J°_v denote the ideal of $\operatorname{alg}\{P_{\mathbb{T}}, \mathbf{PC}\}^\circ$ generated by the maximal ideal v of $\mathbf{C}(\mathbb{T})^\circ$ and let $\operatorname{alg}\{P_{\mathbb{T}}, \mathbf{PC}\}^\circ_v$ be the algebra $\operatorname{alg}\{P_{\mathbb{T}}, \mathbf{PC}\}^\circ/J^\circ_v$. Then (cf. Section 6.13) $\operatorname{alg}\{P_{\mathbb{T}}, \mathbf{PC}\}^\circ_v$ is isometrically star-isomorphic to $\operatorname{alg}(\bar{e}, \bar{p}, \bar{q}) \subset \mathbf{C}\big([0, 1], \mathcal{L}(\mathbb{C}^2)\big)$ and the isomorphism $\varphi^\circ_v\colon \operatorname{alg}\{P_{\mathbb{T}}, \mathbf{PC}\}^\circ_v \to \operatorname{alg}(\bar{e}, \bar{p}, \bar{q})$ takes $\{p_v\}^\circ_v$, $\{P_{\mathbb{T}}\}^\circ_v$ and $\{I\}^\circ_v$ into $\bar{p}, \bar{q}$ and $\bar{e}$, respectively. If $W_{(w,1)}\{A_n\}$ is invertible, then $\varphi^\circ_v\big((W_{(w,1)}\{A_n\})^\circ_v\big)$ is invertible, too.

Hence the proof of Theorem 10.41 will be complete as soon as we have shown that $\varphi^\circ_v\big((W_{(w,1)}\{A_n\})^\circ_v\big) = \varphi^\circ_{(w,v)}(\{A_n\}^\circ_{w,v})$. Since both sides of the last equation are continuous, linear, and multiplicative functions of $\{A_n\}$, it suffices to consider the sequences $\{A_n$

$= \hat{\alpha}_n\}$, $\{A_n = \tilde{c}_n\}$. However, in these special cases our equation is an easy consequence of 10.34(1), 10.40(1) and the definition of φ_v°. ■

Remarks. 1°. The assertion (i) of Theorem 10.3 also follows immediately from Theorem 10.41 and the formulas 10.34(1) and 10.34(3).

2°. Using tensor product arguments, one can easily extend the results of Theorem 10.41 to the system case (i.e., to the case where the functions c, d, α, β are matrix functions).

10.42. Corollary. *Let $\mathfrak{Z}_1' \subset \mathfrak{B}(\mathfrak{Z}_1'' \subset \mathfrak{B})$ denote the smallest closed C^*-algebra containing the set $\left\{\sum_{i=1}^{m} \prod_{j=1}^{l} (c^{ij})_n^{\sim} (\alpha^{ij})_n^{\wedge}\right\}$, where $m, l \in \mathbb{N}$, $c^{ij}, \alpha^{ij} \in \mathbf{PC}(\mathbb{T})$ and $\alpha^{ij} \in \mathbf{C}(\mathbb{T} \setminus \{1\})$ $\big(\alpha^{ij} \in \mathbf{C}(\mathbb{T} \setminus \{1, -1\})\big)$. Put $\mathfrak{Z}' := \mathfrak{Z}_1' + J$ $(\mathfrak{Z}'' = \mathfrak{Z}_1'' + J)$. The sequence $\{A_n\} \in \mathfrak{Z}'$ $(\{A_n\} \in \mathfrak{Z}'')$ is stable if and only if the operators $W_{(w,1)}\{A_n\}$ and $W_{(v,2)}\{A_n\}$ are invertible for all $v \in \mathbb{T}$ and $w \in \{1\}$ $(w \in \{1, -1\})$.*

Proof. The proof of the assertion pertaining to $\mathfrak{Z}''$ is completely analogous to the one concerning $\mathfrak{Z}'$. Therefore, we restrict our consideration to $\{A_n\} \in \mathfrak{Z}'$. In view of Theorem 10.41 it remains to show that the invertibility of the operators $W_{(v,2)}\{A_n\}$ (for all $v \in \mathbb{T}$) implies the invertibility of $W_{(w,1)}\{A_n\}$, where $w \in \mathbb{T} \setminus \{1\}$ is fixed.

First, let $A_n = \sum_{i=1}^{m} \prod_{j=1}^{l} (c^{ij})_n^{\sim} (\alpha^{ij})_n^{\wedge}$, where $c^{ij}, \alpha^{ij} \in \mathbf{PC}(\mathbb{T})$, $\alpha^{ij} \in \mathbf{C}(\mathbb{T} \setminus \{1\})$. Then (cf. 10.34(1))

$$W_{(w,1)}\{A_n\} = \sum_{i=1}^{m} \prod_{j=1}^{l} c^{ij}\alpha^{ij}(w) \tag{1}$$

is the operator of multiplication by a function of $\mathbf{PC}(\mathbb{T})$. Formula 10.34(3) and the proof of Theorem 10.41 yield that

$$W_{(v,2)}\{A_n\} = \sum_{i=1}^{m} \prod_{j=1}^{l} \left(c^{ij}(v+0)\, P_{\mathbb{T}} + c^{ij}(v-0)\, Q_{\mathbb{T}}\right) \tilde{\alpha}^{ij},$$

$$(W_{(v,2)}\{A_n\})_{w^{-1}}^\circ = \left[\sum_{i=1}^{m} \prod_{j=1}^{l} c^{ij}(v+0)\, \alpha^{ij}(w)\, P_{\mathbb{T}} + \sum_{i=1}^{m} \prod_{j=1}^{l} c^{ij}(v-0)\, \alpha^{ij}(w)\, Q_{\mathbb{T}}\right]_{w^{-1}}^\circ$$

$$\varphi_{w^{-1}}^\circ\big((W_{(v,2)}\{A_n\})_{w^{-1}}^\circ\big) = \begin{pmatrix} \sum_{i=1}^{m} \prod_{j=1}^{l} c^{ij}(v+0)\, \alpha^{ij}(w) & 0 \\ 0 & \sum_{i=1}^{m} \prod_{j=1}^{l} c^{ij}(v-0)\, \alpha^{ij}(w) \end{pmatrix} \tag{2}$$

Using a density argument and (1), (2), we conclude that $W_{(w,1)}\{A_n\} \in \mathbf{PC}(\mathbb{T})$ for any $\{A_n\} \in \mathfrak{Z}'$ and, for $v \in \mathbb{T}$,

$$\varphi_{w^{-1}}^\circ\big((W_{(v,2)}\{A_n\})_{w^{-1}}^\circ\big) = \begin{pmatrix} (W_{(w,1)}\{A_n\})\,(v+0) & 0 \\ 0 & (W_{(w,1)}\{A_n\})\,(v-0) \end{pmatrix}.$$

If $W_{(v,2)}\{A_n\}$ is invertible, then so also is $\varphi_{w^{-1}}^\circ\big((W_{(v,2)}\{A_n\})_{w^{-1}}^\circ\big)$, i.e., we have $W_{(w,1)}\{A_n\}\,(v \pm 0) \neq 0$. Hence $W_{(w,1)}\{A_n\}$ is invertible, too. ■

Remark. The assertion (ii) of Theorem 10.3 is an easy consequence of Corollary 10.42 and the formulas 10.34(1) and 10.34(3).

Notes and comments

10.1. The application of the method of piecewise linear collocation to singular integral equations apparently dates back to a paper of LAVRENTYEV [1] that was published as early as 1932. MULTHOPP [1] was probably the first to use the trigonometric collocation for the approximate solution of periodic singular integral equations (his paper appeared in 1938). Only in 1961 IVANOV (see his book [1, pp. 200–205]) showed that, if $\varepsilon = 0$ and $\delta = 1$, then in the case of the Cauchy singular operator $A = S_\Gamma$ the coefficient determinants of the collocation system 10.1(2) vanish with $n \to \infty$. GABDULHAEV [1] proved the convergence of the piecewise linear collocation method for $\varepsilon = 0$ in the case of Hölder continuous coefficients, but under the very restrictive additional assumption that $\|d\|_\infty$ is sufficiently small. The convergence of the quadrature method of discrete whirls for the operator $A = S_\Gamma$ was first investigated by LIFANOV/POLONSKI [1].

10.3.–10.7. Theorem 10.3 is a particular case of the following theorem established in PRÖSSDORF/RATHSFELD [4]. *The sequence* $\{B_n\}$, $B_n \in \mathcal{L}(\mathbf{l}^p(n))$, $1 < p < \infty$, *is stable if and only if the operators* $B_\tau \in \mathcal{L}(\mathbf{L}^p(\mathbb{T}))$ *and* $B^\tau \in \mathcal{L}(\mathbf{L}^q(\mathbb{T}))$ *are invertible for all* $\tau \in \mathbb{T}$, *where* $1/p + 1/q = 1$.

For $\varepsilon = 0$ and $\delta = 1$, Corollary 10.7 was first obtained by PRÖSSDORF/SCHMIDT [2–3] in the case of continuous coefficients and by PRÖSSDORF/RATHSFELD [1] in the case of piecewise continuous coefficients. These authors applied methods which are closely related to our technique. That conditions 10.7(i) and (ii) are sufficient for the convergence of the collocation method with $\varepsilon = 0$ and δ odd in the Sobolev spaces $\mathbf{H}^t$, $t \geqq 0$, as well as the corresponding error estimates in the case of smooth coefficients (see Theorems 10.23 and 10.24) was proved by ARNOLD/WENDLAND [1]. Their analysis rests on an equivalence established between the collocation methods and certain Galerkin methods (see Chapter 13). By applying the same technique and utilizing Fourier expansions of the spline functions, SARANEN and WENDLAND [1] succeeded for $\varepsilon = 1/2$ and even d in the case of constant coefficients. These results were then generalized to the case of variable coefficients independently by PRÖSSDORF [5–6] and by ARNOLD/WENDLAND [2] on the basis of localization techniques (see also Chap. 13). In the general case, Corollary 10.7 was established by SCHMIDT [1, 2, 4]. Note that in the works cited here pseudodifferential operators over closed smooth curves are considered, too. In Sections 10.3–10.7 we essentially follow PRÖSSDORF/RATHSFELD [4, 5].

10.8.–10.11. The material of these sections is taken from PRÖSSDORF/RATHSFELD [3–5]. Theorem 10.9 was first proved in PRÖSSDORF [6] and in SCHMIDT [2].

10.12.–10.30. Theorems 10.13(c) and 10.26(c) were established in PRÖSSDORF/RATHSFELD [6] for the more general case of curves with corners (cf. also Chapter 11). Except for Theorems 10.23 and 10.24, all other results of these sections are due to RATHSFELD [2]. Here we essentially follow RATHSFELD [2] and PRÖSSDORF/RATHSFELD [5].

10.31.–10.42. The material of these sections is taken from HAGEN/SILBERMANN [3]. The methods used there can be regarded as an extension of ideas developed in SILBERMANN [1], PRÖSSDORF/RATHSFELD [4], BÖTTCHER/SILBERMANN [1], HAGEN/SILBERMANN [1] and ROCH/SILBERMANN [4]. They can be applied to the stability analysis of a wide variety of approximation methods such as qualocation (introduced by SLOAN [2] and developed further in SLOAN/WENDLAND [1] and HAGEN/SILBERMANN [4]), and collocation and quadrature methods on curves with corners (see Chapter 11).

Chapter 11
Mellin techniques in the numerical analysis of singular integral equations

11.1. Introduction. Many boundary value problems of elasticity, aerodynamics, hydrodynamics, fluid mechanics, electromagnetics, acoustics, and other engineering applications can be reduced to a singular integral equation of the form

$$A_\Gamma u(t) := c(t)\, u(t) + \frac{d(t)}{\pi \mathrm{i}} \int_\Gamma \frac{u(\tau)}{\tau - t}\, \mathrm{d}\tau + \int_\Gamma k(t, \tau)\, u(\tau)\, \mathrm{d}\tau = f(t) \qquad (t \in \Gamma), \tag{1}$$

where Γ is a closed and piecewise smooth curve in the complex plane or the interval $\mathbb{I} := [0, 1]$, and to the pseudodifferential equation of Mellin type

$$A_{\mathbb{I}} u(t) + \int_0^1 k_t\left(\frac{t}{\tau}\right) u(\tau)\, \frac{\mathrm{d}\tau}{\tau} = f(t) \qquad (t \in \mathbb{I}). \tag{2}$$

In (1) c, d and k are given continuous functions, u is the unknown solution and the first integral is to be interpreted as a Cauchy principal value. The kernel function $k_t(\sigma)$ in (2) is of the form

$$k_t(\sigma) := \frac{1}{2\pi \mathrm{i}} \int_{\mathrm{Re}\, z = 1/2} \sigma^{-z} b(t, z)\, \mathrm{d}z,$$

$b(t, z)$ satisfying suitable smoothness properties with respect to t and analyticity and growth properties with respect to z in a certain strip around the line $\mathrm{Re}\, z = 1/2$.

For the numerical solution of the equations (1) and (2), spline approximation methods are extensively employed. In fact, collocation and quadrature methods are the most widely used numerical procedures for solving the boundary integral equations of the form (1) or (2) arising from exterior or interior boundary value problems of applications. See, e.g., Banerjee/Butterfield [1], Belotserkovski/Lifanov [1], Brebbia/Telles/Wrobel [1], Crouch/Starfield [1], Filippi [1], Mukherjee [1], Brebbia/Kuhn/Wendland [1].

If Γ is a closed smooth curve, a fairly complete stability theory and error analysis of collocation and quadrature methods for (1) using smooth splines and uniform meshes was established in the preceding chapter.

In this chapter we present a stability analysis of quadrature and spline collocation methods for (1) in the case where Γ is a closed curve with a finite number of corners. Furthermore, for (2), we consider quadrature methods using suitably graded meshes.

For these cases, the literature on spline approximation and quadrature methods is much less complete. If Γ is a closed curve with corners, Costabel and Stephan (unpublished) proved the strong ellipticity of the operator A_Γ to be sufficient for the $\mathbf{L}^2$-stability of the piecewise linear collocation for equation (1). In this chapter we establish conditions for the stability of the collocation method with piecewise constant trial functions on uniform partitions. Repeating the reasoning in the corresponding proof one can show that the strong ellipticity is not necessary for the piecewise linear collocation to be stable.

For special cases of equation (2), a convergence analysis of Galerkin and collocation methods with splines has been developed in many papers (see Chapters 5 and 12). Recently Atkinson/Graham [1], Chandler/Graham [4] and Elschner [6] (see also Chap. 5) studied quadrature methods and their iterative versions for (2) in case $c = 1$ and $d = 0$ using simple composite quadrature rules. They proved the stability and established error estimates in the supremum-norm as well as in weighted Sobolev norms. The convergence of the quadrature method of discrete whirls for (1) in case $c = 0$ and $d = 1$ was investigated by Lifanov and Polonski [1].

The discretization of equation (1) via spline collocation is very simple. We take a finite set of collocation points $\{\tau_k^{(n)}: k = 0, \ldots, n-1\} \subset \Gamma$ and choose a space of spline functions $X_n (\dim X_n = n)$ on Γ. For the exact solution $u = A_\Gamma^{-1} f$, we determine an approximation $u_n \in X_n$ by solving the system

$$(A_\Gamma u_n)(\tau_k^{(n)}) = f(\tau_k^{(n)}), \qquad k = 0, \ldots, n-1. \tag{3}$$

If X_n is defined on a suitably graded mesh and the degree of the functions in X_n is sufficiently large, then a high order of convergence is to be expected. However, for the sake of simplicity, we restrict our attention to uniform partitions and piecewise constant splines. Using the arguments of this chapter it is not hard to treat special nonuniform meshes (see, e.g., Section 11.20) and spline functions of higher degree, too.

In order to solve the system of equations (3) one has to compute $(Au_n)(\tau_k^{(n)})$. If this cannot be done analytically, one has to make use of quadrature rules. In this case we recommend the immediate discretization of equations (1) and (2) via quadrature rules. Thereby, the singularity subtraction technique is needed to obtain convergent approximation methods. If suitably graded meshes and quadrature rules with high accuracy are used, then a high order of convergence can be achieved (compare the quadrature methods for the interval studied in Chapter 12). For simplicity's sake, in this chapter we use the rectangle rule. However, in the same way a modified rectangle rule can be considered. In fact, a suitable modification of the quadrature weights in a finite number of knots (in the neighborhood of the corner points or of the end points of the interval) leads to high accuracy of the quadrature rule. It is also possible (though more complicated) to investigate composite Newton-Cotes rules, e.g. the composite Simpson rule.

In order to show the nature of our quadrature methods we discretize the equation

$$A_{\mathbb{R}} u(t) = a(t)\, u(t) + \frac{1}{\pi i} \int_{\mathbb{R}} \frac{u(\tau)}{\tau - t}\, d\tau + \int_{\mathbb{R}} k(t, \tau)\, u(\tau)\, d\tau = f(t), \qquad t \in \mathbb{R}. \tag{4}$$

Although the resulting quadrature methods are not of exciting interest for numerical computation, they are very simple to deduce and give a good motivation for the methods that will be considered. Fix $n \in \mathbb{N}$, $-1 < \varepsilon < 1$ and, for $k \in \mathbb{Z}$, set $t_k^{(n)} := k/n$, $\tau_k^{(n)} := (k + \varepsilon)/n$. Using the rule

$$\int_{\mathbb{R}} g(t)\, dt \sim \sum_{j \in \mathbb{Z}} g(t_j^{(n)}) \frac{1}{n} \tag{5}$$

we obtain

$$\frac{1}{\pi i} \int_{\mathbb{R}} \frac{u(\tau)}{\tau - \tau_k^{(n)}}\, d\tau = \frac{1}{\pi i} \int_{\mathbb{R}} \frac{u(\tau) - u(\tau_k^{(n)})}{\tau - \tau_k^{(n)}}\, d\tau \sim \frac{1}{\pi i} \sum_{j \in \mathbb{Z}} \frac{u(t_j^{(n)}) - u(\tau_k^{(n)})}{t_j^{(n)} - \tau_k^{(n)}} \frac{1}{n} \tag{6}$$

$$\sim \frac{1}{\pi i} \sum_{j \in \mathbb{Z}} \frac{u(t_j^{(n)})}{t_j^{(n)} - \tau_k^{(n)}} \frac{1}{n} - u(\tau_k^{(n)}) \frac{1}{\pi i} \sum_{j \in \mathbb{Z}} \frac{1}{j - k - \varepsilon}.$$

For $\varepsilon \neq 0$, the well-known formula

$$\cot(\pi x) = \frac{1}{\pi} \left\{ \frac{1}{x} + \sum_{j=1}^{\infty} \left[\frac{1}{x - j} + \frac{1}{x + j} \right] \right\}$$

yields

$$\frac{1}{\pi i} \int_{\mathbb{R}} \frac{u(\tau)}{\tau - \tau_k^{(n)}}\, d\tau \sim \frac{1}{\pi i} \sum_{j \in \mathbb{Z}} \frac{u(t_j^{(n)})}{t_j^{(n)} - \tau_k^{(n)}} \frac{1}{n} - u(\tau_k^{(n)})\, i \cot(\pi\varepsilon). \tag{7}$$

Replacing the integrals in (4) by (5), (7) and substituting $u(\tau_k^{(n)})$ by $u(t_k^{(n)})$ we arrive at

$$[a(\tau_k^{(n)}) - i \cot(\pi\varepsilon)]\, u_n(t_k^{(n)}) + \frac{1}{\pi i} \sum_{j \in \mathbb{Z}} \frac{u_n(t_j^{(n)})}{t_j^{(n)} - \tau_k^{(n)}} \frac{1}{n}$$

$$+ \sum_{j \in \mathbb{Z}} k(\tau_k^{(n)}, t_j^{(n)})\, u_n(t_j^{(n)}) \frac{1}{n} = f(\tau_k^{(n)}), \qquad k \in \mathbb{Z}. \tag{8}$$

For $\varepsilon = 1/2$ or $\varepsilon = -1/2$, $\cot(\pi\varepsilon)$ vanishes and the system (8) is called the *method of discrete whirls*.

If $\varepsilon = 0$, then (6) and the equality

$$\sum_{j=1}^{\infty} \left\{ \frac{1}{j} + \frac{1}{-j} \right\} = 0$$

yield

$$\frac{1}{\pi i} \int_{\mathbb{R}} \frac{u(\tau)}{\tau - t_k^{(n)}}\, d\tau \sim \frac{1}{\pi i} \sum_{\substack{j \in \mathbb{Z} \\ j \neq k}} \frac{u(t_j^{(n)})}{t_j^{(n)} - t_k^{(n)}} \frac{1}{n} + \frac{1}{\pi i} \frac{1}{n} u'(t_k^{(n)}). \tag{9}$$

Replacing the integrals in (4) by (5), (9) and neglecting the small term $(1/\pi i)\,(1/n)\, u'(t_k^{(n)})$ we obtain

$$a(t_k^{(n)})\, u_n(t_k^{(n)}) + \frac{1}{\pi i} \sum_{\substack{j \in \mathbb{Z} \\ j \neq k}} \frac{u_n(t_j^{(n)})}{t_j^{(n)} - t_k^{(n)}} \frac{1}{n} + \sum_{j \in \mathbb{Z}} k(t_k^{(n)}, t_j^{(n)})\, u_n(t_j^{(n)}) \frac{1}{n} = f(t_k^{(n)}), \tag{10}$$

$k \in \mathbb{Z}$.

The corresponding quadrature methods to (8) and (10) on smooth curves were considered in Chapter 10. For equation (2), an analysis of these methods is due to RATHSFELD [3], who treated the kernel $k_t(t/\tau)\, 1/\tau$ analogously to the smooth kernel $k(t, \tau)$. In the present chapter we give a simplified proof of Rathsfeld's results. Moreover, we prove them under more general assumptions on the kernel $k_t\big((t/\tau)\big)\, (1/\tau)$ and extend the analysis to the case of curves with corners.

Now let us consider another quadrature method which was studied in Chapter 10 for the case of a closed smooth curve. Namely, in the discretization of $(A_{\mathbb{R}} u)\, (t_k^{(n)}) = f(t_k^{(n)})$ we use

$$\int_{\mathbb{R}} g(t)\, \mathrm{d}t \sim \sum_{\substack{j\in\mathbb{Z} \\ j\equiv k+1 \bmod 2}} g(t_j^{(n)})\, \frac{2}{n}.$$

Similarly to the derivation of (10), we get

$$a(t_k^{(n)})\, u_n(t_k^{(n)}) + \frac{1}{\pi \mathrm{i}} \sum_{\substack{j\in\mathbb{Z} \\ j\equiv k+1 \bmod 2}} \frac{u_n(t_j^{(n)})}{t_j^{(n)} - t_k^{(n)}}\, \frac{2}{n} + \sum_{\substack{j\in\mathbb{Z} \\ j\equiv k+1 \bmod 2}} k(t_k^{(n)}, t_j^{(n)})\, u_n(t_j^{(n)})\, \frac{2}{n}$$

$$= f(t_k^{(n)}), \qquad k \in \mathbb{Z}. \tag{11}$$

Since no substitution $u(\tau_k^{(n)}) \approx u(t_k^{(n)})$ and no neglect of $(1/\pi\mathrm{i})\, (1/n)\, u'(t_k^{(n)})$ is needed, this method converges faster than (8) and (10) in the case of smooth curves. Furthermore, the invertibility of the operator A_Γ will be enough to secure the stability of (11). For the unit circle, method (11) and the method of trigonometric collocation (see Chapter 7) coincide.

For each of the approximation methods under consideration, the problem of stability will be reduced to the same question on the corresponding method for a model problem on an angle or on the half axis, respectively, by utilizing a localization technique. In order to handle the model problems, we shall have recourse to Mellin techniques. This approach is a generalization of the one used in the case of smooth curves (see Chapters 10 and 13). In comparison with proof techniques based on strong ellipticity it is more sophisticated. However, in many situations strong ellipticity arguments do not work. Furthermore, contrary to the strong ellipticity techniques, our proofs yield not only the sufficiency, but also the necessity of the stability conditions. Moreover, by using standard tensor product arguments, one can easily extend the results to the system case. In this chapter we essentially follow PRÖSSDORF/RATHSFELD [6].

Quadrature methods for singular integral equations on curves with corner points

Let us consider quadrature methods for the approximate solution of singular integral equations on curves with corner points. To this end, we shall use simple quadrature methods which are similar to those used in the case of smooth curves (see Chapter 10). Our aim is to establish necessary and sufficient conditions for their stability. Since these conditions will be shown to be of local nature, we start with the simplest situation of an angle. After that we reduce the general case to that one of an angle by using localization techniques.

11.2. Quadrature methods on an angle. Let Γ_ω $(0 < \omega < 2\pi)$ denote the angle $\{t\,\mathrm{e}^{\mathrm{i}\omega}: 0 \leqq t < \infty\} \cup \{t: 0 \leqq t < \infty\}$. Suppose the singular integral operator $A := cI + dS_{\Gamma_\omega}$ with $c, d \in \mathbb{C}$ to be invertible in $\mathbf{L}^2(\Gamma_\omega)$, i.e., $c \pm d \neq 0$. If we seek an approximation u_n for the solution $u \in \mathbf{L}^2(\Gamma_\omega)$ of the equation $Au = f$, $f \in \mathcal{R}(\Gamma_\omega) \cap \mathbf{L}^2(\Gamma_\omega)$ then 10.12(12) to 10.12(16) suggests the following quadrature methods. Choose two different numbers ε, δ $(0 < \varepsilon, \delta < 1)$ and set

$$t_k^{(n)} := \begin{cases} \dfrac{k+\delta}{n} & \text{if } k \geqq 0, \\[2ex] -\dfrac{k+\delta}{n}\,\mathrm{e}^{\mathrm{i}\omega} & \text{if } k < 0, \end{cases} \qquad \tau_k^{(n)} := \begin{cases} \dfrac{k+\varepsilon}{n} & \text{if } k \geqq 0, \\[2ex] -\dfrac{k+\varepsilon}{n}\,\mathrm{e}^{\mathrm{i}\omega} & \text{if } k < 0. \end{cases}$$

Determine approximate values $\xi_j^{(n)}$ of $u(t_j^{(n)})$ $(j \in \mathbb{Z})$ by solving one of the systems

$$\begin{aligned} &\big(c - \mathrm{i}\cot(\pi(\varepsilon - \delta))\,d\big)\,\xi_k^{(n)} + d\,\frac{1}{\pi\mathrm{i}}\left\{\sum_{j=0}^{\infty} \frac{\xi_j^{(n)}}{t_j^{(n)} - \tau_k^{(n)}}\,\frac{1}{n}\right. \\ &\left.+ \sum_{j=-\infty}^{-1} \frac{\xi_j^{(n)}}{t_j^{(n)} - \tau_k^{(n)}}\,\frac{-\mathrm{e}^{\mathrm{i}\omega}}{n}\right\} = f(\tau_k^{(n)}), \qquad k \in \mathbb{Z}, \end{aligned} \tag{1}$$

$$c\xi_k^{(n)} + d\,\frac{1}{\pi\mathrm{i}}\left\{\sum_{\substack{j=0\\ j\neq k}}^{\infty} \frac{\xi_j^{(n)}}{t_j^{(n)} - t_k^{(n)}}\,\frac{1}{n} + \sum_{\substack{j=-\infty\\ j\neq k}}^{-1} \frac{\xi_j^{(n)}}{t_j^{(n)} - t_k^{(n)}}\,\frac{-\mathrm{e}^{\mathrm{i}\omega}}{n}\right\} = f(t_k^{(n)}), \quad k \in \mathbb{Z}, \tag{2}$$

$$\begin{aligned} &c\xi_k^{(n)} + d\,\frac{1}{\pi\mathrm{i}}\left\{\sum_{\substack{j=1\\ j\equiv k+1 \bmod 2}}^{\infty} \frac{\xi_j^{(n)}}{t_j^{(n)} - t_k^{(n)}}\,\frac{2}{n} + \sum_{\substack{j=-\infty\\ j\equiv k+1 \bmod 2}}^{-1} \frac{\xi_j^{(n)}}{t_j^{(n)} - t_k^{(n)}}\,\frac{-2\mathrm{e}^{\mathrm{i}\omega}}{n}\right. \\ &\left.+ \sum_{\substack{j=0\\ j\equiv k+1 \bmod 2}}^{0} \frac{\xi_j^{(n)}}{t_j^{(n)} - t_k^{(n)}}\,\frac{1}{n}\,(1 - \mathrm{e}^{\mathrm{i}\omega})\right\} = f(t_k^{(n)}), \qquad k \in \mathbb{Z}. \end{aligned} \tag{3}$$

If there exists a unique solution $(\xi_k^{(n)})_{k\in\mathbb{Z}}$, then we obtain an approximate solution u_n by setting

$$u_n := \sum_{k\in\mathbb{Z}} \xi_k^{(n)} \chi_k^{(n)}, \quad \chi_k^{(n)}(t) := \begin{cases} 1 \text{ if } \dfrac{k}{n} \leqq t < \dfrac{k+1}{n}, & k = 0, 1, 2, \ldots, \\ 0 \text{ else} \end{cases}$$

$$\chi_k^{(n)}(t) := \begin{cases} 1 \text{ if } \dfrac{k}{n} \leqq -\mathrm{e}^{-\mathrm{i}\omega} t < \dfrac{k+1}{n}, & k = -1, -2, \ldots, \\ 0 \text{ else.} \end{cases}$$

If the methods (1), (2) or (3) are stable, then it is not hard to prove the convergence of u_n to the exact solution u of the equation $Au = f$. However, we consider the quadrature methods (1)–(3) as model schemes for adequate numerical procedures on general curves with corners. From this point of view, it suffices to establish necessary and sufficient conditions for the methods (1), (2) and (3) to be stable.

Let A_n denote the matrix of the system (1), (2) or (3), respectively. We define the interpolation projection K_n by $K_n y := \sum_{k\in\mathbb{Z}} y(\tau_k^{(n)})\,\chi_k^{(n)}$ $\big(y \in \mathcal{R}(\Gamma_\omega)\big)$ and denote the orthogonal projection onto im $K_n \cap \mathbf{L}^2(\Gamma_\omega)$ by L_n. In what follows, we shall identify the operators of $\mathscr{L}(\text{im } L_n)$ with their matrices corresponding to the basis $\{\chi_k^{(n)}: k \in \mathbb{Z}\}$. Because

$$\Big\|\sum_{k\in\mathbb{Z}} \xi_k \chi_k^{(n)}\Big\|_{\mathbf{L}^2(\Gamma_\omega)} = \frac{1}{\sqrt{n}} \|\{\xi_k\}_{k\in\mathbb{Z}}\|_{\tilde{\mathbf{l}}^2}, \tag{4}$$

these matrices may be considered as operators in $\tilde{\mathbf{l}}^2$. In particular, since the matrices $A_n \in \mathscr{L}(\tilde{\mathbf{l}}^2)$ are independent of n, the sequence $\{A_n\}$ $\big(A_n \in \mathscr{L}(\text{im } L_n)\big)$ is stable if and only if $A_1 \in \mathscr{L}(\tilde{\mathbf{l}}^2)$ is invertible.

11.3. Theorem. (a) *The operator $A_1 \in \mathscr{L}(\tilde{\mathbf{l}}^2)$ is a Fredholm operator with index* 0 *if and only if* $(c + d)/(c - d) \notin \Omega$, *where* $\Omega := \{0\}$ *for* 11.2(3), $\Omega := (-\infty, 0]$ *for* 11.2(2) *and* $\Omega := \{e^{-i\pi(\varepsilon-\delta)}t: -\infty < t \leq 0\}$ *for* 11.2(1).

(b) *There is an at most countable subset Φ of $\mathbb{C} \setminus \Omega$ all accumulation points of which belong to Ω such that A_1 is invertible in $\tilde{\mathbf{l}}^2$ if and only if* $(c + d)/(c - d) \notin \Omega \cup \Phi$.

(c) *If $\omega = \pi$, then $\Phi = \emptyset$.*

Assertion (b) of Theorem 11.3 is an easy consequence of part (a). To see this, we denote the approximation matrix for the singular operator S_Γ by S_1 and obtain $A_1 = cI + dS_1$. For $\lambda := (c + d)/(c - d)$, we set $B(\lambda) := (c + d)/(c - d)\,(1/2)\,(I + S_1) + (1/2)\,(I - S_1) = 1/(c - d)A_1$. Obviously, the function $\mathbb{C} \setminus \Omega \ni \lambda \mapsto B(\lambda)$ is analytic (even linear) and its values are Fredholm operators with index 0. Since $\mathbb{C} \setminus \Omega$ is connected and $B(1) = I$, the points of $\Phi := \{\lambda \in \mathbb{C} \setminus \Omega: B(\lambda)$ is not invertible$\}$ must be isolated and so (b) follows.

To show (a) we need some results on Toeplitz operators which are due to GOHBERG and KRUPNIK (see Theorem 6.16). Let $\mathfrak{A} := \text{alg } T(\mathbf{PC})$ denote the smallest algebra containing all Toeplitz operators $T(a)$ with $a \in \mathbf{PC}(\mathbb{T})$. Then $\mathfrak{A}_{n\times n} \subseteq \mathscr{L}(\mathbf{l}^2)_{n\times n}$ $(n \in \mathbb{N})$ is an algebra of continuous operators in $\mathbf{l}_n^2$. There exists a multiplicative linear mapping $\mathfrak{A}_{n\times n} \ni B \mapsto \text{smb } B$ into the algebra of bounded $n \times n$-matrix functions on $\mathbb{T} \times [0, 1]$. The symbol smb B of $B = (B_{kj})_{k,j=1}^n$, $B_{kj} \in \mathfrak{A}$ is equal to $(\text{smb } B_{kj})_{k,j=1}^n$ and the symbol smb $T(a)$ of $T(a)$ with $a \in \mathbf{PC}(\mathbb{T})$ is given by $\text{smb } T(a)\,(\tau, \mu) := \mu a(\tau + 0) + (1 - \mu) \times a(\tau - 0)$, where $(\tau, \mu) \in \mathbb{T} \times [0, 1]$. Furthermore, $B \in \mathfrak{A}_{n\times n}$ is a Fredholm operator if and only if $\det \text{smb } B(\tau, \mu) \neq 0$ for all $\tau \in \mathbb{T}$ and $0 \leq \mu \leq 1$. Suppose $B \in \mathfrak{A}_{n\times n}$ is a Fredholm operator and there exist $\omega_j \in (0, 2\pi)$ $(j = 1, \ldots, m)$, $\omega_0 := 0$, $\omega_{m+1} := 2\pi$ such that $\text{smb } B(\tau, \mu) = \text{smb } B(\tau, \mu')$ for $\tau \neq e^{i\omega_j}$ $(j = 0, \ldots, m)$ and $0 \leq \mu, \mu' \leq 1$. Then the index of B is equal to $-\text{ind det } A_B$, i.e., to the negative index of the curve $\Gamma_0 \cup \Gamma_1 \cup \cdots \cup \Gamma_m$, $\Gamma_j := \{\det \text{smb } B(e^{ix}, 0): \omega_j \leq x \leq \omega_{j+1}\} \cup \{\det \text{smb } B(e^{i\omega_{j+1}}, \mu): 0 \leq \mu \leq 1\}$. Finally, the algebra $\mathfrak{A}_{n\times n}$ contains all compact operators and, moreover, $B \in \mathfrak{A}_{n\times n}$ is compact if and only if $\text{smb } B \equiv 0$.

We identify $\mathbf{l}^2$ with the subspace of $\tilde{\mathbf{l}}^2$ containing all sequences the negatively indexed terms of which vanish. Let $F \in \mathscr{L}(\tilde{\mathbf{l}}^2)$ be defined by

$$F\{\xi_k\}_{k\in\mathbb{Z}} := \{\eta_k\}_{k\in\mathbb{Z}}, \qquad \eta_k := \xi_{-1-k}.$$

Then $\tilde{\mathbf{l}}^2 = F\mathbf{l}^2 \oplus \mathbf{l}^2$, and we denote the orthogonal projection of $\tilde{\mathbf{l}}^2$ onto $\mathbf{l}^2$ by P. Obviously, $\mathscr{L}(\tilde{\mathbf{l}}^2)$ can be identified with $\mathscr{L}(\mathbf{l}^2)_{2\times 2}$ by setting

$$\mathscr{L}(\tilde{\mathbf{l}}^2) \ni A = \begin{pmatrix} PAP & PA(I-P)\,F \\ F(I-P)\,AP & F(I-P)\,A(I-P)\,F \end{pmatrix} \in \mathscr{L}(\mathbf{l}^2)_{2\times 2}. \tag{1}$$

We obtain $\mathfrak{A}_{2\times 2} \subset \mathscr{L}(\tilde{\mathbf{l}}^2)$.

11.4. Lemma. *Let* $z \in \mathbb{C}$, $-1/2 < \operatorname{Re} z < 1/2$, $\Lambda^z := \big((k+1)^z\, \delta_{k,j}\big)_{k,j=0}^{\infty}$ *and* $a \in \mathbf{PC}(\mathbb{T})$. *Suppose there exist* $m \in \mathbb{Z}$, $m \geqq 0$ *and* $\omega_0 := 0$, $\omega_j \in (0, 2\pi)$, $j = 1, \ldots, m$ *such that* a *is continuous on* $\mathbb{T} \setminus \{\exp(\mathrm{i}\omega_j)\colon j = 0, \ldots, m\}$. *Then the following assertions are valid.*

(i) *The matrix* $\Lambda^{-z} T(a)\, \Lambda^z$ *belongs to* $\mathfrak{A}$ *and*

$$\operatorname{smb} \Lambda^{-z} T(a)\, \Lambda^z(\tau, \mu) = \begin{cases} a(\tau) & \text{if } \tau \neq \mathrm{e}^{\mathrm{i}\omega_j},\ j = 0, \ldots, m \\ \dfrac{\mu a(\tau+0) + (1-\mu)\, a(\tau-0)\, \mathrm{e}^{-\mathrm{i}2\pi z}}{\mu + (1-\mu)\, \mathrm{e}^{-\mathrm{i}2\pi z}} & \text{if } \tau = \mathrm{e}^{\mathrm{i}\omega_j} \text{ for } j = 0, \ldots, m. \end{cases} \tag{1}$$

(ii) *The function* $z \mapsto \Lambda^{-z} T(a)\, \Lambda^z$ *is continuous on* $\{-(1/2) < \operatorname{Re} z < (1/2)\}$ *if* a *is twice differentiable at the points of continuity and the second derivative is in* $\mathbf{PC}(\mathbb{T})$.

(iii) *For arbitrary* $D \in \mathfrak{A}$, *the function* $\mathbb{R} \ni \xi \mapsto \Lambda^{-\mathrm{i}\xi} D \Lambda^{\mathrm{i}\xi}$ *takes its values in* $\mathfrak{A}$ *and is continuous in the operator norm.*

Proof. (i) Obviously, it suffices to consider functions $a \in \mathbf{PC}(\mathbb{T})$ having one jump only. Now set $W := (\tau^k \delta_{k,j})_{k,j=0}^{\infty}$ $(\tau \in \mathbb{T})$. Then $A \in \mathfrak{A}$ implies $WAW^{-1} \in \mathfrak{A}$ and $\operatorname{smb} WAW^{-1}(t, \mu) = \operatorname{smb} A(t\tau, \mu)$ for all $t \in \mathbb{T}$ and $0 \leqq \mu \leqq 1$. Consequently, without loss of generality, the point of discontinuity of a can be chosen to equal 1. Hence we may assume $m = 0$.

If, in addition, $a(1+0) = a(1-0)$, then a is continuous and so Remark 4.51, 2° yields that $\Lambda^{-z} T(a)\, \Lambda^z \in \mathfrak{A}$, $\operatorname{smb} \Lambda^{-z} T(a)\, \Lambda^z = \operatorname{smb} T(a)$. Thus (1) holds for continuous a.

Now let $a = \Phi_z$, where $\Phi_z(t)$ denotes the branch of the function t^z which is continuous on $\mathbb{T} \setminus \{1\}$ and satisfies $\Phi_z(1+0) = 1$. Then a formula of Duduchava [1] yields

$$T(\Phi_{-z})^{-1} = \frac{\pi z}{\sin \pi z}\, (\lambda_k \delta_{kj})_{k,j=0}^{\infty}\, T(\Phi_z)\, (\mu_k \delta_{kj})_{k,j=0}^{\infty}, \qquad \lambda_k := \prod_{m=1}^{k} \left(1 - \frac{z}{m}\right), \qquad \mu_k := \prod_{m=1}^{k} \left(1 + \frac{z}{m}\right). \tag{2}$$

(A proof of (2) can also be found in Böttcher/Silbermann [3], Section 6.20. We remark that $\Phi_z = \mathrm{e}^{\mathrm{i}\pi z} \varphi_\alpha^z$, where $\alpha := 1$ and φ_α^z is defined as in Böttcher/Silbermann [3]).

The inverse of $T(\Phi_{-z})$ belongs to the C^*-algebra $\mathfrak{A}$ and its symbol $\operatorname{smb} T(\Phi_{-z})^{-1}$ equals $[\operatorname{smb} T(\Phi_{-z})]^{-1}$. Let d_1 and d_2, respectively, denote the limit of the sequence $\{\lambda_k/(k+1)^{-z}\}$ and $\{\mu_k/(k+1)^z\}$. Taking into account that $d_1 \neq 0 \neq d_2$, $d_1 d_2 = (\sin \pi z)/z$ and using (2), we arrive at

$$\Lambda^{-z} T(\Phi_z)\, \Lambda^z = T(\Phi_{-z})^{-1} + T,$$

$$T := \frac{\pi z}{\sin \pi z} \left(\left\{ d_1 \frac{(k+1)^{-z}}{\lambda_k} - 1 \right\} \delta_{k,j} \right)_{k,j=0}^{\infty} (\lambda_k \delta_{jk})_{k,j=0}^{\infty}\, T(\Phi_z)\, (\mu_k \delta_{jk})_{k,j=0}^{\infty} + \frac{\pi z}{\sin \pi z}\, d_1 \Lambda^{-z} T(\Phi_z)\, (\mu_k \delta_{kj})_{k,j=0}^{\infty} \left(\left\{ d_2 \frac{(k+1)^{-z}}{\mu_k} - 1 \right\} \delta_{kj} \right)_{k,j=0}^{\infty}.$$

Since T is compact, we obtain that $\Lambda^{-z} T(\phi_z)\, \Lambda^z \in \mathfrak{A}$ and thus (1) for $a = \Phi_z$.

If a is an arbitrary function satisfying the assumptions of the lemma with $m = 0$, then a is the linear combination of Φ_z and a continuous function a_0. Thus assertion (i) is valid in this case, too.

(ii) Analogously to part (i), we only have to consider the case of an arbitrary continuous function a and the case of one special discontinuous function which satisfies the hypotheses of the lemma with $m = 0$. First, suppose a is continuous. Then $|a_k| < Ck^{-2}$ implies that

$$\|\Lambda^{-(z+\xi)}T(a)\,\Lambda^{(z+\xi)} - \Lambda^{-z}T(a)\,\Lambda^{z}\|$$
$$= \left\| \sum_{\substack{l\in\mathbb{Z}\\ l\neq 0}} \left(\delta_{k-j,l}a_l \left(\frac{k+1}{j+1}\right)^{-z} \left\{\left(\frac{k+1}{j+1}\right)^{-\xi} - 1\right\}\right)_{k,j=0}^{\infty} \right\|$$
$$\leqq \sum_{\substack{l\in\mathbb{Z}\\ l\neq 0}} |a_l| \sup_{k\geqq\max\{0,l\}} \left|\left(1 - \frac{l}{k+1}\right)^{z} \left\{\left(1 - \frac{l}{k+1}\right)^{\xi} - \left(1 - \frac{l}{k+1}\right)^{0}\right\}\right|$$
$$\leqq C \sum_{\substack{l\in\mathbb{Z}\\ l\neq 0}} |l|^{-2+|\mathrm{Re}\,z|} \log\left(|l| + 1\right) |l|^{|\mathrm{Re}\,\xi|}\, |\xi| \leqq C\,|\xi|$$

and hence $z \mapsto \Lambda^{-z}T(a)\,\Lambda^{z}$ is continuous.

Let $S_{\mathbb{R}^+}$ denote the Cauchy singular operator on the half axis $\mathbb{R}^+ := (0, \infty)$, χ_j the characteristic function of the interval $[j, j+1)$ and $L \in \mathcal{L}(\mathbf{L}^2(\mathbb{R}^+))$ the orthogonal projection onto the span of $\{\chi_j : j = 1, 2, \ldots\}$. If we identify the operators of the space im L with their matrices corresponding to the basis $\{\chi_j : j = 1, 2, \ldots\}$, then

$$LS_{\mathbb{R}^+}|_{\mathrm{im}\,L} = \left(\int_{\mathbb{R}} \overline{\chi}_{k+1}(S_{\mathbb{R}}\chi_{j+1})\right)_{k,j=0}^{\infty}$$

For $x \in \mathbf{L}^2(\mathbb{R})$, we define the Fourier transform $\mathcal{F}x$ by

$$(\mathcal{F}x)\,(t) := \frac{1}{\sqrt{2\pi}} \int_{\mathbb{R}} \mathrm{e}^{\mathrm{i}st} x(s)\,\mathrm{d}s.$$

In particular, we have

$$(\mathcal{F}\chi_{k+1})\,(t) = \sqrt{\frac{2}{\pi}}\, \mathrm{e}^{\mathrm{i}(k+1+(1/2))t}\, \frac{\sin\,(t/2)}{t}.$$

Using the well-known equalities

$$\int_{\mathbb{R}} \overline{f} g\,\mathrm{d}t = \int_{\mathbb{R}} (\overline{\mathcal{F}f})\,(\mathcal{F}g)\,\mathrm{d}t, \qquad S_{\mathbb{R}} = -\mathcal{F}^{-1}\,\mathrm{sign}\,\mathcal{F},$$

we see that

$$\int_{\mathbb{R}} \overline{\chi}_{k+1}(S_{\mathbb{R}}\chi_{j+1})\,\mathrm{d}t = -\int_{\mathbb{R}} \mathrm{e}^{\mathrm{i}(j-k)t}\,\mathrm{sign}\,t \cdot \frac{2}{\pi}\,\frac{\sin^2(t/2)}{t^2}\,\mathrm{d}t$$
$$= -\int_{\mathbb{R}} \mathrm{e}^{-\mathrm{i}(k-j)2\pi s}\,\mathrm{sign}\,s\,\frac{\sin^2\,(\pi s)}{\pi^2 s^2}\,\mathrm{d}s$$
$$= \int_0^1 \mathrm{e}^{-\mathrm{i}(k-j)2\pi s} \left\{-\frac{\sin^2\,(\pi s)}{\pi^2} \sum_{l\in\mathbb{Z}} \frac{\mathrm{sign}\,(l + (1/2))}{(s+l)^2}\right\} \mathrm{d}s.$$

Thus $LS_{\mathbb{R}^+}|_{\operatorname{im} L} = T(\Phi)$, where

$$\Phi(e^{i2\pi x}) := -\frac{\sin^2(\pi x)}{\pi^2} \sum_{k\in\mathbb{Z}} \frac{\operatorname{sign}\big(k + (1/2)\big)}{(x + k)^2}, \quad 0 < x < 1. \tag{3}$$

Obviously, Φ is smooth on $\mathbb{T} \setminus \{1\}$ and $\Phi(1 + 0) = -1$, $\Phi(1 - 0) = 1$. Therefore, we only have to prove the continuity of $z \mapsto \Lambda^{-z} LS_{\mathbb{R}^+}|_{\operatorname{im} L} \Lambda^z \in \mathcal{L}(\operatorname{im} L)$. Setting

$$g(t) := \frac{t}{h(t)}, \quad h(t) := \begin{cases} k & \text{if } k < t < k + 1,\ k \in \mathbb{N}, \\ t & \text{if } 0 < t < 1, \end{cases}$$

we obtain that $\Lambda^{-z} LS_{\mathbb{R}^+}|_{\operatorname{im} L} \Lambda^z = Lg(t)^z\, t^{-z} S_{\mathbb{R}^+} t^z g(t)^{-z}|_{\operatorname{im} L}$. Since the continuity of $z \mapsto g(t)^z \in \mathcal{L}\big(\mathbf{L}^2(\mathbb{R}^+)\big)$ is easy to check and the continuity of $z \mapsto t^{-z} S_{\mathbb{R}^+} t^z \in \mathcal{L}\big(\mathbf{L}^2(\mathbb{R}^+)\big)$ is well-known (if $z \in \mathbb{R}$, $-1/2 < z < 1/2$, then the continuity follows from Theorem 6.6; since Theorem 6.6 remains true for complex γ, the continuity also holds for $z \in \mathbb{C}$, $-1/2 < \operatorname{Re} z < 1/2$), $z \mapsto \Lambda^{-z} LS_{\mathbb{R}^+}|_{\operatorname{im} L} \Lambda^z$ must be continuous.

(iii) If D is a sum or product of Toeplitz operators, then (ii) implies the continuity of $\mathbb{R} \ni \xi \mapsto \Lambda^{-i\xi} D \Lambda^{i\xi} \in \mathfrak{A}$. Using the estimate $\|\Lambda^{-i\xi} D \Lambda^{i\xi}\| = \|D\|$ and a density argument, we get the continuity for arbitrary $D \in \mathfrak{A}$. ■

Now we come to the proof of Theorem 11.3(a). Let us first consider the method 11.2(1). The operator $A_1 \in \mathcal{L}(\mathbf{l}^2)_{2\times 2}$ takes the form

$$A_1 = \begin{pmatrix} A_{11} & A_{12} \\ A_{21} & A_{22} \end{pmatrix},$$

where

$$A_{11} := \big(c - i\cot(\pi(\varepsilon - \delta))\, d\big)\, I + d\left(\frac{1}{\pi i} \frac{1}{-(k - j) - (\varepsilon - \delta)}\right)_{k,j=0}^{\infty},$$

$$A_{21} := d\left(\frac{1}{\pi i} \frac{1}{(j + \delta) + (-k - 1 + \varepsilon)\, e^{i\omega}}\right)_{k,j=0}^{\infty},$$

$$A_{12} := d\left(\frac{1}{\pi i} \frac{-e^{i\omega}}{-(-j - 1 + \delta)\, e^{i\omega} - (k + \varepsilon)}\right)_{k,j=0}^{\infty},$$

$$A_{22} := \big(c - i\cot(\pi(\varepsilon - \delta))\, d\big)\, I + d\left(\frac{1}{\pi i} \frac{1}{(k - j) + (\delta - \varepsilon)}\right)_{k,j=0}^{\infty}.$$

For $-1 < \nu < 1$, $\nu \neq 0$, we set

$$f^\nu(e^{i2\pi x}) := 2\left\{e^{-i\pi\nu(x-1)} \frac{\sin(-\pi\nu x)}{\sin(-\pi\nu)}\right\} - 1, \qquad 0 \leq x < 1.$$

Then a straightforward computation shows that $f^\nu = \sum_{k\in\mathbb{Z}} f_k^\nu t^k$, where $f_k^\nu := (1/\pi i)\big(1/(-k - \nu)\big) - i\cot(\pi\nu)\,\delta_{k0}$. Thus we obtain $A_{1,1} = T(c + df^{(\varepsilon-\delta)})$ and $A_{2,2} = T(c - df^{(\delta-\varepsilon)})$.

Now let us prove that $A_{2,1} \in \mathfrak{A}$. The residue theorem together with the well-known formula

$$\frac{1}{\pi i} \frac{1}{1 - e^{i\omega} x} = \frac{1}{2\pi i} \int\limits_{\operatorname{Re} z = 1/2} x^{-z} \left\{-i \frac{e^{-i(\omega - \pi)z}}{\sin(\pi z)}\right\} dz$$

(see ERDÉLYI et al. [1]) gives

$$\frac{1}{\pi\mathrm{i}}\,\frac{1}{1-\mathrm{e}^{\mathrm{i}\omega}x}=\frac{1}{2\pi\mathrm{i}}\int\limits_{\mathrm{Re}\,z=1/4}x^{-z}\left\{-\mathrm{i}\,\frac{\mathrm{e}^{-\mathrm{i}(\omega-\pi)z}}{\sin(\pi z)}\right\}\mathrm{d}z,$$

$$\frac{1}{\pi\mathrm{i}}\,\frac{x}{1-\mathrm{e}^{\mathrm{i}\omega}x}=\frac{1}{2\pi\mathrm{i}}\int\limits_{\mathrm{Re}\,z=1/2}x^{-(z-1)}\left\{-\mathrm{i}\,\frac{\mathrm{e}^{-\mathrm{i}(\omega-\pi)z}}{\sin(\pi z)}\right\}\mathrm{d}z$$

$$=\frac{1}{2\pi\mathrm{i}}\int\limits_{\mathrm{Re}\,z=5/4}x^{-(z-1)}\left\{-\mathrm{i}\,\frac{\mathrm{e}^{-\mathrm{i}(\omega-\pi)z}}{\sin(\pi z)}\right\}\mathrm{d}z-\mathrm{i}\,\frac{\mathrm{e}^{-\mathrm{i}\omega}}{\pi},$$

$$\frac{1-x}{1-\mathrm{e}^{\mathrm{i}\omega}x}=\frac{1-\mathrm{e}^{-\mathrm{i}\omega}}{2}\int\limits_{\mathrm{Re}\,z=1/4}x^{-z}\left\{-\mathrm{i}\,\frac{\mathrm{e}^{-\mathrm{i}(\omega-\pi)z}}{\sin(\pi z)}\right\}\mathrm{d}z-\mathrm{e}^{-\mathrm{i}\omega}. \tag{4}$$

Hence

$$A_{21}=d\left(\frac{1-\dfrac{k+1-\varepsilon}{j+\delta}}{1-\dfrac{k+1-\varepsilon}{j+\delta}\,\mathrm{e}^{\mathrm{i}\omega}}\,\frac{1}{\pi\mathrm{i}}\,\frac{1}{(j+\delta)-(k+1-\varepsilon)}\right)_{k,j=0}^{\infty}$$

$$=d\left(\frac{1-\mathrm{e}^{-\mathrm{i}\omega}}{2}\int\limits_{\mathrm{Re}\,z=1/4}\left\{-\mathrm{i}\,\frac{\mathrm{e}^{-\mathrm{i}(\omega-\pi)z}}{\sin(\pi z)}\right\}\frac{1}{\pi\mathrm{i}}\,\frac{\left(\dfrac{k+1-\varepsilon}{j+\delta}\right)^{-z}}{-(k-j)-(1-\varepsilon-\delta)}\,\mathrm{d}z\right)_{k,j=0}^{\infty}$$

$$-\,d\,\mathrm{e}^{-\mathrm{i}\omega}\left(\frac{1}{\pi\mathrm{i}}\,\frac{1}{-(k-j)-(1-\varepsilon-\delta)}\right)_{k,j=0}^{\infty}.$$

The last relation and (4) with $x=1$ imply

$$A_{21}=d\,\frac{1-\mathrm{e}^{-\mathrm{i}\omega}}{2}\int\limits_{\mathrm{Re}\,z=1/4}\left\{-\mathrm{i}\,\frac{\mathrm{e}^{-\mathrm{i}(\omega-\pi)z}}{\sin(\pi z)}\right\}\left((k+1-\varepsilon)^{-z}\,\delta_{kj}\right)_{k,j=0}^{\infty}T(f^{(1-\varepsilon-\delta)})$$

$$\times\left((j+\delta)^{z}\,\delta_{kj}\right)_{k,j=0}^{\infty}\mathrm{d}z-d\,\mathrm{e}^{-\mathrm{i}\omega}T(f^{(1-\varepsilon-\delta)})+d\,\frac{1-\mathrm{e}^{-\mathrm{i}\omega}}{2}\int\limits_{\mathrm{Re}\,z=1/4}\left\{-\mathrm{i}\,\frac{\mathrm{e}^{-\mathrm{i}(\omega-\pi)z}}{\sin(\pi z)}\right\}$$

$$\times\,\mathrm{i}\cot\left(\pi(1-\varepsilon-\delta)\right)\left[\left(\left(\frac{k+\delta}{k+1-\varepsilon}\right)^{z}\delta_{kj}\right)_{k,j=0}^{\infty}-I\right]\mathrm{d}z$$

Since the operator function

$$z\mapsto\left\{\Lambda^{-z}T(f^{(1-\varepsilon-\delta)})\,\Lambda^{z}-\left((k+1-\varepsilon)^{-z}\,\delta_{kj}\right)_{k,j=0}^{\infty}T(f^{(1-\varepsilon-\delta)})\left((j+\delta)^{z}\,\delta_{kj}\right)_{k,j=0}^{\infty}\right\}$$

is continuous and bounded on $\{\mathrm{Re}\,z=1/4\}$ and takes compact values only, there exists a compact operator $T\in\mathcal{L}(\mathrm{l}^2)$ such that

$$A_{21}=T+d\,\frac{1-\mathrm{e}^{-\mathrm{i}\omega}}{2}\int\limits_{\mathrm{Re}\,z=1/4}\left\{-\mathrm{i}\,\frac{\mathrm{e}^{-\mathrm{i}(\omega-\pi)z}}{\sin(\pi z)}\right\}\Lambda^{-z}T(f^{(1-\varepsilon-\delta)})\,\Lambda^{z}\,\mathrm{d}z$$

$$-\,d\,\mathrm{e}^{-\mathrm{i}\omega}T(f^{(1-\varepsilon-\delta)}).$$

Thus, by Lemma 11.4 we obtain that $A_{21}\in\mathfrak{A}$ and that

$$\mathrm{smb}\,A_{21}=d\,\frac{1-\mathrm{e}^{-\mathrm{i}\omega}}{2}\int\limits_{\mathrm{Re}\,z=1/4}\left\{-\mathrm{i}\,\frac{\mathrm{e}^{-\mathrm{i}(\omega-\pi)z}}{\sin(\pi z)}\right\}\mathcal{A}^{z}\,\mathrm{d}z-d\,\mathrm{e}^{-\mathrm{i}\omega}\mathcal{A}^{0},$$

where $\mathcal{A}^z := \operatorname{smb} \Lambda^{-z} T(f^{(1-\varepsilon-\delta)}) \Lambda^z$. Extending $z \mapsto \mathcal{A}^z$ to a 1-periodic analytic function, we get

$$\operatorname{smb} A_{21} = -d\, e^{-i\omega}\mathcal{A}^1 + \frac{d}{2}\left[\int_{\operatorname{Re} z = 1/4} \left\{-i\frac{e^{-i(\omega-\pi)z}}{\sin(\pi z)}\right\}\mathcal{A}^z\, dz - \int_{\operatorname{Re} z = 5/4} \left\{-i\frac{e^{-i(\omega-\pi)z}}{\sin(\pi z)}\right\}\mathcal{A}^z\, dz\right].$$

In the strip $\{1/4 < \operatorname{Re} z < 5/4\}$, the function $z \to \mathcal{A}^z(\tau, \mu)$ is constant if $\tau \neq 1$ and has a pole at $z_0 = 1/2 + i\,(1/2\pi)\log\big(\mu/(1-\mu)\big)$ if $\tau = 1$. Consequently, the residue theorem implies that

$$\operatorname{smb} A_{21}(1, \mu) = -d\, e^{-i\omega}\mathcal{A}^1(1, \mu) - 2\pi i\frac{d}{2}\left[\frac{-i\, e^{-i(\omega-\pi)}}{\pi}\mathcal{A}^1(1, \mu) + \left(-\frac{1}{\pi i}\right)(-i)\frac{e^{-i(\omega-\pi)[1/2+(i/2\pi)\log(\mu/(1-\mu))]}}{\sin\left(\pi\left(\frac{1}{2} + \frac{i}{2\pi}\log\left(\frac{\mu}{1-\mu}\right)\right)\right)}\right],$$

$$\operatorname{smb} A_{21}(\tau, \mu) = -d\, e^{-i\omega}\mathcal{A}^1(\tau, \mu) - 2\pi i\frac{d}{2}\left[\frac{-i e^{-i(\omega-\pi)}}{\pi}\mathcal{A}^1(\tau, \mu)\right], \quad \tau \neq 1,$$

$$\operatorname{smb} A_{21}(\tau, \mu) = \begin{cases} 0 & \text{if } \tau \neq 1, \\ d(-i)\dfrac{e^{-i(\omega-\pi)[1/2+(i/2\pi)\log(\mu/(1-\mu))]}}{\sin\left(\pi\left(\frac{1}{2} + \frac{i}{2\pi}\log\left(\frac{\mu}{1-\mu}\right)\right)\right)} & \text{if } \tau = 1,\ 0 \leqq \mu \leqq 1. \end{cases}$$

In a similar manner one can prove that $A_{12} \in \mathfrak{A}$ and compute $\operatorname{smb} A_{12}$. Finally, we obtain

$$\operatorname{smb} A_1(\tau, \mu)$$
$$= \begin{cases} \begin{pmatrix} \{c + df^{(\varepsilon-\delta)}(\tau)\} & 0 \\ 0 & \{c - df^{(\delta-\varepsilon)}(\tau)\} \end{pmatrix} & \text{if } \tau \neq 1,\ 0 \leqq \mu \leqq 1, \\[2ex] \begin{pmatrix} \{c + d(-\mu + (1-\mu))\} & \left\{-d(-i)\dfrac{e^{-i(\pi-\omega)[1/2+(i/2\pi)\log(\mu/(1-\mu))]}}{\sin\left(\pi\left(\frac{1}{2} + \frac{i}{2\pi}\log\left(\frac{\mu}{1-\mu}\right)\right)\right)}\right\} \\ \left\{d(-i)\dfrac{e^{-i(\omega-\pi)[1/2+(i/2\pi)\log(\mu/(1-\mu))]}}{\sin\left(\pi\left(\frac{1}{2} + \frac{i}{2\pi}\log\left(\frac{\mu}{1-\mu}\right)\right)\right)}\right\} & \{c - d(-\mu + (1-\mu))\} \end{pmatrix} & \\ \text{if } \tau = 1,\ 0 \leqq \mu \leqq 1. \end{cases} \quad (5)$$

Since $\det \operatorname{smb} A_1$ is independent of ω, we may suppose $\omega = \pi$. In this case the operator $A_1 \in \mathcal{L}(l^2)_{2\times 2}$ is a Fredholm operator with index 0 if and only if the convolution operator $A_1 = cI + d(f_{k-j}^{(\varepsilon-\delta)})_{k,j\in\mathbb{Z}} \in \mathcal{L}(\tilde{l}^2)$ is Fredholm and its index vanishes, i.e., if and only if $c + df^{(\varepsilon-\delta)}(\tau) \neq 0$ for all $\tau \in \mathbb{T}$. A simple computation shows that the last condition is equivalent to the requirement that $(c + d)/(c - d) \notin \Omega$.

The operator A_1 corresponding to the methods 11.2(2) and 11.2(3) can be treated analogously. We omit the details and remark only that in these cases $f^{(\varepsilon-\delta)}$ has to be replaced by the functions f^0 and $f^{\#}$, respectively, where

$$f^0(e^{i2\pi x}) := 2x - 1, \qquad 0 \leqq x < 1,$$
$$f^{\#}(e^{i2\pi x}) := \begin{cases} -1, & 0 \leqq x < 1/2, \\ 1, & 1/2 \leqq x < 1. \end{cases}$$

If $\omega = \pi$, then $A_1 \in \mathcal{L}(\mathrm{l}^2)$ is a discrete convolution operator. Since the Fredholm property of a convolution operator implies its invertibility, assertion (c) is obvious. This completes the proof of Theorem 11.3.

11.5. Quadrature methods on curves with corners. Let the simple closed curve Γ be given by the 1-periodic continuous parametrization $\gamma\colon \mathbb{R} \to \mathbb{C}$. For a finite subset M of $[0, 1)$, we suppose that γ is twice continuously differentiable on $[0, 1) \setminus M$, that γ' and γ'' have finite limits at the points of M and that $\gamma'(s + 0) \neq -\gamma'(s - 0)$, $|\gamma'(s + 0)| = |\gamma'(s - 0)|$, $s \in M$. Let $c, d \in \mathbf{C}(\Gamma)$, $k \in \mathbf{C}(\Gamma \times \Gamma)$ and define $S_\Gamma, T, A \in \mathcal{L}\big(\mathbf{L}^2(\Gamma)\big)$ by

$$(S_\Gamma x)(t) := \frac{1}{\pi i} \int_\Gamma \frac{x(\tau)}{\tau - t}\, d\tau, \qquad (Tx)(t) := \int_\Gamma k(t, \tau)\, x(\tau)\, d\tau,$$
$$A = cI + dS_\Gamma + T.$$

We seek an approximate solution of the equation $Au = f$, $f \in \mathcal{R}(\Gamma)$.

For the sake of simplicity, let us assume that M is contained in $\{k/N_0\colon k = 0, \ldots, N_0 - 1\}$ $(N_0 \in \mathbb{N})$ and choose n to be a multiple of N_0. The quadrature methods will be defined as follows. Let $t_k^{(n)} := \gamma\big((k + \delta)/n\big)$, $\tau_k^{(n)} := \gamma\big((k + \varepsilon)/n\big)$ $(0 < \varepsilon, \delta < 1, \varepsilon \neq \delta, k \in \mathbb{Z})$ and determine approximate values $\xi_k^{(n)}$ of $u(t_k^{(n)})$ by solving one of the systems

$$\big\{c(\tau_k^{(n)}) - i \cot\big(\pi(\varepsilon - \delta)\big)\, d(\tau_k^{(n)})\big\}\, \xi_k^{(n)} + d(\tau_k^{(n)}) \frac{1}{\pi i} \sum_{j=0}^{n-1} \frac{\xi_j^{(n)}}{t_j^{(n)} - \tau_k^{(n)}} \Delta t_j^{(n)}$$
$$+ \sum_{j=0}^{n-1} k(\tau_k^{(n)}, t_j^{(n)})\, \xi_j^{(n)} \Delta t_j^{(n)} = f(\tau_k^{(n)}), \qquad k = 0, \ldots, n - 1, \tag{1}$$

$$c(t_k^{(n)})\, \xi_k^{(n)} + d(t_k^{(n)}) \frac{1}{\pi i} \sum_{\substack{j=0 \\ j \neq k}}^{n-1} \frac{\xi_j^{(n)}}{t_j^{(n)} - t_k^{(n)}} \Delta t_j^{(n)} + \sum_{j=0}^{n-1} k(t_k^{(n)}, t_j^{(n)})\, \xi_j^{(n)} \Delta t_j^{(n)} = f(t_k^{(n)}),$$
$$k = 0, \ldots, n - 1, \tag{2}$$

$$c(t_k^{(n)})\, \xi_k^{(n)} + d(t_k^{(n)}) \frac{1}{\pi i} \sum_{\substack{j=0 \\ j \equiv k+1 \bmod 2}}^{n-1} \frac{\xi_j^{(n)}}{t_j^{(n)} - t_k^{(n)}} \Delta t_j^{(n)}$$
$$+ \sum_{\substack{j=0 \\ j \equiv k+1 \bmod 2}}^{n-1} k(t_k^{(n)}, t_j^{(n)})\, \xi_j^{(n)} \Delta t_j^{(n)} = f(t_k^{(n)}), \qquad k = 0, \ldots, n - 1, \tag{3}$$

where $\Delta t_j^{(n)} := \gamma\big((j + 1)/n\big) - \gamma(j/n)$ for (1) and (2) and $\Delta t_j^{(n)} := \gamma\big((j + 1)/n\big) - \gamma\big((j - 1)/n\big)$ for (3). The number n appearing in (3) is supposed to be even. If $\chi_j^{(n)}$ denotes the characteristic function of the arc $\big[\gamma(j/n), \gamma((j + 1)/n)\big)$, then the approximate solution will be defined by $u_n := \sum_{j=0}^{n-1} \xi_j^{(n)} \chi_j^{(n)}$.

Before formulating the stability theorem, let us introduce some notation. We set $\tilde{A} := cI - dS_\Gamma - T$. Analogously to the method of freezing the coefficients in the theory of partial differential equations, we shall consider certain model problems. For $\tau \in \Gamma$, let us define $\omega_\tau \in (0, 2\pi)$ by $\omega_\tau := \arg\left(-\gamma'(\tau - 0)/\gamma'(\tau + 0)\right)$ and set $A^\tau := c(\tau) + d(\tau)\, S_{\Gamma_{\omega_\tau}}$. The model problem for the quadrature method (1), (2) or (3), respectively, is the method 11.2(1), 11.2(2) or 11.2(3), respectively, applied to the operator $A^\tau \in \mathcal{L}\left(\mathbf{L}^2(\Gamma_{\omega_\tau})\right)$. The matrix of the corresponding system of equations will be denoted by A_1^τ. In the proof of Theorem 11.3 we have shown that $A_1^\tau \in \mathfrak{A}_{2\times 2}$.

11.6. Theorem. (a) *The method* 11.5(1) (*resp.* 11.5(2)) *is stable if and only if the operators* $A \in \mathcal{L}\left(\mathbf{L}^2(\Gamma)\right)$ *and* $A_1^\tau \in \mathcal{L}(\tilde{\mathbf{l}}^2)$ $(\tau \in \Gamma)$ *are invertible. The method* 11.5(3) *is stable if and only if the operators* $A, \tilde{A} \in \mathcal{L}\left(\mathbf{L}^2(\Gamma)\right)$ *and* $A_1^\tau \in \mathcal{L}(\tilde{\mathbf{l}}^2)$ $(\tau \in \Gamma)$ *are invertible.*

(b) *If the quadrature method is stable and* $f \in \mathcal{R}(\Gamma)$, *then the systems* 11.5(1), 11.5(2) *or* 11.5(3), *respectively, are uniquely solvable for n large enough, and the approximate solutions* u_n *converge to* $u = A^{-1}f$ *as* $n \to \infty$.

Remark. Combining Theorem 11.3 and Theorem 11.6 we get necessary and sufficient conditions for the quadrature methods 11.5(1)–11.5(3) to be stable. In general, the only trouble is that the set Φ in Theorem 11.3 is unknown. We conjecture that it is void in nearly all cases. But now suppose we are out of luck and have the following situation. The operators A and, for 11.5(3), also $\tilde{A}$ are invertible and the operators A_1^τ, $\tau \in \Gamma$, are at least Fredholm with index 0. Moreover, let us assume that in one or more corner points the operators A_1^τ have nontrivial null spaces. Then the quadrature methods only need a little modification in the neighborhood of these points in order to become stable (cf. Remark 11.30).

We shall prove Theorem 11.6 in Section 11.15. To prepare this proof we need a local theory which reduces the stability of approximation methods for an operator $A \in \mathcal{L}\left(\mathbf{L}^2(\Gamma)\right)$ to the stability of the corresponding methods for certain model operators. Let us suppose that we are given a sequence $\{A_n\}$ of approximating operators $A_n \in \mathcal{L}(\operatorname{im} L_n)$, where L_n is the $\mathbf{L}^2(\Gamma)$-orthogonal projection onto the subspace span $\{\chi_k^{(n)}: k = 0, \ldots, n - 1\}$ and the scalar product in $\mathbf{L}^2(\Gamma)$ is given by

$$(f, g) := \int_0^1 f(\gamma(t))\, \overline{g(\gamma(t))}\, dt. \tag{1}$$

Furthermore, let there exist certain model operators $A_1^\tau \in \mathfrak{A}_{2\times 2} \subseteq \mathcal{L}(\tilde{\mathbf{l}}^2)$. In order to describe the connection between $\{A_n\}$ and A_1^τ, $\tau \in \Gamma$, we need some notation.

Let the projections $K_n^\varepsilon, K_n^\delta: \mathcal{R}(\Gamma) \to \mathbf{L}^2(\Gamma)$ and $P_n \in \mathcal{L}(\tilde{\mathbf{l}}^2)$ be defined by

$$K_n^\varepsilon f := \sum_{k=0}^{n-1} f(\tau_k^{(n)})\, \chi_k^{(n)}, \qquad K_n^\delta f := \sum_{k=0}^{n-1} f(t_k^{(n)})\, \chi_k^{(n)}.$$

$$P_n\{\xi_k\}_{k\in\mathbb{Z}} = \{\eta_k\}_{k\in\mathbb{Z}}, \qquad \eta_k := \begin{cases} \xi_k & \text{if } -n/2 < k \leq n/2, \\ 0 & \text{else.} \end{cases}$$

For given $\tau \in \Gamma$ and $n \in \mathbb{N}$, we introduce $E_n^\tau: \operatorname{im} P_n \to \operatorname{im} L_n$ by

$$\operatorname{im} P_n \ni \{\delta_{k,j}\}_{k\in\mathbb{Z}} \mapsto \chi_{j+k_E}$$

where $k_E = k_E(\varepsilon, \tau, n) \in \{0, \ldots, n-1\}$ is defined by $\tau \in (\tau_{k_E-1}^{(n)}, \tau_{k_E}^{(n)}]$. Since E_n^τ is bijective, the mapping $\mathcal{L}(\operatorname{im} L_n) \ni B_n \mapsto B_n^E := E_n^{\tau\,-1} B_n E_n^\tau \in \mathcal{L}(\operatorname{im} P_n)$ is an isomorphism. Moreover, a sequence $\{B_n\}$, $B_n \in \mathcal{L}(\operatorname{im} L_n)$, is uniformly bounded (stable) if and only if $\{B_n^E\}$ has the same property.

Let $\mathbf{M}(\Gamma)$ denote the set of Lipschitz-continuous functions χ on Γ satisfying $0 \leqq \chi \leqq 1$. For $\chi \in \mathbf{C}(\Gamma)$, we set $\tilde{\chi}_n := K_n^\varepsilon \chi|_{\operatorname{im} L_n}$. We shall say that $\{A_n\}$ is equivalent to A_1^τ at τ if, for any $\varepsilon' > 0$, there exist $n_0 \in \mathbb{N}$ and a neighborhood U of τ such that $\chi \in \mathbf{M}(\Gamma)$, $\operatorname{supp} \chi \subset U$ and $n \geqq n_0$ imply $\|\tilde{\chi}_n^E(A_n^E - A_1^\tau)\, \tilde{\chi}_n^E\| < \varepsilon'$. Finally, by $\mathfrak{A}_{2\times 2}^1$ we denote the subalgebra of $\mathcal{L}(\tilde{\mathbf{l}}^2)$ generated by the projection P and the operators $C(\alpha)$ with $\alpha \in \mathbf{PC}(\mathbb{T}) \cap \mathbf{C}(\mathbb{T} \setminus \{1\})$, where $C(\alpha) := (\alpha_{j-k})_{j,k=-\infty}^{\infty}$ if $\alpha(t) = \sum_{k \in \mathbb{Z}} \alpha_k t^k$.

11.7. Theorem. *Suppose $A \in \mathcal{L}(\mathbf{L}^2(\Gamma))$ and $\{A_n\}$, $A_n \in \mathcal{L}(\operatorname{im} L_n)$, satisfy the following conditions.*

(i) *There exists a finite subset $\Gamma' \subset \Gamma$ such that $\tilde{\chi}_n A_n \tilde{\chi}_n L_n \to \chi A \chi$ for all functions $\chi \in \mathbf{C}(\Gamma)$ satisfying* $\operatorname{supp} \chi \cap \Gamma' = \emptyset$.

(ii) *The operator $\chi A - A\chi$ is compact in $\mathbf{L}^2(\Gamma)$ for every $\chi \in \mathbf{C}(\Gamma)$.*

(iii) *The norm $\left\|\tilde{\chi}_n A_n - A_n \tilde{\chi}_n - L_n(\chi A - A\chi)|_{\operatorname{im} L_n}\right\|$ converges to 0 for every $\chi \in \mathbf{C}(\Gamma)$ as $n \to \infty$.*

(iv) *There exist operators $A_1^\tau \in \mathfrak{A}_{2\times 2}^1 \subset \mathcal{L}(\tilde{\mathbf{l}}^2)$ $(\tau \in \Gamma)$ such that $\{A_n\}$ is equivalent to A_1^τ at τ.*

Then $A_n L_n$ converges strongly to A. Moreover, $\{A_n\}$ is stable if and only if the operators $A \in \mathcal{L}(\mathbf{L}^2(\Gamma))$ and $A_1^\tau \in \mathcal{L}(\tilde{\mathbf{l}}^2)$ $(\tau \in \Gamma)$ are invertible.

This theorem will be used in order to prove the stability of the methods 11.5(1) and 11.5(2). For the proof of the stability of 11.5(3), we need a slight modification. Define $W_n \in \mathcal{L}(\operatorname{im} L_n)$ by $W_n \chi_j^{(n)} = (-1)^j \chi_j^{(n)}$, $j = 0, \ldots, n-1$, and set $\tilde{B}_n := W_n B_n W_n$ for $B_n \in \mathcal{L}(\operatorname{im} L_n)$. We suppose the number n to be even. Furthermore, by $\mathfrak{A}_{2\times 2}^{\pm 1}$ we denote the subalgebra of $\mathcal{L}(\tilde{\mathbf{l}}^2)$ generated by P and the operators $C(\alpha)$ with $\alpha \in \mathbf{PC}(\mathbb{T}) \cap \mathbf{C}(\mathbb{T} \setminus \{1, -1\})$.

11.8. Theorem. *Suppose $A \in \mathcal{L}(\mathbf{L}^2(\Gamma))$ and $\{A_n\}$, $A_n \in \mathcal{L}(\operatorname{im} L_n)$ satisfy the assumptions* (i) *and* (ii) *of Theorem* 11.7. *Assume that, in addition, the following properties are satisfied.*

(i)′ *There exists an $\tilde{A} \in \mathcal{L}(\mathbf{L}^2(\Gamma))$ such that $\tilde{\chi}_n \tilde{A}_n \tilde{\chi}_n L_n \to \chi \tilde{A} \chi$ for all $\chi \in \mathbf{C}(\Gamma)$ satisfying* $\operatorname{supp} \chi \cap \Gamma' = \emptyset$.

(ii)′ *The operator $\chi\tilde{A} - \tilde{A}\chi$ is compact for all $\chi \in \mathbf{C}(\Gamma)$.*

(iii)′ *For each $\chi \in \mathbf{C}(\Gamma)$,*

$$\left\|(\tilde{\chi}_n A_n - A_n \tilde{\chi}_n) - L_n(\chi A - A\chi)|_{\operatorname{im} L_n} - W_n L_n(\chi\tilde{A} - \tilde{A}\chi)|_{\operatorname{im} L_n} W_n\right\| \to 0 .$$

(iv)′ *There exist operators $A_1^\tau \in \mathfrak{A}_{2\times 2}^{\pm 1} \subset \mathcal{L}(\tilde{\mathbf{l}}^2)$ $(\tau \in \Gamma)$ such that $\{A_n\}$ is equivalent to A_1^τ at τ.*

Then $A_n L_n$ converges strongly to A, and $\{A_n\}$ is stable if and only if the operators $A, \tilde{A} \in \mathcal{L}(\mathbf{L}^2(\Gamma))$ and $A_1^\tau \in \mathcal{L}(\tilde{\mathbf{l}}^2)$ $(\tau \in \Gamma)$ are invertible.

Since the proof of Theorem 11.7 runs analogously to the one of Theorem 11.8, we only prove Theorem 11.8. This theorem in turn follows from Corollary 10.42. To see this, we shall identify $\mathbb{T}$ and Γ and recall some notation. We define $\Phi\colon \mathbb{T} \to \Gamma$ by $\Phi(e^{i2\pi s}) := \gamma(s)$ and identify the functions f on Γ with the functions $f \circ \Phi$ on $\mathbb{T}$. In view of 11.6(1) the spaces $\mathbf{L}^2(\Gamma)$ and $\mathbf{L}^2(\mathbb{T})$ coincide and L_n is the $\mathbf{L}^2(\mathbb{T})$-orthogonal projection onto the span of $\{\chi_j^{(n)} \circ \Phi : j = 0, \ldots, n-1\}$. Furthermore, identifying $\sum_{j \in \mathbb{Z}} \xi_j t^j \in \mathbf{L}^2(\mathbb{T})$ and $\{\xi_j\}_{j \in \mathbb{Z}} \in \tilde{\mathbf{l}}^2$, the projection $P_n \in \mathcal{L}(\mathbf{L}^2(\mathbb{T}))$ of 9.32(10) is the same as the P_n prevailing in Section 11.7, and $P_{\mathbb{T}}$ coincides with P. Let $\hat{\alpha}_n \in \mathcal{L}(\operatorname{im} L_n)$ $(\alpha \in \mathbf{PC}(\mathbb{T}))$, $\tilde{c}_n \in \mathcal{L}(\operatorname{im} L_n)$ $(c \in \mathbf{PC}(\Gamma))$,

$E_n^{(\tau,1)} \in \mathcal{L}(\operatorname{im} L_n)$ $(\tau \in \mathbb{T})$ and $E_n^{(\tau,2)} \in \mathcal{L}(\operatorname{im} L_n, \operatorname{im} P_n)$ denote the operators

$$\hat{\alpha}_n : \sum_{j=0}^{n-1} \xi_j \chi_j^{(n)} \mapsto \sum_{k=0}^{n-1} \left(\sum_{j=0}^{n-1} \alpha_{n,k-j} \xi_j \right) \chi_k^{(n)},$$

$$\alpha_{n,j} := 1/n \sum_{l=0}^{n-1} \alpha\,(\exp\,[\mathrm{i}2\pi l/n]) \exp\,(\mathrm{i}2\pi lj/n), \tag{1}$$

$$\tilde{c}_n : \sum_{j=0}^{n-1} \xi_j \chi_j^{(n)} \mapsto \sum_{j=0}^{n-1} c(\tau_j^{(n)})\, \xi_j \chi_j^{(n)},$$

$$E_n^{(\tau,1)} : \sum_{j=0}^{n-1} \xi_j \chi_j^{(n)} \mapsto \sum_{j=0}^{n-1} t_E^{-j} \xi_j \chi_j^{(n)},$$

$$E_n^{(\tau,2)} := (E_n^{\Phi(\tau)})^{-1},$$

where $t_E = t_E(\tau, n) := \exp\,(\mathrm{i}2\pi j_E/n)$ and $j_E = j_E(\tau, n) \in \{0, 1, \ldots, n-1\}$ is the number satisfying $\tau \in \left(\exp\,[2\pi\mathrm{i}(j_E - 1)/n], \exp\,[2\pi\mathrm{i}j_E/n]\right] \subset \mathbb{T}$. If we consider the algebra of all sequences $\{B_n\}$ with $B_n \in \mathcal{L}(\operatorname{im} L_n)$ and $\|\{B_n\}\| := \sup\,\{\|B_n L_n\| : n \in \mathbb{N}\} < \infty$, then $\mathfrak{Z}''$ will denote the subalgebra containing all sequences of the form $\{\hat{\alpha}_n\}$, $\{\tilde{c}_n\}$, $\{C_n\}$, $\{(E_n^{(\tau,1)})^{-1} \times L_n T^\tau E_n^{(\tau,1)}\}$ and $\{(E_n^{(\tau,2)})^{-1} P_n K^\tau E_n^{(\tau,2)}\}$ with $\alpha \in \mathbf{PC}(\mathbb{T}) \cap \mathbf{C}(\mathbb{T} \setminus \{-1, 1\})$, $c \in \mathbf{PC}(\Gamma)$, $\|C_n\| \to 0$, $\tau \in \mathbb{T}$, $T^\tau \in \mathcal{K}(\mathbf{L}^2(\Gamma))$ and $K^\tau \in \mathcal{K}(\tilde{\mathbf{l}}^2)$. Finally, for $\{B_n\} \in \mathfrak{Z}''$ and $\tau \in \mathbb{T}$, we set

$$W_{(\tau,1)}\{B_n\} := \lim E_n^{(\tau,1)} B_n (E_n^{(\tau,1)})^{-1} L_n,$$

$$W_{(\tau,2)}\{B_n\} := \lim E_n^{(\tau,2)} B_n (E_n^{(\tau,2)})^{-1} P_n = \lim B_n^E P_n.$$

Now, Corollary 9.42 states that the sequence $\{B_n\} \in \mathfrak{Z}''$ is stable if and only if, for all $\tau \in \mathbb{T}$, the operators $W_{(\tau,2)}\{B_n\}$ and, for $\tau \in \{1, -1\}$, the operators $W_{(\tau,1)}\{B_n\}$ are invertible.

11.9. Lemma. *Suppose* $A \in \mathcal{L}(\mathbf{L}^2(\Gamma))$ *and* $\{A_n\}$ $(A_n \in \mathcal{L}(\operatorname{im} L_n))$ *satisfy the assumptions* (ii), (ii)′, (iii)′, (iv)′ *of the Theorems* 11.7 *and* 11.8. *Then* $A_n \in \mathfrak{Z}''$.

For the proof, we need two further lemmas.

11.10. Lemma. *Let* $\tau \in \Gamma$ *be fixed and* $B^\tau \in \mathfrak{A}_{2\times 2}^{\pm 1} \subset \mathcal{L}(\tilde{\mathbf{l}}^2)$. *For each* $\chi' \in \mathbf{M}(\Gamma)$ *which is identically equal to* 1 *in a neighborhood of* τ *and for each* $\varepsilon' > 0$, *there exists a smaller neighborhood* U *of* τ *such that* $\chi \in \mathbf{M}(\Gamma)$ *and* $\operatorname{supp} \chi \subset U$ *imply* $\|\tilde{\chi}_n^E B^\tau (I - \tilde{\chi}_n'^E)\| < \varepsilon'$.

Proof. *Step 1.* Let $\mathfrak{E}$ denote the set of all $B^\tau \in \mathcal{L}(\tilde{\mathbf{l}}^2)$ such that the assertion of the lemma holds. Since

$$\tilde{\chi}_n^E B^\tau C^\tau (I - \tilde{\chi}_n'^E) = [\tilde{\chi}_n^E B^\tau (I - \tilde{\chi}_n''^E)]\, C^\tau (I - \tilde{\chi}_n'^E) + \tilde{\chi}_n^E B^\tau [\tilde{\chi}_n''^E C^\tau (I - \tilde{\chi}_n'^E)]$$

we observe that $\mathfrak{E}$ is an algebra. It is not hard to show that $\mathfrak{E}$ is closed in the operator norm, i.e. that $\mathfrak{E}$ is a closed subalgebra of $\mathcal{L}(\tilde{\mathbf{l}}^2)$.

Step 2. Now consider a $B^\tau = (b_{jk})_{j,k\in\mathbb{Z}}$ which satisfies $b_{jk} = 0$ for $j \neq \pm k$. Choose $U := \{\gamma(s) : \gamma^{-1}(\tau) - \delta < s < \gamma^{-1}(\tau) + \delta\}$ with $\delta > 0$ being small enough. Then one can check that $U \subseteq \{t \in \Gamma : \chi'(t) = 1\}$ and $\tilde{\chi}_n^E B^\tau (I - \tilde{\chi}_n'^E) = 0$. Thus $B^\tau \in \mathfrak{E}$.

Step 3. Let $B^\tau := C(a) = (a_{k-j})_{k,j\in\mathbb{Z}}$, where a is piecewise continuous. Furthermore, suppose a is twice differentiable at the points of continuity and these derivatives are piecewise continuous. We shall show that $B^\tau \in \mathfrak{E}$.

Let $\tau = \gamma(\sigma)$ and assume $\chi'(\gamma(s)) = 1$ for $\sigma - \delta_1 < s < \sigma + \delta_1$. We choose $U := \{\gamma(s) : \sigma - \delta_2 < s < \sigma + \delta_2\}$ for a suitable $\delta_2 < \delta_1$. Then the absolute value of the element in the j-th row and k-th column of $(I - \tilde{\chi}_n'^E) B^{\tau *}\tilde{\chi}_n^E$ does not exceed Cc_{k-j}, where $c_k := |k|^{-1}$ if $|k| \geqq (\delta_1 - \delta_2) n$ and $c_k := 0$ if $|k| < (\delta_1 - \delta_2) n$. For $\{\xi_j\} \in \tilde{\mathrm{l}}^2$, define $\{\eta_j\} \in \tilde{\mathrm{l}}^2$ by $\eta_j = 0$ if $j \geqq \delta_2 n$ and $\eta_j = |\xi_j|$ if $j < \delta_2 n$. Then we get

$$\|((I - \tilde{\chi}_n'^E) B^{\tau *}\tilde{\chi}_n^E) \{\xi_j\}\|_{\tilde{\mathrm{l}}^2} = \left\| \left\{ \sum_{k\in\mathbb{Z}} ((I - \tilde{\chi}_n'^E) B^{\tau *}\tilde{\chi}_n^E)_{j,k}\, \xi_k \right\}_{j\in\mathbb{Z}} \right\|_{\tilde{\mathrm{l}}^2}$$
$$\leqq C \left\| \left\{ \sum_{k\in\mathbb{Z}} c_{j-k}\eta_k \right\}_{j\in\mathbb{Z}} \right\|_{\tilde{\mathrm{l}}^2}.$$

Young's inequality yields

$$\|((I - \tilde{\chi}_n'^E) B^{\tau *}\tilde{\chi}_n^E) \{\xi_j\}\|_{\tilde{\mathrm{l}}^2} \leqq C\, \|\{\eta_k\}\|_{\tilde{\mathrm{l}}^1}\, \|\{c_j\}\|_{\tilde{\mathrm{l}}^2}.$$

Using that

$$\|\{\eta_k\}\|_{\tilde{\mathrm{l}}^1} \leqq \sqrt{\delta_2 n}\, \|\{\eta_k\}\|_{\mathrm{l}^2} \leqq \sqrt{\delta_2 n}\, \|\{\xi_k\}\|_{\tilde{\mathrm{l}}^2},$$
$$\|\{c_j\}\|_{\tilde{\mathrm{l}}^2} \leqq \left(\sum_{\substack{k\in\mathbb{Z} \\ |k|>(\delta_1-\delta_2)n}} \frac{1}{k^2} \right)^{1/2} \leqq C \frac{1}{\sqrt{(\delta_1 - \delta_2)\, n}},$$

we conclude that

$$\|((I - \tilde{\chi}_n'^E) B^{\tau *}\tilde{\chi}_n^E) \{\xi_j\}\|_{\tilde{\mathrm{l}}^2} \leqq C \sqrt{\frac{\delta_2}{\delta_1 - \delta_2}}\, \|\{\xi_j\}\|_{\tilde{\mathrm{l}}^2}.$$

If we choose δ_2 small enough, then

$$\|\tilde{\chi}_n^E B^\tau (I - \tilde{\chi}_n'^E)\|_{\mathcal{L}(\tilde{\mathrm{l}}^2)} = \|(I - \tilde{\chi}_n'^E) B^{\tau *}\tilde{\chi}_n^E\|_{\mathcal{L}(\tilde{\mathrm{l}}^2)} \leqq C \sqrt{\frac{\delta_2}{\delta_1 - \delta_2}} < \varepsilon'.$$

Step 4. Now the inclusion $\mathfrak{A}^{\pm 1}_{2\times 2} \subseteq \mathfrak{C}$ follows from the fact that $\mathfrak{A}^{\pm 1}_{2\times 2}$ is contained in the smallest closed subalgebra of $\mathcal{L}(\tilde{\mathrm{l}}^2)$ containing the operators B^τ considered in the steps 2 and 3 of this proof. ■

11.11. Lemma. *For $\alpha \in \mathbf{PC}(\mathbb{T})$, $t \in \Gamma$, and given $\varepsilon' > 0$, there exists an open arc U on Γ such that $t \in U$ and $\|\tilde{\chi}_n^E(\hat{\alpha}_n^E - C(\alpha))\, \tilde{\chi}_n^E\| < \varepsilon'$ holds whenever $\chi \in \mathbf{M}(\Gamma)$ satisfies* supp χ $\subset U$.

Proof. *Step 1.* Without loss of generality, suppose α is differentiable between the points of discontinuity and its derivative is in $\mathbf{PC}(\mathbb{T})$. (Otherwise density arguments apply.) By definition $C(\alpha) = (\alpha_{k-j})_{k,j=-\infty}^{\infty}$, and $\hat{\alpha}_n^E$ has the matrix $(\alpha_{n,k-j})_{k,j=-[n/2]}^{[(n-1)/2]}$ corresponding to the basis $\{t^j : j = -[n/2], \ldots, [(n-1)/2]\}$ (cf. 11.8(1)). We consider the difference of the entries $\alpha_l - \alpha_{n,l}$ and show the existence of constants $C > 0$, $\delta' > 0$ such that $|l| \leqq \delta' n$ implies

$$|\alpha_l - \alpha_{n,l}| \leqq Cn^{-1}. \tag{1}$$

If $-kn < s \leqq kn$ and $s \in \mathbb{Z}$, then s is the sum of s' and u, where $-k < u \leqq k$, s' is a multiple of $2k$, and u and s' are uniquely determined. We set

$$\alpha_{n,l,k} := \frac{1}{2kn} \sum_{s=-kn+1}^{kn} \exp\,(\mathrm{i}2\pi sl/2kn)\; \alpha(\exp\,[\mathrm{i}2\pi s/2kn])$$

and obtain

$$\alpha_{n,l} = \frac{1}{2kn} \sum_{s=-kn+1}^{kn} \exp\,(\mathrm{i}2\pi s'l/2kn)\; \alpha(\exp\,[\mathrm{i}2\pi s'/2kn]).$$

Thus we arrive at the equality $\alpha_{n,l,k} - \alpha_{n,l} = \alpha^1_{n,l,k} + \alpha^2_{n,l,k}$, where

$$\alpha^1_{n,l,k} := \sum_{s=-kn+1}^{kn} \frac{\exp(i2\pi sl/2kn)}{2kn} \left[\alpha(\exp[i2\pi s/2kn]) - \alpha(\exp[i2\pi s'/2kn])\right],$$

$$\alpha^2_{n,l,k} := \sum_{s=-kn+1}^{kn} \frac{\alpha(\exp[i2\pi s'/2kn])}{2kn} [\exp(i2\pi sl/2kn) - \exp(i2\pi s'l/2kn)].$$

If α is continuous on the arc between $\exp(i2\pi s/2kn)$ and $\exp(i2\pi s'/2kn)$, then $|\alpha(\exp[i2\pi s/2kn]) - \alpha(\exp[i2\pi s'/2kn])| < Cn^{-1}$. Since there is only a finite number of discontinuities of α, we get $|\alpha^1_{n,l,k}| < Cn^{-1}$ for another positive constant C. For $\alpha^2_{n,l,k}$, we obtain

$$\alpha^2_{n,l,k} = \alpha_{n,l}\beta_{n,l,k}, \qquad \beta_{n,l,k} := \frac{1}{2k} \sum_{u=-k+1}^{k} [\exp(i2\pi lu/2kn) - 1].$$

Using the formula for the sum of the geometric series, we get

$$\beta_{n,l,k} = \frac{\sin(\pi l/n)}{\pi l/n} \frac{i2\pi l/2kn}{\exp(i2\pi l/2kn) - 1} \exp(i2\pi l/2kn) - 1,$$

$$\lim_{k\to\infty} \beta_{n,l,k} = \frac{\sin(\pi l/n)}{\pi l/n} - 1 = (l/n)\left\{\pi \sum_{u=1}^{\infty} \frac{(-1)^u}{(2u+1)!} (\pi l/n)^{2u-1}\right\}.$$

Consequently, we have

$$\alpha_l - \alpha_{n,l} = \lim_{k\to\infty} (\alpha_{n,l,k} - \alpha_{n,l}), \qquad |\alpha_l - \alpha_{n,l}| \leqq Cn^{-1} + C\,|\alpha_{n,l}|\, ln^{-1}.$$

Since $|\alpha_l| \leqq Cl^{-1}$ we obtain $|\alpha_l - \alpha_{n,l}| \leqq Cn^{-1} + C\,|\alpha_l - \alpha_{n,l}|\, ln^{-1}$. Thus (1) holds.

Step 2. If we choose U small enough, then $\tilde{\chi}^E_n(\hat{\alpha}^E_n - C(\alpha))\, \tilde{\chi}^E_n$ is a matrix the entries of which vanish outside a submatrix of dimension $n' \times n'$ with $n' < 2\delta' n$. Since the absolute values of the entries in this submatrix are smaller than Cn^{-1} we get $\|\tilde{\chi}^E_n(\hat{\alpha}^E_n - C(\alpha))\, \tilde{\chi}^E_n\| < C2\delta'$. ■

11.12. Proof of Lemma 9. *Step 1.* In the first step we prove the following assertion: for any $\tau \in \Gamma$ and $\varepsilon' > 0$, there exists an open arc $U_\tau \subset \Gamma$ with $\tau \in U_\tau$ and a sequence $\{A^\tau_n\} \in \mathfrak{Z}''$ such that $\chi \in \mathbf{M}(\Gamma)$ and $\operatorname{supp} \chi \subset U_\tau$ imply

$$\|\tilde{\chi}_n(A_n - A^\tau_n)\, \tilde{\chi}_n\| < \varepsilon'. \tag{1}$$

In view of assumption (iv)' it suffices to show that, for any τ and ε', there exist U_τ and $\{A^\tau_n\}$ such that $\|\tilde{\chi}^E_n(A^{\tau E}_n - A^\tau_1)\, \tilde{\chi}^E_n\| < \varepsilon'$. Furthermore, by density arguments we may suppose that A^τ_1 is a finite sum of finite products of operators of the form P and $C(\alpha)$ with $\alpha \in \mathbf{PC}(\mathbb{T}) \cap \mathbf{C}(\mathbb{T} \setminus \{1, -1\})$.

For the sake of simplicity, we consider $A^\tau_1 = P\, C(\alpha)\, P\, C(\beta)$.

If $\tau = \gamma(\sigma)$, then put $\tau' := \gamma(\sigma + 1/2)$ and let p denote the characteristic function of the arc $[\tau, \tau') \subset \Gamma$. We choose $\chi', \chi'' \in \mathbf{M}(\Gamma)$ satisfying $\chi\chi' = \chi$, $\chi''\chi' = \chi'$ and set $A^\tau_n := \tilde{p}_n \hat{\alpha}_n \tilde{\chi}'_n \tilde{p}_n \tilde{\chi}''_n \hat{\beta}_n$.

Obviously, $\{A^\tau_n\} \in \mathfrak{Z}''$. If $\operatorname{supp} \chi''$ is contained in a small neighborhood of τ, we get $\tilde{p}^n_* \tilde{\chi}^E_n = P\tilde{\chi}^E_n$. Using this, and taking into account that $P\tilde{\chi}''^E_n = \tilde{\chi}''^E_n P$ and $\tilde{\chi}'_n \tilde{\chi}''_n = \tilde{\chi}'_n$, we arrive at the decomposition

$$\begin{aligned}\tilde{\chi}_n^E(A_n^{\tau E} - A_1^\tau)\,\tilde{\chi}_n^E &= P\{\tilde{\chi}_n^E(\hat{\alpha}_n^E - C(\alpha))\,\tilde{\chi}_n'^E\}\,P\tilde{\chi}_n''^E\hat{\beta}_n^E\tilde{\chi}_n^E \\ &\quad + P\tilde{\chi}_n^E C(\alpha)\,\tilde{\chi}_n'^E P\{\tilde{\chi}_n''^E(\hat{\beta}_n^E - C(\beta))\,\tilde{\chi}_n^E\} \\ &\quad - P\{\tilde{\chi}_n^E C(\alpha)\,(I - \tilde{\chi}_n'^E)\}\,PC(\beta)\,\tilde{\chi}_n^E.\end{aligned}$$

The last expression becomes small if we choose χ, χ', χ'' in accordance with the Lemmas 11.10 and 11.11. The proof of (1) is complete.

Step 2. Now we fix $\varepsilon' > 0$ and choose $\tau_1, \ldots, \tau_N \in \Gamma$ such that $\Gamma \subset \bigcup_{i=1}^{N} U_{\tau_i}$. We can find functions χ^i, χ'^i, χ''^i $(i = 1, \ldots, N)$ such that $\operatorname{supp} \chi''^i \subset U_{\tau_i}$, $\chi^i\chi'^i = \chi^i$, $\chi'^i\chi''^i = \chi'^i$ $(i = 1, \ldots, N)$ and $\sum_{i=1}^{N} \chi^i = 1$. Furthermore, we may suppose that $\operatorname{supp} \chi''^i \cap \operatorname{supp} \chi''^j \neq \emptyset$, $i \neq j$ implies $i = j \pm 1$ or $i = 1, j = N$, or $i = N, j = 1$. We set $A_n^i := A_n^{\tau_i}$ and

$$B_n := \sum_{i=1}^{N} \tilde{\chi}_n^i A_n^i \tilde{\chi}_n'^i + \sum_{i=1}^{N} \tilde{\chi}_n^i \{\tilde{\chi}_n'^i A_n - A_n \tilde{\chi}_n'^i\}.$$

Because of the inclusion $\{A_n^i\} \in \mathfrak{Z}''$ and the assumptions (iii)′, (ii) and (ii)′, the sequence $\{B_n\}$ belongs to $\mathfrak{Z}''$. Since $\sum_{i=1}^{N} \tilde{\chi}_n^i = I_n$ and $\tilde{\chi}_n^i\tilde{\chi}_n'^i = \tilde{\chi}_n^i$, we have

$$\begin{aligned}A_n - B_n &= \sum_{i=1}^{N} \tilde{\chi}_n^i\tilde{\chi}_n'^i A_n - B_n \\ &= \sum_{i=1}^{N} \tilde{\chi}_n^i A_n \tilde{\chi}_n'^i + \sum_{i=1}^{N} \tilde{\chi}_n^i \{\tilde{\chi}_n'^i A_n - A_n \tilde{\chi}_n'^i\} - B_n \\ &= \sum_{i=1}^{N} \tilde{\chi}_n^i (A_n - A_n^i)\,\tilde{\chi}_n'^i.\end{aligned}$$

For any $f \in \mathbf{L}^2(\Gamma)$, we get

$$\begin{aligned}\|(A_n - B_n)\,L_n f\| &\leqq \sum_{i=1}^{N} \|\tilde{\chi}_n^i (A_n - A_n^i)\,\tilde{\chi}_n'^i \tilde{\chi}_n''^i L_n f\| \\ &\leqq \sup_{i=1,\ldots,N} \|\tilde{\chi}_n^i (A_n - A_n^i)\,\tilde{\chi}_n'^i\| \sum_{i=1}^{N} \|\tilde{\chi}_n''^i L_n f\|.\end{aligned}$$

Using (1) and the fact that the supports of the functions χ''^i are almost disjoint, we arrive at the estimate

$$\|(A_n - B_n)\,L_n f\| \leqq C\varepsilon' \,\|L_n f\|, \qquad \|A_n - B_n\| \leqq C\varepsilon'.$$

Since ε' is arbitrary and $\mathfrak{Z}''$ is closed, we conclude that $\{A_n\} \in \mathfrak{Z}''$. ■

11.13. Proof of Theorem 11.8. *Step 1.* Theorem 11.8 follows from Corollary 10.42, and Lemma 11.9 once we have shown that $W_{(1,1)}\{A_n\} = A$, $W_{(-1,1)}\{A_n\} = \tilde{A}$ and $W_{(\tau,2)}\{A_n\} = A_1^{\Phi(\tau)}$ $(\tau \in \mathbb{T})$.

Step 2. First, we shall prove that $W_{(-1,1)}\{A_n\} = \tilde{A}$. The identity $W_{(1,1)}\{A_n\} = A$ will follow analogously. Since $\{A_n\} \in \mathfrak{Z}''$, the sequences $\{A_n\}$ and $\{W_n A_n W_n\} = \{E_n^{(-1,1)} \times A_n (E_n^{(-1,1)})^{-1}\}$ are uniformly bounded. Therefore, it suffices to verify that $W_n A_n W_n L_n f$

$\to \tilde{A}f$ for all f from a dense subset of $\mathbf{L}^2(\Gamma)$. We may suppose that $\operatorname{supp} f \cap \Gamma' = \emptyset$ and choose $\chi \in \mathbf{M}(\Gamma)$ satisfying $\operatorname{supp} \chi \cap \Gamma' = \emptyset$ and $\tilde{\chi}_n L_n f = L_n f$.

For another $\chi' \in \mathbf{M}(\Gamma)$ with $\chi'\chi = \chi$ and $\operatorname{supp} \chi' \cap \Gamma' = \emptyset$, we get

$$\begin{aligned} W_n A_n W_n L_n f &= \tilde{\chi}_n' W_n A_n W_n \tilde{\chi}_n L_n f + (I_n - \tilde{\chi}_n')\, W_n A W_n \tilde{\chi}_n L_n f \\ &= \tilde{\chi}_n' W_n A_n W_n \tilde{\chi}_n' L_n f + \{W_n A_n W_n \tilde{\chi}_n' - \tilde{\chi}_n' W_n A_n W_n\}\, L_n f. \end{aligned} \tag{1}$$

Using that $W_n \tilde{\chi}_n' = \tilde{\chi}_n' W_n$, we arrive at the equalities

$$\begin{aligned} &\{W_n A_n W_n \tilde{\chi}_n' - \tilde{\chi}_n' W_n A_n W_n\}\, L_n f = W_n \{A_n \tilde{\chi}_n' - \tilde{\chi}_n' A_n\}\, W_n L_n f \\ &= W_n \big\{A_n \tilde{\chi}_n' - \tilde{\chi}_n' A_n - L_n(A\chi' - \chi' A)|_{\operatorname{im} L_n} - W_n L_n(\tilde{A}\chi' - \chi'\tilde{A})|_{\operatorname{im} L_n} W_n\big\} \\ &\quad \times W_n L_n f + W_n L_n (A\chi' - \chi' A)\, W_n L_n f + L_n(\tilde{A}\chi' - \chi'\tilde{A})\, L_n f. \end{aligned}$$

Assumption (ii) and Lemma 10.34(c) yield that $W_n L_n(A\chi' - \chi' A)\, W_n L_n f \to 0$. This and (iii)$'$ imply that

$$\{W_n A_n W_n \tilde{\chi}_n' - \tilde{\chi}_n' W_n A_n W_n\}\, L_n f \to (\tilde{A}\chi' - \chi'\tilde{A})\, f.$$

Taking into account (1) and (i)$'$, we conclude that $W_n A_n W_n L_n f \to \tilde{A} f$.

Step 3. Now we shall show that $W_{(\tau,2)}\{A_n\} = A_1^t$, where $t = \Phi(\tau)$ and $\tau \in \mathbb{T}$. Since $\{A_n^E\} = \{(E_n^t)^{-1} A_n E_n^t\} = \{E_n^{(\tau,2)} A_n (E_n^{(\tau,2)})^{-1}\}$ is uniformly bounded, it suffices to prove that $A_n^E P_n \{\delta_{jk}\}_{k\in\mathbb{Z}} \to A_1^t \{\delta_{jk}\}_{k\in\mathbb{Z}}$ for any $j \in \mathbb{Z}$. To get this, let $\chi, \chi' \in \mathbf{M}(\Gamma)$ satisfy $\chi \equiv 1$, $\chi' \equiv 1$ in a neighborhood of τ and suppose $\chi'\chi = \chi$. Then

$$\begin{aligned} &A_n^E P_n \{\delta_{jk}\}_{k\in\mathbb{Z}} = A_1^t \{\delta_{jk}\}_{k\in\mathbb{Z}} + t_1 + t_2 + t_3 + t_4 + t_5, \\ &t_1 := A_1^t [\tilde{\chi}_n^E - I]\, \{\delta_{jk}\}_{k\in\mathbb{Z}}, \qquad t_2 := [(\tilde{\chi}_n'^E - I)\, A_1^t \tilde{\chi}_n^E]\, \{\delta_{jk}\}_{k\in\mathbb{Z}}, \\ &t_3 := \tilde{\chi}_n'(A_n^E - A_1^t)\, \tilde{\chi}_n^E \{\delta_{jk}\}_{k\in\mathbb{Z}}, \qquad t_4 := (I - \tilde{\chi}_n'^E)\, A_n^E \tilde{\chi}_n^E \{\delta_{jk}\}_{k\in\mathbb{Z}}, \\ &t_5 := A_n^E (I - \tilde{\chi}_n^E)\, \{\delta_{jk}\}_{k\in\mathbb{Z}}. \end{aligned} \tag{2}$$

If we choose χ, χ' in accordance with Lemma 11.10 (or more exactly, the adjoint version of Lemma 11.10) and assumption (iv)$'$, then the terms t_2 and t_3 become small. For j fixed and n large enough, t_1 and t_5 vanish. Now set

$$\begin{aligned} t_4 &= [A_n^E \tilde{\chi}_n'^E - \tilde{\chi}_n'^E A_n^E]\, \tilde{\chi}_n^E \{\delta_{jk}\}_{k\in\mathbb{Z}} = t_6 + t_7 + t_8, \\ t_6 &:= (E_n^t)^{-1} \big\{A_n \tilde{\chi}_n' - \tilde{\chi}_n' A_n - L_n(A\chi' - \chi' A)|_{\operatorname{im} L_n} - W_n L_n(\tilde{A}\chi' - \chi'\tilde{A})|_{\operatorname{im} L_n} W_n\big\} \\ &\quad \times E_n^t \tilde{\chi}_n \{\delta_{jk}\}_{k\in\mathbb{Z}}, \\ t_7 &:= (E_n^t)^{-1}\, L_n(A\chi' - \chi' A)\, E_n^t \tilde{\chi}_n \{\delta_{jk}\}_{k\in\mathbb{Z}}, \\ t_8 &:= (E_n^t)^{-1}\, W_n L_n(\tilde{A}\chi' - \chi'\tilde{A})\, W_n E_n^t \tilde{\chi}_n \{\delta_{jk}\}_{k\in\mathbb{Z}}. \end{aligned}$$

The terms t_7 and t_8 tend to 0 by Lemma 10.34(c) and (ii), (ii)$'$. Assumption (iii)$'$ shows that $t_6 \to 0$. Thus (2) implies that $A_n^E P_n \{\delta_{jk}\}_{k\in\mathbb{Z}} \to A_1^t \{\delta_{jk}\}_{k\in\mathbb{Z}}$. ∎

11.14. Remark. If we wish to apply Theorem 11.8 (Theorem 11.7), we have to verify that $A_1^t \in \mathfrak{A}_{2\times 2}^{\pm 1}$ ($A_1^t \in \mathfrak{A}_{2\times 2}^1$) for all $t \in \Gamma$. We remark here that an operator $A_1^t \in \mathfrak{A}_{2\times 2}$ belongs to $\mathfrak{A}_{2\times 2}^{\pm 1}$ (to $\mathfrak{A}_{2\times 2}^1$) if and only if the entries $(\operatorname{smb} A_1^t)_{i,j}$ $(i, j = 1, 2)$ of its matrix symbol smb A_1^t satisfy

$$(\operatorname{smb} A_1^t)_{i,j}(\tau,\mu) = \begin{cases} (\operatorname{smb} A_1^t)_{i,j}(\tau,1), & \text{if } i = j, \\ 0, & \text{if } i \neq j, \end{cases} \tag{1}$$

whenever $\tau \in \mathbb{T} \setminus \{1, -1\}$ ($\tau \in \mathbb{T} \setminus \{1\}$) and $i, j = 1, 2$. In the following we shall need the sufficiency of (1) only. Thus we prove that $A_1^t \in \mathfrak{A}_{2\times 2}$ and (1) for $\tau \in \mathbb{T} \setminus \{1, -1\}$ imply that $A_1^t \in \mathfrak{A}_{2\times 2}^{\pm 1}$.

Let $\alpha \in \mathbf{PC}(\mathbb{T}) \cap \mathbf{C}(\mathbb{T} \setminus \{1, -1\})$. It is well-known (cf. Böttcher/Silbermann [3]) that the operators $PC(\alpha)P$, $F(I-P)C(\alpha)(I-P)F$, $PC(\alpha)(I-P)F$, $F(I-P) \times C(\alpha)P$ belong to $\mathfrak{A}$ and their symbols satisfy

$$\begin{aligned} &\operatorname{smb}[PC(\alpha)P](\tau,\mu) = \alpha(\tau), \\ &\operatorname{smb}[F(I-P)C(\alpha)(I-P)F](\tau,\mu) = \alpha(\tau^{-1}), \\ &\operatorname{smb}[PC(\alpha)(I-P)F](\tau,\mu) = 0, \\ &\operatorname{smb}[F(I-P)C(\alpha)P](\tau,\mu) = 0, \qquad \tau \in \mathbb{T} \setminus \{1, -1\}. \end{aligned}$$

Thus $C(\alpha)$ and P belong to $\mathfrak{A}_{2\times 2} \cap \mathfrak{A}_{2\times 2}^{\pm 1}$ and their symbols obey (1). The subalgebra of $\mathfrak{A}_{2\times 2}$-symbols satisfying (1) is the smallest subalgebra of $\mathfrak{A}_{2\times 2}$-symbols containing smb P and smb $C(\alpha)$ with $\alpha \in \mathbf{PC}(\mathbb{T}) \cap \mathbf{C}(\mathbb{T} \setminus \{1, -1\})$, i.e., the algebra of symbols of the operators from $\mathfrak{A}_{2\times 2}^{\pm 1}$. Consequently, if (1) holds for the symbol of $A_1^t \in \mathfrak{A}_{2\times 2}$, then there exists a compact operator $T \in \mathcal{L}(\mathbf{l}^2)_{2\times 2}$ such that $A_1^t - T$ is in $\mathfrak{A}_{2\times 2}^{\pm 1} \cap \mathfrak{A}_{2\times 2}$. Since $\mathfrak{A}_{2\times 2} \cap \mathfrak{A}_{2\times 2}^{\pm 1}$ contains the compact operators, $A_1^t \in \mathfrak{A}_{2\times 2}^{\pm 1}$.

11.15. Proof of Theorem 11.6. Let us consider the quadrature method 11.5(1). We identify the operators of $\mathcal{L}(\operatorname{im} L_n)$ with their matrices corresponding to the basis $\{\chi_j^{(n)}: j = 0, \ldots, n-1\}$ and denote the matrix of the system 11.5(1) by A_n. For the proof of Theorem 11.6 in the case of the method 11.5(1), it suffices to prove the assumptions (i)–(iv) of Theorem 11.7. The validity of (ii) is well known. Let us denote the set of all corner points of Γ by Γ'. Then, while proving assumption (i), the curve Γ can be assumed to be smooth. For smooth curves, the convergence $\tilde{\chi}_n A_n \tilde{\chi}_n L_n \to \chi A \chi$ has already been proved (see Theorem 10.27).

Now we shall investigate the validity of (iii) in Theorem 11.7. Setting

$$F: \mathbf{L}^2(\Gamma) \to \mathbb{C}, \qquad Fx := \int_\Gamma x(\tau)\,\mathrm{d}\tau,$$

we get $(Tx)(\tau) = F\big(k(\tau,\cdot)\,x\big)$. The approximate operator $T_n := \big(k(\tau_k^{(n)}, t_j^{(n)})\,\Delta t_j^{(n)}\big)_{k,j=0}^{n-1} \in \mathcal{L}(\operatorname{im} L_n)$ takes the form

$$T_n x_n = K_{n,\tau}^\varepsilon F K_n^\delta\big(k(\tau,\cdot)\,x_n\big).$$

Thus we obtain

$$\big(T_n - K_n^\varepsilon T|_{\operatorname{im} L_n}\big)\,x_n = K_{n,\tau}^\varepsilon F(I - K_n^\delta)\,k(\tau,\cdot)\,L_n x_n.$$

If ω denotes the modulus of continuity

$$\omega(\delta') := \sup\{|k(\tau,t_1) - k(\tau,t_2)| : \tau \in \Gamma,\ t_1, t_2 \in \Gamma,\ |t_1 - t_2| < \delta' \text{ and there is no corner point between } t_1 \text{ and } t_2\},$$

then

$$\|(I - K_n^\delta)\, k(\tau, \cdot)\, L_n\|_{\mathscr{L}(\mathbf{L}^2(\Gamma))} \leqq C\omega(1/n),$$

$$\left|F\big((I - K_n^\delta)\, k(\tau, \cdot)\, L_n x_n\big)\right| \leqq C\omega(1/n)\, \|x_n\|_{\mathbf{L}^2(\Gamma)}$$

(cf. PRÖSSDORF/SCHMIDT [2, Lemma 4.1]). The latter inequalities imply that

$$\left\|T_n - K_n^\varepsilon T|_{\operatorname{im} L_n}\right\|_{\mathbf{L}^2(\Gamma)\to\mathbf{L}^\infty(\Gamma)} \leqq C\omega(1/n),$$

$$\left\|T_n - K_n^\varepsilon T|_{\operatorname{im} L_n}\right\|_{\mathscr{L}(\mathbf{L}^2(\Gamma))} \to 0 \qquad (n \to \infty).$$

Since $T\colon \mathbf{L}^2(\Gamma) \to \mathbf{C}(\Gamma)$ is compact, we see that $\|(K_n^\varepsilon - L_n)\, T\| \to 0$ and $\left\|T_n - L_n T|_{\operatorname{im} L_n}\right\| \to 0$ $(n \to \infty)$. Replacing T by χT or $T\chi$, respectively, and T_n by $\tilde{\chi}_n T_n$ or $T_n K_n^\delta \chi|_{\operatorname{im} L_n}$, respectively, we obtain that

$$\left\|\tilde{\chi}_n T_n - L_n \chi T|_{\operatorname{im} L_n}\right\| \to 0,$$

$$\left\|T_n \tilde{\chi}_n - L_n T\chi|_{\operatorname{im} L_n}\right\| \leqq \|T_n\| \left\|\tilde{\chi}_n - K_n^\delta \chi|_{\operatorname{im} L_n}\right\| + \left\|T_n K_n^\delta \chi|_{\operatorname{im} L_n} - L_n T\chi|_{\operatorname{im} L_n}\right\| \to 0,$$

$$\left\|\tilde{\chi}_n T_n - T_n \tilde{\chi}_n - L_n(\chi T - T\chi)|_{\operatorname{im} L_n}\right\| \to 0.$$

Since $\tilde{\chi}_n \tilde{c}_n = \tilde{c}_n \tilde{\chi}_n$ and $\chi c = c\chi$ imply that $\tilde{\chi}_n \tilde{c}_n - \tilde{c}_n \tilde{\chi}_n - L_n(\chi c - c\chi)|_{\operatorname{im} L_n} = 0$, it remains to show (iii) for the singular operator S_Γ and $S_n := -\mathrm{i} \cot\big(\pi(\varepsilon - \delta)\big)\, I + \left(\frac{1}{\pi \mathrm{i}} \frac{\Delta t_j^{(n)}}{t_j^{(n)} - \tau_k^{(n)}}\right)_{k,j=0}^{n-1}$. Without loss of generality, we suppose that $\chi \circ \gamma$ is continuously differentiable and set $k'(t, \tau) := -\frac{1}{\pi \mathrm{i}} \frac{\chi(t) - \chi(\tau)}{t - \tau}$. Thus $k'(t, \cdot)$ is continuous on $\Gamma \setminus \Gamma'$ and piecewise continuous on Γ. Consequently,

$$\tilde{\chi}_n S_n - S_n \tilde{\chi}_n = M_n^1 - M_n^2, \qquad M_n^1 := \big(k'(\tau_k^{(n)}, t_j^{(n)})\, \Delta t_j^{(n)}\big)_{k,j=0}^{n-1},$$

$$M_n^2 := \left(\frac{1}{\pi \mathrm{i}} \frac{\Delta t_j^{(n)}}{t_j^{(n)} - \tau_k^{(n)}} \big(\chi(\tau_j^{(n)}) - \chi(t_j^{(n)})\big)\right)_{k,j=0}^{n-1},$$

and the argument showing that $\left\|T_n - L_n T|_{\operatorname{im} L_n}\right\| \to 0$ can be without essential modifications applied to prove that $\left\|M_n^1 - L_n(\chi S_\Gamma - S_\Gamma \chi)|_{\operatorname{im} L_n}\right\| \to 0$. The obvious estimate

$$\left|\frac{1}{\pi \mathrm{i}} \frac{\Delta t_j^{(n)}}{t_j^{(n)} - \tau_k^{(n)}} \big(\chi(\tau_j^{(n)}) - \chi(t_j^{(n)})\big)\right| < C \frac{1}{n} \frac{1}{|j \dot{-} k| + 1}$$

with

$$j \dot{-} k := \begin{cases} j - k & \text{if } -\dfrac{n}{2} < j - k \leqq \dfrac{n}{2}, \\[2mm] j - k + n & \text{if } -\dfrac{3n}{2} < j - k \leqq -\dfrac{n}{2}, \\[2mm] j - k - n & \text{if } \dfrac{n}{2} < j - k < \dfrac{3n}{2} \end{cases}$$

implies that $\|M_n^2\| \leqq C 1/n \log n \to 0$ $(n \to \infty)$. Thus we obtain that $\big\|\tilde{\chi}_n S_n - S_n \tilde{\chi}_n - L_n(\chi S_\Gamma - S_\Gamma \chi)|_{\operatorname{im} L_n}\big\| \to 0$ and assumption (iii) of Theorem 11.7 is so seen to be fulfilled.

Now we prove assumption (iv) of Theorem 11.7. Let us fix $\tau \in \Gamma$ and $\varepsilon' > 0$. The elements of T_n satisfy $|k(\tau_k^{(n)}, t_j^{(n)})\,\Delta t_j^{(n)}| < C1/n$. If we choose $U := \{\gamma(s): \sigma - \varepsilon'/2C < s < \sigma + \varepsilon'/2C\}$ and $\chi \in \mathbf{M}(\Gamma)$ with $\operatorname{supp}\chi \subset U$, then simple estimates show that $\|\tilde{\chi}_n^E T_n^E \times \tilde{\chi}_n^E\| < \varepsilon'$. Thus (iv) is proved with T in the place of A. If we choose U in such a manner that $t \in U$ implies $|c(t) - c(\tau)| < \varepsilon'$, then $\|\tilde{\chi}_n^E(\tilde{c}_n^E - c(\tau)\,I)\| < \varepsilon'$, and (iv) is satisfied for c instead of A. It remains to consider the case $A = S_\Gamma$, $A_n = S_n$.

Without loss of generality, let $\tau = \gamma(0)$ be a corner point and set $\omega := \omega_\tau := \arg\big(-\gamma'(+0)/\gamma'(-0)\big) \in (0,2\pi)$. Choose $\chi' \in \mathbf{M}(\Gamma)$ such that the only corner point in $\operatorname{supp}\chi'$ is τ and $\chi' \equiv 1$ in a neighborhood of τ. We define $v: \mathbb{R} \to \Gamma_\omega$ and $\psi: \Gamma \to \Gamma_\omega$ by

$$\psi\big(\gamma(s)\big) := v(s) \quad \text{if} \quad -\frac{1}{2} < s \leqq \frac{1}{2}, \qquad v(s) := \begin{cases} s & \text{if } s \geqq 0 \\ -\mathrm{e}^{\mathrm{i}\omega} s & \text{if } s \leqq 0, \end{cases}$$

and set $S'x := (S_{\Gamma_\omega}[(\chi'x)\circ\psi^{-1}])\circ\psi$. Then $T' := \chi'(S_\Gamma\chi' - S')$ is a compact integral operator and its kernel k' satisfies (see MIKHLIN/PRÖSSDORF [1], Chap. II, § 2)

$$k'(\tau, t) = \chi'(\tau)\,\frac{1}{\pi\mathrm{i}}\left\{\frac{1}{t-\tau} - \frac{(\mathrm{d}/\mathrm{d}t)\,\psi(t)}{\psi(t) - \psi(\tau)}\right\}\chi'(t),$$

$$k'(\tau, \tau) = \frac{1}{2}\,\chi'(\tau)^2\,\frac{1}{\pi\mathrm{i}}\,\frac{(\mathrm{d}^2/\mathrm{d}t^2)\,\psi(\tau)}{(\mathrm{d}/\mathrm{d}t)\,\psi(\tau)}.$$

Setting $T_n' := \big(k'(\tau_k^{(n)}, t_j^{(n)})\,\Delta t_j^{(n)}\big)_{k,j=0}^{n-1}$, for $\chi \in \mathbf{M}(\Gamma)$ and $\chi\chi' = \chi$, we obtain

$$\tilde{\chi}_n S_n \tilde{\chi}_n = \tilde{\chi}_n U_n \tilde{\chi}_n + \tilde{\chi}_n T_n' \tilde{\chi}_n,$$

$$U_n := -\mathrm{i}\cot\big(\pi(\varepsilon - \delta)\big)\,I_n + \left(\frac{1}{\pi\mathrm{i}}\,\frac{\mathrm{d}/\mathrm{d}t\;\psi(t_j^{(n)})}{\psi(t_j^{(n)}) - \psi(\tau_k^{(n)})}\,\Delta t_j^{(n)}\right)_{k,j=0}^{n-1}.$$

As we have shown above, the operator $\tilde{\chi}_n^E T_n'^E \tilde{\chi}_n^E$ becomes smaller than any prescribed $\varepsilon' > 0$ if $\operatorname{supp}\chi$ is contained in a sufficiently small neighborhood of τ. Therefore, it remains to show that $G_n := \tilde{\chi}_n^E(U_n^E - B_1^\tau)\,\tilde{\chi}_n^E$ is small, where

$$B_1^\tau := -\mathrm{i}\cot\big(\pi(\varepsilon - \delta)\big)\,I + \left(\frac{1}{\pi\mathrm{i}}\,\frac{v\left(\dfrac{j+1}{n}\right) - v\left(\dfrac{j}{n}\right)}{v\left(\dfrac{j+\delta}{n}\right) - v\left(\dfrac{k+\varepsilon}{n}\right)}\right)_{k,j\in\mathbb{Z}} \in \mathfrak{A}_{2\times 2}$$

Because

$$\frac{\dfrac{\mathrm{d}}{\mathrm{d}t}\,\psi(t_j^{(n)})}{\psi(t_j^{(n)}) - \psi(\tau_k^{(n)})}\,\Delta t_j^{(n)}$$

$$= \frac{v\left(\dfrac{j+1}{n}\right) - v\left(\dfrac{j}{n}\right)}{v\left(\dfrac{j+\delta}{n}\right) - v\left(\dfrac{k+\varepsilon}{n}\right)}\;\frac{\dfrac{\mathrm{d}v}{\mathrm{d}s}\left(\dfrac{j+\delta}{n}\right)\dfrac{1}{n}}{v\left(\dfrac{j+1}{n}\right) - v\left(\dfrac{j}{n}\right)}\;\frac{\gamma\left(\dfrac{j+1}{n}\right) - \gamma\left(\dfrac{j}{n}\right)}{\gamma'\left(\dfrac{j+\delta}{n}\right)\dfrac{1}{n}},$$

$$\frac{\dfrac{\mathrm{d}v}{\mathrm{d}s}\left(\dfrac{j+\delta}{n}\right)\dfrac{1}{n}}{v\left(\dfrac{j+1}{n}\right) - v\left(\dfrac{j}{n}\right)} = 1$$

we get

$$G_n = \tilde{\chi}_n^E B_1^\tau \tilde{\chi}_n^E D_n, \qquad D_n := (\delta_{jk} d_j^n)_{j,k \in \mathbb{Z}},$$

$$d_j^n := \begin{cases} 0 & \text{if } |j| \geqq n/2, \\ \dfrac{\gamma\left(\dfrac{j+1}{n}\right) - \gamma\left(\dfrac{j}{n}\right)}{\gamma'\left(\dfrac{j+\delta}{n}\right)\dfrac{1}{n}} - 1 & \text{if } |j| < n/2. \end{cases}$$

Since γ' is piecewise Hölder continuous, we conclude that $\|\tilde{\chi}_n'^E D_n\| \to 0$. Consequently, if $\varepsilon' > 0$ is prescribed, then there exists a number n_0 such that $n \geqq n_0$ implies $\|\tilde{\chi}_n'^E D_n\| < \varepsilon' \, \|B_1^\tau\|^{-1}$ and $\|G_n\| < \varepsilon'$.

The method 11.5(2) can be treated analogously. Let us remark only that M_n^2 has to be replaced by

$$\left(\frac{1}{\pi \mathrm{i}} \, \Delta t_j^{(n)} \, \frac{\mathrm{d}\chi}{\mathrm{d}t} \, (t_j^{(n)}) \, \delta_{jk}\right)_{k,j=0}^{n-1},$$

where the norm of the latter term tends to 0 as $n \to \infty$. The verification of (i), (ii), (ii)′ and (iv)′ (see Theorems 11.7 and 11.8) for the method 11.5(3) is also similar to the preceding proof. To show (i)′, we consider $\tilde{A} := cI - dS_\Gamma - T$ and the corresponding quadrature method for $\tilde{A}$. If $(\tilde{A})_n$ denotes the corresponding approximate operator, then $(\tilde{A})_n = \tilde{A}_n := W_n A_n W_n$. Thus (i)′ follows from (i). It remains to show (iii)′.

For T and S_Γ, define T_n, T_n', S_n and S_n' by

$$T_n := \left(k(t_k^{(n)}, t_j^{(n)}) \left(\gamma\left(\frac{j+1}{n}\right) - \gamma\left(\frac{j-1}{n}\right)\right) \tilde{\delta}_{k,j}\right)_{k,j=0}^{n-1},$$

$$T_n' := \left(k(t_k^{(n)}, t_j^{(n)}) \left(\gamma\left(\frac{j+1}{n}\right) - \gamma\left(\frac{j}{n}\right)\right)\right)_{k,j=0}^{n-1},$$

$$S_n := \left(\frac{1}{\pi \mathrm{i}} \frac{1}{t_j^{(n)} - t_k^{(n)}} \left(\gamma\left(\frac{j+1}{n}\right) - \gamma\left(\frac{j-1}{n}\right)\right) \tilde{\delta}_{k,j}\right)_{k,j=0}^{n-1},$$

$$S_n' := \left(\frac{1}{\pi \mathrm{i}} \frac{1}{t_j^{(n)} - t_k^{(n)}} \left(\gamma\left(\frac{j+1}{n}\right) - \gamma\left(\frac{j}{n}\right)\right)\right)_{k,j=0}^{n-1},$$

where $1/(t_j^{(n)} - t_j^{(n)}) := 0$ and $\tilde{\delta}_{k,j} = 0$ for $k - j$ even and $\tilde{\delta}_{k,j} = 1$ for $k - j$ odd. Then it is easy to prove that

$$\|T_n - (T_n' - W_n T_n' W_n)\| \to 0 \qquad (n \to \infty).$$

Thus we obtain that

$$\|\tilde{\chi}_n T_n - T_n \tilde{\chi}_n - \{(\tilde{\chi}_n T_n' - T_n' \tilde{\chi}_n) - W_n(\tilde{\chi}_n T_n' - T_n' \tilde{\chi}_n) W_n\}\| \to 0,$$

$$\|\tilde{\chi}_n S_n - S_n \tilde{\chi}_n - \{(\tilde{\chi}_n S_n' - S_n' \tilde{\chi}_n) - W_n(\tilde{\chi}_n S_n' - S_n' \tilde{\chi}_n) W_n\}\| \to 0.$$

Since T_n', S_n' are the approximate operators corresponding to the method 11.5(2) and

(iii) is fulfilled for 11.5(2), we get

$$\|\chi_n T_n - T_n\chi_n - \{L_n(\chi T - T\chi)|_{\mathrm{im}\,L_n} - W_n L_n(\chi T - T\chi)|_{\mathrm{im}\,L_n} W_n\}\| \to 0,$$
$$\|\chi_n S_n - S_n\chi_n - \{L_n(\chi S_\Gamma - S_\Gamma\chi)|_{\mathrm{im}\,L_n} - W_n L_n(\chi S_\Gamma - S_\Gamma\chi)|_{\mathrm{im}\,L_n} W_n\}\| \to 0.$$

This completes the proof of Theorem 11.6.

Collocation methods for singular integral equations on curves with corners. Piecewise constant trial functions

Similarly to quadrature methods, one can treat other spline approximation methods, i.e., collocation methods and Galerkin-Petrov methods using splines as test or trial functions. For simplicity, we shall restrict our attention to collocation with piecewise constant trial functions. In the following, we establish the stability of the model problem, viz, the stability of collocation for singular integral equations with constant coefficients on an angle. Using these results, in Section 11.17 we extend our analysis to collocation for equations with continuous coefficients on general curves with corners.

Let us retain the notation of Sections 11.2 and 11.3. The ε-collocation method consists in seeking an approximate solution $u_n = \sum_{k\in\mathbb{Z}} \xi_k^{(n)}\chi_k^{(n)} \in \mathrm{im}\,L_n \subset \mathbf{L}^2(\Gamma_\omega)$ satisfying

$$(Au_n)\,(\tau_k^{(n)}) = f(\tau_k^{(n)}), \qquad k \in \mathbb{Z}.$$

The latter system can be written as $A_n u_n = K_n f$, where $A_n := K_n A|_{\mathrm{im}\,L_n} \in \mathcal{L}(\mathrm{im}\,L_n)$. Here again A_n may be thought of as belonging to $\mathcal{L}(\tilde{\mathbf{l}}^2)$, and these operators do not depend on n. Thus, the sequence $\{A_n\}$ $\big(A_n \in \mathcal{L}(\mathrm{im}\,L_n)\big)$ is stable if and only if $A_1 \in \mathcal{L}(\tilde{\mathbf{l}}^2)$ is invertible.

11.16. Theorem. (a) *The operator $A_1 \in \mathcal{L}(\tilde{\mathbf{l}}^2)$ is Fredholm of index zero if and only if $\dfrac{c+d}{c-d} \notin \Omega$, where $\Omega := \left\{\dfrac{\varphi_\varepsilon(\mu) - 1}{\varphi_\varepsilon(\mu)} : 0 \le \mu \le 1\right\}$ and $\varphi_\varepsilon(\mu)$ is defined as*

$$\int_0^1 \mathrm{e}^{-\mathrm{i}\pi(\varepsilon-\delta)(\mu-1)}\,\frac{\sin\big(-\pi(\varepsilon-\delta)\,\mu\big)}{\sin\big(-\pi(\varepsilon-\delta)\big)}\,\mathrm{d}\delta = \frac{1}{2}\left\{1 + \int_0^1 \frac{2\mathrm{e}^{-\mathrm{i}2\pi(\varepsilon-\delta)\mu}}{\mathrm{e}^{-\mathrm{i}2\pi(\varepsilon-\delta)} - 1}\,\mathrm{d}\delta\right\}.$$

In particular, $\Omega = (-\infty, 0]$ for $\varepsilon = 1/2$.

(b) *There is an at most countable subset Φ of $\mathbb{C}\setminus\Omega$ all accumulation points of which belong to Ω such that A_1 is invertible in $\tilde{\mathbf{l}}^2$ if and only if $(c+d)/(c-d) \,\overline{\in}\, \Omega \cup \Phi$.*

(c) *If $\omega = \pi$, then $\Phi = \emptyset$.*

Proof. Assertions (b) and (c) can be derived analogously to the corresponding assertions of Theorem 11.3. In order to verify (a), we shall prove that $A_1 \in \mathfrak{A}_{2\times 2}$ and that det smb A_1 is independent of ω. Thus it suffices to establish (a) for $\omega = \pi$. In this case, A_1 becomes a discrete convolution operator and (a) will follow easily.

For the sake of brevity, we shall restrict ourselves to the case $\varepsilon = 1/2$. Then the operator A_1 takes the form

$$A_1 = cI + d\left(\frac{1}{\pi\mathrm{i}}\int_{\Gamma_\omega} \frac{\chi_j^{(1)}(\tau)}{\tau - \tau_k^{(1)}}\,d\tau\right)_{k,j\in\mathbb{Z}}. \tag{1}$$

If t_j^δ ($j \in \mathbb{Z}$, $0 < \delta < 1$) denotes the point $k + \delta$ for $k \geqq 0$ and $-(k+\delta)\,\mathrm{e}^{\mathrm{i}\omega}$ for $k < 0$, then

$$\frac{1}{\pi \mathrm{i}} \int\limits_{\Gamma_\omega} \frac{\chi_j^{(1)}(\tau)}{\tau - \tau_k^{(1)}}\, \mathrm{d}\tau = \frac{1}{\pi \mathrm{i}} \int\limits_0^{1/2} \left\{ \frac{1}{t_j^\delta - \tau_k^{(1)}} + \frac{1}{t_j^{1-\delta} - \tau_k^{(1)}} \right\} \mathrm{d}\delta \cdot \begin{cases} 1 & \text{if } j \geqq 0, \\ -\mathrm{e}^{\mathrm{i}\omega} & \text{if } j < 0. \end{cases} \tag{2}$$

Let us set

$$A_1^\delta := \left[c - \mathrm{i} \cot\left(\pi\left(\frac{1}{2} - \delta\right)\right) d \right] I + d \left(\frac{1}{\pi \mathrm{i}} \frac{1}{t_j^\delta - \tau_k^{(1)}}\, \Delta t_j^{(1)} \right)_{k,j \in \mathbb{Z}},$$

$$\Delta t_j := \begin{cases} 1 & \text{if } j \geqq 0, \\ -\mathrm{e}^{\mathrm{i}\omega} & \text{if } j < 0, \end{cases}$$

and consider the operator-valued function $\delta \mapsto A(\delta) := A_1^\delta + A_1^{1-\delta}$ defined on $[0, 1/2]$. The proof of Theorem 11.3 shows that $A(\delta) \in \mathfrak{A}_{2\times 2} \subset \mathcal{L}(\tilde{\mathrm{l}}^2)$. Moreover, the obvious estimates

$$\left| \frac{1}{t_j^\delta - \tau_k^{(1)}} - \frac{1}{t_j^{\delta'} - \tau_k^{(1)}} \right| \leqq |\delta' - \delta| \frac{1}{|j - k|^2}, \qquad j \neq k,\ j, k \in \mathbb{Z},$$

$$\frac{1}{t_k^\delta - \tau_k^{(1)}} + \frac{1}{t_k^{1-\delta} - \tau_k^{(1)}} = 0, \qquad k \in \mathbb{Z}$$

imply the continuity of the function $\delta \mapsto A(\delta)$. The equations (1) and (2) yield

$$A_1 = \int\limits_0^{1/2} \{A_1^\delta + A_1^{1-\delta}\}\, \mathrm{d}\delta \in \mathfrak{A}_{2\times 2}, \qquad \operatorname{smb} A_1 = \int\limits_0^{1/2} \{\operatorname{smb} A_1^\delta + \operatorname{smb} A_1^{1-\delta}\}\, \mathrm{d}\delta .$$

By 11.4(5) we conclude that

$$\operatorname{smb} A_1(\tau, \mu) =$$

$$\begin{cases} \begin{pmatrix} \{c + d(2\psi_\varepsilon(\lambda) - 1)\} & 0 \\ 0 & \{c - d\overline{(2\psi_\varepsilon(\lambda) - 1)}\} \end{pmatrix} & \text{if } \tau = \mathrm{e}^{\mathrm{i}2\pi\lambda},\ 0 < \lambda < 1,\ 0 \leqq \mu \leqq 1, \\[2ex] \begin{pmatrix} \{c + d(-\mu + (1-\mu))\} & \left\{ -d(-\mathrm{i}) \dfrac{\mathrm{e}^{-\mathrm{i}(\pi-\omega)[1/2 + (\mathrm{i}/2\pi)\log(\mu/(1-\mu))]}}{\sin\left(\pi\left(\dfrac{1}{2} + \dfrac{\mathrm{i}}{2\pi}\log\dfrac{\mu}{1-\mu}\right)\right)} \right\} \\ \left\{ d(-\mathrm{i}) \dfrac{\mathrm{e}^{-\mathrm{i}(\omega-\pi)[1/2 + (\mathrm{i}/2\pi)\log(\mu/(1-\mu))]}}{\sin\left(\pi\left(\dfrac{1}{2} + \dfrac{\mathrm{i}}{2\pi}\log\dfrac{\mu}{1-\mu}\right)\right)} \right\} & \{c - d(-\mu + (1-\mu))\} \end{pmatrix} & \text{if } \tau = 1,\ 0 \leqq \mu \leqq 1. \end{cases}$$

Thus det smb A_1 is independent of ω. For $\omega = \pi$, A_1 takes the form $A_1 = cI + d(f_{k-j})_{k,j\in\mathbb{Z}}$, where f_j denotes the j-th Fourier coefficient of the function $f(\mathrm{e}^{\mathrm{i}2\pi\lambda}) := 2\psi_\varepsilon(\lambda) - 1$, $0 < \lambda < 1$. This convolution operator is a Fredholm operator with index 0 if and only if $c + df(\tau) \neq 0$ for all $\tau \in T$, i.e., if $(c + d)/(c - d) \notin \Omega$. ∎

Let us retain the notation introduced in Sections 11.5–11.8. The ε-collocation method determines an approximate solution

$$u_n = \sum_{k=0}^{n-1} \xi_k^{(n)} \chi_k^{(n)} \in \operatorname{im} L_n \subset \mathbf{L}^2(\Gamma)$$

by solving the equations

$$(Au_n)\,(\tau_k^{(n)}) = f(\tau_k^{(n)}), \qquad k = 0, \ldots, n-1. \tag{3}$$

The system (3) can be written as $A_n u_n = K_n^\varepsilon f$, where $A_n := K_n^\varepsilon A|_{\operatorname{im} L_n} \in \mathcal{L}(\operatorname{im} L_n)$. If we fix $\tau \in \Gamma$, then the model problem of the ε-collocation for the operator $A \in \mathcal{L}(\mathbf{L}^2(\Gamma))$ is the ε-collocation for the operator $A^\tau \in \mathcal{L}(\Gamma_{\omega_\tau})$ described before Theorem 11.16. The matrix of the corresponding system will be denoted by A_1^τ. By the proof of Theorem 11.16 we get $A_1^\tau \in \mathfrak{A}_{2\times 2}^1$.

11.17. (a) *The ε-collocation $(0 < \varepsilon < 1)$ for the operator A is stable if and only if the operators $A \in \mathcal{L}(\mathbf{L}^2(\Gamma))$ and $A_1^\tau \in \mathcal{L}(\tilde{\mathbf{l}}^2)$ $(\tau \in \Gamma)$ are invertible.*

(b) *If the collocation method is stable and f is Riemann integrable, then the system* 11.16(3) *is uniquely solvable for n large enough and the approximate solutions u_n converge to $u = A^{-1}f$ as $n \to \infty$.*

Combining Theorems 11.16 and 11.17 we obtain necessary and sufficient conditions for the stability of the collocation method.

Proof. It suffices to show that the assumptions of Theorem 11.7 are fulfilled. The validity of (i) and (ii) can be derived analogously to Section 11.15. Now let us verify property (iii) of Theorem 11.7. Without loss of generality, we suppose $\chi \circ \gamma$ to be continuously differentiable and obtain

$$\begin{aligned}
&\tilde{\chi}_n A_n - A_n \tilde{\chi}_n - L_n(\chi A - A\chi)|_{\operatorname{im} L_n} \\
&= K_n^\varepsilon \chi L_n K_n^\varepsilon A|_{\operatorname{im} L_n} - K_n^\varepsilon A L_n K_n^\varepsilon \chi|_{\operatorname{im} L_n} - L_n(\chi A - A\chi)|_{\operatorname{im} L_n} \\
&= K_n^\varepsilon A(I - K_n^\varepsilon)\,\chi|_{\operatorname{im} L_n} + (K_n^\varepsilon - L_n)\,(\chi A - A\chi)|_{\operatorname{im} L_n}.
\end{aligned}$$

Since $\chi A - A\chi : \mathbf{L}^2(\Gamma) \to \mathbf{C}(\Gamma)$ is compact and $(K_n^\varepsilon - L_n) : \mathbf{C}(\Gamma) \to \mathbf{L}^2(\Gamma)$ converges strongly to 0, we get $\|(K_n^\varepsilon - L_n)\,(\chi A - A\chi)|_{\operatorname{im} L_n}\| \to 0$. By virtue of

$$K_n^\varepsilon A(I - K_n^\varepsilon)\,\chi|_{\operatorname{im} L_n} = K_n^\varepsilon b L_n K_n^\varepsilon S_\Gamma(I - K_n^\varepsilon)\,\chi|_{\operatorname{im} L_n},$$

it remains to show that $\|K_n^\varepsilon S_\Gamma(I - K_n^\varepsilon)\,\chi|_{\operatorname{im} L_n}\| \to 0$. The latter relation is an immediate consequence of the equality

$$K_n^\varepsilon S_\Gamma(I - K_n^\varepsilon)\,\chi|_{\operatorname{im} L_n} = \left(\frac{1}{\pi \mathrm{i}} \int_\Gamma \frac{\chi(\tau) - \chi(\tau_j^{(n)})}{\tau - \tau_k^{(n)}}\, \chi_j^{(n)}(\tau)\, \mathrm{d}\tau\right)_{k,j=0}^{n-1}$$

and of the obvious estimate

$$\left|\frac{1}{\pi \mathrm{i}} \int_\Gamma \frac{\chi(\tau) - \chi(\tau_j^{(n)})}{\tau - \tau_k^{(n)}}\, \chi_j^{(n)}(\tau)\, \mathrm{d}\tau\right| \leqq C\,\frac{1}{n}\,\frac{1}{|k \dot{-} j| + 1}.$$

Thus assumption (iii) is satisfied.

Now we consider property (iv) of Theorem 11.7 retaining the notation S', T', ψ and v introduced in Section 11.15. Repeating the reasoning of Section 11.15 we get the validity of (iv) for $A = T$, $A_n := K_n^\varepsilon T|_{\operatorname{im} L_n}$ and for $A = c$, $A_n = K_n^\varepsilon c|_{\operatorname{im} L_n}$. Therefore, we can assume $A = S_\Gamma$ and $A_n := K_n^\varepsilon S_\Gamma|_{\mathrm{im}\, L_n}$. In this case, A_1^τ takes the form

$$A_1^\tau := \left(\frac{1}{\pi \mathrm{i}} \int\limits_j^{j+1} \frac{1}{v(s) - v(k+\varepsilon)}\, \mathrm{d}v(s)\right)_{k,j\in\mathbb{Z}} .$$

Because

$$\int\limits_j^{j+1} \frac{1}{v(s) - v(k+\varepsilon)}\, \mathrm{d}v(s) = \int\limits_{j/n}^{(j+1)/n} \frac{1}{v(s) - v\big((k+\varepsilon)/n\big)}\, \mathrm{d}v(s)$$

$$= \int\limits_{j/n}^{(j+1)/n} \frac{\mathrm{d}\psi/\mathrm{d}t\big(\gamma(s)\big)}{\psi\circ\gamma(s) - \psi\circ\gamma\big((k+\varepsilon)/n\big)}\, \mathrm{d}\gamma(s) = \int\limits_\Gamma \frac{(\mathrm{d}\psi/\mathrm{d}t)/(t)}{\psi(t) - \psi(\tau_k^{(n)})}\, \chi_j^{(n)}(t)\, \mathrm{d}t,$$

and

$$\tilde{\chi}_n\big(K_n^\varepsilon S'|_{\mathrm{im}\, L_n}\big)\, \tilde{\chi}_n = \tilde{\chi}_n \left(\frac{1}{\pi \mathrm{i}} \int\limits_\Gamma \frac{(\mathrm{d}\psi/\mathrm{d}t)/(t)}{\psi(t) - \psi(\tau_k^{(n)})}\, \chi_j^{(n)}(t)\, \mathrm{d}t\right)_{k,j\in\mathbb{Z}} \tilde{\chi}_n\,.$$

$$\tilde{\chi}_n^E\big(K_n^\varepsilon S'|_{\mathrm{im}\, L_n}\big)^E \tilde{\chi}_n^E = \tilde{\chi}_n^E A_1^\tau \tilde{\chi}_n^E$$

we obtain

$$\tilde{\chi}_n^E\big\{\big(K_n^\varepsilon S_\Gamma|_{\mathrm{im}\, L_n}\big)^E - A_1^\tau\big\}\, \tilde{\chi}_n^E = \tilde{\chi}_n^E\big(K_n^\varepsilon T'|_{\mathrm{im}\, L_n}\big)^E\, \tilde{\chi}_n^E\,.$$

Since assertion (iv) is true for A replaced by the compact operator T' and A_1^τ replaced by 0, the last expression becomes smaller than any prescribed $\varepsilon' > 0$. ■

Quadrature methods for singular integral equations on the interval

11.18. Quadrature methods on the half axis. It is known (see Chapter 9) that there exists a well developed theory on quadrature methods for singular integral equations on the inverval. All these methods use quadrature rules based on polynomial interpolation, where the interpolation (quadrature) knots are the zeros of special orthogonal polynomials. Here we shall consider simple quadrature rules on special non-uniform partitions and prove the stability of the corresponding quadrature methods in analogy to Sections 11.2–11.17. We remark that ε-collocation or Galerkin-Petrov methods with splines can be treated similarly. Moreover, in the same manner, the method of discrete whirls (i.e. the method 11.5(1) for $\varepsilon = 0$ or $\varepsilon = 1$ and $\delta = 1/2$) can be studied.

Our first concern is quadrature methods for a model problem on the half axis. After that we shall consider the corresponding methods on the interval utilizing localization techniques.

Let $\mathbb{R}^+ := (0, \infty)$ and denote the Cauchy singular operator on $\mathbb{R}^+$ by $S_{\mathbb{R}^+}$. For $c, d \in \mathbb{C}$, we set $A := cI + dS_{\mathbb{R}^+} \in \mathcal{L}\big(\mathbf{L}^2_\varrho(\mathbb{R}^+)\big)$, where $-1/2 < \varrho < 1/2$ and the weighted $\mathbf{L}^2$-space $\mathbf{L}^2_\varrho(\mathbb{R}^+)$ is defined by the norm

$$\|f\|_{\mathbf{L}^2_\varrho(\mathbb{R}^+)} = \|t^{-\varrho} f(t)\|_{\mathbf{L}^2(\mathbb{R}^+)}\,.$$

It is well known that A is invertible if and only if $c \pm d \neq 0$ and $(c+d)/(c-d) \notin \{e^{i2\pi\varrho t}: -\infty < t \leqq 0\}$ (see Chapter 6 or GOHBERG/KRUPNIK [4]).

The numerical procedures 11.2(1)–11.2(3) suggest the following quadrature methods for solving the equation $Au = f$, where $f \in \mathscr{R}(\mathbb{R}^+) \cap \mathbf{L}^2_\varrho(\mathbb{R}^+)$. Choose $\alpha \geqq 1$, $0 < \varepsilon < 1$ and set

$$t_k^{(n)} := \left(\frac{k+1/2}{n}\right)^\alpha, \qquad \tau_k^{(n)} := \left(\frac{k+\varepsilon}{n}\right)^\alpha, \qquad k = 0, 1, \ldots$$

We determine approximate values $\xi_k^{(n)}$ for $u(t_k^{(n)})$ by solving one of the systems

$$\left\{c - \mathrm{i}\cot\left(\pi\left(\varepsilon - \frac{1}{2}\right)\right)d\right\}\xi_k^{(n)} + d\frac{1}{\pi \mathrm{i}}\sum_{j=0}^{\infty}\frac{\xi_j^{(n)}}{t_j^{(n)} - \tau_k^{(n)}}\Delta t_j^{(n)} = f(\tau_k^{(n)}),$$
$$k = 0, 1, \ldots, \tag{1}$$

$$c\xi_k^{(n)} + d\frac{1}{\pi \mathrm{i}}\sum_{\substack{j=0\\ j\neq k}}^{\infty}\frac{\xi_j^{(n)}}{t_j^{(n)} - t_k^{(n)}}\Delta t_j^{(n)} = f(t_k^{(n)}), \qquad k = 0, 1, \ldots, \tag{2}$$

$$c\xi_k^{(n)} + d\frac{1}{\pi \mathrm{i}}\sum_{\substack{j=0\\ j\equiv k+1 \bmod 2}}^{\infty}\frac{\xi_j^{(n)}}{t_j^{(n)} - t_k^{(n)}}\Delta t_j^{(n)} = f(t_k^{(n)}), \qquad k = 0, 1, \ldots, \tag{3}$$

where $\Delta t_j^{(n)} := \alpha\left(\frac{j+1/2}{n}\right)^{\alpha-1}\frac{1}{n}$ for (1) or (2) and $\Delta t_j^{(n)} = \alpha\left(\frac{j+1/2}{n}\right)^{\alpha-1}\frac{2}{n}$ for (3). To get an approximate solution, we set

$$u_n := \sum_{k=0}^{\infty}\xi_k^{(n)}\chi_k^{(n)}, \qquad \chi_k^{(n)}(t) := \begin{cases} 1 & \text{if } \left(\frac{k}{n}\right)^\alpha \leqq t < \left(\frac{k+1}{n}\right)^\alpha, \\ 0 & \text{else.}\end{cases}$$

Now denote the $\mathbf{L}^2$-orthogonal projection onto span $\{\chi_k^{(n)}: k = 0, 1, \ldots\}$ by L_n, set $K_n^\varepsilon := \sum_{k=0}^{n-1} f(\tau_k^{(n)})\,\chi_k^{(n)}$ and define $\mathbf{l}^2_v$ $(v \in \mathbb{R})$ to be the Hilbert space of all sequences $\{\xi_k\}_{k=0}^\infty$ satisfying

$$\|\{\xi_k\}_{k=0}^\infty\|_{\mathbf{l}^2_v} := \left(\sum_{k=0}^{\infty}|\xi_k|^2(k+1)^{2v}\right)^{1/2} < \infty.$$

Obviously, there exists a constant $C > 0$ such that

$$\frac{1}{C}\left\|\sum_{k=0}^{\infty}\xi_k\chi_k^{(n)}\right\|_{\mathbf{L}^2_\varrho(\mathbb{R}^+)} \leqq n^{\alpha\varrho-(\alpha/2)}\|\{\xi_k\}_{k=0}^\infty\|_{\mathbf{l}^2_{(\alpha-1)/2-\varrho\alpha}} \leqq C\left\|\sum_{k=0}^{\infty}\xi_k\chi_k^{(n)}\right\|_{\mathbf{L}^2_\varrho(\mathbb{R}^+)}.$$

If we identify the operators of $\mathscr{L}(\operatorname{im} L_n)$ with their matrices corresponding to the basis $\{\chi_k^{(n)}: k = 0, 1, \ldots\}$, then $\mathscr{L}(\operatorname{im} L_n)$ is isomorphic to $\mathscr{L}(\mathbf{l}^2_{(\alpha-1)/2-\varrho\alpha})$. Let $A_n \in \mathscr{L}(\operatorname{im} L_n)$ be defined by the matrix of the system (1), (2) or (3), respectively. It is easy to see that A_n is independent of n. Thus the sequence $\{A_n\}$ $(A_n \in \mathscr{L}(\operatorname{im} L_n))$ is stable if and only if $A_1 \in \mathscr{L}(\mathbf{l}^2_{(\alpha-1)/2-\varrho\alpha})$ is invertible.

11.19. Theorem. *For $c \pm d \neq 0$, $(c+d)/(c-d) \notin \{e^{+i2\pi\varrho t}: -\infty < t \leqq 0\}$, set $\varkappa := 1/(2\pi \mathrm{i}) \times \log(c+d)/(c-d)$, where* log *denotes the branch of the logarithm which is continuous on $\mathbb{C} \setminus \{e^{i2\pi\varrho t}: -\infty < t \leqq 0\}$ and takes real values on $\mathbb{R}^+$. (Note that $|\operatorname{Re}\varkappa - \varrho| < 1/2$.)*

(a) *The operator* $A_1 \in \mathcal{L}(l^2_{(\alpha-1)/2-\varrho\alpha})$ *of method* 11.18(2) *or* 11.18(3), *respectively, is invertible if and only if* $c \pm d \neq 0$, $(c+d)/(c-d) \notin \{e^{+i2\pi\varrho}t\colon -\infty < t \leqq 0\}$ *and if* $|\mathrm{Re}\,\varkappa| < 1/2$ *or* $|\mathrm{Re}\,\varkappa + \varrho| < 1/2$, *respectively.*

(b) *The operator* $A_1 \in \mathcal{L}(l^2_{(\alpha-1)/2-\varrho\alpha})$ *of method* 11.18(1) *is a Fredholm operator with index* 0 *if and only if* $c \pm d \neq 0$, $(c+d)/(c-d) \notin \{e^{+i2\pi\varrho}t\colon -\infty < t \leqq 0\}$ *and* $|\mathrm{Re}\,\varkappa + (\varepsilon - 1/2)/2| < 1/2$. *Let* $\Omega := \{z \in \mathbb{C} \setminus \{e^{+i2\pi\varrho}t\colon -\infty < t \leqq 0\} : |\mathrm{Re}[(1/2\pi i)\log z] + (\varepsilon - 1/2)/2| < 1/2\}$. *Then* A_1 *is invertible if and only if* $(c+d)/(c-d) \in \Omega \setminus \Phi$, *where* Φ *is an at most countable subset of* Ω *whose accumulation points belong to* $\mathbb{C} \setminus \Omega$. *Moreover, if* $\alpha = 1$ *or* $\alpha = 2$, *then* $\Phi = \emptyset$.

Proof. *Step 1.* Let $\alpha = 1$. In this case the proof of Theorem 11.3 implies that $A_1 = T(c + df)$ with $f = f^{(\varepsilon-(1/2))}$ for 11.18(1), $f = f^0$ for 11.18(2) and $f = f^{\#}$ for 11.18(3). By the theory of Toeplitz operators (cf. Theorem 6.17 and Duduchava [2]) we obtain that $A_1 \in \mathcal{L}(l^2_{-\varrho})$ is invertible if and only if the following symbol $\mathcal{A}(\tau, \mu)$ does not vanish and if $\mathrm{ind}\,\mathcal{A} = 0$:

$$\mathcal{A}(\tau, \mu) = \begin{cases} c + df(\tau) & \text{if } \tau \neq 1, \\ \dfrac{1 + f^{2\varrho}(e^{i2\pi\mu})}{2}(c - d) + \dfrac{1 - f^{2\varrho}(e^{i2\pi\mu})}{2}(c + d) & \text{if } \tau = 1,\ 0 \leqq \mu \leqq 1 \end{cases} \tag{1}$$

for 11.18(1) and 11.18(3)

$$\mathcal{A}(\tau, \mu) = \begin{cases} c + df(\tau) & \text{if } \tau \neq 1, -1, \\ \dfrac{1 + f^{2\varrho}(e^{i2\pi\mu})}{2}(c - d) + \dfrac{1 - f^{2\varrho}(e^{i2\pi\mu})}{2}(c + d) & \text{if } \tau = 1,\ 0 \leqq \mu \leqq 1, \\ \dfrac{1 + f^{2\varrho}(e^{i2\pi\mu})}{2}(c + d) + \dfrac{1 - f^{2\varrho}(e^{i2\pi\mu})}{2}(c - d) & \text{if } \tau = -1,\ 0 \leqq \mu \leqq 1 \end{cases}$$

for 11.18(3).

Now we consider the symbol $\mathcal{A}(\tau, \mu)$ for 11.18(1). Then the condition $\mathcal{A}(\tau, \mu) \neq 0$ is equivalent to

$$(c + d)\frac{1 + f}{2} + (c - d)\frac{1 - f}{2} \neq 0$$

or

$$\frac{c + d}{c - d} \neq -\frac{1 - f}{1 + f} = -\frac{1 - \left\{2e^{-i\pi(\varepsilon-1/2)(x-1)}\dfrac{\sin\left(\pi(\varepsilon - 1/2)\,x\right)}{\sin\left(\pi(\varepsilon - 1/2)\right)} - 1\right\}}{1 + \left\{2e^{-i\pi(\varepsilon-1/2)(x-1)}\dfrac{\sin\left(\pi(\varepsilon - 1/2)\,x\right)}{\sin\left(\pi(\varepsilon - 1/2)\right)} - 1\right\}}$$

$$= -e^{-i\pi(\varepsilon-1/2)}\frac{\sin\left(\pi(\varepsilon - 1/2)\,(1 - x)\right)}{\sin\left(\pi(\varepsilon - 1/2)\,x\right)}, \qquad 1 \leqq x \leqq 0,$$

or

$$\frac{c + d}{c - d} \notin \mathcal{S}_\varepsilon := \{e^{-i2\pi(\varepsilon-1/2)/2}t\colon -\infty < t \leqq 0\}$$

and

$$(c-d)\frac{1+f^{2\varrho}}{2}+(c+d)\frac{1-f^{2\varrho}}{2}\neq 0$$

or

$$\frac{c+d}{c-d}\notin \mathcal{S}_\varrho := \{e^{i2\pi\varrho t}: -\infty < t \leqq 0\}.$$

If $\mathcal{C}_\varepsilon$ denotes the convex hull of $\mathcal{A}_\varepsilon := \left\{\frac{c+d}{c-d}\,\frac{1+f(e^{i2\pi x})}{2}+\frac{1-f(e^{i2\pi x})}{2}: 0 \leqq x \leqq 1\right\}$, then $0 \in \mathcal{C}_\varepsilon$ if and only if there exists a number $\mu \in [0, 1]$ such that

$$\mu\left[\left(\frac{c+d}{c-d}\right)\frac{1+f}{2}+\frac{1-f}{2}\right]+(1-\mu)=0$$

or

$$\left\{\mu\,\frac{c+d}{c-d}+(1-\mu)\right\}\frac{1+f}{2}+\frac{1-f}{2}=0.$$

Consequently, $0 \in \mathcal{C}_\varepsilon$ if and only if $\mathcal{S} \cap \mathcal{S}_\varepsilon \neq \emptyset$, where $\mathcal{S} := \left\{\mu\,\frac{c+d}{c-d}+(1-\mu): 0 \leqq \mu \leqq 1\right\}$. If $\varepsilon > 1/2$, then $0 \in \mathcal{C}_\varepsilon$ if and only if $-\frac{\varepsilon-1/2}{2}-1/2 \leqq \operatorname{Re}\varkappa \leqq -1/2$ or $1/2-\frac{\varepsilon-1/2}{2} \leqq \operatorname{Re}\varkappa \leqq 1/2$. If $\varepsilon < 1/2$, then $0 \in \mathcal{C}_\varepsilon$ if and only if $-1/2 \leqq \operatorname{Re}\varkappa \leqq -1/2-\frac{\varepsilon-1/2}{2}$ or $1/2 \leqq \operatorname{Re}\varkappa \leqq 1/2-\frac{\varepsilon-1/2}{2}$.

Similarly, we denote the convex hull of

$$\mathcal{A}_\varrho := \left\{\frac{1+f^{2\varrho}(e^{i2\pi x})}{2}+\frac{1-f^{2\varrho}(e^{i2\pi x})}{2}\,\frac{c+d}{c-d}: 0 \leqq x \leqq 1\right\}$$

by $\mathcal{C}_\varrho$ and obtain that $0 \in \mathcal{C}_\varrho$ if and only if $\varrho < 0$ and $\operatorname{Re}\varkappa \leqq -1/2$ or if $\varrho > 0$ and $1/2 \leqq \operatorname{Re}\varkappa$. The condition $\operatorname{ind}\mathcal{A} = 0$ means that the origin 0 does not belong to the domain included between the circular arcs $\mathcal{A}_\varepsilon$ and $\mathcal{A}_\varrho$. Consequently, $\operatorname{ind}\mathcal{A} = 0$ if and only if $0 \notin \mathcal{C}_\varepsilon \cup \mathcal{C}_\varrho$ or 0 is a common inner point of $\mathcal{C}_\varepsilon$ and $\mathcal{C}_\varrho$ or $0 \in \mathcal{S} \neq \mathcal{C}_\varepsilon \cap \mathcal{C}_\varrho$. Thus the invertibility of A_1 is equivalent to $\left|\operatorname{Re}\varkappa+\frac{\varepsilon-1/2}{2}\right| < 1/2$. The methods 11.18(2) and 11.18(3) can be treated analogously.

Step 2. Now consider the method 11.18(1) in the case $\alpha > 1$ and set $\Lambda^z := \left((k+1)^z \times \delta_{kj}\right)_{k,j=0}^{\infty}$ for $z \in \mathbb{C}$. Since the matrix A_1 of the system 11.18(1) depends on α, let us denote it by A_1^α. The operator $A_1^\alpha \in \mathcal{L}(l^2_{((\alpha-1)/2)-\varrho\alpha})$ has the Fredholm property if and only if $A_{1,\alpha} := \Lambda^{((\alpha-1)/2)-\varrho\alpha}A_1^\alpha\Lambda^{(-(\alpha+1)/2)+\varrho\alpha} \in \mathcal{L}(l^2)$ is Fredholm. In order to show the Fredholm property of A_1^α and that $\operatorname{ind} A_1^\alpha = 0$, we shall prove that $A_{1,\alpha} \in \mathfrak{A}$, $\operatorname{smb} A_{1,\alpha}(\tau, \mu) \neq 0$ and $\operatorname{ind}\operatorname{smb} A_{1,\alpha} = 0$. Lemma 11.4 implies that $A_{1,1} \in \mathfrak{A}$ and

$$\operatorname{smb} A_{1,1}(\tau,\mu) = \begin{cases} c+df(\tau) & \text{if } \tau \neq 1, \\ c+d\left\{-i\cot\left(\pi\left(1/2-\varrho+\frac{i}{2\pi}\log\frac{\mu}{1-\mu}\right)\right)\right\} & \text{if } \tau = 1,\ 0 \leqq \mu \leqq 1. \end{cases} \tag{2}$$

It remains to examine $B = A_{1,\alpha} - A_{1,1}$.

Substituting x by x^α in the well-known formula

$$\frac{1}{\pi i}\frac{1}{1-x} = \frac{1}{2\pi i}\int\limits_{\mathrm{Re}\, z=1/2} x^{-z}\{-i\cot(\pi z)\}\, dz,$$

(see e.g. GOHBERG/KRUPNIK [4]) we get

$$\frac{1}{\pi i}\frac{\alpha x^{(\alpha-1)/2}}{1-x^\alpha} x^{-\varrho\alpha} = \frac{1}{2\pi i}\int\limits_{\mathrm{Re}\, z=1/2} x^{-(\alpha z-(\alpha-1)/2+\varrho\alpha)}\{-i\cot(\pi z)\}\,\alpha\, dz$$
$$= \frac{1}{2\pi i}\int\limits_{\mathrm{Re}\,\zeta=1/2+\varrho\alpha} x^{-\zeta}\left\{-i\cot\left(\pi\left(\frac{\zeta+(\alpha-1)/2}{\alpha}-\varrho\right)\right)\right\} d\zeta.$$

The residue theorem implies

$$\frac{1}{\pi i}\frac{\alpha x^{(\alpha-1)/2}}{1-x^\alpha} x^{-\varrho\alpha} = \frac{1}{2\pi i}\int\limits_{\mathrm{Re}\,\zeta=1/2} x^{-\zeta}\left\{-i\cot\left(\pi\left(\frac{\zeta+(\alpha-1)/2}{\alpha}-\varrho\right)\right)\right\} d\zeta. \quad (3)$$

Subtracting the same formula with $\alpha = 1$ from (3), we obtain

$$\frac{1}{\pi i}\frac{\alpha x^{(\alpha-1)/2}}{1-x^\alpha} x^{-\varrho\alpha} - \frac{1}{\pi i}\frac{x^{-\varrho}}{1-x} = \frac{1}{2\pi i}\int\limits_{\mathrm{Re}\,\zeta=1/2} x^{-\zeta} b(\zeta)\, d\zeta,$$

where

$$b(\zeta) := -i\cot\left(\pi\left(\frac{\zeta+(\alpha-1)/2}{\alpha}-\varrho\right)\right) + i\cot\big(\pi(\zeta-\varrho)\big).$$

Analogously to the derivation of 11.4(4) we conclude that

$$k(x) := (1-x)\left\{\frac{\alpha x^{(\alpha-1)/2-\varrho\alpha}}{1-x^\alpha} - \frac{x^{-\varrho}}{1-x}\right\} = 1/2\int\limits_{\mathrm{Re}\,\zeta=\psi} x^{-\zeta}\{b(\zeta)-b(\zeta+1)\}\, d\zeta - x^{-\varrho}, \quad (4)$$

where $\psi := \max\{-1/2, 1/2+\alpha(\varrho-1/2)\} + \varepsilon'$ and $\varepsilon' > 0$ denotes a small number such that $-1/2 < \psi < 1/2$ and $\psi < \varrho$.

Now the operator $B = A_{1,\alpha} - A_{1,1}$ takes the form

$$B = \frac{d}{\pi i}\left(\frac{\left(\frac{k+1}{j+1}\right)^{(\alpha-1)/2-\varrho\alpha}\alpha\big(j+(1/2)\big)^{\alpha-1}}{(j+1/2)^\alpha-(k+\varepsilon)^\alpha} - \frac{\left(\frac{k+1}{j+1}\right)^{-\varrho}}{(j+1/2)-(k+\varepsilon)}\right)_{k,j=0}^{\infty}$$
$$= \left(\left(\frac{k+1}{k+\varepsilon}\right)^{(\alpha-1)/2-\varrho\alpha}\delta_{k,j}\right)_{k,j=0}^{\infty} B' \left(\left(\frac{j+1}{j+1/2}\right)^{-(\alpha-1)/2+\varrho\alpha}\delta_{k,j}\right)_{k,j=0}^{\infty} - d\Lambda^{-\varrho}T(f)\,\Lambda^{\varrho}$$
$$+ d\left(\left(\frac{k+1}{k+\varepsilon}\right)^{(\alpha-1)/2-\varrho\alpha+\varrho}\delta_{k,j}\right)_{k,j=0}^{\infty}\Lambda^{-\varrho}T(f)\,\Lambda^{\varrho}\left(\left(\frac{j+1}{j+1/2}\right)^{-(\alpha-1)/2+\varrho\alpha-\varrho}\delta_{k,j}\right)_{k,j=0}^{\infty} + T_1,$$

where T_1 is compact and

$$B' := \frac{d}{\pi i}\left(k\left(\frac{k+\varepsilon}{j+1/2}\right)\frac{1}{\big(j+(1/2)\big)-(k+\varepsilon)}\right)_{k,j=0}^{\infty}.$$

If B' is bounded, then it is easy to show $B - B'$ to be compact. Thus the inclusion $B \in \mathfrak{A}$ and the equality $\operatorname{smb} B = \operatorname{smb} B'$ follow from the inclusion $B' \in \mathfrak{A}$. Equation (4) implies

$$B' = \frac{d}{2} \int_{\operatorname{Re} z = \psi} \{b(z) - b(z+1)\} \left(\frac{1}{\pi \mathrm{i}} \frac{((k+\varepsilon)/(j+1/2))^{-z}}{(j+1/2) - (k+\varepsilon)} \right)_{k,j=0}^{\infty} \mathrm{d}z$$

$$- d \left(\frac{1}{\pi \mathrm{i}} \frac{((k+\varepsilon)/(j+1/2))^{-\varrho}}{(j+1/2) - (k+\varepsilon)} \right)_{k,j=0}^{\infty} .$$

Analogously to the proof of the inclusion $A_{2,1} \in \mathfrak{A}$ (see Section 11.4) we conclude that

$$B' = \frac{d}{2} \int_{\operatorname{Re} z = \psi} \{b(z) - b(z+1)\}\, \Lambda^{-z}\, T(f)\, \Lambda^{z}\, \mathrm{d}z - d \Lambda^{-\varrho} T(f)\, \Lambda^{\varrho} + T ,$$

where $T \in \mathscr{L}(\mathbf{l}^2)$ is compact. Thus $B' \in \mathfrak{A}$ and

$$\operatorname{smb} B = \frac{d}{2} \int_{\operatorname{Re} z = \psi} \{b(z) - b(z+1)\}\, \mathscr{A}^{z}\, \mathrm{d}z - d \mathscr{A}^{\varrho}$$

$$= \frac{d}{2} \left\{ \int_{\operatorname{Re} z = \psi} b(z)\, \mathscr{A}^{z}\, \mathrm{d}z - \int_{\operatorname{Re} z = \psi + 1} b(z)\, \mathscr{A}^{z}\, \mathrm{d}z \right\} - d \mathscr{A}^{\varrho} ,$$

$$\mathscr{A}^{z} := \operatorname{smb} \Lambda^{-z} T(f)\, \Lambda^{z} .$$

The residue theorem yields (cf. the computation of $\operatorname{smb} A_{2,1}$ in Section 11.4)

$$\operatorname{smb} B(\tau, \mu) = \begin{cases} 0 & \text{if } \tau \neq 1, \\ db \left(\dfrac{1}{2} + \dfrac{\mathrm{i}}{2\pi} \log \dfrac{\mu}{1-\mu} \right) & \text{if } \tau = 1,\ 0 \leqq \mu \leqq 1 . \end{cases}$$

From this formula and from (2), we conclude that

$$\operatorname{smb} A_{1,\alpha}(\tau, \mu) = \begin{cases} c + d f(\tau) & \text{if } \tau \neq 1, \\ c + d \left\{ -\mathrm{i} \cot \left(\pi \left(\dfrac{1}{2} - \varrho + \dfrac{\mathrm{i}}{2\pi\alpha} \log \dfrac{\mu}{1-\mu} \right) \right) \right\} & \text{if } \tau = 1,\ 0 \leqq \mu \leqq 1 . \end{cases}$$

The same formula holds for the method 11.18(2). For 11.18(3), an analogous computation yields

$$\operatorname{smb} A_{1,\alpha}(\tau, \mu) = \begin{cases} c - d & \text{if } \operatorname{Im} \tau > 0, \\ c + d & \text{if } \operatorname{Im} \tau > 0, \\ c + d \left\{ -\mathrm{i} \cot \left(\pi \left(1/2 - \varrho + \dfrac{\mathrm{i}}{2\pi\alpha} \log \dfrac{\mu}{1-\mu} \right) \right) \right\} & \text{if } \tau = 1,\quad 0 \leqq \mu \leqq 1, \\ c - d \left\{ -\mathrm{i} \cot \left(\pi \left(1/2 - \varrho + \dfrac{\mathrm{i}}{2\pi\alpha} \log \dfrac{\mu}{1-\mu} \right) \right) \right\} & \text{if } \tau = -1,\quad 0 \leqq \mu \leqq 1 . \end{cases}$$

Since the set $\{\operatorname{smb} A_{1,\alpha}(\tau, \mu) : \tau \in \mathbb{T}, 0 \leqq \mu \leqq 1\}$ and ind smb $A_{1,\alpha}$ are independent of α, the Fredholm property of $A_{1,\alpha}$ as well as Ind $A_{1,\alpha}$ do not depend on α. Consequently, the Fredholm property of $A_{1,\alpha}$ and the equation Ind $A_{1,\alpha} = 0$ follow from step 1 of this proof.

Step 3. It remains to show (compare the proof of Theorem 11.3) that the null space of $A_{1,\alpha}$ is trivial if $\alpha = 2$ or if $A_{1,\alpha}$ denotes the approximating operator of 11.18(2) or 11.18(3). First, let $\alpha = 2$ and consider 11.18(1). Choose $\{\xi_j\}_{j=0}^{\infty} \in \mathbf{l}^2_{1/2-2\varrho}$ such that $A_1\{\xi_j\} = 0$. Since

$$A_1 = cI + d\left\{\left(\frac{1}{\pi \mathrm{i}}\,\frac{2(j+1/2)}{(j+1/2)^2-(k+\varepsilon)^2}\right)_{k,j=0}^{\infty} - \mathrm{i}\cot\big(\pi(\varepsilon-1/2)\big)\,I\right\},$$

$$A_1 = \big\{c - d\,\mathrm{i}\cot\big(\pi(\varepsilon-1/2)\big)\big\}\,I + d\left\{\left(\frac{1}{\pi\mathrm{i}}\,\frac{1}{j-k-(\varepsilon-1/2)}\right)_{k,j=0}^{\infty}\right.$$

$$\left. + \left(\frac{1}{\pi\mathrm{i}}\,\frac{1}{j+k+1+\varepsilon-1/2)}\right)_{k,j=0}^{\infty}\right\}$$

we get

$$0 = \sum_{j=0}^{\infty}\left\{\big[c - d\,\mathrm{i}\cot\big(\pi(\varepsilon-1/2)\big)\big]\,\delta_{kj} + d\,\frac{1}{\pi\mathrm{i}}\,\frac{1}{j-k-(\varepsilon-1/2)}\right\}\xi_j$$

$$+ \sum_{j=-\infty}^{-1}\left\{\big[c - d\mathrm{i}\cot\big(\pi(\varepsilon-1/2)\big)\big]\,\delta_{kj} + d\,\frac{1}{\pi\mathrm{i}}\,\frac{1}{j-k-(\varepsilon-1/2)}\right\}(-\xi_{(-1-j)}),$$

$$\text{if}\quad k = 0, 1, 2, \ldots \tag{5}$$

For an arbitrary $f = \sum_{k\in\mathbb{Z}} f_k t^k \in \mathbf{L}^2(\mathbb{T})$, we define

$$\tilde{f}(t) := f(t^{-1}), \qquad (Hf)(t) := t^{-1}\tilde{f}(t) = \sum_{k\in\mathbb{Z}} f_{-1-k}t^k,$$

$$(Pf)(t) := \sum_{k=0}^{\infty} f_k t^k, \qquad Q := I - P.$$

Setting $\eta := \sum_{k=0}^{\infty} \xi_k t^k$ and $g := c + df^{(\varepsilon-(1/2))}$, the equations (5) take the form $Pg(\eta - H\eta) = 0$. From $HPH = Q$, $HgH = \tilde{g}$ and $H^2 = I$, we obtain $Q\tilde{g}(\eta - H\eta) = 0$. Thus the function $\tilde{g}(\eta - H\eta) \in \operatorname{im} P$ satisfies $P(g/\tilde{g})\,\tilde{g}(\eta - H\eta) = 0$, i.e., $\tilde{g}(\eta - H\eta)$ is an element of the null space of the Wiener-Hopf operator $P(g/\tilde{g})|_{\operatorname{im}P}$. Obviously, $(\eta - H\eta)$ belongs to the Sobolev space $\mathbf{H}^{1/2-2\varrho}(\mathbb{T})$. Therefore, it suffices to show $\ker P(g/\tilde{g})|_{\operatorname{im}P} \cap \tilde{g}\mathbf{H}^{1/2-2\varrho}(\mathbb{T}) = \{0\}$, where $\ker P(g/\tilde{g})|_{\operatorname{im}P}$ is the null space of $P(g/\tilde{g})|_{\operatorname{im}P} \in \mathcal{L}\big(\mathbf{H}^v(\mathbb{T})\big)$ with $-1/2 < v < 1/2$, $v < 1/2 - 2\varrho$.

Let χ, $-3/4 < \chi < 3/4$, denote the winding number of the circular arc $\{g(\mathrm{e}^{\mathrm{i}2\pi s}) : 0 \leqq s \leqq 1\}$, i.e., $g(1-0)/g(1+0) = |g(1-0)/g(1+0)|\,\mathrm{e}^{\mathrm{i}2\pi\chi}$. Then the winding number of $\{g/\tilde{g}(\mathrm{e}^{\mathrm{i}2\pi s}), 0 \leqq s \leqq 1\}$ equals 2χ and the dimension of the null space of $P(g/\tilde{g})|_{\operatorname{im}P} \in \mathcal{L}\big(\mathbf{H}^v(\mathbb{T})\big)$ equals $\max\{-\operatorname{ind}_v(g/\tilde{g}), 0\}$ (see Theorem 6.17 and Duduchava [2]), where $\operatorname{ind}_v(g/\tilde{g})$ stands for the winding number of the curve

$$\{g/\tilde{g}(\mathrm{e}^{\mathrm{i}2\pi s}) : 0 \leqq s \leqq 1\}$$

$$\cup \left\{g/\tilde{g}(1+0)\,\frac{1 + f^{(-2v)}(\mathrm{e}^{\mathrm{i}2\pi s})}{2} + g/\tilde{g}(1-0)\,\frac{1 - f^{(-2v)}(\mathrm{e}^{\mathrm{i}2\pi s})}{2} : 0 \leqq s \leqq 1\right\}.$$

If $\chi \geqq 0$, then we get $\operatorname{ind}_v (g/\tilde{g}) \geqq 0$ for all v satisfying $-1/2 < v < 1/2$, $v < 1/2 - 2\varrho$. Thus, in this case, the null space of $P(g/\tilde{g})|_{\mathrm{im}P}$ is trivial and we may suppose $\chi < 0$.

Now we know A_1 to be a Fredholm operator and $\operatorname{Ind} A_1 = 0$. By step 1 and 2 of this proof, this yields that $\operatorname{ind}_{-\varrho} g = 0$, which is equivalent to $-2\pi\chi < 2\pi(1/2 - \varrho)$ for $-1/2 < \chi < 0$ and to $2\pi + 2\pi\chi > 2\pi(1/2 + \varrho)$ for $-3/4 < \chi \leqq -1/2$. In each case we have $-\chi < 1/2 - \varrho$.

Let $-1/2 < \chi < 0$ and choose $v := -2\chi - 1/2 + \delta$, where δ is a suitable small positive number such that $-1/2 < v < 1/2$, $v < 1/2 - 2\varrho$. If $-1/4 \leqq \chi < 0$, then $\operatorname{ind}_v (g/\tilde{g}) = 0$ is equivalent to $-4\pi\chi < 2\pi(1/2 + v)$. For $-1/2 < \chi < -1/4$, the equation $\operatorname{ind}_v (g/\tilde{g}) = 0$ holds if and only if $2\pi + 4\pi\chi > 2\pi(1/2 - v)$. Consequently, for our special choice of v, we get $\operatorname{ind}_v (g/\tilde{g}) = 0$ and $\ker P(g/\tilde{g})|_{\mathrm{im}P} = \{0\}$.

Now let $-3/4 < \chi \leqq -1/2$ and choose $v := -2\chi - 3/2 + \delta$, where δ denotes a small number such that $-1/2 < v < 1/2$, $v < 1/2 - 2\varrho$. Then $\operatorname{ind}_v (g/\tilde{g}) = -1$ is equivalent to the condition $-4\pi\chi - 2\pi < 2\pi(1/2 + v)$, which is fulfilled for the special choice of v. Factorizing the function $g/\tilde{g}$ we can determine the elements of $\ker P(g/\tilde{g})|_{\mathrm{im}P}$ (cf. Gohberg/Krupnik [4] and Mikhlin/Prössdorf [1]). We have $g/\tilde{g}(t) = t^{-1}h(t)\, t^{\beta}$, where $h(t)$ is a certain Hölder continuous function with $\operatorname{ind} h = 0$ and β is defined by $-1/2 - v < \operatorname{Re}\beta < 1/2 - v$, $e^{i2\pi\beta} = (g/\tilde{g})(1-0)/(g/\tilde{g})(1+0)$. From

$$(g/\tilde{g})(1-0)/(g/\tilde{g})(1+0) = \big(g(1-0)/g(1+0)\big)^2 = |g(1-0)/g(1+0)|^2\, e^{i4\pi\chi}$$

we conclude $\operatorname{Re}\beta = 2\chi + k$, $k \in \mathbb{Z}$, and $v = -2\chi - (3/2) + \delta$ yields $\operatorname{Re}\beta = 2\chi + 1$. If h_+h_- is the factorization of h, then $\ker P(g/\tilde{g})|_{\mathrm{im}P}$ is spanned by the function $h_+^{-1}(1-t)^{-\beta}$. Obviously, the intersection $\big[\tilde{g}^{-1} \ker P(g/\tilde{g})|_{\mathrm{im}P}\big] \cap \mathbf{H}^s(\mathbb{T})$ is trivial for $-\operatorname{Re}\beta + 1/2 = -2\chi - 1/2 \leqq s$. Thus $-\chi < 1/2 - \varrho$ implies $-2\chi - 1/2 < (1/2) - 2\varrho$ and $\big[\ker P(g/\tilde{g})|_{\mathrm{im}P}\big] \cap \tilde{g}\mathbf{H}^{1/2-2\varrho}(\mathbb{T}) = \{0\}$.

Step 4. Now we consider the method 11.18(2) assuming $\alpha > 1$. The operator A_1 is invertible if and only if A_1'' is invertible, where

$$A_1'' := \big((k + (1/2))^{(\alpha-1)/2}\, \delta_{kj}\big)_{k,j=0}^{\infty} A_1 \big((j + (1/2))^{-(\alpha-1)/2}\, \delta_{k,j}\big)_{k,j=0}^{\infty} \in \mathscr{L}(\mathbf{l}^2_{-\varrho\alpha}),$$

$$A_1'' = cI + d\left(\frac{1}{\pi i}\, \frac{\alpha(j + (1/2))^{(\alpha-1)/2}\, (k + (1/2))^{(\alpha-1)/2}}{(j + (1/2))^{\alpha} - (k + (1/2))^{\alpha}}\right)_{k,j=0}^{\infty}.$$

By duality arguments we may restrict ourselves to the case $\varrho \leqq 0$. Then $\{\xi_j\}_{j=0}^{\infty} \in \ker A_1''$ $\big(A_1'' \in \mathscr{L}(\mathbf{l}^2_{-\varrho\alpha})\big)$ implies $\{\xi_j\}_{j=0}^{\infty} \in \ker A_1''$ $\big(A_1'' \in \mathscr{L}(\mathbf{l}^2)\big)$. Thus we may suppose $\varrho = 0$. For $\varrho = 0$, we shall show that A_1'' is invertible if and only if $c + \lambda d \neq 0$ $(-1 < \lambda < 1)$. In other words, we shall prove that the spectrum $\sigma(S')$ of

$$S' := \left(\frac{1}{\pi i}\, \frac{\alpha(j + (1/2))^{(\alpha-1)/2}\, (k + (1/2))^{(\alpha-1)/2}}{(j + (1/2))^{\alpha} - (k + (1/2))^{\alpha}}\right)_{k,j=0}^{\infty}$$

is contained in $[-1, 1]$. Since S' is selfadjoint, it suffices to show that $\|S'\| \leqq 1$. Formula (3) and the residue theorem lead to

$$\frac{1}{\pi i}\, \frac{\alpha t^{(\alpha-1)/2}}{1 - t^{\alpha}} = -\frac{1}{2\pi} \int_{\operatorname{Re}\zeta=0} t^{-\zeta} \cot\left(\pi\, \frac{\zeta + (\alpha-1)/2}{\alpha}\right) d\zeta,$$

$$\frac{1}{\pi i}\, \frac{\alpha t^{(\alpha-1)/2}}{1 - t^{\alpha}}\, t = -\frac{1}{2\pi} \int_{\operatorname{Re}\zeta=0} t^{-\zeta} \cot\left(\pi\, \frac{\zeta + (\alpha+1)/2}{\alpha}\right) d\zeta.$$

Consequently,

$$\frac{1}{\pi \mathrm{i}} \frac{\alpha t^{(\alpha-1)/2}}{1-t^\alpha}(1-t) = -\frac{1}{2\pi} \int\limits_{\mathrm{Re}\zeta=0} t^{-\zeta} \left\{\cot\left(\pi \frac{\zeta+(\alpha-1)/2}{\alpha}\right) - \cot\left(\pi \frac{\zeta+(\alpha+1)/2}{\alpha}\right)\right\} \mathrm{d}\zeta$$

$$= \frac{1}{\pi \mathrm{i}} \int\limits_{\mathbb{R}} t^{-\mathrm{i}\xi} \frac{\sin(\pi/\alpha)}{\cos(\pi/\alpha) + \cosh(2\pi\xi/\alpha)} \,\mathrm{d}\xi. \tag{6}$$

Now S' takes the form

$$S' = \left(\frac{1}{\pi \mathrm{i}} \frac{\alpha \left(\frac{k+(1/2)}{j+(1/2)}\right)^{(\alpha-1)/2}}{1-\left(\frac{k+(1/2)}{j+(1/2)}\right)^{\alpha}} \left(1 - \frac{k+(1/2)}{j+(1/2)}\right) \frac{1}{(j+(1/2))-(k+(1/2))}\right)_{k,j=0}^{\infty}$$

$$= \int\limits_{\mathbb{R}} \frac{\sin(\pi/\alpha)}{\cos(\pi/\alpha)+\cosh(2\pi\xi/\alpha)} \left(\frac{1}{\pi \mathrm{i}} \frac{\left(\frac{k+(1/2)}{j+(1/2)}\right)^{-\mathrm{i}\xi}}{(j+(1/2))-(k+(1/2))}\right)_{k,j=0}^{\infty} \mathrm{d}\xi.$$

Using

$$\left\| \left(\frac{1}{\pi \mathrm{i}} \frac{\left(\frac{k+(1/2)}{j+(1/2)}\right)^{-\mathrm{i}\xi}}{(j+(1/2))-(k+(1/2))}\right)_{k,j=0}^{\infty} \right\| = \left\| \left(\frac{1}{\pi \mathrm{i}} \frac{1}{j-k}\right)_{k,j=0}^{\infty} \right\| = \|T(f^0)\|$$

$$= \sup |f^0(\tau)| = 1,$$

$$\frac{\sin(\pi/\alpha)}{\cos(\pi/\alpha)+\cosh(2\pi\xi/\alpha)} \geqq 0$$

and (6) with $t = 1$, we arrive at

$$\|S'\| \leqq \int\limits_{\mathbb{R}} \frac{\sin \pi/\alpha}{\cos \pi/\alpha + \cosh(2\pi\xi/\alpha)} \,\mathrm{d}\xi = 1.$$

The method 11.18(3) can be treated analogously. ■

11.20. Quadrature methods on the interval. Let $A := cI + dS_{\mathbb{I}} + T$, where $\mathbb{I} := [0, 1]$, $c, d \in \mathbf{C}(\mathbb{I})$, $S_{\mathbb{I}}$ denotes the Cauchy singular operator on $\mathbb{I}$ and T is defined by

$$(Tx)(t) = \int_0^1 k(t, \tau)\, x(\tau) \,\mathrm{d}\tau, \qquad k \in \mathbf{C}(\mathbb{I} \times \mathbb{I}).$$

Thus A is a bounded operator in the space $\mathbf{L}^2_{\varrho_0,\varrho_1}$ of all functions x satisfying

$$\|x\|_{L^2_{\varrho_0,\varrho_1}} := \left(\int_0^1 \varrho(\tau)^2 \,|x(\tau)|^2 \,\mathrm{d}\tau\right)^{1/2} < \infty,$$

where $\varrho(\tau) := \tau^{-\varrho_0}(1-\tau)^{-\varrho_1}$, $-1/2 < \varrho_0, \varrho_1 < 1/2$.

The quadrature methods 11.5(1)–11.5(3) suggest the following procedures for the numerical solution of the equation $Au = f$, where $f \in \mathcal{R}(\mathbb{I})$. Choose numbers $\alpha_0, \alpha_1 \geqq 1$

and a monotonically increasing bijective function $\gamma\colon \mathbb{I} \to \mathbb{I}$ such that $\gamma(t) = t^{\alpha_0}$ for $0 \leq t \leq 1/3$, $\gamma(t) = 1 - (1 - t)^{\alpha_1}$ for $2/3 \leq t \leq 1$ and the restriction of γ to $[1/4, 3/4]$ is twice continuously differentiable. Let $\varepsilon_I \in (0, 1)$, $\delta := 1/2$ and set

$$t_k^{(n)} := \gamma\left(\frac{k + (1/2)}{n}\right), \qquad \tau_k^{(n)} := \gamma\left(\frac{k + \varepsilon_I}{n}\right), \qquad n \in \mathbb{N}, k = 0, 1, \ldots, n - 1.$$

We determine approximate values $\xi_k^{(n)}$ of $u(t_k^{(n)})$ by solving one of the systems 11.5(1), 11.5(2) or 11.5(3), where $\Delta t_j^{(n)} := (1/n)\, \gamma'\big((j + 1/2)/n\big)$ for 11.5(1) or 11.5(2), $\Delta t_j^{(n)} := (2/n) \times \gamma'\big((j + 1/2)/n\big)$ for 11.5(3). Finally, we define an approximation of the solution $u = A^{-1}f$ by $u_n := \sum_{k=0}^{n-1} \xi_k^{(n)} \chi_j^{(n)}$, where

$$\chi_k^{(n)}(t) := \begin{cases} 1 & \text{if } \gamma\left(\frac{k}{n}\right) \leq t < \gamma\left(\frac{k+1}{n}\right), \\ 0 & \text{else}. \end{cases}$$

Now, for each $\tau \in \mathbb{I}$, we define a model problem of the quadrature method. If $0 < \tau < 1$, then we set $\omega_\tau = \pi$, $\Gamma_{\omega_\tau} := \mathbb{R}$, $A^\tau := c(\tau) + d(\tau)\, S_{\mathbb{R}} \in \mathcal{L}\big(\mathbf{L}^2(\mathbb{R})\big)$ and consider the method 11.2(1), 11.2(2) or 11.2(3), respectively. The matrix of the corresponding system will be denoted by A_1^τ and belongs to $\mathcal{L}(\tilde{\mathbf{l}}^2)$. For $\tau = 0$, set $A^0 := c(0) + d(0)\, S_{\mathbb{R}^+} \in \mathcal{L}\big(\mathbf{L}^2_{\varrho_0}(\mathbb{R}^+)\big)$, $\alpha := \alpha_0$, $\varepsilon := \varepsilon_I$ and apply the method 11.18(1), 11.18(2) or 11.18(3), respectively. In this case let us denote the matrix of the corresponding system by A_1^0. Then we obtain $A_1^0 \in \mathcal{L}(\mathbf{l}^2_{(\alpha_0 - 1)/2 - \varrho_0\alpha_0})$. In the case $\tau = 1$, it would be natural to choose $A^1 := c(1) + d(1)\, S_{\mathbb{R}^-}$, $\mathbb{R}^- := (-\infty, 0)$. However, we set $A^1 := c(1) - d(1)\, S_{\mathbb{R}^+} \in \mathcal{L}\big(\mathbf{L}^2_{\varrho_1}(\mathbb{R}^+)\big)$ (this operator arises after the transformation of the variable $t \to -t$). Choosing $\alpha := \alpha_1$ and $\varepsilon := -\varepsilon_{II}$, we apply the method 11.18(1), 11.18(2) or 11.18(3), respectively. If A_1^1 denotes the matrix of the corresponding system, then we get $A_1^1 \in \mathcal{L}(\mathbf{l}^2_{(\alpha_1 - 1)/2 - \varrho_1\alpha_1})$.

11.21. Theorem. (a) *The quadrature method* 11.5(1) *or* 11.5(2), *respectively, is stable if and only if the operators* $A \in \mathcal{L}(\mathbf{L}^2_{\varrho_0,\varrho_1})$ *and* A_1^τ $(\tau \in \mathbb{I})$ *are invertible. The method* 11.5(3) *for* A *is stable if and only if the operators* A *and* $\tilde{A} := cI - dS_{\mathbb{I}} - T \in \mathcal{L}(\mathbf{L}^2_{\varrho_0,\varrho_1})$ *as well as* A_1^τ $(\tau \in \mathbb{I})$ *are invertible.*

(b) *If the quadrature method is stable and f is Riemann integrable on* $\mathbb{I}$, *then the system* 11.5(1), 11.5(2) *or* 11.5(3), *respectively, is uniquely solvable for n large enough and the approximate solutions u_n converge to $u = A^{-1}f$ as $n \to \infty$.* ∎

Combining Theorems 11.3, 11.19 and 11.21 we get necessary and sufficient conditions for the afore-mentioned quadrature methods to be stable. For the proof of Theorem 11.21, we need a local theory analogous to the Theorems 11.7 and 11.8. To this end, we first introduce some notation. Let L_n denote the $\mathbf{L}^2$-orthogonal projection onto span $\{\chi_j^{(n)}\colon j = 0, \ldots, n - 1\}$ and define K_n^ε by $K_n^\varepsilon f := \sum_{k=0}^{n-1} f(\tau_k^{(n)})\, \chi_k^{(n)}$. Let A_n denote the matrix of the system 11.5(1), 11.5(2) or 11.5(3), respectively. Identifying the operators of $\mathcal{L}(\operatorname{im} L_n)$ with their matrices corresponding to the base $\{\chi_j^{(n)}\colon j = 0, \ldots, n - 1\}$ we obtain $A_n \in \mathcal{L}(\operatorname{im} L_n)$. Obviously, there exists a $C > 0$ such that

$$\frac{1}{C} \left\| \sum_{j=0}^{n-1} \xi_j \chi_j^{(n)} \right\|_{\mathbf{L}^2_{\varrho_0,\varrho_1}} \leq \frac{1}{\sqrt{n}} \left(\sum_{j=0}^{n-1} |\xi_j|^2\, \varrho_{n,j}^2 \right)^{1/2} \leq C \left\| \sum_{j=0}^{n-1} \xi_j \chi_j^{(n)} \right\|_{\mathbf{L}^2_{\varrho_0,\varrho_1}}, \tag{1}$$

where

$$\varrho_{n,j} := \varrho(t_j^{(n)}) \sqrt{\frac{\mathrm{d}}{\mathrm{d}s}\gamma\left(\frac{j+1/2}{n}\right)}.$$

For $\tau = \gamma(s)$, $0 < s < 1$, we define $k_E = k_E(\varepsilon_I, \tau, n)$ by $\tau \in (\tau_{k_E-1}^{(n)}, \tau_{k_E}^{(n)}]$ and $P_n^\tau \in \mathcal{L}(\tilde{\mathbf{l}}^2)$ by $P_n^\tau\{\xi_k\}_{k\in\mathbb{Z}} = \{\eta_k\}_{k\in\mathbb{Z}}$, where $\eta_k := \xi_k$ for $-k_E \leqq k < n - k_E$ and $\eta_k = 0$ for $k < -k_E$ or $k \geqq n - k_E$. If $\tau \in \{0, 1\}$, we set $P_n^\tau \in \mathcal{L}(\mathbf{l}^2)$ and $P_n^\tau\{\xi_k\}_{k=0}^\infty = \{\eta_k\}_{k=0}^\infty$, where $\eta_k := \xi_k$ for $k = 0, \ldots, n-1$ and $\eta_k := 0$ for $k = n, n+1, \ldots$. Now we introduce E_n^τ: $\operatorname{im} P_n^\tau \to \operatorname{im} L_n$ by

$$\begin{aligned} &\operatorname{im} P_n^\tau \ni \{\delta_{jk}\}_{k\in\mathbb{Z}} \mapsto \varrho_{n,j+k_E}^{-1}\chi_{j+k_E}^{(n)} && \text{for } 0 < \tau < 1, \\ &\operatorname{im} P_n^0 \ni \{\delta_{jk}\}_{k=0}^\infty \mapsto \varrho_{n,j}^{-1}\chi_j^{(n)} && \text{for } \tau = 0, \\ &\operatorname{im} P_n^1 \ni \{\delta_{jk}\}_{k=0}^\infty \mapsto \varrho_{n,n-1-j}^{-1}\chi_{n-1-j}^{(n)} && \text{for } \tau = 1. \end{aligned}$$

Furthermore, we set $B_n^E := (E_n^\tau)^{-1}B_nE_n^\tau \in \mathcal{L}(\operatorname{im} P_n^\tau)$ if $B_n \in \mathcal{L}(\operatorname{im} L_n)$.

Let $\mathbf{M}(\mathbb{I})$ denote the set of Lipschitz continuous functions χ on $\mathbb{I}$ satisfying $0 \leqq \chi \leqq 1$. For $\chi \in \mathbf{C}(\mathbb{I})$, we define $\tilde{\chi}_n \in \mathcal{L}(\operatorname{im} L_n)$ by $\tilde{\chi}_n := K_n^\varepsilon \chi|_{\operatorname{im} L_n}$. We shall say that $\{B_n\}$ $\big(B_n \in \mathcal{L}(\operatorname{im} L_n)\big)$ is equivalent to B_1^τ at $\tau \in \mathbb{I}$ if, for each $\varepsilon' > 0$, there exist $n_0 \in \mathbb{N}$ and a neighborhood U of τ on $\mathbb{I}$ such that $\chi \in \mathbf{M}(\mathbb{I})$, $\operatorname{supp}\chi \subset U$ and $n \geqq n_0$ imply $\|\tilde{\chi}_n^E(B_n^E - B_1^\tau)\,\tilde{\chi}_n^E\| < \varepsilon'$. Here B_1^τ belongs to $\mathcal{L}(\tilde{\mathbf{l}}^2)$ for $\tau \in (0, 1)$ and $B_1^\tau \in \mathcal{L}(\mathbf{l}^2)$ for $\tau = 0$ or $\tau = 1$. In particular, it will be shown that $\{A_n\}$ is equivalent to A_1^τ at $\tau \in (0, 1)$ and to $A_1^{\tau\prime}$ at $\tau \in \{0, 1\}$, where

$$A_1^{\tau\prime} := \left(\sqrt{\alpha_\tau}\left(k + \frac{1}{2}\right)^{(\alpha_\tau - 1)/2 - \varrho_\tau\alpha_\tau} \delta_{kj}\right)_{k,j=0}^\infty A_1^\tau \left(\frac{1}{\sqrt{\alpha_\tau}}\left(j + \frac{1}{2}\right)^{-(\alpha_\tau - 1)/2 + \varrho_\tau\alpha_\tau} \delta_{kj}\right)_{k,j=0}^\infty.$$

In the case of the method 11.5(3), we additionally need the operator $W_n \in \mathcal{L}(\operatorname{im} L_n)$ defined by the matrix $W_n := \big((-1)^j\,\delta_{j,k}\big)_{j,k=0}^{n-1}$. If $B_n \in \mathcal{L}(\operatorname{im} L_n)$, then we set $\tilde{B}_n := W_nB_nW_n$. Finally, by $\mathfrak{A}^1$ ($\mathfrak{A}^{\pm 1}$) we denote the subalgebra of $\mathfrak{A}$ generated by the Toeplitz operators $T(\alpha)$ with $\alpha \in \mathbf{PC}(\mathbb{T}) \cap \mathbf{C}(\mathbb{T} \setminus \{1\})$ $\big(\alpha \in \mathbf{PC}(\mathbb{T}) \cap \mathbf{C}(\mathbb{T} \setminus \{1, -1\})\big)$, and $\mathfrak{A}_{2\times 2}^1$ ($\mathfrak{A}_{2\times 2}^{\pm 1}$) will stand for the algebra introduced in Section 11.6 (Section 11.7).

11.22. Theorem. *Suppose $A \in \mathcal{L}(\mathbf{L}^2_{\varrho_0,\varrho_1})$ and $\{A_n\}$ $\big(A_n \in \mathcal{L}(\operatorname{im} L_n)\big)$ satisfy the following conditions.*

(i) *For all functions $\chi \in \mathbf{C}(\mathbb{I})$ satisfying* $\operatorname{supp}\chi \subset (0, 1)$, *we have $\tilde{\chi}_nA_n\tilde{\chi}_nL_n \to \chi A\chi$.*

(ii) *The operator $\chi A - A\chi \in \mathcal{L}(\mathbf{L}^2_{\varrho_0,\varrho_1})$ is compact for each $\chi \in \mathbf{C}(\mathbb{I})$.*

(iii) *The norms $\big\|\tilde{\chi}_nA_n - A_n\tilde{\chi}_n - L_n(\chi A - A\chi)|_{\operatorname{im} L_n}\big\|$ converge to 0 for each $\chi \in \mathbf{C}(\mathbb{I})$ as $n \to \infty$.*

(iv) *There exist operators $A_1^\tau \in \mathfrak{A}_{2\times 2}^1$ $\big(\tau \in (0, 1)\big)$ and $A_1^{\tau\prime} \in \mathfrak{A}^1$ $(\tau \in \{0, 1\})$ such that $\{A_n\}$ is equivalent to A_1^τ at $\tau \in (0, 1)$ and to $A_1^{\tau\prime}$ at $\tau \in \{0, 1\}$.*

Then A_nL_n converges strongly to A. Moreover, $\{A_n\}$ is stable if and only if the operators $A \in \mathcal{L}(\mathbf{L}^2_{\varrho_0,\varrho_1})$, $A_1^\tau \in \mathcal{L}(\tilde{\mathbf{l}}^2)$ $\big(\tau \in (0, 1)\big)$ and $A_1^{\tau\prime} \in \mathcal{L}(\mathbf{l}^2)$ $(\tau \in \{0, 1\})$ are invertible.

11.23. Theorem. *Suppose $A \in \mathcal{L}(\mathbf{L}^2_{\varrho_0,\varrho_1})$ and $\{A_n\}$ $\big(A_n \in \mathcal{L}(\operatorname{im} L_n)\big)$ satisfy the assumptions* (i) *and* (ii) *of Theorem 11.22. Assume that, in addition, the following hypotheses are satisfied.*

(i)′ *There exists an $\tilde{A} \in \mathcal{L}(\mathbf{L}^2_{\varrho_0,\varrho_1})$ such that $\tilde{\chi}_n\tilde{A}_n\tilde{\chi}_nL_n \to \chi\tilde{A}\chi$ for all $\chi \in \mathbf{C}(\mathbb{I})$ satisfying* $\operatorname{supp}\chi \subset (0, 1)$.

(ii)′ *The operator* $\chi\tilde{A} - \tilde{A}\chi \in \mathcal{L}(\mathbf{L}^2_{\varrho_0,\varrho_1})$ *is compact for each* $\chi \in \mathbf{C}(\mathbb{I})$.
(iii)′ *The norms* $\|\tilde{\chi}_n A_n - A_n\tilde{\chi}_n - L_n(\chi A - A\chi)|_{\operatorname{im} L_n} - W_n L_n(\chi\tilde{A} - \tilde{A}\chi)|_{\operatorname{im} L_n} W_n\|$ *converge to* 0 *for each* $\chi \in \mathbf{C}(\mathbb{I})$ *as* $n \to \infty$.
(iv)′ *There exist operators* $A_1^\tau \in \mathfrak{A}^{\pm 1}_{2\times 2}$ $(\tau \in (0, 1))$ *and* $A_1^{\tau\prime} \in \mathfrak{A}^{\pm 1}$ $(\tau \in \{0, 1\})$ *such that* $\{A_n\}$ *is equivalent to* A_1^τ *at* $\tau \in (0, 1)$ *and to* A_1^τ *at* $\tau \in \{0, 1\}$.

Then $A_n L_n$ *converges strongly to* A. *Furthermore,* $\{A_n\}$ *is stable if and only if the operators* $A, \tilde{A} \in \mathcal{L}(\mathbf{L}^2_{\varrho_0,\varrho_1})$, $A_1^\tau \in \mathcal{L}(\tilde{\mathbf{l}}^2)$ $(\tau \in (0, 1))$ *and* $A_1^{\tau\prime} \in \mathcal{L}(\mathbf{l}^2)$ $(\tau \in \{0, 1\})$ *are invertible.*

The proof of Theorem 11.22 is analogous to that of Theorem 11.23. Hence we only prove Theorem 11.23.

Proof. *Step 1.* First, we consider the case $\varrho_0 = \varrho_1 = 0$ and $\gamma(t) = t$. We choose a smooth curve $\tilde{\Gamma}$ containing $\mathbb{I}$ as a subarc and a parametrization $\tilde{\gamma}$ of $\tilde{\Gamma}$ satisfying $\tilde{\gamma}(s) = 2s$, $0 \leqq s \leqq 1/2$. If $\chi_{\mathbb{I}}$ denotes the operator of multiplication by the characteristic function of $\mathbb{I}$, i.e., the $\mathbf{L}^2$-projection of $\mathbf{L}^2(\tilde{\Gamma})$ onto $\mathbf{L}^2(\mathbb{I})$, then we set $A_{\tilde{\Gamma}} := A\chi_{\mathbb{I}} + (I - \chi_{\mathbb{I}})$. Furthermore, if $\tilde{n} := 2n$ and $L_{\tilde{\Gamma},\tilde{n}}$ is the orthogonal projection onto the piecewise constant functions on $\tilde{\Gamma}$ (cf. Section 11.6), then the projection L_n on $\mathbf{L}^2(\mathbb{I})$ coincides with $(\chi_{\mathbb{I}})_{\tilde{n}}^{\sim} L_{\tilde{\Gamma},\tilde{n}}$, and we set $(A_{\tilde{\Gamma}})_{\tilde{n}} := A_n(\chi_{\mathbb{I}})_{\tilde{n}}^{\sim} + \left(I_n - (\chi_{\mathbb{I}})_{\tilde{n}}^{\sim}\right) \in \mathcal{L}(\operatorname{im} L_{\tilde{\Gamma},\tilde{n}})$. Obviously, $A_n L_n \to A$ if and only if $(A_{\tilde{\Gamma}})_{\tilde{n}} L_{\tilde{\Gamma},\tilde{n}} \to A_{\tilde{\Gamma}}$. Moreover, the stability of $\{A_n\}$ is equivalent to the stability of $\{(A_{\tilde{\Gamma}})_{\tilde{n}}\}$. If, for $\{A_n\}$, the assumptions of Theorem 11.23 are fulfilled, then the assumptions of Theorem 11.8 hold for $\{(A_{\tilde{\Gamma}})_{\tilde{n}}\}$. Thus Theorem 11.23 follows from Theorem 11.8.

Step 2. Now we consider arbitrary ϱ_0, ϱ_1 and γ satisfying the assumptions of Section 11.20. Besides the notation in Section 11.20 we introduce the corresponding functions and operators for $\varrho_0 = \varrho_1 = 0$ and $\gamma(t) = t$ and denote them by the same symbols adding a + (e.g. $\chi_j^{(n)+}$, L_n^+). We define $G\colon \mathbf{L}^2_{\varrho_0,\varrho_1} \to \mathbf{L}^2(\mathbb{I})$ by

$$(Gf)(t) := \varrho(\gamma(t))\, f(\gamma(t)) \sqrt{\gamma'(t)}$$

and set $A^+ := GAG^{-1}$. Furthermore, formula 11.21(1) implies that the operators

$$\varrho_{n+}\colon \mathbf{L}^2_{\varrho_0,\varrho_1} \supset \operatorname{im} L_n \to \operatorname{im} L_n^+ \subset \mathbf{L}^2(\mathbb{I}), \qquad \chi_j^{(n)} \mapsto \varrho_{n,j}\chi_j^{(n)+}$$

and $(\varrho_{n+})^{-1}$ are uniformly bounded. We set $A_n^+ := \varrho_{n+}A_n(\varrho_{n+})^{-1}$ and consider them to be approximate operators for $A^+ \in \mathcal{L}\left(\mathbf{L}^2(\mathbb{I})\right)$ defined on uniform meshes. It is not hard to prove that the assumptions of Theorem 11.23 concerning A and A_n imply the validity of the corresponding assumptions for A^+ and A_n^+. Consequently, the general case of Theorem 11.23 follows from the special case proved in Step 1. ∎

11.24. Proof of Theorem 11.21. Let us first prove the assertions concerning the method 11.5(2). We have to show that the assumptions of Theorem 11.22 are satisfied. Assumption (ii) is well known and (i) follows from the strong $\mathbf{L}^2$-convergence of the corresponding operators on closed curves (see Theorem 10.27).

Now we consider (iii) and set $\Delta t_j^{(n)\prime} := \gamma\left((j+1)/n\right) - \gamma(j/n)$, $T_n := (k(t_k^{(n)}, t_j^{(n)})\, \Delta t_j^{(n)})_{k,j=0}^{n-1}$ and $T'_n := \left(k(t_k^{(n)}, t_j^{(n)})\, \Delta t_j^{(n)\prime}\right)_{k,j=0}^{n-1}$. Analogously to Section 11.15 we obtain that $\|T'_n - L_n T|_{\operatorname{im} L_n}\| \to 0$. By the compactness of T we find $\chi \in \mathbf{M}(\mathbb{I})$ such that $\|T\chi\| < \varepsilon'$ and $\chi \equiv 1$ near the end points of the interval. If $\chi^1 \in \mathbf{M}(\mathbb{I})$ satisfies $\chi^1 \equiv 1$ near 0 and 1 and χ is identically equal to 1 in a neighborhood of $\operatorname{supp} \chi^1$, then we get

$L_nT|_{\mathrm{im}L_n}\,\tilde{\chi}_n^1 = L_nT\chi|_{\mathrm{im}L_n}\,\tilde{\chi}_n^1$ and $\|L_nT|_{\mathrm{im}L_n}\,\tilde{\chi}_n^1\| < C\varepsilon'$. Setting $\chi^2 := 1 - \chi^1$, we obtain

$$T_n - T_n' = \left(T_n' - L_nT|_{\mathrm{im}L_n}\right)\tilde{\chi}_n^1 D_n + L_nT|_{\mathrm{im}L_n}\,\tilde{\chi}_n^1 D_n + T_n'\tilde{\chi}_n^2 D_n,$$

$$D_n := \left(\delta_{kj}\left(\frac{\Delta t_j^{(n)}}{\Delta t_j^{(n)\prime}} - 1\right)\right)_{k,j=0}^{n-1}.$$

Since $\|T_n' - L_nT|_{\mathrm{im}L_n}\| \to 0$, $\|L_nT|_{\mathrm{im}L_n}\,\tilde{\chi}_n^1\| < C\varepsilon'$ and, obviously, $\|\tilde{\chi}_n^2 D_n\| \to 0$ $(n \to \infty)$, we get $\|T_n - T_n'\| < C\varepsilon'$ for sufficiently large n. Thus $\|T_n - T_n'\| \to 0$ and $\|T_n - L_nT|_{\mathrm{im}L_n}\| \to 0$. Repeating the proof from Section 11.15, we conclude that

$$\|\tilde{\chi}_n A_n - A_n\tilde{\chi}_n - L_n(\chi A - A\chi)|_{\mathrm{im}L_n}\| \to 0.$$

Now we fix a point $\tau \in \mathbb{I}$ and show that $\|\tilde{\chi}_n^E T_n^E \tilde{\chi}_n^E\|$ becomes smaller than any prescribed positiv number. We choose $\chi' \in \mathbf{M}(\mathbb{I})$ satisfying $\|T\chi'\| < \varepsilon'$ and $\chi' \equiv 1$ near τ. If $\chi \in \mathbf{M}(\mathbb{I})$ satisfies $\chi \equiv 1$ near τ and χ' is identically equal to 1 in a neighborhood of supp χ, then $L_nT|_{\mathrm{im}L_n}\,\tilde{\chi}_n = L_nT\chi'|_{\mathrm{im}L_n}\,\tilde{\chi}_n$ and $\|L_nT|_{\mathrm{im}L_n}\,\tilde{\chi}_n\| < C\varepsilon'$. The convergence $\|T_n - L_nT|_{\mathrm{im}L_n}\| \to 0$ implies that $\|\tilde{\chi}_n T_n \tilde{\chi}_n\| < C\varepsilon'$ for n large enough. By the argumentation in Section 11.15 it remains to prove assumption (iv) for $A = S_{\mathbb{I}}$ and $A_n = S_n$, $S_n := \left(\frac{1}{\pi\mathrm{i}} \times \frac{\Delta t_j^{(n)}}{t_j^{(n)} - t_k^{(n)}}\right)_{k,j=0}^{n-1}$, where $\frac{1}{t_j^{(n)} - t_j^{(n)}} := 0$.

If $\tau \in (0, 1)$, then assumption (iv) is an immediate consequence of the proof given in Section 11.15. Since the case $\tau = 1$ is analogous to the case $\tau = 0$, we may suppose $\tau = 0$ and choose $\chi \in \mathbf{M}(\mathbb{I})$ such that supp $\chi \subseteq [0, 1/3)$. For

$$S_1^{0\prime} := \left(\sqrt{\alpha_0}\,(k + 1/2)^{(\alpha_0-1)/2-\varrho_0\alpha_0}\,\delta_{kj}\right)_{k,j=0}^{\infty} \times \left(\frac{1}{\pi\mathrm{i}}\,\frac{\alpha_0(j + 1/2)^{\alpha_0-1}}{(j + 1/2)^{\alpha_0} - (k + 1/2)^{\alpha_0}}\right)_{k,j=0}^{\infty}\left(\frac{1}{\sqrt{\alpha_0}}\,(j + 1/2)^{-(\alpha_0-1)/2+\alpha_0\varrho_0}\,\delta_{kj}\right)_{k,j=0}^{\infty}$$

we get

$$\chi_n^E(S_n^E - S_1^{0\prime})\,\chi_n^E = D_n^1 S_n^{0\prime} D_n^2 + \chi_n^E S_1^{0\prime} D_n^3,$$

$$D_n^1 := \left(\chi\left(\gamma\left(\frac{k + 1/2}{n}\right)\right)\frac{\varrho_{n,k} - \sqrt{\alpha_0}\left(\dfrac{k + 1/2}{n}\right)^{(\alpha_0-1)/2-\alpha_0\varrho_0}}{\sqrt{\alpha_0}\left(\dfrac{k + 1/2}{n}\right)^{(\alpha_0-1)/2-\varrho_0\alpha_0}}\,\delta_{kj}\right)_{k,j=0}^{\infty},$$

$$D_n^3 := \left(\chi\left(\gamma\left(\frac{j + 1/2}{n}\right)\right)\frac{\varrho_{n,j}^{-1} - \dfrac{1}{\sqrt{\alpha_0}}\left(\dfrac{j + 1/2}{n}\right)^{-(\alpha_0-1)/2+\varrho_0\alpha_0}}{\dfrac{1}{\sqrt{\alpha_0}}\left(\dfrac{j + 1/2}{n}\right)^{-(\alpha_0-1)/2+\varrho_0\alpha_0}}\,\delta_{kj}\right)_{k,j=0}^{\infty},$$

$$D_n^2 := \left(\chi\left(\gamma\left(\frac{j + 1/2}{n}\right)\right)\frac{\varrho_{n,j}^{-1}}{\dfrac{1}{\sqrt{\alpha_0}}\left(\dfrac{j + 1/2}{n}\right)^{(\alpha_0-1)/2+\varrho_0\alpha_0}}\,\delta_{kj}\right)_{k,j=0}^{\infty}.$$

Since $0 \leqq t_j^{(n)} < 1/3$ implies that

$$\varrho_{n,j} = \left(\frac{j+1/2}{n}\right)^{-\alpha_0\varrho_0} \left(1 - \left(\frac{j+1/2}{n}\right)^{\alpha_0}\right)^{-\varrho_1} \sqrt{\alpha_0} \left(\frac{j+1/2}{n}\right)^{(\alpha_0-1)/2},$$

for any prescribed $\varepsilon' > 0$, we can find a neighborhood $U \subset \mathbb{I}$ of 0 such that $\chi \in \mathbf{M}(\mathbb{I})$ and supp $\chi \subset U$ imply $\|D_n^1\| < \varepsilon'$ and $\|D_n^3\| < \varepsilon'$. By the uniform boundedness of D_n^2 and the boundedness of $S_1^{0\prime} \in \mathfrak{A}$ (see the proof of Theorem 11.19) we conclude that $\|\tilde{\chi}_n^E(S_n^E - S_1^{0\prime})\, \tilde{\chi}_n^E\| \leqq C\varepsilon'$. Thus for the method 11.5(2) assumption (iv) of Theorem 11.22 is satisfied.

The assertions of Theorem 11.21 concerning the methods 11.5(1) and 11.5(3) follow from Section 11.15 and the arguments presented in the first parts of this proof. However, for estimating the norm of

$$M_n^2 := \left(\frac{1}{\pi \mathrm{i}} \frac{\Delta t_j^{(n)}}{t_j^{(n)} - \tau_k^{(n)}} \left(\chi(\tau_j^{(n)}) - \chi(t_j^{(n)})\right)\right)_{k,j=0}^{n-1} \in \mathcal{L}(\operatorname{im} L_n) \qquad \text{(cf. Section 11.15)}$$

we need another method. We write $M_n^2 = F_n D_n$, where

$$F_n := \left(\frac{1}{\pi \mathrm{i}} \frac{\Delta t_j^{(n)}}{t_j^{(n)} - \tau_k^{(n)}}\right)_{k,j=0}^{n-1}, \qquad D_n := \left(\left(\chi(\tau_j^{(n)}) - \chi(t_j^{(n)})\right) \delta_{kj}\right)_{k,j=0}^{n-1}.$$

Since $\|D_n\| \to 0 (n \to \infty)$, it suffices to show that $\{F_n\}_{n\in\mathbb{N}}$ is uniformly bounded. If we put $\chi_n^{\#} := \left(\chi(t_j^{(n)})\, \delta_{kj}\right)_{k,j=0}^{n-1}$, then we obtain that $\left\|\tilde{\chi}_n F_n - F_n \chi_n^{\#} - L_n(\chi S_{\mathbb{I}} - S_{\mathbb{I}}\chi)|_{\operatorname{im} L_n}\right\| \to 0$ in analogy to the method 11.5(2). Now the proof to Lemma 11.9 yields the uniform boundedness of $\{F_n\}$. This completes the proof of Theorem 11.21. ■

Quadrature methods for Mellin operators of order zero

11.25. Mellin operators of order zero. The technique which has been developed in the previous sections can be applied to the stability analysis of quadrature and spline approximation methods for a large class of integral operators usually called pseudodifferential operators of Mellin type or shortly, Mellin operators. This class includes not only singular integral operators on an interval or on a curve with corners but also integral operators with fixed singularities and double layer potential operators on curves with corners.

In this section we introduce the class of Mellin operators, formulate a theorem on the Fredholm property of the operators acting in $L_\varrho^2 := L_{\varrho,0}^2$ $(-1/2 < \varrho < 1/2)$ and give some examples. For the proof of these and further results on this subject, we refer e.g., to the papers of Lewis/Parenti [1] and Elschner [5] (see also Sections 5.1–5.2).

Let x be a function given on $\mathbb{I} = [0, 1]$. Setting $x(t) = 0$ for $1 < t < \infty$, we denote the Mellin transform of x by $\tilde{x} = Mx$, i.e.,

$$\tilde{x}(z) := \int_0^1 t^{z-1} x(t)\, \mathrm{d}t.$$

For a given function $a(t, z)$ $(t \in \mathbb{I}, z \in \{z \in \mathbb{C} : \operatorname{Re} z = 1/2\})$, the Mellin operator $a(t, \partial)$ is defined by $a(t, \partial) = M^{-1} a(\cdot, \cdot)\, M$, i.e.,

$$\left(a(t, \partial)\, x\right)(\tau) := \frac{1}{2\pi \mathrm{i}} \int_{\operatorname{Re} z = 1/2} \tau^{-z} a(\tau, z)\, \tilde{x}(z)\, \mathrm{d}z, \qquad 0 \leqq \tau \leqq 1.$$

We assume that $a(t, z) = c(t) - d(t)\,\mathrm{i}\cot(\pi z) + b(t, z)$, where b, c, d are given functions satisfying the following conditions.

(i) The functions c, d are continuous on $\mathbb{I}$.

(ii) There exist real numbers α, β such that $\alpha < 1/2 < \beta$, $\alpha < 1/2 - \varrho < \beta$ and the function $\mathbb{I} \times \{z\colon \alpha < \operatorname{Re} z < \beta\} \ni (t, z) \mapsto b(t, z)$ is infinitely differentiable with respect to t and analytic with respect to z.

(iii) The function b satisfies

$$\sup\left\{\left|\left(\frac{\partial}{\partial t}\right)^l \left(\frac{\partial}{\partial z}\right)^k b(t, z)\,(1 + |z|)^{1+k}\right| : t \in \mathbb{I},\ \alpha < \operatorname{Re} z < \beta\right\} < \infty$$

for all $l, k \in \mathbb{Z}$, $l, k \geqq 0$.

If the symbol a satisfies (i) – (iii), then $a(t, \partial)$ is a bounded operator in $\mathbf{L}^2_\varrho$. The operator $a(t, \partial)$ is Fredholm if and only if 0 does not belong to the curve

$$\Gamma := \left\{a\left(0, \frac{1}{2} - \varrho + \mathrm{i}t\right) : t \in \mathbb{R}\right\} \cup \{c(t) - d(t) : t \in \mathbb{I}\}$$
$$\cup \left\{c(1) + d(1)\,\mathrm{i}\cot\left(\pi\left(\frac{1}{2} + \mathrm{i}t\right)\right) : t \in \mathbb{R}\right\} \cup \{c(t) + d(t) : t \in \mathbb{I}\}.$$

The index of $a(t, \partial)$ is equal to the negative winding number of Γ about the origin.

Let $S_{\mathbb{I}}$ denote the Cauchy singular operator on $\mathbb{I}$ and set $B := b(t, \partial)$. Then $a(t, \partial)$ takes the form $a(t, \partial) = c(t)\, I + d(t)\, S_{\mathbb{I}} + B$ (see, e.g., Gohberg/Krupnik [4]). The well-known properties of the Mellin transform imply that

$$(Bx)(t) = \int_0^1 k_t\left(\frac{t}{\tau}\right) x(\tau)\, \frac{\mathrm{d}\tau}{\tau}, \qquad k_t(\sigma) := \frac{1}{2\pi \mathrm{i}} \int_{\operatorname{Re} z = 1/2} \sigma^{-z} b(t, z)\, \mathrm{d}z.$$

By virtue of (ii) and (iii), we obtain that

$$k_t(\sigma) = \frac{1}{2\pi \mathrm{i}} \int_{\operatorname{Re} z = (1/2) - \varrho} \sigma^{-z} b(t, z)\, \mathrm{d}z, \qquad k_t(\sigma)\,\sigma^{-\varrho} = \frac{1}{2\pi \mathrm{i}} \int_{\operatorname{Re} z = 1/2} \sigma^{-z} b(t, z - \varrho)\, \mathrm{d}z. \tag{1}$$

Furthermore, the function $\{\operatorname{Re} z = 1/2\} \ni z \mapsto b(t, z - \varrho)$ is square integrable and Plancherel's theorem yields that $\left(\sigma \mapsto k_t(\sigma)\,\sigma^{-\varrho}\right) \in \mathbf{L}^2(\mathbb{R}^+)$, i.e. the function $\tau \mapsto k_t(t/\tau) \times 1/\tau$ belongs to $\mathbf{L}^2_{-\varrho}$ for any $t \in (0, 1]$. Thus we arrive at the formula

$$(Bx)(t) = \int_0^1 k_B(t, \tau)\, x(\tau)\, \mathrm{d}\tau, \qquad k_B(t, \tau) := k_t\left(\frac{t}{\tau}\right) \frac{1}{\tau}, \tag{2}$$

where the integral in (2) converges absolutely.

If the symbol $b(z) = b(t, z)$ is independent of t, then $b(\partial)$ is a Mellin convolution operator. In particular, for $0 < \omega < 2\pi$, $j, k \in \mathbb{N}$, $0 \leqq k \leqq j$ and

$$b(z) = -\mathrm{i}(-1)^j\, \mathrm{e}^{-\mathrm{i}(j+1)\omega} \binom{z + k - 1}{j} \frac{\mathrm{e}^{\mathrm{i}(\omega - \pi)(z+k)}}{\sin \pi(z + k)},$$

we get

$$(Bx)(t) = \frac{1}{\pi \mathrm{i}} \int_0^1 \frac{t^k \tau^{j-k}}{(\tau \, \mathrm{e}^{\mathrm{i}\omega} - t)^{j+1}} \, x(\tau) \, \mathrm{d}\tau . \tag{3}$$

Operators of the form (3) appear in certain boundary integral equations on curves with corners (see COSTABEL [2] and COSTABEL/STEPHAN [3]) or are known to be operators with fixed singularities (see DUDUCHAVA [3]).

11.26. Quadrature methods for Mellin convolution operators on the half axis. In the next sections we shall prove the stability of quadrature methods for solving pseudo-differential equations with Mellin operators on the interval. We begin with considering the following model problem. Let $A := cI + dS_{\mathbb{R}^+} + B_{\mathbb{R}^+} \in \mathscr{L}(\mathbf{L}^2_\varrho(\mathbb{R}^+))$, where $c, d \in \mathbb{C}$, $-1/2 < \varrho < 1/2$,

$$(B_{\mathbb{R}^+}x)(t) := \int_0^\infty k\left(\frac{t}{\tau}\right) \frac{x(\tau)}{\tau} \, \mathrm{d}\tau, \qquad k(\sigma) := \frac{1}{2\pi \mathrm{i}} \int_{\mathrm{Re}\, z = 1/2} \sigma^{-z} b(z) \, \mathrm{d}z .$$

The symbol $b(z) = b(t, z)$ of $B_{\mathbb{R}^+}$ is assumed to satisfy (ii) and (iii) of Section 11.25. It is well known that the Mellin convolution operator A is invertible if and only if

$$c - d\mathrm{i} \cot\big(\pi(1/2) - \varrho + \mathrm{i}t\big) + b\big((1/2) - \varrho + \mathrm{i}t\big) \neq 0, \qquad t \in \mathbb{R} .$$

Retaining the notation of Section 11.18 we obtain the following quadrature method for solving the equation $Au = f$, $f \in \mathscr{R}(\mathbb{R}^+) \cap \mathbf{L}^2_\varrho(\mathbb{R}^+)$. We determine $\xi_k^{(n)}$ $(k = 0, 1, \ldots)$ to be the solution of one of the following systems

$$\left\{c - \mathrm{i} \cot\left(\pi\left(\varepsilon - \frac{1}{2}\right)\right) d\right\} \xi_k^{(n)} + d \frac{1}{\pi \mathrm{i}} \sum_{j=0}^\infty \frac{\xi_j^{(n)}}{t_j^{(n)} - \tau_k^{(n)}} \Delta t_j^{(n)}$$

$$+ \sum_{j=0}^\infty k\left(\frac{\tau_k^{(n)}}{t_j^{(n)}}\right) \frac{\xi_j^{(n)}}{t_j^{(n)}} \Delta t_j^{(n)} = f(\tau_k^{(n)}), \qquad k = 0, 1, \ldots, \tag{1}$$

$$c\xi_k^{(n)} + d \frac{1}{\pi \mathrm{i}} \sum_{\substack{j=0 \\ j \neq k}}^\infty \frac{\xi_j^{(n)}}{t_j^{(n)} - t_k^{(n)}} \Delta t_j^{(n)} + \sum_{\substack{j=0 \\ j \neq k}}^\infty k\left(\frac{t_k^{(n)}}{t_j^{(n)}}\right) \frac{\xi_j^{(n)}}{t_j^{(n)}} \Delta t_j^{(n)} = f(t_k^{(n)}),$$

$$k = 0, 1, \ldots, \tag{2}$$

$$c\xi_k^{(n)} + d \frac{1}{\pi \mathrm{i}} \sum_{\substack{j=0 \\ j \equiv k+1 \bmod 2}}^\infty \frac{\xi_j^{(n)}}{t_j^{(n)} - t_k^{(n)}} \Delta t_j^{(n)} + \sum_{\substack{j=0 \\ j \equiv k+1 \bmod 2}}^\infty k\left(\frac{t_k^{(n)}}{t_j^{(n)}}\right) \frac{\xi_j^{(n)}}{t_j^{(n)}} \Delta t_j^{(n)} = f(t_k^{(n)}),$$

$$k = 0, 1, \ldots, \tag{3}$$

where $\Delta t_j^{(n)} := \alpha \left(\dfrac{j + 1/2}{n}\right)^{\alpha - 1} \dfrac{1}{n}$ for (1) or (2) and $\Delta t_j^{(n)} := 2\alpha \left(\dfrac{j + 1/2}{n}\right)^{\alpha - 1} \dfrac{1}{n}$ for (3).

Setting $u_n := \sum_{k=0}^\infty \xi_k^{(n)} \chi_k^{(n)}$, we get an approximate solution to the equation $Au = f$.

Let $A_n \in \mathscr{L}(\operatorname{im} L_n)$ be defined by the matrix of the system (1), (2) or (3) respectively. Analogously to Section 11.18 the matrix $A_n \in \mathscr{L}(\mathbf{l}^2_{(\alpha-1)/2-\varrho\alpha})$ does not depend on n and $\{A_n\}$ is stable if and only if $A_1 \in \mathscr{L}(\mathbf{l}^2_{(\alpha-1)/2-\varrho\alpha})$ is invertible.

11.27. Theorem. *Let* $a(z) = c - d\,\mathrm{i}\cot(\pi z) + b(z)$ *and set*

$$\mathcal{A}(\tau,\nu) := \begin{cases} a\left(1/2 - \varrho + \mathrm{i}\dfrac{1}{2\pi\alpha}\log\dfrac{\nu}{1-\nu}\right) & \text{for } \tau = 1,\ 0 \leqq \nu \leqq 1, \\ c + df(\tau) & \text{for the methods 11.26(1), 11.26(2) and } \tau \neq 1, \\ & \text{or for 1.26(3) and } \tau \neq 1, -1, \\ c + d\,\mathrm{i}\cot\left(1/2 - \varrho + \dfrac{\mathrm{i}}{2\pi\alpha}\log\dfrac{\nu}{1-\nu}\right) - b\left(1/2 - \varrho + \dfrac{\mathrm{i}}{2\pi\alpha}\log\dfrac{\nu}{1-\nu}\right) & \\ & \text{for 11.26(3) and } \tau = -1, \end{cases} \tag{1}$$

where $f = f^0$ *for* 11.26(2), $f = f^{\varepsilon-(1/2)}$ *for* 11.26(1) *and* $f = f^{\#}$ *for* 11.26(3) *(see Section* 11.4*). The operator* $A_1 \in \mathcal{L}(\mathbf{l}^2_{(\alpha-1)/2-\varrho\alpha})$ *is Fredholm of index zero if and only if* $\mathcal{A}(\tau,\nu) \neq 0$ $(\tau \in \mathbb{T},\ \nu \in [0,1])$ *and the index of the curve*

$$\Gamma_1 := \begin{cases} \{\mathcal{A}(\mathrm{e}^{\mathrm{i}2\pi\lambda}, 0) : 0 < \lambda < 1\} \cup \{\mathcal{A}(1,\mu) : 0 \leqq \mu \leqq 1\} \text{ for 11.26(1) and 11.26(2)} \\ \{\mathcal{A}(\mathrm{e}^{\mathrm{i}2\pi\lambda}, 0) : 0 < \lambda < 1/2\} \cup \{\mathcal{A}(-1,\mu) : 0 \leqq \mu \leqq 1\} \\ \cup \{\mathcal{A}(\mathrm{e}^{\mathrm{i}2\pi\lambda}, 0) : 1/2 < \lambda < 1\} \cup \{\mathcal{A}(1,\mu) : 0 \leqq \mu \leqq 1\} \text{ for 11.26(3)} \end{cases}$$

vanishes. A_1 *is invertible if and only if it is Fredholm of index zero and its null space is trivial.*

Proof. It remains to show that the operator

$$A_1' := \left((k+\varepsilon)^{(\alpha-1)/2-\varrho\alpha}\,\delta_{kj}\right)_{k,j=0}^{\infty} A_1\left((j+1/2)^{-(\alpha-1)/2+\varrho\alpha}\,\delta_{kj}\right)_{k,j=0}^{\infty} \in \mathcal{L}(\mathbf{l}^2)$$

is in $\mathfrak{A}$ and its symbol is given by (1). By step 2 of the proof of Theorem 11.19, we may suppose $c = d = 0$, i.e., $A = B_{\mathbb{R}^+}$.

Step 1. We first consider the method 11.26(1) for a special kernel. Let $\alpha = 1$, $\varrho = 0$, $\xi \in \mathbb{R}$, $0 < \mu < 1$ and set $k(\sigma) = k^{\xi}(\sigma) := \dfrac{1}{\pi\mathrm{i}}\dfrac{\mu\sigma^{(\mu-1)/2}}{1-\sigma^{\mu}}(\sigma^{-\mathrm{i}\xi} - 1)$ i.e., $b(z) = b^{\xi}(z) := -\mathrm{i}\cot\left(\pi\dfrac{z - \mathrm{i}\xi + (\mu-1)/2}{\mu}\right) + \mathrm{i}\cot\left(\pi\dfrac{z + (\mu-1)/2}{\mu}\right)$. Then the residue theorem implies

$$k^{\xi}(\sigma) = \frac{1}{2\pi\mathrm{i}}\int\limits_{\mathrm{Re}\,z=0} \sigma^{-z} b^{\xi}(z)\,\mathrm{d}z + \sum_{0<\mathrm{Re}\,z<1/2} \mathrm{Res}\,\{b^{\xi}(z)\,\sigma^{-z}\},$$

$$k^{\xi}(\sigma)\,\sigma = \frac{1}{2\pi\mathrm{i}}\int\limits_{\mathrm{Re}\,z=1/2} \sigma^{-z+1} b^{\xi}(z)\,\mathrm{d}z$$

$$= \frac{1}{2\pi\mathrm{i}}\int\limits_{\mathrm{Re}\,z=1} \sigma^{-z+1} b^{\xi}(z)\,\mathrm{d}z - \sum_{1/2<\mathrm{Re}\,z<1} \mathrm{Res}\,\{b^{\xi}(z)\,\sigma^{-z+1}\},$$

$$k^{\xi}(\sigma)\,(1-\sigma) = \frac{1}{2\pi\mathrm{i}}\int\limits_{\mathrm{Re}\,z=0} \sigma^{-z}\{b^{\xi}(z) - b^{\xi}(z+1)\}\,\mathrm{d}z + \frac{\mu}{\pi\mathrm{i}}(\sigma^{-\mathrm{i}\xi} - 1)\sum_{(\mu-1)/2\mu<k<(\mu+1)/2\mu} \sigma^{-z_k} \tag{2}$$

where

$$z_k := \begin{cases} k\mu - \dfrac{\mu-1}{2} & \text{if } k < 1/2, \\[2ex] k\mu - \dfrac{\mu+1}{2} & \text{if } k > 1/2. \end{cases}$$

Choosing $\sigma = 1$, we arrive at the equality

$$0 = \frac{1}{2\pi \mathrm{i}} \int\limits_{\operatorname{Re} z = 0} \{b^\xi(z) - b^\xi(z+1)\} \, \mathrm{d}z. \tag{3}$$

For our special choice of k, the operator A_1' takes the form

$$A_1' = \left(k^\xi \left(\frac{\tau_k^{(1)}}{t_j^{(1)}}\right) \frac{\Delta t_j^{(1)}}{t_j^{(1)}}\right)_{k,j=0}^\infty = \left(k^\xi \left(\frac{\tau_k^{(1)}}{t_j^{(1)}}\right) \left(1 - \frac{\tau_k^{(1)}}{t_j^{(1)}}\right) \frac{\Delta t_j^{(1)}}{t_j^{(1)} - \tau_k^{(1)}}\right)_{k,j=0}^\infty .$$

Thus (2) implies that

$$A_1' = \frac{1}{2\pi \mathrm{i}} \int\limits_{\operatorname{Re} z = 0} (\tau_k^{(1)-z}\delta_{kj})_{k,j=0}^\infty \left(\frac{\Delta t_j^{(n)}}{t_j^{(1)} - \tau_k^{(1)}}\right)_{k,j=0}^\infty (t_j^{(1)z}\delta_{kj})_{k,j=0}^\infty \{b^\xi(z) - b^\xi(z+1)\} \, \mathrm{d}z$$

$$+ \frac{\mu}{\pi \mathrm{i}} \sum_{(\mu-1)/2\mu < k < (\mu+1)/2\mu} \left(\left(\frac{\tau_k^{(1)}}{t_j^{(1)}}\right)^{-z_k} \left\{\left(\frac{\tau_k^{(1)}}{t_j^{(1)}}\right)^{-\mathrm{i}\xi} - 1\right\} \frac{\Delta t_j^{(1)}}{t_j^{(1)} - \tau_k^{(1)}}\right)_{k,j=0}^\infty .$$

Analogously to the proof of Theorem 11.3 we conclude that

$$A_1' = T + \frac{1}{2} \int\limits_{\operatorname{Re} z = 0} \{b^\xi(z) - b^\xi(z+1)\} \, \Lambda^{-z} T(f^{(\varepsilon - 1/2)}) \, \Lambda^z \, \mathrm{d}z$$

$$+ \mu \sum_{(\mu-1)/2\mu < k < (\mu+1)/2\mu} \{\Lambda^{-z_k - \mathrm{i}\xi_k} T(f^{(\varepsilon-1/2)}) \, \Lambda^{z_k + \mathrm{i}\xi_k} - \Lambda^{-z_k} T(f^{(\varepsilon-1/2)}) \, \Lambda^{z_k}\},$$

$$A_1' \in \mathfrak{A},$$

$$\operatorname{smb} A_1' = 1/2 \int\limits_{\operatorname{Re} z = 0} \mathcal{A}^z \{b^\xi(z) - b^\xi(z+1)\} \, \mathrm{d}z + \mu \sum_{(\mu-1)/2\mu < k < (\mu+1)/2\mu} \{\mathcal{A}^{z_k + \mathrm{i}\xi_k} - \mathcal{A}^{z_k}\},$$

where $\mathcal{A}^z := \operatorname{smb} \Lambda^{-z} T(f^{(\varepsilon-1/2)}) \, \Lambda^z$. This leads to the formula

$$\operatorname{smb} A_1' = 1/2 \left\{ \int\limits_{\operatorname{Re} z = 0} \mathcal{A}^z b^\xi(z) \, \mathrm{d}z - \int\limits_{\operatorname{Re} z = 1} \mathcal{A}^z b^\xi(z) \, \mathrm{d}z \right\}$$

$$+ \mu \sum_{(\mu-1)/2\mu < k < (\mu+1)/2\mu} \{\mathcal{A}^{z_k + \mathrm{i}\xi_k} - \mathcal{A}^{z_k}\}.$$

Now the residue theorem implies that

$$\operatorname{smb} A_1' = -\pi \mathrm{i} \sum_{0 < \operatorname{Re} z < 1} \operatorname{Res} \{\mathcal{A}^z b^\xi(z)\} + \mu \sum_{(\mu-1)/2\mu < k < (\mu+1)/2\mu} \{\mathcal{A}^{z_k + \mathrm{i}\xi_k} - \mathcal{A}^{z_k}\},$$

$$\operatorname{smb} A_1'(\tau, \nu) = \begin{cases} 0 & \text{if } \tau \neq 1, \\[1ex] b^\xi\left(\dfrac{1}{2} + \dfrac{\mathrm{i}}{2\pi} \log \dfrac{\nu}{1-\nu}\right) & \text{if } \tau = 1, \, 0 \leqq \nu \leqq 1. \end{cases}$$

Thus the symbol smb $A_1'(\tau, \nu)$ is given by (1) for the method 11.26(1) if $\alpha = 1$, $\varrho = 0$, $A = B_{\mathbf{R}^+}$ and $k = k^\xi$.

Step 2. Now take arbitrary $\alpha \geqq 1$, $-1/2 < \varrho < 1/2$ and let k satisfy (ii) and (iii) of Section 11.25. For the Mellin transform of k, we obtain

$$\alpha k(\sigma^\alpha)\,\sigma^{-\alpha\varrho+(\alpha-1)/2} = \frac{1}{2\pi i} \int\limits_{\mathrm{Re}\,z=1/2} \sigma^{-z} b\left(\frac{z+(\alpha-1)/2}{\alpha} - \varrho\right) dz.$$

Choosing a sufficiently small $\mu > 0$ and setting $\varrho' := \varrho + \mu/2\alpha$, $\varrho'' := \varrho - \mu/2\alpha$, we conclude that

$$\alpha k(\sigma^\alpha)\,\sigma^{-\alpha\varrho+(\alpha-1)/2} = \frac{1}{\mu} \frac{\mu\sigma^{\mu/2}}{1-\sigma^\mu} \{\alpha k(\sigma^\alpha)\,\sigma^{-\alpha\varrho'+(\alpha-1)/2} - \alpha k(\sigma^\alpha)\,\sigma^{-\alpha\varrho''+(\alpha-1)/2}\}$$

$$= \frac{1}{2\mu}\frac{1}{\pi i}\frac{\mu\sigma^{(\mu-1)/2}}{1-\sigma^\mu} \int\limits_{\mathrm{Re}\,z=0} \sigma^{-z}\left\{b\left(\frac{1}{2}-\varrho'+\frac{z}{\alpha}\right) - b\left(\frac{1}{2}-\varrho''+\frac{z}{\alpha}\right)\right\} dz.$$

In view of the equality

$$\int\limits_{\mathrm{Re}\,z=0} \left\{b\left(\frac{1}{2}-\varrho'+\frac{z}{\alpha}\right) - b\left(\frac{1}{2}-\varrho''+\frac{z}{\alpha}\right)\right\} dz = 2\pi i \sum \mathrm{Res} = 0$$

we get

$$\alpha k(\sigma^\alpha)\,\sigma^{-\alpha\varrho+(\alpha-1)/2} = \frac{1}{2\mu} \int\limits_{\mathrm{Re}\,z=0} k^{\mathrm{Im}\,z}(\sigma)\left\{b\left(\frac{1}{2}-\varrho'+\frac{z}{\alpha}\right) - b\left(\frac{1}{2}-\varrho''+\frac{z}{\alpha}\right)\right\} dz. \tag{4}$$

Since A_1' takes the form

$$A_1' = \left(k\left(\frac{(k+\varepsilon)^\alpha}{(j+1/2)^\alpha}\right)\left(\frac{k+\varepsilon}{j+1/2}\right)^{-\alpha\varrho+(\alpha-1)/2} \frac{\alpha(j+1/2)^{\alpha-1}}{(j+1/2)^\alpha}\right)_{k,j=0}^{\infty},$$

we arrive at the representation

$$A_1' = \frac{1}{2\mu} \int\limits_{\mathrm{Re}\,z=0} B^z\left\{b\left(\frac{1}{2}-\varrho'+\frac{z}{\alpha}\right) - b\left(\frac{1}{2}-\varrho''+\frac{z}{\alpha}\right)\right\} dz, \tag{5}$$

$$B^z := \left(k^{\mathrm{Im}\,z}\left(\frac{k+\varepsilon}{j+1/2}\right)\frac{1}{j+1/2}\right)_{k,j=0}^{\infty} \in \mathfrak{A}.$$

Note that the operator-valued function $\mathrm{Im}\,z \mapsto B^z \in \mathfrak{A}$ is bounded and continuous with respect to the operator norm (see step 1 of this proof and Lemma 11.4(iii)). Formula (5) yields that $A_1' \in \mathfrak{A}$ and

$$\mathrm{smb}\,A_1' = \frac{1}{2\mu} \int\limits_{\mathrm{Re}\,z=0} \mathrm{smb}\,B^z\left\{b\left(\frac{1}{2}-\varrho'+\frac{z}{\alpha}\right) - b\left(\frac{1}{2}-\varrho''+\frac{z}{\alpha}\right)\right\} dz.$$

Now we define the μ-periodic function $\mathfrak{A}^z$ by

$$\mathcal{A}^z(\tau, \mu) := \begin{cases} 0 & \text{if } \tau \neq 1 \\ -i\cot\left(\pi\left(\frac{1}{2}+\frac{i}{2\pi\mu}\log\frac{\nu}{1-\nu}-\frac{z}{\mu}\right)\right) + i\cot\left(\pi\left(\frac{1}{2}+\frac{i}{2\pi\mu}\log\frac{\nu}{1-\nu}\right)\right) \\ \quad \text{if } \tau = 1. \end{cases}$$

Then $\operatorname{Re} z = 0$ implies $\operatorname{smb} B^z = \mathcal{A}^z$ and we get

$$\operatorname{smb} A_1' = \frac{1}{2\mu}\left\{\int\limits_{\operatorname{Re} z = -\mu/2} \mathcal{A}^{z+\mu/2} b\left(\frac{1}{2} - \varrho + \frac{z}{\alpha}\right) dz - \int\limits_{\operatorname{Re} z = \mu/2} \mathcal{A}^{z+\mu/2} b\left(\frac{1}{2} - \varrho + \frac{z}{\alpha}\right) dz\right\}$$

$$= -\frac{\pi i}{\mu} \sum_{-\mu/2 < \operatorname{Re} z < \mu/2} \operatorname{Res}\left\{\mathcal{A}^{z+\mu/2} b\left(\frac{1}{2} - \varrho + \frac{z}{\alpha}\right)\right\}$$

$$= \begin{cases} b\left(\dfrac{1}{2} - \varrho + \dfrac{i}{2\pi\alpha} \log \dfrac{\nu}{1-\nu}\right) & \text{if } \tau = 1,\ 0 \leqq \nu \leqq 1, \\ 0 & \text{else}. \end{cases}$$

Consequently, the symbol of A_1' is given by (1).

Step 3. Using the functions f^0 and $f^{\#}$ instead of $f^{(\varepsilon-1/2)}$ (see Section 11.4), the methods 11.26(2) and 11.26(3) may be treated analogously. ■

11.28. Quadrature methods for Mellin operators on the interval. In the remainder of this chapter we shall consider operators of the form $A = a(t, \partial) + T$, where $a(t, \partial)$ is defined as in Section 11.25 and T as in Section 11.20. Our aim is to solve the equation $Au = f$ approximately. In the case where $c \equiv 1$, $d \equiv 0$ and the kernel k_B satisfies certain smoothness conditions, recently Chandler and Graham [4] proposed quadrature methods using simple composite quadrature rules. They proved the stability and established error estimates. We shall restrict our considerations to a special case of their methods applying the quadrature methods developed in Section 11.20. Contrary to Chandler and Graham [4], we shall examine the case of arbitrary continuous coefficients c and d. Moreover, their smoothness conditions for the kernel k_B will be replaced by the analiticity of the symbol $b(t, z)$ in a certain strip (cf. (ii) and (iii) of Section 11.25).

Let us retain the notation of Sections 11.20–11.21 and set $\alpha_0 = \alpha$, $\alpha_1 = 1$ and $A = a(t, \partial) + T$. The quadrature methods read as follows. We determine approximate values $\xi_j^{(n)}$ of $u(t_j^{(n)})$ by solving one of the systems

$$\left\{c(\tau_k^{(n)}) - i\cot\left(\pi\left(\varepsilon - \frac{1}{2}\right)\right) d(\tau_k^{(n)})\right\} \xi_k^{(n)} + d(\tau_k^{(n)}) \frac{1}{\pi i} \sum_{j=0}^{n-1} \frac{\xi_j^{(n)}}{t_j^{(n)} - \tau_k^{(n)}} \Delta t_j^{(n)}$$
$$+ \sum_{j=0}^{n-1} \{k_B(\tau_k^{(n)}, t_j^{(n)}) + k(\tau_k^{(n)}, t_j^{(n)})\} \xi_j^{(n)} \Delta t_j^{(n)} = f(\tau_k^{(n)}), \qquad k = 0, \dots, n-1, \tag{1}$$

$$c(t_k^{(n)}) \xi_k^{(n)} + d(t_k^{(n)}) \frac{1}{\pi i} \sum_{\substack{j=0 \\ j \neq k}}^{n-1} \frac{\xi_j^{(n)}}{t_j^{(n)} - t_k^{(n)}} \Delta t_j^{(n)} + \sum_{\substack{j=0 \\ j \neq k}}^{n-1} k_B(t_k^{(n)}, t_j^{(n)}) \xi_j^{(n)} \Delta t_j^{(n)}$$
$$+ \sum_{j=0}^{n-1} k(t_k^{(n)}, t_j^{(n)}) \xi_j^{(n)} \Delta t_j^{(n)} = f(t_k^{(n)}), \qquad k = 0, \dots, n-1, \tag{2}$$

$$c(t_k^{(n)}) \xi_k^{(n)} + d(t_k^{(n)}) \frac{1}{\pi i} \sum_{\substack{j=0 \\ j \equiv k+1 \bmod 2}}^{n-1} \frac{\xi_j^{(n)}}{t_j^{(n)} - t_k^{(n)}} \Delta t_j^{(n)}$$
$$+ \sum_{\substack{j=0 \\ j \equiv k+1 \bmod 2}}^{n-1} \{k_B(t_k^{(n)}, t_j^{(n)}) + k(t_k^{(n)}, t_j^{(n)})\} \xi_j^{(n)} \Delta t_j^{(n)} = f(t_k^{(n)}), \qquad k = 0, \dots, n-1, \tag{3}$$

where $\Delta t_j^{(n)} := \dfrac{1}{n} \gamma'\left(\dfrac{j + 1/2}{n}\right)$ for (1) or (2) and $\Delta t_j^{(n)} := \dfrac{2}{n} \gamma'\left(\dfrac{j + 1/2}{n}\right)$ for

Setting $u_n := \sum_{k=0}^{n-1} \xi_k^{(n)} \chi_k^{(n)}$, we obtain an approximate solution u_n to the equation $Au = f$.

Now, for each $\tau \in \mathbb{I}$, we consider a model problem of the quadrature methods. If $0 < \tau \leqq 1$, then the restriction of k_B to a neighborhood of τ is the kernel of a compact operator. Therefore, we define A_1^τ in the same manner as in Section 11.20. If $\tau = 0$, then we set $A^0 := c(0) + d(0)\, S_{\mathbb{R}^+} + B_{\mathbb{R}^+}$, where $B_{\mathbb{R}^+}$ denotes the Mellin convolution operator (cf. Section 11.26) with the kernel $k = k_0 := M^{-1}b(0, \cdot)$. For this operator A^0, we consider the method 11.26(1), 11.26(2) or 11.26(3), respectively, and define A_1^0 to be the matrix of the corresponding system. Thus we get $A_1^0 \in \mathscr{L}(\mathrm{l}^2_{(\alpha-1)/2-\varrho\alpha})$, $A_1^\tau \in \mathscr{L}(\tilde{\mathrm{l}}^2)$ $(0 < \tau < 1)$ and $A_1^1 \in \mathscr{L}(\mathrm{l}^2)$.

11.29. Theorem. (a) *Suppose the assumptions* (i)—(iii) *of Section* 11.25 *are satisfied. The method* 11.28(1) *or* 11.28(2), *respectively, is stable if and only if the operators* $A \in \mathscr{L}(\mathbf{L}^2_\varrho)$ *and* A_1^τ $(\tau \in \mathbb{I})$ *are invertible. The method* 11.28(3) *is stable if and only if the operators* $A, \tilde{A} := cI - dS_{\mathbb{I}} - B - T \in \mathscr{L}(\mathbf{L}^2_\varrho)$ *and* A_1^τ $(\tau \in \mathbb{I})$ *are invertible.*

(b) *If the quadrature method is stable and f is Riemann integrable, then the system* 11.28(1) 11.28(2) *or* 11.28(3), *respectively, is uniquely solvable for n large enough and the approximate solutions* u_n *converge to* $u = A^{-1}f$ *as* $n \to \infty$.

Proof. We denote the matrix of the system 11.28(1), 11.28(2) or 11.28(3), respectively, by A_n $\big(A_n \in \mathscr{L}(\operatorname{im} L_n)\big)$. Then we have to show that the assumptions of Theorem 11.22 or 11.23, respectively, are satisfied. Since all methods can be treated analogously, we limit ourselves to 11.28(2). The validity of assumption (ii) in Theorem 11.22 is well known and (i) follows from Section 11.15. Furthermore, by the proof of Theorem 11.21 it suffices to show (iii) and (iv) for the case $A = B$,

$$A_n = B_n := \left(k_{t_k^{(n)}}\left(\frac{t_k^{(n)}}{t_j^{(n)}}\right)\frac{\Delta t_j^{(n)}}{t_j^{(n)}}\right)_{k,j=0}^{n-1} \in \mathscr{L}\,(\operatorname{im} L_n)\ \big(k_t(1) := 0\big)$$

only.

To prove that $\|\tilde{\chi}_n B_n - B_n\tilde{\chi}_n - L_n(\chi B - B\chi)|_{\operatorname{im} L_n}\| \to 0$ we may bound our attention to the case $k_B(t, \tau) = k^\xi(t/\tau)\,(t/\tau)^\varrho\, 1/\tau$ (see the proof of Theorem 11.27). To see this, we define B^z by

$$(B^z x)\,(t) := \int_0^1 k^{\operatorname{Im} z}\left(\frac{t}{\tau}\right)\left(\frac{t}{\tau}\right)^\varrho \frac{x(\tau)}{\tau}\,\mathrm{d}\tau$$

and set $\varrho' := \varrho + \mu/2$, $\varrho'' := \varrho - \mu/2$ for a suitable small $\mu > 0$. Since B^z is a Mellin convolution operator whose symbol continuously depends on z, the function $\{\operatorname{Re} z = 0\} \ni z \mapsto B^z \in \mathscr{L}(\mathbf{L}^2_\varrho)$ is continuous. The arguments which led to equation 11.27(4) give

$$k_t(\sigma) = \frac{1}{2\mu}\int_{\operatorname{Re} z=0} k^{\operatorname{Im} z}(\sigma)\,\sigma^\varrho\left\{b\left(t, \frac{1}{2} - \varrho' + z\right) - b\left(t, \frac{1}{2} - \varrho'' + z\right)\right\}\mathrm{d}z,$$

$$(Bx)\,(t) = \int_0^\infty \frac{1}{2\mu}\int_0^\infty \left\{b\left(t, \frac{1}{2} - \varrho' + z\right) - b\left(t, \frac{1}{2} - \varrho'' + z\right)\right\} k^{\operatorname{Im} z}\left(\frac{t}{\tau}\right)\left(\frac{t}{\tau}\right)^\varrho \frac{x(\tau)}{\tau}\,\mathrm{d}z\,\mathrm{d}\tau.$$

Thus we arrive at the representation

$$B = \frac{1}{2\mu} \int_{\operatorname{Re} z = 0} \left\{ b\left(\cdot, \frac{1}{2} - \varrho' + z\right) - b\left(\cdot, \frac{1}{2} - \varrho'' + z\right) \right\} B^z \, dz. \tag{1}$$

An analogous reasoning yields that

$$B_n = \frac{1}{2\mu} \int_{\operatorname{Re} z = 0} \left\{ b\left(\cdot, \frac{1}{2} - \varrho' + z\right) - b\left(\cdot, \frac{1}{2} - \varrho'' + z\right) \right\}_n^{\sim} B_n^z \, dz, \tag{2}$$

where

$$\left\{ b\left(\cdot, \frac{1}{2} - \varrho' + z\right) - b\left(\cdot, \frac{1}{2} - \varrho'' + z\right) \right\}_n^{\sim}$$
$$:= \left(\left\{ b\left(t_j^{(n)}, \frac{1}{2} - \varrho' + z\right) - b\left(t_j^{(n)}, \frac{1}{2} - \varrho'' + z\right) \right\} \delta_{k,j} \right)_{k,j=0}^{n-1},$$
$$B_n^z := \left(k^{\operatorname{Im} z} \left(\frac{t_k^{(n)}}{t_j^{(n)}} \right) \left(\frac{t_k^{(n)}}{t_j^{(n)}} \right)^{\varrho} \frac{\Delta t_j^{(n)}}{t_j^{(n)}} \right)_{k,j=0}^{n-1}.$$

(Note that, for the method 11.28(2), one has to apply Theorem 11.22 with $\varepsilon = 0$. In particular, $\tilde{\chi}_n$ denotes the operator $K_n^0 \chi|_{\operatorname{im} L_n}$ (cf. Section 11.21).)

From (1) and (2) we obtain that

$$\tilde{\chi}_n B_n - B_n \tilde{\chi}_n - L_n(\chi B - B\chi)|_{\operatorname{im} L_n}$$
$$= \frac{1}{2\mu} \int_{\operatorname{Re} z = 0} \left\{ b\left(\cdot, \frac{1}{2} - \varrho' + z\right) - b\left(\cdot, \frac{1}{2} - \varrho'' + z\right) \right\}_n^{\sim} \left[\tilde{\chi}_n B_n^z - B_n^z \tilde{\chi}_n - L_n \right.$$
$$\left. \times (\chi B^z - B^z \chi)|_{\operatorname{im} L_n} \right] dz + \frac{1}{2\mu} \int_{\operatorname{Re} z = 0} \left[\left\{ b\left(\cdot, \frac{1}{2} - \varrho' + z\right) - b\left(\cdot, \frac{1}{2} - \varrho'' + z\right) \right\}_n^{\sim} \right.$$
$$\left. \times L_n - L_n \left\{ b\left(\cdot, \frac{1}{2} - \varrho' + z\right) - b\left(\cdot, \frac{1}{2} - \varrho'' + z\right) \right\} \right] (\chi B^z - B^z \chi)|_{\operatorname{im} L_n} \, dz.$$

The second term on the right-hand side tends to 0 since $\|\tilde{\psi}_n - L_n \psi|_{\operatorname{im} L_n}\| \to 0$ and $\|L_n \psi(I - L_n)\| \to 0$ for arbitrary $\psi \in \mathbf{C}(\mathbb{I})$ as $n \to \infty$. Consequently, to show that $\|\tilde{\chi}_n B_n - B_n \tilde{\chi}_n - L_n(\chi B - B\chi)|_{\operatorname{im} L_n}\| \to 0$ we are left with verifying that $\|\tilde{\chi}_n B_n^z - B_n^z \tilde{\chi}_n - L_n(\chi B^z - B^z \chi)|_{\operatorname{im} L_n}\|$ converges to 0 uniformly for all z from a compact subset of $\{\operatorname{Re} z = 0\}$.

Now equation 11.27(2) implies that

$$(B^z x)(t) = \int_0^1 \frac{1}{2} \int_{\operatorname{Re} \zeta = 0} \{ b^{\operatorname{Im} z}(\zeta) - b^{\operatorname{Im} z}(\zeta + 1) \} \frac{1}{\pi i} \frac{(t/\tau)^{-\zeta + \varrho}}{\tau - t} x(\tau) \, d\zeta \, d\tau$$
$$+ \mu \sum_{(\mu - 1)/2\mu < k < (\mu + 1)/2\mu} \int_0^1 \frac{1}{\pi i} \frac{(t/\tau)^{-z_k + \varrho - z} - (t/\tau)^{-z_k + \varrho}}{\tau - t} x(\tau) \, d\tau,$$

$$B^z = \frac{1}{2} \int_{\operatorname{Re} \zeta = 0} \{ b^{\operatorname{Im} z}(\zeta) - b^{\operatorname{Im} z}(\zeta + 1) \} t^{-\zeta + \varrho} S_{\mathbb{I}} t^{\zeta - \varrho} \, d\zeta$$
$$+ \mu \sum_{(\mu - 1)/2\mu < k < (\mu + 1)/2\mu} \{ t^{-z_k + \varrho - z} S_{\mathbb{I}} t^{z_k - \varrho + z} - t^{-z_k + \varrho} S_{\mathbb{I}} t^{z_k - \varrho} \}. \tag{3}$$

If S_n denotes the operator defined in Section 24 then 11.27(2) yields

$$B_n^z = \frac{1}{2} \int\limits_{\mathrm{Re}\zeta=0} \{b^{\mathrm{Im}z}(\zeta) - b^{\mathrm{Im}z}(\zeta+1)\}\, D_n^{-\zeta+\varrho} S_n D_n^{\zeta-\varrho}\, d\zeta$$
$$+ \mu \sum_{(\mu-1)/2\mu<k<(\mu+1)/2\mu} \{D_n^{-z_k+\varrho-z} S_n D_n^{z_k-\varrho+z} - D_n^{-z_k+\varrho} S_n D_n^{z_k-\varrho}\}, \tag{4}$$
$$D_n^\xi := \left((t_j^{(n)})^\xi\, \delta_{k,j}\right)_{k,j=0}^{n-1}.$$

Taking into account that

$$\tilde{\chi}_n B_n^z - B_n^z \tilde{\chi}_n - L_n(\chi B^z - B^z\chi)|_{\mathrm{im}L_n}$$
$$= \frac{1}{2} \int\limits_{\mathrm{Re}\zeta=0} \{b^{\mathrm{Im}z}(\zeta) - b^{\mathrm{Im}z}(\zeta+1)\} \left[D_n^{-\zeta+\varrho}(\tilde{\chi}_n S_n - S_n\tilde{\chi}_n)\, D_n^{\zeta-\varrho}\right.$$
$$\left. - L_n t^{-\zeta+\varrho}(\chi S_{\mathbb{I}} - S_{\mathbb{I}}\chi)\, t^{\zeta-\varrho}|_{\mathrm{im}L_n}\right] d\zeta$$
$$+ \mu \sum_{(\mu-1)/2\mu<k<(\mu+1)/2\mu} \left[D_n^{-z_k+\varrho-z}(\tilde{\chi}_n S_n - S_n\tilde{\chi}_n)\, D_n^{z_k-\varrho+z}\right.$$
$$\left. - L_n t^{-z_k+\varrho-z}(\chi S_{\mathbb{I}} - S_{\mathbb{I}}\chi)\, t^{z_k-\varrho+z}|_{\mathrm{im}L_n}\right]$$
$$- \mu \sum_{(\mu-1)/2\mu<k<(\mu+1)/2\mu} \left[D_n^{-z_k+\varrho}(\tilde{\chi}_n S_n - S_n\tilde{\chi}_n)\, D_n^{z_k-\varrho}\right.$$
$$\left. - L_n t^{-z_k+\varrho}(\chi S_{\mathbb{I}} - S_{\mathbb{I}}\chi)\, t^{z_k-\varrho}|_{\mathrm{im}L_n}\right],$$

it remains to show that $\|D_n^{-\zeta+\varrho}(\tilde{\chi}_n S_n - S_n\tilde{\chi}_n)\, D_n^{\zeta-\varrho} - L_n t^{-\zeta+\varrho}(\chi S_{\mathbb{I}} - S_{\mathbb{I}}\chi)\, t^{\zeta-\varrho}|_{\mathrm{im}L_n}\|$ converges to 0 uniformly for all ζ from a compact subset of $\{\mathrm{Re}\,\zeta = 0\} \cup \{z_k + z, z_k\colon (\mu-1)/2\mu < k < (\mu+1)/2\mu\}$.

The first part of the proof to Theorem 11.21 implies that

$$\|\tilde{\chi}_n S_n - S_n\tilde{\chi}_n - L_n(\chi S_{\mathbb{I}} - S_{\mathbb{I}}\chi)|_{\mathrm{im}L_n}\|_{\mathscr{L}(\mathbf{L}^2_{\mathrm{Re}\zeta})} \to 0, \ -1/2 < \mathrm{Re}\,\zeta < 1/2,$$
$$\|D_n^{-\zeta+\varrho}(\tilde{\chi}_n S_n - S_n\tilde{\chi}_n - L_n(\chi S_{\mathbb{I}} - S_{\mathbb{I}}\chi)|_{\mathrm{im}L_n})\, D_n^{\zeta-\varrho}\|_{\mathscr{L}(\mathbf{L}^2_\varrho)} \to 0.$$

Therefore, it suffices to show that

$$\|D_n^{-\zeta+\varrho} L_n(\chi S_{\mathbb{I}} - S_{\mathbb{I}}\chi)|_{\mathrm{im}L_n}\, D_n^{\zeta-\varrho} - L_n t^{-\zeta+\varrho}(\chi S_{\mathbb{I}} - S_{\mathbb{I}}\chi)\, t^{\zeta-\varrho}|_{\mathrm{im}L_n}\| \to 0. \tag{5}$$

Since $(\chi S_{\mathbb{I}} - S_{\mathbb{I}}\chi)$ is compact, we may find a characteristic function ψ such that $\mathrm{supp}\,\psi \subset (0, 1)$ and the norm of $\psi(\chi S_{\mathbb{I}} - S_{\mathbb{I}}\chi)\,\psi - (\chi S_{\mathbb{I}} - S_{\mathbb{I}}\chi)$ becomes smaller than an arbitrarily prescribed positive number. For $\varepsilon' > 0$, we obtain

$$\|D_n^{-\zeta+\varrho} L_n\{\psi(\chi S_{\mathbb{I}} - S_{\mathbb{I}}\chi)\,\psi - (\chi S_{\mathbb{I}} - S_{\mathbb{I}}\chi)\}|_{\mathrm{im}L_n}\, D_n^{\zeta-\varrho}\| < \varepsilon',$$
$$\|L_n t^{-\zeta+\varrho}\{\psi(\chi S_{\mathbb{I}} - S_{\mathbb{I}}\chi)\,\psi - (\chi S_{\mathbb{I}} - S_{\mathbb{I}}\chi)\}\, t^{\zeta-\varrho}|_{\mathrm{im}L_n}\| < \varepsilon'.$$

These estimates together with the obvious convergences

$$\|\{D_n^{-\zeta+\varrho} L_n - L_n t^{-\zeta+\varrho}\}\,\psi\|_{\mathscr{L}(\mathbf{L}^2_{\mathrm{Re}\zeta}, \mathbf{L}^2_\varrho)} \to 0, \quad n \to \infty,$$
$$\|\psi\{D_n^{+\zeta-\varrho} L_n - t^{\zeta-\varrho} L_n\}\|_{\mathscr{L}(\mathbf{L}^2_\varrho, \mathbf{L}^2_{\mathrm{Re}\zeta})} \to 0, \quad n \to \infty$$

yield (5). Thus the validity of assumption (iii) is proved.

Now let $\tau = 0$ and consider (iv). The first steps in the proof of (iv) are similar to those ones in the proof of (iii). As $A = B$, we get

$$A_1^0 = \left(k_0\left(\frac{(k+1/2)^\alpha}{(j+1/2)^\alpha}\right)\frac{\alpha(j+1/2)^{\alpha-1}}{(j+1/2)^\alpha}\right)_{k,j=0}^{\infty} \in \mathcal{L}(l^2_{(\alpha-1)/2-\varrho\alpha})$$

and analogously to (2) we arrive at the formulas

$$A_1^0 = \frac{1}{2\mu}\int_{\operatorname{Re}z=0}\{b(0,1/2-\varrho'+z) - b(0,1/2-\varrho''+z)\}\,B_1^z\,\mathrm{d}z, \tag{6}$$

$$B_1^z := \left(k^{\operatorname{Im}z}\left(\frac{(k+1/2)^\alpha}{(j+1/2)^\alpha}\right)\frac{\alpha(j+1/2)^{\alpha-1}}{(j+1/2)^\alpha}\,\frac{(k+1/2)^{\alpha\varrho}}{(j+1/2)^{\alpha\varrho}}\right)_{k,j=0}^{\infty}.$$

Here we have used that the integrand is continuous. The continuity of $z \mapsto B_1^z$ follows from Lemma 11.4(iii), since the proof of Theorem 11.27 yields that $\left((k+1/2)^{(\alpha-1)/2-\alpha\varrho}\,\delta_{k,j}\right) \times B_1^z\left((j+1/2)^{-(\alpha-1)/2+\alpha\varrho}\,\delta_{k,j}\right) \in \mathfrak{A}$. Equations (2) and (6) lead to the representation

$$\begin{aligned}\tilde{\chi}_n^E(A_1^0 - B_n^E)\,\tilde{\chi}_n^E &= \frac{1}{2\mu}\int_{\operatorname{Re}z=0}\Big[\{b(0,1/2-\varrho'+z) - b(0,1/2-\varrho''+z)\}\\ &\quad - \left\{b\left(\cdot,\frac{1}{2}-\varrho'+z\right) - b\left(\cdot,\frac{1}{2}-\varrho''+z\right)\right\}_n^E\Big]\,\tilde{\chi}_n^E(B_n^z)^E\,\tilde{\chi}_n^E\,\mathrm{d}z\\ &\quad + \frac{1}{2\mu}\int_{\operatorname{Re}z=0}\left\{b\left(0,\frac{1}{2}-\varrho'+z\right) - b\left(0,\frac{1}{2}-\varrho''+z\right)\right\}\tilde{\chi}_n^E\\ &\quad \times \left(B_1^z - (B_n^z)^E\right)\tilde{\chi}_n^E\,\mathrm{d}z.\end{aligned}$$

The first term on the right-hand side becomes small since $\left\|\left(\psi_n^E - \psi(0)\,I_n\right)\tilde{\chi}_n^E\right\|$ is smaller than any prescribed positive number for χ vanishing outside a suitable neighborhood of 0. Consequently, to prove that $\|\tilde{\chi}_n^E(A_1^0 - B_n^E)\,\tilde{\chi}_n^E\| < \varepsilon'$ it suffices to show that $\left\|\tilde{\chi}_n^E\left(B_1^z - (B_n^z)^E\right)\tilde{\chi}_n^E\right\| < \varepsilon'$.

Analogously to (4) we obtain

$$\begin{aligned}B_1^z &= \frac{1}{2}\int_{\operatorname{Re}\zeta=0}\{b^{\operatorname{Im}z}(\zeta) - b^{\operatorname{Im}z}(\zeta+1)\}\left((k+1/2)^{-\zeta+\varrho}\,\delta_{k,j}\right)_{k,j=0}^{\infty}\\ &\quad \times R\left((j+1/2)^{\zeta-\varrho}\,\delta_{k,j}\right)_{k,j=0}^{\infty}\,\mathrm{d}\zeta\ +\\ &\quad \mu\sum_{(\mu-1)/2\mu<k<(\mu+1)/2\mu}\left\{\left((k+1/2)^{-z_k+\varrho-z}\,\delta_{k,j}\right)_{k,j=0}^{\infty}R\left((j+1/2)^{z_k-\varrho+z}\,\delta_{k,j}\right)_{k,j=0}^{\infty}\right.\\ &\quad \left. - \left((k+1/2)^{-z_k+\varrho}\,\delta_{k,j}\right)_{k,j=0}^{\infty}R\left((j+1/2)^{z_k-\varrho}\,\delta_{k,j}\right)_{k,j=0}^{\infty}\right\},\end{aligned}$$

$$R := \left(\frac{\alpha(j+1/2)^{\alpha-1}}{(j+1/2)^\alpha - (k+1/2)^\alpha}\,\frac{(k+1/2)^{\alpha\varrho}}{(j+1/2)^{\alpha\varrho}}\right)_{k,j=0}^{\infty}.$$

The latter equation and (4) imply that

$$\tilde{\chi}_n^E\left(B_1^z - (B_n^z)^E\right)\tilde{\chi}_n^E = \frac{1}{2}\int\limits_{\mathrm{Re}\zeta=0}\{b^{\mathrm{Im}z}(\zeta) - b^{\mathrm{Im}z}(\zeta+1)\}\,\tilde{\chi}_n^E\left[\left(\left(k+\frac{1}{2}\right)^{-\zeta+\varrho}\delta_{k,j}\right)_{k,j=0}^{\infty}\right.$$

$$\times R\left(\left(j+\frac{1}{2}\right)^{\zeta-\varrho}\delta_{k,j}\right)_{k,j=0}^{\infty} - \left.\{D_n^{-\zeta+\varrho}S_nD_n^{\zeta-\varrho}\}^E\right]\tilde{\chi}_n^E\,\mathrm{d}\zeta$$

$$+\,\mu\sum_{(\mu-1)/2\mu<k<(\mu+1)/2}\tilde{\chi}_n^E\left[\left(\left(k+\frac{1}{2}\right)^{-z_k+\varrho-z}\delta_{k,j}\right)_{k,j=0}^{\infty}\right.$$

$$\times R\left(\left(j+\frac{1}{2}\right)^{z_k-\varrho+z}\delta_{k,j}\right)_{k,j=0}^{\infty}$$

$$-\left.\{D_n^{-z_k+\varrho-z}S_nD_n^{z_k-\varrho+z}\}^E\right]\tilde{\chi}_n^E$$

$$-\,\mu\sum_{(\mu-1)/2\mu<k<(\mu+1)/2\mu}\tilde{\chi}_n^E\left[\left(\left(k+\frac{1}{2}\right)^{-z_k+\varrho}\delta_{k,j}\right)_{k,j=0}^{\infty}\right.$$

$$\times R\left(\left(j+\frac{1}{2}\right)^{z_k-\varrho}\delta_{k,j}\right)_{k,j=0}^{\infty} - \left.\{D_n^{-z_k+\varrho}S_nD_n^{z_k-\varrho}\}^E\right]\tilde{\chi}_n^E.$$

Thus $\left\|\tilde{\chi}_n^E\left(B_1^z - (B_n^z)^E\right)\tilde{\chi}_n^E\right\| < \varepsilon'$ follows from the inequality

$$\left\|\tilde{\chi}_n^E\left[\left(\left(k+\frac{1}{2}\right)^{-\zeta+\varrho}\delta_{k,j}\right)_{k,j}R\left(\left(j+\frac{1}{2}\right)^{\zeta-\varrho}\delta_{k,j}\right)_{k,j} - \{D_n^{-\zeta+\varrho}S_nD_n^{\zeta-\varrho}\}^E\right]\tilde{\chi}_n^E\right\| < \varepsilon',$$

which can be shown by using the same arguments as in the second part of Sect. 11.24.

If $\tau \in (0, 1]$, then $A_1^\tau = 0$ for $A = B$. By (2) we get

$$\tilde{\chi}_nB_n\tilde{\chi}_n = \frac{1}{2\mu}\int\limits_{\mathrm{Re}z=0}\left\{b\left(\cdot,\frac{1}{2}-\varrho'+z\right) - b\left(\cdot,\frac{1}{2}-\varrho''+z\right)\right\}_n^{\sim}\tilde{\chi}_nB_n^z\tilde{\chi}_n\,\mathrm{d}z. \qquad (7)$$

Obviously, $\chi B^z\chi$ is an integral operator with a continuous kernel. Analogously to the estimate $\|\tilde{\chi}_nT_n\tilde{\chi}_n\| < \varepsilon'$ in the proof of Theorem 11.21, we obtain $\|\tilde{\chi}_n^E(B_n^z)^E\,\tilde{\chi}_n^E\| < \varepsilon'$. Together with (7) this yields

$$\|\tilde{\chi}_n^EB_n^E\tilde{\chi}_n^E\| < \varepsilon'$$

if the support of $\chi \in \mathbf{M}(\mathbb{I})$ is contained in a suitable small neighborhood of $\tau \in (0, 1]$. ■

11.30. Remark. If A_1^0 is Fredholm of index 0 but not invertible, then the quadrature method can be modified so that it becomes stable. This modification was proposed and studied by Chandler and Graham [4] and applies to all methods described in the present paper. For example, let us consider the method 11.28(2).

We choose a natural number i_0 independent of n and determine approximate values $\xi_j^{(n)}$ of $u(t_j^{(n)})$ $(j = i_0, i_0+1, \ldots, n-1)$ by solving the system

$$c(t_k^{(n)})\,\xi_k^{(n)} + d(t_k^{(n)})\,\frac{1}{\pi\mathrm{i}}\sum_{\substack{j=i_0\\ j\neq k}}^{n-1}\frac{\xi_j^{(n)}}{t_j^{(n)}-t_k^{(n)}}\,\Delta t_j^{(n)} + \sum_{\substack{j=i_0\\ j\neq k}}^{n-1}k_B(t_k^{(n)}, t_j^{(n)})\,\xi_j^{(n)}\,\Delta t_j^{(n)}$$

$$+\sum_{j=i_0}^{n-1}k(t_k^{(n)}, t_j^{(n)})\,\xi_j^{(n)}\Delta t_j^{(n)} = f(t_k^{(n)}), \qquad k = i_0, i_0+1, \ldots, n-1. \qquad (1)$$

Setting $u_n := \sum_{k=i_0}^{n-1} \xi_k^{(n)} \chi_k^{(n)}$, we obtain an approximate solution u_n to $Au = f$. If the operators A and A_1^τ, $0 < \tau \leqq 1$, are invertible (cf. Theorems 11.3 and 11.19) and A_1^0 is Fredholm of index 0 (cf. Theorem 11.27) then there exists a natural number i_0 such that the modified method (1) is stable.

The proof of this assertion runs analogously to that one of the stability of the method 11.28(2). The only difference is that $A_1^{0\prime} = (a_{kj})_{k,j=0}^\infty \in \mathfrak{A}$ has to be replaced by $A_1^{0\prime\prime} := (a_{k+i_0,j+i_0})_{k,j=0}^\infty$. The proof of Theorem 11.27 yields that $A_1^{0\prime\prime} \in \mathfrak{A}$ and $\operatorname{smb} A_1^{0\prime\prime} = \operatorname{smb} A_1^{0\prime}$. Therefore, we only have to show the invertibility of $A_1^{0\prime\prime}$ for a suitable i_0.

The invertibility of $A_1^{0\prime\prime}$ is equivalent to the invertibility of $Q_{i_0} A_1^{0\prime} Q_{i_0} \in \mathcal{L}(\operatorname{im} Q_{i_0})$, where $Q_{i_0} := I - P_{i_0}$ and $P_{i_0} \in \mathcal{L}(l^2)$ is defined by

$$P_{i_0}\{\xi_j\}_{j=0}^\infty = \{\eta_j\}_{j=0}^\infty, \qquad \eta_j := \begin{cases} \xi_j & \text{if } j \leqq i_0, \\ 0 & \text{else.} \end{cases}$$

S. Roch proved in his papers [2, 4] that $Q_{i_0} A_1^{0\prime} Q_{i_0}$ is invertible for all i_0 large enough provided $A_1^{0\prime}$ is a Fredholm operator of index zero and some other operators are invertible. These other operators are Mellin convolution operators on the finite interval. Thus the Fredholm property as well as the fact that the index vanishes follow by Sec. 11.25. The invertibility is a consequence of the equivalence of Mellin type convolution operators on the interval and Wiener-Hopf integral operators on the half axis (cf. also Sec. 12.46).

Chapter 12
Spline approximation and quadrature methods for singular integral equations on an interval. Graded meshes

In the present chapter we investigate Galerkin and collocation methods with splines for locally strongly elliptic singular integral equations on an interval. As trial functions we utilize smoothest polynomial splines multiplied by a weight function on arbitrary partitions as well as on graded meshes. We provide error estimates in certain Sobolev norms. Sections 12.36 – 12.57 are concerned with a modification of a quadrature method presented in Chapter 11. The highest rate of convergence achieved by this method is $O(h^4)$ in a weighted $\mathbf{L}^2$-norm. Moreover, we give estimates for the speed of convergence in Sobolev norms.

Galerkin methods

The subject of this section is convergence analysis of Galerkin methods with splines for locally strongly elliptic singular integral equations on an interval. As trial functions we utilize smoothest polynomial splines multiplied by a weight function on arbitrary meshes as well as on special nonuniform partitions. Using the complete asymptotics of solutions at the endpoints of the interval and inequalities of Gårding type for singular integral operators in weighted $\mathbf{L}^2$ spaces, we provide error estimates in certain Sobolev norms. For brevity all results of this section and of the subsequent sections as well will be proved for the simplest singular integral equation, since they easily extend to the complete equation (see 11.1(1)) by standard perturbation theorems (cf. Chap. 1).

12.1. Notation. Consider the singular integral operator

$$A := aI + bS, \qquad (Su)\,(t) := \frac{1}{\pi \mathrm{i}} \int\limits_0^1 \frac{u(\tau)}{\tau - t}\,\mathrm{d}\tau, \tag{1}$$

where a and b are continuous functions on the interval $\mathbb{I} = [0, 1]$. Let $-1/2 < \beta_0$ $\beta_1 < 1/2$ and $\varrho(x) = x^{\beta_0}(1 - x)^{\beta_1}$. Then $\mathbf{L}^2(\varrho) := \mathbf{L}^2(\mathbb{I}, \varrho)$, i.e.

$$\mathbf{L}^2(\varrho) = \varrho^{-1}\mathbf{L}^2, \qquad \mathbf{L}^2 = \mathbf{L}^2(0, 1),$$

is a Hilbert space with the scalar product

$$\langle u, v\rangle := (\varrho u, \varrho v) = \int\limits_0^1 \varrho^2(x)\, u(x)\, \overline{v(x)}\,\mathrm{d}x,$$

and A belongs to $\mathscr{L}(\mathbf{L}^2(\varrho))$.

Assume the condition

$$a(t) + \lambda b(t) \neq 0, \qquad t \in [0, 1], \qquad \lambda \in [-1, 1] \tag{2}$$

to be satisfied. Notice that (2) is equivalent to either of the following conditions (see Corollary 6.23).

(i) There exists a function $\theta \in \mathbf{C}[0, 1]$ satisfying

$$\operatorname{Re} \theta(t) \, [a(t) \pm b(t)] > 0, \qquad t \in [0, 1]. \tag{3}$$

(ii) The operator A is locally strongly elliptic in $\mathbf{L}^2$, i.e. there exists an invertible function $\theta \in \mathbf{C}[0, 1]$ such that $\theta A = D + T$, where $T \in \mathcal{K}(\mathbf{L}^2)$,

$$\operatorname{Re} (Dw, w) \geqq c(w, w), \qquad w \in \mathbf{L}^2, \tag{4}$$

and c is some positive constant independent of w.

Under the hypothesis (2), we have $a(t) \pm b(t) \neq 0$ for all $t \in [0, 1]$ and $-1/2 < \operatorname{Re} \varkappa_j < 1/2$ $(j = 0, 1)$ by virtue of (3), where

$$\varkappa_j = \frac{(-1)^j}{2\pi \mathrm{i}} \log \frac{a(j) + b(j)}{a(j) - b(j)} \qquad (j = 0, 1) \tag{5}$$

and log denotes that branch of the logarithm which is continuous in $\mathbb{C} \setminus (-\infty, 0]$ and takes real values on the positive real axis. Thus A is invertible in $\mathbf{L}^2$. Moreover, A is invertible in $\mathbf{L}^2(\varrho)$ if

$$|\operatorname{Re} \varkappa_j + \beta_j| < 1/2, \qquad j = 0, 1 \tag{6}$$

(see Section 6.11). Both conditions (2) and (6) together are necessary and sufficient for the operator A to be locally strongly elliptic in $\mathbf{L}^2(\varrho)$, i.e. for the existence of an invertible function $\vartheta \in \mathbf{C}[0, 1]$ and an operator $T' \in \mathcal{K}(\mathbf{L}^2(\varrho))$ such that $\vartheta A = D' + T'$, where

$$\operatorname{Re} \langle D'u, u \rangle \geqq c\langle u, u \rangle, \qquad u \in \mathbf{L}^2(\varrho).$$

(see Corollary 6.23).

12.2. Standard Galerkin method with splines. For any real number $k \geqq 0$, let $\mathbf{H}^k$ denote the usual Sobolev space of order k on $[0, 1]$ and $\|\cdot\|_k$ the norm in $\mathbf{H}^k$. Furthermore, let $\tilde{\mathbf{H}}^k$ be the closed subspace $\{u \in \mathbf{H}^k(\mathbb{R}) : \operatorname{supp} u \subset [0, 1]\}$ of $\mathbf{H}^k(\mathbb{R})$. $\mathbf{H}^k$ and $\tilde{\mathbf{H}}^k$ $(k \geqq 0)$ are Hilbert scales (cf. e.g. LIONS and MAGENES [1]). Note that $\tilde{\mathbf{H}}^k = \mathbf{H}^k$ when $k < 1/2$ and that $\tilde{\mathbf{H}}^k$ is a subspace of $\mathbf{H}^k$ when $k \geqq 1/2$. The topology of $\tilde{\mathbf{H}}^k$, however, is stronger than that of $\mathbf{H}^k$ if $k - 1/2 \in \mathbb{N}$. We introduce (by duality with respect to the $\mathbf{L}^2$ scalar product $(\cdot, \cdot)$) the dual spaces $\mathbf{H}^{-k} := (\mathbf{H}^k)'$ and $\tilde{\mathbf{H}}^{-k} := (\tilde{\mathbf{H}}^k)'$, $k \geqq 0$.

Let $\mathcal{S}^\delta(\Delta)$, $\delta \in \mathbb{N}_0$, be the space of $\delta - 1$ times continuously differentiable splines of degree δ subordinate to the partition $\Delta = \{x_0 = 0 < x_1 < \cdots < x_n = 1\}$ of the interval $[0, 1]$. We have $\mathcal{S}^\delta(\Delta) \subset \mathbf{H}^s$ if and only if $s < \delta + 1/2$. Let $h = \max\limits_k (x_k - x_{k-1})$ and $\underline{h} = \min\limits_k (x_k - x_{k-1})$, $1 \leqq k \leqq n$. A mesh Δ is said to be *γ-quasiuniform* $(\gamma > 0)$ if $h < \gamma \underline{h}$, and the set of all γ-quasiuniform meshes is denoted by Part_γ.

Let A be the singular integral operator 12.1(1), and let $\varkappa_j$ $(j = 0, 1)$ be the numbers defined by 12.1(5). The standard Galerkin method with splines for solving the equation $Au = f$ consists in finding a spline $u_\Delta \in \mathcal{S}^\delta(\Delta)$ satisfying the Galerkin equations

$$(Au_\Delta, v_\Delta) = (f, v_\Delta) \quad \text{for all} \quad v_\Delta \in \mathcal{S}^\delta(\Delta). \tag{1}$$

12.3. Theorem. *Assume* 12.1(2). *If h is sufficiently small, then the Galerkin equations* 12.2(1) *are uniquely solvable for every $f \in \mathbf{L}^2$, and the approximate solutions u_Δ converge in $\mathbf{L}^2$ to the exact solution u with the error bound*

$$\|u - u_\Delta\|_0 \leqq c \inf_{v \in \mathcal{S}^\delta(\Delta)} \|u - v\|_0 . \tag{1}$$

If, in addition, $a, b, f \in \mathbf{H}^1$ and, in the case $s > 0$, $\Delta \in \mathrm{Part}_\gamma$, then

$$\|u - u_\Delta\|_s \leqq ch^{t-s} \|u\|_t \tag{2}$$

for any s and t satisfying $0 \leqq t < \min\{\mathrm{Re}\,\varkappa_0, \mathrm{Re}\,\varkappa_1\} + 1/2$, $\max\{\mathrm{Re}\,\varkappa_0, \mathrm{Re}\,\varkappa_1\} - 1/2 < s \leqq t$ and $s < \delta + 1/2$.

Proof. We apply standard techniques for Galerkin's method with finite elements. It follows from the proof of Theorem 10.11, that the Galerkin method 12.2(1) is stable in $\mathbf{L}^2$ provided A is locally strongly elliptic in $\mathbf{L}^2$. Thus, the first part of the theorem is a consequence of Cea's Lemma (cf. Section 1.33).

If $a, b, f \in \mathbf{H}^1$, then Corollary 6.30 implies that $u \in \mathbf{H}^t$. By estimate (1) and the approximation property of $\mathcal{S}^\delta(\Delta)$ (which can be proved similarly to the case of periodic Sobolev spaces, see Secs. 2.6–2.10 and Elschner [1]), we obtain (2) for $s = 0$. Furthermore, (1) and the Aubin-Nitsche Lemma (see Section 1.34) imply estimate (2) in the case $s < 0$, since $A \in \mathcal{L}(\mathbf{H}^s, \overline{\mathbf{H}}^s)$ and $A^* \in \mathcal{L}(\overline{\mathbf{H}}^{-s}, \mathbf{H}^{-s})$ are invertible operators in view of Theorem 6.29. For $s > 0$, estimate (2) is a consequence of (1) and Corollary 1.35. ■

Remark. In formula (2) we always have $t - s < 1$ and, for constant coefficients a and b, even $t - s < 1 - 2\,|\mathrm{Re}\,\varkappa|$, where $\varkappa = \varkappa_0 = -\varkappa_1$. Moreover, for higher $\delta (\delta > 1)$ there is no increase of the order of convergence, in general.

12.4. Galerkin methods with weighted splines. We now derive improved error estimates for Galerkin's method using smoothest splines multiplied by a weight function which reflects the principal term of the asymptotics of the solution at the endpoints of the interval. This Galerkin method is non-standard in the sense that the pairing of test and trial functions is made in the scalar product of $\mathbf{L}^2(\varrho)$. Retaining the notation of the preceding sections, we set $\varrho_0(x) = x^{\varkappa_0}(1 - x)^{\varkappa_1}$ and $\tilde{\mathcal{S}}^\delta(\Delta) = \varrho_0 \mathcal{S}^\delta(\Delta)$. Let further $\varrho(x) = x^{\beta_0}(1 - x)^{\beta_1}$ be a weight function satisfying 12.1(6). Then, under the hypothesis 12.1(2), the singular integral operator 12.1(1) is invertible in $\mathbf{L}^2(\varrho)$.

We now approximate the solution $u \in \mathbf{L}^2(\varrho)$ of the equation $Au = f$, $f \in \mathbf{L}^2(\varrho)$, by the solutions $u_\Delta \in \tilde{\mathcal{S}}^\delta(\Delta)$ of the Galerkin equations

$$\langle Au_\Delta, v_\Delta \rangle = \langle f, v_\Delta \rangle \quad \text{for all} \quad v_\Delta \in \tilde{\mathcal{S}}^\delta(\Delta), \tag{1}$$

i.e.

$$(\varrho A u_\Delta, \varrho v_\Delta) = (\varrho f, \varrho v_\Delta) \quad \text{for all} \quad v_\Delta \in \tilde{\mathcal{S}}^\delta(\Delta). \tag{2}$$

Note that the scalar products in (1) are well-defined, since $u_\Delta, v_\Delta \in \mathbf{L}^2(\varrho)$ by virtue of 12.1(6). Let

$$r = \max\{0, -\mathrm{Re}\,\varkappa_0 - \beta_0, -\mathrm{Re}\,\varkappa_1 - \beta_1\},$$

$$q = \max\{0, \mathrm{Re}\,\varkappa_0 + \beta_0, \mathrm{Re}\,\varkappa_1 + \beta_1\}.$$

12.5. Theorem. *Assume* 12.1(2) *and* 12.1(6). *If h is sufficiently small, then the Galerkin* *equations* 12.4(2) *are uniquely solvable for every* $f \in \mathbf{L}^2(\varrho)$, *and the approximate solutions* u_Δ *converge in* $\mathbf{L}^2(\varrho)$ *to the exact solution* u *with the error bound*

$$\|\varrho(u - u_\Delta)\|_0 \leqq c \inf_{v \in \tilde{\mathscr{S}}^\delta(\Delta)} \|\varrho(u - v)\|_0 . \tag{1}$$

If, in addition, $a, b \in \mathbf{H}^3$ *and* $f \in \mathbf{H}^2$, *then*

$$\|\varrho(u - u_\Delta)\|_0 \leqq ch^{t-r} \|\varrho_0^{-1} u\|_t \tag{2}$$

and, for $\Delta \in \mathrm{Part}_\nu$,

$$\|\varrho_0^{-1}(u - u_\Delta)\|_s \leqq ch^{t-r-s-q} \|\varrho_0^{-1} u\|_t \tag{3}$$

for any s *and* t *satisfying* $r \leqq t < 3/2$, $-q \leqq s \leqq t \leqq \delta + 1$ *and* $s < \delta + 1/2$.

Proof. We first show that the right-hand side of (1) converges to 0 as $h \to 0$ on a dense subset of $\mathbf{L}^2(\varrho)$. By Lemma 6.28 we have

$$\|\varrho(u - v)\|_0 \leqq c \|\varrho_0^{-1}(u - v)\|_r \tag{4}$$

which, together with the approximation property for $\mathscr{S}^\delta(\Delta)$, yields

$$\inf_{v \in \tilde{\mathscr{S}}^\delta(\Delta)} \|\varrho(u - v)\|_0 \leqq c \inf_{w \in \mathscr{S}^\delta(\Delta)} \|\varrho_0^{-1} u - w\|_r \to 0$$

as $h \to 0$ for any $u \in \mathbf{C}_0^\infty(0, 1)$. Note that $\mathbf{C}_0^\infty(0, 1)$ is dense in $\mathbf{L}^2(\varrho)$. Since 12.1(2) and 12.1(6) imply the locally strong ellipticity of the operator $A \in \mathscr{L}(\mathbf{L}^2(\varrho))$, the first part of the theorem is a consequence of Cea's Lemma (cf. Section 1.33).

If $a, b \in \mathbf{H}^3$ and $f \in \mathbf{H}^2$, then Corollary 6.37 for $k = 2$ entails that $\varrho_0^{-1} u \in \mathbf{H}^t$ for $t < 3/2$. Therefore, by the estimates (1) and (4) and the approximation property, inequality (2) holds.

It remains to prove (3). Using Lemma 6.28 and the inverse property for $\mathscr{S}^\delta(\Delta)$ (this can be demonstrated in the same way as in the periodic case, see Section 2.11), we obtain for arbitrary $v \in \tilde{\mathscr{S}}^\delta(\Delta)$

$$\|\varrho_0^{-1}(u_\Delta - v)\|_s \leqq ch^{-s-q} \|\varrho_0^{-1}(u_\Delta - v)\|_{-q} \leqq c_1 h^{-s-q} \|\varrho(u_\Delta - v)\|_0 . \tag{5}$$

From (1) and the triangle inequality it follows that

$$\|\varrho(u_\Delta - v)\|_0 \leqq c_2 \|\varrho(u - v)\|_0 ,$$

which implies that

$$\|\varrho(u_\Delta - v)\|_0 \leqq c_3 \|\varrho_0^{-1}(u - v)\|_r , \qquad v \in \tilde{\mathscr{S}}^\delta(\Delta) \tag{6}$$

in view of (4). Furthermore, by the approximation property, we can choose an element $v \in \tilde{\mathscr{S}}^\delta(\Delta)$ such that

$$\|\varrho_0^{-1}(u - v)\|_r \leqq c_4 h^{t-r} \|\varrho_0^{-1} u\|_t , \tag{7}$$

$$\|\varrho_0^{-1}(u - v)\|_s \leqq c_4 h^{t-s} \|\varrho_0^{-1} u\|_t . \tag{8}$$

Finally, combining the estimates

$$\|\varrho_0^{-1}(u - u_\Delta)\|_s \leqq \|\varrho_0^{-1}(u - v)\|_s + \|\varrho_0^{-1}(v - u_\Delta)\|_s$$

and (5) through (8), we get (3). ∎

Remarks 1°. In (2) and (3) the best asymptotic error estimates are obtained for $\beta_j = -\text{Re}\,\varkappa_j$ ($j = 0, 1$), since then $r = q = 0$. If $\varkappa_j \in \mathbb{R}$ and $\beta_j = -\varkappa_j/2$ ($j = 0, 1$), the Galerkin method 12.4(2) coincides with the *method of weighted residuals* proposed by Washizu and Ikegawa [1].

2°. For higher $\delta(\delta > 1)$, we have no increase of the order of convergence in Theorem 12.5, in general. However, in the case of constant coefficients a and b, we obtain from Corollary 6.38 that the estimates (2) and (3) hold for any s and t satisfying $r \leqq t \leqq k + 1 - |\text{Re}\,\varkappa|$, $-q \leqq s \leqq t \leqq \delta + 1$ and $s < \delta + 1/2$ if $f \in \mathbf{H}^{k+1}$, $k \in \mathbb{N}$, where $\varkappa = \varkappa_0 = -\varkappa_1 = (2\pi i)^{-1} \log [(a + b)/(a - b)]$.

12.6. Galerkin's method with splines on special graded meshes. For each fixed $n \in \mathbb{N}$, we define a partition $\Delta_n = \{0 = \sigma_0 < \sigma_1 < \cdots < \sigma_{2n} = 1\}$ of the interval $[0, 1]$ by

$$\sigma_k = 2^{-1}(k/n)^{\beta_0}, \qquad \sigma_{2n-k} = 1 - 2^{-1}(k/n)^{\beta_1}, \quad k = 0, \ldots, n, \quad (1)$$

where $\beta_j \geqq 1$ ($j = 0, 1$), are given real numbers. Note that the partitions Δ_n are not γ-quasiuniform for any $\gamma > 0$.

12.7. Theorem. *Assume* 12.1(2), $a, b \in \mathbf{H}^{r+2}$ *and* $f \in \mathbf{H}^s$, *where* $\text{Re}\,\varkappa_j + 1/2 + r < s < \text{Re}\,\varkappa_j + 3/2 + r$ ($j = 0, 1$). *Let* u *and* u_{Δ_n} *be the solutions of the equations* $Au = f$ *and* 12.2(1), *respectively. Then*

$$\|u - u_{\Delta_n}\|_0 \leqq C(1/n)^\mu (\log n)^\nu,$$

where C is a constant and

$$\mu = \min\{s, \delta + 1, \beta_0(\text{Re}\,\varkappa_0 + 1/2), \beta_1(\text{Re}\,\varkappa_1 + 1/2)\},$$

$$\nu = \begin{cases} 1/2 & \text{if } s \geqq \delta + 1 \text{ and } \min\limits_{j=0,1} \{\beta_j(\text{Re}\,\varkappa_j + 1/2)\} = \delta + 1, \\ 0 & \text{else.} \end{cases}$$

12.8. Corollary. *Let the assumptions of Theorem* 12.7 *be fulfilled. If* $s \geqq \delta + 1$ *and*

$$\beta_j > (\delta + 1)/(\text{Re}\,\varkappa_j + 1/2), \qquad j = 0, 1$$

then $\|u - u_{\Delta_n}\|_0 = O(n^{-\delta-1})$ *as* $n \to \infty$.

By Theorem 12.3, the proof of Theorem 12.7 requires estimating

$$E_n(u) := \inf_{v \in \mathcal{S}^\delta(\Delta_n)} \|u - v\|_0.$$

We first determine the asymptotic behavior of $u = A^{-1}f$ at the points 0 and 1. From the proofs in Sections 6.28–6.37 we conclude that u has the representation

$$u(t) = u_0(t) + t^{\varkappa_0} \sum_{m=0}^{r} \sum_{s=0}^{m} \alpha_{m,s} t^m (\log t)^s$$
$$+ (1 - t)^{\varkappa_1} \sum_{m=0}^{r} \sum_{s=0}^{m} \beta_{m,s} (1 - t)^m \big(\log (1 - t)\big)^s, \quad (1)$$

where $u_0 \in \mathbf{H}^s$ and $\alpha_{m,s}, \beta_{m,s} \in \mathbb{C}$. As a consequence of the approximation property, we have

$$E_n(u_0) \leqq C(1/n)^{\min\{s,\delta+1\}}. \quad (2)$$

For further estimates, we need the quasi-interpolant Q_{Δ_n} introduced by DE BOOR [1],

$$Q_{\Delta_n} g := \sum_{i=-\delta}^{2n-1} \lambda_i(g)\, \psi_i \sqrt{\frac{\sigma_{i+\delta+1} - \sigma_1}{\delta + 1}},$$

where

$$\begin{aligned}
&\sigma_{-r} := -1/2r(\sigma_1 - \sigma_0), && r = 1, 2, \ldots, \delta,\\
&\sigma_{2n+r} := 1 + 1/2r(\sigma_{2n} - \sigma_{2n-1}), && r = 1, 2, \ldots, \delta,\\
&\lambda_i(g) := \sum_{j=0}^{\delta} (-1)^{\delta-j}\, \omega_i^{(\delta-j)}(\tau_i)\, g^{(j)}(\tau_i), && i = -\delta, \ldots, 2n-1,\\
&\omega_i(t) := (t - \sigma_{i+1}) \ldots (t - \sigma_{i+\delta})/\delta!, && \tau_i \in (\sigma_i, \sigma_{i+\delta+1}) \cap [\sigma_1/2, 1],
\end{aligned} \tag{3}$$

and ψ_i is defined as in Section 2.14. The operator Q_{Δ_n} is a projection onto $\mathscr{S}^\delta(\Delta_n)$.

Note that $\{\psi_i : i = -\delta, \ldots, 2n-1\}$ is a basis of $\mathscr{S}^\delta(\Delta_n)$. Moreover, there exists a positive constant D such that (see DE BOOR [1, 2])

$$D^{-1} \left\| \sum_{i=-\delta}^{2n-1} \xi_i \psi_i \right\|_0 \leqq \left(\sum_{i=-\delta}^{2n-1} |\xi_i|^2 \right)^{1/2} \leqq D \left\| \sum_{i=-\delta}^{2n-1} \xi_i \psi_i \right\|_0 \tag{4}$$

for arbitrary $\xi_i \in \mathbb{C}$ $(i = -\delta, \ldots, 2n-1)$.

12.9. Lemma. *For any* $(\delta + 1)$*-times continuously differentiable function* f *on* $(0, 1)$, *the estimate*

$$\int_{\sigma_i}^{\sigma_{i+1}} |f - Q_{\Delta_n} f|^2 \leqq C(\sigma_{i+\delta+1} - \sigma_{i-\delta-1})^{2(\delta+1)+1} \max_{t \in [\sigma_i - \delta - 1, \sigma_i + \delta + 1]} |f^{(\delta+1)}(t)|^2 \tag{1}$$

$(i = 0, \ldots, 2n-1)$

holds, where C *is a constant independent of* f *and* n.

Proof. For $t \in [\sigma_i, \sigma_{i+1}]$, we obtain

$$\left| f(t) - \sum_{j=0}^{\delta} f^{(j)}(\sigma_i) \frac{(t - \sigma_i)^j}{j!} \right| \leqq C(\sigma_{i+1} - \sigma_i)^{(\delta+1)} \max_{t \in [\sigma_i, \sigma_{i+1}]} |f^{(\delta+1)}(t)|,$$

whence

$$\int_{\sigma_i}^{\sigma_{i+1}} \left| f(t) - \sum_{j=0}^{\delta} f^{(j)}(\sigma_i) \frac{(t - \sigma_i)^j}{j!} \right|^2 dt \leqq C(\sigma_{i+1} - \sigma_i)^{2(\delta+1)+1} \max_{t \in [\sigma_i, \sigma_{i+1}]} |f^{(\delta+1)}(t)|^2. \tag{2}$$

On the other hand,

$$\begin{aligned}
\left| \sum_{j=0}^{\delta} f^{(j)}(\sigma_i) \frac{(t - \sigma_i)^j}{j!} - Q_{\Delta_n} f(t) \right| &= \left| Q_{\Delta_n} \left(\sum_{j=0}^{\delta} f^{(j)}(\sigma_i) \frac{(\cdot - \sigma_i)^j}{j!} - f \right)(t) \right|\\
&= \left| \sum_{s=i-\delta}^{i+1} \lambda_s \left(\sum_{j=0}^{\delta} f^{(j)}(\sigma_i) \frac{(\cdot - \sigma_i)^j}{j!} - f \right) \sqrt{\frac{(\sigma_{s+\delta+1} - \sigma_s)}{\delta + 1}}\, \psi_s(t) \right|,
\end{aligned}$$

and 12.8(4) yields

$$\int_{\sigma_i}^{\sigma_{i+1}} \left| \sum_{j=0}^{\delta} f^{(j)}(\sigma_i) \frac{(t-\sigma_i)^j}{j!} - Q_{\Delta_n} f(t) \right|^2 \mathrm{d}t$$

$$\leqq C \sum_{s=i-\delta}^{i+1} (\sigma_{i+\delta+1} - \sigma_{i-\delta-1}) \left| \lambda_s \left(\sum_{j=0}^{\delta} f^{(j)}(\sigma_i) \frac{(\cdot - \sigma_i)^j}{j!} - f \right) \right|^2. \tag{3}$$

Setting $g(t) = f(t) - \sum_{j=0}^{\delta} f^{(j)}(\sigma_i) \frac{(t-\sigma_i)^j}{j!}$, we obtain

$$|g^{(j)}(\tau_s)| \leqq C(\sigma_{i+\delta+1} - \sigma_{i-\delta-1})^{\delta+1-j} \max_{t \in [\sigma_{i-\delta-1}, \sigma_{i+\delta+1}]} |f^{(\delta+1)}(t)|$$

and

$$|\omega_s^{(\delta-j)}(\tau_s)| \leqq C(\sigma_{i+\delta+1} - \sigma_{i-\delta-1})^j \qquad (s = i-\delta, \ldots, i+1).$$

Therefore, by means of 12.8(3) we have

$$|\lambda_s(g)| \leqq C(\sigma_{i+\delta+1} - \sigma_{i-\delta-1})^{\delta+1} \max_{t \in [\sigma_{i-\delta-1}, \sigma_{i+\delta+1}]} |f^{(\delta+1)}(t)|.$$

From the last inequality and (3) we infer

$$\int_{\sigma_i}^{\sigma_{i+1}} \left| \sum_{j=0}^{\delta} f^{(j)}(\sigma_i) \frac{(t-\sigma_i)^j}{j!} - Q_{\Delta_n} f(t) \right|^2 \mathrm{d}t$$

$$\leqq C(\sigma_{i+\delta+1} - \sigma_{i-\delta-1})^{2(\delta+1)+1} \max_{t \in [\sigma_{i-\delta-1}, \sigma_{i+\delta+1}]} |f^{(\delta+1)}(t)|^2$$

which together with (2) implies (1). ■

12.10. Lemma. *Let* $f(t) := t^{\alpha+i\beta}(\log t)^r$ $(\alpha, \beta \in \mathbb{R}, \alpha > -1/2, r \in \mathbb{N})$. *Then there exists a constant* C *such that*

$$E_n(f) \leqq \|(I - Q_{\Delta_n}) f\| \leqq C \begin{cases} \left(\frac{1}{n}\right)^{\beta_0(\alpha+1/2)} (\log n)^r & \text{if } \beta_0\left(\alpha + \frac{1}{2}\right) < \delta + 1, \\ \left(\frac{1}{n}\right)^{\delta+1} (\log n)^{2r+1/2} & \text{if } \beta_0\left(\alpha + \frac{1}{2}\right) = \delta + 1, \\ \left(\frac{1}{n}\right)^{\delta+1} (\log n)^{2r} & \text{if } \beta_0\left(\alpha + \frac{1}{2}\right) > \delta + 1. \end{cases} \tag{1}$$

Proof. Fix an arbitrary positive number $\varepsilon < 1$. From Lemma 12.9 and the inequalities $\sup_{\varepsilon \leqq t \leqq 1} |f^{(\delta+1)}(t)| < \infty$ and $|\sigma_{i+\delta+1} - \sigma_{i-\delta-1}| \leqq C/n$ one can easily see that

$$\int_{\varepsilon}^{1} |f - Q_{\Delta_n} f|^2 \leqq C 1/n^{2(\delta+1)}.$$

Consequently, we may assume without loss of generality that $\sigma_i = (i/2n)^{\beta_0}$, $i = 0, 1, \ldots, 2n$.

Let $i \geqq 2(\delta + 1)$. Then

$$(\sigma_{i+\delta+1} - \sigma_{i-\delta-1}) \leqq C(i/n)^{\beta_0 - 1}\, 1/n,$$
$$\max_{t \in [\sigma_i - \delta - 1, \sigma_i + \delta + 1]} |f^{(\delta+1)}(t)| \leqq C(i/n)^{\beta_0(\alpha - \delta - 1)} (\log i)^r (\log n)^r + C,$$

and Lemma 12.9 yields

$$\int_{\sigma_{2(\delta+1)}}^{1} |f - Q_{\Delta_n} f|^2 \leqq C \sum_{i=2(\delta+1)}^{2n} (1/n)^{2(\delta+1)+1} (i/n)^{(\beta_0 - 1)(2(\delta+1)+1)}$$
$$\times \{(i/n)^{2\beta_0(\alpha-\delta-1)} (\log i)^{2r} (\log n)^{2r} + 1\},$$

$$\left(\int_{\sigma_{2(\delta+1)}}^{1} |f - Q_{\Delta_n} f|^2\right)^{1/2} \leqq C \begin{cases} (1/n)^{\beta_0(\alpha+1/2)} (\log n)^r & \text{if } \beta_0(\alpha + 1/2) < \delta + 1, \\ (1/n)^{\delta+1} (\log n)^{2r+1/2} & \text{if } \beta_0(\alpha + 1/2) = \delta + 1, \\ (1/n)^{\delta+1} (\log n)^{2r} & \text{if } \beta_0(\alpha + 1/2) > \delta + 1. \end{cases} \tag{2}$$

For $i \leqq 2(\delta + 1)$, we have

$$|\omega_s^{(\delta-j)}(\tau_s)| \leqq C(1/n)^{\beta_0 j}, \qquad |f^{(j)}(\tau_s)| \leqq C(1/n)^{\beta_0(\alpha-j)} (\log n)^r.$$

Now, 12.8(3) and 12.8(4) together yield

$$\int_0^{\sigma_{2(\delta+1)}} |Q_{\Delta_n} f|^2 \leqq (\sigma_{3(\delta+1)} - \sigma_0) \sum_{s=0}^{2(\delta+1)} |\lambda_s(f)|^2 \leqq C(1/n)^{\beta_0(2\alpha+1)} (\log n)^{2r},$$
$$\left(\int_0^{\sigma_{2(\delta+1)}} |Q_{\Delta_n} f|^2\right)^{1/2} \leqq C(1/n)^{\beta_0(\alpha+1/2)} (\log n)^r, \tag{3}$$

whereas

$$\left(\int_0^{\sigma_{2(\delta+1)}} |f|^2\right)^{1/2} \leqq C(1/n)^{\beta_0(\alpha+1/2)} (\log n)^r. \tag{4}$$

Finally, (1) follows from (2), (3) and (4). ■

Theorem 12.7 is an immediate consequence of 12.8(1), 12.8(2), Lemma 12.10 and Theorem 12.3. ■

12.11. Remarks. 1°. Theorem 12.7 remains true for the operator $A = aI + Sb$ and functions $f \in \mathbf{H}^s$ satisfying

$$f^{(k)}(0) = f^{(k)}(1) = 0, \qquad k = 0, 1, \ldots, [s - 1/2].$$

2°. Furthermore, Theorem 12.7 can be generalized to the case of singular integral equations with piecewise smooth coefficients on open or closed Lyapunov curves (see Prössdorf and Rathsfeld [3]).

3°. Condition 12.1(2) can be proved to be necessary for the $\mathbf{L}^2$-stability of the standard Galerkin method in the case of equidistant partitions (see Prössdorf and Rathsfeld [3] and Sec. 10.11). The necessity of conditions 12.1(2) and 12.1(6) in Theorem 12.5 may be shown via the techniques developed in Chap. 11.

Collocation methods

Spline collocation methods for solving singular integral equations on an interval are not yet developed as far as their counterparts for equations on closed curves. In this section we study collocation methods utilizing smooth polynomial splines of degree $2\delta + 1$ multiplied by a weight function such that the elements of the spline spaces are polynomials of degree δ on the two grid segments at the endpoints of the interval $[0, 1]$. Following a basic idea of ARNOLD and WENDLAND [1] (see also Chapter 13), we use the fact that nodal collocation is equivalent to a certain Galerkin method and, under condition 12.1(2), we obtain convergence and optimal error estimates in appropriate weighted Sobolev norms in the case of graded meshes. We begin with formulating two convergence theorems for the collocation with piecewise linear trial functions which are immediate consequences of Corollary 10.7 (see also PRÖSSDORF and RATHSFELD [1]).

12.12. Collocation by piecewise linear splines. Let us seek an approximate solution as a piecewise linear spline vanishing at the endpoints on the uniform partition $\{x_k = k/n: k = 0, 1, \ldots, n\}$, $n \in \mathbb{N}$, i.e.

$$u_n = \sum_{k=1}^{n-1} \xi_k \varphi_k^{(n)},$$

where

$$\varphi_k^{(n)}(x) = \begin{cases} (x - x_{k-1})/(x_k - x_{k-1}) & \text{for } x_{k-1} \leqq x \leqq x_k, \\ (x_{k+1} - x)/(x_{k+1} - x_k) & \text{for } x_k \leqq x \leqq x_{k+1}, \\ 0 & \text{otherwise}. \end{cases}$$

The coefficients $\xi_k = \xi_k^{(n)}$ must be determined so that

$$(Au_n)(x_k) = f(x_k), \qquad k = 1, \ldots, n-1, \tag{1}$$

where A stands for the operator 12.1(1).

12.13. Theorem. *Let a and b be continuous functions on* $[0, 1]$. *Then the collocation method* 12.12(1) *is stable in* $\mathbf{L}^2$ *if and only if condition* 12.1(2) *is satisfied.*

12.14. Theorem. *Let a and b be piecewise continuous functions on* $[0, 1]$. *Suppose the following three conditions are fulfilled:*

(i) $d(x) := a(x) - b(x) \neq 0$, $x \in [0, 1]$;

(ii) $\mu(c/d)(x + 0) + (1 - \mu)(c/d)(x - 0) \notin (-\infty, 0]$
for all $0 < x < 1$ *and all* $\mu \in [0, 1]$, *where* $c := a + b$;

(iii) $\mu c(0+) + (1 - \mu) d(0+) \neq 0$,
$\mu c(1-) + (1 - \mu) d(1-) \neq 0$, $\mu \in [0, 1]$.

Then the collocation method 12.12(1) *converges in* $\mathbf{L}^2$ *for every bounded Riemann integrable function f on* $[0, 1]$. *The conditions* (i) *through* (iii) *are necessary and sufficient for the* $\mathbf{L}^2$-*stability of the method* 12.12(1).

12.15. Remark. Theorems 12.13 and 12.14 remain valid for the midpoint collocation with piecewise constant trial functions on uniform partitions (see SCHMIDT [6], where the more general case of ε-collocation is studied; cf. also Chapter 13).

For our further analysis of collocation methods we have to extend the results of Chapter 6 on smoothness and asymptotics of solutions of singular integral equations to a scale of weighted Sobolev spaces which we are going to define now.

12.16. Notation. For $l \in \mathbb{N}$, we introduce the weighted Sobolev space

$$\mathbf{H}_l^k = \{u \in \mathbf{H}^k : x^j(1-x)^j \,\mathrm{D}^j u \in \mathbf{H}^k,\ j = 1, \ldots, l\}$$

with the canonical norm

$$\|u\|_{k,l} = \sum_{0 \leq j \leq l} \|x^j(1-x)^j \,\mathrm{D}^j u\|_k,$$

where $\mathrm{D} = \mathrm{d}/\mathrm{d}x$. We use the convention that $\mathbf{H}_l^0 = \mathbf{L}_l^2$. Analogously we define

$$\overline{\mathbf{H}}_l^k = \{u \in \overline{\mathbf{H}}^k : x^j(1-x)^j \,\mathrm{D}^j u \in \overline{\mathbf{H}}^k,\ j = 1, \ldots, l\}.$$

Note that $\overline{\mathbf{H}}_l^k = \mathbf{H}_l^k \cap \overline{\mathbf{H}}^k$. Consider the Fuchsian differential operator $\mathcal{D}_l = \mathrm{D}^l x^l (1-x)^l$. The following lemma allows us to define an equivalent Hilbert space structure on $\mathbf{L}_l^2$ which is crucial for our further considerations.

12.17. Lemma. *For any $k \geq 0$, $\mathcal{D}_l$ is a continuous Fredholm operator of $\mathbf{H}_l^k$ into $\mathbf{H}^k$ with trivial kernel and index $-l$. Moreover, an element $f \in \mathbf{H}^k$ belongs to the image $\mathcal{D}_l(\mathbf{H}_l^k)$ if and only if*

$$(f, x^j) = 0, \qquad j = 0, \ldots, l-1. \tag{1}$$

Proof. Let first $k \in \mathbb{N}$. Note that then $\mathcal{D}_l$ is a Fredholm operator of $\mathbf{H}_l^k$ into $\mathbf{H}^k$ with index $-l$ (cf. ELSCHNER [2, Th. 3.2.1]). It remains to verify the second assertion of the lemma. Let $u \in \mathbf{H}_l^k$. It is easily seen that, for $i = 0, \ldots, l-1$, $\mathrm{D}^i x^l (1-x)^l u$ are continuous functions on $[0, 1]$ vanishing at the endpoints. Setting $f = \mathcal{D}_l u$, we therefore obtain (1) by integration by parts. Conversely, suppose $f \in \mathbf{H}^k$ satisfies (1). Putting $v = J^l f$, where J denotes the operator $Jf(x) = \int_0^x f(t)\,\mathrm{d}t$, we observe that $v \in \mathbf{H}^{k+l}$ and, by virtue of (1), $\mathrm{D}^j v(0) = \mathrm{D}^j v(1) = 0$ for $j = 0, \ldots, l-1$. Hence $u = x^{-l}(1-x)^{-l}\, v \in \mathbf{H}_l^k$ by Lemma 6.28, and we have $\mathcal{D}_l u = f$. Finally, for arbitrary $k \geq 0$, one gets the result by interpolation. ∎

Lemma 12.7 implies that $\|\mathcal{D}_l u\|_k$ is an equivalent norm on $\mathbf{H}_l^k$ and the corresponding scalar product on $\mathbf{L}_l^2$ is given by $(u, v)_l := (\mathcal{D}_l u, \mathcal{D}_l v)$.

Consider the singular integral operator

$$A = a(x)\, I + b(x)\, S, \qquad Su(x) = (\pi \mathrm{i})^{-1} \int_0^1 u(y)\,(y-x)^{-1}\,\mathrm{d}y, \tag{2}$$

where I is the identity operator and, for the sake of simplicity, the coefficients are assumed to be infinitely differentiable on $[0, 1]$:

$$a, b \in \mathbf{C}^\infty[0, 1]. \tag{3}$$

The next lemma concerns the commutator $T = [\mathcal{D}_l, A]$ of $\mathcal{D}_l$ and A.

12.18. Lemma. *Assume* 12.17(3). *Then, for any $l \geq 1$ and $k \geq 0$, T is a continuous operator of $\overline{\mathbf{H}}_{l-1}^k$ and $\overline{\mathbf{H}}_l^k$ into $\mathbf{H}^k$ and H^{k+1}, respectively.*

Proof. Let $u \in \mathbf{C}_0^\infty(0, 1)$. We have $T = [\mathcal{D}_l, aI] + [\mathcal{D}_l, bS]$ and

$$[\mathcal{D}_l, aI]\, u = \sum_{j=0}^{l-1} a_j x^{j+1}(1-x)^{j+1}\, \mathrm{D}^j u \tag{1}$$

with suitable functions $a_j \in \mathbf{C}^\infty[0, 1]$. Analogously, since $[\mathrm{D}, S]\, u = 0$ (cf. Mikhlin and Prössdorf [1, Chap. II, Lemma 6.1]) and $[bI, S]$ is an integral operator with smooth kernel for any $b \in \mathbf{C}^\infty[0, 1]$, one obtains the relation

$$[\mathcal{D}_l, bS]\, u = \sum_{j=0}^{l-1} Sb_j x^{j+1}(1-x)^{j+1}\, \mathrm{D}^j u + Ku, \tag{2}$$

where $b_j \in \mathbf{C}^\infty[0, 1]$ and K is a continuous map of $\mathbf{L}^2$ into $\mathbf{H}^k$ for all $k \geqq 0$. Let first $k \in \mathbb{N}$. Note that $\mathbf{C}_0^\infty(0, 1)$ is dense in $\overline{\mathbf{H}}_l^k$ (cf. Elschner [2, Lemma 3.2.3]). Then it follows from (1) that $[\mathcal{D}_l, aI]$ is a continuous map of $\overline{\mathbf{H}}_{l-1}^k$ into $\overline{\mathbf{H}}^k$ and of $\overline{\mathbf{H}}_l^k$ into $\overline{\mathbf{H}}^{k+1}$. Furthermore, since $S: \overline{\mathbf{H}}^k \to \mathbf{H}^k$ is bounded for every $k \in \mathbb{N}$, (2) implies the continuity of the maps $[\mathcal{D}_l, bS]: \overline{\mathbf{H}}_{l-1}^k \to \mathbf{H}^k$ and $[\mathcal{D}_l, bS]: \overline{\mathbf{H}}_l^k \to \mathbf{H}^{k+1}$, which completes the proof. By interpolation we obtain the assertion for all $k \geqq 0$. ■

As a consequence of the preceding lemma and the continuity of the operator S: $\overline{\mathbf{H}}^k \to \mathbf{H}^k$ ($k \geqq 0$), we get the following result.

12.19. Corollary. *Assume* 12.17(3), $k \geqq 0$ *and* $l \in \mathbb{N}$. *Then the singular integral operator* A *defined by* 12.17(2) *is a continuous map of* $\overline{\mathbf{H}}_l^k$ *into* $\mathbf{H}_l^k$.

If 12.1(2) and 12.17(3) hold, then the operator A is invertible in $\mathbf{L}^2$ (cf. Sec. 6.27) as well as in the spaces $\mathbf{L}_l^2$ (by virtue of Lemma 12.17).

The following lemma generalizes Corollary 6.30 to the case of weighted Sobolev spaces.

12.20. Lemma. *If the hypotheses* 12.1(2), 12.17(3) *are satisfied and if* $0 < k < \operatorname{Re} \varkappa_j + 1/2$ $(j = 0, 1)$, *then* $u \in \mathbf{L}_l^2$ *and* $Au \in \mathbf{H}_l^k$ *imply that* $u \in \mathbf{H}_l^k$.

Proof. Because $\mathcal{D}_l u \in \mathbf{L}^2$, $\mathcal{D}_l Au \in \mathbf{H}^k$ and $k < 1$, Lemma 12.18 yields that $A\mathcal{D}_l u \in \mathbf{H}^k$. Thus $\mathcal{D}_l u \in \mathbf{H}^k$ by Corollary 2.30 ,whence $u \in \mathbf{H}_l^k$. ■

Choosing a function $\varphi_0 \in \mathbf{C}_0^\infty(\mathbb{R})$ such that $\varphi_0(x) = 1$ for $x \in [0, 1/2]$ and $\varphi_0(x) = 0$ for $x \geqq 2/3$, we set $\varphi_1(x) = \varphi_0(1-x)$. Our aim now is to extend Theorem 6.36 to the spaces $\mathbf{H}_l^k$.

12.21. Lemma. *Assume* 12.1(2), 12.17(3) *and*

$$\operatorname{Re} \varkappa_j + 1/2 + r - 1 < k < \operatorname{Re} \varkappa_j + 1/2 + r \qquad (j = 0, 1)$$

for some positive integer r. *If* $u \in \mathbf{L}_l^2$ *and* $Au \in \mathbf{H}_l^k$, *then* u *is of the form*

$$u = u_0 + \varphi_0 \sum_{m=0}^{r-1} \sum_{j=0}^{m} c_{mj} x^{\varkappa_0 + m} \log^j x + \varphi_1 \sum_{m=0}^{r-1} \sum_{j=0}^{m} d_{mj}(1-x)^{\varkappa_1 + m} \log^j (1-x) \tag{1}$$

with certain complex numbers c_{mj}, d_{mj} *and some* $u_0 \in \overline{\mathbf{H}}_l^k$.

Proof. We proceed by induction on l. For $l = 0$, the assertion was proved in Chapter 6. Let $l \geqq 1$. By the induction hypothesis, u admits the representation (1) with $u_0 \in \overline{\mathbf{H}}_{l-1}^k$. Putting $u_1 = u - u_0$, it follows from Lemmas 6.32 and 6.33 that

$$Au_1 = v + \varphi_0 \sum_{m=0}^{l+r-1} \sum_{j=0}^{m+1} e_{mj} x^{\varkappa_0 + m} \log^j x + \varphi_1 \sum_{m=0}^{l+r-1} \sum_{j=0}^{m+1} f_{mj}(1-x)^{\varkappa_1 + m} \log^j (1-x) \tag{2}$$

with complex numbers e_{mj}, f_{mj} and $v \in \overline{\mathbf{H}}^{k+l}$. Now $Au_0 \in \mathbf{H}^k$ and $Au \in \mathbf{H}_l^k$ imply that $Au_1 \in \mathbf{H}^k$. Hence $Au_1 \in \mathbf{H}_l^k$, since each term on the right-hand side of (2) is either zero or belongs to H_l^k. Therefore $Au_0 \in \mathbf{H}_l^k$, and by Lemma 12.18 and $u_0 \in \overline{\mathbf{H}}_{l-1}^k$, one obtains that $A\mathcal{D}_l u_0 \in \mathbf{H}^k$. Because $\mathcal{D}_l u_0 \in \mathbf{L}^2$, Theorem 6.36 yields that $\mathcal{D}_l u_0$ is modulo $\overline{\mathbf{H}}^k$ a linear combination of the functions

$$\begin{aligned} &\varphi_0 x^{\varkappa_0+m} \log^j x, \qquad \varphi_1(1-x)^{\varkappa_1+m} \log^j (1-x), \\ &0 \leqq j \leqq m, \qquad 0 \leqq m \leqq r-1. \end{aligned} \tag{3}$$

It is not difficult to check that there is another linear combination w of the functions (3) such that $\mathcal{D}_l(u_0 + w) \in \overline{\mathbf{H}}^k$. Therefore, $u_0 + w \in \mathbf{H}_l^k$ by the injectivity of $\mathcal{D}_l$ (see Lemma 12.17), and hence $w \in \mathbf{H}^k$. Consequently, $w \in \mathbf{H}_l^k$ by the argument we have used above, whence $u_0 \in \mathbf{H}_l^k \cap \overline{\mathbf{H}}^k = \overline{\mathbf{H}}_l^k$ which completes the proof. ■

Finally, in the case $\varkappa_0 = \varkappa_1 = 0$ we obtain the following extension of Lemma 12.20.

12.22. Corollary. *Under the hypotheses* 12.1(2), 12.17(3), *if* $b(0) = b(1) = 0$ *and* $1/2 < k < 3/2$, *then* $u \in \mathbf{L}_l^2$ *and* $Au \in \mathbf{H}_l^k$ *imply that* $u \in \mathbf{H}_l^k$.

Proof. Note that $\varkappa_0 = \varkappa_1 = 0$ by the third assumption. Thus Lemma 12.21 gives $u = \varphi_0 c_{00} + \varphi_1 d_{00} + u_0$ with $u_0 \in \overline{\mathbf{H}}_l^k$, which proves the assertion. ■

12.23. Collocation by splines of odd degree. Let $\mathcal{S}^\delta(\Delta)$ ($\delta \in \mathbb{N}$) be the space of $\delta - 1$ times continuously differentiable polynomial splines of degree δ subordinate to the partition $\Delta = \{0 = x_0 < x_1 < \cdots < x_n = 1\}$ of the interval $[0, 1]$. We have $\mathcal{S}^\delta(\Delta) \subset \mathbf{H}^k$ if and only if $k < \delta + 1/2$. Let $h = \max\limits_{1 \leqq i \leqq n} (x_i - x_{i-1})$ be the mesh size of Δ. Setting

$$\mathring{\mathcal{S}}^{2\delta+1}(\Delta) = \{u \in \mathcal{S}^{2\delta+1}(\Delta) \colon \mathrm{D}^j u(0) = \mathrm{D}^j u(1) = 0,\, j = 0, \ldots, \delta\},$$

we define the weighted spline spaces

$$\overline{\mathcal{S}}^{2\delta+1}(\Delta) = x^{-\delta-1}(1-x)^{-\delta-1}\, \mathring{\mathcal{S}}^{2\delta+1}(\Delta).$$

Note that $\dim \mathring{\mathcal{S}}^{2\delta+1}(\Delta) = \dim \overline{\mathcal{S}}^{2\delta+1}(\Delta) = n - 1$. Let A be the singular integral operator 12.17(2). Our collocation method for solving the equation $Au = f$ defines the approximate solution $u_\Delta \in \overline{\mathcal{S}}^{2\delta+1}(\Delta)$ by

$$Au_\Delta(x_i) = f(x_i), \qquad i = 1, \ldots, n-1. \tag{1}$$

In order to formulate (1) as a Galerkin method we need the following key lemma.

12.24. Lemma. *Let* $f \in \mathbf{L}_{\delta+1}^2$. *Then* $f(x_i) = 0$, $i = 1, \ldots, n-1$, *if and only if*

$$(f, u_\Delta)_{\delta+1} = 0 \quad \textit{for all} \quad u_\Delta \in \overline{\mathcal{S}}^{2\delta+1}(\Delta). \tag{1}$$

Proof. Setting $v_\Delta = x^{\delta+1}(1-x)^{\delta+1} u_\Delta$, integration by parts gives

$$(f, u_\Delta)_{\delta+1} = (\mathcal{D}_{\delta+1} f, \mathrm{D}^{\delta+1} v_\Delta) = (-1)^\delta \left(\mathrm{D} x^{\delta+1}(1-x)^{\delta+1} f, \mathrm{D}^{2\delta+1} v_\Delta\right) \tag{2}$$

for all $u_\Delta \in \overline{\mathcal{S}}^{2\delta+1}(\Delta)$. Next we observe that $\mathrm{D}^{2\delta+1}$ is an injective map of $\mathring{\mathcal{S}}^{2\delta+1}(\Delta)$ into $\mathcal{S}^0(\Delta)$, since $v \in \mathring{\mathcal{S}}^{2\delta+1}(\Delta)$ and $\mathrm{D}^{2\delta+1} v = 0$ imply that v is a polynomial of degree 2δ having the zeros 0 and 1 of multiplicity $\delta + 1$, whence $v = 0$. Furthermore, we have

$$\mathrm{D}^{2\delta+1}\left(\mathring{\mathcal{S}}^{2\delta+1}(\Delta)\right) \dotplus \mathbb{C} = \mathcal{S}^0(\Delta)$$

since $\dim S^0(\Delta) = n$ and every $\mathbf{L}^2$ solution of the equation $\mathrm{D}^{2\delta+1}v = 1$ is a polynomial of degree $2\delta + 1$ with leading term $x^{2\delta+1}/(2\delta + 1)!$, which cannot belong to $\mathring{\mathcal{S}}^{2\delta+1}(\Delta)$. Therefore, because of (2) and $\left(1, \mathrm{D}x^{\delta+1}(1 - x)^{\delta+1} f\right) = 0$, (1) is equivalent to

$$\left(\mathrm{D}x^{\delta+1}(1 - x)^{\delta+1} f, w_\Delta\right) = 0 \quad \text{for all} \quad w_\Delta \in \mathcal{S}^0(\Delta). \tag{3}$$

Let θ_i be the characteristic function of the interval $[x_{i-1}, x_i]$. Putting $w_\Delta = \theta_i$ in (3), we obtain

$$[x(1 - x)]^{\delta+1} f\big|_{x=x_{i-1}} = [x(1 - x)]^{\delta+1} f\big|_{x=x_i}, \qquad i = 1, \ldots, n.$$

Consequently, (3) implies that $f(x_i) = 0$ $(i = 1, \ldots, n - 1)$ since $[x(1 - x)]^{\delta+1} f\big|_{x=0} = 0$. Conversely, if $f(x_i) = 0$ $(i = 1, \ldots, n - 1)$, then (3) holds for all elements of the basis $\{\theta_i : i = 1, \ldots, n\}$ of $\mathcal{S}^0(\Delta)$. ∎

Now it follows from the preceding lemma that $u_\Delta \in \bar{\mathcal{S}}^{2\delta+1}(\Delta)$ satisfies the collocation equations 12.23(1) if and only if

$$\bar{P}_\Delta A u_\Delta = \bar{P}_\Delta f, \tag{4}$$

where $\bar{P}_\Delta$ denotes the orthogonal projection of $\mathbf{L}^2_{d+1}$ onto $\bar{\mathcal{S}}^{2\delta+1}(\Delta)$. The next lemma, which is an extension of Corollary 6.23 implies the stability of the Galerkin method (4) if A satisfies the assumption

$$\operatorname{Re}\,[a(x) \pm b(x)] > 0, \qquad x \in [0, 1]. \tag{5}$$

12.25. Lemma. *Assume* 12.17(3) *and* 12.24(5). *Then there exist a constant* $c > 0$ *and a compact operator* K *on* $\mathbf{L}^2_l$ *such that*

$$\operatorname{Re}\left((A + K)\, u, u\right)_l \geq c\, \|u\|^2_{0,l} \quad \textit{for all} \quad u \in \mathbf{L}^2_l. \tag{1}$$

Proof. Since the commutator of $\mathcal{D}_l$ and A is a continuous mapping of $\mathbf{L}^2_l$ into $\mathbf{H}^1$ by Lemma 12.18, we have the representation $\mathcal{D}_l A = A\mathcal{D}_l + K_1$ with some compact operator $K_1 : \mathbf{L}^2_l \to \mathbf{L}^2$. Furthermore, by the proof of Corollary 6.23 there is a constant $c > 0$ and a compact operator K_2 in $\mathbf{L}^2$ such that

$$\operatorname{Re}\left((A + K_2)\, \mathcal{D}_l u, \mathcal{D}_l u\right) \geq c\, \|\mathcal{D}_l u\|^2_0, \qquad u \in \mathbf{L}^2_l. \tag{2}$$

Let π_l be the orthogonal projection of $\mathbf{L}^2$ onto $\mathcal{D}_l(\mathbf{L}^2_l)$ and $\mathcal{D}_l^{-1} : \mathcal{D}_l(\mathbf{L}^2_l) \to \mathbf{L}^2_l$ the inverse of $\mathcal{D}_l$. Note that

$$(K_1 u, \mathcal{D}_l u) = (\mathcal{D}_l^{-1} \pi_l K_1 u, u)_l, \qquad (K_2 \mathcal{D}_l u, \mathcal{D}_l u) = (\mathcal{D}_l^{-1} \pi_l K_2 \mathcal{D}_l u, u)_l$$

and

$$(Au, u)_l = (\mathcal{D}_l A u, \mathcal{D}_l u) = (A \mathcal{D}_l u, \mathcal{D}_l u) + (K_1 u, \mathcal{D}_l u).$$

Setting

$$K = \mathcal{D}_l^{-1} \pi_l K_2 \mathcal{D}_l - \mathcal{D}_l^{-1} \pi_l K_1,$$

from (2) we thus obtain (1). ∎

We are now able to prove the main result of this section. In the sequel c denotes a generic constant independent of u and the mesh size h.

12.26. Theorem. *Assume* 12.1(2) *and* 12.17(3). *If h is sufficiently small, then the collocation equations* 12.23(1) *are uniquely solvable for every* $f \in \mathbf{L}^2_{\delta+1}$, *and the approximate* *solutions u_Δ converge in $\mathbf{L}^2_{\delta+1}$ to the exact solution u with the error bound*

$$\|u - u_\Delta\|_{0,\delta+1} \leqq c \min_{v \in \mathcal{S}^\delta(\Delta)} \|\mathcal{D}_{\delta+1} u - v\|_0 . \tag{1}$$

Proof. Because of 12.1(3) we may replace 12.1(2) by condition 12.24(5), since the collocation equations 12.23(1) and the equations

$$(\theta A u_\Delta)(x_i) = (\theta f)(x_i), \qquad i = 1, \ldots, n-1, \qquad \theta \in \mathbf{C}^\infty[0, 1],$$

are obviously equivalent. Note that $\overline{P}_\Delta$ converges strongly to the identity operator in $\mathbf{L}^2_{\delta+1}$ as $h \to 0$; see (3) below. Therefore, using the equivalence of 12.23(1) and 12.24(4), Lemma 12.25 and standard theory of Galerkin methods (cf. Chapter 1), we obtain the convergence of the collocation method in $\mathbf{L}^2_{\delta+1}$ and the error estimate

$$\|u - u_\Delta\|_{0,\delta+1} \leqq c \min_{w \in \overline{\mathcal{F}}^{2\delta+1}(\Delta)} \|u - w\|_{0,\delta+1} . \tag{2}$$

The proof will be complete once we have shown that

$$\min_{w \in \overline{\mathcal{F}}^{2\delta+1}(\Delta)} \|u - w\|_{0,\delta+1} = \min_{v \in \mathcal{S}^\delta(\Delta)} \|\mathcal{D}_{\delta+1} u - v\|_0 \tag{3}$$

since, in view of the approximation property of the spline spaces $\mathcal{S}^\delta(\Delta)$, the right-hand side of (3) tends to zero as $h \to 0$.

Let P_Δ be the orthogonal projection of $\mathbf{L}^2$ onto $\mathcal{S}^\delta(\Delta)$. Then (3) is a consequence of the equality $\mathcal{D}_{\delta+1}\overline{P}_\Delta = P_\Delta \mathcal{D}_{\delta+1}$. Indeed, we first observe that $\mathcal{D}_{\delta+1}$ is an injective map of $\overline{\mathcal{F}}^{2\delta+1}(\Delta)$ into $\mathcal{S}^\delta(\Delta)$ (cf. the proof of Lemma 12.24). Because of Lemma 12.17 and $\dim \mathcal{S}^\delta(\Delta) = n + d$, $\mathcal{S}^\delta(\Delta)$ is the orthogonal sum (with respect to the $\mathbf{L}^2$ scalar product) of $\mathcal{D}_{\delta+1}\big(\overline{\mathcal{F}}^{2\delta+1}(\Delta)\big)$ and the space of all polynomials of degree $\leqq d$. Therefore

$$(I - \pi_{\delta+1}) P_\Delta = P_\Delta (I - \pi_{\delta+1});$$

recall that $\pi_{\delta+1}$ is the orthogonal projection of $\mathbf{L}^2$ onto $\mathcal{D}_{\delta+1}(\mathbf{L}^2_{\delta+1})$ whose complementary projection, $I - \pi_{\delta+1}$, projects onto the polynomials of degree $\leqq \delta$. Consequently,

$$\mathcal{D}^{-1}_{\delta+1} \pi_{\delta+1} P_\Delta \mathcal{D}_{\delta+1} = \mathcal{D}^{-1}_{\delta+1} P_\Delta \mathcal{D}_{\delta+1}$$

is a projection of $\mathbf{L}^2_{\delta+1}$ onto $\overline{\mathcal{F}}^{2\delta+1}(\Delta)$ with norm 1, which proves the assertion. ∎

We remark that (1) implies the uniform convergence of the approximate solutions u_Δ on every compact subinterval of (0, 1). Let $\varkappa_j$ $(j = 0, 1)$ be the numbers defined by 12.1(5).

12.27. Corollary. *Under the hypotheses* 12.1(2) *and* 12.17(3), *if* $f \in \mathbf{H}^k_{\delta+1}$, $0 < k < \operatorname{Re} \varkappa_j + 1/2$ $(j = 0, 1)$ *and* $k < \delta + 1/2$, *we have the error estimate*

$$\|u - u_\Delta\|_{0,\delta+1} \leqq c h^k \|u\|_{k,\delta+1} . \tag{1}$$

Proof. Lemma 12.20 ensures that $\mathcal{D}_{\delta+1} u \in \mathbf{H}^k$. Furthermore, by the approximation property of $\mathcal{S}^\delta(\Delta)$ and since $k < \delta + 1/2$, the inequality $\|(I - P_\Delta)\, \mathcal{D}_{\delta+1} u\|_0 \leqq c h^k \times \|\mathcal{D}_{\delta+1} u\|_k$ holds, so that (1) follows from 12.26(1). ∎

Note that in formula (1) we always have $k < 1$, and for constant coefficients a and b, even $k < 1/2 - |\operatorname{Re} \varkappa|$, where $\varkappa = \varkappa_0 = -\varkappa_1$. Moreover, for higher δ $(\delta > 1)$ there is no increase of the order of convergence, in general. In the special case $\varkappa_0 = \varkappa_1 = 0$, from Theorem 12.26 and Corollary 12.22 we obtain the following result.

12.28. Corollary. *Suppose* 12.1(2) *and* 12.17(3) *are in force and let* $f \in \mathbf{H}^k_{\delta+1}$, $b(0) = b(1) = 0$, $1/2 < k < 3/2$ *and* $k < \delta + 1/2$. *Then estimate* 12.27(1) *is valid again.*

Using special nonuniform partitions, we finally get the optimal order of convergence in the weighted Sobolev space $\mathbf{L}^2_{\delta+1}$. For every fixed $n \in \mathbb{N}$, define the grid $\Delta_n = \{0 = x_0 < x_1 < \cdots < x_{2n} = 1\}$

$$x_i = 2^{-1}(i/n)^{q_0}, \qquad x_{2n-i} = 1 - 2^{-1}(i/n)^{q_1}, \qquad i = 0, \ldots, n, \tag{1}$$

where $q_j > (\delta + 1)/(\operatorname{Re} \varkappa_j + 1/2)$, $j = 0, 1$.

12.29. Corollary. *Assume* 12.1(2), 12.17(3) *and* $f \in \mathbf{H}^{\delta+1}_{\delta+1}$. *Then the collocation method* 12.23(1) *converges in* $\mathbf{L}^2_{\delta+1}$ *with the error bound*

$$\|u - u_{\Delta_n}\|_{0,\delta+1} = O(n^{-\delta-1}) \quad as \quad n \to \infty. \tag{1}$$

Proof. Applying Lemma 12.21 with $r = l = \delta + 1$, we see that $\mathcal{D}_{\delta+1}u$ has the representation

$$\mathcal{D}_{\delta+1}u = v + \varphi_0 \sum_{m=0}^{\delta} \sum_{j=0}^{m} c_{mj} x^{\varkappa_0+m} \log^j x + \varphi_1 \sum_{m=0}^{\delta} \sum_{j=0}^{m} d_{mj}(1-x)^{\varkappa_1+m} \log^j (1-x),$$

where $c_{mj}, d_{mj} \in \mathbb{C}$ and $v \in \overline{\mathbf{H}}^{\delta+1}$. Now the results of Secs. 12.8–12.10 imply that

$$\min_{v \in \mathcal{S}^{\delta}(\Delta_n)} \|\mathcal{D}_{\delta+1}u - v\|_0 = O(n^{-\delta-1}),$$

which gives (1) in view of 12.26(1). ■

12.30. Remark. The results of Secs. 12.16–12.29 are due to Elschner [7]. By using similar techniques, Elschner [3] also established the convergence of nodal collocation with splines of even degree in suitable weighted Sobolev spaces when condition 12.1(2) is replaced by the requirement that

$$b(t) + \lambda a(t) \neq 0, \qquad t \in [0, 1], \lambda \in [-1, 1].$$

Note that the practical important case of the generalized airfoil equation (cf. Golberg [3]) is covered by these results.

Region method with splines of even degree

We now study a variant of the so-called region method, which is a Galerkin-Petrov method using splines of even degree on arbitrary grids as trial functions and piecewise constant splines as test functions.

12.31. Prologue. Retaining the notation of the preceding sections, we set

$$\mathring{\mathcal{S}}^{2\delta}(\Delta) = \{u \in \mathcal{S}^{2\delta}(\Delta) \colon \mathrm{D}^j u(0) = \mathrm{D}^j u(1) = 0, j = 0, \ldots, \delta - 1\},$$

$\delta \geq 1$ and $\tilde{\mathcal{S}}^{2\delta}(\Delta) = x^{-\delta}(1-x)^{-\delta} \mathring{\mathcal{S}}^{2\delta}(\Delta)$. Note that $\dim \tilde{\mathcal{S}}^{2\delta}(\Delta) = n$. Let A be the singular integral operator defined by 12.1(1). Our variant of the region method for the approximate solution of the equation $Au = f$ defines $u_\Delta \in \tilde{\mathcal{S}}^{2\delta}(\Delta)$ by

$$\int_{x_{i-1}}^{x_i} x^\delta (1-x)^\delta A u_\Delta(x)\, \mathrm{d}x = \int_{x_{i-1}}^{x_i} x^\delta (1-x)^\delta f(x)\, \mathrm{d}x, \qquad i = 1, \ldots, n. \tag{1}$$

The following lemma is crucial for proving the stability of the method in $\mathbf{L}^2_\delta$.

12.32. Lemma. *Let $f \in \mathbf{L}^2_\delta$. Then the equations*

$$\int\limits_{x_{i-1}}^{x_i} x^\delta (1-x)^\delta f(x)\,\mathrm{d}x = 0, \qquad i = 1, \dots, n \tag{1}$$

hold if and only if

$$(f, u_\Delta)_\delta = 0 \quad \textit{for all} \quad u_\Delta \in \mathring{\mathcal{S}}^{2\delta}(\Delta). \tag{2}$$

Proof. It is easy to verify that $\mathrm{D}^{2\delta}$ is an isomorphism of $\mathring{\mathcal{S}}^{2\delta}(\Delta)$ onto $\mathcal{S}^0(\Delta)$. Therefore, integrating by parts, we infer that (1) is equivalent to

$$\left(x^\delta (1-x)^\delta f, w_\Delta\right) = 0 \quad \text{for all} \quad w_\Delta \in \mathcal{S}^0(\Delta). \tag{3}$$

On the other hand, (3) is valid if and only if equations 12.31(1) hold. ■

Let $\overline{P}_\Delta$ and P_Δ be the orthogonal projections of $\mathbf{L}^2_\delta$ and $\mathbf{L}^2$ onto $\mathring{\mathcal{S}}^{2\delta}(\Delta)$ and $\mathcal{S}^\delta(\Delta)$, respectively. By the preceding lemma, 12.31(1) is equivalent to the Galerkin method

$$\overline{P}_\Delta A u_\Delta = \overline{P}_\Delta f, \qquad u_\Delta \in \mathring{\mathcal{S}}^{2\delta}(\Delta). \tag{4}$$

The next lemma enables us to replace 12.1(2) by condition 12.24(5) in the stability proof.

12.33. Lemma. *If $\theta \in \mathbf{C}^\infty[0, 1]$, then the norm of the operators $(I - \overline{P}_\Delta)\,\theta \overline{P}_\Delta$ in $\mathbf{L}^2_\delta$ tends to zero as $h \to 0$.*

Proof. As in Sect. 12.26 one can check that

$$\mathcal{D}_\delta \overline{P}_\Delta = P_\Delta \mathcal{D}_\delta. \tag{1}$$

Together with Lemma 12.18 this implies

$$\mathcal{D}_\delta (I - \overline{P}_\Delta)\,\theta \overline{P}_\Delta = (I - P_\Delta)\,\mathcal{D}_\delta \theta \overline{P}_\Delta = (I - P_\Delta)\,\theta P_\Delta \mathcal{D}_\delta + (I - P_\Delta)\,K \overline{P}_\Delta$$

with a compact operator K of $\mathbf{L}^2_\delta$ into $\mathbf{L}^2$. Since the norm of the operators $(I - P_\Delta)\,\theta P_\Delta$ in $\mathbf{L}^2$ tends to zero (see Theorem 2.14) and P_Δ converges strongly to the identity as $h \to 0$, we obtain the result. ■

12.34. Theorem. *Assume* 12.1(2) *and* 12.17(3). *If h is sufficiently small, then equations* 12.31(1) *are uniquely solvable for every $f \in \mathbf{L}^2_\delta$, and the approximate solutions u_Δ converge in $\mathbf{L}^2_\delta$ to the exact solution u with the error bound*

$$\|u - u_\Delta\|_{0,\delta} \le c \min_{v \in \mathcal{S}^\delta(\Delta)} \|\mathcal{D}_\delta u - v\|_0. \tag{1}$$

Proof. Suppose first that condition 12.24(5) is satisfied. Applying Lemma 12.25 and standard theory of Galerkin methods to 12.32(4), we get the convergence of the region method in $\mathbf{L}^2_\delta$ and the error estimate

$$\|u - u_\Delta\|_{0,\delta} \le c \min_{w \in \mathring{\mathcal{S}}^{2\delta}(\Delta)} \|u - w\|_{0,\delta}. \tag{2}$$

Now (2) and 12.33(1) imply (1). In the general case, we deduce (2) with the help of 12.1(3) and Lemma 12.33; see Sec. 10.11. ■

In the same manner as in Secs. 12.27–12.29, one can derive the following results on the rate of convergence for the region method.

12.35. Corollary. *Suppose* 12.1(2) *and* 12.17(3) *are in force.*

(i) *Under the hypotheses* $f \in \mathbf{H}^k_\delta$, $0 < k < \operatorname{Re} \varkappa_j + 1/2$ $(j = 0, 1)$ *and* $k < \delta + 1/2$ *the error estimate*

$$\|u - u_\Delta\|_{0,\delta} \leqq ch^k \|u\|_{k,\delta}$$

holds.

(ii) *Let* Δ_n $(n \in \mathbb{N})$ *be the graded meshes defined by* 12.28(1). *If* $f \in \mathbf{H}^{\delta+1}_\delta$, *then the region method* 12.31(1) *converges in* $\mathbf{L}^2_\delta$ *with the error bound*

$$\|u - u_{\Delta_n}\|_{0,\delta} = O(n^{-\delta-1}) \quad as \quad n \to \infty.$$

Moreover, Corollary 12.35(i) continues to hold when the fourth assumption is replaced by $\varkappa_0 = \varkappa_1 = 0$ and $1/2 < k < 3/2$.

Quadrature methods for strongly elliptic Cauchy singular integral equations on an interval

12.36. Introduction. The results of Sections 12.12–12.30 can be viewed as only a first step in establishing the stability for adequate spline collocation methods. Furthermore, in order to implement collocation procedures, one has to compute the singular integrals of splines. If this cannot be done analytically, then one uses quadrature rules. Thus, it is natural and easier to discretize a singular integral equation of the form

$$\mathsf{a}(t)\,\mathsf{x}(t) + \frac{\mathsf{b}(t)}{\pi \mathrm{i}} \int_0^1 \frac{\mathsf{x}(\tau)}{\tau - t}\, \mathrm{d}\tau + \int_0^1 \mathsf{k}(t, \tau)\, \mathsf{x}(\tau)\, \mathrm{d}\tau = \mathsf{y}(t), \qquad 0 \leqq t \leqq 1, \tag{1}$$

directly using quadrature rules. Here a, b and k are continuous functions, y is at least Riemann integrable and x is the unknown function.

In the following sections we consider a modification of a quadrature method (cf. 12.37(15)) exhibited in Chapter 11. In order to obtain the system of equations 12.37(15) for this modified quadrature method, we proceed as follows (cf. Sec. 12.37). First, we transform (1) substituting $\tau = \sigma^\alpha$, $t = s^\alpha$ in a neighborhood of 0 and $\tau = 1 - (1 - \sigma)^\alpha$, $t = 1 - (1 - s)^\alpha$ in a neighborhood of 1, where $\alpha \geqq 1$ (cf. the transformation technique for quadrature methods in the case of Fredholm integral equations of the second kind, for which see e.g. Kress [1, 2] and Micke [1]). Then we set up the usual quadrature method using the subtraction technique (cf. the modified quadrature methods in Baker [1]) and Simpson's rule.

In Sec. 12.50 we prove the method under consideration to be stable if and only if Eq. (1) is locally strongly elliptic. For the stable quadrature method, the rates of convergence are derived in Secs. 12.40–12.43. Sec. 12.51 contains numerical examples. Notice that all the results of Secs. 12.37–12.50 apply also to singular integral equations with fixed singularities (cf. Secs. 12.52–12.56 and, for equations of the second kind, Chandler/Graham [4]). Furthermore, analogous modifications carried out in the case of other quadrature methods by Rathsfeld [3] lead to numerical procedures which are stable for certain classes of non-strongly elliptic equations (cf. Secs. 12.52–12.57).

12.37. Derivation of the quadrature method. For the sake of uniqueness, the solution is sought in the form $\mathsf{x} = \varrho \mathsf{z}$, where $\varrho(t) = t^{\varrho_0}(1-t)^{\varrho_1}$, $-1/2 \leqq \operatorname{Re} \varrho_0, \operatorname{Re} \varrho_1 \leqq 1/2$ and $\mathsf{z} \in \mathbf{L}^2 := \mathbf{L}^2(0, 1)$. By ${}^0\log$ and by ${}^1\log$ we denote the branch of the logarithm which takes real values on the positive real axis and is continuous on $\mathbb{C} \setminus \{\exp(\mathrm{i}2\pi\varrho_0)\, t: -\infty < t \leqq 0\}$ and $\mathbb{C} \setminus \{\exp(\mathrm{i}2\pi\varrho_1)\, t, -\infty < t \leqq 0\}$, respectively. If we set

$$\varkappa_j := \frac{(-1)^j}{2\pi\mathrm{i}}\, {}^j\log \frac{\mathsf{a}(j) + \mathsf{b}(j)}{\mathsf{a}(j) - \mathsf{b}(j)}, \qquad j = 0, 1, \tag{1}$$

then (cf. Chapter 6 and Sec. 12.38) there exist $\lambda, \mu \in \mathbb{C}$ such that the first term in the asymptotics of x is $\lambda t^{\varkappa_0}$ for $t \to 0$ and $\mu(1-t)^{\varkappa_1}$ for $t \to 1$.

In order to set up our quadrature methods, we begin with transforming Eq. 12.36(1). Choose $\alpha \geqq 1$ and $\gamma: [0, 1] \to [0, 1]$ such that $\gamma(\sigma) = \sigma^\alpha$ for $0 \leqq \sigma \leqq 1/3$, $\gamma(\sigma) = 1 - (1-\sigma)^\alpha$ for $2/3 \leqq \sigma \leqq 1$ and γ is infinitely differentiable on $[1/4, 3/4]$. Substituting $\mathsf{x}(\tau) = \varrho(\gamma(\sigma))\, \mathsf{z}(\gamma(\sigma))$ in 12.36(1), we arrive at the equation

$$\mathsf{a}(\gamma(s))\, \varrho(\gamma(s))\, \mathsf{z}(\gamma(s)) + \frac{\mathsf{b}(\gamma(s))}{\pi\mathrm{i}} \int_0^1 \frac{\varrho(\gamma(\sigma))\, \mathsf{z}(\gamma(\sigma))}{\gamma(\sigma) - \gamma(s)}\, \gamma'(\sigma)\, \mathrm{d}\sigma$$

$$+ \int_0^1 \mathsf{k}(\gamma(s), \gamma(\sigma))\, \varrho(\gamma(\sigma))\, \mathsf{z}(\gamma(\sigma))\, \gamma'(\sigma)\, \mathrm{d}\sigma = \mathsf{y}(\gamma(s)), \qquad 0 \leqq s \leqq 1. \tag{2}$$

Taking into account that

$$\int_0^1 |\mathsf{z}(\tau)|^2\, \mathrm{d}\tau = \int_0^1 |\mathsf{z}(\gamma(\sigma))\, \gamma'(\sigma)^{1/2}|^2\, \mathrm{d}\sigma, \tag{3}$$

we introduce the new unknown function $z(\sigma) := \mathsf{z}(\gamma(\sigma))\, \gamma'(\sigma)^{1/2}$. We now multiply (2) by $\gamma'(s)^{1/2}/\varrho(\gamma(s))$ to obtain

$$a(s)\, z(s) + \frac{b(s)}{\pi\mathrm{i}} \int_0^1 \frac{r(\sigma)}{r(s)} \frac{\gamma'(s)^{1/2} \gamma'(\sigma)^{1/2}}{\gamma(\sigma) - \gamma(s)}\, z(\sigma)\, \mathrm{d}\sigma + \int_0^1 k(s, \sigma)\, z(\sigma)\, \mathrm{d}\sigma = y(s),$$

$$0 \leqq s \leqq 1, \tag{4}$$

where $a(s) := \mathsf{a}(\gamma(s))$, $b(s) := \mathsf{b}(\gamma(s))$, $r(\sigma) := \varrho(\gamma(\sigma))$, $y(s) := \mathsf{y}(\gamma(s))\, \gamma'(s)^{1/2}/\varrho(\gamma(s))$ and

$$k(s, \sigma) := \frac{\varrho(\gamma(\sigma))}{\varrho(\gamma(s))}\, \mathsf{k}(\gamma(s), \gamma(\sigma))\, \gamma'(s)^{1/2}\, \gamma'(\sigma)^{1/2}. \tag{5}$$

Note that the operator on the left-hand side of (4) is a pseudodifferential operator of Mellin type (cf. Chap. 11).

We write (4) in the form

$$a(s)\, z(s) + \frac{b(s)}{\pi\mathrm{i}} \int_0^1 \frac{z(\sigma)\, \mathrm{d}\sigma}{\sigma - s} + \int_0^1 \left[\frac{b(s)}{\pi\mathrm{i}}\, l(s, \sigma) + k(s, \sigma)\right] z(\sigma)\, \mathrm{d}\sigma = y(s),$$

$$0 \leqq s \leqq 1, \tag{6}$$

$$l(s, \sigma) := \left\{\frac{r(\sigma)}{r(s)} \frac{\gamma'(s)^{1/2}\, \gamma'(\sigma)^{1/2}}{\gamma(\sigma) - \gamma(s)} - \frac{1}{\sigma - s}\right\}$$

and observe that the second term is the only integral containing a singular function. However, for α large enough, $z(0) = z(1) = 0$. Setting $z(\sigma) := 0$ for $\sigma \notin [0, 1]$, the subtraction method leads to

$$\int_0^1 \frac{z(\sigma)}{\sigma - s}\, d\sigma = \int_{-1}^{1+2s} \frac{z(\sigma)}{\sigma - s}\, d\sigma = \int_{-1}^{1+2s} \frac{z(\sigma) - z(s)}{\sigma - s}\, d\sigma . \tag{7}$$

Now we discretize the integrals in (6), (7). We choose an even integer $n > 0$, set $t_k := k/n$ $(k \in \mathbb{Z})$, $\omega_j := 4/3n$ if $j = 1, 3, 5, \ldots, n-1$, $\omega_j := 1/3n$ if $j = 0$ or $j = n$ and $\omega_j := 2/3n$ if $j = 2, 4, \ldots, n-2$. Then Simpson's rule gives

$$\int_0^1 f(t)\, dt \sim \sum_{j=0}^{n} f(t_j)\, \omega_j . \tag{8}$$

Furthermore, for fixed k, we put $\nu_j := 4/3n$ if j is odd, $\nu_j := 2/3n$ if $j = -n+2, -n+4$ $\ldots, n+2k-2$ and $\nu_j := 1/3n$ if $j = -n$ or $j = n+2k$ to get

$$\int_{-1}^{1+2t_k} f(t)\, dt \sim \sum_{j=-n}^{n+2k} f(t_j)\, \nu_j , \qquad k = 0, \ldots, n . \tag{9}$$

Obviously, $\nu_j = 4/3n$ for $j = 1, 3, \ldots, n-1$ and $\nu_j = 2/3n$ for $j = 0, 2, \ldots, n-1$ independently of k. For $s = t_k$, formulas (7), (9) yield

$$\int_0^1 \frac{z(\sigma)}{\sigma - t_k}\, d\sigma \sim \sum_{j=-n}^{n+2k} \frac{z(t_j) - z(t_k)}{t_j - t_k}\, \nu_j$$

$$= \sum_{\substack{j=0 \\ j\neq k}}^{n} \frac{z(t_j)}{t_j - t_k}\, \nu_j - z(t_k) \sum_{\substack{j=-n \\ j\neq k}}^{n+2k} \frac{\nu_j}{t_j - t_k} + z'(t_k)\, \nu_k .$$

Using the equality $\sum_j \nu_j (t_j - t_k)^{-1} = 0$ and taking into consideration the difference formula $z'(t_k) \sim n\big[8\big(z(t_{k+1}) - z(t_{k-1})\big) - \big(z(t_{k+2}) - z(t_{k-2})\big)\big]/12$ as well as $\nu_j z(t_j) = \omega_j z(t_j)$, we obtain

$$\int_0^1 \frac{z(\sigma)}{\sigma - t_k}\, d\sigma \sim \sum_{\substack{j=0 \\ j\neq k}}^{n} \frac{z(t_j)}{t_j - t_k}\, \omega_j + \nu_k n \left\{8\, \frac{z(t_{k+1}) - z(t_{k-1})}{12} - \frac{z(t_{k+2}) - z(t_{k-2})}{12}\right\}, \tag{10}$$

where $z(t_l) := 0$ for $l < 0$ and $l > n$.

Applying (8) to the integral in the third term of (6), we get

$$\int_0^1 k(t_k, \sigma)\, z(\sigma)\, d\sigma \sim \sum_{j=0}^{n} k(t_k, t_j)\, z(t_j)\, \omega_j , \tag{11}$$

$$\int_0^1 l(t_k, \sigma)\, z(\sigma)\, d\sigma \sim \sum_{j=0}^{n} l(t_k, t_j)\, z(t_j)\, \omega_j . \tag{12}$$

Here, for $0 < k < n$, $n \in \mathbb{Z}$, the value $l(t_k, t_k)$ is given by

$$l(t,t) := \lim_{\sigma \to t} \left\{ \frac{\varrho(\gamma(\sigma))}{\varrho(\gamma(t))} \frac{\gamma'(\sigma)^{1/2}\,\gamma'(t)^{1/2}}{\gamma(\sigma) - \gamma(t)} - \frac{1}{\sigma - t} \right\} = I_1 + I_2,$$

$$I_1 := \lim_{\sigma \to t} \left\{ \frac{\varrho(\gamma(\sigma)) - \varrho(\gamma(t))}{\gamma(\sigma) - \gamma(t)} \frac{\gamma'(\sigma)^{1/2}\,\gamma'(t)^{1/2}}{\varrho(\gamma(t))} \right\},$$

$$I_2 := \lim_{\sigma \to t} \left\{ \frac{\gamma'(\sigma)^{1/2}\,\gamma'(t)^{1/2}}{\gamma(\sigma) - \gamma(t)} - \frac{1}{\sigma - t} \right\}.$$

Obviously, $I_1 = \varrho'(\gamma(t))\,\gamma'(t)/\varrho(\gamma(t))$. An easy computation shows that $I_2 = 0$, i.e.

$$l(t_k, t_k) = (\varrho \circ \gamma)'\,(t_k)/\varrho(\gamma(t_k)), \qquad k = 1, \ldots, n-1. \tag{13}$$

If $k = 0$ or $k = n$, then we set $l(t_k, t_k) := (\varrho \circ \gamma)'\,(t_k)/\varrho(\gamma(t_k)) := 0$ in (12). This neglect of one term is justified by the equalities

$$\lim_{\sigma \to 0} l(\sigma, 0)\, z(\sigma) = \lim_{\sigma \to 0} z(\sigma)/\sigma = 0, \qquad \lim_{\sigma \to 1} l(\sigma, 1)\, z(\sigma) = 0$$

which hold provided $\alpha > 3$. Thus (12) and (13) lead to

$$\int_0^1 l(t_k, \sigma)\, z(\sigma)\, \mathrm{d}\sigma \sim \sum_{\substack{j=0 \\ j \neq k}}^{n} \left\{ \frac{r(t_j)}{r(t_k)} \frac{\gamma'(t_j)^{1/2}\,\gamma'(t_k)^{1/2}}{\gamma(t_j) - \gamma(t_k)} - \frac{1}{t_j - t_k} \right\} z(t_j)\, \omega_j + \begin{cases} 0 & \text{if } k = 0 \text{ or } k = n, \\ \dfrac{(\varrho \circ \gamma)'\,(t_k)}{\varrho(\gamma(t_k))}\, z(t_k)\, \omega_k & \text{else}. \end{cases} \tag{14}$$

Eq. (6) for $s = t_k$, $k = 0, \ldots, n$ together with (10), (11) and (14) suggests the following quadrature method: set $u := 0$, $v := 0$ and determine approximate values ξ_k ($k = u, \ldots, n - v$) for $z(t_k)$ by solving the system

$$\begin{aligned} a(t_k)\, \xi_k &+ \frac{b(t_k)}{\pi \mathrm{i}} \left\{ \sum_{\substack{j=u \\ j \neq k}}^{n-v} \frac{\varrho(\gamma(t_j))}{\varrho(\gamma(t_k))} \frac{\gamma'(t_k)^{1/2}\,\gamma'(t_j)^{1/2}}{\gamma(t_j) - \gamma(t_k)}\, \omega_j \xi_j \right. \\ &\left. + \nu_k n\, \frac{8(\xi_{k+1} - \xi_{k-1}) - (\xi_{k+2} - \xi_{k-2})}{12} + \omega_k \frac{(\varrho \circ \gamma)'\,(t_k)}{\varrho(\gamma(t_k))}\, \xi_k \right\} \\ &+ \sum_{j=u}^{n-v} k(t_k, t_j)\, \omega_j \xi_j = y(t_k), \qquad k = u, \ldots, n - v, \end{aligned} \tag{15}$$

where $\xi_k = 0$ for $k < u$ and $k > n - v$. An approximation for $\mathbf{z}(\gamma(k/n))$ and $\mathbf{x}(\gamma(k/n))$, respectively, is given by

$$\mathbf{z}(\gamma(k/n)) \sim \xi_k \gamma'(k/n)^{-1/2}, \qquad \mathbf{x}(\gamma(k/n)) = \mathbf{z}(\gamma(k/n))\, \varrho(\gamma(k/n)). \tag{16}$$

The number $\gamma'(k/n)^{1/2}$ is small provided k or $n - k$ is small. Hence, division by $\gamma'(k/n)^{1/2}$ leads to larger errors near the end-points of the interval. However, in Sec. 12.42 it will be shown that our transformation technique improves the rate of convergence not only away from the end-points, but also near them.

If α is large, then $z(t)$ tends to zero very rapidly as $t \to 0$ or $t \to 1$. For small integers $u, v \geqq 0$, the neglect of the first u and the last v terms in the sums of formulas (10), (11) and (14) leads to small errors only. Consequently, the system (15) yields a natural method also in the case $u, v \neq 0$. In the subsequent sections we shall fix u and v and consider the stability of (15) for $n \to \infty$. It turns out that if Eq. 12.36 (1) is locally strongly elliptic, then there exist u and v such that the method (15) is stable.

Now let us propose a slight modification of the method (15). For the sake of simlicity, let $\alpha = 3$ and $\mathsf{k}(t, \tau) \equiv 0$. Set $\varrho_j = \varkappa_j$ $(j = 0, 1)$, $\varrho(t) := t^{\varkappa_0}(1 - t)^{\varkappa_1}$ and seek $\mathbf{z}$ and $\mathbf{x} = \varrho\mathbf{z}$ in the form

$$\begin{aligned} \mathbf{z}(t) &= \mathbf{z}(0)\,(1 - t)^{2-\varkappa_0-\varkappa_1} + \mathbf{z}(1)\, t^{2-\varkappa_0-\varkappa_1} + \mathbf{z}_0(t), \\ \mathbf{x}(t) &= \mathbf{z}(0)\, t^{\varkappa_0}(1 - t)^{2-\varkappa_0} + \mathbf{z}(1)\, t^{2-\varkappa_1}\,(1 - t)^{\varkappa_1} + \varrho(t)\, \mathbf{z}_0(t), \end{aligned} \tag{17}$$

where $\mathbf{z}_0(t)$ vanishes at 0 and 1. The formulas

$$\begin{aligned} p_0(s) &:= \int_0^1 \frac{(1 - \sigma)^{2-\varkappa_0}\, \sigma^{\varkappa_0}}{\sigma - s}\, d\sigma = -\pi \cot(\pi\varkappa_0)\,(1 - s)^{2-\varkappa_0}\, s^{\varkappa_0} \\ &\quad + \pi/\sin(\pi\varkappa_0) \left\{s^2 + (\varkappa_0 - 2)\, s + \frac{1}{2}\,(\varkappa_0 - 1)\,(\varkappa_0 - 2)\right\}, \\ p_1(s) &:= \int_0^1 \frac{(1 - \sigma)^{\varkappa_1}\, \sigma^{2-\varkappa_1}}{\sigma - s}\, d\sigma = \pi \cot(\pi\varkappa_1)\,(1 - s)^{\varkappa_1}\, s^{2-\varkappa_1} \\ &\quad + \pi/\sin(\pi\varkappa_1) \left\{s^2 - \varkappa_1 s + \frac{1}{2}\,\varkappa_1(\varkappa_1 - 1)\right\} \end{aligned} \tag{18}$$

(cf. Chap. 9) yield

$$\begin{aligned} &\int_0^1 \frac{\varrho(\gamma(\sigma))\, \mathbf{z}(\gamma(\sigma))}{\gamma(\sigma) - \gamma(s)}\, \gamma'(\sigma)\, d\sigma \\ &= \mathbf{z}(0)\, p_0(\gamma(s)) + \mathbf{z}(1)\, p_1(\gamma(s)) + \int_0^1 \frac{\varrho(\gamma(\sigma))\, \mathbf{z}_0(\gamma(\sigma))}{\gamma(\sigma) - \gamma(s)}\, \gamma'(\sigma)\, d\sigma. \end{aligned} \tag{19}$$

Introducing $z_0(\sigma) := \mathbf{z}_0(\gamma(\sigma))\, \gamma'(\sigma)^{1/2}$ and substituting (19) in (2), we proceed analogously to the derivation of (15). For the approximate values η_0 of $\mathbf{z}(0)$, η_n of $\mathbf{z}(1)$ as well as ξ_k of $z_0(t_k)$ $(k = 1, \ldots, n - 1)$, we obtain

$$\begin{aligned} &a(t_k)\left[\xi_k + \gamma'(t_k)^{1/2}\left\{\left(1 - \gamma(t_k)\right)^{2-\varkappa_0-\varkappa_1} \eta_0 + \gamma(t_k)^{2-\varkappa_0-\varkappa_1}\, \eta_n\right\}\right] \\ &+ \frac{b(t_k)}{\pi i} \left\{ \sum_{\substack{j=1 \\ j \neq k}}^{n-1} \frac{\varrho(\gamma(t_j))}{\varrho(\gamma(t_k))}\, \frac{\gamma'(t_j)^{1/2}\, \gamma'(t_k)^{1/2}}{\gamma(t_j) - \gamma(t_k)}\, \omega_j \xi_j \right. \\ &+ \nu_k n\, \frac{8(\xi_{k+1} - \xi_{k-1}) - (\xi_{k+2} - \xi_{k-2})}{12} + \omega_k\, \frac{(\varrho \circ \gamma)'\,(t_k)}{\varrho(\gamma(t_k))}\, \xi_k \\ &\left. + \frac{\gamma'(t_k)^{1/2}}{\varrho(\gamma(t_k))}\left[p_0(\gamma(t_k))\, \eta_0 + p_1(\gamma(t_k))\, \eta_n\right] \right\} = y(t_k), \qquad k = 1, \ldots, n - 1, \end{aligned} \tag{20}$$

where $\xi_k := 0$ for $k < 1$ and $k > n - 1$. Furthermore, since $\alpha = 3$ the derivative $z_0'(0)$ is equal to $\mathsf{z}_0(0)$. Therefore, we find

$$0 = z_0'(0) \sim \frac{z_0(t_1) - z_0(0)}{t_1} = n z_0(t_1),$$

and analogously that $n z_0(t_{n-1}) \sim 0$. Thus the approximate values η_0, η_n, ξ_j ($j = 1, \ldots, n - 1$) are to be determined by (20) and the equations

$$n\xi_1 = 0, \qquad n\xi_{n-1} = 0. \tag{21}$$

An approximation for $\mathsf{z}(\gamma(k/n))$ and $\mathsf{x}(\gamma(k/n))$, respectively, is given by

$$\mathsf{z}(\gamma(k/n)) \sim \begin{cases} \eta_0 & \text{if } k = 0, \\ \eta_n & \text{if } k = n, \\ \eta_0(1 - \gamma(k/n))^{2-\varkappa_0-\varkappa_1} + \eta_n\gamma(k/n)^{2-\varkappa_0-\varkappa_1} + \dfrac{\xi_k}{\gamma'(k/n)^{1/2}} & \text{else}, \end{cases} \tag{22}$$

$$\mathsf{x}(\gamma(k/n)) = \mathsf{z}(\gamma(k/n))\, \varrho(\gamma(k/n)).$$

The advantage of the method (20)–(21) is that the singular integrals of the terms $\mathsf{z}(0)\,(1 - t)^{2-\varkappa_0-\varkappa_1}$ and $\mathsf{z}(1)\,t^{2-\varkappa_0-\varkappa_1}$ (cf. (17)) are evaluated exactly. Furthermore, z_0 vanishes faster at the end-points of the interval. Hence, the functions in the integrals of (6), (7) become smoother if z is replaced by z_0. Consequently, Simpson's rule converges faster. For the method (20)–(21), we renounce proving convergence or stability. However, in Sec. 12.51 numerical results for this method are given and compared with the corresponding results of (15).

12.38. Rates of convergence. For the sake of simplicity, we make some assumptions. Let $\mathbf{H}^s$ and $\mathbf{H}^s(\mathbb{R})$ denote the usual Sobolev space of order s on $[0, 1]$ and on $\mathbb{R}$, respectively. Suppose $\mathsf{a}, \mathsf{b}, \mathsf{y} \in \mathbf{H}^s$ for arbitrary $s > 0$. Furthermore, we restrict our analysis to the case $\mathsf{k}(t, \tau) \equiv 0$ since an arbitrary $\mathsf{k}(t, \tau)$ can be treated analogously to the theory of Fredholm integral equations of the second kind (cf. Chapter 3). We set $u = v = 0$ and define $\varkappa_0, \varkappa_1$ by (37.1), where ${}^j\!\log$ denotes the branch of the logarithm which is continuous on $\mathbb{C} \setminus (-\infty, 0]$ and takes real values on $(0, \infty)$. Choosing $\varrho_0 := \varkappa_0$, $\varrho_1 := \varkappa_1$, we consider the method 12.37(15). If the operator on the left-hand side of 12.36(1) is invertible in $\mathbf{L}^2_\varrho := \{\varrho f : f \in \mathbf{L}^2\}$ and x is the solution of 12.36(1), then (cf. 12.21(1)) x takes the form

$$\mathsf{x}(t) = g_x(t) + \sum_{m=0}^{k-2} \sum_{j=0}^{m} c_{m,j,x} t^{\varkappa_0+m} \log^j t, \qquad 0 \leqq t \leqq 2/3,$$

$$\mathsf{x}(t) = h_x(t) + \sum_{m=0}^{k-2} \sum_{j=0}^{m} d_{m,j,x}(1 - t)^{\varkappa_1+m} \log^j (1 - t), \qquad 1/3 \leqq t \leqq 1,$$

where $k \geqq 2$ is an arbitrary integer, $c_{m,j,x}, d_{m,j,x} \in \mathbb{C}$, $g_x \in \mathbf{H}^s$ $(0 \leqq s < \varkappa_0 + k - 1/2)$, $g_x^{(l)}(0) = 0$ $(l \in \mathbb{Z},\ 0 \leqq l < \varkappa_0 + k - 1)$, $h_x \in \mathbf{H}^s$ $(0 \leqq s < \varkappa_0 + k - 1/2)$ and $h_x^{(l)}(1) = 0$ $(l \in \mathbb{Z}, 0 \leqq l < \varkappa_1 + k - 1)$. Consequently,

$$\begin{aligned} z(t) &= g_z(t) + \sum_{m=0}^{k-2} \sum_{j=0}^{m} c_{m,j,z}\, t^{(\alpha-1)/2+m\alpha} \log^j t, \qquad 0 \leqq t \leqq 2/3, \\ z(t) &= h_z(t) + \sum_{m=0}^{k-2} \sum_{j=0}^{m} d_{m,j,z}(1 - t)^{(\alpha-1)/2+m\alpha} \log^j (1 - t), \qquad 1/3 \leqq t \leqq 1, \end{aligned} \tag{1}$$

where $k \geqq 2$ is an integer, $c_{m,j,z}, d_{m,j,z} \in \mathbb{C}$, $g_z, h_z \in \mathbf{H}^s$ $\big(0 \leqq s < (\alpha - 1)/2 + (k - 1) \times \alpha + 1/2\big)$ and $g_z^{(l)}(0) = 0$, $h_z^{(l)}(1) = 0$ $\big(l \in \mathbb{Z}, 0 \leqq l < (\alpha - 1)/2 + (k - 1)\,\alpha\big)$.

In order to deduce convergence or stability, it is useful to introduce an interpolation z_n of the approximate values ξ_j $(j = 0, \ldots, n)$ satisfying $z_n(t_j) = \xi_j$. For numerical computation, one would choose piecewise polynomial interpolation, i.e. a local interpolation procedure. However, to obtain error estimates also in higher Sobolev norms let us introduce the set $\mathcal{S}_n^{\mathbb{R}}$ of smoothest splines of order 5, i.e., $\mathcal{S}_n^{\mathbb{R}}$ is the collection of all $\varphi \in \mathbf{C}^4(\mathbb{R})$ such that the restrictions $\varphi|_{(t_j,t_{j+1})}$, $j \in \mathbb{Z}$ are polynomials of degree less than or equal to 5. Furthermore, let the basis $\{\varphi_j, j \in \mathbb{Z}\}$ be defined by $\varphi_j(t_k) = \delta_{jk}$, $j, k \in \mathbb{Z}$, and let $L_n^{\mathbb{R}}$ denote the orthogonal projection of $\mathbf{L}^2(\mathbb{R})$ onto $\mathcal{S}_n^{\mathbb{R}}$. For a function f on $\mathbb{R}$, we set $K_n^{\mathbb{R}} f := \sum_j f(t_j)\,\varphi_j$, and, for a function g defined on $[0, 1]$, we put $K_n g := \sum_{j=0}^{n} g(t_j)\,\varphi_j$. If $L_n^{\mathbb{R}} f = \sum_{j \in \mathbb{Z}} \lambda_j(f)\,\varphi_j$, then we define L_n by $L_n g := \sum_{j=0}^{n} \lambda_j(g)\,\varphi_j$. Let $A_n \in \mathcal{L}(\operatorname{im} L_n)$ be the operator the matrix of which corresponding to the basis $\{\varphi_j : j = 0, \ldots, n\}$ is the matrix of the system 12.37(15). Then 12.37(15) can be written in the form $A_n z_n = K_n y$, where $z_n := \sum_{j=0}^{n} \xi_j \varphi_j$. Because

$$C^{-1} \frac{\Big(\sum_{j \in \mathbb{Z}} |\xi_j|^2\Big)^{1/2}}{n^{1/2}} \leqq \Big\| \sum_{j \in \mathbb{Z}} \xi_j \varphi_j \Big\|_{\mathbf{L}^2(\mathbb{R})} \leqq C \frac{\Big(\sum_{j \in \mathbb{Z}} |\xi_j|^2\Big)^{1/2}}{n^{1/2}},$$
$$C^{-1} \frac{\Big(\sum_{j=0}^{n} |\xi_j|^2\Big)^{1/2}}{n^{1/2}} \leqq \Big\| \sum_{j=0}^{n} \xi_j \varphi_j \Big\|_{\mathbf{L}^2} \leqq C \frac{\Big(\sum_{j=0}^{n} |\xi_j|^2\Big)^{1/2}}{n^{1/2}}, \tag{2}$$

the $\mathcal{L}(\operatorname{im} L_n)$-norm of A_n is equivalent to the Euclidean norm of its matrix corresponding to $\{\varphi_j\}$. (Here and in the following $C > 0$ denotes a generic constant the value of which varies from instance to instance.) In this section let us suppose the method 12.37 (15) to be stable (cf. Corollary 12.50), i.e., let there exist an integer $n_0 \geqq 1$ such that A_n is invertible for $n \geqq n_0$ and $\sup \|A_n^{-1}\| < \infty$.

We start our error analysis by estimating $\|z_n - z\|_{\mathbf{L}^2}$. Let $A \in \mathcal{L}(\mathbf{L}^2)$ be defined by the left-hand side of 12.37(6), i.e.,

$$(Az)(t) := a(t)\,z(t) + b(t)\,(Sz)(t) + \frac{b(t)}{\pi \mathrm{i}}\,(K_0 z)(t),$$

$$(Sz)(t) := \frac{1}{\pi \mathrm{i}} \int_0^1 \frac{z(\tau)}{\tau - t}\,\mathrm{d}\tau, \qquad (K_0 z)(t) := \int_0^1 l(t, \tau)\,z(\tau)\,\mathrm{d}\tau.$$

Then the formula

$$z_n - z = (K_n - I)\,z + A_n^{-1}\{K_n A z - A_n K_n z\}$$

leads to the inequality

$$\|z_n - z\|_{\mathbf{L}^2} \leqq C\{\|(K_n - I)\,z\|_{\mathbf{L}^2} + \|K_n A z - A_n K_n z\|_{\mathbf{L}^2}\}. \tag{3}$$

Here the first term on the right-hand side can be estimated utilizing the approximation property of $\mathcal{S}_n^{\mathbb{R}}$. In fact, for $0 \leqq s \leqq u \leqq r \leqq 6$, $s < 5.5$, $1/2 < r$ and any $f \in \mathbf{H}^r(\mathbb{R})$,

$g \in \mathbf{H}^u(\mathbb{R})$ we have (see Chapter 2)

$$\begin{aligned} \|(I - L_n^{\mathbb{R}})\, g\|_{\mathbf{H}^s(\mathbb{R})} &\leqq C n^{s-u}\, \|g\|_{\mathbf{H}^u(\mathbb{R})}, \\ \|(I - K_n^{\mathbb{R}})\, f\|_{\mathbf{H}^s(\mathbb{R})} &\leqq C n^{s-r}\, \|f\|_{\mathbf{H}^r(\mathbb{R})}. \end{aligned} \tag{4}$$

In particular, if $f \in \mathbf{H}^r(\mathbb{R})$ and f vanishes outside the interval $[0, 1]$, then $K_n f = K_n^{\mathbb{R}} f$ and we get

$$\|(I - K_n)\, f\|_{\mathbf{H}^s} \leqq C n^{s-r}\, \|f\|_{\mathbf{H}^r}. \tag{5}$$

For the second term on the right-hand side of (3), we obtain

$$K_n A z - A_n K_n z = K_n b|_{\operatorname{im} L_n} \left\{ [K_n S z - S_n K_n z] + \frac{1}{\pi \mathrm{i}}\, [K_n K_0 z - K_{0,n} K_n z] \right\}, \tag{6}$$

where the operators S_n and $K_{0,n}$ are defined by the matrices

$$S_n := \left(\frac{1}{\pi \mathrm{i}} \left[\frac{(1 - \delta_{jk})}{t_j - t_k}\, \omega_j + \nu_k n\, \frac{8(\delta_{j,k+1} - \delta_{j,k-1}) - (\delta_{j,k+2} - \delta_{j,k-2})}{12} \right] \right)_{k,j=0}^{n},$$

$$K_{0,n} := \left(l(t_k, t_j)\, \omega_j \right)_{k,j=0}^{n},$$

respectively. Since $\|K_n b|_{\operatorname{im} L_n}\| \leqq \sup |b(t)|$, it remains to estimate $\|K_n S z - S_n K_n z\|_{\mathbf{L}^2}$ and $\|K_n K_0 z - K_{0,n} K_n z\|_{\mathbf{L}^2}$.

The first step in estimating $\|K_n S z - S_n K_n z\|$ will be Lemma 12.39, where the rate of convergence for a more general approximate operator on the real axis is established. Thus set $K_n^0 f := \sum f(t_{2j})\, \varphi_{2j}$, $K_n^1 f := \sum f(t_{2j+1})\, \varphi_{2j+1}$, i.e., $K_n^{\mathbb{R}} = K_n^0 + K_n^1$. If $L_n^{\mathbb{R}} f = \sum \lambda_j(f)\, \varphi_j$, then set $L_n^0 f := \sum \lambda_{2j}(f)\, \varphi_{2j}$, $L_n^1 f := \sum \lambda_{2j+1}(f)\, \varphi_{2j+1}$. For a 2×2-matrix function $\boldsymbol{a} = (\boldsymbol{a}^{r,s})_{r,s=0}^1$ on the unit circle $\mathbb{T}$, denote the k-th Fourier coefficient by $\boldsymbol{a}_k := (\boldsymbol{a}_k^{r,s})_{r,s=0}^1$ and the block convolution matrix $(\boldsymbol{a}_{k-j})_{k,j \in \mathbb{Z}}$ by $C(\boldsymbol{a})$. We identify this matrix with the operator $C(\boldsymbol{a}) \in \mathscr{L}(\operatorname{im} L_n^{\mathbb{R}})$ defined by

$$\begin{aligned} C(\boldsymbol{a})\, \varphi_{2j} &= \sum_{k \in \mathbb{Z}} \{\boldsymbol{a}_{k-j}^{0,0} \varphi_{2k} + \boldsymbol{a}_{k-j}^{1,0} \varphi_{2k+1}\}, \\ C(\boldsymbol{a})\, \varphi_{2j+1} &= \sum_{k \in \mathbb{Z}} \{\boldsymbol{a}_{k-j}^{0,1} \varphi_{2k} + \boldsymbol{a}_{k-j}^{1,1} \varphi_{2k+1}\}. \end{aligned} \tag{7}$$

Let $S_{\mathbb{R}}$ denote the Cauchy singular operator on $\mathbb{R}$ and set $P_{\mathbb{R}} := 1/2(I + S_{\mathbb{R}})$, $Q_{\mathbb{R}} := I - P_{\mathbb{R}}$.

12.39. Lemma. *Let $\boldsymbol{a}$ be piecewise continuous and suppose that the functions $p(\mathrm{e}^{\mathrm{i}2\pi\lambda}) := \boldsymbol{a}^{0,0}(\mathrm{e}^{\mathrm{i}2\pi\lambda}) + \boldsymbol{a}^{0,1}(\mathrm{e}^{\mathrm{i}2\pi\lambda})\, \mathrm{e}^{-\mathrm{i}\pi\lambda}$, $q(\mathrm{e}^{\mathrm{i}2\pi\lambda}) := \boldsymbol{a}^{1,0}(\mathrm{e}^{\mathrm{i}2\pi\lambda})\, \mathrm{e}^{\mathrm{i}\pi\lambda} + \boldsymbol{a}^{1,1}(\mathrm{e}^{\mathrm{i}2\pi\lambda})$ $(-1/2 < \lambda < 1/2)$ satisfy $p(1 \pm 0) = q(1 \pm 0)$ and $p^{(j)}(1 \pm 0) = q^{(j)}(1 \pm 0) = 0$ for $j = 1, 2, \ldots, l - 1$, where l is an integer and $0 < l \leqq 6$. Let $1/2 < s \leqq l$. Then there exists a constant $C > 0$ such that, for any $f \in \mathbf{H}^s(\mathbb{R})$,*

$$\|C(\boldsymbol{a})\, K_n^{\mathbb{R}} f - K_n^{\mathbb{R}}[p(1 - 0)\, P_{\mathbb{R}} + q(1 + 0)\, Q_{\mathbb{R}}]\, f\|_{\mathbf{L}^2(\mathbb{R})} \leqq C n^{-s}\, \|f\|_{\mathbf{H}^s(\mathbb{R})}. \tag{1}$$

Proof. It is well known that the vector $(\mathrm{e}^{-\mathrm{i}2\pi\lambda j})_{j \in \mathbb{Z}}$ $(-1/2 < \lambda \leqq 1/2)$ is an eigenvector of the convolution matrix $(\boldsymbol{a}_{k-j}^{r,s})_{k,j \in \mathbb{Z}}$ corresponding to the eigenvalue $\boldsymbol{a}^{r,s}(\mathrm{e}^{\mathrm{i}2\pi\lambda})$. Furthermore, for $g^\lambda(t) := \mathrm{e}^{-\mathrm{i}2\pi\lambda t}$, we obtain

$$K_n^0 g^\lambda = \sum_j \mathrm{e}^{-\mathrm{i}2\pi(2\lambda/n)j} \varphi_{2j}, \qquad K_n^1 g^\lambda = \mathrm{e}^{-\mathrm{i}2\pi\lambda/n} \sum_j \mathrm{e}^{-\mathrm{i}2\pi(2\lambda/n)j} \varphi_{2j+1}$$

and

$$C(\boldsymbol{a})\, K_n^0 g^\lambda = \boldsymbol{a}^{0,0}(\mathrm{e}^{\mathrm{i}2\pi(2\lambda/n)})\, K_n^0 g^\lambda + \boldsymbol{a}^{1,0}(\mathrm{e}^{\mathrm{i}2\pi(2\lambda/n)})\, \mathrm{e}^{\mathrm{i}2\pi\lambda/n} K_n^1 g^\lambda,$$
$$C(\boldsymbol{a})\, K_n^1 g^\lambda = \boldsymbol{a}^{0,1}(\mathrm{e}^{\mathrm{i}2\pi(2\lambda/n)})\, \mathrm{e}^{-\mathrm{i}2\pi\lambda/n} K_n^0 g^\lambda + \boldsymbol{a}^{1,1}(\mathrm{e}^{\mathrm{i}2\pi(2\lambda/n)})\, K_n^1 g^\lambda,$$
$$C(\boldsymbol{a})\, K_n^{\mathbf{R}} g^\lambda = p(\mathrm{e}^{\mathrm{i}2\pi(2\lambda/n)})\, K_n^0 g^\lambda + q(\mathrm{e}^{\mathrm{i}2\pi(2\lambda/n)})\, K_n^1 g^\lambda.$$

Hence,

$$C(\boldsymbol{a})\, K_n^{\mathbf{R}} g^\lambda = r(\mathrm{e}^{\mathrm{i}2\pi(2\lambda/n)})\, K_n^0 g^\lambda + s(\mathrm{e}^{\mathrm{i}2\pi(2\lambda/n)})\, K_n^1 g^\lambda + \begin{cases} p(1+0)\, K_n^{\mathbf{R}} g^\lambda & \text{if} \quad \lambda > 0, \\ p(1-0)\, K_n^{\mathbf{R}} g^\lambda & \text{if} \quad \lambda < 0, \end{cases} \tag{2}$$

where r and s satisfy

$$|r(\mathrm{e}^{\mathrm{i}2\pi\lambda})| \leqq \min\{C, |\lambda|^l\}, \qquad |s(\mathrm{e}^{\mathrm{i}2\pi\lambda})| \leqq \min\{C, |\lambda|^l\} \tag{3}$$

and $C > 0$ is a constant. If $\mathcal{F}f$ denotes the Fourier transform,

$$\mathcal{F}f(t) := \int_{\mathbf{R}} f(\tau)\, \mathrm{e}^{\mathrm{i}2\pi\tau t}\, \mathrm{d}\tau,$$

then

$$f = \int_{\mathbf{R}} \mathcal{F}f(\lambda)\, g^\lambda\, \mathrm{d}\lambda, \qquad Q_{\mathbf{R}} f = \int_0^\infty \mathcal{F}f(\lambda)\, g^\lambda\, \mathrm{d}\lambda, \qquad P_{\mathbf{R}} f = \int_{-\infty}^0 \mathcal{F}f(\lambda)\, g^\lambda\, \mathrm{d}\lambda, \tag{4}$$

and we arrive at

$$C(\boldsymbol{a})\, K_n^{\mathbf{R}} f = \int_{\mathbf{R}} \mathcal{F}f(\lambda)\, [C(\boldsymbol{a})\, K_n^{\mathbf{R}} g^\lambda]\, \mathrm{d}\lambda.$$

Eqs. (2), (4) lead to

$$\begin{aligned} C(\boldsymbol{a})\, K_n^{\mathbf{R}} f &= \int_0^\infty \mathcal{F}f(\lambda)\, p(1+0)\, K_n^{\mathbf{R}} g^\lambda\, \mathrm{d}\lambda + \int_{-\infty}^0 \mathcal{F}f(\lambda)\, p(1-0)\, K_n^{\mathbf{R}} g^\lambda\, \mathrm{d}\lambda \\ &\quad + \int_{\mathbf{R}} \mathcal{F}f(\lambda)\, [r(\mathrm{e}^{\mathrm{i}2\pi(2\lambda/n)})\, K_n^0 g^\lambda + s(\mathrm{e}^{\mathrm{i}2\pi(2\lambda/n)})\, K_n^1 g^\lambda]\, \mathrm{d}\lambda \\ &= K_n^{\mathbf{R}}\{p(1-0)\, P_{\mathbf{R}} + q(1+0)\, Q_{\mathbf{R}}\}\, f + I_1 + I_2, \end{aligned} \tag{5}$$
$$I_1 := \int_{\mathbf{R}} \mathcal{F}f(\lambda)\, r(\mathrm{e}^{\mathrm{i}2\pi(2\lambda/n)})\, K_n^0 g^\lambda\, \mathrm{d}\lambda, \qquad I_2 := \int_{\mathbf{R}} \mathcal{F}f(\lambda)\, s(\mathrm{e}^{\mathrm{i}2\pi(2\lambda/n)})\, K_n^1 g^\lambda\, \mathrm{d}\lambda.$$

Setting $I_3 := L_n^0 \int_{\mathbf{R}} \mathcal{F}f(\lambda)\, r(\mathrm{e}^{\mathrm{i}2\pi(2\lambda/n)})\, g^\lambda\, \mathrm{d}\lambda$ and $I_4 := I_1 - I_3$, we get

$$\begin{aligned} I_1 &= I_3 + I_4, \\ \|I_3\|_{\mathbf{L}^2(\mathbf{R})} &\leqq C \left\| \int_{\mathbf{R}} \mathcal{F}f(\lambda)\, r(\mathrm{e}^{\mathrm{i}2\pi(2\lambda/n)})\, g^\lambda\, \mathrm{d}\lambda \right\|_{\mathbf{L}^2(\mathbf{R})} = C\, \|\mathcal{F}f(\lambda)\, r(\mathrm{e}^{\mathrm{i}2\pi(2\lambda/n)})\|_{\mathbf{L}^2(\mathbf{R})}. \end{aligned} \tag{6}$$

Using (3), we obtain that

$$\|I_3\|_{\mathbf{L}^2(\mathbf{R})} \leqq Cn^{-l}\, \|\mathcal{F}f(\lambda)\, \lambda^l\|_{\mathbf{L}^2(\mathbf{R})} \leqq Cn^{-l}\, \|f\|_{\mathbf{H}^l(\mathbf{R})}. \tag{7}$$

The definition of I_4 and formulas 12.38(2) and 12.38(4) yield

$$\begin{aligned}\|I_4\|_{\mathbf{L}^2(\mathbf{R})} &= \left\|(K_n^0 - L_n^0)\int_{\mathbf{R}} \mathcal{F}f(\lambda)\, r(e^{i2\pi(2\lambda/n)})\, g^\lambda\, d\lambda\right\|_{\mathbf{L}^2(\mathbf{R})} \\ &\leqq \left\|(K_n^{\mathbf{R}} - L_n^{\mathbf{R}})\int_{\mathbf{R}} \mathcal{F}f(\lambda)\, r(e^{i2\pi(2\lambda/n)})\, g^\lambda\, d\lambda\right\|_{\mathbf{L}^2(\mathbf{R})} \\ &\leqq Cn^{-l}\left\|\int_{\mathbf{R}} \mathcal{F}f(\lambda)\, r(e^{i2\pi(2\lambda/n)})\, g^\lambda\, d\lambda\right\|_{\mathbf{H}^l(\mathbf{R})} \leqq Cn^{-l}\,\|f\|_{\mathbf{H}^l(\mathbf{R})}. \end{aligned} \tag{7'}$$

Formulas (5)–(7′) together with analogous arguments for I_2 prove (1) in case $s = l$ is an integer. Thus Lemma 12.39 follows by interpolation. ■

We now consider the operator $S_{\mathbf{R},n} \in \mathcal{L}(\text{im } L_n^{\mathbf{R}})$ given by its matrix representation corresponding to the basis $\{\varphi_j : j \in \mathbb{Z}\}$, i.e.

$$S_{\mathbf{R},n} := \left(\frac{1}{\pi i}\left[\frac{(1-\delta_{jk})}{t_j - t_k}\,\theta_j + \theta_k n\,\frac{8(\delta_{j,k+1} - \delta_{j,k-1}) - (\delta_{j,k+2} - \delta_{j,k-2})}{12}\right]\right)_{k,j\in\mathbb{Z}},$$

where $\theta_j := 4/3n$ if j is odd and $\theta_j := 2/3n$ if j is even. Define $z(t) := 0$ for $t \notin [0, 1]$. Since $K_nSz - S_nK_nz$ is a projection of $K_n^{\mathbf{R}}S_{\mathbf{R}}z - S_{\mathbf{R},n}K_n^{\mathbf{R}}z$ into im L_n, we get (cf. 12.38(2))

$$\|K_nSz - S_nK_nz\|_{\mathbf{L}^2} \leqq \|K_n^{\mathbf{R}}S_{\mathbf{R}}z - S_{\mathbf{R},n}K_n^{\mathbf{R}}z\|_{\mathbf{L}^2(\mathbf{R})}. \tag{8}$$

On the other hand, the equality $S_{\mathbf{R},n} = C(\boldsymbol{a})$ holds, where

$$\begin{aligned}(\boldsymbol{a}_{k-j}^{00})_{k,j\in\mathbb{Z}} &= \left(\frac{2}{3n}\,\frac{1}{\pi i}\left[\frac{1-\delta_{2j,2k}}{t_{2j} - t_{2k}} - n\,\frac{\delta_{2j,2k+2} - \delta_{2j,2k-2}}{12}\right]\right)_{k,j\in\mathbb{Z}} \\ &= \left(\frac{1}{3}\,\frac{1}{\pi i}\,\frac{1-\delta_{jk}}{j-k} - \frac{1}{18}\,\frac{1}{\pi i}\,(\delta_{j,k+1} - \delta_{j,k-1})\right)_{k,j\in\mathbb{Z}},\end{aligned}$$

$$\begin{aligned}(\boldsymbol{a}_{k-j}^{11})_{k,j\in\mathbb{Z}} &= \left(\frac{4}{3n}\,\frac{1}{\pi i}\left[\frac{1-\delta_{2j+1,2k+1}}{t_{2j+1} - t_{2k+1}} - n\,\frac{\delta_{2j+1,2k+2} - \delta_{2j+1,2k-1}}{12}\right]\right)_{k,j\in\mathbb{Z}} \\ &= \left(\frac{2}{3}\,\frac{1}{\pi i}\,\frac{1-\delta_{jk}}{j-k} - \frac{1}{9}\,\frac{1}{\pi i}\,(\delta_{j,k+1} - \delta_{j,k-1})\right)_{k,j\in\mathbb{Z}},\end{aligned}$$

$$(\boldsymbol{a}_{k-j}^{01})_{k,j\in\mathbb{Z}} = \left(\frac{2}{3}\,\frac{1}{\pi i}\,\frac{1}{(j-k)+1/2} + \frac{4}{9}\,\frac{1}{\pi i}\,(\delta_{jk} - \delta_{j,k-1})\right)_{k,j\in\mathbb{Z}},$$

$$(\boldsymbol{a}_{k-j}^{10})_{k,j\in\mathbb{Z}} = \left(\frac{1}{3}\,\frac{1}{\pi i}\,\frac{1}{(j-k)-1/2} + \frac{8}{9}\,\frac{1}{\pi i}\,(\delta_{j,k+1} - \delta_{jk})\right)_{k,j\in\mathbb{Z}}.$$

An elementary computation of the Fourier coefficients shows that

$$\boldsymbol{a}^{00}(e^{i2\pi\lambda}) = \frac{1}{3}\,(2\lambda - 1) - \frac{1}{18}\,\frac{1}{\pi i}\,(e^{-i2\pi\lambda} - e^{i2\pi\lambda}),$$

$$\boldsymbol{a}^{11}(e^{i2\pi\lambda}) = \frac{2}{3}\,(2\lambda - 1) - \frac{1}{9}\,\frac{1}{\pi i}\,(e^{-i2\pi\lambda} - e^{i2\pi\lambda}),$$

$$\boldsymbol{a}^{01}(e^{i2\pi\lambda}) = \frac{2}{3}\,(-e^{i\pi\lambda}) + \frac{4}{9}\,\frac{1}{\pi i}\,(1 - e^{i2\pi\lambda}),$$

$$\boldsymbol{a}^{10}(e^{i2\pi\lambda}) = \frac{1}{3}\,(-e^{-i\pi\lambda}) + \frac{8}{9}\,\frac{1}{\pi i}\,(e^{-i2\pi\lambda} - 1).$$

Consequently, Lemma 12.39 applies with $l = 5$ and we find

$$\|K_n^{\mathbb{R}} S_{\mathbb{R}} f - S_{\mathbb{R},n} K_n^{\mathbb{R}} f\|_{\mathbf{L}^2(\mathbb{R})} \leqq C n^{-s} \|f\|_{\mathbf{H}^s(\mathbb{R})}. \tag{9}$$

By 12.38(1) we conclude that $z \in \mathbf{H}^s(\mathbb{R})$ for $0 \leqq s < \alpha/2$. Therefore, if $1/2 < s \leqq 5$, $s < \alpha/2$, then formulas (8), (9) imply

$$\|K_n S z - S_n K_n z\|_{\mathbf{L}^2} \leqq C n^{-s}. \tag{10}$$

In order to simplify our considerations, we shall give coarse estimates for $\|K_n K_0 z - K_{0,n} K_n z\|_{\mathbf{L}^2}$. By 12.38(2) we see that

$$\begin{aligned} \|K_n K_0 z - K_{0,n} K_n z\|_{\mathbf{L}^2} &\leqq \sup_{k=0,\dots,n} |K_0 z(t_k) - K_{0,n} K_n z(t_k)| \\ &\leqq \sup_{0 \leqq t \leqq 1} \left| \int_0^1 l(t,\tau)\, z(\tau)\, \mathrm{d}\tau - \sum_{j=0}^{n} l(t, t_k)\, z(t_k)\, \omega_k \right| \\ &\leqq C n^{-s} \sup_{0 \leqq t \leqq 1} \|l(t,\cdot)\, z\|_{\mathbf{W}_1^s}, \end{aligned} \tag{11}$$

where $0 \leqq s \leqq 4$ and $\mathbf{W}_1^s$ denotes the Sobolev-Slobodetski space of power 1 and order s (cf. Triebel [1], Secs. 2.3, 2.5). It remains to examine whether the norms $\|l(t,\cdot)\, z\|_{\mathbf{W}_1^s}$ are uniformly bounded or not.

In case $\tau, t < 1/3$, we have

$$l(t,\tau) = \begin{cases} -1/\tau & \text{if } t = 0, \\ \dfrac{\alpha t^{(\alpha-1)/2-\alpha\varkappa_0} \tau^{(\alpha-1)/2+\alpha\varkappa_0}}{\tau^\alpha - t^\alpha} - \dfrac{1}{\tau - t} & \text{if } t \neq 0. \end{cases}$$

Setting $x = t/\tau$ $(0 < x < \infty)$, we may write

$$l(t,\tau) = \frac{1}{\tau} \begin{cases} -1 & \text{if } t = 0, \\ \dfrac{\alpha x^{(\alpha-1)/2-\alpha\varkappa_0}}{1 - x^\alpha} - \dfrac{1}{1-x} & \text{if } t \neq 0, \end{cases} \tag{12}$$

$$l(t,\tau) = \frac{1}{\tau} \begin{cases} -1 & \text{if } t = 0, \\ \dfrac{I_1 - I_2}{\int_0^1 \big(x + \mu(1-x)\big)^{\alpha-1} \mathrm{d}\mu} & \text{if } t \neq 0, \end{cases} \tag{13}$$

$$\begin{aligned} I_1 := \left[\frac{\alpha-1}{2} + \alpha\varkappa_0\right] \int_0^1 \int_0^1 (1-\mu)\, [x + \mu(1-x)]^{(\alpha-1)/2-\alpha\varkappa_0} \\ \times [1 - (1-\nu)(1-\mu)(1-x)]^{(\alpha-1)/2+\alpha\varkappa_0-1}\, \mathrm{d}\nu\, \mathrm{d}\mu, \end{aligned}$$

$$I_2 := \left[\frac{\alpha-1}{2} - \alpha\varkappa_0\right] \int_0^1 \int_0^1 \mu [x + \mu(1-\nu)(1-x)]^{(\alpha-1)/2-\alpha\varkappa_0-1}\, \mathrm{d}\nu\, \mathrm{d}\mu.$$

Consequently, the function $\tau \to \tau l(t,\tau)$ is uniformly bounded with respect to t. Differentiating (12) and (13), we conclude that the function $\tau \mapsto \tau^{l+1} (\partial/\partial\tau)^l\, l(t,\tau)$ is uniformly bounded if l is a positive integer. In case $\tau > 2/3$, $t < 1/3$, we get

$$l(t,\tau) = \begin{cases} -1/\tau & \text{if } t = 0, \\ \dfrac{\alpha t^{(\alpha-1)/2-\alpha\varkappa_0}(1-\tau)^{(\alpha-1)/2+\alpha\varkappa_0}}{1-(1-\tau)^\alpha - t^\alpha} - \dfrac{1}{\tau - t} & \text{if } t \neq 0. \end{cases}$$

We now have to suppose $(\alpha - 1)/2 - \alpha\varkappa_0 \geqq 0$ in order to guarantee the uniform boundedness with respect to t. If this assumption is fulfilled, then the function $\tau \mapsto (1-\tau)^{l+1} \times (\partial/\partial\tau)^l\, l(t,\tau)$ is uniformly bounded for $\tau \geqq 2/3$, $t \leqq 1/3$. The other cases can be treated analogously. Thus, assuming $\alpha \geqq 1/(1 - 2\varkappa_i)$ $(i = 0, 1)$, we obtain that the function $\tau \mapsto \tau^{l+1}(1-\tau)^{l+1}\,(\partial/\partial\tau)^l\, l(t,\tau)$ is uniformly bounded with respect to t, where l is an arbitrary non-negative integer.

This fact together with 12.38(1) implies that

$$\sup_{0 \leqq t \leqq 1} \|l(t,\cdot)\, z\|_{\mathbf{W}_1^s} < \infty \tag{14}$$

for any s satisfying $0 \leqq s < (\alpha - 1)/2$. Indeed, consider for example, $z(\tau) = \tau^{(\alpha-1)/2}$ (cf. 12.38(1)), $0 \leqq \tau \leqq 1/3$ and $s \in \mathbb{Z}$, $s > 0$. Then, for certain constants C_j,

$$\begin{aligned}(\partial/\partial\tau)^s \{l(t,\tau)\,\tau^{(\alpha-1)/2}\} &= \sum_{j=0}^{s} C_j (\partial/\partial\tau)^j\, l(t,\tau)\,\tau^{(\alpha-1)/2-(s-j)} \\ &= \left\{\sum_{j=0}^{s} C_j \tau^{j+1} (\partial/\partial\tau)^j\, l(t,\tau)\right\} \tau^{(\alpha-1)/2-s-1}.\end{aligned}$$

The last function is integrable if $(\alpha - 1)/2 - s - 1 > -1$, i.e., $s < (\alpha - 1)/2$. If $s \notin \mathbb{Z}$, $s > 0$, then another straightforward argumentation including the special definition of the norm in $\mathbf{W}_1^s$ leads to the same result.

12.40. Theorem. *Suppose all the assumptions of Sections* 12.38–12.39 *are fulfilled. Then* $\|z_n - z\|_{\mathbf{L}^2} \leqq Cn^{-s}$, *where* $1/2 < s < (\alpha - 1)/2$, $s \leqq 4$.

Proof. This theorem follows from 12.38(3), 12.38(5), 12.38(6), 12.39(10), 12.39(11) and 12.39(14). ■

We now define $\mathbf{z}_n$ by $\mathbf{z}_n(\gamma(\tau))\,\gamma'(\tau)^{1/2} = z_n(\tau)$, $0 \leqq \tau \leqq 1$, i.e., we set $\mathbf{z}_n(t) := z_n(\gamma^{-1}(t))/\gamma'(\gamma^{-1}(t))^{1/2}$. In view of 12.37(3) we obtain $\|\mathbf{z}_n - \mathbf{z}\|_{\mathbf{L}^2} = \|z_n - z\|_{\mathbf{L}^2}$.

12.41. Corollary. *Suppose all the assumptions of Sections* 12.38–12.39 *are fulfilled,* $1/2 < s < (\alpha - 1)/2$ *and* $s \leqq 4$. *Then the following estimates hold.*

(i) $\|\mathbf{z}_n - \mathbf{z}\|_{\mathbf{L}^2} \leqq Cn^{-s}$.

(ii) *If* $0 \leqq r \leqq s$, *then* $\|z_n - z\|_{\mathbf{H}^r} \leqq Cn^{r-s}$.

(iii) *Suppose* $0 < \varepsilon < 1/2$ *is fixed and* $s < 4$. *Then*

$$\sup_{\varepsilon \leqq t \leqq 1-\varepsilon} |\mathbf{z}_n(t) - \mathbf{z}(t)| \leqq Cn^{1/2-s}.$$

Proof. Estimate (i) follows from Theorem 12.40. Furthermore, an arbitrary function φ from im $L_n^{\mathbb{R}}$ satisfies the inverse property (cf. Sec. 2.11), i.e. $0 \leqq s \leqq r < 5.5$ implies that

$$\|\varphi\|_{\mathbf{H}^r(\mathbb{R})} \leqq Cn^{r-s}\,\|\varphi\|_{\mathbf{H}^s(\mathbb{R})}. \tag{1}$$

Thus we obtain

$$\begin{aligned}\|z_n - z\|_{\mathbf{H}^r} &\leqq \|z_n - L_n^{\mathbb{R}} z\|_{\mathbf{H}^r} + \|z - L_n^{\mathbb{R}} z\|_{\mathbf{H}^r} \leqq Cn^r\,\|z_n - L_n^{\mathbb{R}} z\|_{\mathbf{L}^2(\mathbb{R})} + \|z - L_n^{\mathbb{R}} z\|_{\mathbf{H}^r} \\ &\leqq Cn^r\,\|z_n - z\|_{\mathbf{L}^2} + Cn^r\,\|z - L_n^{\mathbb{R}} z\|_{\mathbf{L}^2(\mathbb{R})} + \|z - L_n^{\mathbb{R}} z\|_{\mathbf{H}^r(\mathbb{R})}.\end{aligned}$$

Together with 12.38(4) and Theorem 12.40, this implies (ii). Assertion (iii) follows from

$$\sup_{\varepsilon\le 1\le t-\varepsilon} |\mathbf{z}_n(t) - \mathbf{z}(t)| \le C \sup_{0\le t\le 1} |z_n(t) - z(t)|,$$

the embedding $\mathbb{C}[0, 1] \supset \mathbf{H}^{1/2+\delta}[0, 1]$ $(\delta > 0)$ and (ii). ∎

12.42. Remark. Corollary 12.41 states that $\|z_n - z\|_{\mathbf{H}^r}$ is of order n^{r-s}, $s := \min\{4, (\alpha - 1)/2\}$. In view of the coarse estimate 12.39(11), we conjecture $\|z_n - z\|_{\mathbf{H}^r} \le Cn^{r-s}$, $s := \min\{4, \alpha/2\}$. Furthermore, the use of the embedding $\mathbb{C}[0, 1] \supset \mathbf{H}^{1/2+\delta}[0, 1]$ in the proof of (iii) leads to a similar effect. Thus we conjecture that $\sup_{\varepsilon\le t\le 1-\varepsilon} |\mathbf{z}_n(t) - \mathbf{z}(t)|$ is of order n^{-s}, where $s := \min\{4, \alpha/2\}$.

Now let us consider the convergence near the end-points of the interval. By Corollary 12.41(ii) this problem is solved for z_n. However, when passing from $z_n(t_k)$ to $\mathbf{z}_n(\gamma(t_k))$, we divide by the small number $\gamma'(t_k)^{1/2}$. Multiplication by an unbounded function is an archetypal example of an ill-posed problem. The corresponding theory proposes regularization. More exactly, we approximate $\mathbf{z}(0)$ not by $\mathbf{z}_n(0)$, but by $\mathbf{z}_n(\gamma(\psi_n))$, where ψ_n is a small number yet to be specified. From the asymptotics of $\mathbf{z}$ (cf. 12.38(1)) we conclude that

$$|\mathbf{z}(0) - \mathbf{z}(\gamma(\psi_n))| \le C\gamma(\psi_n) \log \gamma(\psi_n). \tag{1}$$

Using the definition of γ, z and $\mathbf{z}_n$ as well as Corollary 12.41 (ii), we get

$$|\mathbf{z}(\gamma(\psi_n)) - \mathbf{z}_n(\gamma(\psi_n))| \le C \frac{|z(\psi_n) - z_n(\psi_n)|}{\gamma'(\psi_n)^{1/2}} \le C \frac{n^{-s}}{\psi_n^{(\alpha-1)/2}}, \tag{2}$$

where $0 < s < \min\{3.5, (\alpha - 2)/2\}$. The sum of the left-hand sides in (1), (2) becomes small if we choose ψ_n such that $\psi_n^\alpha \log \psi_n \approx n^{-s}/\psi_n^{(\alpha-1)/2}$, i.e. $\psi_n := Cn^{-2s/(3\alpha-1)}$. Due to this choice and formulas (1), (2) we have

$$\gamma(\psi_n) = n^{-2\alpha s/(3\alpha-1)}, \qquad |\mathbf{z}(0) - \mathbf{z}_n(\gamma(\psi_n))| \le Cn^{-2\alpha s/(3\alpha-1)} \log n.$$

12.43. Corollary. *Suppose all the assumptions of the preceding sections are fulfilled,* $\alpha > 2$, $0 < s < \min\{7\alpha/(3\alpha - 1), (\alpha(\alpha - 2))/(3\alpha - 1)\}$, *and choose* $\gamma(\psi_n) = Cn^{-s}$. *Then*

$$|\mathbf{z}(0) - \mathbf{z}_n(\gamma(\psi_n))| \le Cn^{-s} \log n.$$

Remark. If our conjectures in Remark 12.42 are true, then Corollary 12.43 holds with $s < \min\{8\alpha/(3\alpha - 1), \alpha^2/(3\alpha - 1)\}$.

12.44. Stability of the quadrature method. In the following sections we shall give necessary and sufficient conditions for the stability of method 12.37(15). These conditions are of local nature (cf. Theorem 12.49). Therefore, in a first step we analyse the quadrature method for a singular integral operator with constant coefficients on the half axis or on the entire real axis. Analogously to the Fredholm theory of singular integral operators, we introduce the operator $A^0 := \mathsf{a}(0)I + \mathsf{b}(0)S_{\mathbb{R}^+} + (\mathsf{b}(0)/\pi\mathrm{i})K_0^0 \in \mathcal{L}(\mathbf{L}^2(\mathbb{R}^+))$, where $S_{\mathbb{R}^+}$ is the Cauchy singular integral operator on the positive real half axis and K_0^0 is defined by

$$K_0^0 z(t) := \int_0^\infty k_0^0(t, \tau)\, z(\tau)\, \mathrm{d}\tau,$$

$$k_0^0(t, \tau) := \left\{\alpha \frac{t^{(\alpha-1)/2-\alpha\varrho_0}\tau^{(\alpha-1)/2+\alpha\varrho_0}}{\tau^\alpha - t^\alpha} - \frac{1}{\tau - t}\right\}. \tag{1}$$

In a sense, A^0 is locally equivalent at 0 to the operator A defined by the left-hand side of 12.37(6) (cf. also the operator A introduced in Sec. 12.38). From

$$k_0^0(t,\tau) = \frac{1}{\tau}\, k^0\left(\frac{t}{\tau}\right), \qquad k^0(x) := \left\{\frac{\alpha x^{(\alpha-1)/2-\alpha\varrho_0}}{1-x^\alpha} - \frac{1}{1-x}\right\} \tag{2}$$

it follows that A^0 is a Mellin convolution operator, i.e. $A^0 := M^{-1}mM$, where M is the Mellin transform

$$Mz(\zeta) := \int_0^\infty t^{\zeta-1}\, z(t)\,\mathrm{d}t, \qquad M: \mathbf{L}^2(\mathbb{R}^+) \to \mathbf{L}^2(\{\zeta \in \mathbb{C}: \operatorname{Re}\zeta = 1/2\}),$$

and m denotes the operator of multiplication by the symbol function

$$m(\zeta) := \mathsf{a}(0) + \mathsf{b}(0)\,(-\mathrm{i})\cot\left[\pi\left(\frac{\zeta + (\alpha-1)/2}{\alpha} - \varrho_0\right)\right], \qquad \zeta \in \mathbb{C}, \qquad \operatorname{Re}\zeta = 1/2. \tag{3}$$

Obviously, A^0 is invertible if and only if $m(\zeta)$ does not vanish on the line $\{\zeta \in \mathbb{C}: \operatorname{Re}\zeta = 1/2\}$, i.e., if and only if

$$\frac{\mathsf{a}(0)+\mathsf{b}(0)}{\mathsf{a}(0)-\mathsf{b}(0)} \notin \{t\,\mathrm{e}^{\mathrm{i}2\pi\varrho_0}: -\infty < t \leqq 0\}. \tag{4}$$

The "local representative" of the method 12.37(15) at the point $\tau = 0$ will be shown to be the following quadrature method for the equation $A^0 z = y$. Let $y \in \mathbf{L}^2(\mathbb{R}^+)$ be Riemann integrable. Determine approximate values ξ_j for $z(t_j)$ $(k = u, u+1, \ldots)$ by solving the system (cf. 12.37(15))

$$\begin{aligned} y(t_k) = \mathsf{a}(0)\,\xi_k + \frac{\mathsf{b}(0)}{\pi\mathrm{i}} \Bigg\{ &\sum_{\substack{j=u\\ j\neq k}}^{\infty} \alpha\, \frac{t_k^{(\alpha-1)/2-\alpha\varrho_0} t_j^{(\alpha-1)/2+\alpha\varrho_0}}{t_j^\alpha - t_k^\alpha}\, \omega_j \xi_j \\ &+ \eta_k n\, \frac{8(\xi_{k+1}-\xi_{k-1}) - (\xi_{k+2}-\xi_{k-2})}{12} + \omega_k \alpha\varrho_0 \frac{1}{t_k}\,\xi_k \Bigg\}, \quad k = u, u+1, \ldots \end{aligned} \tag{5}$$

Here we set $\xi_k := 0$ for $k < u$ and

$$\eta_j := \begin{cases} 4/3n & \text{if } j \text{ is odd}, \\ 2/3n & \text{if } j \text{ is even}, \end{cases} \qquad \omega_j := \begin{cases} 4/3n & \text{if } j \text{ is odd}, \\ 2/3n & \text{if } j = 2, 4, \ldots, \\ 1/3n & \text{if } j = 0. \end{cases}$$

We start our analysis of method (5) in the special case $u = 0$. Analogously to Sec. 12.38 we introduce

$$K_n^{\mathbb{R}^+} f := \sum_{k=0}^\infty f(t_k)\,\varphi_k, \qquad L_n^{\mathbb{R}^+} f := \sum_{k=0}^\infty \lambda_k(f)\,\varphi_k, \qquad z_n := \sum_{k=0}^\infty \xi_k \varphi_k.$$

We define $A_n^0 \in \mathscr{L}(\operatorname{im} L_n^{\mathbb{R}^+})$ to be the operator whose matrix with respect to the basis $\{\varphi_j\}$ is the matrix of the system (5). Then (5) holds if and only if $A_n^0 z_n = K_n^{\mathbb{R}^+} y$. Since the norm of an operator in $\mathscr{L}(\operatorname{im} L_n)$ is equivalent to the $\mathbf{l}^2$-operator norm of its matrix, we have to prove the stability of $\{A_n^0\}_{n=1}^\infty$ with respect to this norm. However, it is easy to check that the matrix A_n^0 is independent of n. Hence, $\{A_n^0\}_{n=1}^\infty$ (i.e. the method (5)) is stable if and only if $A_1^0 \in \mathscr{L}(\mathbf{l}^2)$ is invertible.

12.45. Theorem. *The operator* $A_1^0 \in \mathscr{L}(\mathbf{l}^2)$ *is Fredholm with index zero if and only if* 12.44(4) *is fulfilled and*

$$|\mathrm{Re}\, \varkappa_0| < 1/2, \tag{1}$$

where $\varkappa_0$ *is given by* 12.37(1) *and* $^0\log$ *is defined as in Sect.* 12.37.

Proof. To show this we need some results on Toeplitz operators which are due to Gohberg and Krupnik (see e.g. Böttcher/Silbermann [3]; cf. also Chapter 6). Let $\mathfrak{A} = \mathrm{alg}\, \mathscr{T}(\mathbf{PC}) \subset \mathscr{L}(\mathbf{l}^2)$ denote the smallest Banach algebra containing all Toeplitz operators $T(\boldsymbol{a}) = (\boldsymbol{a}_{k-j})_{k,j=0}^\infty$ with piecewise continuous $\boldsymbol{a} = \sum \boldsymbol{a}_k t^k$ on the unit circle $\mathbb{T}$. Then the set of 2×2-matrix operators $\mathfrak{A}_{2\times 2} \subset \mathscr{L}(\mathbf{l}^2)_{2\times 2}$ is an algebra of continuous operators in $\mathbf{l}_2^2 := \mathbf{l}^2 + \mathbf{l}^2$. There exists a multiplicative linear mapping $\mathfrak{A}_{2\times 2} \ni B \mapsto \mathrm{smb}\, B$ into the algebra of bounded 2×2-matrix functions over $\mathbb{T} \times [0, 1]$. The symbol $\mathrm{smb}\, T(\boldsymbol{a})$ is given by $\mathrm{smb}\, T(a)\,(\tau, \mu) = \mu \boldsymbol{a}(\tau + 0) + (1 - \mu)\, \boldsymbol{a}(\tau - 0)$, where $\tau \in \mathbb{T}$, $0 \leqq \mu \leqq 1$. Furthermore, $B \in \mathfrak{A}_{2\times 2}$ is a Fredholm operator if and only if $\det \mathrm{smb}\, B(\tau, \mu) \neq 0$ for all $(\tau, \mu) \in \mathbb{T} \times [0, 1]$. If $\mathrm{smb}\, B(\tau, \mu) = \mathrm{smb}\, B(\tau, 1)$ for $\tau \in \mathbb{T} \setminus \{1\}$ and $0 \leqq \mu \leqq 1$, then the index of B is minus the index of the curve $\{\det \mathrm{smb}\, B(e^{i2\pi\lambda}, 1) : 0 < \lambda < 1\} \cup \{\det \mathrm{smb}\, B(1, \mu) : 0 \leqq \mu \leqq 1\}$. Finally, $\mathfrak{A}_{2\times 2}$ contains all compact operators on $\mathbf{l}_2^2$, and $B \in \mathfrak{A}_{2\times 2}$ is compact if and only if $\mathrm{smb}\, B \equiv 0$.

For $C \in \left(\mathscr{L}(\mathbf{l}^2)\right)_{2\times 2}$, $C = (C^{rs})_{r,s=0}^1$, $C^{rs} = (C_{kl}^{rs})_{k,l=0}^\infty$, we define $C \in \mathscr{L}(\mathrm{im}\, L_n)$ by

$$C\varphi_{2j} := \sum_{k=0}^\infty \{C_{kj}^{00}\varphi_{2k} + C_{kj}^{10}\varphi_{2k+1}\},$$

$$C\varphi_{2j+1} := \sum_{k=0}^\infty \{C_{kj}^{01}\varphi_{2k} + C_{kj}^{11}\varphi_{2k+1}\}.$$

(cf. the definition of $C(\boldsymbol{a})$ in 12.38(7)). Thus $\mathfrak{A}_{2\times 2} \subset \mathscr{L}(\mathrm{im}\, L_n)$. On the other hand, there exists a compact $T \in \mathscr{L}(\mathbf{l}^2)_{2\times 2}$ such that $A_1^0 = C + T$, where

$$\begin{aligned} C^{00} &= \left(\mathsf{a}(0)\, \delta_{2k,2j} + \frac{\mathsf{b}(0)}{\pi i} \frac{t_{2k}^{(\alpha-1)/2-\alpha\varrho_0} t_{2j}^{(\alpha-1)/2+\alpha\varrho_0}}{t_{2j}^\alpha - t_{2k}^\alpha} \frac{2}{3} \right. \\ &\quad \left. - \frac{2}{3} \frac{\mathsf{b}(0)}{\pi i} \frac{\delta_{2j,2k+2} - \delta_{2j,2k-2}}{12} \right)_{k,j=0}^\infty \\ &= \left(\mathsf{a}(0)\, \delta_{kj} + \frac{\mathsf{b}(0)}{3\pi i} \frac{k^{(\alpha-1)/2-\alpha\varrho_0} j^{(\alpha-1)/2+\alpha\varrho_0}}{j^\alpha - k^\alpha} - \frac{\mathsf{b}(0)}{18\pi i} (\delta_{j,k+1} - \delta_{j,k-1}) \right)_{k,j=0}^\infty, \\ C^{11} &= \left(\mathsf{a}(0)\, \delta_{2k+1,2j+1} + \frac{\mathsf{b}(0)}{\pi i} \frac{t_{2k+1}^{(\alpha-1)/2-\alpha\varrho_0} t_{2j+1}^{(\alpha-1)/2+\alpha\varrho_0}}{t_{2j+1}^\alpha - t_{2k+1}^\alpha} \frac{4}{3} \right. \\ &\quad \left. - \frac{4}{3} \frac{\mathsf{b}(0)}{\pi i} \frac{\delta_{2j+1,2k+3} - \delta_{2j+1,2k-1}}{12} \right)_{k,j=0}^\infty \\ &= \left(\mathsf{a}(0)\, \delta_{kj} + \frac{2}{3} \frac{\mathsf{b}(0)}{\pi i} \frac{(k + 1/2)^{(\alpha-1)/2-\alpha\varrho_0} (j + 1/2)^{(\alpha-1)/2+\alpha\varrho_0}}{(j + 1/2)^\alpha - (k + 1/2)^\alpha} \right. \\ &\quad \left. - \frac{\mathsf{b}(0)}{9\pi i} (\delta_{j,k+1} - \delta_{j,k-1}) \right)_{k,j=0}^\infty, \end{aligned}$$

$$C^{01} = \left(\frac{2}{3}\,\frac{\mathsf{b}(0)}{\pi \mathrm{i}}\,\frac{k^{(\alpha-1)/2-\alpha\varrho_0}(j+1/2)^{(\alpha-1)/2+\alpha\varrho_0}}{(j+1/2)^\alpha - k^\alpha} + \frac{4}{9}\,\frac{\mathsf{b}(0)}{\pi \mathrm{i}}\,(\delta_{jk} - \delta_{j,k-1})\right)_{k,j=0}^{\infty},$$

$$C^{10} = \left(\frac{\mathsf{b}(0)}{3\pi \mathrm{i}}\,\frac{(k+1/2)^{(\alpha-1)/2-\alpha\varrho_0}\,j^{(\alpha-1)/2+\alpha\varrho_0}}{j^\alpha - (k+1/2)^\alpha} + \frac{8}{9}\,\frac{\mathsf{b}(0)}{\pi \mathrm{i}}\,(\delta_{j,k+1} - \delta_{j,k})\right)_{k,j=0}^{\infty}.$$

From this we conclude (cf. Sec. 11.19) that $C \in \mathfrak{A}_{2\times 2}$, $A_1^0 = C + T \in \mathfrak{A}_{2\times 2}$, $\operatorname{smb} A_1^0 = \operatorname{smb} C$ and

$$\operatorname{smb} C(\tau, \mu) = \mathsf{a}(0) \begin{pmatrix} 1 & 0 \\ 0 & 1 \end{pmatrix} + \mathsf{b}(0)\, F(\tau, \mu), \tag{2}$$

$$F(\tau, \mu) := \begin{cases} (-\mathrm{i}) \cot \left[\pi \left(\frac{1}{2} - \varrho_0 + \frac{\mathrm{i}}{2\pi\alpha} \log \frac{\mu}{1-\mu}\right)\right] \begin{pmatrix} 1/3 & 2/3 \\ 1/3 & 2/3 \end{pmatrix} \\ \qquad\qquad \text{if } \tau = 1,\ 0 \leqq \mu \leqq 1, \\ \begin{pmatrix} \left\{\frac{1}{3}(2\lambda - 1) + \frac{1}{9\pi}\sin(2\pi\lambda)\right\} & \left\{-e^{\mathrm{i}\pi\lambda}\left[\frac{2}{3} + \frac{8}{9\pi}\sin(\pi\lambda)\right]\right\} \\ \left\{-e^{-\mathrm{i}\pi\lambda}\left[\frac{1}{3} + \frac{16}{9\pi}\sin(\pi\lambda)\right]\right\} & \left\{\frac{2}{3}(2\lambda - 1) + \frac{2}{9\pi}\sin(2\pi\lambda)\right\} \end{pmatrix} \\ \qquad\qquad \text{if } \tau = e^{\mathrm{i}2\pi\lambda},\ 0 < \lambda < 1,\ 0 \leqq \mu \leqq 1. \end{cases}$$

The eigenvalues $\lambda_0(\tau, \mu)$ and $\lambda_1(\tau, \mu)$ of $F(\tau, \mu)$ are given by

$$\lambda_k(\tau, \mu) = \begin{cases} (-\mathrm{i}) \cot \left[\pi \left(\frac{1}{2} - \varrho_0 + \frac{\mathrm{i}}{2\pi\alpha} \log \frac{\mu}{1-\mu}\right)\right] & \text{if } \tau = 1, 0 \leqq \mu \leqq 1, k = 0, \\ 0 & \text{if } \tau = 1, 0 \leqq \mu \leqq 1, k = 1, \\ \frac{1}{2}\left\{(2\lambda - 1) + \frac{1}{3\pi}\sin(2\pi\lambda)\right\} + (-1)^k & \\ \times \left[\frac{1}{36}\left\{(2\lambda - 1) + \frac{1}{3\pi}\sin(2\pi\lambda)\right\}^2 + \left\{\frac{1}{3} + \frac{16}{9\pi}\sin(\pi\lambda)\right\}\right. & \\ \left.\times \left\{\frac{2}{3} + \frac{8}{9\pi}\sin(\pi\lambda)\right\}\right]^{1/2} & \text{if } \tau = e^{\mathrm{i}2\pi\lambda},\ 0 < \lambda < 1,\ 0 \leqq \mu \leqq 1. \end{cases}$$

Obviously, $\lambda_0(e^{\mathrm{i}2\pi\lambda}, \mu)$ varies continuously from 0 to 1 if λ varies from 0 to 1. Furthermore, $\lambda_1(e^{\mathrm{i}2\pi\lambda}, \mu) = -\lambda_0(e^{\mathrm{i}2\pi(1-\lambda)}, \mu)$ varies from -1 to 0. We set $F(\tau, \mu) = \left(F^{rs}(\tau, \mu)\right)_{r,s=0}^1$ and introduce the matrix norm $\|F(\tau, \mu)\|_\infty := \max\{|F^{r0}(\tau, \mu)| + |F^{r1}(\tau, \mu)| : r = 0, 1\}$. Then an easy computation yields that $\|F(e^{\mathrm{i}2\pi\lambda}, \mu)\|_\infty \leqq 1$, $0 < \lambda < 1$, and so we obtain that $|\lambda_k(e^{\mathrm{i}2\pi\lambda}, \mu)| \leqq 1$, $0 < \lambda < 1$, $k = 0, 1$. On the other hand,

$$\det \operatorname{smb} A_1^0(\tau, \mu) = \begin{cases} \mathsf{a}(0)\left\{\mathsf{a}(0) + \mathsf{b}(0)\,(-\mathrm{i}) \cot\left[\pi\left(\frac{1}{2} - \varrho_0 + \frac{\mathrm{i}}{2\pi\alpha}\log\frac{\mu}{1-\mu}\right)\right]\right\} \\ \qquad\qquad \text{if } \tau = 1,\ 0 \leqq \mu \leqq 1, \\ \{\mathsf{a}(0) + \mathsf{b}(0)\,\lambda_0(e^{\mathrm{i}2\pi\lambda}, \mu)\}\,\{\mathsf{a}(0) + \mathsf{b}(0)\,\lambda_1(e^{\mathrm{i}2\pi\lambda}, \mu)\} \\ \qquad\qquad \text{if } \tau = e^{\mathrm{i}2\pi\lambda},\ 0 < \lambda < 1,\ 0 \leqq \mu \leqq 1. \end{cases}$$

Using the properties of $\lambda_k(\tau, \mu)$, we see that det smb A_1^0 does not vanish if and only if $\mathcal{A}$ has no zeros, and in this case we have ind det smb $A_1^0 = \operatorname{ind} \mathcal{A}$, where

$$\mathcal{A}(\tau, \mu) := \begin{cases} \mathsf{a}(0) + \mathsf{b}(0)\,(-\mathrm{i}) \cot \left[\pi \left(\dfrac{1}{2} - \varrho_0 + \dfrac{\mathrm{i}}{2\pi\alpha} \log \dfrac{\mu}{1-\mu}\right)\right] & \\ \qquad\qquad\qquad\qquad\qquad\qquad\qquad\qquad \text{if } \tau = 1,\ 0 \leqq \mu \leqq 1, & \\ \mathsf{a}(0) + (2\lambda - 1)\,\mathsf{b}(0) \qquad\qquad \text{if } \tau = \mathrm{e}^{\mathrm{i}2\pi\lambda},\ 0 < \lambda < 1,\ 0 \leqq \mu \leqq 1. & \end{cases}$$

However, the relations $\mathcal{A} \neq 0$ and ind $\mathcal{A} = 0$ are equivalent to 12.44(4) and 12.45(1) (cf. Sec. 11.19). ■

Now, let $u \geqq 0$. We define A_n^{0u} to be the operator in $\mathcal{L}(\operatorname{span}\{\varphi_j : j = u, u+1, \ldots\})$ the matrix of which corresponding to the basis $\{\varphi_j : j = u, \ldots\}$ is the matrix of 12.44(5). Again, method 12.44(5) is stable if and only if A_1^{0u} is invertible.

12.46. Theorem. *If, for $u \in \mathbb{Z}$, $u \geqq 0$, the operator $A_1^{0u} \in \mathcal{L}(\mathrm{l}^2)$ is invertible, then* 12.44(4) *and* 12.45(1) *are satisfied. On the other hand, if* 12.44(4) *and* 12.45(1) *hold, then there exists a $u \in \mathbb{Z}$, $u \geqq 0$ such that $A_1^{0,u}$ is invertible.*

Proof. Let $Q_u \in \mathcal{L}(\operatorname{im} L_n^{\mathbb{R}^+})$ be defined by $Q_u\varphi_k := 0$ if $k < u$ and $Q_u\varphi_k := \varphi_k$ if $k \geqq u$. We thus have $A_1^{0u} = Q_u A_1^{00}|_{\operatorname{im} Q_u}$. If A_1^{0u} is invertible, then so also is $D := (I - Q_u) + Q_u A_1^{00} Q_u$. Since $A_1^{00} - D$ is an operator of finite rank, A_1^{00} is Fredholm with index zero. Hence Theorem 12.45 implies 12.44(4) and 12.45(1).

Vice versa, let 12.44(4) and 12.45(1) be satisfied. We then apply a theorem due to Roch ([4], Secs. 2 and 3). For $C \in \mathfrak{A}_{2\times 2}$, this theorem states the equivalence of the following two assertions.

(i) The operator C is Fredholm with index zero. For all $\tau \in \mathbb{T}$, the restriction of the Mellin convolution operator M^{-1} smb $C\big(\tau, (1 + \mathrm{i}\cot(\pi\cdot))/2\big)\, M \in \mathcal{L}\big(\mathbf{L}^2(\mathbb{R}_+)\big)_{2\times 2}$ to the interval $[1, \infty)$ is invertible. The Toeplitz operator $T(\mathbf{c}) \in \mathcal{L}(\mathrm{l}^2)_{2\times 2}$ is invertible, where $\mathbf{c}(t) :=$ smb $C(t, 1)$, $t \in \mathbb{T}$.

(ii) There exists $u_0 \in \mathbb{Z}$, $u_0 \geqq 0$ such that $Q_{2u} C|_{\operatorname{im} Q_{2u}}$ is invertible for $u \geqq u_0$ and $\sup_{u \geqq u_0} \left\|(Q_{2u} C|_{\operatorname{im} Q_{2u}})^{-1}\right\| < \infty$.

Setting $C = A_1^{00}$, it remains to verify (i). By our assumption and Theorem 12.45 we get the Fredholm property of A_1^{00} and ind $A_1^{00} = 0$. Furthermore, if $\tau \neq 1$, then M^{-1} smb $A_1^{00}\big(\tau, (1 + \mathrm{i}\cot(\pi\cdot))/2\big)\, M$ is a scalar multiple of the identity and, consequently, it is invertible. In the case $\tau = 1$, we choose a 2×2-matrix G satisfying

$$\begin{pmatrix} 1/3 & 2/3 \\ 1/3 & 2/3 \end{pmatrix} = G^{-1} \begin{pmatrix} 1 & 0 \\ 0 & 0 \end{pmatrix} G$$

to obtain

$$M^{-1} \operatorname{smb} A_1^{00} \left(\tau, \frac{1 + \mathrm{i}\cot(\pi\cdot)}{2}\right) M = G^{-1} \begin{pmatrix} A^0 & 0 \\ 0 & \{\mathsf{a}(0)\, I\} \end{pmatrix} G.$$

It is well known that the restriction of operator A^0 to $[1, \infty)$ is equivalent to the Wiener-Hopf integral operator on the half axis which has the symbol

$$\mathbb{R} \ni t \mapsto \mathsf{a}(0) + \mathsf{b}(0)\,(-\mathrm{i}) \cot \left[\pi \left(\frac{1}{2} - \varrho_0 + \mathrm{i}\frac{t}{\alpha}\right)\right].$$

Hence, it is invertible if and only if $\mathcal{A}(\tau, \mu) \neq 0$ (for the definition of $\mathcal{A}$ we refer to the proof of Theorem 12.45) for all $(\tau, \mu) \in \mathbb{T} \times [0, 1]$ and $\operatorname{ind} \mathcal{A} = 0$, i.e., if and only if 12.44(1) and 12.45(1) are satisfied.

Finally, we write $\boldsymbol{c} = a(0)\,(I_{2\times 2} + \boldsymbol{d})$, where $\boldsymbol{d}(\tau) := b(0)/a(0)\, F(\tau, 1)$, $\tau \in \mathbb{T}$. Then 12.44(4), 12.45(1) and the inequality $\|F(\tau, 1)\|_\infty \leq 1$ yield that $|b(0)/a(0)| < 1$ and $\|\boldsymbol{d}\|_\infty < 1$. Hence, $T(\boldsymbol{c}) = a(0)\,\big(I + T(\boldsymbol{d})\big)$ is invertible. ■

In the case $0 < \tau < 1$ we define $A^\tau := \mathsf{a}(\tau) + \mathsf{b}(\tau)\, S_{\mathbb{R}} \in \mathcal{L}\big(\mathbf{L}^2(\mathbb{R})\big)$. (Notice that the restriction of $l(t, \tau)$ to a neighborhood of τ is the kernel of a compact operator.) This operator is locally equivalent to A at the point τ. It is invertible if and only if $\mathsf{a}(\tau) \pm \mathsf{b}(\tau) \neq 0$. We consider the equation $A^\tau z = y$, where $y \in \mathbf{L}^2(\mathbb{R})$ and y is Riemann integrable. Analogously to 12.37(15), we determine approximate values ξ_j of $z(t_j)$ $(j \in \mathbb{Z})$ by solving the system

$$\mathsf{a}(\tau)\, \xi_k + \frac{\mathsf{b}(\tau)}{\pi \mathrm{i}} \left\{ \sum_{\substack{j \in \mathbb{Z} \\ j \neq k}} \frac{\eta_j}{t_j - t_k}\, \xi_j + \eta_k\, \frac{8(\xi_{k+1} - \xi_{k-1}) - (\xi_{k+2} - \xi_{k-2})}{12} \right\} = y(t_k),$$

$$k \in \mathbb{Z}, \qquad \eta_j := \begin{cases} 4/3n & \text{if } j \text{ is odd}, \\ 2/3n & \text{if } j \text{ is even}. \end{cases} \tag{1}$$

This quadrature method is the "local representative" of 12.37(15) at the point τ, $0 < \tau < 1$.

Retaining the definitions of $S_n^{\mathbb{R}}$, $\{\varphi_j\}$, $K_n^{\mathbb{R}}$, $L_n^{\mathbb{R}}$, $C(\boldsymbol{a})$ from Sec. 12.38, we set $z_n := \sum \xi_j \times \varphi_j$. Furthermore, let A_n^τ denote the operator whose matrix with respect to the basis $\{\varphi_j\}$ is the matrix of the system (1). Hence, (1) is equivalent to $A_n^\tau z_n = K_n^{\mathbb{R}} y$. Again, the matrix A_n^τ is independent of n, and so method (1) is stable if and only if A_1^τ is invertible.

12.47. Theorem. *The operator $A_1^\tau \in \mathcal{L}(\mathrm{l}^2)$ is invertible if and only if*

$$\frac{\mathsf{a}(\tau) + \mathsf{b}(\tau)}{\mathsf{a}(\tau) - \mathsf{b}(\tau)} \notin (-\infty, 0]. \tag{1}$$

Proof. Section 12.38 yields that $A_1^\tau = \mathsf{a}(\tau) + \mathsf{b}(\tau)\, C(\boldsymbol{a})$, where $\boldsymbol{a} = (\boldsymbol{a}^{rs})_{r,s=0}^1$ is the matrix function defined in Sec. 12.39. Thus A_1^τ is a discrete convolution matrix with the symbol $\mathbb{T} \ni \tau \mapsto \operatorname{smb} C(\tau, 1)$ (cf. 12.45(2)). It is invertible if and only if this symbol does not degenerate, i.e., $\mathsf{a}(\tau) + \mathsf{b}(\tau)\, \lambda_k(\tau, 1) \neq 0$ for $\tau \in \mathbb{T}$ and $k = 0, 1$. Since $\{\lambda_k(\tau, 1), \tau \in \mathbb{T}, k = 0, 1\} = [-1, 1]$, Theorem 12.47 follows. ■

The case $\tau = 1$ is akin to the case $\tau = 0$. However, instead of considering $\mathsf{a}(1) + \mathsf{b}(1) \times S_{\mathbb{R}_-} + \mathsf{b}(1)/\pi\mathrm{i}\, K_0^1$ we transform this operator by setting $t' = -t$. Thus we get $A^1 = \mathsf{a}(1) - \mathsf{b}(1)\, S_{\mathbb{R}_+} - \mathsf{b}(1)/\pi\mathrm{i}\, K_0^1$, where

$$K_0^1 z(t) = \int_0^\infty k_0^1(t, \tau)\, z(\tau)\, \mathrm{d}\tau,$$

$$k_0^1(t, \tau) := \left\{ \alpha\, \frac{t^{(\alpha-1)/2 - \alpha\varrho_1}\, \tau^{(\alpha-1)/2 + \alpha\varrho_1}}{\tau^\alpha - t^\alpha} - \frac{1}{\tau - t} \right\}.$$

We define the quadrature method for A^1 in the same way as in (44.5) for A^0. For this method, we introduce the operators A_n^1, A_n^{1v} which correspond to A_n^0 and A_n^{0u}. Hence, the quadrature method for A^1 is stable if and only if $A_1^{1v} \in \mathcal{L}(\operatorname{span}\{\varphi_j : j = v, \ldots\})$ is invertible. Thus Theorem 12.46 applies.

12.48. Theorem. *Let $v \in \mathbb{Z}$, $v \geqq 0$. If A_1^{1v} is invertible, then*

$$\frac{\mathsf{a}(1) - \mathsf{b}(1)}{\mathsf{a}(1) + \mathsf{b}(1)} \notin \{t\, \mathrm{e}^{\mathrm{i}2\pi\varrho_1}\colon -\infty < t \leqq 0\}, \tag{1}$$

$$|\mathrm{Re}\, \varkappa_1| < 1/2, \tag{2}$$

where $\varkappa_1$ is given by 12.37(1) *and* $^1\log$ *is defined as in Sec.* 12.37. *On the other hand, if* (1) *and* (2) *are satisfied, then there exists a $v \in \mathbb{Z}$, $v \geqq 0$ such that A_1^{1v} is invertible.*

Now, the stability of method 12.37(15) can be deduced from Theorems 12.46–12.48 and the following localization principle.

12.49. Theorem. (a) *The method* 12.37(15) *is stable in* $\mathbf{L}^2$ *if and only if the following conditions are satisfied.*

(i) *A is invertible.*
(ii) *Method* 12.44(5) *for A^0 is stable.*
(iii) *For all $\tau \in (0, 1)$, method* 12.46(1) *for A^τ is stable.*
(iv) *The quadrature method* 12.44(5) *applied to the operator A^1 is stable.*

(b) *If the quadrature method* 12.37(15) *is stable, then the linear system* 12.37(15) *is uniquely solvable for n large enough. Furthermore, if the right-hand side y is Riemann integrable, then z_n converges to the exact solution z in the L^2-norm.*

The proof of this theorem is analogous to the proof of Theorem 11.21. (Setting $B_n := A_n\Omega_n$, $\Omega_n := (\delta_{kj}n/\omega_j)_{k,j=0}^n$, the sequence $\{A_n\}$ is stable if and only if $\{B_n\}$ is stable. For the sequence $\{B_n\}$, Theorem 11.22 applies. The validity of the assumptions of Theorem 12.22 can be shown analogously to Sec. 11.24.)

12.50. Corollary. *Suppose $\varkappa_j$, $j = 0, 1$ is given by* 12.37(1), *where now* $^j\log$ *denotes the branch of the logarithm which is continuous on $\mathbb{C} \setminus (-\infty, 0]$ and takes real values on $\mathbb{R}^+$ (i.e.* $^1\log = {}^2\log$). *Furthermore,* set $\varrho_0 := \varkappa_0$, $\varrho_1 := \varkappa_1$ *and let the operator A on the left-hand side of* 12.37(6) *be invertible in* $\mathbf{L}^2$. *Then the stability of* 12.37(15) *for some u and v implies the locally strong ellipticity, i.e.*

$$\mathsf{a}(t) + \lambda \mathsf{b}(t) \neq 0, \qquad 0 \leqq t \leqq 1, \qquad -1 \leqq \lambda \leqq 1.$$

Vice versa, if the latter condition is fulfilled, then there exist $u, v \in \mathbb{Z}$, $u, v \geqq 0$ such that the quadrature method 12.37(15) *is stable.*

12.51. Numerical results. When applying method 12.37(15), the first question is how to choose the parameters u, v and α. The number $u(v)$ has been introduced in order to guarantee stability in case $A_1^{00}(A_1^{10})$ is Fredholm with index zero, but its null space is not trivial (cf. Secs. 12.46–12.49). Actually, it is not clear whether this situation is possible. However, if it should happen, then it is an exceptional case. (Let us consider A_1^{00} as a function of $\mathsf{a}(0) \in \mathbb{C}$. Then it is a Fredholm operator with index zero and nontrivial null space only for $\mathsf{a}(0)$ belonging to an at most countable subset of $\mathbb{C}$. Moreover, at the accumulation points of this subset A_1^{00} is either not Fredholm or its index differs from zero. For the proof of these claims, we refer to Sec. 11.3.) Therefore, we propose $u = v = 0$. If it would turn out that the resulting method is not stable, then we would choose $u = v = 1$ or $u, v \geqq 1$ (cf. also the corresponding parameters in Chandler/Graham [4], where another class of equations is considered). In all the examples presented here the choice $u = v = 0$ is satisfactory.

Sect. 12.40 suggests to choose α large, at least greater than 9. Then we have $(\alpha - 1)/2 > 4$ and Theorem 12.40 insures a convergence of order n^{-4}. However, α must not be too large. Namely, to obtain 12.37(15) we have discretized 12.36(1) using the non-uniform mesh $\{\gamma(j/n): j = 0, \ldots, n\}$. If $n < 100$ and α is very large, then the mesh points are concentrated near the end-point of the interval, and in the inner part of $[0, 1]$ the mesh is coarse. Hence, though the rate of convergence may be high, a large error is to be expected. For this reason, we have set $\alpha \leq 5$. Moreover, we have carried out our computations using $\gamma(t) := t^\alpha/\{t^\alpha + (1 - t)^\alpha\}$. This transformation of the interval behaves like t^α as $t \to 0$ and like $1 - (1 - t)^\alpha$ as $t \to 1$. It is not hard to extend all the results to the case of this function γ.

The *first example* we have considered is that of THOMAS [1]. If $\mathsf{a}(t) \equiv 1$, $\mathsf{b}(t) \equiv -i$, $\mathsf{k}(t, \tau) \equiv 0$ and $\mathsf{y}(t) \equiv \sqrt{2}$, then the solution x of 12.36(1) takes the form $\mathsf{x}(t) = t^{-1/4} (1 - t)^{1/4}$. Choosing $\varrho_0 = \varkappa_0 = -1/4$, $\varrho_1 = \varkappa_1 = 1/4$, we get $\mathsf{z}(t) \equiv 1$. The error $\sup_{0.1 \leq t \leq 0.9} |\mathsf{z}(t) - \mathsf{z}_n(t)|$ for $\alpha = 1, 3, 4, 5$ is given in Table 1:

Table 1. Example 1. The error for the method 12.37(15)

n ↓ $\alpha \to$	1	3	4	5
12	0.08	0.0017	0.008	0.013
24	0.053	0.00013	0.00036	0.0010
48		0.00005	0.000012	0.000041
96		0.000005	0.0000046	0.0000014

For our special choice, equation 12.36(1) is strongly elliptic. In accordance with the stability theorem the condition numbers of the systems 12.37(15) are bounded. Table 2 gives the approximate values $\mathsf{z}_n(t)$ for $\alpha = 5$.

Table 2. Example 1. The approximate values for the method 12.37(15), $\alpha = 5$

t ↓ $n \to$	12	24	48	96
0.5	1.011	1.00091	1.000035	1.0000011
0.8432	1.013	0.99940	0.999977	0.9999993
0.9959	1.006	1.00023	1.000008	1.0000002
1	1.013	0.99978	0.999982	1.0000002

The approximate value for $\mathsf{z}(1)$ was not determined by the way described in Sec. 12.42. We have set $\mathsf{z}(1) \sim \mathsf{z}_n(\gamma(t_{j_0}))$ and

$$|\mathsf{z}_n(\gamma(t_{j_0+1})) - \mathsf{z}_n(\gamma(t_{j_0}))| = \min_{j=n/2, n/2+1, \ldots, n-2} |\mathsf{z}_n(\gamma(t_{j+1})) - \mathsf{z}_n(\gamma(t_j))|.$$

This principle of choosing an approximation for $\mathsf{z}(1)$ seems to be good if $\mathsf{z}(1)$ does not vanish. Note that the choice $\mathsf{z}(1) \sim \mathsf{z}_n(\gamma(\varphi_n))$ is practicable only if the constants in 12.42(1) and 12.42(2) are known explicitly.

For the case of the method 12.37(20) – 12.37(21), we have put $\alpha = 3$, and the approximate values for $\mathsf{z}(t)$ (cf. 12.37(22)) are given in Table 3. It turns out that 12.37(20) to 12.37(21) provides better approximations near the endpoints of the interval. The condition numbers of the systems of equations 12.37(20)—12.37(21) seem to tend to infinity as $n \to \infty$. However, they are still acceptable if $n < 100$.

Table 3. Example 1. The approximate values for the method 12.37(20)—(21), $\alpha = 3$

t ↓ $n \to$	12	24	48	96
0.5	1.006	1.00034	1.000012	1.0000004
0.8889	0.993	0.99983	0.999996	0.9999999
0.9920	1.001	1.00006	1.000002	1.0000001
1	1.003	1.00023	1.000022	1.0000024

The results presented in the Tables 1–3 show that quadrature methods lead, roughly speaking, to the same order of convergence as the computations in Thomas [1], where a Galerkin method was proposed and the solution was approximated by continuous piecewise linear functions on non-uniform meshes. (Moreover, the spline functions were modified near the end-points of the interval.) Gerasoulis [2] and Jen/Srivastav [1] considered the case $\mathsf{a} \equiv 0$, $\mathsf{b} \equiv 1$ only (where 12.36(1) is not strongly elliptic). The approximation obtained by spline collocation seems to be better. However, if the methods of Gerasoulis [2] and Jen/Srivastav [1] are applied to the case of variable a and b, then logarithmic terms appear in the asymptotics of the solution (cf. Sec. 12.38), and the convergence will be slower. For our quadrature methods, the rate of convergence remains the same. Tables 4 and 5 give the approximate values for $\mathsf{z}(t)$ obtained by the methods 12.37(15), $\alpha = 5$ and 12.37(20) – 12.37(21), $\alpha = 3$ in the case $\mathsf{a}(t) = 1 + \sin(\pi t)$, $\mathsf{b}(t) = i(-1 + t^2(1-t))$, $\mathsf{y}(t) = \sin(t)$ and $\varrho_0 = -1/4$, $\varrho_1 = 1/4$ (*Example* 2).

Table 4. Example 2. The approximate values for the method 12.37(15), $\alpha = 5$

t ↓ $n \to$	12	24	48	96
0.5	0.275	0.27859	0.278737	0.278739
0.8432	0.643	0.64436	0.643071	0.643005
0.9959	0.802	0.83839	0.838294	0.838282
1	0.802	0.83839	0.839237	0.833825

Table 5. Example 2. The approximate values for the method 12.37(20)—(21), $\alpha = 3$

t ↓ $n \to$	12	24	48	96
0.5	0.280	0.27881	0.278742	0.278739
0.8889	0.720	0.71778	0.717624	0.717619
0.9920	0.838	0.83937	0.839382	0.839382
1	0.838	0.83426	0.833872	0.833819

Finally, we have tested method 12.37(15) in the case where the stability conditions are violated. Then the condition numbers are larger, but still acceptable for $n < 100$. However, the errors are much larger than in the case of stable methods. For instance, if $\mathsf{a}(t) = 1 - 4t(1-t)$, $\mathsf{b}(t) = \mathrm{i}(-1 + 3t(1-t))$, $\mathsf{y}(t) = \cos(t)$ and $\varrho_0 = -1/4$, $\varrho_1 = 1/4$ (*Example* 3), then the stability condition fails to be true at $t = 1/2$. The approximate values of $\mathsf{z}(t)$ are given in Table 6.

Table 6. Example 3. The approximate values for the method 12.37(15), $\alpha = 5$

t $\downarrow$ $n \rightarrow$	12	24	48	96
0.5	5.594	3.129	1.875	1.949
0.8432	−3.517	0.434	−0.406	−0.406

If $\mathsf{a}, \mathsf{b}, \mathsf{y}$ are the same as in our first example and $\varrho_0 = 0.4$, $\varrho_1 = -0.4$, $\alpha = 5$ (*Example* 4), then the condition numbers of 12.37(15) slowly turn to infinity. The approximate values $\mathsf{z}_n(t)$ seem to converge to $f(t) := t^{-0.65}(1-t)^{0.65}$ (cf. Table 7). However, although $g = \varrho f$ is a solution of 12.36(1), it is not the one sought in the class $\{\varrho \mathbf{z} : \mathbf{z} \in \mathbf{L}^2\}$.

Table 7. Example 4. The approximate values for the method 12.37(15), $\alpha = 5$

t $\downarrow$ $n \rightarrow$	12	24	48	96	$f(t)$
0.5	0.676	0.99720	0.999987	0.9999996	1.0
0.8432	0.529	0.33415	0.335024	0.3350301	0.3350301
0.9959	0.228	0.02824	0.028144	0.0281421	0.0281421

A quadrature method for a Cauchy singular integral equation with a fixed singularity

12.52. Notation. In Section 12.53 we present a quadrature method for the numerical solution of the singular integral equation of the first kind

$$\int_0^1 \mathsf{k}(t, \tau)\, \mathsf{x}(\tau)\, d\tau = \mathsf{y}(t), \qquad 0 \leqq t \leqq 1, \tag{1}$$

where y is a given Riemann integrable function, the kernel k is of the form

$$\mathsf{k}(t, \tau) = \frac{1}{\pi \mathrm{i}} \frac{1}{\tau - t} + \mathsf{k}_0\left(\frac{t}{\tau}\right)\frac{1}{\tau} + \mathsf{k}_1(t, \tau), \qquad 0 \leqq t, \tau \leqq 1,$$

the function k_1 is continuous on $[0, 1] \times [0, 1]$ and the Mellin transform of k_0 (cf. 12.53(1)) satisfies certain analiticity and growth conditions (cf. 12.53(2)). This quadrature method (cf. 12.53(9)) is a slight modification of the so called *method of discrete whirls* (cf. Polonski/Lifanov [1]) and a modification of a method presented in 11.28(1).

We shall derive the corresponding system of equations for the numerical procedure in Sec. 12.53. In Secs. 12.55–12.56 we formulate necessary and sufficient conditions for the method to be stable and consider the speed of convergence. Since the proofs of these results run analogously to those given in Secs. 12.38–12.50, we only give some hints. Finally, in Sec. 12.57 we discuss numerical examples. We solve (1) for $\mathsf{k}(t, \tau) = 1/\pi\mathrm{i} \times 1/(\tau - t)$ and for two special kernels k arising in the analysis of crack problems.

12.53. Derivation of the quadrature method. Let us consider 12.52(1) and suppose the function

$$b(\zeta) = \int_0^\infty \mathsf{k}_0(t)\, t^{\zeta-1}\, \mathrm{d}t \tag{1}$$

is analytic in the strip $\mu < \operatorname{Re}\zeta < \nu$, where $\mu < 1/2 < \nu$. Furthermore, let

$$\sup_{\mu<\operatorname{Re}\zeta<\nu} (1 + |\zeta|)^{k+1} |(\mathrm{d}/\mathrm{d}\zeta)^k\, b(\zeta)| < \infty, \qquad k = 0, 1, \dots \tag{2}$$

and choose ϱ_0, ϱ_1 such that $-1/2 < \varrho_0, \varrho_1 < 1/2$ and $\mu < 1/2 - \varrho_0 < \nu$. We seek a solution $\mathbf{x}$ of 12.52(1) in the form $\mathbf{x} = \varrho\mathbf{z}$, where $\mathbf{z} \in \mathbf{L}^2(0, 1)$ and $\varrho(t) = t^{\varrho_0}(1 - t)^{\varrho_1}$.

In order to obtain an equation the solution of which vanishes at the end-points of the interval, we take $\alpha \geqq 1$ and choose a bijective map $\gamma\colon [0, 1] \to [0, 1]$ such that $\gamma(t) = t^\alpha$ for $t < 1/3$, $\gamma(t) = 1 - (1 - t)^\alpha$ for $t > 2/3$ and $\gamma \in \mathbf{C}^\infty(1/4, 3/4)$. We transform 12.52(1) by substituting $t = \gamma(s)$, $\tau = \gamma(\sigma)$, $\mathrm{d}\tau = \gamma'(\sigma)\, \mathrm{d}\sigma$. After multiplying the resulting equation by $\gamma'(s)^{1/2}/\varrho(\gamma(s))$ we arrive at the equation

$$\int_0^1 \frac{\varrho(\gamma(\sigma))}{\varrho(\gamma(s))}\, \gamma'(\sigma)^{1/2}\, \gamma'(s)^{1/2}\, \mathsf{k}(\gamma(s), \gamma(\sigma))\, z(\sigma)\, \mathrm{d}\sigma = y(s), \qquad 0 \leqq s \leqq 1. \tag{3}$$

Here we have set

$$z(\sigma) := \mathbf{z}(\gamma(\sigma))\, \gamma'(\sigma)^{1/2}, \qquad y(s) := \mathbf{y}(\gamma(s))\, \gamma'(s)^{1/2}/\varrho(\gamma(s)). \tag{4}$$

Denoting the operator on the left-hand side of (3) by A, we have $Az = y$. Note that A is a pseudodifferential operator of Mellin type and that its symbol is given by

$$\operatorname{smb} A(t, \zeta) := \begin{cases} -1 & \text{if } 0 \leqq t \leqq 1,\ \zeta = 1/2 + \mathrm{i}\infty, \\ 1 & \text{if } 0 \leqq t \leqq 1,\ \zeta = 1/2 - \mathrm{i}\infty, \\ -(-\mathrm{i}) \cot\left[\pi\left(1/2 - \varrho_1 - \mathrm{i}\dfrac{\xi}{\alpha}\right)\right] & \text{if } t = 1,\ \zeta = 1/2 + \mathrm{i}\xi,\ \xi \in \mathbb{R}, \\ (-\mathrm{i}) \cot\left[\pi\left(1/2 - \varrho_0 + \mathrm{i}\dfrac{\xi}{\alpha}\right)\right] + b\left(1/2 - \varrho_0 + \mathrm{i}\dfrac{\xi}{\alpha}\right) & \\ & \text{if } t = 0,\ \zeta = 1/2 + \mathrm{i}\xi,\ \xi \in \mathbb{R}. \end{cases}$$

Hence, $A \in \mathscr{L}(\mathbf{L}^2(0, 1))$ is a Fredholm operator if and only if $\operatorname{smb} A$ does not vanish. The index of A is equal to the negative index of the curve

$$\{(-\mathrm{i}) \cot [\pi(1/2 - \varrho_0 + \mathrm{i}\xi)] + b(1/2 - \varrho_0 + \mathrm{i}\xi) \colon \xi \in \mathbb{R}\}$$
$$\cup\ \{\mathrm{i} \cot [\pi(1/2 - \varrho_1 + \mathrm{i}\xi)] \colon \xi \in \mathbb{R}\}.$$

We write (3) in the form

$$\frac{1}{\pi \mathrm{i}} \int_0^1 \frac{z(\sigma)}{\sigma - s}\, \mathrm{d}\sigma + \int_0^1 k(s, \sigma)\, z(\sigma)\, \mathrm{d}\sigma = y(s),$$

$$k(s, \sigma) := \frac{1}{\pi \mathrm{i}} \left\{ \frac{\varrho(\gamma(\sigma))}{\varrho(\gamma(s))} \frac{\gamma'(\sigma)^{1/2}\, \gamma'(s)^{1/2}}{\gamma(\sigma) - \gamma(s)} - \frac{1}{\sigma - s} \right\} \tag{5}$$

$$+ \frac{\varrho(\gamma(\sigma))}{\varrho(\gamma(s))} \gamma'(\sigma)^{1/2}\, \gamma'(s)^{1/2} \left\{ \mathrm{k}_0 \left(\frac{\gamma(s)}{\gamma(\sigma)} \right) \frac{1}{\gamma(\sigma)} + \mathrm{k}_1(\gamma(s), \gamma(\sigma)) \right\}.$$

Finally, we set $z(\sigma) = 0$ for $\sigma \leqq 0$ or $\sigma \geqq 1$. Using that $\int_{-1}^{1+2s} \mathrm{d}\sigma/(\sigma - s) = 0$, we arrive at the equations

$$\int_0^1 \frac{z(\sigma)}{\sigma - s}\, \mathrm{d}\sigma = \int_{-1}^{1+2s} \frac{z(\sigma)}{\sigma - s}\, \mathrm{d}\sigma,$$

$$\frac{1}{\pi \mathrm{i}} \int_{-1}^{1+2s} \frac{z(\sigma) - z(s)}{\sigma - s}\, \mathrm{d}\sigma + \int_0^1 k(s, \sigma)\, z(\sigma)\, \mathrm{d}\sigma = y(s), \qquad 0 \leqq s \leqq 1. \tag{6}$$

Now we set $t_k := (k + 1/2)/n$, $\tau_k := k/n$. We shall approximate the integrals in (6) applying the modified rectangle rules. (Note that these quadrature rules are of order n^{-4}.) So

$$\int_0^1 f(\sigma)\, \mathrm{d}\sigma \sim \sum_{j=0}^{n-1} f(t_j)\, \omega_j, \qquad \int_{-1}^{1+2\tau_k} f(\sigma)\, \mathrm{d}\sigma \sim \sum_{j=-n}^{n+2k-1} f(t_j)\, \omega_j^k, \tag{7}$$

where

$$\omega_j := \begin{cases} 1/n & \text{for } j = 3, 4, \ldots, n-4 \\ 9/8n & \text{for } j = 0,\ j = n-1 \\ 19/24n & \text{for } j = 1,\ j = n-2 \\ 13/12n & \text{for } j = 2,\ j = n-3, \end{cases}$$

$$\omega_j^k := \begin{cases} 1/n & \text{for } j = -n+3, 4, \ldots, n+2k-4 \\ 9/8n & \text{for } j = -n,\ j = n+2k-1 \\ 19/24n & \text{for } j = -n+1,\ j = n+2k-2 \\ 13/12n & \text{for } j = -n+2,\ j = n+2k-3 \end{cases}$$

and we obtain

$$y(\tau_k) \sim \frac{1}{\pi \mathrm{i}} \sum_{j=-n}^{n+2k-1} \frac{z(t_j) - z(\tau_k)}{t_j - \tau_k}\, \omega_j^k + \sum_{j=0}^{n-1} k(\tau_k, t_j)\, \omega_j z(t_j), \qquad k = 0, \ldots, n.$$

We observe that $\sum 1/(t_j - \tau_k)\, \omega_j^k = 0$ and $z(t_j) = 0$ for $j \leqq 0$ and $j \geqq n$ provided that α is large enough. Thus

$$y(\tau_k) \sim \frac{1}{\pi \mathrm{i}} \sum_{j=0}^{n-1} \frac{z(t_j)}{t_j - \tau_k} \frac{1}{n} + \sum_{j=0}^{n-1} k(\tau_k, t_j)\, \omega_j z(t_j), \qquad k = 0, \ldots, n. \tag{8}$$

This suggests the following quadrature method (cf. (3), (5), (8)). We determine approximate values ξ_j for $z(t_j)$ ($j = 0, \ldots, n-1$) by solving

$$y(\tau_k) = \sum_{j=0}^{n-1} \frac{\varrho(\gamma(t_j))}{\varrho(\gamma(\tau_k))} \gamma'(t_j)^{1/2} \gamma'(\tau_k)^{1/2} \mathsf{k}(\gamma(\tau_k), \gamma(t_j)) \omega_j \xi_j + \frac{1}{\pi \mathrm{i}} \left\{ \sum_{j=0}^{2} + \sum_{j=n-3}^{n-1} \right\} \frac{1}{t_j - \tau_k} \left[\frac{1}{n} - \omega_j \right] \xi_j, \qquad k = 0, 1, \ldots, n-1. \tag{9}$$

Finally, approximate values for the functions z and x are given by

$$\mathsf{z}(\gamma(t_j)) \sim \xi_j/\gamma'(t_j)^{1/2}, \qquad \mathsf{x}(\gamma(t_j)) \sim \varrho(\gamma(t_j))\, \xi_j/\gamma'(t_j)^{1/2}. \tag{10}$$

12.54. Stability and convergence. Set

$$\varphi_j(t) := \begin{cases} \dfrac{t - \tau_{j+1}}{\tau_j - \tau_{j+1}} & \text{if } \tau_j \leqq t \leqq \tau_{j+1}, \\ \dfrac{t - \tau_{j-1}}{\tau_j - \tau_{j-1}} & \text{if } \tau_j \geqq t \geqq \tau_{j-1}, \\ 0 & \text{else}, \end{cases} \tag{1}$$

$$\psi_j(t) := \begin{cases} \dfrac{t - t_{j+1}}{t_j - t_{j+1}} & \text{if } t_j \leqq t \leqq t_{j+1}, \\ \dfrac{t - t_{j-1}}{t_j - t_{j-1}} & \text{if } t_j \geqq t \geqq t_{j-1}, \\ 0 & \text{else}, \end{cases}$$

$j \in \mathbb{Z}$, and define $K_n f := \sum_{j=0}^{n-1} f(\tau_j)\, \varphi_j$. We denote the orthogonal projection of $\mathbf{L}^2(0, 1)$ onto span $\{\psi_j : j = 0, \ldots, n-1\}$ by L_n and identify the operators of $\mathscr{L}(\operatorname{im} L_n, \operatorname{im} K_n)$ with their matrices corresponding to the bases $\{\psi_j : j = 0, \ldots, n-1\}$ and $\{\varphi_j : j = 0, \ldots, n-1\}$. If $A_n \in \mathscr{L}(\operatorname{im} L_n, \operatorname{im} K_n)$ is given by the matrix of the system of equations 12.53(9), then 12.53(9) is equivalent to $A_n z_n = K_n y$, where $z_n := \sum_{j=0}^{n-1} \xi_j \psi_j$. Furthermore, the operator norm in $\mathscr{L}(\operatorname{im} L_n, \operatorname{im} K_n)$ is equivalent to the Euclidean norm of the corresponding matrices.

Before stating the stability conditions, let us consider two familiar integral equations as well as the corresponding quadrature methods. These numerical procedures serve as local representatives of the method 12.53(9) and play an important role in proving the stability of 12.53(9). We define the Mellin convolution operators $A^i \in \mathscr{L}(\mathbf{L}^2(0, \infty))$, $i = 0, 1$, by

$$(A^0 z)(s) = \frac{1}{\pi \mathrm{i}} \int_0^\infty \frac{\sigma^{\alpha\varrho_0}}{s^{\alpha\varrho_0}} \frac{\alpha \sigma^{(\alpha-1)/2} s^{(\alpha-1)/2}}{\sigma^\alpha - s^\alpha} z(\sigma)\, \mathrm{d}\sigma + \int_0^\infty \alpha k_0 \left(\left(\frac{s}{\sigma} \right)^\alpha \right) \left(\frac{s}{\sigma} \right)^{(\alpha-1)/2 - \alpha\varrho_0} \frac{1}{\sigma} z(\sigma)\, \mathrm{d}\sigma, \tag{2}$$

$$(A^1 z)(s) = -\frac{1}{\pi \mathrm{i}} \int_0^\infty \frac{\sigma^{\alpha\varrho_1}}{s^{\alpha\varrho_1}} \frac{\alpha \sigma^{(\alpha-1)/2} s^{(\alpha-1)/2}}{\sigma^\alpha - s^\alpha} z(\sigma)\, \mathrm{d}\sigma, \qquad 0 < s < \infty.$$

For $A^0 z = y$ and $A^1 z = y$, we get the following quadrature methods. Set $\eta_j := 1/n$ for $n > 2$, $\eta_0 := 9/8n$, $\eta_1 := 19/24n$, $\eta_2 := 13/12n$ and determine approximate values ξ_j of $z(t_j)$, $j = 0, 1, \dots$ by solving

$$y(\tau_k) = \frac{1}{\pi i} \sum_{j=0}^{\infty} \frac{t_j^{\alpha\varrho_0}}{\tau_k^{\alpha\varrho_0}} \frac{\alpha t_j^{(\alpha-1)/2} \tau_k^{(\alpha-1)/2}}{t_j^\alpha - \tau_k^\alpha} \eta_j \xi_j$$
$$+ \sum_{j=0}^{\infty} \alpha k_0 \left(\frac{\tau_k^\alpha}{t_j^\alpha}\right) \left(\frac{\tau_k}{t_j}\right)^{(\alpha-1)/2-\alpha\varrho_0} \frac{1}{t_j} \eta_j \xi_j + \frac{1}{\pi i} \sum_{j=0}^{2} \frac{1}{t_j - \tau_k} \left[\frac{1}{n} - \eta_j\right] \xi_j, \tag{3}$$

$k = 0, 1, \dots,$

or

$$y(\tau_k) = -\frac{1}{\pi i} \sum_{j=0}^{\infty} \frac{t_j^{\alpha\varrho_1}}{\tau_k^{\alpha\varrho_1}} \frac{\alpha t_j^{(\alpha-1)/2} \tau_k^{(\alpha-1)/2}}{t_j^\alpha - \tau_k^\alpha} \eta_j \xi_j - \frac{1}{\pi i} \sum_{j=0}^{2} \frac{1}{t_j - \tau_k} \left[\frac{1}{n} - \eta_j\right] \xi_j,$$

$$k = 1, 2, \dots \tag{4}$$

Let us denote the matrices of (3) and (4) by A_n^0 and A_n^1, respectively. We regard them as operators mapping the space $\mathbf{l}^2 := \left\{\{\xi_j\}_{j=0}^\infty : \sum |\xi_j|^2 < \infty\right\}$ into $\mathbf{l}^2$ or $\mathbf{l}^{2,0} := \left\{\{\xi_j\}_{j=1}^\infty : \sum |\xi_j|^2 < \infty\right\}$, respectively. Remark that A_n^i $(i = 0, 1)$ is independent of n.

12.55. Theorem. *Suppose the kernel* k *satisfies the assumptions of Sec.* 12.53. *Let the function* $\mathcal{A}$ *on* $\{(t, \mu) : t \in \mathbb{T}, 0 \le \mu \le 1\}$ *be defined by*

$$\mathcal{A}(t, \mu) := \begin{cases} (-i) \cot \left[\pi \left(1/2 - \varrho_0 + i \dfrac{1}{2\pi\alpha} \log \dfrac{\mu}{1-\mu}\right)\right] & \\ \quad + b\left(1/2 - \varrho_0 + i \dfrac{1}{2\pi\alpha} \log \dfrac{\mu}{1-\mu}\right) & \text{if } t = 1, \\ -e^{i\pi\lambda} & \text{if } t = e^{i2\pi\lambda}, 0 < \lambda < 1. \end{cases}$$

Then the method 12.53(9) *is stable if and only if*
(i) $A \in \mathcal{L}(\mathbf{L}^2(0, 1))$ *is invertible,*
(ii) $\varrho_1 < 0$,
(iii) $\mathcal{A}(t, \mu) \ne 0$, $0 \le \mu \le 1$, $|t| = 1$,
(iv) *the index of the curve* $\{\mathcal{A}(e^{i2\pi\lambda}, 1) : 0 < \lambda < 1\} \cup \{\mathcal{A}(1, \mu) : 0 \le \mu \le 1\}$ *vanishes,*
(v) *the null spaces of the operators* $A_n^0 \in \mathcal{L}(\mathbf{l}^2)$ *and* $A_n^1 \in \mathcal{L}(\mathbf{l}^2, \mathbf{l}^{2,0})$ *are trivial.*

The proof of Theorem 12.55 resembles the arguments given in Sections 12.46–12.49 and Section 11.24 (see also Rathsfeld [3]). Condition (v) is hardly verifiable. However, the case where (v) is violated is exceptional and may be treated by a little modification of 12.53(9) (cf. Secs. 12.44–12.49 and Sec. 11.30).

In order to get asymptotic error estimates, we need further assumptions.

(A.1) Suppose that, for all $k \in \mathbb{Z}$, $k \ge 2$, the solution x of 12.52(1) takes the form

$$\mathsf{x}(t) = c_0 t^{\varkappa_0} + \sum_{i=1}^{L} \sum_{j=0}^{M} c_{ij} t^{\mu_i} (\log t)^j + f(t),$$

$$\mathsf{x}(t) = \sum_{m=0}^{k-2} \sum_{j=0}^{m} d_{mj} [1 - t]^{\varkappa_1 + m} (\log [1 - t])^j + g(t),$$

where $L = L(k)$, $M = M(k)$ are positive integers, $\varkappa_1 := -1/2$, c_0, c_{ij}, $d_{mj} \in \mathbb{C}$ and $-1 < \varkappa_0 < \mu_i \leqq k - 1/2$ $(i = 1, 2, \ldots, L)$. The functions f and g are supposed to belong to the Sobolev space $\mathbf{H}^k[0, 2/3]$ and $\mathbf{H}^s[1/3, 1]$, $0 \leqq s < \varkappa_1 + k - 1/2$, respectively.

(A.2) Let $\alpha \geqq (1 \pm 2\varrho_i)^{-1}$, $i = 0, 1$.

(A.3) Suppose that, for $0 \leqq t \leqq 1$, the functions $\tau \mapsto \mathsf{k}_1(t, \tau)$ belong to $\mathbf{H}^4[0, 1]$ and

$$\sup_{0 \leqq t \leqq 1} \|\mathsf{k}_1(t, \cdot)\|_{\mathbf{H}^4[0,1]} < \infty .$$

(A.4) Suppose $\tau \mapsto \mathsf{k}_0(t/\tau)\, 1/\tau$ is in $\mathbf{H}^4[0, 1]$ for $1/3 \leqq t \leqq 1$ and in $\mathbf{H}^4[1/3, 1]$ for $0 \leqq t \leqq 1$. Let

$$\sup_{0 \leqq t \leqq 1} \left\| \mathsf{k}_0\left(\frac{t}{\cdot}\right)\frac{1}{\cdot} \right\|_{\mathbf{H}^4[1/3,1]} < \infty, \qquad \sup_{1/3 \leqq t \leqq 1} \left\| \mathsf{k}_0\left(\frac{t}{\cdot}\right)\frac{1}{\cdot} \right\|_{\mathbf{H}^4[0,1]} < \infty .$$

(A.5) For $j = 0, \ldots, 4$, let

$$\sup_{0 < t < \infty} |(\mathrm{d}/\mathrm{d}t)^j\, \mathsf{k}_0(t)\, t^{(\alpha-1)/2\alpha - \varrho_0 + j}| < \infty .$$

We remark that (A.1) can be verified using the theory by Elschner [5]. Furthermore, if (A.2) holds, then (A.4) and (A.5) follow for any k_0 of the form $\mathsf{k}_0(t) = p(t)/q(t)$, where $p(t)$ and $q(t)$ are polynomials such that $q(t)$ does not vanish on $(0, \infty)$ and the degree of q is greater than that of p. If (A.5) is satisfied, then

$$K(s, \sigma) := \frac{\sigma^{\alpha\varrho_0}}{s^{\alpha\varrho_0}}\, \alpha\sigma^{(\alpha-1)/2}\, s^{(\alpha-1)/2}\mathsf{k}_0\left(\frac{s^\alpha}{\sigma^\alpha}\right)\frac{1}{\sigma^\alpha} = \mathsf{k}_0\left(\left(\frac{s}{\sigma}\right)^\alpha\right)\left(\frac{s}{\sigma}\right)^{(\alpha-1)/2 - \alpha\varrho_0}\frac{1}{\sigma},$$

$$\sup_{\substack{0 \leqq \sigma \leqq 1/3 \\ 0 \leqq s \leqq 1/3}} |\sigma^{l+1}(\partial/\partial\sigma)^l\, K(s, \sigma)| < \infty, \qquad l = 0, \ldots, 4,$$

which leads to an estimate analogous to 12.39(14).

12.56. Theorem. *Suppose $A \in \mathcal{L}(\mathbf{L}^2[0, 1])$ is invertible and 12.53(9) is stable. Then we have the following.*

(i) *If r and s satisfy $1/2 < s \leqq 4$, $s < \alpha(\varkappa_i - \varrho_i) + (\alpha - 1)/2$, $i = 0, 1$ and $0 \leqq r \leqq s$, then $\|z_n - z\|_{\mathbf{H}^r[0,1]} \leqq Cn^{r-s}$. (Note that, for $s > 2$ or $r > 1.5$, one has to change the notion of z_n. The piecewise linear interpolation z_n of the values ξ_k, $k = 0, \ldots, n - 1$ must be replaced by spline interpolation of higher order.)*

(ii) *Let $\mathbf{z}_n$ denote the function $\mathbf{z}_n(t) := z_n(\tau)/\gamma'(\tau)^{1/2}$, where $t := \gamma(\tau)$. If s satisfies $1/2 < s \leqq 4$, $s < \alpha(\varkappa_i - \varrho_i) + (\alpha - 1)/2$, $i = 0, 1$, then $\|\mathbf{z}_n - \mathbf{z}\|_{\mathbf{L}^2(0,1)} \leqq Cn^{-s}$.*

(iii) *If $s < 4$ and $s < \alpha(\varkappa_i - \varrho_i) + (\alpha - 1)/2$, $i = 0, 1$, then*

$$\sup_{k = 0, \ldots, n-1} |\xi_k - z(t_k)| \leqq Cn^{1/2 - s} .$$

(iv) *Let $s < 4$, $s < \alpha(\varkappa_i - \varrho_i) + (\alpha - 1)/2$, $i = 0, 1$ and choose $\varepsilon \in (0, 1/2)$. Define $n_i := n_i(n)$, $i = 0, 1$ by $\varepsilon n \leqq n_0 < \varepsilon n + 1$ and $(1 - \varepsilon)\, n - 1 < n_1 \leqq (1 - \varepsilon)\, n$. Then*

$$\sup_{k = n_0, n_0+1, \ldots, n_1} \left| \frac{\xi_k}{\gamma'(t_k)^{1/2}} - \mathbf{z}(\gamma(t_k)) \right| \leqq Cn^{1/2 - s} .$$

(v) *Let s satisfy* $0 \leq s < \min\left\{7\alpha/(3\alpha-1), \left(2\alpha^2(\varkappa_i - \varrho_i) + \alpha(\alpha-2)\right)/(3\alpha-1), i = 0, 1\right\}$ *and determine* $n_3 := n_3(n)$ *by* $n^{-s/\alpha} \leq n_3 < n^{-s/\alpha} + 1$. *Then*

$$\left|\frac{\xi_{n_3}}{\gamma'(t_{n_3})^{1/2}} - \mathsf{z}(0)\right| \leq Cn^{-s} \log n\,.$$

Theorem 12.56 is an easy consequence of Secs. 12.38–12.43. Since there coarse estimates were used, we conjecture that the results of the theorem can be improved.

12.57. Numerical examples. Let $\alpha = 4$. For the sake of simplicity, choose $\gamma(t) := t^4/[t^4 + (1-t)^4]$. As a *first example* we consider the equation

$$\frac{1}{\pi}\int_0^1 \frac{\mathsf{x}(\tau)}{\tau - t}\, d\tau = 1, \qquad 0 \leq t \leq 1,$$

whose solution x is given by $\mathsf{x}(t) = t^{1/2}(1-t)^{-1/2}$. We set $\varrho_0 := 1/4$ and $\varrho_1 := -1/4$. Table 8 gives the approximate values $\mathsf{z}_n(t)\, t^{1/4}\,(1-t)^{1/4}$ for the function $\mathsf{x}(t)\,(1-t)^{1/2} =: t^{1/2}$.

Table 8. Example 1. The approximate values for method 12.53(9), $\alpha = 4$, and the exact values

t	approximate values	exact values
$n = 6$		
0.2065	0.449	0.454
0.7934	0.882	0.891
0.9878	0.973	0.993
$n = 24$		
0.2679	0.51724	0.51759
0.5826	0.76320	0.76329
0.9171	0.95746	0.95763
0.9922	0.99558	0.99611
$n = 96$		
0.2848	0.533671	0.533679
0.5208	0.721671	0.721680
0.9099	0.953867	0.953882
0.9913	0.995601	0.995637

Our *second example* is

$$\frac{1}{\pi i}\int_0^1 \left\{\frac{1}{\tau - t} - \frac{1}{\tau + t} - \frac{2t}{(\tau+t)^2} + \frac{4t^2}{(\tau+t)^3}\right\} \mathsf{x}(\tau)\, d\tau = -i4t, \qquad 0 \leq t \leq 1,$$

(cf. WIGGELSWORTH [1]). The asymptotics of the solution x can be found in ELSCHNER [5]. We apply 12.53(9) with $\tau_k := (k + 1)/n$, $k = 0, \ldots, n - 1$. For this case, Theorem 12.55 holds if condition (ii) is replaced by $\varrho_1 > 0$ and if $\mathcal{A}(t, \mu)$ is defined by

$$\mathcal{A}(t, \mu) := \begin{cases} (-\mathrm{i}) \cot \left[\pi \left(1/2 - \varrho_0 + \mathrm{i} \dfrac{1}{2\pi\alpha} \log \dfrac{\mu}{1 - \mu}\right)\right] & \\ \quad + b \left(1/2 - \varrho_0 + \mathrm{i} \dfrac{1}{2\pi\alpha} \log \dfrac{\mu}{1 - \mu}\right) & \text{if } t = 1, \\ -\mathrm{e}^{-\mathrm{i}\pi\lambda} & \text{if } t = \mathrm{e}^{\mathrm{i}2\pi\lambda}, \quad 0 < \lambda < 1. \end{cases}$$

Choosing $\varrho_0 := -1/4$ and $\varrho_1 := 1/4$, we get $\mathcal{A}(t, \mu) \neq 0$ and $\operatorname{Im} \mathcal{A}(t, \mu) \geqq 0$. Thus, if condition (v) in Theorem 12.55 is satisfied, then 12.53(9) is stable. In Table 9 the approximate values of the function x are given and compared with $\mathsf{x}_w(t)$, where $\mathsf{x}_w(t)$ denotes the sum

$$\mathsf{x}_w(t) := \sum_{j=0}^{8} a_j (1 - t)^{j+1/2}$$

of the first terms of a series derived in WIGGELSWORTH [1]. (Note that the a_j are known only approximately.)

Table 9. Example 2. The approximate values of the method 12.53(9), $\alpha = 4$ and of Wiggelsworth's method

t	approximate values	x_w
$n = 12$		
0.0048	−5.6623	−5.8123
0.1147	−5.4593	−5.4610
0.6611	−3.5243	−3.5254
0.8853	−2.1130	−2.1141
$n = 24$		
0.0078	−5.79419	−5.80252
0.1556	−5.33336	−5.33430
0.5826	−3.87482	−3.87485
0.8444	−2.44788	−2.44793
$n = 48$		
0.0098	−5.78658	−5.79590
0.1339	−5.39997	−5.40128
0.5415	−4.04161	−4.04161
0.8202	−2.62246	−2.62246

As a *third example* we consider the integral equation

$$\frac{1}{\pi} \int_{-1}^{1} \frac{u(\tau)}{\tau - t} \, \mathrm{d}\tau + \frac{1}{\pi} \int_{-1}^{1} \frac{\tau(\tau^2 - t^2)}{(\tau^2 + t^2)^2} u(\tau) \, \mathrm{d}\tau = 1, \qquad -1 < t < 1, \tag{1}$$

subject to the condition

$$\int_{-1}^{1} u(\tau)\, d\tau = 0. \tag{2}$$

The problem (1)–(2) arises in the study of a cruciform crack in an infinite isotropic elastic medium under constant load along its four branches (see e.g. Panasyuk/Savruk/Dacyshin [1]).

We choose α and β satisfying $-1/2 < \beta < 1/2$, $-1/2 < \alpha < 0$ and define $\mathbf{L}^2_{\alpha,\beta,\alpha}$ as the space of all functions u on $(-1, 1)$ with finite norm

$$\|u\|_{\mathbf{L}^2_{\alpha,\beta,\alpha}} := \left(\int_{-1}^{1} |t|^{-2\beta} (1 - t^2)^{-2\alpha} |u(t)|^2 \, dt \right)^{1/2}.$$

Then the following holds (see Prössdorf/Rathsfeld [9]). There exists a unique solution $u \in \mathbf{L}^2_{\alpha,\beta,\alpha}$ to the problem (1)–(2). This solution u is odd and $u(t) = \mathsf{x}(t^2)$ for $0 < t < 1$, where $\mathsf{x} \in \mathbf{L}^2_{\gamma,\alpha}$, $\gamma := (\beta + 1/2)/2$, is the unique solution of the equation

$$\frac{1}{\pi} \int_0^1 \frac{\mathsf{x}(\tau)}{\tau - t} \, d\tau + \frac{1}{\pi} \int_0^1 \frac{\tau - t}{(\tau + t)^2} \mathsf{x}(\tau) \, d\tau = 1, \qquad 0 < t < 1. \tag{3}$$

Moreover, the solution u behaves like $C(1 \pm t)^{-1/2}$ as $t \to \mp 1$ and $u(0) = 0$.

We set $\varrho_0 = 1/4$ and $\varrho_1 = -1/4$ and apply the quadrature method 12.53(9) with $\tau_k = (k + 1/4)/n$, $t_k = (k + 3/4)/n$ and $\omega_j = 1/n$ to the equation (3). We list the computed approximate values of the function $(1 - t^2)^{1/2} u(t)$ in Table 10.

Table 10. Example 3. The approximate values of the method 12.53(9), $\alpha = 4$

n ↓	$t \to$ 0.0204	0.5821	0.999998
6	−0.0007	0.296484448	0.843
30	−0.0002915	0.297013265	0.866
54	−0.000291466	0.297013258	0.8638
78	−0.000291467	0.297013257	0.86360
102	−0.000291527	0.297013257	0.86356

We observe that the rate of convergence is lower near the end points 0 and 1 (cf. also the theoretical rates in Sec. 12.56). This is a consequence of the multiplication by large numbers in 12.53(10). Hence, to get good approximations at the point $t = 1$, we must not take the approximate value at the point nearest to 1. In Table 11 we list three points t_0, t_1 and t_2 as well as the approximate values of $(1 - t^2)^{1/2} u(t)$ at $t = 1$ obtained by quadratic extrapolation of the approximate values at t_0, t_1 and t_2. Finally, we remark that better rates of convergence at $t = 1$ are presented in Gerasoulis [1] and Jen/Srivastav [1]. However, method 12.53(9) is very easy to implement and is more general. (E.g., it works also if the solution $\mathsf{x}(t)$ has a representation of the form $\mathsf{x}(t) = c_0 t^{\varkappa_0} + c_1 t^{\varkappa_1} + \cdots$ as $t \to 0$, where $-1/2 < \varkappa_0 < \varkappa_1 < 1/2$.)

Table 11. Example 3. Quadratic extrapolation at the point $t = 1$

n	t_0	t_1	t_2	approximate values
6	0.58	0.94	0.99	0.8666
18	0.67	0.91	0.99	0.8637
30	0.90	0.94	0.99	0.8636
42	0.90	0.93	0.99	0.8636

Notes and comments

12.1.—12.5. Here we follow Elschner [1].

12.6.—12.11. This material is taken from Prössdorf/Rathsfeld [3, Sec. 3].

12.12.—12.35. In these sections we follow Elschner [7]. A somewhat different presentation of these results can be found in Elschner [3]. Costabel and Stephan [2] treated essentially the same collocation method with splines of first degree applied to the operator of the single layer potential on a polygon. For collocation with special weighted spline functions, good rates of convergence have been obtained in Gerasoulis [2] and Jen/Srivastav [1].

12.37.—12.51. The material of these sections is taken from Prössdorf/Rathsfeld [8].

The first convergence analysis of simple quadrature methods was apparently given by Lifanov and Polonski [1] in the case of the method of discrete whirls. A generalization of this method can be found in Rathsfeld [3] and in Prössdorf/Rathsfeld [5]. The method 12.37(15) represents a modification of a quadrature method studied by Rathsfeld [3].

12.53.—12.57. This material is adapted from Rathsfeld [4]. The third example in Sec. 12.57 is taken from Prössdorf/Rathsfeld [9].

Notice that a simple method for obtaining the approximate solution of the equation 12.36(1) was suggested in Fabrikant/Hoa/Sankar [1]. The solution is represented in the form of a power series with undetermined coefficients, multiplied by a function which matches the singular behaviour of the solution. The collocation method is used to determine the unknown coefficients. However, for this method no convergence analysis is known.

Chapter 13
Spline approximation methods for pseudodifferential equations on closed curves

In the first part of this chapter we study ε-collocation of pseudodifferential equations on closed curves in a scale of Sobolev spaces by smooth polynomial splines subordinate to uniform meshes. We give necessary and sufficient stability conditions essentially depending on the choice of the collocation points and on the degree of the splines used as trial functions. We shall present two different and independent approaches to the stability theory. The first one has its origin in the idea of reducing the pseudodifferential equation to a singular integral equation and applying the corresponding results of Chapter 10. The crucial tool of the second approach is a localization technique for spline approximation of pseudodifferential equations, which relies upon locally freezing coefficients and using perturbation techniques familiar from the existence theory for partial differential equations. Moreover, we prove optimal order error estimates in Sobolev norms. In the second part we investigate the nodal collocation of pseudodifferential equations by polynomial splines on arbitrary meshes. The corresponding error analysis is based on the equivalence of the collocation method to certain Galerkin methods and on the use of the Aubin-Nitsche lemma. Finally, we compare the collocation method with a standard Galerkin method and with the region method using splines of the same degree.

Spline collocation on uniform meshes

13.1. Notation. Let Γ be a simple closed $\mathbf{C}^\infty$-curve in the complex plane given by a regular parametric representation $\Gamma := \{z = \gamma(x) : 0 \leqq x \leqq 1\}$, where γ is a 1-periodic function of the real variable x and $|\mathrm{d}\gamma/\mathrm{d}x| \neq 0$. Via the parametrization we have a one-to-one correspondence between functions on Γ and 1-periodic functions. Throughout this chapter we identify a function u defined on Γ with the 1-periodic function $u(x) = u(\gamma(x))$. Let $\mathbf{H}^s$ ($s \in \mathbb{R}$) denote the periodic Sobolev space of order s. For fixed δ and $n \in \mathbb{N}$, we set $h = 1/n$ and denote by $\mathcal{S}_h^\delta := \mathcal{S}^\delta(\Delta)$ the space of all 1-periodic, $\delta - 1$ times continuously differentiable splines of degree δ on the uniform mesh $\Delta_h := \{kh : k = 0, \ldots, n-1\}$; $\mathcal{S}_h^0$ stands for the space of piecewise constant functions.

Let A be a classical elliptic pseudodifferential operator of order $\alpha \in \mathbb{R}$ on Γ with the principal symbol $a_0(x, \xi) \in \mathbf{C}^\infty(\mathbb{R} \times \mathbb{R} \setminus \{0\})$ (see Chapter 6). Thus $a_0(x, \xi)$ is a positively homogeneous function with respect to $\xi \in \mathbb{R} \setminus \{0\}$ of degree α, and $a_0(x, \xi) \neq 0$ for all x, ξ. Let ε, $0 \leqq \varepsilon < 1$, be a fixed number, put $x_k := (k + \varepsilon) h$, $k = 0, \ldots, n-1$, and let $f \in \mathbf{H}^s$, $s > 1/2$. The ε-collocation method determines an approximate solution $u_h \in \mathcal{S}_h^\delta$

to the operator equation

$$Au = f \quad \text{on} \quad \mathbb{R} \tag{1}$$

by the equations

$$Au_h(x_k) = f(x_k), \qquad k = 0, \ldots, n-1. \tag{2}$$

Once a basis for $\mathscr{S}_h^\delta$ is chosen, (2) is easily reduced to an $n \times n$ linear system for the unknown coefficients of u_h.

It follows from Section 2.20 that the left-hand side of Eq. (2) makes sense if and only if either

$$\delta - \alpha > 0 \quad \text{for} \quad \varepsilon = 0 \tag{3}$$

or

$$\delta - \alpha > -1 \quad \text{for} \quad \varepsilon \in (0, 1). \tag{4}$$

In what follows the conditions (3) or (4) are assumed to be satisfied.

In particular, 0-collocation (i.e. nodal collocation) and 1/2-collocation (i.e. midpoint collocation) are frequently used for integral equations of the first kind with a logarithmic kernel, which define a strongly elliptic ψdo of order -1. However, it is often more advantageous to apply ε-collocation with $\varepsilon \neq 0$ because of the relations (3) and (4) and the fact that, generally speaking, the numerical computation of singular integrals, e.g. at the nodes of splines, is rather complicated. Moreover, by appropriately choosing the parameter ε in dependence on the principal symbol of A, spline collocation is stable and converges with optimal order for a class larger than strongly elliptic pseudodifferential equations (see Theorems 13.2 and 13.14).

In order to formulate stability conditions for the ε-collocation (2) we define the functions

$$a(x) := a_0(x, +1), \qquad b(x) := a_0(x, -1) \; (x \in \mathbb{R}). \tag{5}$$

Since $a_0(x, \xi)$ is a positively homogeneous function with respect to ξ of degree α, we have

$$a_0(x, \xi) = \frac{|\xi|^\alpha}{2} \left[a(x)\,(1 + \xi/|\xi|) + b(x)\,(1 - \xi/|\xi|)\right]. \tag{6}$$

Furthermore, we introduce the "symbol" of the ε-collocation (2) by

$$\sigma_A^{\delta,\varepsilon}(x, s) := \sum_{k \in \mathbb{Z}} a_0(x, k + s)\,(k + s)^{-\delta-1}\, e^{2\pi i k \varepsilon} \tag{7}$$

for $x \in \mathbb{R}$ and $0 < s < 1$. By virtue of (6), Eq. (7) can be rewritten as

$$\sigma_A^{\delta,\varepsilon}(x, s) = a(x) \sum_{k=0}^{\infty} (k + s)^{\alpha-\delta-1}\, e^{2\pi i k \varepsilon} + b(x) \sum_{k=-1}^{-\infty} |k + s|^\alpha\, (k + s)^{-\delta-1}\, e^{2\pi i k \varepsilon} \tag{8}$$

or

$$\sigma_A^{\delta,\varepsilon}(x, s) = a(x) \sum_{k=0}^{\infty} (k + s)^{\alpha-\delta-1}\, e^{2\pi i k \varepsilon} + (-1)^{\delta+1}\, b(x) \sum_{k=-1}^{-\infty} |k + s|^{\alpha-\delta-1}\, e^{2\pi i k \varepsilon}.$$

13.2. Theorem. *Assume* 13.1(3) *and* 13.1(4). *The* ε*-collocation* 13.1(2) *is stable in* $(\mathbf{H}^{t+\alpha}, \mathbf{H}^t)$, $-1/2 < t < \delta - \alpha + 1/2$, *if and only if* (i) $\dim \ker A = 0$ *and*

$$\text{(ii)} \quad \begin{cases} a(x)\, b(x) \neq 0,\; x \in \mathbb{R}, \\ \sigma_A^{\delta,\varepsilon}(x, s) \neq 0,\; x \in \mathbb{R},\; 0 < s < 1. \end{cases}$$

Notice that if $\varepsilon = 0$ or $\varepsilon = 1/2$, then condition (ii) is equivalent to the requirement that

$$\mu a(x) + (-1)^{\delta+1+2\varepsilon} (1 - \mu) b(x) \neq 0 \tag{1}$$

for all $\mu \in [0, 1]$ and $x \in \mathbb{R}$. This is an immediate consequence of the equalities

$$\text{im } \sigma^-_{0,\beta}/\sigma^+_{0,\beta} = \text{im } \sigma^+_{1/2,\beta}/\sigma^-_{1/2,\beta} = [-1, 1]$$

(see Section 2.21; cf. also Section 10.7).

In the case of integer order α and $t \geqq 0$, Theorem 13.2 easily follows from Corollary 10.7. Indeed, A admits the representation

$$A = A_0 \Lambda^\alpha + T, \qquad T \in \mathscr{L}(\mathbf{H}^{t+\alpha}, \mathbf{H}^{t+1}), \tag{2}$$

where $A_0 = aP_{\mathbb{T}} + bQ_{\mathbb{T}}$ is the singular integral operator on the unit circle $\mathbb{T}$ with the coefficients $a, b \in \mathbf{C}^\infty$ defined by 13.1(5) (see Corollary 6.63). Since (see 2.20(4))

$$\mathscr{S}_h^\delta = \begin{cases} \Lambda^{-\delta}(\mathscr{S}_h^0) & \text{if } \delta \text{ is even,} \\ S\Lambda^{-\delta}(\mathscr{S}_h^0) & \text{if } \delta \text{ is odd,} \end{cases} \tag{3}$$

where $S := S_{\mathbb{T}}$ and $S^2 = I$, we obtain

$$\Lambda^\alpha(\mathscr{S}_h^\delta) = \begin{cases} \mathscr{S}_h^{\delta-\alpha} & \text{if } \alpha \text{ is even,} \\ S(\mathscr{S}_h^{\delta-\alpha}) & \text{if } \alpha \text{ is odd.} \end{cases} \tag{4}$$

We now define the singular integral operator $A_1 \in \mathscr{L}(\mathbf{H}^t)$ by

$$A_1 = \begin{cases} A_0 + T\Lambda^{-\alpha} & \text{if } \alpha \text{ is even,} \\ A\Lambda^{-\alpha}S = aP_{\mathbb{T}} - bQ_{\mathbb{T}} + T\Lambda^{-\alpha}S & \text{if } \alpha \text{ is odd.} \end{cases}$$

So 13.1(2) is equivalent to the following ε-collocation for the operator A_1: find $v_h \in \mathscr{S}_h^{\delta-\alpha}$ such that

$$A_1 v_h(x_k) = f(x_k), \qquad k = 0, \ldots, n-1. \tag{5}$$

In view of Theorem 1.37 and the estimate 10.23(2) the collocation (5) is stable in $\mathbf{H}^t$ i and only if it is stable in $\mathbf{L}^2$. Since $T\Lambda^{-\alpha}$, $T\Lambda^{-\alpha}S \in \mathscr{K}(\mathbf{L}^2, \mathbf{C})$, the result follows from Corollary 10.7. ∎

In what follows we give another, independent proof of Theorem 13.2, which applies to the case of arbitrary real order α. The crucial tool of this proof is a localization technique for spline approximation methods which we are going to formulate now. Moreover, we prove optimal order estimates in Sobolev norms for the ε-collocation 13.1(2).

Since for given ε, $0 \leqq \varepsilon < 1$, 13.1(3) resp. 13.1(4) is satisfied, in one of the spaces $\mathscr{M}_h^{\delta-\alpha} := \Lambda^{\alpha-\delta}(\mathscr{S}_h^0)$ or $\mathscr{N}_h^{\delta-\alpha} := S\Lambda^{\alpha-\delta}(\mathscr{S}_h^0)$ we can choose the interpolation basis $\{\chi_k\}_{k=0,\ldots,n-1}$ i.e., $\chi_k(x_j) = \delta_{kj}$ $(k, j = 0, \ldots, n-1)$ (see Theorem 2.22). By Q_h we denote the interpolation projection corresponding to the basis $\{\chi_k\}_{k=0,\ldots,n-1}$, i.e.

$$Q_h u = \sum_{k=0}^{n-1} u(x_k)\, \chi_k \tag{6}$$

if $\delta - \alpha > 0$. For $-1 < \delta - \alpha \leqq 0$, Q_h is defined by (6), where $\chi_k = \theta_k$ stands for the characteristic function of $[x_k, x_{k+1})$ (notice that in this case $\varepsilon > 0$, since 13.1(4) holds). Now Eq. 13.1(2) can be written as

$$Q_h A u_h = Q_h f, \qquad u_h \in \mathscr{S}_h^\delta. \tag{7}$$

Let M_y, $y \in [0, 1)$, denote the set of all 1-periodic $\mathbf{C}^\infty$-functions g_y satisfying $0 \leqq g_y \leqq 1$, $\operatorname{supp} g_y \subset U$ and $g_y \equiv 1$ in V for some neighborhoods U and $V \subset U$ of y. For fixed $y \in [0, 1)$ and given $a, b \in \mathbf{C}^\infty$, with the ψdo of the form

$$A = (aP_{\mathbf{T}} + bQ_{\mathbf{T}})\,\Lambda^\alpha + T, \qquad T \in \mathscr{L}(\mathbf{H}^{t+\alpha}, \mathbf{H}^{t+1}) \tag{8}$$

we shall associate the operator with constant coefficients

$$A_y := \big(a(y)\,P_{\mathbf{T}} + b(y)\,Q_{\mathbf{T}}\big)\,\Lambda^\alpha. \tag{9}$$

Suppose the following conditions are satisfied:

I . There exist projections P_h of $\mathbf{H}^{t+\alpha}$ onto $\mathscr{S}_h^\delta$ such that $\|P_h\|_{t+\alpha} \leqq C$. Furthermore, for all $g \in \mathbf{C}^\infty$, $\big\|(I - P_h)\,g|_{\mathscr{S}_h^\delta}\big\|_{t+\alpha} \to 0$ as $h \to 0$.

II . For all $g \in \mathbf{C}^\infty$, we have $\big\|Q_hg|_{\operatorname{im} Q_h}\big\|_t \leqq C$, and there exists a $\beta > 0$ such that $\|u - Q_hu\|_t \to 0$ as $h \to 0$ for every $u \in \mathbf{H}^{t+1-\beta}$.

III. For all $g \in \mathbf{C}^\infty$, $\big\|Q_hA(I - P_h)\,g|_{\mathscr{S}_h^\delta}\big\|_{t+\alpha\to t}$ tends to zero as $h \to 0$.

IV . The convergence $\|v - v_h\|_{t+\alpha} \to 0$ $(h \to 0)$ implies $\|Q_hAv_h - Av\|_t \to 0$ for all $v \in \mathbf{H}^{t+\alpha}$, $v_h \in \mathscr{S}_h^\delta$.

V . For all $y \in [0, 1)$ and $\eta > 0$, there exists a $g_y \in M_y$ such that

$$\inf_T \big\|Q_h\big(g_y(A - A_y) - T\big)|_{\mathscr{S}_h^\delta}\big\|_{t+\alpha\to t} < \eta \tag{10}$$

for all sufficiently small $h = 1/n$. Here the infimum is taken over the set of all bounded operators $T\colon \mathbf{H}^{t+\alpha} \to \mathbf{H}^{t+1}$.

Notice that condition I implies the strong convergence $P_h \to I$ on $\mathbf{H}^{t+\alpha}$, since $\|g - P_hg\|_{t+\alpha} \to 0$ as $h \to 0$ for all g of the dense subset $\mathbf{C}^\infty \subset \mathbf{H}^{t+\alpha}$.

Moreover, (10) is equivalent to

$$\inf_T \big\|Q_h[(A - A_y)\,g_y - T]|_{\mathscr{S}_h^\delta}\big\|_{t+\alpha\to t} < \eta,$$

since $g_yA - Ag_y \in \mathscr{L}(\mathbf{H}^{t+\alpha}, \mathbf{H}^{t+1})$.

The sequence $\{Q_hA|_{\mathscr{S}_h^\delta}\}$ is said to be *stable* in $(\mathbf{H}^{t+\alpha}, \mathbf{H}^t)$ if $\|Q_hAv_h\|_t \geqq c\,\|v_h\|_{t+\alpha}$ for all sufficiently small h and all $v_h \in \mathscr{S}_h^\delta$. We call $\{Q_hA|_{\mathscr{S}_h^\delta}\}$ *locally stable at* y in $(\mathbf{H}^{t+\alpha}, \mathbf{H}^t)$ if there exist $g_y \in M_y$ and operators $T_y, T_y' \in \mathscr{L}(\mathbf{H}^{t+\alpha}, \mathbf{H}^{t+1})$ and $C_{y,h}, D_{y,h} \in \mathscr{L}(\operatorname{im} Q_h, \operatorname{im} P_h)$ such that

$$Q_hg_y(A_y + T_y)\,C_{y,h} \cong Q_hg_yP_h, \qquad D_{y,h}Q_h(A_y + T_y')\,g_yP_h \cong P_hg_yP_h, \tag{11}$$
$$\sup_h \|C_{y,h}\|_{t\to t+\alpha} < \infty, \qquad \sup_h \|D_{y,h}\|_{t\to t+\alpha} < \infty.$$

Here and in the sequel $B_h \cong C_h$ stands for two sequences of operators B_h and C_h satisfying $\|B_h - C_h\| \to 0$ as $h \to 0$.

13.3. Theorem. *Suppose the conditions* I *through* V *are satisfied, where A and A_y are the ψdo's defined by* 13.2(8) *and* 13.2(9), *respectively. If A is invertible then the sequence* $\{Q_hA|_{\mathscr{S}_h^\delta}\}$ *is stable in* $(\mathbf{H}^{t+\alpha}, \mathbf{H}^t)$ *if and only if* $\{Q_hA|_{\mathscr{S}_h^\delta}\}$ *is locally stable at y in* $(\mathbf{H}^{t+\alpha}, \mathbf{H}^t)$ *for all* $y \in [0, 1)$.

Proof. For the sake of brevity we put $A_h := Q_hAP_h$ and

$$f_h^P := P_hfP_h, \qquad f_h^Q := Q_hfQ_h = Q_hf \qquad (f \in \mathbf{C}^\infty).$$

First suppose $\{A_h\}$ is stable in $(\mathbf{H}^{t+\alpha}, \mathbf{H}^t)$, i.e.

$$\|A_h v_h\|_t \geqq c\,\|v_h\|_{t+\alpha} \qquad \forall h < h_0 \quad \forall v_h \in \mathscr{S}_h^\delta.$$

In view of conditions III and V, for each $y \in [0, 1)$ and $0 < q < 1$, there exist $f_y \in M_y$ and $T_y \in \mathscr{L}(\mathbf{H}^{t+\alpha}, \mathbf{H}^{t+1})$ such that

$$\|[A_h - (A_y)_h]\,(f_y)_h^P - (T_y)_h\|_{t+\alpha\to t} \leqq q/c \qquad \forall h < h_0.$$

Choose $g_y \in M_y$ so that $g_y = g_y f_y$ to get

$$\begin{aligned} A_h^{-1}(A_y)_h\,(g_y)_h^P &= (g_y)_h^P + A_h^{-1}[(A_y)_h - A_h]\,(g_y)_h^P \\ &\cong (I + B_{y,h})\,(g_y)_h^P - A_h^{-1}(T_y)_h\,(g_y)_h^P, \end{aligned} \tag{1}$$

where $B_{y,h} := A_h^{-1}\{[(A_y)_h - A_h]\,(f_y)_h^P + (T_y)_h\}$ (recall condition I). Since $\|B_{y,h}\|_{t+\alpha} \leqq q < 1$, the operators $I + B_{y,h} \in \mathscr{L}(\mathscr{S}_h^\delta)$ are invertible and $\|(I + B_{y,h})^{-1}\|_{t+\alpha} \leqq (1 - q)^{-1}$ for all $h < h_0$. Consequently, the second relation in 13.2(11) follows from (1). Analogously the first relation in 13.2(11) can be derived.

We now prove the converse. Thus, assume 13.2(11) is fulfilled for all $y \in [0, 1)$ and $h < h_0$. Then

$$D_{y,h}(A_y + T_y')_h\,(g_y)_h^P \cong (g_y)_h^P \tag{2}$$

and

$$(g_y)_h^Q\,(A_y + T_y)_h\,C_{y,h} \cong (g_y)_h^Q. \tag{3}$$

From condition V we infer that, for each $y \in [0, 1)$, there are $a_y \in M_y$, $T_y \in \mathscr{L}(\mathbf{H}^{t+\alpha}, \mathbf{H}^{t+1})$ and $B_{y,h} \in \mathscr{L}(\operatorname{im} P_h, \operatorname{im} Q_h)$ satisfying

$$(a_y)_h^Q\,[A_h - (A_y)_h] \cong B_{y,h} + (T_y)_h \tag{4}$$

and $\|B_{y,h}C_{y,h}\|_t \leqq q < 1$. Choosing $f_y \in M_y$ such that $f_y a_y = f_y g_y = f_y$, we see from (4) that

$$(f_y)_h^Q\,[A_h - (A_y)_h] \cong (f_y)_h^Q\,[B_{y,h} + (T_y)_h]. \tag{5}$$

Combining (3) and (5), we find that

$$(f_y)_h^Q\,A_h C_{y,h} \cong (f_y)_h^Q\,[(I + B_{y,h}C_{y,h}) + (T_y)_h\,C_{y,h}].$$

Thus

$$(f_y)_h^Q\,A_h G_{y,h} \cong (f_y)_h^Q\,[I + (T_y)_h\,G_{y,h}], \tag{6}$$

where $G_{y,h} := C_{y,h}(I + B_{y,h}C_{y,h})^{-1}$. Obviously, $\sup\limits_h \|G_{y,h}\|_{t\to t+\alpha} < \infty$.

We now select a finite number of functions $f_{y_1}, \ldots, f_{y_N}$ such that the function $f := \sum\limits_{k=1}^N f_{y_k} \in \mathbf{C}^\infty$ is invertible. Setting $C_h := \sum\limits_{k=1}^N P_h f_{y_k} G_{y_k,h}$ and taking into account condition III, we obtain that

$$A_h C_h \cong \sum_{k=1}^N (f_{y_k})_h^Q\,A_h G_{y_k,h} + \sum_{k=1}^N Q_h T_k G_{y_k,h},$$

where $T_k := Af_{y_k} - f_{y_k}A \in \mathscr{L}(\mathbf{H}^{t+\alpha}, \mathbf{H}^{t+1})$. Apply (6) to get

$$A_h C_h \simeq f_h^Q + \sum_{k=1}^{N} Q_h T_k' G_{y_k,h}, \tag{7}$$

where $T_k' = T_k + f_{y_k} T_{y_k} \in \mathscr{L}(\mathbf{H}^{t+\alpha}, \mathbf{H}^{t+1})$. Finally, for $\tilde{C}_h := C_h - \sum_{k=1}^{N} P_h A^{-1} T_k' G_{y_k,h}$, it follows from (7) that $A_h \tilde{C}_h \simeq f_h^Q + W_h$, where $W_h := \sum_{k=1}^{N} (Q_h - A_h A^{-1})\, T_k' G_{y_k,h}$. In view of conditions I, II and IV, $Q_h - A_h A^{-1}$ converges strongly to zero on $\mathbf{H}^t$. Since $T_k' \in \mathscr{K}(\mathbf{H}^{t+\alpha}, \mathbf{H}^t)$ assertion 1.1(h) yields that $\|W_h\|_t \to 0$ as $h \to 0$. Hence $A_h \tilde{C}_h \simeq f_h^Q$. Since $(f_h^Q)^{-1} = (f^{-1})_h^Q$ and $\sup_h \|(f^{-1})_h^Q\| < \infty$ (see condition II), there exist operators $D_h \simeq \tilde{C}_h (f^{-1})_h^Q$ such that

$$A_h D_h = I\big|_{\operatorname{im} Q_h}, \qquad \sup_h \|D_h\|_{t \to t+\alpha} < \infty.$$

Thus, because $\dim \operatorname{im} Q_h = \dim \operatorname{im} P_h < \infty$, A_h is invertible and $(A_h)^{-1} = D_h$. This proves the stability of the sequence $\{A_h\}$ in $(\mathbf{H}^{t+\alpha}, \mathbf{H}^t)$. ■

Note that the invertibility of A has been used only in the second part of the preceding proof.

13.4. The case of constant coefficients. By means of Theorem 13.3, all considerations may be reduced to the case of ψdo's with constant coefficients. To this end we first prove Theorem 13.2 under the additional assumption that A has constant coefficients,

$$A = (cI + dS)\, \Lambda^\alpha; \qquad c, d \in \mathbb{C},\ \alpha \in \mathbb{R}. \tag{1}$$

This operator will be analyzed using only the results on periodic spline interpolation established in Chapter 2.

Retaining the notation of Section 13.2 we introduce a basis $\{\vartheta_k\}_{k=0}^{n-1}$ in $\mathscr{S}_h^\delta$ by

$$\vartheta_k = \begin{cases} \Lambda^{-\alpha}\chi_k & \text{if } \chi_k \in \mathscr{M}_h^{\delta-\alpha}(\mathscr{N}_h^{\delta-\alpha}) \text{ and } \delta \text{ is even (odd)},\\ S\Lambda^{-\alpha}\chi_k & \text{if } \chi_k \in \mathscr{N}_h^{\delta-\alpha}(\mathscr{M}_h^{\delta-\alpha}) \text{ and } \delta \text{ is even (odd)}. \end{cases}$$

Obviously,

$$\left\| \sum_{k=0}^{n-1} y_k \vartheta_k \right\|_{t+\alpha} = \left\| \sum_{k=0}^{n-1} y_k \chi_k \right\|_t \tag{2}$$

for any $y_k \in \mathbb{C}$ $(k = 0, \ldots, n-1)$. If we seek the solution of Eq. 13.1(2) in the form $u_h = \sum_{k=0}^{n-1} \xi_k \vartheta_k$, then 13.1(2) is a linear algebraic system with the matrix

$$A_n := \begin{cases} cI_n + d\big(S\chi_k(x_r)\big)_{k,r=0}^{n-1} & \text{if } \chi_k \in \mathscr{M}_h^{\delta-\alpha}(\mathscr{N}_h^{\delta-\alpha}) \text{ and } \delta \text{ is even (odd)},\\ c\big(S\chi_k(x_r)\big)_{k,r=0}^{n-1} + dI_n & \text{if } \chi_k \in \mathscr{N}_h^{\delta-\alpha}(\mathscr{M}_h^{\delta-\alpha}) \text{ and } \delta \text{ is even (odd)}, \end{cases}$$

I_n being the $n \times n$ identity matrix. From 2.22(2) it follows that

$$S\chi_k(x_r) = \begin{cases} \sum_{j=0}^{n-1} a_{kj} S\Lambda^{\alpha-\delta}\theta_j(x_r) \quad \text{with} \quad (a_{kj})_{k,j=0}^{n-1} = \big((\Lambda^{\alpha-\delta}\theta_j(x_r))_{j,r=0}^{n-1}\big)^{-1} \\ \qquad\qquad\qquad\qquad\qquad\qquad\qquad\qquad \text{if} \quad \chi_k \in \mathscr{M}_h^{\delta-\alpha}, \\ \sum_{j=0}^{n-1} a_{kj} \Lambda^{\alpha-\delta}\theta_j(x_r) \quad \text{with} \quad (a_{kj})_{k,j=0}^{n-1} = \big((S\Lambda^{\alpha-\delta}\theta_j(x_r))_{j,r=0}^{n-1}\big)^{-1} \\ \qquad\qquad\qquad\qquad\qquad\qquad\qquad\qquad \text{if} \quad \chi_k \in \mathscr{N}_h^{\delta-\alpha}. \end{cases}$$

As a product of two circulants, the matrix $\left(S\chi_k(x_r)\right)_{k,r=0}^{n-1}$ possesses the eigenvalues ν_l^n/μ_l^n if $\chi_k \in \mathcal{M}_h^{\delta-\alpha}$ resp. μ_l^n/ν_l^n if $\chi_k \in \mathcal{N}_h^{\delta-\alpha}$, $l = 0, \ldots, n-1$. Hence from Lemma 2.21 we conclude that A_n has the eigenvalues $\lambda_n^l = \varrho(lh)$, $l = 0, \ldots, n-1$, where

$$\varrho(s) := \begin{cases} c + d\Phi_{\varepsilon,\delta-\alpha}(s),\ \chi_k \in \mathcal{M}_h^{\delta-\alpha} \text{ and } \delta \text{ even}, \\ c + d\left(\Phi_{\varepsilon,\delta-\alpha}(s)\right)^{-1},\ \chi_k \in \mathcal{N}_h^{\delta-\alpha} \text{ and } \delta \text{ odd}, \\ c\left(\Phi_{\varepsilon,\delta-\alpha}(s)\right)^{-1} + d,\ \chi_k \in \mathcal{N}_h^{\delta-\alpha} \text{ and } \delta \text{ even}, \\ c\Phi_{\varepsilon,\delta-\alpha}(s) + d,\ \chi_k \in \mathcal{M}_h^{\delta-\alpha} \text{ and } \delta \text{ odd}, \end{cases} \tag{3}$$

$s \in [0, 1]$. Here $\Phi_{\varepsilon,\beta}$ denotes the function defined by

$$\Phi_{\varepsilon,\beta}(s) := \begin{cases} +1 & \text{if } s = 0, \\ \sigma_{\varepsilon,\beta}^+(s)/\sigma_{\varepsilon,\beta}^-(s) & \text{if } 0 < s < 1, \\ -1 & \text{if } s = 1 \end{cases}$$

with

$$\sigma_{\varepsilon,\beta}^{\pm}(s) := \sum_{k=0}^{\infty} (k+s)^{-\beta-1}\, e^{2\pi i k\varepsilon} \pm \sum_{k=1}^{\infty} (k-s)^{-\beta-1}\, e^{-2\pi i k\varepsilon}.$$

Notice that under the assumptions ensuring the existence of an interpolation basis the functions $\Phi_{\varepsilon,\delta-\alpha}$ resp. $\Phi_{\varepsilon,\delta-\alpha}^{-1}$ are $\mathbf{C}^\infty$ on [0,1].

Thus the eigenvalues of A_n lie on the non-closed $\mathbf{C}^\infty$-curve ϱ given by (3), where $\varrho'(s) \neq 0$, $0 < s < 1$. Furthermore, the union over n of the eigenvalues λ_n^l is a dense subset of the image of the function ϱ.

Since A_n is a circulant, we have (see Remark 2.4)

$$\|A_n\|_{t,h\to t,h} \leqq \max_{0\leqq s\leqq 1} |\varrho(s)|, \qquad t \in \mathbb{R},$$

where $\|\cdot\|_{t,h\to t,h}$ denotes the operator norm in $\mathbb{C}^n$, the latter space endowed with the discrete Sobolev norm $\|\cdot\|_{t,h}$. Furthermore, the inverses of the matrices A_n exist and are uniformly bounded with respect to the norm $\|\cdot\|_{t,h\to t,h}$ if and only if $\varrho(s) \neq 0$, $s \in [0, 1]$, and then

$$\|A_n^{-1}\|_{t,h\to t,h} \leqq \max_{0\leqq s\leqq 1} |\varrho(s)|^{-1}, \qquad t \in \mathbb{R}.$$

Notice that $\operatorname{im} \Phi_{1/2,\delta-\alpha} = \operatorname{im} \Phi_{0,\delta-\alpha}^{-1} = [-1, 1]$.

Now combine the results of this section with Theorem 2.23 to get the following lemma.

13.5. Lemma. *For all $n \in \mathbb{N}$, there exist bases $\{\vartheta_k : k = 0, \ldots, n-1\}$ of $\mathcal{S}_h^\delta$ such that, for all $\boldsymbol{y} = (y_k)_{k=0}^{n-1} \in \mathbb{C}^n$ and $t < \delta - \alpha + 1/2$, $c_1 \|\boldsymbol{y}\|_{t,h} \leqq \left\|\sum_{k=0}^{n-1} y_k\vartheta_k\right\|_{t+\alpha} \leqq c_2 \|\boldsymbol{y}\|_{t,h}$ and $\|A_n\|_{t,h\to t,h} \leqq c_3$. Moreover, $\|A_n^{-1}\|_{t,h\to t,h} \leqq c_4$ if and only if*

$$\begin{aligned} &\textit{either } c + \lambda d \neq 0,\ \lambda \in [-1, 1] \textit{ for } \varepsilon = 1/2 \textit{ and } \delta \textit{ even or } \varepsilon = 0 \textit{ and } \delta \textit{ odd}, \\ &\textit{or } \lambda c + d \neq 0,\ \lambda \in [-1, 1] \textit{ for } \varepsilon = 1/2 \textit{ and } \delta \textit{ odd or } \varepsilon = 0 \textit{ and } \delta \textit{ even}, \\ &\textit{or } c + d\left(\Phi_{\varepsilon,\delta-\alpha}(s)\right)^{(-1)^\delta} \neq 0,\ s \in [0, 1], \textit{ in all other cases.} \end{aligned} \tag{1}$$

Here A_n denotes the matrix of 13.1(2) corresponding to $\{\vartheta_k\}$, where A is defined by 13.4(1), and the constants c_j $(j = 1, \ldots, 4)$ are independent of y and n.

13.6. Remarks. 1°. Condition 13.5(1) is equivalent to condition (ii) of Theorem 13.2, where $\sigma_A^{\delta,\varepsilon}$ is defined by 13.1(8) with $a = c + d$, $b = c - d$. This follows easily from the remark to Theorem 13.2.

2°. If condition 13.5(1) is violated, but A_n^{-1} exists for some n, then

$$\|A_n^{-1}\|_{t+\delta,h\to t,h} \geqq \gamma h^{\delta-1} \tag{1}$$

for $0 \leqq \delta$ 1, where γ is a constant independent of n.

Proof. Assuming $\varrho(\tau) = 0$, $\tau \in [0, 1]$, we choose $l_n \in \{0, \ldots, n-1\}$ satisfying $|\tau - l_n h| = \min\limits_l |\tau - lh|$, where the minimum is taken over $l \in \{0, \ldots, n-1\}$. Let $v_n \in \mathbb{C}^n$ be an eigenvector of A_n corresponding to the eigenvalue $\varrho(l_n h)$, i.e. $A_n v_n = \varrho(l_n h)\, v_n$. Since $|\varrho(l_n h)| \leqq \gamma h$, we have $\|A_n v_n\|_{t,h} \leqq \gamma h\, \|v_n\|_{t,h}$. Thus, by Corollary 2.24 $h^{\delta-1} \|A_n v_n\|_{t+\delta,h} \leqq \gamma\, \|v_n\|_{t,h}$ for $0 < \delta < 1$, which implies (1). ■

13.7. Lemma. *Let A be defined by* 13.4(1). *Then the sequence $\{Q_h A|_{\mathscr{S}_h^\delta}\}$ is uniformly bounded in $(\mathbf{H}^{t+\alpha}, \mathbf{H}^t)$ for $t < \delta - \alpha + 1/2$. It is stable if and only if condition* 13.5(1) *is fulfilled.*

Proof. Let $u_h = \sum\limits_{k=0}^{n-1} \xi_k \vartheta_k$, where $\vartheta_0, \ldots, \vartheta_{n-1}$ is the basis of $\mathscr{S}_h^\delta$ whose existence was established in Lemma 13.5. Then

$$Q_h A u_h = \sum_{k=0}^{n-1} (A_n \bar{\xi})_k\, \chi_k,$$

where $(A_n\bar{\xi})_k$ denotes the k-th coordinate of $A_n\bar{\xi} \in \mathbb{C}^n$. Hence Lemma 13.5 and Theorem 2.23 imply that

$$\|Q_h A u_h\|_t \leqq c_1\, \|A_n \bar{\xi}\|_{t,h} \leqq c_2\, \|\bar{\xi}\|_{t,h} \leqq c_3\, \|u_h\|_{t+\alpha}.$$

Analogously, if 13.5(1) is satisfied and $v_h = \sum\limits_{k=0}^{n-1} \eta_k \chi_k \in \operatorname{im} Q_h$, then

$$(Q_h A|_{\mathscr{S}_h^\delta})^{-1}\, v_h = \sum_{k=0}^{n-1} (A_n^{-1}\bar{\eta})_k\, \vartheta_k.$$

Consequently,

$$\left\|(Q_h A|_{\mathscr{S}_h^\delta})^{-1}\, v_h\right\|_{t+\alpha} \leqq c_1\, \|A_n^{-1}\bar{\eta}\|_{t,h} \leqq c_2\, \|\bar{\eta}\|_{t,h} \leqq c_3\, \|v_h\|_t,$$

i.e. the sequence $\{Q_h A|_{\mathscr{S}_h^\delta}\}$ is stable in $(\mathbf{H}^{t+\alpha}, \mathbf{H}^t)$.

Conversely, assuming the stability of $\{Q_h A|_{\mathscr{S}_h^\delta}\}$, it follows from Lemma 13.5 and Theorem 2.23 that

$$\|A_n^{-1}\bar{\eta}\|_{t,h} \leqq c_1^{-1} \left\|(Q_h A|_{\mathscr{S}_h^\delta})^{-1}\, v_h\right\|_{t+\alpha} \leqq c_5\, \|v_h\|_t \leqq c_6\, \|\bar{\eta}\|_{t,h}.$$

Thus 13.5(1) is satisfied by Lemma 13.5. ■

13.8. Remarks. 1°. If $\beta \geqq \delta - \alpha$ and $\tilde{Q}_h$ denotes any interpolation projection satisfying $\tilde{Q}_h f(x_k) = f(x_k)$, $k = 0, \ldots, n-1$, $\operatorname{im} \tilde{Q}_h = \mathscr{M}_h^\beta$ (or $\mathscr{N}_h^\beta$), then Lemma 13.7 remains valid for the sequence $\{\tilde{Q}_h A|_{\mathscr{S}_h^\delta}\}$ (cf. Theorem 2.23). Moreover, in this case 13.1(2) is equivalent to

$$\tilde{Q}_h A u_h = \tilde{Q}_h f, \qquad u_h \in \mathscr{S}_h^\delta. \tag{1}$$

For example, in the case $\beta = 0 \geqq \delta - \alpha$ we can choose the interpolation projection onto $\mathcal{S}_h^\delta$ defined by $\tilde{Q}_h f = \sum_{k=0}^{n-1} f(x_k)\,\theta_k$, where θ_k stands for the characteristic function of $[x_k, x_{k+1})$ (notice that in this case $\varepsilon > 0$ in view of 13.1(4)).

2°. For im $\tilde{Q}_h = \Lambda^{-1}(\text{im } Q_h)$, the representation (1) is also useful in the degenerated case, i.e. when $\varrho(\tau) = 0$ for some $\tau \in [0, 1]$. Then it follows from 13.6(1) that there exists a sequence $\{v_h\}$, $v_h \in \mathcal{S}_h^\delta$, $h \to 0$, satisfying

$$\|\tilde{Q}_h A v_h\|_{t+\sigma} \leqq \gamma h^{1-\sigma} \|v_h\|_{t+\alpha}, \qquad t < \delta - \alpha + 1/2 \tag{2}$$

for $0 \leqq \sigma < 1$. Consequently, if condition 13.5(1) is violated then the ε-collocation 13.1(2) is not stable in $(\mathbf{H}^{t+\alpha}, \mathbf{H}^{t+\sigma})$ for $0 \leqq \sigma < 1$.

In order to check the conditions of Theorem 13.3 and to prove optimal error estimates in the scale of Sobolev norms we need some auxiliary estimates.

13.9. Lemma. *Let $v_h \in \mathcal{S}_h^0$, $\varphi_h \in \mathcal{S}_h^1$, $c, d \in \mathbb{C}$, $-1 < \alpha \leqq 0$, and let Q_h be the interpolation projection*

$$Q_h u = \sum_{k=0}^{n-1} u(x_k)\,\theta_k, \qquad x_k = (k + \varepsilon)\,h, \qquad 0 < \varepsilon < 1.$$

Then there exists a constant γ independent of $h = 1/n$ such that

$$\|Q_h(cI + dS)\,\Lambda^{-\alpha}(v_h + \varphi_h)\|_\alpha \leqq \gamma\,\|v_h + \varphi_h\|_0\,.$$

To prove this lemma we first introduce the n-dimensional space $C_n := \text{lin }\{\psi_k\}_{k=0}^{n-1}$, where ψ_k stands for the piecewise linear function defined by

$$\psi_k(x) = \begin{cases} 2xh^{-1} - 2k - 1, & x \in [kh, (k+1)\,h), \\ 0 & \text{otherwise}, \end{cases}$$

$k = 0, \ldots, n - 1$. Let C_n^0 denote the $(n - 1)$-dimensional subspace

$$C_n^0 := \left\{\sum_{k=0}^{n-1} \beta_k \psi_k \in C_n : \sum_{k=0}^{n-1} \beta_k = 0\right\}.$$

13.10. Lemma. *There exists a constant γ independent of h such that*

$$\left\|Q_h(cI + dS)\,\Lambda^{-\alpha}\big|_{C_n}\right\|_{0\to\alpha} \leqq \gamma h^\alpha, \qquad \left\|Q_h(cI + dS)\,\Lambda^{-\alpha}\big|_{C_n^0}\right\|_{0\to\alpha} \leqq \gamma\,.$$

Proof. We introduce in C_n the basis

$$\varphi_j(x) = \sum_{l=0}^{n-1} e^{2\pi i j l h}\,\psi_l(x), \qquad j = 0, \ldots, n - 1.$$

Obviously, $\langle \varphi_j, \varphi_l \rangle_0 := (\varphi_j, \varphi_l) = \delta_{jl}/3$, and $\{\varphi_j\}_{j=0}^{n-1}$ is a basis in C_n^0. From the Fourier expansion of ψ_l it results that

$$\Lambda^{-\alpha}\psi_l(x) = i \sum_{0 \neq k \in \mathbb{Z}} \frac{\exp\big(2\pi i k x - \pi i k(2l + 1)\,h\big)}{|2\pi k|^\alpha\,\pi k}\,m_k,$$

$$S\Lambda^{-\alpha}\psi_l(x) = i \sum_{0 \neq k \in \mathbb{Z}} \frac{\exp\big(2\pi i k x - \pi i k(2l + 1)\,h\big)}{|2\pi k|^\alpha\,|\pi k|}\,m_k,$$

where $m_k = \cos \pi k h - \sin \pi k h/(\pi k h)$.

The series converge uniformly on each closed subinterval which does not contain the points $x = lh$ and $x = (l + 1)\,h$. Thus

$$\sum_{l=0}^{n-1} e^{2\pi i j l h} \Lambda^{-\alpha} \psi_l(x_r)$$
$$= i \sum_{l=0}^{n-1} e^{2\pi i j l h} \sum_{k \neq 0} \frac{\exp\big(2\pi i k(r - l)\,h - \pi i k h + 2\pi i k \varepsilon h\big)}{|2\pi k|^\alpha\, \pi k}\, m_k = e^{2\pi i j r h} \varrho_j^n .$$

Here

$$\varrho_j^n = \frac{i h^\alpha}{2^\alpha \pi^{\alpha+1}} \exp\,(2\pi i \varepsilon j h - \pi i j h) \sum_{k \in \mathbb{Z}} \frac{\exp\,(2\pi i k \varepsilon)}{|k + jh|^\alpha\,(k + jh)}$$
$$\times \left(\cos \pi j h - \frac{\sin \pi j h}{\pi(k + jh)}\right)$$

with $\sum_{k \in \mathbb{Z}}$ interpreted as $\sum_{0 \neq k \in \mathbb{Z}}$ for $j = 0$.

By analogy, we find

$$\sum_{l=0}^{n-1} e^{2\pi i j l h} S \Lambda^{-\alpha} \psi_l(x_r) = e^{2\pi i j r h} \varkappa_j^n ,$$

where

$$\varkappa_j^n = \frac{i h^\alpha}{2^\alpha \pi^{\alpha+1}} \exp\,(2\pi i \varepsilon j h - \pi i j h) \sum_{k \in \mathbb{Z}} \frac{\exp\,(2\pi i k \varepsilon)}{|k + jh|^{\alpha+1}} \left(\cos \pi j h - \frac{\sin \pi j h}{\pi(k + jh)}\right).$$

Hence

$$\{(cI + dS)\, \Lambda^{-\alpha} \varphi_j(x_r)\}_{r=0}^{n-1} = (c\varrho_j^n + d\varkappa_j^n)\, e_j$$

with $e_j = \{\exp\,(2\pi i j r h)\}_{r=0}^{n-1}$. Since $(e_j, e_k)_{\alpha,h} = (1 + 4h^{-2} \sin^2 \pi j h)^\alpha\, \delta_{jk}$, we deduce from Theorem 2.23 that

$$\|Q_h(cI + dS)\, \Lambda^{-\alpha} \sum_{j=0}^{n-1} \beta_j \varphi_j\|_\alpha^2 \leqq \gamma \sum_{j=0}^{n-1} |\beta_j|^2\, |c\varrho_j^n + d\varkappa_j^n|^2\, (1 + 4h^{-2} \sin^2 \pi j h)^\alpha .$$

Utilize the uniform boundedness of the functions

$$\left|\sum_{k \in \mathbb{Z}} \frac{\exp\,(2\pi i k \varepsilon)}{|k + s|^{\alpha+1}} \left(\cos \pi s - \frac{\sin \pi s}{\pi(k + s)}\right)\right|^2 |\sin \pi s|^{2\alpha}$$

for $0 < \varepsilon, s < 1$ to obtain

$$\left\|Q_h(cI + dS)\, \Lambda^{-\alpha} \sum_{j=1}^{n-1} \beta_j \varphi_j\right\|_\alpha^2 \leqq \gamma \sum_{j=1}^{n-1} |\beta_j|^2 \leqq 3\gamma \left\|\sum_{j=1}^{n-1} \beta_j \varphi_j\right\|_0^2 .$$

Since $|c\varrho_0^n + d\varkappa_0^n|^2 = O(h^{2\alpha})$, we have

$$\|Q_h(cI + dS)\, \Lambda^{-\alpha} \varphi_0\|_\alpha \leqq \gamma h^\alpha . \quad \blacksquare$$

13.11. Proof of Lemma 13.9. For $v_h \in \mathcal{S}_h^0$, $\varphi_h \in \mathcal{S}_h^1$, we have the representation

$$v_h + \varphi_h = \sum_{k=0}^{n-1} (\alpha_k \theta_k + \beta_k \psi_k) .$$

Now $\int\limits_0^1 \varphi_h'(x)\,dx = 0$ implies $\sum\limits_{k=0}^{n-1} \beta_k = 0$. Since $\mathcal{S}_h^0 \cap \mathcal{S}_h^1$ contains the constant functions only, we find $\mathcal{S}_h^0 + \mathcal{S}_h^1 = \mathcal{S}_h^0 + C_n^0$, the sum being orthogonal with respect to the $\mathbf{L}^2$ scalar product $(\cdot, \cdot)$. Hence $v_h + \varphi_h = w_h + \psi_n$, where $w_h \in \mathcal{S}_h^0$, $\psi_n \in C_n^0$ and $(w_h, \psi_n) = 0$. Since by Lemma 13.5

$$\|Q_h(cI + dS)\,\Lambda^{-\alpha} w_h\|_\alpha \leqq c\,\|w_h\|_0,$$

it follows from Lemma 13.10 that

$$\begin{aligned}\|Q_h(cI + dS)\,\Lambda^{-\alpha}(v_h + \varphi_h)\|_\alpha &\leqq \gamma(\|w_h\|_0 + \|\psi_n\|_0)\\ &\leqq \gamma\sqrt{2}\,(\|w_h\|_0^2 + \|\psi_n\|_0^2)^{1/2} = \gamma\sqrt{2}\,\|v_h + \varphi_h\|_0. \quad \blacksquare\end{aligned}$$

Now we are in a position to prove the properties I–V formulated in Section 13.2.

13.12. Theorem. *The conditions* I *through* V *of Sec.* 13.2 *are satisfied for every operator A of the form* 13.2(8) *with coefficients $a, b \in \mathbf{C}^\infty$ and for $-1/2 < t < \delta - \alpha + 1/2$.*

Proof. Property I follows from Theorem 2.13, which yields $\|P_h\|_{t+\alpha} \leqq c$ and

$$\left\|(I - P_h)\,g|\mathcal{S}_h^\delta\right\|_{t+\alpha} \leqq ch^r, \qquad r = \min\{1, 1 + \delta - \alpha - t\}. \tag{1}$$

Notice that, for $\delta - \alpha \leqq 0$, condition I is satisfied with $P_h := \mathrm{D}^{-\delta}\tilde{P}_{h,0}\mathrm{D}^\delta$, where

$$\tilde{P}_{h,0}u(x) = h^{-1}\sum_{k=0}^{n-1} \theta_k(x) \int\limits_{kh}^{(k+1)h} u(y)\,dy\colon \mathbf{L}^2 \to \mathcal{S}_h^0 \tag{2}$$

is the orthogonal projection (see Sec. 2.7).

For $\delta - \alpha > 0$, condition II is a consequence of Theorem 2.26 and Lemma 2.28. If $\delta - \alpha \leqq 0$ then $\operatorname{im} Q_h = \mathcal{S}_h^0$ and condition II follows from Theorem 2.30 and the estimate

$$\left\|(I - Q_h)\,g|\mathcal{S}_h^0\right\|_t \leqq ch^{\min\{1-t,1+t\}},$$

which can easily be verified.

III. We first assume $\delta - \alpha > 0$. For $1/2 < t < \delta - \alpha + 1/2$, condition III results from (1), since the projections Q_h are uniformly bounded in $\mathbf{H}^t$. For $0 \leqq t \leqq 1/2$ and $v_h \in \mathcal{S}_h^\delta$, we find from (1) that

$$\begin{aligned}\|Q_hA(I - P_h)\,gv_h\|_t &\leqq \|(I - Q_h)\,A(I - P_h)\,gv_h\|_t + \|A(I - P_h)\,gv_h\|_t\\ &\leqq c_1h^{s-t}\,\|A(I - P_h)\,gv_h\|_s + c_2h^{\min\{1,1+\delta-\alpha-t\}}\,\|v_h\|_{t+\alpha}\\ &\leqq c_3(h^{s-t+\min\{1,1+\delta-\alpha-s\}}\,\|v\|_{s+\alpha} + h^{\min\{1,1+\delta-\alpha-t\}}\,\|v\|_{t+\alpha})\\ &\leqq c_4h^{\min\{1,1+\delta-\alpha-s\}}\,\|v_h\|_{t+\alpha},\end{aligned}$$

where $1/2 < s < \delta - \alpha + 1/2$. Hence, for $-1/2 < t < 0$,

$$\|Q_hA(I - P_h)\,gv_h\|_t \leqq \|Q_hA(I - P_h)\,gv_h\|_0 \leqq ch^q\,\|v_h\|_{t+x},$$

where $q := t + \min\{1, 1 + \delta - \alpha - s\} > 0$.

We now consider the case $\delta - \alpha \leqq 0$. Using 13.2(8) we see that

$$\left\|Q_h T(I - P_h)\, g\big|_{\mathcal{S}_h^\delta}\right\|_{t+\alpha \to t} \to 0 .$$

Since $Q_h a = Q_h a Q_h$ and condition II is satisfied, it remains to prove that

$$\left\|Q_h \Lambda^\alpha(I - P_h)\, g\big|_{\mathcal{S}_h^\delta}\right\|_{t+\alpha \to t} \to 0 \quad \text{as} \quad h \to 0 \tag{3}$$

and

$$\left\|Q_h S \Lambda^\alpha(I - P_h)\, g\big|_{\mathcal{S}_h^\delta}\right\|_{t+\alpha \to t} \to 0 \quad \text{as} \quad h \to 0 . \tag{4}$$

For example, let δ be even and $v_h \in \mathcal{S}_h^\delta$. Utilize the orthogonal projection (2) and $P_h := \mathrm{D}^{-\delta} \tilde{P}_{h,0} \mathrm{D}^\delta$ to get

$$\begin{aligned} Q_h \Lambda^\alpha(I - P_h)\, g v_h &= Q_h \Lambda^{\alpha-\delta}(I - \tilde{P}_{h,0})\, \mathrm{D}^\delta g v_h \\ &= Q_h \Lambda^{\alpha-\delta}(I - \tilde{P}_{h,0})\, g D^\delta v_h + Q_h \Lambda^{\alpha-\delta}(I - \tilde{P}_{h,0})\, (\mathrm{D}^\delta g - g \mathrm{D}^\delta)\, v_h . \end{aligned} \tag{5}$$

Notice that $\mathrm{D}^\delta v_h =: \tilde{v}_h \in \mathcal{S}_h^0$ and $(I - \tilde{P}_{h,0})\, g \tilde{v}_h(x) = \tilde{v}_h(x)\, (I - \tilde{P}_{h,0})\, g(x)$. Choosing $g_h \in \mathcal{S}_h^1$ such that $g_h(kh) = g(kh)$ we have

$$Q_h \Lambda^{\alpha-\delta}(I - \tilde{P}_{h,0})\, g \tilde{v}_h = Q_h \Lambda^{\alpha-\delta} \tilde{v}_h(g - g_h) + Q_h \Lambda^{\alpha-\delta} \tilde{v}_h(g_h - \tilde{P}_{h,0} g) . \tag{6}$$

It can easily be verified that $\tilde{v}_h(x)\, \big(g(x) - g_h(x)\big) \in \mathbf{H}^r$ for $r < 3/2$ and $\|\tilde{v}_h(g - g_h)\|_s \leqq c h^{3/2-s}\, \|\tilde{v}_h\|_0\, \|g\|_2$ for $0 \leqq s < 3/2$. Since $\delta - \alpha > -1$, we infer from Theorem 2.30 that

$$\begin{aligned} \|Q_h \Lambda^{\alpha-\delta} \tilde{v}_h(g - g_h)\|_t &\leqq \|(I - Q_h)\, \Lambda^{\alpha-\delta} \tilde{v}_h(g - g_h)\|_t + \|\Lambda^{\alpha-\delta} \tilde{v}_h(g - g_h)\|_t \\ &\leqq c h^{3/2-\alpha+\delta+\min\{0,-t\}}\, \|\tilde{v}_h\|_0 \leqq c h^q\, \|\tilde{v}_h\|_{t+\alpha-\delta} , \end{aligned}$$

where

$$q = \begin{cases} 3/2 - \alpha + \delta + \min\{0, -t\}, & t + \alpha - \delta \geqq 0 , \\ 3/2 + t + \min\{0, -t\}, & t + \alpha - \delta < 0 . \end{cases}$$

Moreover, $\tilde{v}_h(g_h - \tilde{P}_{h,0} g) \in \mathcal{S}_h^0 + C_n^0$. Thus Lemma 13.10 and arguments like in the proof of Lemma 2.16 yield that

$$\|Q_h \Lambda^{\alpha-\delta} \tilde{v}_h(g_h - \tilde{P}_{h,0} g)\|_{\delta-\alpha} \leqq c h^{\delta-\alpha}\, \|\tilde{v}_h(g_h - \tilde{P}_{h,0} g)\|_0 \leqq c h^{3/2+\delta-\alpha}\, \|\tilde{v}_h\|_0 .$$

Hence, for $-1/2 < t < \delta - \alpha$,

$$\|Q_h \Lambda^{\alpha-\delta} \tilde{v}_h(g_h - \tilde{P}_{h,0} g)\|_t \leqq c h^{3/2+\delta-\alpha}\, \|\tilde{v}_h\|_0 \leqq c h^{3/2+t}\, \|\tilde{v}_h\|_{t+\alpha-\delta} .$$

For $\delta - \alpha \leqq t \leqq \delta - \alpha + 1/2$, it follows from the inverse property IE (1/2) for $\mathcal{S}_h^0$ that

$$\|Q_h \Lambda^{\alpha-\delta} \tilde{v}_h(g_h - \tilde{P}_{h,0} g)\|_t \leqq c h^{3/2+2(\delta-\alpha)-t}\, \|\tilde{v}_h\|_0 \leqq c h^{3/2+2(\delta-\alpha)-t}\, \|\tilde{v}_h\|_{t+\alpha-\delta} .$$

Since all exponents of h in the preceding estimates are positive, we conclude from (6) that

$$\|Q_h \Lambda^{\alpha-\delta}(I - \tilde{P}_{h,0})\, g \mathrm{D}^\delta v_h\|_t \leqq c h^r\, \|v_h\|_{t+\alpha} , \qquad r > 0 .$$

To estimate $\|Q_h \Lambda^{\alpha-\delta}(I - \tilde{P}_{h,0})\, (\mathrm{D}^\delta g - g \mathrm{D}^\delta)\, v_h\|_t$ we proceed similarly. Let $\tilde{P}_{h,1} : \mathbf{L}^2 \to \mathcal{S}_h^1$ denote the orthogonal projection. Since $K := \mathrm{D}^\delta g - g \mathrm{D}^\delta$ is an operator of order $\delta - 1$, we find

$$\begin{aligned} \|Q_h \Lambda^{\alpha-\delta}(I - \tilde{P}_{h,0})\, K v_h\|_t &\leqq \|Q_h \Lambda^{\alpha-\delta}(I - \tilde{P}_{h,1})\, K v_h\|_t \\ &\quad + \|Q_h \Lambda^{\alpha-\delta}(\tilde{P}_{h,1} - \tilde{P}_{h,0})\, K v_h\|_t . \end{aligned}$$

Since $(I - \tilde{P}_{h,1}) Kv_h \in \mathbf{H}^s$, $s < 3/2$, we have

$$\|(I - Q_h) \Lambda^{\alpha-\delta}(I - \tilde{P}_{h,1}) Kv_h\|_t \leqq ch^{1+\min\{t,0\}} \|v_h\|_{t+\alpha},$$

and hence

$$\|Q_h\Lambda^{\alpha-\delta}(I - \tilde{P}_{h,1}) Kv_h\|_t \leqq c(h^{1+\min\{t,0\}} + h^1) \|v_h\|_{t+\alpha}.$$

Apply Lemma 13.9 to get

$$\|Q_h\Lambda^{\alpha-\delta}(\tilde{P}_{h,1} - \tilde{P}_{h,0}) Kv_h\|_{\delta-\alpha} \leqq c \|(\tilde{P}_{h,1} - \tilde{P}_{h,0}) Kv_h\|_0 \leqq ch \|v_h\|_\delta.$$

Consequently,

$$\|Q_h\Lambda^{\alpha-\delta}(\tilde{P}_{h,1} - \tilde{P}_{h,0}) Kv_h\|_t \leqq ch^{1+t+\alpha-\delta} \|v_h\|_{t+\alpha}$$

for $-1/2 < t \leqq \delta - \alpha$ and

$$\|Q_h\Lambda^{\alpha-\delta}(\tilde{P}_{h,1} - \tilde{P}_{h,0}) Kv_h\|_t \leqq ch^{1+\delta-\alpha-t} \|v_h\|_\delta \leqq ch^{1+\delta-\alpha+t} \|v_h\|_{t+\alpha}$$

for $\delta - \alpha < t \leqq \delta - \alpha + 1/2$. Combining all the preceding estimates and taking into account (5) we find

$$\|Q_h\Lambda^\alpha(I - P_h) gv_h\|_t \leqq ch^r \|v_h\|_{t+\alpha}, \qquad r > 0$$

for $-1/2 < t < \delta - \alpha + 1/2$ and any $v_h \in \mathcal{S}_h^\delta$. Thus we have proved (3). Replacing Λ^α by $S\Lambda^\alpha$ in the preceding considerations we get (4). Analogously one can proceed in the case where δ is odd.

IV. From 13.2(8), Lemma 13.7, Theorem 2.23 and condition II we deduce that the operators

$$Q_hA\big|_{\mathcal{S}_h^\delta} : \mathbf{H}^{t+\alpha} \to \mathbf{H}^t, \qquad -1/2 < t < \delta - \alpha + 1/2$$

are uniformly bounded. In view of conditions II and III, we have

$$\|Q_hAP_hg - Ag\|_t \to 0 \quad \text{as} \quad h \to 0$$

for all $g \in \mathbf{C}^\infty$. Since $\mathbf{C}^\infty$ is a dense subset of $\mathbf{H}^{t+\alpha}$, condition IV follows.

V. Since $A - A_y = \left[\big(a(x) - a(y)\big) P_{\mathbf{T}} + \big(b(x) - b(y)\big) Q_{\mathbf{T}}\right] \Lambda^\alpha + T$, there is a sufficiently small neighborhood $U \ni y$ such that, for every $g_y \in M_y$ satisfying $\operatorname{supp} g_y \subset U$, we have $\|B\|_{\alpha\to 0} < \eta$, where $B := g_y(A - A_y) - g_yT$. Because

$$\|B\|_{\alpha\to 0} = \|\Lambda^{-t}B\Lambda^t\|_{t+\alpha\to t} = \|B + (\Lambda^{-t}B\Lambda^t - B)\|_{t+\alpha\to t}$$

and $\Lambda^{-t}B\Lambda^t - B \in \mathcal{L}(\mathbf{H}^{t+\alpha}, \mathbf{H}^{t+1})$, we get

$$\inf_T \|g_y(A - A_y) + T\|_{t+\alpha\to t} \leqq \|B\|_{\alpha\to 0} < \eta. \tag{7}$$

If $1/2 < t < \delta - \alpha + 1/2$, then (7) implies condition V, since the projections Q_h are uniformly bounded.

For $-1/2 < t \leqq 1/2 \leqq \delta - \alpha + 1/2$ or $-1/2 < t < \delta - \alpha + 1/2 \leqq 1/2$ we proceed as follows. For any $g \in M_y$ and $f \in \mathbf{C}^\infty$ satisfying $f(y) = 0$, we have $\|g_kf\|_1 \to 0$ as $k \to \infty$, where $g_k(x) := g\big(y + k(x - y)\big)$. Since

$$Q_h[g_k(A - A_y) - g_kT] v_h(x) = Q_hg_k(x) [c(x) - c(y)] Q_h\Lambda^\alpha v_h(x) \\ + Q_hg_k(x) [d(x) - d(y)] Q_hS\Lambda^\alpha v_h(x),$$

where $c = (a + b)/2$, $d = (a - b)/2$, condition II together with the uniform boundedness of $Q_h A^\alpha|_{\mathscr{S}_h^\delta}$ and $Q_h S A^\alpha|_{\mathscr{S}_h^\delta}$ implies that

$$\left\| Q_h(g_k(A - A_y) - g_k T)|_{\mathscr{S}_h^\delta} \right\|_{t+\alpha\to t} < \eta$$

for some k and for all sufficiently small h. ■

13.13. Second proof of Theorem 13.2. Assume first $\{Q_h A|_{\mathscr{S}_h^\delta}\}$ to be stable in $(\mathbf{H}^{t+\alpha}, \mathbf{H}^t)$, $-1/2 < t < \delta - \alpha + 1/2$, i.e. $\|Q_h A v_h\|_t \geqq c \|v_h\|_{t+\alpha}$ for all $v_h \in \mathscr{S}_h^\delta$ and all sufficiently small h. Given $v \in \mathbf{H}^{t+\alpha}$, choose $v_h \in \mathscr{S}_h^\delta$ such that $\|v - v_h\|_{t+\alpha} \to 0$ as $h \to 0$. Passing to the limit in the last inequality, condition IV yields $\|Av\|_t \geqq c \|v\|_{t+\alpha}$. Thus dim ker $A = 0$. Consequently, $a(y)\, b(y) \neq 0$ and the operators $A_y \in \mathscr{L}(\mathbf{H}^{t+\alpha}, \mathbf{H}^t)$ are invertible for all $y \in [0, 1)$. Moreover, by Theorem 13.3, $\{Q_h A|_{\mathscr{S}_h^\delta}\}$ is locally stable at x for all $x \in [0, 1)$. Hence $\{Q_h A_x|_{\mathscr{S}_h^\delta}\}$ is locally stable at y for all $y \in [0, 1)$, since A_x is an operator with constant coefficients. Applying Theorem 13.3 again we find that $\{Q_h A_x|_{\mathscr{S}_h^\delta}\}$ is stable in $(\mathbf{H}^{t+\alpha}, \mathbf{H}^t)$ for each fixed $x \in [0, 1)$. Now condition (ii) of Theorem 13.2 follows from Lemma 13.7, Remark 13.8.1° and Remark 13.6.1°.

Conversely, suppose the conditions (i) and (ii) of Theorem 13.2 are satisfied. Then there is a homotopy connecting the curves given by a and b and consequently, ind $A = 0$. Hence A is invertible. By Lemma 13.7 and Remark 13.6.1°, $\{Q_h A_x|_{\mathscr{S}_h^\delta}\}$ is stable in $(\mathbf{H}^{t+\alpha}, \mathbf{H}^t)$ for any $x \in [0, 1)$. Applying Theorem 13.3 to the operator A_x we observe that $\{Q_h A_x|_{\mathscr{S}_h^\delta}\}$ is locally stable at y for all $y \in [0, 1)$. Thus the stability of $\{Q_h A|_{\mathscr{S}_h^\delta}\}$ results from Theorem 13.3. ■

13.14. Theorem. *Assume* 13.1(3) *or* 13.1(4). *Let A be an elliptic ψdo of the form* 13.2(8) *satisfying* dim ker $A = 0$ *and*

$$\sigma_A^{\delta,\varepsilon}(x, s) \neq 0, \qquad x \in \mathbb{R}, \qquad 0 < s < 1. \tag{1}$$

Then there exists $h_0 > 0$ such that for $0 < h \leqq h_0$ and any function $f \in \mathbf{H}^s$, $s > 1/2$, the collocation equations 13.1(2) *are uniquely solvable. Furthermore, for $t < \min\{s, \delta - \alpha + 1/2\}$, we have $\|u - u_h\|_{t+\alpha} \to 0$ as $h \to 0$, where $u \in \mathbf{H}^{t+\alpha}$ is the solution of* 13.1(1). *Moreover, we have the optimal error estimate*

$$\|u - u_h\|_{t+\alpha} \leqq c h^{s-t} \|f\|_s$$

for $0 \leqq t \leqq s \leqq \delta - \alpha + 1$, $t < \delta - \alpha + 1/2$, $s > 1/2$, *and*

$$\|u - u_h\|_{t+\alpha} \leqq c h^{\min\{s, \delta-\alpha+1-t\}} \|f\|_s$$

for $-1/2 < t < \delta - \alpha + 1/2 \leqq 0$, $s > 1/2$.

Proof. We again regard the ε-collocation 13.1(2) as the projection method 13.2(7). Notice that $A \in \mathscr{L}(\mathbf{H}^{t+\alpha}, \mathbf{H}^t)$ is invertible (cf. Sec. 13.13). Consider first the case $\delta - \alpha > 0$. Then, by Remark 1.27.3°, Theorem 2.26 and Theorem 13.2, $A \in \Pi\{\mathscr{S}_h^\delta, Q_h\}$ and the optimal error estimate

$$\|u - u_h\|_{t+\alpha} \leqq c \inf_{v_h \in \mathscr{S}_h^\delta} \|u - v_h\|_{t+\alpha} \leqq c h^{s-t} \|u\|_{s+\alpha}$$

holds for $1/2 < t < \delta - \alpha + 1/2$. If $0 \leqq t \leqq 1/2$, then Proposition 1.23 and Theorem 13.2 yield that $\mathbf{H}^s \subset \mathfrak{R}\big(A, \{Q_h A|_{\mathcal{S}_h^\delta}, Q_h\}\big)$ and

$$\|u - u_h\|_{t+\alpha} \leqq \inf_{v_h \in \mathcal{S}_h^\delta} \big(\|u - v_h\|_{t+\alpha} + c\,\|Q_h A(u - v_h)\|_t\big). \tag{2}$$

Choose $v_h \in \mathcal{S}_h^\delta$ from the approximation property AP($\delta + 1/2$) and apply Theorem 2.26 to obtain

$$\begin{aligned}\|Q_h A(u - v_h)\|_t &\leqq \|(I - Q_h)\, A(u - v_h)\|_t + \|A(u - v_h)\|_t \\ &\leqq c_1 h^{r-t}\, \|A(u - v_h)\|_r + c_2\, \|u - v_h\|_{t+\alpha} \leqq c h^{s-t}\, \|u\|_{s+\alpha}\end{aligned}$$

where $1/2 < r < \delta - \alpha + 1/2$.

Assume now $-1 < \delta - \alpha \leqq 0$. Then again $\mathbf{H}^s \subset \mathfrak{R}\big(A, \{Q_h A|_{\mathcal{S}_h^\delta}, Q_h\}\big)$ and (2) holds for $s > 1/2$, $-1/2 < t < \delta - \alpha + 1/2$. Obviously,

$$\|Q_h A(u - v_h)\|_t \leqq \inf_{w_h \in \mathcal{S}_h^{\delta+1}} \big(\|Q_h A(u - w_h)\|_t + \|Q_h A(v_h - w_h)\|_t\big).$$

Taking into account that $u \in \mathbf{H}^{s+\alpha}$ and choosing w_h from the approximation property AP($\delta + 3/2$), we get for $s + \alpha \leqq \delta + 2$,

$$\begin{aligned}\|Q_h A(u - w_h)\|_t &\leqq \|(I - Q_h)\, A(u - w_h)\|_t + \|A(u - w_h)\|_t \\ &\leqq c_1 h^{s-\max\{t,0\}}\, \|u\|_{s+\alpha} + c_2 h^{s-t}\, \|u\|_{s+\alpha}.\end{aligned}$$

By 13.2(8),

$$Q_h A(v_h - w_h) = Q_h(cI + dS)\, \Lambda^\alpha(v_h - w_h) + Q_h T(v_h - w_h),$$

where $c := (a + b)/2$, $d := (a - b)/2$. Since $v_h - w_h = \mathrm{D}^{-\delta}(\tilde{v}_h - \tilde{w}_h)$, $\tilde{v}_h \in \mathcal{S}_h^0$, $\tilde{w}_h \in \mathcal{S}_h^1$ and $\big\|Q_h g|_{\mathcal{S}_h^0}\big\|_t \leqq c_1$ for $g \in \mathbf{C}^\infty$, we need estimates for $Q_h \Lambda^{\alpha-\delta}(\tilde{v}_h - \tilde{w}_h)$ and $Q_h S \Lambda^{\alpha-\delta}(\tilde{v}_h - \tilde{w}_h)$. By Lemma 13.9,

$$\|Q_h(cI + dS)\, \Lambda^\alpha(v_h - w_h)\|_{\delta-\alpha} \leqq c_2\, \|v_h - w_h\|_\delta.$$

From the inverse property IP(1/2) for $\mathcal{S}_h^0$ it follows that, for $s \geqq t \geqq \delta - \alpha$,

$$\begin{aligned}\|Q_h(cI + dS)\, \Lambda^\alpha(v_h - w_h)\|_t &\leqq c_3 h^{\delta-\alpha-t}(\|u - v_h\|_\delta + \|u - w_h\|_\delta) \\ &\leqq c_4 h^{\delta-\alpha-t+\min\{1,s+\alpha-\delta\}}\, \|u\|_{s+\alpha} \\ &= c_4 h^{\min\{s-t,\delta-\alpha-t+1\}}\, \|u\|_{s+\alpha}.\end{aligned}$$

It remains to estimate $\|Q_h T(v_h - w_h)\|_t$. Since $T \in \mathcal{L}(\mathbf{H}^{t+\alpha}, \mathbf{H}^{t+1})$, we have $T(v_h - w_h) \in \mathbf{H}^r$ for some $r \in (1/2, 1]$. Hence

$$\begin{aligned}\|Q_h T(v_h - w_h)\|_t &\leqq \|(I - Q_h)\, T(v_h - w_h)\|_t + \|T(v_h - w_h)\|_t \\ &\leqq c_5 h^{r-\max\{t,0\}}\, \|T(v_h - w_h)\|_r + \|T(v_h - w_h)\|_t \\ &\leqq c_6 h^{r-\max\{t,0\}+\min\{s,\delta-\alpha+1\}-r+1}\, \|u\|_{s+\alpha} \\ &= c_6 h^{1+\min\{s,\delta-\alpha+1\}-\max\{t,0\}}\, \|u\|_{s+\alpha}.\end{aligned}$$

Combining the preceding estimates we find that

$$\|u - u_h\|_{t+\alpha} \leqq \text{const}\; h^\mu\, \|f\|_s$$

for $-1/2 < t < \delta - \alpha + 1/2$, $1/2 < s \leqq \delta - \alpha + 1$ and $\delta - \alpha \leqq t \leqq s$, where

$$\mu = \min \{s - t, s - \max \{t, 0\}, \min \{s - t, \delta - \alpha - t + 1\}, \\ 1 + \min \{s, \delta - \alpha + 1\} - \max \{t, 0\}\}.$$

Obviously, $\mu = s - t$ for $0 \leqq t < \delta - \alpha + 1/2$, $1/2 < s \leqq \delta - \alpha + 1$ and $\mu = \min \{s, \delta - \alpha + 1 - t\}$ for $-1/2 < t < \delta - \alpha + 1/2 \leqq 0$, $s > 1/2$. Notice that $\mu = \min \{s, \delta - \alpha + 1 - t\}$ for $-1/2 < \delta - \alpha \leqq t \leqq 0$, $s > 1/2$. ■

The following theorem may be viewed as a converse of Theorem 13.14.

13.15. Theorem. *Assume* 13.1(3) *or* 13.1(4). *Suppose A is an invertible ψdo of the form* 13.2(8) *and the ε-collocation* 13.1(2) *converges in $\mathbf{H}^{t+\alpha}$ with optimal order for all $f \in \mathbf{H}^s$, $s > 1/2$, where $0 \leqq t \leqq s < \delta - \alpha + 1/2$ if $\delta - \alpha > 0$ or $0 \leqq t < \delta - \alpha + 1/2$ if $-1/2 < \delta - \alpha \leqq 0$. Then condition* 13.14(1) *is satisfied.*

Proof. Set $f = Q_hAv_h$, $u_h = v_h \in \mathcal{S}_h^\delta$ in the optimal error estimate

$$\|u_h - u\|_{t+\alpha} \leqq ch^{s-t} \|f\|_s$$

to get

$$\|v_h - A^{-1}Q_hAv_h\|_{t+\alpha} \leqq ch^{s-t} \|Q_hAv_h\|_s.$$

Applying the inverse property IP$(\delta + 1/2)$ we find that

$$\|v_h\|_{t+\alpha} \leqq ch^{s-t} \|Q_hAv_h\|_s + \|A^{-1}Q_hAv_h\|_{t+\alpha} \leqq c_1 \|Q_hAv_h\|_t.$$

Hence the ε-collocation is stable in $(\mathbf{H}^{t+\alpha}, \mathbf{H}^t)$ and the assertion follows from Theorem 13.2. ■

13.16. Remarks. 1°. Suppose A is an invertible ψdo of the form 13.2(8) with constant coefficients $a, b \in \mathbb{C}$ (i.e. the principal symbol $a_0(x, \xi)$ is independent of x). If the ε-collocation converges in $\mathbf{H}^{t+\alpha}$, $t < \min \{s, \delta - \alpha + 1/2\}$, for all $f \in \mathbf{H}^s$, $1/2 < s \leqq \delta - \alpha + 1$, $\delta - \alpha > -1$ and $s < t + 1$, then condition 13.14(1) is satisfied.

Proof. From the Banach-Steinhaus theorem we infer that there exists a constant c independent of h and $f \in \mathbf{H}^s$ such that

$$\left\|\left(Q_hA\big|_{\mathcal{S}_h^\delta}\right)^{-1} Q_hf\right\|_{t+\alpha} \leqq c \|f\|_s.$$

Thus the ε-collocation 13.1(2) for the operator A and, consequently, for $A - T$ (cf. Corollary 1.26) is stable in $(\mathbf{H}^{t+\alpha}, \mathbf{H}^s)$ which contradicts Remark 13.8.2°. ■

2°. The results of the preceding sections and the methods for proving them extend to the case of arbitrary ψdo's of the form 13.2(8), where T is an operator of order β, $\alpha - 1 < \beta < \alpha$. A proof of this merely requires a corresponding modification of Theorem 13.3, which leads to a variation of the bounds for t. In particular, if $\delta - \alpha > 0$, then the ε-collocation 13.1(2) is stable in $(\mathbf{H}^{t+\alpha}, \mathbf{H}^t)$, at least for $1/2 < t < \delta - \alpha + 1/2$, if and only if the conditions (i) and (ii) of Theorem 13.2. are satisfied.

3°. Assume 13.1(3) or 13.1(4). Suppose A is an invertible ψdo of the form 13.2(8) satisfying condition 13.14(1). Then

$$R \subset \mathfrak{R}\left(A, \left\{Q_hA\big|_{\mathcal{S}_h^\delta}, Q_h\right\}\right), \qquad A \in \mathcal{L}(\mathbf{H}^{t+\alpha}, \mathbf{H}^t), \qquad t \leqq 0, t < \delta - \alpha + 1/2,$$

where R stands for the set of all 1-periodic bounded and Riemann-integrable functions on $\mathbb{R}$.

Proof. We think of the ε-collocation 13.1(2) as the projection method 13.2(7), where

$$Q_h f = \sum_{k=0}^{n-1} f(x_k)\, \theta_k, \qquad x_k = (k+\varepsilon)\, h, \qquad 0 \leqq \varepsilon < 1, \tag{1}$$

θ_k being the characteristic function of $[kh, (k+1)\, h)$, $k = 0, \ldots, n-1$. Then Theorem 2.23 and Theorem 13.2 give that $\{Q_h A|_{\mathcal{S}_h^\delta}\}$ is stable in $(\mathbf{H}^{t+\alpha}, \mathbf{H}^t)$, $-1/2 < t < \min\{1/2, \delta - \alpha + 1/2\}$. Thus the assertion is a consequence of Proposition 1.23, since for all $f \in R$

$$\begin{aligned}\|f - Q_h f\|_0^2 &= \sum_{k=0}^{n-1} \int_{kh}^{(k+1)h} |f(x) - f(x_k)|^2 \,\mathrm{d}x \\ &\leqq 2 \sup_x |f(x)|\, h \sum_{k=0}^{n-1} \sup_{kh \leqq x,y \leqq (k+1)h} |f(x) - f(y)| \to 0 \quad \text{as} \quad h \to 0. \ \blacksquare\end{aligned}$$

Replacing the right-hand sides of the equations 13.1(2) by the mean-value of f on $[kh, (k+1)\, h)$, we get the following modification of the ε-collocation: find $u_h \in \mathcal{S}_h^\delta$ such that

$$Au_h(x_k) = h \int_{kh}^{(k+1)h} f(x) \,\mathrm{d}x, \qquad k = 0, \ldots, n-1. \tag{2}$$

13.17. Theorem. *Assume A to be an invertible ψdo of the form 13.2(8) satisfying condition 13.14(1). Then there exists $h_0 > 0$ such that for $0 < h \leqq h_0$ and any function $f \in \mathbf{H}^s$, $s > -1/2$, the equations 13.16(2) are uniquely solvable. Furthermore, for $-1/2 < t \leqq s$, $t < \delta - \alpha + 1/2$, we have $\|u - u_h\|_{t+\alpha} \to 0$ as $h \to 0$, where $u \in \mathbf{H}^{t+\alpha}$ is the solution of 13.1(1). Moreover, the error estimate*

$$\|u - u_h\|_{t+\alpha} \leqq ch^{s-t} \|f\|_s \tag{1}$$

holds for $0 \leqq t \leqq s \leqq 1$, $t < \delta - \alpha + 1/2$, $s \leqq \delta - \alpha + 1$.

Proof. Let Q_h be as in 13.16(1) and let $P_h\colon \mathbf{L}^2 \to \mathcal{S}_h^0$ denote the orthogonal projection defined by

$$P_h f = \sum_{k=0}^{n-1} \theta_k h \int_{kh}^{(k+1)h} f(y) \,\mathrm{d}y.$$

Then 13.16(2) is equivalent to the equation

$$Q_h A u_h = P_h f.$$

Thus, for all $f \in \mathbf{H}^s$, $s > -1/2$, the convergence $\|u - u_h\|_{t+\alpha} \to 0$ follows from Remark 1.27.3° and Theorem 2.12. Moreover, in view of Proposition 1.23,

$$\|u - u_h\|_{t+\alpha} \leqq c \left[\inf_{v_h \in \mathcal{S}_h^\delta} (\|u - v_h\|_{t+\alpha} + \|Q_h A v_h - Au\|_t) + \|P_h f - f\|_t \right].$$

If $\delta - \alpha > 0$, then choose $v_h \in \mathcal{S}_h^\delta$ from the approximation property AP$(\delta + 1/2)$ to get

$$\begin{aligned}\|Q_h A v_h - Au\|_t &\leqq \|(I - Q_h)\, A v_h\|_t + \|A(v_h - u)\|_t \\ &\leqq c_1 h^{r - \max\{t,0\}} \|v_h\|_{r+\alpha} + c_2 h^{s-t} \|u\|_{s+\alpha} \\ &\leqq c_3 h^{s - \max\{t,0\}} \|v_h\|_{s+\alpha} + c_2 h^{s-t} \|u\|_{s+\alpha} \leqq ch^{s - \max\{t,0\}} \|u\|_{s+\alpha},\end{aligned}$$

where $1/2 < r < \min\{1, \delta - \alpha + 1/2\}$, $r \geqq s$.

Suppose now $-1 < \delta - \alpha \leqq 0$. Given $u \in \mathbf{H}^{s+\alpha}$, we choose $v_h \in \mathcal{S}_h^\delta$, $w_h \in \mathcal{S}_h^{\delta+1}$ according to the approximation property and estimate each term on the right-hand side of

$$\|(I - Q_h) Av_h\|_t \leqq \|(I - Q_h) Aw_h\|_t + \|A(v_h - w_h)\|_t + \|Q_hA(v_h - w_h)\|_t .$$

In analogy to the proof of Theorem 13.14 we find that

$$\|(I - Q_h) Aw_h\|_t \leqq ch^{s-\max\{t,0\}} \|u\|_{s+\alpha},$$

$$\|A(v_h - w_h)\|_t \leqq ch^{s-t} \|u\|_{s+\alpha}$$

and, for $s \geqq t \geqq \delta - \alpha$,

$$\|Q_hA(v_h - w_h)\|_t \leqq ch^{\min\{s-t,\delta-\alpha-t+1\}} \|u\|_{s+\alpha} .$$

Putting all preceding estimates together we arrive at (1) for $\delta - \alpha > -1/2$ and $0 \leqq t \leqq 1$, $t < \delta - \alpha + 1/2$. ■

Spline collocation on arbitrary meshes

13.18. Notation. Let $\mathcal{S}^\delta(\Delta)$, $\delta \in \mathbb{N}_0$, be the space of all 1-periodic, $\delta - 1$ times continuously differentiable splines of degree δ subordinate to the partition $\Delta = \{x_0 = 0 < x_1 < \dots < x_n = 1\}$ with the meshwidth $h = \max (x_k - x_{k-1})$. Suppose A is a classical pseudodifferential operator of order $\alpha \in \mathbb{R}$ on Γ with the principal symbol $a_0(x, \xi) \in \mathbf{C}^\infty(\mathbb{R} \times \mathbb{R} \setminus \{0\})$ and $f \in \mathbf{H}^s$, $s > 1/2$.

We consider the collocation method which determines an approximate solution $u_\Delta \in \mathcal{S}^\delta(\Delta)$ to Eq. 13.1(1) by the equations

$$Au_\Delta(x_k) = f(x_k), \qquad k = 0, \dots, n - 1. \tag{1}$$

Throughout what follows the condition $\delta - \alpha > 0$ is assumed to be satisfied (cf. also Sec. 13.1).

We now reduce the collocation equations (1) to equivalent Galerkin equations. To this end we define mappings J and J_Δ by the definite integral and its numerical counterpart, the trapezoidal rule, namely

$$Jf := \int_0^1 f(x)\, dx \quad \text{and} \quad J_\Delta f := \sum_{j=0}^{n-1} f(x_j)\, (x_{j+1} - x_{j-1})/2 .$$

Moreover, we introduce the operator $I_\Delta := I - J + J_\Delta \in \mathcal{L}(\mathbf{H}^s)$, $s > 1/2$.

13.19. Lemma. *Let $f \in \mathbf{H}^t$, $t > 1/2$ and $s \in \mathbb{R}$. Then the equations*

$$f(x_k) = 0, \qquad k = 0, \dots, n - 1,$$

hold if and only if

$$\langle I_\Delta f, v_\Delta\rangle_s = 0 \quad \forall\, v_\Delta \in \mathcal{N}_\Delta^{2s-1} := \Lambda^{-2s+1}S(\mathcal{S}^0(\Delta)) .$$

Proof. Since $\Lambda S = D$, the definition of the scalar product $\langle\cdot, \cdot\rangle_s$ implies that $\langle DI_\Delta f, v\rangle_0 = \langle I_\Delta f, \Lambda^{-2s+1}Sv\rangle_s$. So the assertion is a consequence of Lemma 2.17. ■

13.20. Corollary. *Let $\mu > 0$. For an arbitrary partition Δ, there exists an interpolation projection $Q_\Delta : \mathbf{H}^s \to \mathcal{N}_\Delta^\mu$, $s > 1/2$, such that*

$$Q_\Delta u(x_k) = u(x_k), \qquad k = 0, \ldots, n-1.$$

The projections Q_Δ are uniformly bounded in $\mathbf{H}^{(\mu+1)/2}$. Moreover,

$$\|u - Q_\Delta u\|_t \leqq ch^{s-t} \|u\|_s \tag{1}$$

for all $u \in \mathbf{H}^s$ and $0 \leqq t \leqq (\mu+1)/2 \leqq s \leqq \mu + 1$, c being a constant independent of u and h.

If $\Delta \in \mathrm{Part}_\sigma$ for fixed $\sigma \geqq 1$, then the Q_Δ's are uniformly bounded in $\mathbf{H}^t$, $1/2 < t < \mu + 1/2$, and (1) holds for all $0 \leqq t \leqq s \leqq \mu + 1$, $t < \mu + 1/2$ and $s > 1/2$.

Proof. If the projection Q_Δ exists, then by Lemma 13.19

$$\langle I_\Delta Q_\Delta u, v_\Delta\rangle_\varrho = \langle I_\Delta u, v_\Delta\rangle_\varrho \qquad \forall\, v_\Delta \in \mathcal{N}_\Delta^\mu, \tag{2}$$

where $\varrho = (\mu+1)/2$. Since $\mu > 0$, we observe that $\mathcal{N}_\Delta^\mu \subset \mathbf{H}^\varrho$. Introducing the orthogonal projection $P_\Delta^{\varrho,\mu} : \mathbf{H}^\varrho \to \mathcal{N}_\Delta^\mu$, we find that (2) can be written as

$$P_\Delta^{\varrho,\mu} I_\Delta Q_\Delta u = P_\Delta^{\varrho,\mu} I_\Delta u.$$

Since $\mathcal{N}_\Delta^\mu$ contains all constant functions, we have $I_\Delta : \mathcal{N}_\Delta^\mu \to \mathcal{N}_\Delta^\mu$. Now it follows from Section 2.16 that

$$Q_\Delta = I_\Delta^{-1} P_\Delta^{\varrho,\mu} I_\Delta = (I + J - J_\Delta)\, P_\Delta^{\varrho,\mu} I_\Delta = P_\Delta^{\varrho,\mu} + (J - J_\Delta)\,(P_\Delta^{\varrho,\mu} - I). \tag{3}$$

Hence, the projections Q_Δ exist and are uniformly bounded in $\mathbf{H}^\varrho$. Noticing that $P_\Delta^{\varrho,\mu} = \Lambda^{-\mu} S P_\Delta^{\varrho-\mu,0} S \Lambda^\mu$, where $P_\Delta^{\varrho-\mu,0} : \mathbf{H}^{\varrho-\mu} \to \mathcal{S}^0(\Delta)$ stands for the orthogonal projection, we conclude from Theorem 2.12 that

$$\|u - P_\Delta^{\varrho,\mu} u\|_t \leqq ch^{s-t} \|u\|_s, \qquad u \in \mathbf{H}^s \tag{4}$$

for $2(\varrho - \mu) - 1 + \mu = 0 \leqq t \leqq (\varrho - \mu) + \mu = (\mu+1)/2 \leqq s \leqq \mu + 1$, if Δ is an arbitrary partition and for $0 \leqq t \leqq s \leqq \mu + 1$, $t < \mu + 1/2$, $s > 1/2$, if $\Delta \in \mathrm{Part}_\sigma$. Finally, combining (3) and (4) with the estimate for the trapezoidal rule (see Lemma 2.16), we get (1). ∎

In what follows we denote by Q_Δ the interpolation projection $Q_\Delta : \mathbf{H}^s \to \mathcal{N}_\Delta^{\delta-\alpha}$, $s > 1/2$, which exists in view of Corollary 13.20, since $\delta - \alpha > 0$. Then the collocation 13.18(1) is nothing else than the projection method $\{Q_\Delta A|_{\mathcal{S}^\delta(\Delta)}\}$. Moreover, it follows from (3), (4) and Lemma 2.16 that

$$\|(J - J_\Delta)\,(P_\Delta^{(\delta-\alpha+1)/2,\delta-\alpha} - I)\|_{s\to t} = \|Q_\Delta - P_\Delta^{(\delta-\alpha+1)/2,\delta-\alpha}\|_{s\to t} \leqq ch^s \tag{5}$$

for $1/2 < s < \delta - \alpha + 1/2$ and $t < \delta - \alpha + 1/2$. Hence, for all $v_\Delta \in \mathcal{S}^\delta(\Delta)$ and $1/2 < t < \delta - \alpha + 1/2$,

$$\|(Q_\Delta - P_\Delta^{(\delta-\alpha+1)/2,\delta-\alpha})\, A v_\Delta\|_t \leqq ch^t \|v_\Delta\|_{t+\alpha}. \tag{6}$$

As an immediate consequence of (6), we get the following theorem.

13.21. Theorem. *The projection method $\{Q_\Delta A|_{\mathcal{S}^\delta(\Delta)}\}$ is stable in $(\mathbf{H}^{t+\alpha}, \mathbf{H}^t)$, $1/2 < t < \delta - \alpha + 1/2$, if and only if the projection method $\{P_\Delta^{(\delta-\alpha+1)/2,\delta-\alpha} A|_{\mathcal{S}^\delta(\Delta)}\}$ is stable in $(\mathbf{H}^{t+\alpha}, \mathbf{H}^t)$.*

The projection method $\{P_\Delta^{(\delta-\alpha+1)/2,\delta-\alpha}A|_{\mathcal{S}^\delta(\Delta)}\}$ determines an approximate solution $v_\Delta \in \mathcal{S}^\delta(\Delta)$ to Eq. 13.1(1) by the equations

$$\langle Av_\Delta, \varphi_\Delta\rangle_{(\delta-\alpha+1)/2} = \langle f, \varphi_\Delta\rangle_{(\delta-\alpha+1)/2}$$

for all $\varphi_\Delta \in \mathcal{N}_\Delta^{\delta-\alpha}$, which can be rewritten as

$$\langle Av_\Delta, \psi_\Delta\rangle_{(\delta+1)/2} = \langle f, \psi_\Delta\rangle_{(\delta+1)/2} \qquad \forall\, \psi_\Delta \in \mathcal{N}_\Delta^\delta, \tag{1}$$

since $\langle u, w\rangle_{r+s} = \langle u, \Lambda^{2s}w\rangle_r$ and $\Lambda^{-\alpha}(\mathcal{N}_\Delta^{\delta-\alpha}) = \mathcal{N}_\Delta^\delta$. Because $\mathcal{N}_\Delta^\delta = \mathcal{S}^\delta(\Delta)$ for odd δ and $\mathcal{N}_\Delta^\delta = S(\mathcal{S}^\delta(\Delta))$ for even δ, we see that (1) can be written in the form

$$(\Lambda^{\delta+1}Av_\Delta, \psi_\Delta) = (\Lambda^{\delta+1}f, \psi_\Delta) \qquad \forall\, \psi_\Delta \in \mathcal{S}^\delta(\Delta) \tag{2}$$

if δ is odd or

$$(\Lambda^{\delta+1}Av_\Delta, S\psi_\Delta) = (\Lambda^{\delta+1}f, S\psi_\Delta) \qquad \forall\, \psi_\Delta \in \mathcal{S}^\delta(\Delta) \tag{3}$$

is δ is even. Here $(\cdot, \cdot) = \langle\cdot, \cdot\rangle_0$ is the scalar product in $\mathbf{L}^2$. Thus the projection method $\{P_\Delta^{(\delta-\alpha+1)/2,\delta-\alpha}A|_{\mathcal{S}^\delta(\Delta)}\}$ is nothing but Galerkin's method for the operator equation

$$\Lambda^{\delta+1}Au = \Lambda^{\delta+1}f \tag{4}$$

if δ is odd and

$$\Lambda^{\delta+1}ASu = \Lambda^{\delta+1}f \tag{5}$$

if δ is even, where both the trial and test functions are choosen in $\mathcal{S}^\delta(\Delta)$ and $S(\mathcal{S}^\delta(\Delta))$, respectively.

The projection method (1) is also referred to as a *j-Galerkin method*, $j = (\delta + 1)/2$. It follows easily from the definition of the orthogonal projection

$$P_\Delta^{(\delta+\alpha+1)/2,\delta}\colon \mathbf{H}^{(\delta+\alpha+1)/2} \to \mathcal{N}_\Delta^\delta$$

that its $\mathbf{L}^2$-adjoint has the form

$$(P_\Delta^{(\delta+\alpha+1)/2,\delta})^* = \Lambda^{\delta+1}P_\Delta^{(\delta-\alpha+1)/2,\delta-\alpha}\Lambda^{-\delta-1}.$$

Consequently, the Galerkin equations (2), (3) for the operator equations (4), (5), respectively, take the form

$$P_\Delta^{(\delta-\alpha+1)/2,\delta-\alpha}Av_\Delta = P_\Delta^{(\delta-\alpha+1)/2,\delta-\alpha}f, \qquad v_\Delta \in \mathcal{S}^\delta(\Delta) \tag{6}$$

if δ is odd or

$$P_\Delta^{(\delta-\alpha+1)/2,\delta-\alpha}ASw_\Delta = P_\Delta^{(\delta-\alpha+1)/2,\delta-\alpha}f, \qquad w_\Delta \in S(\mathcal{S}^\delta(\Delta)) \tag{7}$$

if δ is even.

As a corollary of Remark 1.27.3°, Theorem 13.21 and the properties of the corresponding projections we get the following proposition.

13.22. Proposition. *Let* $1/2 < t < \delta - \alpha + 1/2$ *and* $A\colon \mathbf{H}^{t+\alpha} \to \mathbf{H}^t$. *Then* $A \in \Pi\{Q_\Delta A|_{\mathcal{S}^\delta(\Delta)}\}$ *if and only if* $A \in \Pi\{P_\Delta^{(\delta-\alpha+1)/2,\delta-\alpha}A|_{\mathcal{S}^\delta(\Delta)}\}$ *and* δ *is odd or* $AS \in \Pi\{P_\Delta^{(\delta-\alpha+1)/2,\delta-\alpha}AS|_{S(\mathcal{S}^\delta(\Delta))}\}$ *and* δ *is even.*

We recall that the ψdo A with the principal symbol $a_0(x, \xi)$ is said to be *locally strongly elliptic* if there exists a function $\theta \in \mathbf{C}^\infty$ such that

$$\operatorname{Re} \theta(x)\, a_0(x, \xi) > 0, \qquad (x, \xi) \in \mathbb{R} \times \{\pm 1\}. \tag{1}$$

Note that if (1) holds, then θA satisfies the Gårding inequality

$$\langle \theta Au, u\rangle_s \geqq \gamma \|u\|^2_{s+\alpha/2} - \langle Ku, u\rangle_s \tag{2}$$

for all $u \in \mathbf{H}^{s+\alpha/2}$ and $s \in \mathbb{R}$, where $\gamma > 0$ and $K: \mathbf{H}^{s+\alpha/2} \to \mathbf{H}^{s-\alpha/2}$ is a compact linear operator (see Corollary 6.23). Choosing $s = (\delta + 1)/2$, (2) assumes the form

$$(\Lambda^{\delta+1}\theta Au, u) \geqq \gamma \|u\|^2_{(\delta+\alpha+1)/2} - (\tilde{K}u, u), \tag{3}$$

where $\tilde{K} := \Lambda^{\delta+1}K: \mathbf{H}^{(\delta+\alpha+1)/2} \to \mathbf{H}^{-(\delta+\alpha+1)/2}$ is compact, i.e. the operator $\tilde{A} := \Lambda^{\delta+1}\theta A \in \mathcal{L}(\mathbf{H}^{(\delta+\alpha+1)/2}, \mathbf{H}^{-(\delta+\alpha+1)/2})$ satisfies a Gårding inequality.

13.23. Theorem. *Assume* $\delta > \alpha$ *and* $\dim \ker A = 0$. *If* δ *is odd (resp. even) and* $A \in \mathcal{L}(\mathbf{H}^{(\delta+\alpha+1)/2}, \mathbf{H}^{(\delta-\alpha+1)/2})$ *(resp.* AS*) is locally strongly elliptic, then* $A \in \Pi\{Q_\Delta A|_{\mathcal{S}^\delta(\Delta)}\}$, *i.e., for any function* $f \in \mathbf{H}^{(\delta-\alpha+1)/2}$ *and all sufficiently small* $h > 0$, *the collocation equations* 13.18(1) *are uniquely solvable and* $\|u - u_\Delta\|_{(\delta+\alpha+1)/2} \to 0$ *as* $h \to 0$, *where* $u \in \mathbf{H}^{(\delta+\alpha+1)/2}$ *is the solution of* 13.1(1).

Proof. Since $Q_\Delta\theta = Q_\Delta\theta Q_\Delta$, the projection method 13.18(1), i.e. $\{Q_\Delta A|_{\mathcal{S}^\delta(\Delta)}\}$, is stable in $(\mathbf{H}^{(\delta+\alpha+1)/2}, \mathbf{H}^{(\delta-\alpha+1)/2})$ if and only if $\{Q_\Delta\theta A|_{\mathcal{S}^\delta(\Delta)}\}$ is stable. Hence the assertion follows from Proposition 13.22 and Cea's lemma (Sec. 1.33). ■

13.24. Theorem. *Suppose the hypotheses of Theorem* 13.23 *are satisfied. Then the optimal error estimate*

$$\|u - u_\Delta\|_{t+\alpha} \leqq ch^{s-t}\|f\|_s, \qquad f \in \mathbf{H}^s \tag{1}$$

holds for $0 \leqq t \leqq (\delta - \alpha + 1)/2 \leqq s \leqq \delta - \alpha + 1$.

If, in addition, $\Delta \in \mathrm{Part}_\gamma$ *for fixed* $\gamma \geqq 1$, *then* (1) *is true for* $0 \leqq t \leqq s \leqq \delta - \alpha + 1$, $t < \delta - \alpha + 1/2$, $s > 1/2$.

Proof. From 13.20(3) we see that u_Δ satisfies the equation

$$[P_\Delta^{\varrho,\delta-\alpha} + (J - J_\Delta)(P_\Delta^{\varrho,\delta-\alpha} - I)]\, Au_\Delta = [P_\Delta^{\varrho,\delta-\alpha} + (J - J_\Delta)(P_\Delta^{\varrho,\delta-\alpha} - I)]\, f, \tag{2}$$

where $\varrho = (\delta - \alpha + 1)/2$. In view of Theorem 13.23 and Proposition 13.22, for all sufficiently small h, there exists a unique solution $v_\Delta \in \mathcal{S}^\delta(\Delta)$ to Eq. 13.21(6) (resp. $w_\Delta \in S(\mathcal{S}^\delta(\Delta))$ to Eq. 13.21(7)).

If δ is odd, then apply the Aubin-Nitsche lemma to the operator $\Lambda^{\delta+1}A \in \mathcal{L}(\mathbf{H}^{(\delta+\alpha+1)/2}, \mathbf{H}^{-(\delta+\alpha+1)/2})$ and the Galerkin method 13.21(2) (or, equivalently, 13.21(6)) to get

$$\|u - v_\Delta\|_t \leqq ch^{s-t}\|u\|_s$$

for $\delta + \alpha + 1 - (\delta + 1) \leqq t \leqq (\delta + \alpha + 1)/2 \leqq s \leqq \delta + 1$, i.e.

$$\|u - v_\Delta\|_{t+\alpha} \leqq ch^{s-t}\|f\|_s \tag{3}$$

for $0 \leqq t \leqq (\delta - \alpha + 1)/2 \leqq s \leqq \delta - \alpha + 1$. If δ is even, then applying the Aubin-Nitsche lemma to the operator $\Lambda^{\delta+1}AS \in \mathcal{L}(\mathbf{H}^{(\delta+\alpha+1)/2}, \mathbf{H}^{-(\delta+\alpha+1)/2})$ and the Galerkin method 13.21(3) (or, equivalently, 13.21(7)) and setting $u = Sw$, $v_\Delta = Sw_\Delta \in \mathcal{S}^\delta(\Delta)$, $w_\Delta \in S(\mathcal{S}^\delta(\Delta))$ gives

$$\|u - v_\Delta\|_{t+\alpha} = \|w - w_\Delta\|_{t+\alpha} \leqq ch^{s-t}\|f\|_s \tag{4}$$

for $0 \leqq t \leqq (\delta - \alpha + 1)/2 \leqq s \leqq \delta - \alpha + 1$.

As a consequence of (2), 13.21(6) and 13.21(7), $u_\Delta - v_\Delta$ satisfies the relation

$$P_\Delta^{\varrho,\delta-\alpha} A(u_\Delta - v_\Delta) = (J - J_\Delta)\,(P_\Delta^{\varrho,\delta-\alpha} - I)\,(f - Au_\Delta). \tag{5}$$

Since the Galerkin method 13.21(6) (resp. 13.21(7)) is stable, it follows from (5) and 13.20(5) that

$$\begin{aligned}\|u_\Delta - v_\Delta\|_{(\delta+\alpha+1)/2} &\leqq c\,\|P_\Delta^{\varrho,\delta-\alpha} A(u_\Delta - v_\Delta)\|_\varrho \leqq ch^\varrho\,\|f - Au_\Delta\|_\varrho \\ &\leqq c_1 h^\varrho\,\|u - u_\Delta\|_{(\delta+\alpha+1)/2}.\end{aligned}$$

Using Theorem 13.23, Remark 1.27.3° and the approximation property AP$(\delta + 1/2)$ for $\mathcal{S}^\delta(\Delta)$, we find that

$$\|u - u_\Delta\|_{(\delta+\alpha+1)/2} \leqq c_2 h^{s-\varrho}\,\|u\|_{s+\alpha} \tag{6}$$

for $\varrho = (\delta - \alpha + 1)/2 \leqq s \leqq \delta - \alpha + 1$. Hence

$$\|u_\Delta - v_\Delta\|_{t+\alpha} \leqq \|u_\Delta - v_\Delta\|_{(\delta+\alpha+1)/2} \leqq c_3 h^s\,\|f\|_s. \tag{7}$$

Combining (3), (4) and (7), we get (1).

If $\Delta \in \mathrm{Part}_\gamma$, then Proposition 13.23, Corollary 1.36, and Theorem 2.11 show that $A \in \Pi\{Q_\Delta A|_{\mathcal{S}^\delta(\Delta)}\}, A: \mathbf{H}^{t+\alpha} \to \mathbf{H}^t$, for all t satisfying $1/2 < t < \delta - \alpha + 1/2$. Hence, in view of Remark 1.27.3°, and the approximation property AP$(\delta + 1/2)$ for $\mathcal{S}^\delta(\Delta)$, the optimal error estimate

$$\|u - u_\Delta\|_{t+\alpha} \leqq ch^{s-t}\,\|f\|_s, \qquad f \in \mathbf{H}^s \tag{8}$$

holds for $1/2 < t \leqq s \leqq \delta - \alpha + 1$, $t < \delta - \alpha + 1/2$. Repeating the preceding argument and using (8) instead of (6), we get (1) for the remaining case $0 \leqq t \leqq 1/2$, $s > 1/2$. ■

13.25. Remarks. 1°. The locally strong ellipticity of the ψdo A is equivalent to each of the following two conditions:

(i) $\mu a(x) + (1 - \mu)\,b(x) \neq 0$, $x \in \mathbb{R}$, $\mu \in [0, 1]$;
(i′) $c(x) + \lambda d(x) \neq 0$, $x \in \mathbb{R}$, $\lambda \in [-1, 1]$,

where

$$a(x) = a_0(x, +1), \qquad b(x) = a_0(x, -1), \qquad c = (a + b)/2, \qquad d = (a - b)/2.$$

The locally strong ellipticity of AS is equivalent to each of the subsequent two conditions:

(ii) $\mu a(x) - (1 - \mu)\,b(x) \neq 0$, $x \in \mathbb{R}$, $\mu \in [0, 1]$;
(ii′) $d(x) + \lambda c(x) \neq 0$, $x \in \mathbb{R}$, $\lambda \in [-1, 1]$.

Condition (i) (resp. (ii)) coincides with 13.2(1) for $\varepsilon = 0$ and δ odd (resp. $\varepsilon = 0$ and δ even). (See also Secs. 6.21—23).

2°. It follows from Theorem 13.2 that if Δ is a uniform mesh and $A \in \Pi\{Q_\Delta A|_{\mathcal{S}^\delta(\Delta)}\}$, $A: \mathbf{H}^{t+\alpha} \to \mathbf{H}^t$, for some $t \in (1/2, \delta - \alpha + 1/2)$ and $\delta > \alpha$, then the conditions of Theorem 13.23, in particular (i) or (ii), respectively, are satisfied. We do not know whether this is true for an arbitrary fixed mesh Δ.

Spline Galerkin method on arbitrary meshes

13.26. Notation. The standard spline Galerkin method determines an approximate solution $u_\Delta \in \mathcal{S}^\delta(\Delta)$ to Eq. 13.1(1) by

$$(Au_\Delta, \psi_\Delta) = (f, \psi_\Delta), \qquad \forall\, \psi_\Delta \in \mathcal{S}^\delta(\Delta). \tag{1}$$

Notice that (1) makes sense for $\delta > (\alpha - 1)/2$ and $f \in \mathbf{H}^s$, $s > -\delta - 1/2$, where $\alpha \in \mathbb{R}$ denotes the order of the ψdo A.

Setting $B := \Lambda^{-\delta-1} A$ and $g := \Lambda^{-\delta-1} f$, we find that (1) can be written in the form (see 13.21(2))

$$(\Lambda^{\delta+1} Bu_\Delta, \psi_\Delta) = (\Lambda^{\delta+1} g, \psi_\Delta) \qquad \forall\, \psi_\Delta \in \mathcal{S}^\delta(\Delta). \tag{2}$$

Since $S\Lambda = \Lambda S$ and $(f, \psi) = (Sf, S\psi)$ for $\psi \in \mathbf{H}^{-s}$, (2) is equivalent to (see 13.21(3))

$$(\Lambda^{\delta+1} SBu_\Delta, S\psi_\Delta) = (\Lambda^{\delta+1} Sg, S\psi_\Delta) \qquad \forall\, \psi_\Delta \in \mathcal{S}^\delta(\Delta).$$

Obviously, B is a ψdo of order $\beta = \alpha - \delta - 1$, and the operators B and SBS are locally strongly elliptic exactly if A is so. Hence, the following theorem is an immediate consequence of Theorems 13.23 and 13.24.

13.27. Theorem. *Assume* $\delta > (\alpha - 1)/2$ *and* $\dim \ker A = 0$. *If* $A \in \mathcal{L}(\mathbf{H}^{\alpha/2}, \mathbf{H}^{-\alpha/2})$ *is locally strongly elliptic and* $f \in \mathbf{H}^s$, $-\alpha/2 \leqq s \leqq \delta - \alpha + 1$, *then the Galerkin equations* 13.26(1) *are uniquely solvable for* $h \leqq h_0$, *and*

$$\|u - u_\Delta\|_{t+\alpha} \leqq c h^{s-t} \|f\|_s \tag{1}$$

for $-\delta - 1 \leqq t \leqq -\alpha/2$.

If, in addition, $\Delta \in \mathrm{Part}_\gamma$ *for fixed* $\gamma \geqq 1$, *then* (1) *holds for* $-\delta - 1 \leqq t \leqq s \leqq \delta - \alpha + 1$, $t < \delta - \alpha + 1/2$, $s > -\delta - 1/2$.

Remark. In view of Remark 13.25.2°, the locally strong ellipticity of the ψdo A is even necessary for the first assertion of Theorem 13.27 to be true.

13.28. Region method. The region method defines $u_\Delta \in \mathcal{S}^\delta(\Delta)$ by

$$\int_{x_k}^{x_{k+1}} Au_\Delta(x)\, dx = \int_{x_k}^{x_{k+1}} f(x)\, dx, \qquad k = 0, \ldots, n - 1. \tag{1}$$

Notice that (1) makes sense for every integrable function f.

Let θ_k denote the characteristic function of the interval $[x_k, x_{k+1})$. Then (1) can be written in the form

$$(Au_\Delta, \theta_k) = (f, \theta_k), \qquad k = 0, \ldots, n - 1.$$

Recalling that $(f, \theta_k) = (\Lambda^\delta Sf, \Lambda^{-\delta} S\theta_k)$ and that $\Lambda^{-\delta} S\theta_k$ (resp. $\Lambda^{-\delta}\theta_k$), $k = 0, \ldots, n - 1$, is a basis in $\mathcal{S}^\delta(\Delta)$ if δ is odd (resp. even) (see Sec. 2.20), we find that (1) is equivalent to the Galerkin method

$$(\Lambda^\delta SAu_\Delta, \psi_\Delta) = (\Lambda^\delta Sf, \psi_\Delta) \qquad \forall\, \psi_\Delta \in \mathcal{S}^\delta(\Delta)$$

if δ is odd and equivalent to the Galerkin method

$$(\Lambda^\delta Au_\Delta, \psi_\Delta) = (\Lambda^\delta f, \psi_\Delta) \qquad \forall\, \psi_\Delta \in \mathcal{S}^\delta(\Delta)$$

in case δ is even. As a corollary of Theorem 13.27, we obtain the following result.

13.29. Theorem. *Assume* $\delta > \alpha - 1$ *and* $\dim \ker A = 0$. *If* δ *is odd (resp. even) and* $AS \in \mathcal{L}(\mathbf{H}^{(\delta+\alpha)/2}, \mathbf{H}^{(\delta-\alpha)/2})$ *(resp. A) is locally strongly elliptic, then the equations* 13.28(1) *are uniquely solvable for* $h < h_0$, *and*

$$\|u - u_\Delta\|_{t+\alpha} \leqq ch^{s-t} \|f\|_s \tag{1}$$

for $f \in \mathbf{H}^s$ *and* $-1 \leqq t \leqq (\delta - \alpha)/2 \leqq s \leqq \delta - \alpha + 1$.

If, in addition, $\Delta \in \mathrm{Part}_\gamma$ *for fixed* $\gamma \geqq 1$, *then* (1) *holds for* $-1 \leqq t \leqq s \leqq \delta - \alpha + 1$, $t < \delta - \alpha + 1/2$, $s > -1/2$.

13.30. Comparison of the rates of convergence. The error estimates of the present chapter can be used to compare the ε-collocation method (ε — C) with the Galerkin method (GM) and with the region method (RM).

Let us first consider the case in which the same degree splines $\mathcal{S}^\delta(\Delta)$ are used for all methods. Further let us restrict ourselves to smooth solutions and quasiuniform partitions (equidistant partitions in the case of ε-collocation for $0 < \varepsilon < 1$). Moreover, the ψdo A of order $\alpha \in \mathbb{R}$ is assumed to satisfy the conditions of the corresponding convergence theorems.

Table 1

Method	Restrictions	Optimal convergence in $H^{t+\alpha}$	Highest order of convergence
GM	$\delta > (\alpha - 1)/2$	$-\delta - 1 \leqq t < \delta - \alpha + 1/2$	$2\delta + 2 - \alpha$
$0 - \mathrm{C}$	$\delta > \alpha$	$0 \leqq t < \delta - \alpha + 1/2$	$\delta + 1 - \alpha$
RM	$\delta > \alpha - 1$	$-1 \leqq t < \delta - \alpha + 1/2$	$\delta + 2 - \alpha$
$\varepsilon - \mathrm{C}$	$\delta \geqq \alpha$	$0 \leqq t < \delta - \alpha + 1/2$	$\delta + 1 - \alpha$
$\varepsilon - \mathrm{C}$	$-1/2 < \delta - \alpha < 0$	$\delta - \alpha \leqq t < \delta - \alpha + 1/2$	1
$\varepsilon - \mathrm{C}$	$-1 < \delta - \alpha \leqq -1/2$	$-1/2 < t < \delta - \alpha + 1/2$	$< \delta - \alpha + 3/2$

When working with boundary integral methods for plane problems in mathematical physics and mechanics, the curve Γ is the boundary of a two-dimensional domain, and the physical field u is defined by integrals of the form

$$\tilde{u}(z) = \int_\Gamma k(z, \tau)\, u(\tau)\, d\tau,$$

where u is the solution of a pseudodifferential equation $Au = f$ on Γ and $k(z, \tau)$ is a smooth kernel for $z \notin \Gamma$. Hence, for points outside Γ, the approximate value $\tilde{u}_h(z)$ of the field will converge with a rate equal to the highest rate of convergence achieved by the approximate solution u_h to u.

Table 1 shows that the highest order of convergence gained by the 0-collocation method is $\delta + 1 - \alpha$ in $\mathbf{H}^\alpha$, while the Galerkin method converges with rate $2\delta + 2 - \alpha$ in $\mathbf{H}^{\alpha-\delta-1}$. Hence, to obtain the same order of convergence for both methods, one has to use splines of different orders δ_G for the Galerkin method and δ_C for the 0-collocation method which are related by the equality

$$\delta_C = 2\delta_G + 1.$$

On the other hand, the construction of the stiffness matrix for the Galerkin method requires the evaluation of double integrals whilst the collocation method only needs one integration. Moreover, contrary to Galerkin's method, the ε-collocation methods and the region method converge for classes of elliptic equations essentially larger than strongly elliptic pseudodifferential equations.

Systems of pseudodifferential equations

Let $A = (A_{ij})_{i,j=1}^N$ be a matrix of classical pseudodifferential operators A_{ij} of order $\alpha \in \mathbb{R}$ on the simple closed curve Γ with principal symbols $a_{ij}^0(x, \xi) \in C^\infty(\mathbb{R} \times \mathbb{R} \setminus \{0\})$. We consider the system

$$Au = f \quad \text{on} \quad \mathbb{R}, \tag{1}$$

where $f = (f_1, \ldots, f_N)$ is a given 1-periodic vector valued function and $u = (u_1, \ldots, u_N)$ is the desired unknown vector valued function. Note that each pseudodifferential equation on a system of mutually disjoint Jordan curves $\Gamma = \bigcup_{j=1}^N \Gamma_j$ may be reduced to a system (1) by parametrizing each curve Γ_j and identifying functions on Γ with N-vector valued 1-periodic functions.

The results of this chapter can easily be extended to the systems (1). Set $a_0 = (a_{ij}^0)_{i,j=1}^N$ and define the matrix $\sigma_A^{\delta,\varepsilon}$ by 13.1(7).

13.31. Theorem. *Let* $0 \leqq \varepsilon < 1$. *Assume* 13.1(3) *or* 13.1(4). *The* ε*-collocation* 13.1(2) *is stable in* $(\mathbf{H}_N^{t+\alpha}, \mathbf{H}_N^t)$, $-1/2 < t < \delta - \alpha + 1/2$, *if and only if the operator* $A \in \mathcal{L}(\mathbf{H}_N^{t+\alpha}, \mathbf{H}_N^t)$ *is invertible and*

$$\det \sigma_A^{\delta,\varepsilon}(x, s) \neq 0, \qquad x \in \mathbb{R}, \qquad 0 < s < 1. \tag{1}$$

Moreover, if these conditions are satisfied, then the error estimates of Theorem 13.14 *hold.*

This follows from the proofs of Theorem 13.2 (see also Remark 10.30.3°) and of Theorem 13.14.

Notice that condition (1) is equivalent to

$$\det\big(c(x) + \lambda d(x)\big) \neq 0, \qquad x \in \mathbb{R}, \qquad \lambda \in [-1, 1] \tag{2}$$

in case $\varepsilon = 0$ and δ is odd or $\varepsilon = 1/2$ and δ is even, where

$$c(x) := \frac{1}{2}\,[a_0(x, 1) + a_0(x, -1)], \qquad d(x) := \frac{1}{2}\,[a_0(x, 1) - a_0(x, -1)].$$

If $\varepsilon = 0$ and δ is even or if $\varepsilon = 1/2$ and δ is odd, then (1) is equivalent to

$$\det\big(d(x) + \lambda c(x)\big) \neq 0, \qquad x \in \mathbb{R}, \qquad \lambda \in [-1, 1]. \tag{3}$$

Further, condition (2) is satisfied if and only if the operator A is locally strongly elliptic, i.e. if there is a $\theta \in C_{N\times N}^\infty$ such that

$$\operatorname{Re} \theta(x)\, a_0(x, \xi) > 0, \qquad (x, \xi) \in \mathbb{R} \times \{\pm 1\}$$

(see Corollary 6.23, and Prössdorf/Rathsfeld [2], Prössdorf/Schmidt [3]). Hence, (3) is equivalent to the locally strong ellipticity of the operator AS.

13.32. Remark. The assertions of Theorems 13.23, 13.24, 13.27 and 13.29 remain true for an invertible operator $A \in \mathcal{L}(\mathbf{H}_N^{t+\alpha}, \mathbf{H}_N^t)$.

This easily follows from the proofs of these theorems.

Notes and comments

13.2.—13.17. For $\alpha = \varepsilon = t = 0$ and $\delta = 1$, Theorem 13.2 was first proved in Prössdorf/Schmidt [2]. In the case where $\varepsilon = 0$ and δ is odd, Theorem 13.14 was established by Arnold and Wendland [1]. Saranen and Wendland [1] succeeded for $\varepsilon = 1/2$, δ even and $\delta > \alpha$ in the case of constant coefficients. Their results were then generalized to the case of variable coefficients independently by Prössdorf [5—6] and by Arnold and Wendland [2] on the basis of localization techniques which are similar to those of Sec. 13.3 (cf. "Notes and comments" to Secs. 10.3—10.7). In the general case, Theorems 13.2 and 13.14 were established in Schmidt [2—6] by utilizing the techniques presented in Secs. 13.3—13.14. Theorem 13.3 is due to Prössdorf [5—6]. In Secs. 13.4—13.17 we essentially follow Schmidt [4, Secs. 3.4—3.8].

13.19.—13.24. Lemma 13.19 is due to Arnold/Wendland [1]. For odd δ and under additional hypotheses, Theorems 13.21 and 13.24 were established by Arnold and Wendland [1] (cf. also Prössdorf [11]). In the general case, the results of Secs. 13.20—13.24 were proved by Schmidt [1—3].

13.27.—13.29. In particular cases, Theorem 13.27 was established by many authors, see e.g. Arnold/Wendland [1, 2], Hsiao/Wendland [1, 2], Nedelec [1], Prössdorf [6, 11] and Stephan/Wendland [1]. Schmidt [1, 2] succeeded in the general case. Theorem 13.29 goes back to Schmidt [4].

Notice that the equations to which the analysis of the present chapter applies include Fredholm integral equations of the second kind, certain first kind integral equations with logarithmic kernels, singular integral equations involving Cauchy or Hilbert kernels, a variety of integro-differential equations, and two-point boundary value problems for ordinary differential equations (see the examples considered in Secs. 6.54—6.71). Finally, we briefly mention some results which are closely related to the subject of this chapter.

Following the first proof of Theorem 13.2 (see Sec. 13.2) one can easily extend the results of Secs. 10.12—10.30 concerning quadrature methods to pseudodifferential equations of integer order α (cf. Bühring [1]).

All the results of the present chapter remain true in the case where the subspace $\mathcal{S}^\delta(\Delta)$ of periodic splines is replaced by the corresponding subspace of complex splines (i.e. piecewise polynomials with respect to the complex variable $z = \gamma(x)$, $0 \leqq x \leqq 1$, where γ is the parametrization of the curve Γ), see G. Schmidt [4, Sec. 5]. For some equations, the use of complex splines as trial functions provides essential advantages since the computational expense for determination of the stiffness matrix is less.

Rannacher and Wendland [1, 2] proved pointwise error estimates of almost optimal order for the spline collocation method. In the case where trigonometric polynomials are used as the space of trial functions, Hölder-Zygmund norm error estimates and pointwise rates of convergence were considered in McLean/Wendland [1] for the trigonometric Galerkin method and in McLean/Prössdorf/Wendland [1] for the collocation method (see also Chapter 8). Elschner [2] succeeded in establishing suboptimal orders of convergence for spline Galerkin methods applied to non strongly elliptic pseudodifferential equations.

Wendland [4] and Saranen [1] derived error estimates for the main spline approximation methods such as the Galerkin method, collocation method and the least-squares method applied to locally strongly elliptic pseudodifferential equations taking into account the effect of numerical quadratures. Lamp, Schleicher and Wendland [1] proved $\mathbf{H}^t$ norm error estimates for a fully discrete trigonometric Galerkin method applied to elliptic pseudodifferential equations on closed curves. For the case of the first kind integral equation with logarithmic kernel, $\mathbf{L}^p$ Sobolev norm error estimates for the exact trigonometric Galerkin method were obtained in McLean [1, 2]. Amosov [1] studied an approximation method for solving elliptic pseudodifferential equations on smooth closed curves which is based on the use of the parametrix of the corresponding ψdo and is combined with the trigonometric collocation. This numerical procedure requires $O(N \log N)$

significant operations for determining N coefficients of the approximate trigonometric polynomial. Moreover, asymptotic Sobolev norm error estimates with optimal order including quadrature errors are proved. The papers Hsiao [1, 2] and Hsiao/Prössdorf [1] are concerned with estimates of the variation in approximate solutions due to small perturbations of the data for the spline Galerkin method and the spline collocation method, respectively, applied to a class of integral equations of the first kind with logarithmic kernels. It is shown that a proper choice of the mesh size can be made in the numerical computation so that one will obtain an optimal rate of convergence for the approximate solutions.

Applying Theorems 13.3, 13.23 and 13.24, Prössdorf and Szyszka [1] proved that, for the singular integral equation with continuous coefficients on a smooth closed curve, the spline collocation method with quasiuniform meshes is stable in $\mathbf{L}^2$ provided the corresponding singular operator is locally strongly elliptic. Szyszka [1] investigated collocation methods for pseudodifferential equations on smooth closed curves in the case where odd degree splines with defect r, $r > 1$, are used as the space of trial functions.

Notice that Theorems 13.2, 13.14 and 13.24 have been confirmed for Cauchy singular integral equations in Niessner [1] and Niessner/Ribault [1] and for the integral equation of the first kind with logarithmic kernel in Hoidn [1].

Many of the discretization methods including collocation can be viewed as special cases of the *Galerkin-Petrov method*: Find a trial function $u_h \in V_h$ such that $(Au_h, v_h) = (f, v_h)$ for all test functions $v_h \in T_h$. Here V_h denotes the trial space and T_h stands for the test space with $\dim V_h = \dim T_h$, but in general V_h and T_h can be different. For strongly elliptic boundary integral equations, the choice of spline trial spaces V_h and trigonometric test spaces T_h yields an exponentially convergent procedure (see Arnold [1]). In Ruotsalainen/Saranen [1, 2] the trial space V_h consists of Dirac functions and the test space T_h of splines (see also Ruotsalainen [1]). The *qualocation method* introduced and studied by Sloan [2] and further developed in Sloan/Wendland [1] and Chandler/Sloan [1] is a Galerkin-Petrov method in which the outer integrals (i.e. the scalar products) are performed numerically by special quadrature rules. In particular cases involving ψdo's with even symbols (such as the first kind integral equation with logarithmic kernel), this method, with appropriate choice of the quadrature points and weights, has a maximal order of convergence two powers of h higher than the collocation method (cf. also Sloan/Burn [1], Saranen/Sloan [1] and Prössdorf/Saranen/Sloan [1]).

References

Agranovich, M. S.

[1] Elliptic singular integro-differential operators. Uspehi Mat. Nauk **20** (1965) 5 (125), 3–120 (Russian); *also in:* Russ. Math. Surveys **20** (1965) 1–121.

[2] Spectral properties of elliptic pseudodifferential operators on a closed curve. Funkts. Anal. Prilozh. **13** (1979) 4, 54–56 (Russian); *also in:* Funct. Anal. Appl. **13** (1979) 279–281.

[3] On elliptic pseudodifferential operators on a closed curve. Trudy Moskov. Mat. O.-va **47** (1984) 22–67 (Russian); *also in:* Trans. Moscow Math. Soc. **47** (1985) 23–74.

Akhiezer, N. I.

[1] Some formulas for the inversion of singular integrals. Izv. Akad. Nauk SSSR, Ser. Mat. **9** (1945) 4, 275–290 (Russian).

[2] Vorlesungen über Approximationstheorie. Akademie-Verlag, Berlin 1967.

Alexits, G.

[1] Einige Beiträge zur Approximationstheorie. Acta Scient. Math. **26** (1965) 212–224.

Amosov, B. A.

[1] On the approximate solution of elliptic pseudodifferential equations on a smooth closed curve. Z. Anal. Anw., to appear.

Anselone, P. M.

[1] Collectively compact operator approximation theory and applications to integral equations. Prentice-Hall, INC., Englewood Cliffs, New Jersey 1971.

Anselone, P. M., and I. H. Sloan

[1] Integral equations on the half-line, J. Integral Equations **9** (Suppl.) (1985) 3–23.

[2] Numerical solution of integral equations on the half-line II. The Wiener-Hopf case. J. Integral Equations and Applications **1** (1988) 203–225.

Arnold, D. N.

[1] A spline-trigonometric Galerkin method and an exponentially convergent boundary integral method. Math. Comp. **41** (1983) 383–397.

Arnold, D. N., and W. L. Wendland

[1] On the asymptotic convergence of collocation methods. Math. Comp. **41** (1983) 349–381.

[2] The convergence of spline collocation for strongly elliptic equations on curves. Numer. Math. **47** (1985) 317–341.

Atkinson, K. E.

[1] A survey of numerical methods for the solution of Fredholm integral equations of the second kind. SIAM, Philadelphia 1976.

[2] A discrete Galerkin method for first kind integral equations with a logarithmic kernel. J. Integral Equations and Applications, to appear.

Atkinson, K. E., and A. Bogomolny

[1] The discrete Galerkin method for integral equations. Math. Comp. **48** (1987) 595–616.

Atkinson, K. E., and F. De Hoog

[1] The numerical solution of Laplace's equation on a wedge. IMA J. Numer. Anal. **4** (1984) 19–41.

Atkinson, K. E., and I. G. Graham
[1] An iterative variant of the Nyström method for boundary integral equations on non-smooth boundaries. Proc. MAFELAP, Academic Press, New York 1988.
Atkinson, K. E., and I. H. Sloan
[1] The numerical solution of first-kind logarithmic-kernel integral equations on smooth open arcs. Math. Comp., to appear.
Atkinson, K. E., I. G. Graham, and I. H. Sloan
[1] Piecewise continuous collocation for integral equations. SIAM J. Numer. Anal. **20** (1983) 172–186.
Aubin, J. P.
[1] Approximation of elliptic boundary value problems. Wiley Interscience, New York 1972.
Babuska, I., and A. K. Aziz
[1] Survey lectures on the mathematical foundations of the finite element method. *In*: A. K. Aziz (ed.), The mathematical foundation of the finite element method with applications to partial differential equations, Academic Press, New York 1972, 3–35.
Baker, C. T. H.
[1] The numerical treatment of integral equations. Clarendon Press, Oxford 1977.
Banerjee, P. K., and R. Butterfield
[1] Boundary element methods in engineering science. McGraw-Hill, London 1981.
Banerjee, P. K., and R. P. Shaw
[1] Developments in boundary element methods. Appl. Sci. Publ., London 1982.
Bari, N. K., and S. B. Stechkin
[1] Best approximations and the differential properties of two conjugate functions. Trudy Moskov. Mat. O.-va **5** (1956) 483–522 (Russian).
Belotserkovski, S. M., and I. K. Lifanov
[1] Numerical methods for singular integral equations. Nauka, Moscow 1985 (Russian).
Berezanski, Yu. M.
[1] Expansions with respect to eigenfunctions of selfadjoint operators. Naukova Dumka, Kiev 1965 (Russian).
Berthold, D.
[1] Punktweise Konvergenz der Quadraturformelmethode für singuläre Integralgleichungen, Diss. A, Fakultät für Math. und Naturwiss. TU Karl-Marx-Stadt, 1989.
Boikov, I. B.
[1] On the approximate solution of singular integral equations. Dokl. Akad. Nauk SSSR **203** (1972) 3, 511–514 (Russian); *also in:* Sov. Math. Dokl. **13** (1972), 400–404.
Borja, M., and H. Brakhage
[1] Über die numerische Behandlung der Tragflächengleichung. Z. angew. Math. Mech. **47** (1967), T102–T103. Braunschweig, Oktober 1968.
Böttcher, A., and B. Silbermann
[1] The finite section method for Toeplitz operators on the quarter-plane with piecewise continuous symbols. Math. Nachr. **110** (1983) 279–291.
[2] Invertibility and asymptotics of Toeplitz matrices, Akademie-Verlag, Berlin 1983.
[3] Analysis of Toeplitz operators. Akademie-Verlag, Berlin 1989, Springer-Verlag, New York, Heidelberg, Berlin 1990.
Böttcher, A., N. Krupnik, and B. Silbermann
[1] A general look at local principles with special emphasis on the norm computation aspect. Integral Equations Oper. Theory **11** (1988) 4, 455–479.
Brakhage, H.
[1] Über die numerische Behandlung von Integralgleichungen nach der Quadraturformelmethode. Numer. Math. **2** (1960) 183–196.
[2] Zur Fehlerabschätzung für die numerische Eigenwertbestimmung bei Integralgleichungen. Numer. Math. **3** (1961) 174–179.

BRAMBLE, J. H., and R. SCOTT
[1] Simultaneous approximation in scales of Banach spaces. Math. Comp. **32** (1978) 947–954.
BREBBIA, C. A.
[1] Boundary element methods. Springer-Verlag, Berlin, Heidelberg, New York 1981.
[2] Progress in boundary element methods. Plymouth: Pentech Press, London 1981.
[3] Boundary element techniques, methods in engineering. Springer-Verlag, Berlin, Heidelberg, New York 1983.
BREBBIA, C. A., T. FUTAGAMI, and M. TANAKA
[1] Boundary elements. Springer-Verlag, Berlin, Heidelberg, New York 1983.
BREBBIA, C. A., G. KUHN, and W. L. WENDLAND (eds.)
[1] Boundary elements IX. Springer-Verlag, Berlin, Heidelberg, New York 1987.
BREBBIA, C. A., J. C. F. TELLES, and L. C. WROBEL
[1] Boundary element techniques. Springer-Verlag, Berlin, Heidelberg, New York 1984.
BRUHN, G., and W. WENDLAND
[1] Über die näherungsweise Lösung von linearen Funktionalgleichungen. *In:* L. COLLATZ, G. MEINARDUS, H. UNGER (eds.), Funktionalanalysis, Approximationstheorie, Numerische Mathematik, Birkhäuser Verlag, Basel 1967, 136–164.
BRUNNER, H., and P. J. VAN DER HOUWEN
[1] The numerical solution of Volterra equations. North-Holland Publ. Comp., Amsterdam, New York, Oxford 1986.
BÜCKNER, H.
[1] Konvergenzuntersuchungen bei einem algebraischen Verfahren zur näherungsweisen Lösung von Integralgleichungen. Math. Nachr. **3** (1950) 358–372.
[2] Die praktische Behandlung von Integralgleichungen. Ergebnisse der Angewandten Mathematik, Band 1, Springer-Verlag, Berlin 1952.
[3] Numerical methods for integral equations. *In:* J. TODD (ed.), A survey of numerical analysis, McGraw-Hill Book Comp., New York 1962.
BUTZER, P. L., and R. J. NESSEL
[1] Fourier analysis and approximation, Vol. 1. Birkhäuser Verlag, Basel, Stuttgart 1971.
CALDERON, A. P., and A. ZYGMUND
[1] On the existence of certain singular integrals. Acta Math. **88** (1952) (1–2), 85–139.
[2] On singular integrals. Amer. J. Math. **78** (1956) 2, 289–309.
CHANDLER, G. A.
[1] Superconvergence for second kind integral equations. *In:* R. S. ANDERSSEN, F. DE HOOG, M. LUCAS (eds.), The application and numerical solution of integral equations. Sijthoff and Noordhoff, Groningen 1980.
[2] Galerkin's method for boundary integral equations on polygonal domains. J. Austral. Math. Soc. **B 26** (1984) 1–13.
CHANDLER, G. A., and I. G. GRAHAM
[1] Uniform convergence of Galerkin solutions to non-compact integral operator equations, IMA J. Numer. Anal. **7** (1987) 327–334.
[2] Product integration-collocation methods for non-compact integral operator equations. Math. Comp. **50** (1988) 125–138.
[3] The convergence of Nyström methods for Wiener-Hopf equations. Numer. Math. **52** (1988) 345–364.
[4] Higher order methods for linear functionals of solutions of second kind integral equations. SIAM J. Numer. Anal. **25** (1988) 1118–1137.
CHANDLER, G. A., and I. H. SLOAN
[1] Spline qualocation methods for boundary integral equations. Numer. Math., to appear.
CHANDRA, P.
[1] On the degree of approximation of functions belonging to Lipschitz class. Nanta Math. **8** (1970) 88–91.

[2] On the generalized Fejer means in the metric of Hölder space. Math. Nachr. **109** (1982) 39–45.
[3] Degree of approximation of functions in the Hölder metric by Borel's means. J. Math. Anal. Appl. **149** (1990) 236–248.

Çiarlet, P. G.
[1] The finite element method for elliptic problems. North-Holland, Amsterdam, New York, Oxford 1978.

Clancey, K., and I. Gohberg
[1] Factorization of matrix functions and singular integral operators. Birkhäuser Verlag, Basel 1981.

Cordes, H. O.
[1] The algebra of singular integral operators in $\mathbb{R}^n$. J. Math. Mech. **14** (1965) 1007–1032.
[2] Elliptic pseudodifferential operators — an abstract theory. Springer Lecture Notes in Mathematics, Vol. 756, Springer-Verlag, Berlin, Heidelberg, New York 1979.

Costabel, M.
[1] An inverse for the Gohberg-Krupnik symbol map. Proc. of the Royal Soc. of Edinburgh **87** A (1980) 153–165.
[2] Boundary integral operators on curved polygons. Ann. Mat. Pura Appl. **133** (1983) 305–326.
[3] Singular integral operators on curves with corners. Integral Equations Oper. Theory **5** (1983) 353–371.

Costabel, M., and E. P. Stephan
[1] Boundary integral equations for mixed boundary value problems in polygonal domains and Galerkin approximation. Banach Center Publications, Vol. 15, pp. 175–251, PWN, Warszawa 1985.
[2] On the convergence of collocation methods for boundary integral equations on polygons. Math. Comp. **49** (1987) 461–478.
[3] The method of Mellin transformation for boundary integral equations on curves with corners. Preprint No 761, Fachbereich Mathematik, Technische Hochschule Darmstadt 1983.

Costabel, M., V. J. Ervin, and E. P. Stephan
[1] On the convergence of collocation methods for Symm's integral equation on open curves. Math. Comp. **51** (1988) 167–179.

Costabel, M., E. P. Stephan, and W. L. Wendland
[1] On boundary integral equations of the first kind for the bi-Laplacian in a polygonal plane domain. Ann. Scuola Norm. Sup. Pisa **10** (1983) 197–241.

Crouch, S. L., and A. H. Starfield
[1] Boundary element methods in solid mechanics. George Allen & Unwin, London 1983.

Crouzeix, M., and V. Thomée
[1] The stability in $\mathbf{L}_p$ and $\mathbf{W}_p$ of the $\mathbf{L}_2$-projection onto finite element function spaces. Math. Comp. **48** (1987) 521–532.

Dahmen, W.
[1] On the best approximation and de la Vallee Poussin sums. Matem. Zametki **23** (1978) 671–683 (Russian).

De Boor, C.
[1] The quasi-interpolant as a tool in elementary polynomial spline theory. *In*: G. G. Lorentz (ed.). Approximation theory. Academic Press Inc., New York, London 1973, 269–276. Proceedings of an International Symposium, Austin 1973.
[2] A bound on the $\mathbf{L}^\infty$-norm of $\mathbf{L}_2$-approximation by splines in terms of a global mesh ratio. Math. Comp. **30** (1976) 765–771.
[3] A practical guide to splines. Springer-Verlag, New York, Heidelberg, Berlin 1978.

De Hoog, F., and I. H. Sloan
[1] The finite-section approximation for integral equations on the half-line. J. Austral. Math. Soc. Series B, **28** (1987) 415–434.

Delves, L. M., and J. Mohammed
[1] Computational methods for integral equations. Cambridge University Press, Cambridge 1985.

DEVINATZ, A., and M. SHINBROT
[1] General Wiener-Hopf operators. Trans. Amer. Math. Soc. **145** (1969) 467–494.
DIDENKO, V. D.
[1] On a direct method for solving the Hilbert problem. Math. Nachr. **133** (1987) 317–342 (Russian).
DOUGLAS, R. G.
[1] Banach algebra techniques in operator theory. Academic Press, New York 1972.
DOW, M. L., and D. ELLIOTT
[1] The numerical solution of singular integral equations over (−1, 1). SIAM J. Numer. Anal. **16** (1979) 115–134.
DUBEAU, F., and J. SAVOIE
[1] Periodic quadratic spline interpolation. J. Approximation Theory **39** (1983) 77–88.
DUDUCHAVA, R. V.
[1] Discrete Wiener-Hopf equations in spaces $\mathbf{l}_p$ with weight. Soobshzh. Akad. Nauk Gruz. SSR **67** (1972) 1, 17–20 (Russian).
[2] On discrete Wiener-Hopf equations. Trudy Tbilissk. Mat. Inst. **50** (1975) 42–59 (Russian).
[3] Integral equations with fixed singularities. Teubner-Texte zur Mathematik, Bd. 24, BSB B. G. Teubner Verlagsges., Leipzig 1979.
DZISKARIANI, A. V.
[1] On the solution of singular integral equations by means of approximate projection methods. Zhurnal vychisl. matem. fiz. 19, Nr. **5** (1979) 1149–1161 (Russian).
[2] On the solution of singular integral equations by means of collocation methods. Zhurnal vychisl. matem. i matem. fiz. 21, Nr. **2** (1981) 355–362 (Russian).
EHLICH, H., and K. ZELLER
[1] Auswertung der Normen von Interpolationsoperatoren. Math. Ann. **164** (1966) 105–112.
ELLIOTT, D.
[1] The approximate solution of singular integral equations. *In:* M. A. GOLBERG (ed.), Solution methods for integral equations. Plenum Publishing Corporation, New York, London 1979, 83–107.
[2] A convergence theorem for singular integral equations. J. Austr. Math. Soc., Series B, **22** (1981) 539–552.
[3] The classical collocation method for singular integral equations. SIAM J. Numer. Anal. **19** (1982) 816–832.
[4] Orthogonal polynomials associated with singular integral equations having a Cauchy kernel. SIAM J. Math. Anal. **13** (1982), 1041–1052.
[5] Rates of convergence for the method of classical collocation for solving singular integral equations. SIAM J. Numer. Anal. **21** (1984) 136–148.
[6] Convergence theorems for singular integral equations. *In:* M. A. GOLBERG (ed.), Solution methods for integral equations, Plenum Press 1988.
ELSCHNER, J.
[1] Galerkin methods with splines for singular integral equations over (0, 1). Numer. Math. **43** (1984) 265–281.
[2] Singular ordinary differential operators and pseudodifferential equations. Math. Research, Vol. 22, Akademie-Verlag, Berlin 1985 and Lecture Notes in Mathematics, Vol. 1128, Springer-Verlag, New York, Heidelberg, Berlin 1985.
[3] On spline collocation for singular integral equations on an interval. Seminar Analysis. Operator equat. and numer. anal. 1985/86, Karl-Weierstraß-Inst. Math., Akad. Wiss. DDR, Berlin (1986) 31–54.
[4] On spline approximation for a class of integral equations III. Collocation methods with piecewise linear splines. Seminar Analysis. Operator equat. and numer. anal. 1986/87, Karl-Weierstraß-Inst. Math., Akad. Wiss. DDR, Berlin (1987) 25–40.
[5] Asymptotics of solutions to pseudodifferential equations of Mellin type. Math. Nachr. **130** (1987) 267–305.

[6] On spline approximation for a class of non-compact integral equations. Report R-MATH-09/88, Karl-Weierstraß-Inst. Math., Akad. Wiss. DDR, Berlin 1988.
[7] On spline approximation for singular integral equations on an interval. Math. Nachr. **139** (1988) 309–319.
[8] On spline approximation for a class of integral equations I. Galerkin and collocation methods with piecewise polynomials. Math. Meth. in the Appl. Sci. **10** (1988) 543–559.
[9] On spline collocation for convolution equations. Integral Equations Oper. Theory **12** (1989) 486–510.
[10] On spline approximation for a class of integral equations II. Galerkin's method with smooth splines. Math. Nachr. **140** (1989) 273–283.

ELSCHNER, J., S. PRÖSSDORF, A. RATHSFELD, and G. SCHMIDT
[1] Spline approximation of singular integral equations. Demonstratio Math. **18** (1985) 3, 661 to 672.

ELSCHNER, J., and G. SCHMIDT
[1] On spline interpolation in periodic Sobolev spaces. Preprint P-MATH-01/83, Inst. Math., Akad. Wiss. DDR, Berlin, 1983.

ERDÉLYI, A., W. MAGNUS, F. OBERHETTINGER, and F. TRICOMI
[1] Tables of Integral Transforms, Vol. 1, McGraw-Hill Book Company Inc., New York, Toronto, London 1954.

ERDOGAN, F., G. D. GUPTA and T. S. COOK
[1] Numerical solution of singular integral equations. Mech. of Fracture **1** (1973) 368–425.

ESKIN, G. I.
[1] Boundary value problems for elliptic pseudodifferential equations. Nauka, Moscow 1973 (Russian); *English transl.:* Amer. Math. Soc. Transl. of Math. Monographs 52, Providence, R. I., 1981.

FABRIKANT, V., S. V. HOA, and T. S. SANKAR
[1] On the approximate solution of singular integral equations. Computer Methods in Applied Mechanics and Engineering **29** (1981) 19–33.

FENYÖ, S., and H. W. STOLLE
[1] Theorie und Praxis der linearen Integralgleichungen. Bände 1–4, VEB Deutscher Verlag der Wissenschaften, Berlin 1982–1984.

FILIPPI, P.
[1] Theoretical acoustics and numerical techniques. CISM Courses and Lectures 277, Springer-Verlag, Wien, New York 1983.

FOX, L., and E. T. GOODWIN
[1] The numerical solution of non-singular linear integral equations. Phil. Trans. Roy. Soc. London A **245** (1953) 501–534.

FREDHOLM, I.
[1] Sur une classe d'equations fonctionelles. Acta Math. **27** (1903) 365–390.

FREUD, G.
[1] Orthogonale Polynome, Akademie-Verlag, Berlin 1969.

GABDULKHAEV, B. G.
[1] On a direct method for the solution of integral equations. Izv. Vyssh. Uchebn. Zaved., Mat. **3** (1965) 51–60 (Russian).
[2] Approximate solution of singular integral equations by the method of mechanical quadratures. Dokl. Akad. Nauk SSSR **179** (1968) 2, 260–263 (Russian).

GÄHLER, S., and W. GÄHLER
[1] Quadrature methods for the solution of Fredholm integral equations on the half-line, Math. Nachr. **140** (1989) 321–346.

GAIER, D.
[1] Saturation bei Spline-Approximation und Quadratur. Numer. Math. **16** (1970) 129 to 140.

GAZEWSKI, R., and W. LENSKI
[1] On the approximation by the generalized de la Vallee Poussin means of Fourier series in the space $\mathbf{W}^{\gamma}\mathbf{H}^{\alpha}$. Math. Nachr. **120** (1985) 63—71.
GELFAND, I. M., and N. J. VILENKIN
[1] Verallgemeinerte Funktionen (Distributionen). Bd. IV. VEB Deutscher Verlag der Wissenschaften, Berlin 1964.
GERASOULIS, A.
[1] Singular integral equations — The convergence of the Nyström interpolant of the Gauss-Chebyshev methods. BIT **22** (1982) 200—210.
[2] The use of piecewise quadratic polynomials for the solution of singular integral equations of Cauchy type. Comp. Math. Appl. **8** (1982) 15—22.
[3] On the existence of approximate solutions for singular integral equations of Cauchy type discretized by Gauss-Chebyshev quadrature formulae. BIT **21** (1981) 377—380.
GERASOULIS, A., and R. P. SRIVASTAV
[1] A method for the numerical solution of singular integral equations with a principal value integral. Int. J. Engng. Sci. **19** (1981) 1293—1298.
[2] On the solvability of singular integral equations via Gauss-Jacobi quadratures. Int. J. Comp. Math. **12** (1982) 59—75.
GERASOULIS, A., and R. VICHNEVETSKY (eds.)
[1] Numerical solution of singular integral equations. Proceedings of an IMACS International Symposium held at Lehigh University Bethlehem, Pennsylvania, USA, June 21—22, 1984.
GOHBERG, I.
[1] The factorization problem in normed rings, functions of isometric and symmetric operators and singular integral equations. Uspehi Mat. Nauk **19** (1964) 1, 71—124 (Russian).
GOHBERG, I., and I. A. FELDMAN (GOCHBERG, I., and I. A. FELDMANN)
[1] Convolution equations and projection method for their solution. Nauka, Moscow 1971 (Russian); *Engl. transl.:* Amer. Math. Soc. Transl. of Math. Monographs 41, Providence, R. I., 1974; *German transl.:* Akademie-Verlag, Berlin 1974.
GOHBERG, I., and N. YA. KRUPNIK
[1] On the algebra generated by Toeplitz matrices. Funkts. Anal. Prilozh. **3** (1969) 2, 46—56 (Russian); *also in:* Funct. Anal. Appl. **3** (1969) 119—127.
[2] On the algebra generated by one-dimensional singular integral operators with piecewise continuous coefficients. Funkts. Anal. Prilozh. **4** (1970) 3, 26—36 (Russian); *also in:* Funct. Anal. Appl. **4** (1970) 193—201.
[3] Singular integral operators with piecewise continuous coefficients and their symbols. Izv. Akad. Nauk SSSR, Ser. Mat. **35** (1971) 4, 940—964 (Russian); *also in:* Math. USSR Izv. **5** (1971), 955—979 (1972).
[4] Introduction to the theory of one-dimensional singular integral operators. Shtiintsa, Kishinev 1973 (Russian); *Germ. transl.:* Birkhäuser Verlag, Basel, Boston, Stuttgart 1979.
GOHBERG, I., and V. I. LEVCHENKO
[1] On the projection method for the degenerate discrete Wiener-Hopf equation. Mat. Issled. **VII** (1972) 3, 238—253 (Russian).
GOLBERG, M. A.
[1] Projection methods for Cauchy singular integral equations with constant coefficients on [—1, 1]. *In:* C. T. H. BAKER and G. F. MILLER (eds.), Treatment of integral equations by numerical methods, Academic Press, London 1982, 261—272.
[2] The convergence of a collocation method for a class of Cauchy singular integral equations. J. Math. Anal. Appl. **100** (1984) 500—512.
[3] The numerical solution of Cauchy singular integral equations with constant coefficients. J. Integral Equations **9** (1985) (Suppl.), 127—151.

Golberg, M. A., and J. A. Fromme
[1] On the L_2-convergence of collocation for the generalized airfoil equation. J. Math. Anal. Appl. **71** (1979) 271–286.
Golberg, M. A., M. Lea, and G. Miel
[1] A superconvergence result for the generalized airfoil equation with application to the Hap problem. J. Integral Equations **5** (1983) 175–186.
Graham, I. G.
[1] Galerkin methods for second kind integral equations with singularities, Math. Comp. **39** (1982) 519–533.
Grubb, G.
[1] Functional calculus of pseudodifferential boundary problems. Birkhäuser Verlag, Basel, Boston, Stuttgart 1986.
Hackbusch, W.
[1] Theorie und Numerik der Integralgleichungen. Teubner, Stuttgart 1989.
Haftmann, R.
[1] Numerische Behandlung einer singulären Integralgleichung aus der Strömungsmechanik. Diss. A, Fakultät für Math. und Naturwiss., TH Karl-Marx-Stadt, 1979.
[2] Über die Quadraturformelmethode zur Lösung singulärer Integralgleichungen. Beiträge zur Numer. Math. **10** (1982) 47–56.
Hagen, R., and B. Silbermann
[1] A finite element collocation method for bisingular integral equations, Appl. Anal. **19** (1985) 117–135.
[2] Local theory of the collocation method for the approximate solution of singular integral equations, II. Seminar Analysis. Operator equat. and numer. anal. 1986/87, Karl-Weierstraß-Inst. Math., Akad. Wiss. DDR, Berlin 1987, 41–56.
[3] A Banach algebra approach to the stability of projection methods for singular integral equations, Math. Nachr. **140** (1989) 285–297.
[4] On the stability of the qualocation method. Seminar Analysis. Operator equat. and numer. anal. 1988/89, Karl-Weierstraß-Inst. Math., Akad. Wiss. DDR, Berlin 1989, 43–52.
Halmos, P. R.
[1] Two subspaces. Trans. Amer. Math. Soc. **144** (1969) 381–389.
Hardy, G. H.
[1] Note on Lebesgue's constants in the theory of Fourier series. J. London Math. Soc. **17** (1942) 4–13.
Havin, V. P., S. V. Khrushchev, and N. K. Nikol'ski (eds.)
[1] Linear and complex analysis problem book. Lect. Notes Math., Vol. **1043**, Springer-Verlag, Heidelberg, 1984.
Helfrich, H. P.
[1] Simultaneous approximation in negative norms of arbitrary order. RAIRO Numer. Anal. **15** (1981) 231–235.
Hilbert, D.
[1] Grundzüge einer allgemeinen Theorie der linearen Integralgleichungen. Verlag B. G. Teubner, Leipzig, Berlin 1912.
Hille, E.
[1] Analytic function theory. Ginn and Company, Boston 1962.
Hoidn, H. P.
[1] Die Kollokationsmethode angewandt auf die Symmsche Integralgleichung. Doctoral Thesis, ETH Zürich, 1983.
Hörmander, L.
[1] Pseudodifferential operators. Commun. Pure Appl. Math. **18** (1965) 3, 501–517.
[2] The analysis of linear partial differential operators. III. Pseudodifferential operators. Springer-Verlag, Berlin, Heidelberg, New York, Tokyo 1985.

HSIAO, G. C.
[1] The finite element method for a class of improperly posed integral equations. *In*: G. HÄMMERLIN and K. H. HOFFMANN (eds.), Improperly Posed Problems and Their Numerical Treatment, 117–131, Birkhäuser Verlag, Basel 1983.
[2] On the stability of integral equations of the first kind with logarithmic kernels. Arch. Rat. Mech. Anal. **94** (1986) 179–192.
HSIAO, G. C., and R. C. MAC CAMY
[1] Solution of boundary value problems by integral equations of the first kind. SIAM Review **15** (1973) 687–705.
HSIAO, G. C., and S. PRÖSSDORF
[1] On the stability of the spline collocation method for a class of integral equations of the first kind. Applicable Analysis **30** (1988) 249–261.
HSIAO, G. C., and W. L. WENDLAND
[1] A finite element method for some integral equations of the first kind. J. Math. Anal. Appl. **58** (1977) 449–481.
[2] The Aubin-Nitsche lemma for integral equations. J. Integral Equations **3** (1981) 299 to 315.
HSIAO, G. C., P. KOPP, and W. L. WENDLAND
[1] A Galerkin collocation method for some integral equations of the first kind. Computing **25** (1980) 89–130.
[2] Some applications of a Galerkin collocation method for integral equations of the first kind. Math. Meth. Appl. Sci. **6** (1984) 280–335.
IOAKIMIDIS, N. I.
[1] Application of the method of singular integral equations to elasticity problems with concentrated loads. Acta Mech. **40** (1981) 159–168.
[2] A new method for the numerical solution of singular integral equations appearing in crack and other elasticity problems. Acta Mech. **39** (1981) 117–125.
[3] An iterative algorithm for the numerical solution of singular integral equations. J. Comp. Phys. **43** (1981) 164–176.
IOAKIMIDIS, N. I., and P. S. THEOCARIS
[1] Numerical solution of Cauchy type singular integral equations by use of Lobatto-Jacobi numerical integration rule. Aplikace Matematiky **23** (1978) 439–452.
[2] A comparison between the direct and classical methods for the solution of Cauchy-type singular integral equations. SIAM J. Numer. Anal. **17** (1980) 115–118.
[3] On convergence of two direct methods for solution of Cauchy type singular integral equations of the first kind. BIT **20** (1980) 83–87.
IVANOV, V. V.
[1] The theory of approximate methods and their application to the numerical solution of singular integral equations. Naukova Dumka, Kiev 1968 (Russian); *Engl. transl.:* Noordhoff Int. Publ., Leyden 1976.
JEN, E., and R. P. SRIVASTAV
[1] Cubic splines and approximate solution of singular integral equations. Math. Comp. **37** (1981) 417–423.
JOE, S.
[1] Discrete collocation methods for second kind Fredholm integral equations, SIAM J. Numer. Anal. **22** (1985) 1167–1177.
[2] Discrete Galerkin methods for Fredholm integral equations of the second kind. IMA J. Numer. Anal. **7** (1987) 149–164.
JOURNÉ, J.-L.
[1] Calderon-Zygmund operators, pseudodifferential operators and the Cauchy integral of Calderon. Springer Lecture Notes in Mathematics, Vol. 994, Springer-Verlag, Berlin, Heidelberg, New York, Tokyo 1983.

JUNGHANNS, P.
[1] Kollokationsverfahren zur näherungsweisen Lösung singulärer Integralgleichungen mit unstetigen Koeffizienten. Math. Nachr. **102** (1981) 17–24.
[2] Kollokations- und Quadraturformelverfahren zur näherungsweisen Lösung einer Klasse singulärer Integralgleichungen mit einer festen Singularität, Wiss. Zeitschr. d. TH Karl-Marx-Stadt **24** (1982) 295–303.
[3] Polynomiale Näherungsverfahren für singuläre Integralgleichungen über beschränkten Intervallen, Diss. B, Fakultät für Math. und Naturwiss., TH Karl-Marx-Stadt 1983.
[4] Uniform convergence of approximate methods for Cauchy type singular integral equations over (−1, 1), Wiss. Zeitschr. d. TH Karl-Marx-Stadt **26** (1984) 251–256.
[5] Some remarks on the zero distribution of pairs of polynomials associated with singular integral operators. A convergence theorem for the quadrature method. Wiss. Zeitschr. d. TH Karl-Marx-Stadt **27** (1985) 88–93.
[6] Effective solution of systems of algebraic equations occuring in the approximate solution of singular integral equations by means of the method of quadrature formulae. Wiss. Zeitschr. d. TH Karl-Marx-Stadt **27** (1985) 94–96.
[7] A new convergence rate for the quadrature method for solving singular integral equations. Banach Centre Publications, Vol. 22, PWN, Warszawa 1989, 183–191.
JUNGHANNS, P., and D. BERTHOLD
[1] Direct multiple grid methods for Cauchy type singular integral equations. Wiss. Zeitschr. d. TU Karl-Marx-Stadt **29** (1987) 180–186.
JUNGHANNS, P., and D. ÖSTREICH
[1] Numerische Lösung des Staudammproblems mit Drainage. ZAMM **69** (1989) 2, 83–92.
JUNGHANNS, P., and B. SILBERMANN
[1] Zur Theorie der Näherungsverfahren für singuläre Integralgleichungen auf Intervallen. Math. Nachr. **103** (1981) 199–244.
[2] Numerical analysis for one-dimensional Cauchy type singular integral equations. *In*: Probleme und Methoden der Mathematischen Physik, 8. Tagung in Karl-Marx-Stadt 1983, 122–129, Teubner-Texte zur Mathematik, Bd. 63, BSB Teubner Verlagsges., Leipzig 1984.
[3] Local theory of the collocation method for the approximate solution of singular integral equations. I. Integral Equations Oper. Theory **7** (1984) 6, 791–807.
[4] The numerical treatment of singular integral equations by means of polynomial approximations, I, Preprint, P-Math-35/86, Karl-Weierstraß-Inst. Math., Akad. Wiss. DDR, Berlin 1986.
[5] Numerical analysis of the quadrature method for solving linear and nonlinear singular integral equations. Wiss. Schriftenreihe der TU Karl-Marx-Stadt 10/1988, Karl-Marx-Stadt 1988.
KALANDIYA, A. I.
[1] Mathematical methods of two-dimensional elasticity. Mir Publishers, Moscow 1975.
KANTOROVICH, L. V., and G. P. AKILOV (KANTOROWITSCH/AKILOW)
[1] Funktionalanalysis in normierten Räumen. Akademie-Verlag, Berlin 1964.
KANTOROVICH, L. V., and V. I. KRYLOV
[1] Approximate methods of higher analysis. Interscience Publishers, Inc., New York 1958.
KARPENKO, L. N.
[1] The approximate solution of a singular integral equation by means of Jacobi polynomials. Prikl. Mat. and Mech. **30** (1966) 564–569.
KESLER, S. SH., and N. YA. KRUPNIK
[1] On the invertibility of matrices with entries from a ring. Uchebn. Zap. Kishinev Gos. Univ. **91** (1967) 51–54 (Russian).
KÖHLER, U., and B. SILBERMANN
[1] Einige Ergebnisse über Φ_+-Operatoren in lokal konvexen topologischen Vektorräumen. Math. Nachr. **56** (1973) 145–153.
[2] Über algebraische Eigenschaften einer Klasse von Operatormatrizen und eine Anwendung auf singuläre Integraloperatoren. Math. Nachr. **57** (1973) 245–258.

KOHN, J. J., and L. NIRENBERG
[1] An algebra of pseudodifferential operators. Commun. Pure Appl. Math. **18** (1965) (1–2), 269–305.
KOZAK, A. V., and I. B. SIMONENKO
[1] Projection methods for the solution of multidimensional discrete convolution equations. Sibir. Math. Zh. **21** (1980) 2, 119–127 (Russian).
KRASNOSELSKI, M. A., G. M. VAINIKKO, et al.
[1] Näherungsverfahren zur Lösung von Operatorgleichungen. Akademie-Verlag, Berlin 1973.
KREIN, M. G.
[1] Integral equations on a half-line with kernel depending upon the difference of the arguments. Uspehi Mat. Nauk **13** (1958) 3–120 (Russian); *also in:* Amer. Math. Soc. Transl. **22** (1962) (2) 163–288.
KRENK, S.
[1] On quadrature formulas for singular integral equations of the first and second kind. Quarterly of Appl. Math. **33** (1975/76) 225–232.
[2] Polynomial solutions to singular integral equations with applications to elasticity theory. Lyngby 1981.
KRESS, R.
[1] Linear integral equations. Springer-Verlag, Berlin, Heidelberg, New York, Tokyo 1989.
[2] Boundary integral equations in time-harmonic acoustic scattering. NAM-Bericht Nr.66, Institut für Numerische und Angewandte Mathematik, Universität Göttingen 1989.
KROTOV, V. G.
[1] Note on the convergence of Fourier series in the spaces Λ_{ω}^{p}. Acta Sci. Math. **41** (1979) 335 to 338.
KRUPNIK, N. YA.
[1] Banach algebras with a symbol and singular integral operators. Shtiintsa, Kishinev 1984 (Russian); *Engl. transl.:* Birkhäuser Verlag, Basel 1987.
KUMANO-GO, H.
[1] Pseudodifferential operators. M.I.T. Press, Cambridge 1981.
LAMP, U., T. SCHLEICHER, and W. L. WENDLAND
[1] The fast Fourier transform and the numerical solution of one-dimensional boundary integral equations. Numer. Math. **47** (1985) 15–38.
LAVRENTYEV, M. A.
[1] On the building up of the flow past an arc of given shape. Trudy ZAGI **118** (1932) 3–56 (Russian).
LEINDLER, L.
[1] Generalizations of Prössdorf's theorems. Studia Sci. Math. **14** (1979) 431–439.
LE ROUX, M. N.
[1] Resolution numerique du probleme du potential dans le plan par une methode variationuelle d'elements finis. These, L'Universite de Rennes, Serie A, No. d'ordre 347, No. de serie 38 (1974).
LEWIS, J. E., and C. PARENTI
[1] Pseudodifferential operators of Mellin type. Comm. Part. Diff. Eq. **8** (1983) 477–544.
LIFANOV, I. K., and YA. E. POLONSKI
[1] Foundation of the numerical method of discrete whirls for the solution of singular integral equations. Prikl. Math. Mech. **39** (1975) 742–746 (Russian).
LINZ, P.
[1] An analysis of a method for solving singular integral equations. BIT **17** (1977) 329–337.
LIONS, J. L., and E. MAGENES
[1] Non-homogeneous boundary value problems and applications. Vol. I. Springer-Verlag, Berlin, Heidelberg, New York 1972.
LITVINCHUK, G. S., and I. M. SPITKOVSKI
[1] Factorization of measurable matrix functions. Akademie-Verlag, Berlin 1987, Birkhäuser Verlag, Basel 1987.

MacCamy, R. C.
[1] On singular integral equations with logarithmic or Cauchy kernels. J. Math. Mech. **7** (1958) 355–376.
Marcus, M., and H. Minc
[1] A survey of matrix theory and matrix inequalities. Allyn and Bacon, Boston 1964.
Markus, A. S., and I. A. Feldman
[1] On the index of an operator matrix. Funkts. Anal. Prilozh. **11** (1977) 2, 83–84 (Russian); *also in:* Funct. Anal. Appl. **11** (1977) 149–151.
[2] On the connection between certain properties of an operator matrix and its determinant. *In:* Lin. Operatory (Mat. Issled., Vyp. 54), 110–120, Shtiintsa, Kishinev 1980 (Russian).
McLean, W.
[1] Boundary integral methods for the Laplace equation. Thesis, Australian National University, Canberra 1985.
[2] A spectral Galerkin method for a boundary integral equation, Math. Comp. **47** (1986) 597–607.
McLean, W., and W. L. Wendland
[1] Trigonometric approximation of solutions of periodic pseudodifferential equations. *In:* H. Dym et al. (eds): The Gohberg anniversary Collection. Vol. 2, Topics in analysis and operator theory. Birkhäuser Verlag, Basel, Boston, Berlin 1989, 359–383.
McLean, W., S. Prössdorf, and W. L. Wendland
[1] Pointwise error estimates for the trigonometric collocation method applied to singular integral equations and periodic pseudodifferential equations. J. Integral Equations and Applications **2** (1989) 1, 125–146.
Meyer, Ch.
[1] Reduktionsverfahren für singuläre Integralgleichungen mit speziellen Carleman'schen Verschiebungen. Wiss. Zeitschr. d. TH Karl-Marx-Stadt **25** (1983) 381–387.
Micke, A.
[1] Die Auflösung schwach singulärer Fredholmscher Integralgleichungen mit angepaßten Quadraturen. ZAMM **70** (1990) 1, 49–62.
Mikhlin, S. G. (Michlin, S. G.)
[1] Singular integral equations. Uspehi Mat. Nauk **3** (1948) 3 (25), 29–112 (Russian).
[2] Singular integral equations with two independent variables. Mat. Sb. **1** (4), 535–550 and **1** (6) (1936) 963–964 (Russian).
[3] Multidimensional singular integrals and integral equations. Fizmatgiz, Moscow 1962 (Russian); *Engl. transl.:* Pergamon Press, New York 1965.
[4] Approximation auf dem kubischen Gitter. Akademie-Verlag, Berlin 1976 und Birkhäuser Verlag, Basel, Stuttgart 1976.
[5] Fehler in numerischen Prozessen. Akademie-Verlag, Berlin 1985; *Engl. transl.:* Wiley, New York 1990.
Mikhlin, S. G., and S. Prössdorf
[1] Singuläre Integraloperatoren. Akademie-Verlag, Berlin 1980; *Engl. transl.:* Akademie-Verlag, Berlin 1986, Springer-Verlag, Berlin, Heidelberg, New York, Tokyo 1986.
Mikhlin, S. G., and Ch. L. Smolitski
[1] Näherungsverfahren zur Lösung von Differential- und Integralgleichungen. BSB B. G. Teubner Verlagsges., Leipzig 1969.
Mohapatra, R. N., and P. Chandra
[1] Hölder continuous functions and their Euler, Borel and Taylor means. Math. Chronicle **11** (1982) 81–96.
[2] Degree of approximation of functions in the Hölder metric. Acta Math. Hung. **41** (1983) 67–76.
Mohapatra, R. N., and R. S. Rodriguez
[1] On the rate of convergence of singular integrals for Hölder continuous functions. Math. Nachr. **149** (1990) 117–124.

MONEGATO, G., and S. PRÖSSDORF
[1] On the numerical treatment of an integral equation arising from a cruciform crack problem. Math. Methods Appl. Sci. **12** (1990) 489–502.
MUKHERJEE, S.
[1] Boundary element methods in creep and fracture. Appl. Sci. Publ. London, New York 1982.
MULTHOPP, H.
[1] Die Berechnung der Auftriebsverteilung von Tragflügeln. Luftfahrt-Forschung **XV** (1938) 4, 153–169.
MUSAEV, B. I.
[1] On approximate solution of singular integral equations. *In:* Singular Integral operators. Izd. Azerb. Univ., Baku 1986, 33–61 (Russian).
MUSKHELISHVILI, N. I.
[1] Singular integral equations. Noordhoff, Groningen 1953; *German transl.:* Akademie-Verlag, Berlin 1965.
MYSOVSKIKH, I. P.
[1] Estimation of error arising in the solution of an integral equation by the method of mechanical quadratures. Vestnik Leningrad Univ. **11** (1956) 66–72 (Russian).
[2] An error estimate for the numerical solution of a linear integral equation. Dokl. Akad. Nauk SSSR **140** (1961) 763–765 (Russian).
[3] On the method of mechanical quadrature for the solution of integral equations. Vestnik Leningrad Univ. **17** (1962) 78–88 (Russian).
NAIMARK, M. A.
[1] Normierte Algebren. VEB Deutscher Verlag der Wissenschaften, Berlin 1959; *Engl. transl.:* Noordhoff, Groningen 1959 *and* Hafner, New York 1964.
NATANSON, I. P.
[1] Konstruktive Funktionentheorie. Akademie-Verlag, Berlin 1955.
NEDELEC, J. C.
[1] Approximation des equations integrales en mecanique et en physique. Lecture Notes, Centre de Mathematiques Appliquees, Ecole Polytechnique, Palaiseau 1977.
NEUMANN, C.
[1] Untersuchungen über das logarithmische und Newtonsche Potential. Teubner, Leipzig 1877.
NIESSNER, H.
[1] Significance of kernel singularities for the numerical solution of Fredholm integral equations. *In:* C. A. BREBBIA, G. KUHN, and W. L. WENDLAND (eds.) [1], 213–227.
NIESSNER, H., and H. RIBAULT
[1] Condition of boundary integral equations arising from flow computations. Comp. and Appl. Math. **12–13** (1985) 491–503.
NIKOL'SKI, S. M.
[1] Approximation of functions of several variables and imbedding theorems. Springer-Verlag, Berlin, Heidelberg, New York 1975.
NITSCHE, J. A.
[1] Ein Kriterium für die Quasi-Optimalität des Ritzschen Verfahrens. Numer. Math. **11** (1968) 346–348.
[2] Zur Konvergenz von Näherungsverfahren bezüglich verschiedener Normen. Numer. Math. **15** (1970) 224–227.
NITSCHE, J. A., and A. SCHATZ
[1] On local approximation properties of L_2-projections on spline subspaces. Appl. Anal. **2** (1972) 161–168.
[2] Interior estimates for Ritz-Galerkin methods. Math. Comp. **28** (1974) 937–958.
NOBLE, B.
[1] Methods based on the Wiener-Hopf technique for the solution of partial differential equations. Pergamon, London 1958.

[2] The numerical solution of integral equations. *In:* The state of the art in numerical analysis. Proc. Conf. Univ. York, Heslington 1976, Academic Press, London 1977, 915–966.

Nyström, E. J.

[1] Über die praktische Auflösung von Integralgleichungen mit Anwendungen auf Randwertaufgaben. Acta Math. **54** (1930) 185–204.

Obreshkov, N.

[1] Verteilung und Berechnung der Nullstellen reeller Polynome. VEB Deutscher Verlag der Wissenschaften, Berlin 1963.

Östreich, D.

[1] Zum Staudammproblem mit Drainage. Z. angew. Math. Mech. **67** (1987) 7, 293–300.

Panasyuk, V. V., M. Savruk, and A. P. Dacyshin

[1] Stress distribution on crakcs in plates and shells. Naukova Dumka, Kiev 1976 (Russian).

Parton, V. Z., and P. I. Perlin

[1] Integral equations in elasticity theory. Nauka, Moscow 1977 (Russian).

Peetre, J.

[1] New thoughts on Besov spaces. Durham, Math. Depart. Duke Univ., 1976.

Peterson, G. K.

[1] Measure theory for C^*-algebras II. Math. Scand. **22** (1968) 63–74.

Pietsch, A.

[1] Operator ideals. VEB Deutscher Verlag der Wissenschaften. Berlin 1978.

Plamenevski, B. A.

[1] Algebras of pseudodifferential operators. Nauka, Moscow 1986 (Russian).

Plamenevski, B. A., and V. N. Senichkin

[1] On C^*-algebras of singular integral operators with discontinuous coefficients on composed contours. I. Izv. Vyssh. Uchebn. Zaved., Mat. **260** (1984) 1, 25–33 (Russian).

[2] On C^*-algebras of singular integral operators with discontinuous coefficients on composed contours. II. Izv. Vyssh. Uchebn. Zaved., Mat. **262** (1984) 4, 37–46 (Russian).

Pöltz, R.

[1] Operators of local type in spaces of Hölder continuous functions. Seminar Analysis. Operator equat. and numer. anal. 1986/87, Karl-Weierstraß-Inst. Math., Akad. Wiss. DDR, Berlin 1987, 107–122 (Russian).

Power, S. C.

[1] Hankel operators on Hilbert spaces. Pitman Research Notes, No. 64, Pitman, Boston, London, Melbourne 1982.

Prestin, J.

[1] Best approximation in Lipschitz spaces. *In:* J. Szabados and K. Tandori (eds.), Coll. Math. Soc. Jan. Bolyai 49: Alfred Haar Memorial Conference. Proc. Conf. Budapest (Hungary) 1985, Tandori, 753–759, Jan. Bol. Math. Soc., Budapest 1987 and North-Holland Publ. Comp., Amsterdam, Oxford, New York 1987.

[2] On the approximation by de la Vallee Poussin sums and interpolatory polynomials in Lipschitz norms. Anal. Math. **13** (1987) 251–259.

[3] Trigonometric interpolation in Hölder spaces. J. Approximation Theory **53** (1988) 145–154.

[4] Lagrange interpolation for functions of bounded variation. Acta Math. Hung., to appear.

Prestin, J., and S. Prössdorf

[1] Error estimates in generalized trigonometric Hölder-Zygmund norms. Z. Anal. Anw. **9** (1990) 4, 343–349.

Prössdorf, S.

[1] Zur Konvergenz der Fourierreihen hölder-stetiger Funktionen. Math. Nachr. **69** (1975) 7–14.

[2] Systeme einiger singulärer Gleichungen vom nicht normalen Typ und Projektionsverfahren zu ihrer Lösung. Studia Math. LIII (1975) 225–252.

[3] Einige Klassen singulärer Gleichungen. Akademie-Verlag, Berlin 1974: *English transl.:* North-Holland Publ. Comp., Amsterdam, New York, Oxford 1978; *Russian transl.:* Mir, Moscow 1979.

[4] Approximation methods for solving singular integral equations. *In:* E. LANCKAU and W. TUTSCHKE (eds.), Complex Analysis, Methods, Trends, and Applications, Akademie-Verlag, Berlin 1983, 131–141.

[5] A localization principle in the theory of finite element methods. *In:* Probleme und Methoden der Mathematischen Physik, 8. Tagung in Karl-Marx-Stadt 1983, 169–177, Teubner-Texte zur Mathematik, Bd. 63, BSB B. G. Teubner Verlagsges., Leipzig 1984.

[6] Ein Lokalisierungsprinzip in der Theorie der Splineapproximationen und einige Anwendungen. Math. Nachr. **119** (1984) 239–255.

[7] Recent results in numerical analysis for singular integral equations. *In:* Problems and methods in mathematical physics, 9. TMP, Karl-Marx-Stadt 1988, Proceedings, Teubner-Texte zur Mathematik, BSB B. G. Teubner Verlagsges., Leipzig 1989, 224–234.

[8] Linear integral equations. *In:* V. G. MAZYA and S. M. NIKOL'SKI (eds.), Encyclopaedia of Mathematical Sciences, Vol. 27, VINITI, Moscow 1988 and Springer-Verlag, Berlin, Heidelberg, New York, Tokyo 1989.

[9] Numerische Behandlung singulärer Integralgleichungen. ZAMM **69** (1989) T5–T13.

[10] On the super-approximation property of Galerkin's method with finite elements. Numer. Math., to appear.

[11] Zur Splinekollokation für lineare Operatoren in Sobolewräumen. *In:* Recent Trends in Mathematics, Reinhardsbrunn 1982, Teubner-Texte zur Mathematik, Bd. 50, BSB B. G. Teubner Verlagsges., Leipzig 1983, 251–262.

PRÖSSDORF, S., and J. ELSCHNER

[1] Finite element methods for singular integral equations on an interval. Engineering Analysis **1** (1984) 2, 83–87.

PRÖSSDORF, S., and A. RATHSFELD

[1] A spline collocation method for singular integral equations with piecewise continuous coefficients. Integral Equations Oper. Theory **7** (1984) 4, 536–560.

[2] On strongly elliptic singular integral operators with piecewise continuous coefficients. Integral Equations Oper. Theory **8** (1985) 6, 825–841.

[3] On spline Galerkin methods for singular integral equations with piecewise continuous coefficients. Numer. Math. **48** (1986) 99–118.

[4] Stabilitätskriterien für Näherungsverfahren bei singulären Integralgleichungen in L^p. Z. Anal. Anw. **6** (1987) 6, 539–558.

[5] On quadrature methods and spline approximation of singular integral equations. *In:* C. A. BREBBIA, G. KUHN and W. L. WENDLAND (eds.) [1], 193–211.

[6] Mellin techniques in the numerical analysis for one-dimensional singular integral equations. Report R-MATH-06/88, Karl-Weierstraß-Inst. Math., Akad. Wiss. DDR, Berlin 1988.

[7] Quadrature and collocation methods for singular integral equations on curves with corners. Z. Anal. Anw. **8** (1989) 3, 197–220.

[8] Quadrature methods for strongly elliptic Cauchy singular integral equations on an interval. *In:* H. DYM et al. (eds.), The Gohberg anniversary collection. Vol. 2, Topics in analysis and operator theory. Birkhäuser Verlag, Basel, Boston, Berlin 1989, 435–471.

[9] On an integral equation of the first kind arising from a cruciform crack problem: *In:* Bl. SENDOV (ed.), Integral Equations and Inverse Problems. Proc. Intern. Conf. Varna (Bulgaria) 1989, Longman, Coventry 1990, 210–219.

PRÖSSDORF, S., and G. SCHMIDT

[1] Notwendige und hinreichende Bedingungen für die Konvergenz des Kollokationsverfahrens bei singulären Integralgleichungen. Math. Nachr. **89** (1979) 203–215.

[2] A finite element collocation method for singular integral equations. Math. Nachr. **100** (1981) 33–60.

[3] A finite element collocation method for systems of singular integral equations. *In:* L. ILIEV and V. ANDREEV (eds.), Complex Analysis and Applications 81, 428–439, Publishing House Bulgarian Acad. Sci., Sofia 1984.

PRÖSSDORF, S., and B. SILBERMANN

[1] Ein Projektionsverfahren zur Lösung abstrakter singulärer Gleichungen vom nicht normalen Typ und einige seiner Anwendungen. Math. Nachr. **61** (1974) 133–155.

[2] Verallgemeinerte Projektionsverfahren zur Lösung singulärer Gleichungen vom nicht normalen Typ. Math. Nachr. **68** (1975) 7–28.

[3] Zur Kollokations- und Reduktionsmethode für Systeme singulärer Integralgleichungen. *In:* VII. Internationaler Kongreß über Anwendungen der Mathematik in den Ingenieurwissenschaften, Weimar 1975, Berichte, 289–293, VEB Verlag für Bauwesen, Berlin 1975.

[4] Einige allgemeine Sätze zur Theorie der Projektionsverfahren für lineare Operatorgleichungen in Banachräumen. Math. Nachr. **75** (1976) 61–72.

[5] On the convergence of the reduction and collocation methods for systems of singular integral equations. Dokl. Akad. Nauk SSSR **226** (1976) 3, 516–519 (Russian); *also in:* Sov. Math. Dokl. **17** (1976) 140–143.

[6] General theorems on the convergence of projection methods for operator equations in Banach spaces. Dokl. Akad. Nauk SSSR **230** (1976) 3, 527–529 (Russian); *also in:* Sov. Math. Dokl. **17** (1977) 1347–1349.

[7] Projektionsverfahren zur Lösung von Systemen singulärer Gleichungen vom nicht normalen Typ. Rev. Roum. Pures Appl. **XXII** (1977) 7, 965–991.

[8] Projektionsverfahren und die näherungsweise Lösung singulärer Gleichungen. Teubner-Texte zur Mathematik, BSB B. G. Teubner Verlagsges., Leipzig 1977.

[9] Über Näherungsverfahren zur Lösung singulärer Gleichungen. *In:* 7. Tagung über Probleme und Methoden der Mathematischen Physik in Karl-Marx-Stadt 1979, Tagungsberichte, Teil II, 95–114, Wiss. Schriftenreihe der TH Karl-Marx-Stadt, 1979.

PRÖSSDORF, S., and I. H. SLOAN

[1] A quadrature method for singular integral equations on closed curves. SIAM J. Numer. Anal., to appear.

PRÖSSDORF, S., and U. SZYSZKA

[1] On spline collocation of singular integral equations on nonuniform meshes. Seminar Analysis. Operator equat. and numer. anal. 1986/87, Karl-Weierstraß-Inst. Math., Akad. Wiss. DDR, Berlin 1987, 123–137.

RANNACHER, R., and W. L. WENDLAND

[1] On the order of pointwise convergence of some boundary element methods. Part I. Operators of negative and zero order. Math. Modelling and Numer. Anal. **19** (1985) 65–88.

[2] On the order of pointwise convergence of some boundary element methods. Part II. Operators of positive order. Math. Modelling and Numer. Anal. **22** (1988) 343–362.

RATHSFELD, A.

[1] Über das Reduktionsverfahren für singuläre Integralgleichungen mit stückweise stetigen Koeffizienten. Math. Nachr. **127** (1986) 125–143.

[2] Quadraturformelverfahren für eindimensionale singuläre Integralgleichungen. Seminar Analysis. Operator equat. and numer. anal. 1985/86, Karl-Weierstraß-Inst. Math., Akad. Wiss. DDR, Berlin 1986, 147–186.

[3] Eine Quadraturformelmethode für Mellin-Operatoren nullter Ordnung. Math. Nachr. **137** (1988) 321–354.

[4] A quadrature method for a Cauchy singular integral equation with a fixed singularity. Seminar Analysis. Operator equat. and numer. anal. 1987/88, Karl-Weierstraß-Inst. Math., Akad. Wiss. DDR, Berlin 1988, 107–117.

RICE, J. R.

[1] On the degree of convergence of nonlinear spline approximation. *In:* I. J. SCHOENBERG (ed.), Approximation with special emphasis on spline functions, Academic Press, New York 1969, 349–365.

ROCH, S.

[1] Locally strongly elliptic singular integral operators, Wiss. Zeitschr. d. TH Karl-Marx-Stadt **29** (1987) 224–229.

[2] Finite sections of operators belonging to the closed algebra of singular integral operators. Seminar Analysis. Operator equat. and numer. anal. 1986/87, Karl-Weierstraß-Inst. Math., Akad. Wiss. DDR, Berlin 1987, 139–148.

[3] Finite sections of operators generated by singular integrals with Carleman shift, Preprint 52/87, TU Karl-Marx-Stadt, 1987.

[4] Lokale Theorie des Reduktionsverfahrens für singuläre Integraloperatoren mit Carlemanschen Verschiebungen, Dissertation A, TU Karl-Marx-Stadt 1988.

[5] Finite sections of operators generated by convolutions. Seminar Analysis. Operator equat. and numer. anal. 1987/88, Karl-Weierstraß-Inst. Math., Akad. Wiss. DDR, Berlin 1988, 118–138.

[6] Finite sections of singular integral operators with measurable coefficients, Wiss. Zeitschr. d. TH Karl-Marx-Stadt **31** (1989) 2, 236–242.

ROCH, S., and B. SILBERMANN

[1] Toeplitz-like operators, quasicommutator ideals, numerical analysis. I. Math. Nachr. **120** (1985) 141–173.

[2] Functions of shifts on Banach spaces — invertibility, dilations, numerical analysis, Preprint P-MATH-11/87, Karl-Weierstraß-Inst. Math., Akad. Wiss. DDR, Berlin, 1987.

[3] Toeplitz-like operators, quasicommutator ideals, numerical analysis. II. Math. Nachr. **134** (1987) 245–255.

[4] Finite sections of singular integral operators with Carleman shift. Seminar Analysis. Operator equat. and numer. anal. 1986/87, Karl-Weierstraß-Inst. Math., Akad. Wiss. DDR, Berlin 1987, 149–180.

[5] A symbol calculus for finite sections of singular integral operators with flip and piecewise continuous coefficients, J. Funct. Anal. **78** (1988) 365–389.

[6] Algebras generated by idempotents and the symbol calculus for singular integral operators Integral Equations Oper. Theory **11** (1988) 3, 385–419.

[7] Algebras of convolution operators and their image in the Calkin algebra. Report R-MATH-05/90, Karl-Weierstraß-Inst. Math., Akad. Wiss. DDR, Berlin, 1990

[8] The Calkin image of algebras of singular integral operators. Integral Equations Oper. Theory **12** (1989) 855–897.

[9] A symbol calculus for the algebra generated by shift operators. Z. Anal. Anw., 8 (1989) 4, 293–306.

[10] Non-strongly converging approximation methods. Demonstratio Math. **22** (1989) 3, 651–676

ROOKE, D. P., and I. N. SNEDDON

[1] The crack energy and the stress intensity factor for a cruciform crack deformed by internal pressure. Int. J. Engng. Sci. **7** (1969) 1079–1089.

ROSENBLUM, M., and J. ROVNYAK

[1] Hardy classes and operator theory. Oxford University Press, New York, Clarendon Press, Oxford 1985.

RUDIN, W.

[1] Functional analysis. Mc Graw-Hill Book Company, New York 1973.

RUOTSALAINEN, K.

[1] On the convergence of some boundary element methods in the plane. Report 37, Department of Mathematics, University of Jyväskylä 1987.

RUOTSALAINEN, K., and J. SARANEN

[1] A dual method to the collocation method. Math. Methods Appl. Sci. **10** (1988) 439 to 445.

[2] Some boundary element methods using Dirac's distributions as trial functions. SIAM J. Numer. Anal. **24** (1987) 816–827.

SARANEN, J.

[1] On the effect of numerical quadratures in solving boundary integral equations. *In:* Notes on Numerical Fluid Mechanics, Verlag Vieweg, Braunschweig (to appear).

SARANEN, J., and W. L. WENDLAND

[1] On the asymptotic convergence of collocation methods with spline functions of even degree. Math. Comp. **45** (1985) 91–108.

[2] The Fourier series representation of pseudodifferential operators on closed curves. Complex Anal. **8** (1987) 55–64.

SCHECHTER, M.

[1] Quantities related to strictly singular operators. Indiana Univ. Math. J. **21** (1972) 1061–1071.

SCHLEIFF, M.

[1] Über Näherungsverfahren zur Lösung einer singulären linearen Integrodifferentialgleichung. ZAMM **48** (1968) 477–483.

[2] Singuläre Integraloperatoren in Hilberträumen mit Gewichtsfunktion. Math. Nachr. **42** (1969) 145–155.

[3] Zur numerischen Lösung einer linearen Integrodifferentialgleichung mit der Kollokationsmethode. Wiss. Zeitschr. d. TH Karl-Marx-Stadt **11** (1969) 3, 375–379.

SCHMIDT, E.

[1] Auflösung der allgemeinen linearen Integralgleichung. Math. Ann. **64** (1907) 161–174.

SCHMIDT, G.

[1] On spline collocation for singular integral equations. Math. Nachr. **111** (1983) 177–196.

[2] The convergence of Galerkin and collocation methods with splines for pseudodifferential equations on closed curves. Z. Anal. Anw. **3** (1984) 4, 371–384.

[3] On spline collocation methods for boundary integral equations in the plane. Math. Methods Appl. Sci. **7** (1985) 74–89.

[4] Splines und die näherungsweise Lösung von Pseudodifferentialgleichungen auf geschlossenen Kurven. Report R-MATH-09/86, Karl-Weierstraß-Inst. Math., Akad. Wiss. DDR, Berlin 1986.

[5] On ε-collocation for singular integro-differential equations over (0, 1). Numer. Math. **50** (1987) 337–352.

SCHNEIDER, C.

[1] Product integration for weakly singular integral equations. Math. Comp. **36** (1981) 207–213.

SCHOENBERG, I. J.

[1] On interpolation by spline functions and its minimal properties. *In:* P. L. BUTZER and J. KOREVAR (eds.), On approximation theory. Birkhäuser Verlag, Basel 1972, 109–129.

SCHULZE, H., and B. SILBERMANN

[1] One-dimensional singular integral operators on Hölder-Zygmund spaces. Seminar Analysis. Operator equat. and numer. anal. 1988/89, Karl-Weierstraß-Inst. Math., Akad. Wiss. DDR, Berlin 1989, 129–139.

SCHUMAKER, L. L.

[1] Spline functions: basic theory. Wiley, New York 1981.

SEELEY, R. T.

[1] Integro-differential operators on vector bundles. Trans. Amer. Math. Soc. **117** (1965) 5, 167 to 204.

[2] Topics in pseudodifferential operators. *In:* L. NIRENBERG (ed.), Pseudodifferential operators, CIME, Cremonese, Roma 1969, 169–305.

SHUBIN, M. A.

[1] Pseudodifferential operators and spectral theory. Nauka, Moscow 1978 (Russian); *Engl. transl.:* Springer-Verlag, Berlin, Heidelberg, New York, Tokyo 1987.

SICKEL, W.

[1] Some observations on equivalent quasi-norms in generalized Besov spaces and approximation of functions in Lipschitz norms. Preprint N/86/49, Friedrich-Schiller-Universität, Jena 1986.

SILBERMANN, B.

[1] Lokale Theorie des Reduktionsverfahrens für Toeplitzoperatoren. Math. Nachr. **104** (1981) 137–146.

[2] Lokale Theorie des Reduktionsverfahrens für singuläre Integraloperatoren. Z. Anal. Anw. **1** (1982) 6, 45–56.
[3] Numerical analysis for Wiener-Hopf integral equations in spaces of measurable functions. Seminar Analysis. Operator equat. and numer. anal. 1985/86, Karl-Weierstraß-Inst. Math., Akad. Wiss. DDR, Berlin 1986, 187–203.
[4] Local objects in the theory of Toeplitz operators. Integral Equations Oper. Theory **9** (1986) 5, 706–738.
[5] The Banach algebra approach to the reduction method for Toeplitz operators. In: Havin/Khrushchev/Nikolski [1], 293–297.
[6] The C^*-algebra generated by Toeplitz and Hankel operators with piecewise quasicontinuous symbols. Integral Equations Oper. Theory **10** (1987) 5, 730–738.
[7] Symbol constructions and numerical analysis. Preprint 1989, Sektion Mathematik, TU Karl-Marx-Stadt.
[8] Approximation of periodic pseudodifferential equations and optimal error estimates. Seminar Analysis. Operator equat. and numer. anal. 1988/89, Karl-Weierstraß-Inst. Math., Akad. Wiss. DDR, Berlin 1989, 141–152.

Simonenko, I. B., and Chin Ngok Minh
[1] A local method in the theory of one-dimensional singular integral equations with piecewise continuous coefficients: Fredholmness. Izd. Rostovsk. Univ., Rostov-na-Donu 1986 (Russian).

Sloan, I. H.
[1] Improvement by iteration for compact operator equations. Math. Comp. **30** (1976) 758 to 764.
[2] A quadrature-based approach to improving the collocation method. Numer. Math. **54** (1988), 41–56.

Sloan, I. H., and A. Spence
[1] Projection methods for integral equations on the half line. IMA J. Numer. Anal. **6** (1986) 153–172.

Sloan, I. H., and V. Thomée
[1] Superconvergence of the Galerkin iterates for integral equations of the second kind. J. Integral Equations **9** (1985) 1–23.

Sloan, I. H., and W. L. Wendland
[1] A quadrature-based approach to improving the collocation method for splines of even degree. Z. Anal. Anw. **8** (1989) 4, 361–376.

Sloan, I. H., and Y. Yan
[1] On integral equations of the first kind with logarithmic kernels. J. Integral Equations and Applications **1** (1988) 549–579.

Spitkovski, I. M., and A. M. Tashbaev
[1] Factorization of certain piecewise continuous matrix functions and its applications. Math. Nachr., in print.

Srivastav, R. P.
[1] On the numerical solution of singular integral equations using Gauss-type formulae — I, quadrature and collocation on Chebyshev nodes. IMA J. Numer. Anal. **3** (1983) 305–318.

Srivastav, R. P., and E. Jen
[1] Numerical solutions of singular integral equations using Gauss type formulae — II, Lobatto-Chebyshev quadrature and collocation on Chebyshev nodes. IMA J. Numer. Anal. **3** (1983), 319–325.

Stallybrass, M. B.
[1] A pressurized crack in the form of a cross. Quart. J. Mech. and Appl. Math. **23** (1970) 35 to 48.

Stephan, E., and W. L. Wendland
[1] Remarks to Galerkin and least squares methods with finite elements for general elliptic problems. Manuscripta Geodaetica **1** (1976) 93–123.

STOLLE, H.-W.
[1] Zur Nachrechnung freifahrender Propeller unter angenäherter Berücksichtigung der starken Belastung. Wiss. Zeitschr. Univ. Rostock, Math.-nat. Reihe **1** (1952) 15–28.
STYPINSKI, Z.
[1] Generalization of the theorem of Prössdorf. Funct. Approx. Comment. Math. **7** (1979) 101 to 104.
SUBBOTIN, YU. N.
[1] Interpolating splines. *In:* Z. CIESIELSKI and J. MUSIELAK (eds.), Approximation theory. Warszawa, Dordrecht, Boston, 1975, 221–234.
SZYSZKA, U.
[1] Spline-Kollokation mit nicht glatten Splines für singuläre Integralgleichungen auf geschlossenen Kurven. Dissertation A, Wilhelm-Pieck-Universität Rostock 1989.
TAYLOR, M. E.
[1] Pseudodifferential operators. Princeton University Press, Princeton, New Jersey 1981.
THEOCARIS, P. S., and N. I. IOKIMIDIS
[1] Numerical integration methods for the solution of singular integral equations. Quarterly of Appl. Math. **35** (1977/78) 173–183.
[2] The V-notched elastic half-plane problem. Acta Mech. **32** (1979) 125–140.
[3] A remark on the numerical solution of singular integral equations and the determination of stress-intensity factors. J. Eng. Math. **13** (1979) 213–222.
THOMAS, K. S.
[1] Galerkin methods for singular integral equations. Math. Comp. **36** (1981) 193–205.
TIMAN, A. F.
[1] Theory of approximation of functions of a real variable. Fizmatgiz, Moscow 1960 (Russian); *English transl.:* Pergamon Press, Oxford 1963.
TREVES, F.
[1] Introduction to pseudodifferential and Fourier integral operators I. Plenum Press, New York, London 1980.
TRICOMI, F. G.
[1] Integral equations. Interscience, New York 1957.
TRIEBEL, H.
[1] Interpolation theory, function spaces, differential operators. VEB Deutscher Verlag der Wissenschaften, Berlin, 1978, North-Holland, Publ. Comp., Amsterdam, New York, Oxford 1978.
TSAMASHYROS, G., and P. S. THEOCARIS
[1] Equivalence and convergence of direct and indirect methods for the numerical solution of singular integral equations. Computing **27** (1981) 71–80.
VAINIKKO, G. M.
[1] Funktionalanalysis der Diskretisierungsmethoden. Teubner-Texte zur Mathematik, BSB B. G. Teubner Verlagsges., Leipzig 1976.
VAINIKKO, G. M., A. PEDAS, and P. UBA
[1] Methods for solving weakly singular integral equations. Gos. Univ. Tartu, 1984 (Russian).
VERBITSKI, I. E.
[1] Projection methods for the solution of singular integral equations with piecewise continuous coefficients. In: Oper. v. Banach. Prostr. (Mat. Issled., Vyp. 47), 12–24, Shtiintsa, Kishinev 1978 (Russian).
VOLTERRA, V.
[1] Lecons sur le equations integrales et les equations integro-differentielles. Gauthier-Villars, Paris 1913.
WASHIZU, K., and M. IKEGAWA
[1] Finite element technique in lifting surface problems. *In:* International symposium on finite element methods in flow problems. University of Wales, Swansea 1974.

WENDLAND, W. L.
[1] Elliptic systems in the plane. Pitman, London 1979.
[2] Boundary element methods and their asymptotic convergence. *In:* P. FILIPPI (ed.), Theoretical Acoustics and Numerical Treatments, CISM Courses and Lectures, No. 277, Springer-Verlag, Wien, New York 1983, 135–216.
[3] On the spline approximation of singular integral equations and one-dimensional pseudodifferential equations on closed curves. *In:* A. GERASOULIS and R. VICHNEVETSKY [1], 113–119.
[4] On some mathematical aspects of boundary element methods for elliptic problems. *In:* J. R. WHITEMAN (ed.), The mathematics of finite elements and applications V. Academic Press, London 1985, 193–227.
[5] Asymptotic accuracy and convergence for point collocation methods, *In:* C. A. BREBBIA (ed.), Topics in Boundary Element Research, Vol. 2, Springer-Verlag, Berlin 1985, 230–257.
[6] Strongly elliptic boundary integral equations. *In:* A. ISERLES and M. J. D. POWELL (eds.), The State of the Art in Numerical Analysis, Clarendon Press, Oxford 1987, 511–562.
WHEELER, J. C.
[1] Modified moments and Gaussian quadratures. Rocky Mountain J. Math. **4** (1974) 287–296.
WIGGELSWORTH, L. A.
[1] Stress distributions in a notched plate. Matematika **4** (1957) 76–96.
WOLFERSDORF, L. v.
[1] On the theory of nonlinear singular integral equations of Cauchy type. Math. Meth. Appl. Sci. **7** (1985) 493–517.
ZACHARIAS, K.
[1] Eine Bemerkung zur trigonometrischen Interpolation. Beiträge zur Numerischen Mathematik **9** (1981) 195–200.
ZHUK, V. V.
[1] On the accuracy of the representation of a continuous periodic function by linearapproximation methods. Izv. Vyssh. Uchebn. Zavad., Math. **8** (1972) 46–59 (Russian).
ZOLOTAREVSKI, V. A.
[1] On the convergence of the collocation method for systems of singular integral equations. Mat. Issled. **X** (1974) 1, 56–69 (Russian).
[2] The solution of singular integral equations by the reduction method. Mat. Issled. **X** (1974) 2, 38–52 (Russian).
[3] On the approximate solution of singular integral equations. Mat. Issled. **X** (1974) 3, 82–94 (Russian).
[4] On the method of mechanical quadratures for systems of singular integral equations. Izv. Vyssh. Ucheb. Zaved., Math. **4** (1976) 47–55 (Russian).
ZYGMUND, A.
[1] Trigonometric series, Vol. I–II. Cambridge Univ. Press, London 1959.

Supplemental references

BÜHRING, K.
[1] Quadraturformelverfahren für die approximative Lösung von Pseudodifferentialgleichungen über geschlossenen Kurven. Seminar Analysis. Operator equat. and numer. anal. 1989/90, Karl-Weierstraß-Inst. Math., Akad. Wiss. DDR, Berlin 1990, 45–68.
MASTROIANNI, G., and S. PRÖSSDORF
[1] A quadrature method for Cauchy integral equations with weakly singular perturbation kernel. J. Integral Equations and Applications, submitted.
PRÖSSDORF, S., J. SARANEN and I. H. SLOAN
[1] A discrete method for the logarithmic-kernel integral equation on an open arc. J. Integral Equations and Applications, submitted.

Saranen, J., and I. H. Sloan
[1] Quadrature methods for logarithmic-kernel integral equations on closed curves. Applied Mathematics Preprint AM 90/28. The University of New South Wales, Australia.
Sloan, I. H., and B. J. Burn
[1] An unconventional quadrature method for logarithmic kernel integral equations on closed curves. J. Integral Equations and Applications, to appear.

Notation index

Operators

$[A, B]$	commutator of A and B, i.e. $AB - BA$
coker A	$=$ codim im A (if one of them is finite)
$\mathcal{D}(A)$	domain of the operator A
det A	determinant of a matrix A
$\Phi(X, Y)$	collection of all Fredholm operators mapping X into Y
$\Phi_{\pm}(X, Y)$	collection of all semi-Fredholm operators
F	Fourier transform
ind A	index of a Fredholm operator A
im A	image of the operator A
ker A	kernel of the operator A
$\mathfrak{R}(A;\{A_n,P_n\})$	convergence manifold
$\mathcal{K}(X, Y)$	collection of all compact operators mapping X into Y
$\mathcal{L}(X, Y)$	space of all bounded linear operators mapping X into Y
P_Γ	Riesz projection, i.e. $1/2(I + S_\Gamma)$
Q_Γ	$= I - P_\Gamma$
smb A	symbol of an operator A
sp A	spectrum of the operator (element) A
S_Γ	singular integral operator with Cauchy kernel along Γ
D	differential operator $\mathrm{d}/\mathrm{d}x$
I	identity operator

Spaces and sets

alg A	smallest closed subalgebra of an algebra B containing the set $A \subset B$.
cen A	center of the algebra A
codim M	codimension of a linear set M in a linear space X, i.e. dim X/M
$\mathbf{C}(X)$	space of all continuous (complex-valued) functions on X
$\mathbf{C}^m(X)$	space of all m-times continuously differentiable functions on X
$\mathbf{C}^{m,\lambda}(X)$	set of all $f \in \mathbf{C}^m(X)$ for which $f^{(m)} \in \mathcal{H}^\lambda(X)$ $(0 < \lambda < 1)$
$\mathbf{C}^\infty(X)$	space of all infinitely differentiable functions
dim M	dimension of the linear set M
$\mathcal{H}_p^s$	Hölder-Zygmund space

$\mathscr{H}^s$	classical Hölder-Zygmund space, i.e. $\mathscr{H}^s = \mathscr{H}^s_\infty$
$\mathbf{H}^s$	periodic Sobolev space
$\mathbf{H}^s(X)$	Sobolev space of functions defined on X
$\mathbf{L}^p(X)$	Lebesgue space on X $(1 \leqq p \leqq \infty)$
$\mathbf{L}^p(X, \varrho)$	weighted Lebesgue space
$\mathbf{l}^p$, $\tilde{\mathbf{l}}^p$	spaces of p-summable sequences
$\mathbf{PC}(\Gamma)$	space of all piecewise continuous functions on Γ
$\mathscr{R}(X)$	class of all bounded Riemann integrable functions
$\mathscr{S}^\delta_n$, $\mathscr{S}^\delta(\Delta)$	spaces of splines
$\mathbb{T}$	complex unit circle
$\mathbf{W}$	Wiener algebra
$\mathbf{W}^{\alpha,\beta}$, $\mathbf{W}(v)$	weighted Wiener algebras
$X_{n\times n}$	set of square matrices of order n with entries from X

General symbols

$\mathbb{C}$	complex field
δ_{ij}	Kronecker symbol
$\mathbb{N}$	set of all natural numbers
$\mathbb{N}_0 = \mathbb{Z}^+$	set of all nonnegative integers
$\mathbb{R}$	real field
$\mathbb{R}^+$	$= \{x \in \mathbb{R} : x \geqq 0\}$
$\mathbb{I}$	$= [0, 1]$
$\mathbb{Z}$	set of all integers
supp f	support of the function f
wind f	winding number of a continuous complex-valued function f
χ_E	characteristic function of the set E

Name index

Subject index

W. G. Mazja/S. A. Nasarow/B. A. Plamenewski

Asymptotische Theorie elliptischer Randwertaufgaben in singulär gestörten Gebieten

I. Störungen isolierter Randsingularitäten
Mathematische Lehrbücher und Monographien, Abt. II, Band 82

1991. 432 Seiten — 18 Abb. — 17,0 × 24,0 cm — Hardcover — 148,— DM
ISBN 3-05-500693-3

Zum ersten Mal wird hier ein einheitlicher und allgemeiner Zugang zur asymptotischen Analysis elliptischer Randwertprobleme in singulär gestörten Gebieten gegeben. Im ersten Band werden Gebiete betrachtet, deren Rand in der Umgebung endlich vieler konischer Punkte „geglättet" ist, insbesondere ist der wichtige Spezialfall eines Gebietes mit kleinen Löchern enthalten.

Im 4. Kapitel — dem Kernstück des 1. Bandes — werden vollständigen asymptotischen Entwicklungen von Lösungen allgemeiner elliptischer Randwertaufgaben nach Potenzen eines kleinen Parameters, welcher die Störung des Gebietes charakterisiert, konstruiert und begründet. Dabei wird die in den letzten zwanzig Jahren entstandene Lösbarkeitstheorie elliptischer Differentialgleichungen in Gebieten mit nicht glattem Rand entscheidend benutzt.

Konkreten Problemen der mathematischen Physik, insbesondere der Elastizitätstheorie, wird große Aufmerksamkeit geschenkt. Hervorzuheben sind in diesem Zusammenhang die Kapitel 6—10, in denen die Asymptotik von Spannungsintensitätsfaktoren, Energieintegralen und Eigenwerten untersucht wird.

Bitte umblättern ...

II. Nichtlokale Störungen
Mathematische Lehrbücher und Monographien, Abt. II, Band 83

1991. 319 Seiten – 34 Abb. – 17,0 × 24,0 cm – Hardcover – 148,– DM
ISBN 3-05-501332-8

Der zweite Band untersucht Randsingularitäten höherer Dimension und gewisse nichtlokale Störungen.

Es werden asymptotische Entwicklungen von Lösungen elliptischer Randwertaufgaben nach Potenzen eines kleinen Parameters für die folgenden wichtigen Spezialfälle singulär gestörter Gebiete konstruiert und begründet: dünne Gebiete, Gebiete mit schmalen Hohlräumen, Einschlüssen oder Verbindungsstücken, in der Nähe von Kanten geglättete Gebiete, schnelle Oszillationen des Randes sowie der Koeffizienten der Gleichungen. Dabei wird die in den letzten zwanzig Jahren entstandene Lösbarkeitstheorie elliptischer Differentialgleichungen in Gebieten mit nicht glattem Rand entscheidend benutzt. Des weiteren wird die Mittelung von Differential- und Differenzengleichungen auf feinen periodischen Netzen sowie diskreten Gittern studiert. Konkreten Problemen der mathematischen Physik, insbesondere der Elastizitätstheorie und der Elektrostatik, wird große Aufmerksamkeit geschenkt.

Beide Bände beruhen zum größten Teil auf Resultaten der Autoren und weisen keine wesentlichen Überschneidungen mit anderen Monographien zur Theorie elliptischer Randwertaufgaben auf.

9783055006968